100 YEARS OF ENGINE DEVELOPMENTS

Other related publications available by SAE, include:

Hands-On Race Car Engineering
By John H. Glimmerveen
Order No. R-323

Racing Chassis and Suspension Design
By Carroll Smith
Order No. PT-90

The Ford GT: New Vehicle Engineering and Technical History of the GT-40
Order No. PT-113

Ferrari Formula 1: Under the Skin of the Championship-Winning F1-2000
By Peter G. Wright
Order No. R-356

For more information or to order these publications, contact SAE Customer Service at:

400 Commonwealth Drive, Warrendale, PA 15096-0001 USA
Web site: http://store.sae.org
E-mail: CustomerService@sae.org
Phone: Toll-free 1-877-606-7323 (U.S./Canada) or 1-724-776-4970

100 Years of Engine Developments

PT-115

Edited by
Daniel J. Holt

Published by
Society of Automotive Engineers, Inc.
400 Commonwealth Drive
Warrendale, PA 15096-0001
U.S.A.
Phone: (724) 776-4841
Fax: (724) 776-5760
www.sae.org
April 2005

All rights reserved. No part of this publication may be reproduced, stored in a retrieval system, or transmitted, in any form or by any means, electronic, mechanical, photocopying, recording, or otherwise, without the prior written permission of SAE.

For permission and licensing requests contact:

SAE Permissions
400 Commonwealth Drive
Warrendale, PA 15096-0001-USA
Email: permissions@sae.org
Fax: 724-772-4891
Tel: 724-772-4028

All SAE papers, standards, and selected books are abstracted and indexed in the Global Mobility Database.

For multiple print copies contact:

SAE Customer Service
Tel: 877-606-7323 (inside USA and Canada)
Tel: 724-776-4970 (outside USA)
Fax: 724-776-1615
Email: CustomerService@sae.org

ISBN 0-7680-1498-0
Library of Congress Catalog Card Number: 2005921721
SAE/PT-115
Copyright © 2005 SAE International

Positions and opinions advanced in this publication are those of the author(s) and not necessarily those of SAE. The author is solely responsible for the content of the book.

SAE Order No. PT-115

Printed in USA

Preface

Engine development – over 100 years and still going strong

By Daniel J. Holt

Historians sometimes differ as to who is credited with the invention of the first automobile. Some suggest that the very first self-powered vehicles used a steam engine, which gives credit for the first automobile to Nicolas Joseph Cugnot for his 1769 invention. Others suggest that the 1885 invention of Karl Benz, which used an internal combustion engine (ICE), should be given the credit.

The situation is also complicated if one looks at who is credited with the invention of the internal combustion engine. An internal combustion engine is loosely defined as any engine that uses an explosive combustion of fuel to push a piston that then moves a crankshaft, which eventually moves the wheels. As early as 1680 people such as Christian Huygens were trying to develop such an engine, using combustible substances such as gunpowder. However, it was Nikolaus August Otto that finally got a patent in 1876 on what he called the Otto cycle, which was a gasoline-powered four stroke engine. In 1887 Gottlieb Daimler patented a carburetor that he used to build his 1885 prototype of the modern day engine. In 1886 Karl Benz obtained a patent for an automobile that used gasoline in an internal combustion engine. From that time on, the internal combustion powerplant began its journey though history.

Throughout the past century many changes have been made to powerplants to power the automobile. During the 20th century the spark ignition engine had competition from the compression ignition engine, solar-powered electric motors, battery-powered electric motors, and steam-powered engines.

Halfway into the first decade of 21st century we find that gasoline spark ignition engines are still king among the choice of powerplants. Compression ignition engines powered by diesel fuel are popular in Europe and seem to be making their way, albeit slowly, into the United States. The high cost of fuel has caused a new entry to take the limelight--the hybrid vehicle. With these vehicles combining an internal combustion engine and electric motor, hybrids are becoming popular and nearly all vehicle manufacturers are developing their own versions or buying the powerplant from another vehicle manufacturer. How long the hybrid trend will last is up to discussion; some prognosticators predict that half the vehicles of the future will be hybrids, others think that they may come and go as the battery-powered and/or solar-powered electric vehicle has over the years. Fuel cell vehicles have been creating interest because they are vehicles powered via an electric motor driven by electricity. The electric-motor-driven vehicle is still the only powerplant that can meet the zero emissions criteria since they do not have any combustion of fuel associated with them. The fuel cell vehicle has a number of challenges to overcome including a fuel infrastructure, cost, range, and consumer acceptance. The hydrogen fuel infrastructure is being addressed because some internal combustion engine developers contend that an ICE running on hydrogen may be the answer to some of the emissions challenges. The best guess is that fuel cell powered vehicles will not be manufactured in mass until about 2020. In the meantime, engineers are faced with producing vehicles that meet the needs and desires of the consumer, the regulations imposed upon them by the governments, and cost restraints set by their respective companies.

The very first single cylinder engines barely created enough power to propel the horseless carriage to a speed of 3 mph. Today, even with the cost of gasoline, consumers are buying V-8 powered engines and some are interested in V10 and V12 engines. At the 2005 North American International Auto Show, BMW was showing a 500 hp, 5-L, V10; BMW was also showing a 6-L, V12 engine; Chevrolet had a 500 hp, V8, 7-L engine; and Cadillac was displaying a turbocharged 4.4-L, 440 hp, V8 engine.

Horsepower and displacement are terms that are discussed more than fuel economy and emissions by the buying public. Drop the engine into a vehicle and the consumer asks what are the 0-60 times? Fuel economy figures are clearly displayed on the sticker of the new vehicles but these numbers are not the deal maker or breaker. It should be noted that as the price of gasoline increases, more consumers are taking an interest in the fuel economy numbers, but will they continue to do so if the price of gasoline drops to a price that they deem acceptable?

The objective of this Progress in Technology (PT) book is to look at some of the key technologies that were used to mold the internal combustion engine of today. This PT examines some of the papers presented at SAE technical meetings over SAE's 100 year history (1905-2005).

From 1905 through the 1960s it seemed that the development of the internal combustion engine was progressing with numerous advancements and refinements being made in the area of spark plugs, carburetion, valve timing, combustion advances, etc. Engines were becoming more powerful and larger, and in fact, a 50-year anniversary of the small block V8 engine will be celebrated in 2005. After the second World War, consumers wanted faster, bigger cars, which meant more powerful engines. All seemed to be going well until the 1960s when the engineers were being faced with automotive emissions regulations. In the 1960s and early 1970s, the California Air Resources Board was created, the Clean Air Act was introduced in the United States, and the Environmental Protection Agency was created in the United States. CAFE or Corporate Average Fuel Economy combined with emissions regulations caused the industry to step back and examine the situation. If that wasn't enough, the oil embargo of the 1970s drove consumers to seek out smaller, more fuel-efficient vehicles, which caused the engineers to develop smaller powerplants.

In the 1970s unleaded gasoline was developed, emissions regulations were introduced in Europe, and Japan introduced its version of emissions testing. Engineers reacted to the new emissions developments and created new systems to meet the new regulations, such as:
Positive crankcase ventilation (PCV) valve
Catalytic converter
Open loop control
Electronic engine control
Evaporative emissions
Exhaust gas recirculation
On-board diagnostics

It seemed that during the 1970s the day of the high performance vehicle was gone. The triple threat of CAFE, emissions regulations, and the oil embargo (which forced gasoline prices to begin a steady climb) caused many people to wonder if the love affair with the automobile was over. Consumers were turning down the bigger, more powerful cars for smaller, fuel-efficient vehicles. Imported vehicles with better fuel economy started to take a foothold in the United States.

However, the pessimists underestimated the ingenuity and ability of the engineers. Today, the 2005 vehicles are faster, more responsive, more fuel efficient, and create fewer emissions than previous models.

This Progress in Technology book contains 100 papers on the technical developments of the internal combustion engine divided into 6 chapters:
Chapter 1 – Spark plugs
Chapter 2 – Carburetors
Chapter 3 – Ignition timing
Chapter 4 – Valve timing
Chapter 5 – Fuel injection
Chapter 6 -- Emissions

Although this PT is limited in just how many papers it can contain, the following is a list of associated SAE products that can give a more detailed, targeted look at engines:

PT-53 : Design of Racing and High Performance Engines
PT-80 : Direct Fuel Injection for Gasoline Engines
PT-94 : Homogeneous Charge Compression Ignition (HCCI) Engines: Key Research and Development Issues
PT-100 : Design of Racing and High-Performance Engines -- 1998-2003
PT-109 : The Diesel Engine
PT-110 : Electronic Engine Control Technologies, 2nd Edition
Transactions Journal of Engines--annual collection of the best SAE papers on engines
Diesel Engines Technology Collection on CD-ROM
Emissions Technology Collection on CD-ROM
Spark-Ignition Engines Technology Collection on CD-ROM.

It is important to remember that the internal combustion engine is constantly evolving. Engineers have perfected the cylinder cutout mechanism so that a V8 powered vehicle can operate on four cylinders to improve fuel economy. Diesel engines are cleaner and more efficient. Hydrogen refueling centers are starting to be developed in various parts of the United States to service both a hydrogen powered ICE and fuel cell powered vehicles. Engineers are standing up to the task and are developing vehicles that are better in every way. The rest of this decade and the next should reveal many exciting new technological developments.

TABLE OF CONTENTS

Preface v

Chapter 1 - Spark Plugs

170045	Spark-Plugs for High-Speed Engines	3
200015	Preignition and Spark-Plugs	13
400146	Ceramic Insulators for Spark Plugs	29
460016	Spark Plugs	37
480194	Factors Affecting Functioning of Spark Plugs	55
490001	Ignition and Spark Plug Design	69
510190	How Large Can the Gaps Be	75
520065	Operation of Spark Plugs in Present Day Engines	85
560125	New Engines Demand Tailored Spark Plugs	89
670149	Review of Spark-Plug Technical Developments	103
700081	Spark-Plug Design Factors and Their Effect on Engine Performance	123
760264	The Performance of A Multigap Spark Plug Designed for Automotive Applications	135
770853	New Aspects on Spark Ignition	147

Chapter 2 – Carburetors

060003	Some Requirements of Carburetor Design	161
070009	The Carburetor and Its Functions	165
120020	Carburetors	179
120032	Carburetors and Fuels	191
130033	A Consideration of Certain Problems of Carburetion	195
160035	Carburetor Investigations	213
170018	A Standard of Carburetor Performance	247
190069	Heavy-Fuel Carburetor-Type Engines for Vehicles	263
200060	Factors Involved In Fuel Utilization	279
210047	Fundamental Points of Carburetor Action	295
250007	Best Location for Carburetor Intake	325
290020	Dual Carburetion and Manifold Design	337
310025	Influence of Carburetor Setting and Spark Timing on Knock Ratings	349
660114	Control of Auto Exhaust Emission By Modifications of The Carburetion System	355
670485	Vehicle Exhaust Emission Experiments Using a Premixed and Preheated Air-Fuel Charge	371
690137	A Study of Factors Affecting Carburetor Performance at Low Air Flows	385
690138	Process and Control for Producing Anti-Emission Carburetors	395
750371	Closed Loop Carburetor Emission Control System	401
760226	Emission Control With Lean Mixtures	411
760288	A Vaporized Gasoline Metering System for Internal Combustion Engines	427
770352	A Feedback Controlled Carburetion System Using Air Bleeds	443
854244	Engine Ignition and Carburettor Electronic Control	451
871293	Electronic Gaseous Fuel Carburetion System	461

Chapter 3 - Ignition Timing

500025	Full Vacuum Controlled Ignition System	473
670115	Capacitor Discharge Ignition - A Design Approach	483
700083	Some Factors to Consider In the Design and Application of Automotive Ignition Systems	491
750346	HEI - A New Ignition System Through New Technology	507
770854	A Production Computerized Engine Timing Control System	519
780118	Microprocessor Engine Controller	527
780655	Electronic Spark Timing Control for Motor Vehicles	537
785133	Electronic Spark Timing Control for Motor Vehicles	555
820004	Electronic Control of Ignition Timing	593
854217	Evolution of Ignition Systems	599
860249	General Motors Computer Controlled Coil Ignition	609
870082	A Compact Microcomputer Based Distributorless Ignition System	615
871914	Crankshaft Position Measurement With Applications to Ignition Timing, Diagnostics and Performance Measurement	623

Chapter 4 - Valve Timing

740102	Design and Development of A Variable Valve Timing (VVT) Camshaft	631
825056	Variable Valve Timing as a Means to Control Engine Load	641
850074	Effects of Intake-Valve Closing Timing on Spark-Ignition Engine Combustion	647
854241	Electronics ~ The Key To Engine Management	657
860537	Development of a Variable Valve Timed Engine to Eliminate the Pumping Losses Associated With Throttled Operation	661
885065	Adaptive Ignition and Knock Control	669
890673	Variable Valve Actuation Mechanisms and the Potential for Their Application	677
890674	A Review and Classification of Variable Valve Timing Mechanisms	695
910445	Perspectives on Applications of Variable Valve Timing	709
914173	A High Speed Variable Valve Timing Mechanism for Engines	723
960496	Spark-Ignition Engine Knock Control and Threshold Value Determination	731
960584	Comparison of Variable Camshaft Timing Strategies at Part Load	741
980769	VAST: A New Variable Valve Timing System for Vehicle Engines	761
890681	Application of A Valve Lift and Timing Control System to an Automotive Engine	769
2004-01-1869	Study on Variable Valve Timing System Using Electromagnetic Mechanism	779

Chapter 5 - Fuel Injection

680043	New Fuel-Injection System and Its Potential for Reducing Exhaust Gas Emissions	789
741224	Electronic Fuel Injection in the U.S.A.	799
874345	The Ford Central Fuel Injection System	813
891330	Current Status of New Development in Electronic Fuel Injection System	819
921496	From the Simple Carburetor to the Electronic Fuel Injection SPI and MPFI	825
950228	Development of First Volume Production Thermoplastic Throttle Body	839
1999-01-0171	A Comparison of Gasoline Direct Injection Systems and Discussion of Development Techniques	849
1999-01-0174	Influence of Fuel Injection Timing Over the Performances of Direct Injection Spark Ignition Engine	863

Chapter 6 – Emissions

620154	The General Motors Catalytic Converter	875
620397	Development and Evaluation of Automobile Exhaust Catalytic Converter Systems	891
620398	Catalytic Converter Development Problems	957
660106	Design and Development of the General Motors Air Injection Reactor System	965
680110	Exhaust Emission Control by Engine Design and Development	983
680123	Potentialities of Further Emissions Reduction by Engine Modifications	997
680242	Automobiles and Air Pollution	1013
690503	Performance of a Catalytic Converter That Operates With Nonleaded Fuel	1021
700151	Chrysler Cleaner Air System for 1970	1033
710013	Effectiveness of Exhaust Gas Recirculation With Extended Use	1045
720481	Methods for Fast Catalytic System Warm-Up During Vehicle Cold Starts	1079
720487	Toyota Status Report on Low Emission Concept Vehicles	1103
725029	Predicting Performance of Future Catalytic Exhaust Emission Control Systems	1125
730014	Gasoline Lead Additive and Cost Effects of Potential 1975-1976 Emission Control Systems	1137
730569	Thermal Response and Emission Breakthrough of Platinum Monolithic Catalytic Converters	1157
741159	Pre-Chamber Stratified Charge Engine Combustion Studies	1171
750414	The Effects of Exhaust Gas Recirculation and Residual Gas on Engine Emissions and Fuel Economy	1191
750178	Reliability Analysis of Catalytic Converter as an Automotive Emission Control System	1225
760199	NO X Catalytic Converter Development	1237
780006	The Fast Burn With Heavy EGR, New Approach for Low NO_x and Improved Fuel Economy	1243
780203	Ford Three-Way Catalyst and Feedback Fuel Control System	1259
780624	Effect of Catalytic Emission Control on Exhaust Hydrocarbon Composition and Reactivity	1275
780672	Performance and Emission Predictions for A Multi-Cylinder Spark Ignition Engine With Catalytic Converter	1299
780843	Emission Control at GM	1321
790387	The Effects of Varying Combustion Rate in Spark Ignited Engines	1327
800053	GM Micro-Computer Engine Control System	1339
865033	Economy Increase for Otto Engines By Introducing Exhausts Into Carburettor	1355
874354	Implications of the Future European Emissions Scene for the Motor Industry ~ Effect on Overall Industry Costs and Complexity	1359

About the Editor 1363

CHAPTER 1

SPARK PLUGS

INDIANA SECTION PAPER

SPARK-PLUGS FOR HIGH-SPEED ENGINES

By Albert Champion
(*Member of the Society*)

Abstract

The author first mentions hot-tube ignition, which preceded the spark-plug, then the low-tension type of make-and-break ignition, following with the development of spark-plugs. He compares the conical and petticoat types of porcelain and discusses at some length the insulating materials used in spark-plugs, such as porcelain, mica, steatite, glass and quartz. In this connection he covers the composition of the material, its dielectric strength, carbon-absorbing ability, heat-conductivity, mechanical characteristics, and the difficulties of glazing. Under the heading of general design of spark-plugs the author takes up such points as assembly of insulator and center electrode, electrodes, gaskets, and separable and integral plugs.

At various meetings of the Society the question of standardizing spark-plug sizes has been discussed. The functions, shortcomings and close relationship of the spark-plug to the efficiency of the engine have received, however, but scant attention, and the engineer will find it highly important to give more consideration to this very essential part.

I want to go back further than spark-plugs, as far as automobile engines are concerned, to the first ignition used. I believe that most engineers have never had any experience with the hot-tube ignition that preceded the spark-plug. Some of the younger engineers do not realize the difficulties in the old days when we did not have electrical ignition. A thin platinum tube was heated by a gasoline burner and when the gas was compressed in this hot tube the explosion would be produced. That was the first means of ignition for automobile engines. Then we had the low-tension type of ignition in connection with the make-and-break system. De Dion was the first manufacturer to use spark-plugs successfully. The plugs were very much like the article we make today. The porcelain was of the conical type and had a center wire cemented in and a single ground electrode.

In the early days of the industry in this country there were no spark-plug makers, as there was no spark-plug business. The few manufacturers who were really putting out experimental cars had to make their own plugs. Some of them used mica for insulation, others porcelain; but all of the porcelain came from France. As the business grew, the majority of spark-plugs were imported from France. Then we had spark-plug manufacturers in the United States, but those who used porcelain insulation brought it from France.

Impossible Claims for Plugs

Up to a few years ago all spark-plug manufacturers thought that if they could make a spark-plug, the insulation of which would not crack or soot too easily, they would have accomplished all that was expected of them. Innumerable claims were made about soot-proof plugs, but to date no spark-plugs are made that will not soot under certain conditions; neither are there porcelains that will not finally crack. We used to have a good deal more porcelain breakage on plugs than we have today but the reduction of the breakage is not due to the better porcelain but to the improved design of the plugs as developed by some of the spark-plug manufacturers. The porcelain maker was blamed for the breakage, but later on we found that the design had more to do with that trouble than the quality of the porcelain itself.

It is sometimes thought that as regards sooting, the design known as the petticoat type of porcelain is better than the conical type, but this is a mistake. The conical type porcelains are now made in such a way that the porcelains are of a sufficient length and are small enough to get hot. This type of plug will not soot so easily as the petticoat type, which does not project far enough into the combustion chamber, resulting in a comparatively low temperature which causes it to soot and short-circuit more rapidly than the conical porcelain. The disadvantage of the petticoat type as with high-speed engines today is that the center wire is very long, and being away from the cooling points it becomes overheated and will cause preignition.

INSULATING MATERIALS

Porcelain.—Porcelain is a mixture of different clays such as feldspar, flint, kaolin and ball clay. The composition used by the different manufacturers are very similar but the variation in the quality of the several materials is great; to date it has been found that the combination of English and French clays really constitutes the best body for porcelain for use in the manufacture of spark-plugs. After all of this clay has been ground and mixed together it is ready for what is called vitrifying, for what we really have is a multitude of molecules that we simply weld together. The firing is a very careful process. There is no such thing as the highest heat for porcelain. What is required is the proper heat, just as in treating steel. If we under-heat, it will not stand heating and cooling. If we over-heat, it will blister.

The first quality needed in porcelain for spark-plugs is the greatest dielectric strength, which means the least amount of electric leakage. Any porcelain weak dielectrically will let an appreciable flow of current go through when subjected to heat, and the higher the temperature the greater the leakage. We also found that under pressure and low temperatures, which would be the condition when starting an engine, the plug with the porcelain having the greatest dielectric strength will produce a spark under high pressure, while the one with low dielectric strength will not give a good spark even under a comparatively low pressure. A porcelain with good dielectric strength not only means more power, smoother running, better pick-up, but also easier starting.

Many engineers have probably wondered why, with a certain make of plug, they can throttle their engine down lower than with some other make. If an insulator in the spark-plug permits so much electric leakage that most of the current goes through the insulation little is left to ignite the gas. This will not only prevent the engine's running smoothly, but on a quick pick-up the car will be jerked, owing to missed explosions, as at that time the mixture of gas is not the best and the spark is so weak that the mixture will not ignite until it has adjusted itself.

Many years ago, after having made some great improvements in our spark-plugs, we decided that we ought to find some way of putting them out of business. We had an engine with a detachable head so that we could vary the compression and when we ran it so as to get to a fairly good heat we found that in about forty minutes the engine would stop. We began to blame the ignition, which was by a well-known battery system. We also tried about every make of plug we could find on the market. Some of them we found would make the engine stop in a very short time, so we realized that a good deal depended on the spark-plug itself. On the other hand, we thought the trouble was partly due to the ignition. We made tests with a larger coil and found that it helped. Then we made a special high-tension switch, so that when the engine was ready to quit we could shift to the other coil; the engine would at once pick-up again. This showed us that after a coil became hot leakage developed. The combined electric leakage of the coil and the spark-plug insulation was enough to reduce the current until the spark became so weak that it could not ignite the gas. Further tests that we made demonstrated that if there had been less leakage in the insulation we could have run continuously with the same coil.

One of the greatest troubles on most insulators is the fact that they will change form. All American-made porcelain will absorb carbon. Some engines must be operated a couple of thousand miles, others less, depending upon their carbonization. It will be worse in the winter when the choker is used, especially if the engine pumps a little oil, for with the gas and the oil a combination is formed such as is found in a carburizing furnace. This change of form in the porcelain baffles many people. If their engines do not run smoothly they clean the plugs. If there is no break or crack in the porcelain they blame anything in the engine but the spark-plugs. They adjust the carbureter, they may grind the valves and do other things; then finally they decide to try a new set of plugs and just as soon as the new plugs are put in the engine runs properly again. If they do not use a plug with the right kind of insulation it is right only for a short period. This is why so many plugs are made for resale. One spark-plug manufacturer has claimed to make at least sixteen million plugs per year; we make about twelve million ourselves and there are about two hundred more plug makers and assemblers who, combined, make at least as many plugs as the two larger manufacturers, so that about six sets of plugs per year are used on every car that is running in this country today. If that is the condition, we will have to admit that the spark-plugs made today need much improvement.

Effect of Conductivity

In making a spark-plug for high-speed engines, we must figure the conductivity of heat. Porcelain, or any other refractory material, is a very poor conductor of heat and the center wire, which is in the center of the insulator, should be kept comparatively cool so as not to result in preignition. The insulator really gives us an ideal condition for a heat-retaining furnace, it being w' known that all furnaces are lined with a refractory m..erial to hold the heat.

As far as the sooting of porcelain is concerned, there is no doubt that the quality which is non-absorbent, such ɛ that made in France or Bohemia, will not become sooted or short-circuited as rapidly as the other. As long as porcelains of this quality do not break or crack the engine will run smoothly, as they do not change form. If oil or soot should result in a short-circuit they will, after being cleaned, be as good as they were before. But the trouble with all American-made porcelain is that after the plug starts to become short-circuited we might just as well buy a new plug, for although cleaned on the surface they have absorbed so much carbon that they will soon become short-circuited again.

Steatite.—Steatite is a commonly known material, used in the Bosch spark-plug, the manufacturers of which were the first to use it. This insulator is made in Germany.

The principal material entering into the manufacture of steatite is what is known as soapstone, or lava, or talc, which are all of the same family. A great variation is found in talc; some that would be very fine for talcum powder would not be good for steatite. In this country today we have many varieties of soapstone used for different purposes. The German soapstone, which is found near Nuremberg, has proved to be the best to date. This material has great mechanical strength and will stand much more abuse than porcelain.

Difficulties with Glazing

The only real trouble with steatite is that so far it is not possible to apply a glaze at a high heat. Therefore it is necessary to vitrify the material first and then use a low-fusing glaze. Because of the difference between the temperature at which the steatite is vitrified and the temperature at which the glaze is fused, there is a great difference of expansion so that after a short time the glaze cracks like a very poor quality of china. When this happens, it gives the impression that the steatite is porous and that the spark is jumping through, but this is not the case, as carbon is lodged in the multitude of cracks and the spark jumps from one to the other, instead of taking place at the gap. At high speed a steatite glaze has been frequently known to melt and for high-speed engines I thing that much better results would be obtained if steatite were used without any glaze and simply given a high polish.

Mica.—Mica is good dielectrically before being subjected to oil and the heat of an engine. Its main quality is that it will not break under stress of a heavy bl v. The best quality of this material is what is known as india-ruby mica and the reason it is found very good, especially for spark-plugs, is that it will split evenly. The process may be still continued after a piece of it has been split down to a few thousandths of an inch. We are very well posted on the mica-insulator proposition, having built in our experimental department plugs with this insulation and also having tested all makes of mica spark-plugs.

We may first mention the mica core, which consists of a certain thickness of mica wrapped on a metallic center wire. A large number of mica washers are slipped over this mica tube, then tightened together, giving what might be termed a mica cylinder, which is turned to the form desired and given a high polish. In the experiments we made we always found that with any degree of pressure we might use we never could compress the mica enough to make the core gas-tight. Shellac has been used between the mica washers when assembling them; this will make the insulation tight when cold, but when subjected to heat it will leak very badly. Some of the latest mica plugs, as made for airplane engines, have layers of mica on the inner part of the shell and then a certain number of mica washers to cover the layers of mica. Actually, the other construction has been reversed in order to keep the insulating material as cool as possible.

Plugs made of this design are successful for aviation engines, but on account of the large number rejected in manufacture they cost several dollars apiece. Furthermore, they are only good for a limited number of hours, and one of their greatest troubles is that they will become short-circuited with oil very easily.

A great difficulty in the manufacture of plugs with mica insulation is to duplicate; in other words, make two good plugs in succession. Not only does it take much skill, but if the operator should happen to have moist hands as he wraps the mica, moisture will get between the layers, and this is enough to make a bad plug.

Glass and Quartz.—Glass has been used in spark-plugs. These were on the market for a short time, but have disappeared. The reason for it is that glass is dielectrically inferior to the poorest kind of porcelain, and consequently with glass plugs the engine will not run very smoothly, which is enough to condemn them. Quartz, dielectrically, is just as bad as glass, and is also bad mechanically. We tested quartz at one time, knowing its ability to stand heating and cooling, but when we found the amount of electric leakage we concluded it was unsatisfactory.

GENERAL DESIGN

Assembly of Insulator and Center Electrode.—We now have the porcelain through which a center wire must be inserted. It is imperative for high-speed work that the plug be gas-tight.

A valve cap in an engine fitted with a copper-asbestos gasket and tightened with all possible force will, upon being tested, still show some slight leakage. Further, this being metal to metal, the coefficient of expansion and contraction is the same, and the greatest pressure can be applied. Yet the valve cap cannot be made gas-tight.

Considering these facts, it will probably be realized what a problem the spark-plug manufacturer has to solve to make a gas-tight plug with a porcelain assembled in a steel shell and a metallic core running through the porcelain.

Electrodes and Materials.—The best practice is to cement to the center wire in the porcelain. As the cement dries the evaporation leaves it porous, and this porosity means leakage.

Another design of center wire consists of a long screw headed at one end. A packing is placed between the head and the porcelain at one end, and the other end is tightened with a nut. This is only used on a small number of plugs, as not only is the leakage great but the strain on the porcelain will crack it.

These two methods, cement and packing, are the most commonly used, but the first design mentioned is the best practice known today with the insulators we have to work with.

At the same time we are limited as to the size of the electrodes, owing to the fact that they are made, in most cases, of an alloy with a high percentage of nickel. Nickel alloys have large coefficients of expansion, so that a heavy wire will crack the porcelain; therefore, we must endeavor to make them small enough to preclude this. This is one of the things that handicaps the spark-plug maker using porcelain insulators. If a larger electrode could be used the plug would be better for aviation and racing purposes so far as the center electrode is concerned, as it would assist considerably in conducting the heat away.

We have heard of all kinds of materials used for center electrodes, but all of them are nickel alloys. The material most commonly used with nickel is manganese. The proportions vary, but run about 96 or 97 per cent nickel, about one and one-half per cent manganese, and a little foreign matter, like cobalt. The manganese is really used to help flux and rolling, and my opinion is that pure nickel is better. Pure nickel, which is used by some makers, has a very much better conductivity of heat, but its expansion is greater, consequently it will crack the porcelain much more easily.

The reducing flame of the explosion will carbonize any material now used in spark-plugs. At times, after using a spark-plug for a certain period, one attempting to adjust the electrode will find that it has become very brittle, and will break, and will probably blame the material, not realizing that the principal trouble is due to carbonization. For instance, an engine may be equipped with plugs of the same material throughout, and on examination electrodes in three cylinders will be found to be still good, while in the fourth one, owing to pumping of oil in the cylinder which creates a reducing condition, the electrode will be absolutely gone. So far we find that pure nickel is the best for the purpose.

Gaskets.—The design of the gasket is a very important factor in spark-plug efficiency. Porcelain is a very poor conductor of heat, consequently anything we can do to conduct the heat away will improve the plug, because the cooler we keep the insulation the less electric leakage we have. Asbestos as a cushion is good, but it should be inclosed in copper so the porcelain will be in contact with the copper and the copper with the body, in order to carry the heat away. Otherwise, it only helps to insulate it from the cooling points and hold the heat in the porcelain, in the same manner as is done with pipes and boilers, where asbestos is used to retain the heat.

Shells and Bushings.—I do not know of anything but cold-rolled steel being used for the shell itself.

Separable Plugs.—On the plugs that come apart a brass bushing is good, as it will help to carry a certain amount of heat away from the porcelain. The principal difficulty with separable plugs, however, is that it is really impossible not only to make them gas-tight, but to keep them so when subjected to heat. When the parts are put together and the tightening is done on the bushing, either the porcelain will turn on the packing or the porcelain and packing will turn against the seat. This will impair the packing, and cause leakage. The other great trouble is that on account of variation in the size of porcelains it is seldom possible to get them tightened exactly as they should be, so that many of these porcelains will break while in use.

Integral Plugs.—It is difficult to make a plug of the separable type, but on the other hand it is much more difficult to make a good integral-type plug. It was only after a good many years of research work that our engineering department did develop a plug that has been successful.

AUTHOR'S CONCLUSION

Poor throttling, poor pick up, missing on hard pulls and at high speed are caused mostly by the spark-plugs. These things happen because of poor insulation. We must have and are getting good design in spark-plugs, and it is quite possible to have a gas-tight plug. But the most needed thing today is better insulation. Efficient plugs will mean a great deal to the manufacturers and users. Inferior plugs will certainly spoil a good engine and cause trouble to the service man, and I know that one of the biggest costs of service today is caused by owners continually going into service stations for various adjustments because their engines are not running properly, when nearly all of the trouble is owing to faulty spark-plugs.

PREIGNITION AND SPARK-PLUGS

By Stanwood W Sparrow[1]

The author proposes to determine what features of spark-plug construction cause preignition and how this preignition manifests itself. To this end observed conditions on an Hispano-Suiza aviation engine following 4 hr. of an intended 6-hr. run are reported, with supplementary tests and observations. This resulted in experiments made to determine the cause of preignition, using spark-plugs constructed so that different features of their design were exaggerated. Illustrations of these plugs are shown and the results obtained from their tests are described. The different observed peculiarities are then stated, analyzed and compared with normal spark-plug performance. The experiments serve as a means of identification of special forms of preignition and as an indication of the abnormally high temperatures to which valves and combustion-chamber walls are thus subjected. They also serve as a basis for predicting the performance of a given design of spark-plug or for designing a spark-plug to meet known conditions.

What features of spark-plug construction cause preignition and how does this preignition manifest itself? These two questions this paper proposes to answer. There seems to be no better way of describing the particular phases to be treated than to tell of the events that led us to undertake the work.

After about 4 hr. of what was intended to be a 6-hr. test of a 180-hp. Hispano-Suiza aviation engine, the flame from one of the exhaust ports was noticed to be about twice its usual length and of that light-yellow color which usually indicates a rich mixture. At the same time the power of the engine decreased about 10 per cent, vibration was much increased and the exhaust valve appeared to be extremely hot. The engine was stopped and inspected as thoroughly as possible without dismantling, but except that the spark-plugs were badly carbonized no trouble was found. With new spark-plugs the engine not only completed this run, but made several later runs without any recurrence of this trouble. It should be borne in mind that no adjustments had been

---

[1]Associate engineer, Bureau of Standards, Washington.

made of mixture ratio or throttle during the 4 hr. prior to the appearance of this long yellow flame and that the engine had been developing about 140 hp. during this time. Some weeks later the phenomena again appeared. Temperature measurements were made at this time with a thermocouple inserted in one of the spark-plug holes. This couple showed an average temperature of 150 deg. cent. (392 deg. fahr.) with normal operation of the engine and 400 deg. cent. (752 deg. fahr.) with preignition. The couples used were not constructed for these tests and the actual temperatures have no particular significance as they are temperatures of a part of the cylinder wall thermally insulated from the jacket. It is the relative temperature that is important, noting that at the time the flame appeared and the power decreased the temperature was much greater than during normal operation. This was confirmed by the fact that when the engine was stopped after a few minutes operation, the exhaust valve in a cylinder that had been preigniting would remain red for nearly a minute after the valves in all the other cylinders had become black. Experiments were then begun to determine the cause of this occurrence and the results of this work are here summarized.

What Is Preignition

Whenever a spark-plug is referred to in this paper as causing preignition, it is the effect of preignition just described that is meant, the long yellow flame, engine vibration, drop in power and extreme heat. The first task is then to show that preignition does produce this effect and to define just what is meant by preignition. Although preignition is quite commonly defined as "self-ignition from the heat of compression," in this case the literal meaning of the word has been more closely adhered to and it has been called ignition from any cause before the proper time, or the time of ignition which will result in maximum power production. It was not difficult to determine that preignition would cause the above noted effects. This was accomplished experimentally by setting one of the two magnetos successively ahead by 15-deg. steps, with the high-tension wire from this magneto disconnected from all but one cylinder. This magneto was not switched in until the engine had been brought up to speed on the other set of plugs. By changing the wires the effect of the advance could be studied in any cylinder. The magnetos are set for an advance of about 20 deg. ordinarily and it was found that firing from about 65 to 215 deg. before center did produce the effects I have described, including the long yellow flame. The intensity of the flame seemed greatest at an advance about midway between these two points and was about the same for all cylinders.

It is evident then that this trouble is caused by preignition. It remains to determine just how the sparkplugs cause preignition. The line of investigation followed was to construct plugs in which different features were exaggerated. The accompanying sketches illustrate some of these plugs. No special significance attaches to the exact shape of the porcelain or the shell. The plugs in which the special features were incorporated were of two types that have been used with satisfaction in the laboratory and in actual service; and, since both would produce the results noted, the drawings have been made to show approximately the average proportions of these two plugs.

The first definite indication as to the cause of the trouble came with a plug having the so-called petticoat type of porcelain, the lower end of the petticoat in this case having been broken off so that it rested near the center electrode in about the position shown in Fig. 2. This plug produced preignition when used successively in four different cylinders. Moreover, removal of the piece of porcelain broken off cured the trouble and its substitution on a standard one-piece plug, as illustrated in Fig. 2, again produced it. It would be assumed from this that either the porcelain becomes hot and fires the incoming charge, or that it is instrumental in heating the center electrode to the igniting temperature. A happy accident at this point gave considerable aid. The bit of porcelain slipped into the position shown in Fig. 1, while being inserted in the cylinder and in that position operated without preignition. This pointed to the heating of the center electrode as being the root of the trouble, even though this heating effect be of no great magnitude. Further experiments have supported this supposition, as regards both the cause of the trouble and the narrow range existing between normal operation and preignition. Thus, the plug shown in Fig. 3, with a piece of porcelain wedged between the center and the side electrodes will persistently cause trouble; while with the piece removed the plug will operate satisfactorily. The breaking away of a large portion of the porcelain from around the center electrode will not change the operation of the plug.

If a plug porcelain be cracked as shown in Fig. 4, so that the break can be detected only by the fact that the center electrode can be moved with the finger, preignition will result. If now the porcelain be broken away, leaving the center electrode bare down to the crack, the plug will again operate without trouble.

Fig. 5 shows a plug upon the center electrode of which has been placed a metal disk of about 5/16-in. diameter and 7/64 in. thick; the spark jumping the annular gap. The plug so constructed caused preignition, but with the disk removed or with one side of the disk flattened as

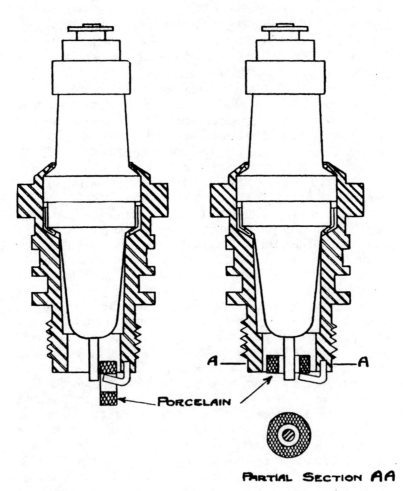

Fig. 1   Fig. 2

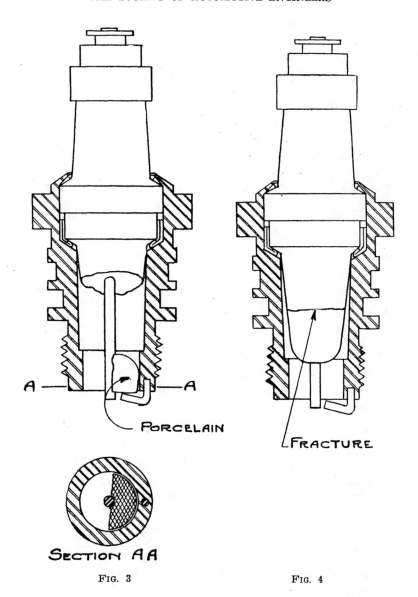

Fig. 3 Fig. 4

in Fig. 6, it operated normally. Fig. 7 illustrates another case where the line dividing preignition from satisfactory operation is extremely narrow. The plug is unusual only in the respect that the side electrode is replaced by a circular bushing of brass. Several were made leaving the bushing a slip-fit to the shell, all of them causing trouble, while others having the bushing a drive-fit to the shell operated satisfactorily, apparently because of the better heat conduction in the latter case.

What Causes Preignition

Thus far the types have been selected that were of most aid in fixing the cause of the trouble. That cause is an overheated center electrode which preignites the incoming charge and soon causes continuous burning throughout the cycle, extremely high average cylinder-temperature and loss in power. Further examination of the distinctive features of the plugs mentioned thus far shows in each case that these features do cause heating of the center electrode. It would be more direct,

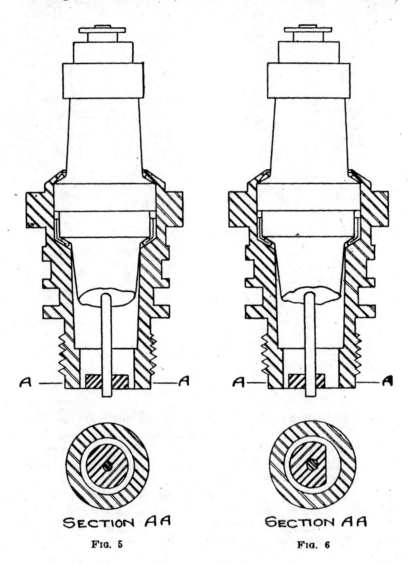

SECTION AA SECTION AA
FIG. 5   FIG. 6

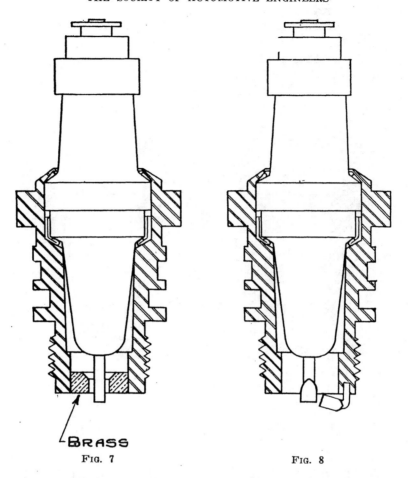

Fig. 7   Fig. 8

to say that they prevent the cooling of the center electrode, through preventing its free contact with the cool incoming charge. It is not possible, however, to state whether the trouble produced by the fracture in Fig. 4 is due to the decrease in the heat conducted away by the porcelain or to the free access given to the hot gases by the burning away of the cement around the center electrode.

It is of interest to study a few of the other freak plugs tried in this series. Fig. 8 shows a plug with the electrodes flattened nearly to a knife-edge without producing trouble. Fig. 9 shows an ordinary plug to which has been added a loop of 1/32-in. steel wire extending into the cylinder 2 3/16 in. from the shell. Here again, although there is a wire that doubtless becomes extremely hot, the fact that it is so freely exposed to the incoming cool gases prevents it from attaining an average temper-

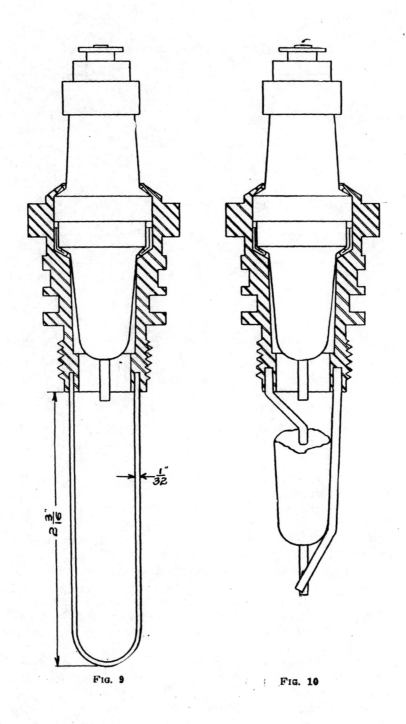

Fig. 9 Fig. 10

ature which is sufficient to cause preignition. Fig. 10 is another illustration of the same fact. In this case a section of a porcelain from a spark-plug is supported on wires extending into the cylinder. Again there is no preignition. Metal discs fitted on electrodes as shown in Figs. 11 and 12 cause no trouble.

The question now arises: How could there be continuous burning in the cylinder throughout the cycle without "popping back" in the carbureter? The explanation is that the rate of flame propagation is less than the velocity of the incoming charge, when the engine operates with the wide-open throttle at a speed of 1800 r.p.m., as in these tests. However, when preignition was present, if the engine was slowed down to 1400 r.p.m. "popping" in the carbureter would appear.

To prove definitely that preignition did produce this continuous burning, a window was constructed in the inlet manifold so that, with the inlet valve open, a portion of the interior of the cylinder could be observed. With a plug such as is shown in Fig. 4, a small section of the side electrode could be seen. When there was "popping" in the carbureter there would be a series of flashes visible through the window, but with the preignition at the speed of 1800 r.p.m. the observed phenomena were of continuous and rather gradual growth. Thus, with the engine operating normally everything would appear black; then the electrode would in turn become brighter and soon be surrounded by a small spot of reddish flame. As soon as the preignition trouble became pronounced, all of the cylinder that could be observed would appear to be filled with bright yellow flame. The special electrodes shown in Fig. 13 caused preignition and were of sufficient length that they could be easily observed through the window in the manifold. At about 1600 r.p.m. and 80 per cent full load the center electrode would be seen to glow. Under these conditions the engine would operate without trouble, but with a slight increase of load there would be a continuous flame around the electrode and all the evidences of preignition, namely, the long yellow flame, vibration, drop in power and excessive heat would appear. The plug shown in Fig. 14 gave no trouble itself, but when used in a cylinder in which the other plug was one that caused preignition, the nichrome wire, which melts at a temperature of 1500 deg. cent. (2732 deg. fahr.), fused. This is several hundred degrees higher than the average

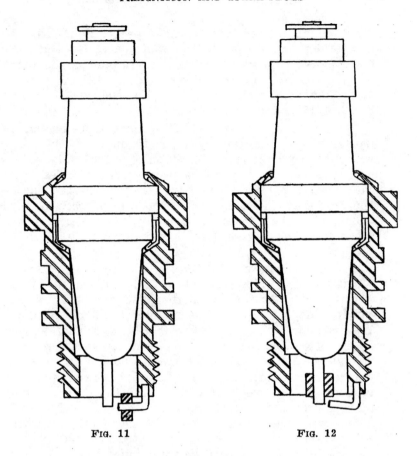

Fig. 11 Fig. 12

temperature of a freely exposed wire in a cylinder under normal conditions.

Thus far these remarks have been confined to the "freak" plugs. The question now arises as to why the apparently normal plugs gave trouble. While most plugs that have given trouble have been found to leak around the center electrode, by no means all the plugs that leak here give trouble. Of twelve plugs known to have caused preignition, four were found to leak badly; while of the twelve that operated satisfactorily at the same time, none showed excessive leakage. While it seems reasonable that a leak should cause higher electrode temperatures, the leakage in the above mentioned four plugs seems as likely to have been the effect of the high temperatures, as that it was the cause of the preignition. There is a rather remote possibility that there exists sufficient difference in the heat conductivity of different porcelains to alter the electrode temperature enough to bring about these results. It seems plausible that in many cases the trouble is initiated not by the plug itself but by bits of incandescent carbon or other hot points within the cylinder. The design of the plug determines how long this can happen before the plug attains a sufficiently high temperature to continue the preignition temperature of its own accord. Another point that tends to show this is that some cylinders seem more prolific trouble-makers than others, and that often an adjustment of the mixture ratio away from that of maximum power will prevent the trouble; but once the trouble appears no adjustment will cure it.

Discussion of Results

These then are the outstanding features of this comparatively short series of tests. They serve as a means of identification of this form of preignition and as an indication of the abnormally high temperatures to which valves and combustion-chamber walls are subjected because of this trouble. Finally, they serve as a basis for predicting the performance of a given design of sparkplug or for designing a spark-plug to meet known conditions. Of chief importance in this regard is the fact that, to attain cooling of a spark-plug element, it is more effective to give free access to the cooling action of the incoming charge than to sacrifice this effect to protect the element from the hot explosive gases.

The contradictory demands made upon the spark-plug are realized. Electrodes must be run hot enough to burn clean and prevent fouling at light loads and at the same time be cool enough to prevent preignition at full load. A test similar to this suggests itself as a means for proportioning electrode sizes. It is to design the plug so that each electrode will just produce preignition trouble, at a compression ratio slightly higher than that with which the plug will be used. Care must be taken to avoid badly carbonized cylinders or anything other than the spark-plug that might cause preignition. It appears that in many cases the side electrode can be made smaller and still be less likely to produce preignition than the center electrode, or the latter could be made larger and still maintain a factor of safety against fouling greater than that of the side electrode. That, however, is the problem of makers and users of sparkplugs. It is as a supplement to their own experiments

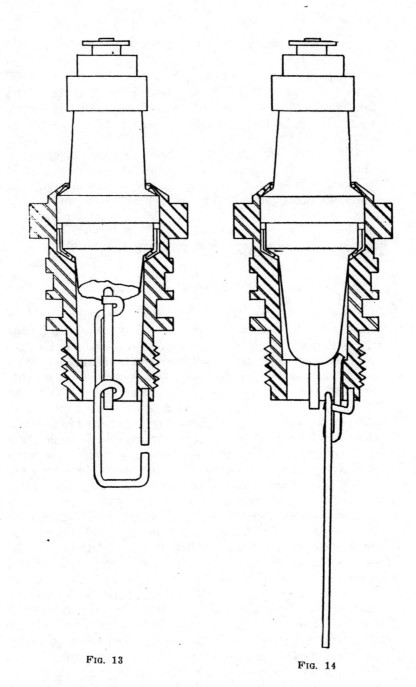

Fig. 13   Fig. 14

and as a suggestion for their further efforts that these results are presented.

THE DISCUSSION

THOMAS S. KEMBLE:—In 1915 the Curtiss VX engine was developed with a rather high compression-ratio for the sake of the advantages to be obtained at great altitudes. Inspection requirements, however, demanded full-throttle dynamometer tests of considerable duration and preignition developed. This was reduced to passable dimensions for dynamometer tests by reducing the compression somewhat and the air performance was very good for that period. Later, the volumetric efficiency was improved to such an extent that serious preignition again developed. This time repeated tests demonstrated that preignition occurred only in conjunction with leaky spark-plugs. The trouble was eliminated by using gas-tight plugs only. At that time gas-tight plugs were very difficult to obtain. Since then the situation has altered materially through the cooperation of spark-plug manufacturers and good plugs are easily obtainable. An interesting feature developed in that on at least three occasions a hole appeared in the aluminum intake manifold near one of the intake ports. This hole looked as if it had been caused by intense heat. We had been unable to get these engines to pop back through the carbureter intake without holding one of the intake valves well open. We therefore attempted to explain the manifold failure as being due to continuous popping back which failed to reach the carbureter intake. In view of Mr. Sparrow's observations through the intake manifold, I am convinced that our explanation was probably the correct one.

R. CHAUVEAU:—The tests we made during the war on various types of engines confirmed entirely the statements made by Mr. Sparrow. I will add an assumed explanation of the causes of the preignition. In the spark plug shown in Fig. 2 it is, to my mind, due to the fact that there is an opening between what I would call the inner chamber of the spark-plug and the cylinder. The volume of this chamber is several times larger than the opening. During the explosion the pressure in the chamber is raised to the same degree as the pressure in the cylinder. This necessitates the passing of an amount of hot gases through the opening into the chamber. With these gases passing through a small opening alongside the electrode, it is natural to assume that they will trans-

PREIGNITION AND SPARK-PLUGS

fer a certain amount of heat to the electrode itself, and this amount of heat is proportional to the speed and pressure of the gases.

The cooling of the center electrode is, in my opinion, effected mainly through the top of the electrode outside of the porcelain and only to a very small extent through incoming gases from the intake valve. That theory explains why we have preignition on the plug shown in Fig. 2 and not with the plug shown in Fig. 1, as the opening of the chamber is not restricted to any large extent. This theory was confirmed by another type of plug, of which I submit a sketch. As shown in Fig. 15, it

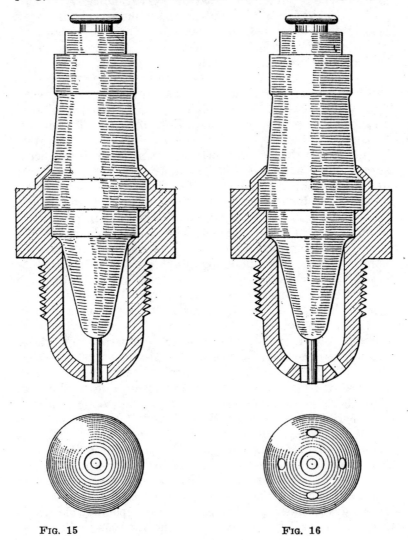

FIG. 15   FIG. 16

would consistently cause preignition. By drilling four holes, as shown in Fig. 16, in the shell, we would deflect a certain amount of the gases from the annular opening around the center electrode and eliminate the preignition. The plug shown in Fig. 3 in Mr. Sparrow's paper was claimed to cause preignition. This can be explained by the fact that while the opening is not restricted as much as in Fig. 2, at the same time the chamber in the shell has been increased 150 to 200 per cent, so that the proportion between these two remains about the same and, in addition, a larger area of the center electrode has been exposed. It is further proved in Figs. 5 and 6 that preignition is eliminated by increasing the opening to the chamber.

In conclusion, I would say that we have found the open-type plug the most satisfactory, and that the larger the center electrode is the less is the danger of preignition, provided the free end which extends beyond the porcelain inside the shell is short.

Another advantage in running the electrode as cool as possible, accomplished by the use of a heavy electrode, is the fact that it will cut down the leakage of the high-tension current and provide a more concentrated spark. The ground electrode is not likely to cause preignition, in conformity with Mr. Sparrow's statement. It is preferable always to use the single-ground electrode to concentrate the high-tension energy in one strong spark rather than distribute it among a number of weaker sparks. The single-ground electrode also cuts down electrical leakage.

In designing spark-plugs for aviation engines, I have found, however, that the engineer is limited as long as he has to use the materials now available for spark-plugs, and that it is impossible on this account to design one plug that will be universally good for all types of engines. In fact, a new engine and an older one of the same make, type and construction, require as a rule differently designed plugs, as the spark-plug in a new engine must be designed for high temperatures and a small amount of oil, whereas in an old engine the spark-plug must be designed for lower temperatures but for an excess of oil. It is only in rare cases that these two features can be combined into one spark-plug.

S. W. SPARROW:—Spark-plug leakage would be very apt to produce the preignition troubles described by Mr. Kemble. I think his explanation as to the cause of the hole in the intake manifold is undoubtedly correct. It is a pleasure to know that Mr. Chauveau's experience confirms so completely the results obtained in the tests at the Bureau of Standards. His explanation of the cause of the heating of the center electrode is of decided interest. I feel sure, however, that when an electrode is exposed to the incoming gases, much of its cooling is due to the heat taken from the electrode to vaporize the liquid fuel or oil that is thrown against it.

Ceramic Insulators for Spark Plugs

By Frank H. Riddle

Director of Research, Champion Spark Plug Co., Ceramic Division

THE word "ceramic" is derived from the Greek word "keramos" meaning "burned stuff," thus indicating its inorganic character, Mr. Riddle explains. Although the word is commonly associated with the substance known as clay, in modern usage it includes a large number of minerals and rocks and a correspondingly large number of mineral products, such as the many clay products, glass in its many forms, enamels, and the cements. The chief actors on the stage of ceramic insulators, he indicates, are the silicates, especially the silicates of aluminum.

Taking up clays first in his discussion of ceramics, he shows that clays differ enormously in their physical properties chiefly because of differences in mineral structure, particle size, and the impurities associated with them because of their geological history. Quartz, feldspar, porcelain and spark-plug porcelain also are treated in his presentation. He shows that spark-plug glaze has a surprising effect on the physical properties of the insulator.

Mr. Riddle then lists and discusses some of the properties which are involved in the performance of spark plugs: density, porosity, and refractoriness; mechanical strength; thermal expansion; electrical resistance and other electrical properties; thermal conductivity; resistance to heat shock; and resistance to chemical agencies.

THIS paper deals with the materials and properties of the ceramic insulators employed in spark plugs. By way of explanation, the word "ceramic" is derived from the Greek word "keramos," meaning "burned stuff," thus indicating its inorganic character. Commonly, the word is associated with the substance known as clay but, in modern usage, it includes a large number of minerals and rocks, and a correspondingly large number of mineral products, such as the many clay products, glass in its many forms, enamels, and the cements.

Chemically speaking, we deal in this field with minerals that have as important components the SiO_4 tetrahedron and the SiO_4 ring, as well as the hexagonal rings of the AlO_6 groups. We are thus concerned with the world of Si ions, and the characteristic properties of each material depend upon two things: the arrangement of the atoms within the component crystals, and the size and distribution of the crystals.[1] The ionic structure of the crystals with the relatively small ionic radius of Si, the moderately small radius of Al, and the large one of O, tends to build atomic complexes of many kinds, in the form of many minerals. Commonly speaking, the chief actors on this stage are the silicates, and especially the silicates of aluminum.

The first ceramic insulators were made of porcelain which consists essentially of three components – clay, quartz and feldspar. Thus, the well-known hard porcelain tableware of Europe is composed of 50% clay, 25% quartz, and 25% feldspar. By molding the desired articles from this mixture, drying and firing them to a sufficiently high temperature, we obtain the vitreous and translucent product known as porcelain.

Clays

Clays are essentially silicates of aluminum and correspond to the chemical formula $Al_2O_3.2SiO_2.2H_2O$. They differ enormously in their physical properties chiefly because of differences in mineral structure, particle size, and the impurities associated with them because of their geological history.

For our purposes we need to consider only the purer grades of clay such as are used in making the finer grades of pottery. These may be roughly divided into two classes – the kaolins and the plastic bond clays. The former are relatively larger grained, less plastic but, when fired, assume a more or less nearly white color. The latter clays have a very fine structure, with a large percentage of particles below 0.5 microns in size and, as a result, possess highly developed plasticity and bonding power. They fire to a somewhat cream color on account of impurities which are lacking in the kaolins. Both types of clay consist essentially of a mineral base which corresponds to the typical kaolin formula but may differ in crystal structure, so that we may have kaolinite, anauxite, nacrite, dickite, halloysite, and allophane. The first named is, however, by far the most common. All types show certain colloidal properties, depending upon their fineness of grain and the impurities present. Thus, all of them may be dispersed or deflocculated, through the addition of alkali, and coagulated by acids or salts.

[This paper was presented at the Annual Meeting of the Society, Detroit, Mich., Jan. 16, 1940.]

[1] See the *Journal of the American Ceramic Society*, Vol. 20, February, 1937, pp. 31-42: "Contribution of Mineralogy to Ceramic Technology," by W. J. McCaughey.

The outstanding properties of the clays are their plasticity, their shrinkage in drying, their condensation and shrinkage in firing, and their growth in hardness and mechanical strength as the firing temperature is increased. By determining the porosity of clay specimens fired at definite temperatures, and plotting the porosity values in per cent against the temperatures, curves are obtained which show the pyrophysical behavior of the material. In this manner, the tangent of the curve shows the rate at which the clay contracts under the heat treatment given it in that, with increased temperature, the porosity becomes lower and lower until it approaches zero. The more rapid the descent of the curve, the more impurities are present in the clay. The contraction of the clay volume is due to the effect of surface tension which tends to contract the material to a minimum volume. The temperature at which zero porosity is approached is known as the vitrification point and is characteristic of the clay in question. The higher this temperature, the purer is the material. Upon further heating, the porosity may again rise due to the evolution of gases, a stage known as overfiring which indicates

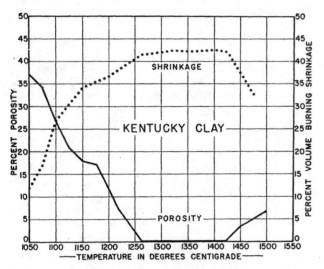

Fig. 1 — Change in porosity and shrinkage of a clay fired at different temperatures

deterioration or a breakdown of the clay. Such a curve is shown in the diagram of Fig. 1. Finally, the clay melts to a slag-like mass, at a temperature depending upon the amount of impurities it contains. For the purer clays it varies from 1704 C (3100 F) to 1760 C (3200 F).

The outstanding points during the heating cycle of clays are as follows: They dry completely at 104 C (220 F); at 500 C (932 F), they lose their chemically combined water in an endothermic reaction requiring 10.8 cal per gram-molecule; at this temperature they also undergo an increase in volume; at 800 C (1472 F), they pass through an exothermic reaction which may be due to molecular dissociation; above this temperature the action of any fluxes present begins to be effective as shown by the contraction of the mass and a decrease in the true density. At still higher temperatures the clays begin to dissociate according to the reaction: $3(Al_2O_3.2SiO_2) \rightarrow 3Al_2O_3.2SiO_2 + 4SiO_2$, where the new aluminum silicate formed is mullite, a compound which at first is amorphous or cryptocrystalline, but later develops into needle-like crystals that become larger as the temperature continues to rise.

The clays also carry smaller percentages of accessory minerals of which mica is probably the most persistent. Several types of this mineral may occur. In mica the structure consists of ions which are arranged in sheets, each layer forming a hexagonal network. As a component of ceramic mixtures the micas are not desirable.

Quartz

This common form of silica is an important constituent of most ceramic products. Simple as quartz in the form of sand, quartzite, or sandstone may appear to us, it offers an unsuspected, complicated story when it is subjected to heat. While stable at ordinary temperatures, its characteristic is to change its structure as soon as it is heated to higher temperatures. At 575 C (1067 F) it changes sharply from its original crystalline structure to a new one, accompanied by an increase in volume. We say that it has inverted from alpha to beta quartz. Above 870 C (1598 F) we have another inversion to the modification known as tridymite, but this inversion, however, is so sluggish that ordinarily a third form, cristobalite, first may be produced above 1470 C (2678 F). The cristobalite has two forms, alpha and beta cristobalite, and the tridymite has three — alpha, beta, and gamma tridymite. At each inversion a volume change takes place. It is thus seen readily that, unless every portion of a piece of quartz rock, be it large or small, reaches an inversion temperature at the same time, enormous stresses are set up. The tridymite has two low-temperature inversions at 117 C (243 F) and 163 C (325 F), respectively. The cristobalite has one which is variable and may be as high as 277 C (531 F), or as low as 198 C (388 F). Above 1710 C (3110 F) the silica melts to a glass (fused quartz) which may become transparent and is noted for two of its very important properties — its transparency to ultraviolet light, and its exceedingly low coefficient of thermal expansion.

The original quartz thus may exist in eight modifications. The volume changes involved in the inversions are greatest for the cristobalite, followed by the change from alpha to beta quartz, and least for the tridymite.

The various silica inversions are shown in the diagram of Fig. 2. For ceramic purposes the quartz is ground quite fine, to pass the 200-mesh sieve and blended with the other constituents of the body.

Feldspar

The minerals chiefly to be considered in this group are the potash feldspars, microcline and orthoclase, both of which have the typical formula: $K_2O.Al_2O_3.6SiO_2$. They are thus silicates of alumina and potash and may be considered to be alkali salts. They do not exist in the pure form but invariably carry smaller percentages of soda.

This crystalline component is a hard mineral with well defined cleavage. It is introduced in the finely ground state. When it is heated to a sufficiently high temperature, it gradually loses its crystalline character and becomes a glass. It is

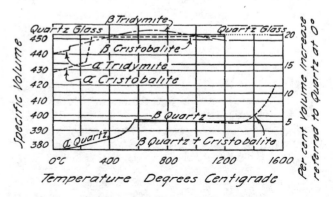

Fig. 2 — Quartz inversion — Specific volumes of the silica minerals and quartz glass (arranged by McDowell)

more or less transparent but shows an extraordinary reluctance to flow as a glass should. It is this characteristic behavior which makes it so valuable a flux in ceramics since it brings about the consolidation and vitrification of a porcelain without undue distortion and deformation of shape. Even at temperatures well above the so-called fusion point, a potash feldspar has been found to have a viscosity of 1,104,000 poises

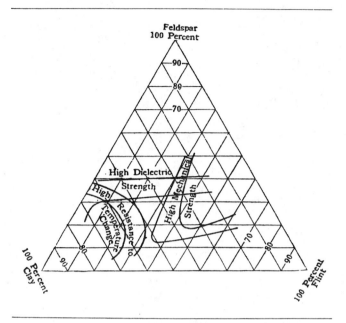

Fig. 3 – Tri-axial diagram showing the field of possible porcelain compositions (by Kleinfelter and Gilchrist[4])

at 1310 C (2390 F), 602,000 at 1366 C (2491 F) and 525,000 poises at 1394 C (2541 F). This explains the remarkably sluggish flow of this flux and which performs its function only because of the great surface tension force to which all ceramic bodies are subjected at higher temperatures. Yet, at temperatures sufficiently high, the mobility of the feldspars becomes great enough to assume more of the properties of glass and behave as the alkali salts it really is.

From conductivity measurements[2,3] it has been found that glass possesses the properties of an electrolyte, and that the alkalies, especially the soda, are most active in promoting the conductivity. Other bases, like the oxides of the alkaline earth group, bring about a distinct increase in the electrical resistance. In this respect feldspar is analogous to glass and is subject to similar laws at the temperatures with which we are concerned.

Porcelain

The ceramic bodies known as porcelain consist essentially of clay, chiefly kaolin, quartz, and feldspar. The clay is the plastic agent which permits the wet mixture to be molded and shaped. The quartz produces the rigid skeleton which lowers both the drying and firing shrinkage and imparts to the mass the necessary resistance to distortion at the high temperatures which are involved. The feldspar functions as the glassy cement which unites the whole to form a vitrified structure that approaches zero porosity and produces a certain degree of translucency. Porcelain is thus a vitreous, hard mass which is impermeable to liquids, even under pressure. But the internal structure varies according to the temperature and the time of exposure to the maximum temperature range. Rapid firing may cause the porcelain to consist of poorly developed mullite, $3Al_2O_3.2SiO_2$, some undissociated kaolin, with quartz grains in practically their original size and shape, with feldspar glass surrounding the particles. With slower firing, the kaolin will be dissociated more completely to mullite and the quartz grains more vigorously attacked and partially dissolved by the fusing feldspar. In the extreme case the mullite formation will be carried to its maximum, producing a well developed network of crystals with the quartz completely dissolved by the feldspar. Here then, we would have left only the two principal phases, mullite and glass. It is evident that the physical properties of the porcelain must vary according to the kind of heat-treatment it has received.

Spark-Plug Porcelain

It is from these feldspathic porcelains that the present spark-plug compositions have been derived. The field of possible porcelain compositions has been outlined by tri-axial diagrams, of which one is shown by the diagram of Fig. 3. But about 1918[4] it was realized that at least two basic defects were inherent in this type of porcelain. One defect was the peculiarity of quartz to invert to its different crystalline modifications, and the resulting volume changes cause stresses to be produced within the porcelain structure. The other was the electrolytic conductance of the feldspar at higher temperatures which reduced the insulating efficiency of the insulator.

The manufacturers of these insulators promptly began a vigorous research which soon resulted in the replacement of quartz by inert components like mullite (previously known as sillimanite) through the introduction of minerals of the sillimanite group, namely, andalusite, kyanite, and natural sillimanite. All have the same formula $Al_2O_3.SiO_2$. Other minerals included a mineral richer in alumina, dumortierite ($8Al_2O_3.6SiO_3.B_2O_3.H_2O$) and also zircon (zirconium silicate). The advantages of aluminum oxide were known at the time, but lack of a suitable source prevented its use. In the reference just cited it states that extensive work with aluminum oxide was prohibited by lack of material of suitable quality. Likewise, the feldspar was replaced by other fluxes, like the oxides of magnesium and calcium. Improvements were carried still further by the insistence upon fine grinding

---
[2] See the *Journal of the American Ceramic Society*, Vol. 7, February, 1924, pp. 86-104: "The Electrical Conductivity of Sodium Chloride in Molten Glass," by W. J. Sutton and A. Silverman.
[3] See *Zeitschrift für Technische Physik*, Vol. 6, 1925, pp. 544-554, by G. Gehlhoff and M. Thomas.
[4] See the *Journal of the American Ceramic Society*, Vol. 2, July, 1919, pp. 564-575: "Special Spark Plug Porcelains," by A. V. Bleininger and F. H. Riddle.

Fig. 4 – Photomicrograph of a section of a typical kaolin-quartz-feldspar body – 150X

and the employment of high temperatures which made it possible to reduce the volume of the glassy matrix in the insulator to a minimum. Strict inspection methods and control were initiated.

The results of these changes were striking since the mechanical strength and the dielectric resistance of the insulators were improved greatly, so that two investigators, Kraner and Snyder[5] (Westinghouse), could say: "American spark-plug porcelains of the mullite type are vastly superior to all other ceramic materials, as to mechanical strength, thermal-shock resistance and glaze-fit."

But the manufacturers were not satisfied with the progress made and continued their researches in the direction of further replacements of the quartz by oxides like those of zirconium, aluminum, and titanium. Developments are under way also in the direction of producing insulators consisting of single compounds or oxides, in the virtual absence of glass. Satisfactory insulator bodies have been formed from single oxides or from a single component which is the satisfied crystalline combination of two or more oxides.

In this connection the study of the microstructure of the

Fig. 5 – Photomicrograph of a section of an American alumina body – 450X

insulators by means of the petrographic microscope has become increasingly important and evidence by means of X-ray analysis is often necessary for the confirmation of observations. Photomicrographs of two sections, one of a typical kaolin-quartz-feldspar body, and one of an American alumina composition are shown in Figs. 4 and 5.

Glaze

Spark plugs are covered with a glaze, which is virtually a glass, and this thin coating, often only 0.1 mm in thickness, has a surprising effect upon the physical properties of the insulator. The two determining factors which govern this relation are the thermal expansion and the modulus of elasticity of body and glaze. The coefficient of expansion of the glaze should be lower than that of the body so that the glaze is in a state of compression while the body is in tension. Should, by chance, this condition be reversed, it would affect the status of both glaze and body. The adhesion of the glaze would be reduced and the insulator as a whole would show a distinct loss in mechanical strength.

Between the glaze and the body there usually forms an

Fig. 6 – Microsection showing the junction of body and glaze with mullite crystals growing into the glaze

intermediate layer resulting from the solution attack of the glaze upon the body which frequently gives rise to a crystalline growth into the glaze as is shown by the photomicrograph of Fig. 6. This layer tends to moderate somewhat the physical differences between the two interfaces.

Processing

The accuracy of the various steps in the manufacturing process of the spark-plug insulators is today under very careful control. The milling operations are subject to constant testing, and enormous strides have been made in the preparation of the body material and in the shaping of the spark-plug insulators. Some methods of forming insulators may be said to be revolutionary in character, especially since at times it is necessary to form bodies entirely lacking in plasticity.

Also, the firing process is controlled not only with respect to the accurate measurement of temperature and its distribution but also as regards the kiln gases, the composition of which has a profound influence upon the character of the product.

Much progress also has been made in connection with the type of the electrodes, the assembly of the insulators, and the fitting of the metal parts.

Properties of Insulators

Whether or not a spark-plug insulator serves the purpose for which it is intended depends after all only upon the physical properties of the final product. If they satisfy the requirements that are exacted, all is well, but failure to meet even one of the specifications means failure of the insulator.

Some of the properties which are involved in the performance of spark plugs are:

Density, porosity, and refractoriness.
Mechanical strength.
Thermal expansion.
Electrical resistance and other electrical properties.
Thermal conductivity.
Resistance to heat shock.
Resistance to chemical agencies.

Density and Refractoriness

Density, per se, is not of much significance, except as it affects other properties, and obviously must vary with the specific gravity of the body constituents. The density of mullite-type insulators fluctuates around 2.5 to 2.9; that of alumina bodies, 3.3 to 3.9; magnesium bodies, 3.0 to 3.6. The porosity must approach zero so that the permeability is vir-

---

[5] See the *Journal of the American Ceramic Society*, Vol. 14, September, 1931, pp. 617-623: "Mechanical and Thermal Shock Tests on Ceramic Insulating Materials," by H. M. Kraner and R. A. Snyder.

tually nil, even to liquids of low surface tension, and under pressure. The spark-plug insulators, in a sense, are refractories and must be able to withstand high temperatures. In this respect no difficulties have been experienced, and practically no insulators have fusion points below 1649 C (3000 F).

Mechanical Strength

Spark plugs are subjected to severe mechanical stresses, both internal and external, which must be resisted. It is for this reason that insulator bodies are tested by all possible means, for compressive, tensile, transverse strength, and resistance to impact.

Some characteristic values for the principal mechanical qualities are given in Table 1.

Table 1 – Mechanical Qualities of Insulator Bodies

Tensile strength	10,000- 30,000 lb per sq in. (Area tested, 0.2 sq in.)
Compressive strength	60,000-200,000 lb per sq in. (Area tested, 0.114 sq in.)
Modulus of rupture	15,000- 25,000 lb per sq in. (Rods ½ in. dia. x 2 in. on 2½ in. span)

Tests determining the resistance to impact usually are made part of the daily control program.

While such strengths as are attained may not be required in actual use, the tests afford an excellent means of checking the structure of the porcelain. These types of bodies are several times as strong as porcelain tableware in ordinary use.

Thermal Expansion

The thermal expansion of the insulators is an important physical constant on account of the frequent and wide temperature changes which are involved and the relation between the metal parts and the ceramic. At the same time, the coefficient of thermal expansion curve affords an excellent indication of the type of insulator material and the presence or absence of any tendency toward any molecular transformation or phase changes through the prevailing temperature range.

If the component parts of a spark plug had the same thermal expansion throughout their entire operating temperature range and all parts were heated uniformly, the conditions obviously would be ideal. As these conditions are not possible, the engineers must make compromises based on the use of the best available materials and their own experience over a period of years. In general, ceramic insulators have lower expansions than have the metals with which they are assembled, and also have lower thermal conductivities so that clearances are necessarily required where the ceramic surrounds metal parts. Some insulators have been developed that have quite high expansions. See Table 2.

In this table A is an ordinary electrical porcelain; B is the mullite type of spark-plug insulator; C is an alumina body; D is of the spinel type; E is a substantially pure zirconium-oxide body; F is substantially pure zirconium silicate; G is magnesium oxide.

It is evident from these values that there is a wide variation among the various compositions, and this condition makes possible a classification of the different types of insulators. This classifying has been done in the diagram of Fig. 7 where the several fields are shown graphically.

Electrical Resistance

In the case of ordinary electrical porcelains we need only be concerned with their resistance at atmospheric tempera-

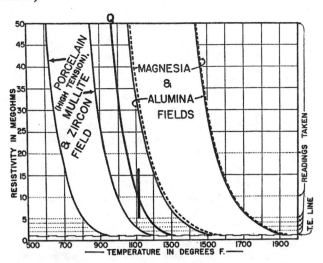

Fig. 8 – Resistivity fields above 1 megohm per cm³ (Te value) of different types of ceramic insulator bodies – The line marked Q represents fused quartz

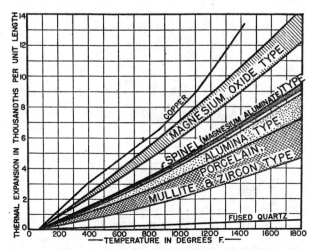

Fig. 7 – Thermal expansion fields of different types of ceramic insulator bodies

Table 2 – Coefficients of Linear Thermal Expansion $\times 10^{-6}$ per deg C

Temperature, C	Temperature, F	Various Insulator Compositions						
		A	B	C	D	E	F	G
25-200	77-392	5.87	3.45	6.26	7.88	6.60	3.68	12.17
25-400	77-752	6.38	3.81	7.13	8.43	6.84	4.01	12.90
25-500	77-932	6.58	3.94	7.38	8.78	6.95	4.10	13.16
25-600	77-1112	7.02	4.09	7.53	9.12	7.12	4.30	13.68
25-800	77-1472	6.44	4.60	7.73	9.68	7.47	4.56	14.20
25-1000	77-1832	6.52	5.06	7.91	10.04	7.98	4.59	14.56
25-200	77-392	5.87	3.45	6.26	7.88	6.60	3.68	12.17
200-400	392-752	6.82	4.13	7.88	8.93	7.06	4.30	13.53
400-500	752-932	7.33	4.44	8.34	10.04	7.33	4.52	14.17
500-600	932-1112	9.12	4.75	8.31	10.75	7.89	5.17	16.19
600-800	1112-1472	4.78	6.08	8.32	11.31	8.51	5.30	15.68
800-1000	1472-1832	6.82	6.84	8.55	11.46	9.98	4.70	15.91

For comparison, the approximate coefficients for the range 20 – 100 C for some metals follow: iron – 11.7; copper – 16.6; silver – 19.0; brass – 19.5; steel – 13.0; stainless steel – 17.3.

tures but, in spark plugs, higher temperatures are involved. Due to the ionic activity at higher heats we must expect a certain decrease in resistance. Thus, while the resistivity in ohms per cm of an insulator, at 100 C (212 F) is 3×10^{13}, it becomes 9×10^{11}, at 150 C (302 F), and 6×10^{8} at 300 C (572 F).

Again, when we measure the temperature and resistance of 1 cc of insulating body which is being slowly heated and we note the temperature at which the resistance remains at one megohm, we obtain a value of considerable interest. This is known as the Te value and shows to what extent the electrolytic effect has progressed.

Electrical measurements are compiled in Table 3.

A curve giving the relation between the temperature and the electrical resistivity of various types of bodies is shown in Fig. 8.

Thermal Conductivity

The thermal conductivity of spark-plug insulators is an important quality and concerns operating conditions since it determines very largely what the plug temperature will be.

Table 3 – The Effect of Temperatures Upon the Electrical Resistivity of Various Types of Spark-Plug Insulator Compositions

Resistivity, Megohms	High-Tension Porcelain, C	Mullite and Zircon Types, Range in C	Magnesia Types, Range in C	Alumina Types, Range in C	Fused Quartz, C
50	321	441-471	576- 782	571- 782	524
20	360	482-513	637- 844	632- 846	569
10	396	516-541	688- 894	682- 896	599
5	424	546-577	738- 944	727- 944	630
4	435	560-591	755- 963	743- 963	641
3	449	574-604	777- 985	763- 985	654
2	466	591-621	805-1010	785-1010	677
Te-1	499	632-663	860-1060	838-1066	721

| | | Converted to Fahrenheit | | | |
Meghoms	F	Range in F	Range in F	Range in F	F
50	610	825- 880	1070-1440	1060-1440	975
20	680	900- 955	1180-1550	1170-1555	1055
10	745	960-1005	1270-1640	1260-1645	1110
5	795	1015-1070	1360-1730	1340-1730	1165
4	815	1040-1095	1390-1765	1370-1765	1185
3	840	1065-1120	1430-1805	1405-1805	1210
2	870	1095-1150	1480-1850	1445-1850	1250
Te-1	930	1170-1225	1580-1940	1540-1950	1330

These determinations were made in the Champion Laboratories on centimeter cubes using a 240-v megohmer for measuring resistance and a platinum thermocouple with a potentiometer for determining temperature.

The Bureau of Standards reports (1918) tests using 60 cycles at 500 v giving Te values as follows: fused quartz, 890 C (1634 F); high-tension porcelain, 490 C (1914 F); mica (phlogopite), 720 C (1328 F). The difference between the Te value for quartz obtained by the Bureau of Standards and the Champion Laboratories is unexplained. The values for quartz which are shown in the table, however, were obtained recently (and duplicated on a second sample) on a sample suitable for spark plugs and, since the method was the same, these values can be compared safely with the others shown in the table.

	Mullite and Zircon Types	Magnesia Type	Magnesium Aluminate Type	Alumina Type
Dielectric Constant	6.2 -6.8	10.0-11.0	7.5	8.4
Per Cent Power Factor	0.40-0.47	4.0-12.0	0.10	0.10-0.18
Loss Factor	2.73-2.90	41.0-136.0	0.75	0.85-1.5

These tests were made at 2000 cycles using a Leeds and Northrup Capacitance and Inductance Bridge in the Champion Laboratories.

An insufficient number of compositions of any one type has been tested to permit giving the complete range of values. Those shown in the table, however, are typical of their class.

Table 4 – Range of Thermal Conductivities for Various Types of Spark-Plug Insulator Compositions

Temperature, C	Mullite and Zircon Types, range	Magnesium Aluminate Types, range	Alumina Types, range	Magnesia Types, range
38	0.0037-.0045[a]	0.0069-.0073	0.0047-.0076	0.0083-.0090
204	0.0039-.0048	0.0070-.0074	0.0065-.0091	0.0094-.0105
427	0.0041-.0052	0.0071-.0076	0.0083-.0101	0.0107-.0119
649	0.0043-.0054	0.0073-.0077	0.0092-.0107	0.0114-.0125
871	0.0046-.0055	0.0074-.0079	0.0093-.0105	0.0114-.0123
Temperature, F				
100	10.8-13.0[b]	19.9-21.2	13.8-22.2	24.2-26.2
400	11.3-14.0	20.2-21.6	19.0-26.5	27.4-30.4
800	12.0-15.0	20.6-22.1	24.0-29.5	31.0-34.7
1200	12.6-15.7	21.2-22.5	26.7-31.0	33.0-36.4
1600	13.5-15.9	21.5-23.0	27.0-30.6	33.0-35.8

[a] CGS Unit – Gram-calories per square centimeter per second per degree centigrade per centimeter thickness.

[b] English Unit – Btu per square foot per hour per degree fahrenheit per inch thickness.

These determinations were made by J. L. Finck Laboratories on special discs 1 in. thick x 4 in. diameter made for this purpose.

| | Thermal Conductivity | |
	CGS Units at 38C	English Units 100F
*Fused Quartz	0.0036	10.4
*High Tension Porcelain	0.0003	7.3
*Mica (Phlogopite)	0.0001	3.5
**Aluminum	0.5	1450.0
**Copper	0.9	2610.0
**Steel	0.1	290.0

* From "Electrical Engineer's Handbook," Volume V, by Pendar and McIlwain.

** From "Handbook of Physics and Chemistry," by Hodgman and Lange.

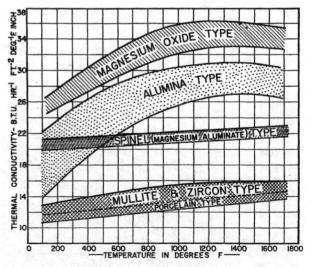

Fig. 9 – Thermal conductivity fields of different types of ceramic insulator bodies

At the same time it governs to a considerable extent the design of the insulator which must make allowance for the heat conductivity. In general, the higher the thermal conductivity of the insulator, the more easily it will carry heat away and remain cool. This permits the use of insulators with longer firing ends and, therefore, longer electrical leakage paths.

In Table 4, the thermal conductivity expresses the heat in Btu which will flow in one hour through an area of 1 sq ft, a thickness of 1 in. and with a temperature gradient of 5/9 C (1 F). Here again we find it possible to map the different fields of insulator types with reference to their thermal conductivities, as is shown in Fig. 9.

Resistance to Thermal Shock

It goes without saying that spark plugs must show good resistance to heat shock. Numerous control tests are made daily in the laboratories of the manufacturers by alternatingly heating and cooling the firing-end tips of the insulators.

Resistance to Chemical Agencies

Considerable attention must be paid also to possible chemical reactions taking place in the engine cylinder which involve carbon or its compounds; vapors of various sorts, including water vapor; and chemicals introduced, including those used for antiknock purposes. These conditions must be considered with respect to the various temperatures of operation. The most widely used antiknock agent today contains lead tetraethyl. This burns to lead oxide and other salts which are active basic fluxes that react with most silicates at engine temperatures. Insulators made of pure silica or bodies containing high percentages of silica are attacked much more readily by lead fluxes than bodies low in silica compounds, neutral or basic in composition. The unfortunate effect of lead is emphasized further by the fact that an objectionable coating or deposit forms more readily on the silicious bodies. It is difficult and sometimes impossible to remove such deposits from the silicious bodies as a glass has been formed on the surface. This glass has a different coefficient of expansion from the body proper, and cleaning results in chipping the surface of the insulators. With the less silicious and the basic bodies the deposit is of a dusty nature, forms more slowly, and can be removed more easily.

In this paper the attempt has been made to trace the origin of the American spark-plug insulators and to follow their development. Much could be said on the subject of the engine and practical working tests which have followed the results of the laboratory findings. But, finally, these insulators must be judged by their actual performance in spark plugs in the field. Of real value must be considered the performances under specially severe conditions, as in aviation, racing boats, certain trucks, and other punishing services. The ceramic should be considered as a raw material to be used by the spark-plug engineer the same as he would use any other raw material. It has been shown that the characteristics of the ceramic vary widely and the spark-plug designer must necessarily know and use insulators best suited for his requirements.

Acknowledgment

The writer wishes to thank the officials of the Champion Spark Plug Co. for permission to publish this work. He also expresses his appreciation to his associates who have compiled the critical data and particularly to Dr. A. V. Bleininger for his cooperation. All critical data determinations were made in the Champion Laboratories unless references are given.

Shortcomings of Mica Insulation for Aviation Spark Plugs

(Concluded from page 235)

of "fouling" is aggravated by the use of mica as an insulator material. As has been stated, mica is a silicate, and all silicates are very reactive with inorganic lead compounds. The products of combustion, when lead as an antiknock ingredient is present in the fuel, deposit lead bromide and lead sulphate on the spark-plug surfaces. The sulphate does not, as a rule, react further, but the bromide oxidizes to litharge (PbO). Many silicates react vigorously with litharge, forming low-melting lead-silicates and these will often drip off a badly attacked insulator nose.

Cleaning a mica spark plug to remove these deposits, an operation which is necessary to restore the thermal characteristics of the plug, gives at the same time a fresh and particularly susceptible surface for attack by the compounds already mentioned. Removal of deposit accumulations from the comparatively soft surface of the mica washers, with the laminations exposed edgewise, is a difficult and hazardous operation.

The extensive and excellent mechanical engineering work done on the ground and center electrodes has resulted in fair service with respect to electrode erosion, and instances are on record where mica spark plugs have been operated for approximately 350 hr in a modern engine and still had gap clearance that would allow fairly uniform firing of the charge. If electrode erosion becomes excessive, missing is likely to occur (mostly at take-off conditions), and, in addition, the voltage may rise to such an extent as to cause damage to coils and high-tension wires. Any substitute for mica that may be proposed must contemplate the use of electrode designs that give similar service. Thus it would be well to take advantage of the tremendous store of experience available. To repeat in essence what Arthur Nutt[3] so ably said: people who propose new insulating materials know too little about spark plugs and their requirements, and the spark plug people too little about insulators. Here is an excellent opportunity for two specialists to get together.

Finds "Spark-Plug Complex"

When appraising the seriousness of the present aviation spark plug situation, a word of caution may be in order. The field mechanics responsible for the maintenance of the airline schedules, of which we are justly proud, have undoubtedly a spark-plug complex. As mentioned before, they immediately blame spark plugs for faulty engine operation, even though other units may be, and often are, the cause of the trouble. As a result, the mechanical wear and tear that the spark plug gets from continual removal from the cylinder, replacement, and incessant taking apart, is likely to damage the plug to a degree equal to an extended service period. It is not unusual to see a spark plug that has been overhauled to death. Until mica is replaced by a more inert and stable material, the author does not believe that spark plugs will become available which will operate consistently for the overhaul period of the engine (350-600 hr) without being removed from the cylinders. With the adoption of new and more suitable insulating materials it would appear wise to use, as far as possible, the wide background of engineering and operating experience gained from the mica plug. However, future engines and future fuels may require radical developments in spark plugs and ignition systems, and these developments should not be allowed to be retarded by a slavish regard for the findings of the past.

[3] See SAE Transactions, Vol. 34, December, 1939, pp. 501-512: "Aircraft Engines and Their Lubrication," by Arthur Nutt.

Thermal Conductivities Quoted Incorrectly in Spark-Plug Paper

In the latter part of Table 4 of the paper: "Ceramic Insulators for Spark Plugs," published on p. 241 of the June, 1940, SAE Transactions, the thermal conductivities in cgs units at 38 C of high-tension porcelain and mica (phlogopite) were given erroneously as 0.0003 and 0.0001, respectively. The correct values, as quoted in the "Electrical Engineer's Handbook," Vol. 5, by Pendlar and McIlwain, are 0.0025 for high-tension porcelain and 0.0012 for mica.

REPRINT.--Paper was presented at a meeting
of the Northwest Section of the Society of
Automotive Engineers at Tacoma, on Jan. 4, 1946.
Subject to revision. All papers presented
at meeting of the Society are the exclusive
property of the Society, from which permission
to publish this paper, in full or in part, with
credit to the author and the Society, may be
obtained upon request. The Society is not responsible for statements or opinions advanced
in papers or discussions at its meetings.

460016

SPARK PLUGS

by
Robert M. Ward
Champion Spark Plug Company

Consider the rhythmic, pulsing energy of the spark-ignition internal combustion engine. The resonant whirr of passenger cars; the powerful drone of heavy-laden trucks and busses; the lumbering energy of tractors and industrial equipment; the muffled beat of marine engines; the powerful roar of aircraft - symbols of the vitality that is America......And the pulse of that vitality, hinging on the life-giving spark across the gaps of the spark plugs in those engines.

As we reflect on this, and on the many mechanized units performing their assigned tasks for our well being, it is well for us to pause occasionally and think - how are the spark plugs performing? And what about that important engine part? What are they? What do they do? What do they require?

Source of great power development, these engines, after all, are really inanimate things until the vitalizing sprak, biting into the compressed mixture with flashing energy, starts and sustains the process which transforms inanimation into powerful action.

Sometimes neglected; sometimes banished in an effort to utilize an alternative means to the same end; and sometimes misunderstood; the spark plug, we must admit, then finds itself ignored or in the midst of very hectic experience. We eventually find they cannot be ignored, and in the rush of concern on their importance and condition we sometimes overlook what might be causing their improper performance.

This talk, tonight, cannot possibly cover all the technical reasons and proof of those things that are overlooked in seeking the cause of improper spark plug performance; nor is it intended to excuse spark plug design and progress; the latter of which is considerably further ahead than is sometimes realized. But it is hoped by the focusing of our mutual attention on the fundamentals of spark plugs and engine operation, that helpful and constructive thinking, will encourage a better understanding of spark plugs and the establishment of their service on the best possible plane.

THE ENGINE

To understand spark plugs better, let's slow up the engine action for a moment and reflect on each of the four strokes in the cycle of an engine:

1. The downward stroke of the piston draws into the cylinder a charge of highly inflammable gas, produced by the proper uniting of gasoline and air in the carburetor and intake manifold.

2. The upward stroke of the piston compresses this gas and at a point just before the piston reaches the top of the stroke, the compressed gas is ignited by a timed spark at the points of the spark plug.

3. The ignited gas burns very rapidly and the more complete the combustion the better the engine operates, the almost instantaneous expansion of the rapidly burning gas driving the piston down and thus developing power.

4. On the next upward stroke of the piston, the exhaust valve opens and the burned gas is expelled from the cylinder preparatory to the drawing in of another fresh charge, as explained in the intake stroke.

Occurring hundreds or thousands of times per min., it is obvious that the rapidity of the instantaneous ignition and the completion of the combustion help to determine the efficiency and economy of engine performance - the aim being to convert the highest possible percentage of the fuel used into power.

Along with spark plug, however, there are other engine factors which must be in proper condition to assure the best performance. For example:

Intake Stroke

Condition of the cylinder; piston; rings; and valves.
Condition of the carburetor and manifolding.
Volumn of the cylinder charge.

Compression Stroke

Condition of the cylinder; piston; rings; ahd valves.
The factor of compression ratio.

Firing Stroke

>Condition of the cylinder; piston; rings; and valves.
>Condition of the fuel; air-fuel ratio; and octane values.
>Condition of the combustion chamber; and deposits therein.
>Positioning of the spark plugs.
>Condition of the ignition system.
>Condition of the spark plug.
>Spark timing.

Exhaust Stroke

>Condition of the cylinder; piston; rings; and valves.
>Condition of the manifolding.
>Condition of the muffler.
>Degree of back pressure and heat remaining in the cylinder, retarding cooling.
>Degree of exhaust gases remaining to dilute the oncoming fresh fuel charge.

The malfunctioning of the engine, in any of these phases, can affect the spark plug in occasioning any premature plug fouling or burning tendencies noted.

It will be noted in observing the processes of the engine and spark plug that the plug cooling period is during the intake stroke; both from the lowering of cylinder temperature and the cooling effect of the incoming fuel charge.

The internal combustion engine is essentially a heat engine and as such they have a range of best operating temperatures. Considering the fact that this one factor has so much to do with good engine performance, it would appear that more attention would be given to engine operating temperatures. The fact that there is a glaring lack of this attention to engine service by motorists, and some operators, is evidenced by the complaints on fuels; depreciation of the mechanical condition of engines; and the use of hotter operating spark plugs; during the war, with its slow speed, start and stop, short trip, engine operation.

Certain engine parts are designed with the use of metals best suited to the function and stresses with which they are involved during engine operation. Because of varying expansion co-efficients of these metals, clearances are planned between these parts to provide for the best positioning and operation of the moving parts compatible with their function in the engine. Provision is made, of course, for lubrication of such moving parts which includes the pistons and valves.

These clearances are important in their effect on spark plugs, because too little thought is given to the need for any excessive clearances to be closed to proper limits, in engine warm-up.

Pending full warm-up, excessive oil can be permitted to pass into the combustion chamber, resulting in an abundance of elements in the mixture to be burned and causing residue products to coat the combustion chamber surfaces, including the depositing of fouling matter on the insulator of the spark plug which had not reached proper operating temperature before subsequent engine use.

These conditions also occur, sometimes quite severely, during long idling period.

When the engine is in sustained operation, the continuous rapidity of combustion at elevated temperatures, are applied with blow-torch effect on the spark plugs.

Therefore, during engine operation, it will be noted that great stresses and a wide range of changes in the combustion chamber are imposed on the spark plug. Rapid change in temperatures; presence of oil; hammering shocks of compression and combustion; terrific heat and pressures; severe chemical and electrical attack; all of which require the spark plug to be designed to withstand such stresses and retain operating dependability.

This operation requires the spark plug to perform not enough to avoid fouling during warm-up or cool engine periods, or in the presence of unfavorable mixture conditions; and perform cool enough to prevent pre-ignition of burning during sustained or full power operation; while always delivering a properly timed, intense spark, dependably and well in every cycle of the engine, hundreds or thousands of times per minute for thousands of miles or extended periods of time.

One of the interesting things about a spark plug, often lost sight of, is that a portion of the plug operates inside the inferno of the combustion chamber, and another portion operates outside the engine - a tough job for any device in being subjected to such extremities.

Sometimes, the positioning of the spark plug seems to be one of the last things considered in designing the engine; and its location is not always of the best for advantages in installation; removal; or operation. In such cases, abuse sometimes results and the spark plug is the innocent victim of explosive and uncomplimentary expletives.

Yet, considering the entire picture, the spark plug does its job amazingly well and dependable, we believe you will agree, asking only reasonable attention to enable it to maintain its essential service efficiently.

THE SPARK PLUG - DEFINITION AND IMPORTANT ELEMENTS

What is a spark plug? And what is its function?

A spark plug is a device made of two wires, carefully insulated from each other. A gap is spaced between these two wires, or electrodes, and the device is assembled in a metal shell by which it is fitted to the engine. Its function is to bring the high tension current into the cylinder and across the gap in a spark and thereby ignite the compressed gases and thus provide power to the engine.

Its important elements are:

1. The insulator; which must retain its efficiency under all temperatures and pressures in the firing chamber or cylinder of the engine. It must prevent the current from flowing in any other direction than across the gap of the electrodes, and should run hot enough to avoid fouling and cool enough to avoid burning within the assigned heat range position of the spark plug to which it is fitted.

2. The electrodes; the center electrode which conveys the current through the insulator, and the side electrode, which is brought up adjacent to the inner end of the center electrode to form the

gap over which the high tension current passes in the form of a spark.

3. The metal parts; such as the shell and gaskets which introduce the insulator and electrodes into the cylinder without loss of compression; and the terminal, by which the spark plug cable is fitted to the spark plug.

4. The gap; which should be properly formed to assure maximum intensity and distribution of the spark, and with the proper setting to provide smoothest idle and maximum performance values throughout the engine power range.

5. The gas-tight construction of all component parts, to prevent troublesome internal hot-gas leakage between the insulator and center electrode; and between the insulator and spark plug shell; thus assuring that the spark plug will retain its assigned heat range position.

CAUSES OF UNSATISFACTORY SPARK PLUG PERFORMANCE

On a study of the aforementioned elements, it will be noted what may happen to the spark plug during engine service life; or during spark plug servicing; which may cause unsatisfactory spark plug performance and to understand why and how aprk pulgs need periodic inspection and replacement when worn-out. For example:

Insulator

Fouled from wet gas; oil; or products of combustion.
Fouled from dirt; paint; grime; or other accumulation on the top half of insulator.
Burning because of wrong heat range; lean mixture; improperly installed; hot gas-leakage within the plug itself; poor quality of materials and design.
Pre-ignition because of wrong heat range; improper installation; incandescence of encrusted deposits; hot gas-leakage; or unable to withstand the temperatures encountered in maintaining a balanced operating temperature.
Broken by rough handling; dropping; slipping wrenches; sttempting to set gaps by bending center electrode; or sudden high temperature change causing firing end fracture.
Altered firing end design caused by too many or too excessive sand blast cleanings; resulting in chronic burning or fouling because the heat range shaping of the insulator has been critically altered from its original design.

Worn-out in prolonged service; imbedded conductive deposits; and burned and blistered surfaces.

Electrodes

Oxidized and corroded causing increased resistance to the passage of current.
Honeycombed by chemical and electrical attack.
Burned excessively because of wrong heat range; lean mixtures; improper plug installation; or hot gas-leakage within the plug itself.

Damaged by improper gap setting.
Worn-out by constant application of heat, electrical and chemical attack in prolonged service; whereby the electrodes are worn to excessive thinness and causing chronic gap opening and added gap resistance to the forming of the spark.

Metal Parts

Damaged and distorted shells by use of wrong, slipping or cocked wrenches.
Damaged threads by being dropped; cross-threaded; or inefficient by being filled with carbon, combustion deposits, or bushing metal; burned from exposure to cylinder gases and temperatures; or sheared-off from bushing seizure.
Torn, cocked, burned, or excessively flattened gaskets; resulting in blow-by and poor heat transfer from insulator to shell, or from shell to cylinder.
Terminals not properly fitted or damaged.

Gaps

Incorrectly spaced and shaped; resulting in poor idling, acceleration, or full power performance.

Gas-Tightness

Expansion and contraction of metal parts, in relation to the insulator, permitting leakage of hot gases between the center electrode and insulator; and between the insulator and spark plug shell: resulting in pre-ignition, service troubles, early electrode destruction, premature plug burning, or a complete failure of the entire plug.

SPARK PLUG DESIGN AND HEAT RANGE

Engines are designed to give their performance values with certain factors at the disposal of the engineers and designers to utilize in accomplishing their desired results. Among these are:

        Liquid Cooled or Air Cooled Design
        Dimensions of Bore & Stroke
        Fuels Used
        Compression Ratio
        Valve Design & Timing
        Cylinder Head Construction
        Combustion Chamber Design
        Spark Plug Positioning & Size
        Carburetion
        Cylinder Load
        Air-Fuel Ratio
        Spark Plug Gap
        Spark Timing
        Engine Speeds
        Ignition System Design

These bring about operating characteriestics which vary according to the respective uses of these factors by the different engine designers and manufacturers; and these operating characteristics are reflected in the cylinder and combustion chambers by varying temperatures, pressures, and mixture conditions, for which it is necessary to design spark plugs to operate within specific temperature ranges.

Where high temperatures are encountered it is necessary to provide a cool operating spark plug to avoid plug burning or pre-ignition dangers; while in the opposite extreme a hot operating spark plug type must be provided to prevent fouling dangers.

Between the two operating extremes, spark plugs must be provided with operating characteristics suitable to the respective engine requirements as to the combustion chamber temperature and mixture conditions; and which require spark plugs in modifications of a very cool or very hot operating type.

The length and shape of the spark plug insulator at the firing end; the type of insulating material used; spark plug shell design; and gas-tight construction; are factors controlling the heat range of spark plugs.

Generally speaking, spark plugs can be identified as to their operating characteristics, as to their being a colder or hotter operating type by the following:

Cold Plug

A cold operating spark plug is one where the insulator projection at the firing end is relatively short in exposure to combustion chamber; thus providing a short heat path from the insulator tip to the insulator gasket seat in the spark plug shell. This affords a fast transfer of heat from the insulator firing end to the plug cooling medium - either the circulating water, or other liquid, around the spark plug boss; or to the air - and thereby assists the spark plug to operate at a temperature low enough to avoid burning or pre-ignition, while still running hot enough to prevent fouling in the proper balance to suit the engine requirements for which it is recommended.

Hot Plug

A hot operating spark plug is one where the insulator projection at the firing end is relatively long, thus providing a long heat path from the insulator tip to the insulator gasket seat in the spark plug shell This affords a delay to the passage of heat, so as to retain a temperature balance within the insulator to burn off the fouling matter which may have a tendency to form from low engine operating temperature or from the prevalence of wet gas and oil in the combustion chamber; while at the same time operating cool enough to avoid burning or pre-ignition dangers in suiting the engine service for which it is recommended.

Please note that the aforementioned descriptions are general. We should like to digress here, for a moment, and point out that these descriptions will not be a true indicator in comparing the plug types of different manufacturers; or even in comparing the plug types of the manufacturer the speaker represents. Like the factors available to engine designers and manufacturers, in designing an engine; there are also factors, as mentioned earlier for example, available to spark plug designers and manufacturers to utilize in accomplishing their desired performance results. The descriptions mentioned in generally identifying a plug heat range can be helpful, but the safest practice in coming to any conclusions is to refer to the spark plug manufacturer's charts, or consult with the manufacturer or its representatives.

SPARK PLUG SIZES

In general, as cylinders were made smaller in bores for the high compression, high speed automotive engines, the evolution of spark plugs sizes in smaller diameters followed in like trend. Sizes are available as small as 6 mm at the present time, if they should be necessary. However, as spark plug diameters are reduced in thread size, it naturally follows that insulators, center electrodes, and shell bores, must likewise be restricted in diameters. There then becomes present, certain complexities and limitations in the smaller sizes that must be considered in engine design and operating characteristics. Results secured by manufacturers using the 14 mm or 10 mm sizes, can be equally well secured by another engine manufacturer who may elect to use the 18 mm size. Examples of this are in the almost universal use of the 18 mm spark plug size for aircraft and racing engines. However, the selection of sizes or types in spark plugs should be considered in the light that such are factors as additional engineering elements that may be utilized by the engineer in accomplishing the desired objective -- much the same as a certain selection of dimensions as to cylinder bore and piston stroke may be utilized.

Advantages of the smaller sizes are:

1. Easier location in the engine in the presence of and when any problems of castings or other restrictions on spark plug location; for installation, removal, or operation, make them preferable.

2. Providing for the cooling medium to be closer to the insulator firing end, because of the shortened distance.

3. A slightly faster warm-up characteristic because of the smaller mass exposed to the combustion; and a slightly faster levelling off temperature in operation.

Advantages of the larger sizes are:

1. The use of larger and heavier center electrodes.

2. A greater bore area in which fouling matter can accumulate before fouling would occur.

3. Additional ruggedness of construction and more latitude in spark plug design.

4. A satisfactory warm-up characteristic and levelling off temperature in operation.

SPARK PLUG THREAD LENGTHS

The threads on the spark plug have the following functions:

1. They provide the means to install and remove the plug readily for servicing the engine.

2. They provide the means whereby the plug is held securely to the engine to provide the ground in the electrical circuit; and in assuring transfer of heat within the plug to the plug cooling medium.

3. The thread length, or reach, provides the means whereby the gap of the plug can be positioned for best ignition of the compressed gases.

Some cylinder heads are designed and constructed of certain metals to provide advantages in rapid cooling or light weight; and are often of a material where the threads in the spark plug boss, being softer than the spark plug threads, would wear excessively in the torque applied during successive spark plug installations and removals. To compensate for this, some manufacturers make their spark plug boss (or bushing) of greater depth and having more threads, so that the torque applied on installation and removal of spark plugs is distributed over a greater thread area - thereby assuring longer thread life for the spark plug boss (or bushing) threads. In such installations it is important that spark plugs with proper thread lengths should be used.

HOW THE SPARK PLUG WORKS

When the electric current arrives at the end of the center electrode, it doesn't automatically jump the gap as is sometimes supposed. It must wait, temporarily, at this point because there is a purposely designed gap void in the circuit due to the separation of the electrodes, which means resistance to the current passage and which resistance is further increased by compression pressure. In the presence of this resistance, and in order to complete its circuit to the ground electrode, the electric current harnesses the gas mixture in the gap area to provide the conductive agency necessary to get across the gap - as the spark plug insulator and design has the function of preventing the current from flowing in any other direction.

In this action a phenomena of ionization takes place, (the charging of gaseous molecules so that the gas conducts an electric current) which might be compared to the current creating its own bridge by arranging the ions of the gaseous elements present in the gas mixture, to go through this ionized field in reaching the side electrode.

Great friction is generated in the ionized mixture, by this action, and a showering of sparks and a veritable bombardment of the mixture elements occurs, and the further phenomena of burning takes place in igniting the compressed mixture of great expansion powers.

This is the point where the spark plug releases, or unlocks so to speak, the great energy stored in the gasoline - in each cylinder charge.

SIDE ELECTRODES, MULTIPLE or SINGLE?

Relative to the merits of multiple side electrodes as compared to single side electrodes, here again the purposes desired must be considered. For example; the aircraft types have multiple side electrodes to provide a greater gap area and the longer gap life needed in such engines which operate mostly at constant speed.

In the case of the automotive or marine types of plugs, the engine must have flexibility in quick starting, acceleration, stop and start idling service - wherein these are best ensured by concentrating the spark at a single gap.

In the case of multiple gaps, the current would have the tendency to split its energy by ionizing the gases in all the gaps to spark at any one or several points, resulting in a delayed action of the spark through certain engine ranges to cause poor idling or acceleration.

In the case of the aircraft engine, constant engine speed and maximum sparking strength from the magneto at those speeds assures positive action at one of the gaps so that the foregoing is no factor in such operation. It should be remembered, however, that the spark occurs principally at only one gap of a multiple gap spark plug; as it would be a very rare occasion for the conditions of the electrodes and the gases present in the gap area to be so identical in values that the spark would occur at more than one of the multiple gaps.

The actual reason multiple gap spark plugs are used in aircraft, or other constant speed engines, is so when one gap burns open, the spark occurs at the next closest gap, and so on as the sparking action takes place at the various points until all the gaps are opened to the point where service hour recommendations provide for plug servicing and gap re-setting.

Long experience and many tests have shown that the single gap has more values for the automotive or marine type engines wherein operation requires considerable flexibility through the various engine speeds selected.

GAP SETTINGS USED

Every engine has a gap tolerance range which its design and ignition current strength will permit. This can be classed as the minimum and maximum permissible setting. The minimum setting is that point where the gap can be closed on the spark plug and the engine will idle although the idle will be very rough. Any closer setting, would cause the engine to stop at the recommended engine idling speed; so to keep it running, at that close gap setting, would require an increase in engine speed. The maximum setting is that point where the gap can be opened on the spark plug and the engine will still give full power on a heavy load or hard pull. Any wider setting would occasion outright missing under full power or heavy load.

Between these two points, we can then set the spark plug gaps, according to what we would desire in engine performance. For example; if the operation was consistently at low power and idling, we would prefer the widest practical setting to gain engine smoothness; if the operation was for long periods at more sustained power, we would prefer the closest possible setting, commensurate with idling requirements, in order to secure the longest gap life before re-setting would be necessary.

It is this gap tolerance range which permits the engine to operate properly even though there may be a few thousandths variation in the accuracy of the gap setting performed. However, when the gaps grow out to the maximum permissible setting, those plugs which started with a wider gap will naturally be more susceptible to occasional missing under acceleration or a full power operation.

The reason wide gaps are used on some types of automotive engines, is calculated on engineering the travel of the spark through more of the mixture, so that if lean mixture stratifications are encountered there is a better chance of the spark contacting some portion of a richer mixture stratification in or near the gap zone. This requires the use of ignition system capable of adequate current strength and designed for such problems as may be encountered in a high compression engine according to the resistances occurring. All this engineering is coupled with the air-fuel mixture ratios in use by the engine, which mixture condition is an important factor in spark efficiency. Such engineering would not necessarily imply that an .033, .039, or .040 setting would be better than an .025 setting, as many other factors of engine design and characteristics must be studied to determine relative values and methods of securing dependable economy and performance. Gap settings used, are another factor with which the automotive engineer may work in designing an engine for certain objectives desired. One engineer may use one setting; and another engineer may elect to use a different setting in accomplishing parallel results.

SELECTION OF SPARK PLUG TYPE

The spark plug manufacturer's charts provide recommendations which apply in the majority of cases and only unusual engine operating conditions will require a hotter or colder operating plug type. There are, however, differences in the merits of spark plugs in their abilities to perform and these must also be studied in deciding on the spark plug to be used.

When cases arise where it is necessary to handle a vexing or chronic spark plug problem it will be found that there are four factors involved:

1. A mal-functioning of the engine due to mechanical or other deficiencies.

2. The engine's operating characteristics are not within its normal operating range.

3. The installation and gap setting of the plug may be incorrect.

4. The wrong type or quality of plug is in use.

Referring back to the operation of the engine in the four strokes of the cycle, it will be recalled that there are many factors which can occasion premature plug fouling or burning; and before coming to the conclusion that a hotter or colder plug type should be used it would be wise practice to ascertain whether corrections could be better made in the engine.

There are, however, times when the use of a hotter or colder spark plug is the better and more practical course; or mandatory for the service required.

Conditions which may require a hotter plug type are: (chronic occurences)

Fouling	Low Compression Engine
Sooting	Start and Stop Work
Oily Motor	Governed Engine Speed
Cold Operating Engine	Cold Climate Operation
Slow Speed Operation	Engine Mechanical Condition

Conditions which may require a colder plug type are: (chronic occurences)

Rapid Gap Growth	High Compressions
Pre-ignition, Ping	Heavy Load
Overheating	Long Haul
Core Fracture	Long Upgrades, Low Gear Work
Hot Type Engine	Hot Climate Operation
High Speeds	Engine Mechanical Condition

Whenever a colder or hotter plug type is used to handle an aggravated condition pending later engine overhaul, or a change in engine operating conditions; it naturally follows that the spark plug type regularly recommended for the engine should be again installed, in lieu of the colder or hotter type previously used.

ESTIMATING THE NEED FOR CHANGE OF SPARK PLUG HEAT RANGE BY OPERATING COLOR

In our opinion there is no specific spark plug insulator operating color that you can standardize on in determining the need for hotter or colder plugs. Engines are purposely designed to predetermined standards at various engine speeds. Cylinder loads vary. Some engines are designed to operate with fuel mixtures on the "lean" side; and others on the "rich" side. Spark plugs used in one engine may be operating normally and yet have an entirely different operating color than the set coming out of another engine of the same kind. This is equally true in cinsidering driver's habits with the same engine, and where the drivers are subject to changes. Gasolines and oils have different chemical constituents and structure, and are purposely prepared for certain specific values. Spark plugs, themselves, have insulators of different properties for certain specific values, which take on an operating color entirely different than another insulator whould take on in the same engine.

If you were to remove a set of plugs from a given engine, directly after a hard run, the plugs may appear to be operating on the "hot" side - and yet this would be a perfectly normal appearance. If you checked this same engine, after a period of stop and start, cool engine and long idling service - may then find the same set appearing to run on the "cool" side - or dark in color. Suppose, in the case of the hot appearing set, you installed a colder running plug type. You may then find a fouling situation developing because the plugs were too cold in operating range for stop and start, and cool engine operation. Suppose we installed a hotter running plug type, when we noted the set appearing on the dark side in the earlier reference. We may then find such a set would cause burning, pre-ignition, and short plug life, when the engine was subjected to hard service. Because of these extremes, spark plugs have to have extra range properties - and you can readily understand just what the term implies.

Unless you are in complete knowledge of the plug operating color resulting from the use of the various fuels; the engine operating conditions; the engine air-fuel mixture ratio; and the properties of the spark plug insulating material in operating appearance characteristics - the safest rule to apply in checking plugs by appearance or color, is to look for either of two definite conditions:

1. Are the plugs fouled by a sooty or oily deposit in such accumulation as to be indicative of a low plug operating temperature, or engine maladjustment?

2. Are they so burned in appearance so as to be indicative that a cooler plug type, or engine correction, is necessary?

The simple check by comparison with a new or good plug, will quickly enable you to determine relative conditions.

SPARK PLUG SERVICING AND INSTALLATION

To cover these aspects of spark plugs and their service, it is well for us to pause now so that we may view a film on these subjects, which will cover the matters far better than we can discuss them here.

This film, "Ignition and Spark Plugs", was prepared for the training of personnel in our Armed Forces, and in their libraries for the purpose. Adhering to the fundamentals, the subjects are covered in such a way as to provide excellent pictorial reference to the subjects thus far discussed:

RE-GAPPING SPARK PLUGS - ROUND WIRE GAGES or FLAT GAGES?

In connection with re-gapping, we should remember that electrode corrosion and oxides at the gap area vitally affect spark efficiency. The cleaner can remove oxides and deposits from the core, but because of gap location, the cleaner stream cannot always reach this area with full effect. Therefore, when plugs are worth of further use, it is a good practice to dress the gap area, on both center and side electrodes, with a small file of the distributor point type. When dressing the gap area, sucure flat parallel surfaces so that on regapping to required spacing you can approximate the flat gap of the new plug.

There is no objection to the use of a round wire gage in measuring the spark plug gap in automotive or other types having the flat gap, though better practice is to use the flat gage in such instances so as to secure as near parallel gap you will note is present on such new plugs.

The real purpose of measuring spark plug gaps with a round wire gage is for those types where gaps may be inaccessible for a flat gage, such as in aircraft types where the gaps are formed from a three or four prong "clover-leaf" type setting, and where inaccessibility is the only factor. With the aircraft type setting, this gap area is thoroughly cleaned by the cleaner blast stream, however, and gap setting is accomplished by a special gap forming tool, which reforms the side electrode in a parallel relationship to the center electrode.

The automotive types are purposely designed with the flat gap for maximum parallel areas so as to provide a hotter, spreading, and more effective sparking action and longer gap life.

When electrodes of the flat gap type are burned in such a manner as to present a curvature of electrode surfaces at the gap area - it should be remembered that such surfaces may be highly corroded and offer resistance to the passage of the spark - even if set at the correct gap spacing with a round wire gage. This corrosion, etc., is generally a few thousandths thick on both the center and side electrodes, so that if the correct physical measurement of the gap checked out as desired - we would still have a slightly wider gap in considering the actual spacing between the clean, virgin metal areas of each electrode which provide the better sparking action. It should also be remembered that the current must fight its way through this scaled and corroded area, further hampering its efficiency. It is for this reason that the earlier recommendation is made, so that on presenting clean metal surfaces, we approximate the desired flat gap setting for a renewal of efficient sparking action and longer life for the gap setting.

As mentioned previously, there is no objection to the use of the round wire gage, so long as proper care is taken to provide a gap setting compatible with new plug design. Very definitely, the round wire gage should be used, and of the proper type, for those gaps requiring such gage accessibility.

AIRCRAFT SPARK PLUGS

Airplane engines are remarkably high in efficiency and dependability. This progress has been due to the close attention by the aircraft industry to all engine values which can be developed to create more power.

This high degree of efficiency; coupled with sustained high cylinder temperature and relatively continuous engine operation; requires that aircraft spark plugs be designed and built to withstand the rigors of such service. The basic requirements of an aircraft spark plug, in engines of high output, are:

1. Rugged construction.
2. Gas-tightness.
3. Good fouling range for idling operation.
4. Ability to remain cool enough at high operating temperatures to avoid burning or pre-ignition.
5. Long gap life.
6. Shielding.

In essence, these are exactly the same as required for all types of spark plugs, commensurate with their service intents and costs, excepting for the factor of shielding.

Facilities for communication from ship to airport and from ship to ship are rapidly becoming a necessity in avaition, not only for military and commercial aircraft, but also for private aircraft as well. The trend indicates its almost universal use in due time.

Because the aircraft engine's ignition system is in itself a type of miniature radio station; its wave radiations during operation will interfere with radio reception by the presence of noises emanating from the ignition system and conveyed to the airplane's radio receiving apparatus. To prevent this interference, the entire ignition system is enclosed in a metal shield. This shielding picks up the noises enamating from the engine's ignition system and carries them to ground, groviding clarity in reception. Spark plugs are a part of the ignition system so affected and are produced in both unshielded and shielded types for aircraft use.

These charts will illustrate their general design:

The factor of gap erosion is extremely important on types used in aircraft engines of high output for the following reasons:

1. As the aircraft ascends to higher altitudes the atmospheric pressure reduces and with this comes a reduction of the insulating value of this air surrounding the exterior of the ignition system. Consequently, flashover dangers would be present with increased resistance as would be present with a gap too wide on the spark plugs.

2. With the shielded types, the presence of trapped air, on assembly of the cable terminal sleeve in the shielding well, eventually becomes sufficiently ionized to increase the danger of internal flashover or current leakage.

3. The airplane, particularly those in commercial operation, accumulate so many hours in a relatively short time, it would therefore be necessary to remove spark plugs frequently for servicing and re-gapping. This has its effects both in delay to the airplane in service, and in costs for spark plug removal and reinstallation. With multiple cylinder engines, and dual ignition, you can appreciate the time factor necessary for removal and installation of spark plugs if this is done too often.

For these reasons, considerable research has been devoted to methods for the control of gap erosion on aircraft types. This has shaped along two lines; one, in the metallurgy of the electrode design and construction; and two, by the incorporation of a resistor within the spark plug itself.

The chief reason for the use of the resistor is to reduce gap growth. This is accomplished in the following manner. The normal ignition spark consists of a series of individual discharges or pulses of gradually diminishing intensity. Ignition of the gaseous mixture in a combustion chamber is accomplished by the first two or three pulses, and the remaining discharges occur after ignition has taken place, and therefore serves no useful purpose. However, these succeeding discharges, which increase electrode wear, can be stopped or cut off by the use of a suitable resistor, and the gap growth is thereby reduced.

The servicing and overhaul of aircraft spark plugs does not differ greatly from the servicing and overhaul of regular types. The steps of degreasing; sand-blast cleaning; inspection; gap setting and testing being the important steps.

IN WHAT WAY DO PROPERLY OPERATING SPARK PLUGS SAVE GASLOINE?

Since the electric current travels to the gap by means of the center electrode coming out of the insulator, and in view of the resistances at the gap, whenever carbon deposits are present and other conductive materials are coated on or burned into the insulator, these form conductive paths whereby part of the electric current can easily leak to ground without performing its function in assisting to ignite the mixture. When the electrodes are scaled, oxidized, and corroded, these poorly conducting conditions retard the proper releasing of the spark which must endeavor to overcome this added resistance, tending to increase the rate of current leakage over and through the conductive deposits on and in the insulator, resulting in a weak, wavering spark traveling across the gap zone. This reduction of spark intensity, and the spark intensity, and the smallness of its volume, hampers proper ignition of the compressed gas and burning is slow, sluggish, and incomplete.

Throughout the ranges of engine operating demands, pressures mount in the cylinder due to power requirements and the conditions described plus the widening of the gap due to sparking action in prolonged service, so that there often occurs an occasional misfiring or complete failure of the spark plug to properly function during certain engine power ranges needed in the operation of the car.

When the spark plugs fire, the mixture does not explode on ignition, but burns with great rapidity. The rate of burning, however, is in direct proportion to the intensity of ignition and the size of the area of the gas mixture originally ignited. At the start of burning, the flame front of the original ignition, ignites the adjacent layer of gas mixture, and each successive flame front in the burning process ignites successive layers, until the entire mass is burning. Then, the pressure developed from the great expansion of the burning mixture forces the piston downward with the energy of irresistable power.

Let's study the details of this burning action a moment, by a comparison, in order to picture the importance of intensity of ignition and the size of the area originally ignited as very definite factors affecting the time necessary to burn anything. Suppose we took two pieces of paper of exactly the same size and material and we lit one of these with a match and the other with a torch of larger initial flame. We would then note, if we clocked their burning, that the piece of paper ignited by the larger flame of the torch starting the paper burning on a wider flame front, and more intensely, so that the ever widening successive flame fronts consumed the paper a great deal more rapidly - and consequently we received the heat values a great deal more quickly and more efficiently withing the time element noted.

The gasoline mixture compressed into the combustion chambers of the engine cylinders, contains a very definite quantity of heat units and expansion values in the ability of each charge to generate considerable power. Once in the combustion chamber, it's up to the spark plug to unleash the great energy stored in the gas mixture by starting an intense and efficient burning within the time the engine allows.

The piston, however, cannot wait at firing position very long for this burning and expansion to take place, and it's even more acute when the engine is turning at several thousand revolutions per minute - as other spark plugs will be firing in sequence in the engine, and engine momentum tends to keep the engine revolving. So, when the spark at the gap of the spark plug is weak, the bruning of the gas mixture is slowed and, because of the events just described, the piston is compelled to move downward on its power stroke prematurely, in ratio to the burning of the gasoline, but on time according to engine design - with only a partial power development occasioning its travel with the slowly burning mixture only partially expanded. Thus, the piston reaches bottom center and starts upward on its exhaust strike, the exhaust valve opens and the valuable heat units and power qualities of the still burning gasoline is expelled from the cylinder and is wasted out through the exhaust system.

In such cases, the back pressure exercised on the piston moving upward, by the still expanding mixture, further reduces engine efficiency - attacks cylinder walls, piston, rings, and valves - and accentuates general wear on the engine; while the engine requires a further opening of the throttle, for added volumes of gasoline, to overcome this braking action so as to provide the power that should have come in the proper ignition of the original gas mixture provided.

That properly operating spark plugs can save gasoline cannot be denied. How much they will save is dependent upon how bad the old ones are, and whether they are all firing. In function the automotive engine is in effect a miniature chemical factory and a type of pump. It pumps in a mixture of gasoline and air - burns it - and pumps out the residue product of gas the energy qualities of which the engine had previously harnessed into power before its expulsion from the cylinder. It will be readily noted that the efficient firing and thorough burning of this gas and air mixture are all important if the engine is to be powerful, dependable, and economical. Dependable ignition and efficient burning are essential, otherwise the engine becomes more of a wasteful pump than an efficient source of power. Remember, if oné plug is consistently mis-firing, in the case of a six cylinder engine, this means that one sixth of the engine is not delivering power and consequently all the gas going into that particular cylinder is completely wasted. In the rapidly occuring firing sequence of our modern engines, we are not always conscious of occasionally missing spark plugs and here, too, considerable power is lost and gas wasted.

By their operation in this "chemical factory" we should consider that spark plugs are expendable, in the very nature of the work they must do. They cannot go on indefinitely and their cycle of service must be renewed with fresh elements at regular intervals if we are to have a dependable and efficient engine.

So have your spark plugs maintained in good condition.

By so little, so much is gained.

Factors Affecting FUNCTIONING OF SPARK PLUGS

by W. A. BYCHINSKY
AC Spark Plug Division, GMC

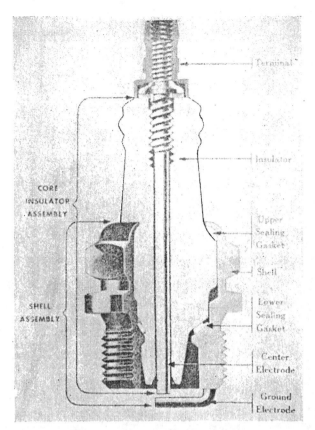

Fig. 1—Section showing principal components of spark plug

A SPARK plug must do two things. It must: (1) conduct high-tension current from the outside of the combustion chamber to the inside, and (2) furnish a pair of electrodes between which the high-tension voltage can spark. These requirements are simple and are easily met.

A spark plug, in performing these two functions must *not* do many things. It must not interfere with the functioning of the engine in any way. It must not break or fracture or disintegrate. It must not run at a temperature sufficiently high to cause preignition of the gaseous mixture. It must not occupy so much space that it cannot be located at the point in the chamber where ignition should be initiated, nor should it interfere with the normal cooling of the engine. It should not create any extraneous load on the ignition system, such as

[This paper was presented at a meeting of the Pittsburgh Section of the SAE, Pittsburgh, Pa., March 25, 1947.]

from fouling of the surface of the insulation by electrically conducting coatings, or from electrical loss in the body of the insulator, or from flash-over along the upper part of the insulator. It should not change in its ability to produce a spark with a given high-tension voltage, that is, the gap should not change or wear, and the sparking voltage should not vary. Furthermore, the spark plug should not be affected by the engine. The materials of the plug should not be attacked by the fuels and products of combustion at the high temperatures that exist in an engine; nor should they be affected by the rapid changes in temperature to which they are exposed.

The negative requirements impose very severe limitations on the materials and design of spark plugs. In order to meet them the advances in many fields of engineering and science have been utilized. Research, investigation, and development in the

---

The spark plug must satisfy two simple and easily met requirements: it must conduct high-tension current into the combustion chamber, and it must furnish a pair of electrodes between which the voltage can spark.

Much more difficult to satisfy are the negative requirements, which impose severe limitations on both materials and design, as discussed here by Mr. Bychinsky.

In addition, he reports that the high operating speeds, wide speed range, high compression ratio, high specific output, and economy of the present-day engine have made the job of overcoming these limitations a particularly tough one, requiring continued research, investigation, and development.

April 1948 Vol. 2 No. 2

The Author

W. A. BYCHINSKY (M '44) is chief ignition engineer of GMC's AC Spark Plug Division. He joined the Spark Plug Engineering Department in 1930 shortly after his graduation from the University of Michigan. After completing graduate work in electrical engineering, he received his MSE in 1931 and Ph.D. in 1933.

field of spark plugs have been aggressively pursued and refinements in design have been continuously applied.

Historical Survey

In addition to the familiar high-tension spark, many means have been proposed and put into practice for the ignition of the fuel in an internal-combustion engine. Most have been discarded either because of inherent unsatisfactoriness, high cost, or unadaptability, to such changing trends in an engine as, for example, higher speeds. The open flame pilot light is an example of the latter; compression ignition aided by a hot spot or glowing surface cyclically exposed to the gases is another example. The diesel engine, of course, has had a separate and distinct history. Another old type which, for all its attractive characteristics, has been superseded by the spark, is the interrupter type wherein an arc is created by separating a pair of contact points carrying a current flowing through an inductive circuit.

The application of a high-tension spark to the ignition of the fuel in an internal-combustion engine seems to have been made before 1860. It is interesting to note that one of the first spark-plug insulators was made by Blin at a place near Limoges, France, from a pottery clay he was then using for making artists' paint cups. The first insulators made in the United States were formed from a porcelain of the type used by the electrical industry for general-purpose insulation.

The inadequacy of the early porcelain insulators soon became apparent. Failures due to thermal shock and shunting due to electrical conductivity at high operating temperatures became major problems.

Early insulators were made from compositions containing (1) a plastic clay, hydrated aluminum silicate ($Al_2O_3 \cdot 2SiO_2 \cdot 2H_2O$), such as ball clay or kaolin, to permit the wet mixture to be molded and shaped, that is, to impart forming characteristics, (2) a so-called inert material such as flint (SiO_2), which forms the rigid skeleton of the body and which reduces shrinkage in drying and firing, and improves the mechanical strength, and (3) a flux or glass-forming agent such as feldspar, potassium-aluminum silicate ($K_2O \cdot Al_2O_3 \cdot 6SiO_2$), which acts as a cement or bond on the crystals of the fired body.

The first step in counteracting the low resistivity was to eliminate or to reduce the alkali content by replacing the potash by lime or magnesia.

The cause of low thermal shock was traced to the presence of the crystalline silica, which changes phases within the operating range of a plug with resultant sudden changes in volume and hence discontinuities in the expansion curve. This produces strains in the body of the insulator which impair the heat-shock resistance of the structure.

Methods of reducing the silica content were devised. The first method of doing so was to replace the flint, that is, silica, by calcined clay which contains mullite. Mullite crystals ($3Al_2O_3 \cdot SiO_2$) form a desirable component of porcelain. Thus, the skeleton structure of the insulator body had a decreased silica content and an increased mullite content. Plugs made by AC during World War I were manufactured from this body.

The next step in the removal of the silica was to add alumina to the calcined clay, which alumina would react with some of the excess silica to form mullite. Finally, the silica (flint) was entirely replaced by andalusite ($Al_2O_3 \cdot SiO_2$), which is a naturally occurring mineral, or by artificially made mullite. Natural mullite is not commercially available although some deposits have been discovered on the island of Mull off the coast of Africa. These mullite bodies still are in general use for spark-plug insulators. Their limitations are well understood and the possibility of their improvement is quite restricted.

Much effort, therefore, has been expended in research and investigation of ceramic compositions other than porcelain for spark-plug insulator use. The alumina insulator, patented by AC, represents a very successful solution to this quest. Until this development it was necessary to use clay in the body as a bond to permit forming. Clay is a natural product and never occurs in a pure state but is always contaminated with various other inorganic compounds, which in the firing produce undesirable glasses in the insulator. An AC development was to replace clay by an organic bonding agent which would permit forming and yet which, in the firing of the body, would burn out without leaving a trace. By these processes it was possible to work with minerals entirely devoid of plasticity. Instead, therefore, of being limited to compositions containing a certain minimum percentage (20 to 50%) of natural clay in every body developed, the whole periodic table was opened for use. The criterion now used in the development of a ceramic insulator body is only concerned with the end or fired product.

The mullite bodies are being gradually displaced by alumina bodies. During World War II all ceramic aircraft plugs and substantially all high-output plugs were made of alumina. The improvement over the mullite is very marked. For example, electrical resistance at elevated temperatures is many times greater than that of the mullite body. The thermal conductivity is 15 times greater, and the mechanical strength in compression doubled, and the modulus of rupture is three times as great.

Another important characteristic, which will be more fully discussed later, is the inertness of alumina to attack by the lead compounds in the fuel.

Early plugs were made in a variety of sizes in both straight and taper threads. One-half inch pipe threads and 7/8-18 were common types. The prior European development in spark plugs is reflected in the early introduction and standardization of a metric thread size, namely the 18-mm plug.

Current usage is almost entirely limited to three thread sizes, the 18 mm, the 14 mm, and the 10 mm, with aircraft and heavy-duty equipment favoring the 18 mm and passenger-car engines specifying 14 mm and 10 mm.

Present-Day Requirements

Current internal-combustion engines are characterized by their higher operating speeds and wider speed range, higher compression ratios, higher specific outputs, and greater economy.

From the spark-plug point of view this resolves itself into operation over wider ranges of speed, pressures, and temperatures, and in leaner or less easily ignitible mixtures.

Furthermore, the addition of detonation inhibitors to the fuel, notably tetraethyl lead, has greatly influenced spark-plug behavior. Another requirement, by no means new, is concerned with the adaptability of the plug to servicing or reconditioning. The advantages inherent in the one-piece nondemountable plug, that is, one in which the insulator could not be removed for cleaning, were made possible by the development and introduction of spark-plug cleaning devices. Other needs, now limited to special applications, but probably of more universal future concern, are moisture proofing, radio suppressing, and radio shielding. In addition to the technical considerations, mention must be made of the very important item of cost.

Selecting Plug Components

A spark plug has three principal components, the insulator, the electrodes, and the shell. (See Fig. 1.)

The service loads imposed upon the shell are generally sufficiently low to permit the use of low-carbon free-machining steel. The punishment to which a shell is subjected in the normal process of manufacturing, such as crimping and heat sealing, is usually much more severe than that encountered in service. To meet these manufacturing requirements it is usually necessary to use specially selected steel of special analysis. Rustproofing beyond that obtained with a conventional blackening process is seldom warranted.

The electrodes are subjected to and must resist high temperatures, corrosive gases, and erosive electrical discharges. Nevertheless, they should be reasonably good thermal conductors, they should permit the formation of a spark discharge at a low and constant voltage, and finally they should be easily fabricated and should be available at low cost. These specifications are met, in their entirety, by no known metal or alloy; but much work has been done in devising and testing numerous alloys and metals in an effort to approximate those qualities which are of prime importance in certain applications. The nickel-chromium-barium alloy currently used on many plug types is the result of such research. The corrosion resistance of high nickel-chromium alloys is well known. The reason underlying the addition of the barium is more obscure. It stems from an investigation made some years ago into the factors which affect the sparking voltage of a spark plug. It was discovered that the so-called work function of materials exposed on the surface of the electrodes greatly influenced the sparking voltage of the electrodes. The nickel alloys in use at the time were found to be very erratic in their sparking voltage because of the nonuniform distribution in the metal of the electrode of magnesium, which is customarily added to de-oxidize the melt. Magnesium is known to have a low work function, which greatly affects sparking voltage. A quest was, therefore, made for a low work function metal which could be alloyed with the electrode material to ensure uniformity of distribution. Barium was found to be especially well suited to this purpose but, because of the lack of commercial usage, it was obtainable only in small quantities and at a very high cost. Spark-plug electrode requirements were sufficient to warrant commercial scale refining, and the cost was accordingly brought down to within economic limits. This alloy is known of as isovolt alloy–the name emphasizing the fact that the sparking voltage of electrodes made from the alloy remains constant.

In selecting the insulator material, consideration must be given to the mechanical, thermal, electrical, and chemical conditions under which the insulator is to operate. The direct mechanical loads imposed by the engine, even at extreme output, are small compared to the loads (imposed by the plug assembly) which are required to ensure gastight operation. Mechanical loads often encountered during installation or removal of plugs due to wrench slippage or other accidental mistreatment are by no means negligible and on their own account merit the use of the highest strength insulators available. In this regard, the alumina-type insulator is outstanding among insulator compositions.

Resistance to heat shock, which has been the subject of intensive investigation, was greatly enhanced when the mullite body, now in general use, replaced the old type porcelain. The new alumina body which is making its appearance represents a further marked improvement in this characteristic. In addition, the alumina body possesses high thermal conductivity which permits the use of long tips on high-output (or cold) plugs.

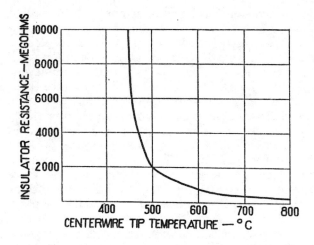

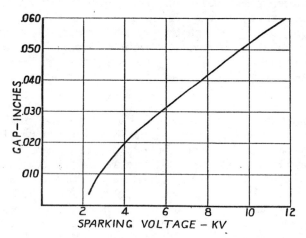

Fig. 2 (top)—Effect of temperature on electrical resistance of spark-plug insulators

Fig. 3 (bottom)—Variation of sparking voltage with gap size

The electrical strength characteristics of most spark-plug insulating materials have been adequate for the needs. The electrical stresses involved are not excessive. The electrical conductivity of the early porcelains is too high for current engine use. The electrical conductivity of the alumina insulator at elevated temperatures is very low. In fact, the alumina composition forms one of the best high-temperature insulators known today. The effect of temperature on the electrical resistance of spark-plug insulators is shown in Fig. 2.

The chemical reactions of the insulator tip with the products of fuel combustion are of prime importance. The lead compounds added to many fuels as an antiknock have a particularly deleterious effect on some ceramic materials. They react with silica to form a low melting point lead silicate glass. This glass is chemically quite active at high temperatures. It is also a good electrical conductor at elevated temperatures. At one time most plugs had glazed tips, that is, the tips were coated with glass, which has a high silica content. This was done to retard the attachment of carbon. Upon the introduction of tetraethyl lead it was found very desirable to eliminate this glaze in order to retard the reaction of the lead salts.

Fitting Plug to Engine

To perform with the optimum degree of satisfactoriness a spark plug must fit the engine in which it is used. It must fit the engine mechanically; it must fit the engine electrically; and it must fit the engine thermally. These requirements will now be considered in more detail.

The mechanical needs of any given engine are fixed or constant. The mechanical misfits are usually quite obvious. Due to standardization the number of thread sizes in use is quite limited. Only an insignificant percentage have threads other than 10, 14, 18 mm or 7/8 in. More variation, however, is present in the reach of the thread. The use of a long-reach plug in a short-reach hole may, obviously, lead to difficulty due to interference with some engine part, such as the valves or piston. The reverse, that is, the use of a short-reach plug in a long-reach hole may also, but less obviously, lead to improper functioning. In such a case two troubles may be encountered. First, the uncovered threads of the engine head will be exposed to corrosion and encrustation, thereby making it difficult to insert the proper reach of plug at a later time. Second, the location of the spark gap will be substantially changed from the normal location fixed by the long-reach plug. Under critical engine conditions such a change in location could lead to engine roughness or, in extreme cases, to miss.

A mechanical misfit at the terminal end of a plug can also lead to a difficulty which, when it does occur, is often difficult to trace. The trouble usually arises due to bringing the plug terminal into proximity with some grounded engine part, thus encouraging flash-over. This can happen when the overall length of the plug is too great as well as when it is too short. This condition is especially prevalent when a cover plate is mounted over the spark plugs, as it is in some valve-in-head engines.

The electrical needs of the engine are not, by any means, constant in value. We are required to furnish a voltage sufficient to jump the gap of the plug. The amount of voltage required, as a function of gap size, is shown in Fig. 3. The measurements were made in a conventional automotive engine at road load. This voltage is dependent on the size of the gap, its geometry, the temperature of the electrodes, and the pressure temperature and air-fuel ratio of the gas mixture in the region of the gap. Furthermore, it is also influenced by the rate of voltage rise applied to the plug. To furnish this voltage we have available the ignition coil. The magnitude of the voltage available to the plug depends on the current flowing through the coil primary and breaker points at the moment of breaker opening. This current depends on speed, decreasing as the engine speed increases. The voltage available at the plug also depends on the amount of shunting, that is, degree of fouling, of the plug. For instance, an ignition system which is capable of furnishing 20,000 v across the electrodes of a new clean plug may only be able to furnish 10,000 v across the electrodes of a moderately fouled plug. To understand what causes this condition, let us consider the action which takes place when the high-tension voltage is applied to the plug. When the breaker points are closed energy is stored in the ignition coil in the form of a magnetic field. When the breaker points are suddenly opened this energy is transformed to electrostatic energy and appears as a high voltage across the terminals of the coil. This surge of voltage is not produced instantaneously but builds up gradually to a peak value, and then gradually decays. By gradually is meant, in this case, in about 1/20,000 of a sec. Thus the voltage applied to the spark plug starts at zero and gradually increases in value until the peak is reached, or if the sparking voltage is below this value, until the plug sparks over. If the plug is fouled, then, while the voltage applied to it is building up in magnitude, some current flows across the fouled coating. This loss of current, which means a loss of energy, cuts down the peak voltage to which the high tension can build. If the plug is badly fouled the drain in current can be so high that the high-tension voltage never builds up to the required sparking voltage and the plug fails to spark. The mechanism of the process thus is seen to be a race; on the one hand, the high-tension voltage tries to build up to a high enough value to cause spark over; on the other hand, the fouling coating of the plug tries to drain off energy as fast as it builds up.

It is, therefore, apparent that the faster we can build up the high-tension voltage, the less effect will the fouling have. This is precisely the principle used in high-frequency ignition systems. In such systems, instead of taking 1/20,000 of a sec for the high-tension voltage to reach its peak value, it may take less than 1/1,000,000 of a sec. In practice, high-frequency ignition systems have demonstrated their ability to fire plugs so badly fouled that their electrical conductivity is of the order of 2000 micromhos, whereas conventional systems may fail with only 5 micromhos. Although high-frequency systems have been used on aircraft there is little likelihood, because of cost considerations, of their use on transportation vehicles in the near future.

Another type of high-frequency ignition system is closer at hand. This type, which also has seen service on aircraft, is the well-known series gap or, as it is sometimes called—intensifier gap. In Fig. 4 is shown a sectioned view of the LC-36, which is such a plug. If certain complex critical relationships between the series gap and the spark-plug gap are fulfilled, the equivalent of a very-high-frequency ignition system is obtained. Great difficulty has been encountered in building reliable series gaps, and much work has been done on them during the years. The problems involved in the production of such gaps with the desired stable characteristics are well on their way to an economically sound solution. It is not unlikely that spark plugs containing built-in series gaps will soon be considered for certain vehicle applications.

In discussing the voltage required by a plug we have said that it must be sufficient to spark the gap and we have noted that the gap size is not a constant.

In the first place it is not constant because it wears. Part of the wear is due to the corrosive action of the active high-temperature gases to which the electrodes are exposed. However, this effect accounts for but a small portion of the gap wear. Part of the wear is due to the action of the spark and subsequent electrical discharge in tearing off particles of the electrode. At ordinary temperatures, material loss and hence electrode wear due to this cause would be small. At the elevated temperatures and in the presence of corrosive gases, the loss of electrode material due to the electrical discharge is the major cause of gap wear. The rate of wear is influenced not only by the characteristics and operating conditions of the electrodes but also by the characteristics of the ignition system. A factor of great importance is the secondary capacity of the ignition coil and harness. Shielding of the high-tension leads, which tends to increase the

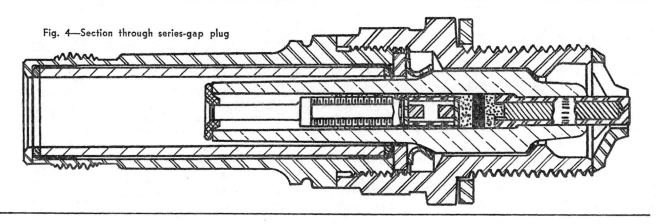

Fig. 4—Section through series-gap plug

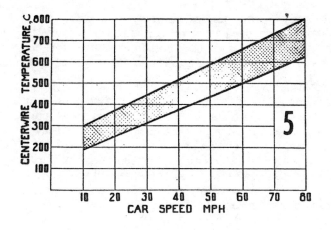

Fig. 5—Variation in spark-plug-tip temperature with car speed

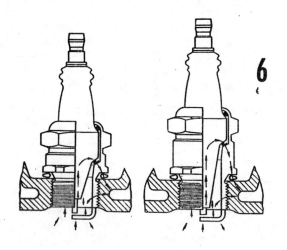

Fig. 6—Path of heat travel with long (right) and short (left) insulator tip

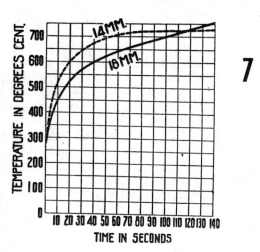

Fig. 7—Temperature-time curve at wide-open throttle and 4000 rpm

discharge current accounts for a large part of the electrode material loss. A resistor having a nominal 100-ohm value sealed in the insulator of the plug was in general usage on aircraft during World War II. Gap wear was reduced through this expedient to 1/3 or less of what it would have been without it. Such gains are not obtained, however, with ignition systems having low secondary capacities.

Not only is the gap size variable because of wear, it is variable because it must be changed to fit the engine in which it is used. Some engines operate satisfactorily with 0.025 gap. Other engines will not operate satisfactorily with gaps less than 0.040 in. The reason for this phenomenon has been the object of intensive study. It is related to the problem of the fat spark versus the thin spark. Many laboratory experiments have been run over ranges of air-fuel ratio greatly exceeding that encountered in practice and they have invariably shown that if a spark is obtained across the gap of a plug, regardless of how thin it is, it will ignite the mixture, and regardless of how fat it is it will not influence the burning after the mixture is ignited. In other words, the combustion process in an engine has been shown to be independent of the igniting spark, and the minimum amount of energy required in the spark to initiate combustion has been shown to be exceedingly small compared to the amount available. These are test results, but the fact remains that some engines operating with a reasonable air-fuel ratio require wide gaps in order to function properly. The explanation must lie in the nature of the air-gasoline mixture in the vicinity of the gap at the time of the spark. Although the overall air-fuel ratio may be nominal, the mixture in the combustion chamber may be highly stratified and the mixture in the gap, which because of turbulence changes rapidly with time, may at one moment be normal and at another moment be excessively lean or rich; alternatively, at any instant the mixture in one small region may be normal and in an adjacent small region it may be excessively lean or rich or contaminated with residual exhaust gas. With a wide gap the chances of a fortuitous mixture being in the gap region at the moment of sparking is greater than for a narrow gap simply because the wide gap includes more volume of mixture. The same effect might be expected from narrow gaps with increased energy sparks that is obtained from wide gaps with normal energy. This is borne out by laboratory experiments on single-cylinder engines operated under very critical and very abnormal conditions.

capacity of the system, results in increased gap wear. In aircraft ignition systems long-shielded high-tension leads are employed having capacities of up to 250 mmf. In such cases the high rate of electrode wear can be counteracted by the inclusion of a resistor in the high-tension circuit, preferably as near to the sparking points of the plug as possible. The resistor reduces the peak current which flows through the electrodes at the time the capacity of the system is being discharged. The high

SAE *Quarterly* Transactions

From indicator cards taken under conditions which duplicate the performance of engines requiring wide gaps, it is interesting to note the effect of either gap size or ignition coil energy on the engine output. With narrow gaps it was observed that the cycle-to-cycle variation was great with a large percentage of late or slow burns, whereas with either the wide gap or the high energy system a more uniform cycle was obtained with a sizeable reduction in the percentage of late burns. In neither case was actual missing recorded, although the loss of power with the narrow gap was significant. In ordinary service an effect such as this would be difficult to detect but can be counteracted by the use of the proper gap setting.

Since wide gaps will ignite more critically unfavorable mixtures than narrow gaps, we may be tempted to gap all plugs at the highest setting. To do so, however, would increase the sparking voltage requirements and would decrease the time interval between gap adjustments. In addition, the higher sparking voltage requirement would decrease the amount of fouling which could be tolerated without engine miss. These are undesirable consequences of the use of wide gaps.

Consideration has been given, so far, to selecting a plug which fits the engine mechanically and electrically. The somewhat more complicated problem of the thermal or heat range fitting of the plug to the engine will now be discussed.

For optimum operation the surface temperature of the firing end of a spark plug must be high enough to burn off or retard the deposition of carbon and low melting point electrically conductive salts, but it must never get so hot that it becomes a source of preignition, or that it glazes or fuses or promotes chemical reaction with any of the high melting point oxides deposited on the insulator.

The range of spark-plug-tip temperatures encountered in conventional automotive service is shown in Fig. 5.

The temperature of the spark-plug insulator is a function not only of the spark plug but also of the temperature of the combustion gases to which it is exposed. This latter temperature is itself a function of the engine design and of the engine operating conditions. It is, therefore, necessary to fit the plug not only to the engine type but also to the particular mode of operation to which the engine is subjected. Let us deal in more detail with these factors.

The tip or electrode end of a spark plug absorbs heat from the hot combustion gases and it dissipates heat to the cooling system of the engine and to the air surrounding the exposed or terminal end of the plug.

The path of heat travel is shown in Fig. 6. The temperature which is reached by the plug, or any component of it, is that value at which the amount of heat absorbed is equal to the amount of heat dissipated.

The amount of heat absorbed by the insulator tip increases as the average gas temperature increases, that is, as the engine output is increased. It also increases as the area of the insulator tip exposed to the hot gases is increased. Thus, other things being unchanged, more heat will be transferred to a long insulator tip than to a short one. The difference in length of heat path in a plug with a long and a short insulator is shown in Fig. 6.

For a given amount of heat absorbed, the temperature which the insulator tip will assume depends on the ease with which heat can be transferred from the tip to the cooler engine parts, such as the cylinder head, and the cooling air. If the insulator material is a poor heat conductor, the tip will operate at a higher temperature than, other things being the same, it would reach if the insulator were a good heat conductor; or, for the same insulator material, if the path the heat must follow to reach the cooling medium is short, the tip will assume a lower temperature than if the path is long. A plug of the former type would be said to have a short insulation length, the latter type, a long insulation length; or, referring to the temperature, the tip of the plug would reach under the same engine operating conditions, the former, short tip, would be called a cold plug and the latter, long tip, would be called a hot plug.

The tip length is thus a measure of the temperature at which a plug will operate, but only if all other variables, such as material and design, are not changed.

With the introduction of new insulator compositions having markedly different heat-conduction characteristics and with the development of new insulator shapes, made possible by the releasing of processing limitations inherent in the old type of insulator bodies, better methods of characterizing the operating temperatures of spark-plug tips were needed.

One scheme, which has been used to some extent, is to run the plug in a certain type and make of single-cylinder sleeve-valve engine under fixed load, speed, and fuel conditions and to observe, by means of an optical pyrometer synchronized with the engine, the temperature of the tip of the insulator. This temperature, expressed in hundreds of degrees centigrade, was used as the heat rating of the plug. For instance, a plug which rated 7.5 would attain a maximum tip temperature of 750 C when operated in the particular test engine and under the particular operating conditions chosen for the rating test. Although this method is attractive in that a direct measurement of the insulator tip temperature is obtained, the accuracy of the reading is low and, of more importance, the results so obtained are susceptible of misinterpretation. The reason for this lies in the fact that the thermal behavior of a plug at a low engine gas temperature (low output) and at a high gas temperature (high output) are not necessarily identical. Two plugs may have the same tip temperature at the engine output corresponding to the temperature rating test

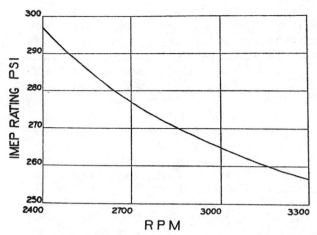

Fig. 8—Effect of engine rpm on preignition rating

and yet one may be operable at very much greater engine output without preignition than the other. In fact, one of the important objectives of the spark-plug design is to maintain the spark-plug-tip temperature as nearly independent of engine temperature as possible.

Another high-temperature rating test was extensively used at one time under the name of "efficiency test." In this test a high-speed high-output single-cylinder engine was used with the spark timing adjusted so that spark advance beyond the setting resulted in a loss in load which, under the dynamometer setup employed, appeared as a drop in speed. The engine output was chosen sufficiently high to cause preignition of all the plugs tested. With a fixed engine starting temperature, the engine speed was rapidly brought up to 4000 rpm, at which point the load was set. The engine was then allowed to run until the speed dropped to 3500 rpm because of spark advance due to preignition. The time, in minutes, required to reach this speed was termed the efficiency rating. Since this test is carried out under unstable engine conditions, the absolute value of the ratings varied greatly from day to day, so that its main usefulness was to furnish a means of comparing different plugs on the same day. Since the time required to reach preignition temperatures is influenced by plug design, the proper interpretation of the results of the efficiency test required much experience. This is shown in Fig. 7 for a 14-mm and an 18-mm plug.

The rating method in widest usage today is the imep (indicated mean effective pressure) preignition rating. By this method a plug is run at successively higher engine outputs until preignition is encountered. The maximum indicated engine output, expressed in terms of imep, which the plug can withstand without preigniting is the imep preignition rating of the plug. Thus, the measurement of the high-temperature characteristic of the plug is made at the maximum temperature at which the plug can, or at least should, be operated. It is, therefore, not necessary to extrapolate performance at high temperatures from data obtained at some lower temperature.

The variables which enter into the determination of preignition rating are numerous. Consistent and reliable values require very complete control. In the first place the test engine must be designed to eliminate the possibility of preignition from no matter what engine part at the highest output at which plugs are to be rated. Sodium-cooled valves and, in some cases, sodium-cooled pistons are used. In practice the engine is checked with a calibration plug known to have a rating substantially above that of the plugs to be run. If preignition is observed with the extremely cold calibration plug, the cause must be due to some engine part other than the plug.

Since the average combustion gas temperature depends on the air-fuel ratio, means are provided to vary the air-fuel ratio while measuring the average combustion gas temperature by means of a thermocouple mounted in the combustion chamber. The fuel mixture is set for maximum temperature for each engine setting by adjusting the stroke of the fuel injector pump by which means the fuel is supplied to the engine. The intake air is held at a constant temperature and humidity.

The fuel used is specially selected with an octane rating greater than 100 so that detonation-free operation is obtainable at the highest output. This is essential, since temperature rise due to detonation would present an uncontrollable factor in the heating of the plug.

Preignition ratings are affected by engine speed, as shown by Fig. 8.

It is desirable to standardize on a single, representative speed and to extend the results to other speed ranges. Speeds of 2700 or 3000 have been found useful.

The output level of the engine is varied by varying the intake pressure from below atmospheric to a high degree of supercharge. In an actual rating test the amount of boost is increased in increments of ½ in. of hg, adjusting the fuel for maximum temperature at each setting and allowing the plug and engine to stabilize before each increase. When preignition is obtained it is made evident by a drop in load and by a sudden change in the rate of rise of combustion-chamber temperature. This latter indication is more reliable in the early stages of preignition.

The high-temperature thermal performance of a plug is adequately characterized by the imep preignition rating. No satisfactory means has been developed for measuring the low-temperature performance.

The resistance of a plug to carbon fouling is closely associated with its behavior at low temperatures, that is, at low engine outputs.

Innumerable tests and testing cycles, both on the bench and in test engines, have been devised in an effort to predetermine the ability of a plug to resist carbon fouling. Although much has been learned from such work the problem is by no means solved.

One of the directions in which to work is fairly apparent. Carbon fouling of the insulator tip, when it is excessive, occurs because of the load or shunt which it produces on the ignition coil. For a given thickness of coating, the smaller the diameter of the tip and the longer the tip, the lower will be the shunting value of the coating. Therefore, it is usually desirable to use as long an insulator tip as possible. But the length for the conventional plug is fixed by the preignition limitation. Insulators with higher thermal conductivities are required if longer tips are to be tolerated. This possibility was realized with the introduction of the alumina body, the thermal conductivity of which is substantially greater than that of the older mullite bodies.

Another avenue of attack on the problem of low-temperature fouling is through the geometry of the insulator tip.

New methods of forming the insulator have been developed that remove the limitations formerly encountered with clay bodies. As a result, it is now possible to manufacture insulators having very thin recessed tip sections which, because of their low mass can more readily follow temperature changes in the combustion chamber and so operate for a greater percentage of the time at temperatures sufficiently high to burn off or, at least, to retard the deposition of carbon. It must be mentioned that, along with the ability to form and fire such insulator shapes, it was necessary to have at hand an insulating material of the requisite strength, heat shock resistance, and imperviousness to attack by fuel components at elevated temperatures. Alumina is such a material.

Having considered the factors involved in determining the operating temperature of a spark plug, let us now apply them to the selection of the optimum heat range plug for a particular engine application. The temperature which the insulator tip should reach to prevent or retard the deposition of carbon is the same for all plugs in all engines. The temperature which must be reached by the plug before preignition is incurred is similarly independent, for practical purposes, of the engine and

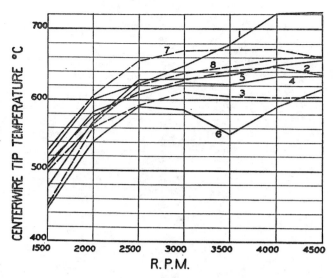

Fig. 10—Variations in temperature for different cylinders of 8-cyl engine

plug. Therefore, regardless of the engine type and plug type chosen, the plug temperature over the maximum possible range should lie between these two limits.

If the engine output is high, for instance, if we are dealing with a high-compression high-speed engine run at full load, when the combustion gas temperature is very high, then we must use a cold plug, that is, one with short insulation length or, more scientifically, one having a high imep rating in order that it can more readily carry away the heat it absorbs from the engine and so will not overheat. If, on the other hand, we are dealing with a low-output engine, that is, low compression ratio, low speed, run at light load, when the combustion gas temperatures are low, then we would choose a hot plug, that is, one with a long insulation length, in other words, one having a low imep rating in order that the heat imparted to the insulator tip is restricted in its flow, thereby permitting the temperature of the tip to build up to the required value.

In practice the gradations in imep rating or heat range are small enough to permit the accurate selection of plugs for all types of operation from the lowest-output kerosene-burning stationary engine to the highest-output triptane-fueled water-injection aircraft engine. In terms of imep ratings this represents a range of from 70 to greater than 500. Typical ratings for common plug type are shown in Fig. 9.

Checking Plug from Engine

Much can be learned from an examination of a set of plugs removed from an engine not only regarding the behavior of the plugs but also regarding the operation of the engine. Spark plugs have often been likened to clinical thermometers in that they are indicators of many engine malfunctionings. Unfortunately, this useful role has one un-

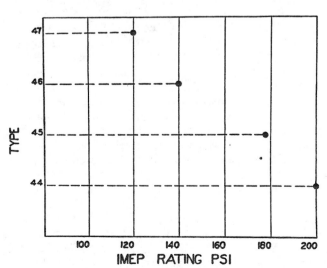

Fig. 9—Imep ratings for several plugs

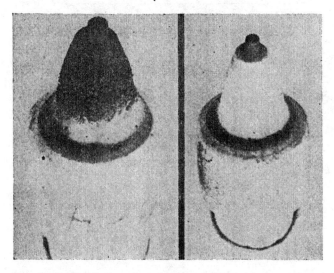

Fig. 11 (left)—Black carbon fouling usually indicates missing or roughness

Fig. 12 (right)—Insulator shown in Fig. 11 after having been subjected to engine operation at slightly higher output than that under which it became fouled

desirable aspect, namely that spark plugs are often blamed for the engine troubles which they reflect. In reporting spark plug difficulties it is important to include all other pertinent indications of the trouble so that an accurate diagnosis of the underlying cause can be made.

There are three general types of spark-plug trouble. These are:

1. Effects of operation at excessive temperature.
2. Missing or roughness due to low speed or carbon fouling.
3. Missing or roughness due to high speed or lead fouling.

Let us consider these problems in more detail.

Excessive temperature can be due to several causes and can manifest itself in several ways. Broken insulator tips, discolored shells, and badly corroded electrodes are the usual indications of excessive temperature operation, although the latter, badly corroded electrodes, may also be due to corrosive agents in the gasoline, such as high sulfur.

While the obvious remedy may be to use a colder, that is, higher rating plug, this may not be the correct remedy. The original plug may have been in the correct heat range but it could have operated at a higher temperature than normal because of leakage past the engine seat gasket. Such leakage could be due to improper tightening or to the use of worn or distorted gaskets or to defective gasket seats on the cylinder head. A good seal at this point is important if normal plug operation is to be insured. Another cause of excessive temperature is through lean mixture operation occasioned by leaks in the intake manifolding.

Abnormally high temperatures can also be induced by operating the engine under conditions of excessive detonation or knock. This can be attributed to the antiknock quality of the fuel used, or to the presence of hot spots in the combustion chamber due to faulty circulation of the coolant, or to the buildup of heavy carbon layers on the walls of the combustion chamber and dome of the piston or, finally, to incorrect spark advance. Some of these faults may even lead to destructive preignition. These possibilities should, therefore, be considered before changing to a plug of a colder heat range. The operating temperature varies from cylinder to cylinder and such effects may be expected to be more pronounced in certain cylinders. Cylinder-to-cylinder variation in temperature in an 8-cyl automotive engine operated at road load is shown in Fig. 10.

Missing or roughness at low speed is usually indicated on a plug by black carbon fouling (Fig. 11). The carbon fouling is *the fact*. That the carbon fouling caused the noted missing is *the conjecture*. If the conjecture is correct the remedy is (1) to go to a hotter plug, or (2) to eliminate the source of carbon fouling either by reconditioning the engine, if due to high oil consumption, or by adjusting the carburetor, if too rich, or (3) to change the mode of operation to include sufficient high-output runs to burn off any carbon accumulations before they become troublesome.

Fig. 12 is a photograph of the insulator shown in Fig. 11 after having been subjected to engine operation at a slightly higher output than that under which it became fouled.

But if the conjecture is wrong, then the observed fouling is due to the noted missing, that is, the engine intermittently fails to fire, allowing the plug to cool off and to collect gasoline and oil vapor, which carbonizes and eventually shorts out the plug. In this case the heat range of the plug is not at fault and the remedies just mentioned are not indicated. Instead, we should investigate the ignition system to make sure that sufficient voltage to spark the plug gap is delivered to the plug. We should check the battery, condenser, and points. We should inspect the high-tension cable and the distributor for leaks or breaks in the insulation.

We should check for places where the high tension current could bypass the plug either as a spark through a break in the insulator surface or as a spark from the terminal to some projection or boss on the cylinder head. We should check the plug gap. If it is too wide the available high-tension voltage may not be sufficient to fire it regularly. If it is too small it may not always ignite the mixture when it does spark because of inherently bad mixture due to imperfect carburetion. Just as in the case of excessive temperature, we should expect cylinder-to-cylinder variation in the engine. This is illustrated in Figs. 13 and 14, which show insulators from plugs operated in cylinders 1, 2, 3, 4, 5, 6, 7, and 8 of an 8-cyl automotive engine operated at low output. The corresponding insulators, from plugs operated at slightly higher outputs, are shown in Figs. 15 and 16.

SAE *Quarterly* Transactions

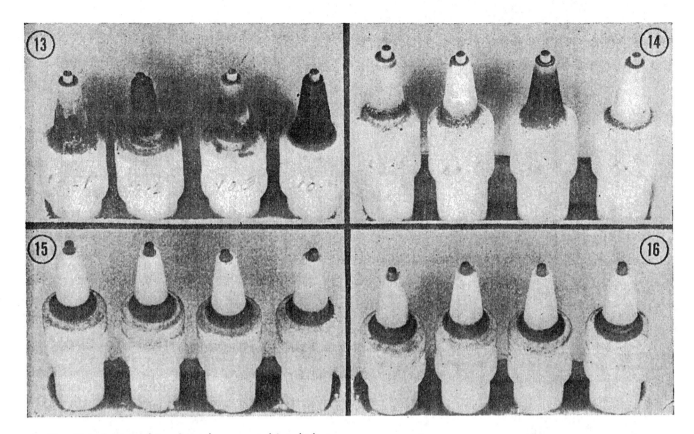

Figs. 13 and 14—Insulators from plugs operated in cylinders of engine run at low output

Figs. 15 and 16—Corresponding insulators from plugs operated at slightly higher output than that used to obtain conditions shown in Figs. 13 and 14

Figs. 17, 18, and 19—Plugs of different heat ranges operated under identical engine conditions for three different cylinders of 8-cyl engine

Fig. 20—Badly lead-fouled spark plug

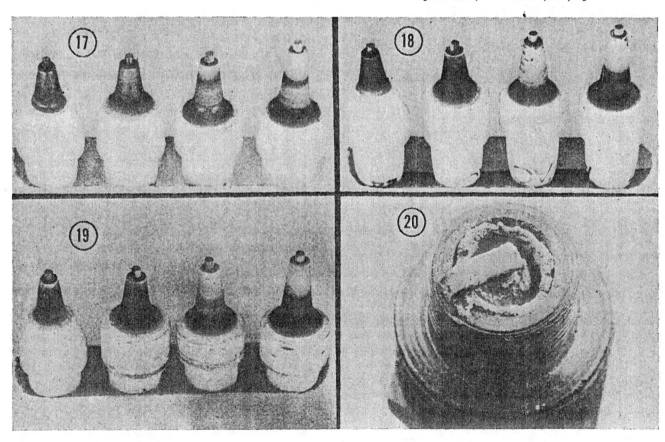

April 1948   Vol. 2 No. 2

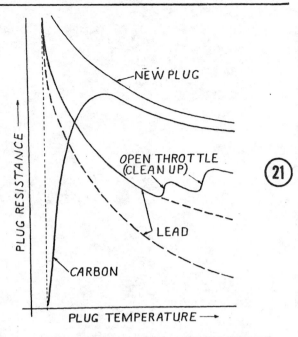

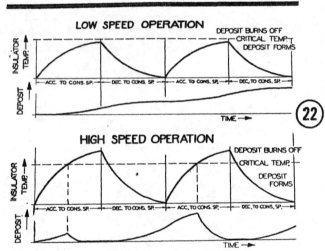

Fig. 21—Behavior of fouled plug when it is subjected to gradually increased operating temperatures

Fig. 22—Effect of level of operating temperature on buildup of carbon deposit

Figs. 17, 18, and 19 show plugs of different heat ranges operated under identical engine conditions for three different cylinders of an 8-cyl engine.

Missing or roughness at high speed or on acceleration is usually an indication of the presence of a coating of lead oxide and other lead salts on the tip of the insulator. Fig. 20 shows a badly lead-fouled spark plug. These compounds have little effect on the performance of a plug at the temperatures encountered in part-throttle and low-speed operation. At high speeds and loads, however, when the temperatures also are high, they form good electrical conductors which tend to short out the plug.

In normal operation the oxides condense as a loose powder on the cooler engine parts, such as the roof of the combustion chamber, shell of the spark plug, and regions of the insulator. As the temperature is increased the powder fuses, forming a continuous glaze which reacts with silica, if it is present in the insulator, to form a low-melting point lead silicate glass. The deleterious effect of the lead silicate glaze is due to its high electrical conductivity at high temperatures which, like the lead salts and oxides, tends to promote ignition miss. Such glazes, once formed, are difficult to remove except by mechanical abrasion in the spark-plug cleaner. The lower melting point lead salts, such as lead oxychloride and lead oxybromide, can often be burned off in the engine by slowly increasing the lead, at all times remaining below the missing point, until the plug is free of miss at all loads.

The process may be better illustrated by an example. In a certain car all the plugs were so badly fouled that it was not possible to run above 50 mph. At this speed (and temperature) the resistance of the coating was measured and found to be below 50,000 ohms. At 40 mph no miss was noted. The plugs were accordingly run at 40 mph. The speed was then slowly increased; if the rate of increase was too rapid and miss occurred, the speed was dropped back slightly. In this way the car speed was gradually brought up until it could be operated at top speed without miss. The resistance, after the test, was over 20 megohms and the plugs were free of deposit. This is an extreme case, and is mentioned here merely to indicate the nature of high-speed miss.

In Fig. 21 is shown the general behavior of a fouled plug when it is subjected to gradually increased operating temperature.

In Fig. 22 is shown the effect of the level of operating temperature on the buildup of carbon deposit.

In checking the plugs which have been removed from an engine, the question of the useful life of a plug often arises. Although it is reasonable to expect about 10,000 miles of service or 250 hr of operation, a precise figure cannot be given, since the life is governed by the operating conditions and by the service expected. Plugs with electrodes which have been worn thin must be regapped at more frequent intervals than plugs with new electrodes, so that the economically useful life would be determined by the expense of servicing and the cost of the interruption in service caused by removing the plugs for service.

The cost, in terms of wasted gasolene, of operating fouled plugs in an engine is well known and even if the poorer response of an engine with chronic miss were unobjectionable, the extra gasoline consumption should not be tolerated.

The abrasive blasting technique represents the most satisfactory method of removing the coatings deposited on the insulator tips and shells.

The operation of the cleaner is simple and, unless the compound blast is directed at the plug for an excessively long period of time, the plug is not injured in any way. The abrasive compound is selected to have sharp cutting edges rather than the rounded ones associated with ordinary sand. In addition, the grain-size distribution is controlled to give the optimum proportions of the various sizes of particles.

An important part of the reconditioning operation is the cleaning of the terminal end of the insulator and the electrodes. The former can usually be wiped clean with the aid of a little grease solvent, the latter can be cleaned with sandpaper.

Another item of the reconditioning is the gap setting which, because of irregularities of worn electrodes, should be checked with round wire feeler gages rather than with flat, leaf-type feelers.

Although methods of reconditioning spark plugs have been well developed, methods of bench testing plugs that are completely reliable have not been discovered. The difficulty lies in the inability to reproduce on a plug on the bench, the thermal conditions which prevail when the plug is operated in an engine. Because of this, certain defects, namely oxide coatings, which would cause malfunctioning in a hot engine have no deleterious effect in a cold compression box or similar tester. The difficulty in evaluating the effect of such coatings on a cold bench test makes it appear desirable to eliminate the coating by cleaning, as determined by visual inspection.

The Electric Auto-Lite Company

TOLEDO 1, OHIO

February 1, 1950

Mr. Norman G. Shidle, Editor
Society of Automotive Engineers, Inc.
29 West 39th St.,
New York 18, New York

Dear Mr. Shidle:

I am enclosing a copy of a condensation of a talk given by C. C. Cipriani before the Toronto section of the S.A.E.

This condensation, plus glossy prints of two illustrative charts is provided in the hope that it will be of some value to you.

Sincerely yours,

Zeke Cook
Steve Hannagan Associates

ZC: rs

enclosures (3)

IGNITION AND SPARK PLUG DESIGN

Condensation of a talk given by C. C. Cipriani, chief of the spark plug engineering division of The Electric Auto-Lite Company, at the November meeting of the Toronto section of the Society of Automotive Engineers.

Most SAE members will agree that the modern internal combustion engine as we know it today profoundly affects our everyday lives because of its wide and varied usage. However, not all of us appreciate that the electric ignition system as used on these engines is responsible to a large degree for the efficiency, low weight per horsepower, and convenience of operation which makes this form of power predominant in modern times.

Why does the electric ignition system contribute so much to the modern otto cycle engine? Principally because it provides the means for rapid combustion, and rapid combustion makes the high speed engine possible with its attendant advantages in efficiency and low weight per horsepower. Also, the electric ignition system in combination with the electric starting motor provides a convenience of operation unequalled by any other form of motive power.

Auto-Lite, as manufacturers of all components of the ignition system has always stressed the inter-dependency of these units for most satisfactory performance, and Figure #1 illustrates why this is the case. From Figure #1 it will be seen that as the spark plug firing tips become fouled and provide a lowered resistance to ground, the voltage of which the coil is capable depreciates markedly, also that the voltage required by the spark plugs at road loads and wide open throttle for new plugs and for plugs which have operated the equivalent of 10,000 miles is quite different. It will be observed that the used plugs require increased voltage because electrode erosion has increased

—more—

the gap width. It is obvious that spark plugs with good fouling characteristics will contribute to satisfactory performance by keeping the voltage available as high as possible, and spark plugs with good electrode erosion characteristics will contribute to satisfactory performance by keeping the voltage required as low as possible.

The modern spark plug with built in non-inductive ceramically bonded resistor not only reduces ignition interference with radio, short wave communications, and television, to acceptable levels, but also reduces electrode erosion to the point where wider gap settings and their attendant contributions to improved engine performance become practical.

Wide gap settings are capable of igniting lean or stratified mixtures which cannot be ignited by smaller settings. Therefore, under any condition of operation where the mixture at the spark plug gap is lean or stratified, wide gaps will minimize the missing condition which would exist were plugs with smaller settings in use. Because the initial volume of mixture ignited has a reduced surface to volume ratio, and less spark energy in the form of heat is lost under these conditions to the unburned mixture.

Heat range determinations on modern spark plugs are now much more precise than in past years. Practice at Auto-Lite is to rate plugs in terms of I.M.E.P. in a 17.6 inch supercharged single-cylinder engine operating on a fuel with an extrapolated rating of 150 octane. During these tests all variables of engine operation are maintained constant with the exception of engine

-more-power, which is increased in small increments by a means of increased air flow from a separately driven compressor. Pre-ignition is indicated by thermo-couple within the engine cylinder.

Lead fouling characteristics of spark plugs are determined in a similar engine equipped with a means of measuring the spark plug shunt resistance to ground during operation. Figure #2 illustrates the performance of plugs with good fouling characteristics and with poor fouling characteristics in this type of test. In this case, the engine is operated at very light loads with a heavily leaded fuel for a specified period of time. Following this a regular pre-ignition test is run using the same fuel and the I.M.E.P. that a plug will obtain expressed as a percentage of its normal rating without the shunt resistance dropping to 1 megohm is taken as the measure of its fouling rating.

-o-

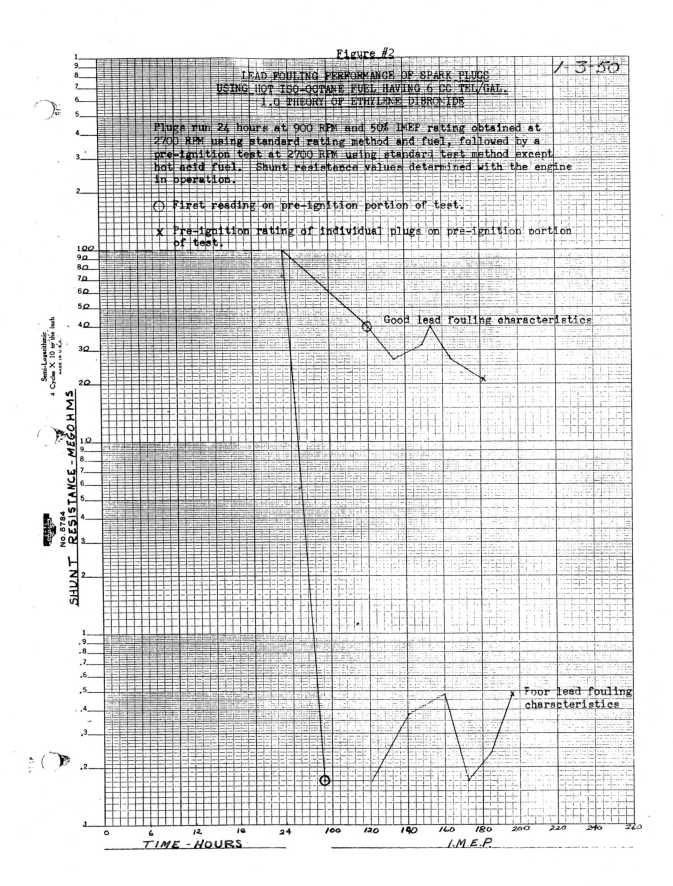

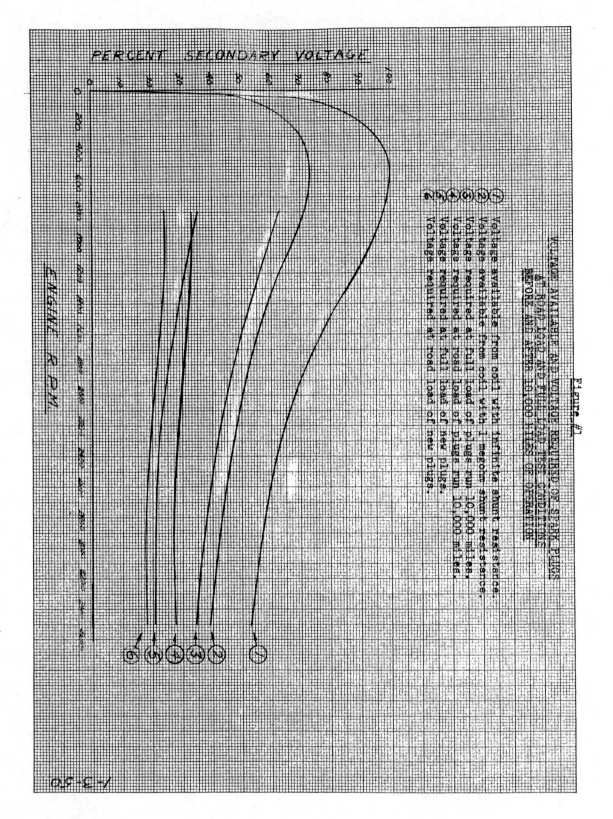

Figure #1

How Large Can the Gaps Be?

C. E. Swanson and J. W. Miller, Northwest Airlines, Inc.

This paper was presented by title at the SAE Annual Meeting, Detroit, Jan. 8, 1951.

ALLOWING spark-plug gaps to erode to a larger dimension can mean substantial savings for airlines by reducing the number of spark plugs overhauled for a given period of flying time. However, as the sparking gap increases the voltage required to fire the gap also increases, thus placing higher electrical stresses on the ignition generating and transmission system. Available space in magnetos and the continuity of insulating material in the cable to the spark plug have permitted designs that withstand voltages higher than the spark plug can transmit. The space limitations of the spark plug itself, coupled with the necessity for a connection to the cable, have limited the voltage that can be transmitted to the spark gap. This investigation is concerned with determining the maximum permissible spark-plug gap that can be fired by voltages transmitted through this critical region, the spark-plug barrel.

The first step in the investigation was to determine a satisfactory way to simulate engine operating conditions in the laboratory, this to consist of a correlation of spark-plug bomb pressures to various conditions of engine power. The second step was to determine sparking voltages required to fire various gaps at the different bomb pressures corresponding to various engine power configurations. The third step was to determine experimentally the breakdown voltage for various spark-plug barrel configurations at different altitudes and, if possible, to express this relationship analytically. The final step in the procedure was to construct a set of curves assembling the information of the first three steps into a directly usable form.

By pumping down the barrel interior while the engine was in operation, we attempted to eliminate all intermediate steps and go directly from engine power settings to maximum permissible altitude for a given gap size, but this method of attack was abandoned because of the severe carbon tracking it caused. We also attempted to eliminate the intermediate steps involving voltage measurements by simply observing the barrel pressures at which barrel breakdown occurred for various gap dimensions and bomb pressures. This procedure was also abandoned because of inadequate means to detect the advent of the first sparks in the barrel.

To obtain the correlation between engine power and bomb pressure, a spark plug was placed in a bomb and wired in parallel with a plug in an engine. (See Fig. 1.) With a given engine power setting, the CO_2 pressure in the bomb was gradually lowered until the first spark fired in the bomb. The CO_2 pressure was then lowered still further until the sparking was quite regular in the bomb and then raised until all sparks in the bomb were extinguished. The pressure at which the first spark occurred and the pressure at which the sparks were extinguished were recorded.

A SET of curves has been developed that makes it possible to determine the maximum permissible spark-plug gap that can be fired by voltages transmitted through this critical region, the spark-plug barrel.

The investigation that led to the development of these curves consisted of the following steps:

1. To determine a satisfactory way to simulate engine operating conditions in the laboratory, this to consist of a correlation of spark-plug bomb pressures to various conditions of engine power.

2. To determine the sparking voltages required to fire various gaps at the different bomb pressures corresponding to various engine power configurations.

3. To determine experimentally the breakdown voltage for various spark-plug barrel configurations at different altitudes and to express this relationship analytically.

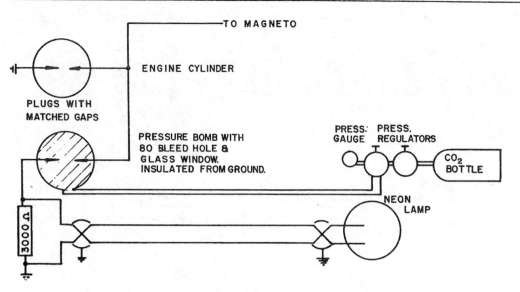

Fig. 1 – Schematic diagram of apparatus employed to determine CO_2 bomb pressures equivalent to various engine power configurations

A spark in the bomb was detected by means of a neon lamp in parallel with a 3000-ohm resistor, the combination being inserted in the ground path for the bomb. (See Fig. 1.) This method of detection was previously tried in the laboratory and proved satisfactory if the resistance was of a large enough value to eliminate interference from other plugs firing in the same vicinity. A window was placed in the bomb to permit a visual check on the indicator when it was installed on the engine test stand. The check proved satisfactory but the window could not be used throughout the test because of its proximity to the propeller.

A Sperry ignition analyzer was also installed to see if it could be used to detect the transfer of the spark from the electrodes of the plug located in the combustion chamber to the barrel of this plug as the barrel pressure was reduced to simulate increased altitude. It did detect a complete breakdown of the harness or barrel but would not permit detection of the transfer of the first spark. The predominant pattern of the very frequently repeated transient associated with the sparking in the combustion chamber masked the single trace produced by the relatively isolated first spark occurring in the barrel.

The engine used was a P&W R-2800-CA18 equipped with water injection. The operating conditions used were held as closely as possible to those shown in Table 1.

Table 1 – Engine Operating Conditions

Rpm	Map	Torque (Bmep + 1.79)	Mixture
2100	34	81	Rich
2100	34	81	Lean
2270	36.5	93	Rich
2270	36.5	93	Lean
2500	45	110	Rich
2500	45	110	Lean
2590	48.5	118	Rich
2680	50	Max (126)	Rich-wet

Lower power settings were attempted but the results were too erratic to use and not of enough importance to continue running.

Matched sets of plugs with gaps ranging from 0.010 to 0.045 in. were used for the test. The gaps for these pairs were eroded to the proper width and the two plugs fired in the same bomb and from the same source of voltage until they would fire alternately through a range of air pressures from 0-95 psi.

The results of this part of the test are plotted in Fig. 2.

In choosing an abscissa for the plot a unit proportional to charge density was desired. Preferred abscissas in their respective order would be: (1) Gaseous charge per power stroke, (2) indicated mean effective pressure, (3) brake mean effective pressure, (4) manifold pressure. Gaseous charge per power stroke and indicated mean effective pressure were not used because suitable instrumentation was not available. Since the ratio of bmep to imep is approximately the same for most currently operated reciprocating aircraft engines and since imep is nearly proportional to gaseous charge, it follows that bmep is approximately proportional to gaseous charge. Accordingly, it seems reasonable to presume that if the quantities determined are correlated to bmep they will be applicable to other engines without great error or without great difficulty, since bmep values for various operating conditions are readily available for most engines. The engine used was equipped with a torquemeter so bmep could be obtained by applying a simple multiplier to the torque readings. Because of these factors of convenience bmep was chosen as the preferred abscissa.

Differences between the sparking characteristics of the engine charge and CO_2, and differences in temperature of the electrodes, make a detailed analysis of the general curves difficult. Equivalent

CO_2 pressures would, however, be expected to rise with bmep at an increasing rate because the rapidly increasing friction losses require a larger than proportionate charge. The upward concavity of the curves in Fig. 2 show this trend.

In plotting the curves, averages of all readings were used. The large number of readings taken and the resultant smooth curve of averages serve to give assurance that the influence of experimental errors has been reduced to relative insignificance. However, because of the large departures of the individual readings from the averages, these departures being attributable to inherent characteristics of the reciprocating engine more than to experimental errors, it was decided to use the highest average values plus the mean absolute departure from the averages as the spark-plug bomb pressures for determining maximum permissible gaps. Very few points actually lie above this curve.

It is interesting to note that the equivalent bomb pressures are very little higher for lean mixtures than for rich mixtures, although a difference definitely exists. The results obtained using water injection are also very interesting. They are not, however, considered as reliable as the rest of the data because power settings could not be held constant and readings had to be taken in a very short time to avoid injury to the engine. Six sets of readings with an average bmep of 221 psi had an average equivalent CO_2 bomb pressure of 63.5 psi. Since only two of these sets of readings were higher than the average, it is expected that the true equivalent CO_2 bomb pressure would be about 55-60. Additional tests with water injection should be run, but it is believed that an extrapolation of the curves obtained without water injection will give reason-

The Authors

C. E. SWANSON has been manager of general and aircraft engineering with Northwest Airlines, Inc., since 1939. He obtained his degrees of B.S. and M.S. in electrical engineering from the University of Minnesota.

J. W. MILLER is an aircraft electrical engineer with Northwest Airlines, Inc. He studied electrical engineering at the University of Minnesota, graduating in 1948.

ably accurate figures for equivalent bomb pressures.

The equipment used to accomplish the second step of obtaining curves of sparking voltage versus CO_2 bomb pressure, for various spark plugs and gaps, consisted of a variable 60-cycle a-c supply (see Figs. 3 and 4) connected to a spark plug in a CO_2 bomb. The bomb had a glass window to observe the spark. The spark plugs used were Bendix type 7KLS-2 and Champion type R37S-1. The R37S-1 gaps were eroded to desired dimensions by actual service in a P&W R-2000 engine. The 7KLS plug gaps were eroded to dimension by fast cycle sparking in the laboratory.

The procedure used to conduct the test was to set a given pressure of CO_2 in the bomb and raise the voltage until it fired the gap in the plug continuously, or nearly continuously. The tests on the Bendix 7KLS-2 plugs were run at Northwest Airlines, Inc., in 1941 and the tests on the Champion R37S-1 plugs were run at Northwest Airlines, Inc., in 1950.

The results of the two tests are plotted in Fig. 5. In spite of considerable differences in the con-

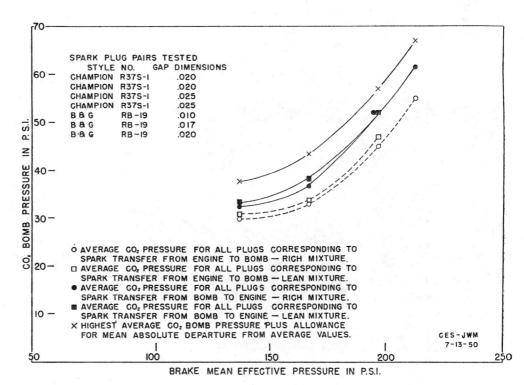

Fig. 2 – CO_2 bomb pressure equivalent to various engine power configurations – P & W R-2800-CA18 engine

April 1951 Vol. 5 No. 2

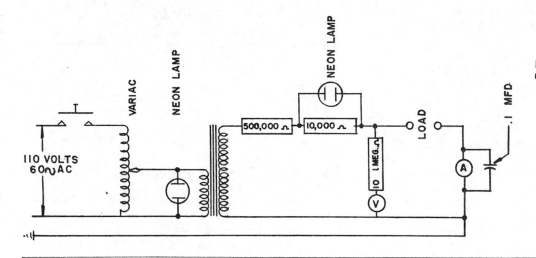

Fig. 3 – Schematic diagram of 60-cycle high-voltage source and breakdown indicator

struction of the two types of plugs, the voltages required to fire equal gaps are practically identical. Several other curves of Champion R37S-1 plugs were run at gaps between 0.025 and 0.035 in., and the curves fell almost exactly in their proper place as interpolated between these curves. It is quite difficult to get an exact measurement of a sparkplug gap with errors not exceeding 0.001 in. Inspection of the curves of Fig. 5 seems to indicate that an error was made in measuring the gap designated as 0.030 in. Probably a dimension of about 0.029 in. existed. A fine wire plug was also tested in a similar manner but a lesser degree of accuracy existed in determining the gap dimensions. Ninety per cent of the points taken had less than 5% departure from these curves. These results lead to the conclusion that sparking voltages for a given CO_2 pressure are primarily a function of gap dimension and that existing differences in sparking electrode construction do not change it appreciably.

The family of curves of Fig. 5 clearly shows that the potential gradient required to produce sparkover increases as the gap spacing is reduced.

As one would expect, these curves conform with the general statement of Paschen's law, which states that the sparkover voltage is a function of δ_x, the product of the relative density and the gap spacing. Values of sparkover voltage were read from each of the curves while holding constant the product of absolute pressure in pounds per square inch and the gap spacing in inches. In one check this product was arbitrarily chosen as 1 and each curve showed a sparking voltage within the range of $6333 ^{+167}_{-283}$ v, with a mean absolute departure of only 1.66% from the average. In a second check the product was arbitrarily chosen as 2 and each curve showed a sparking voltage within the range of $9791 ^{+209}_{-191}$ v, with a mean absolute departure of only 0.93% from the average. Further, it is to be observed that every member of this family of curves exhibits the concavity downward which characterizes the curve of Paschen's law in the regions corresponding to high values of the product δ_x.

The step of comparing barrel breakdown voltage versus air density was accomplished with test equipment consisting of a voltage supply and current indicator, a vacuum pump, and gage, and spark plugs with the electrodes cut away. (See Figs. 3, 4, and 6.) A high air pressure was held in the bomb to make certain sparks were suppressed at the electrode end of the plug, even though the electrodes were removed.

The procedure used was to set a given barrel pressure and raise the voltage until the neon lamp flashed at intervals of less than 5 sec. This procedure was checked by looking into the interior of glass insulated barrels to make sure that a flash of the neon lamp did indicate a flashover in the barrel. Barrel pressure readings were taken in inches of mercury and the air density ratios were computed by dividing the absolute pressures ob-

Fig. 4 – Equipment used to determine variation in sparking voltage with CO_2 bomb pressure

SAE *Quarterly* Transactions

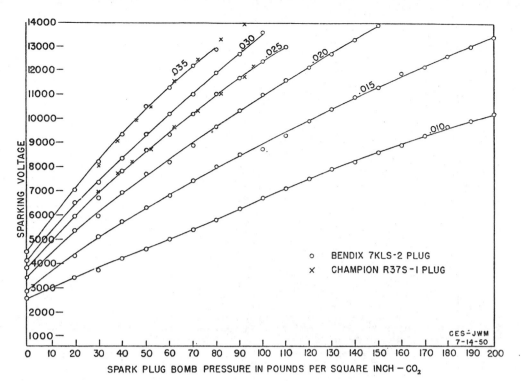

Fig. 5 – Sparking voltage versus CO_2 bomb pressure for various spark plugs and gaps

tained by the standard absolute air pressure. The temperature throughout the test was not controlled but room temperature did not vary appreciably during the tests. Cable and insulators were changed with the same spark plugs to get a comparison of 5-mm and 7-mm cable and of different types of insulators.

Equivalent altitudes with a constant barrel temperature were calculated by the following equation:

$$\delta = \frac{PT_0}{P_0T} \quad (1)$$

where:

δ = Air density ratio at altitude in question
P = Standard air pressure at altitude in question
P_0 = Standard air pressure (29.92 in. of Hg) at sea level
T = Standard air temperature (15C or 288K) at sea level

For a constant barrel temperature of 100C:

$$\delta_{100C} = \frac{288P}{373 \times 29.92} = 0.0258P \quad (2)$$

Results of this test are plotted in Fig. 7.

The curve of all the points plotted was expected to follow the general form of Peek's[1] equation for the advent of corona between two concentric cylinders:

$$g_v = 31{,}000\delta \left(1 + \frac{0.308}{\sqrt{\delta r}}\right) \quad (3)$$

where:

g_v = Potential gradient for advent of visual corona, v per cm
δ = Air density ratio
r = Radius of inner cylinder, cm

Two points were chosen and an equation of that general form fitted to them and plotted. The curve fits the points very well in the lower and middle extremities but in the upper portion it breaks and levels off. The reason for this radical and abrupt departure is not known, neither is it consistently reproducible. Because of this departure and because of inability definitely to segregate curves corresponding to individual barrel configurations, the lower boundary curve of all points obtained was chosen as a representative curve to determine actual limitations. It is on the safe side but holds to the points obtained very closely.

Two theories have been considered in attempting to arrive at an analytical expression for the breakdown voltage of the spark-plug barrel. The first of these is based on the well-known fact that visual corona is evidence of ionized air and on the premise

Fig. 6 – Spark plugs used in test

[1] See "Dielectric Phenomena in High Voltage Engineering," by F. W. Peek. Pub. by McGraw-Hill, New York City, 1929.

April 1951 Vol. 5 No. 2

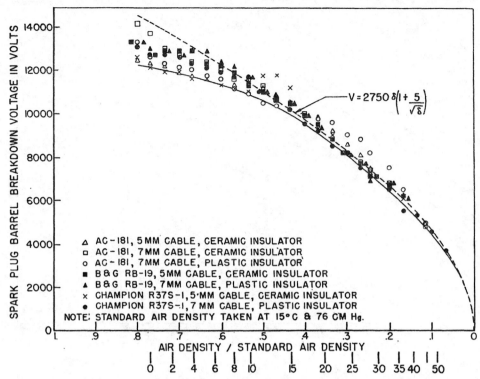

Fig. 7 – Breakdown voltages for various spark-plug barrel configurations and various air density ratios

that the existence of corona for the full length of a possible breakdown path will cause disruptive breakdown along that path. The second theory is that sparkover forms progressively by following successive local breakdowns, thus forming its own breakdown path as it progresses.

It is very important to determine which theory predominates because design changes that will effect improvement may be radically different for each. The efficaciousness of such changes cannot be predicted unless the basic governing theory is known.

Calculations of existing potential gradients for a given applied voltage can be made as explained in Table 2 and Fig. 8. If it is assumed that the first theory predominates, it would be expected that the potential gradients required to cause breakdown could be calculated by equation 3. How-

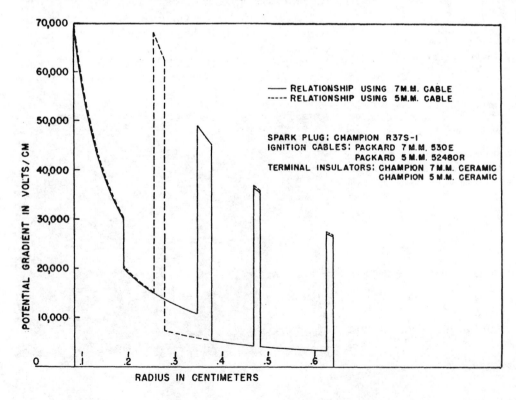

Fig. 8 – Computed potential gradients in spark-plug barrel – assuming electric field to be radial and voltage equals 10,000 v.

SAE *Quarterly* Transactions

ever, a review of Peek's experimental development of the equation disclosed that it did not necessarily apply for very small spacings between concentric cylinders. He did not make an investigation of this phenomenon for concentric cylinders but did make a limited investigation of breakdown gradients for small sphere gaps. As an approximation, his conclusions regarding potential gradients required to break down small gaps between spheres can be applied to the advent of corona in small concentric cylindrical gaps. This is justifiable because the electric intensity vectors in both cases are very nearly parallel and constant and because with these spacings and this electric field configuration, breakdown and the advent of corona are simultaneous. The critical spacing, below which Peek's empirical equation no longer applies to sphere gaps, is 0.54 $\sqrt{r}$, where r is the radius of the spheres. Applying this to 5-mm wire the critical spacing would be $0.54\sqrt{2.5} = 0.855$ mm. The spacing that exists in a spark-plug barrel between the wire and terminal sleeve, where it has been found that breakdown occurs, is about 0.24 mm, which is considerably smaller than the critical spacing.

Although Peek does not give an exact relation for the advent of corona or sparkover for these small spacings, he does present data which show that a potential gradient of 110 kv per cm would be required to cause breakdown in a gap of this size when the air density ratio is 1. At the voltage causing spark-plug barrel breakdown adjacent to 5-mm cable the potential gradient calculated to exist adjacent to the cable insulation is 88 kv per cm. For 7-mm cable it is about 62 kv per cm. It appears improbable that this large difference between breakdown gradients in the air space adjacent to 5-mm and 7-mm cable can be solely attributed to the difference in radii. That barrel breakdown occurs in this region in spark plugs employing one-piece core construction was conclusively established by direct observation using the special AC plug illustrated in Fig. 6. The findings obtained in the investigations reported herein constitute considerable support to the applicability of the first of these theories in that the sparkover occurs in the regions predicted by the theory and by the fact that substantial departures from the standard barrel length produced only relatively small changes in barrel sparkover voltage. (See Figs. 9 and 10.)

In order to ascertain more conclusively the applicability of this theory to spark-plug barrel problems and design, it will be necessary to conduct further experimental investigations to determine the critical potential gradients of radial fields required to produce visual corona in very thin air films of various densities and having the form of cylindrical shells of various thicknesses and radii. In addition, experiments will have to be conducted with longitudinal fields superposed on the aforementioned radial fields to determine the circumstances under which sparkover occurs longitudi-

Table 2 – Determination of Potentials and Potential Gradients Existing within Spark-Plug Barrel under Postulate of Radial Electric Field

Let:

a = Radius of conductor (assumed to be circular in cross-section)

r_1 = Radius of outer surface of first insulation layer

r_2 = Radius of outer surface of second insulation layer and so on

r_{n-1} = Radius of outer surface of last insulation layer

b = Radius of inner surface of grounded shield

x_i = Distance from center of conductor to an interior point of i th insulation layer

$$r_{i-1} \quad x_i \quad r_i \qquad i = 1, 2, 3 \ldots\ldots\ldots n$$

E_i = Electric intensity at a point distant x_i from center of conductor

V_i = Potential at a point x_i from center of conductor

k_i = Specific inductive capacity of material comprising i th insulation layer

λ = Electrostatic unit charge per unit length of conductor (chosen positive)

The electric intensity E at all points of a cylindrical surface concentric with the conductor may be computed by the use of Gauss' Law:

$$\int_S E \cos\alpha\, dS = \frac{4\pi}{k}\sum q \qquad (1)$$

where S is a surface enclosing the free charges Σq, and α is the angle between the lines of force and the outward drawn normal to the surface.

In this particular case the lines of force are everywhere radial, so for a unit length equation (1) yields:

$$2\pi x_i E_i = \frac{4\pi\lambda}{k_i}$$

$$E_i = \frac{2\lambda}{k_i x_i} \qquad (2)$$

The potential at any point in the system is obtained by computing the work required to move a unit positive charge from the inner surface of the shield to that point. It is thus determined that V_o, the potential of the conductor, may be expressed as:

$$V_o = \lim_{\epsilon \to 0} \left[\int_{a+\epsilon}^{r_1-\epsilon} E_1 dx_1 + \int_{r_1+\epsilon}^{r_2-\epsilon} E_2 dx_2 + \cdots \int_{r_{n-1}+\epsilon}^{b-\epsilon} E_n dx_n \right] \qquad (3)$$

where $o < \epsilon$

The potential gradients $\dfrac{dV_i}{dx_i}$ are given by the relationship:

$$\frac{dV_i}{dx_i} = -E_i \qquad (4)$$

Substituting equation (2) in (3) and (4) and eliminating λ between the resulting equations it is determined that

$$\left|\frac{dV_i}{dx_i}\right| = \frac{1}{k_i x_i} \cdot \frac{V_o}{\left[\dfrac{1}{k_1}\log_e\dfrac{r_1}{a} + \dfrac{1}{k_2}\log_e\dfrac{r_2}{r_1} + \cdots \dfrac{1}{k_n}\log_e\dfrac{b}{r_{n-1}}\right]} \qquad (5)$$

A graph of these relationships applied to a Champion R37S-1 spark plug used in conjunction with Champion style NW-1 ceramic terminal insulator and Packard style No. 52480-R 5 mm copper conductor cable appears in Fig. 8.

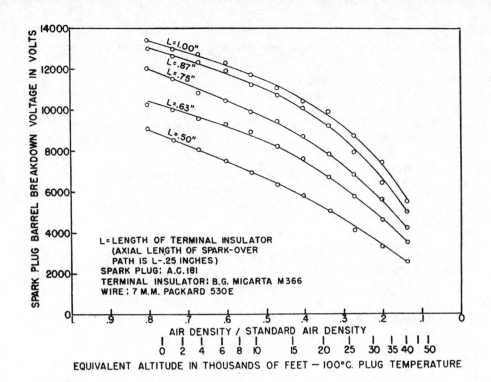

Fig. 9 — Breakdown voltages for various spark-plug barrel lengths and various air density ratios

nally through these air layers. For these tests the curved surfaces bounding the thin cylindrical air films will, of course, have to be of nonconducting material in order to prevent radial sparkover.

If it is assumed that the second theory of a self-propagating sparkover is true, barrel length should be a prime factor in determining the breakdown voltage. The curves of Figs. 9 and 10, plotted from data taken with various terminal sleeve lengths, show that barrel length does affect the breakdown voltage but at a definitely diminishing rate near the existing length and at high air density ratios. At the lower air density ratios, or higher altitudes, it appears that further increases in barrel length would be of some benefit. This lends some support to the second theory, but it appears that a combination of the two theories may actually govern the phenomenon of sparkover.

The final results of the step of applying the information previously obtained to the actual determination of the maximum permissible spark-plug gap are presented in the family of curves appearing in Fig. 11.

To develop this family of curves, a given altitude temperature and barrel temperature were chosen and the breakdown voltage corresponding to the air density determined by them was taken from Fig. 7. Applying this breakdown voltage to Fig. 5, maximum permissible spark gaps for various bomb pressures were determined. These bomb pressures were then converted to bmep by Fig. 2 and the results plotted. This procedure was repeated for

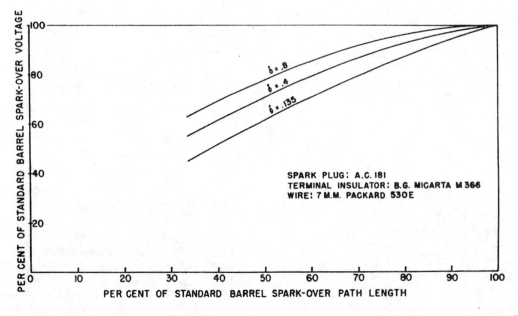

Fig. 10 — Relation between sparkover voltage and barrel length

various altitudes and barrel temperatures. Since exact temperatures existing in spark-plug barrels were not known, curves for 100 C and 150 C were plotted.

Interpolation can be employed to ascertain the maximum permissible gap for spark plugs having barrels operating at intermediate temperatures. If necessary, additional curves can be plotted using the procedure heretofore described.

If supercharged harnesses are employed the altitude curve to be applied would be the one having a standard atmospheric pressure equal to the actual pressure existing in the spark-plug barrel.

The inflection of these curves from concavity upwards to concavity downwards as bmep is reduced is attributable to the diminishing slope of the curve of Fig. 2 in the corresponding region of bmep.

Possible sources of error in performing experiments of this character are: temperature influence on the barrel atmosphere, dirty barrels and cable, and impulse effect errors due to using 60-cycle a-c supply voltage in the laboratory.

Throughout the conduction of the test, the room temperature did not depart more than 15 C from the arbitrary aeronautical standard air temperature of 15 C. The maximum possible error produced by this, in calculations of air density ratio for a given altitude, would be $\frac{288-303}{288} = 5\%$. About half of this is a usual error but is of little consequence to final results.

Barrels, insulators, and cable were kept as clean as possible throughout the tests. This condition may or may not exist in actual service depending on harness configuration and care in handling of components.

Since the breakdown voltage for a given gap configuration is not necessarily the same for a rapidly increasing voltage as for a slowly increasing voltage, it was considered necessary to determine if the altitude limitations of a spark-plug barrel would be different when the source of high voltage was a magneto instead of an alternating-current transformer. To accomplish this a glass barreled spark-plug with a window was supplied with adequate a-c 60-cycle voltage to fire its spark gap for a given bomb pressure. The barrel was then pumped down until the spark began to transfer from the bomb into the barrel. This barrel pressure was recorded. The spark transfer test was then repeated with the same bomb pressures using a magneto for the power source. The results are shown in Table 3.

The points obtained using a magneto voltage source agree very well with the points obtained using a 60-cycle a-c voltage source. This shows that the impulse ratio for breakdown in the barrel is approximately equal to the impulse ratio for breakdown at the electrodes.

Since the spark transfer in the above test was detected directly by visual observation, the results obtained afford an excellent check on the data plotted in Fig. 7. If the points defined by the voltages and barrel pressures given in Table 3 are superposed on Fig. 7, it will be seen that they agree very well with the plotted data.

Table 3 – Test Results

Bomb Pressure, psi	Barrel Pressure Causing Spark Transfer, in. of Hg		60-Cycle Peak Voltage
	60-Cycle A-C Source	Magneto Source	
92	22.5	23.5	13,300
72	17.5	17.5	12,000
49	12.9	13.0	10,000
22	7.0	7.0	7,000
0	3.5	3.25	4,530

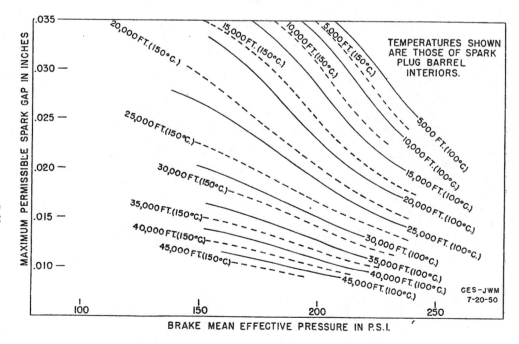

Fig. 11 – Maximum permissible spark gap for various operating conditions

Operation of Spark Plugs in Present Day Engines

In the present passenger car engine, the old adage, that a cold plug was needed for a hot engine, is not holding true. It has been the feeling for years that high compression engines being driven at high speeds should have a cold operating plug. Today in our modern engines, which have compression ratios between 7.2 and 8 to 1, and which are equipped with over-drive or fourth gear, we find that the horsepower required to move this vehicle over the road at 70 to 80 miles per hour is only a small part of the horsepower that is available from the engine. Consequently, the spark plug situation has to be taken into account.

In the early days of motoring, all spark plugs had a glaze on the firing end of the insulator to better resist carbon fouling which was then the main problem. In the early 20's, the introduction of tetraethyl lead required the glaze to be eliminated due to attack by lead deposits. About 1935 to 1937, the fuel situation was starting to change again with the increasing use of leaded fuels. Up to that time, the insulating material used in spark plugs was of such a nature that it would be affected by the tetraethyl lead in the fuel, and tests were run on new Champion insulator materials which would resist the attacks of the lead. After many hours and miles of testing, it was found that insulators incorporating a large percentage of aluminum oxide had the greatest resistance to the attack of tetraethyl lead. In 1937 this new improved insulator, incorporating aluminum oxide bodies, was placed in Champion production, and as fast as the transfer could be made all the plugs were changed over from our original Sillimanite Body to our improved ceramic, which was basically aluminum oxide.

With the change to the improved ceramic, it was necessary to redesign many of the plugs, since the thermal conductivity of the new insulator body was higher than the original Sillimanite Body. So to keep preignition rating the same, it was necessary to lengthen the insulator nose from 1/16" to 1/8", depending on the type of the plug. This meant that the fouling range of the plug was improved while keeping the heat range of the plug constant. Also, the tetraethyl lead had no effect on the new insulator body.

At the present time, we are again having changes - this time the engines being changed to a more efficient and higher specific output. Owners of cars are also conscious of the fact that with the advent of improved driving comforts, such as automatic transmissions, the fuel economy of many units is not what it used to be when they shifted gears and had over-drive on their cars. Also, fuel conditions are changing slightly and in certain localities the sulphur content of the fuel has increased, with the octane rating of the fuel dropping slightly, thus creating a problem which is slightly different than previously encountered back in 1937, and which must be handled in a slightly different manner. Along with this, the use of certain types of detergent oils or oil additives has created a spark plug problem which to date is not entirely solved.

One of the main problems which faces all car manufacturers is the fact that when new models are brought out on the market, the original spark plugs recommended for many of these cars are the plugs which have been used on the dynamometer stands; and engineers who set up the engines on the stand are trying to find out how much horsepower the engine develops, along with life tests of the engine components. Many times with full throttle, full power ratings, the standard spark plugs which they had been using for some time may be slightly on the hot side and may cause a slight pre-ignition. To fight this problem, a colder plug is used which will eliminate this situation from their dynamometer test.

Many times these spark plugs would be approved, production released, and then after the customer drives the car it is found the spark plug fouls. This condition may be due to the customer not driving the car with as high a percentage of the available power as he did with his lower compression engines. This is born out in the fact that city driving has created the greatest fouling problems.

Along with this fouling problem, there arises a fuel deposit problem which creates a skip or miss in the engine after approximately 1500 to 3000 miles of driving. The deposits on the insulators vary considerably in different localities, but we found that the majority of these deposits are broken down into about four basic types - lead sulphate, lead oxide, barium sulphate and lead phosphate.

We believe that the barium sulphate and the lead phosphate may come from some of the detergent oils. It has been found from our tests that when the fuel deposit is of a fluffy nature on the insulator nose, shorting out does not occur as quickly as when the deposit is of a vitreous nature. In actual resistance tests, we find that as the temperature increases, the resistance across the insulator nose seems to drop very suddenly after the insulator nose had attained a temperature of around 500 to 700°F. From the results of our examination, a slight skip or miss will be felt in the engine on a hard pull or at high speeds. We also find that of the fuel deposits which contain a heavy lead phosphate content, the material has a tendency to attack the insulator under certain operating conditions.

At the present time, the general solution for curing the skip or miss has been to clean the plugs at intervals of about 3000 miles, file the sparking surfaces of the electrodes, and the car will then operate satisfactorily. It has also been noted that some of these conditions exist mainly when the car is new but after it is broken in, the deposit situation seems to lessen slightly.

We have also found in many cases that the problem of spark plug installation has not been seriously considered by the car manufacturers nor by the service departments at the different dealers. Many times spark plugs are installed properly at the factories, and after delivery of the cars to the dealers, the mechanic will remove the plugs to check them, and then not properly reinstall them. With improper installation of the correct heat range plug, we find that the plug will tend to overheat and the deposit problem becomes more serious. It is also true that a hotter operating plug, say one step hotter, can be used very satisfactorily in many of the high compression engines if properly installed.

It is not very often that people drive over 70 to 75 miles per hour on the road, since state laws and driving conditions are a factor in the high speed operation of the vehicle. The hotter type spark plug will operate satisfactorily and will provide a longer fouling path, thereby lengthening the period of time when the spark plug should be serviced.

In conjunction with proper installation of spark plugs, we also find that the elimination of induced current between ignition cables is becoming more and more of a factor with the use of higher compressions. It has come to our attention time and time again that the skip or miss in an automobile can be eliminated with the use of a new spark plug, but after a couple of hundred miles of operation, the same skip is back in the engine. People have gone so far as to grind valves, install new distributor points and completely overhauled carburetors when the only thing necessary to eliminate the trouble was the separation of the ignition cables.

We are carrying on extensive research toward the development of plugs which will give broader fouling ranges, together with the elimination of fuel deposit problems which are now crowding into the picture differently than when tetraethyl lead was first introduced. We believe that the time is not too far off when these problems will be overcome, but until that time arrives, it will be necessary to clean and test our plugs more frequently so that we all can enjoy the pleasure of driving a car which will give us the driving comfort desired by everyone.

REPRINT---Paper was presented at a meeting of the Mid-Continent Section of the Society of Automotive Engineers on October 5, 1956. Subject to revision. Permission to publish the paper, in full or in part, after its presentation and with credit to the author and the Society, may be obtained upon request. The Society is not responsible for statements or opinions advanced in papers or discussions at its meetings.

NEW ENGINES DEMAND TAILORED SPARK PLUGS

By

RICHARD C. TEASEL

Director of Research
Champion Spark Plug Co.

Society of Automotive Engineers, Inc., 485 Lexington Avenue, New York 17, N. Y.

NEW ENGINES DEMAND TAILORED SPARK PLUGS

Our production spark plug is a simple little thing. It has only three main parts --- an insulator, a shell and electrodes. Like the weather, the spark plug is taken for granted, and nothing much is ever done about it. Once in awhile somebody tries a new material. For example, in a French patent issued on January 24, the shell is brass - we use steel; the electrodes are platinum - we use platinum in a few plugs, but usually nickel alloy; the insulator is ceramic - so is ours. In fact, our present spark plug is quite well described in that French patent of January 24 ----- January 24, 1860.

So what has been done for 95 years since the brilliant Jean Lenoir patented his celebrated engine and described our spark plug in a single sentence? Of course, for the first 30 years we had a good excuse ... we didn't have an automobile to use it in. But then things began to happen, and the automobile has been the subject of more research, development and testing than anything man has ever built. Change after change, improvement and more improvement ... Clutch, gearshift, choke, spark advance ... all automatic and out of sight ... Radiator and gas tank have gone "underground". From a rubber-bulb toy, the horn has become a finely-tuned electric organ. The crank --- well, you just can't hardly get them no more. The only crank many of you have ever seen is the one behind the wheel!

But you lift the hood and there is the spark plug --- still pretty much as Lenoir described it before the Civil War.

Still, it looks as though the lawyers have been busy, for in the United States alone spark plugs have been the subject for more than two thousand eight hundred patents.

Why all these patents? The plug has only one job; to ignite the fuel charge. Even in more technical terms, it still seems simple: The spark plug must conduct into the combustion chamber the high-voltage current and discharge this current across the electrode gap so as to ignite the fuel-air mixture.

Yet, in the last two-and-a-half years alone, our Research personnel have spent more than 65,000 man-hours trying to improve it.

Actually, we have two sets of problems. We have some hardy perennials that are as old as the spark plug itself, and a crop of new ones that come in with every change in engine design. First, let's take a brief look at some of the old problems.

As you may know, the engine imposes upon the spark plug three main requirements. (See Fig. 1.) One, the plug must operate below preignition under the most severe conditions. Two, the insulator tip must operate sufficiently hot under low-speed conditions to resist fouling deposit. And three, the electrode materials must withstand corrosion and erosion.

To assign values to these factors is the function of the test equipment. By so doing we are able not only to solve a current problem in the least time, but to some extent, anticipate a problem and correct it before the engines and plugs reach the field.

LET'S LOOK AT THE WORKING CONDITIONS. The spark plug must operate at temperatures that may cycle from sub-zero to 4000-degrees Fahrenheit. In fact, in cold-weather starting, these two extremes can exist at opposite ends of the plug at the same time. Despite the widely different expansion characteristics of its components, the spark plug must remain not only intact but also gas-tight. Pressures may range from sub-atmospheric up to 2000 lbs. per square inch; voltage may go as high as 18,000 volts - yet flashover or electrical leakage must be avoided. The fuel-air mixture may fluctuate from lean to rich; fuel and oil may contain various amounts of anti-knock compounds, sulfur, scavengers, and metallic additives; the spark plug must resist chemical attack by these substances. It must also prevent their accumulating in amounts that will cause fouling. It must do these things throughout the entire output range of the engine and under all habits of driving.

The newer problems are no easier.

Perhaps some of you have wondered why, after so many pre-war years of relative freedom from spark plug fouling, it became so serious with the advent of the redesigned post-war cars.

Many concurrent changes brought on this situation. (Fig. 2)

Engine horsepowers were sharply increased, and to improve, or in many cases merely to maintain fuel economy, the engine builders increased the compression ratios. Higher compression ratios necessitated an increase in the octane number of the gasoline to prevent detonation; that was accomplished in part by increasing the amount of Tetraethyl lead.

Not only was the total power output increased, but also the horsepower per cubic inch of engine displacement. As only a very small percentage of all this power would be used at low speeds, during break-in and city driving these engines had to be operated with the throttle nearly closed. This of course produced a high vacuum in the intake manifold. At the same time, because of the very low-power operation, combustion chamber temperatures were extremely low.

The high manifold vacuum aggravated the oil leakage problem, (Fig. 3) particularly around the intake valve guides in some of the early overhead-valve V-8 engines. Of course, added to this was the old problem of oil leakage past the piston rings during the engine break-in period.

Because of the low temperatures in the combustion chamber, all this oil could not be completely burned. The result was excessive deposition of carbonaceous material on the spark plug. In plain English --- FOULING.

The throttling down of the large engines had other bad effects. It caused uneven distribution of the fuel. Some cylinders ran rich and others ran lean, causing wide temperature variations between cylinders. It even affected the distribution of the tetra-ethyl lead within the fuel. Some cylinders could be getting twice their share of lead while others got practically none at all.

The introduction of four-barrel carburetion in the large-displacement engines further complicated the distribution problem. Spark plug temperature at high speed or large throttle opening is an important factor in controlling spark plug fouling. We have found spark plug temperature variations between cylinders, under certain operating conditions, greater than the temperature difference between a J-8 and a J-14 spark plug. This was, and still is, a major fouling factor.

I do not mean to criticize the engine and fuel changes I have just mentioned. The post-war period was and continues to be a time of change. These changes, all or in part, are common to all car makes. It is true, however, and is becoming more apparent all the time, that every engine model has its own fouling peculiarities.

For a moment let's have a look at the magnitude of some of the things I have just mentioned.

Here we have charted the average concentration of tetra-ethyl lead in U. S. passenger car fuels and its increase over the years. (See Fig. 4.) Currently, the legal maximum is 3.0 ml/gal and from 1940 to the present time, we have had a 50% increase.

In this chart, Fig. 5, are shown the trends of engine factors that primarily determine the spark plug requirement. Note that between 1934 and 1950, though there was a gradual increase in the horsepower per cubic inch of engine displacement, the maximum horsepower remained practically constant. However, note the sudden increase in power after 1950. The compression ratio has followed the octane improvement. It would seem these curves are going to have to level off sometime but it won't be for awhile.

To translate these late engine changes into spark plug performance is a difficult job. However, this chart pretty well represents the fouling problem, Fig. 6. Car A may be considered representative of the cars from 1933 to 1950; Car B - of high horsepower cars since 1950. The high horsepower, particularly the high horsepower per cubic inch, requires a colder plug to prevent preignition under high output. Such a plug of course runs quite cold under low output. The large engine does very little work in moving the car at low speeds. This results in lower horsepower per cubic inch and lowers the combustion chamber temperature. All of this encourages lead and carbonaceous deposits ... result ... fouling.

It became very apparent that we needed new standards and new techniques so that our spark plug designs could be literally tailored, or matched, to these highly critical new engines.

To do this, and retain our leadership in the industry, it was necessary to learn more about engines, and more than we know, or thought we knew, about spark plugs.

It was decided to tackle the problem from some new angles which, as far as we know, no one else in the industry had tried. This would call for some special new instrumentation that simply did not exist, and there was no choice but to develop and build it ourselves.

Sometimes we were lucky enough to find something from which certain parts could be used, but usually it was a case of starting from scratch.

To make things more hectic, we did not always know that an instrument would work, or that it would tell us anything useful even if it did. Often we found ourselves right back where we started. But let's skip that, and tell you a little about some of the results.

For example, thermocouple spark plugs. The job may sound simple: measure the temperature at various points on the spark plug while it is operating in an engine. But the measurements had to be accurate, and the thermometer had to be able to take the terrific beating of 4700-degree gas temperatures and 2000-pound pressures. Still it did not dare affect the operation or the temperature of the spark plug. So anything that went into the business end of the spark plug had to be very, very small.

It is just that ----- two wires, welded together, one platinum and the other a platinum alloy, each wire slightly thicker than a human hair, .005-inch in diameter.

As you know, a thermocouple operates on the principle that two dissimilar metals, when joined together, give off an electrical voltage. This voltage is of course very small, but it is measurable, and it is proportional to the temperature existing at the point of junction.

Special techniques had to be developed for making and drilling the insulators, and a machine was designed and built for electrically welding the fine wires together.

Probably the most remarkable feature of this thermocouple instrumentation is the method of measuring the infinitesimal voltage of the element itself (about eight one-thousandths of one volt). We virtually sift it out of the ten to fifteen thousand-volt ignition potential that surges through the thermocouple. That measurement is about as fine as laying a trans-Atlantic cable and having one inch of cable left over.

Here in Fig. 7. is a typical thermocouple spark plug as used in these programs.

This one has two thermocouples -- one on the center electrode and one on the insulator tip. The leads are brought up along the center electrode, but they do not make electrical contact with it. They terminate in a special connector at the top. This construction makes possible a thermocouple plug having the same heat range as the standard spark plug.

Our thermocouple temperature measurement work has been used in the aircraft and automotive industries. American Airlines used our thermocouple equipment in their first DC-6-B's, and it was aboard one of the B-36's on the history-making non-stop flight from Carswell Air Force Base to Arabia.

Another important piece of equipment is the voltage wave form analyzer. See Fig. 8. Most of you know it as the ignition analyzer, used in aircraft ignition work. In fact, this is only one of several cases where we took instrumentation and techniques that had been used only in the aviation field, and adapted them to this program.

Our analyzer is a basic oscilloscope. We use the secondary voltage at the spark plug for the pattern on the screen. This tells us a great deal about normal and abnormal spark plug performance better than primary voltages. Three typical secondary voltage patterns are shown in the next figure.

Each of these patterns is actually a slow-motion picture of an event that lasted just about one-thousandth of a second.

On the left is a typical normal pattern. A pattern is read from left to right. The vertical pulse at the left indicates the voltage required to fire the gap -- the longer the vertical pulse, the higher the voltage. This one is downward because this was a negative polarity spark. The series of shorter lines following are re-firings. They are shorter because the gap is already ionized from the initial pulse.

The second pattern is that of a fouled spark plug. You can see that there is no sharp pulse or break in the pattern to show the gap firing. The vertical pulse is not as long as the normal pattern.

The third pattern is that of a resistor spark plug, or a spark plug with an external suppressor. There is the initial vertical pulse, like the normal pattern, but no re-firings. The resistor has eliminated them. You will note that the resistor has not affected the voltage required to initially fire the gap. So resistor spark plugs or external suppressors do not put a heavier load on the cables or other parts of the ignition system. However, the elimination of the re-firings can in some cases aggravate a fouling condition.

We also developed a Shunt Resistance Meter, Fig. 9. The term "shunt resistance" confuses many people. What it amounts to is this: the voltage at the spark plug must build up to a certain level or the spark will not jump the gap. The deposits that form on the insulator nose can- and do - drain off some of this voltage. If they drain off too much, the spark plug does not fire. How much they drain off depends on their electrical resistance. This in turn depends on their composition and temperature. In general, shunt resistance is used to measure the ability of a spark plug to resist electrical losses through the deposits on the insulator nose.

(Incidentally, here the 12-volt system does have some slight advantage over the 6-volt system - - at a given shunt resistance, it produced a higher output.)

By measuring the resistance of the deposits during actual engine operation, the Shunt Resistance Meter has told us a great deal that could not have been learned in any other way, about what goes on in the combustion chamber.

Still another interesting instrument is the Misfire Indicator, Fig. 10. This is a modification of a development by the Standard Oil Company of Ohio.

When a spark plug misfires, the unburned fuel charge ignites in the exhaust manifold and produces a change in the normal gas pressure pulse.

By recording all the pressure pulses on a tape which we can examine late, the misfire indicator can detect a single misfire among thousands of normal firings. It is better by far than the sharpest mechanic in the business.

Also worthy of mention is our Preignition Detector, Fig. 11. This instrument indicates preignition caused by overheated combustion chamber components, spark plugs, engine deposits, and the like.

It operates on the principle of pre-ionization of the spark plug gap by a flame front prior to the normal ignition spark. Here again an oscilloscope is used, but this one shows the ignition patterns in a circular form.

Now suppose you are an engine manufacturer with a new engine and you want to determine the best spark plug design for the engine. How do we apply these pieces of instrumentation?

Generally we first run a temperature survey using the thermocouple spark plugs. We determine what the temperature distribution is throughout the engine over a range of part and full throttle operation. An example of the results of such a survey is shown Fig. 12. You can observe that there is a slightly smaller spread in temperatures at the lower speed than at the higher speeds. This is due, we believe, to better distribution of fuel and air to the cylinders at low speed.

Occasionally we find that the maximum insulator tip temperature for a given cylinder is higher at some condition less than full throttle maximum RPM. Such an example is shown in Fig. 13.

From this temperature data we can determine what the temperature requirements are for the spark plug to get optimum performance. As indicated earlier we need to operate at 30 miles per hour with the insulator tip at a temperature above 750° F. At the same time, under maximum temperature conditions, the insulator tip must remain below preignition. It is not always possible to accomplish these two requirements in a "conventional" spark plug design.

As a result we may have to develop new spark plug designs. Such an example is the Turbo-action series as used in the 1956 Ford, Lincoln and Mercury vehicles. A comparison of the firing end of Turbo-action and conventional plugs is shown in Fig. 14. This T. A. design raised the low speed driving temperatures by 115°F., and yet gave a three times greater margin of safety from preignition than the more conventional type spark plugs.

On the basis of the temperature data, experimental designs are fabricated for further testing. One important factor to be considered is that the spark plug must have some leeway from preignition due to slight mal-adjustments of the engine. One of the variables having the greatest effect is spark advance.

Therefore, we generally install the preignition detector and a complete set of the proposed spark plugs. We then advance the spark at various operating conditions until either preignition is detected or detonation becomes too severe. We believe a well-matched spark plug should have a spark advance leeway of at least 5-10° above the spark advance-manifold pressure curve.

On the basis of this spark advance leeway and the temperature data, a spark plug design is proposed for further tests. While this may sound somewhat complicated it actually is less time-consuming and considerably more accurate than the older methods of running large numbers of a given design, and attempting to evaluate the suitability of the design on the basis of appearance and dynamometer performance.

After a spark plug is acceptable on the basis of the temperature and preignition surveys, the engine manufacturer, or in some cases Champion, sets up a fouling test to further evaluate the design before release.

Because these tests vary in some details I will only describe them in broad terms. The spark plugs are installed in the prototype engine and operated for a given period at constant speed equivalent to road load city driving.

The spark plugs are periodically checked during an acceleration from low speed to high speed. During this acceleration period we examine the shunt resistance, the voltage waveform and the misfires, using the pieces of equipment previously described. From this data an accurate comparison of the proposed spark plug design can be made.

Such a comparison is shown in Fig. 15. This is a comparison of a conventional spark plug and a Turbo-action plug on the basis of misfires during the acceleration.

On this illustration (Fig. 16) are the same two spark plugs compared on the basis of shunt resistance. You will notice the marked improvement of the Turbo-action design.

While it is not always possible to make such marked improvement as just shown, these techniques do provide an accurate pre-evaluation of the spark plug before actual service. In some cases the engine manufacturer, on the basis of the temperature surveys, has altered engine combustion chamber configuration and spark plug locations to gain in spark plug performance.

To date we have conducted surveys at Ford, White, Studebaker, Packard, International Harvester, Chrysler, Dodge and American Motors as well.

So this program, though designed to solve our _own_ problems, provides the engine manufacturer and the consumer with the best-matched spark plug for the engine - _before_ the engine reaches service.

I want to thank many of my associates who assisted in the preparation of the talk and particularly Bob Nostrant, Champion Engineering Department, who is directing these survey programs in the field.

ENGINE IMPOSED SPARK PLUG REQUIREMENTS

1. Operate Below Preignition.
2. Resist Fouling Deposits.
3. Resist Corrosion and Erosion.

FIGURE I

INCREASED COMPRESSION RATIO

INCREASED OCTANE NUMBER

INCREASED QUANTITIES LEAD

FIGURE 2

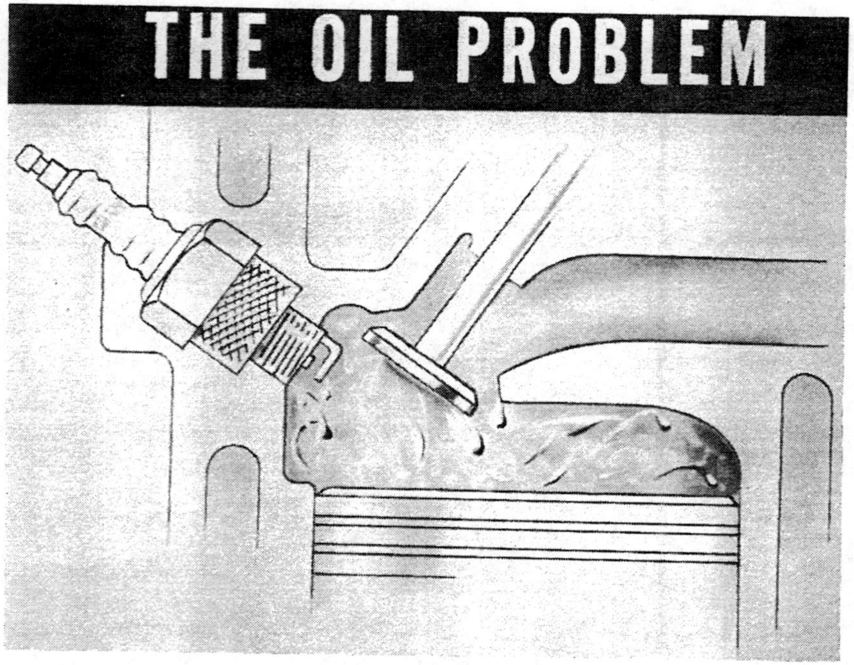

THE OIL PROBLEM

FIGURE 3.

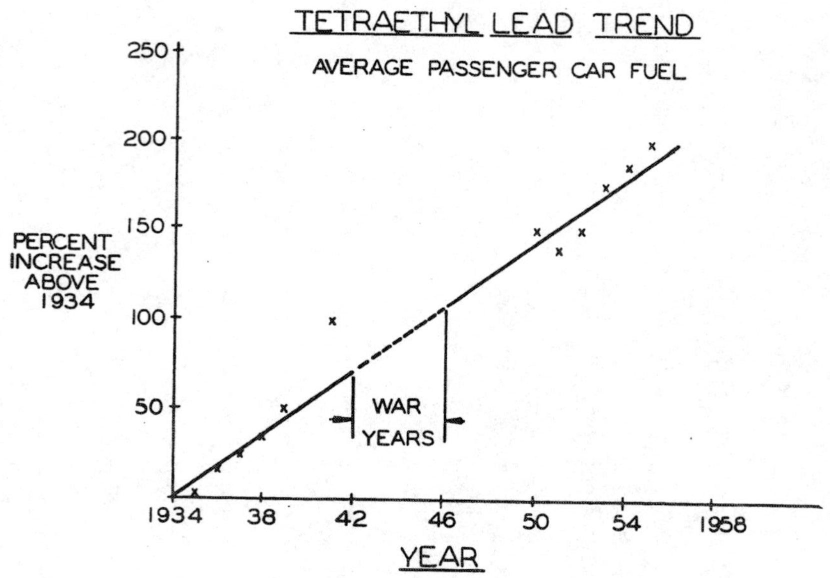

TETRAETHYL LEAD TREND
AVERAGE PASSENGER CAR FUEL

PERCENT INCREASE ABOVE 1934 vs YEAR (1934–1958), with WAR YEARS indicated.

FIGURE 4

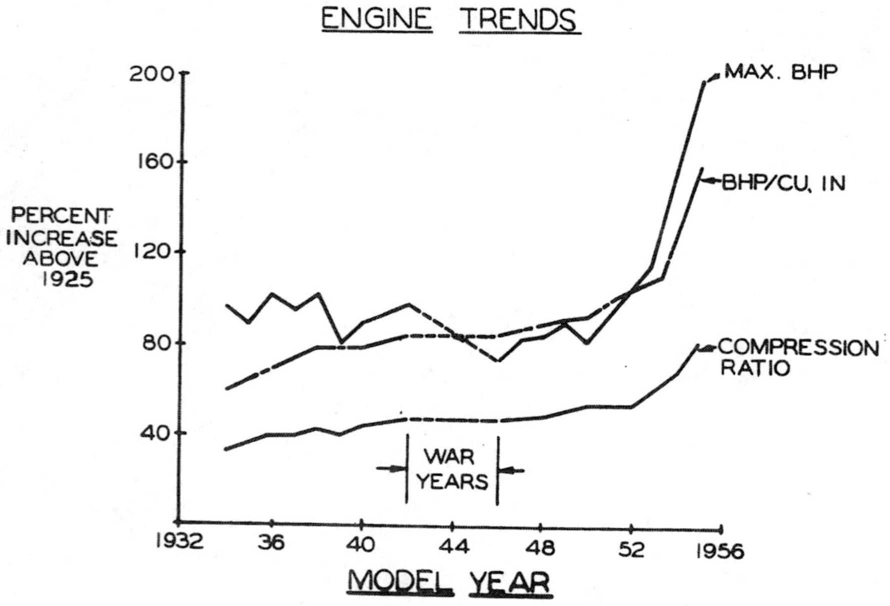

ENGINE TRENDS

PERCENT INCREASE ABOVE 1925 vs MODEL YEAR (1932–1956), showing MAX. BHP, BHP/CU. IN., and COMPRESSION RATIO, with WAR YEARS indicated.

FIGURE 5

SPARK PLUG OPERATING TEMP.
LEVEL ROAD LOAD CONDITIONS

SPARK PLUG INSULATOR TIP TEMP. vs CAR SPEED – M.P.H., showing CAR "A" and CAR "B", with PREIGNITION and LOW TEMPERATURE FOULING regions.

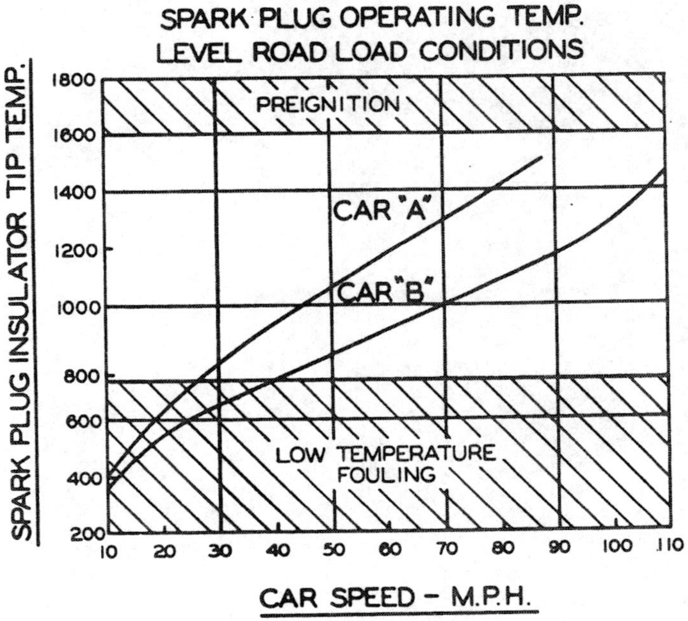

FIGURE 6

98

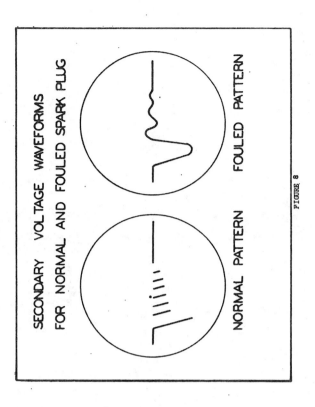

FIGURE 8

SECONDARY VOLTAGE WAVEFORMS FOR NORMAL AND FOULED SPARK PLUG

NORMAL PATTERN   FOULED PATTERN

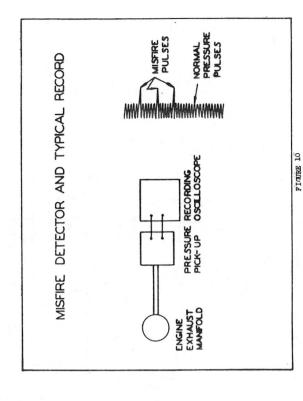

FIGURE 10

MISFIRE DETECTOR AND TYPICAL RECORD

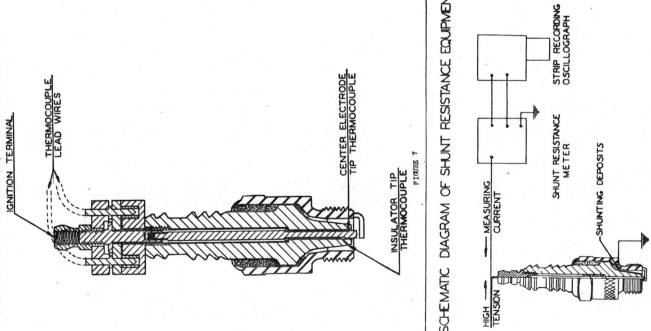

FIGURE 7

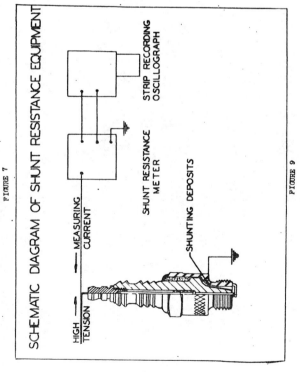

FIGURE 9

SCHEMATIC DIAGRAM OF SHUNT RESISTANCE EQUIPMENT

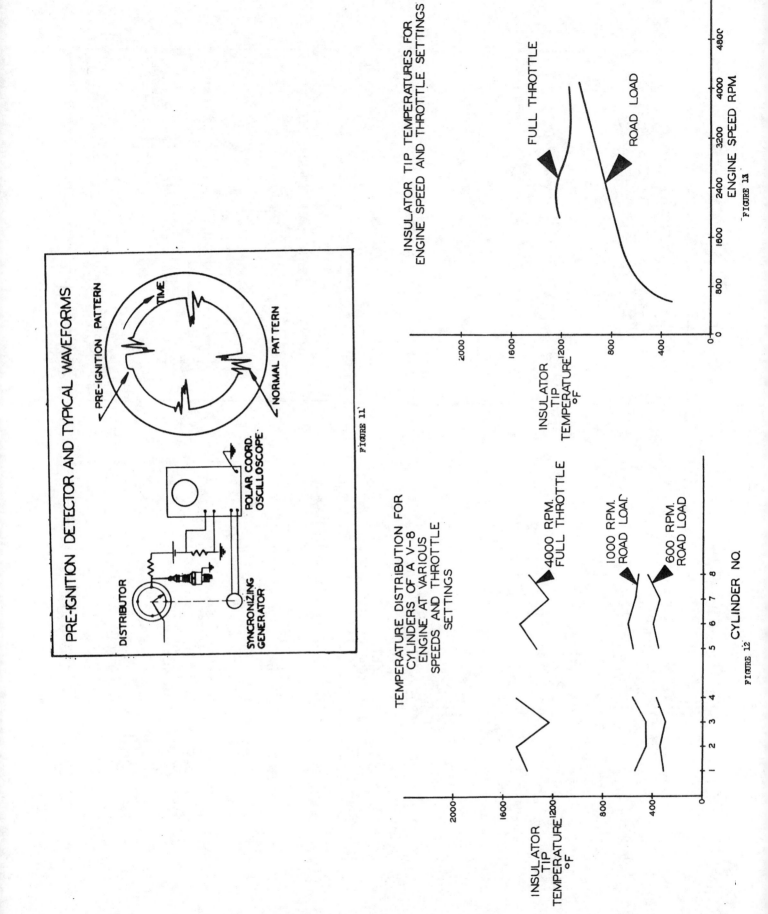

FIGURE 14

COMPARISON OF THE NUMBER OF SPARK PLUGS MISFIRING DUE TO LEAD FOULING

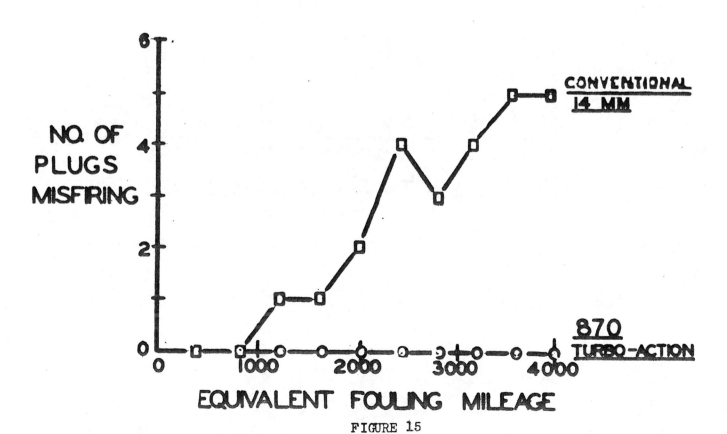

FIGURE 15

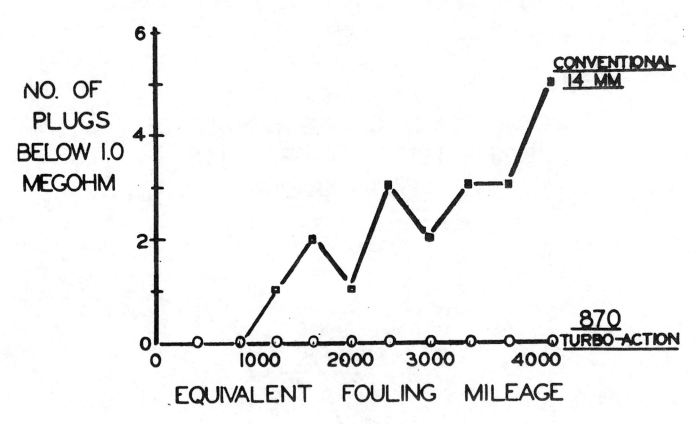

FIGURE 16

It Has All Been Done Before

I. B. K. Gough
Smiths Industries Ltd.

SPARK PLUG DEVELOPMENT has probably included more patents covering weird and wonderful as well as useful ideas than any other engine component; anybody may think himself qualified to invent a new plug!

When the author first joined his present company, the development engineer (of some 30 year's service), who had collaborated with its founder, said at their first meeting, "You'll soon find there's nothing new in spark plugs; it has all been done before."

This comment has remained firmly implanted in the author's mind. When dealing with problems as they arise, it has become apparent that they tend to come round in cycles and that similar solutions are generally found, although perhaps more highly developed, as knowledge progresses. Many of the earliest inventions were impractical at the time, owing often to material or ancillary equipment limitations. The history of spark plugs is almost that of materials applied to a relatively few basic design principles surviving from the welter. Whenever a particular problem arises, it may be that subsequent material improvements will permit use of an old idea and make it appear to be brand new. With all respect to the engineers who report apparently recent developments, this report shows how it looks on historical analysis.

HISTORICAL BACKGROUND

Because of the vast number of plug designs made since the spark plug came into existence, this paper could not possibly cover spark plug development exhaustively. What is attempted is a definition of ignition problems as they apply to internal combustion engines today and a trackback to the origins of the solutions now being offered.

Before tackling this, however, and with apologies to those who may think "it has all been said before," brief reference is made to the main phases of development of ignition principles. This is necessary to the theme chosen because of reference back to early origins of some of the apparently modern solutions.

IGNITION DEVELOPMENT - Man has been trying for at least three centuries to utilize explosions to provide motive power. One of the earliest records is of the Dutch scientist Huygens, who pumped water from the Seine, although his engine depended upon the pressure differential below atmospheric created by cooling gases in a cylinder after the explosion of a charge of gunpowder. Ignition was by fuse!

The basic principles on which are based the automobile engine as we know it today are approximately one century old. They are generally attributed to Lenoir, working in France, and Otto in Germany. Lenoir's engine, patented in 1860, employed a Ruhmkorff trembler coil ignition system and spark plug, the latter being virtually the progenitor of those in use today. Fig. 1A shows the plug and Fig. 1B is a reconstruction from various recorded data of the ignition system.

Otto, at about the same time, developed engines using the four stroke cycle which originally used an external flame as the source of ignition; quite an ingenious valve mechanism (even by today's standards) was employed to admit first the fuel into the cylinder and then the flame for ignition, with both accomplished against compression pressure! Later, in 1876, Otto patented an engine that was virtually the prototype of today's reciprocating internal combustion engine.

---------------ABSTRACT---------------

The history of spark plug development proves this author's contention that an astute and imaginative designer should look to the past as well as to the future for solutions to plug problems. Many failures of the past can become successful applications today through the use of modern-day materials and techniques. Several adaptations of old ideas to new designs are described here as examples of this approach.

Clearly, flame ignition had many shortcomings, and other workers in the same field around this time used a hot tube ignition source. The first traceable patent was No. 562, in 1855, in the name of A. V. Newton. A metal or ceramic tube, closed at the outer end, was heated by an external flame and was attached to the cylinder head so as to have its open end in communication with the combustion space. In other designs the tube was housed in an antechamber, which was opened to the combustion space at an appropriate time by a valve; this arrangement was particularly popular for gas engines. Although hot tube ignition was better than the direct flame system it was still a limitation on engine development, for which electric ignition held most promise.

As far as can be ascertained, the first recorded attempt to apply electricity to the ignition function in an internal combustion engine was a United Kingdom patent by J. Reynolds in 1844. It described a system whereby a platinum element was heated by current drawn from primary batteries.

In 1857, Barsanti and Matteucci proposed an atmospheric engine in which the charge was to be ignited by sparks produced by an induction coil and battery. It did not apparently reach practical form, but the Lenoir engine (which did) had obvious similarities to it.

The Ruhmkorff coil employed by Lenoir was invented by a Paris optician in 1851, originally as a scientific device solely to produce high tension discharges, but Lenoir succeeded in harnessing the energy to ignite gas/air mixtures in his engine.

Oddly enough, despite the lead given by Lenoir, early vehicle engines with electric ignition employed a low tension system in which a device something like the modern contact breaker was located in the combustion space and ignition was effected by the spark across the contacts when they were opened. Fig. 2 shows one such plug. Apparently the high tension systems were regarded as too complicated or insufficiently reliable.

However, any system which relied on moving parts in a hot corrosive atmosphere had obvious limitations. In 1887, electrostatic friction machines were tried out, but advances in electrical engineering, in particular the introduction of secondary batteries, gave high tension ignition a boost. It was employed on a De Dion Bouton car, which was one of the first successful passenger vehicles, introducing a system that eventually became world wide standard automobile practice, once electric lighting and starting became established.

The De Dion Bouton system originally employed a specially developed Ruhmkorff coil, of robust construction to withstand the arduous conditions endured by motor vehicles

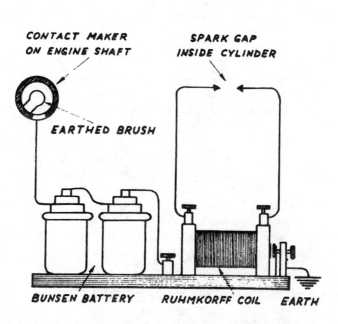

Fig. 1 - Lenoir plug and ignition circuit

of that era. Later, in 1895, the electromagnetic interrupter was replaced by a mechanical "make-and-break" operated by the engine. For some years, the merits and demerits of the two systems were contested, many drivers preferring the trembler system because, mounted within sight and earshot, it produced both a loud buzz and vivid stream of sparks as proof that the system was working properly! It is believed that the last example of the trembler system being used as standard in a series production car was the famous Ford model T, lasting until well after World War I.

THE MAGNETO - The next important step was the magneto, particularly welcomed by some engineers because batteries were not then very reliable; it enabled engines to be built as compact self-contained units that were not dependent upon any external accessories. In many cases it superseded coil systems for a long time, surviving until this day for aviation purposes and for industrial, agricultural, and marine engine applications where there are no auxiliary systems such as lighting that require a battery. Its decline in automobiles occurred simply because a battery was required for lighting and starting and a coil/contact breaker/distributor system could be made more cheaply than a magneto.

Electric lighting and starting steadily progressed from about 1912 onward. By 1930, coil ignition like that basically used in De Dion Bouton cars had again become the standard for most production automobiles. Many drivers regretted the passing of the magneto because batteries do go dead sometimes and, with magneto ignition, one could always get started even if only by pushing or towing. Starting handles were eventually sacrificed on the altar of cost, and the battery does have an awful lot dependent upon it. However, we are stuck with the situation in this highly competitive automobile world.

However, to return to the magneto: Its origin as a practical device appears to be due to work done by F. R. Simms in England and Robert Bosch in Germany, around 1895. The first Bosch machines were heavy devices, having low tension armatures oscillating at about 150 times per minute between the pole pieces. The actuating mechanism was a powerful back spring working off the engine camshaft; the interrupting device was a low tension spark plug exposed to the combustion. Simms saw the necessity for a much lighter machine capable of running at higher speeds, and by working jointly with Bosch, a family of progressively more successful low tension machines was made and marketed in both their names.

There was still the drawback of the mechanical action of the low tension plug, however. In 1898, André Boudeville produced the first high tension magneto, although it was not a practical success because of poor mechanical design and construction. Simms, in 1899, fed a high tension coil from a low tension magneto, after which Honold, a Bosch engineer, improved it by combining the low (primary) and high (secondary) tension windings on the one armature.

This was the practical prototype of the classic high tension magneto that has survived until today. It first appeared in 1900, going into production in 1901.

There were many disagreements between Simms and Bosch, and they eventually severed business connections. There is no doubt, however, that these men, together with some clever engineers employed by them, made tremendous contributions to electric ignition, apart from other electrical devices. The Bosch designs have been firmly established in the European field of ignition for many years, and those of Simms, both directly and indirectly, in the United Kingdom and in the United States. It may be of interest to know that Simms built a factory in New Jersey in 1910, that the American Simms Magneto Co. was established against Bosch competition, and that these activities later passed into the hands of the Bendix Aviation Corp.

Because of their position in the automotive industry, it was natural that Bosch in Europe became the leader in coil ignition development, with Simms leading in the United Kingdom. Lucas later attained leadership in the United Kingdom for general automotive ignition equipment and in the United States the Autolite and AC/Delco companies have played major roles in this development.

SPARK PLUG PROTOTYPES - A few of the many interesting developments around the turn of the century, a very fertile phase of the internal combustion engine development, were:

1. In 1889, a U.K. Patent was granted to Edward Butler for an electrostatic generator followed by a Ruhmkorff coil and battery, the plug being formed by a side gap between an insulated electrode in the cylinder head and a protuber-

Fig. 2 - "Hilo" low tension plug

ance on the top of the piston. This idea was revived in the United Kingdom during the scooter boom in the 1950's as a way of eliminating plug fouling of the "feathering" or "whiskering" variety.

2. In 1892, J. D. Roots brought out a tricycle in which the engine employed hot tube ignition.

3. In the 1890's, the first practical motor cycles were produced by Werner Freres of Paris. They had platinum tube ignition heated externally at the closed end by a petrol burner in a perforated box, the open end projecting into the cylinder. The result of a spill was a brisk fire and no more motorcycling that day! Henrich and Wilhelm Hildebrand of Munich also used hot tube ignition on their production motor cycles.

4. In 1896, Colonel H. Capel-Holden started producing motorcycles using coil, battery, and a high tension commutator type of distributor. Also in this year, an American, E. J. Pennington, produced a motorcycle using a coil and battery system with a single pole plug in the cylinder.

5. Plugs with in-built transformers fed from a low tension magneto developed their own high tension. See Fig. 3. This idea was revived experimentally at the beginning of World War II as a means of overcoming radio interference.

For the purposes of this paper, it is proposed to deal mainly with plugs that, in one form or another, are in use today.

Consequently, the make and break type of low tension plug will not be pursued further than the mention already made and the illustration (Fig. 2).

In conjunction with coil ignition and magneto systems, high tension plugs have been used. Starting with Lenoir's plug in 1860 (Fig. 1), all those plugs associated with the various phases of ignition system development have consisted fundamentally of a body with means of attachment to the cylinder head and, usually, carrying one of the sparking electrodes. Into this body was fitted an insulated center containing the other electrode.

The original Lenoir plug carried its own thread, a principle which, oddly enough, was more like those of today but which was not used in the early vehicle engines. Although combustion pressures were low by today's standards, the plug bodies were formed with massive flanges and held down with nuts and studs reminiscent of high pressure steam joints (Fig. 4). What factors of safety were used before engineers bothered much about stressing!

Around the turn of the century, however, plugs reverted to threaded mountings, and it is from this point on that the history of plugs becomes the history of materials as suggested earlier in this paper, for despite a vast amount of invention, research, and development, today's plugs have a great similarity to Lenoir's except for dimensions and materials.

SPARK PLUG MATERIALS - The heart of a spark plug is the insulator. It has to withstand severe thermal and mechanical stresses, operating for most of its life at a red heat while one end is under high pressures and the other is in the atmosphere and may even at times be sprayed with water. Yet, ideally, it is required to ensure that there is never an

Fig. 3 - Transformer plug

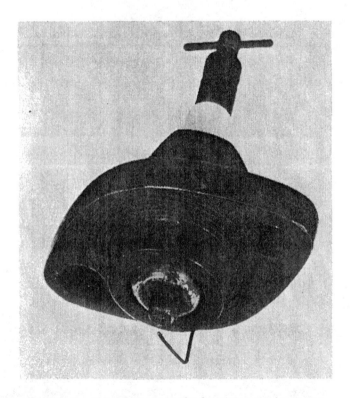

Fig. 4 - Flange mounting high tension spark plug

easier path for the electric discharge than across the gap to form a combustion promoting spark. Ceramic materials in the broadest sense have predominated in plug history, but at various times glass, quartz, mica, and even wood have been used.

The first plugs employed porcelain clay-based ceramic insulators, which were excellent electrically at room temperature and pressure when unstressed, but of very poor strength when subjected to mechanical and thermal shock. To be robust enough to resist room temperature and pressure, they were made in thick section, which made them even more susceptible to mechanical and thermal shock. In consequence, the threads of early self-mounting plugs were of large diameter: 1, 7/8, 1/2 in. taper, gas, etc. As cylinders were then of comparatively large bore and low compression ratio, this did not matter too much. However, the subsequent trend of engine development was toward smaller bores and higher compression ratios, linked with larger valves to let more combustible mixture in and the resultant residues out. Hence it was necessary to reduce the size of the plug proportionally, often to the despair of the plug designer who was left with the impression that the engine designer had forgotten he needed a plug until after all the other engine features had been designed.

Alternative insulating materials made size reduction possible; with better mechanical strength, small sections could be used and the insulators housed in smaller bodies. Metric thread sizes were introduced, starting with 22 (similar to the older 7/8 in. ALAM) and then 18, 14, 12, and even 10 mm was reached in the late 1930's. However, the short insulating lengths and small electrode sizes permissible in the smallest sizes gave the plugs rather narrow operating range and short life, so their use today is virtually limited to high performance engines of small capacities where limitations of space are more important. Today, 14 mm seems the nearest to a universal standard possible to achieve, although an uninformed observer could be forgiven for appearing perplexed by the number of different thread lengths in use around the world. This is not the choice of the plug makers, it should be noted, and a plea for some common agreement among engineering organizations would earn their undying gratitude. For an example of conflicting standards, we have the case in 1966 of two very similar engines produced by related companies in the United Kingdom and on the Continent wherein one has a 3/4 in. and the other 1/2 in. reach plug of otherwise identical specification.

But to return to our theme: A sad feature of early porcelain insulators was that pieces would break off and fall into cylinders, with generally disastrous results. The development of the internal combustion engine has always been closely linked with racing, competitions, record breaking, and so on, and many engines in that fertile, first decade of the twentieth century were wrecked from this cause.

Kenelm Lee Guiness, the founder of the author's company, whose initials form the name of the plug (similar to AC, Champion, Lodge, BG, and others), and who raced Sunbeam cars with considerable success, pioneered in 1911 the use of mica to replace porcelain. This involved a completely different method of construction and the plugs were virtually built up by hand, stage by stage, instead of assembled from a number of separate components.

Starting with the central electrode, thin leaves of mica 0.0015 in. thick, previously cut to size and split down from mineral lumps, were wrapped around and stuck with shellac; a gland or even the body was pressed on and swaged to produce a strong, pressure tight assembly. Finally, the lower end of the mica was machined to the desired shape and the upper end covered by a stack of mica washers subsequently held tight by means of a washer and nut secured to the terminal post. These washers were finally turned and even ground to a super, moisture resisting finish that gave the plug the attractive appearance of grained wood. The electrodes were also finally finished off as required. For special purposes (for example, racing or record breaking and aviation plugs), the washers were inspected on both sides for flaws and conductive inclusions; this was quite a costly process. Fig. 5 shows a typical plug.

Mica insulation was adopted by many plug makers and enjoyed a quarter of a century of reliable use until the advent of tetraethyllead. The combustion residues of lead containing fuels attacked the mica, breaking down its structure.

So plugs were once again at the mercy of available materials. Many and varied have been those tried, but we can

Fig. 5 - Typical mica insulated plug

be concerned here only with those that endured as practical production solutions. The next phase was "back to ceramics," but with a difference. Steatite, sillimanite, and finally alumina were tried, bringing at each stage an improvement in thermal and mechanical strength and other desired qualities. Siemens in Germany, with Bosch of magneto fame, contributed greatly to the development of these ceramics, as did Turner of Lodge in the United Kingdom and Albert Champion in the United States, through both companies with whom he was associated. One company, it is reported, bought an entire mine or quarry (according to one's terminology) to ensure its supply of raw materials.

The return thereby of individual components capable of being mechanized in production was in keeping with the quantity outlook developing with the expanding car industry; the era of mass production had begun and is here to stay. This is where we are now at an advanced stage today!

Before passing on to more detailed consideration of modern ignition problems, it is thought worthy of mention that one of the features of spark plugs, which has produced the most prolific collection of ideas and patents over their history, has been electrode geometry. Internal combustion engines and their ignition systems developed contemporarily with electrical technology generally. It is perhaps the more understandable, therefore, that there should be many theories of spark discharge phenomena which led to a vast array of electrode formations, each with its claimed advantages. Fig. 6 shows a very small selection among the many. Today, however, they all fall into two or three categories, of which one (described as single point overhead earth or ground, electrode) probably accounts for 90% of the world's output, which is estimated at about 1500 million per annum. Lenoir's plug (Fig. 1) and others at the dawn of high tension ignition used just such a simple geometry that has survived right through the jungle of alternatives to become the modern vogue. Merely by using up-to-date materials and techniques, the original spark plug cannot be bettered for the vast majority of the world's automotive engines.

As a final comment in this section, it is interesting to reflect that the high tension plugs used with the various coil and magneto systems in the vital period around the turn of the century were mostly of French origin and cost about 17 cents. French engineers, of course, contributed a great deal to the early history of the automotive vehicle.

CURRENT PROBLEMS

Spark plug engineers today face two principal problems with normal passenger vehicle engines. In addition, others with specialized engine applications will be mentioned.

1. Vehicles tend toward greater powers and higher potential performance, but road conditions and speed limits compel them to spend a great percentages of road time at low power.

2. The modern operational trend is toward longer periods between servicing and toward component life equal to the first life of the car. One major U.S. car manufacturer writes a 40,000 mile plug life into his specification, for example, although it may not yet be obtainable.

Spark plug performance to meet these requirements may not seem too difficult of attainment, but in fact it is. To facilitate appreciation of the problems, it is proposed to discuss briefly the various factors governing spark plug operation.

The job of the spark plug is to receive an electric impulse from the ignition system and to discharge it across the gap between the plug electrode to produce a spark to ignite the gasoline/air mixture. So, what are the problems?

1. Taking a conventional coil system (Fig. 7) we have, in electrical terms, the spark plugs representing shunt loads

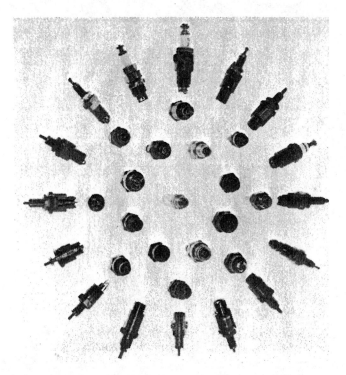

Fig. 6 - Various electrode arrangements

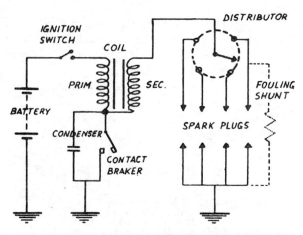

Fig. 7 - Conventional coil ignition circuit

across the secondary winding. It might be expected that the coil would always produce its full output, providing, of course, the system is in a good state of maintenance, and that whether or not the plugs sparked is entirely their concern. In fact, the system will only produce such high tension as the shunt load of the spark plugs will permit. New plugs have very high resistance (10^5 megohms or higher). The ignition system will develop maximum voltage on open circuit (infinite shunt resistance, one might say) and new plugs represent the nearest practical approach to this.

Plug insulation resistance drops as the ceramic gets hot, an inescapable property of most materials, and still more so as combustion deposits accumulate on that part of the insulator exposed to the combustion chamber. Within limits, which will be discussed later, these deposits are burnt off by the operating temperature of the ceramic, but the fact remains that steady reduction of plug insulation resistance occurs with time. So, the coil output falls in consequence, irrespective of the condition of the rest of the system.

2. To produce a spark between the plug electrodes requires a particular voltage, which varies with many other things but principally with:

(a) The spark gap width.
(b) The shape of the electrodes.
(c) The temperature of the electrodes.
(d) The compression pressure at the instant of firing.

In practice, this covers a range of approximately 20 kv at starting or snap acceleration down to 3 kv at light throttle cruising, with various stages between depending upon the instantaneous engine conditions.

The easiest condition is when the plugs are new, the gap width is at the basic setting, and the electrodes still bear the sharp, sheared edges of initial manufacture. This concentrates ionization and facilitates gap breakdown, and once started, the engine will bring up the electrode temperature, which further facilitates gap breakdown.

However, these ideal conditions are short lived. The following happens, usually in that order, to make the plugs more difficult to spark and the ignition coil less able to provide the necessary voltage:

(a) the electrodes lose their sharp edges.
(b) The insulator accumulates deposit.
(c) The electrodes erode away, increasing the gap width.

Factors (a) and (c) require more voltage, yet (b) makes less available. Fig. 8 shows graphically what is happening. This is diagrammatic only and does not relate to any specific vehicle but illustrates only the principle. What happens is that the plug voltage increases, owing to loss of the sharp edges and general gap erosion. While this is going on, the coil output is falling because of the reduced insulation resistance of the insulator when hot and the shunting effect of insulator deposits. It may also be worsened by contact breaker deterioration. Eventually, the lines cross, at which point misfiring or even failure to start will set in.

It will be noted that, for purposes of illustration on the graph, maximum coil output has been shown at 100% and the new plug requirement at 50% and the crossover point at 6000 miles. These figures will vary with different engines, but the same principle applies. The difference in voltages represents ignition reserve. Clearly, the greater the reserve, the longer the vehicle can go before the crossover point is reached. Unfortunately, one can do very little to reduce the basic plug voltage, since this is dependent upon so many factors external to the plug. Even the spark gap width is determined by the engine maker and is largely dependent upon the carburetion balance employed.

For slow smooth idling, important with automatic transmissions, and regular firing at low throttle lean mixture cruising, quite wide gaps are used. Thus in the United States, the average gap is around 0.035 in., whereas it is nearer 0.025 in. in Europe. Leaning off mixture strength to reduce atmospheric pollution certainly will not bring any reduction of gap width; rather, it might be increased.

It is beyond the scope of this paper to comment in any detail on ignition supply systems, which have in any case been adequately covered in other SAE papers and elsewhere. Suffice it to say that practical limitations prevent the indefinite raising of the coil output level as a means of increasing the margin of reserve. For one thing, contact breakers already suffer from burning, even welding together at very low cranking speeds, as in subzero winter weather because of the high primary currents they already have to handle with prevailing secondary voltages.

So we are stuck with the situation that, sooner or later, what is generally regarded as "plug trouble" sets in, although clearly the plugs are not solely to blame.

Traditionally, the cure for this condition has been to maintain the ignition system properly and to clean and reset the plugs periodically or replace them with new ones. Unfortunately, the worsening contrast between open road and city congestion, coupled with the increasing complication of removing and refitting plugs in today's multicylinder engines, with their hoods cluttered with filters, heaters, and other equipment, makes this an unpopular chore, so the demand for plugs that will last the first life of the car is understandable.

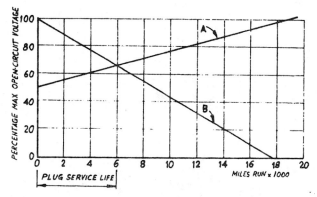

Fig. 8 - Graph showing coil output and plug voltage demand

SOLUTIONS

How can we tackle these problems? Let it be said immediately that improvement of the ignition system can be made. The advent of solid-state electronics has opened up new possibilities of making the coil output less dependent upon the condition of the plug and contact breaker, and of making the whole ignition system capable of operating for longer periods without attention. In its simplest form, electronic aid replaces the heavy current breaking function of the contact breaker and eliminates the need for frequent cleaning, resetting, or replacement of this component. More sophisticated arrangements provide for generation of the ignition pulse by different means; for example, condenser discharge into the coil primary. The sensitivity of the coil output to secondary shunting by the plugs is related to the secondary rise time, that is, the time taken from the change of primary condition until the secondary voltage reaches its operational peak (when the plug fires).

The significance of rise time is that the shorter it can be made, the lower the secondary shunt resistance (greater amount of plug fouling) the coil can tolerate and yet still operate satisfactorily. To quote some approximate figures as a guide: A typical coil/contact breaker system may have a rise time of 10 μsec to 20 kv and will probably stop working properly when plug shunt resistances fall to about 1 megohm. An electronic system can reach 20 kv in 1-1/2 to 2 μsec and will continue working down to a proportionately lower plug shunt resistance in consequence. It is convenient to refer to the former as low frequency and the latter as high frequency, but strictly speaking, the system never develops a full wave form. Beyond the plug gap breakdown point, surplus energy simply dissipates in a manner determined by the ringing characteristics of the coil circuit (Fig. 9).

So far, reference has been made only to high tension systems, which work in conjunction with orthodox spark plugs. Later, mention will be made of so-called low voltage high frequency systems and the plugs that go with them.

Notwithstanding the advantages offered by more sophisticated systems, however, the bulk of the world's production vehicles still employ the coil/contact breaker system with all its shortcomings. Through its many years of use it has been engineered down to remarkably low cost and anything else is almost bound to be more expensive. So, apart from after-sales replacement or partial adoption of a few of the more expensive line models, the new systems do not seem to be a likely aid to the plug designer for a good many years yet.

Let us now take a look at what the plug makers have been doing about the problems. We begin with a recapitulation of plug operating conditions.

PLUG TEMPERATURE - It has been stated that over a period of operation, lead products of combustion and, of course, carbon become deposited on insulator noses in the same manner as on other components in the combustion chamber. Because of its lower thermal conductivity and its relative isolation in the gas space within the plug, the insulator runs at relatively high temperatures, similar to the exhaust valve. The characteristics of the combustion deposit are such that most of it burns away if the insulator temperature is in the range 500-600 C. However, insulator temperature varies over the operating range of the engine by very virtue of the fact that power is a function of heat produced by the amount of combustible mixture admitted to the cylinders at any given throttle setting.

Therefore a top limit must be put on plug temperature; otherwise, ignition by hot surface (akin to the early hot tubes) takes place instead of by spark. As this is uncontrollable timing-wise, it leads to preignition and engine damage. It is generally agreed that preignition results when plug temperatures exceed approximately 950 C. Hence, for any given state of engine tune, the maker must select a plug whose insulator temperature is safely below this figure at the condition that gives the highest temperature (full throttle and maximum power or maximum bmep). It is for this reason that plug makers have to produce ranges of different heat values within a given thread size and reach. Each plug has its own rate of dissipating excess heat to the engine cooling system, so that appropriate matches for various stages of engine tune can be made to ensure that temperatures are kept below the danger point.

Having established this upper operational limit, one must look at the other end. If plug temperatures fall below about 350 C for any considerable length of time, excessive combustion deposition takes place, misfiring sets in, and eventually the fouled plug oils up as well and is incapable of being revived without removal from the engine. Yet, at idling, insulator temperatures can fall as low as 150 C. Clearly, then, it remains largely a matter of luck related to chosen (or enforced) driving conditions as to how much of their time the plugs operate in the "self-cleaning" temperature range.

INSULATORS - The great majority of American cars today, and an increasing number in the United Kingdom and on the Continent, employ plugs where insulator noses project beyond their bodies. Described variously as projecting core, extended nose, self-cleaning plugs, and so forth, they make use of the greater exposure in the flame to broaden the proportion of the engine operating range over which self-cleaning temperatures are maintained. They do not become hot

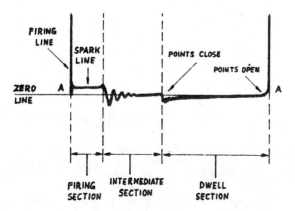

Fig. 9 - Typical ignition pulse trace

tube igniters at the top end, however, because large volumes of relatively cool mixtures impinge on them during the induction stroke and cool them down. What happens is diagrammatically shown in Fig. 10.

But is this a new idea? By no means. Even the flanged plug shown in Fig. 4, made before this century, had an extended insulator nose. In those days they were concerned only with carbon and oil, not lead as well, but it was difficult in the low power engines to keep the plug hot enough to burn off carbon. So, in effect, there was the same problem. The principle has been widely used for many years in, for example, tractor plugs where carbon fouling and oiling have always been problems. Even the mica plug for the Model T Ford (Fig. 11), with the added feature of a stepped surface to break up the fouling line by a series of hot edges, employed the idea. This was made about the end of World War I.

Also shown in Fig. 11 are:

1. A Molla plug made very early in this century, with the added feature of adjustable (by rotating the central electrode) twin gaps.

2. A mica plug of unknown origin and with duplex gaps.

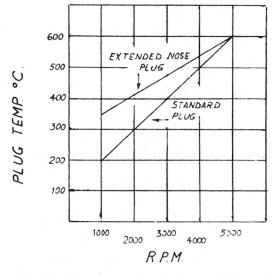

Fig. 10 - Temperature comparison of orthodox and extended nose plugs

3. A KLG plug for racing Guzzi motorcycles similar to the Model T Ford plug and made in 1932.

4. A model extended nose plug.

ELECTRODES - While it is obviously desirable to keep the insulation as free as possible from deposits, to minimize the shunting effect on the coil, anything that keeps down the rate of plug voltage increase will also contribute to lengthening the serviceability period. Spark polarity has an effect, and it has been established practice with coil systems for many years to make the central (hotter) electrode negative, which minimizes erosion from this cause.

One obvious solution in the plug itself is to use electrode materials that erode less rapidly. Erosion is a complex process between chemical corrosion and electrical volatization. The combined effect of high temperature and corrosive attack (oxygen and sulfur predominating) enlarges the grain structure and corrodes along the boundaries. The sparks nibble out the corroded material and (on a microscopic scale, of course) pieces of the electrode fall off; it is rather like chipping out the mortar in a stone wall. Fig. 12 shows a microsection of a typical electrode after many thousands of miles of service.

The earliest electrodes were of platinum or copper. Later, nickel was used, progressively alloyed with maganese and silicon to increase resistance to sulfur and oxygen attack. More recently, various other elements have been alloyed with what had virtually become the classic manganese-silicon-nickel plug alloy; these include titanium, zirconium, barium, thorium, columbium, and chromium, to name a few. Some were intended to improve corrosion resistance, either directly or by grain refining; others, to reduce the work function in an attempt to influence breakdown voltage.

Chromium has always held an attraction as it has such a good high temperature performance. In combination with nickel it has been used for electrical heating elements over many years. In similar combination and also including iron and other elements, it has literally made possible the jet age by producing such alloys as the Nimonic and Inconel.

To meet the modern equipment requirement of low gap erosion, an increasing number of plug makers are turning to these high chromium content materials. It is unfortunate that they have considerably lower thermal conductivities than the classic manganese-silicon-nickel materials, and

Fig. 11 - Historical collection of extended nose plugs

design must circumvent this drawback. However, nothing better has been found for regular price plugs.

Is this a brand new jet age idea? Not a bit. Brightray, a high chromium/nickel alloy was used as long ago as 1929, but was discontinued owing to its low thermal conductivity. Availability of the Inconel range of materials and ceramics with somewhat better conductivity now permits the better use of chromium. Again the right materials at the right time!

The more rapidly heat can be transferred out of the central firing point and back to the engine cooling system via the plug insulator, body shell, thread, and gasket, the longer can the insulator nose be made, for a given plug heat factor. This is an obvious advantage from the fouling point of view. Although copper has been used in many plugs as the actual electrode right through to the firing point, it could be used successfully only with nonleaded fuels. Lead by-products attack it and life is seriously reduced. Silver has also been used for the same purpose, but it is a very soft metal of relatively low melting point and has not been a very practical material other than as a backing to platinum or other precious metal firing points.

However, for some applications, for example, military vehicles and aircraft, plugs are often made with nickel alloy electrodes having copper cores to improve heat transfer. Indeed some Japanese plugs embody a nickel alloy tipped copper electrode in their normal automotive range (probably the only country that can afford to do so!). But this bi-metal arrangement, again, is as old as the hills. Kenelm Lee Guiness, as long ago as the early 1920's, produced (mica) plugs having copper electrode stems carrying inserted nickel alloy firing points. In 1929 a range of plugs designed for mass production embodied nickel alloy central electrodes having copper tube swages on them. Two or three years later, production was assisted by the advent of copper sheathed steel telephone cable, which was bought in lengths and applied to the purpose. Another variant was the insertion of copper rod into drilled electrodes. Fig. 13 shows a selection.

METHODS OF SEALING SPARK PLUGS TO CYLINDER HEADS - Over the years the classic method of achieving a gastight seal has been to use a gasket. Early components were composed of copper containing asbestos string. Largely, because the asbestos represented an insulating sandwich against heat flow, this was later omitted and gaskets of solid copper, rolled (hollow) copper, and steel have variously been used. Because they often fell off the plug during the fitment or removal, various means of locking the gaskets to the plug have been used in the last decade or so. Is this new?

The Mosler platinum pointed plug (shown in Fig. 6) which was made at the turn of the century and carries patent dates October 16, 1898, April 22, 1902, and September 15, 1903, fired into the insulator glaze, has a retained external copper gasket. Every original plug also had a strong label attached to the terminal which said "Centre firing point guaranteed platinum."

Fig. 13 - Selection of plugs having copper cored and sheathed electrodes

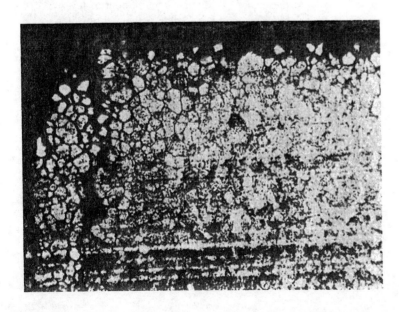

Fig. 12 - Photomicrograph of nickel alloy electrode

TAPER SEAT PLUGS - An American car manufacturer has been using a taper seat for some years, as a means of obtaining a gas seal, instead of the more conventional gasket. A new idea? In 1928 the author's company produced plugs, shown in Fig. 14, that were held down by cams operated by a common shaft, the idea being to give a quick change in racing engines. With engines having 8, 12, and 16 cylinders in use today, sometimes with two plugs per head, think what racing mechanics would be involved in if we were to revert to this old arrangement! Nevertheless, the principle was right then - in 1928.

AVIATION SPARK PLUGS - While piston engines are in their decline for major airline transport, their use in light aircraft will undoubtedly persist for many years yet. American and United Kingdom practices in general have sharply contrasted until fairly recently. American practice has favored a cheap throwaway plug with limited servicing and ultimate life. These plugs employ, in the main, heavy multiple tongue earth electrodes in conjunction with correspondingly heavy central firing points, often with copper core to improve heat flow. We have already seen how telephone cables might be said to have been the father of these.

However, in the United Kingdom, where longer operating periods have been a design requirement, precious metals have been the vogue for many years. Back to de Dion or earlier? In England we are proud to possess a message of gratitude from your late, revered President Roosevelt, saying that British (platinum pointed) plugs made it possible for World War II bombers to safely fly the Atlantic to the European theater. Today such plugs use iridium tips, which erode much less and are immune to lead contamination. By using iridium, operating lives up to 800 flying hours are approved and over 1000 hr without attention in certain instances has been achieved. A new mid-twentieth century idea? Oh no! The first evidence of its possible use, traced by the author, was U.K. patent 18654 taken out in 1904 by one Theodore Rene! Only the inability of the metallurgists to process is suitably for mass production having the required ductility prevented its use until 1958 in pure form. The previous use of iridium as an alloyed element with platinum, while of some value, was nowhere near so effective. Again, we have a development related to a material.

Fig. 15 shows comparisons between plugs having platinum alloy and iridium electrodes, made after the stated test times. That with iridium ultimately completed 1014 hr before the test was stopped, during which time the single cylinder test engine consumed 22,388 gal of high octane fuel containing 729 lb TEL (468 lb of lead as metal). At the end of the test the plug was still functioning satisfactorily, although the electrodes were reduced to some 50% of their original thickness in the sparking area. Fig. 16 shows photomicrographs of the platinum and iridium after test.

ALTERNATIVE SOLUTIONS

Plugs described variously as low tension, surface gap, or surface discharge have attracted plug engineers (and some engine designers) for automobile, military, and aviation use for many years. The principles of operation appear to make it possible theoretically to get away from the "heat value" problem, reducing plug types to one for each thread size and reach; also to provide other benefits such as complete freedom from fouling and preignition. This development so gripped the imagination of ignition engineers that, soon after

Fig. 14 - Early taper seat plug

Fig. 15 - Comparison of plugs having platinum and iridium electrodes

World War II, the general manager of a well-known U.K. ignition system company, with a plug engineer colleague of the author's on the pillion, could be seen riding a motor cycle around the streets of Coventry enveloped in thick blue smoke made by a considerable overdose of oil. Neither the plugs nor the hearts of the rider and passenger missed a beat! However, the system only worked for a thousand miles, so there was obviously more work to be done.

SURFACE DISCHARGE PLUGS - Fig. 17 illustrates the principle of such plugs. The normal spark plug practice of having the insulator free-standing in a gas space and caused to run at high temperature to burn off fouling is abandoned. Instead, the gap is formed across the end of the insulator between the central electrode and the body, the whole thing being solidly composed of the metal and ceramic materials. The gap is circumferential, or polar, and one current and practical but special purpose application of such plugs describes them as "polar gap plugs." Their potential benefits are:

1. Being of solid construction they operate at much lower temperatures than conventional air gap plugs, virtually that of the cylinder head face, thus eliminating the risk of pre-ignition.

2. The presence of a dielectric material other than air in the gap reduces the voltage required to create a spark. If that part of the insulator which is in the gap is treated with a suitable material in the semiconductor class, the sparking voltage is lowered still more. Moreover, and this is the theory of the unfoulable plug, products of combustion depositing on the surface will not (as with a high tension spark plug) shunt the current and prevent a spark occurring, but will merely modify the value of the semiconductor surface resistance. Since the only way for the spark to go is across this surface it will still spark. It would then be possible to operate at much lower voltages than with conventional spark plugs.

It is an obvious advantage to have ignition systems operating at low voltage; less primary current is required to generate it, insulation is easier, and there is less sensitivity to dampness and contamination.

It has been found in practice that the voltage reducing effect of a pure insulant is not so great as might be supposed. Although combustion deposits perform the duty of semiconducting layer to a degree, there is none present to assist the first start when the plugs are new and certain combustion conditions can heat the end of the plug sufficiently to clean off the deposits. Thus, to preserve the low voltage qualification, pure insulant surface discharge plugs can be used only with small gaps.

Unfortunately, such plugs have a restricted burning range of mixture strengths in gasoline engines, owing to flame nucleus quenching by their relatively cold, solid ends. Some improvement was made by partway drilling of the central electrode (Fig. 17), but in general, larger gaps than would be required by a conventional air gap plug have to be employed in engine applications where smooth running at a variety of mixture conditions is required. Then the choice is whether to return to the high voltage class, or to employ semiconductors.

In any case, there is a further problem. Whether the plugs work themselves into the semiconductor class by starting off with insulated gaps that collect deposits, or work as proper semiconductors by design, they will not operate properly with conventional ignition systems. To such systems they stand in the same relationship as fouled high tension plugs; that is, they load or short circuit the secondary circuit and prevent its building up the necessary voltage. To use surface

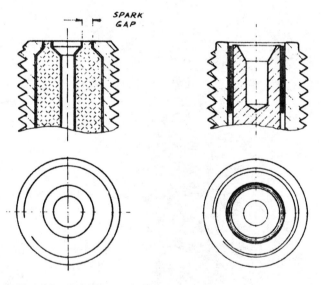

Fig. 17 - Schematic diagram of surface discharge plug

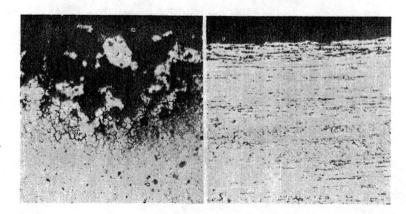

Fig. 16 - Photomicrographs of electrode in Fig. 15

discharge plugs, therefore, it is necessary to have an ignition pulse of very short rise time and, preferably, greater energy so as to cause the spark to arc away from the surface; this improves its incendivity by getting it away from the relatively cold plug end.

A system was envisaged in which a condenser was charged to a few thousand volts, as controlled by a gap (Fig. 18), and then applied impulsively to the plug as the gap broke down and the condenser discharged. Very short rise times less than 1 μsec could be achieved for sparking across the gap. There were two theories of how this occurred:

1. The initial current across the deposit covered surface, or semiconductor, heats it, ionizing the air immediately above it and also causing its resistance to fall. This further increases ionization until it is sufficient to allow a spark to occur.

2. If the semiconducting surface is regarded as a dispersal of semiconducting particles among the insulating ones, the total discharge is formed by a collection of small sparks hopping across the surface like on stepping stones.

Despite these attractions, such systems have not come into general use except for aviation jet and turbine engines. Here, some of the problems described as applying to spark plugs in gasoline engines are the same, but there are others. Because of the altitude that modern aircraft reach, low voltage is obviously desirable, since flashover in the harness system becomes easier as atmospheric pressure falls. The plugs are required only to initiate combustion, which becomes self-sustaining, and the ignition is then switched off. Hence the plug will collect fouling and will likely not operate on the next occasion.

The major additional problem, however, is that the fuel is relatively difficult to ignite, compared with gasoline, and needs to be volatilized by the spark. To meet all these requirements, surface discharge plugs, or igniters as they are generally called, are the modern solution in conjunction with high energy levels in the range 2-12 joules at usually about 2 kv.

Historically, turbines have been ignited by (in turn) wax tapers, oily waste, spark plugs, glow plugs, torch igniters wherein a jet of fuel is lit by a spark plug, glow plug or SD ignitor, and then, finally, by surface discharge plugs on their own, the latest stage of which is the multiple or cascade gap that gives in effect a longer duration spark for the same voltage.

As to why the system has not been generally applied to automobile engines, in view of its obvious theoretical attractions, one finds:

1. The plugs and ignition systems are more expensive than the conventional high tension system.

2. The system cannot operate at anywhere near the energy levels used for aviation purposes because

(a) Such levels are difficult to generate at high repetition rates.

(b) The plugs would have very short lives under continuous sparking conditions, owing to rapid erosion.

(c) If the energy is scaled down to acceptable levels, there is a fouling problem of another sort. Under some conditions, particularly cold starting, oil or even ice can accumulate across the plug face and the permissible energy from the other considerations is too low to blast it off. Instead, the discharge fizzles across below the fouling and cannot get at the mixture to light it.

A possible cure for this problem is to arrange for two energy levels: high for cold starting, low for normal running. Unfortunately, cost then becomes prohibitive.

(d) The mixture burning range is restricted, compared to that of air gap plugs, owing to flame nucleus quenching by the relatively cold plug face. The associated problems of high voltage cum-air gaps or low voltage cum-semiconductor gaps and ignition system energy levels have already been covered.

So, until further work is done, we do not seem likely to see such systems on production cars.

In any case, the shorter rise times being obtained from solid state ignition systems enable conventional spark plugs to function more satisfactorily, so there may be no need to go as far as surface discharge ignition for other than special applications.

One such application, which must be mentioned, is in the sphere of high power outboard marine engines. Outboards, particularly under racing conditions, run perilously near to preignition, overspeeding being inescapable as the propellers leave the water momentarily. The benefit to be derived from the relatively cold operating end of a surface discharge plug is therefore obvious, and the development of a suitable, short rise time ignition system (albeit a high voltage one with insulated surface gap) has enabled this benefit to be realized while still permitting satisfactory ignition over the remainder of the engine power range. The plug's business end is virtually identical to the schematic diagram (Fig. 17).

DEVELOPMENT OF GAP PLUGS - But let us examine the matter historically. It is generally assumed that the application of low voltage surface discharge ignition dates from patents taken out by a Dutchman named W. B. Smits (later

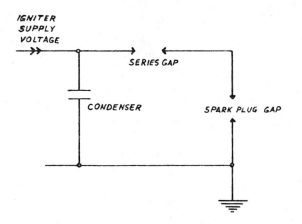

Fig. 18 - Ignition circuit for use with surface discharge plug

operating as Smitsvork) immediately before and after World War II.

French patent 811908, granted in 1937, seems to have started it all. The essence of the principle was to charge a condenser and subsequently discharge it across a plug of the surface type which was improved, not deteriorated in performance from the effect of the combustion deposits. Smits seems to have had a vast number of patents assigned to him in various countries and to have covered the theoretical ground of insulated and semiconducting gaps pretty thoroughly, but there is little evidence of any practical systems having been made and marketed.

However, research into earlier work has revealed that:

1. An engine was run by the KLG Co. at the turn of the 1920's, wherein the surface gap was formed by interleaving the mica insulator wrapping with single turns of metallic foil, thus forming a series of concentric circular gaps of gradually increasing diameter. In 1932 one Frank Watson took out U.K. Patent 385208, employing a ceramic insulator, the end of which was stepped to carry a series of metallic washers spaced apart on the steps by insulation.

The effect of both the KLG and Watson arrangements (shown in Fig. 19) was to produce a rudimentary, mechanical semiconductor as a precurser to the modern arrangement of dispersed conductive particles.

2. U.S. Patent 1,537,903, applied for in 1922 by Egbert von Kepel (and referring to earlier work on the subject), concerned the charging of a condenser under the control of a spark gap in series with the spark plug.

3. U.K. Patent 2162 was applied for in 1903 and granted in 1904 to Sir Oliver Lodge. Sir Oliver, of course, was a famous British scientist who, among his many achievements, founded the spark plug company which bore his name, now amalgamated with that founded by Kenelm Lee Guiness to form the author's company. It is believed that Sir Oliver's patent really is the origin of this type of ignition system (used then with high tension spark plugs) and is of such interest in this connection that the author begs leave to quote from it. The patent reads:

"We, Sir Oliver Joseph Lodge F.R.S. Principal of the University of Birmingham, and Alexander Marshall Lodge, Engineer, both of Mariemont, Edgbaston, in the City of Birmingham, do hereby declare the nature of this invention to be as follows:

"This invention relates to electric ignition apparatus for use with internal combustion engines or motors and for blasting or other purposes, the object being to provide for the production of an effective spark in places or under circumstances where insulation is difficult, uncertain or non-existent, as for instance under water or oil or between terminals joined by charred fragments or otherwise partially short-circuited.

"The invention comprises the application to the purpose aforesaid of a phenomenon not previously unknown to science but emphasized by the first named applicant in the year 1888 and described in his published book on 'Lightning Conductors etc.' Such phenomenon, as set forth in the stated work, is the exciting and precipitation by a spark termed an A spark of another or B spark in what is described as an 'impulsive rush'.

"To carry the invention into effect, a pair of small Leydens or other condensers are employed with four terminals, one to each coating. Two of the said terminals are connected to the exciting coil and the exciting or A spark gap, whilst the other two are led to another or secondary or B spark gap (which is arranged within the motor cylinder or in the position required for the desired ignition) by leads which must not be too perfectly insulated from each other throughout their whole length. The said leads remain at zero, or at the same potential, until the A spark occurs, when the contents of the jars are suddenly liberated and with an impulsive rush surge along the leads and jump across the B gap, thus producing the required igniting spark.

"The A spark gap and the B spark gap are arranged in separate circuits, the former being in circuit with the small Ruhmkorff or exciting coil and one pair of condenser coatings (one of each jar). The other pair of coatings and the aforesaid leads from the B gap circuit which, though completed through the knobs or terminals at the A gap, is independent of the coil.

"With the aforesaid arrangement defective insulation in the leads of the B gap circuit is advantageous, as it ensures the charging of the condensers. A complete insulation in this circuit must be avoided and to ensure that it shall not occur an imperfect conductor or "leak" is placed in the leads between the jars and the gap, preferably adjacent to the former. The same "leak" serves also for the A discharge, or the completion of the A circuit.

"The B spark can be produced under water or in any circumstances of ordinary defective insulation. Any charred fragments or particles of moisture or oil that may collect between the terminals will be blown away by the B spark which is as effective for ignition purposes as any ordinary spark.

"The knobs or terminals of the A spark gap are conveniently arranged near the Ruhmkorff coil which is provided with any ordinary make and break or other exciter, and the complete A circuit can be so placed as to permit of ready inspection and control. The knobs or terminals of the B spark gap, which may be fixed inside a cylinder or, in the case of blasting, submarine mining, and for other purposes, at some distant or submerged station, will be unaffected by dirt,

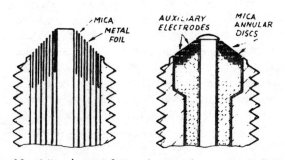

Fig. 19 - Mica/metal foil and stepped ceramic surface discharge plugs

grease, water or other substances which would place an ordinary sparking device out of action.

"The Leyden jars or other condensers, of any desired shape, are enclosed in a box or casing.

"This invention relates to electric ignition for use in internal combustion engines, or for blasting or other similar purposes. For example, in internal combustion engines with ordinary electric ignition, and especially in oil engines, it is common for a charred oil deposit to take place on the plug, practically short-circuiting the spark gap and preventing ignition.

"Now the object of the invention is to render the effectiveness of the spark employed for ignition purposes practically independent of the insulation of the points between which it is caused to occur."

Fig. 20 illustrates the basic circuit. Further in the text of the patent it also states that the "leak" V or V^1 is not absolutely essential because "charred particles" (which today we call fouling) form on the spark plug insulation and form an automatic leak.

Unlike many patentees, Sir Oliver Lodge actually made and sold many products based on his patents. The ignition system just described was indeed marketed. The patent referred to the A (series gap) and B (condenser discharge) sparks, and the author feels, without wishing to prolong this particular section of the paper unnecessarily, that the following quotation from the Lodge catalogue of 1912 is well worth noting. Incidentally, Sir Oliver was awarded the Gold Medal in 1908 by the Societe D'encouragement pour L'industrie Nationale, Paris, for his ignition system.

"The diagram [Fig. 20] shows displayed the general principle of the arrangement stowed away in the case of a Lodge igniter. The terminals of the coil are not connected to the sparking plug directly as usual, but are led to it through the intervention of a pair of coated insulators, or Leyden jars with their outer coatings short-circuited by a leak or imperfect conductor, the object of which is to keep them always at the same potential except at the instant of a sudden electric rush. Accordingly there is no strain thrown upon the leads or the sparking plug, whose terminals remain at the same potential up to the last moment when the two jars are full and over-flow at A. At this instant everything is liberated, and with a rush of inconceivable rapidity, the jars empty themselves across A and around the complete circuit through the sparking plug. No leak or imperfect conductor has time to exert any influence on the rush, which is over in the millionth of a second but not before it has ignited the combustible mixture exposed to the sparking plug.

"The rush is so violent that not only is dirt in the path blown away, but the electric momentum overshoots the mark, and the jars are charged up in the reverse direction by the impetus. They then discharge again, and again are charged in the ordinary way and so on many times, without the coil taking any further part in the action; its function is over when it has filled the jars to overflowing. The jar spark is a noisy white-hot spark of extreme suddenness all over in the millionth of a second, which can be timed to occur with great accuracy.

"By this arrangement a discharge is brought about quite suddenly between points which except during the rush are completely inert, so that they might be handled with impunity, or placed under water, or clogged with dirt. But during the violence of the rush the dirt in the path is flung away, the water is burst through, the circuit is bound to be completed when the full discharge occurs at A. the A spark is therefore a pioneer spark, which precipitates the sudden rush and causes the B spark. The A spark is in the box under glass, so as to be easily open to inspection, and where it can be kept quite clean. The charge is prepared or generated by the coil, as usual, all the strain being thrown upon the clean gap at A, and directly this gap gives way the whole accumulated contents of the jars are suddenly emptied without warning through the combustible mixture, the time of firing being accurately adjustable by the primary cam which regulates the charging action of the coil.

"A rather extraordinary result can be obtained in connection with the adjustable spark gap on the igniter. Should the sparking plugs become foul from over-lubrication or other cause, the points of the plugs can be cleaned from the driving seat. This is done by simply opening out the adjustable spark gap, whereby the intensity of the sparks at the plugs is so increased that any accumulations are blown off the points. The spark at the adjustable gap can be seen through the glass window in the lid of the igniter. This spark acts as an indicator and shows clearly whether the ignition is in good order or not."

What a charming contrast to the sort of language employed in promotional literature today! There could surely not be a better interpretation of what was to become the main problem for ignition engineers over half a century later. Fig. 21 shows the equipment in polished mahogany case.

SURFACE GAP PLUGS - In an attempt to achieve a halfway house situation between the high tension and surface discharge systems, various surface gap plugs have appeared

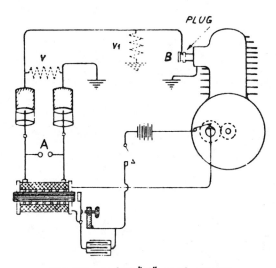

Fig. 20 - Lodge "B" spark circuit

on the market, in recent years. These still employ a measure of heat value effect to prevent complete fouling-out, but make partial use of surface sparking. In effect, the plugs are of side gap high tension design but with the ceramic insulator brought right forward until its end literally forms part of the gap (Fig. 22). Thus the spark discharge takes place partly across a surface and partly across air. Two useful features derive from this:

1. As already mentioned, the voltage to break down a given gap width across a surface is less than the same gap of air; it is therefore possible to employ a greater combination surface/air gap width than air for a given voltage. The largest practicable spark gap is always desirable in order to ignite the weakest mixtures, but with plain high tension plugs a limitation is placed by the maximum the ignition system can produce, bearing in mind the need for reserve (Fig. 8).

2. The part of the spark that occurs across the insulator surface burns away deposits, breaking the fouling line, and increases the period of serviceability.

Plugs known variously as Fire Injectors, Fuel Igniters, and Golden Lodge operate on this principle. More recently in Europe, a plug has appeared of this form but with a normal high tension gap as well, the theory being that the spark will appear at either the air or surface gap, depending upon the instantaneous values of the various parameters that control sparking. Surface gap plugs suffer two major problems:

(a) Ignition of the weaker mixtures occurs chiefly from the part of the spark that takes place across the air gap. Generally, the air gap is made small compared with the ceramic and if means are not provided for gap adjustments (which is tricky anyway with the particular gap geometry employed), one either has to produce a vast range of plugs identical except for gap or accept a compromise arrangement that is bound to lead to cases where gaps are wrongly matched and misfiring will occur.

(b) The surface gap is vulnerable to oil blobbing.

Plugs of this design are without doubt potentially capable of giving longer service and total lives, but to give of their best they really need to be fed from modified ignition systems. It may well be that with the more general application of solid state techniques to ignition systems, taking advantage of the more favorable rise times they make possible, surface gap plugs may come into wider use. Although commercially available plugs of this sort are relatively recent, we find once again that the principle is not.

The principle is certainly covered in the Von Lepel patent previously mentioned, which is generally taken as the origin, although earlier workers may well have covered it. The revival in recent years is covered principally by F. P. Dollenberg in U.S. Patent 2,899,585 dated 1957.

SERIES GAPS - Cyclicly every decade or so somebody comes up with a device variously described as an ignition supercharger, spark booster, high frequency converter, and so on. On the Continent there was even a "Funken-strecher" The promoters undoubtedly make some money for a time, but then the gadgets quietly disappear. When public memory has faded, they appear again in a new guise, and this could well go on ad infinitum.

These devices are, basically, series gaps. They may, in simplest form, be merely air gaps between two rods or cascade gaps formed by a stack of alternative metal and in-

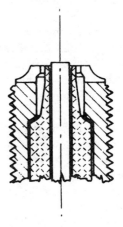

Fig. 22 - Surface gap plug

Fig. 21 - Lodge "B" equipment

sulating washers. They have been found with values as low as 0.5 kv up to 10 or 12 kv. They can also be built into spark plugs.

The intended functions of these gaps are:

1. To isolate the ignition generating source from the shunt load of the spark plugs so that, however low this may become due to fouling, the voltage can build up at least to the breakdown value of the series gap. If this is sufficient to fire the plugs at their instantaneous conditions, then ignition will take place where otherwise it would not.

2. To shorten the pulse rise time when the series gap breaks down by virtue of the capacitively stored energy in the system, further favoring the firing of low shunt resistance plugs.

Function 2 is similar to that of the system used with the low tension surface discharge plugs, but as the self-capacity of the ignition system is low, the energy stored is also small. It could be increased by placing a condenser in parallel with the coil secondary, but any useful capacity value places such a further load on the coil that its output is seriously reduced (Fig. 23), unfortunately making it impracticable with existing production systems.

To be of real value, the gap breakdown voltage needs to be greater than the maximum spark plug voltage, for clearly it will be shunted along the fouled insulator after jumping the series gap if the plug gap demand is significantly greater than the series gap setting.

Fig. 24 shows the electric circuit, where:

V_{in} = Coil output voltage (input to the plug)
V_p = Plug gap voltage
C_1 = Coil self-capacitance
C_2 = Plug self-capacitance
R Shunt = Plug shunt resistance (fouling)
A = Series gap
B = Plug gap
S_1, S_2 = Switch analogies for series and plug gaps

To facilitate understanding of the system it is useful to introduce the concept of a preset switch analogy of spark discharges; when a gap has broken down, it becomes conducting, with a very low voltage drop once the arc is established. Such a circuit is shown.

If the contact breaker has just opened, a voltage V_{in} will be impressed across the points X and Z. Until S_1 closes (series gap fires), this voltage will be available across X and Y because of R shunt. There is zero voltage across Y and Z because no current can flow through R shunt.

At some stage, the voltage across S_1 will rise to a level sufficient to close this switch (series gap fires). As a result, a pulse of short rise time and of amplitude V_p will occur across Y and Z, where

$$V_p = \frac{C_1}{C_1 + C_2} V_{in} \qquad (1)$$

The rise time of this pulse is short, being only that corresponding to a frequency of 100 Mc/sec or higher. It is from this feature that the high frequency aspect of series gaps is derived.

Arranged another way, Eq. 1 becomes

$$V_{in} = \frac{C_1 + C_2}{C_1} V_p$$

In other words, the voltage the coil is permitted to generate is only that much greater than the series gap setting by the ratio of the system and plug capacitances. Typically, C_1 may be 50 pf and C_2 may be 5-10 pf, so the two volt-

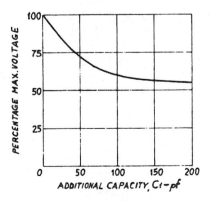

Fig. 23 - Effect of extra capacity on coil ignition voltage

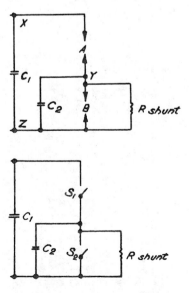

Fig. 24 - Series gap circuit and switch analogy

ages are not very different. Let us now consider two extreme conditions, one where R shunt is very high (say, infinity, as might be said to correspond to a new plug).

As the contact breaker opens, the voltage rises and distributes itself between S_1 and S_2 in a manner determined by the stray capacities of the system. In these circumstances the voltage across X and Z reaches the sum of the closing voltages of S_1 and S_2 (the series and plug gap voltages).

Since the sparking voltage of a plug gap in an engine may read a maximum of 20 kv or even higher under some conditions, it would be necessary to provide at least twice this from the coil if the series gap is to be effective when the plugs are fouled yet need their maximum voltage to ignite the charge.

In practice, having regard to practical limitations experienced by the ignition system, the maximum, safe, series gap setting to use is about 2-4 kv.

Thus, returning to the circuit diagram, if we consider the other extreme case where R shunt is low (say, approaching zero), V_{in} can be only slightly greater than the series gap settings by the ratio $(C_1 + C_2)/C_1$, and in the event of a plug's needing more than this to produce a spark (which will be a considerable proportion of the engine operating range), it will not fire.

Hence the series gap is of very limited use with present ignition system voltage levels. It can only be said that its use will increase the possibility of a fouled plug starting to fire again, and perhaps clean up by flame action if engine conditions (probably by accident, as the average driver will be quite unaware of the significance of all the electrical parameters involved) are held long enough corresponding to plug voltage values below the series gap setting.

Again, this a principle that may enjoy greater chances of success with solid state ignition systems.

The historical aspect (Fig. 25) shows a selection of series gap devices, commencing with (yes!) a nonmetallic trouser button! The brass bodied spark plug with a tubular, ventilated mica insulator cannot, unfortunately, be dated, but it must be very early in the century and is probably the father of the modern series gap plug.

The writer's first association with series gaps goes back some 30 years, before he had the least idea that he would ultimately work as a spark plug engineer. Passing a field in the country on a bicycle, he was amazed to see what appeared to be a blue oil smoke cloud, in the center of which was the vague outline of a motor vehicle, approaching the gate. Fascinated, he dismounted and awaited its arrival. It turned out to be a bull-nosed Morris with modified body from which food was being distributed for cattle. Clearly, it was burning as much oil as gasoline and the author who, although not old enough for a license, was rebuilding an old motorcycle and fancied he knew a few things about engines, asked the farmer how he managed to keep his spark plugs firing. Throwing open the bonnet (hood, of course, in American parlance), the farmer pointed at the ignition system saying "Ah! that be the secret. Gent from Town fixed it that a way back last summer; ain't 'ad plugs out since. Used to 'ave em out most every day."

Fig. 26 shows the arrangement which, one might say, is self-explanatory. A lot cheaper than the commercial gadgets! It probably worked very well on an old, low compression engine whose plug voltage demand was low at all conditions compared with the output of the magneto.

As far as it has been possible to trace, the first reference to series gaps goes back to the Sir Oliver Lodge patent of 1903, but it could be earlier. There are also International Patent 16916, 1906, assigned to Luis Keller Leahy of Los Angeles and U.K. Patent 3403, 1909, to Henry Herbert Mattoret covering series gaps, although the purposes are more to indicate whether or not the ignition system is working and, in the case of the former, to "regulate" the spark and "permit changing of a plug without stopping the engine" than to have any effect in dealing with fouled plugs electrically. U.K. Patent 1288, 1908, granted to Richard Forbes Collum, describes a gap and condenser system fed from an electrostatic generator.

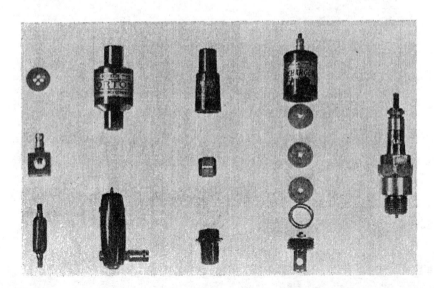

Fig. 25 - Selection of series gap devices

To a small extent, all jump gap distributors confer a measure of series gap effect; however, to make best use of the stray capacities, the gap should be as close as possible to the ignition spark gap. An ad hoc application of the principle employed for many years is to disconnect the high tension lead from a fouled plug and hold it about 1/4 in. away from the terminal while the engine runs on the other cylinders until the plug fires again. A bit complicated on modern automobile engines (and difficult if you are moving!) but widely used 30 years ago.

GLOW PLUGS - Finally in the few fields of ignition that have been considered, to illustrate the author's theme, it is thought to be not out of place to consider glow ignition in its modern context, bearing in mind that it was one of the earliest methods. Two types are:

1. Hot Wire - This does not have serious place in gasoline engines unless one includes miniature units used for model boats and aircraft. It is, of course, widely employed as a starting aid for diesels with certain combustion systems.

It might interest the reader to know, however, that in 1939 we had a passenger car in England operating on platinum glow plugs. Starting was achieved by battery energization, which was switched off once running. Ignition timing was varied by feeding in a trickle of current under the control of a rheostat attached to the accelerator, to change the element temperature.

2. Hot Tube - Under this heading there is something of interest. In the early development phase of the aircraft gas turbine, some were found to suffer from "flame-out." Due to certain peculiarities of air flow in conjunction with particular pressures and temperatures, particularly under certain tropical conditions, the normally self-sustaining combustion would extinguish. To restart, the crew had first of all to recognize the condition and then operate the igniters with an engine full of hot gas, with frightening results for passengers particularly when flames up to 12 ft long belched from the tail pipes!

The solution, a hot tube, goes back almost to the very dawn of engine history. Originally of platinum but later of refractory material for cost reasons, a glow plug was made and inserted in the combustion system at a strategic position determined by test. Under normal conditions it simply stood in the flame and glowed red hot. If the flame extinguished, the glow plug reignited it and delays up to as long as 11 sec have been recorded. Fig. 27 shows such a plug.

Later engines had their combustion systems modified in an attempt to overcome the problem and obviate the need for the glow plugs. However, there is evidence to suggest that this has been only partly successful. Conditions such as thick slush on runways can still produce unstable combustion, and it is understood that some pilots operate the ignition systems continuously to guard against flame-outs. This obviously has an effect on igniter life, and "it has all been done before" may well bring glow plugs back to act as silent sentinels in their modern role.

CONCLUSION

The author has often heard it said that there is nothing in spark plugs; they are only nails through pieces of china with means of attachment to the cylinder! This can scarcely be true when one considers the wealth of work and knowledge

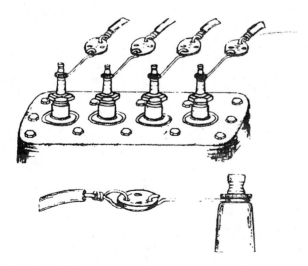

Fig. 26 - "Trouser button" series gap installation

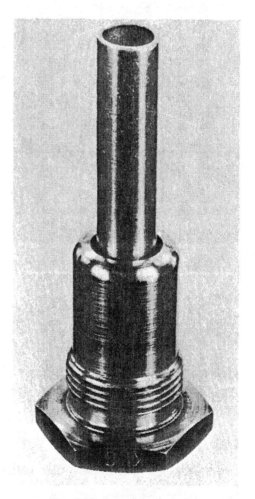

Fig. 27 - Aviation glow plug

that has been put into their history by contributions from very eminent people. As an example, in the U.K. Patent Library alone there are 1884 patents on record covering spark plugs, so on a world wide basis the figure must be enormous.

While, as was said at the beginning of this discussion, this paper could not be an exhaustive catalogue of a century of plug design, it is hoped that it has established how little that is really novel has been applied in our present decade to solve our current spark plug problems: old ideas, new materials, the right combination at the time, and then give the Advertising Department its head!

This paper lists a number of references for those sufficiently interested to study details if they wish to follow up any points presented here. However, the author has not made his journey simply to "point the finger" at the present generation of engineers, which in any case includes himself.

Many years ago, a sales slogan of the author's company was "fit and forget." Even a slogan can be equally applicable to the modern situation. The thought with which this author leaves you is that there is something we as engineers, facing problems, should not forget and that is to look back on what has passed before, just as the good automobile driver constantly watches his mirror to solve the problems of finding the best way ahead to the end of his particular journey. He observes not only what is immediately behind him, but whatever is as far back as he can see.

ACKNOWLEDGMENTS

The author wishes to thank his company for permission to write the paper and use the various items in illustration; also, various present and past members of the staff, in particular G.W. Shoobert whose memory of 43 years of spark plug engineering has been invaluable in preparing the manuscript.

Additionally, the British Science Museum and libraries of the Institution of Mechanical Engineers and U.K. Patents Office have provided useful contributions.

REFERENCES

1. E.T. Westbury, "Ignition equipment." U.K.: Marshall, 1948.

2. F.R. Simms, "The History of the Magneto," 1940.

3. "Bibliography on Ignition," U.S. Dept. of Commerce, 1963; N.B.S. Publication 251. (This lists 161 reports specifically on spark plug research and development dating from 1916, including a bibliography listing many books and reports on the ignition function and ignition systems, exclusive of spark plugs, and a miscellaneous section covering approximately 730 references in all.)

700081

Spark Plug Design Factors and Their Effect on Engine Performance

Robert J. Craver, Richard S. Podiak, and Reginald D. Miller
Champion Spark Plug Co.

THE PURPOSE of this paper is to provide to those who are relatively unfamiliar with spark plugs, information as to the variables affecting their design, application, and performance. In performing this assignment, basic spark plug theory will be presented as well as test data relating the effect of various spark plug configurations on engine performance.

With the improvement of existing and the introduction of new spark ignited engines (1)* over the past twenty years, several new spark plug features have also been introduced. A review of designs such as projected insulator, projected electrode gap, resistor, and surface gap spark plugs will reveal that the principles are not new (2). However, the proper application and refinement of these features have resulted in improved spark plug and engine performance.

EFFECT OF ENGINE DESIGN ON SPARK PLUG DESIGN

While the spark plug manufacturer has the responsibility of the plug design, the engine designer also has considerable in-

*Numbers in parentheses designate References at end of paper.

fluence on its final configuration and performance. The engine designer accomplishes this by influencing the physical dimensional characteristics of the spark plug. The choice of thread diameter and reach are generally determined by engine features such as displacement, valve size and arrangement, combustion chamber configuration, etc. Spark plug thread diameters of 10 and 12 mm are desirable from the engine designer viewpoint but are usually objectionable to the spark plug manufacturer because of their small size. As a result, their use has been confined to the smaller cylinder bore, high performance engines such as motorcycles. The 14 mm diameter designs are in common use throughout the world in both two and 4-cycle engines. The 18 mm design is also extensively used in automotive and tractor engines and is in general usage in aviation and industrial natural gas fueled applications. Other thread diameters such as 7/8, 1-1/8, and 1/2 and 3/4 in. pipe threads are also produced but have limited application.

Different reaches or thread lengths are generally available in any given thread diameter. This is illustrated in the 14 mm diameter gasket seated designs shown in Fig. 1 where standard lengths of 3/8, 7/16, 1/2, and 3/4 in. are available.

ABSTRACT

Spark plug design features are established by both the engine designer and the spark plug engineer to obtain the optimum engine and spark plug performance for the specific application. The paper describes the elements which influence spark plug design, discusses factors affecting the spark plug voltage requirement, operating temperature and heat range, and presents test data showing the effect of various firing end configurations on engine performance.

The engine tests were performed in a medium displacement V-8 engine at a simulated 30 mph road load operating condition. The results suggest that the ignition of lean fuel/air ratios can be extended by projection of the spark plug gap deeper into the combustion chamber and that there exists an optimum relationship between electrode size and gap spacing. Further tests, with ignition systems having widely varying spark discharge characteristics, indicate the very short spark durations to be detrimental in igniting lean fuel/air ratios.

In the past, the majority of the spark plugs have been of the gasket seated design as just shown in Fig. 1. However, in recent years, there has been a strong trend to the gasketless conical-seated configuration (3) as illustrated in Fig. 2 for the 14 mm and 18 mm thread diameters. The spark plug hex size can thereby be decreased from 13/16 to 5/8 in. in the 14 mm spark plug and from 7/8 to 13/16 in. in the 18 mm design. This reduction in hex size has been found to be advantageous in some engine designs.

In addition to selecting the spark plug thread diameter, reach, and seating configuration, the engine designer has also exercised control over the installation height of spark plugs. This control has been very apparent in small engine applications, where overall physical size and weight of the engine package is of utmost importance (4). The small 14 mm compact design spark plug shown in Fig. 3 is typical of a chain saw engine application. Also shown is a conventional automotive design and an extreme in installation height, as represented by a spark plug for a large natural gas industrial engine. The insulator flashover distance has been reduced in the compact designs which somewhat limits their applications. However, this limitation can be extended with the use of an insulating type of terminal or cover as part of the ignition lead.

SHIELDED SPARK PLUGS

The majority of the spark ignited engines use the unshielded type of spark plug shown in Figs. 1-3. Completely shielded ignition systems are primarily used for the suppression of ignition noise (5) in two-way radio equipped military vehicles, marine, and aircraft applications. In these applications they may also provide for a waterproof ignition system. Shielded plugs are also used in industrial engine applications, but more for safety reasons in potentially explosive atmospheres (6), than for radio interference suppression. The two types of 5/8-24 barrel configurations and the single 3/4-20 termination are shown in Fig. 4. All require their own type of ignition harness connectors with the two on the right having the advantage of more positive moisture-proof seals.

After the engine designer has established the physical criteria for the spark plug (7), preferably in cooperation with the spark plug engineer, the spark plug manufacturer can incorporate certain features into the product for optimum performance. These features are selected on the basis of engine usage and the environment in which it will operate.

RESISTOR SPARK PLUGS

Resistors are commonly incorporated in a spark plug design for either suppression of the high frequency electromagnetic

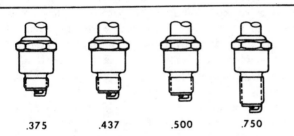

.375   .437   .500   .750

Fig. 1 - Spark plug thread reach

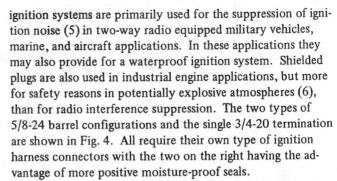

COMPACT   AUTOMOTIVE   INDUSTRIAL

Fig. 3 - Spark plug installation configuration

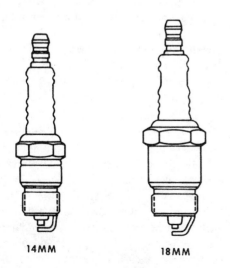

14MM   18MM

Fig. 2 - 14 and 18 mm conical-seated spark plugs

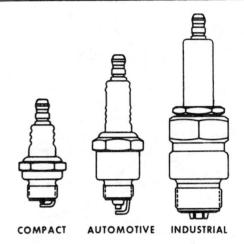

5/8-24   5/8-24   3/4-20

Fig. 4 - Shielded spark plug barrel configuration

radiation emanating from the spark discharge (8), or for the control of electrode erosion (9). In general automotive use where the vehicles are now equipped with resistance spark plug cable, resistor plugs are being used to obtain greater suppression of radio interference. In aviation and industrial engine usage, the resistor spark plug is primarily used for control of electrode erosion.

SPARK PLUG GAP CONFIGURATIONS

No one spark plug firing end configuration is best for all engine applications. The selection of this design feature will depend upon factors such as the type of engine, type of service, fuel, ignition system, and economics. A photograph of the more popular configurations is shown in Fig. 5. A brief description of each is as follows:

1. High-Performance - This is a very cold operating spark plug in which the insulator and gap are retracted into the shell bore. The insulator nose is short and the ground electrode is reduced in length by the so-called pushwire configuration.

2. Conventional Automotive - Typical electrode configuration for the majority of two and four-cycle gasoline engines.

3. Massive Electrodes - This may take the form of two, three, or four prong ground electrodes and is in common usage in aircraft engines and industrial natural gas engines.

4. Fine Wire - This is a relatively expensive spark plug having small electrodes of platinum, iridium, or other related precious metal alloys. These are primarily used in the aircraft and industrial engine field where long life is required.

5. Surface Gap - In conjunction with the proper ignition system, this design provides good service in two-cycle outboard engines.

SPARK PLUG HEAT RANGE

All spark plugs have a characteristic known as heat range. This is a relative indication or rating of the ability of the spark plug to reject heat. It is determined in accordance with SAE procedures (10) in a specially designed, carefully controlled single cylinder 17.6 cu in. engine. Results are expressed as the maximum indicated mean effective pressure at which the engine can operate without spark plug induced preignition (11).

Ratings range from 100-500 and are measured at a fixed speed and spark advance. These values are useful only for comparing spark plugs and are not directly related to the output of other engines because of differences in combustion chamber design, engine speeds, etc.

The simplified heat flow drawing shown in Fig. 6 illustrates the heat rejection path through the insulator of two spark plugs. A "cold" plug will have a short insulator nose, a short heat rejection path, and a high imep number. A "hot" plug will generally have a long insulator nose, thereby a long heat rejection path and a low imep number.

SPARK PLUG OPERATING TEMPERATURES

A specially constructed spark plug, as shown in Fig. 7, having a thermocouple located on the external surface and near the tip of the insulator, is used in determining spark plug operating temperatures. The actual insulator tip temperature, under various modes of engine operation, can be determined through the use of this laboratory tool. Thermocouple spark plugs are also constructed with the thermocouple junction within the ceramic insulator body or on either of the electrodes.

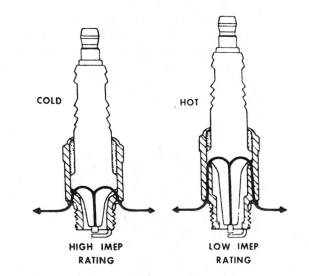

Fig. 6 - Spark plug heat flow drawing

Fig. 5 - Typical spark plug gap configurations

There is a desired spark plug operating temperature range for each engine application. The operating temperature of three different heat range spark plugs in a typical automotive application is shown in Fig. 8. It is desirable to have the spark plug insulator tip temperature in excess of 700 F under low speed city driving conditions to prevent carbon or cold fouling of plugs. It is also necessary to limit the insulator tip temperature under high speed, high power operation. Continuous operation in a 1400-1600 F range is undesirable for good electrode gap life as chemical corrosion begins to significantly contribute to electrode deterioration. Temperatures of 1750 F and greater can promote preignition of the fuel charge and cause engine damage as a result of excessive combustion chamber temperatures. Control of these temperatures is accomplished by varying such design features as insulator nose length, insulator and gap position, and electrode material.

SPARK PLUG FOULING

As just indicated, the operating temperature of the spark plug has an effect on spark plug fouling. Carbon or cold fouling is aggravated by rich fuel/air ratios and high oil consumption, particularly when combined with continuous low power operation. Low spark plug operating temperatures cannot burn the conductive carbonaceous deposits from the insulator and as a result the output of the ignition system, with this shunt loading (12), is reduced until there is insufficient voltage available to break down the plug gap.

Spark plug lead fouling is generally observed under higher speed operation. During low speed, low power city type service, the lead anti-knock compound from the fuel is deposited upon the spark plug insulator nose. These deposits are nonconductive under driving conditions where the insulator tip temperature is low. Upon a heavy acceleration to a higher power operation, the temperature of the insulator and its deposits is increased. This increase in temperature converts the nonconductive compounds into new compounds which vary in resistivity (13). As the resistivity of these deposits decreases, the output of the ignition system is decreased. The engine, therefore, misfires and spark plug fouling is said to occur.

The fuel refiners have placed various lead scavenger additives in the gasolines (14) to reduce spark plug fouling but they are not 100% effective under all conditions. Spark plug engineers also attempt to reduce this tendency through design but are often limited as to alternatives (15). Ignition system manufacturers can also exercise control over spark plug performance (16). While the lead additive could be reduced in the fuel and ignition systems designed to decrease the tendency for spark plug fouling, these modifications would undoubtedly result in increased cost to the motorist.

SPARK PLUG REQUIRED VOLTAGE

In order to develop an electrical discharge across the spark plug electrodes and ignite the fuel mixture (17), the available voltage of the ignition system must exceed that required by the spark plug. This is illustrated in Fig. 9, a plot of available and required voltages versus engine speed for new and used spark plugs. Under steady state operation, the ignition system available voltage is shown to exceed the spark plug required voltage for both new and used plugs throughout the speed range. However, under a sudden acceleration, the increased requirements of used, marginally fouled spark plugs exceed the voltage available from the ignition system and misfiring occurs.

Many other factors also affect the spark plug voltage requirements, some of which are shown in Fig. 10. The breakdown voltage is noted to increase with an increase in both engine compression pressure and spark plug gap spacing in accordance with Paschen's law.

Fig. 7 - Thermocouple spark plug

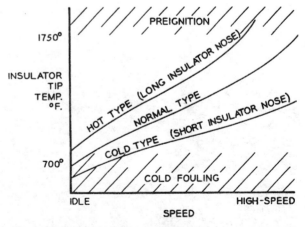

Fig. 8 - Spark plug operating temperature versus engine speed

The effect of lean mixtures on the voltage requirement in engines, equipped with exhaust emission devices, is further illustrated in Fig. 11. At road load speeds above 2000 rpm, and under wide open throttle operation, there was no appreciable change in requirement between the standard engine and the engine equipped with the emission control devices incorporating changes in both carburetion and timing. However, at road loads below 2000 rpm, the difference in voltage requirements becomes significant as we approach an engine idling condition. At idle, the voltage required to fire the spark plug with the lean carburetor is almost doubled that of the standard carburetor. It is of particular interest to note that the increase in voltage requirement is due almost entirely to the leaner fuel/air ratio, since the effect of spark advance is almost negligible. The spark plug performance could be expected to deteriorate more rapidly under such conditions. Not only would the rate of gap erosion become more critical, but the maintenance of the ignition system itself would become more important.

In Fig. 10, we also note that the required voltage of a gap is reduced when the hotter of the two electrodes is of a negative polarity. Since the center electrode generally operates at a temperature greater than the ground electrode, it is desirable to supply a negative polarity voltage to this electrode for minimum spark plug voltage requirement.

The voltage required to break down a gap is also found to decrease as the electrodes become small or sharper in configuration. New or filed electrodes having sharp corners concentrate the electron emission and reduce the voltage requirements. Rounding of the electrodes, even without appreciable increase in gap spacing, may increase the firing voltage 25%.

The spark plug operating temperature also affects the voltage requirements by changing the rate of electrode emission and the temperature differential between the two electrodes.

Having presented the factors affecting the physical design of the spark plug and some of the requirements for optimum plug performance, a series of engine tests were performed to determine the effect of spark plug firing end design on engine operation. Features such as gap spacing, ground electrode orientation, gap location, surface gap design, ignition system

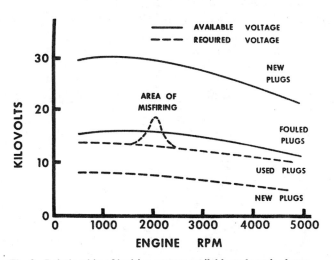

Fig. 9 - Relationship of ignition system available and spark plug required voltages

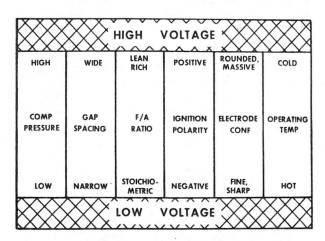

Fig. 10 - Factors affecting spark plug required voltage

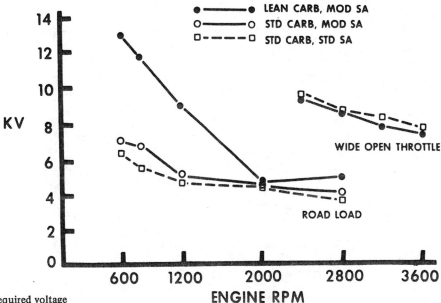

Fig. 11 - Effect of fuel-air ratio on spark plug required voltage spark discharge, and insulator projection on engine performance were investigated in carefully controlled dynamometer tests.

EFFECT OF SPARK PLUG GAP SPACING AND GROUND ELECTRODE ORIENTATION ON ENGINE PERFORMANCE

The following engine test was conducted in a medium displacement V-8 engine using spark plugs having the basic automotive electrode configuration shown in Fig. 2. The results were obtained under steady state 30 mph road load operation, since previous investigations had shown this to be a sensitive area of operation for observing the effect of small changes in spark plug design. The data shown in Fig. 12 indicates that the electrode gap for optimum brake specific fuel consumption must be increased as the physical size of electrodes is increased. A 0.030 in. gap is satisfactory for center electrode diameters less than approximately 0.050 in. diameter, and 0.035 in. spacing for electrodes less than 0.110 in. diameter. Increasing the gap spacing improves the specific fuel consumption of the engine, until with a 0.040 in. gap, the effect of electrode sizes up to 0.130 in. diameter can be neglected. These results are in agreement with present practice as spark plugs, for the majority of automotive usage, have electrode diameters of 0.090-0.110 in. and a gap spacing of 0.035 in.

Our tests indicated that under this 30 mph road load operating condition, the orientation of the ground electrode in the combustion chamber can also have an effect on the combustion process. In this testing, the area around the spark plug position was arbitrarily divided into four segments as shown in Fig. 13. Four different sets of spark plugs, in each of two electrode dimensions, were constructed in which the position of the ground electrode was located in each of the individual segments.

In assuming a clockwise motion of the fuel mixture within the chamber, spark plugs constructed with the heel of the ground electrode in Segment No. 2 would hinder the entry of the fuel mixture into the gap area. Spark plugs constructed with the heel of the ground wire in Segment No. 4 would thereby allow easy access of the mixture into the gap area.

The results of tests with a 0.090 in. center electrode, 0.050 X 0.100 in. ground electrode and 0.035 in. gap spacing are shown in Fig. 14, and indicate that the position of the ground electrode had no significant effect on the brake specific fuel consumption of the engine. However, in spark plugs constructed with a 0.125 in. diameter center electrode, 0.072 X 0.125 in. ground electrode and 0.030 in. gap, the ground electrode orientation was noted to affect the fuel consumption. With the heel of the ground wire oriented in Segment No. 2, an increase in the fuel consumption was observed. With the ground wire of this spark plug oriented in Segment No. 4, the brake specific fuel consumption was minimum. When the heel of the ground wires were located in positions No. 1 and 3, the fuel consumption was equal and just slightly greater than that obtained for the most optimum location. Under this light load engine operating condition, the spark plugs with the heavy duty electrodes and smaller gap spacings apparently do not permit easy flow of the fuel charge into the electrode area. The larger electrodes also tend to shield the gap area and cause occasional fuel charge misfires or delayed ignition of the charge. However, small gap spacings in combination with large electrodes are successfully used in continuous high power applications, such as gaseous fueled engines and heavy duty gasoline engines where a good homogenous fuel mixture is

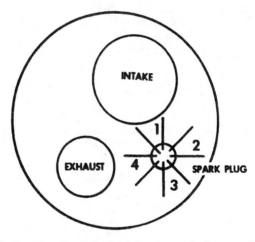

Fig. 13 - Top view of cylinder head showing spark plug ground electrode orientation

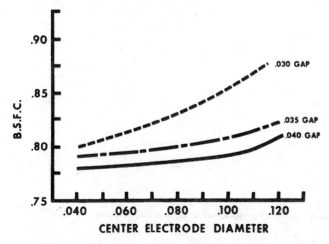

Fig. 12 - Effect of spark plug electrode size on fuel economy

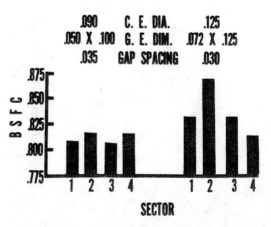

Fig. 14 - Effect of gap orientation on fuel economy

available, and where a high degree of combustion chamber turbulence is present. Intentional orientation of the spark plug ground electrodes in an engine would, of course, be undesirable for both the engine and spark plug manufacturers.

EFFECT OF SPARK PLUG GAP LOCATION ON ENGINE PERFORMANCE

The position of the spark plug gap in the combustion chamber has been noted to have an appreciable effect on engine performance in areas of lean fuel/air ratios. Five sets of spark plugs, with the gaps located as shown in Fig. 15, were also tested at the simulated 30 mph road load operating condition over a wide range of fuel/air ratios. The results of this testing are plotted in Fig. 16. It will be noted that spark plugs having gaps projected into the combustion chamber greater than 0.062 in. have very little effect on the minimum fuel consumption of the engine. However, there is a strong indication that locations retracted from the 0.062 in. projected position, will result in an increase in the brake specific fuel consumption, as evidenced by the performance of the 0.078 in. retracted gap spark plug. Of particular interest, is the continuous improvement in the ability to ignite lean fuel/air ratios as the gap was projected into the combustion chamber; optimum performance being obtained with the gap projected 0.531 in. beyond the end of the spark plug. The identification character on each curve indicates the fuel/air ratio at the point of consistant engine misfiring. This data, therefore, suggests the application of a 0.531 in. gap projection. However, a 0.215 in. gap projection is the approximate limit in which a gap can be extended into the combustion chamber of modern day engines, and still satisfy the spark plug operating temperature and life requirements. Greater projections result in higher electrode operating temperatures, causing more rapid electrode erosion, and the danger of engine preignition.

EFFECT OF SURFACE GAP SPARK PLUGS ON ENGINE PERFORMANCE

In recent years, considerable interest has been shown in the surface gap type of spark plug configuration (18). This design has been very successful in many two-cycle engine applications in combination with a capacitor discharge type ignition system. A view of the firing ends of a typical surface gap plug and a conventional spark plug, comparing insulator nose length, heat range, and gap configuration, is shown in Fig. 17. Because of its restricted nose length and cold operating temperature, the surface gap design is very susceptible to fouling, and thereby requires a capacitor discharge ignition system having a fast secondary voltage rise time (19). Fig. 18 is a plot of spark plug operating temperatures comparing a particular sur-

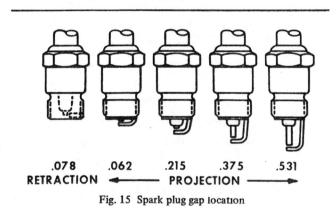

Fig. 15 Spark plug gap location

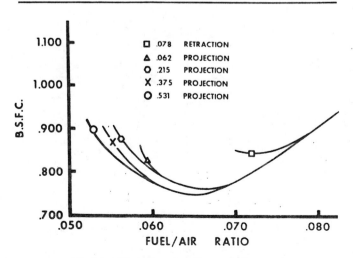

Fig. 16 - Effect of gap location on fuel economy

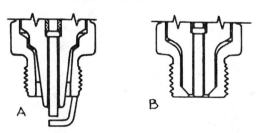

Fig. 17 - A-conventional and B-surface gap spark plug firing end configuration

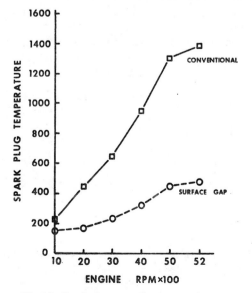

Fig. 18 - Spark plug operating temperature

face gap design to a conventional spark plug in a 50 hp outboard engine. A maximum wide open throttle temperature of only 450 F is noted for the surface gap plug as compared to 1400 F for the conventional plug. There has been only limited success with the surface gap plug in four-cycle engines, subject to the many modes of automotive type operation. The graph of brake specific fuel consumption versus fuel/air ratio in Fig. 19, shows a significant loss in performance in the leaner operating range for the surface gap spark plug and capacitor discharge ignition system. This is attributed to two major factors: poor physical exposure of the spark discharge as compared to a conventional spark plug and retracted position of the gap in the combustion chamber. The electrical discharge takes place over the surface of the insulator in the surface gap plug which does not result in the optimum exposure of the spark energy to the fuel charge. The relationship of gap position and brake specific fuel consumption has been previously noted.

EFFECT OF SPARK DISCHARGE CHARACTERISTICS ON ENGINE PERFORMANCE

While ignition system performance was not originally meant to be a part of this paper, it is believed necessary to show that a certain relationship exists between spark plugs and the spark discharge characteristics of the ignition system. As a result, further test work was undertaken to determine the effect of spark discharge characteristics on engine performance. In order to determine the effect of the electrical discharge, four ignition systems having widely varying spark discharge characteristics were selected for testing. The voltage waveform which appears across the spark plug at the time of the electrical discharge for each of these systems is shown in Fig. 20. The duration of the conventional ignition system shown as "A", is approximately 1600 μ-sec. Transistorized systems having durations of 3000, 300, and 12 μ-sec respectively, are shown as B, C, and D and were selected since they represent extremes in output characteristics.

System "B" is representative of the more common inductive type of transistorized system. The discharge is similar to that of the conventional system except the inductive portion or the arc duration is about 3000 μ-sec. Systems "C" and "D" are capacitor discharge type systems in which a charged capacitor is discharged into the primary of the ignition coil. In System "C", the overall duration is 300 μ-sec and unique, in that a refire and inductive arc of a positive polarity occurs after about 150 μ-sec. System "D" has a very fast rise time, is 12 μ-sec in duration, and has only a negative polarity arc.

A simulated 30 mph road load operating condition was again selected for this investigation. The testing was performed with the engine equipped with standard suppressive ignition cable and projected gap type of spark plugs. Measurements of horsepower were made with each type ignition system at various fuel/air ratios. Only very minor adjustments in throttle setting were found necessary to maintain an approximate 100 lb/hr of dry air flow through the engine.

Fig. 21 is a plot of horsepower versus spark advance for each of the ignition systems at a fuel/air ratio of 0.072 with 0.035 in. gap spark plugs. It will be noted that in the ranges

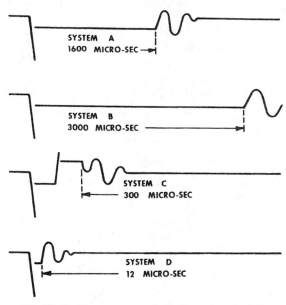

Fig. 20 - Ignition system spark discharge characteristics

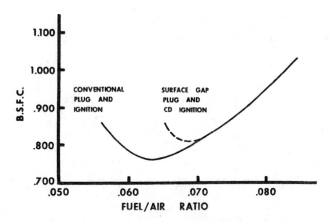

Fig. 19 - Effect of surface gap spark plug on fuel economy

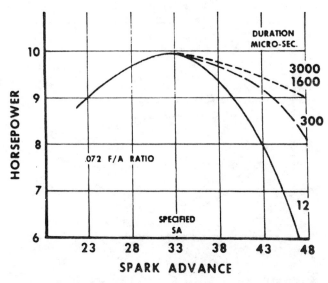

Fig. 21 - Effect of arc duration on horsepower at 0.072 fuel-air ratio of spark advance from 23 deg to slightly beyond the specified ignition timing of 33 deg, the spark discharge characteristic had no appreciable effect on engine output. However, at the higher spark advances, the longer duration type discharge appeared advantageous although no appreciable difference was noted between the 3000 and 1600 μ-sec systems. The spark advance for maximum horsepower was approximately the same for all systems.

In Fig. 22 at the considerably leaner fuel/air ratio of 0.061, the engine output at the same initial throttle setting, decreased from 10 to 6.5 hp. Contrary to the previous results, the engine performance at the specified spark advance was affected by the spark discharge characteristics as evidenced by the loss in horsepower with the 12 μ-sec, System "D". The horsepower developed with the 300 μ-sec system was again found to decrease immediately beyond the specified spark advance, while a distinct difference between the 1600 and 3000 μ-sec systems was apparent at the higher spark advance.

These results suggest a relationship between combustion chamber turbulence and spark duration. Under conditions of high spark advance where the turbulence in the area of the spark plug was probably reduced, a long duration type of discharge was required to initiate combustion and minimize misfiring. With a more turbulent and homogenous fuel charge, as would be expected when the piston approached top dead center, a discharge of shorter duration provided for satisfactory ignition.

In the past, the manufacturers of several of the shorter-duration capacitor discharge ignition systems have recommended the use of larger gap spark plugs, to eliminate a slight miss which may be experienced under certain light load conditions. In order to determine the effect of a larger gap spacing with shorter duration systems, several tests were conducted using the 1600 and 300 μ-sec duration systems. As shown in Fig. 23, investigation of the 1600 μ-sec system revealed no appreciable difference in engine output with either 0.035 in. or 0.050 in. gap plugs with a 0.072 in fuel/air ratio. However, at a 0.061 fuel/air ratio and greater than specified spark advance, a slight improvement in horsepower was noted with the 0.050 in gap plugs.

With the shorter duration 300 μ-sec system, improvements in performance with the use of larger gaps were clearly demonstrated as shown in Fig. 24. A slight improvement was evident at the 0.072 fuel/air ratio at the high spark advance, while a more noticeable difference was observed at a 0.061 fuel/air ratio.

The improvement in engine performance as a result of 0.050 in. spark plug gaps, was believed to be due to the increased volume between the electrodes for the accumulation of a combustible fuel charge. The increase in voltage requirement resulting from a larger electrode spacing had also increased the energy released in the capacitance portion of the discharge, but the increase in volume between the electrodes for the ac-

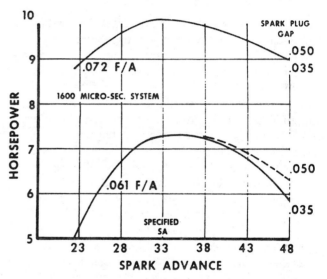

Fig. 23 - Effect of spark plug gap spacing at different fuel-air ratios with a 1600 μ-sec duration ignition system

Fig. 22 - Effect of arc duration on horsepower at 0.061 fuel-air ratio

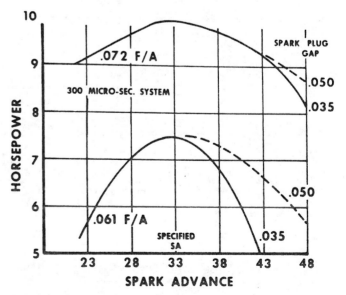

Fig. 24 - Effect of spark plug gap spacing at different fuel-air ratios with a 300 μ-sec duration ignition system

cumulation of a combustible fuel charge was believed to be the more important factor.

Because of the current interest by the engine manufacturers in using leaner mixtures to reduce the unburned hydrocarbons in the engine exhaust (20), the ability of an ignition system to ignite the fuel mixture is important. As just shown, the test results at lean fuel/air ratios indicated that the longer duration spark discharges gave the best overall engine performance. Some of the loss in performance at lean fuel/air ratios with the extreme short duration spark discharge was recovered by using larger spark plug gap spacings. However, these larger spark plug gaps have the disadvantages associated with the generating and distributing of greater spark plug required voltages, as well as increasing the intensity of the radiated ignition noise.

EFFECT OF INSULATOR PROJECTION ON SPARK PLUG PERFORMANCE

Spark plugs in which the insulator is projected beyond the end of the shell, while not directly affecting engine performance, have shown desirable spark plug temperature characteristics. The spark plug shown on the right in Fig. 25 has the insulator projected 0.060 in. beyond the end of the shell, as compared to an insulator retraction of 0.030 in. for the plug on the left. The projected insulator spark plug is said to have a broader heat range as a result of its operating temperature characteristics shown in Fig. 26. It will be noted that the insulator tip temperature has a more desirable, flatter part throttle temperature response, as well as a reduction in wide open throttle temperature. Under low power operating conditions, the projected insulator is more exposed to the combustion process and absorbs greater heat than the retracted insulator design. However, under the higher power conditions the projected insulator, while still more exposed to the combustion process, is likewise exposed to the "charge cooling" effect of the incoming fuel mixture. This "charge cooling" effect is the predominate factor in the reduction in insulator tip temperature under high power operation. This feature can be used to advantage in many automotive engines. A few engine designs do not show this appreciation of the projected insulator and, therefore, it is not of particular advantage. However, this design is accompanied with the projected gap feature which showed improved performance in regard to ignition of lean fuel/air mixtures. The projected insulator spark plug by virtue of its increased low power operating temperature is also beneficial in reducing cold fouling tendencies, Although this design has significant performance advantages in many engines, the projected insulator spark plug should not be used indiscriminately. Under continuous high power type applications, undesirable insulator tip corrosion and more rapid electrode erosion could be expected.

SUMMARY

It has been shown that many of the physical characteristics of the spark plugs are influenced by the engine designer and the engine application. The individual design features must be considered for each application if optimum spark plug and engine performance are to be obtained. The spark plug must operate within a specified temperature range in order to resist fouling and insure adequate electrode gap life.

The 30 mph road load engine testing has shown that spark plugs having increased gap spacings and gap projections are beneficial in improving the brake specific fuel consumption of the engine, and the ability to ignite lean fuel/air ratios. While surface gap spark plugs have been successfully used in two-cycle engines equipped with capacitor discharge ignition, their success in automotive engines has been limited. The test results have also shown that normal engine performance is obtained with conventional spark plugs over a wide range of ignition system spark discharge characteristics. However, adverse effects have been noted with a very short duration in regard to brake specific fuel consumption and the ability to ignite lean fuel/air ratios.

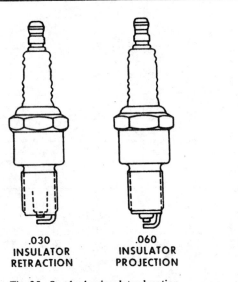

Fig. 25 - Spark plug insulator location

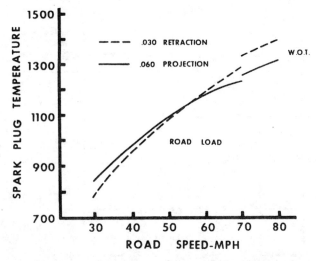

Fig. 26 - Effect of insulator location on spark plug operating temperature
Nose length: A-0.561, B-0.050; A-200, B-400+; Gap spacing: A-0.035, B-0.050

REFERENCES

1. C. G. Studaker, "Design and Development of New Buick/400 and 300 cu in. V-8 Engines." Paper 670083 presented at SAE Automotive Engineering Congress, Detroit, January 1967.
2. I. B. K. Gough, "It Has All Been Done Before." Paper 670149 presented at SAE Automotive Engineering Congress, Detroit, January 1967.
3. W. R. Edwards and G. E. Roy, "Development of a New Automotive Spark Plug." Paper 670148 presented at SAE Automotive Engineering Congress, Detroit, January 1967.
4. H. I. Hazzard, "A Modern 2 Cu. In. 6.5 Lb. Chain Saw." Paper 690599 presented at SAE Farm, Const., and Industrial Machinery Meeting, Milwaukee, September 1969.
5. "Radio Shielding of Aircraft Engine Ignition Systems." United States Army Air Force, 1944.
6. "A Discussion of Industrial Low-Fire-Hazard Ignition." Scintilla Div. of Bendix Corp., January 1963.
7. SAE Standard, Spark Plugs—SAE J548a, SAE Handbook.
8. P. J. Kent, "The Automotive Industry's Participation in Reduction of Radio and Television Interference." Paper presented at SAE Summer Meeting, French Lick, Indiana, June 1948.
9. "Effect of Internal Resistors on Spark Plug Electrode Erosion." Report for Army Air Forces, Ethyl Corp., 1943.
10. SAE Recommended Practice, Preignition Rating of Spark Plugs for Ground Vehicles—SAE J549, SAE Handbook.
11. J. R. Sabina, J. J. Mikita, and M. H. Campbell, "Preignition in Automotive Engines." Paper presented at American Petroleum Institute Meeting, New York, May 1953.
12. E. T. Jones, "Induction Coil Theory and Applications." London: Sir Isaac Pitman & Sons, Ltd., 1932.
13. F. W. Lamb, "Postulated Mechanism of Spark Plug Fouling." Paper presented at 1954 Champion Spark Plug Aircraft Conference, October 1954.
14. "Motor Gasoline Additives and Their Functions." Ethyl Technical Notes.
15. C. A. Hall, R. C. Beaubier, E. C. Marckwardt, and R. L. Courtney, "Spark Plug Fouling." Paper presented at SAE Annual Meeting, Detroit, January 1957.
16. L. R. Hetzler and P. C. Kline, "Engineering C. D. Ignition for Modern Engines." SAE Transections, Vol. 76, paper 670116.
17. D. H. Holkeboer, "The Effect of Compression Ratio on the Ability to Burn Lean Fuel-Air Mixtures in a Spark-Ignition Engine." Univ. of Michigan, Industry Program of College of Engineering, March 1961.
18. G. H. Millar, "Surface Gap Ignition as Applied to Large High Performance Outboard Motors." Paper 724-B presented at SAE West Coast Meeting, Seattle, Washington, 1963.
19. W. R. Eason, "Voltage Risetime - A New Ignition Criterion?" Paper 652C presented at SAE Automotive Engineering Congress, Detroit, January 1963.
20. J. S. Minomiya and A. Golovoy, "Effects of Air-Fuel Ratio on Composition of Hydrocarbon Exhaust from Isooctane, Diisobutylene, toluene, and toluene-M-Heptane Mixture." Paper 690504 presented at SAE Mid-Year Meeting, Detroit, May 1969.

Society of Automotive Engineers, Inc.
TWO PENNSYLVANIA PLAZA, NEW YORK, N.Y. 10001

This paper is subject to revision. Statements and opinions advanced in papers or discussion are the author's and are his responsibility, not the Society's; however, the paper has been edited by SAE for uniform styling and format. Discussion will be printed with the paper if it is published in SAE Transactions. For permission to publish this paper in full or in part, contact the SAE Publications Division and the authors.

12 page booklet.   Printed in U.S.A.

The Performance of a Multigap Spark Plug Designed for Automotive Applications

W. G. Rado, J. E. Amey, B. Bates, and A. H. Turner
Ford Motor Co.

AS THE EMISSION control requirements for passenger cars become more and more stringent, the demand for burning marginally ignitable mixtures as defined by previously accustomed to modes of engine operation continuously increases. Some of the new ignition systems currently being developed in order to improve the initiation of combustion use spark plugs with larger gap sizes combined with substantially higher available ignition energies.

Similar improvements in the initiation of combustion are expected to be achieved by the use of multigap spark plugs and current production ignition systems. The multiplicity of gaps is a way to produce an enlarged flame kernel. This is expected to facilitate achievement of performance comparable to that of wide gap plugs, especially under light load or exhaust gas recirculating conditions. Such plugs, however, would require no additional available ignition voltage to initiate combustion reliably under heavy load, high cylinder pressure operation.

The purpose of this paper is to describe the concept underlying the operation, the geometrical and electrical circuit constraints on the design of, and the results of evaluating the electrical and in-engine performance of multigap spark plugs intended for automotive applications.

PRINCIPLE OF OPERATION

The concept underlying the operation

---ABSTRACT

The electrical principle of operation, the geometrical and electrical circuit constraints on the design of, and the electrical and in-engine performance of a multigap spark plug developed for automotive applications are described.

The electrical principle of operation is based on successively breaking down an array of spark gaps through the use of a resistive ladder network.

The measurements evaluating the electrical performance of various multigap designs indicate that these plugs can deliver up to twice the energy of a single gap plug to the arcs, using the same ignition system. The increased amount of energy is also delivered in a shorter time than for single gap plugs.

The measurements evaluating the in-engine performance of these plugs further indicate that improvements of up to 6% in fuel economy for a simulated CVS test run, the extension of the lean misfire limit by several A/F ratio numbers under various engine operating conditions, and improvements in driveability can be achieved compared to the performance of standard plugs under the same operating conditions.

of the multigap spark plugs to be reported on here makes use of a ladder type impedance-spark gap network. The basic function of this network is to apply very nearly the full available ignition voltage in sequence to the individual gaps of an array of gaps not yet electrically broken down. This sequential breakdown of an array of gaps then allows the formation of long arc lengths without the need for increased available ignition voltage.

A model of a ladder type resistor-spark-gap-network multigap spark plug with 3 gaps is shown in Figure 1. The corresponding ideal electrical equivalent circuit and the salient voltage relationships describing this spark plug's

CONCEPTUAL MODEL

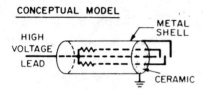

IDEAL EQUIVALENT CIRCUIT

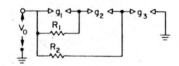

VOLTAGE RELATIONSHIPS

$$V_{g_3} = V_o \left(\frac{R_{g_3}}{R_{g_3} + R_2} \right)$$

$$V_{g_2} = (V_o - V_{g_3}^*) \frac{R_{g_2}}{R_{g_2} + R_1}$$

$$V_{g_1} = V_o - V_{g_3}^* - V_{g_2}^*$$

Fig. 1-The conceptual model of a 3 gap spark plug using a ladder type resistor-spark-gap-network with the corresponding ideal electrical equivalent circuit and salient voltage relationships

operation are also depicted. The available ignition voltage, applied to the spark plug, first divides across gap g3 and resistor R_2. If the resistance of gap g3 before electrical breakdown is substantially higher than the value of R_2, most of the available voltage appears across gap g3. When the gap breaks down electrically, the voltage appearing across R_2 will be very nearly equal to V_o, since the arc sustaining voltage is generally much smaller than the value of V_o (.5 to 1 Kv for sustaining versus 10-26 Kv for electrical breakdown). The high voltage across R_2 now divides across the resistor R_1 and the electrically unbroken down gap g2. Again, if the resistance of the gap is substantially higher than the value of R_2, most of the voltage will appear across the gap g2 until it breaks down electrically. When both gaps g2 and g3 are electrically broken down, the voltage across R_1 very nearly equals V_o, it being reduced by the sum of the sustaining voltages of gaps g2 and g3. This finally leads to the breakdown of gap g1, after which the stored energy of the ignition system is dissipated in the arcs as though the resistor-coupled intermediate electrodes were not there.

The same idea can be implemented by using capacitors (1)* instead of resistors in the ladder type network. However, due to the constraints on spark plug geometry and the achievable dielectric strength of ceramics used in production, multigap spark plugs of practical size using a capacitive ladder type network appear difficult to be fabricated.

On the other hand, resistive network multigap spark plugs can be fabricated directly into the ceramic inserts currently used in production spark plugs. Since the resistance of air or of the combustible gases in the spark plug gaps before electrical breakdown is well into the tens of megohms, the resistor-coupled electrodes require the use of resistors of a few megohms. The encasing of such resistors in the ceramic shell does not present a technologically unsolvable problem.

CONSTRAINTS ON THE DESIGN OF RESISTIVE-NETWORK MULTIGAP PLUGS

The previous discussion of the operation of the multigap spark plug using a resistance-spark gap network was based on an ideal equivalent circuit. A physical realization of this concept and the real equivalent electrical circuit is shown in Figure 2. This equivalent circuit reveals the existence of two types of problems that have to be faced before a usable prototype plug can be fabricated. There are problems related to current flow and to interelectrode capacitance. There are two current flow problems. First, there is a limitation as to how low the value of the resistor can be, because the ignition system is instantaneously loaded down by that resistor as soon as the first gap is broken down electrically. This is equivalent to having the ignition system connected to a fouled plug with low leakage resistance. This then requires a

---

*Numbers in parentheses designate reference at end of paper.

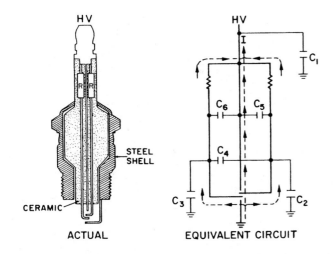

Fig. 2-An actual 3 gap spark plug using a resistor-spark-gap network with its equivalent electrical circuit

larger available ignition voltage to insure that the yet electrically unbroken down gaps will break down.

The second current flow related problem arises when the arcs are already established. If the value of the resistors is low enough, the sustaining voltage associated with some segment of the arc array will produce a current flow through the resistors which could amount to a substantial fraction of the total arc current. This current flow constitutes a waste of the stored available ignition energy. Since the sustaining voltages across the arc segments are generally independent of the arc currents for spark plugs used with the conventional ignition systems (they are only functions of gas pressure and gap size), the larger the value of resistance is, the less important this current drain will be.

Therefore, both limitations on the design of a resistive-network multigap spark plug due to current flow require the use of resistance values as high as other constraints permit.

The most significant remaining constraint is due to interelectrode capacitance. A feel for the extent of this effect can be developed by analyzing the voltage variation at various points in the equivalent circuit of Figure 2, as the plug is undergoing the process of successfully breaking down electrically the array of gaps. When a voltage first appears at the high voltage lead of the spark plug, it immediately begins to charge capacitors C_2 and C_3 with respective time constants RC_2 and RC_3 to voltages of V_{C_2} and V_{C_3}, where R is the value of the resistors embedded in the spark plug's ceramic insert. Depending upon the differences in the values of C_2 and C_3, capacitor C_4 will be charged anywhere from zero volts (when $C_3 = C_2$) to some value less than V_{C_2} or V_{C_3}, depending upon which is greater. As the arc forms in the first gap, it immediately shorts out capacitor C_3. This leads to an instantaneous redistribution of charge on C_4 and C_2, lowering the voltage V_{C_2} depending on the capacitance value of C_4 and C_2 and the voltages which these capacitors were charged to. For the next gap to break down, the voltage on C_2 has to rise to the breakdown value. This is accomplished by the ignition system charging C_2 through the resistor R again. When the second gap breaks down, the remaining available voltage now appears directly across the last electrically unbroken down gap with no capacitance charging effects influencing its breakdown. Thus, for a 3 gap multigap spark plug, depending on the values of the electrode capacitances and the values of the resistors coupling the gaps to the high voltage lead, either all three gaps will break down electrically or only one will. Since the risetimes increase with an increase in the values of both resistors and capacitances, it is of prime importance to keep electrode capacitances as small as possible and use resistor values which are the lowest allowed by other constraints.

Combining the demands of these constraints, then, resistor values of 1-2 megohms and wires for electrodes with the smallest usable diameters are recommended.

THE EXPERIMENTAL TECHNIQUE USED TO EVALUATE THE ELECTRICAL PERFORMANCE OF RESISTIVE-NETWORK MULTIGAP SPARK PLUGS

The electrical test facility used in evaluating the performance of multigap spark plug designs is shown in Figure 3. A complete ignition system was assembled with a variable speed electric motor driving the distributor. The facility

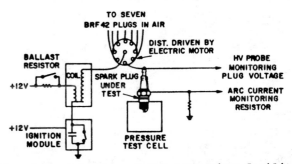

Fig. 3-Description of the measuring facility used to evaluate the electrical performance of various multigap designs

permitted the use of several types of ignition systems, such as the Ford production breakerless ignition system (SBI) and the experiemental Ford breakerless high energy ignition system (HEI). Production wiring, coils and electronic modules were used except for the experimental HEI units. Seven of the 8 voltage leads from the distributor were connected to production BRF42 Autolite spark plugs firing in air. The eighth high voltage lead using a standard length ignition cable was routed to a spark plug screwed into a pressurizable test cell. Dry air was used in all of the evaluations because dry air requires a higher voltage to break down electrically in the pressure range of interest than N_2, O_2, CO_2 and other gases normally found in operating automotive engines at the time of ignition. A high voltage probe monitored the voltage waveform, appearing at the high voltage connection of the spark plug under test, while a resistor placed between the steel shell of the spark plug and ground was used to monitor the arc current waveform of the plug.

Two parameters derived from the above electrical waveforms were used as the basis for comparing the electrical performance of multigap spark plugs to that of standard spark plugs. These were the total energy dissipated in the arcs, considered to represent the actual ignition energy a spark plug is delivering to the gases to be ignited and a parameter reflecting the relationship between available ignition voltage and the time it takes to reach complete electrical breakdown of the gaps after the ignition process is initiated in the ignition module. The typical traces from which this information was derived is shown in Figure 4. The total energy delivered was calculated from traces represented by the upper oscilloscope display. An average value of arc current was derived from the triangular shaped current waveform. This value was then multiplied by the value of the sustaining voltage, corresponding to the constant part of the upper voltage trace. This gave a value for the average power in the arc. Multiplying this value by the duration of the arcs, corresponding to the time for which there was current flow in the arcs, the value of the energy delivered by the arcs to the surrounding gases was obtained.

The measure of a spark plug's ability to electrically break down a set of gaps was determined from the trace of the lower oscilloscope display. It is a time magnified part of the upper display's voltage trace. It shows the voltage V_B,

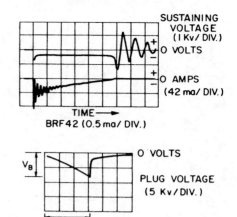

Fig. 4-Typical voltage-current traces associated with the arc generated in the gap of a spark plug. Upper oscilloscope display depicts the sustaining voltage and the corresponding arc current phase. Lower oscilloscope display depicts the voltage rise on the electrode to the electrical breakdown level

that had to be reached at the plug's electrode to produce breakdown and the time, τ, it took to have the voltage rise to this value. The parameter of comparison was defined as $V_B\tau$, the product of V_B and τ.

Neither V_B, nor V_B/τ was found to be a reliable measure of performance when multigap spark plugs were compared to production spark plugs because for a multigap spark plug V_B was not related to the gap size directly due to the capacitance charging effects of the resistive-ladder-type network built into them.

For the purposes of later making comparisons between the electrical

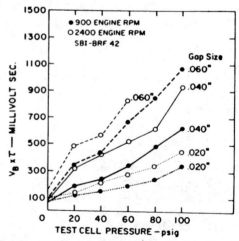

Fig. 5-The $V_B\tau$ performance of the reference BRF42 spark plug used with a SBI system as a function of cell pressure, gap size, and engine speed

performances of multigap and production spark plugs, first the electrical performance of standard production plugs was mapped for both the SBI and HEI system. Figures 5 and 6 represent the variation of the parameter $V_B\tau$ and the energy dissipated in the arc of a BRF42 spark plug used with a SBI system as a function of dry air pressure and gap size. Figures 7 and 8 represent the same data for a BRF42 spark plug used with a HEI system.

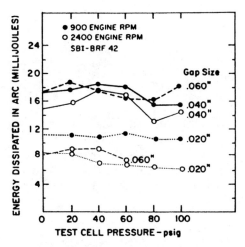

Fig. 6-The variation of the energy delivered through the arc of the reference BRF42 spark plug used with a SBI system as a function of test cell pressure, gap size, and engine speed

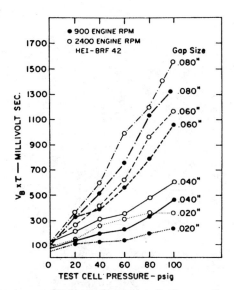

Fig. 7-The $V_B\tau$ performance of the reference BRF42 spark plug used with a HEI system as a function of test cell pressure, gap size, and engine speed

THE ELECTRICAL PERFORMANCE OF RESISTIVE-NETWORK MULTIGAP SPARK PLUGS

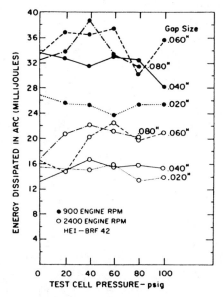

Fig. 8-The variation of the energy delivered through the arc of the reference BRF42 spark plug used with a HEI system as a function of test cell pressure, gap size, and engine speed

Two basic designs of the resistive-network multigap spark plug were evaluated using the previously described technique. One used "thin" electrodes and another design used "thick" electrodes. The values of resistances and the values of the various electrode capacitances (refer to Figure 2 for the capacitance designations) for all of the spark plugs tested are tabulated in Table I. The thin electrode spark plugs used .032 inches thick diameter nichrome wire while the thick electrode plugs used 0.040 inches by .070 inches rectangular shaped inconel ribbon. The "thick" electrode spark plugs also had copper-glass seals in the ceramic insert which increased the electrode diameter to about .140 inches for a length of .5 inches, within the steel shell part of the ceramic insert. This contributed substantially to the electrode capacitances of the "thick" electrode spark plug. As can be seen from Table I, the "thick" electrode spark plug is expected to have two or three times the voltage rise time of a thin electrode spark plug for the same value of resistor. The resistors in these experimental plugs were epoxy coated 0.5 watt carbon resistors, generally two of them in series making up the specific value of resistance. For any one spark plug both resistors were always chosen to be of the same value.

First, the existence of the correct scaling of $V_B\tau$ with the value of resistance was checked for both types of

Table 1 - Resistance and Capacitance* Values of Multigap Spark Plugs

Thick Electrode (.070" x .040") Plugs

Resistance	C_1	C_2	C_3
0.5MΩ	10.5	10.0	10.5
1.0MΩ	10.5	10.0	10.0
2.0MΩ	10.0	9.5	10.0
4.0MΩ	12.0	10.5	10.5

Resistance	C_4	C_5	C_6
0.5MΩ	8.5	8.0	9.0
1.0MΩ	8.5	7.5	8.5
2.0MΩ	9.0	7.5	8.0
4.0MΩ	8.5	8.0	9.0

Thin Electrode (.032" diam.) Plugs

Resistance	C_1	C_2	C_3
0.5MΩ	4.0	3.0	3.5
1.0MΩ	4.0	5.0	5.0
2.0MΩ	4.5	5.5	4.0
4.0MΩ	4.5	6.0	6.5

Resistance	C_4	C_5	C_6
0.5MΩ	4.0	4.5	4.5
1.0MΩ	4.0	2.5	2.5
2.0MΩ	4.0	3.0	3.0
4.0MΩ	4.0	3.0	3.0

*All values in picofarads

plugs. That is, the larger the value of the resistor, the larger should the product $V_B\tau$ be for any one value of test cell pressure.

As can be seen from Table I, the capacitance values of the "thick" electrode plugs are almost independent of the values of the resistors in contrast to the "thin" electrode spark plugs. As a result, a better scaling is expected for the "thick" electrode spark plugs.

This was, in fact, the case as indicated by the data in Figures 9, 10, and 11. Looking at the data in Figure 9, for example, at a test cell pressure of 60 psig, the 4MΩ resistor plug had the highest value of $V_B\tau$, the 2MΩ had a lower, the 1MΩ had an every lower value, and the 0.5MΩ resistor plug had the lowest value of $V_B\tau$. A similar effect is indicated in Figure 10, looking again at the data corresponding to a test call pressure of 60 psig. At lower pressures the scaling is not that obvious due to the experimental errors in the measurements and the fact that the spark plugs are electrically broken down at a voltage well

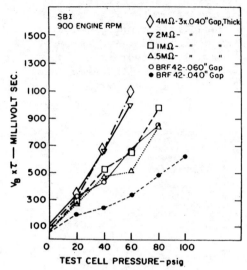

Fig. 9-The $V_B\tau$ performance of "thick" electrode 3 gap spark plugs used with a SBI system as a function of test cell pressure at 900 engine rpm

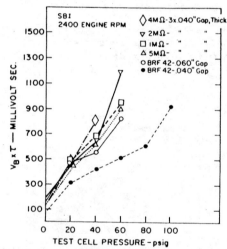

Fig. 10-The $V_B\tau$ performance of "thick" electrode 3 gap spark plugs used with a SBI system as a function of test cell pressure at 2400 engine rpm

below the system's maximum capability. The effect is more pronounced near breakdown voltages corresponding to the maximum available ignition voltage values. These data then qualitatively support the expected scaling of $V_B\tau$ with R for "thick" electrode multigap plugs. A quantitative agreement cannot be expected because these resistors can severely degrade in value under the application of high voltages.

In contrast to Figures 9 and 10, the data of Figure 11 representing the

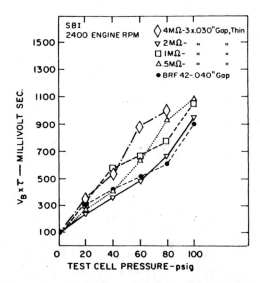

Fig. 11-The $V_B\tau$ performance of "thin" electrode 3 gap spark plugs used with a SBI system as a function of test cell pressure at 2400 engine rpm

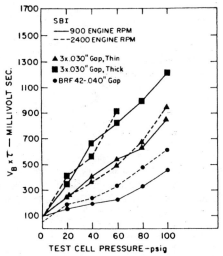

Fig. 12-A comparison of the $V_B\tau$ performance of a BRF42 reference plug and a "thin" and a "thick" electrode 3 gap plug, all used with a SBI system, as a function of test cell pressure and engine speed

performance of "thin" electrode plugs indicate a value of $V_B\tau$ for a 0.5MΩ resistor plug which is larger than for either the 1MΩ or 2MΩ resistor plug at a test cell pressure of 80 psig. This is due to the fact that the value of the capacitance C_4 is greater than those of the capacitors C_2 and C_3, and after the first gap is broken down, the charge redistribution from C_2 to C_4 is more severe than for the other "thin" electrode plugs. This requires a longer time to recharge C_2 to the voltage level needed for electrically breaking down the second gap.

For purposes of comparison, the $V_B\tau$ performances of production BRF42 spark plugs are also included in Figures 9, 10, and 11. It is seen that using the Ford standard production breakerless ignition, SBI, system, of all of the multigap spark plug designs, only the "thin" electrode 2MΩ resistor plug performed close to the production BRF42 spark plug with a gap size of .040 inches. Most of the other multigap spark plugs performed more like a BRF42 spark plug with a gap size of .060 inches.

Further evaluations of the electrical performance of 3 gap spark plugs having individual gap sizes of .030 and .040 inches were carried out using both the SBI and the HEI systems. Typical results, using the SBI system, are shown in Figures 12, 13, and 14. Figure 12 presents a comparison between the values of $V_B\tau$ for "thin" and "thick" electrode 3 gap plugs using 2MΩ resistors and a BRF42 spark plug with a .040 inch gap size. These data

clearly demonstrate the large variation of $V_B\tau$ with the various designs. These substantially larger $V_B\tau$ values associated with the multigap spark plugs are expected

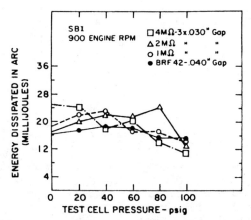

Fig. 13-The variation of the energy delivered through the arcs of "thin" electrode 3 gap spark plugs used with a SBI system as a function of test cell pressure at 900 engine rpm. For reference, the values corresponding to a BRF42 spark plug are also included

to produce a sizable deficiency, even for the "thin" electrode 3 gap spark plug, in the amount of energy dissipated in the arcs compared to the energy dissipated in the single arc of a BRF42 plug. This is seen in Figures 13 and 14. In fact, the effect is especially severe for the high speed (2400 engine RPM) data, Figure 14, where the peak available ignition voltage is reduced due to speed.

A more favorable performance is demonstrated by the multigap plugs when

141

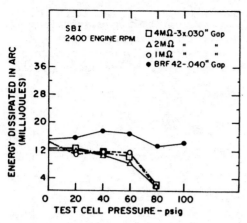

Fig. 14-The variation of the energy delivered through the arcs of "thin" electrode 3 gap spark plugs used with a SBI system as a function of test cell pressure at 2400 engine rpm. For reference, the values corresponding to a BRF42 spark plug are also included

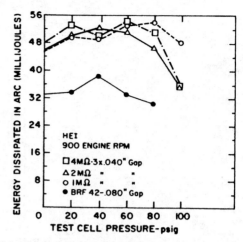

Fig. 16-The variation of the energy delivered through the arcs of "thin" electrode 3 gap spark plugs used with a HEI system as a function of test cell pressure at 900 engine rpm. For reference, the values corresponding to a BRF42 spark plug are also included

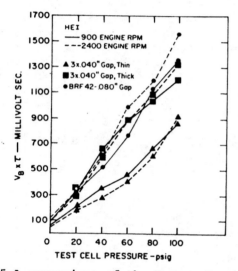

Fig. 15-A comparison of the $V_B\tau$ performance of a BRF42 reference plug, a "thin" and a "thick" electrode 3 gap plug, all used with a HEI system, as a function of test cell pressure and engine speed. All resistors were 2MΩ in value

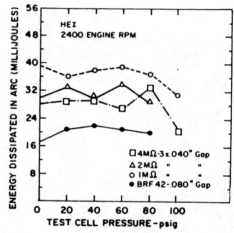

Fig. 17-The variation of the energy delivered through the arcs of "thin" electrode 3 gap spark plugs used with a HEI system as a function of test cell pressure at 2400 engine rpm. For reference, the values corresponding to a BRF42 spark plug are also included

the HEI system is used. Here the comparison is to a BRF42 production spark plug with a gap size of .080 inches. The variation of $V_B\tau$ with pressure for the two experimental and the one reference plug is shown in Figure 15. As seen in the figure, the "thin" electrode multigap plug is far superior to the BRF42 reference plug. As suggested by the favorable range of $V_B\tau$ values, the "thin" electrode plug was found to deliver substantially more energy through its arcs than the BRF42 plug with a gap size of .080 inches through its single arc. This improvement is shown in Figures 16 and 17. Again, as the peak available ignition voltage decreases with speed, the lower $V_B\tau$ characteristic of the 3 gap plug is so favorable that for the speed of 2400 engine RPM, the thin electrode plug delivers about twice the energy of a single gap plug.

The above data indicate that, as previously proposed, fabricating a 3 gap multigap spark plug using a ladder type resistive network is possible with electrical performance superior to present day production spark plugs. Not only can these plugs provide substantially longer arc lengths, but they can also dissipate in their arcs more of the available ignition energy stored in the primary winding of the ignition coil than single gap plugs can. Experimental observations indicate that the increased amount of energy was delivered in a shorter time and the breakdown voltage required to establish the long arc length could be made substantially less than that for a single gap plug.

THE IN-ENGINE PERFORMANCE OF RESISTIVE-NETWORK MULTIGAP SPARK PLUGS

Since only "thin" electrode multigap spark plugs approached the electrical performance of production BRF42 spark plugs with .040 inches gap size when used with the standard production ignition, SBI, system, only this type of of spark plug was used in evaluating the in-engine performance of multigap spark plugs. Both in-car and dynamometer mounted engine experiments were carried out. In-car experiments were made to evaluate the extent to which driveability could be improved. Cars with low driveability ratings were evaluated, back to back, with regular spark plugs first and then with 3 gap plugs. Typical results are shown in Table II. In both cases the idle, light load, part throttle performance of the cars was improved. To check on the effect of the thinness of the electrodes, special BRF42 single gap spark plugs were fabricated having the same .035 inches gap size as the reference plug. As the data indicate, no improvement in driveability was achieved when thin electrode single gap plugs were used (car 2), in contrast to the benefits obtained with 3 gap plugs.

Dynamometer tests on engines using multigap spark plugs were carried out to determine the plugs' effect on emissions and fuel economy under various characteristic engine operating conditions. Data corresponding to the operation of a fuel injected 351 cubic inch displacement, otherwise production engine, using 3 types of spark plugs, are shown in Figures 18, 19, 20, and 21. Figure 18 represents data on idle, Figure 19 on acceleration, Figure 20 on cruise, while Figure 21 represents data on deceleration performance. The three types of plugs compared were the standard production BRF42 spark plug with a gap size of .035 inches, a special BRF42 spark plug of the same gap size with thin electrodes (diameter of .032 inches) and a "thin" electrode multigap spark plug with 3 gaps of .035 inches each using 2MΩ resistors. All spark plugs were used with the same production SBI system.

For every one of the 4 engine operating conditions investigated as a function of A/F ratio, the multigap spark plugs produced an extension of the lean misfire limit and allowed the engine to maintain the appropriate speed and load condition with lower fuel consumption. The thin electrode single gap BRF42 spark plug also produced an improvement in fuel economy and an extension of the lean misfire limit over that of the reference BRF42 plug. However, the effect was substantially smaller than that produced by the 3 gap experimental spark plug.

Additional tests carried out on the dynamometer using the same engine allowed a computer calculation to be made of the plug's projected performance operating the engine over a CVS test cycle. By finding the best fuel economy operating point for

Table 2 - Driveability Improvements Produced by the Use of Multigap Spark Plugs

Car 1

Operating Condition	Standard Plug	Experimental Plug
Crowds below 40 MPH	4.0	5.0
Crowds above 40 MPH	6.0	6.5
Road loads below 40 MPH	5.0	5.0
Road loads above 40 MPH	5.0	6.0
W.O.T. accel. to 30 MPH	6.0	6.0
Part Throttle accel. to 30 MPH	4.0	5.0
Tip-in 0-30 MPH	7.0	7.0
Idle Quality	5.0	5.0

Car 2

Operating Condition	Standard Plug	Thin Electrode Standard Plug	Experimental Plug
Crowds below 40 MPH	5.0	5.0	5.0
Crowds above 40 MPH	5.0	5.0	6.0
Road loads below 40 MPH	5.0	5.0	6.0
Road loads above 40 MPH	6.0	6.0	6.0
W.O.T. accel. to 30 MPH	6.0	6.0	6.0
Part Throttle accel. to 30 MPH	5.0	5.0	6.0
Tip-in 0-30 MPH	6.0	6.0	6.0
Idle Quality	5.0	5.0	6.0

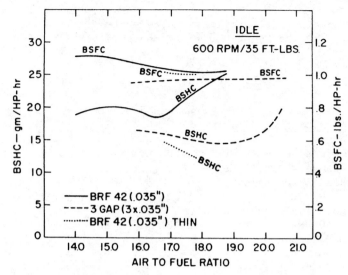

Fig. 18-The in-engine idle performance of a "thin" electrode 3 gap spark plug with 2M resistors, of a "thin" electrode BRF42, and a reference BRF42 spark plug using a SBI system as a function of A/F ratio. All data points were taken at MBT spark timing

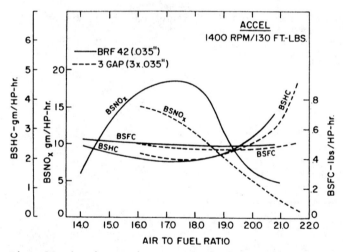

Fig. 19-The in-engine acceleration performance of a "thin" electrode 3 gap spark plug with 2MΩ resistors, of a "thin" electrode BRF42, and of a reference BRF42 spark plug using a SBI system as a function of A/F ratio. All data points were taken at MBT spark timing

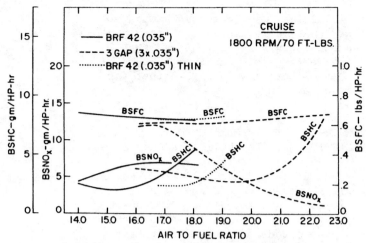

Fig. 20-The in-engine cruise performance of a "thin" electrode 3 gap spark plug with 2MΩ resistors, of a "thin" electrode BRF42 and of a reference BRF42 spark plug using a SBI system as a function of A/F ratio. All points were taken at MBT spark timing

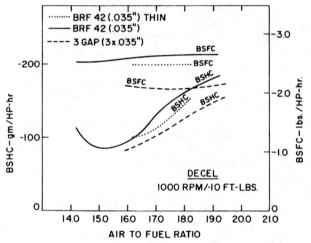

Fig. 21-The in-engine deceleration performance of a "thin" electrode 3 gap spark plug with 2MΩ resistors, of a "thin" electrode BRF42 and of a reference BRF42 spark plug using a SBI system as a function of A/F ratio. All points were taken at MBT spark timing

9 speed/load engine operating points representative of the modes of engine operation encountered during the CVS test cycle, always operating at the maximum brake torque, MBT, spark advance, the performance of the three types of plugs was calculated. The results are shown in Table III. As before, some improvement in fuel economy was achieved by just using thin electrodes, but a more substantial improvement was achieved when a multigap spark plug was used.

CONCLUSIONS

A concept has been presented for the design of multigap spark plugs in which an array of spark gaps are broken down in sequence by the use of a resistive ladder type network. This approach allows the fabrication of spark plugs with arc lengths substantially larger than the single gaps of current production spark plugs, without the need for an increase in the available ignition system voltage at the spark plug.

The evaluation of the electrical

Table 3 - CVS Test Run Performance of Various Types of Spark Plugs Based on Computer Calculation from Model Results

Type of Spark Plug	HC gm/mile	CO gm/mile	NO_x gm/mile
BRF42 Reference	4.0	5.4	4.2
BRF42 Thin Electrode 3 Gap	3.0	3.5	5.2
Thin Electrode	3.7	4.9	5.2

Type of Spark Plug	Fuel Economy miles/gal	% Improvement
BRF42 Reference	14.0	0
BRF42 Thin Electrode 3 Gap	14.3	2.1
Thin Electrode	14.8	5.7

performance of several multigap spark plug designs indicates that there are constraints on the values of the resistors making up the ladder type network in combination with the effects of electrode capacitances. In order to insure that there be no loading of the ignition system and that there be no excessive current flow through the ladder type network, the resistors have to be 2 megohms or more in value. This value can lead to excessive slowing of voltage rise on the electrodes, forming the array of gaps, which problem can only be solved by decreasing electrode capacitance. That can be achieved by using thin wires at least in that part of the ceramic insert located inside the outer steel shell of the spark plug or by using distributed resistance in the form of a ceramic "wire" replacing the metal leads.

The evaluation of the in-engine performance of 3 gap multigap spark plugs indicates that the lean misfire limit of the engine under any specific engine operating condition can be extended and an improvement in fuel economy can be achieved when 3 gap spark plugs are used. Both of these indications verify the earlier suggestion that increasing the burn-initiating flame kernel's size improves the reliability of ignition. In addition, the driveability of vehicles with poor ratings can be improved.

REFERENCES

1. G. Pratt, MIT and D. Moyer, Ford Motor Company (private communication)

New Aspects on Spark Ignition

H. Albrecht
Research Div., Daimler-Benz AG (Germany)

W. H. Bloss, W. Herden, R. Maly, B. Saggau, and E. Wagner
Institut für Physikalische Elektronik, Universität Stuttgart (Germany)

FROM A HISTORICAL point of view ignition raised one of the biggest problems in the development of the combustion engine and Carl BENZ once called it in this connection "the problem of problems". At that time many different methods of ignition were tested. At the turn of the century the high voltage magnetic ignitor was invented and shortly afterwards the battery ignition system similar in principle followed. Its fundamental conception has been maintained until now. Such conventional ignition systems supply energy between 10 and 50 mWs per spark event of a duration of approximately 1 ms and ignite homogeneous and stoichiometric ($\lambda = 1$) or not very rich ($\lambda < 1$) air-fuel-mixtures without difficulty. Lean mixtures ($\lambda > 1$), however, generally require additional precautions, such as increased electrode distances in the spark plug, prolonged duration, higher energy etc. Obviously, such relatively small changes in ignition considerably affect the performance of the combustion engine. Therefore special attention should be paid to the starting phase of the combustion.

Nowadays modern semiconductor electronics can deliver ignition pulses with great variety. Energy, voltage, rise time and duration can be tailored to suit many demands. Not "how" but rather "what" to design is the main issue to be considered by the electrical engineer. In this decision the main aims are maximum gain and minimum costs.

The task of the spark is to fire the chemically exothermal reaction by as exact a timing as possible. For this purpose a certain volume of gas must be activated. What is the significance in this activation process of maximum temperature, emitted radiation and the produced radicals in or around the spark channel? What is the mechanism of converting electrical energy into activation energy?

Research on spark discharges with electrode distances in the mm-range has not been done very much in the past. Especially, processes in the very early phases of breakdown have been extra-

---— ABSTRACT ———

Fundamental investigations have been made on processes in common spark plugs useful for ignition of fuel-air-mixtures. Development of spark discharge from the early beginnings as well as the initiation of chemical reaction have been studied experimentally by using very high frequency measuring technique, radiation measurements, spectroscopic methods and interferometry by a nitrogen laser (~ 300 ps). The measurements have been carried out in non-reacting gases (N_2, air) as well as in inflammable mixtures. A complete description will be given of the total process from the spark breakdown to the arc and glow-phase, resp., in the millisecondrange including temperature, local dimensions, energy distribution and radicals which can be formed.

polated from measurements in later stages of the discharge (1)*. Such work requires the use of ultra high frequency technique as can be realized from the schematic current time dependence of Fig. 1 without considering any details. In praxi there are often additional transient currents caused by unavoidable inductivities and capacities. Definition of the different stages during the spark event is not very clear as yet and use of these definitions therefore differs widely.

EXPERIMENTAL ARRANGEMENTS

For investigations on spark plasma the spark plug or at least the electrodes, have to be mounted in an appropriate apparatus device to allow UHF-measurements of current and voltage as well as of emitted radiation to be made. On the other hand one must ensure that the sparks in the measuring set-up work under conditions which are comparable with those in the engine.

Fig. 2 shows a cross-section through the device which we have developed and used. The sparks were fed by different ignition systems (like e.g. conventional coil- or CD-ignitor) or cable generators. The latter has the advantage of producing defined pulses for easier interpretations of the fundamental processes. We built a special current shunt consisting of a 5 μm metallic resistor foil to guarantee a low rise-time (≈100 ps). The working pressure can be varied up to about 8 bars.

---

* Numbers in parantheses designate References at end of paper

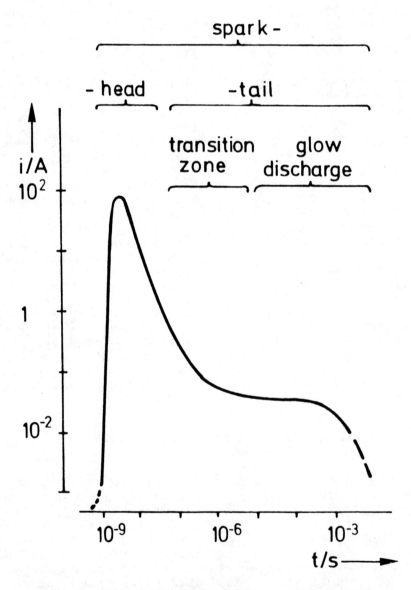

Fig. 1 - Idealized spark current as function of time with definition of different spark regions

The windows for radiation measurements are made from silica, MgF_2, for wavelengths below approx. 220 nm (vacuum-ultra-violet range). Investigation of the emitted radiation is useful only if an adequate spectral resolution is used. For this purpose the spark channel can be projected onto the slit of a spectrograph (focus length 0,5 m). Magnified projection made possible an investigation of geometry of the spark.

In order to measure in inflammable fuel-air-mixtures a small combustion chamber has been built (see Fig. 3) which can be fed by premixed gas-mixtures. Both devices enable repetitive working to use signal averaging measuring techniques. Because of low signal levels long measuring intervals are often required and investigations were carried out automatically under computer control. For interferometric investigations there has been developed in the institute, an atmospheric pressure nitrogen laser (2) of high power ($\sim$1 MW with a pulse length of 300 ps). The principal set-up is shown in Fig. 4.

RESULTS

I-V- AND RADIATION MEASUREMENTS - The transience of spark current and of voltage, resp., is shown in Fig. 5 with increasing time resolution. The first picture at the top is well-known and has been chosen for orientation. With low time resolution the spark head cannot be seen but only the decreasing glow discharge current. Due to the high time resolution in the lower picture it is possible to distinguish between different spark plugs. Small differences in their construction determine the first current pulse which has a rise of up to $> 10^{11}$ A/s. The power input to the spark plasma during this phase is immense

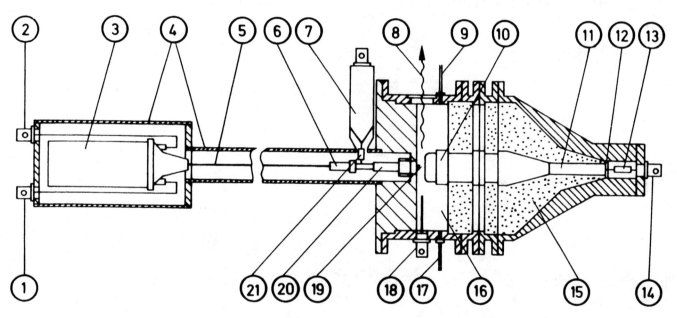

1,2	primary current	11,13	termination: 50 Ω
3	ignition coil	12	foil shunt 42 mΩ, Nichromium
4	shielding (brass tube)		15 μm, τ < 0,1 ns
5	high voltage cable, 800 mm long	14	BNC connector
6	spark plug connector, suppression resistor 1 kΩ	15	coaxial termination
		16	coaxial spark chamber
7	high voltage probe, spark voltage	18	trigger signal
8	LiF window for spectroscopic measurements	19	spark gap
9,17	gas inlet	20	spark plug, ground electrode removed
10	variable electrode: brass	21	damping resistor: 270 Ω

Fig. 2 - Cross section of measuring device for current, voltage, and emitted radiation

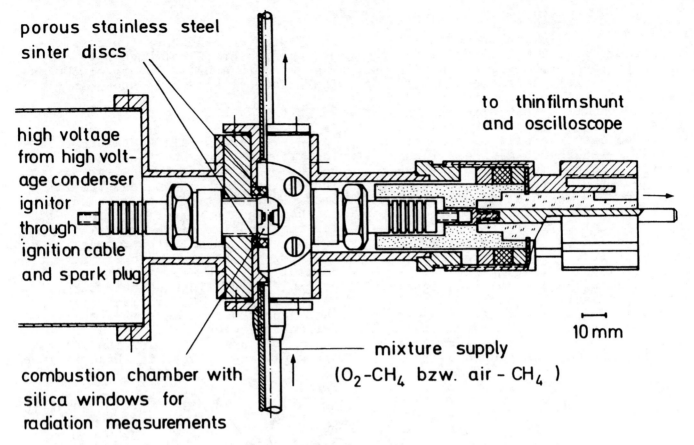

Fig. 3 - Special combustion chamber for measurements in inflammable gas-mixtures (cross-section)

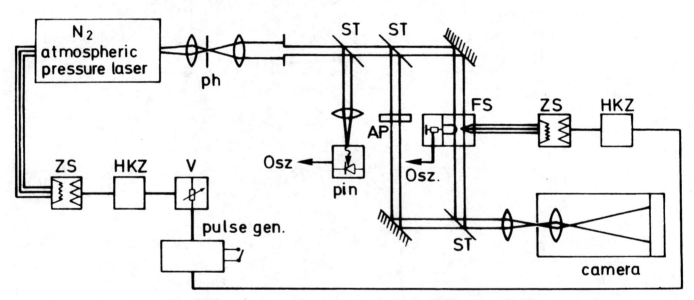

N$_2$ atmospheric pressure laser   Osz: oscilloscope
ph: pin hole   pin: pin diode
ZS: ignition coil   ST: beam splitter
HKZ: CDI   AP: compensation plate
V: delay   FS: spark chamber
Pulsg.: pulse generator

Fig. 4 - MACH-ZEHNDER, set-up for pick-up short-time interferograms, radiation source - atmospheric pressure N$_2$-Laser (300 ps, 2MW) (order of magnitude some 100 kW) and the results show that a sizeable amount of total energy (about 10 % with a conventional ignition system) is being fed into the plasma during a relatively small time period (less than 10 ns) of the spark head.

We will see later on that this energy is very effective because it is mainly stored in the plasma as potential energy (dissociation, excitation, and multiple ionization) and feeds the expanding spark channel at later times. The correlation between spark current and different characteristic wavelengths of emitted radiation is shown in Fig. 6 whereas a cable generator as power supply has been used (10 m, puls length t ≈ 100 ns). Wavelengths emitted by molecules (example in Fig. 6 belongs to the 2. positive system of N_2) are intense only at the very first beginning of breakdown and disappear already during current rise because the molecules become dissociated quickly. Also early, one obtains wavelengths emitted by ions of up to the 4th ionization stage. The behaviour of a single-ion-line ("NII") shows the best correlation with the current. Therefore it should be possible to use it for an indirect measuring method instead of the current where a suitable UHF-shunt cannot be mounted.

Radiation typical for electrode materials rises slower and may be used in investigations of electrode materials.

Radiation in the vacuum-uv-range (< 200 nm) was measured with adequate precautions (evacuated spectrograph, MgF_2-optics etc.). In pure N_2 we measured intense emission also during the starting part of the spark head. Because of the high photon energy this radiation may be photochemically active, and for this reason it could be of high importance in the ignition process (3). However, little oxygen absorbs this radiation on short distances due to the high absorption coefficient of the O_2-molecule. This has to be taken into account with regard to a possible direct activating action of the radiation. The radial distribution at different wavelengths has been measured with a resolution of 2 µm. The results obtained after Abel-inversion with a molecule-line and a single-ion-line are shown in Fig. 7 (4). Because we were interested in the spark head a cable generator (1 m, pulse length ti ≈ 13 ns) was used for these investigations. It was proved that the spark behaviour in these time regions could be compared with that produced by conventional ignition systems.

It can be seen that very early (~ 3 ns) the discharge is concentrated on a small cylinder (r ≈ 20 µm). About 12 ns later there is no molecule radiation out of the inner core of the channel because the plasma is dissocia-

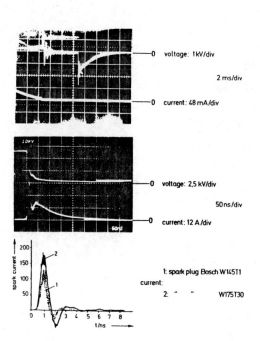

Fig. 5 - Transiences of current and voltage, resp., with increasing time resolution

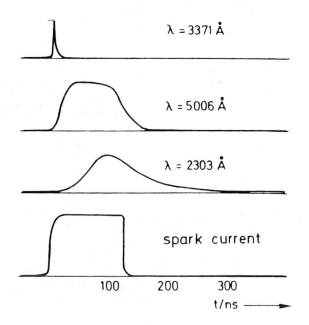

Fig. 6 - Correlation between current and different characteristic wavelengths of emitted radiation - cable generator pulse: 100 ns 337,1 nm molecule line (N_2) 500,5 single-ion line (N^+) 230,3 nm ion line (Mn^+, electrode material)

ted already. At this time (current is decreasing) the ion line shows a maximum.

SPECTROSCOPIC ANALYSIS - Electro-magnetic radiation, particularly in the optical range can be used as an excellent probe for investigations of plasmas. Intensity ratios of different emitted lines or shapes of bands due to vibration and rotation of molecules can be used as an indirect method to measure temperatures in a plasma. Measuring of line width and appropriate evaluation makes it possible to determine electron densities because there is a strong dependence on acting electric fields due to the charge carrier (5).

The feasibility often depends on thermal equilibrium to interpret the results in a correct way. Of course broadening of lines due to the quadratic Stark-effect (6) does not depend on this demand. Therefore this method has been used to determine the average electron density in the channel. In the current rise a density of $n_e \approx 10^{19}$ cm^{-3} is reached (see Fig. 8). The evaluation of different NII-line intensities by a Boltzmann-plot results in an excitation temperature (which can be set equal to an electron temperature excitating those lines) of more than 60.000 K also in the start phase of the spark. As can be seen, a high electron density is reached soon and due to the large cross section of Coulombcollisions as well as efficient collisions between the heavy particles (ions and neutrals) the gas will be heated up fast (7) soon reaching thermal equilibrium and full ionization. These high temperatures cause a high pressure in the thin spark channel which starts to expand with supersonic velocity. The temperature decreases continuously with time as can be seen in Fig. 8 and with it the intensity of the emitted radiation. Therefore, measurements of emitted radiation become more and more difficult at lower temperatures and are nearly impossible when the temperature at the channel has dropped below about 15.000 K. We found two ways to overcome these problems. First, we made use of the interferometric measuring technique described in Fig. 3 by which above all it is possible to get information on density gradients and so on volume-temperature-progresses of the spark channel. Second, if temperature comes down to values at which molecules are existing again, their rotation temperature could be determined by evaluation of the band radiation. With excitation by electron collision of such bands with an additional weak spark at an arbitrary point of time and measuring the radiation emitted in its new start phase, the gas temperature (unchanged by this time) can be deduced from it. We took for this, specially, the 2. pos. system of the nitrogen molecules because it shows a

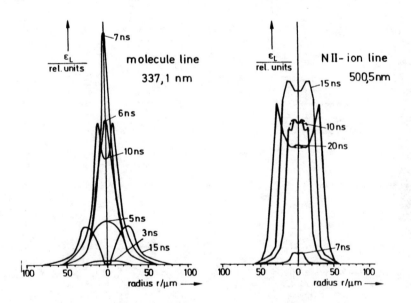

Fig. 7 - Radial distribution of emission coefficient in a spark discharge fed by a cable generator pulse - 13ns (current and voltage as in Fig. 8); gap distance, 1 mm; pressure, 1 bar (N_2)

large cross section for electron collision.

In combustable mixtures many other rotation bands are present (like CN- or OH-bands) which we also used for temperature determinations. Fig. 9 shows a section of a measured spectrum with CN-bands during the inflammation phase if CH_4-air-mixture ($\lambda \approx 1$) ignited by a common CDI.

INTERFEROMETRIC MEASUREMENTS - Examples of short time interferograms are shown in Fig. 10. From pictures like this dimensions of the expanding spark channel can be taken. The shock wave being in advance of the hot plasma channel is easy to distinguish. Thus the diameters of shock wave and channel as well as their propagation velocities depending on time can be determined (see Fig. 11). The investigations show that the shock wave velocity is the same in both air and air-methane-mixture. The channel diameter is unchanged up to this point of time at which combustion starts (about 20 μs), which can be easily determined. Common ignition systems (e.g. CD-Ignitor) produce a hot volume of about 3 mm³ after 20 μs. If this occurs in a reactive mixture, the combustion will start. Results of trials to increase this start volume are demonstrated in Fig. 12 with air-methane-mixture ($\lambda \approx 1$, electrode spacing 1 mm). We found that it is possible to increase this volume just by temporal

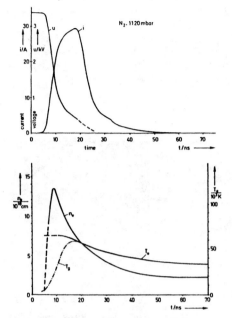

Fig. 8 - Current i, voltage u, electron density n_e, temperature (of electrons T_e, gas T_g) in a spark discharge fed by a cable generator pulse ≈ 13 ns, gap distance, 1 mm; pressure, 1 bar N_2

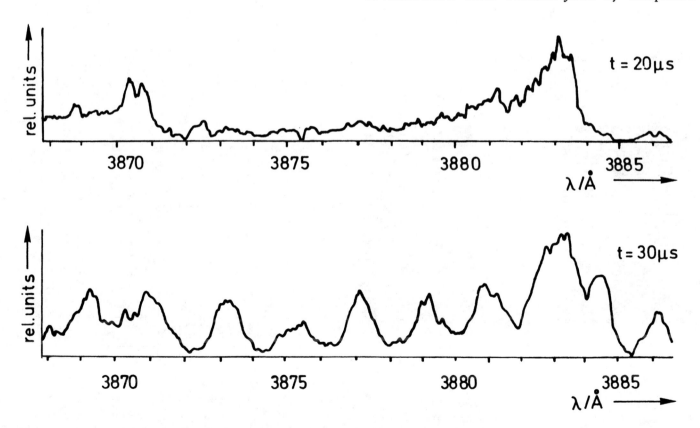

Fig. 9 - Section of the spectrum with CN-bands during inflammation in CH_4-air-mixture ($\lambda \approx 1$)

redistribution of the energy of the CD-ignition without changing the total energy supplied to the spark. This has been done by setting a greater portion of the total energy into the spark head. The third curve in Fig. 12 has been obtained by 10-fold increased energy compared to the CDI. It has to be stated that the influence of a forced spark head is longstanding up to far in the ms-range. Also, the combustion, measured with a pressure probe, shows an accelerated burning rate in the ms-range in the case of a forced spark head. The burning velocity in flowing mixtures can be influenced (increased) in the ms-range, too, as has been demonstrated in the small combustion chamber of Fig. 3.

Summarizing, there can now be given a complete description of the spark discharge of Fig. 8 (spark head) by application of the mentioned methods. Putting together the results,

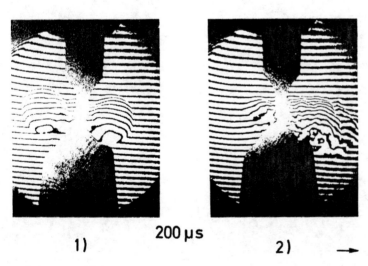

Fig. 10 - Short time interferograms of spark discharges (CD-ignitor), (air, electrode distance 2,3 mm)

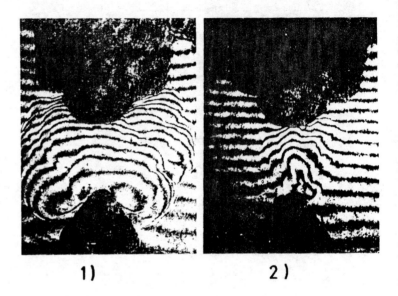

Fig. 10A - Short time interferograms in a mixture of CH_4-O_2, 50 μs after spark breakdown 1) $\lambda \approx 1$, 2) $\lambda = 0$ (electrode distance 2,3 mm)

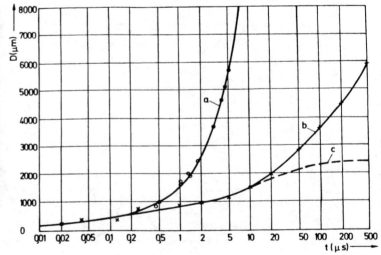

Fig. 10B - Short time interferograms in a mixture of CH_4-O_2 200 μs after spark breakdown 1) gas resting, 2) gas flowing (electrode distance 1 mm)

Fig. 11 - Diameters of shockwave (a) and plasma channel (b,c) in air (c) and CH_4-air-mixture (a,b; $\lambda \approx 1$) of a spark discharge (CD-ignitor), gap spacing, 1 mm

Fig. 13 represents the time dependence of temperature, volume and pressure. The volume has been obtained by interferometry, and during the earlier times (below 50 ns), by radial radiation measurements. The temperature has been deduced from spectroscopic analysis of N^+-lines and N_2-rotation bands, resp. (the bands were excited by a tiny second spark discharge as described above). The pressure has been calculated from the shock wave velocity (8) and the Saha equation, resp.

The temperature curve shows two plateaus at which cooling down of the expanding spark is delayed. During these times it can be shown that the spark is being fed by the potential energy stored during the breakdown phase by the particle which becomes free due to recombination of the ions (1st plateau) into atoms, and of atoms to molecules, respectively (2nd plateau). A detailed analysis of the distribution of energy on the different possible energy forms will be published in a separate paper shortly.

Following, we only give the most important results. The total energy converted in the spark discharge results from integration power over time. The question is which portion stays in the plasma and which one will be lost? Loss processes during early times are caused by radiation, by electrodes and by shock wave:

Reasonable estimations and measurements show that the radiation losses are below 1 % of the total energy. Short wave radiation (VUV) will be re-

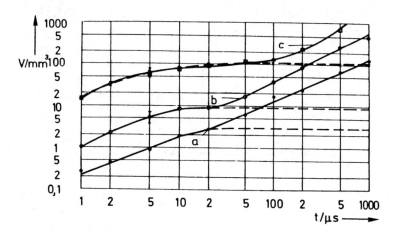

Fig. 12 - Plasma volumes in CH_4-air-mixture $\lambda \approx 1$ of a spark discharge after Fig. 11 (dotted lines in air), a) common DC-ignition; b) total energy as in a), but redistributed in favour of the spark head; c) discharge of 2 nF-HF-condenser (10 times the total energy of a,b)

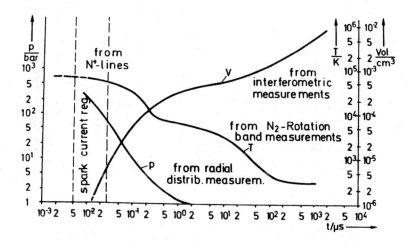

Fig. 13 - Time dependence of volume, pressure, and temperature in a plasma of a spark head (discharge after Fig. 8) in N_2, 1 bar

absorbed and remains in the spark plasma. Electrode losses have been estimated to amount to about 5 %. The energy loss in the shock disturbed zone (8) has been calculated to be about 13 % of the electrically supplied energy.

The results (temperature and density of electrons, total pressure and energy density in the plasma) and understanding permit one to calculate the conditions of the plasma in the spark head with the assumption of thermodynamic eqilibrium. By differentiation between potential (ionization and dissociation) and kinetic (thermal) energy the time dependence of each portion has been determined (see Fig. 14). The first maximum of the kinetic energy at about 100 ns is caused by neutralization of ionized particles and a short time later, when the ionization portion is reduced drastically, the dissociation increases, even at the cost of thermal energy in the expanding channel. After this time the thermal energy portion grows continuously and at times around 50 μs all the energy is converted to thermal energy.

However, it should be pointed out that the fundamental investigations last mentioned were done with spark heads produced by a cable pulse. But measurements for comparison with conventional ignition systems show that these findings can be transferred without problems.

Investigations during the glow phase of the spark are just going on in the institute and will be reported at later times. First results show that the temperature seems to be in this region more than 1.000 K. However loss processes during these relatively long times become more important. It has been found that about 50 % of the supplied energy gets lost in the cathode, and 5 % in the anode. All thing considered, it seems to us that the glow phase of the spark cannot be as efficient as the spark head with regard to energy transfer to the gas.

SUMMARY AND CONCLUSIONS

In this paper we report on detailed research being done on spark discharges

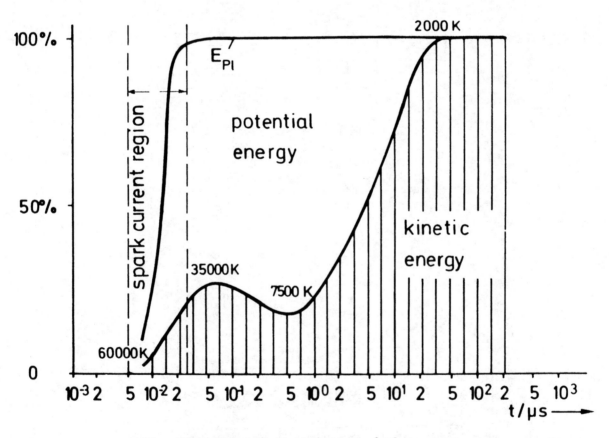

Fig. 14 - Temporal distribution of the energy conducted into the plasma by a spark head after Fig. 8 subdivided into potential energy (ionization, dissociation) and kinetic (thermal) energy in N_2, 1 bar

relevant to ignition in combustion engines. There have been applied electrical and radiation measurements in the ultra high frequency range, spectroscopic analysis and interferometry both in air and in air-methane-mixtures. Ingeniously, the spark event should be divided into spark head and glow phase and in between a transition region. The starting diameter of the spark channel is determined by the breakdown mechanism and corresponds to the diameter of an avalanche.

The current rise is enormous and in nano-seconds full ionization of the thin channel is achieved. Due to intense interaction of the charged particles the gas is heated up to temperatures above 60.000 K quickly leading to pressures between 100 and 200 bars. The spark channel expands with supersonic velocity and becomes cooled down. Short wave radiation which could be photochemically of interest, is emitted only during the start phase but will be absorbed after very short distances (μm-range) if oxygen and hydrocarbons are present. Eventually, stimulated activation will be swept over by the hot expanding spark channel a little later.

In inflammable mixtures radicals (CN, NH) are detectable at an early stage already (10 ns). An independent flame front, however, can be identified as soon as the expansion velocity of the spark becomes equal to or smaller than the possible flame velocity. The starting point of the combustion has been determined to exist between 10 and 20 μs. The heated volume produced with conventional ignition systems and arrangements is about 3 mm^3 (channel diameter 1 to 2 mm). It could be shown that the activated volume and the following inflammation can be influenced by redistribution of the temporal power supply in favour of the spark head. Hence, it follows that a spark head could have a particular meaning for the process of inflammation. Nevertheless, research is going on to clear up the tasks of the glow phase in which sufficiently high temperatures have been detected, too.

A consideration of energy distribution in the spark head is given.

Glossary:

E - energy
i - current
n_e - electron density
p - pressure
r - radius
t - time
T - temperature
u - voltage
V - volume
λ/Å - wavelength in Ångstrom units
λ - $\dfrac{\text{air mass for stoichiometric mixture}}{\text{actual air mass}}$

REFERENCES

1. G. Conzelmann, "Über die Entflammung des Kraftstoffluftgemisches im Ottomotor". Bosch Techn. Berichte 1 Heft 6. p. 297 - 304. Sept. 1966
2. W. Herden, "Compact high power subnanosecond nitrogen and "open-air" Lasers at 760 Torr". Physics Letters A, 54 A1, p. 96. 1975
3. A.E. Cerkanowicz et.al., "Photochemical ignition of gaseous fuel oxydizer mixtures at subatmospheric pressures", Report VL 2482-3-0-US Air Force, April 1970
4. B. Saggau, "Spektroskopisch untersuchte zeitliche Entwicklung kurzer Stickstoffunken". Spring Meeting of the German Physical Society, Hannovia, March 1976
5. W. Lochte-Holtgreven, "Plasma Diagnostics", North-Holland Co., 1968
6. H.R. Griem, "Spectral line broadening by plasmas". Academic Press, New York, 1974
7. L. Spitzer, "Physics of Fully Ionized Gases". Interscience Publishers Inc., New York, 1956
8. A.G. Gaydon, I.R. Hurle, "The Shock tube in high Temperature Chemical Physics". Chapman and Hall Ltd., London, 1963

CHAPTER 2

CARBURETORS

SOME REQUIREMENTS OF CARBURETER DESIGN.

BY EDWARD T. BIRDSALL, M.E.

Ever since the idea of using a vaporized or gasified liquid fuel in the internal combustion engine was suggested, the device for preparing the fuel for use in the engine has been a subject for much thought and study. Numberless designers more or less insufficiently armed with the proper experience, knowledge and data for the task, have undertaken to solve the problem with varying results.

As long as the principal requirement was to furnish fuel to engines working under a practically constant load and speed and fuel was cheap, the defects of the early carbureters were not such as to interfere seriously with the operation of the engine. Other troubles, such, for example, as ignition, occupied so much of the operator's time, that the carbureter, so long as it worked at all, was neglected.

In the following remarks it is assumed that the engine used has a sufficient number of cylinders to produce a steady flow of mixture and that the carbureter is of the modern float feed type, with a fuel jet and main and auxiliary air inlets. The fuel is assumed to be gasoline, although in the main alcohol or heavier oils require the same general conditions. The object to be attained is a mixture that will develop a maximum of power from a given size of motor with a minimum of fuel, not an average or "good-enough" result.

With the use of the internal combustion engine under extreme variations of load and speed, as demanded by the modern automobile, with the perfection of the ignition and other features, and with the rapid rise in price of the lighter oils, the subject of carbureter design becomes one of great interest and importance. Again, in a few years, when the commercial wagon will demand a low fuel cost combined with great certainty and flexibility of engine operation, the carbureter will probably determine the ex-

CARBURETER DESIGN

tent of the development of this, the most important branch of the automobile industry.

The function of a carbureter is to supply the proper mixture of air and fuel to the engine, under all conditions of speed and power. The four essential conditions under which carbureters must work are:

First. Wide open throttle and high engine speed, as when climbing hills or running fast on the level.

Second. Wide open throttle and slow engine speed, as when traveling slowly on the high gear or picking up from standstill.

Third. Partly closed throttle and high engine speed, as when running fast down grade or on a low gear.

Fourth. Nearly closed throttle and low engine speed, as with engine running idle when car is standing.

For some time it was thought that the best carbureter was one that gave a constant mixture under all conditions. But we now know that a constant mixture is not the best from either the standpoint of best operation or full economy. It was also thought that the best mixture contained just sufficient oxygen to entirely consume the carbon and hydrogen. It was found, however, that a mixture with a slight excess of fuel gave the best results. These facts being demonstrated, it becomes almost obvious that the different engine speeds will demand different mixtures for maximum results. Thus at slow speeds, the mixture should be richer than at high. This is due to the fact that at low speeds more heat is lost to the cylinder walls, more compression pressure is lost by leakage and the combustion can therefore be slower, thus sustaining the pressure. At high speeds the compression is higher due to less leakage and less loss of heat. Therefore unless the mixture was leaner at high speed there might be danger of pre-ignition. A lean and highly compressed charge also burns faster and hence gives better pressures and fuel economy than a richer one.

The quantity of mixture that an engine will take varies greatly with the speed. At slow speeds the quantity is equal to the cubic contents of the cylinders multiplied by the number of power strokes. At high speeds of one thousand revolutions and over the quantity may drop to less than one-half the theoretical amount, depending on the design of the valves, inlet piping and carbureter passages. This peculiarity reacts upon the compression and hence on the mixture desired for best results. It will thus be seen that the design of the engine has a great deal to do with the carbureter design, which explains the well-known but seemingly mysterious fact that a carbureter that gives good results one engine fails to maintain its reputation when applied to one of different design.

The design and class of ignition used have also a marked influence. Poorer mixtures can be used as the spark is hotter, the throttle can be more nearly closed, resulting in increased engine capacity and fuel economy.

To get the maximum power out of a given sized engine the fuel should be introduced into the cylinders as cold as possible consistent with complete evaporation, intimacy of mixture and completeness of combustion. To provide for the heat absorbed by the evaporation of the fuel, hot air is drawn in to form the mixture, the entire apparatus is heated by means of a hot water or the general heat of the engine compartment under a closed bonnet is relied upon. The adjustment of this heat is an important matter, but exact knowledge on the subject is apparently non-existent.

The ever varying density and compositions of the fuels used and obtainable introduce many and very serious complications into the problem. These differences demand different sizes of jets, different float levels, different amounts of heat to be supplied, and different proportions of air for combustion.

Different densities and temperatures of the fuel affect to a very appreciable extent the flow of the fuel from the jet. Between extremes this has been found to vary as much as 40 per cent. Thus a carbureter exposed to atmospheric temperatures in this latitude would seem to require a wide range of adjustment.

Owing to the absence of a ready means—like the pressure gauge on the water circulation, or the voltmeter on the accumulators—of ascertaining the quality of the mixture being delivered by a carbureter, the majority of the motors in use are operating under more or less disadvantageous conditions, even if carefully and properly regulated at the outset.

The amount of reliable data and facts concerning the action of air and gasoline in a carbureter at the command of designers and students is remarkably small. Of no other part of the automobile is so little known. What is badly needed is a series of carefully-planned and exhaustive experiments with data so arranged that it can be analyzed and deductions made.

CARBURETER DESIGN

DISCUSSION.

PRESIDENT RIKER: Gentlemen, you have heard Mr. Birdsall's paper on Carburators and it is open for discussion. I might state that the perfect carburator has been discovered and is on exhibition at the Armory; it consists of the old surface carburator rejuvenated.

JOSEPH TRACY: Mr. Birdsall says under certain conditions one carburator might give satisfaction on one car and poor satisfaction on another.

E. T. BIRDSALL: I had a carburator on a Mercedes in Paris and brought it home and put it on another machine of quite different design, and the carburator behaved very badly until we reconstructed it to accord with changed conditions. We altered the auxiliary air supply to correspond to the slightly different engine. I finally got the carburator to work on the other engine as well as it did on the Mercedes.

JOSEPH TRACY: You said something about different float levels of the fuels.

E. T. BIRDSALL: The levels seem to make a difference, although, as we all know, the ordinary float feed carburator will draw every drop of gasoline out of the float chamber.

JOSEPH TRACY: Can you imagine that rasing or lowering the level would give more gasoline?

E. T. BIRDSALL: I do not see why, but it does seem to.

JOSEPH TRACY: If you carried the level very low you would not get as much gasoline, would you.

E. T. BIRDSALL: If you are running an engine you do not know that the gasoline is low until it gives out, there are no symptoms that it is going to give out.

JOSEPH TRACY: Do you think there would be any advantage in carrying the level very low?

E. T. BIRDSALL: It makes it very difficult to start.

JOSEPH TRACY: Did the motor that you tried have automatic valves?

E. T. BIRDSALL: It was the same general type of motor as the Mercedes. With four-cylinder engines, the flow is fairly steady. The design of the engine has a great effect on the working of the carburator.

THE CARBURETER AND ITS FUNCTIONS

By CHARLES E. DURYEA, Reading, Pa.

MEMBER.

The carbureter is the lungs of the engine, and large power, long service and efficient action depend upon this device. Many varieties have been offered, and the road to the patent office is busier to-day than ever with people who think they have carbureters superior to former designs. With a fixed gas no carbureter is necessary, but air and gas adjustably proportioned are permitted to mix on their way to the cylinder with good results. The earlier inventors generally attempted to provide the gas by drawing air in some manner through a tank containing gasoline, which permitted the air to absorb gasoline vapor and issue from the tank practically saturated with vapor. This over-rich mixture was then diluted by the admission of air to form the proper mixture. In one form wicks of cotton, or even excelsior, served to distribute the vapor through the air. In another form the air was drawn down into the liquid and, bubbling up through it, became saturated. Other inventors seeking simplicity admitted the gasoline directly into the air passage, trusting that it would be sprayed or vaporized and mixed with the air before the end of the compression stroke. Still others provided a spray nozzle, past which the air is drawn with sufficient velocity to break the liquid into a spray. This form is now in almost general use to the exclusion of other forms. Each is usually called a carbureter, but properly the gas tanks only are entitled to this name, and the present form is more appropriately an atomizer, or, since its essential service is to mix liquid fuel and air to form what is universally called a "mixture," I prefer the short, simple, expressive word, *mixer*.

A number of facts concerning gasoline engines must be kept in mind when considering the mixing device if the engine is to give superior results:

First, the mixer must perform its function to the fullest possible extent and intimately mix the air and liquid. It is not enough that it should provide a proper mixture at high speed only, for, although this will cause the engine to show a high power, it will not give smooth running or great power at slow speeds. If the mixture is not intimately mixed, some parts are too poor to burn, others burn slowly because lean, while other parts are too fat to burn, or burn very slowly because overfat. The result is little power, a hot engine, much deposit of soot and an ill-smelling exhaust.

Second, in order to have full power and give the best results the liquid fuel must be properly proportioned to the air. Too much or too little liquid produces slow-burning mixtures and undesirable results. Further, although during each cycle the engine may receive the proper amount of air and liquid for the perfect mixture, if the early portion is air and the latter portion largely liquid, it is quite evident that a homogeneous mixture will not be produced and that proper ignition with perfect engine behavior cannot follow. It is therefore necessary that the air and liquid be proportioned constantly in a proper manner, and this may be rightly termed the second great requirement.

Third, it is also evident that different sized engines will have different requirements, and that a mixing device suited to one may not be suitable when fitted to another. The same is true in connection with speed. A proper mixture at one speed may be completely thrown out of proportion, or may be improperly mixed at another speed. Engines nowadays run at rotative speeds from 200 to 2,000, and the perfect mixer must meet these requirements. Since at high speeds full charges are usually used, while at the low speeds the throttle reduces the charge admitted, it is quite evident that the service required of a mixing device is not adequately represented by the proportion 10 to 1, but that it is probably more nearly 20 to 1, and possibly may vary as much as 50 to 1. Such wide variation increases the difficulty of maintaining proper proportions and making a perfect mixture, and renders it necessary that the mixer should automatically adjust itself to the varying requirements.

These three features are the basic ones which must be kept in mind while considering the minor but important points of the perfect mixer. Most mixing devices heretofore constructed have aimed to provide for these three points, but more often than not each provision has been an imperfect one and the results not of superior quality. The typical mixer of to-day takes air from the atmosphere at practically constant pressure, and liquid from a float chamber presumably having a constant level. Since, however, the quantity of air required is about fifteen hundred times greater in volume than of the liquid, and since the speed under a given suction is much greater than the speed of the liquid, it will be seen that wide opportunity for improper proportion exists. Add to this the facts that at very slow speeds the liquid may not be sprayed, but may be simply drawn from its nozzle in large drops, or even in a stream running down the outer walls of the nozzle, while at very high speeds the air inlet may be too small to admit a sufficient quantity of air, and the difficulty of maintaining a proper proportion under such wide variation will become apparent. To meet this difficulty the perfect mixer must automatically enlarge the supply of air and vary the liquid to maintain it proportionate to the air as the needs of the engine grow greater. To do this with certainty it should have a diaphragm acted upon by the suction of the engine, which diaphragm should be large enough to respond to slight variations, and thus prevent high vacuums, with consequent reduced power at high speeds. This method of providing for wide range is the only correct method. The mere opening of the usual auxiliary air port cannot perform this service, for the suction must increase considerably before the air port will open, and there is seldom or never provision made for securing either intimate mixture or proper admission of proportionate amounts of air and liquid with this auxiliary device.

This necessary automatic adjustment should not only be operated by the suction, but it should be sensitive enough to prevent much variation in vacuum between high and low speeds, and the mixer at high speeds should have openings large enough to admit the fullest possible charges, while at low speeds the opening should be so small as to secure sufficient air velocity to make a perfect mixture, that is, a fine spray of the liquid properly proportioned and intimately mixed, even when turning the engine over by hand. This can only be attained by permitting a large diaphragm to vary the size of the passage or passages under increased suction, and consequently proportionate to the speed.

The float chamber should be concentric with the liquid inlet, so that inclination in any direction will not cause more or less liquid to be admitted. The float should be surrounded by a substantially concentric volume of liquid that will support and balance the float, with the result that sudden vertical movements, such as jolts, are without effect. This arrangement is superior to floats balanced by weights, in addition to the column of liquid, for, owing to their different densities, the liquid and the weight may interfere in their duties and destroy the perfect balance sought for. The float should be a single piece preferably without working joints, and particularly without frictional contacts with levers, which may sooner or later wear through its thin metal and cause it to leak. The float should be constant in weight and buoyancy, and is therefore preferably of metal, since few cork floats can be depended upon to remain impervious to gasoline and retain their buoyancy. The float point should be adjustable, so that the level of the liquid may be maintained at the most advantageous point to suit the vacuum necessary to make the proper spray, and also to overcome the effect of different heads of gasoline which may be used. It should be quite evident that the mixer, giving excellent satisfaction when attached to the bottom of a tank three or four inches deep, may fail when piped to a tank several feet deep in the bow of the boat, which in rough weather may rise several feet above the level of the mixer. The float point may be of such taper and size as to in some degree vary the gasoline level in action, giving a higher level and better mixture at slow speeds.

The float point should be easily ground so that it may be kept tight and in perfect working order. Further, the motion of the vehicle should tend to move the float point to some degree, even though slight, which movement serves to force away any particles of dirt that may lodge on the point during the passage of liquid. On this account it is best if the float and point are fixed one to the other so that the point partakes of the motion of the float and liquid in the chamber. The gasoline should enter the float chamber from a single direction, either up or down, so that no pockets exist in which water or dirt may gather. It is best to feed the chamber by gravity from a tank above the float chamber and with downwardly extending pipe, without pockets, leading into the chamber near the top, with float point upward, such point being attached directly to the float without levers, weights or other unnecessary parts. The float chamber should open at the bottom for automobile use. This facilitates removal of any water, ice or dirt, and removal of float itself, without opening the top and permitting dirt to fall in from above. The float and removable bottom can be replaced with a stream of gasoline flowing upon them, which will wash away particles of dirt, if any accidentally get on the parts while being replaced. With top opening, ice in the bottom of the chamber may not only support the float and prevent its falling to admit gasoline, but may also bind the float so firmly that it cannot be removed to permit removal of ice, which may prove an unpleasant predicament if away from means of warming the mixer. The float chamber should have an air vent to permit proper action, and this vent should preferably terminate above the gasoline tank, so that if for any reason the float fails in its duty the gasoline rising in the vent tube will not rise higher than the tank level, and so cannot escape. Where convenient, the tickler, or device for depressing the float and flooding the mixer, should pass down this vent tube. This arrangement, in connection with a needle that closes the nozzle when the motor is stopped, prevents danger from leaking gasoline and possibly fire. It is more reliable than a stop cock, for the operator will grow careless about the stop cock, but will, if needed, adjust the nozzle daily to secure best results under prevailing weather conditions for that day.

All gasoline entering the mixer should be strained through ample gauze, so that particles likely to clog the nozzle may be kept from entering. Such gauzes are usually provided at the opening of the tank or in the funnel, but this is not sufficiently certain, for the best results and the perfect mixer should be self-protected from this certain cause of trouble.

The outlet from the float chamber, usually termed the nozzle, should be nearly concentric with the chamber. If centrally located, variations in angle do not affect the level at this point, but it is some advantage to have this point slightly behind the center, so that going up hill or accelerating the action of the vehicle automatically raises the level of the liquid, and thus slightly increases the flow, making the mixture slightly more fat and powerful. This arrangement permits the normal mixture to be lean, insures perfect combustion, great economy, and no odor, yet automatically brings the mixture to maximum fatness and power when power is needed.

Since liquid has considerable weight, and consequent inertia, the passage to the nozzle should be both short and large, for large passages do not clog easily, and, if short, the liquid can flow quickly and will likewise cause flowing without delay when the suction ceases. If large, the friction is less, and no particle of liquid need acquire high momentum. If, on the other hand, this passage is long, the liquid does not get started until a large volume of air has passed the nozzle, making the early part of the charge too lean, while as the suction decreases and the air flow ceases, the inertia of the liquid causes it to continue to flow, making the latter portion of the charge overfat, and leaves between charges probably unsprayed drops of liquid, which fall upon the walls or are drawn into the motor.

Such liquid as remains in the passage unsprayed should be retained and not permitted to run into the motor or upon the ground. This liquid should also, by the shape of the passage, or by other suitable means provided, be broken up, sprayed or finally divided at the next suction stroke, so that it may properly serve its purpose within the engine. If, because of a faulty float, the nozzle should flood, the air passage should not fill with gasoline, for, when attempting to start the engine, this would result in a large volume of liquid being drawn into the cylinder, making its contents too fat to ignite. To prevent such flooding, the air passage should have an opening at a proper distance above the bottom, to permit the escape of excess liquid in case such exists.

The nozzle should be closed from above by an adjustable needle, for the inverted conical point of such a needle assists in making a fine spray. This needle-adjusting handle should terminate near the operator and permit him, while operating the vehicle, to vary the proportion of the mixture, and thus secure the greatest power by trial, as well as accommodate the device to the temperature and humidity of different days, and also to the gravity and composition of different fuels. No adjustment while the vehicle is standing can compare with adjustments in actual road service in point of accuracy. Further, the mixer should be adjustable at low speeds to secure certain ignition and steady running. Gas engines are particularly prone to misfire at their limits, and the perfect mixing device for automobiles will provide superior conditions at these limits in order to secure the most satisfactory range of service. This necessitates provision also for adjustment at normal or high speeds, and by inference the device should automatically compensate at intermediate speeds. Most present-day devices are adjustable for one speed only, and depend for automatic adjustment upon considerable variation in the suction vacuum, and so cannot give good results at widely varied speeds from that to which they are adjusted. This defect need not, and most certainly should not, exist.

That the largest possible charges may be drawn into the motor at high speeds, it is self-evident no needless friction should be caused the air as it passes toward the engine. On this account a single air passage is better than several, because there is less wall surface and friction. It is also evident that the air passage should be easy and not tortuous or broken. It is undoubtedly true that the tortuous passage will break up the particles of gasoline and help to form a homogeneous mixture, but this is done at the cost of increased suction and of some loss of volume and consequent needless loss of power from the motor, particularly at high speed.

Since most engines may occasionally back fire through their inlet valves, the mixer should be provided with escape for such explosion, for if this is not done the pressure may force into the float chamber and will more certainly interfere with the next succeeding charges, than if allowed to escape into the atmosphere freely and promptly. To prevent such explosions from igniting anything on the outside, the pipe entrance should be provided with a gauze strainer, which mainly serves to keep out particles of dirt that otherwise would enter the engine and likely stick to the walls and cause rapid wear and pre-ignition. Much of the carbon deposit, so common in automobile engines, is caused by road dust with enough oil to bind it together.

The rapid evaporation of the liquid not only takes heat from the passages in which the evaporation takes place, but frequently causes a deposit of moisture, which in the presence of low atmospheric temperature becomes ice and clogs the passage. This freezing may be prevented and a more perfect evaporation, with consequent intimate mixture, secured by heating the passage where the mixture is taking place. I therefore favor a heater jacket outside the mixture passage, through which hot gas from the exhaust or hot water from the circulating system may flow, and I consider it advisable to place within the mixture passage at this point one or more gauzes of large area to positively intercept large particles of liquid and prevent their being carried into the cylinder. All gauzes should be removable for cleaning purposes, and frequent attention to the various details of this most necessary part of the vehicle is necessary to insure perfect work.

We may get a better understanding of the features necessary

in a perfect mixer by considering a typical present-day carbureter. This consists of a float chamber usually at one side of the air passage, and with a long, small nozzle for gasoline reaching into the air passage, which at this point is strangled or contracted to increase the velocity of the air past the nozzle. Between the nozzle and the engine an auxiliary opening is provided, closed by a spring valve, which, when the suction is increased sufficiently, opens more or less, admitting a quantity of pure air with which to dilute the over-rich mixture coming from the strangled passage. The action of this device is about as follows: At extremely slow engine speeds, say under 200, the mixture is imperfect because the air passage is not small enough to give proper air velocity for a suitable spray. This is one of the reasons why the gas engine is regarded as inflexible and why many engines fail to develop power as soon as their speeds are reduced. If this passage is small enough for perfect running at very slow engine speeds, say 50 and 100 with throttle practically closed, it is too small to admit a practical amount of air at higher speeds, so the gasoline by itself, or badly mixed with air, is drawn from this passage, while the greater portion of air, with imperfect provision for mixing, enters at the auxiliary valve. Clearly this cannot give a proper mixture or proper proportion. Next it must be remembered that, while the strangled passage is constantly open, the auxiliary passage is closed, except when sucked open. Further, the auxiliary valve flutters, and the result may often be that in the early part of a stroke the mixture is exceedingly rich, because it all comes from the strangled passage, while later, the auxiliary having been sucked open, a large quantity of air enters (larger than necessary), with resultant poor mixture, followed by closing of the valve as the suction decreases near the end of the stroke, with consequent rich mixture at this time. Add to this the fact that with a long, slim nozzle the gasoline will continue to flow for some time after the suction stops, because of its momentum, and it will be seen that the beginning and end of each charge are probably overfat, while the center of the charge is very lean.

There is also a wide range of suction, because at the beginning and end of the stroke there is little or no vacuum, and the strangled tube offers a free passage, while at the center of the stroke there must be, and is, enough vacuum to open the auxiliary, so it is quite evident that the engine is not drawing uniformly and is not free from that negative pressure or vacuum necessary to get the largest charges and to avoid needless loss of power. The ideal carbureter will avoid this irreg ly by opening a passage proportionate to the amount of mixture required, and it will not only open the air passage, but it will adjust the gasoline to suit. If, for example, a piston or diaphragm is provided, operated by the suction of the engine in one direction and by gravity in the other, with a dash pot so that it cannot flutter, it may be made to open the air passage and to adjust the gasoline, so that with little or no increase of suction the proper amount of air and liquid is admitted. With such arrangement the vacuum need only be sufficient to give the air the necessary velocity required to make a proper spray, and higher speeds will not starve the engine because of higher vacuum. The dash pot insures average openings, so that at the beginning and end of the stroke the velocity will be low, at the middle high, but with an average somewhat higher than the least practical velocity, while good results will be obtained, even during the slow portions of the stroke.

Many typical carbureters have quite abrupt corners. This decreases the amount of air that can enter and thus impairs the efficiency. Some provide for complete vaporization within the carbureter, or very close thereto, with the result that in wet weather the moisture of the atmosphere is condensed, and in cold weather frozen, thus choking the device with ice. It is better practice to carry the spray some distance, and thus distribute this refrigerating effect with less likelihood of ice formation.

The typical carbureter has but a single adjustment for the gasoline. It is argued that the gasoline may be adjusted for low speeds when the auxiliary air valve is shut, but this very frequently does not give the proper quantity of gasoline for high speeds, so it usually becomes necessary to adjust by gases, and after a trial adjust again, until that adjustment which gives fairly good results at high speeds, and permits getting along at low speeds, is found. That this is not ideal is readily seen. The ideal method would vary the air passage so as to supply the requisite amount of air with the least possible variation in vacuum and would also vary the amount of gasoline to suit this amount of air. The ideal mixer should be adjustable at low speeds for starting or running the engine idle, and it should also be adjustable at maximum or normal speeds, so that the best possible condition can be had at this time. It should automatically vary this normal or running gasoline adjustment as the proportion of air is varied. In short, it should not have less than four adjustments, two of which (*i.e.*, gasoline and air) are automatic, and two of which are manually operated, as indicated by the behavior of the motor. The typical mixer has but half this number, and these badly deranged. The writer patented more than a half-dozen years ago the first automatic air inlet applied to automobile carbureters, but because of the defects of this method did not use it to any great extent, although by careful adjustment of the auxiliary valve springs it may be made to serve better than most carbureters will serve without this auxiliary valve.

In conclusion, the requirements of the perfect mixer may be summed up as follows: It must intimately mix, properly proportion and satisfactorily adjust, and also have the following specifications: Float chamber concentric with inlet and nearly concentric with outlet. Float of metal with point adjustable to different heads, different liquids and different weights of float. Float point easily ground, and moved by any motion of the float. The float should be free from balance-weights or levers. The mixer should be adjustable by the operator while driving. It should have adjustments for very low speeds and also for normal or high speeds, and should automatically adjust between these speeds. It should have a short gasoline passage for quick action and a large gasoline passage to prevent clogging or ramming. It should retain in the air passage unsprayed liquid, but have provision to let out any excess. A gauze strainer at the gasoline inlet, and also at the air inlet, are strongly advised. The gasoline should flow in a single direction, either up or down, to the float chamber from the tank. The float chamber must have vent at the top, which should, if possible, open higher than the tank. It should have removable bottom and a means for daily use to shut off the gasoline. The air passage should be easy and single, rather than multiple, and have a removable gauze to prevent unsprayed liquid reaching the engine. This passage should be adjustable to the engine speed by the amount of suction, and should open freely in a reverse direction to permit back explosion to escape. A dash pot must prevent fluttering with change of opening, so that the suction vacuum is closely constant. Provision for heating is necessary in cold weather or with low gravity liquids. A mixing device which meets these requirements leaves little room for improvement.

THE CARBURETER AND ITS FUNCTIONS

The Duryea mixer has been designed after a very long experience with stationary, automobile and marine engines of all varieties. As shown in the sectional sketch herewith, it has an air passage nearly horizontal, curved easily, and provided at its forward end with gauze screen to exclude dirt. At the opposite end a disk or other throttle is provided, while midway is located the gasoline inlet with adjusting needle and an adjustable air gate, spring-mounted, so that it may open freely to let out any explosions or back pressure from the engine, but which ordinarily

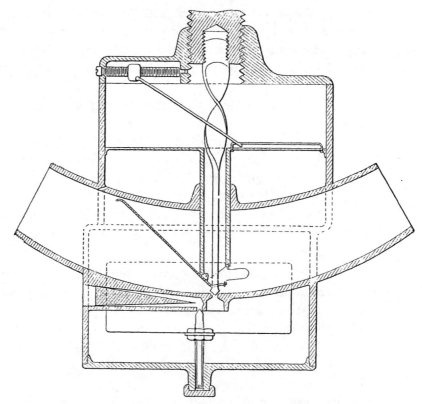

Fig. 1.—Section of the Duryea Carbureter.

remains closed and causes the major portion of the air to pass under its lower end and intimately mix with the gasoline. The air passage is of necessity rectangular at the portion where the air gate is placed. This air gate is carried by a diaphragm and regulated by the suction of the engine, which withdraws the air from above the diaphragm, raising it and the gate support. This diaphragm is large, so a very slight variation in the amount of suction suffices to change the area of the air inlet. The gasoline passage is extremely short, with level adjustable by the float needle, so that the gasoline can be drawn with very slight suction, which also contributes to full charges and large power. The float and its conical point act as one piece without pivots, levers or weights, but capable of a slight twisting motion, which twisting permits the point to grind itself upon its seat, and thus remain tight and clean. The gasoline inlet is protected by a gauze cone, which catches any dirt from the pipes or tank, and which is removed readily for cleaning. The float chamber is closed at th bottom by a cap, screw-threaded, and with a lead gasket, making it absolutely tight. The float needle is adjustable without removing the float or this main cap, by unscrewing a smaller cap and reaching the needle stem with a small screwdriver, the slight leak of gasoline doing no harm. The gasoline-adjusting needle is adjusted by a differential screw on top of the mixer, which may be provided with a stem carried to a point on the dash or elsewhere accessible by the operator while driving the vehicle. This enables the mixer to be adjusted at any time in order to secure the best results. Ordinarily the screw is set to secure easy starting and certain running at very low speeds. The needle is spirally flattened, forming a sort of screw which passes through a long slot in a radius bar pivoted at one end to an adjustable nut in the cover and at the other end, sliding in a radial slot in the diaphragm.

This arrangement causes the needle to be partly revolved as the diaphragm rises or falls, thus varying its adjustment and the amount of this adjustment may be greater or less, as the radius bar is, by its nut and screw, caused to be farther from or nearer to the needle. This radius bar is adjusted for nominal or high speeds, as may be desired, and the adjustment is found by trial either by adjusting the differential screw to produce the desired result, and then changing the radius bar screw accordingly, or by adjusting the radius bar nut and screw while the differential remains in a fixed position. This arrangement provides for this mixer superior adjusting facilities not found in others, in that the mixer may be perfectly adjusted for low speeds and for high speeds by the two separate adjusting devices, and because of the diaphragm, radius bar and twist of the needle it will automatically adjust itself in a reasonably proportionate manner for intermediate speeds. Since it depends upon suction for this automatic adjustment, it secures the same quality of

THE CARBURETER AND ITS FUNCTIONS

mixture, regardless of the action of the throttle or the size of the engine, and is therefore more nearly universal than previous devices. It has no working parts exposed to mud, and may, therefore, be placed in any desired position. Its gasoline outlet is slightly to the rear of the center of the float chamber, which slightly increases the gasoline flow when great power is needed, as on hills.

This device may be used without a heating chamber if it is supplied with warm air, or if the quality of fuel is such that heat is not needed, but a heating chamber can be furnished which is attached immediately after the throttle where the evaporation is greatest and heat most needed, and this chamber is arranged for either water or hot gas, as may be preferred.

Not many power tests have been made of this device, but such as have been made indicate five to ten per cent. more power than other carburetors gave on the same motor.

SOME PAPERS PRESENTED AT MEETINGS OF SECTIONS OF THE SOCIETY DURING 1911

DETROIT SECTION

CARBURETERS

Present and Future. Starting Troubles, Causes and Remedies. A New Calibrated System Described

By George M. Holley

(Member of the Society)

This is not intended to be a historical record, but a simple record which tends to show what our future designs will be, as we are still in the development stage on motors and carbureters.

Carbureter manufacturers are forced to meet severe and extreme conditions on the high-speed automobile motor. These extremes of "speed range," "flexibility," "starting," extreme hot and cold weather, etc., are too wellknown to take up in detail.

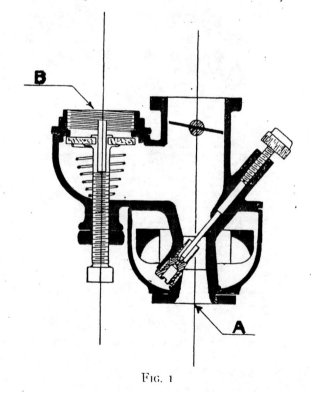

Fig. 1

On the majority of cars, the conditions are becoming more rational, as designers are building slower-speed and longer-stroke motors, which tend to give more uniform carbureter action.

Practically all carbureters in use at the present time are constructed as shown in Fig. 1 with constant air opening at A and a spring actuated auxiliary air valve at B.

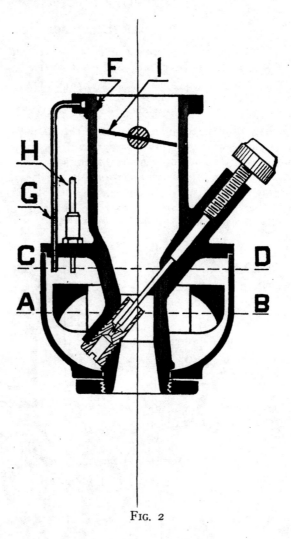

Fig. 2

Typical Carbureter Construction

These valves are extremely sensitive. It must be perfectly clear that a very fine adjustment is required to cover a starting suction as well as a slow-running and high-speed suction.

In cold weather a great deal of difficulty has been experienced when first starting up, due to two causes.

First. The gasoline being supplied is getting heavier. The refiners cannot supply enough of the high test gasoline. Only 15 per cent. of gasoline is obtained from a barrel of Pennsylvania crude oil; while 41 per cent. is kerosene. To raise this percentage of commercial gasoline, the cut is made lower in gravity, which increases the quantity of gasoline obtained from each barrel of crude oil; but at the same time the fuel is of course heavier.

Second. 'This heavy gasoline does not vaporize as readily as the lighter, especially in cold weather, as even if it is lifted from the nozzle (necessarily at low velocities) it is liable to lie in the *bottom of the carbureter*, while undiluted air simply passes direct to the throttle.

Various devices are being provided to overcome this difficulty. Several makers have added a small tank on the dash for introducing gasoline directly into the manifold.

Starting Device

A simple and positive starting device which is easily attached to any carbureter is shown in Fig. 2.

The gasoline level is normally on the line A B.

When the flusher H is depressed, it presses down on float E, which

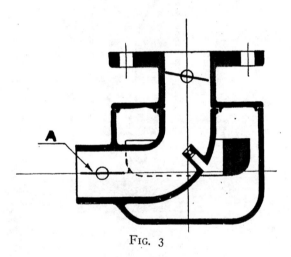

Fig. 3

lifts the gasoline inlet needle and causes the gasoline level to rise to the line C-D.

When the motor is cranked with throttle I opened *only* two or three notches on the throttle lever, there is a high vacuum on the upper side of the throttle plate I, causing the gasoline to rush up through the tube G, spraying through the measuring hole F and mixing with the incoming air past the throttle plate, both coming in under high velocity.

After the first few revolutions, the gasoline drops back away from the starting tube G to its normal level.

A butterfly valve in the air intake as shown at A in Fig. 3 is being used to a considerable extent, and causes the gasoline to come out of the nozzle with great velocity, literally sousing the inside of the carbureter.

But this is not really effective unless all the air has one entrance or the auxiliary air valve is held closed as well as the constant air supply, which in most cases causes mechanical complication.

Valve in Constant Air Intake

The modern carbureter is not only a measuring device, but a vaporizing device as well; the latter function becoming harder as the fuel gets heavier. The gasoline is vaporized in two different ways: First, by the high velocity of fuel flow into the inrushing air, which might be termed mechanical vaporization; second, by heat from the motor, water-jacket and hot air tube.

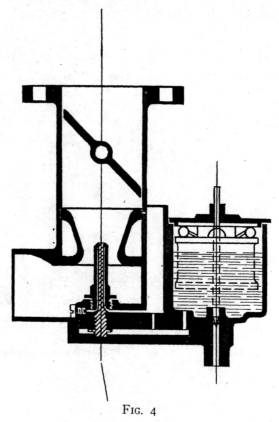

FIG. 4

Even with all these methods, vaporization is becoming more and more difficult as the gravity of the fuel drops, and it is of vital importance from now on that we should have *all the air entering through one opening and that all the air should pass the fuel* nozzle as in Fig. 4. That the fuel and air are better mixed is self-evident. This is common practice on practically all foreign carbureters, most of which have dropped the spring-controlled air valve and have no adjustments.

The difficulty of producing an automatic carbureter without auxiliary air valves to compensate for fluctuations in vacuum produced in the carbureter, is largely due to the fact that we are handling fluids which vary greatly in density. The rate of increased flow due to vacuum produced

Construction in Which All Air Passes Fuel Nozzle

by the piston in a plain tube carbureter, as shown in Fig. 4, is not the same for gasoline and air. The vacuum rapidly increases as the motor speed increases. This causes a freer flow of gasoline and a *decreasing flow of air per power stroke when certain speeds are reached.* This decrease of air flow is largely due to restrictions on the motor and to the small amount of time given to suck in the charge on each stroke. Added to this is the fact that the motor speeds do not increase uniformly according to throttle openings.

Vacuum Tests

To ascertain certain difficulties as to fuel flow necessary to perfect a carbureter without moving parts, vacuum tests were taken 1½" below the throttle plate on a plain tube carbureter shown in Fig. 9. Tests were made

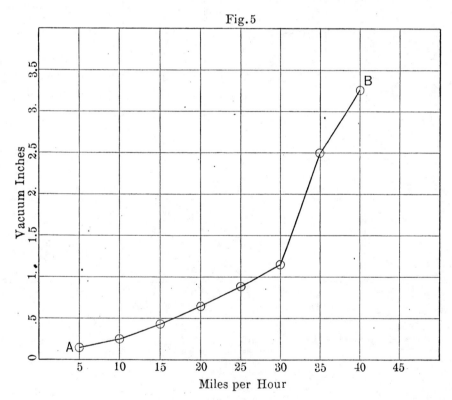

Spark retarded. Car on level asphalt road. Throttle opened notch by notch

on the car with the vacuum gage attached to the dash. The results of these tests are shown in Figs. 5, 6, 7 and 8.

The curve A-B in Fig. 5 shows the vacuum in the carbureter 1½" below the throttle plate, the curve being for from 5 to 40 miles per hour, with spark retarded.

C-D in Fig. 6 represents vacuum curve from 5 to 45 miles per hour, with spark advanced.

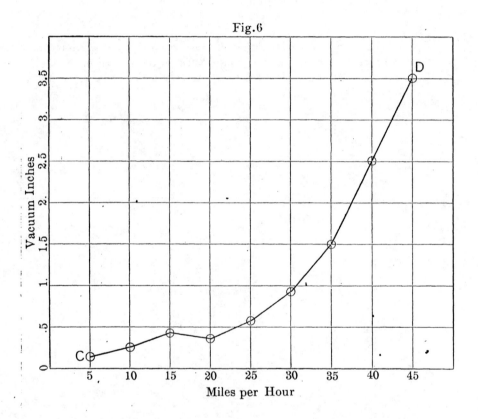

Spark advanced. Car on level asphalt road. Throttle opened notch by notch

The vacuum tests in Figs. 5 and 6 were taken by opening the throttle notch by notch on high gear, with the car running on level asphalt road.

Curve E F in Fig. 7 represents the vacuum with the motor pulling very hard through sand from 1 to 25 miles per hour, with the spark advanced to the most advantageous position with the throttle wide open.

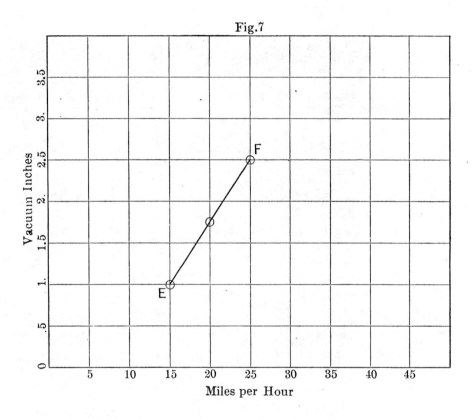

Motor pulling hard. Car running through sand. Spark advanced to most advantageous position. Wide-open throttle

Fig. 8 clearly shows the different vacuums in the carbureter, caused by varying loads under the same throttle openings as used on a high-speed automobile. This figure merely represents the three different vacuum curves 5, 6 and 7, all conditions of which have to be covered in an automatic carbureter without hand adjustment. We might say that the dotted line shows the vacuum range area under these particular tests which are covered by A, E, F, B, D.

It is a wellknown fact that for motor car work, a richer mixture is more reliable for starting, slow running and hard pulling, but for average driving from 10 miles per hour and up, a leaner mixture is more economical of fuel, heats the motor least and causes less carbonization in the

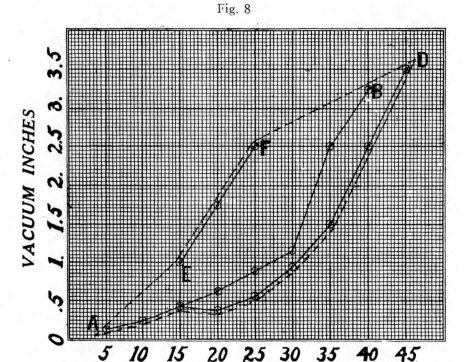

Fig. 8

Dotted line shows vacuum range area

cylinders. This has heretofore been a hard combination to obtain without considerable complication, including moving parts, which easily get out of adjustment due to delicacy and natural wear which arises in service.

Fig. 9 illustrates a new design to overcome the difficulties encountered on the constructions described and to fulfill the conditions mentioned in the previous paragraph.

This problem has been solved (after considerable experimentation) without a single moving part, except the float, by a special construction of the spray nozzle which is fully shown by the sectional cut.

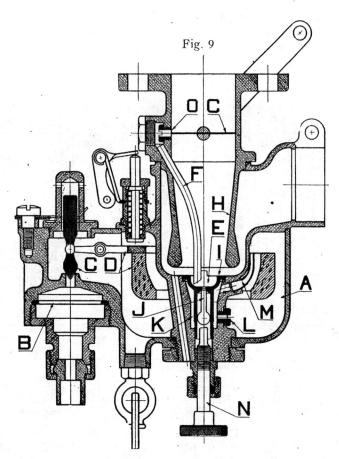

Design producing rich starting-mixture, and fuel economy above ten miles an hour

All float feed carbureters depend on a partial vacuum at the spray nozzle while the full atmospheric pressure is exerted upon the top of the fuel in the float chamber, this causing the fuel to be ejected from the spray nozzle.

In the sectional cut, the fuel from the tank enters the float chamber (A) through the strainer (B) and the level is regulated by the inlet needle (C) actuated by float and lever (D).

When the motor is at rest, the fuel level is half way up in the cup (E). This submerges the lower end of low speed tube (F) in the fuel. When starting the motor, the throttle (G) being nearly closed, fuel and air are drawn through the low speed tube (F) with a very high velocity, owing to the high vacuum above the throttle plate (G), which forms a rich mixture, thus making starting very easy.

The low speed tube continues its supply of mixture to the motor for low speeds (thereby allowing the motor to be throttled exceptionally low), but as the throttle opens this low speed tube gradually merges its mixture into the large tube (H), all the mixture being supplied through (H) above a motor speed of about 300 RPM.

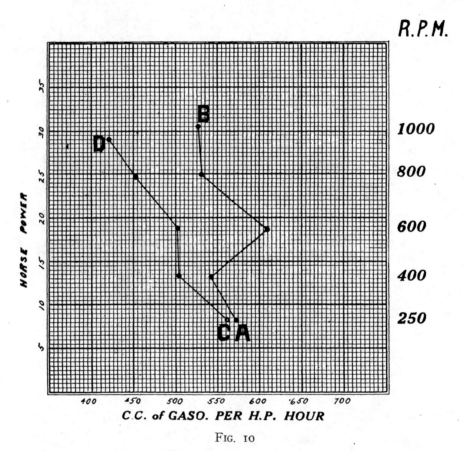

Fig. 10

Curve A B—Fuel consumption without slot J shown in Fig. 9. Curve C D—Fuel consumption with slot J

The spray nozzle (I), having a slot (J), and being in the reservoir (K), is supplied with fuel through two separate channels; first, through the series of holes (M); second, through plug (L) having a limiting hole. At low speeds both supply the fuel, the row of holes (M) predominating. But as the motor speed increases, the fuel level automatically drops, due to the fact that the needle (C) has to lift higher and higher as the fuel supplied increases; the leverage being about three

CARBURETERS

to one, the level drops three times the movement of the needle, and the holes (M) are uncovered to the air at atmospheric pressure above the fuel level. As this occurs, the air passes through the slot (J) in spray nozzle and maintains the uniformity of the mixture.

The fuel supplied at low speeds is adjusted by the low speed plug (O) at extreme high speeds by the plug (L) and at intermediate speeds through the automatic action of the series of holes (M), the slot (J) in the nozzle (I) and adjusting needle (N).

Fuel Consumption Tests to Show Effect of Nozzle Slot

To verify the results which were obtained on actual road tests, as to the action of the slot (J), fuel consumption tests were run on a motor of standard design in connection with an electric dynamometer. The

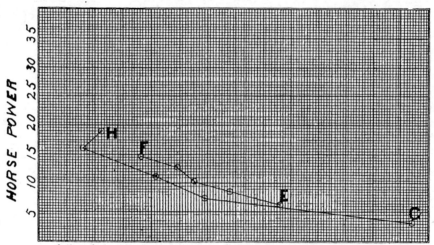

FIG. 11

E F, without slot.   G H, with slot

results of these tests are shown in the fuel curves, Figs. 10 and 11. The fuel curve (10) was run at various speeds. The curve A B shows the increased consumption when the slot is not used and the curve C D with the slot in use. The two tests were made with exactly the same needle setting, but with the vent (M) plugged in A B. The fuel curve (11) was run at constant speed, but with an increasing load. This also shows the difference between using the slot and not. E F represents the fuel curve without the slot and G H with the slot in action.

The slot in these was slightly large, which is shown by the horsepower falling off slightly and the mixture being slightly lean.

By the construction described, we obtain the following advantages:

Once adjusted, always adjusted; as there are no moving parts to wear out or get out of adjustment, which gives absolute reliability.

Positive starting *due to high vacuum and air velocity directly applied to the source of fuel supply.*

High horse-power due to large tubes machine-finished containing no obstructions.

Positive action at low and idling speeds due to rich mixture.

Most economical mixture for driving speeds.

Rapid acceleration, proven by actual road tests, owing to the more homogeneous and better proportioned mixtures.

A homogeneous mixture and low fuel consumption due to the atomizing effect of the air passing through the nozzle with the fuel, and the fact that all the air passes the source of fuel supply.

Richer mixture automatically obtained for hill climbing and hard pulling, due to the action of the fuel level rising with slower motor speeds and thus feeding the spray nozzle through both channels.

CARBURETERS AND FUELS

By E. J. Stoddard

(Member of the Society)

I will try to digest the important facts relating to carburetion as follows:

By way of laying a foundation for and illustrating what I wish to present for your consideration let us look at our materials in the first place. For our purpose we may say that air is a mixture of

	Oxygen.	Nitrogen.
By volume	20.81	79.19
By weight	23.10	76.90

We may neglect the other gases as inconsiderable.

But we think the amount of water vapor in the air does make a difference. It has been said that the engine acts quite differently on a rainy day. If the temperature is 62° a cubic foot of air should weigh .0761 and a cubic foot of saturated air .075581. That is, the air is a little over .7 of a per cent. heavier than the mixture. I have figured that only .0119, or a little over 1 per cent. of the air would be displaced. Even with the air at 82° it would be only $\frac{.0733}{.072267} = 1.014+$, a little over one and four-tenths per cent. heavier than the saturated mixture, and only about .0236 + or a little over 2 per cent. of the air would be displaced.

To me the cause of the observed difference does not appear here. Through the courtesy of the Detroit Weather Bureau I have been given the maximum variations. This indicates a variation of about ½ pound as the largest in the barometer. The active constituent of the air is oxygen. In a pound of air there is .231 of a pound of oxygen; and at 62° .228 pound of oxygen in a pound of mixture and .01179 pound of water vapor, a little over 5 per cent.; and at 82° there is .2276 pound of oxygen and .02361 of water vapor, or something over 10 per cent.

We are accustomed to think of water as opposed to combustion. Perhaps this biases our judgment so that we see more effects in damp weather; or perhaps it is due to our depressed spirits. If the water is in the form of vapor it can not carry away the heat in the usual way; that is, by vaporization.

Then comes in the question of dissociation. Does the initial heat of combustion change the vapor into its elements and thus dissipate its energies? I do not know. We seem to need data as to how and when heat is developed and what, when and how the resulting chemical combinations are formed. But it seems to me there are more immediately accessible facts which can be obtained with less trouble.

GASOLINE

Gasoline is said to be a mixture of the lighter distillates of petroleum, mainly of the methane or paraffine series, having the general chemical formula C_nH_{2n+c}. The following table contains data relative to some of the liquid distillates from Pennsylvania petroleum, compiled from the books of Redwood and Crew on petroleum.

Name.	Chemical Formula.	Per Cent. of Carbon.	Per Cent. Hydrogen.	Boiling Point, F.	Specific Gravity.	Vapor Density.	
Pentane	C_5H_{12}	12064 83.20	10365 16.71	87 100	.64	2.538	22430
Hexane	C_6H_{14}	9984 83.68	60156 16.32	143 158	.676	3.053	159996 air
Heptane	C_7H_{16}	83.90	16.04	165 210	.718	3.547	
Octane	C_8H_{18}	12204 84.17	9820 15.83	247 255	.737	3.992	22025
Nonane	C_9H_{20}			303	.756	4.8	14792 air

If we assume that the liquid is a mixture of the above-named substances, there is still a good deal of room for variation in the proportions. If we take the heat value of its constituent parts, and my figures are correct, pentane would have a value of 22,430 B.T.U. per pound and octane 22,025, not a very great difference. When we come to figure the quantity of air required, there is apparently a greater difference, 16 pounds of air being required for pentane and 14.79 for octane. We would expect a difference in readiness to ignite, and a considerable difference as to the rate of diffusion and readiness with which a perfect mixture is formed. We need experimental data I think on these points.

As to vaporization we notice that pentane is nearly a gas and octane nearly a solid. We would expect difference in viscosity and readiness of flow; besides there is a distinct difference in specific gravity which would probably affect the action of the carbureter. The quality of gasoline is usually tested by its specific gravity and the Baume gage. This is obviously a very crude method. I have advocated testing the vapor tension also, which may be done very simply. If the quantity of gasoline to the volume of enclosed air were measured carefully and standardized I think we would have more information.

EXPERIMENTS

It is desirable in testing carbureters to draw a charge of the mixture from the intake pipe and find out of what it is composed. I have tried taking a given volume of mixture and dissolving out the hydro-carbon with olive oil noting the diminution of volume. I have not had satisfactory results so far, perhaps because of imperfect apparatus. I believe the chemists use alcohol to dissolve out the heavy hydrocarbons, afterward removing the alcohol vapor with water (Hempel's Gas Analysis). If gasoline vapor is three times as heavy as air, we may estimate roughly the relative volumes of gasoline vapor and air in a 16 to 1 mixture as

$$\frac{13 \times \frac{1}{3}}{16 \times 13} = \frac{1}{16 \times 3} = \frac{1}{48},$$

or about 2 per cent. Now I think we would want to distinguish a one-tenth variation anyway, and therefore with this method, if it worked perfectly, we would want to distinguish clearly a variation of one-fifth per cent., a fineness of measurement that makes me distrust it. I have thought we might measure the degree of saturation thus: Suppose we put some pure air into a receptacle, introduce sufficient gasoline to saturate it with vapor, and note that the water column rises and stops at 60" of water. We know that this is the saturation pressure. Now if we put into a similar vessel the mixture to be estimated and then introduce gasoline as before and note that the water column now rises to 44" instead of 60 we would think that the mixture was $\frac{60-44}{60} = \frac{4}{15}$ saturated. It would seem that this was somewhat more hopeful. Now suppose we take a cubic foot of gas from the intake pipe during the operation of the engine. This might weigh .0777 pound and one-sixteenth of it would be about .00486 of a pound. If we take 22,000 B.T.U. as the heat value of gasoline, $22{,}000 \times .00486 = 106.92$ B.T.U., a value that could be easily read to a small fraction of a per cent. Even a much smaller quantity of mixture might be taken. For instance, a gallon, 231 cubic inches, which would contain about 14 B.T.U., is sufficiently large to measure. It seems to me that by this latter method the quantity of gasoline in any mixture may be measured with sufficient accuracy. If we could measure the variations in the mixture supplied, it seems to me we would have mastered the most immediate and accessible data. And this seems to me entirely possible.

What occurs in the cylinder would require deeper study, including I think the study of radiation of the gases at high temperatures.

The careful analysis of the exhaust gases seems to promise results as to the effectiveness of the combustion. I have not had satisfactory results, however, nor have I seen reports of results that satisfied me.

A CONSIDERATION OF CERTAIN PROBLEMS OF CARBURETION

By Arthur B. Browne

(Member of the Society)

The lack of uniformity exhibited by the great and ever-increasing variety of carbureters on the market, suggests that as yet, no comprehensive principle of automatic regulation of the gas to air ratio has been generally recognized. The simplest form of carbureting device consists of a fuel jet introduced into the moving air column within the intake pipe. If the velocity of the fuel flow were directly proportional to the velocity of the air flow, the mixture from such a device would be of constant composition under all conditions and the principal problem of carburetion would be resolved at once to its simplest terms. Unfortunately the relation between the air and fuel velocities is not a direct proportion, but, as will be demonstrated, it is none the less definite. Once recognized, its application to practical carburetion not only eliminates the necessity for most of the mechanical complications now in use but it explains clearly the errors which are introduced by their use.

CARBURETION

THE LAW OF GRAVITY APPLIED TO CARBURETION

The law of the flow of fluids, including gases within certain limits of pressure differences, is expressed

$$v = \sqrt{2gh} \qquad (1)$$

where
- v = velocity in feet per second.
- g = acceleration of gravity (32.2 feet per second).
- h = head, or height in feet of the fluid, required to produce the pressure necessary to cause the flow.

The velocity of the air (Va) in a carbureter will be expressed

$$Va = \sqrt{2gh} \qquad (2)$$

whence

$$Va^2 = 2gh \qquad (3)$$

and

$$h = \frac{Va^2}{2g} \qquad (4)$$

In this case, h is the height in feet, of a column of air, the weight of which will exert the pressure necessary to cause a flow of air at the velocity Va, or conversely, the loss of head caused by the air flowing at the velocity Va.

The head of fuel caused by air passing at a velocity Va, will be

$$h \times \frac{Wa}{Wf}$$

where
- Wa = weight 1 cubic foot of air (.076 pounds at 62° F.).
- Wf = weight 1 cubic foot of fuel (weight 1 cubic foot of water [62.355 pounds] × Sp. Gr. of fuel).

Applying equation (2) to the fuel velocity, Vf, we have

$$Vf = \sqrt{2gh \frac{Wa}{Wf}}$$

But as, before actual discharge commences, the fuel must rise from the level in the float chamber to the mouth of the fuel nozzle, a distance of h' feet, subject to the retardation of gravity, we must deduct the value of $2gh'$ and hence

$$Vf = \sqrt{2gh \frac{Wa}{Wf} - 2gh'}$$

Substituting the value of $2gh$ as determined by equation (3), the velocity of the fuel is expressed in terms of air velocity as follows:

$$Vf = \sqrt{\frac{Wa}{Wf} Va^2 - 2gh'} \qquad (5)$$

Wimperis (The Internal Combustion Engine, page 268), arrives at the same relation between air and fuel velocities by methods of the calculus.

EFFECT OF TEMPERATURE

The density of both fuel and air is, of course, modified by temperature. The density of the air varies inversely as the absolute temperature, while the density of gasoline is shown by Clerk and Burls (The Gas, Petrol and Oil Engine, Vol. II, page 623) to be modified by temperature as follows:

$$\text{Sp. Gr.} = 0.72 \{1 - .0007(t-60)\}$$

whence

$$Wf = W \times s \{1 - .0007(t-60)\} \qquad (6)$$

where

Wf = weight of 1 cu. ft. of gasoline.
W = weight of 1 cu. ft. of water.
s = specific gravity of gasoline at 60° F.
t = temperature of the gasoline in F.°.
t' = temperature of the air in F.°.

Substituting these values in equation (5) we have:

$$Vf = \sqrt{\frac{\frac{(460+62)\,0.076}{460+t'}}{62.355 \times s\{1-.0007(t-60)\}}(Va^2 - 2gh')} \qquad (7)$$

The range of values for t and t' to be used in equation (7) is so small that it will be readily seen that the effect of temperature is negligible.

WORKING FORMULÆ

Omitting the temperature correction, a simple working equation for gasoline of a Sp. Gr. of 0.72, may be expressed

$$Vf = \sqrt{(.00169\,Va^2) - 2gh'} \qquad (8)$$

For fuel of any other gravity, equation (5) becomes

$$Vf = \sqrt{\left(\frac{.076}{62.355\,s}Va^2\right) - 2gh'}$$

which reduces to

$$Vf = \sqrt{\left(\frac{.00122}{s}Va^2\right) - 2gh'} \qquad (9)$$

APPLICATION OF THE LAW TO VARIOUS TYPES

In order to obtain a clear understanding of the application of the law, let us consider the action of various types of carbureting devices in view of the relation of air and fuel velocities as expressed in equation (8).

Assume (A) that a unit quantity of air is passing each device with a given velocity and then (B) that a greater quantity of air is demanded. For the sake of uniformity let us assume that each device maintains a constant level of fuel 0.5 inch. (0.0416 feet) below the mouth of the fuel nozzle and that the fuel employed is gasoline of a specific gravity of 0.72.

CARBURETION

I. THE SIMPLE CARBURETER

(A) In this device, the velocity of the fuel discharge for an air velocity of say 90 feet per second will be, by equation (8)

$$Vf = \sqrt{(.00169 \times 90^2) - (64.4 \times .0416 \times 0.72)} = 3.43 \text{ ft. per sec.}$$

(B) As the area of air admission is constant, four times the air will pass at four times the velocity. By equation (8) this will induce a fuel flow of

$$Vf = \sqrt{(.00169 \times 360^2) - 1.9} = 14.73 \text{ ft. per sec.}$$

Hence, while the quantity of air has been increased four times, the quantity of fuel has increased 4.4 times and the resulting mixture is 10.4 per cent. richer than formerly.

II. THE MIXING VALVE

In this device, head, pressure on the valve, amount of valve opening, admission area exposed by said opening and the quantity of air admitted are in direct proportion to each other, if friction is disregarded. It follows therefore, that, as the head varies with the square of the velocity (Equation 4), the quantity of air bears the same relationship. Conversely we may state that the velocity varies as the square root of the quantity of air admitted.

(A) The fuel flow, for an air velocity of 90 feet per second will be 3.43 feet per second as in (I-a).

(B) By the proportion stated above, four times the initial quantity of air will pass the apparatus at twice the initial velocity. Hence the fuel flow induced by the increased quantity will be, by equation (8)

$$Vf = \sqrt{(.00169 \times 180^2) - 1.9} = 7.27 \text{ ft. per sec.}$$

showing that while the air quantity has increased four times, the fuel quantity has increased only 2.17 times, or but 54 per cent. of the fuel is present that is necessary for a constant mixture. It is thus readily seen why the mixing valve cannot be used for carburetion where any material degree of flexibility is desired.

III. THE COMPENSATING CARBURETER

Attempts to correct the tendency to over-richness exhibited by the simple carbureter, led to the early adoption of the auxiliary air-valve. The popular conception of the auxiliary-air inlet is that the air thus admitted serves to dilute the necessarily over-rich mixture formed at the mouth of the fuel nozzle. As all the air entering the carbureter, through either the primary or auxiliary inlet, finally reaches the cylinders as part of the explosive mixture, the foregoing statement is obviously true, but the most important function of the auxiliary inlet is likely to be lost sight of in such an explanation of its purpose.

The area of the auxiliary opening modifies the velocity of all the incoming air and hence exercises a direct influence upon the amount of fuel inspirated. This function will be better understood if the primary and auxiliary inlets are considered as a divided unit. Any enlargement of the auxiliary area increases the *total* area of admission and hence modifies both quantity and velocity.

In a carbureter of this type let
Q = the quantity of air.
V = velocity of the air.
a = auxiliary area.
c = the primary area.
A = total admission area = $a + c$.
g = acceleration of gravity.

Disregarding friction, the quantity of fluid discharged by an orifice is expressed

$$Q = VA \qquad (10)$$

Hence the quantity of air passing the carbureter will be

$$Q = V(a+c)$$

which, by substituting the value of V from equation (1) may be written

$$Q = (a+c)\sqrt{2gh} \qquad (11)$$

In this equation h is the height of a column of air necessary to cause a unit deflection of the spring governing the auxiliary valve; therefore the velocity of a given quantity of air is directly dependent upon spring tension and deflection, as well as upon the relative areas of both primary and auxiliary openings. As these variables are fixed by construction, determination of the quantity and velocity may be effected by simple substitution of the known values in equation (11).

For instance, assume that in a carbureter of this type, provided with a primary inlet 5-8 inch in diameter (0.3 square inch area), a vacuum of 1 inch of water causes an auxiliary area of 0.05 square **inch to be opened**

(A) A head of 1 inch of water is equivalent to a head of 68.284 ft of air at normal pressure and temperature.

By equation (1)

$V = \sqrt{64.4 \times 68.28} = 66.31$ feet (or 796 inches) per second.
$A = 0.3 + .05 = 0.35$ square inch.
$Q = 796 \times 0.35 = 278.6$ cubic inches per second.

By equation (8)

$Vf = \sqrt{(.00169 \times 66.31^2) - 1.9} = 2.35$ feet per second.

(B) Assume now that, on open throttle, the vacuum within the carbureter is 20 inches of water. The head of air would be

$20 \times 68.28 = 1,365.6$ feet.
$V = \sqrt{64.4 \times 1,365.6} = 296.5$ feet (or 3,559 inches) per second.
$A = 0.3 + (0.5 \times 20) = 1.3$ square inches.
$Q = 3,559 \times 1.3 = 4,616.7$ cubic inches per second.
$Vf = \sqrt{(.00169 \times 296.5^2) - 1.9} = 12.1$ feet per second.

Therefore, the air flow has increased $\frac{4,617}{279} = 16.5$ times while the fuel flow has increased only $\frac{12.1}{2.35} = 5.15$ times; or but 31 per cent of the former proportion of fuel is present. In other words, had the original mixture in (A) been in the air/gas ratio of say 10/1, the high-speed mixture of (B) would be in the ratio of 32/1, which is far beyond the limits of combustibility.

As may be readily determined, no adjustment of spring tension can do more than very slightly modify this tendency toward impoverishment of the mixture, while the addition of various fo. of subsidiary springs, becoming operative only at some point of the valve-opening, can do no more than correct the error at one given point and then start, as it were, merely a new scale of errors.

The inherent error of the auxiliary valve is by no means of theoretical interest only. It still remains a factor of so intensely practical effect, despite the remarkable ingenuity that has been displayed in various attempts to correct it, that its elimination would effect an annual saving of thousands of dollars to both manufacturer and user of motor cars through the increased efficiency of the liquid fuel engine.

IV. THE MULTIPLE-JET CARBURETER

Attempts to correct the error in mixture composition introduced by the increasing air flow have been confined largely to two principal channels. Abroad, the tendency is toward the use of multiple fuel jets, while in this country more attention has perhaps been given to the direct mechanical regulation of the area of the orifice in the fuel nozzle.

It will be apparent from the foregoing treatment of the subject that, in multiple-jet practice, the flow from each succeeding jet is, in turn, amenable to the law of fluid flow as expressed in equation (8). Hence, each succeeding jet, like the subsidiary spring on the auxiliary valve, merely corrects the error at the point where its own discharge commences and then the flow suffers a cumulative error until corrected by the introduction of the flow from still another jet.

It is evident that the use of a sufficient number of jets might be made to reduce the error to very small proportions, and in fact good results have been obtained from such construction. Mechanical complications and the nicety of constructional detail have proven serious disadvantages, however.

V. THE VARIABLE FUEL ORIFICE

Inspection of equation (8) and the substitution of values therein in the examples cited, disclose that the fuel velocity is in constantly decreasing proportion to the air velocity. In example III, the quantity of fuel discharge has been treated of in terms of fuel velocity. It is evident, however, from equation (10) that the actual fuel discharge is the product of its velocity and the area of the fuel orifice. Hence, it will be recognized that variation of the area of the fuel orifice may be made to compensate for the increasing ratio between the fuel and air velocities. In III-B, for instance, while the quantity of air was increased 16.6 times, the fuel velocity increased only 5.15 times; therefore, to maintain constancy of mixture, the area of the fuel orifice should have been increased

$$\frac{16.5}{5.15} = 3.2 \text{ times.}$$

The withdrawal of a straight tapered pin from the fuel nozzle increases the area of discharge in direct proportion to the lift of the pin; consequently delicate mechanical complications are resorted to in effecting the desired decrease in the proportional area opened. Properly designed and properly adjusted, there is no reason why this method should not give results approaching accuracy but when we consider the almost microscopic nicety of adjustment necessary to effect accurate sub-division of the minute fuel stream, we realize the practical difficulty of both making and maintaining such adjustments. When we remember, too, that the volume of liquid gasoline is less than 1/8000 of the volume of the air with which it is mixed, it is apparent that regulation of the 8,000 parts would be much more practical than any attempt to subdivide the 1 part.

VI. A NEW METHOD OF COMPENSATION

It has been shown in Case V that compensation can be effected by the variation of the area of the fuel nozzle. It is equally true that automatic variation of the total air admission area will accomplish the same result with much greater accuracy and without adjustments or mechanical complications of any kind. For this purpose it is necessary to determine the velocity of the air corresponding to any given fuel velocity.

If, by equation (8)
$$Vf = \sqrt{(.00169 \, Va^2) - 2gh'}$$
$$Va = \sqrt{\frac{1}{.00169}(Vf^2 + 2gh')}$$

or, more conveniently
$$Va = \sqrt{591.71 \, (Vf^2 + 2gh')} \qquad (12)$$

The practical application of these formulæ is, perhaps, best made clear by a concrete example. Let us consider a carbureter with a primary inlet 5/16 inch in diameter (area, 0.077 square inch). Let us assume the auxiliary valve to be governed by a spring that will deflect 0.01 inch for a vacuum in the carbureter of 1 inch of water.

(A) Assume that 230 cubic inches of air per second is passing through this carbureter at a velocity of 90 feet per second. By equation (8) the fuel velocity will be

$$Vf = \sqrt{(.00169 \times 90^2) - 1.9} = 3.43 \text{ feet per second,}$$

the vacuum will be $\dfrac{90^2}{64.4 \times 68.28} = 1.84$ inches of water.

The deflection of the valve will be
$$0.01 \times 1.84 = 0.0184 \text{ inches.}$$
The total admission area will be
$$\frac{230}{90 \times 12} = 0.213 \text{ square inch.}$$
The auxiliary area will be
$$0.213 - 0.077 = 0.136 \text{ square inch.}$$

CARBURETION

(B) Assume now that ten times the original quantity of air is demanded.

The quantity of air would be

$$230 \times 10 = 2{,}300 \text{ cubic inches per second.}$$

This air must pass the fuel jet with a velocity sufficient to induce a flow 10 times the initial quantity of the fuel.

As, by equation (12)

$$Va = \sqrt{591.71\,(Vf^2 + 2gh')}$$

the air velocity that will increase the fuel flow 10 times may be expressed

$$Va_{10} = \sqrt{591.71\,\{10(Vf^2) + 2gh'\}}$$

Substituting the valves of the present example

$$Va_{10} = \sqrt{591.71\,\{10(3.4^2) + 1.9\}} = 263.67 \text{ feet per second.}$$

The vacuum will be $\dfrac{264^2}{64.4 \times 62.28} = 15.85$ inches of water.

The deflection of the valve

$$15.85 \times 0.01 = 0.158 \text{ inch.}$$

The total admission area

$$\frac{2300}{264 \times 12} = 0.73 \text{ square inch.}$$

The auxiliary area

$$0.73 - 0.077 = 0.653 \text{ square inch.}$$

As a practical convenience these equations may be simplified and expressed in terms of fuel velocity as follows:

$$\text{Velocity of the air} = 24.32\sqrt{Vf^2 + 2gh'} \tag{13}$$

$$\text{Total admission area} = \frac{Qa}{292\sqrt{Vf^2 + 2gh'}} \tag{14}$$

$$\text{Vacuum in inches of water} = \frac{Vf^2 + 2gh'}{7.44} \tag{15}$$

$$\text{Total spring deflection} = \frac{(Vf^2 + 2gh')d}{7.44} \tag{16}$$

where $d =$ the spring deflection for a vacuum of 1 inch of water.

By the use of these formulæ the auxiliary-air admission area may be determined for any number of points in the travel of the valve and the walls surrounding the valve may be made to conform to the curve so plotted, thus assuring the permanent maintenance of any desired air/gas ratio without adjustments of any kind.

RELATION OF VELOCITY TO VACUUM

In all the foregoing calculations the influence of friction and other factors modifying the flow of liquids in a carbureter have been omitted for the purpose of permitting simplified statements of fundamental principles. These modifications are, however, of prime importance, none the less so because their variant values are undetermined. They affect the flow of both fuel and air to such an extent that without giving them due consideration, the application of any formulæ expressing the relationship of actual flow of fuel and air would be impossible.

Thus the formulæ herein expressed have, so far, tentatively assumed that the drop in pressure or "vacuum" at the mouth of the fuel nozzle was the same as that within the mixing chamber. Repeated experiments have demonstrated the fallacy of such an assumption, to which indeed must be attributed the failure of many otherwise meritorious devices. Solution of the intricate problems existing between the mouth of the fuel nozzle and the mixing chamber, involving marked physical changes in both the liquid fuel and the air, would be interesting theoretically, but, from a practical standpoint, we are fortunately able to eliminate the effect of these modifying influences instrumentally. This can be accomplished by two structural modifications. First, the control of the auxiliary area directly by the vacuum at the mouth of the fuel nozzle, which construction also presents the further practical advantage of rendering the action of the instrument practically insusceptible to barometric changes. Second, by a slight modification of the curve of auxiliary-admission areas, so that the air velocities are increased a sufficient amount, determined experimentally, to compensate for the frictional resistance offered by the nozzle to the flow of the fuel. Instruments constructed in accordance with the foregoing principles have been found to maintain a constancy of mixture in strict accord with the theory, and it has been determined that the slightest departure from the theoretical curve of admission areas, produces negative results in constancy of composition.

VARIABLE MIXTURES

If, however, it were desirable to vary the mixture composition for different operating conditions, the proposed method lends itself readily to that end. Thus, the auxiliary areas may be diminished at and near the starting end of the curve, resulting in the richer mixture so often claimed to be necessary for easy starting. At ordinary road speeds the areas may be so calculated that a mixture of high fuel economy will result, while at extreme open-throttle for high speed, contraction of the admission curve will increase the richness of the mixture for the development of maximum power. In other words, the designer has but to determine the range of mixture composition which he considers most satisfactory and construct the admission curve in accordance therewith, knowing that whatever action has been selected will be repeated with invariable exactitude.

CONSTANT MIXTURE

The results obtained from many different engines by the use of gasoline mixtures of really constant composition have been so pronounced as to be in the nature of a revelation, particularly as regards certain details not ordinarily considered as primary functions of carburetion. Among these it is worthy of note that without exception every engine developed its maximum torque, both at high and low speeds, with a fixed spark ad nce. There is noticeable also a marked quietness of operation not easily explained, unless, possibly, the uniform rate of flame propagation establishes a rhythmical vibratory effect. The objectionable features of fluctuating mixtures are, naturally, minimized. After a full season's running the cylinders of several cars were found free from carbon, while the spark-plug points were clean and the porcelains discolored by heat only. Exhaust gas analysis shows practically no loss through incomplete combustion. The average of 44 samples taken from several different cars under all sorts of road conditions gave 0.43 per cent. CO, while 29 samples yielded no CO.

DISCUSSION.

C. P. Grimes.—I think Mr. Browne covered the subject very well in that he omitted at the start the variation of temperature before he attempted his discussion. The problem of temperature and pressure in the carbureter to-day is the most complicated we have to deal with. I think every one knows that a small variation in the temperature of kerosene or gasoline will make a very wide variation in the flow of the fuel in relation to the air. I have found in my experiments that the pressure in the carbureter has far more to do with the economy of carburetion to-day than seems to be generally recognized. I at one time designed a carbureter with a variable throat. I varied the amount of gasoline flowing from the nozzle by varying the area of the throat into which the gasoline was drawn by placing a bob in the center capable of movement along its axis. This carbureter worked out nicely, but I found it was altogether too sensitive to temperature change, and further that the absolute pressure in the carbureter at low speeds was too high to effect proper breaking up of the fuel.

E. B. Wood.—Mr. Browne will correct me if I misunderstand the paper, which I possibly have. I think his whole theory of carburetion and the peculiar behavior of the carbureter is founded on the assumption that the gasoline level is half an inch lower than the jet, because if you take his first example and cut out the correction for level, which you can do for experimental purposes by simply letting the jet flood, according to his figures, it would give a constant mixture, whereas it very certainly does not. The fault, I think, arises in this manner—it is probable that the air flow does follow the law $V = K\sqrt{2gh}$, but the difficulty is we cannot measure the height H in feet of air. If we measure it with the water gage it is apparently not the real height. Dr. Watson did some experiments in that line. He compared the mean pressure in an induction pipe taken from the manograph (indicator) diagram with that of the water gage, taken at the same point and they were entirely different. I think the fuel follows the $V = K\sqrt{2gh}$ law over a fairly wide range and is under very similar conditions to the water gage which measures the suction.

Thomas S. Kemble.—I have been very much interested in this paper and am glad that we are going to have it in the records of our Society. I have done considerable work with an air valve which compensated according to the principle which Mr. Browne has described, except that the valve was operated by the pressure from the mixing chamber and not by the pressure at the fuel orifice. This worked very satisfactorily. It does not seem to me, though, that the real difficult problem which now confronts us in carburetion is the obtaining of the proper ratio between air and fuel, but rather the proper breaking up of the fuel and the proper application of heat so as to approximate as nearly as possible real carburetion.

I cannot understand some of the results that are given in the paper as being due simply to the proper proportions of the mixture, but believe that there must have been some very happy designing which brought good carburetion as well as proper mixing.

The matter of the best results being obtained with a fixed spark, I am unable to understand. A time lag exists in the magneto and at high speed we have a lower volumetric efficiency and consequent lower rate of flame propagation. In view of these facts I cannot see how the same spark setting can be proper for both high and low speed.

C. P. GRIMES.—This is not very technical, but I wish to suggest that the real problem in this commercial enterprise is to provide an instrument that will do so much. An ammeter to measure current will cost $25, a voltmeter about $50, a venturi meter $90. And these meters vary greatly. My grocer paid $250 for some new scales and when I questioned the weight of a purchase his scales were wrong, because, he said, they had to be set for each ten degrees temperature.

A carbureter must measure air and gasoline under all conditions of humidity, temperature and pressure, and yet the manufacturer demands such an instrument for about $10, which must include proper overhead, research and maintenance of the goods sold.

JOHN WILKINSON.—I would like to make a little comment on that spark question. I do not agree with the gentleman at all. We have made experiments for years on the question of power in relation to spark advance. We make motors with small combustion chambers where the distance of the spark point from the flame propagation is about as small as you could get it, and have never been able to find any conditions wherein the increasing speed did not require an increasing spark advance to obtain the maximum power, no matter what the condition of the mixture.

ARTHUR B. BROWNE.—I am aware that I am open to the criticism, in this paper, of touching only one of the problems, and the one that may be considered the minor problem in carburetion. Personally I think that idea has become prevalent, but I believe it to be a mistake that is holding us back in the production of efficient gaseous mixtures and obtaining efficiency from engines.

I did not bring the matter up in the discussion of the paper read by Mr. Chase, but if the gentlemen who have copies of his data sheet will turn to the table they will notice that the chemical composition, the quality of the mixture—I prefer that term—varied from 12.2 ratio to 7.8 during this test as the speed increased. The curves which Mr. Chase showed indicate, to my mind, without carefully calculating them, not only a considerable loss in volumetric efficiency, but in addition to that a tremendous loss from imperfect combustion caused by overrich mixtures. We cannot maintain, in my opinion, mixtures of varying richness, mixtures richer than the point of chemical composition, without sustaining serious power losses.

To state the problem a little bit differently from what I did in my remarks prefacing the paper, I believe that the problem of carburetion as a whole consists, first, of quantitatively mixing a liquid and a gas, and second, in producing complete vaporization of the liquid in the gas. Then only can you have proper distribution and equal power in the cylinders.

The gentleman spoke of the many variables in the problem of carburetion. It is true; and as I have attempted to state in my paper, they would make an exceedingly interesting study from a theoretical standpoint, if it were not possible to eliminate those variables instrumentally, except the ones that I have mentioned.

Regarding the fixed spark position, I can assure the gentleman that no one has been more surprised at my experience than I have myself. I do not attempt to defend it. I do not even attempt to claim that it is true. I will cite a single illustration of a certain car with a certain carbureter where the spark could not be advanced more than twenty-three degrees on the quadrant without a serious knock. A carbureter embodying the principles outlined in my paper was installed on this car. The spark lever was placed at full advance and the car driven from three miles an hour to its limit of speed with no indication of knocking. Now on that car it was not an accident. Had this condition occurred but once or twice, it might have been accidental. But as a matter of fact it has been observed on every one of fifteen or twenty different makes of cars that I have experimented on. Consequently, while I do not attempt to define its cause, it is certainly an apparent fact up to date.

I have failed to comprehend the criticism made by Mr. Wood that the elimination of the head or vertical distance between the level of the float chamber and the mouth of the fuel jet would give rise to a constant mixture.

The head (h' in equation 8) is not, in reality, simply the vertical distance between the level of the fuel in the float chamber and at the mouth of the fuel jet. To this must be added the "friction head" imposed on the fuel by its passage through the nozzle. This quantity is subject to constant variation and depends for its value upon the velocity, density and viscosity of the fuel.

It by no means follows, therefore, that a constant mixture would result by maintaining the level in the float chamber identical with the level at the mouth of the fuel nozzle. Could some means be devised whereby the true value of h' would become zero I would certainly expect a constant mixture to result.

The effect of friction head is so well recognized in hydraulics and so difficult of accurate determination under the conditions of carburetion that I refrained from complicating the formulæ by its consideration and hence throughout my paper have referred to the quantity h′ as a mere difference of level.

I have been experimenting with approximately constant mixtures. I say approximately because my work has been conducted very largely on the road, without adequate laboratory facilities, and I have depended for my knowledge of the correct mixtures entering the cylinders on gas analysis of the exhaust. So I say that I have been experimenting with mixtures of reasonable constancy. The results have been very apparent in increased power, which results agree quite closely with pre-determination of gaseous mixtures. That is, the greatest power has been obtained with approximately a twelve- to thirteen-to-one mixture, at the same time securing extreme flexibility and extreme quietness, not on one engine but on many operating with this mixture. The subject has not undergone such development as I propose to give it, and much desirable data are lacking, but it seems that we are opening a new field and that many discrepancies that we now find in motor car operation can be explained and possibly remedied by carbureter modification.

C. P. Grimes.—I was going to say that we have found in our carbureter economy readings on the road, that very often you could tell whether one carbureter is more economical than another by the position of the spark on the quadrant at which the mixture would fire. The chemical analyses that we have made show that no two mixtures will ignite at the same rate.

P. S. Tice.—Practically the whole of the experimental work of my firm has been conducted with a view to producing a simple carbureter, the type alluded to by Mr. Browne—a single air passage and single jet—and we have subordinated our structure entirely to that. But we have found that compensation can be secured by a simple rearrangement of the fuel passage with reference to the air passage; that is, a rearrangement of the simple carbureter, as one ordinarily thinks of it. We use for our air passage a form commonly termed the "reëntrant nozzle" and locate the fuel conveying air passage within that one.

I would like to touch upon the matter of mixture formation as distinguished from compensation; that is, as distinguished from mixture proportioning. We find it of the utmost importance to secure a thorough breaking up of the fuel, a thorough mixing, and as thorough vaporization as can be had. Practically the whole of carburetion is bound up in vaporization; and in order to secure it we have to apply, or have to put into the fuel, a certain quantity of heat, depending on the fuel. There is only one way of getting heat into the fuel. But we can assist in getting heat into the fuel by securing a thorough mechanical breaking up. This can be secured by high air velocities, as the most simple way to begin, and, of course, permits of transferring very rapidly what heat is available.

I would like to allude to certain things brought out in some kerosene work which we have done. They will illustrate the point, although the difficulties are greater with kerosene than with gasoline. A carbureter of elementary form was used and the heat input necessary to secure complete vaporization of the fuel determined. The motor was run under throttle, but the throttling was identical at all points in all the runs. Now this motor was maintained at the predetermined speed and torque, which were the maxima for the elementary carbureter in question, over a very wide range in fuel proportion. The normal consumption rate of the motor under conditions of the test, with complete vaporization, was around one and a quarter pounds of kerosene per brake-horsepower-hour. This consumption was very low. This motor could be made to deliver with the same carbureter, the same power, at the same speed, with a consumption of over four pounds of fuel per brake-horsepower-hour. Of course, the higher consumption was obtained, or rather was necessitated, when the heat input to the fuel was lowered, and vaporization was only very partially secured.

Now then, does it not seem, at this time, if we put the greater part of our attention to the securing of vaporization, that the other end of it, the compensation, will practically take care of itself? Some of us may be able to build the most perfectly compensating carbureter that can be made right now. I do not doubt that it can be done. But even if we do accomplish this result under a given set of conditions, we cannot do it with, say, a temperature change of fifteen or twenty degrees without a corresponding change in adjustment. Suppose we are running on the block with certain temperature and other conditions, and that the temperature alone changes; the whole of the carburetion, so far as *exact* compensation is concerned, will be thrown out by a change of ten to fifteen degrees.

L. V. CRAM.—One problem Mr. Browne did not touch on I have found difficult, when it comes to the motor on commercial work, and that is what the customer will demand in the line of acceleration. Mr. Grimes in a paper at the winter meeting discussed this subject and I wish to call particular attention to the truth of his last paragraph. In this country you must greatly increase gasoline consumption over the best possible economy in order to take care of your customer's requirements in acceleration. On the block I have often adjusted a carbureter for very good work, fair acceleration, and put it in the hands of a user who did not like it because it would not "get away" fast enough.

RICH MIXTURE FOR STARTING.

ARTHUR B. BROWNE.—Mr. President, in the matter of acceleration I would like to have an expression of opinion. If there is anything that is definitely fixed in the research of the authority of the world, it is the fact that the chemical composition of the mixture plays a part in every function of that mixture. The greatest speed range, the greatest power, the greatest economy, have been determined by various authorities who agree very closely. As Mr. Chase has said, it is noticeable that power, possibly maximum power on the block or on the road, if we had any means of measuring it—can be produced throughout a considerable fuel range. But in these days of soaring prices of gasoline can we afford to use any richer mixture than is necessary to give maximum power? I think President Browne will bear me out in the statement that the Royal

Automobile Club have determined that about ninety-five per cent. of the power and about ninety-five per cent. of the maximum thermal efficiency occur with a mixture of about 14 to 1. I may misquote the exact figures.

T. B. BROWNE.—15 to 1.

ARTHUR B. BROWNE.—Any departure from that composition towards the side of richness means a loss of fuel. The power curve will probably remain fairly level down to a mixture approximating 10 to 1, and I doubt if much difficulty will be encountered even a little lower than that.

That it is necessary to produce a rich mixture for either starting or acceleration I cannot understand. A mixture of perfect chemical composition is the most inflammable, therefore it is the easier starting; it gives the most powerful explosion, therefore it produces the greatest power effect; it is the quickest burning, therefore we have the greatest speed. Now why has this notion become prevalent that we must have a richer mixture for acceleration? To my mind it seems absurd. I admit the force of the argument of those who bear down particularly on the gaseous character of the mixture—it is the amount of *gas* we get in the mixture that produces results. We may inject liquid gasoline without practical effect, unless it be in the form of exceptionally minute particles constituting a spray which vaporizes in the heat of the cylinders. *Liquid* particles are so slow burning as to be almost wholly ineffective and the *gas* content of the mixture is alone available for the development of power. In starting an engine the failure of the carbureter to give to the cylinders a sufficiently rich mixture is because ordinarily it is cold, the vaporization temperature has not been reached and consequently but a small proportion of the fuel is vaporized and becomes effective as gas. While admitting the necessity for increased temperature and high velocities in effecting vaporization, I really believe that most of our carbureter troubles are due to inconstancy in the relative proportion of the true gases constituting the mixture.

C. P. GRIMES.—I wish to say, Mr. President, that the practical everyday men about me who have tried know that it is an absolute necessity to have a rich mixture for starting on a cold morning and for getting away in a hurry.

The graph of the relation existing between the volume of the mixture before and after combustion starts in a straight line from a ratio of 20 lbs. of air per pound of gasoline up to fourteen. At that point there is a very marked change. The curve continues upward at a greater angle to a ratio of nine and I dare say continues thus to the limits of combustion. This change in volume ratios I believe accounts for the increased power in acceleration given by the richer mixture.

I recall a very interesting experiment that was completed within twenty minutes that will illustrate the folly of maximum power.

N	Lb.	H.P.	Lb./H.P.-Hr.
1,000	50.5	28.9	.5665
1,000	51.4	29.35	.840

In both cases the engine ran fine; trained men could not tell the difference by the sound or action. The figures show that there was expended 48.4 per cent. more fuel for an increase of horsepower of 1.557 per cent.

T. B. BROWNE.—I do not carry the exact figures in my head, but Dr. Watson has in his paper to the Institution shown us that we do get with the rich mixture a certain increase of power, beyond that obtained from a mixture producing perfect combustion. And I think it is that rich mixture which you are using in this country, as denoted by the analysis of the exhaust gases, to get rapid acceleration.

I would like to say a word about acceleration. Some of you here seem to be very much more keen on rapid acceleration than we are. Most of our makers set their carbureters on cars for ordinary uses, as apart from those turned out for racing purposes or high-speed work, so that very rapid acceleration is not capable of being attained, with a view to prolonging the life of the tires and also to reduce the fuel consumption. The figures are given in Dr. Watson's papers in our Institution Proceedings, and I would refer Mr. Browne to those proceedings for the exact proportions of the rich mixture giving the greatest power.

J. G. VINCENT.—I would like to say one word about this matter. It is one I have been very much interested in. I think carburetion is blamed a great many times for things it is not responsible for. We all know that valve-timing is more or less of a compromise, especially in the six-cylinder motor. I have run some tests which have proved to me that valve-timing has a tremendous effect on acceleration. I have been working along the line of increasing the torque of the motor below 1,200, letting it take care of itself above 1,200. I conducted one interesting experiment with two motors. One showed nine horsepower more above 1,200 than the other, and the one that had nine horsepower less above 1,200 r.p.m. was ahead of the game always, because the overlap of the suction strokes in getting the high power and great speed absolutely killed the motor for low torque, around 400, the motor speed you usually have when you are accelerating. So that I think investigation along the lines of closing the inlet valves early, especially on the six-cylinder motor, and properly taking care of the intake header, and having the passages as small as you can have them, would get you away from a great many difficulties that you are ordinarily up against.

HERBERT CHASE.—I think Mr. Browne is absolutely right in the matter of a constant mixture actually giving the best results, except possibly at very low speeds, where the quantity of gas remaining in the clearance chamber dilutes greatly the incoming charge. The charge needs to be slightly richer in some cases, but much less rich than some people imagine. I believe that at low speed, and in cranking in particular, a large proportion of the fuel actually goes through the motor without entering at all into combustion. The trouble is that vaporization does not take place; with the low gas velocities atomization is not accomplished. The mixture is supposedly rich but in reality a great deal of the gasoline present remains inert so far as effective combustion is concerned.

T. B. BROWNE.—I would say in connection with that that Dr. Watson found that the rich mixture caused the excessively greater amount of carbon monoxide in the exhaust, proving that it was really a rich mixture, and not because of the gasoline globules passing through. It was proved by the proportion of carbon monoxide in the exhaust as obtained by actual analysis.

PAPERS PRESENTED AT MEETINGS OF SECTIONS OF THE SOCIETY

CLEVELAND SECTION

CARBURETER INVESTIGATIONS
By Frank H. Ball and Frederick O. Ball

(Members of the Society)

Abstract

The results are given of laboratory investigations made of a number of different types of carbureters, showing the relation between their gasoline and air consumptions over a wide range. This relation is plotted on so-called quality diagrams, on which is indicated the range between which high power and high efficiency can be expected.

A description is given of a carbureter arranged in two stages, the first being used at light load and the second coming into action when the throttle is nearly open, thereby more than doubling the carbureter capacity. Engine performance curves are presented showing the result when only one or both stages of this carbureter are used.

The laboratory used in the investigations reported in this paper contained a gasoline engine and electrical equipment for loading the engine; also a metering outfit for measuring the gasoline and air used by the carbureter under all conditions. A steam injector was used as another means for drawing air and gasoline through the carbureter.

A water column provided the means for observing the suction or head that impelled the flow, thus furnishing the data necessary for plotting curves showing the quantities of gasoline and air discharged at each end.

FLOW OF AIR AND LIQUIDS THROUGH ORIFICE

Contrary to prevalent opinion, the same law governs the flow of air and of gasoline through a fixed orifice. The quantities discharged in both cases vary as the square root of the suction or head. This law does not hold good at extremely high velocities, but can be accepted as approximately correct throughout the range covered in carbureter practice.

It follows therefore that in a fixed-orifice carbureter the ratio of the flow of gasoline and air at any point will be the ratio throughout the whole range of working capacity.

In practice it is not desirable to maintain the gasoline level at the overflow point in the discharge nozzle; therefore gasoline and air flow do not coincide. This is illustrated in Fig. 2. These curves indicate that with a suction of 20 in. of water the capacity of the respective orifices was such that 10 oz. of gasoline and 100 cu. ft. of air were discharged per minute. This ratio has been found to give a strong clean-burning mixture. It represents approximately 12 lb. of air to 1 lb. of gasoline. The air and gasoline curves in Fig. 2 do not have a common zero and therefore the ratio of gasoline to air is not constant.

The actual ratio of gasoline to air, under the conditions shown in Fig. 2, is represented in Fig. 1. The right-hand end of the quality

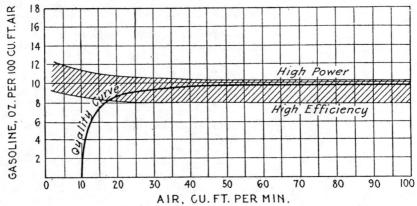

Fig. 1—Quality Diagram for Elemental Fixed-Orifice Carbureter

curve indicates 10 oz. of gasoline per 100 cu. ft. of air and corresponds with the right-hand end (where they coincide) of the quantity curves in Fig. 2. Toward zero, the quality curve of Fig. 1 shows the effect of the difference in the zeros of the two quality curves of Fig. 2.

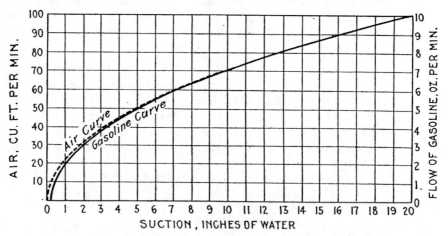

Fig. 2—Air and Gasoline Flow for Elemental Fixed-Orifice Carbureter

The shaded zone in Fig. 1 covers the range within which it has been found desirable that the quality should be maintained by the carbureter. The boundaries of this zone may not be accepted universally as the best, but they are approximately the best. The zone is entirely satisfactory as a field in which to plot the characteristic quality diagrams of various types of carbureter.

ELEMENTAL FIXED-ORIFICE CARBURETERS

When less than 18 cu. ft. of air per minute is drawn through the carbureter, the quality curve, Fig. 1, drops below the desirable zone. This makes it impossible to enrich small quantities of air. Therefore,

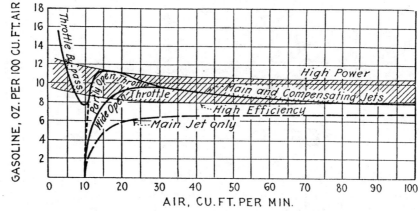

FIG. 3—QUALITY DIAGRAM FOR COMPENSATED FIXED-ORIFICE CARBURETER

the elemental fixed-orifice carbureter cannot be used with success, and it becomes necessary to embody in the carbureter some means for augmenting the gasoline supply where it is deficient. Various devices are used for this purpose. Some effect the necessary compensation by

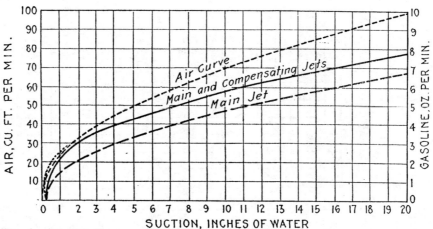

FIG. 4—AIR AND GASOLINE FLOW FOR COMPENSATED FIXED-ORIFICE CARBURETER

acting on the gasoline supply, others act on the air and still others on both.

It is proposed to analyze the effect of the most important of these devices, selecting those which have been used most extensively in high-grade instruments. The analysis is made solely to show the effect of the various devices in maintaining a normal quality at a lower minimum quantity than is possible with the elemental fixed-orifice carbureter. No attention is paid to the relative maximum

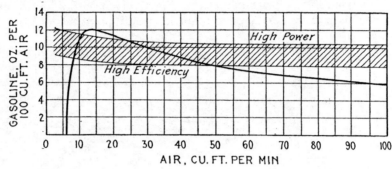

FIG. 5—DIAGRAM FOR ELEMENTAL AIR-VALVE CARBURETER WITHOUT FIXED ORIFICE

capacities or to any other questions relating to the several types. These will be considered later in connection with quality diagrams for the actual carbureters.

COMPENSATED FIXED-ORIFICE CARBURETER

This differs from the elemental fixed-orifice carbureter in that it is provided with compensating devices that increase the supply of

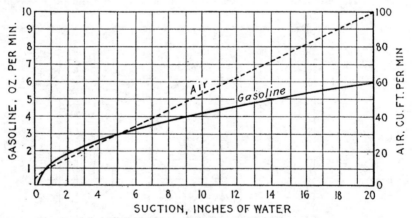

FIG. 6—FLOWS FOR ELEMENTAL FIXED-ORIFICE CARBURETER WITHOUT FIXED ORIFICE

gasoline, particularly where it is most deficient. Figs. 3 and 4 illustrate the effect of these devices. The quantities discharged by the main jet under various amounts of suction are shown by the main-jet quantity curve in Fig. 4. The quality produced by these quantities is shown in Fig. 3 by the curve marked "main jet only."

A second supply is drawn from a well open to the atmosphere and supplied with a constant quantity of gasoline from the float-chamber by gravity. The total quantities of gasoline from this double supply are indicated by the curve marked "main and compensating jets" in Fig. 4; the effect on the quality is indicated in Fig. 3 by the curve marked "main and compensating jets." The two branches of the latter marked "partly-open throttle" and "wide-open throttle" come together at a common zero. This will be explained later. A third supply of

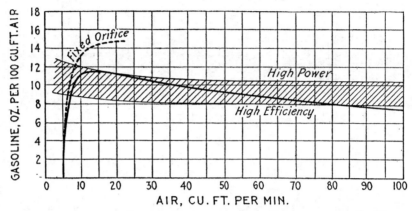

Fig. 7—Diagram for Simple Carbureter with Compensating Air Valve

gasoline is drawn from the gravity well already referred to and is discharged at a point on the engine side of the throttle. This is also essentially a constant quantity and its effect on the quality is indicated by the curve marked "throttle by-pass" in Fig. 3.

When the throttle is nearly closed the by-pass comes into action and draws gasoline from the gravity well, but when the throttle is wide-open and the engine is slowed down by additional load the

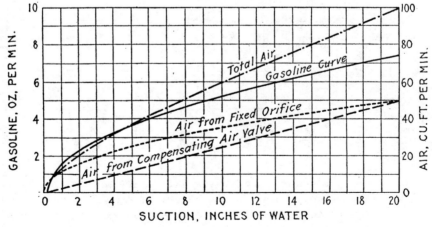

Fig. 8—Flows for Simple Carbureter with Compensating Air Valve

decreasing suction of the engine permits the well to gradually fill with gasoline, thus decreasing the gravity head in the float-chamber and causing the quantity discharged into the well to decrease. The effect

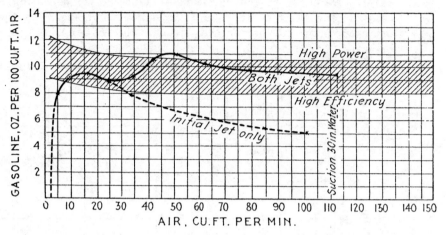

Fig. 9—Diagram for Actual Air-Valve Carbureter with Supplemental Gasoline Jet

of this smaller quantity of gasoline drawn from the atmospheric well is shown on that part of the quality curve marked "wide-open throttle." The other branch, marked "partly-open throttle," indicates the

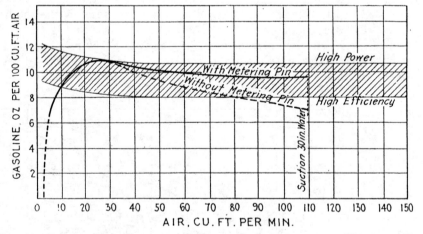

Fig. 10—Diagram for Actual Air-Valve Carbureter with Metering Pin Attached to Air Valve

quality when, by progressively closing the throttle, the by-pass comes into action and prevents the level in the well from rising.

With this arrangement then the carbureter has two distinct quality curves as shown by the full lines. This completes the analysis of this class of carbureter. It is a small class and no further types will be considered.

AIR-VALVE CARBURETERS

This class is much larger than all others combined and is represented by a large number of types. Only a few of the most important will be considered.

The air-valve carbureter will be best understood by first making an analysis of the simplest form, in which all the gasoline is admitted through a fixed orifice and all the air through a valve closed by a spring. This is an imaginary type, which could not be used in practice.

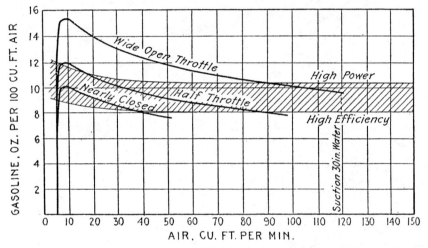

FIG. 11—DIAGRAM FOR ACTUAL AIR-VALVE CARBURETER WITH NEEDLE VALVE CONTROLLED BY THROTTLE

In Fig. 5 the quality curve does not follow the desirable zone, but crosses it at an angle. The range of capacity through which the mixture would fire is thus very limited. The analysis of this quality diagram will be found in Fig. 6.

Comparing the quality diagram, Fig. 5, with that of Fig. 1, it is evident that a combination of the two systems of control would make a better quality diagram than either. This suggests the type of air-valve carbureter that **will** be **next** considered.

Simple Carbureter with Compensating Air-Valve and Fixed Orifice

This type is well known in a variety of forms. The fixed air-orifice is often called the "strangle tube" or the "Venturi choke," and ordinarily the gasoline nozzle terminates in this passage. It will be assumed in this, as in the preceding cases, that the gasoline is regulated by a fixed orifice, or in other words, that the area of the gasoline orifice remains unchanged. A second air passage is closed by a spring-opposed air-valve, which opens as the suction increases.

The three dotted curves in Fig. 8 represent quantities of air passing through the air passages with different degrees of vacuum in the manifold. The lower dotted curve indicates the quantities admitted by the progressively-opening air-valve. The next represents the quantities admitted through the fixed air-orifice and the upper dotted curve is the sum of these quantities and is therefore the total air. The full-

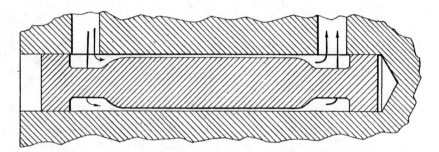

FIG. 12—DEVICE DESIGNED FOR MAGNIFYING LIQUID FRICTION

line curve indicates the quantities of gasoline admitted through the fixed gasoline orifice.

In Fig. 7 the dotted line shows what the quality would have been if the air-valve had been prevented from opening, and the full line indicates the quality when the air-valve is free to open in response to increasing suction.

Air-Valve Carbureter With Supplemental Gasoline Jet

Various means have been used to cause a second gasoline jet to come into action before reaching maximum capacity, thus augmenting the supply when it has become deficient. If the second jet comes into service abruptly, it causes the quality diagram to make abrupt changes of quality. This is shown in Fig. 9, which actually represents the action of a well known carbureter of this type. The dotted line shows what the quality would have been with the first jet only, and the added richness caused by the second jet is shown by the full line.

Carbureter With Metering Pin Attached to Air-Valve

Fig. 10 shows how the effect sought with the double jet (shown in Fig. 9) is produced by a taper metering pin attached to the air-valve and arranged so that when the air orifice is increased by opening the air-valve, the gasoline orifice is simultaneously increased by withdrawing the tapering pin.

With this device a fluid dashpot attached to the valve is necessary to prevent objectionable reciprocating motion of the metering pin. These moving parts attached to the air-valve prevent a delicacy of action that is desirable when small quantities are being used. The forces acting on the valve are consequently weak. To overcome this difficulty and give stability of action it has been found necessary to make the fixed air-orifice larger than that in the simple carbureter, thus making it possible to keep the air-valve in contact with the seat until the actuating forces shall have become of some magnitude. The effect of this is to make it impossible (see Fig. 10) to enrich a small quantity of air. Under these conditions the closing of the throttle for slow speed reduces the engine speed, partly by reducing the quantity

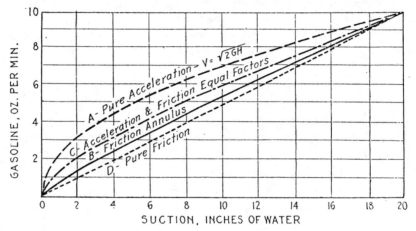

Fig. 13—Relation of Liquid Friction to Flow of Gasoline

of mixture entering the cylinder and partly by decreasing the richness of the mixture. This results in an extremely lean mixture for idling and slow running, which fires with uncertainty and is susceptible to the cold.

Carbureter With Needle-Valve Controlled by Cam on Throttle

Fig. 11 shows quality diagrams for the best known carbureter of this type. It also illustrates the effect produced by this method of control. These curves show that in each case the needle-valve remained in one position throughout, which means that the throttle-valve also remained in one position and the speed was changed by changing the load.

If the load at each speed is such that the necessary position of the throttle causes the needle-valve to supply the amount of fuel required by the air that is being used, then a normal quality of mixture is supplied, but under all other conditions the mixture is either too lean or too rich.

To understand this clearly, refer to Fig. 11, which shows that when 100 cu. ft. of air is being used, the quality of the mixture is 10. Assuming now that the throttle remains wide-open, and the speed is reduced by increasing the load (as it would be in climbing a hill) the quality grows steadily richer and the engine presently ceases to fire because of excessive richness. This condition is always found when the gasoline orifice is varied by opening and closing the throttle.

FRICTION IN CONTROLLING FLOW OF LIQUIDS

So far the quantity of gasoline flowing through a fixed orifice has been assumed to vary as the square root of the suction or head. The curve representing these quantities is also the curve representing the velocity of a mass in equal increments of time when acted on by a constant force, such as the force of gravitation, and the law is called the "law of acceleration." The friction of liquid flowing through an orifice is therefore a negligible quantity. The inertia of the liquid opposing acceleration through the orifice is practically the only factor to be considered and the flow can be called an "acceleration flow." Liquid friction can, however, be so magnified relatively to the inertia of the liquid as to become the dominating factor. The shape of the curve is

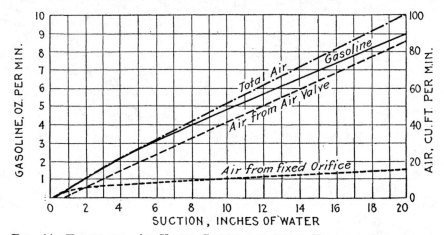

Fig. 14—Flows for Air-Valve Carbureter with Friction Control of Gasoline

then greatly changed because the resistance of friction varies directly as the velocity and not as the square of the velocity. Fig. 12 represents a device for magnifying liquid friction relatively to the resistance of its inertia. The liquid passes through a thin annulus of relatively large diameter. The thin liquid tube encounters friction on a large surface, both inside and outside. Because of the high friction thus developed, the cross-sectional area of the annulus must be much larger for a given quantity than would be required with an orifice flow, and this decreases the velocity and consequently the effect of inertia. The rounded corners of the entrance to the annulus eliminate a resistance called "entrance head."

The flow of gasoline is represented by the curves in Fig. 13. The curve A represents the discharge through an orifice. The curve D connecting the ends of A represents the discharge that would take place if friction were the only resistance to the flow. The flow would then increase directly as the head. This condition is, of course, impracticable. Curve B represents the actual discharge when the flow is resisted by a device such as is shown in Fig. 12. The curve C passes midway between curves A and D. All curves between A and C can be called curves of "acceleration flow," because in every case acceleration is the dominating factor; curves between C and D can be called curves of "friction flow," because in every case friction is the dominating factor.

Air-Valve Carbureter With Friction Gasoline Control

Fig. 14 shows the relation of air and gasoline flow to suction for a carbureter using friction control of gasoline. This analysis follows the same general plan that was followed on the preceding pages, and the resulting quality diagram is given in Fig. 15.

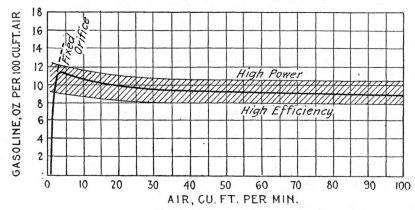

FIG. 15—DIAGRAM FOR AIR-VALVE CARBURETER WITH FRICTION CONTROL OF GASOLINE

It is evident from Fig. 14 that for use with a straight-line air curve, a gasoline flow represented by curve B, Fig. 13, is more desirable than that represented by curve A. This is true because the quantities of gasoline discharged by a fixed orifice are found (Fig. 6) along a curve, while the quantities of air discharged by a spring-opposed valve are found along a straight line, or nearly so. As a result a constant ratio of air and gasoline cannot be maintained.

COMPARISONS OF QUALITY DIAGRAMS FROM ACTUAL CARBURETERS

The preceding pages have been devoted mainly to the theoretical diagrams relating to several types of carbureter. It is now proposed to investigate the diagrams showing results actually produced by these instruments. Comparisons will be made as to: uniformity of quality;

maximum capacity (for high speed); minimum capacity (for slow speed); and working range.

A desirable diagram is one that runs parallel to and does not cross the zone of best quality. If the quality diagram indicates rich or lean mixtures, trouble will result under varying atmospheric conditions.

The diagrams were all made from 1¼-in. carbureters. This is the size generally used on engines of from 275 to 300 cu. in. piston displacement. Such engines, if properly designed, require for maximum power at high speed a carbureter capacity of about 150 cu. ft. of air per minute with a manifold vacuum of 30 in. of water.

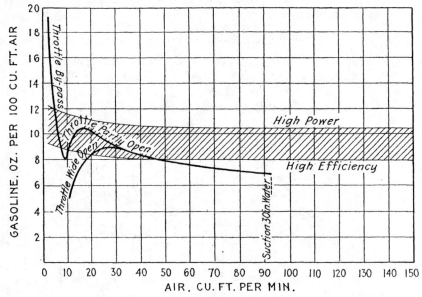

Fig. 16—Diagram for Actual Compensated Fixed-Orifice Carbureter

The maximum capacity means little unless considered in connection with the minimum capacity because a carbureter with limited working range can show large capacity, when good performance is sacrificed at slow speed.

To obtain the best slow running and idling, these engines should be supplied with less than 4 cu. ft. of air per minute, and this must be enriched nearly to the maximum. A lean mixture is not desirable when idling as the engine is then susceptible to changes of temperature; an excessively rich mixture is also faulty in that it makes soot on the valves and spark-plugs and loads up the intake passages.

For the purpose of comparison a uniform standard has been adopted for the working range. In each case the maximum capacity is determined by measuring the quantity of air the carbureter delivers with a manifold vacuum of 30 in. of water. The minimum capacity is located at the point where the curve passes out of the zone of best quality.

Compensated Fixed-Orifice Carbureter

The quality curve, Fig. 16, crosses the zone three times. At high speed the quality is too lean for high power. At 16 to 20 ft. the quality is too rich for high mileage. At 8 ft. there is a lean spot, and at 3 to 4 ft. the mixture is too rich for good idling. The working range is from 4½ to 92 ft., or a ratio of 1 to 20. The capacity is too small for high power.

Simple Carbureter With Compensating Air-Valve

The quality curve, Fig. 17, crosses the zone twice. At 90 ft. the quality is too lean for high power. From 10 to 30 ft. it is too rich for high mileage. The working range is from 6 to 90, or 1 to 15. The capacity is too little for high power (even if the quality be correct).

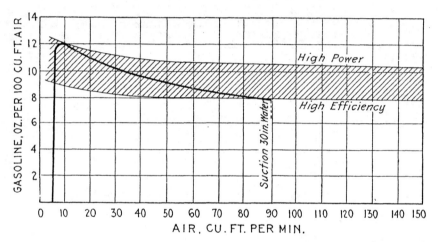

Fig. 17—Diagram for Actual Simple Carbureter with Compensating Air Valve

Air-Valve Carbureter With Supplemental Gasoline Jet

The quality, Fig. 9, zigzags badly, and while good at 114 ft., is much too rich at 50 ft. and too lean at anything less than 8 ft. The power would be fairly good at high speed, but when using from 40 to 60 ft. of air, the mixture would be much too rich for good performance. The mileage would be low and there would undoubtedly be "loading" in the intake passages. The idling and slow running could not be good as the minimum amount of air enriched is not small enough. The working range is from 8 to 114 ft. or 1 to 14.

Air-Valve Carbureter With Metering Pin Attached to Air Valve

The quality, Fig. 10, does not, in this case, stagger across the zone and for part of its range is correct for high power. At from 20 to 40 ft. (where most of the ordinary running occurs) the quality is too rich for good mileage and "loading" will occur. This can be corrected by a different adjustment, but it must be done at a further sacrifice at slow speed, which is already faulty.

The mixture goes out of the zone at 10 ft. and would go out at 15 to 20 ft. if the quality were made efficient at 25 ft. The working range is from 10 to 110 ft., or 1 to 11, and would be much less if an efficient mixture were used at ordinary speeds.

Air-Valve Carbureter With Needle-Valve Controlled By Throttle

The quality, Fig. 11, of the mixture with this carbureter depends on the position of the throttle-valve. If the throttle is skillfully handled, the quality stays well within the zone. If, however, the throttle

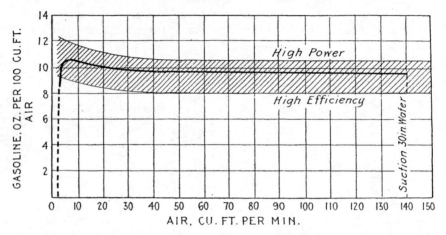

Fig. 18—Diagram for Actual Air-Valve Carbureter with Friction Control of Gasoline

is not handled well, the quality will go outside the zone, being both too rich and too lean.

The working range is fairly good—5½ to 120 ft., or 1 to 22. The maximum power will be good and the idling fairly good, but this instrument will load badly when pulling at medium or slow speeds with open throttle.

Air-Valve Carbureter With Friction Control of Gasoline

The quality curve in Fig. 18 follows parallel to the zone of quality throughout its entire range. The working range is from 3 to 140 ft., or 1 to 47. In view of this remarkable showing, it is unfortunate that fluctuations in temperature make so great a variation in the flow of gasoline that the device is impracticable.

Two-Stage Carbureter

(The description of this carbureter is given later.)

The quality curve, Fig. 19, is divided into two parts. Most of the running is done on the part marked "primary stage." The curve follows along the efficient edge of the zone, giving high mileage, and has a low minimum capacity, giving good idling and slow running. When both stages are in use the quality is rich enough for maximum power and speed. The two stages are used only for maximum power and will give that power without affecting the mileage obtained at ordinary speeds with the primary stage.

The working range is from 3 to 140 ft., or 1 to 47, and a still greater range can be used successfully, if desired, by increasing the capacity of the second stage. Since the primary stage is independent of the second stage, it is not affected by adjusting the latter.

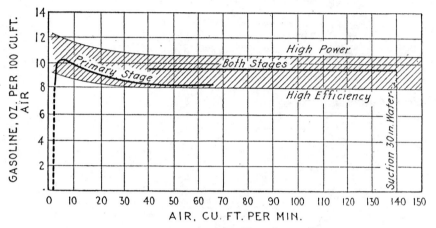

Fig. 19—Diagram for Actual Two-Stage Carbureter

CONSTRUCTION OF TWO-STAGE CARBURETER

Many of the difficulties inherent in the types described can be minimized by dividing the carbureter into two stages. Such a device is illustrated in Fig. 20, in which *2* is the primary fixed-orifice for air, and *3* is the gasoline jet located as usual in this passage. A spring-opposed valve *4* controls the flow of air through the valve passage to the mixing chamber. These parts, when connected to a gasoline supply and an outlet to an engine, constitute a simple air-valve and fixed-orifice carbureter, called the "primary stage" of the instrument. Associated with this in the structure is the air passage *5* containing the gasoline jet *6*. These parts, when in action, constitute an elemental fixed-orifice carbureter, such as has already been described. This fixed-orifice carbureter is the "second stage" and generally has more than half the total capacity. Normally the second stage is closed by the butterfly valve *7* and held closed by a spring. A connection to the throttle is so arranged that when the throttle is nearly open, the final full opening throws the valve *7* wide-open, thereby more than doubling the carbureter capacity.

Vacuum Device Gives Quick Acceleration at Slow Speeds

This carbureter has a device, actuated by fluctuations of the vacuum in the manifold, that discharges a predetermined quantity of gasoline whenever the throttle is suddenly opened at slow speed. This consists of the plunger *8* having on its upper end an extension *9* which acts as a piston to move the plunger under the influence of the fluctuations of the manifold vacuum communicated through the passage *10*. The plunger is fitted loosely in a cylindrical chamber having a restricted passage at the bottom communicating with the float-chamber so that the level of gasoline in the plunger-chamber is maintained at the float-chamber level. The plunger-chamber has an atmospheric opening *11*, and a passage *12*, to the mixing-chamber.

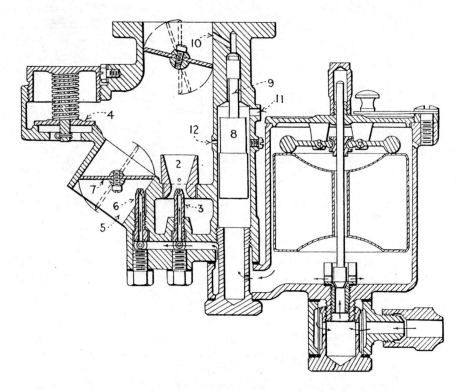

Fig. 20—Sectional View of Two-Stage Carbureter

The operation is as follows: When the throttle is nearly closed, the vacuum in the manifold lifts the plunger to the position shown, and the space below the plunger fills with gasoline. The device is now ready for action. A sudden opening of the throttle breaks the vacuum in the manifold, which releases the plunger and it drops by gravity, causing the gasoline to pass up to the space above the plunger, where it is swept into the mixing-chamber by the air entering through the passages *11* and *12*. This operation is repeated as often as the throttle is suddenly opened from a nearly closed position.

Advantages of Two-Stage Instrument

The working range of a two-stage instrument should be greater than that of a single stage because it is the summing up of the ranges of both stages. The atomizing of the primary stage must necessarily be high, even when the total capacity of the instrument is great, because the dividing of a carbureter into two equal stages makes the

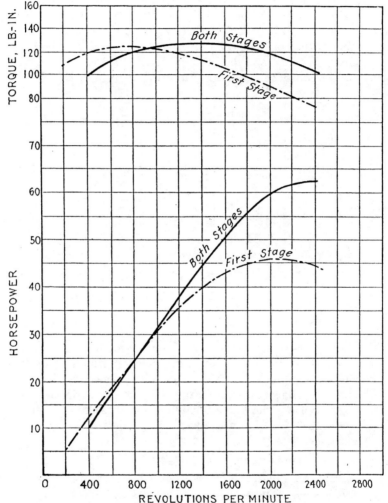

FIG. 21—HORSEPOWER AND TORQUE CURVES FOR TWO-STAGE CARBURETER

atomizing energy of the primary stage four times as great as with the full capacity. This high atomizing facilitates the pick-up to such an extent that it is fairly good practice to make the capacity of the primary stage so small that no pick-up device is necessary. All things considered, however, it is better to divide the two stages more evenly and incorporate a pick-up device.

The two-stage instrument is peculiarly adapted to obtaining high fuel efficiency, as will be seen by referring to Fig. 19. In this case the setting of the primary stage is such that the mixture is lean. The high atomizing of this stage and the pick-up discharge make this practicable, the result being high fuel efficiency.

Under these conditions, the setting of the second stage is made for a rich mixture so that when it is brought into service for maximum power, the resulting quality from the combined jets will be rich enough for maximum torque.

Inasmuch as more than 90 per cent of the running of the engine is done with the primary stage, the fuel efficiency is practically determined by the quality of mixture used in this stage. The quality diagram of the primary stage differs somewhat from those of the

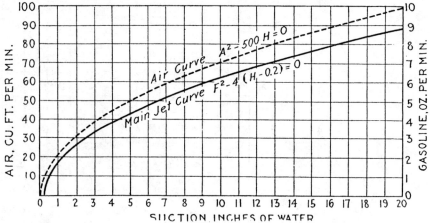

FIG. 22—AIR AND GASOLINE FLOWS WITH DIFFERENT MAIN JETS

simple air-valve carbureters. This is explained by the restricted air passage, or throat, under the air-valve, which becomes a fixed-orifice control of air when the valve has opened slightly. The air-valve simply bridges over from the primary Venturi throat to the largest fixed orifice under the air-valve. This arrangement limits the capacity and could not be used with a single-stage carbureter of normal capacity.

Power and Torque With Two-Stage Carbureter

Fig. 21 illustrates one of the effects of dividing a carbureter into two stages. The torque curve shows this most clearly. At about 900 r.p.m. the torque is the same whether the single stage or both stages are in service. At this speed the loss of atomizing of the large capacity offsets the gain due to low resistance. With the primary stage the gain due to higher atomizing is offset by the higher resistance. Below this speed the primary stage shows the higher torque because of the superior atomizing, and above that speed, the lower resistance of the two stages results in the higher torque. A proper use of the two stages produces in effect a flat torque curve.

DISCUSSION

V. R. HEFTLER:—In the first part of the paper certain theoretical assumptions are made, which, as stated, are not generally held. The opinion* was presented by Prof. K. Rummel and was contrary to the view of A. Krebs.† Krebs held that the air flow follows a parabolic law, and that the gasoline flow folows the same law, with a slight lag at the start. He attributed the lag to the slight difference in head between the constant level and the tip of the nozzle. Krebs determined the lag experimentally and found the corrective term to be 21 mm. when the nozzle was not 21 mm. above the level.

If the law, upon which all the curves in the paper are based, is true it would be interesting to know how it has been established. The new opinion could easily be tested by eliminating the disturbing influence caused by the nozzle being too high above the gasoline level. We could readily make an experimental carbureter with the nozzle, say ⅛ in. below the level. Such a carbureter would be impractical

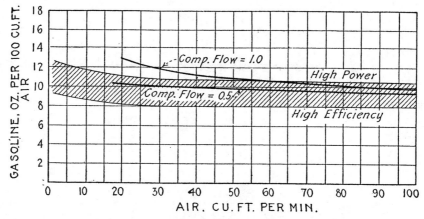

FIG. 23—QUALITY DIAGRAM PRODUCED WITH DIFFERENT COMPENSATING JETS

for an automobile, because it would spill gasoline while at rest, but it could be used to determine why the gasoline curve does not follow the air curve.

JET DESIGN BETTERS PERFORMANCE

But even taking the premises used in the paper in analyzing the various carbureters, it is possible, at least in one case, to arrive at a different conclusion. I refer now to Figs. 3 and 4 for the compensated fixed-orifice carbureter, which is described as follows: "A second supply is drawn from a well open to the atmosphere and supplied with a constant quantity of gasoline from the float-chamber by gravity." This is quoted from a patent granted to Baverey, and therefore refers to only one carbureter (the Zenith).

---
*See *Horseless Age*, Apr. 7, 1915, p. 474.
†See *Horseless Age*, Apr. 14, 1915, p. 508.

I have considered the parabolas used in making the curves in Fig. 4 and find that the air curve is accurately represented by the equation $A^2 - 500 H = 0$, where A is the air flow and H the suction. The fuel curve of the main jet is represented by the equation $F^2 - 2.316 (H - 0.2) = 0$, where F is the fuel flow. These equations (plotted in Fig. 22) can be verified easily.

I have assumed that the flow through the compensating jet follows a more complex law; that it is constant and gives a quantity of fuel of 1 oz. per minute, provided the suction is greater than 2 in. With lower suctions the flow through the jet is not constant. We shall see later whether these assumptions are justified.

Fig. 4 of the paper shows a tendency for the mixture to become too lean at high suctions, while it is nearly correct at low suctions. The remedy for this is obviously to increase the size of the main jet,

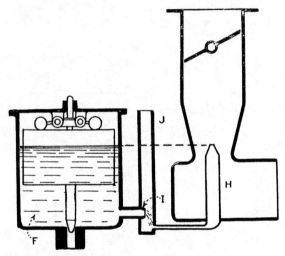

FIG. 24—ILLUSTRATING OPERATING PRINCIPLE OF ZENITH CARBURETER

that is, change the parameter in the main jet equation. For instance, make $F^2 - 4 (H - 0.2) = 0$. With the same compensating jet at 20 cu. ft. of air per minute, the quality would be .13 oz. gasoline per 100 cu. ft. air. At 100 cu. ft. per minute it would be 9.9, so I must use a smaller compensating jet. I assume one represented by $C = 0.5$, giving only ½ oz. gasoline per minute. The quality varies from 10.5 for 20 cu. ft. to 9.4 for 100 cu. ft. What a difference! The curves in Fig. 23 show the quality curves with $C = 1.0$ and with $C = 0.5$, the flow of fuel through the compensating jet then being respectively 1.0 and 0.5 oz. gasoline per 100 cu. ft. of air.

The assumption has been made that the compensating flow is constant only when the suction is 2 in. of water. It is possible, however, to have constant flow at much lower suctions. This can be made clear by referring to Fig. 24, which shows a constant level fuel chamber F, a well J open to the atmosphere, a fuel orifice I below the level, and a connecting channel terminating in this nozzle H. The suction at which the flow through the compensating jet I becomes constant depends on the areas of H and I and on the height between H and the connecting channel. At exceedingly low suctions, the fuel drops in the well J, and a suction is soon reached at which the air passes through the channel from I to H. From that time on the flow through the compensating orifice is practically constant. The value of this constant-flow suction depends on the instrument. In the Zenith carbureter constant flow is reached at a suction much lower than 2 in. of water. The fair average would be 1.1 in. of water, so that we have there one reason why it is legitimate to continue the quality curve of the main and compensated jets nearer the origin of coordinates.

There is another reason why it can be continued still nearer. It is stated in the paper that "when the throttle is wide-open and the engine is slowed down by additional load, the decreasing suction of the engine permits the well to gradually fill with gasoline, thus decreasing the gravity head in the float-chamber and causing the quantity discharged into the well to decrease." This is a peculiar point, hardly known outside our organization, and I do not wonder that an error has been made. After the suction is reached at which air passes from I to H, we have no gasoline in the channel, but instead an emulsion, which is much lighter than gasoline. Less suction is required to hold and carry this emulsion up in the well, so that if the critical suction is 1.1 in. of water, going up, it does not follow that the well again fills when the same suction is reached, going down; this kind of hysteresis effect causes the flow from the compensating jet to remain constant for suctions much lower than have evidently been assumed.

To sum up, I have taken the premises under which the theoretical analysis has been made in the paper, and have shown that, by simply changing the parameters, which is equivalent to changing the sizes of the jets, I arrive at results quite different. I question, therefore, whether by changing the jets it would not have been possible to obtain with the actual carbureter entirely different and much more satisfactory results.

FRANK H. BALL:—If the curves were based on theory, it would be debatable whether the same law governs the flow of gasoline and air through fixed orifices, but inasmuch as the flow of both have been repeatedly measured through the range covered in carbureter practice and have always been found to correspond approximately to the law in question, it cannot be attacked without refusing to admit the accuracy of the measurements that have been made.

Mr. Heftler suggests that as an experiment the level might be carried at the overflow point in the discharge nozzle, or even higher. That is just what has been done, and the result absolutely confirms the theory regarding such a condition. When the level was at the overflow point the quality diagram was a straight horizontal line through the whole length of the quality zone. When the level was slightly higher than the discharge nozzle, the resulting quality diagram curved upward at the slow-speed end; the diagram curved downward at that end when the level was slightly below the overflow point. This, of course, was a fixed-orifice carbureter and had no air-valve.

A successful valveless carbureter was made on this plan by having the discharge nozzle terminate in a shallow cup that prevented the gasoline from wasting when not in service. When in service an air-stream was made to sweep the gasoline from this cup. With an adjustable gasoline level, a perfect quality diagram could be obtained and the quality was under full control down to zero quantity.

Referring now to Mr. Heftler's discussion of the quality diagram, Fig. 3, it is quite true that a smaller compensating jet and a larger main jet would make the diagram more nearly horizontal between 20 and 100 cu. ft. of air, but they would also make the lean spot at 10 ft. still leaner. If this was corrected by changing the by-pass adjust-

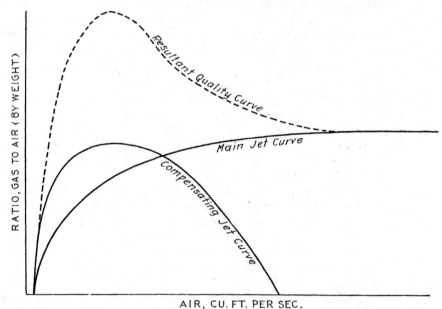

Fig. 25—Quality Curves for Carbureters with Two Variables in Parallel

ment, the by-pass curve would be moved bodily in a horizontal direction, so that the quality would be too rich for idling. The adjustment from which the diagram in Fig. 16 is made is assumed to be a fair compromise, particularly as it is the setting used regularly by a prominent automobile manufacturer with this type of carbureter.

ACTION OF PLAIN-TUBE INSTRUMENT WITH SIMPLE JET

L. ARNSON:—I would like to call attention to a so-called plain-tube carbureter with a simple jet that reverses the usual action of the simple jet mentioned in the paper; I refer to the carbureter developed in the last eight years by George Longuemare, Paris.

The correct quality curve of a carbureter is in the form of a rectangular hyperbola. Mr. Heftler and Mr. Ball in his paper have

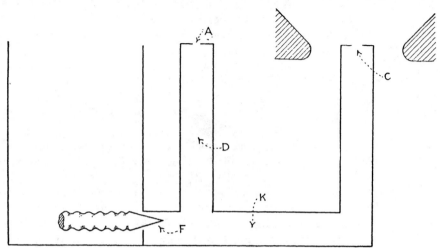

Fig. 26—Illustrating Operating Principle of Longuemare Carbureter

shown this with two variables (main jet and compensating jet) in parallel, producing the upper curve in Fig. 25. In the Zenith carbureter two variables are in parallel. Longuemare has worked on a different principle in that two variables are in series. This principle is illustrated in Fig. 26. At the end of the jet is a restriction or

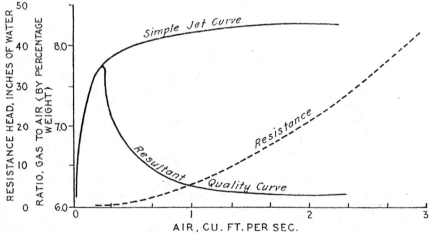

Fig. 27—Quality Curves for Longuemare Carbureter (Two Variables in Series)

calibrated orifice C, and also a fuel orifice F, in connection with a large channel K and air orifice A. The action of this carbureter is that when the suction below begins to act on the jet F the chamber D empties into the channel K. Air bubbles then pass through the channel into the jet C. The amount of the infusion of air into the channel is approximately proportional to the speed. The two variables in series are the suction head produced in the channel and the resistance at C, varying with the velocity of the fuel. With increased speed the resistance is increased, because instead of a perfect gas an emulsion lighter than gasoline and heavier than air is passing through K. The resulting velocities are abnormal and do not increase in proportion to the resistance, but at a slightly greater rate.

The equation for the flow of fuel from a jet emptied upon a column of air is

$$V_f^2 = V_a^2 \frac{d_a}{d_f} - 2gh$$

The square of the fuel velocity is equal to the square of the air velocity times the air density divided by the fuel density minus $2gh$, the head effect due to the difference of the jet and fluid level, viscosity, etc. If d_f is reduced as we are reducing it by the emulsion of air, we can get extremely high velocities, and consequently an increased resistance, the tendency of which is to decrease the flow of fuel from the nozzle and at the same time produce an emulsion in the passages. In that way, the mixture is enriched in a simple jet at starting (which is absolutely the reverse of the action of an ordinary simple jet) and becomes leaner with increased velocity. We have, therefore, a simple jet that is the reverse of the jet mentioned by Mr. Ball and that in addition has the necessary characteristics for pick-up without the use of a moving part.

FUEL CONSUMPTION OF TWO-STAGE CARBURETER

The high efficiency limit, Fig. 18, is on normal speeds from about 50 to 140 cu. ft. of air per minute. The air/gas ratio is apparently 15.3 by weight. In Fig. 19 Mr. Ball shows what appears to be a quality curve of a two-stage carbureter. According to this curve when both stages are working the air/gas ratio is 12 at speeds of from 40 to 140 cu. ft. of air per minute. Any design produced in our laboratory showing such a consumption of fuel would be speedily abandoned. We would not permit an instrument to go out at any stage of these speeds with an air/gas ratio less than 15; it must be more than 16.8 with the commercial fuel now available.

FREDERICK O. BALL:—A year or so ago we tested the Longuemare type of compensation. It does compensate, but has the same fault as all fixed-orifice carbureters; the quality falls off to zero at too great a quantity of air. The minimum is not low enough and, as is true of all fixed-orifice carbureters, other means must be used to obtain idling and slow running.

MATHEMATICS UNTRUSTWORTHY

C. P. GRIMES:—We should feel indebted to Mr. Ball for producing a carbureter that, according to his paper, has a working range of from 1 to 47. Previous to my present professional connection I spent almost 3 years in what I think to be the finest carbureter research laboratory in the United States, and never found, in testing some 40 carbureters, one with such a range. As a rule they had not half that range.

I have yet to find a case where the truth regarding carburetion is expressed by a mathematical statement. My experiments covered Venturis of various shapes, exits of 8 or 10 different angles, 6 or 8 entrance radii and different exit lengths. I found that each time the Venturi, the exit, or the approach was changed the curve of the air flowing through this Venturi was also changed. I tried jets up to $5/8$ in. long and could not get consistent results. As the gasoline comes up in the nozzle and enters the zone of high suction, the more volatile parts of the hydrocarbons are vaporized. When perhaps half way through the nozzle they burst and fly apart and cause a disagreeable hissing noise. This flying apart, I believe, upsets all mathematical theories relating to the flow.

I have obtained curves similar to those shown in Fig. 2 of the paper. These seem to show that the making of a carbureter is a very simple matter; in reality a curve showing the area necessary to maintain a uniform mixture ratio is very irregular. With a jet $5/8$ in. long, for instance, the curve will indicate lean mixtures at lower speeds. But if the jet is shortened the curve will be much more uniform. I made some interesting experiments on the height of the gasoline in the carbureter, and found that the depth of the gasoline in the nozzle has a great deal to do with keeping the curve uniform.

I believe that a carbureter giving a little richer mixture at low speeds and a leaner one at high speed is preferable. This point is illustrated in Fig. 3. The intensity of combustion depends upon the closeness of the association of the particles. To obtain uniform explosion pressure a less rich mixture is used at the higher speeds. By compressing fewer molecules closer together, a good power curve is obtained.

FREDERICK O. BALL:—We have endeavored to avoid mathematics as much as possible, because a formula does not mean much unless it corresponds to actual practice. The tests we have made show that in many cases the flow of gasoline and air follow practically the same law. As the fuel gets heavier the curve of flow will be different from the air curve. The different carbureters tested have been taken just as they were used on cars by the manufacturers, or by the individual users.

QUALITY CURVES HAVE SAWTOOTH CONTOUR

F. C. MOCK:—The mathematics and the theory used in connection with carburetion have been founded mostly upon research made years ago with the fuel then common. They are based on the assumption that a uniform mixture of fuel and air gives ideal performance. As

DISCUSSION OF CARBURETER INVESTIGATIONS

a matter of fact, the days have gone by, probably never to return, when a uniform mixture can operate an automobile engine under all the temperatures of customary use. If it were possible to take a moving picture record of the mixture proportion, in the same form as the curves in the paper, of one of the best carbureters, the resulting curve would have a sawtooth contour, with variations much greater than any shown by Mr. Ball. These would be necessary in order that the car might give its best performance. This is because the present type of motor-car engine, with ignition by single electric spark, and high speed with consequent short time for flame propagation, will not use efficiently any part of the fuel except that which is in vapor form. The same economic causes that have resulted in the present high price of gasoline have so changed its nature that not all of it evaporates; and further the amount that can evaporate changes continually during the varying conditions of average driving.

TABLE I—PROPERTIES OF AUTOMOBILE-ENGINE FUELS*

Name	Chemical Composition	Density, Deg. Baume	Boiling Point, Deg. Fahr.	Specific Heat	Latent Heat of Evaporation, Cal. per Kg.
Pentane	C_5H_{12}	87	97	0.52	85
Hexane	C_6H_{14}	81	156	0.50	80
Heptane	C_7H_{16}	70	208	0.48	80
Octane	C_8H_{18}	65	257	0.51	81
Nonane	C_9H_{20}	61	302	0.50	......
Decane	$C_{10}H_{22}$	58	342	0.50	84
Undecane	$C_{11}H_{24}$	55	383	0.50	......
Tetradecane	$C_{14}H_{30}$	46	485	0.50	......
		Sp. Gr.			
Benzol	C_6H_6	0.880	176	0.34	93
Ethyl alcohol	C_2H_6O	0.792	173	0.60	220
Methyl alcohol	CH_4O	0.796	150	0.60	284

*Computed values are based on Landolt and Bornstein Physico-Chemical Tables and nternational Annual Physical Tables.

VAPORIZING DEPENDS ON FUEL PROPERTIES

In general we can say: 1. More of the fuel is vaporized at closed than at open throttle; 2. More of the fuel is vaporized when the engine is warm than when it is cold; 3. More of the fuel reaches the cylinders in vapor form when the air velocities are high in the carbureter and in the manifold. The vaporizing conditions change constantly during the operation of a motor car, so that the quality curve of a carbureter giving flexibility and power, along with maximum fuel efficiency, must have an irregular contour.

To find the underlying reasons and the laws governing these conditions we must examine the properties of the fuel that concern its use in the engine. These properties, as given in Tables I and II, cover completely, I believe, the requirements of successful engine operation, so far as a fuel supply system is concerned. Further, they explain definitely many peculiar things noticed in engine and carbureter performance.

Petroleum is an irregular natural blend of a number of elements of similar molecular structure, which grade in order as to density, volatility, etc. Table I covers the range of these elements that have been used in gasoline, past and present. The gasoline of the early days was simply the liquid portion of the petroleum that was too volatile to be incorporated safely in the mixture sold as kerosene, and therefore was to a considerable extent a by-product. It consisted mainly of pentane, hexane and heptane. Before the war the fuel used in Europe was also largely of this character. As the demand

TABLE II—PROPERTIES OF AUTOMOBILE-ENGINE FUELS

Name	Chemical Composition	Percentage in Gasoline		Proportion, Vapor to Air, for Perfect Combustion		Critical Temp., Deg. F., of Vapor Saturation			Initial Temp.* Deg. F. of Air to Vaporize Fuel Charge
		Present (1915)	Early	By Weights	By Volumes	Full Cyl. Charge	2/3 Cyl. Charge	1/3 Cyl. Charge	
Pentane	C_5H_{12}	....	10	0.065	0.0272	0	0	0	....
Hexane	C_6H_{14}	3	45	0.066	0.0220	0	0	0	....
Heptane	C_7H_{16}	12	35	0.066	....	....	....	....	....
Octane	C_8H_{18}	30	10	0.066	0.0170	68	54	33	98
Nonane	C_9H_{20}	35	....	....	....	....	....	....	....
Decane†	$C_{10}H_{22}$	15	....	0.066	0.0136	109	95	72	152
Undecane	$C_{11}H_{24}$	5	....	....	....	....	....	....	....
Benzol	C_6H_6	....	....	0.075	0.028	61	59	56	114
Ethyl alcohol	C_2H_6O	....	....	0.111	0.070	57	41	24	257
Methyl alcohol	CH_4O	....	....	0.155	0.140	65	52	36	384

*Initial temperature of gasoline 90 deg. F.

†Loss of density of air charge for complete vaporization equals 10 per cent (90 deg. F. equals 100 per cent).

for motor-car fuel increased it was necessary to use, year by year, more and more of the petroleum elements, with the result that to-day gasoline runs from octane to undecane and contains scarcely a trace of its earlier ingredients. Tetradecane is an element typical of what is now sold as kerosene; its molecule is heavy and complex.

A molecule of tetradecane can be broken to form one molecule of hexane and one of heptane, with one atom of carbon left over. This is practically what is done by the Burton and Rittman cracking processes, which, under the combination of high temperature and high pressure, with no oxygen present, turn the heavier petroleum elements into lighter ones, with a carbon residue. This process can ta, place in cylinders, when we try to run the engine cold with a rich and largely unvaporized mixture and is often the cause of the troublesome carbon deposits on cylinder walls and spark-plugs.

In Table II are shown some properties of the fuel elements that concern more closely their adaptation to use in the engine. The percentage of the different elements varies throughout the country, but those given are typical of the fuel available in the central and eastern States. The early gasoline was almost entirely different in composition from that of the present day. European gasoline (petrol) is similar to our early gasoline, but contains more octane.

The proportions *by weight* of fuel vapor to air for perfect chemical combustion vary but little, as the actual quantities of hydrocarbon must be the same in all cases. The proportion *by volume* of vapor to air, however, decreases markedly as we come to the heavier gasoline elements. This is because the molecules are heavier and therefore fewer are needed to make a given weight; according to an elementary law of physics equal volumes of gases (at the same temperature) contain equal numbers of molecules, so that for the same weight of the heavier elements the volumes must be less as the molecules are heavier.

A recent bulletin of the Bureau of Mines on Inflammability of Mixtures of Gasoline Vapor and Air indicates that the limits of inflammability with an electric spark depend more upon the ratio of volumes than upon the ratio of weights and that the lower limit is around 0.014. This limit would exclude decane and undecane, which are important components in practically all the gasoline sold to-day. It was found, of course, that under the heat of compression the limits of inflammability were extended. This simply emphasizes the fact we know from experience, that the proportion of fuel needed is based on the amount vaporized in the manifold rather than upon the amount fed from the carbureter.

ACTION OF CRITICAL TEMPERATURES

The most important part of these tables is that giving the critical temperature necessary to allow the existence of enough vapor for a running mixture. The maximum density of the vapor of any liquid is determined and limited by the temperature. This density is called the saturation point and beyond it no more vapor can be gotten into a given space; any compression will result in the condensation of enough of the vapor so that the density is the same as before. Thus each temperature gives a different density of saturation; conversely, there is a different saturation temperature for each density.

What has happened to engine fuel is this: A point has been reached where at ordinary operating temperatures the vapors of some of the elements of gasoline cannot be made dense enough to furnish their full share of the running mixture. The proportion of the running mixture that any element can furnish in vapor form depends on the temperature and the density of charge required. For instance, referring to octane, to furnish a vapor sufficiently dense to carburet a full charge requires a vapor or mixture temperature of 68 deg. F. If the throttle is partly closed, however, so that only two-thirds of an air charge is taken in, of course only two-thirds of the vapor charge is needed, or two-thirds the density, which can then be obtained at 54 deg. F. At one-third of a cylinder-charge, which corresponds to a speed of 20 m.p.h. on a smooth city street, the necessary vapor can be obtained at 33 deg. F.

A certain drop in temperature takes place while the gasoline changes from a liquid to a vapor state. This is usually compensated for by preheating the air. With octane the air charge must be preheated 30 deg. F. What actually happens under driving conditions is this: Driving at 20 m.p.h. (closed-throttle or nearly closed-throttle position) the 30-per cent octane in the gasoline becomes vaporized if the air taken in is at 63 (33 + 30) deg. F. But if the driver opens the throttle to accelerate, although more gasoline will be admitted through the carbureter, no more octane can be vaporized at the temperature, and it requires instead 98 (68 + 30) deg. temperature before it will become vaporized in sufficient quantity to give the full (wide-open) throttle mixture.

For decane much higher temperatures are required. The mixture temperature for wide-open throttle is 109 deg. F. The drop of temperature for vaporization is 43 deg. F., so that a wide-open throttle charge requires an initial air temperature of 152 deg, or a closed-throttle temperature of 115 deg. In many cars the air-intake temperature during the eight colder months of the year is not much over 120 deg. As a result the drivers find that although they can run on an economical setting, that is, using all the components of the gasoline, at closed throttle, at wide-open throttle the engine will not fire. The gasoline orifice must then be opened so that at wide-open throttle enough fuel for a firing mixture is obtained from the heptane, octane and part of the nonane elements of the gasoline. The remainder of the nonane, decane and undecane, although fed from the carbureter, is in the form of liquid or drops in the intake manifold and is practically wasted.

These are the reasons for what is generally observed in practice, that we have to use a richer mixture for wide-open than for closed-throttle positions, and why wide-open throttle operation is much more sensitive to temperature than closed-throttle operation.

OPERATING DIFFICULTIES WITH HEAVY FUELS

There are other disadvantages in using heavier fuels. Preheating the air involves expanding it and decreasing its density, with the result that for a given suction less air charge is taken into the engine. Even if the necessary vapor temperature be maintained without preheating the air, there will be a certain loss in density as long as the air is taken in along with and through the same passage as the gasoline vapor. In the case of decane the loss from preheating the air amounts to 10 per cent, which shows that we have, even with our present fuel, reached a point where to get flexible operation (with a uniform-mixture carbureter at least) it is necessary to sacrifice volumetric efficiency. With kerosene the mixture temperature is so high that this loss in charge volume is serious indeed, making necessary larger piston displacements for the same horsepower output. I believe that this fact alone, exclusive of the difficulty of maintaining a sufficiently high temperature, makes it impossible to use kerosene below tetradecane in motor cars of the present type.

Tables I and II show the fuel properties for benzol, ethyl (grain or sugar beet), alcohol, and methyl or wood alcohol. Benzol grades in quality slightly better than octane and would be a good substitute for gasoline for motor-car use, if it could be obtained. It has the disadvantage that it freezes.

The alcohols will run on low mixture-temperatures, but their latent heat of vaporization is high. This, in conjunction with the fact that more than twice as much alcohol as gasoline is needed for a running mixture, makes it necessary to preheat the air to a high temperature; in fact, to a point hardly obtainable in a motor car. Even an exhaust-jacketed intake-manifold will not furnish sufficient heat under average closed-throttle driving. A high thermal efficiency is obtainable with alcohol on account of the fact that high compressions can be used with a corresponding increase in the efficiency of the heat cycle. At all times the volumetric proportion required of fuel to air is more than double that of the gasoline hydrocarbons, and the consumption per horsepower hour is always greater.

C. P. GRIMES:—At present engine manufacturers are confronted with a serious problem, caused by some carbureting devices. It has been found that after putting clear lubricating oil in the engines and running them for several hundred miles, perhaps 10 per cent of the fluid in the crankcase is gasoline or kerosene. Mr. Mock explained this action. The heat is not sufficient to fracture the fuel or at least to maintain it as a vapor.

ATOMIZING AND VAPORIZING FUELS

E. H. SHERBONDY:—We ordinarily think of a carbureter as a device separate from the complete engine. As a matter of fact, the intake manifold and what we call the carbureter serve separate functions in preparing the combustible mixture for use in the engine cylinder. Fundamentally, all the carbureters in this day of heavy fuels act only as weighing devices, giving a certain weight of fuel to a certain weight of air. This fuel must be atomized. The atomization requires a certain amount of energy. In almost all carbureters the arrangement for atomizing is very poor. A butterfly throttle-valve is placed immediately above the fuel nozzle and is partly closed. The fuel is thereby thrown out of the mixture, resulting in bad distribution.

Following atomization, we attempt to vaporize. Some vaporization starts with light fuel immediately at the fuel nozzle, but about 80 per cent of the fuel that goes into the cylinders is vaporized not in the manifolds but in the engine cylinders.

The engines as designed to-day are arranged so that the fuel has to be carried up through the valves. A system arranged so that the fuel travels from the fuel nozzle through the intake manifold and valves in a downward direction would assist the travel of the fuel greatly. Then if the engine itself were arranged to perform the function of atomization, the fuel would be ready, after passing into the cylinder, to be vaporized by the heat of compression. In the Diesel engine, which does not require an ignition device, the fuel is introduced by spraying under high pressure into the cylinder and is immediately vaporized.

VAPORIZATION INSTANTANEOUS

Vaporization is often thought of as requiring a relatively long time to take place. For instance, if we take a square foot of surface and wet it with gasoline, we observe that the gasoline apparently vaporizes slowly until it is all gone. Vaporization appears then to require a time interval, but this is not the case. The particles on the top of the surface change spontaneously from a liquid to a vapor condition. Absolute proof that vaporization is instantaneous is found in the indicator card of the Diesel engine. If any period were required for vaporization, the curve would drop.

I think that the future of the automobile industry hinges on the development not of carbureters but engines. So far a weighing device has been developed, but the carbureter engineer has dealt with only a small portion of the problem of carburetion. He has been greatly handicapped, largely by the construction of the engine.

LAW OF FLOW THROUGH AIR-VALVE

GEO. W. SMITH, JR:—Fig. 8 of Mr. Ball's paper shows that the air supplied by an air-valve weighted with a spring is directly proportional to the head. The area of the opening varies directly as the head and the velocity varies directly as the square root of the head. Therefore, the flow should be represented by a curve instead of by a straight line. According to my explanation, the flow of air varies as the three-halves power.

FREDERICK O. BALL:—We have obtained with certain carbureters a curve of air that was made to coincide with the curve of gasoline, the two being identical throughout the full range. We have also succeeded in getting an annulus with enough friction that the curve of the flow of gasoline is almost a straight line.

DISCUSSION OF CARBURETER INVESTIGATIONS

ANDREW AUBLE:—On one particular economy run a normal mileage was obtained on the first half gallon of gasoline, and on the next or last half gallon more than double the mileage was obtained. I believe this was due to the fact that the engine became hot enough to evaporate the vapor before the explosion took place in the cylinder. When the engine was heated to a point where I felt it was receiving fuel in an evaporated form, it was much more efficient, the economy being brought to a point hardly thought possible. When the gas reaches a high temperature the engine temperature decreases automatically. My experience with an air-cooled engine was that it would become heated to a certain point, and then with continuous running eventually become cooler and remain so in mile after mile of low-gear testing. I believe this was due to the fuel being evaporated instead of vaporized.

FUEL TRAVEL IN TWO-STAGE CARBURETER

E. H. SHERBONDY:—The device shown in Fig. 20 of Mr. Ball's paper has a gasoline passage from the float-chamber feeding to the two nozzles. When operating normally under low throttle, the fuel would be discharged through nozzle 3. For fuel to be discharged from that nozzle compression must take place in the gasoline passage, thus tending to throw the fuel through the second nozzle, even though the throttle valve 7 be closed.

FREDERICK O. BALL:—The gasoline level is lower in the second nozzle than in the float-chamber, owing to the resistance of the passage between the primary nozzle and the float-chamber. If this resistance were not kept large, all the gasoline would be drawn from the second-stage nozzle and air would go in that way, destroying the action of the primary carbureter.

We recently made a demonstration of the flow of gasoline through an orifice and a friction annulus. The apparatus used consisted of a chamber with a float to maintain a constant level. The chamber was mounted so that it could be raised or lowered, thus changing the head on two discharge openings. The discharge openings were at the same level and connected with the float-chamber. One discharge opening had a fixed orifice and the other a friction annulus similar to that shown in Fig. 12.

The float-chamber was located so that the gasoline level was about 1 in. above the discharge nozzles. The amount of gasoline flowing through both nozzles was then adjusted so that the flow was equal as measured by graduates arranged to catch the discharge. Next the float-chamber was raised so as to make the head about 20 in. If the same law of flow applied the discharge from both openings would be equal at the higher head, but the discharge from the friction annulus was really more than twice that from the orifice.

The amounts of gasoline discharged at 1 in. head and 20 in. head are in the same ratio as the air the carbureter must deliver to keep the mixture correct. If two carbureters, one with a regular orifice and the other with the friction annulus, have the same capacity at 1 in. suction in the mixing chamber, then at 20 in. suction, since the friction annulus delivers twice as much gasoline as the orifice, it can also deliver twice as much air; in other words, the friction-annulus carbureter covers a wide range of capacity with small range of suction compared to the orifice carbureter.

SERVICE CONDITIONS NOT CONSIDERED

P. S. TICE:—The authors are to be congratulated on the ingenious graphic method employed to present the outstanding characteristics of compensation in the several well known types of carbureters. The diagrams, however, have seemingly been constructed largely from data obtained under conditions materially different from those obtaining in service on engines. Therefore the diagrams are apt to be misleading. It is regrettable that more space was not devoted to a description of the apparatus and a summary of the methods employed in its use. A case in point is presented by Figs. 1 and 2 of the paper and the statements that accompany them. It is assumed apparently that the flows of air and fuel are caused by identical heads or differences of pressure. In the type of carbureter (elemental fixed-orifice) here considered, this condition is impossible of realization. Not only is the mean head causing efflux of liquid always greater than the corresponding mean head causing the flow of air, but also the two do not bear a fixed ratio, in that the head upon the fuel orifice increases at a higher rate than does that impelling the air flow. A further distinction arises in a consideration of the relative quantities of each fluid delivered under given mean heads in that the densities of the two are widely separated, and the discharges do not therefore bear the same relationship under different conditions of installation, even though the mean heads remain identical in value or relationship. Thus it follows that if any carbureter, of any type, be set or fitted to any engine, an investigation of that carbureter when removed from the conditions of its installation on that particular engine will be valueless as a statement of its characteristics, even for purposes of comparison with other carbureters investigated with it under like experimental conditions.

A practical carbureter is not merely one of its given nominal size with hard and fast characteristics displayed under any and all conditions; but it is a Packard carbureter, or a Ford carbureter, or a Hupmobile carbureter, and will display its true characteristics only upon investigation conducted on its home lot, as it were. If the depressions and velocities induced in the carbureter were constant at a given engine speed, the comparison of characteristics would be simple and not involve the engine and its type. Then and then only could the admirably simple and controllable ejector be made to give usable readings of heads and quantities.

DISCUSSION OF CARBURETER INVESTIGATIONS

DISCHARGE THROUGH ORIFICES

In those portions of the paper dealing with the flow of fuel through an orifice reference is had to the factors acceleration, inertia and friction. In carbureter practice it is insufficient to call the fuel passage an orifice, simply, and to assume that the discharge from it will or can vary directly as the square root of the head. With the sizes of fuel orifice permissible in common carbureter practice, it is, practically speaking, a physical impossibility even to approach a discharge directly proportional to the square root of the head. With any permissible orifice having length, that is other than a true sharp edge without approach, the coefficient in the expression $V = C \sqrt{2gh}$ increases with the head, the rate of increase being controlled in a general way by the ratio of the length to the diameter of the orifice.

I am thoroughly in accord with the authors in the matter of the importance of atomization of the fuel. It is safe to say that it is impossible, under the conditions and with the values to be worked with in carburetion, to carry this phase of the carbureter's performances too far—provided, of course, that its other functions are not curtailed.

The atomizing at other than full throttle, or before the second-stage butterfly valve is well opened, can be assisted by the air drawn through nozzle *6* and discharged with fuel through nozzle *3*. This action would take place unless the common fuel passage to both nozzles were extremely large and the outlet orifice of nozzle *3* correspondingly small. The pick-up device is clever, susceptible of control in fitting, and should be of great value with certain types of engines.

INDIANA SECTION PAPERS

A STANDARD OF CARBURETER PERFORMANCE

By Prof. O. C. Berry

(Non-Member)

Carbureter performance is often reported in terms of engine performance. This may serve the purpose when it is desired to compare the merits of different carbureters for use on a given engine, but it does not throw much light upon the performance of the carbureter itself. As a standard of carburetion, it is far from satisfactory. We ought to establish a standard of carbureter performance that would be expressed in terms of the ability of a carbureter to perform those functions for which it was designed.

Opinions may differ as to the proper division of the requirements placed upon the carbureter and those placed upon the intake manifold of the engine, as both influence the quality of the carburetion. For this reason no attempt will be made to enumerate all of the functions that a carbureter should perform. We will probably agree that usually: (1) It must be a good mixing device; (2) It must avoid decreasing the volumetric efficiency of the engine; (3) It must supply the engine with the best mixture of fuel and air for each speed and load at which the engine is capable of running.

(1) To be a good mixing device, the carbureter should cause the proportion of fuel and air to be the same in every part of the charge supplied to the engine. In order to do this to the best advantage as much of the fuel as possible must be converted into a gas, and the remainder must be so finely divided that it will float along in the air in the form of a fog. To gasify the fuel, heat must be added. This can be done by heating the charge of air, or by causing the fuel to impinge against a hot plate. Atomizing the fuel also helps in gasifying it. One of the best ways to accomplish this is to introduce it into a stream of rapidly moving air so that the directions of flow of the fuel and air will be nearly at right angles to each other. After the fuel has been properly atomized and gasified, it must be thoroughly and uniformly mixed with the charge of air.

(2) An ideally perfect carbureter will supply the mixture at the engine throttle at atmospheric pressure and unheated. Under practical working conditions it is obviously impossible to produce a carbureter for our present commercial gasoline that can meet this requirement perfectly, as condition (1) must also be met. There must be an appreciable vacuum in the body of the carbureter in order to produce the air velocities necessary to atomize the fuel, and

the addition of heat to the mixture is required in order to gasify the fuel sufficiently to get satisfactory performance, especially in the winter season. Because of both the vacuum in the carbureter and the heated mixture, the volumetric efficiency of the engine will be decreased.

The best carbureter that can be produced from this point of view will therefore be the one maintaining a sufficient vacuum to atomize the fuel properly and at the same time adding just enough heat to gasify the mixture. The amount of vacuum required will depend

FIG. 1—VIEW OF TESTING LABORATORY AT PURDUE UNIVERSITY

partly upon the scheme used in atomizing and partly upon the character of the fuel. The amount of heat required will also depend partly upon the way in which it is applied and partly upon the character of the fuel. More and better experimental data are needed in order to determine how to atomize the fuel to the best advantage and how to apply the heat; also the pressure difference and the amount of heat necessary to produce the desired results with any given fuel.

PROPER FUEL AND AIR MIXTURES

(3) There is some difference of opinion as to just what the character of the mixture should be, and how it should vary with the speed and load of the engine. Is it true that a comparatively rich

mixture is needed for proper performance when the engine is running slowly and unloaded, and that a lean mixture is needed for speeding? Or does the engine require the same mixture at all speeds and all loads? Is the mixture that will give the best efficiency the same as the one that will give the best power? If so, what is this mixture? If not, how much do the two mixtures differ, what are they in each case, and how much will the powers and efficiencies differ for these two mixtures? What is the maximum range of mixtures that can be used successfully in an engine to produce regular firing, and how will this range vary with the speed and load of the engine?

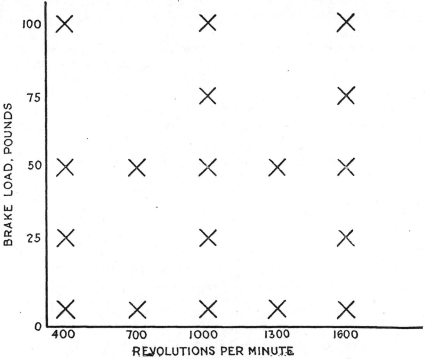

Fig. 2—Speeds and Loads at Which Tests Were Made

How will the power and efficiency vary as the character of the mixture changes?

Some experiments have been completed recently in the laboratories of Purdue University in an attempt to answer these questions. The tests were run on a Haynes "Light Six" engine connected to a Diehl electric dynamometer. The set-up is shown in Fig. 1. The gasoline was piped from the gasoline tank A into a 2-qt. vessel placed on one of the scale pans. In order to give the balances freedom of motion, the gasoline was siphoned from this vessel to the carbureter. The balances B were capable of weighing the gasoline to the one-hundredth part of an ounce, and were equipped with electrical contacts so that when they came to a balance they would start a stopwatch, a revolution-counter on the end of the engine shaft and would ring a gong on the air meter. This made it possible to take all of the readings at the same time.

DESCRIPTION OF TESTS

The air was measured through an Emco No. 4 gas meter C, reading in cubic feet, so that tenths of a cubic foot could be estimated. The meter was connected to the carbureter by rubber tubing, which was kept from collapsing by a coil of wire inside of it. The barometer was read, together with the temperature of the air to and from the meter, the pressure drop through the meter in inches of water, and the wet and dry bulb thermometers on a hygrometer. In this way the amount of air could be measured to within 1 per cent. The air leaving the meter was maintained at a temperature of about 90 deg. F., thus insuring sufficient heat in the mixture.

The supply of cooling water for the engine in tank D was kept between 125 and 135 deg. F. It was caused to circulate through the engine jackets by the pump regularly supplied with and driven by the engine.

The speed of the engine was read on a tachometer, as well as being computed from the stop-watch and revolution-counter readings, giving a good check on this important factor. The brake load on the engine was read by means of a sensitive set of Fairbanks scales, thus making it possible to compute the power developed by the engine to a satisfactory degree of accuracy.

Tests were run at the different speeds and loads indicated by the crosses on Fig. 2. The maximum brake load that the engine could maintain is about 100 lb., so that the loads were about zero, one-quarter, one-half, three-quarters and full. At the speed and load represented by each one of the crosses a set of tests was run in which the mixture of fuel and air was varied while the throttle opening was kept constant, and the brake load was so varied as to keep the speed constant. This was accomplished as follows: The carbureter was adjusted so as to give a mixture yielding good power and the dynamometer (F) and engine throttle were then adjusted to give the desired load and speed. For example, a load of 50 lb. and a speed of 1300 r.p.m. were selected to obtain the curves shown in Fig. 3. A test was then run at this speed, load and mixture, and the weight of gasoline per pound of dry air in the mixture, the power developed by the engine and efficiency computed. This gave the first point on the curves in Fig. 3.

The mixture was then made slightly richer by opening the gasoline needle, the throttle opening remaining the same, and the brake load being adjusted to bring the speed back to 1300 r.p.m. A second test was made under these conditions, and a point on each of these curves determined. This was repeated until the mixture finally became so rich that the engine would miss and could carry but a fraction of its original load. The mixture was then brought back to approximately the original point and a set of tests run with increasingly leaner mixtures until the engine could no longer perform properly. In this way the power and efficiency curves, Fig. 3, were completed. A similar set of tests was run at each of the speeds and loads indicated in Fig. 2, and curves similar to Fig. 3 were drawn in each case.

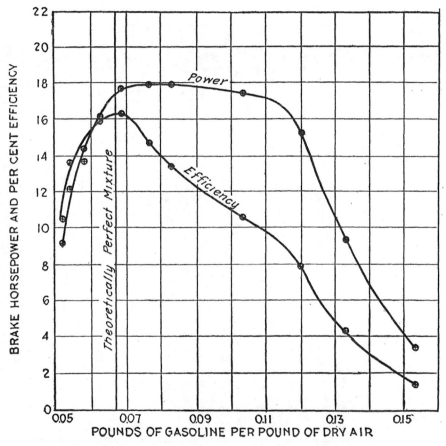

Fig. 3—Engine Performance Curves at 1300 R.P.M.

The vertical line in Fig. 3 (at 0.0671 lb.) represents the theoretically perfect mixture of fuel and air, or the one in which there is just enough oxygen in the air to burn the fuel and no excess of either fuel or air exists. The curve shows that the engine will run with a mixture of less than 0.055 lb. of gasoline per pound of air, but will not pull well with so lean a mixture. As more fuel is added the power will increase rapidly until nearly full power is reached, when the curve becomes almost horizontal, increasing slowly to a maximum, then decreasing slowly for a time, but finally reaching a point where it falls off rapidly. The richest mixture with which we were able to run the engine was 0.155 lb., or nearly three times as rich as the leanest mixture. The engine will carry nearly full load with a mixture as lean as 0.065, or as rich as 0.115 lb. In other words, a carbureter can be adjusted with as lean a mixture as can be used to carry full load, and the amount of gasoline can be nearly doubled, without greatly affecting the power capacity of the engine.

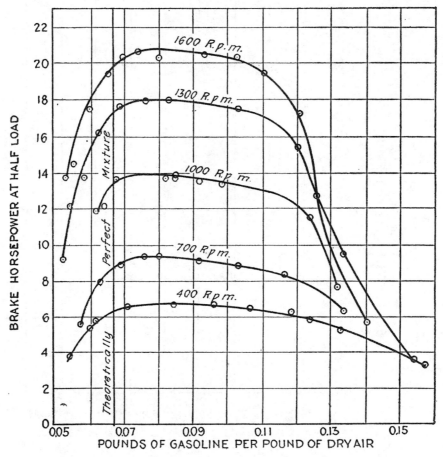

FIG. 4—PERFORMANCE CURVES AT HALF LOAD WITH DIFFERENT SPEEDS

POINTS OF HIGHEST EFFICIENCY AND POWER

It is practically impossible to stand by the side of an engine mounted on a test block and distinguish any difference whatsoever in its performance as the mixture is being changed through this range. A listener standing with his back to the engine could not tell at any point within these limits whether the mixture was being made richer or leaner. The effect of this change upon the efficiency of the engine is very different. The point of highest efficiency seems to fall almost exactly on the point of the theoretically perfect mixture. As the mixture is made richer than this the efficiency will decrease even while the power is increasing slightly, and will decrease rapidly after the point is reached when the power is also decreasing. Thus the point of highest efficiency comes with a mixture of 0.0671 lb., while the point of highest power is found near 0.08 lb. The efficiency of 16.2 per cent will be recognized as good indeed for a gasoline engine running at half load.

In Fig. 4 is shown a set of these curves taken at about half load, but at different speeds from 400 to 1600 r.p.m. They show that the mixture for maximum power is not noticeably affected by the speed, but that at high speeds the engine cannot hold up its power with quite so much excess fuel as at lower speeds. With leaner mixtures the power holds up about equally well at all speeds. The efficiency curves are not given in this figure but the tests show that the highest efficiency is developed with the theoretically perfect mixture in each case.

Fig. 5 shows a set of these curves for a speed of 1000 r.p.m., but for about one-quarter, one-half, three-quarters and full load. These curves show that at full load a slightly wider range of mixtures can be used than at light loads. The point of best power comes at about 0.08 lb. in each case, however, so that the curves simply show that with an open throttle, and a consequently higher compression, a mixture can be exploded which is a little further removed from normal than is the case with the closed throttle. The efficiency curves are not shown in this figure, but the point of highest efficiency comes at about 0.0671 lb. in each case.

EFFECT OF MANIFOLD CONSTRUCTION

The point may well be raised that the engine used had a wet manifold, and that the curves here shown are therefore not theoretically perfect. As a matter of fact the engine was chosen for its fine reputation for being able to fire regularly under adverse conditions in service. It upheld its reputation with us in a satisfactory manner. The intake manifold was entirely inside of the block casting, and was heated by the jacket water, as well as by the exhaust gases as they passed out to the exhaust manifold. We felt that the manifold construction was good, and that even a perfectly dry manifold could not have changed our results appreciably.

These curves offer an explanation of why it is that in two cars of the same make, both having good power, good acceleration and smooth running engines, one can travel 10 miles on a gallon of gasoline and the other 15 miles. One will have the carbureter adjusted so as to get a powerful, but lean and therefore efficient mixture, while the other carbureter will be adjusted to give too rich a mixture. This fact will suggest a rule to follow when adjusting a carbureter. Decrease the quantity of gasoline until the engine loses power and then increase it slowly until good power is restored, but not a notch beyond this point.

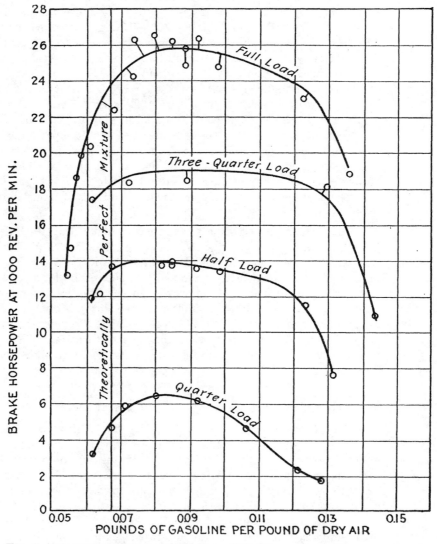

Fig. 5—Brake Horsepower Curves at 1000 R.P.M. with Different Loads

There are numerous carbureters on the market which are not capable of much adjustment, but which will run on cold or on warm days equally well. They may enable the engine to start easily and still hit every cylinder after it is warmed up. They accomplish this in spite of the fact that a greater weight of cold than of warm air can pass through a given orifice, while the increased viscosity of the gasoline will cause less cold than warm gasoline to pass. Thus with the same carbureter adjustment the mixture is necessarily leaner when the fuel and air are cold than it is after they have become warmed up. This seemingly good practical performance of the simple or so-called fool-proof carbureters is due to the wide limits between which the mixture is capable of giving good power. The efficiency of these carbureters is low, for when they are adjusted so that the mixture is rich enough to start the engine readily, it will necessarily be too rich for the best efficiency after the engine becomes warm. This same line of reasoning shows the value of having a good means of adjusting the carbureter placed where it can be reached from the driving seat while the car is in motion.

AUTHOR'S CONCLUSIONS

A wide range of mixtures of gasoline vapor and air can be used in an engine to give regular firing in all of the cylinders. At half load and mid-speed on an automobile engine any mixture between 0.055 and 0.155 lb. of gasoline per pound of dry air can be fired, and practically full power will be developed between the limits of 0.065 and 0.115 lb. The highest power will be developed with a mixture of about 0.080, while the best efficiency will accompany the theoretically perfect mixture of 0.0671 lb. A change of engine speed apparently does not affect the mixture at which the best power or the best efficiency will be realized, or the ability of the engine to use the leaner mixture, but at the higher speeds a smaller amount of excess gasoline can be used successfully. A change in load does not noticeably affect the points of highest power with reference to mixture ratio or efficiency, but at the higher loads a wider range of extreme mixtures can be used. The engine can run idle with as wide a range of mixtures as it can use when pulling a good load.

The set of tests described is only one step in the large task of mastering the subject of the proper carburetion of petroleum fuels. Purdue University plans to continue its investigations, and it is hoped that other experimental laboratories of the country will join in publishing their accumulation of information upon this subject and enable us as a nation to get the most out of the use of our limited supply of fuel oils.

DISCUSSION

A. P. BRUSH:—Will Professor Berry state why the maximum power output is not coincident with the theoretically perfect mixture? Also, at what point in the enriching of the mixture does soot begin to appear on the spark-plug?

PROF. O. C. BERRY:—Two factors tend to enable the richer mixtures to give higher power than does the theoretically perfect mixture. In the first place the fuel is probably never quite uniformly mixed

DISCUSSION OF PERFORMANCE

with the air, so that when the total air and gasoline are in proper proportion, some parts of the mixture are too lean. Then when a rich mixture is used the hydrogen tends to burn in preference to the carbon, leaving a deposit of carbon in the cylinder. The effect of this upon the power of the engine may be understood by analogy to the case where methane, CH_4, is burned. To burn 1 cu. ft. of methane requires 9.56 cu. ft. of dry standard air, and will produce 1072 B.t.u. To burn the hydrogen out of 2 cu. ft. of methane will require the same amount of air, but will produce 1392 B.t.u. Thus more power can be produced by the second method of burning.

My opinion is that the sooting is in almost exact proportion to the excess of fuel used, but that the effect in depositing carbon on the plugs will be much more noticeable when the mixture is rich enough to cause the engine to "miss" occasionally.

A. P. BRUSH:—All of us have seen engines that run in service and on the dynamometer without visible carbon deposit on the spark-plugs. I would like to determine the point in the mixture at which discoloration from carbon begins to occur. It would then show us the practical running range of mixtures.

We might define the practical running range of mixtures as from the theoretically perfect mixture to that point at which carbon begins to be deposited. It would be a valuable contribution if such a point could be determined before Professor Berry's apparatus is dismantled.

USE OF LOW-GRADE FUEL

J. W. ESTERLINE:—We have made an extensive investigation dealing with the use of low-grade fuels in automobiles, and that part of the investigation which was assigned to me was the burning of fuels in the form of a gas, rather than in the form of a vapor. We ran several cars on kerosene and fuel oil, however, and had the novel experience of going around the Speedway at 50 m.p.h. on some crankcase oil that a racing driver left in one of the cans in the pit.

We succeeded in obtaining a device that gave sufficient gas at a uniform pressure, at sufficiently low speed and at sufficiently high speed to operate the engine under practically all conditions of running. The device was simple, devoid of moving parts and could be regulated in an excellent manner. One car was run 3200 miles and the cylinders were clean at the expiration of the test and appeared to be well polished inside. There was no residue left either in the cylinders or in the gas generators.

We had no difficulty in getting the mixture to ignite. There was no smoke and no smell. The engine efficiency was higher when running on gas mixed with air than when running on a vapor from a modern carbureter. On one test we selected a sample of gasoline with a given number of thermal units per pound and a sample of kerosene with the same heat value. The gasoline and the kerosene were both used in the same engine, the gasoline being admitted through a modern carbureter and the kerosene through the gas generator. Tests showed that the efficiency on the kerosene was from 15 to 20 per cent higher than on the same engine running on gasoline and with the carbureter most advantageously adjusted.

In most of our work we utilized the exhaust heat from the engine for producing the gas. The only problem met was that of getting a production of gas at sufficiently low speed and an amount of gas proportional to the requirements of the engine at all times. This was successfully done, however, and the engine operated without difficulty.

The temperatures of the gas used were quite high, and our results showed that no doubt it would be necessary to design the engine to burn fuels at such temperatures.

We drove four different cars a total of seven or eight thousand miles, and reduced the operation of the apparatus to a point where it was quite reliable. An engine running on a gas produced in this manner is powerful, snappy and flexible and is at all times clean, provided the proper amount of air is mixed with the gas.

F. E. MOSKOVICS:—Was any attempt made to analyze the exhaust for various mixtures?

PROF. O. C. BERRY:—They were not analyzed, as these chemical analyses require considerable time, which was not available.

HEAT IS BEST REMEDY

V. R. HEFTLER:—Mr. Mock started his paper by recommending heat and small manifolds. The question of small manifolds is one that is very interesting. In our work we have found it advisable to have a ready means of determining gas velocities. We have found what we call the nomograph a convenient means of making calculations. This is based on the slide rule and the scales are slide-rule scales. They are extended or contracted according to what exponent of the quantity is used and are shifted one way or another according to what factor is to be used in multiplying that particular quantity.

I do not know what Mr. Mock's idea of a small manifold is—that is, how small can it be? That is a matter of compromise, but I would like to know where in his opinion it is wise to stop. I suppose the gas velocities depend on the sort of manifold. With a dry manifold the size can be greater. A wet manifold must be a very small one if we want to get good action. We have found in our experience that heat is a remedy for most carburetion ills.

If heat is applied to the manifold, the mixture range can be much closer to the ideal of smaller range, or we have high power and economy. If heat is not applied the mixture is wet. If we do not want to adjust the carbureter constantly, we must keep the mixture rich enough so that the lighter elements will be present in large enough quantity to keep the engine running under all conditions. Heat will therefore help economy. It will have its effect on the condition that exists when the throttle is opened. If a manifold is heated, the wet condition described will not obtain to such an extent; on opening the throttle, there will be more gas in the manifold. With a good manifold the throttle can be opened as quickly as desired and there will be no bark.

This explanation of what happens when the throttle is opened disposes of the wrong theory that for acceleration a rich mixture is needed. I believe Mr. Mock's explanation sets forth what we need. We may need more gasoline from the carbureter but we need the same mixture in the engine. That we need more gasoline at the carbureter to keep the same mixture at the cylinder is due to the time necessary to pass from a dry to a wet manifold. This will not happen if the manifold is properly heated. With such a manifold and with one not too large, acceleration can be obtained without any sacrifice of economy.

FREDERICK PURDY:—A manifold takes the mixture from the carbureter, which we assume has delivered at the throttle, or a little back of it, a proper mixture for the load at which the engine is running. It is my impression that what we really want all the time is a theoretically perfect mixture or as near it as we can get. It is quite absurd to assume that the mixture that gives the highest mean effective pressure is the mixture that we want all the time.

There are many curves inside the manifold. We cannot make them straight and feed all the cylinders. They cannot always have an even rise and an even length. If the manifold is not uniform throughout its length, the gas velocities must change rapidly, no matter whether the car speed is accelerated or retarded.

Mr. Esterline called attention to the fact that in a carbureter or in the device he used he got higher efficiency. I attribute that to the fact a gas was used. I saw some beautiful experiments made in Joseph Tracy's laboratory a couple of years ago on the burning of fuel oils; a portion of the oil was burned and the flame put out, but this portion raised the temperature of the remaining fuel to a point where a gas resulted, because the chemical composition of the material was changed.

NECESSITY OF ATOMIZATION

It has been said that a vapor was intended to be used in the original combustion engine. I believe the original one was developed to use a manufactured gas. A vapor is substantially a gas in which the particles are exceedingly small; the smaller they are the more nearly possible it will be to carry them when air is not present. For the same reason, fine dust will float, because of its own surface tension and the quantity of air it carries on its surface; the finer the gasoline is broken up, the more it will be subject to the movement of the air. The larger they are, the less subject they are to the movement of the air, so the more they will be affected by inertia. The best atomizing carbureter is desirable, largely for this reason, it seems to me. The finer the atomization, the more nearly gas will respond to the changes air is subject to, and that is the crux of this situation.

I am not an advocate of extreme temperature, because I do not like to reduce the volumetric efficiency of the engine. In 1906, the bigger the engine, the better the car was considered to be, and the more we got for it, but conditions are different to-day. We use the smallest engine that will drive the car. The engine must be made just as efficient as possible. There is no advantage in reducing the volumetric efficiency in order to use gas at a high temperature. The economy will be higher when a higher temperature is used, because the gas will be vaporized and will be in almost a molecular condition.

We can only add a certain quantity of gasoline to the air and have it evaporate. That quantity at moderate temperatures is not sufficient to make a satisfactory running mixture so that a wet manifold is required. The object of a small manifold is to keep the speed so high that gasoline will be carried along the walls of the manifold at a uniform rate. The only reason for using a smaller manifold is because it is wet. It will not take up any more of the gasoline because it is running fast. If the air is passing fast enough over a body it will carry the gasoline along in small units, so that it will feed into the cylinder at a uniform rate.

The smaller manifold is not altogether the answer to the problem. I cannot say what the quantity of gasoline is that will be carried to the manifold in suspension. When the engine is cold, the air going into it is colder and heavier. The volume taken in is not greater, but the weight is, and it is pounds that count, not cubic feet. When the engine is cold, more fuel is needed. The air weighs more, so that the mixture is actually thin when it gets to the engine. In going over a mountain, the volume or cubic feet of air is the same, but the air is of lighter weight and the fuel supply must be decreased to balance it. The situation is reversed when the air and engine are cold. If the manifold is smooth enough, the mixture will be carried along the side more uniformly. Vents in the carbureter provide means for changing the speed of air slowly so that the inertia of the mixture will not stop it. If the mixture in the manifold is subject to changes in speed, and the elements are not thoroughly evaporated, the speeds must be changed so slowly that each element of the mixture can be kept up to the other. The heavy element will lag behind, otherwise. In some carbureters this is taken care of with a dashpot and some take care of it in other ways. If the direction of movement of the mixture is changed less rapidly, there will be a good deal less engine trouble. All the good engines that seldom miss are built with manifolds so that there is not such a rapid change in the speed or direction of movement of the gas.

MORE INTIMATE MIXTURE REQUIRED

A. P. BRUSH:—We have been told manifolds should be kept small, so they can handle liquid; that they should be heated, so there will not be any liquid. Is one or the other, or a little of both necessary? We have been told manifolds should not have corners, because that would make the fuel separate from the air; that the corners, which we should not have, should be heated, so that the fuel will be evaporated. Is it all of one, or a little of both?

I would remind you, in analyzing this subject, it must be remembered that liquid fuel is never burned; that a gas is never burned. The only thing that is ever burned in an internal combustion engine is a mixture of one gas with another; some hydrocarbon gas combined with two gases, nitrogen and oxygen. After all, we are building gas engines, and all the fuel is gasified before it begins to burn, and is mixed with sufficient intimacy with the oxygen to burn.

My belief is that we should do everything possible to secure segregation as promptly as possible after the fuel leaves the nozzle. It should be secured before the first bend of the manifold has been met or at that point. Two forces can be used to induce segregation—inertia due to a change in direction of the flow, and gravity. A drop of liquid is affected by gravity and will fall out of the air-stream if the velocity is sufficiently low. It seems to me, in place of trying to use small manifolds with few bends and putting the liquid into the engine and vaporizing it too late to secure a sufficiently intimate mixture with the air, so that it can all be burned, we should use this early segregation and throw the fuel into the air-stream in the shape of a fixed gas. Thereafter each bend and every change in flow produces a more intimate mixture between gasified hydrocarbon and the air.

The question of supplying heat in connection with carburetion should be studied from the standpoint of what we desire to accomplish, which is the vaporization of the liquid fuel with the least possible addition of heat to the entire mixture.

If all the manifold is heated enough to vaporize fully the fuel, the air is also heated to such an extent that the power output of the engine is reduced an undesirable amount. The same thing is true if the air is heated enough to vaporize fully the fuel within the time of its passage to the manifold.

The proportion of vaporized fuel to air should be only about 1 to 15; assuming that the entire amount of the fuel must be heated to 400 deg. F. to vaporize it completely, and that the air is not heated at all, then complete vaporization of the fuel will increase the temperature of the entire mixture only about one-sixteenth of 400, or about 25 to 30 deg.

The suggestion I make may and probably will prove to be the method of doing what we have been told we ought to do, but we have not been told how to do, namely, keep the density high enough to obtain good volumetric efficiency and not be in trouble winter or summer. If we can keep the air away from the vaporizing surface and make the fuel lie against that surface until vaporized, we are sure of a mixture which follows the flow of the manifold throughout its length and it is impossible to get too much heat into the mixture.

CARBURETER SHOULD USE ALL FUELS

I would like to raise the question as to whether a kerosene carbureter would be a desirable instrument if it could be secured. If we should originate a carbureter that would run on kerosene and would not run on gasoline, I wonder if we would not reverse the present condition between kerosene and gasoline. Is not that carbureter and manifold arrangement which will handle a combination of all the non-viscous products of petroleum distillation the ideal device so far as petroleum-derived fuels are concerned?

FREDERICK PURDY:—Some time ago a carbureter was developed in which the primary mixture was carried in a small tube that was heated, alongside of which a great quantity of air was carried. This design did not prove to be very successful. I think Mr. Brush's idea is excellent and that it can be carried out successfully if we can find a way of doing it without having to use too massive a construction.

A. P. BRUSH:—My judgment is only that the manifold should be large enough. I believe that a horizontal portion of the header might be provided with a heated trough narrow and small enough to have small effect on the major air-stream, but so located as to take care of any condensation of the gasified fuel. I doubt that any perceptible condensation would occur in the largest manifold with the lowest-speed engine between the time of evaporization and delivery into the cylinders.

I believe the ponderousness of such a carbureter is not the problem of the carbureter man, but of the engine man. The former is through when he gives us an instrument that delivers a correctly proportioned mixture of hydrocarbon fuel and air. From there on, it devolves upon the designer of the engine. All of the energy and money and time that the carbureter companies have spent in devising means for giving a temporary excess feed of gasoline to compensate for defective manifolds has been due to our neglect or misunderstanding of the problem; the carbureter designer has been forced by the engine designers to make his instruments less perfect than they were before that line of development was started.

VARIATION OF SUCTION

C. P. GRIMES:—I would like to substantiate Mr. Mock's statement regarding the variation of suction. On a recent hill-climbing test, if the suction in the manifold fell below 2½ in. Hg., we could not climb the ascent, regardless of the make of carbureter or the size of the manifold. In our case it resolved itself into the question of whether we wished to close the throttle or build a manifold with a choke of the proper size, thus necessitating a compromise with speedway performance.

V. R. HEFTLER:—One of the most successful airplane engines used in France not only has the heated manifold, but also has hot air introduced in this way: all the exhaust passes around the intake. The intake is in the center of the exhaust muffler. At the high altitude at which these engines are designed to operate the temperature is so low that no heating arrangement can be too efficient.

A MEMBER:—I would like to ask Mr. Brush if the gasified fuel will be precipitated out at 150 deg. if it has been mixed properly with the air, or if the affinity will keep the two in suspension in a saturated form. The mixture is precipitated out, but only the excess leaves as the air temperature changes. Is the temperature of the air in the manifold low enough so that the proper amount of gas to give a proper mixture will precipitate?

A. P. BRUSH:—Taking both air and fuel at running temperatures, there is not time enough to vaporize the fuel from the delivery at the jet to the ignition at the spark-plug. The air in this room is probably so much below the saturation point with water at this temperature that it would take up a pitcherful of water. At the same time, I might throw a glass of water into the air, and the major portion of it would come down. Some of it would evaporate, but a very small part. I might spread the same amount of water over a sheet of cheese cloth, spreading it over a large surface, and that cloth would dry and the water would be in the air. There is no question that the time is too short for natural evaporation to diffuse the fuel. I am sure that even if the air is given heat enough, in time, to vaporize the fuel, it does not get to the fuel fast enough and that considerable-size drops of fuel pass into the engine as such.

A MEMBER:—I am not sure that a proper amount of gas will precipitate, but I thought it might from the fact that precipitation occurs from mixtures that have also been highly atomized. When a mixture leaves the jet, it certainly cannot form much of a stream; it is at least nebulous, but it does condense, even after that. The vapor is almost in a molecular state, so it is not likely that it will condense as rapidly as if merely nebulous.

F. C. MOCK:—The subject of carburetion is so broad and practical in every detail, that it is hard to tell where to start. Our present engines have not changed from the original design, which was intended to operate on gasoline vapor and air, but the fuel available has changed so we cannot always have gasoline vapor. I think that a great part of our trouble with engine operation is due to inefficiency in the gas charge. When we do not have proper fuel, the engine does not run right. When particles gather in liquid form, we cannot further control the flow, regardless of the care taken in design.

HEAVY-FUEL CARBURETER-TYPE ENGINES FOR VEHICLES

By J H Hunt[1]

Manufacturers of carbureters and ignition devices are called upon to assist in overcoming troubles caused by the inclusion of too many heavy fractions in automobile fuels. So far as completely satisfactory running is concerned, the difficulty of the problem with straight petroleum distillates is caused by the heaviest fraction present in appreciable quantity.

The problems are involved in the starting, carburetion, distribution and combustion. An engine is really started only when all its parts have the same temperatures as exist in normal running, and when it accelerates in a normal manner. Two available methods, (a) installing a two-fuel carbureter, using a very volatile fuel to start and warm-up the engine, and (b) heating the engine before cranking by a burner designed to use the heavier fuel, are described and discussed.

It is pointed out more effort has been made to mix the fuel with air in correct proportion and distribute it among the cylinders than to handle it properly in the cylinders. It appears that knocking occurs later than has been supposed, and is caused by exceedingly high pressure in the engine. The meeting of the new conditions in the cylinders has not had proper attention. The fuel situation makes the problem pressing. A heavy-fuel carbureter-type engine could probably handle some of the lighter fractions of gas and fuel oils.

It should, however, be capable of handling all the fractions, beginning with the gasolines and on down through the fuel oils.

Rising fuel prices in the face of an approaching shortage of fuels suitable for the present type of motor-car engine have led to considerable experimental work with kerosene and lower-grade petroleum-distillates with the idea of using them to supplement the supply of gasoline. Many of these attempts have been made with kerosene or distillates having no lighter fractions. This work has been prompted by the present discrepancy in the selling

---

[1]Research engineer, Dayton Engineering Laboratories Co., Dayton, Ohio.

price of gasoline and kerosene, a difference which will certainly vanish as soon as there are any appreciable number of motor-car engines equipped to handle the lower-grade fuel. There has also been considerable work with fuels obtained by mixing a certain percentage of heavier fuels with ordinary or casinghead gasolines. Most of this latter work has been done by the fuel manufacturers. They have carried out experiments of enormous extent in an effort to find how much of the heavier fuels can be used in an engine of ordinary design. They have made and sold the fuels, and the user has carried out the road tests. Manufacturers of carbureters and ignition devices have been called upon to help in overcoming troubles caused primarily by the introduction of too many heavy fractions into automobile fuels. There has also been a limited amount of laboratory work with fuels consisting of cuts taken from the heaviest fractions that it is considered possible to use, through and including the lighter fractions called gasoline 10 years ago.

As far as completely satisfactory running is concerned, the difficulty of the problem is determined by the heaviest fraction present in an appreciable quantity. This statement applies to straight petroleum distillates. For the purpose of this paper the only fuels under consideration are straight petroleum distillates which have not been subjected to any cracking process and to which no other materials have been added. When lighter fuels are present in quantity it is possible to obtain deceiving results and secure what seems to be fair operation for a time, the fact that the result is not the final solution not becoming evident until the condition of the engine is examined after a service test of some length.

The problems connected directly with the use of heavier fuels may be classified as starting, carburetion, distribution and combustion problems.

Starting

If starting is defined as the complete process of getting the engine into a condition where it is satisfactorily flexible and operating with its normal economy, it is obvious that much more is involved than simply obtaining a few explosions. The engine is really started only after all parts are at the same temperature conditions that exist in normal running and when it responds to the accelerator in the normal manner.

With fuels containing a fair percentage of light hydrocarbons like our present gasoline, getting a few explosions is fairly simple even in cold weather. All that is necessary is to provide a good "choking" device aided by a "priming" device, if the inlet manifolds are long and of exceptionally large area, or if the lift is great. An excessive amount of fuel is supplied of which only a small part is vaporized to produce an explosive mixture. Much of the remainder is deposited on the walls of the combustion-chamber where it is cracked by the heat of the explosion, forming tarry carbon deposits, or it flows by the piston-rings into the crankcase, cutting the oil off the cylinder walls, polluting the crankcase oil and reducing its lubricating qualities. The inevitable result is excessive wear of piston-rings, cylinder walls and bearings in cold-weather service. As a result of the poor fit of the rings the products of combustion leak by during the explosion and the working stroke. The vapor formed by the burning of the hydrogen in the fuel condenses in the crankcase. The water freezes whenever the temperature of the crankcase falls low enough, with the result of broken oil-pump drives or obstruction of the oil supply and sticking of the engine if it is driven before the ice has time to thaw.

The addition of more heavy hydrocarbons to our present fuel will simply increase these troubles with our present type of engine. The longer an engine takes in reaching its final operating temperature, the worse these difficulties become. Anything that can be done to accelerate the warming-up of the engine will tend to reduce them and the present prevalence of thermostatic controls on the water circulation, radiator shutters, etc., shows the increasing appreciation of this fact on the part of the manufacturer. In spite of all that has been done to date, the situation with even the present grade of fuel is not satisfactory in cold weather. It is no unusual thing to remove from the crankcase of an automobile after only a few hundred miles' service, a mixture which on distillation gives from 30 to 60 per cent of hydrocarbons that belong in kerosene or gasoline and not in the lubricating oil, and where excessive priming occurs it is quite possible to have a mixture in the crankcase which will give off explosive vapors when the engine heats up. If the fit of a set of rings is so bad that the flame gets by during the explosion, a wrecked engine results. The ordinary driver today is a sufficiently good judge of car performance to insist on repairs before this condition arises; nevertheless, one hears rumors of occasional crankcase explosions.

The only complete remedy for this condition would be to bring the engine before cranking into a condition such that it would start and run with perfect distribution without any choking or priming attachment whatever. The only way to accomplish this would be to bring the temperature of all parts of the carbureter, inlet manifold and cylinder walls to the final operating temperature before cranking or at least before using a heavy fuel.

There are two possible methods of getting the engine to the proper condition for running. One is to employ a two-fuel carbureter, using a very volatile fuel to start and warm-up the engine. This gives perfectly satisfactory results as far as the engine is concerned. There are several disadvantages in the use of two fuels. The first is the nuisance of keeping two fuel tanks filled. The second is the difficulty of keeping a float-valve, which is subjected to vibration on the road, tight. The ordinary float-valve is not tight under these conditions, but as the slight amount of leakage is removed by the normal action of the carbureter we do not recognize the leakage. There is also a possibility of some loss of the light fuel by evaporation due to the effect of engine heat.

The other method of starting is to heat the engine before cranking by a burner designed to use the heavier fuel. As far as the carbureter and manifold are concerned it has been demonstrated that this is fairly easy as burners have been developed which will deliver 40,000 to 60,000 B.t.u. per hr., a rate which is sufficient to heat a heavy cast-iron manifold in less than 2 min. in zero weather to the point where starting is possible without any choking or priming, and if the manifold is made of thin welded steel, or, better, of aluminum, good running conditions are obtained in 20 sec. These burners use an airblast from a motor-driven fan and are electrically ignited by a spark-plug. They can thus be controlled by a simple switch on the dashboard.

Some tests were carried out just before our entry into the war, which will suffice to give some idea of the conditions existing with our present fuels. A six-cylinder engine, having practically the entire distribution system external to the cylinder block, was selected for test as it permitted easy modification of the distribution system. The manifold first used was not equipped with any provision for heating, although both the throat of the carbureter and the incoming air were heated. The engine was mounted in a frame with a radiator in the normal position and refrigerated air was blown through the radiator. The arrangement is shown in the upper illustration on page 535. A series of thirty-six tests was made by cranking the engine when the incoming air was at 0 deg. fahr. and when the radiator water was from 20 to 32 deg. fahr., representing conditions occurring when the engine has been standing for some time in zero weather. The standard equipment was used as far as carburetion and distribution were concerned. After starting it was necessary to use the choker for a considerable period to keep the engine running. The engine was run at about 600 r.p.m. with a load of from 3 to 4 hp., corresponding to driving within the traffic laws on a cold day on good pavement. It was occasionally slowed down to about 300 r.p.m. and the throttle opened. When the engine had warmed to the point where it would accelerate without the use of the choker and without missing, the test was discontinued. The test runs were about 15 min. long. From time to time during the test samples of the crankcase oil were removed for distillation tests. The percentage of hydrocarbons belonging in kerosene and gasoline which distilled off below 450 deg. steadily increased and amounted to about 63 per cent at the end of the thirty-six starts. Fresh oil was put in and a run of an equivalent length of time, that is for 9 hr., was made at a constant speed of 600 r. p. m. and with the same load as before, the engine being cranked only twice in this time. At the end of the running it was found that there had been no increase in the lighter hydrocarbons. About 400 cu. ft. of air, cooled down to zero, was being blown through the radiator each minute of the test run.

The inlet header was then provided with a sheet metal jacket, being arranged so that the carbureter also was heated. An electrically-heated burner was fitted to this jacket and mounted at the rear of the dashboard, the arrangement of the jacket being as shown at the bottom of page 535. The burner was run long enough before each test so that the engine would start without any choking. A series of thirty-six starts and runs of about 15 min. each was made as before. In the tests the lighter hydrocarbons in the crankcase increased about 23 per cent. As no provision was made for maintaining the temperature of the inlet header after

TEST ARRANGEMENT FOR BLOWING REFRIGERATED AIR THROUGH THE RADIATOR OF AN INTERNAL-COMBUSTION ENGINE TO SIMULATE WINTER CONDITIONS

starting it soon cooled down to where the distribution was unsatisfactory under the test conditions. Another series of tests was then made in which the burner was kept running long enough after cranking so that no

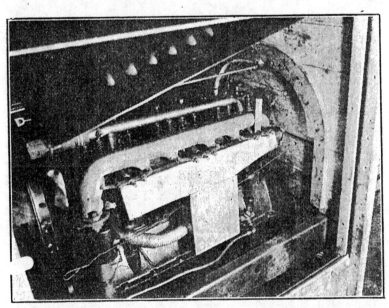

HEATING THE CARBURETER OF AN INTERNAL-COMBUSTION ENGINE TO ELIMINATE CHOKING AT START

evidence of faulty distribution serious enough to cause missing developed during the 15-min. run. Under these conditions the increase in light hydrocarbons during the thirty-six starts and runs was reduced to about 13 per cent.

It is my opinion that the crankcase pollution met with in the last test was due to the fact that the cylinder walls were entirely too cold in the early part of this run. In fact, they were never hot enough as no radiator shutters were fitted and there was no thermostatic control of the water circulation. The cold air was blown through the radiator and over the engine all the time. All crankcase pollution, therefore, cannot be removed unless the combustion-chamber and cylinder walls are properly heated before cranking. This will require very large burners or a much lower thermal capacity of the cylinder block than at present or a compromise. A larger burner alone does not seem desirable. The burner will either be very expensive or an undesirable length of time will be required. The difficulty is increased because means must be provided whereby the cylinder block can absorb the heat efficiently at the high rate required. The conditions discussed will, of course, be aggravated by increasing the percentage of heavy fractions and any engine designed to use heavier fuels than at present will require preheating by one or the other of the two possible means before cranking, or serious lubrication troubles will develop, particularly in cold weather.

In some of the experimental work a modification of the burner described in the paper[2] entitled "Kerosene versus Gasoline in Standard Automobile Engines," which was presented by Dr. C. E. Lucke at the 1916 Semi-Annual Meeting of the Society, describing the Good kerosene equipment, was used. The motor-driven fan supplies air at 9 to 12 in. water pressure to the chamber a. Some of this air passes through the opening b, atomizing the fuel issuing from the nozzle c. The fuel is forced from the chamber d by the air under pressure which comes through the hole e, that is provided with an adjusting screw f. The combined action of the vent-hole g and the restriction caused by the screw f permit any desired fraction of the pressure in the chamber a to be effective on the fuel in d. The atomized fuel from c is ignited by the spark-plug at h and mixed with the proper amount of air through the openings at the right in the central

---

[2] See S. A. E. TRANSACTIONS, vol. XI, part 2, p. 118.

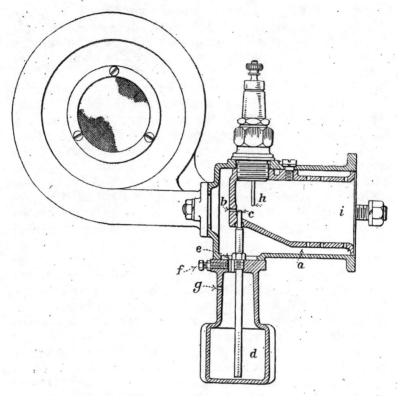

The Electrically-Heated Burner Used in the Tests

casting i. Apparently practically perfect combustion is obtained, either kerosene or gasoline being used as a fuel.

How the burner was applied to the distributing system of a six-cylinder marine gasoline engine is shown diagrammatically on page 539. In this case an exhaust gas-heated venturi vaporizer was combined with the burner installation. A very great improvement in starting conditions in winter weather was noticed. The application of the starting burner to a six-cylinder automobile engine, a special cast inlet manifold being fitted to permit the burner installation, is also illustrated on page 541. With this installation the car has been run all winter in Ohio without using the choker.

Carburetion and Distribution

The problems of carburetion and distribution can hardly be separated as it is possible to go a considerable distance on the road toward securing good distribution in the carbureting device, and on the other hand an ordinary gasoline carbureter can be used with possibly slight modifications in the jet and in the idling device, if the distributing system is properly designed. If carburetion is defined simply as the metering of the fuel into the incoming air in the proper proportions for all ranges of speed and load, including the conditions where these are changing rapidly, there is really no great difficulty involved with fuels no heavier than the present-day kerosene. As the heavier fuels have a higher viscosity than gasoline, it is undoubtedly necessary to arrange conditions so that the temperature of the carbureter is practically independent of the weather. There have been a great many attempts in the past to handle the heavier fuels in the carbureting device alone, usually with the aid of heat from the exhaust. So much heat, however, is required that it is not possible to transfer it from the exhaust to the incoming mixture in the carbureter without making this entirely too bulky. The present indication is that it will be better to have that part of the equipment which has most to do with the vaporization of the mixture built into that part of the engine usually manufactured by the engine builders rather than to attempt to incorporate it in the carbureter.

Assuming that the fuel has been properly metered into the incoming air, the problem of handling it so that it will be distributed to the cylinders by some means which will not cause too great a loss in volumetric efficiency is one of the two real problems involved in the use of the heavy fuels. The other problem is the proper control of the combustion, once the mixture is in the cylinder. Past experience has shown that it is comparatively easy to distribute the mixture when the fuel is perfectly vaporized. It is therefore natural that considerable effort has been applied to devices which produce a dry mixture with heavy fuels and then maintain this mixture hot enough through the remaining part of the inlet manifold so that no condensation can occur. Two losses in volumetric efficiency are involved in this process. The first is due to the increased temperature of the mixture and the resulting smaller charge in the combustion-chamber; the second is caused by the need of considerable suction to draw the mixture through the restriction caused by the heating device. The heating surfaces must be of considerable area to transmit the amount of heat required. Part of the problem, of course, is the collection of this heat from the exhaust. To keep the surfaces from being entirely too large the mixture must be drawn by

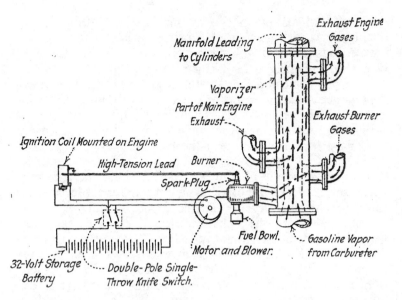

DIAGRAM SHOWING THE APPLICATION OF THE ELECTRIC BURNER TO THE DISTRIBUTING SYSTEM

them at high velocity, resulting in great friction losses. In order to make these surfaces effective at comparatively light load and speed, the designs are usually such that the suction head required to draw the mixture through at high speed involves very serious losses in volumetric efficiency. Some of the devices now developed maintain very uniform temperatures of the mixture at varying loads and speeds but many permit too great an increase with load. This results in still further losses in volumetric efficiency at high speeds.

The inevitable result is to cause the engine not only to have a reduced torque at all speeds but to reach the peak of its power curve at much lower speeds than normal. If the same car performance is to be obtained the piston displacement must be considerably greater than at present, as the low peak speed of the power curve will require a change in the gear ratio. The increased piston displacement would result in decreased economy under normal driving conditions. Means of avoiding these troubles are needed if we are to use heavier fuels without a very undesirable increase in engine dimensions. There is no doubt, however, that a very large percentage of the automobile users probably would be well satisfied with cars whose performance, as far as ultimate speed is concerned, is much below that of the average car supplied today, providing, of course, that satisfactory flexibility and power were available at speeds below 40 miles per hr.

The Good venturi vaporizer, which Dr. Lucke described in the paper previously referred to, has, in addition to the bench testing, had a very considerable amount of service work on the road which indicated that the cars of drivers who would be satisfied with a comparatively moderate maximum speed could be taken care of in the immediate future. One of these equipments has been in service for about 2½ yr. on a Ford engine, during which time the car has been run 20,000 miles in all kinds of weather. In the first 2 yr. there was no mechanical trouble whatever with the engine. It has been found necessary to change the oil in the crankcase more frequently than if gasoline were being used. It has also been necessary to clean out the carbon more frequently than in the case of the average engine using gasoline. In general, however, the service of the engine might be considered very satisfactory.

A Buick D-45 six-cylinder car has been equipped with such a device and run a little over 10,000 miles in a period of about 14 months, including the larger part of two winters. No mechanical trouble at all developed in this engine. About half the crankcase oil was removed every 1000 miles and fresh oil added. Both of these cars were run as single-fuel cars, the manifolds being preheated by a burner preparatory to starting. Both cars were run with the standard equipment except for the vaporizer and the burner, no provision being made to control the temperature in the radiator to any value higher than normal and no radiator shutters being fitted. The temperature of the combustion-chamber was undoubtedly too low for the best results with heavy fuels. The compression was reduced in both engines to avoid excessive knocking.

There has also been a great deal of development work on devices where only part of the air is heated with the fuel to a point above vaporizing temperatures, colder air being added later. It is probable, however, that in most of these as now installed a really dry mixture is not secured but a sort of fog when the colder air is added for tempering the overheated, over-rich mixture. I have had no opportunity to test such devices on engines that could be called high-speed. It seems probable, however, that at high speeds distribution troubles would be encountered unless the manifolds were very carefully de-

The Starting Burner Applied to a Six-Cylinder Automobile Engine Enabled the Car to Be Run an Entire Winter Without a Choker

signed. It will perhaps not be out of place to mention in passing that where the attempt is made to handle wet mixtures or use devices which on some engines give dry mixtures, the data available are not yet sufficiently definite to permit going from one engine to another and obtaining first-class results with the first effort. Frequently a device which seems very promising on one engine will be a complete failure on another that appears to be very similar.

Combustion

When the mixture finally reaches the cylinder, the heavy fuels do not burn in the same manner as the lighter hydrocarbons. When auxiliary equipment giving proper carburetion and distribution is fitted to a normal design of motor-car engine, exceedingly severe knocking occurs when fuels of the type of kerosene are used at anything like full load, that is, at normal compressions. Investigation has shown that this knocking is due to the momentary existence of exceedingly high pressure in the engine. At the 1919 Annual Meeting of the Society, C. F. Kettering gave a report covering some experimental work with a special type of indicator which has demonstrated conclusively not only the existence of these pressures but that they come at a point in the revolution of the engine later than the ignition and are not in any way due to what is known as preignition. This discussion by Mr. Kettering was along the same lines as an informal report before the S. A. E. Council at a meeting in Dayton nearly a year ago.

At present the only commercial way of eliminating this knock is to reduce the compression which, of course, has an injurious effect upon the power and economy. Introducing a certain amount of water with the fuel will permit the compression to be somewhat higher than without water but less than with gasoline. Apparently the effect of the water is to hold down the maximum temperatures reached during combustion due to the heat required for its evaporation.

In the January, 1919, number of *The Automobile Engineer* of London, Harry R. Ricardo discusses the probable nature of the fuel knock and some methods of handling it in engines of compressions such as are used today for vehicle work. Mr. Ricardo's scheme is, briefly, to mix with the secondary air of the carbureter a certain amount of cooled exhaust gases, a method which has found previous application in engines operated with producer gas. This idea is very suggestive and deserves to be very carefully investigated. Some investigators have told me that they tried this method without success several years ago. Mr. Ricardo's test report covers the work on a single-cylinder engine at speeds up to only 1400 r. p. m. Where an attempt is made to distribute to a multiple-cylinder engine, temperatures in the mixture must be used which tend to make the knock more severe and require a greater reduction of compression than is necessary in a single-cylinder engine. The Ensign fuel converter, which burns a part of the fuel to vaporize the remainder and which permits the products of this partial combustion to enter the cylinders with the remainder of the fuel, evidently will permit higher compressions than would be possible with a vaporized kerosene mixture at the same temperature. This fact seems significant in connection with the results reported by Mr. Ricardo. I have not had sufficient experience with this latter device to justify a surmise as to the reasons for the results obtained.

The most important problem to be solved in handling the heavy fuels in the motor-car engine is to find a means of eliminating the fuel knock. The decrease of compression is very undesirable from the standpoint of efficiency and power. Anything that can be done to the engine or to the fuel which will permit the use of the usual gasoline compression will go far toward solving the ultimate problem. A vehicle engine for motor-car work must, of course, be very flexible and do the most of its running at comparatively light loads. While we cannot consider that the point is as yet definitely settled it seems pretty certain that it is desirable if not necessary to increase the temperature of the combustion chamber for light-load work above that obtainable with water as a cooling fluid, certainly above that which can be obtained with a cooling mixture of water and alcohol so much used in the winter. Experience covering some thousands of engines in service all over the United States has shown that it is possible to handle a wet mixture in an air-cooled single-cylinder engine running under steady load conditions without pollution of the crankcase oil or the formation of injurious carbon deposits and when obtaining very satisfactory economy. This indicated that a high temperature in the combustion-chamber will take care of the effects of previous lack of vaporization. The hot combustion-chamber walls undoubtedly tend to prevent the condensation and depositing of fuel.

A theoretical reason for using higher temperatures with the heavy fuels is supplied by the fact that a saturated petroleum vapor tends to condense when compressed unless additional heat is supplied. If the temperature of the uncompressed mixture is above that of the cylinder walls the mixture cannot draw any heat from the walls during compression and may remove heat from the compressed air adjacent to it to the point where the fuel mixed with these layers will condense. In the *Report of the French Academy of Science* for March 4, 1918, there is a short paper by M. Jean Rey giving data for the entropy curve of petroleum. If it proves absolutely necessary to raise the temperature of the combustion-chamber, naturally a very complete redesign of the engine will be necessary.

A trouble which is usually met in engines running on kerosene is the dilution of the crankcase oil with the heavier portions of the fuel. A great portion of this trouble is undoubtedly due to the faulty distribution and vaporization of the fuel by the devices with which the engines have been equipped in the past. As previously mentioned, it is pretty well established that this trouble is practically negligible in engines which are run under constant speed and at approximately full load where the temperature conditions in the combustion-chamber are such that complete vaporization is obtained. It is, therefore, probable that more complete investigation will show that this trouble can be eliminated by proper design of the distribution system and the engine itself. If not, means must be found to remove the condensed hydrocarbons from the lubricating oil, or to prevent them from reaching the oil.

Mr. Ricardo has developed for the British Tank Service a special type of engine having a guide for the upper end of the connecting-rod, independent of that part of the piston which carries the ring. This construction has the incidental benefit of preventing any leakage past the rings from reaching the crankcase and would undoubtedly be a very complete method of preventing the pollution of the lubricant. However, the construction requires numerous engine changes and it is very likely that less expensive means will be found to obtain the required result.

To date very much more effort has been expended in attempting to mix the fuel with the air in the proper proportions and distribute it among the cylinders of the engine than has been made to handle it properly in the cylinders after it once reaches them. This has undoubtedly been due to the fact that commercial conditions have led the engine manufacturer to concentrate on production and on detailed improvements in his standard product, leaving the development work along the line of handling heavier fuels to accessory manufacturers or to investigators working with very limited facilities for experiments.

The problem has not been properly understood by a great many workers in this field, with the result that the necessity of meeting new conditions in the cylinder has not received the proper attention. The present fuel situation makes the problem very pressing. If manufacturers are to continue their present schedules they must have engines to handle heavier fuels in the immediate future. The only possible contingency that could prevent this need would be the development of commercial cracking processes to the point where an ample supply of fuels that are satisfactory when burned in present-type engines will be available. Failing this, the industry will soon be compelled to slow down owing to lack of sufficient fuel of the quality required at present.

HEAVY-FUEL CARBURETER-TYPE ENGINES

When we examine the data that are available today, we must admit that there is no demonstrated means by which the heavy-fuel carbureter engine will improve the fuel situation to any great extent. The equipments which have shown promising results in service have handled nothing heavier than kerosene. At present only about half as much kerosene as so-called gasoline is being produced. If the outfits now available were marketed in quantity and applied to truck and tractor engines and to the moderate-speed car engines where their use would not sacrifice performance unduly, only about 50 per cent more fuel would be available. Fifty per cent increases, while highly desirable, do not begin to meet the situation before us.

Nearly half the petroleum is marketed as gas and fuel oil, a quantity over twice that marketed as gasoline, averaging all crudes. The heavy-fuel carbureter-type engine can probably handle some of the lighter fractions of gas and fuel oils, particularly in the case of petroleum from some fields. The probable result of putting heavy-fuel engines on the market in quantities would be, first, a rise in the price of kerosene and next a fall in volatility and an increase in density such as we have been having with gasoline. In 2 years the engine pronounced satisfactory today would sound like a 1910 model on a cold morning. To prevent this condition from developing, it will be necessary to make the heavier-fuel engine capable of handling all of the fractions, beginning with the gasolines and down through the fuel oils. We are not yet ready to do this with the carbureter type of engine. We must remember also that the motor-car industry may not use all of this fuel oil. A large part of it will be required for marine and power engines of the Diesel and semi-Diesel type. For the motor-car industry to take proper steps to meet the fuel situation a fuel priority board would be needed to tell us what part of the total supply of the petroleum we could use. With this information the development work could be pushed. As it is some work may be wasted as the fuel on which it is based may not be available in the very near future.

FACTORS INVOLVED IN FUEL UTILIZATION

By P S Tice[1]

From a laboratory examination of the controlling relationships between carburetion and engine performance still in progress, the general conclusions so far reached include fuel metering characteristics, the physical structure of the charge, fuel combustion factors and details of engine design and manufacture.

In every throttle-controlled engine, the variation in fuel metering for best utilization is inversely functional with the relative loading and with the compression ratio, but the nature of the fuel leaves these general relationships undisturbed. The physical structure of the charge influences largely the net engine performance and the order of variation of the best metering with change in load. Perfect homogeneity in the charge is theoretically desirable but entails losses in performance. With the least perfectly formed charge that will operate an engine, increased outputs result from a more nearly uniform distribution of fuel in the charge, even though accompanied by a reduction in charge weight. Considering available petroleum fuels, those requiring the higher charge temperatures for maximum utilization also possess the ability to ignite at relatively lower temperatures when mixtures suitable for engine use are made.

It should be possible to operate an engine continuously at open throttle, without detonation, on a properly constructed charge of any petroleum fuel proposed to date, with a compression ratio resulting in gage pressures including 75 lb. per sq. in., but an analysis of the failure to do this shows that preignition is the direct result of local overheating in the charge. Augmenting turbulence decreases local overheating and so appreciable inequalities of combustion-chamber surfaces must exist, and high piston-head and exhaust-valve temperatures contribute to a condition favoring detonation. Piston design, exhaust-valve and spark-plug electrode cooling are then considered, mention being made also of piston gas-tightness as being very important. Net engine performance is the square root of the brake mean effective pressure divided by the brake specific consumption, combining the cost of unit

---

[1] M.S.A.E.—Engineer in charge of carbureter division, Stewart-Warner Speedometer Corporation, Chicago.

power and the amount of power obtained. Its value attains a maximum with mixtures only slightly richer than those resulting in the true minimum specific consumption. With manual throttle control, steady and consistent operation obtains throughout the throttling range when realizing the minimum possible specific fuel consumption but, so far, possible maximum utilizations cannot be realized in operation under governor control. These considerations are illustrated by charts, showing curves resulting from the experiments made.

My laboratory activities have been concentrated for the past seven months upon an examination of the controlling relationships between carburetion and engine performance. This work was undertaken to provide a summary of fundamental data for carburetion system design, and to develop an effective short program of work to be done in fitting an engine for its particular class of service. While this program of investigation is still going forward, a sufficient number of engines, varying in size, speed, general design, type and class of service, have been examined to permit arrival at some interesting and important conclusions. These are quite general in their scope, and involve (a) fuel-metering characteristics, (b) the physical structure of the charge, (c) combustion characteristics of the fuel and (d) several details of engine design and manufacture.

The first item can be treated very briefly for the present purpose, since, as between typical engines with respect to size and speed range, the variation in requirements is small. In every throttle-controlled engine the variation in fuel metering for optimum utilization is inversely functional with the relative loading and with the compression ratio. The nature of the fuel leaves these general relationships undisturbed. On the other hand, the physical structure of the charge, by which is meant both the degree of distribution of the fuel content and the state in which it exists, is of great importance, since upon this factor depends largely the net performance of the engine, and also, to some extent, the order of variation of the optimum metering with any change in load. It will be granted that perfect homogeneity in the charge is theoretically the desired condition, but it was early found in a progression toward this result that it cannot be attained without inducing losses in performance. In order that each minute space occupied by the charge may contain the same proportions of air and fuel as every other like space, which is the condition of perfect homogeneity, it is obvious that the fuel must exist in the vapor rather than the liquid state, and that admixture must have been carefully attended to in forming the charge. But the condition that the whole of the fuel be present as vapor presupposes that the temperature of the charge be high enough to maintain it in this state.

Effect of Charge Structure on Engine Performance

Since the output of an engine is limited solely by the amount of heat made available in it in unit time, which in turn is directly as the relative oxygen content of the charge, with any one fuel, it is seen that as structural perfection in the entering charge is approached the possible useful output of the engine is reduced. Starting with the least perfectly formed entering charge that will operate an engine, increased outputs result from a more nearly uniform distribution of fuel in the charge, even though this last is accompanied by a reduction in charge weight. It is true also that materially reduced specific fuel consumptions result from further equalization of the fuel distribution in the charge to a point where much less than maximum possible output is obtainable. It is equally true that when the goal of minimum specific fuel consumption is attained in practice the charge is far from being a truly homogeneous one, as it passes the intake valves. The distribution of fuel in the air of the charge under this condition is relatively very good, but the major art of it *must* exist as liquid; and anything done to augment the vapor content at this point brings about a net loss in performance. What happens in the charge structure within the cylinder during the aspiration and compression strokes is in some respects an open question. It is hoped that certain much-needed direct experimental evidence will be available shortly in this matter.

The outstanding charactertistic differences among our fuels are their relative volatilities and their behaviors at and during combustion. Unfortunately, one cannot consider these items separately under service conditions. The less volatile the fuel the more extensive the initial preparation that must be devoted to it before entry to the cylinders, to insure that it be most usefully distributed, and this most useful charge structure is attained only through means which raise the charge temperature. Experimental evidence shows that to have the same charge structure with two fuels, the temperatures in the two cases will stand somewhat as the total heats of vaporization of the fuels.

Considering available petroleum fuels, those requiring the higher charge temperatures for maximum utilization also possess the ability to ignite at relatively lower temperatures, when worked up into mixtures suitable for engine use. It is no difficult thing to visualize what happens when fuels are substituted in an engine. A fixed compression ratio will result in higher terminal compression temperatures, the higher the initial charge temperature. At the same time, the charge having the higher initial temperature is often inflammable at a lower temperature than is the other. The result in such a case must be at least a nearer approach to autoignition upon compression of the higher temperature charge.

In Fig. 1 are curves of terminal compression temperature against compression ratio for several initial charge temperatures. If the act of compression alone were responsible for heating of the charge, it would be possible to use ratios resulting in a close approach to the temperature indicated as that of autoignition, but, because of such other sources of heating, we are limited to compression ratios which in themselves would not take the charge temperature beyond that value indicated by the line marked "usable compression ratios." Experience shows that there is incipient autoignition along this latter line, proving that autoignition temperatures are attained. Hence this graph may be taken as approximately expressing the reduction in the compression ratio, and therefore in performance, imposed by our present imperfect temperature distribution in the combustion-chamber walls, and by the initial charge temperatures imposed by the nature of our fuels.

If autoignition does occur before passage of the spark, it is a case of preignition, but if the autoignition temperature is not attained before the passage of the spark, one of two things results. Either the combustion is sufficiently near normal rate to be usable, or the ignition by spark and the resulting pressure wave through the charge cause the temperature of the major portion to rise to that of autoignition. When this happens, nearly the whole charge burns at once, giving detonation, the evidences of which are violent knocking, loss of power, sooty exhaust and a reduced exhaust temperature.

What data we have show that it should be possible to operate an engine continuously at open throttle, without detonation, on a properly constructed charge of any petroleum fuel proposed to date, with a compression ratio resulting in gage pressures up to and including 75 lb. per sq. in. Actually, we are not able to do this as yet, in service. The analysis of our failure does not proceed very far before it becomes evident that the detonation or preignition experienced is the direct result of local overheating in the charge. Everyone is familiar with cases

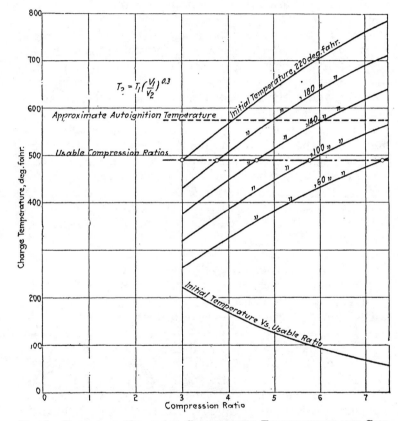

FIG. 1—CURVES OF TERMINAL COMPRESSION TEMPERATURE AND COMPRESSION RATIOS

where a great improvement has been effected by augmenting the turbulence. Now, the important effect of increased turbulence in a charge having an initially suitable structure is to equalize the temperature distribution in it. It is very likely that greater turbulence results in a higher mean temperature in the charge, rather than a lower one, but no one will gainsay its advantages in such a case.

Conditions in the Cylinder

This brings us to a consideration of the conditions surrounding the charge in a cylinder. If turbulence prevents local overheating, it must be that appreciable inequalities of wall temperature exist. This is evidently the case. There are the piston, the exhaust valve and the plug electrodes, as representing combustion-chamber surfaces that are not in direct contact with an external cooling or temperature equalizing medium. As yet it is impossible to state with complete accuracy with which of these hotter surfaces the greater responsibility rests. However, this much is certain: high piston-head and exhaust-valve temperatures contribute much to the condition of detonation, while very rarely if ever causing preignition; and preignition is almost always initiated at the spark-plug electrodes.

Piston design is a subject that has had much attention, but from our present point of view, this attention has not until recently been most usefully directed. It is safe to say that the most advanced thought in piston design, which treats it as a thermal rather than a mechanical member, has hardly reached those who can make the best use of it. For a brief statement of some of the very much worth-while things the piston can be made to accomplish, one cannot do better than read the paper[2] by Dr. A. H. Gibson, on air-cooled cylinders.

The exhaust valve, it seems, is an old offender. Beyond question, this is the most grossly overheated part in the engine and requires very effective treatment of design and environment if active trouble is to be avoided. Of the two parts by which heat can be removed from the valve-head, the seat can conveniently be made most useful. At the same time, the path afforded by the stem must not be overlooked, since it is always an open one. Housing of valve-stems and springs is not conducive to cooler valve-heads. Piston-head surface is relatively large, and exhaust-valve-head surface relatively small, but the former is much cooler. The piston-head can be responsible for a considerable rise in temperature of a large mass of charge, while the exhaust-valve head may account for an excessive rise in a comparatively small mass. Both these surfaces emit much heat by radiation, as well as by convection and conduction.

With the higher initial charge temperatures employed

---

[2] See *Automotive Industries*, May 20, 1920, p. 1156.

with the heavier grades of fuel, all working faces of the combustion space acquire higher temperatures, and therefore require that some advance be made in their cooling provisions. While it is unthinkable that internal cooling can be employed to control the piston and exhaust-valve temperatures, the evidence all points to that means as most suitable in the case of the spark-plug electrodes. If the electrode masses are made as small as possible consistent with complete failure to fuse at full load, and they are given relatively large surfaces freely exposed in the main combustion space, the flow of the fresh charge about them can ordinarily be depended upon to keep their temperatures sufficiently low.

Before leaving engine design details, it is desired to mention briefly one other phase of piston design, its gastightness. Not only does piston leakage reduce output and thereby increase the specific consumption of fuel, but it is responsible for dilution of the lubricating oil. Appreciable dilutions occur, we know, with even the most volatile fuels. Several possible remedies can be employed, and probably one of the best of these is to draw the carbureter air through the crankcase, thereby reducing the partial pressures of the vapors in that space to values well below those of saturation. Much of the diluent will then evaporate and be removed with the air. If the case is kept hot enough, with this method, average dilution can be made almost negligible. The real stumbling block here is that the air-cleaners of today are not to be trusted to this extent. But after considering everything, we can hardly fail to conclude that the best treatment is prevention rather than cure. The oil diluent is getting past the pistons.

Criterion of Engine Performance

At this time there is no generally accepted working definition of fuel utilization, or of engine performance in its relation to utilization. In most minds utilization has meant merely the inverse of consumption; while performance is synonymous with power. It seems that when we say "fuel utilization" we mean primarily the cost of power, rather than the fuel cost of operation. Likewise, by the expression "engine performance" we must mean the relation between the power obtained and the fuel outlay. Viewed in this way, the one is correlative to the other.

In the foregoing the term "net performance" has been

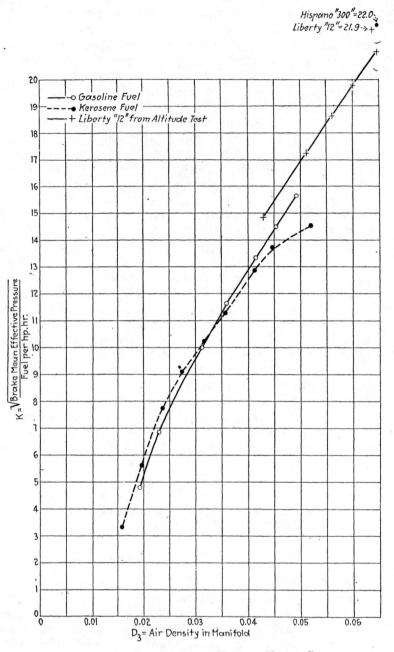

Fig. 2—Comparison of a Tractor Engine Using Gasoline and Kerosene Fuel with a Liberty Aircraft Engine

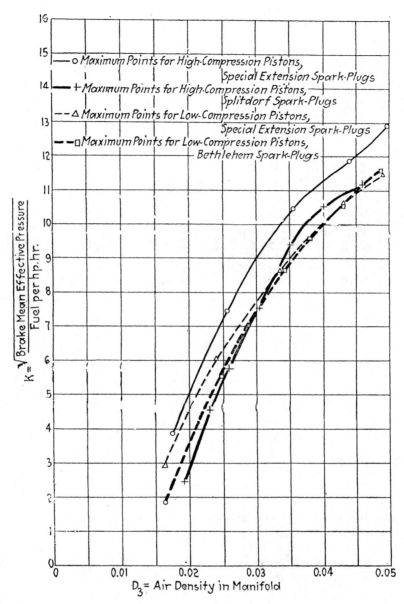

Fig. 3—Effect of Changing the Compression Ratio and Redesigning and Relocating the Spark-Plug Electrodes in a Kerosene-Burning Tractor Engine

used, the purpose being to embrace in a single expression the two ideas, the cost of unit power and the amount of power obtained. It seems that the first is best expressed

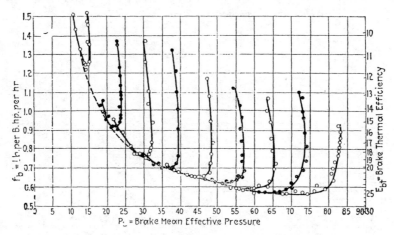

FIG. 4—TEST RESULTS UPON WHICH THE CRITERION VALUES GIVEN ARE BASED

by the brake specific consumption and the second in terms of the brake mean effective pressure. Consideration given this matter indicates that a useful criterion of "net engine performance" is established in the quotient resulting from the division of the square root of the brake mean effective pressure by the brake specific consumption. The use of this quantity as a criterion of performance expresses an easily workable and just relationship between cost and value received, without exaggerating the importance of either item. That it is not a wholly illogical expression is seen when it is written in terms of the engine constants and of the quantities directly observed in a test.

$$\text{Criterion} = \frac{\sqrt{P}}{f} \times \frac{NQ^{1.5}}{F\sqrt{La}}, \text{ where}$$

$P =$ Brake mean effective pressure
$f =$ Fuel consumption in pounds per brake horsepower per hour
$N =$ Engine speed in revolutions per minute
$Q =$ Torque
$F =$ Fuel consumed in a unit time
$L =$ Length of stroke in feet
$a =$ Area of piston in square inches

For a given engine it is proportional to $NQ^{1.5}/F$; and

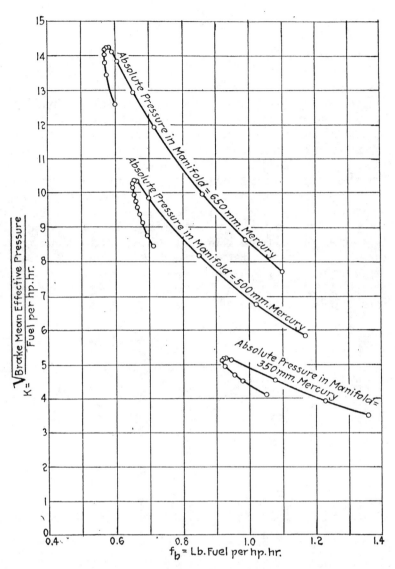

Fig. 5.—Curves of Criterion Values Plotted Against Brake Specific Fuel Consumption

for a given speed is simply $Q^{1.5}/F$. The value of this criterion, under best general conditions of engine design and carburetion, attains its maximum with mixtures only slightly richer than those resulting in the true minimum specific consumption. It shows in good perspective the results of carburetion, compression ratio or other modifications in an engine, it permits of a truthful interpretation of part-throttle operation in terms of that at open throttle and affords a means of comparing directly the performances of engines, regardless of type of design or service. A very good tractor engine, operating on kerosene, can have an open-throttle performance rating under this criterion of from 13.5 to 14.0; while the same engine on gasoline can rate at from 14.0 to 14.5. The accepted type of higher-speed car engine will very rarely exceed 15.5, and highly developed aeronautic engines rate at between 21 and 24 at their best speeds.

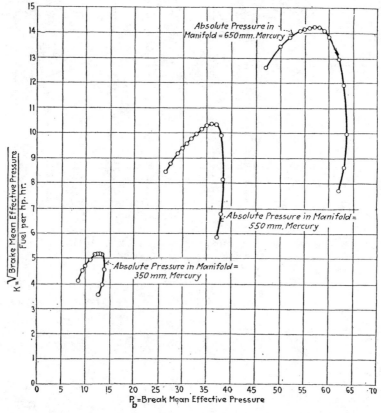

Fig. 6—Curves of Criterion Values and Brake Mean Effective Pressures

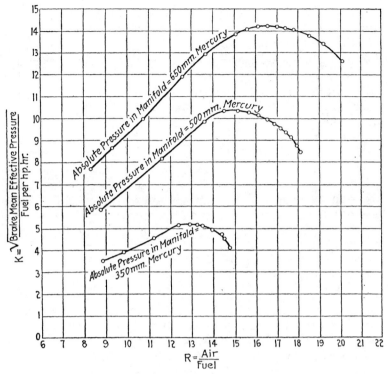

Fig. 7—Relation Between Criterion Values and the Mixture Ratio

In the curves of Fig. 2 is a direct comparison of a highly developed tractor engine, operating on both gasoline and kerosene, with the Liberty-twelve aviation engine. In the case of the former the effect of incipient detonation with kerosene fuel is well shown by the abrupt change in slope between the densities of 0.045 and 0.052. Points for the Liberty twelve are taken from published data in altitude performance tests, the criterion values at the smallest densities being omitted because of the self-evidently too rich mixtures used.

The failure of the tractor engine curve to join up with that for the Liberty engine may be accounted for by the higher mechanical efficiency of the latter, and also to some extent by possible errors in the density values assigned to the Liberty points. Judging from the close agreement in momentary slopes and the straightness of the curves for gasoline in these two engines, it may be taken that the net performance as expressed by this criterion is directly as the air density of the charge, at densities greater than half that of the standard atmosphere. This points to the importance of securing high volumetric efficiency and of maintaining the lowest permissible charge temperature.

Fig. 3 shows kerosene performance in another tractor engine and directly compares the effects of compression-ratio change and of the redesign and relocation of the spark-plug electrodes. A comparison of the triangles and squares shows the effect of substituting long, small mass electrodes for short heavy ones, with a compression pressure somewhat under 70 lb. gage. Open-throttle performance is but little affected, while a marked gain is shown at reduced loads. The effect of raising the compression to about 75 lb. gage is shown, with these two sets of plugs, by the circles and the crosses. The direct effect of this compression-ratio change, using the more favorable plug design, is shown by comparison of the circles with the triangles.

Fig. 4, from tests of a four-cylinder 4½ by 5½-in. tractor engine shows the nature of the test results from which the foregoing maximum criterion values were obtained. Each member of this family of curves represents a fixed manifold pressure, with the charge proportions varied over nearly the whole of the steady operating range. The broken line drawn tangent to the members of the group expresses the maximum possible utilization in this engine under the conditions of the test. That this performance is very creditable is seen by consulting the scale of brake thermal efficiency which shows 25 per cent at full load and just over 20 per cent at half load.

It has been stated that with suitable carburetion the criterion $\sqrt{P}/f$, attains its maximum value on mixtures only slightly richer than those giving minimum brake specific consumption. The truth of this is shown in Fig. 5, where the value of the criterion, from the data of Fig. 4, is plotted against the brake specific consumption for the three absolute intake pressures of 350, 500 and 650 mm. of mercury.

The corresponding curves of the criterion plotted against the brake mean effective pressure are given in Fig. 6. It shows that the maximum criterion occurs not far from the maximum pressure at and below half load, and that as the load approaches its full value, the peak net performance increasingly departs from the maximum possible output.

Fig. 7 shows the relationship between this criterion

and the mixture ratio for the same values of the intake pressure and indicates clearly the required variation in the fuel metering to obtain the greatest net performance from the engine.

With manual throttle control, perfectly steady and consistent operation is had throughout the throttling range when realizing the minimum possible specific fuel consumption, but it has so far been found without fail that possible maximum utilizations cannot be realized in operation under governor control. To eliminate excessive hunting, the mixtures must be enriched by an amount which results in from 5 to 10 per cent increase in the specific consumption.

FUNDAMENTAL POINTS OF CARBURETER ACTION*

By F C Mock[1]

The author selects and sets forth some of the main laws and basic considerations influencing carbureter action. A brief defense of the carbureter as a means of supplying fuel to an engine is made, as compared with the fuel-injection method, and conditions in the cylinder, the manifold and the carbureter during normal operation are stated. The relations of throttle positions, manifold vacuum and engine torque are discussed, followed by an exposition of the effect of manifold vacuum upon vaporization.

The subject of air-flow in carbureters is treated at some length and the venturi-tube form of air-passage is commented upon in considerable detail. The flow of air through air-valves, fuel-flow and mixture-proportion requirements are given detailed consideration, the last being inclusive of passenger-car, motor-truck, tractor, motorboat and airplane needs. The essentials for obtaining accurate information with regard to carbureter and engine tests are outlined. Drawings and charts supplement the text.

It is always difficult to determine exactly, and define accurately, the actions that enter into the functioning of a carbureter because these are so intimately interconnected with many others equally complex and equally vital, all uniting in the operation of the modern internal-combustion engine. In carbureter work one must hold fast to an absolute faith in physical science and the immutability of Nature's laws; many things seem to happen that we can reject, at the time, only because we know they cannot be and yet we must be always alert to find the errors of our previous conceptions and ready to accept the existence of forces and phenomena that were not known before. We must steer a mean course between the attitude of the unskilled service-man, who believes that the carbureter, if properly built, can compensate for all deficiencies in the 600-lb. mass of metal beside it under the hood, and that of the theorist, who would plan it all out on paper without soiling his hands in experimental tests.

---

*Buffalo Section paper.

[1] M. S. A. E.—Research engineer, Stromberg Motor Devices Co., Chicago.

As the subject of carbureter action is broad, it seemed possible in such a paper as this only to select and set forth some of the main laws and basic considerations that apparently exist and which we use as guides in the engineering work of our own organization. I believe that an understanding of these laws will be of great assistance to all automotive engineers, particularly those engaged in engine and motor-car development.

Carbureter versus Fuel Injection

Before proceeding further with the subject I want to justify the effort thus involved by making a brief defense of the carbureter as a means of supplying fuel to an engine, particularly as compared with the fuel-injection method. In the modern carbureter, which if my memory is correct has been cruelly characterized by a fuel-injection advocate as an "uncertain and complicated device suitable only for laboratory use," rotation of the engine draws in the air, the inflow of air draws its charge of gasoline in a fixed proportion and the engine cannot rotate without getting its predetermined mixture proportion. Further, the mixture is *stable* under changes of load and speed, a point that we appreciate only after trying other methods of fuel feed; for instance, a gravity, needle controlled, flow of gasoline into the mouth of an upturned intake-pipe.

In the carbureter the fuel is fed as a steady stream through a metering orifice and the effect of surface tension becomes negligible. For fuel injection each charge must be an individual minute drop or globule. Under the range of temperatures encountered the change in metering due to change of capillarity and viscosity will be very high and as soon as the walls of the injection duct get hot the globule, with the range of fuel volatility we have to handle today, will tend to disintegrate due to the boiling of its light elements.

Furthermore, the present motor-car fuel system operates on a range of quantity of mixture, from light load low speed to high speed full load, of over 30 to 1, and goes smoothly and positively from any one point in the range to any other. The very essence of the usefulness of the motor car and motor truck lies in this flexibility, in traffic service, under adverse road conditions and the like. The tractor and the airplane were satisfactory at first with a limited speed-range and little flexibility, but as they have developed much greater flexibility has been demanded. Is there a small high-speed multi-cylinder fuel-injection engine in existence today that has a speed and load range of 3 to 1 and will make the change smoothly and without danger of stalling?

With the modern type of carbureter there is very little trouble due to wear, deterioration or change of the instrument in service. We know now that most of the difficulties formerly charged up to the carbureter were due to the non-volatility of the fuel and could be overcome only by changes in other parts than the carbureter.

The fuel situation has been steadily changing in the last 7 years. In each engine-building department of our large factories the value of the production equipment is hundreds of thousands of dollars. Is it not sane economics to do as much as possible in adaptation to the changing fuel situation with a small easily modified device such as the carbureter, rather than worry along under continuous and uncertain, but always costly, modifications in engine design? If we can make the car operate better by introducing new elements into the carbureter and these elements are positive and permanent in their operation, is there any reason whatsoever why this should not be done? In fact, there are some carbureters in use today whose construction is not complicated enough. If the characteristic of a given type of carbureter is such, for instance, as to give an unduly rich mixture in certain parts of the average driving range, or if some other equally important defect is present, we can say, I think, that the design of this carbureter is not complete until these faults have been remedied.

Conditions in Cylinder, Manifold and Carbureter During Normal Operation

The first thing necessary in a consideration of this subject is a clear and definite understanding of the conditions existing in the carbureter and the intake-manifold. As every carbureter man knows to his sorrow, the ordinary dynamometer "horsepower curve" is not at all representative of general engine performance. The engine can operate at all speeds between idle and full load; that is, the total range of performance possible, if the conditions of operation permit, can be designated by the area between the power curve and its baseline. Very often, however, the conditions of operation are confined, more or less definitely, to portions only of this area. I have endeavored to show these limitations by the curves of Fig.

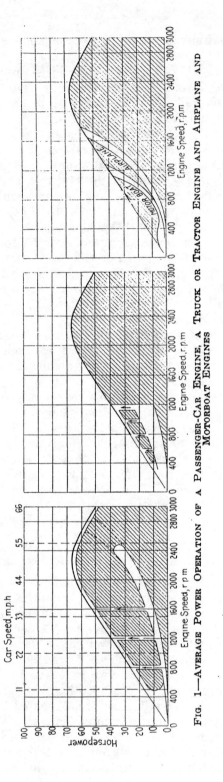

Fig. 1—Average Power Operation of a Passenger-Car Engine, a Truck or Tractor Engine and Airplane and Motorboat Engines

1. A more intelligent representation might be obtained perhaps with curves showing the torque at different speeds.

The specific conditions of engine performance will be considered in more complete detail in later portions of this paper relating to mixture requirements. It is sufficient to point out here that for passenger cars the full power of the engine is seldom used at ordinary speeds, but is used very often at low engine speeds because only small throttle-openings will give full torque at this time. As we all know, the full power of the engine at high speeds is used very seldom indeed. The motor truck and the tractor have considerably less range of action than the passenger-car engines, while motorboat and airplane engines, working under propeller load, are very definitely limited as to range of torque and speed.

We all understand how at wide-open throttle the rate of mixture delivery from the carbureter depends upon the engine speed and the volumetric efficiency. At part throttle and part load, however, these relations are less clear, the quantity of mixture passed being governed partly by the throttle position and partly by the engine speed. It is very important that we understand the conditions existing in the intake-manifold of these different throttle positions, loads and speeds, as well as under changes from one condition to another.

Relation of Throttle Position, Manifold Vacuum and Engine Torque

The throttle, I might point out, is not necessarily a part of the carbureter, although usually so considered. There are some who think of the throttle as controlling the volume of air fed to the cylinders; instead, the volume of charge per cylinder is constant and the power of the engine is regulated by the change of the mixture density in this fixed and continuous volume of cylinder charge. Now, if we select any engine speed and vary the throttle position under changing load with the correct mixture proportion, we find that the charge density in the manifold, as shown by barometer readings, bears a definite relation to the torque developed; also that this relation is almost identical for all engine speeds. Fig. 2 contains a set of curves illustrating this, and Fig. 3 gives a general average of the results of our record tests. We believe that the curving over of the lines of Fig. 2 is an indication of the intake-valve restriction and that it is

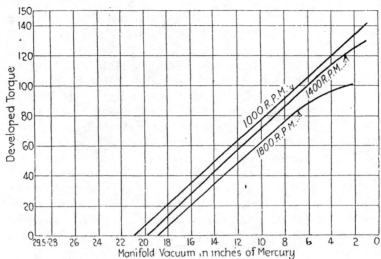

Fig. 2—Curves Showing the Relation between the Vacuum and the Torque Developed by an Engine

possible to determine the valve limitations by this method of analysis.

The maximum torque is usually obtained at not over 1½ in. of vacuum. This cannot be given as an exact figure because the vacuum registered is partially due to air velocity and partially to kinetic energy losses, and for other reasons, but 1½ in. is a fair average for all speeds with engines of four or more cylinders on one manifold. One of the reasons for not securing an air charge at atmospheric density, or 100 per cent volumetric efficiency, is the expansion of the air charge when it enters the hot combustion-chamber. It is obvious that under such conditions a slight initial rarefaction of the air charge, as expressed by this 1½ in. of mercury vacuum, will have little or no effect upon the charge density at the end of the intake stroke.

The developed torque decreases with an increasing vacuum or a decreasing barometer down to zero with the engine idling at about 18-in. vacuum. It is very difficult to find what charge density corresponds with zero indicated torque and we can only guess at these points, with the volume of exhaust residue as a guide. For any given cylinder temperature and compression-ratio it should be possible to construct a line such as that dotted on Fig. 2, from which the density of air charge corresponding with any given torque can be estimated, according to the horizontal distance between the torque and zero consumption line. The space at the left of the zero consumption line represents the volume of exhaust residue in the cylinder at the beginning of the intake stroke.

The left portion of Fig. 4 shows the gain of torque with decreasing speed, for a given size of air-opening. The line a indicates the torque that would be obtained with a throttle-opening which would give a 16-in. vacuum at 1000 r.p.m. This is about the torque required to drive the average car on a not too smooth country road. If the road becomes rougher or a hill is encountered so that the speed of the engine is reduced to 800 r.p.m., this will decrease the air velocity through the throttle orifice, which will in turn reduce the manifold vacuum with an increase of about one-third in the torque as shown. A further increase of the load will slow down the engine and raise the torque still more, until, with this same throttle-opening, a full air-charge will be delivered to the engine, and full torque developed at about 130 r.p.m.

A similar condition exists with regard to the resistance of the entire carbureter at wide-open throttle and high speed. The lines b and c at the right of Fig. 4 show a comparison of the power developed by two carbureters, one of which, b, has 2 in. of mercury vacuum at 2400 r.p.m., and the other, c, considerably smaller, 4 in. The

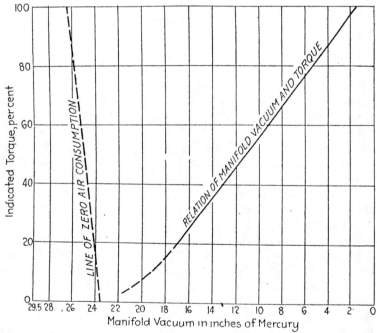

FIG. 3—MANIFOLD VACUUM-TORQUE CURVES BASED ON ACTUAL TESTS

dotted line is a rough approximation of the power required to drive a 3400-lb. touring car, at road speeds in the relation of 22 m.p.h. to 1000 r.p.m. of the engine. It will be noticed that the difference between the carbureters b and c is insignificant at 2000 and disappears entirely at 1800 r.p.m. This is the reason carbureters of widely differing air-capacity may show so little difference in steady pulling torque at ordinary driving speeds. Nearly every engineer remembers that sometime in his experience he expected to get a tremendous gain in power from a car by using a larger carbureter, and how very slight a gain, if any, was shown. Occasional instances have been reported in which the larger carbureters, and vacuum of less than 1 in. of mercury, gave better torque than carbureters more nearly the customary size, but I think that in each case careful investigation would have shown that the gain was due to something else than an increased delivery of the air charge.

Effect of Manifold Vacuum Upon Vaporization

As we have seen, decreasing the manifold vacuum tends in general to increase the torque, but another opposite effect is possible, due to changes in the rate of evaporation. Uniform mixture quality requires that the fuel-vapor density maintain a fixed proportion with the air-charge density. Also, we know that with the fuels used today the vapor density obtainable at ordinary atmospheric temperatures and pressures is much less than necessary to furnish a combustible mixture with air. Giving concrete estimates, at 80 deg. fahr. the vapor density of our average gasoline might average only about 1/45 of that of air, in which case, if you feed from the carbureter a fuel-to-air mixture of 1 to 15, only one-third of the fuel could and would vaporize. If, without change of mixture temperature, we close the throttle and cut down the density of the air in the manifold to one-third that of the outside atmosphere, a fuel-air ratio from the carbureter of 15 to 1 in this reduced charge will give a vapor density of 1/45 that of atmosphere and under such a condition the manifold will be substantially dry. This results from the fact that for average temperatures and grades of fuel there is a torque limit below which the mixture will be substantially dry. Further throttle-opening may not increase the torque in proportion to the increased air-charge because of incomplete vaporization, a condition that can be remedied somewhat

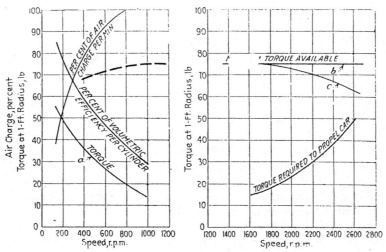

Fig. 4—Curves Showing the Relation between the Torque and the Speed of an Engine

by enriching the mixture so as to get a greater vapor-density from the light elements of the fuel. Stated in another way, for every change of temperature, below a certain critical point, there may be a change in the mixture proportion needed for maximum power.

Another effect of this manifold vacuum condition, less evident in a theoretical consideration of the subject than in actual experimental work, is the invitation to air leakage into the intake system. This tendency to leakage is the highest when the weight of air-charge per cylinder is least, so that a very small leak at the exhaust-valves, valve-guides or manifold joints may seriously disturb the operation of the engine under light loads, but show no effect on full-load pulling.

This relation of vacuum and torque explains the failure of the multitude of "economizing" devices put on the market in recent years, which consisted of suction valves designed to be mounted on the intake-manifold, opening inwardly under the vacuum. Their opening was greatest when the throttle-opening and the air delivery of the carbureter was least, and in no possible way could they be made to exert anything like a uniform corrective effect upon the mixture at different engine speeds.

Air-Flow

A study of the air-flow in carbureters presents continuous difficulty because this flow, while in reality following known physical laws, is often deceptively disturbed by factors whose presence is not evident. As it is difficult for many to understand how air-flow in the carbureter can generate "suction," it may be best to begin our consideration of this subject with an analysis of the conditions of energy existing in the airstream.

According to our present conception of matter, a gas under pressure is made up of molecular particles in rapid motion, individually irregular as to direction and velocity, their average impact against the retaining walls developing what we observe as the gas pressure. The amount of physical energy present in this body of gas will be proportionate to the number and mass of these molecular particles and to the square of their average velocity, the manifestations of this velocity also appearing in what we call the temperature. If the side of a vessel containing such a gas be opened into a passage of lower pressure, the gas particles will issue forth, but without any addition or subtraction of energy. The total amount of energy present in their motion must be the same as before, so that if this stream of particles has a definite direction, instead of being balanced in all directions as before, whatever velocity is gained in this one direction of flow, means a reduction in the relative molecular velocities normal and opposite thereto. That is, when a fluid enters upon a condition of flow, there is a corresponding reduction in its pressure. If we can get this conception of air at rest consisting of molecules in motion in all directions at a tremendous velocity, and of the air as it flows through the carbureter to the intake-manifold having this molecular velocity slightly increased in the direction of flow to the engine and correspondingly decreased in other directions, we will then be able to understand many of the apparent anomalies which our experimental determinations seem to disclose.

Much of our experimental uncertainty arises from the difficulty of measuring average pressures in the airstream. In a carbureter or intake-manifold the stream is nearly always turbulent and non-uniform; also nearly any sort of pressure-measuring tube is likely to generate eddies around its measuring orifice, so that the greatest care is necessary for any reliable determination of the air-pressure distribution. If we consider the action of air-flow past a resistance concentrated at one point of the passage, such as the venturi tube in a carbureter, an equation based on the laws of conservation of energy takes the following form:

$$V = \sqrt{([2gP_1U_1] [N \div (N-1)] [1-(P_2/P_1)^{(N-1)/N}])}$$

where

    g = the acceleration due to gravity
    $N = 1.41$
    P_1 = the external air-pressure which is assumed to be 2102 lb. per sq. ft.
    P_2 = the air-pressure in the orifice
    U_1 = the specific volume of the external air or 13.13 cu. ft. per lb.
    V = the air velocity in the orifice in feet per second

The above, plotted in Fig. 5, is the form commonly incorporated in textbooks and gives the velocity at the throat in terms involving the pressure at the throat, which is variable. I have found that the application of this equation is more convenient if the velocity term is modified to give, instead of the actual velocity, the velocity that would exist if the density were the same at the restriction as in the air entering the carbureter. With such a velocity term it is possible to compute much more readily the weight of gas flowing under known areas and venturi throat pressures.

$$V_d = \sqrt{([2gP_1U_1] [N \div (N-1)] [(P_2/P_1)^{2/N} - (P_2P_1)^{(N+1)/N}]}$$

The notation in this equation is the same as in the previous one except that

    V_d = the velocity at the initial density or external air-pressure P_1 that would give the same weight of flow as the velocity V at the pressure P_2

These values also are plotted in Fig. 5. In both curves the rate of air-flow in feet per second is given in terms

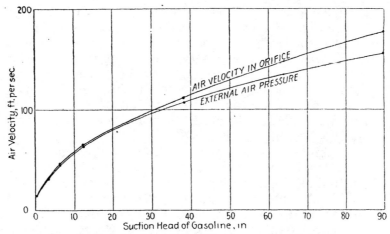

FIG. 5—CURVES OF THE AIR-FLOW AND AIR VELOCITY

of the corresponding reduction of pressure in inches of head of gasoline of 0.75 specific gravity.

As is shown by an inspection of these equations, a peculiar condition arises when the air discharges into a vacuum or pressure of less than half the original absolute pressure. In general, the rate of flow increases with the drop of final pressure, but at a certain point, about 14½-in. vacuum, a limiting velocity is reached, which is, by the way, the velocity of sound. This has a bearing upon the flow of air past small throttle-openings. At all vacuums above 14½ in. of mercury the velocity of flow is constant and the quantity of air-flow going to the engine depends only upon the effective area of the throttle orifice.

The Venturi Tube

From the viewpoint of the carbureter engineer we are interested in the air-flow only as a means of obtaining the desired fuel-flow; otherwise we would not interfere with it except at the throttle to control the power output of the engine. We obtain the fuel-flow by locating the fuel-discharge orifice in a region of suction or depression in pressure and naturally we will have the greatest range of action by getting the greatest possible depression for a given rate of flow. In other words, we desire equally an unrestricted air-flow, which usually implies low velocity, and a high suction. These two apparently opposing requirements have led to the general adoption in carbureter design of what is commonly known as the venturi-tube form of air passage, shown at the left of Fig. 6. The action of the molecular forces in passing through this shape of channel is obvious upon consideration. The rounding or converging entrance gives a free air-supply right up to the point of greatest contraction. Upon passing this the components of pressure against the expanding taper of the walls yield an onward reaction that permits a rise of pressure. This rise or recovery of pressure at the outlet is what distinguishes the venturi-tube action from that of a bushing or choke.

One thing I do not understand, and for which I have never seen an explanation, is the fact that there is a definite limit upon the taper of the discharge part of the venturi tube; above 7 deg. on a side the action seems to decrease markedly, regardless of the velocity or, so far as we have measured, of the density of the medium. At smaller angles than this the efficiency falls off slightly, to an extent that might be due to the wall friction. The angle of entrance seems to be of minor importance provided there is a smooth transition at the throat to the diverging contour. One point in venturi-tube design which is seldom emphasized is that the outer or collecting part of the entrance should be definitely larger in area than the outlet, in the same way that it is necessary to obtain a high coefficient of discharge for a given size of plain orifice. The dotted line at the left of Fig. 6 may show more plainly what I mean.

Another thing not generally understood is that the flow of gasoline in a venturi tube greatly disturbs its action,

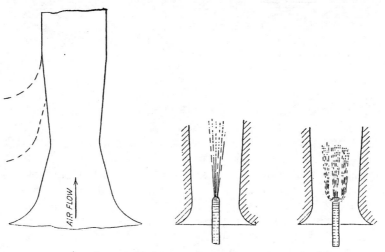

Fig. 6—Types of Carbureter Jets

the derangement increasing with the rate at which gasoline is fed and the extent to which it is atomized. Some of the loss in energy of air-flow is occasioned in accelerating the gasoline from the jet, as shown in the views in the center and at the right of Fig. 6. In the latter, which is the jet with the side outlets, the fuel is shown by its parabolic line of travel, which we have often observed, to accelerate to the speed of the air in a very short distance of travel; this generates a fine atomization, but considerable resistance to air-flow. A jet such as that in the center does not exercise so great a restriction on the air-flow, but the fuel is not so finely broken up nor will the spray spread as widely in the same shape of passage as with the one shown at the right.

The general conditions vary so widely with the design of carbureter that no general analysis of the flow of air through air-valves is possible. It should be mentioned, however, that very few carbureter air-valves give a constant pressure-regulation for the reason that there is a diminution of pressure on the surface of the valve near the approach to the opening orifice where the velocity is high. The area of this reduced-pressure section increases with the valve-opening. For instance, in the case of the

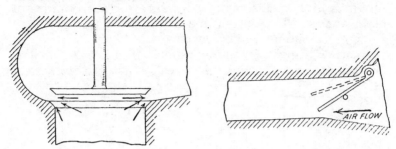

FIG. 7—WEIGHTED AND HINGED TYPES OF AIR-VALVE

weighted air-valve at the left of Fig. 7, if the area of the valve were 4 in. and the weight 2 lb., the valve would begin to open at a suction of ½ lb. per sq. in. and maintain this pressure above it for the first small opening. By the time the crack of opening became appreciable there would be a diminution of pressure around the circumference of the valve near the crack, so that the vacuum above it might perhaps reach ⅝ lb. As the opening increased, this action would increase still further. Therefore a valve of this type cannot be depended upon to give anything like a constant vacuum in the carbureter.

This action is still more marked in a hinged type of valve inclined slantwise across the air-opening. I have experimented with a valve such as that shown in Fig. 7 at the right and found that, with no spring at all, if I held it wide-open, it would suck to the position shown by the dotted line at a high engine speed. It will be noticed that the valve constitutes in effect a nozzle-shaped entrance and the fall of pressure within it is very similar to that in the entrance to the venturi tube shown at the left of Fig. .

FUEL-FLOW

The rate of liquid-flow through an orifice of fixed size is customarily and most conveniently gaged according to the pressure causing the flow. This pressure may be that of gravity as shown in Fig. 8, or a difference of pressure measured in terms of the gravity-head. In either case, under an equality of forces, the rate of flow should be the same. In our work we usually refer to the pressure-difference causing the flow in terms of gasoline-head. Disregarding the fluid friction in the passage leading to the jet of the gravity feed, Fig. 8, the velocity of the liquid discharged from the orifice should equal the velocity of a falling body which has passed through the distance h. Consequently, the rate of flow V should equal $\sqrt{2gh}$.

Fig 9 shows approximately the flow, under gravity-head, of a No. 52, 0.0635-in. diameter jet and a No. 60, 0.0400-in. diameter jet, plotted against the square root of the head. On such a diagram, a flow corresponding to the equation just mentioned would intersect the corner, as shown. If we assume that the line corresponding to this flow is an asymptote of the straight-line curve of the No. 60 jet, and would give a uniform mixture proportion through a considerable range of air-flow, the actual mixture proportion of the jets would plot as shown in Fig. 10.

The relation of the actual flow to the mathematical one, which we may call the coefficient of efflux, is in my opinion due mainly to surface tension, for the following reasons:

(1) At times when this coefficient is low the fuel can be seen very definitely clinging to the tip of the jet
(2) Submerging the jet and providing means for drawing off the fuel as fast as it is supplied to a normal

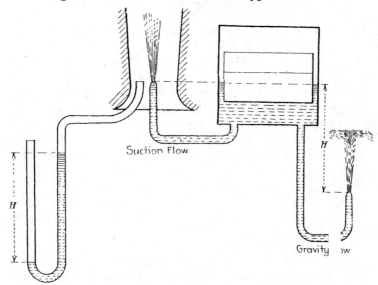

Fig. 8—Suction and Gravity Forms of Fuel-F w

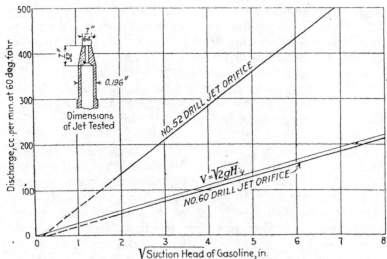

Fig. 9—Curves of Fuel-Flow through Two Carbureter Jets Plotted against the Suction-Head

level, without imposing any additional suction on the jet, will give very close to a uniform ratio of flow to the square root of the head

(3) Introduction of air near the outlet of a submerged jet, with sufficient definite bubbling within and a splashing discharge from the jet will give a quantity to give a substantially straight-line mixture curve

It has been suggested that the height of the jet above the level is responsible for this deficiency. This height is to a certain extent responsible, but inspection of curves similar to those of Fig. 9 shows that the jet would need to be about $3/8$ in. below the level to give an approximately straight line.

It will be noticed that the flow from a plain jet has varying characteristics. If the jet can be used under suctions whe its range of action extends from a to b, it will give a substantially uniform mixture. If, on the other hand, as in the traditional type of air-valve carbureters, its a ion begins at c, from c to a its coefficient of discharge i constantly increasing and an air-valve or similar regulation will be necessary to keep the mixture from becoming unduly rich. Some years ago, when we were working with air-valve carbureters, we found by experience that it was scarcely ever wise to go below $1\frac{1}{2}$ to 2 in. of suction on the plain jet for idling, that is, the lowest part of the delivery range. Below this point the action was very uncertain, the discharge being particularly affected by the temperature, which seems natural when we consider how largely the coefficient of efflux under these heads is governed by the surface tension. From time to time efforts have been made to get a large high-to-low speed-range performance from a plain jet by coming down to very low metering heads at the lower end. As nearly as we can find out, this invariably results in the difficulties just mentioned, very uncertain action and undue sensitiveness to temperature change, as well as pronounced differences of action with fuels having different viscosity values.

It may be worthwhile at this time to analyze the matter of carbureter range. It seems to be the general belief that the range of a carbureter is governed by its range of air-opening sizes. This, is, of course, one limitation, but the size and capacity of the fuel orifice is equally responsible. As a sample computation, the No. 60 plain jet above described, at a minimum operating suction-head of 1½ in. will pass 23 cc. of gasoline per min. If we assume that the limiting manifold depression at high speed is 2 in. of mercury and that the efficiency of the venturi system employed in the carbureter is such that a 2-in. manifold depression gives 4 in. of mercury depression on the jet or 64 in. of gasoline-head, the flow from the jet at this time would be 214 cc. per min., a ratio from high to low speed of 9.3 to 1. A greater range than this from such a jet can be obtained only by increasing the efficiency of the venturi by in-

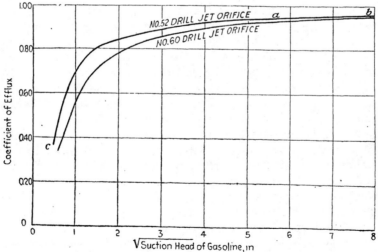

FIG. 10—THE ACTUAL MIXTURE PROPORTIONS OF THE TWO JETS WHOSE FUEL-FLOW IS GIVEN IN FIG. 9

creasing the manifold vacuum, or by going lower on the low end of the metering curve.

A jet structure whose coefficient of efflux is free from the retarding influence of surface tension and gravity will meter down to very much lower heads, with the result of a considerably greater range. For example, with a minimum metering-head of 7/16 in., a value easily obtained, and a maximum suction of 64 in. at a constant coefficient of efflux, the range will be equal to the square root of this suction ratio, or 12.1 to 1. In modern carbureter practice the venturi efficiency gives considerably higher suction than this, so that with the help of an auxiliary idling jet a ratio of 50 to 1 from minimum idling to maximum wide-open throttle capacity is easily obtained with fixed-size air and fuel openings. It is obvious that a further marked increase of range would necessarily involve a proportional increase of both air openings and fuel jets, with considerable mechanical complication of the carbureter structure and a loss of the inevitable precision of metering characteristic of the fixed-opening type. In dealing with the question of carbureter range, it should be borne in mind that the maximum air demand of the engine will be reached at about the point of maximum high-speed indicated horsepower; also that the air capacity of a carbureter system is not a definite amount but increases considerably with a slight increase in the manifold vacuum.

There are a few miscellaneous points, which may be of interest, concerning the behavior of the fuel in the air column in the carbureter. The most noticeable characteristic of the gasoline, under observation of glass models, is the extent to which its action is governed by the surface tension. It clings to every surface it touches and would always much rather run along the surface than fly off. Very many jets that have been proposed as having fine spraying and atomizing qualities prove exactly the opposite, on account of this clinging effect. The fuel spray tends always to seek out and deposit at the points of low velocity. One of the most difficult things in developing a carbureter of very high speed-range is the problem of keeping the spray in suspension. To vary the fuel and the air openings in proportion is much less difficult than to assure the gasoline rising steadily from the small low-speed air opening up to the throttle and the intake-manifold through the large body space necessary for the full ir-opening. The tendency of the fuel to gather and load in this space is always very pronounced. In fact, much of the difficulty ascribed to lack of compensation in the jet delivery is really a failure to get the fuel from the jet to the throttle

MIXTURE-PROPORTION REQUIREMENTS

Since practically all of our automotive applications of power have a reserve capacity, that is, their full power-range is in excess of their normal operating power, it follows that the mixture-proportion requirements can to some extent be gaged as dependent upon the position of the throttle. At partially closed throttle maximum power is not a consideration, and a mixture of maximum economy is desirable. At wide-open throttle, however, the maximum-power mixture is desired in nearly every case, provided this can be obtained without too much sacrifice in fuel consumption. The records of the Society contain many experiments to determine the exact mixture proportion required to give these two characteristics, maximum power and economical fuel-consumption and, as might be expected, the values obtained vary with every engine tested. It is obvious, first, that the fuel supply must be increased whenever there is any unequal distribution of the fuel, as the weakest cylinder must be supplied. Low fuel-consumption, therefore, requires either a partial vaporization or fogging of the mixture, or else a high air-velocity in the manifold along with the proper contour. Since the vaporization is so much more complete at closed than at open throttle, our distribution troubles and the manifold contour are important almost only at wide-open throttle and for this reason, with the average motor-car engine, we can run on a thinner mixture at part throttle, or above about 12-in. manifold vacuum, than at wide-open throttle. That is, the mixture of maximum power at part throttle is nearly always leaner than the mixture of maximum power at full throttle. Under circumstances where the fuel evaporates completely in the intake-manifold, the leanness of mixture that can be used is limited by the ability to ignite and propagate combustion, for which conditions are more favorable at the full compression of a wide-open throttle than at the lower compression of part throttle, with also a higher percentage of exhaust dilution.

We do not find any reliable evidence that an increase in the fineness of atomization beyond a certain point, reached in nearly all carbureters, will give any effect upon power or fuel consumption, provided the discharge from the jet is steady and not intermittent. It seems that vaporization reaches its limit at 100-per cent atomization. It is, of course, worthwhile to keep the spray off the manifold walls so far as possible.

In illustration of the foregoing, Fig. 11 is a typical mixture-proportion curve. The wide-open throttle setting, being that of maximum power, will usually be uniform for intermediate and high speeds but somewhat richer at low speeds, both because of poorer manifold conditions at this time and because detonation is less likely to occur with such a mixture. The closed-throttle range may be leaner than that at wide-open throttle at ordinary driving speeds and such a setting will give exceedingly smooth operation when the engine is in good condition. When the valves begin to leak, a richer mixture may be needed to cover up partly the irregularity of firing charge. At idling the very thin air-charge, the high percentage of exhaust residue and the uncertain leakage of air through the valve-guides make a very rich mixture necessary. The exact mixture-proportion required under each of these conditions will depend, as can readily be understood, upon the individual characteristics of the engine and the fuel used. Also, they vary quite definitely with the service for which the engine is used. An effort will be made to analyze these requirements under the different divisions of passenger cars, motor trucks, tractors, motorboats, airplanes, racing cars and dynamometer tests.

Passenger-Car Mixture Requirements

For passenger-car service the predominant requirement today is ability to accelerate from closed to open throttle at the moderate or legal range of speeds. Practically all carbureter adjustments are made to give this acceleration, most drivers gage the ability of their car by its response to the throttle, and a large part of our factory demonstration work on the road is directed toward the attainment of this important point. This is natural and proper because the full power of the engine in present-day driving is nearly always used to give a gain in speed, except in a few hilly localities, and many times when it is called upon, the safety of the driver depends upon its response. At present, country roads and city streets alike are so generally crowded that the time made on either a long or short trip depends more upon the

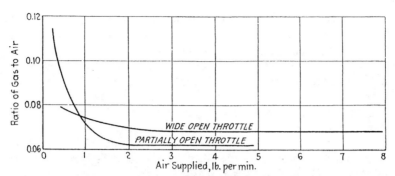

Fig. 11—A Typical Mixture-Proportion Curve

ability of the car to thread its way through traffic, to slow down and pick up quickly, than upon the maximum speed. This matter of acceleration is so important that nearly any carbureter will receive favor, regardless of faults in other details, provided it fulfills the function of giving prompt and positive acceleration.

So far as the action in the carbureter is concerned, acceleration requires, first, a mixture-proportion and condition of vaporization that will pull full power at wide-open throttle and low speeds. I might comment on the fact that, while it is customary nowadays to take power curves from 3500 to 800 r.p.m., it is very seldom that the dynamometer horsepower curves are carried down much lower. But just as soon as we make a carbureter test on the road we try to get acceleration from something like 1 m.p.h., which means full power from 60 r.p.m. up.

We further find that in addition to a powerful steady running condition, acceleration is greatly improved by a temporary excess of gasoline. The need of this excess feed was very clearly explained by the recent experiments of the Bureau of Standards with glass intake-manifolds. It was found that at wide-open throttle, with ordinary fuels and the ordinary range of temperature available, a large part of the gasoline travels from the carbureter to the cylinder along the manifold walls at a rate of speed relatively much lower than the air-speed. At closed throttle so much greater a percentage of the fuel evaporated that almost none was on the walls. Now, in throwing the throttle open from the closed position, the wall-flow took a considerable time to beecome established, during which period the cylinders were receiving only the percentage of gasoline that evaporated. Under this condition, with a normally lean mixture from the carbureter, the mixture reaching the cylinders was very much too lean for power. By raising the proportion of the gasoline fed to the air, the amount of evaporable elements could be increased to a point where they alone would supply a firing mixture without the aid of the heavier elements. It is obvious that continuous use of such a mixture would mean a considerable waste of fuel, but the actual amount needed to fill the temporary deficiency on acceleration is small. We find from experiments that a thin steady running mixture, with means for temporarily enriching the mixture as necessary for acceleration, will give better acceleration than a carbureter without such an accelerating device, set so rich that the engine lopes in driving along the street at closed throttle. It is rather remarkable that the application of different means of heating the walls of the manifold does not do away with the need of this accelerating extra fuel-supply, although it cuts down the amount required. In comparison, the use of hot air on the carbureter comes nearer giving acceleration on a normal mixture. Visible observation shows that when such conditions are attained the fuel evaporates almost entirely within a few inches from the jet and that there is much less wall-flow. According to the analysis just given, the extra charge on acceleration should be delivered at the moment of the change from high to low manifold-vacuum, and we find in practice that this is the case. This requirement cannot be fulfilled with precision by accelerating means dependent upon the throttle opening nor upon an increase of the carbureter suction, since neither of these coordinates invariably with the fall of the manifold-vacuum; on the other hand, some precautions are necessary to prevent the large accelerating charge that is necessary for some engines from all going into one or two cylinder ports. However, too much is far less objectionable than too little and there are a number of satisfactory accelerating devices in use which do not conform entirely to the above statement of requirements.

Fuel Economy.—We all try to obtain low fuel-consumption as a characteristic of passenger-car operation. Since the average speed of a car in most sections of the country lies between 10 and 35 m.p.h. and at closed throttle, the normal driving range indicated in Fig. 1 should be set for best economy, although in practice drivers will not use such a mixture setting unless it is possible at the same time to get good acceleration.

There are two factors, really wholly outside the scope of this paper, that have much to do with the fuel consumption. The first is the mechanical efficiency of the engine. With the average 3500-lb. car having a 4½ to 1 gear-ratio about 5 hp. is required to drive the car at a speed of 20 m.p.h., about 1¾ hp. to overcome the loss of pumping against the manifold-vacuum on the intake stroke and about 3½ hp. to overcome the friction of the engine, a mechanical efficiency of 48 per cent. As many cars are delivered with their engines very stiff, for the first 2000 miles the mechanical efficiency is less than 35 per cent. With, for instance, a 2¼ to 1 gear-ratio in a four-speed gearbox, the pumping loss would be about 1 hp. and the engine friction about 1¾ hp., so that at the same power output the mechanical efficiency would be about 65 per cent. If the 4½ to 1 gear-ratio gave 16 miles per gal., the 2¼ to 1 ratio should give 22. A trial of a well designed four-speed gearbox with a flowmeter indicating the fuel consumption, will show what a tremendous effect this factor exerts. An average value of the torque required to propel a car at moderate speeds on smooth level roads is 50 lb. per ton of car weight. At a fuel consumption of 0.65 lb. per hp-hr. a 3500-lb. car would travel 41 miles per gal. and a 2000-lb. small car, 72 miles. The difference between these figures and what we actually do make is due less to deficiencies in our engine cycle than to the fact that we operate at so low a percentage of the normal engine torque. A fuel-injection engine would probably be even less efficient under these same conditions, on account of the extra power required for driving the various auxiliaries not found on our present type of engine.

Another factor that we find in our service-stations raises the passenger-car fuel-consumption is the tendency of the average driver to retard the spark to prevent the engine from knocking at wide-open throttle, and leave it that way for steady driving at closed throttle. That is very common with cars having high-compression engines and poor manifold systems which invite the accumulation of carbon in the combustion-chambers. For ordinary legal-speed driving at from 14 to 16-in. manifold-vacuum, the spark should be advanced almost to the point necessary for maximum car-speed.

Hill-climbing at ordinary speeds is largely a question of proper mixture proportion and favorable conditions of vaporization. The capacity of the carbureter does not enter much into this because nearly all carbureters, as has been brought out, are adequate in size for the speeds at which hills are climbed. At low manifold-velocities and wide-open throttle with the present heavy fuels there is an accumulation in the manifold that increases rapidly and if it goes into the cylinders in occasional slugs will give the action known as "loading." It is not generally appreciated, but a hill test can be made a very delicate demonstration of the nature of the torque curve of a car, particularly with engines whose torque drops off at low speeds. Referring to the dotted-line curve at the left of Fig. 4, if the inclination of the hill is such that it just equals the torque of the engine at this speed, any slight increase that could be made in the torque would increase the engine speed to a region of increased torque whereby the engine would accelerate and continue to gain speed, provided the inclination of the hill were constant. On the other hand, the merest diminution in torque, even a single misfire in one cylinder, would cause the engine speed to drop slightly to a region of decreased torque, under which the car would start to lose speed still further and it would probably be necessary to shift gears within 60 ft. In the hill testing that I have witnessed, particularly before proper methods of manifold heating came into vogue, this point was a source of very great waste of time and much useless work. The skill of the driver and the natural variations of manifold temperature made big differences in the hill performance and all engaged in the test used to rack their imagination to find explanations for the different results we seemed to find.

Speed, in the average driver's estimation, is what we might characterize more definitely as good acceleration up to 45 m.p.h. Such determinations as we have been able to make show that the power required for high speeds crosses the curve of engine torque at such an angle that considerable variations of torque at the very high end make only small changes in car speed and such apparently small factors as wind direction and wheel-traction road-surface make very considerable differences. For long stretches and speedway work speed is simply a question of having enough air and the right amount of fuel to go with any type of carbureter which I know of that is positive in its operation. Our service experience indicates maximum car-speed is to the average car-owner the least important of all requirements; not even 1 per cent of our complaints are directed to this point. It is not so much that the drivers do not think they want speed as that they do not use it often enough to know whether they have it, and it takes a very considerable loss in engine power to make a change in maximum speed that the average driver can detect.

A carbureter for any service should, in my opinion, be reasonably simple and positive in operation. It should be possible for production carbureters to be made alike and so that after any reasonable overhauling any one can be restored easily to its original condition. The carbureter should be as wearproof and dustproof as possible, so that it will not deteriorate or change adjustment during the life of the car. This is rather difficult to insure with carbureters of the variable gasoline orifice or metering-needle type. In every case this type of carbureter should have a dashpot piston to keep the attendant air-valve from hammering and wearing the seat of the metering-needle seat. It is advantageous to have the idling mixture adjustable separately from the remainder of the mixture range, because slight manifold leakages, valve-stem leakages and the like affect this proportionately much more than other parts of the mixture range. The action of the carbureter should be unaffected by the jolting of rough roads.

Motor Trucks

The range of operation of motor trucks is such that the engine works through a lesser range of speeds than in the case of passenger cars, and the manifolds are usually designed so that this speed range corresponds with the higher range of manifold-velocities, with the result that carbureting conditions are much easier. Also, four-cylinder engines are as a rule easier to handle than those of larger numbers of cylinders, due mainly to conditions of manifold distribution. These advantages are offset, however, by the practice, which is extensive in motor-truck construction, of dropping the carbureter very low down to get gravity fuel-feed from a gasoline tank under the seat. This necessitates a long manifold, and very few trucks indeed are equipped for proper heating of the fuel charge after it leaves the carbureter, with the result that the fuel condenses in the manifold and the difficulties of acceleration are considerable. The governor valve and its chamber also act as intercepting and loading points for the fuel spray. I believe that experience will show that the life of the engine in a motor truck can be very greatly prolonged by proper design of the intake-manifold and correct selection of the carbureter and that this consideration, even more than that of fuel consumption, will soon lead to the use of more effective provisions for vaporization on motor-truck engines.

TRACTORS

The duty of a carbureter in tractor work is considered easy because the load is supposedly steady. Actually, however, there are periods in service when flexibility is needed badly. With kerosene as fuel it is possible with high manifold-velocities and by use of a carbureter having good atomizing properties and a charge well diffused through the airstream, to perform pretty well with a low mixture-temperature, provided the cylinder jackets are fairly warm. Even so, we can scarcely say that successful general operation on kerosene has been attained as yet. The tractor carbureter should be of the simplest construction and entirely dustproof.

MOTORBOATS

As will be noted from the diagram at the right of Fig. 1, the range of performance of an engine attached to a propeller falls within very definite limits; also the condition of unfavorable evaporation at wide-open throttle is encountered only under the correspondingly favorable condition of high manifold-velocities. The requirements of motorboat carburetion are, therefore, the simplest of any automotive branch. Due to this condition, however, only recently has there been any pressure upon the designers to provide for adequate heating of the intake-manifold, and very many engines are in use today whose performance suffers on this account. Nearly all the later designs show that these considerations are now being given attention.

AIRPLANE CARBURETION

In airplane service the propeller load and the fact that relatively highly volatile fuels are used make the engine typical of this field the easiest on which to demonstrate carbureter performance on the dynamometer test-stand. But in the air there are a number of special requirements which render the airplane field perhaps the most difficult of all the branches of carbureter work. First, the mixture tends to grow richer with increase of altitude, in the proportion of the square root of the change in the air density. Second, it is desirable that the carbureter should function properly in every position that the airplane may assume. Third, the carbureters are usually more or less subject to the deranging influence of the propeller blast, which sometimes almost equals in magnitude the velocity of the air past the metering jets of the carbureter. It is nearly always possible to compensate for this by balancing the float-chamber of the carbureter on the air entrance. Fourth, the high mechanical efficiency of airplane engines and the fact that a number of carbureter units are usually employed on one engine, make it very difficult to obtain a satisfactory idle and low-speed range, and, on this account, positive and certain acceleration. When we add to these points the requirements that the carbureter should give at all times the smoothest possible firing of the engine, that effective precautions must be taken against any stoppage of the jets or mechanical failure and that every effort should be exerted to prevent backfires, it is evident that much work is necessary before a perfect airplane carbureter can be produced.

Carbureter and Engine Tests

In our demonstration work throughout the country we encounter widely different methods of engine testing to determine carbureter merit, some good and some, in my estimation, not so good. In view of the importance of this work it seems worthwhile to set forth, for discussion at least, what seem to be the essentials for obtaining accurate and trustworthy information.

Dynamometer tests should obviously reproduce conditions of actual operation, not only as to the temperature of cooling water, the air around the engine, and the air entering the carbureter, but also as to speeds and loads. The full power-range of a passenger-car engine should be brought down to the speed demanded in road operation, while the fuel consumption should be taken at loads corresponding to normal driving. Most important of all, in my estimation, the power should be plotted as torque against speed. The method in general use at present of plotting the horsepower, which is the torque times the speed, against the speed, results in a form of curve in which undue prominence is given to a relatively unimportant and rarely used part of engine operation. When one looks at such a power curve the eye instinctively goes to the peak and gives this consideration to the exclusion of everything else. Yet nearly every car makes its maximum road-speed at a higher number of engine revolutions than that of the peak of the power curve, so that it is the power beyond the peak that determines the maximum car-speed. Also the horsepower-speed curve is very unsatisfactory as a means of indicating low-speed pulling ability, or as a basis for fuel-consumption tests at part load. In fact, it does not at all depict the power performance or ability of a car, except possibly for advertising purposes.

Since the power of the engine at full air-charge is dependent upon the density of the air entering the carbureter, this should be recorded, and corrections made for temperature and barometer in all tests of maximum torque. Care should be taken to have the air-supply clean and not contaminated by the exhaust. If the intake-manifold is near the exhaust-manifold, at high speed and full load the radiation from the latter may heat the intake unduly and cause the power to fall off.

For fuel-consumption tests the friction horsepower of the engine should always be recorded and comparisons made upon the basis of indicated horsepower. This is particularly important with part-load tests.

Since present-day requirements are such that the carbureter adjustment used in service is determined by the ability to accelerate, it would be of great value to have a dynamometer equipped with a flywheel disc to give an inertia equivalent to that of the finished car. A disc of this sort is in use at the Bureau of Standards.[2]

There should by all means be in plain view a thermometer giving the intake-manifold temperature, without which intelligent comparisons cannot be made. It is advisable also to have a gage reading the intake-manifold vacuum, as this, along with the temperature, governs the rate of vaporization and the extent to which the mixture delivered by the carbureter will approximate that received by the cylinders. A flowmeter is very useful to give a general indication of the fuel consumption under different conditions of driving.

An easily read accelerometer is of great value. It not only gives an idea of the power developed but can also be used to read the road and wind-resistance, from the rate of deceleration with the clutch out, thus removing two otherwise unknown variables in a fuel-consumption test.

---

[2] See TRANSACTIONS, vol. 15, part 2, p. 173.

ELEMENTS OF AUTOMOBILE FUEL ECONOMY

It is believed that the foregoing covers the main points to be considered in the development of proper carbureter action. It will be noticed that this work falls definitely into two distinct parts; the first, a joint responsibility of the engine designer and the carbureter engineer, being to determine the specific mixture and vaporization requirements of an engine; and the second, for which the carbureter engineer alone is responsible, to produce an instrument that will deliver reliably and consistently such a fuel-charge with as favorable vaporization conditions as can be obtained. Successful results can be obtained only by close and cordial cooperation. I am glad to say that in the past few years this cooperation has been general and that our work has been more efficient and much more pleasant on this account.

BEST LOCATION FOR CARBURETER INTAKE[1]

By A H Hoffman[2]

ABSTRACT

Tests to determine the location under the hood of a motor vehicle where the air-intake of the carbureter will be exposed to the least dust were made by the agricultural engineering division of the University of California at Davis, Cal., and the results are given in the hope that they will serve a useful purpose. Of three types of dust-screen devised to catch the dust at different locations so that it could be photographed, and still would present little hindrance to passage of the air from point to point under the hood, the most effective was one of coarse hospital gauze stretched over frames set in transverse vertical positions on either side of and above the engine.

The tests were made on two phaetons and a speed truck, run for less than 3 miles and following another car on a dusty road. Photographs show the screens in position in the vehicles and removed to show the distribution of dust as collected under various operating conditions, as with the radiator fan idle and with it revolving with the fan-belt normally tight.

The conclusion reached as a result of the tests is that, in normal cases, the best carbureter-intake position would seem to be (a) on the side of the engine on which the fan-blades have a descending motion, (b) about midway between the radiator and the dash, (c) about midway between the side of the hood and the center line of the engine, and (d) about one-third of the distance down from the top of the hood toward the top of the cylinder-head.

---

[1] Los Angeles Group paper.
[2] Agricultural engineering division, University of California, Davis, Cal.

Where it would encounter the least dust is, evidently, the best place under the hood to locate the air inlet of a carbureter. Several tests just made at Davis, Cal., by the agricultural engineering division of the University of California had as their object to determine the relative dustiness of various locations under the hoods of three machines. The first efforts were not aimed at an exact quantitative determination and may be supplemented by more exact work. However, since the determination of the best available place is probably all that usually would be required in practice, it is hoped that the methods here indicated may serve a useful purpose.

A Ford phaeton, a Studebaker Light-Six phaeton and an I. H. C. 1-ton speed truck were fitted successively with devices for indicating to the eye or on the photographic plate the relative dustiness at different places under the hood. The machines were then run for less than 3 miles, following another car on a dusty road. The dust-holding devices were designed so as to offer little hindrance to the passage of the air from point to point under the hood,

FIG. 1—INDICATOR SCREENS OF WHITE COTTON TWINE IN POSITION

These Gave a Fair Indication of Relative Quantities of Dust at Different Locations under the Hood but Were Not Satisfactory for Making Photographs of the Results. The Picture Was Taken Before the Test Began

Fig. 2—Screens of Coarse-Mesh Hospital Gauze
These Were Used in Transverse Vertical Planes Only, in the Second Attempt To Secure Records

especially when the object was to determine as between front and back positions. Hence the metal frames for supporting the dust-catching materials and these dust catchers themselves were placed with their longer dimensions approximately parallel to the air stream as the illustrations indicate.

In the first attempt, ordinary white cotton wrapping twine, moistened with Household Lubricant diluted with four volumes of high-test gasoline, was wound around the frames, as in Fig. 1. The spacing for adjacent turns of the twine was about $3/4$ in. While this gave positive indications, it was not sufficiently definite to give good photographs.

In the second attempt, coarse-mesh white hospital gauze, similarly moistened with oil, was employed, as in Fig. 2. It was used only for determinations in transverse vertical planes, since its presence in longitudinal planes might conceivably have changed the air-flow under the hood from its normal direction.

In the third attempt, screens of No. 22 iron wire, with meshes $1\frac{1}{4}$ in. square, were woven on frames of

Fig. 3—Wire-Mesh and White-Felt Screens in Vertical Longitudinal Position

1/8 x 3/4-in. iron built to fit the space where the determinations were to be made. At alternate intersection points of the wires 1/3-in. squares of white felt, 1/16 in. thick and oiled as before indicated, were pushed in and held by friction between the wires, as shown in Figs. 3 and 4. These gave satisfactory indications.

It appears from the photographs reproduced in Fig. 5 that when the fan-belt is loose the dust distribution is more uniform on both sides of the engine and is considerably less above the midplane than below, as revealed by the darker portions of the screens shown in the photograph at the left. When fan-belt is normally tight, the fan tends to throw the dust up from the rising blades, thus increasing the dustiness in the upper spaces on that side of the engine, as indicated by the screens in the central picture.

In the case of normal fan-belts and no shroud around the fan, it was found that there was less difference in dustiness between the two sides of the engine but that an upper current of air and dust passed back to the fan just under the top of the hood.

For the normal case, it would seem that the best inlet-position would be (a) on the side of the engine on which the fan-blade tips descend, (b) about midway between the radiator and the dash, (c) about midway between the engine center-line and the side of the hood, and (d) about one-third of the distance down from the top of the hood toward the top of the cylinder-head. The presence of accessories, shrouds and the like would necessitate making a special test on the given machine to determine exactly what is the best location for the inlet under that particular set of conditions.

THE DISCUSSION

EMMETT F. ANNIS[3]:—We can hardly overestimate the value of the work of Professor Hoffman. He has taken hold of the air-cleaner problem as probably no other man has done before in the United States and has systematized

---

[3] M.S.A.E.—Consulting engineer, Los Angeles.

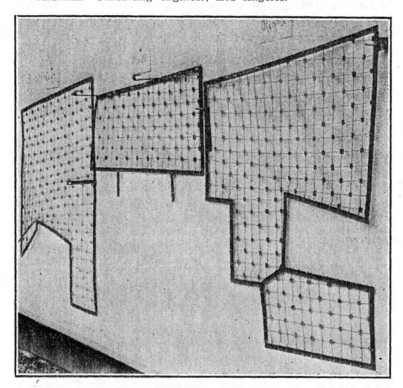

FIG. 4—SCREENS WITH SQUARE WHITE-FELT INSERTS

The Felt Squares Were Inserted at Alternate Intersections of the Wires and Moistened With Oil. Satisfactory Indications Were Secured with These Screens. This View Is From the Left Side, Rear Quarter. Concentration of Dust Just in Front of the Dash Is Evident. The Rear Edges of the Upper Felts Are Also Dusty. The Fan-Belt Was Tight in This Test

his methods of tests in such a way that now his methods are considered practically standard throughout the Country.

The question of air-cleaners is just beginning to receive the necessary attention it deserves. Heretofore the automobile engineer has devoted his attention to the designing and production of cars, without giving much attention to performance after they have had the demonstration tests. I know of one engineer who is designing a very high-class motorcoach engine and specifies as his objective a powerplant that will operate 100,000 miles without an overhaul. One of the first things he took up was the installation of air, oil and gasoline-cleaning devices.

The General Motors Corporation has appointed a board of 20 engineers to investigate air-cleaner methods and types of devices and to make recommendations to its factory heads on types of air-cleaner for all of the General Motors cars and trucks. Mack Trucks has also made an exhaustive investigation of this subject and many other companies are considering air-cleaners.

The most difficult problem we have to solve is that of education. The opportunity to sell air-cleaners, if we can convince the user of their real value, exists. Professor Hoffman gave his paper[4] on the final results of his 1924 series of tests at the last Annual Meeting of the Society and the principal discussion at that meeting was on air-cleaners and crankcase-oil dilution. Available data on the actual performance of air-cleaners in service are lacking to a great extent; the data are hard to secure and they vary considerably with different operatio s and methods of operation and also with the men wh are conducting the tests. An attempt was made last ummer to obtain some valuable data on air-cleaners fitted o a number of trucks operated by the California State ghway Commission but, unfortunately, when the trucks ere put into operation, the main objective was to hur y through with the jobs, and the control of the apparatus and the drivers was not vested entirely in those who were making the tests. As a consequence, much useful information was lost; cleaners were removed and not put back properly, and, when the results of the tests were analyzed, it was surprising how few reliable data were secured.

The Quartermaster Corps made one test several years

---

[4] See p. 161.

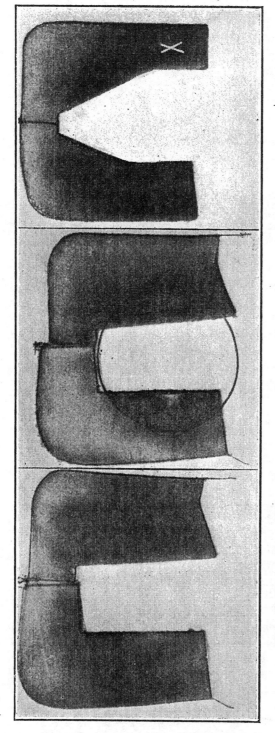

Fig. 5—Results of Tests Obtained with Transverse Vertical Screens of Hospital Gauze

Left View Shows Uniformity of Dust Distribution on Both Sides of Engine with Fan-Belt Off. Center View Shows Greater Dustiness in Upper Regions on the Side on Which the Fan Blades Were Ascending When the Fan-Belt Was Normally Tight. In Both of These Views the Screen Was Placed 9 In. Back of the Radiator on an I. H. C. Speed Truck. Right View Shows Little Difference in Dustiness of the Two Sides of the Engine with the Fan-Belt Normal but Notable Predominance of Dust in the Lower Regions. This Screen Was Located 8 In. Back of the Radiator on a Studebaker Light-Six Phaeton. The Position of the Carbureter Air-Intake Is Indicated by X

ago on two Liberty 4¾ x 6-in. engines, one with and the other without an air-cleaner. I do not have available the exact mileage of that test but it was conducted along the Mexican border and the engine without an air-cleaner was completely worn out after approximately 3000 miles, while the one with the cleaner showed so little wear that it could be continued in operation with very little overhauling.

The objectives are to (a) remove, as nearly as possible, all dirt from the air; (b) do it with the least possible restriction of air-flow and the least possible increase in fuel consumption and (c) have as little mechanical trouble as possible with the cleaner. The third objective is difficult of attainment with most air-cleaner constructions if high efficiency and low vacuum are maintained.

One designer of air-cleaners states that he set out to design a filter to fulfill the three objectives mentioned. He had the benefit of experience with the first types of centrifugal oil-cleaners and of the dry-cleaners having a felt-covered wire cylinder which had been on the market for a number of years, and realized that if he was to succeed in putting on the market a filter that would not choke he would need to have a filter sufficiently large to pass air freely and still not take too much space. He recalled that precipitation in fluids varies as the sixth power of velocity and believed that certainly in air travel the precipitation would be to the sixth root. I was skeptical of the results to be obtained with a filter-type cleaner but, after going over the tests, I was surprised to find an efficiency of 99.8 per cent and that the cleaner would not choke; it actually worked in practice, and, by slowing down the air current, it really did precipitate the dirt out of the filter. It did not depend entirely on the velocity; the direction of the air current was altered by lugs and it was also necessary to pass the air through a felt filter. The felt used had to be strong enough to withstand hard usage and moreover of a kind through which air would pass freely without restriction but which would take out the dust. The search for the felt cloth took more than a year. Now the cleaner has high-test dust extraction with freedom of air passage and a very low vacuum. I have yet to see any car burning more gasoline with the cleaner on than it used before. We never have changed a carbureter on account of installation of the cleaner, although in some instances we have had to redesign the fittings so they would work better. We make the cleaner in larger units for tractors and shovels and have established a service department and are watching installations carefully.

R. K. HAVIGHORST[4]:—I wish to emphasize Professor Hoffman's remarks about giving air-cleaners a chance by placing the intake where it will get the cleanest air possible. At the Denver tractor exhibition in 1919 the tractor representatives always removed the air-cleaners, evidently being willing to ruin the engines for the sake of giving a satisfactory 2 to 4-hr. demonstration. In one case, after a run of 3 hr., the top piston-ring was missing, the middle one looked like a safety razor blade and the third ring was very thin, though still measurable. I examined some other engines and found the experience almost identical. That drew my attention to the necessity of having efficient air-cleaners. Experiments were carried along in an unsatisfactory way until the Samson Company finally developed a cleaner of its own. During the summer of 1922 I had the privilege of viewing Professor Hoffman's experiments and obtained some ideas that were applied later. The air should be taken from a point where the dust is the least, if possible. In a passenger car it can be taken from under the seat. If the air is taken from a point where it is cleanest and if a cleaner of high efficiency is used and is connected so that no leaks of dusty air occur, and if the lubricating oil is kept clean, the well-designed engine will wear slowly.

EUGENE POWER[5]:—We are trying two or three makes of air-cleaner but they have not been on the trucks long enough to indicate beneficial results positively. We are convinced that they will do all that we expect of them but I have no information based on actual performance. It is essential that care be given them by the operators and that the cleaners be given a chance to do good work. After we had one in use for 2 months, we wrote to our district office to find out what the men there thought of it and the office replied that it never had been opened. Then they began to complain of carbureter trouble. It should be impressed on users that the air-cleaners must be kept clean.

T. O. DUGGAN[6]:—As a result of considerable experi

---

[4] United Parcel Service, Los Angeles.

[5] M.S.A.E.—Superintendent of automotive equipment, Union Oil Co. of California, Los Angeles.

[6] A.S.A.E.—Sales manager, Chanslor & Lyon Co., Los Angeles.

ence we have found that the human element is a serious problem in the air-cleaner situation. We could not get the operators to clean the air-clarifiers and there was restriction due to carbon. We would like to know what is to be done about the human element in actual operation. If the operator does not take care of the air-cleaner, the efficiency of the engine is reduced, gasoline consumption increases, coke is formed and carbon is deposited in the engine. What we would like to know, when recommending air-cleaner installations, is how the equipment for which money is spent is to do the work expected of it if the operator is negligent and does not look after it. It seems that the problem is how to make the device automatic and foolproof and to induce the user to operate it intelligently.

E. B. Moore[7]:—Two years ago we were putting-in 800 acres of alfalfa and after the tractor had been in service less than 15 days, following a complete rebuilding, it would not pull. It was equipped with an air-cleaner and we found that it had not been taken care of. We overhauled the engine, put it back on the job and told the foreman to clean the air-cleaner regularly. They finished planting the alfalfa without further engine trouble. If an air-cleaner can ever be built that will be automatic in its operation, such a device will be the solution of this vexing problem.

W. H. Fairbanks[8]:—We have in service two types of air-cleaner but they have not operated long enough to prove much. I have heard almost no complaint of bad effects from fog or atmospheric moisture. However, we are checking-up this condition closely. We operate a long-distance telephone line across the Colorado Desert to Yuma, Ariz., and we have a specially equipped Ford for the use of our trouble men. Dust is plentiful and sand-storms frequent. Among other things, the car is fitted with an air-cleaner. We brought this Ford in for overhaul after about 5 months' service and 5500 miles of travel. The engine was fouled with dirt, grit and sand on the outside, but the pistons, rings and cylinder-walls were in good condition and showed no appreciable wear. We are not worrying about the human element in regard to the use or abuse of air-cleaners. If a man will not obey orders and give his equipment

---

[7] M.S.A.E.—Superintendent and manager, L. A. Automotive works, Los Angeles.

[8] Supervisor of shops and vehicles, Southern California Telephone Co., Los Angeles.

fair usage, we let him go and get someone on the job who will.

Mr. Powers:—I have been told that in certain types of dry air-cleaner an appreciable increase of carbureter trouble due to lean mixture is experienced if they are subjected to heavy fog. Has anyone had experience along that line?

Professor Hoffman:—Reports of certain cleaners choking-up have come to me a number of times and I have tried to trace them down but found the evidence inconclusive. Here are a few questions handed me: (a) How often does a cleaner need attention? Answer: That varies tremendously; no definite period can be given; it should be cleaned whenever necessary to maintain efficiency and avoid undue restriction of air. (b) What quantity of dust is collected per mile of travel? Answer: A list of six vehicles was given in the report on the 1924 series of air-cleaner tests[D] with the quantity collected by each. A Dodge car, in 1 year's use, driven 3600 miles, about 50 per cent of which was on paved roads, showed 0.00277 grams of dust per mile. An International Harvester truck, driven over the State of California for 4 months and run over 6000 miles, of which 87 per cent was on paved highways, collected 0.00106 grams per mile. I do not see any reason why, under the average conditions in California, an air-cleaner could not be put on any ordinary automobile or truck and forgotten for considerable periods, given attention only once or twice a year. How often it should be cleaned depends largely upon the nature of the work and the weather conditions and the amount of dust met with. (c) How great is the wear of the top piston-ring and the upper portion of the cylinder? Answer: The top ring wears several times as fast as the second and third rings, because the dust gets first chance at it, and the oil at the top ring, being warmer, is much less viscous than it is at the lower rings. In addition, the top ring must stand direct explosion pressures. The film of hot oil can be squeezed out readily and, being thin, allows more particles of dust to rest against and cut the bearing surfaces. (d) How much dust gets into the crank-case oil? Answer: Some idea can be obtained by referring to the paper presented at the 1925 Annual Meeting by G. A. Round and the one that I presented at the same meeting covering the results obtained with

---

[D] See p. 165.

air-cleaners on trucks operated by the California State Highway Commission.[10] Carelessness in permitting the use of dirty funnels and vessels in refilling crankcases is often to blame for rapid bearing-wear. Anything that will cause high vacuum in the crankcase should be carefully avoided. Any chance leak between a shaft and its bushing or past the relieved edges of a two-piece bearing will permit the dust to be sucked directly into the bearings. Hence, using flap-valves to cover filler tubes and connecting the breather to the carbureter air-cleaner outlet are usually not considered as being representative of the best practice.

Dual Carburetion and Manifold Design

By F. C. Mock[1]

ANNUAL MEETING PAPER — *Illustrated with* CHARTS AND DRAWINGS

DUAL carbureters, as equipment for eight-cylinder passenger-car engines, have recently come into special prominence and, compared with a single carbureter, give a gain in power in the middle-speed range, between 1400 and 2800 r.p.m. This is an adaptation from airplane-engine practice, in which greater power-output and better distribution have been realized by multiplying carbureter units as the number of cylinders is increased. An absence of overlapping and interfering suction-strokes and the use of larger manifold-passages are apparently responsible for this gain.

Tests made on a number of eight-cylinder engines, of both the in-line and the V-types confirmed this gain, which was, however, unaccompanied by any particular gain in fuel economy. With dual manifolding, the weight and heat-capacity of the manifold mass are high in proportion to the piston displacement, with a resultant lag in mixture temperature, as compared with changes in the manifold temperatures, the extremes of the former being greater than in other engines.

While the dual carbureter has been blamed for inability of the engine to idle smoothly and for poor fuel-economy at light loads, the substitution of a single carbureter on the selfsame engine failed, in virtually every case, to improve either condition.

Carburetion problems are relatively simple with the dual system. Low velocities can be used, and a nearly uniform mixture gives good results except at low speeds. Some minor variations between the suction and the airflow of the two carbureter barrels result from the difference in the length and volume of the two manifold members. In service, some difficulty was experienced at first in securing a satisfactory setting of the two idling adjustments.

Dual carbureters have not successfully replaced the single system on six-cylinder engines because of "blowback of fuel spray and air charge from the mouth of the carbureter." This condition probably can be remedied by a slight change in the valve timing.

The discussion covers the necessity for overcoming spiralling of the airstream, the increase in aircraft-engine power-output following the change from a six-cylinder manifold to a three-cylinder grouping and the "ramming" and blowback effects in six and eight-cylinder engines.

MULTIPLYING carbureter units with increase in the number of cylinders is not new; in airplane-engine practice, in particular, greater power-output and better fuel-charge distribution have been realized by increasing the number of carbureter units as required to avoid overlapping suction-strokes and yet maintain even suction intervals. In motor-car practice the multiple carbureter has come into special prominence as equipment for the eight-cylinder engine, to which, as compared with a single carbureter, it seems to contribute a definite gain in power in the middle speed-range, from 1000 to 3000 r.p.m.; a less marked gain in power at high speed; little if any saving in fuel consumption; and, in the opinion of nearly every driver, a "feeling" of smooth free pulling at full load and more willing response to the throttle that can be ascribed to better distribution and naturally more favorable conditions of carburetion.

In the preparation of this paper I wrote to a number of American manufacturers who had been experimenting with single and dual-manifold systems on eight-cylinder engines, requesting power comparisons and permission to make them public. Each company addressed responded generously, for which I wish to express my appreciation and thanks.

The Effect on Power

Fig. 1 shows a typical comparison of power curves obtained with 1½-in. single and 1¼-in. dual carbureter systems on one of our leading eight-cylinder-in-line cars. The setting of each carbureter had been developed from thorough tests, which involved also changes of manifold diameter and shape. It will be noted that the torques were equal at 1200 r.p.m., that the dual system gained about 10 per cent at 2400 r.p.m., and about 26 per cent at 3600 r.p.m. This differs from the comparison obtained with some other engines in that the gain increases with the engine-speed. In the car, the single system felt a little more "solid" below 7 m.p.h., both were about alike from 7 to 20, and above 20 m.p.h. the dual carbureter gave the feeling of considerably more power, just the difference between an ordinary automobile and a rather superior one. The second-gear acceleration of the dual system was much faster. I believe the dual system gave 4 or 5 m.p.h. greater maximum speed; exact comparisons of this character are always difficult to obtain.

Fig 2 shows power curves obtained on another well-known eight-cylinder-in-line engine with a 1¾-in. single and a 1¼-in. dual carbureter. These curves show that the latter is better all along the line; 3½ per cent at 400 r.p.m., 7 per cent at 1200 to 2400 r.p.m., and 8½ per cent at 3600 r.p.m. The improvement in car performance was about that described in the first instance, except that not more than 3 m.p.h. was gained in maximum speed.

Fig. 3 shows curves of two engines that have slightly different characteristics. Here the gain in torque is confined to the region between 1000 and 3200 r.p.m., and in working with these engines we found the rather peculiar fact that, whereas we could bring the power of the engine equipped with the single carbureter up to that of the engine with the dual carbureter at 3600 r.p.m. by putting a large air passage in the carbureter,

---

[1] M.S.A.E.—Research engineer, Stromberg Motor Devices Co., Chicago.

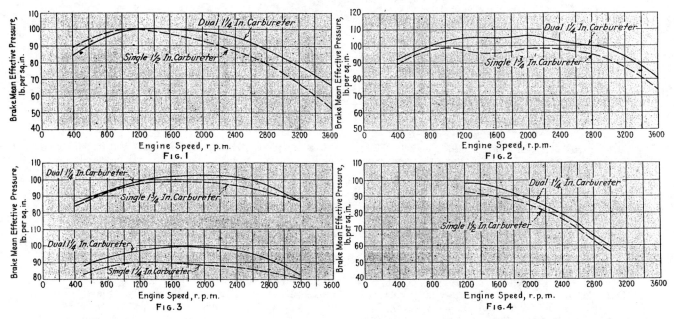

POWER COMPARISONS OF EIGHT-CYLINDER ENGINES EQUIPPED WITH SINGLE AND DOUBLE-CARBURETER SYSTEMS SHOWING A GAIN WITH THE LATTER AT INTERMEDIATE AND HIGH SPEEDS

Fig. 1—A Well-Known 1929 In-Line Engine
Fig. 2—Another In-Line Engine
Fig. 3—Two More In-Line Engines
Fig. 4—A V-Engine with Four Crank Throws Spaced at 90 Deg.

no such enlargement, within practical limits, would bring the former up to the latter at 2400 r.p.m. We have experienced this a number of times, and, making allowance for the natural difference to be expected between engines, it seems that, while the gain in power with the dual system at maximum speed may be ascribed to the freer air-delivery, the gain at intermediate speeds must be due to something else, probably the elimination of overlapping suction-strokes.

Gain in Intermediate Car-Speed

A rather interesting test of actual performance was obtained with two cars having engines whose curves are shown in the lower part of Fig. 3. With the single manifold and carbureter on one car and the dual system on the other, the cars were driven side by side on a level concrete road at 5 m.p.h., and at a given signal both throttles were opened wide. Neither gained a yard until a 20-m.p.h. speed was reached, but at this point the dual-manifold car began to draw ahead, gaining about four car-lengths by the time a speed of 40 m.p.h. was attained. Beyond this speed it gained less rapidly, and finally the two cars settled to a constant distance of about six car-lengths apart. The carbureters and manifolds were then interchanged and the test repeated, with substantially the same result. This checks with what we have found from other tests, that a given gain of torque is more advantageous to car performance if it comes in the intermediate speed-range than if it comes only at maximum speed.

Fig. 4 is a comparison similar to the preceding, made on an eight-cylinder V-type engine with four crankshaft-throws 90 deg. apart. The firing order was such that one 90-deg. overlap of suction strokes occurred in the sequence of the four cylinders of each bank, as fed by one barrel of the 1¼-in. dual carbureter. A 5/16 x 5/8-in. balancing port, located at the junction of the manifold and carbureter, connected the two manifolds of the dual system. The engine is reported to have shown better power, smoother running and slightly better idling than with a single manifold and ¼-in. larger single carbureter of the same type.

Altogether, out of nine different eight-in-line engines tested with single and dual carbureters, eight showed definite gains in power, seven showing values such that the dual equipment was adopted. One engine, however, with a very creditable high-speed performance both on the dynamometer and in the car, seemed to show a little less power at high speed with a 1¼-in. dual carbureter than with a 1¼-in. single carbureter, while the gain with the dual in the intermediate range was only 4 or 5 per cent. The size of valves and the timing on this engine were such that with the dual carbureter a very marked "blowback" of fuel spray out of the air entrance of the carbureter occurred, to the extent that its use would have been dangerous. Trial of balancing ports between the dual manifolds reduced the blowback, brought the power up to that developed with the single carbureter at maximum speed and down to that at intermediate speeds. It seems probable that the dual system would have performed better with a different valve timing, with perhaps earlier closing of the intake valve.

Application to Six-Cylinder Engines

With dual manifolds on six-cylinder engines, a definite interval of absolutely no induction between suction strokes occurs, which develops or permits a blowback of fuel spray out of the entrance to each carbureter barrel, if the customary high-speed intake-valve timing be used. On slower-speed marine and heavy-duty engines, with intake valves closing not later than 35 deg. past bottom center, this blowback has been barely noticeable. In the few efforts we have made to fit dual carbureters and manifolds on high-speed six-cylinder motor-car engines, very little gain in torque has been realized. The violent blowback suggests that power would be gained if this charge rejection were stopped, for instance by closing the intake valve earlier.

One carbureter barrel for each three cylinders, in either 6 or 12-cylinder engines, with intake valve closing at about 42 deg. after bottom center, is almost the universal practice in aircraft engines. Under these conditions blowback is dealt with by turning the entrance of the carbureter toward the propeller so that its blast keeps the fuel spray from blowing out.

Reasons for Power Gain

Considering the energy of airflow in the intake manifold during the suction stroke of one cylinder alone, the induction of air may be termed to some extent potential in the first part of the stroke and more nearly kinetic in the latter part of the intake-valve opening. That is, during the first part of the intake diagram the velocity through the intake manifold is below that corresponding to the depression at the intake-valve port, while during the closing of the valve the airflow, due to inertia, is greater than would be proportional to the depression at the valve port. This inertia or ramming effect should be constant as to the time required for its swing or oscillation from maximum depression to maximum momentum, so that its duration in degrees of

[2] See Air Service Engineering Division Report No. 2608, p. 160.

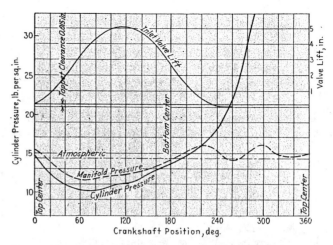

Fig. 6—Indicator Diagram Showing Pulsation Waves of Pressure in the Intake Manifold

crankshaft rotation changes with the engine speed; though it must be initiated more quickly under the more rapid piston-motion of higher speeds. To obtain the best power, we close the intake valve at the time this inertia surge is at a peak, or at least on the rise, and when the piston is rising on the compression stroke; and the greatest volumetric efficiency will be obtained if the valve can be closed at or just beyond the surge peak and when the pressure in the cylinder is at or above atmospheric.

If, however, the forced depression of some other cylinder occurs in the manifold at the time corresponding to the surge pressure-rise of the cylinder we are discussing, the surge peak will not be reached, the energy inducing inward flow will remain in potential form, as depression in the manifold, and the volumetric efficiency and power will be lowered by this interference. Fig. 5 shows the angular time of mutual interference of four, six and eight-cylinder engines with a single carbureter, alternate or interfering suctions being placed on opposite sides of a median line. It will be noted that the interference is little in the case of the four, more with the six and considerable with the eight.

If this surge or ram of the air column occupies a definite duration of time, its effect on the volumetric efficiency will be favorable through a certain range of speed and unfavorable at other speeds above and below. This is borne out by what we find on test; that, in the neighborhood of 2000 r.p.m., the power for a given manifold-vacuum is usually higher with the dual carbureter than the single; but in several instances at 3000 r.p.m. the situation was, if anything, reversed. To move the surge effect up in the speed range would require experimenting with both intake-pipe length and valve timing with the aid of an indicator.

Pressure Pulsations and Valve Timing

Pressure pulsations appear on all the indicator diagrams of intake pressures that I have ever seen, but few efforts have been made to analyze them. One careful study of this particular subject is contained in a report by F. Glen Shoemaker, entitled Cylinder and Manifold Pressures During the Induction and Exhaust Strokes. Fig. 6, which is reproduced from this report[2], clearly shows the existence of these pressure waves; it will be noticed that 40 deg. after bottom center the pressure in

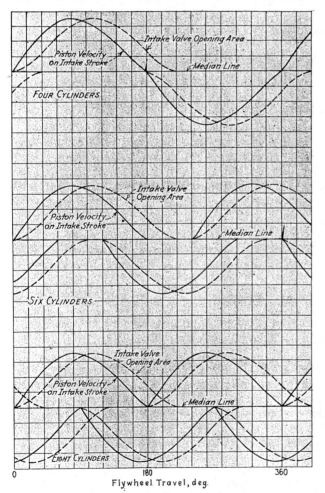

Fig. 5—Superposed Diagrams of Intake-Valve Opening for Four, Six and Eight-Cylinder Engines, Showing Periods of Overlapping and Interfering Suctions

the intake manifold reaches 1¾ lb. above atmospheric. These readings were taken with a single carbureter feeding three cylinders equally spaced in firing order. Apparently, the pulsation characteristics of the air column in the intake manifold are superposed upon the pressure change at the valve port, as forced by the piston travel on the intake stroke. The single run here reproduced does not indicate whether the manifold pulsations have the frequency of sound waves, although their periodicity corresponds roughly with that of a closed-end pipe of the length of the intake manifold; nor is it possible to deduce the point at which the manifold pulsations begin to appear as a deviation from the forced-pulsation curve. With the aid of the successful high-speed indicators now available, investigating this phenomenon fully and utilizing to full advantage the intake-stroke energy represented by the peaks of these pulsations should be easy. In the report referred to, a material gain in volumetric efficiency and power is recorded as a result of changing the valve timing of the engine to take full advantage of this inertia effect.

Manifold Arrangement, Shape and Size

We have made some effort to ascertain the sound-pulsation characteristics of different intake-manifolds, by directing a jet of air into them and noting the character and pitch of the resulting reverberation. With a manifold feeding four cylinders, such as the half of an eight-cylinder dual-manifold, a definite pitch is easily noted, which is constant regardless of which cylinder has its intake valve open. With an eight-cylinder single-manifold, or six-cylinder manifolds of either the two-port or three-port type, a definite pitch is less easily found; sometimes two different notes can be distinguished, and the frequency changes considerably as the crankshaft is rotated into different positions. I hope that later we shall be able to find a relation between the natural resonant pitch of an intake manifold and the pulsation waves existing therein; this will, of course, depend on whether the pulsation waves at all times attain the velocity and other characteristics of sound waves. In any case it seems likely that best utilization of this inertia effect to gain power will require symmetrically formed manifolds arranged to avoid overlap of suction strokes.

It has been found, I believe, that suction-stroke interference was considerably worse with a long column of air between the point of junction and the air entrance, as shown in Fig. 7. This indicates that, with a single carbureter on an eight-cylinder engine, the lower manifold at the left of Fig. 8 would give better power than the one directly above, provided the fuel distribution were equally good; I believe this has been found true. This also explains why large free air entrances are necessary to the dual carbureter.

The chief consideration in dual-manifold arrangement is, from the foregoing, to have each carbureter barrel feed an evenly spaced order of non-overlapping suction-strokes; also, this will later be shown to be necessary for smooth idling and light-load operation. With an eight-in-line engine and 2-4-2 crankshaft, this requires one of the arrangements shown at the right, the lower being perhaps preferable on account of the more equal lengths. With a 4-4 crankshaft, something like the arrangement in Fig. 9 will be needed.

The V-eights with 180-deg. crankshaft simply require one barrel to each bank. The V-eight with 90-deg. shaft, however, introduces a more difficult problem. Fig. 10 shows an arrangement that will give even firing but look rather intricate in the V; I believe, however, that the advantages of this arrangement will justify the apparent complication if the details of design can be worked out.

The dual manifold is less sensitive to shape than the single manifold. The reason for this becomes obvious when one watches the flow of water through a model passage of similar shape. Very slight causes, in the entrance of the carbureter or beyond it, are sufficient to throw the stream of mixture slightly to one side. As in Fig. 11 with simultaneous flow from both branches of

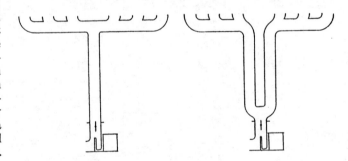

FIG. 7—TWO FORMS OF SIX-CYLINDER MANIFOLD
The Construction at the Left Gives Marked Suction Interference, While That at the Right Was Found To Have Less

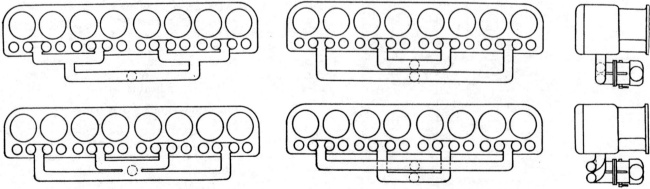

FIG. 8—MANIFOLD ARRANGEMENTS FOR TWO EIGHT-CYLINDER-IN-LINE ENGINES
The Arrangements Shown at the Left Are for Single Carbureter and the Lower of These Should Give Less Suction Interference and Better Power than the Upper One. The Arrangements at the Right Are the Customary Types for an Engine with a 2-4-2 Crankshaft. The Construction Shown in the Lower View Gives Less Difference in Length and Volume of the Two Manifolds with a Resulting Smaller Difference in Inertia Effect and Idling Vacua

a T, any unevenness of stream-borne content at the entrance gives highly uneven distribution at the outlets; whereas, with flow from only one branch at any one time, an eddy blocks the flow into the other branch, and the distribution becomes a matter of chronology rather than of contour. With the ordinary dual-manifold, the directions of flow do not regularly alternate at the T, but the times of flow barely overlap, and I have seen no dual manifold with marked signs of bad distribution.

The better distributional qualities of the dual manifold make it possible to use velocities lower by far than with any other type of engine with which we have worked. The areas in square inches of manifold passage now in use are approximately 1.20 per cent of the piston displacement in cubic inches as compared with 0.75 per cent for eight-cylinder single-carbureter and 0.65 per cent for six-cylinder single-carbureter manifolds. For some reason, cutting down the area of a dual

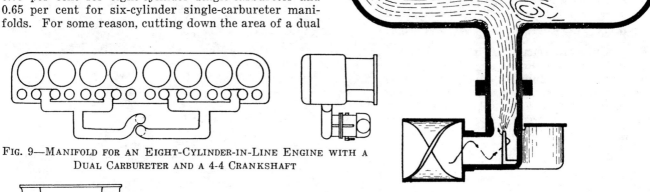

Fig. 9—Manifold for an Eight-Cylinder-in-Line Engine with a Dual Carbureter and a 4-4 Crankshaft

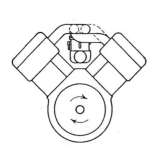

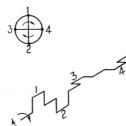

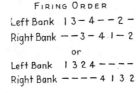

Fig. 10—Manifold Arrangement and Firing Order for a Dual Carbureter on an Eight-Cylinder V-Engine with the Cylinder-Block Axes at 90 Deg. and a Four-Throw 90-Deg. Crankshaft

manifold seems to cut about the same percentage off the power at 2400 r.p.m. as it does at maximum speed, which is seldom true of single-carbureter manifolds.

Importance of Adequate Mixture-Temperature

In my judgment the different elements of manifold design rank about as follows in importance: (a) adequate provision in either cold or warm weather for fairly complete vaporization of the fuel before it reaches the first division point in the manifold passage; (b) velocity as high as high-speed considerations will permit; and (c) a little common sense as regards the contour of the passages. This opinion is based on the following observations:

(1) I have seen no form of manifold that would enable an engine to run satisfactorily on definitely wet mixtures at average intake-manifold air-velocities below 37 ft. per sec.

(2) Some engines that can scarcely be made to run without heat at manifold velocities around 100

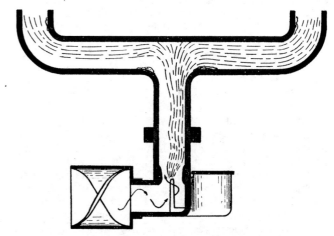

Fig. 11—Fuel Flow in Two Dual Manifolds

If the Fuel Flow Is Unequally Distributed in the Airstream at the Entrance to the Intake Manifold and Flow Exists Simultaneously through Both Branches, the Inequality Is Continued at the T-Point of Division in the Design Shown in the Lower Drawing. If Flow Exists in But One Branch of the Manifold at Any Given Time, the Conditions of Distribution Are More Favorable, as Shown in the Upper Drawing

ft. per sec., figured on a basis of 75-per cent volumetric efficiency, seem to have almost perfect charge-distribution above a mixture temperature of 130 deg. fahr.

(3) Such engines as I have seen that could fire smoothly when cold at high intake-velocities did not show, in a car, more power or pickup at these speeds than after they had warmed up enough to pull steadily at low speeds.

(4) Under conditions of reasonably complete vaporization and mixing between the carbureter and division point of the manifold, indicated in

practice by the ability to pull at low velocities without "loading," the contour of the manifold has not seemed to matter very much, and a number of different shapes have been equally satisfactory. In actual use of a car, of course, to have the engine run as well as possible during the warming-up period is desirable; but the best selection I have seen of contour and firing order is still entirely dependent upon mixture temperature as to its satisfactory operation.

These observations point to the conclusion that the *chief* consideration in a passenger-car intake-system is the maintenance of the mixture temperature between practical limits of, say, 120 and 160 deg. fahr. This may not be true of other types of engine service, for instance those of continuous load at a fixed fairly high speed; but I have never seen an engine operate well in any service with a "stone cold" intake manifold.

Reasons for Mixture-Temperature Variation

Maintenance of efficient mixture-temperature is largely a matter of the ability to apply heat in varying rate and quantity under different atmospheric temperatures. With any given design and dimension of hot-spot, the mixture temperature is almost constant for different steady loads and speeds, provided the temperature of the air and fuel in the carbureter do not change. In actual car operation, however, the air and fuel temperatures vary widely, and with fixed hot-spot action the mixture temperature follows. This is better understood if one appreciates the nature and conditions of the heat transfer. A thin metal plate exposed on one side to the flame of a blowtorch will not rise to a high temperature if a small spray of liquid is discharged against the other side; the high heat-capacity of the liquid and its intimate contact with the metal will carry heat out as fast as it can be transmitted into the plate from the hot gases of the flame. Similar conditions exist in the heating wall of a hot-spot, so that its temperature varies in opposite direction, and through a wide range, with the quantity of liquid fuel striking the wall.

Assume that a manifold of the type shown in Fig. 12 is used on a cold day with air and fuel entering the carbureter at 40 deg. fahr. Approximately 8 per cent of the fuel will evaporate at the jet, and the remainder will strike the heating walls; if the exhaust temperature be 900 deg. fahr., the wall temperature will be perhaps 120 deg., and the fuel will flow along it at slightly below this temperature, evaporating at its surface as the airstream "scrubs" it. The final mixture will be low in temperature and decidedly wet.

Next assume a spring day, with air and fuel entering the carbureter at 100 deg. fahr.; 40 per cent of the fuel may evaporate in the carbureter, and the remainder may cool the hot-spot wall to about 380 deg.; the liquid fuel on the walls will be much less in quantity than before, and probably it will evaporate before leaving the manifold, reaching the cylinders as a thin fog at a temperature between 130 and 150 deg. fahr.

Finally, assume entering air and fuel temperatures of 150 deg. fahr., which I have actually observed on a hot summer day. Under these conditions the fuel has probably disappeared into vapor 6 in. beyond the jet, and no liquid is on the hot-spot walls, the temperature of which may now rise above 600 deg. fahr., hot enough to discolor brass and to char any but asbestos gaskets. The radiation effect from the walls will be considerable at this temperature, and the mixture temperature at the cylinder port may reach 200 deg. fahr.

Perhaps it is worthwhile to interrupt here to point out the very considerable change that these conditions make in the accelerating charge required from the carbureter. In the first instance, any increased rate of mixture delivery at the carbureter will give an immediate increase of air at the cylinder ports, but several seconds might elapse before the corresponding fuel delivery could arrive at these same ports; and unless the mixture proportion at the carbureter were temporarily enriched, the engine probably would be unable to fire just after the throttle is opened from the idling position. In the case of summer operation, any change in mixture delivery at the carbureter will instantly carry up air and fuel in equal proportions to the cylinder port; and any temporary extra fuel-feed, as needed to avoid too lean a mixture in the cylinder port in the first instance, would give a mixture temporarily too rich to fire.

"Heat-Flywheel" Effect

Even greater extremes of temperatures are experienced, due to what may perhaps be graphically described as a "heat-flywheel" effect. The actual temperature-distribution during steady high-speed running in the common form of intake manifold may be about as shown in the upper drawing of Fig. 13. Upon starting, some time is required to reach this temperature; and upon stopping, or even upon the diminution of liquid feed against the walls following closing the throttle to the idling position, the temperature will tend to equalize throughout the metal of the manifold as in the lower drawing, resulting in an unusually high temperature on the intake-manifold walls. I have seen iron and steel discolored in this way. Upon starting or opening the throttle soon afterward, an augmented heating-

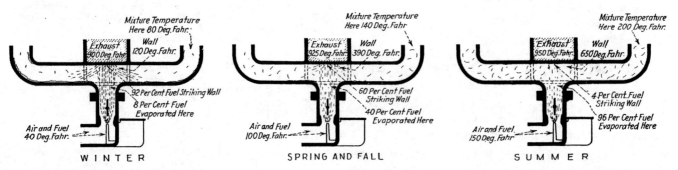

Fig. 12—Differing Air and Fuel Temperatures at the Carbureter Cause a Temperature Variation in the Intake Manifold

effect will be experienced, the magnitude and duration of which will be proportional to the manifold mass, and specific heat, in relation to the piston displacement. That the time of warming up, and the heating of the intake after slowing down, will be much greater with a manifold of exaggerated wall mass, is apparent.

Means of Heat Control

Doubling the number of intake manifolds on an engine, as in the dual system, is unfavorable in the way described, both because of the greater mass of the casting and the greater sectional area available for heat conduction. Engines of different make vary greatly in this respect according to their design; but with intake and exhaust manifolds cast in one piece, the residual-heat effects, loss of power, decreased speed, fuel boiling in carbureter, and sensitiveness to too much accelerating charge are definitely more marked in the engine with dual manifolds than in one with a single-intake system.

The most effective steps in dealing with this problem seem to be (a) to make the exhaust manifold, including the heat-control valve, in a casting separate from that of the intake; (b) to use a gasket spacer with effective heat-insulating qualities between the two; (c) to use aluminum for the intake manifold; (d) to design the heat jacket of the intake casting with more surface exposed to the exhaust than is now common practice, taking care that ribs or boundary surfaces are disposed to conduct to the inner or charge-heating wall the heat picked up from the exhaust by the outer wall of the jacket; and, (e) most important, to design the passages and control valve so that all the exhaust can circulate through the jacket during the warming-up period in cold weather, that part of the exhaust can be utilized for steady running in cold weather, and that virtually all exhaust circulation can be cut-off on a hot summer day. These changes should be easily made by a control on the dash; in any position of the control the exhaust passages should be large enough not to restrict the power by back-pressure.

The chief need of full exhaust-circulation through the heater is for warming up, so as to overcome the heat inertia of the manifolds; and it is remarkable how little heat application need be used once the manifolds are warm. Considerable heat is necessary for city driving in spring and fall, so that intermediate positions of the heat-control valve are necessary.

In a few years a construction may be devised whereby the temperature of the air and fuel in the carbureter, and of the carbureter itself, can be held at about 100 deg. fahr. This would obviate the need of hot-spot and accelerating-charge adjustments, and also eliminate the trouble of boiling gasoline in the carbureter.

In the eight-cylinder dual system, when the manifold feeding cylinders Nos. 3, 4, 5 and 6 is made short and next the cylinder-block, as in the arrangement shown at the upper right of Fig. 8, it usually carries a higher mixture-temperature than the other manifold by reason of the shorter distance and greater mass of metal for heat conduction; but apparently the longer passage to cylinders Nos. 1, 2, 7 and 8 has better distribution, or better heating of the charge before it reaches the valve ports, because, after starting, in two engines I have observed, normal firing was attained in these cylinders some time before those fed by the shorter manifold. One attempt to bring the mixture temperature of the short

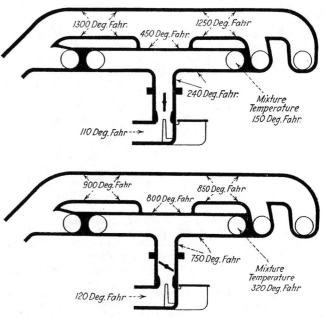

Fig. 13—"Heat-Flywheel" Effect

The Upper View Shows the Distribution of Temperatures in the Manifold Wall During Steady High-Speed Running across Country. The Lower View Shows the Temperatures in the Same System a Few Minutes After Speed Was Reduced

branch down to that of the long one did not give satisfactory operation.

Idling and Light-Load Performance

Ever since the introduction of the eight-cylinder engine, its idling has been subject to criticism. I do not mean that the fours and sixes, as classes, idled satisfactorily; as a matter of fact, our demonstrating engineers probably spend as much time on all engines trying to lower the speed of safe idling, to overcome stalling and to eliminate misfiring and an evil smell from the exhaust while the car or the motorcoach is coasting down from high speed with clutch engaged and throttle closed, as they do on any other phase of engine operation. Furthermore, they spend this time unprofitably, because the fault seldom lies in the carbureter, which, though popularly supposed to control entirely the ability of an engine to idle, actually can do no more than supply a mixture of any desired strength.

A low idling-speed is more important in sales demonstrations than in average service; few drivers will complain of a positive non-stalling road idling-speed of 6 m.p.h. More serious is the fact that, with increasing proportion of reserve torque, acceleration and hill-climbing ability, each year our engines approach more nearly an idling condition as they travel along at legal city-driving speeds. In particular, the fuel economy suffers. Yet, so far as I know, no systematic research has been made regarding the factors that govern the ability to fire smoothly and on thin mixtures at light loads; no authoritative articles on this subject are to be found in the journals of the different automobile engineering societies; and so little is known of the requirements of light-load efficiency that they are seldom considered in designing an engine.

In our own organization we have for some time had a program of research before us on this subject, but more urgent matters have continually taken precedence. It may be of service, however, to state the two main phases of the problem as we see them.

(1) To fire regularly at light load and on attenuated air-charges, the average engine seems to demand increasing richness of mixture as the manifold vacuum increases above 14 in. of mercury; at about 18 to 20 in. the firing becomes uncertain with any mixture strength; and above 21 in. the engine scarcely fires at all. Conditions are worse at the low idling-speeds, in that it is harder to avoid an occasional miss.

(2) Applying to idling only is the tendency toward variation of the engine speed. When an attempt is made to lower the idling speed of an engine, a point is reached at which the rate of angular rotation begins to oscillate or "roll," as this is commonly called; if the roll augments, the engine will soon stop in one of the phases of slow travel. Sometimes an engine will not idle on account of definite missing, regular or intermittent, but more often the minimum speed of safe idling is that just above the rolling point. If we could reduce this tendency to roll, as by a heavier flywheel, or keep the conditions of firing stable, so as to dampen out the roll once it starts, a lower steadier idling-speed would result.

Idling Stability

Considering the latter phase first, for reasons of convenience, it is necessary to understand the relation, or rather lack of relation, between airflow and engine speed at the high manifold-vacua attending idling. When a gas at pressure P_1 flows through an orifice into a region of lower pressure P_2, the rate of flow is greater as the pressure P_2 is lowered, until a certain limiting velocity is reached, which is the velocity of sound in the gas at the condition P_1. For air at sea-level pressure flowing into the intake manifold through the throttle orifice, this limiting velocity is reached at a manifold vacuum of about 14.4 in. of mercury. This means that, at vacua greater than this, engines usually idle at vacua over 16 in., the airflow, and with most types of carbureter the fuel flow, past the throttle into the engine are constant per unit of time, independent of engine speed, and *the quantity of charge per cylinder is directly inverse to the engine speed.* This has several results worth studying. These are

(1) If the throttle stop-screw of an engine is set to give an idle of 5 m.p.h. and, while driving 50 m.p.h. the throttle be allowed to return to the same closed position, the mixture charge *per cylinder* at 50 m.p.h. will be one-tenth the weight of the 5-mile idling charge and entirely too weak to fire. The unburned mixture passing through the hot cylinder develops a pungent disagreeable smell that is annoying when it enters the car or motorcoach body. As the vehicle slows down, the charge per cylinder will increase in strength, reaching a richness at which occasional ignition will occur, but the rate of flame propagation will be so slow that burning will exist when the exhaust valve opens. Under certain conditions the unburned gas reaches an ignitible strength in the muffler, may ignite from the delayed-burning exhaust of some cylinder that has ignited and give a muffler explosion. As the vehicle slows down still more, a charge density is reached at which ignition will generally occur but will not be entirely regular; if this condition is marked, the universal-joints and other loose places in the drive will "chuckle" until firing becomes regular. About all we have been able to do so far in the way of cure is to make firing conditions as uniform as possible in the cylinders, so that they all come in at once when they do start to fire. Careful adjustments of the mixture and spark-advance help a little, and usually a certain throttle opening is better than either a faster or slower idling-opening. Enough trouble has been experienced from this source to warrant a thorough investigation, but I do not know that anyone has found time for one.

(2) Another result of this idling-feed constancy per unit of time is that *the charge per cylinder varies if the time between suction strokes is irregular*. Instances of this sort are (See Fig. 14) a twin V-type motorcycle-engine having both connecting-rods on the same crankpin; an eight-cylinder V-engine with four crankshaft throws at 90 deg. and one carbureter barrel feeding each bank; and the experiment sometimes tried of putting three carbureters, or at least three throttles, on successive pairs of cylinders of a six-cylinder engine having the standard form of crankshaft. In any of these instances, flow of mixture past the idling throttle-opening will go on whether an intake valve is open or not, and the cylinder whose intake valve opens after the long interval will receive a definitely stronger charge than one drawing after a short interval.

(3) Lastly, this constant flow furnishes the explanation of the tendency of an idling engine to roll. If for any reason one cylinder fails to fire with average strength, the flywheel will slow down a little, and the cylinder at that time drawing will receive a greater-than-normal idling-charge, which will then tend to increase the angular speed of the flywheel to above that of normal idle; if it does so increase, the next cylinder will receive less than the regular idling-charge, the flywheel will slow down again, and so on. Whether the roll dampens out or increases seems to depend upon the relation of the flywheel moment of inertia to the friction drag of the engine, which in turn is equal to the idling indicated torque, and upon the uniformity of firing of the cylinders. During the slowing-down phase of the oscillation, certain essential parts of the engine operation, such as the spark break or valve openings, may get "out of time," which would definitely account for the engine dying on the slow phase of the roll. Why a slight increase of flywheel moment of inertia is so often effective in improving the idle is easily understandable.

Irregular Miss When Idling

Even though the roll be overcome, we encounter many instances in which engines show a definite irregular miss on idle, and this tendency usually carries up into the level-road driving-range. Sometimes we have found the cause of the trouble and its cure, sometimes it has disappeared during our work without our knowing just which step was responsible, and sometimes we have found no remedy. The chief known cause is disturbance of the normal process of fuel and air proportioning through the carbureter, by manifold or valve-guide air-leaks, intake or exhaust valves not seating tightly, or non-uniform overlap of exhaust and intake-valve openings in the different cylinders (See Fig. 15). Inadequate mixture-temperature would account for irregular firing, but in our observation the mixture is usually dry under sustained idling, even though wet at full throttle. Different forms of equalizer duct have been used on eight-cylinder engines, but after long trial I could not see that they did any good, except in the case of a V-engine with 90-deg. cranks, where they were necessary, to compensate for uneven suction-intervals.

Work on the ignition system has often improved the idle. The beneficial effect of widening the spark-plug gaps is well known. This is effective even though no spark is being induced at some plug, which is out of time, through proximity of the high-tension paths. Producing a spark longer either in path or in duration seems likely to make ignition more certain. I have witnessed no positive demonstration that changing the shape of the cylinder-head or the location of the spark-plug made a definite improvement in the light-load operation, although this has often been reported.

Some engines we have had idled well, others would not idle at all steadily, at 5 m.p.h. In no case did the substitution of a single-carbureter system for the dual or vice versa, on the same identical engine, affect appreciably the idling performance.

Fuel Consumption at Light Loads

This tendency toward irregularity of firing has extended up into the legal city-driving speeds, where it can often be cured by the use of a slightly rich mixture. The dual system seems in this respect a little more sensitive than the single system on some engines, possibly because of the lower intake-velocities. Another

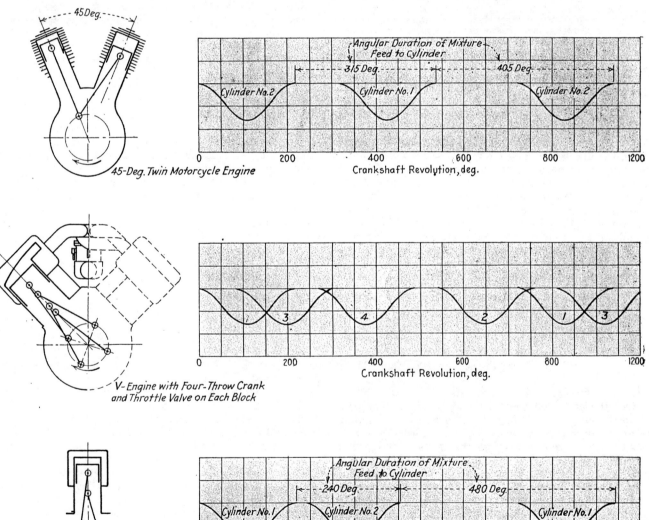

Fig. 14—Diagrams Showing How Irregular Spacing of Suction Strokes Results in Unequal Strengths of Air and Fuel Charges

During Light-Load Running the Flow of Mixture past the Partly Closed Throttle-Valve into the Vacuum of the Intake Manifold Continues After an Intake Valve Is Closed, and the Quantity of Charge Received per Cylinder Is Greater after the Longer Interval

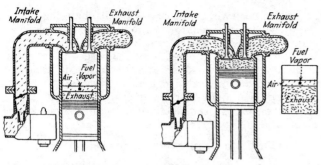

FIG. 15—COMPARISON OF THE CONTENTS OF THE CYLINDER UNDER DIFFERENT CONDITIONS OF OPERATION

The View at the Left Shows Conditions at the Beginning of the Compression Stroke under Normal Idling-Conditions. The Larger View at the Right Shows Conditions in the Cylinder at Light Load When the Closing of the Exhaust Valve Overlaps the Opening of the Intake Valve. With Atmospheric Pressure in the Exhaust Manifold and an 8-Lb. Vacuum in the Intake Manifold, a Very Slight Simultaneous Lift of the Exhaust and Intake Valves Will Permit Cross Flow as Shown. The Small View at the Extreme Right Shows the Resulting Proportions of Air, Gasoline Vapor and Exhaust in the Cylinder When Compression Begins

difficulty with regard to fuel consumption on the dual system comes from its increased tendency to detonate, on the same compression-ratio, because of its higher compression and greater power. When the spark is set back to cure this, we often find that it is too far retarded for best economy at closed throttle. In other words, the difference between the proper spark-advance for closed throttle and for open throttle seems to be greater on the dual than on the single system. Also, the peculiarity in the torque curves illustrated in Fig. 3 seems to develop a special requirement in the spark-advance needed for best power at different speeds, that is difficult to fulfill by automatic means. I trust our ignition engineers will correct me if I am not right.

Carburetion problems are in general easier of solution with a dual carbureter and manifold on an eight-cylinder engine than with a single carbureter on either an eight or a six. As previously stated, we can use larger areas and lower velocities. Fewer departures from a uniform mixture-proportion are required, enrichment at light loads and for idling being the only pronounced essential. Almost no accelerating charge is required; in fact, the sensitiveness to too much accelerating charge is perhaps the most outstanding characteristic. The warming-up period is short and taken care of without much enrichment of the mixture.

One structural difficulty is the need of exact fitting of the throttle valves for idling. Instructing the service men how to handle twin idle adjustments, though this is not difficult, has taken a little time. We have had some difficulty overcoming a tendency to "load" when pulling at full-load wide-open throttle at very low speeds, say below 8 m.p.h. This is perhaps explained by the very low intake-velocities used and is not of much importance in the actual use of a car. No such tendency was experienced on one engine which had the heat jacket on the manifold brought right down to the carbureter flange; with such a construction, however, a control must be provided to cut off nearly all the exhaust heating in summer, otherwise much trouble will be experienced from fuel boiling in the carbureter.

Altogether, the dual manifold and carbureter seem necessary to the full development of the superior qualities of the eight-cylinder engine. I have seen good carburetion and low power with the single-carbureter system, also less satisfactory carburetion and high power; but never as good all-round performance, judged by the standards of either the engineer or the ordinary motorist, as is easily obtained with the dual system.

THE DISCUSSION

CLAUDE S. KEGERREIS[3]:—Our work has checked Mr. Mock's eight-cylinder observations and we agree mainly in the six-cylinder work. We have found some differences but these are too minor to enumerate. I absolutely agree with Mr. Mock in the four requirements he listed.

We are building air-cleaners, because we want to protect the carbureter metering and power. We have found, in our development work on air-cleaners, that if we have a high resistance, or even a low resistance with considerable spiralling through the cleaner, the results suffered. Consequently, we took great pains in the design of our cleaner to straighten out that airflow and have found, on block tests, a slight supercharging effect at high-torque points instead of a loss in power. It is nothing but the straightening effect in the airstream coupled with the minimum pressure-loss in the device.

Several times in the last 6 or 8 months engineers of designers and manufacturers have asked us to recommend intake-manifold temperatures. I have given them as 120 to 155 deg. Mr. Mock says that he goes up to 160. Apparently the carbureter men are agreeing.

F. G. SHOEMAKER[4]:—The tests that Mr. Mock mentioned in regard to intake-manifold pressures were made in 1922 at McCook Field on a Packard 1237 aircraft engine. The diagrams were made with a Bureau of Standards balanced-pressure diaphragm-indicator of an early type. Each card took about 45 min. of continuous full-throttle operation, the results being a panoramic picture of the conditions in this period. Due to the location of the pressure connections in the manifolds, the observed pressures are the resultant of both static and impact effects which cannot be separated and hence cannot be used for precise calculations. However, when considered only as fairly accurate pictures, they reveal some very interesting characteristics.

In contrast to the reports about not obtaining this ramming effect on six-cylinder automobiles, it has been shown repeatedly on aircraft engines that a gain in power of at least 5 per cent is obtainable by going from a six-cylinder to a three-cylinder manifold grouping. It is obvious from a study of the pressure cards mentioned that the valve timing must be different with the two types of manifold to take full advantage of the ramming effect. This may account for the negative results obtained on some six and eight-cylinder automobile engines.

I should like to emphasize the importance of knowing the actual instantaneous pressures in the manifold and

---

[3] M.S.A.E.—Chief engineer, Tillotson Mfg. Co., Toledo.
[4] M.S.A.E.—Research engineer, powerplant section, General Motors Corp. Research Laboratories, Detroit.

cylinder before any intelligent improvements can be made in valve timing and manifold arrangement. The customary cut-and-try method is particularly hopeless in this case on account of the great number of factors that can be varied without being able to evaluate the effects of each change. The number of experimental camshafts in the engine laboratories is ample proof of this point. The electrical indicators now being developed should make such pressure studies entirely feasible, and a basic study along these lines would be of great benefit to the industry, particularly in the design of camshafts and manifolds.

Diagrams of Blowback and Ramming

H. M. JACKLIN[5]:—Mr. Mock's remarks about blowback through the carbureter and about ramming effects have caused me to think that the members might be interested in some indicator diagrams taken to show these effects. These diagrams were taken with a 20-lb. spring, as indicated alongside the upper diagram in Fig. 16, on cylinder No. 6 of a six-cylinder engine operating with full throttle at 500, 1000 and 2000 r.p.m. The small loops at the end of the inlet or suction stroke and the beginning of the compression before the inlet valve closed at about 35 to 40 deg. were very clear in the original diagrams and have lost little in tracing, while the ramming effect, totalling 3 lb. per sq. in., is very clear at the higher speed. This blowback, or more accurately this drawback, was not at all apparent at the carbureter inlet, as it would extend only to the laterals of the manifold and be absorbed or cured by the succeeding cylinder. With a dual manifold this blowback would probably be very apparent at low speed, since no overlapping of suction strokes in each bank of three cylinders would occur. The remedy for the fuel losses because of such blowback would be to use a fairly long inlet-horn on the carbureter. If this blowback were eliminated by a change in inlet-valve closing, the engine would certainly suffer at the higher speeds. Therefore, we must compromise in the valve timing as well as in other things.

The exhaust-valve timing and the manifolding seem to be especially well chosen in this particular engine, as the exhaust gases actually help the engine at 500 r.p.m. At 1000 r.p.m. the exhaust pressure drops to the atmospheric line and does not rise until near the end when the exhaust valve is closing. At 2000 r.p.m. the exhaust back-pressure does not appear excessive at any time. The high velocity induced in the inlet manifold results in some ramming at all three speeds, but the inertia of the mixture column is not great enough to prevent blowback at the two lower speeds. This inertia actually

[5] M.S.A.E.—Associate professor of automotive engineering, Purdue University, West Lafayette, Ind.

results in a pressure increase of 3 lb. above atmospheric at the start of compression at 2000 r.p.m.

In some tests on an eight-cylinder engine, with a dual manifold and using a 2-4-2 crankshaft, I have found that the ramming effect on cylinders Nos. 1, 2, 7 and 8 is between 1½ and 2 lb. higher than that for the center four cylinders. This is due entirely to the longer

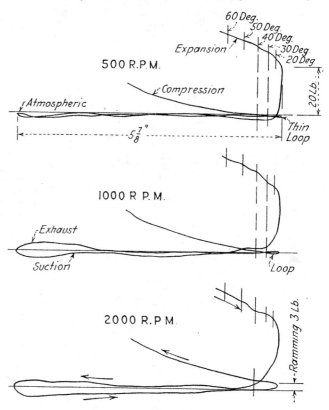

Fig. 16—Indicator Diagrams Taken To Show Blowback Through the Carbureter and Ramming Effects

These Diagrams Were Taken with a 20-Lb. Spring on Cylinder No. 6 of a Six-Cylinder Engine Operating with Full Throttle at 500, 1000 and 2000 R.P.M. The Increase in the Back-Pressure and the Early Suction-Pressure Lines with Increased Speed Should Be Noted as Well as the Blowback Loops at the End of the Inlet and Early Compression Lines at the Lower Speeds and the Ramming Effect at the Higher Speed

laterals to the outer cylinders, which gives a greater mass of air-fuel mixture acting to cause such ramming.

I must also point out the greatly increased "pumping losses" at higher speeds as indicated in these diagrams by the increasing area of the diagram between the exhaust and suction lines with increased speed. This area is a direct measure of the work done in removing the exhaust gases and drawing in a new charge.

Influence of Carbureter Setting and Spark Timing on Knock Ratings

Annual Meeting Paper

By John M. Campbell,[1] Wheeler G. Lovell[2] and T. A. Boyd[3]

THAT the antiknock quality of a motor fuel is best determined by comparison with a standard or reference fuel is generally agreed. The results of the investigation described in this paper show how this comparison may be affected either by the carbureter adjustment or by the spark timing or by both. Therefore, in establishing a standard method of comparing one fuel with another, some definite specification covering each of these two engine-variables must be drawn up. This specification should correspond reasonably well with conventional conditions of engine operation and at the same time with convenient laboratory technique. From the results of this study these conditions would appear to be fulfilled best by using the carbureter setting for maximum knock and the spark timing for maximum power.

THE knock rating of a motor fuel may be defined as the measure of the tendency of the fuel to knock in an engine with respect to that of some standard fuel. A knock rating, therefore, involves a comparison between one fuel and another as it is observed in an engine. This investigation was concerned primarily with the determination of whether two fuels that knock just alike at one carbureter setting or at one spark timing will knock just alike at some other adjustment of the carbureter or of the spark timing. For this purpose several pairs of fuels were selected for comparison.

(1) A commercial gasoline containing benzene and the same gasoline containing lead tetraethyl
(2) A straight-run gasoline and a reference fuel consisting either of benzene in normal heptane or of iso-octane in normal heptane
(3) Two reference fuels, one consisting of benzene in normal heptane and the other of iso-octane in normal heptane.

These fuels were compared by the bouncing-pin indicator, and we assumed that fuels that gave equal bouncing-pin readings during the same period of test were equivalent in tendency to knock.

Description of Experimental Equipment

The experiments were carried out with one of the single-cylinder, variable-compression test-engines being developed in connection with the work of the Subcommittee on Methods of Measuring Detonation of the Cooperative Fuel-Research Steering Committee[4]. The engine was connected to an electric dynamometer for absorbing the output, and the speed was held constant at 600 r.p.m. by a 2-hp. synchronous converter that was connected by V-belts to the flywheel of the engine. An evaporative cooling-system provided with forced circulation maintained the temperature of the jacket water at about 100 deg. cent. (212 deg. fahr.) The engine was fitted with a bouncing-pin indicator for comparing the knock intensity of different fuels.

The carbureter was one that was built at the General Motors Corp. Research Laboratories and has been described before[5]. It had a fixed jet and two float-bowls that were independently adjustable for height so that the mixture ratio for each of the two fuels could be independently controlled as desired by adjusting the height of each float-bowl with respect to the common fuel-jet. A three-way cock inserted in the fuel line between the float-bowls and the jet enabled a change from one fuel to another to be made quickly without interrupting the operation of the engine. The venturi tube of the carbureter was in a horizontal position and discharged the fuel mixture directly into the intake port of the engine. No heat was applied to the mixture before it entered the cylinder-block. The carbureter had no throttle; hence the engine was always operated with the same intake-opening.

The timing of the spark was observed by the flash in a small neon tube that was attached to the crankshaft and that passed a stationary protractor insulated from the ground and connected to the high-tension terminal of the spark-plug. Samples of the exhaust gases were taken at different carbureter-settings during some of the tests, and the carbon-dioxide content was determined to estimate approximately the magnitude of the effect of unit changes in carbureter setting upon the air-fuel ratio.

Benzene in Gasoline Compared with Lead Tetraethyl in Gasoline

Several experiments were made to determine the effect of carbureter setting and of spark timing respectively upon the comparison of a gasoline containing benzene with the same gasoline containing lead tetraethyl. In these experiments a fuel consisting of 40 per cent by volume of benzene in gasoline was compared with fuels containing various concentrations of lead

---

[1] Research chemist, General Motors Corp. Research Laboratories, Detroit.
[2] Assistant head of fuel section, General Motors Corp. Research Laboratories, Detroit.
[3] M.S.A.E.—Head of fuel section, General Motors Corp. Research Laboratories, Detroit.
[4] See *Bulletin of the American Petroleum Institute*, Jan. 2, 1930, section 3, p. 32.
[5] See p. 128.

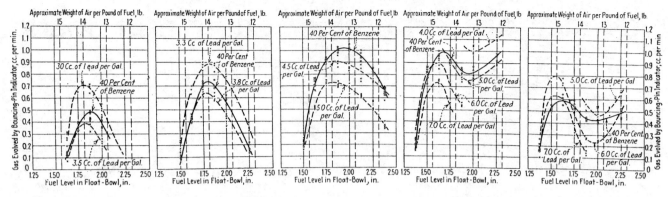

Fig. 1—Spark Timing 7½ Deg. before Top-Dead-Center; Compression Ratio 6.5 to 1

Fig. 2—Spark Timing 20 Deg. before Top-Dead-Center; Compression Ratio 5.7 to 1

Fig. 3—Spark Timing 28 Deg. before Top-Dead-Center; Compression Ratio 5.65 to 1

Fig. 4—Spark Timing 37½ Deg. before Top-Dead-Center; Compression Ratio 5.7 to 1

Fig. 5—Spark Timing 42 Deg. before Top-Dead-Center; Compression Ratio 5.8 to 1

Effect of Carbureter Setting on Knock Rating

In Making These Tests the Carbureter Setting Which Is Expressed Arbitrarily in Terms of the Fuel Level in the Float-Bowl, Was Varied, While the Spark Timing Remained Fixed. Each Chart Represents a Series of Knock Ratings of 40 Per Cent of Benzene in Gasoline Compared to Various Quantities of Lead Tetraethyl. The Engine Used Was a Single-Cylinder, Variable-Compression Engine Running at a Constant Speed of 600 R.P.M., Having a Jacket-Water Temperature of Approximately 100 Deg. Cent. (212 Deg. Fahr.) and Fitted with a Bouncing-Pin Indicator

tetraethyl in the same gasoline. The object of this comparison was to ascertain the concentration of lead tetraethyl in gasoline which would be equivalent in antiknock effect to 40 per cent of benzene in gasoline at various carbureter-settings and spark timings.

In making this comparison the spark timing was held constant while the carbureter setting was varied. At each carbureter-setting bouncing-pin readings were taken alternately until four readings had been obtained from each fuel. Then the carbureter setting was changed and the bouncing-pin readings continued. After completing a series of bouncing-pin readings taken in this way at different carbureter-settings but at some fixed timing of the spark, the spark timing was changed and the whole process was repeated. In this way five runs were made at spark timings of 7½, 20, 28, 37½ and 42 deg. before top dead-center respectively. The data are presented in Figs. 1 to 5.

In these figures, bouncing-pin readings, expressed as cubic centimeters of gas generated per minute, are plotted on the vertical axis, and carbureter settings, which are expressed arbitrarily in terms of the fuel level in the float-bowls, are plotted along the lower horizontal axis. The approximate mixture-ratios corresponding to different fuel-levels, as determined by referring the percentage of carbon dioxide in the exhaust gases to a representative plot of exhaust-gas composition at different mixture-ratios, are plotted along the upper horizontal axis. These mixture-ratio values apply in a quantitative way only to the gasoline containing lead tetraethyl and not to the mixtures of benzene in gasoline. The composition of the exhaust gases was not appreciably affected by changes in spark timing.

Each of these charts represents a series of knock ratings of 40 per cent of benzene in gasoline with respect to lead tetraethyl in gasoline at different carbureter-setting but at a fixed spark-timing. For example, by interpolation of the data in Fig. 1, at the carbureter settings for maximum knock and at a spark timing of 7½ deg. before top dead-center, 40 per cent of benzene in gasoline would give the same bouncing-pin readings as about 3.4 cc. of lead tetraethyl per gal. in the same gasoline; and, since the bouncing-pin readings would be the same for these two fuels, they are considered to have an equivalent tendency to knock. It will be observed in Fig. 1 that this relative relationship between benzene and lead tetraethyl in gasoline was not seriously affected by changes in carbureter settings corresponding to changes between 13 and 15 lb. of air per lb. of fuel at this spark timing. On the other hand, Fig. 4 shows that at a spark setting of 37½ deg. before top dead-center the same mixture of 40 per cent of benzene in gasoline was equivalent to as much as 6.0 cc. of lead tetraethyl per gal. at a carbureter setting corresponding to about 15 lb. of air per lb. of fuel, but only to about 4.8 cc. of lead tetraethyl per gal. at a carbureter setting corresponding to 13 lb. of air per lb. of fuel.

An analysis of the data in Figs. 1 to 5 shows that, as the timing of the spark is advanced, two important changes take place in the relative knock-ratings of these two types of fuel. First, greater concentrations of lead tetraethyl are necessary to match a given concentration of benzene in the same gasoline; and, second, the magnitude of the influence of carbureter setting upon the relationship between the antiknock effect of lead tetraethyl and of benzene respectively increases.

The specific effect of spark timing, as distinguished from that of the carbureter setting, upon the concentration of lead tetraethyl that would be found equivalent to 40 per cent of benzene is shown in Fig. 6. In this chart the concentration of lead tetraethyl equivalent to 40 per cent of benzene in gasoline, as determined from the data in Figs. 1 to 5 at the carbureter setting for maximum knock, is plotted on the vertical axis and the particular spark-timing at which the determination was made is plotted along the horizontal axis. This chart shows clearly how the concentration of lead tetraethyl that is equivalent to 40 per cent of benzene increases as the spark timing is advanced. Furthermore, from the shape of this curve the inference can be drawn that, at spark timings before that for maximum power, the effect of spark timing on the comparison between fuels containing lead tetraethyl and those containing benzene is more pronounced than at spark timings after that for maximum power.

In obtaining the data presented in Figs. 1 to 5, the compression ratio, of course, had to be changed when the spark timing was changed, to keep the knock intensity within the range of the bouncing-pin instrument. Consequently, the variations in knock rating that thus far have been ascribed to spark timing conceivably might have been due to the changes in compression ratio that were made at the same time. To preclude this possibility, an experiment was made comparing fuels containing benzene and lead tetraethyl respectively at different spark-timings but at a fixed compression-ratio and at practically constant carbureter-setting, that for maximum knock. This comparison was made by taking bouncing-pin readings at different spark-timings with all other engine-variables remaining constant.

The data for this experiment are plotted in Fig. 7, where bouncing-pin readings for the different fuels are plotted against spark timing. These data show that, even at a fixed compression-ratio and at a substantially constant carbureter-setting, the concentration of lead tetraethyl that is required to match a given concentration of benzene in gasoline depends to a large extent upon the spark timing at which the comparison is made. Thus, in this gasoline at a spark-advance of 20 deg., 40 per cent of benzene was equivalent to less than 4.0 cc. of lead tetraethyl per gal., and at a spark-advance of 40 deg. the same concentration of benzene was equivalent in antiknock effect to about 6.0 cc. of lead tetraethyl per gal.

Although the data in Fig. 7 represent different knock-intensities at each spark-timing, no indication is given that this comparison between lead tetraethyl and benzene was affected by changes in knock intensity. These data obtained at different knock-intensities agree with the data in Fig. 6, which represent data obtained at substantially constant knock-intensity. Furthermore, unpublished data that we have obtained in comparing the relationship between the antiknock properties of gasoline solutions of benzene and of lead tetraethyl, as determined by different methods, have shown that this relationship was not affected beyond the experimental error whether the determination was made under conditions of incipient knock or under conditions of fairly severe knock. This was true provided that the only change made to increase the knock intensity was to increase the compression by about one ratio and simultaneously to retard the spark so that it remained at that for maximum power.

A Straight-Run Gasoline Compared with Benzene in Heptane or with Iso-Octane in Heptane

In making further studies of the effect of spark timing on knock ratings, one of the test fuels of the Detonation Subcommittee, designated as gasoline *AAA*-1, was compared with two different reference-fuels. One of the reference fuels consisted of a mixture of normal heptane and iso-octane—2, 2, 4-trimethyl pentane—and the other of a mixture of normal heptane and benzene.

The comparisons were made at two different spark-timings—20 deg. and 40 deg. before top dead-center. All comparisons were made at the carbureter setting for

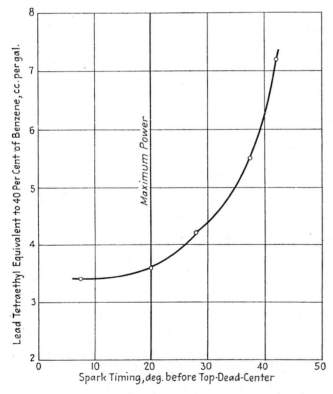

FIG. 6—EFFECT OF SPARK TIMING ON KNOCK RATING

As the Spark Timing Is Advanced, the Concentration of Lead Tetraethyl That Is Equivalent to 40 Per Cent of Benzene Increases. From the Shape of the Curve the Inference Can Be Drawn That the Effect of Spark Timing on the Comparison between Fuels Containing Lead Tetraethyl and Those Containing Benzene Is More Pronounced at Spark Timings before That for Maximum Power Than at Those after

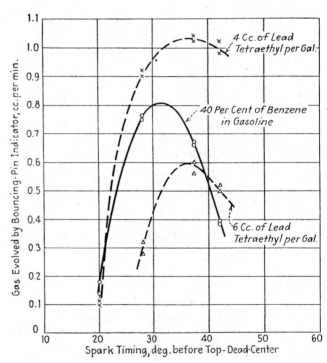

FIG. 7—EFFECT OF SPARK TIMING ON BOUNCING-PIN READINGS

In This Test, Which Was Made with a Compression Ratio of 5.5 to 1 and a Carbureter Setting for Maximum Knock, the Concentration of Lead Tetraethyl That Is Required To Match a Given Concentration of Benzene in Gasoline Was Found To Depend to a Large Extent on the Spark Timing at Which the Comparison Was Made

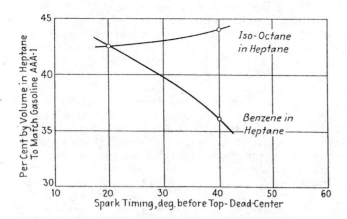

FIG. 8—EFFECT OF SPARK TIMING ON KNOCK RATING

The Results Plotted Were Obtained by Comparing the Knocking Tendency of a Straight-Run Gasoline with Those of Mixtures of Iso-Octane in Heptane and Benzene in Heptane at the Carbureter Setting for Maximum Knock. The Bouncing-Pin Indicator Was Used To Make This Comparison, Which Showed That the Concentrations of Benzene or of Iso-Octane in Heptane Which Were Equivalent to the Straight-Run Gasoline Depend to Some Extent on the Spark Timing at Which the Comparison Was Made

maximum knock. The fuels were matched by the bouncing-pin indicator, and the results are presented in Fig. 8. In this chart the concentrations of benzene or of iso-octane in heptane which were found to be equivalent to gasoline AAA-1 depend somewhat upon the timing of the spark at the time the comparison is made.

An explanation for this may be had from a study of Fig. 9. The data plotted in this chart were obtained by comparing gasoline AAA-1 with two different mixtures of benzene in heptane at constant compression-ratio. Bouncing-pin readings were taken alternately on all three fuels at the carbureter setting for maximum knock; first at 20 deg., then at 30 deg. and finally at 40 deg. before top dead-center. As the spark timing was advanced the knock intensity increased with all three fuels, as shown by the increase in bouncing-pin readings. But the important thing to observe is that as the spark was advanced the bouncing-pin readings obtained while the engine was running on gasoline AAA-1 increased faster than those obtained with mixtures of benzene in heptane under the same conditions. In other words, the rate of increase in knock intensity with respect to spark-advance was greater with gasoline AAA-1 than with mixtures of benzene and heptane. Thus, according to Fig. 9, at a spark timing of 23 deg., gasoline AAA-1 knocked just like a mixture of 40 per cent of benzene in heptane; and, at a spark timing of 30 deg., gasoline AAA-1 knocked appreciably more than the same mixture of benzene in heptane.

Benzene in Heptane Compared with Iso-Octane in Heptane

The effect of spark timing upon the composition of mixtures of benzene in heptane and iso-octane in heptane which have equivalent tendencies to knock can be determined from the data presented in Fig. 8. At a spark-advance of 20 deg., a mixture containing 42.5 per cent by volume of benzene in heptane would be equivalent to 42.5 per cent by volume of iso-octane in heptane since each of these mixtures was equivalent in knocking tendency to gasoline AAA-1. On the other hand, if the comparison was made at a spark-advance of 40 deg., only

[6] See p. 127.

36 per cent of benzene in heptane would be required to match 44 per cent of iso-octane in heptane since each of these mixtures was equivalent to gasoline AAA-1.

The effect of carbureter setting on the comparison between benzene and iso-octane in heptane is shown by the data plotted in Fig. 10. Although these data were obtained from another engine[6], it was similar to the Co-operative Fuel-Research engine and was operated under conditions comparable with those used in obtaining the other data reported in this paper. In Fig. 10 bouncing-pin readings are plotted on the vertical axis against carbureter settings at a fixed spark-timing on the horizontal axis. This chart shows that, at a carbureter setting corresponding to about 13.1 lb. of air per lb. of fuel, based on the exhaust-gas composition of the mixture of iso-octane in heptane, 79 per cent of iso-octane in heptane was equivalent in knocking tendency to 68 per cent of benzene in heptane. But, at a carbureter setting corresponding to about 12.2 lb. of air per lb. of fuel, the same 79 per cent of iso-octane in heptane was equivalent to 68.5 per cent of benzene in heptane. Thus, the knock rating of one reference-fuel with respect to another appears to depend both upon the spark timing and upon the carbureter setting at which the comparison is made.

General Discussion of Results

Methods so far proposed for determining the knocking property of a fuel depend fundamentally upon finding, by experiment in an engine, what mixture of known pure substances knocks like the fuel under investigation. In ordinary work, of course, pure substances are not always used, but gasolines are used which have been compared with the definite pure substances.

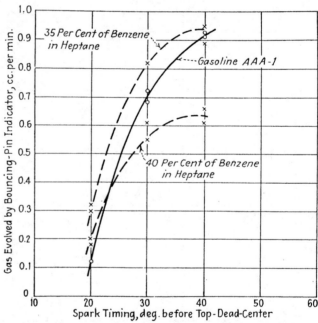

FIG. 9—EFFECT OF SPARK TIMING ON THE BOUNCING-PIN READINGS

In This Test Bouncing-Pin Readings Were Taken Alternately on a Straight-Run Gasoline and Two Different Benzene-Heptane Mixtures at the Carbureter Setting for Maximum Knock and at Spark Timings of 20, 30 and 40 Deg. before Top Dead-Center. The Knock Intensity Increased for All Three Fuels as the Spark Was Advanced, but the Increase for the Straight-Run Gasoline, as Indicated by the Change in the Bouncing-Pin Readings, Was Greater than for the Benzene-Heptane Mixtures under the Same Conditions

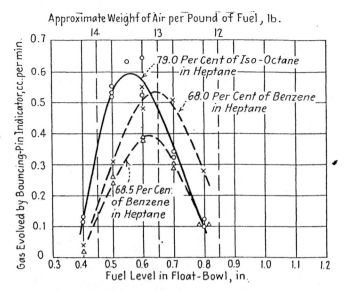

Fig. 10—Effect of Carbureter Setting on the Knock Rating of Benzene in Heptane Compared to Iso-Octane in Heptane

These Results Were Obtained on a Different Engine from That Used in the Other Tests, Although the Two Were Similar. The Compression Ratio Was 5.8 to 1; the Spark-Advance, 26 Deg.; the Speed, 600 R.P.M., and the Jacket-Water Temperature, 212 Deg. Fahr. Both the Spark Timing and the Carbureter Setting at Which the Comparison Was Made Apparently Affect the Knock Rating of One Reference-Fuel with Respect to Another

The data presented here show that fuels equivalent under one set of conditions may not be equivalent under another set, and in particular when the conditions varied are the spark timing and the mixture ratio. Consequently, in making comparisons that are to be definite and that can be standardized and reproduced, specifying accurately the carbureter setting and spark timing to be used in the making of measurements is necessary.

What carbureter setting and what spark timing are most suitable for a standard method of rating fuels for knock? With respect to carbureter adjustment, the setting for maximum knock is probably the most suitable. The three main reasons why this setting appears to be the best are, first, the adjustment for maximum knock is definite and can be reproduced; second, it is an adjustment that can be made by the same instrumentation as is used for making the knock rating itself; and, third, the carbureter setting for maximum knock is a practical one, because it lies within the range of ordinary engine-operation. The precision with which this adjustment is made is thus suited to any particular method of comparing the antiknock qualities of fuels. Although the mixture ratio for maximum knock is not necessarily the same as that for maximum power, the two adjustments usually are so close together that commercial carbureters designed to compromise between maximum power and maximum fuel-economy will vary enough in their metering to operate at a mixture ratio for maximum knock at some time.

With respect to spark timing the best setting to use apparently is that for maximum power. This setting is not materially affected by ordinary changes in fuel composition at compression ratios below the detonation point. When knock-rating tests are carried out at such a degree of knock intensity that the knock will affect the spark timing for maximum power, it would be preferable not to adjust the spark for maximum power with the knocking fuel but rather to maintain the spark timing for the maximum power that would be obtained under the same engine-conditions but with a fuel that did not knock. If this procedure is followed, determining the spark setting for maximum power independently for every knock-rating is not necessary. Once the spark setting for maximum power has been determined for a particular engine under all the conditions at which it might be used for knock-rating work, future adjustments can be made immediately by referring to the previously determined data.

Making knock ratings at the spark timing for maximum power possesses a number of advantages. First, with such an adjustment the conditions for laboratory measurements will correspond as closely as possible, in respect to spark timing, with those prevailing in actual practice. Second, this adjustment is usually such that the effects of variations in carbureter setting are less than at some of the more advanced spark-timings. Third, the possible effect of small changes in the spark timing appears to be less at the spark timing for maximum power than at more advanced spark-settings. Consequently, we suggest that in making knock measurements, the necessary engine comparisons of the fuels be made at the carbureter setting for maximum knock and at the spark timing for maximum power.

660114

Control of Automotive Exhaust Emission by Modifications of the Carburetion System

Richard D. Kopa
Dept. of Engineering, University of California, Los Angeles

CONTROL OF EXHAUST emission from automotive engines can be effectively approached from two directions:

1. At the exhaust manifold by treating the exhaust gas already contaminated.
2. At the inlet manifold by treating the air-fuel mixture prior to combustion in the engine cylinder.

The first approach does not require any modification of the existing engine powerplant; however, it does require that some additional equipment be added to the exhaust system, such as an air injector, afterburner, convertor, and so forth. The second approach requires certain modifications of the engine, particularly of that part which involves the induction system.

The effects of engine operating parameters on the emission of unburned hydrocarbons, carbon monoxide, and nitrogen oxides from the exhaust of internal combustion engines have been a matter of study by several investigators (1-5)*. Some parameters, such as spark advance, compression ratio, manifold vacuum, engine speed, and so forth, affect the emission moderately. A parameter affecting exhaust emission most profoundly is the air-fuel ratio of the combustible mixture.

The carburetion system of an internal combustion engine is designed to supply to the cylinders a combustible mixture with an air-fuel ratio predetermined to provide the best engine performance or economy. The engine designer has to consider the carburetor and inlet manifold as an integral system which, because of the transient flow phenomena occurring in the manifold as well as the heterogeneity of the combustible mixture delivered by the carburetor, imposes quite narrow limits to the range of air-fuel ratios which will provide satisfactory engine operation. Unfortunately, the range of air-fuel ratios which gives best engine performance does not coincide with that which results in minimum emission of contaminants from the engine exhaust system. A minimum emission of all exhaust contaminants results when the engine operates with about 20-30% excess of air (Fig. 1), which is definitely above the lean limit of satisfactory operation of an engine equipped with a conventional carburetor.

The first part of this paper describes studies of exhaust-

---

*Numbers in parentheses designate References at end of paper.

ABSTRACT

Carburetion of a lean air-fuel mixture is proposed for reduction of all major contaminants emitted from automotive exhaust. Since the mixture lean limits tolerable by an internal combustion engine equipped with conventional carburetor make this approach unsuitable for nitrogen oxides control, exhaust gas recirculation as a supplemental control method is considered.

Good fuel atomization and homogeneous mixing with air prior to admission into the inlet manifold eliminate "power surging" and extend the mixture lean limit of satisfactory engine operation. These tasks can be accomplished by pneumatic fuel atomization carburetion. Experimental results indicate effective reduction of unburned hydrocarbons, carbon monoxide, and nitrogen oxides as well as good engine performance at air-fuel ratio settings over 18:1.

contaminant control by the induction method, the tests having been performed on internal combustion engines equipped with conventional carburetors. The scope of the investigation was limited to nitrogen-oxides control by dilution of the air-fuel mixture with inert gas. Exhaust-gas recycling appeared to be the most practical method; however, the resulting unstable operation of the engine presented a problem. The outcome of these studies led to investigation of the principle of pneumatic-fuel-atomization carburetion as a means of producing homogeneous air-fuel and recycled-exhaust-gas mixtures. This method is described in the second part of this paper.

EXPERIMENTAL PROCEDURE

One test engine and four automobiles were tested during this research program. The test engine was a 1955 Mercury, V-8 cylinder, with compression ratio 9:1, rated 180 bhp at 4800 rpm, connected to a General Electric dynamometer. Most of the preliminary investigation of exhaust gas recycling and pneumatic fuel atomization was performed on this engine test setup.

Airflow was measured by means of a calibrated venturi flowmeter and fuel flow by a calibrated rotameter. Shielded thermocouples were installed in the vicinity of each exhaust valve and the output recorded on a multichannel high-speed recorder. Two Kistler SLM pressure transducers built into spark plugs were employed to determine the maximum cylinder-to-cylinder and cycle-to-cycle pressure variation. Recorded exhaust gas temperatures and observed pressure variations served as the basis for estimation of the relative fuel distribution to individual cylinders. A 2 in. thick, transparent plastic spacer was mounted between the carburetor and the flange of the inlet manifold to allow for visual observation of the wetness of the air-fuel mixture. A Dumont Engine-Scope served for periodic checks of the ignition system. Engine rpm, cooling water temperature, oil pressure, temperature, and mixture temperature were measured on portable instruments during laboratory experiments as well as in test automobiles during road tests.

Exhaust gas was analyzed for carbon monoxide, unburned hydrocarbons, and nitrogen oxides. One M.S.A.-LIRA, Model 300, infrared, nondispersive instrument was sensitized for a carbon monoxide 10% full-scale reading. Another M.S.A.-LIRA-300 infrared, dual-range instrument was sensitized for hexane 1500 ppm and 15,000 ppm full-scale readings, respectively. Saltzman and phenoldisulfonic acid methods, and in the latter part of this research program a Davis Nitrous Fumes Analyzer, were employed for determination of nitrogen oxides.

The first automobile tested was a 1959 Ford equipped with an automatic transmission and 292 cu in. V-8 engine (the car will be hereafter referred to as test car A). The following automobiles were also tested: a 1951 Ford with standard shift and a 6-cyl engine (test car B), a 1958 Dodge with automatic transmission and a V-8 engine (test car C), and a 1965 Ford with automatic transmission and a 352 cu in. V-8 engine (test car D). All automobiles were tested on a Clayton-Research type chassis dynamometer. All tests were conducted with commercial grade gasoline. California Standard "Driving Cycle" tests were performed by the California Vehicle Pollution Laboratory for the Motor Vehicle Pollution Control Board, Los Angeles.

EXHAUST EMISSION CONTROL BY INDUCTION METHOD

As mentioned in the introduction, the most important parameter affecting auto exhaust emission is the air-fuel ratio of the combustible mixture. Fig. 1 presents a typical concentration of unburned hydrocarbons, carbon monoxide, and nitrogen oxides in automotive exhaust gas as the function of air-fuel ratio at a car cruising speed of 60 mph. Shown also is the typical air-fuel-ratio range and misfire lean limit of operation of an engine equipped with a conventional carburetor. (Misfire lean limit is indicated by the beginning of the broken line.)

Slight leaning of the air-fuel mixture, still tolerable with regard to satisfactory engine performance, would result in a significant decrease of carbon monoxide and a noticeable decrease of unburned hydrocarbons. However, there would

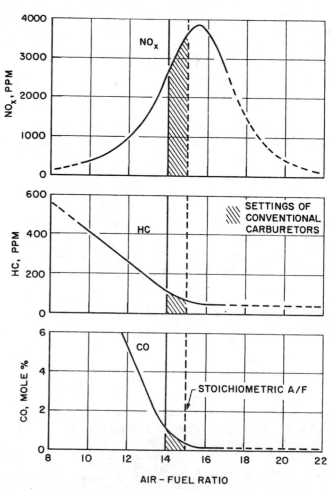

Fig. 1 - Typical concentration of unburned hydrocarbons, carbon monoxide, and nitrogen oxides as function of air-fuel ratio. Car speed 60 mph

be an increase rather than a reduction of nitrogen oxides. Obviously, only a leaning of the air-fuel mixture to a ratio range over 19:1 would accomplish the desired significant decrease of all three contaminants. Unfortunately, for reasons discussed later in this paper, this approach is not practicable for engines equipped with conventional carburetors. Nevertheless, nitrogen oxides can be controlled independently of air-fuel-ratio change. Exhaust gas recirculation as a method of nitrogen oxides control has been already reported (6-10). In a previous paper (6), the theoretical background, some experimental data, and a description of an exhaust gas recycling system were presented.

EXHAUST GAS RECYCLING

By summarizing from earlier publications, the effect of exhaust recycling on the emission of nitrogen oxides can be described as follows: Addition of exhaust gas, or any inert gas, such as carbon dioxide, water vapor, or nitrogen, to the combustible mixture will lower the combustion temperature during the combustion process in the engine cylinder. Since the amount of nitric oxide produced in the engine cylinder from atmospheric nitrogen and oxygen is an exponential function of combustion temperature, even a moderate decrease in the combustion temperature will result in a significant decrease in nitric oxide production. A theoretical analysis based on computation of chemical equilibrium indicated that a 16% lowering of peak combustion temperature would cause about 85% reduction of nitric oxide concentration in the combusted gases. This affect was experimentally accomplished by recycling about 15% of the exhaust gas back to the inlet manifold.

In this paper, additional experimental data and findings are published in order to offer a more complete picture of the potentials of and problems inherent in exhaust gas recycling. Fig. 2 presents schematically the exhaust gas recycling system. Exhaust gas is tapped, preferably from the exhaust pipe between the exhaust manifold and muffler. A 1/2 in. brass or stainless steel tubing conducts the exhaust gas to the flow control valve actuated by the carburetor throttle (the tubing should be efficiently cooled by air-blast from the radiator fan). Gas passing through the control valve is admitted to the inlet manifold through a sandwich mounted between the carburetor and inlet manifold flange.

The effect of exhaust recycling on nitrogen oxides emission, engine power, and economy is shown in Fig. 3. Reduction of nitrogen oxides in the engine exhaust, decrease in power, and increase in brake specific fuel consumption is plotted against the percentage of exhaust recycling. The tests from which these data were obtained were performed on a 1955 Mercury V-8 engine at constant throttle setting, air-fuel ratio, and spark advance. The initial load was 30 hp at 2400 rpm. At 15% exhaust recycling, nitrogen oxides were reduced by about 88%, power output by 16%, and economy by about 14%. Power loss and reduced economy can be balanced by change of the spark timing. By advancing the spark timing 19 deg, the power and economy drop was compensated for; however, nitrogen oxides emission was raised considerably. The net effect of exhaust recycling at balanced power was approximately 60% reduction of nitrogen oxides.

In the subsequent testing program, three cars were equipped with exhaust gas recycling devices. The percentage of exhaust gas recycling and spark timing change was experimentally determined as a compromise between maximum nitrogen oxides control and minimum increase in specific fuel consumption. A typical characteristic of the exhaust gas recycling rate as, for example, determined for test car A is presented in Fig. 4. The per cent of exhaust gas recycling is plotted as a function of car speed at road load, medium load, and heavy load conditions. Up to 16 mph and at wide open throttle, the exhaust recycling is shut off in order to prevent rough idling and to obtain full power at full throttle operation. The average reduction of nitrogen oxides at cruising speeds from 20-70 mph was 75% on test car A, 87% on test car B, and 70% on test car C. The specific fuel consumption as determined by road testing remained essentially unchanged on test cars A and C. On car B, where no change of spark timing was made, about 8% decrease in fuel economy was recorded.

The experimental data presented here probably do not reflect the best results obtainable. No further study in this direction was conducted; rather, attention was given to a problem of power surging which became apparent at this stage of experimentation with exhaust gas recycling.

At this point, it is well to mention that one quite unexpected observation was made at the conclusion of the above tests. When the recycling devices were dismounted from cars A, B, and C, the emission of nitrogen oxides was rechecked.

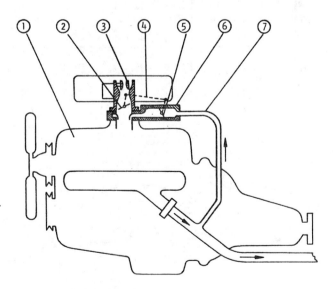

1 ENGINE
2 THROTTLE VALVE
3 CARBURETOR
4 THROTTLE AND RECYCLING VALVE CONNECTING LINKAGE
5 EXH. GAS RECYCLING FLOW CONTROL VALVE
6 RECYCLING DEVICE
7 CONNECTING TUBING TO EXHAUST SYSTEM

Fig. 2 - Exhaust gas recycling system

As compared to emissions prior to exhaust gas recycling operation, the concentration of nitrogen oxides reached only about 65% of the original level. After an extended period of operation without recycling, a progressive return to the original concentration was observed. A similar observation was also orally communicated to the author by H. Daigh(9). It could be speculated that during exhaust gas recycling a new kind of deposit is being formed in the combustion chamber with some catalytic effects on nitric oxide decomposition, and it appears that these effects deserve thorough investigation.

In Refs. 7, 9, and 10 are reported the effects of exhaust gas recycling on other exhaust contaminants, on engine wear and performance, and so forth. In the sutdies reported here,

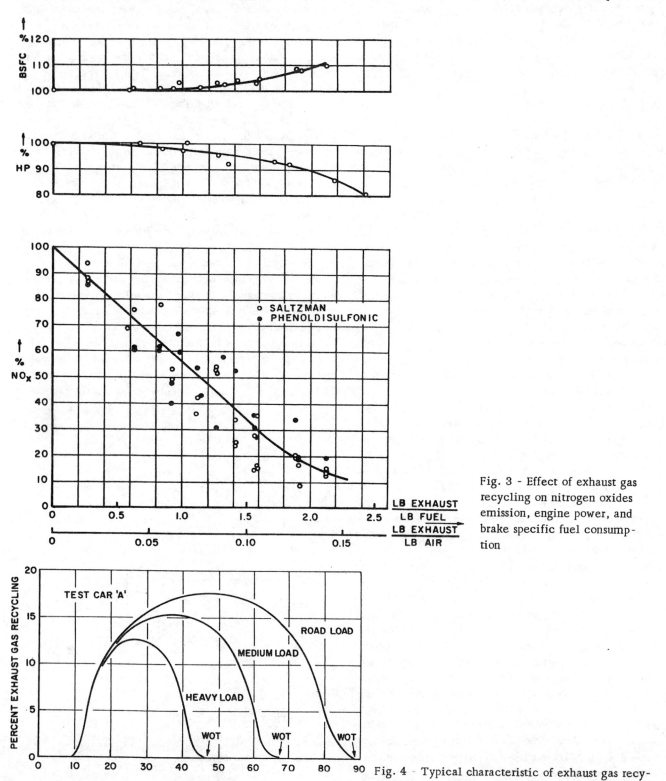

Fig. 3 - Effect of exhaust gas recycling on nitrogen oxides emission, engine power, and brake specific fuel consumption

Fig. 4 - Typical characteristic of exhaust gas recycling rate as function of car cruising speed

no effect on emission of unburned hydrocarbons was observed as long as all engine parameters were kept constant and the amount of exhaust gas recycling did not reach a level which would cause the engine to misfire.

ELIMINATION OF POWER SURGING BY ENRICHMENT OF MIXTURE

During road testing of cars equipped with exhaust gas recycling devices, it was noted that at certain cruising speeds the engines operated in an unstable way commonly called "power surging." This effect was most noticeable at around 50 mph cruising speed and about 15 in. Hg manifold vacuum. The severity of power surging depended on the rate of exhaust recycling and was more evident on older cars.

Since the exhaust was recycled into the inlet manifold below the carburetor, the air-fuel ratio, or rather oxygen-fuel ratio, remained unaffected. However, the ratio of total gas flow (air plus exhaust) to fuel flow was increased. For example, the carburetor of test car A at 50 mph delivered the mixture at an air-fuel ratio of 14.0:1. Recycling of 17.5% of exhaust gas represents the addition of 2.60 lb of gases, which brings the total gas-fuel ratio to 16.6:1 and results in heavy surge. It was therefore anticipated that balancing the total gas-fuel ratio by increasing fuel flow to the original ratio of 14:1 would eliminate power surging.

Fig. 5 presents a comparison of standard carburetion without exhaust gas recycling and of enriched mixture carburetion with exhaust gas recycling. Emission of nitrogen oxides, unburned hydrocarbons, and carbon monoxide is plotted as a function of car cruising speed. Because of the combined effect of exhaust recycling and mixture enriching (see Fig. 1), quite impressive reduction of nitrogen oxides resulted. However, as expected, emission of unburned hydrocarbons and carbon monoxide had increased. About 10% increase in fuel consumption was recorded. Yet car road performance was very good, with acceleration even better than that with standard carburetion.

Bishop and Nebel (3) came to the conclusion that a 90% reduction in emission of nitrogen oxides as accomplished by enriching the mixture to maximum performance air-fuel ratio would cost the driver about a 26% loss in fuel economy. The results obtainable with exhaust gas recycling and total gas-fuel balancing would compare quite favorably with these figures.

CARBURETOR AND INLET MANIFOLD

With the occurrence of power surging as induced by exhaust gas recycling, the problem of the air-fuel-ratio-range limits of a conventional carburetor become even more challenging. Among the three cars tested, an appreciable difference in the tendency toward power surging was observed. The question arose as to whether this difference was due only to variation in age and condition of the test cars or to differences in engine design. For example, car C, one year older than car A but in comparable mechanical condition, had shown a much greater tendency to power surge when operated under road conditions identical to those to which car A was subjected. Both cars were equipped with a V-8 engine and automatic transmission; the striking difference was in the design of the carburetors.

The carburetor of car A had a booster venturi designed according to Fig. 6 (c). In this design, fuel from the float chamber enters an annular cavity in the venturi, is discharged through a circumferential slot, spreads over the internal surface of the venturi, and is evenly atomized by the inlet air at the lower edge of the booster venturi. The carburetor of car C, designed according to Fig. 6 (a), had a booster venturi with a single spout discharging the fuel in an uneven spray. Observed by stroboscopic light, this spray appeared to be composed mostly of a single trickle consisting of large fuel droplets (compare Ref. 11). The trickle continuously changed shape, shifting its position around the circumference of the throttle plate. During the test, the measured manifold vacuum was over 15 in. Hg, and accordingly the pressure drop across the throttle valve was critical, causing the inlet air to flow at sonic velocity. Since the large fuel droplets of the trickle did not pass through the clearance between the throttle plate and carburetor barrel in a steady

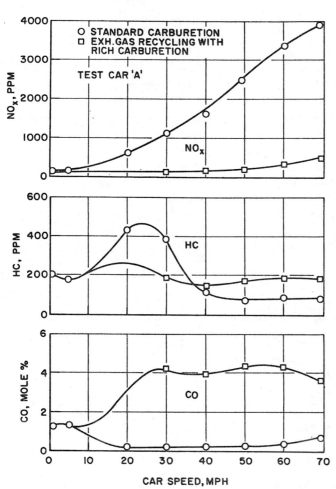

Fig. 5 - Comparison of emission resulting from standard carburetion and enriched mixture carburetion with exhaust gas recycling

stream, it is conceivable that during the progressive evaporation of fuel droplets in the space downstream of the throttle valve, an aggravated stratification of the air-fuel mixture may have taken place (12).

At a certain engine speed and corresponding frequency of pulsation in the inlet manifold, the stratification of the mixture could be sustained or even augmented by gas dynamical resonance effects. The resulting maldistribution of fuel to individual cylinders might cause an instantaneous drop in power and engine speed. A drop in engine speed, in turn, might offset the gas dynamic resonance effects, with the result of temporarily improved fuel distribution. Better fuel distribution would cause the engine to speed up again until inlet manifold resonance frequency is reached, and thus the cycle would be repeated. Consequently, power surging could be pictured as an unstable engine operation resulting from fluctuations in fuel distribution caused by the transient stratification of the air-fuel mixture.

To test this hypothesis, the following experiment was performed. The speed of car C on the chassis dynamometer was set at 50 mph, the load to 15 in. Hg manifold vacuum, and the exhaust gas recycling rate at 18%. Heavy power surging was observed. Then the fuel flow to the carburetor was shut off and a hypodermic needle N with a stop cock S, connected to the fuel supply, was inserted into the carburetor barrel, as shown in Fig. 6 (b). The flow of fuel was adjusted by means of the stop cock S to the same rate as the flow through the carburetor. The engine was now operating at the same speed, load, exhaust recycling rate, and air-fuel ratio as before. When the needle was positioned so that the tip reached downstream of the throttle plate (Fig. 6 (b)), power surging ceased. It was therefore concluded that minute variations in delivery of fuel to the inlet manifold, as for example can be caused by the varying size of the fuel droplets produced by conventional carburetors, may promote stratification of the air-fuel mixture in the inlet manifold and maldistribution of fuel to individual cylinders. To prevent this occurrence, fuel should be finely atomized and evenly mixed with air before admission into the inlet manifold, or directly injected and atomized into the inlet manifold.

The amount of heat supplied to the inlet manifold and the location of hot spots also has effect on the stratification of the air-fuel mixture. Increased mixture stratification and fuel maldistribution to individual cylinders may, in turn, result in increased emission of unburned hydrocarbons and of carbon monoxide (13). From Fig. 7, it is apparent that a change in heat input to the inlet manifold can effect an increase in fuel maldistribution and consequent increase in emission of hydrocarbons and carbon monoxide. Furthermore, lowering of mixture temperature will result in lower emission and raising of mixture temperature in higher emission of nitrogen oxides. Data on emission presented in Fig. 7 resulted from a test performed on test car A equipped with a standard carburetor. For operation with a cold inlet manifold, the cross-over heating passage was blocked by means of brass shims inserted below the manifold gasket. The resultant effect on mixture temperature in shown in Fig. 8.

FUEL ATOMIZATION CARBURETION

The investigation described in the previous section led to the conclusion that the lean limit of the air-fuel ratio can be considerably extended and satisfactory engine operation maintained if the engine is supplied with a homogeneous air-fuel mixture. This known fact was corroborated in a comprehensive study by Robison and Brehob (13). The goal of the investigation presented in this paper was to obtain an overall picture of the merits as well as feasibility of this approach to automotive emission control.

Generally, homogenation of the air-fuel mixture can be

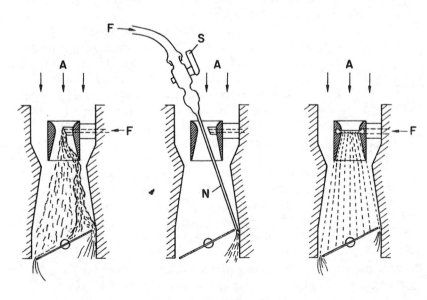

Fig. 6 - Effect of carburetor design on fuel spray

accomplished in two ways:

1. Vaporization of fuel prior to mixing it with air.
2. Atomization of liquid fuel into tiny droplets injected in the stream of air supplied to the engine.

The first principle is commonly used in L-P gas (liquefied petroleum) carburetion systems. Vaporization of fuel in this case can be readily accomplished because propane at atmospheric conditions is in a gaseous state. With commercial gasoline, however, a considerable amount of heat and time is consumed for complete vaporization. This fact presents difficult problems for the design of a carburetion system which has to satisfy the requirements of good engine performance, economy, and low emissions as well. If the second means of homogenation is chosen, the gasoline can be atomized into droplets a few microns in size, and mixing with air will result in either a mixture of air and air-suspended, partially vaporized, minute fuel droplets (nonequilibrium state), or a mixture of air and gaseous fuel from the completely vaporized fuel droplets (equilibrium state).

Whichever state will be reached depends on the initial heat content of the system and the amount of heat supplied before or during the mixing process. In view of power, economy, and emission considerations, the nonequilibrium state is preferable. From Refs. 14 and 15, the time needed for complete vaporization of the fuel droplet can be computed and the state of the mixture admitted into the engine cylinders estimated.

Several methods of liquid atomization are described in technical literature (16). The smallest droplet size can be conveniently obtained by means of pneumatic atomization as it is commonly practiced in spray drying or spray painting. Fig. 9 presents schematically one of the possible arrangements of pneumatic fuel atomization as applied to engine carburetion. Fuel supplied from the engine fuel pump passes through a fuel pressure regulator and is fed to a pneumatic atomizing nozzle of the external mixing type. The pressure of the fuel and the opening position of the needle valve built into the atomizing nozzle determine the rate of flow of the fuel. The suppression in the inlet manifold and the opening position of the throttle determine the rate of flow of the air, if the throttle is located upstream of the atomizing nozzle (Fig. 17), or the rate of flow of the air-fuel mixture, if the throttle is located downstream of the atomizing nozzle (Fig. 9). The needle valve is mechanically connected to the throttle so as to maintain a constant ratio between the throttle and needle valve flow areas. The fuel pressure regulator responds to inlet manifold suppression and modifies the fuel pressure according to a predetermined characteristic so that an optimal air-fuel ratio for minimum emissions is maintained over the entire operating range of the engine.

Fig. 10 presents the airflow characteristic of a conventional carburetor throttle. Airflow is plotted as a function of inlet manifold vacuum, with the angle α of the throttle opening as a parameter. For a given position α of the throttle, the airflow remains constant, independent of the manifold vacuum until this drops below a critical value corresponding to 14.2 in. Hg vacuum. A further drop in the manifold vacuum causes a decrease of airflow according to a flow function, which can be computed from the energy and polytropic change of the state of the gas. In Fig. 10 the computed flow functions are compared with measured flow rates. It is obvious that the pressure of the fuel can be kept constant for all engine operating conditions which produce

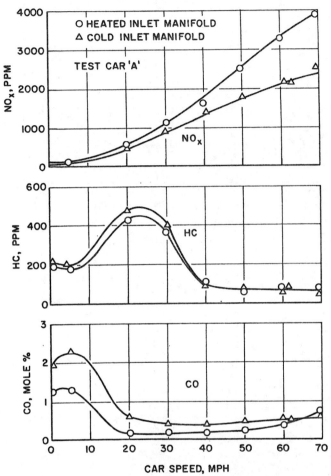

Fig. 7 - Comparison of emission resulting from car operation with standard heated and with cold (unheated) inlet manifold

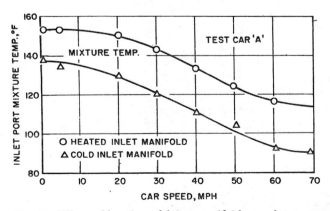

Fig. 8 - Effect of heating of inlet manifold on mixture temperature

an inlet manifold vacuum equal to or higher than 14.2 in. Hg.

Other functions of the fuel atomizing carburetor, Type MV (manifold vacuum controlled), are apparent from Fig. 9. The carburetor incorporates exhaust gas recycling and positive crankcase ventilation. The amount of exhaust gas recycled into the carburetor is controlled by a valve, mechanically actuated by the throttle. Compressed air for fuel atomization is supplied under a pressure of 20-30 psi from an air compressor driven by a fan belt. Induction air enters the mixing chamber through a swirl cage and mixes in a cyclonic motion with the injected spray of the atomized fuel. Larger fuel droplets which are thrown against the wall of the mixing chamber are instantly vaporized by the heat supplied through the wall from a heating jacket. To this end, exhaust gas or engine cooling water is bypassed through the heating jacket. Fig. 11 shows a fuel atomizing carburetor Type MV installed on the V-8 engine of test car A.

During the preliminary tests of car A equipped with a fuel atomizing carburetor, a transparent plastic spacer was installed between the carburetor and the flange of the inlet manifold. The flow of exhaust gas through the heating jacket of the mixing chamber was adjusted to a level at which no droplets of liquid fuel could be observed in the mixture or on the walls of the plastic spacer. In order to prevent excessive heating of the air-fuel mixture, heat input from the inlet manifold hot-spot was eliminated by inserting two brass shims below the manifold gasket over the opening of the

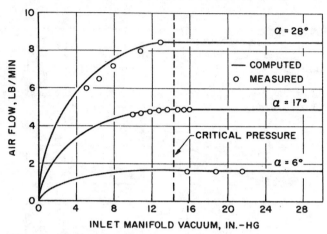

Fig. 10 - Airflow through standard carburetor as function of manifold vacuum at constant throttle opening α

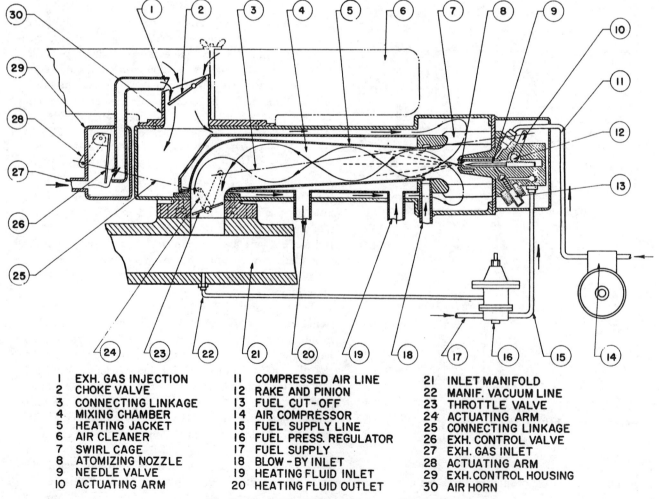

1	EXH. GAS INJECTION	11	COMPRESSED AIR LINE	21	INLET MANIFOLD
2	CHOKE VALVE	12	RAKE AND PINION	22	MANIF. VACUUM LINE
3	CONNECTING LINKAGE	13	FUEL CUT-OFF	23	THROTTLE VALVE
4	MIXING CHAMBER	14	AIR COMPRESSOR	24	ACTUATING ARM
5	HEATING JACKET	15	FUEL SUPPLY LINE	25	CONNECTING LINKAGE
6	AIR CLEANER	16	FUEL PRESS. REGULATOR	26	EXH. CONTROL VALVE
7	SWIRL CAGE	17	FUEL SUPPLY	27	EXH. GAS INLET
8	ATOMIZING NOZZLE	18	BLOW-BY INLET	28	ACTUATING ARM
9	NEEDLE VALVE	19	HEATING FLUID INLET	29	EXH. CONTROL HOUSING
10	ACTUATING ARM	20	HEATING FLUID OUTLET	30	AIR HORN

Fig. 9 - Fuel atomizing carburetor - Type MV

crossover heating passage. The temperature of the mixture as measured by a bare-wire thermocouple in the inlet manifold varied from 140 F at high engine speeds and loads to 160 F at low speeds and loads. A series of tests were performed at operating conditions corresponding to car cruising speeds of 0-70 mph and air-fuel ratio settings from 11:1-21:1. Emission of unburned hydrocarbons, carbon monoxide, and nitrogen oxides was recorded during every run. The test car did not show signs of power surging up to an air-fuel ratio setting of 20:1. At higher ratios, slight roughness of engine operation was observed.

To take full advantage of excess oxygen at lean air-fuel ratio settings for the oxidation of unburned hydrocarbons, both exhaust manifolds and exhaust pipes down to the muffler were insulated with asbestos tape. Fig. 12 presents a comparison of test results of hydrocarbon emission with and without an insulated exhaust manifold as a function of car cruising speed. The air-fuel ratio was 17:1 at all speeds and 16:1 at idling. The emission of unburned hydrocarbons was appreciably reduced with the insulated exhaust manifold, especially at engine operation corresponding to higher car cruising speeds resulting in higher exhaust temperatures. At transient engine operating conditions, the reduction was also more pronounced at lower speeds and idle, probably because of the carry-over effect of accumulated heat in the insulated exhaust system. The lower emission data in Table 1, obtained from the California Cycle Test Procedure, suggest this possibility.

Figs. 13-15 present the results of tests performed at constant car cruising speeds at 30, 45, and 60 mph with air-fuel ratio as a variable. All tests were run with cold inlet manifold and insulated exhaust manifold. Exhaust gas recycling rates were 12% at 30 mph, 13% at 45 mph, and 8% at 60 mph. These results indicate that at engine operating conditions corresponding to car cruising speeds up to 70 mph appreciable reduction in emission of carbon monoxide, unburned hydrocarbons, as well as nitrogen oxides can be accomplished with fuel atomization carburetion at air-fuel ratio settings of 17:1-18:1 and exhaust gas recycling rates of 8-12%. Fig. 16 presents a comparison of emissions as measured on test car A equipped with a conventional carburetor (engine according to factory standards) and on the same car with a fuel atomizing carburetor Type MV, and cold inlet and insulated exhaust manifolds except at idling, where it was retarded to 0 deg btc. Ignition spark advance was within the limits of factory standards. For the control of exhaust emission during car deceleration, the fuel atomizing carburetor was equipped with a throttle cracker and an ignition over-advance device as described in Ref. 2.

Table 1 presents the results of the California Standard "Driving Cycle" test, performed by the California Vehicle Pollution Laboratory in Los Angeles. The average emission

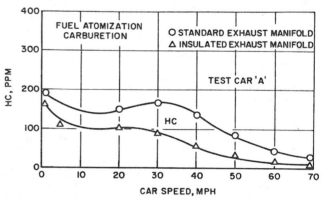

Fig. 12 - Comparison of emission of hydrocarbons with and without insulation of exhaust manifold

Fig. 11 - Fuel atomizing carburetor - Type MV, installed on test car A

from the sixth and seventh hot cycle was 95 ppm of unburned hydrocarbons and 0.3 mole per cent of carbon monoxide.

The experimental results were encouraging enough to be the stimulus for one more inquiry, namely, could the principle of fuel atomization carburetion be simplified so as to provide a basis for the development of a system comparable in cost to the conventional high-performance carburetor. Fig. 17 presents schematically an arrangement resulting from this inquiry.

Two principal changes distinguish later systems from the former:

1. The throttle valve is located upstream of the atomizing nozzle.
2. The mixing chamber is designed as a double-cyclonic flow inductor.

The fuel atomizing nozzle needle valve is mechanically connected to the throttle in the same manner as in the former system. Since the nozzle is subjected to manifold vacuum, the pneumatic atomization is accomplished by bleeding ambient air through the nozzle (see Fig. 17), and consequently no air compressor is necessary. The function of the double-cyclonic flow inductor is to promote and complete homogeneous intermixing of the fuel (in droplets or vapor form), of the induction air, and of the recirculated exhaust gas. The mixing process can be visualized in two steps:

1. Larger unvaporized fuel droplets are driven by centrifugal force against the walls of the flow inductor, which are heated to effect instantaneous vaporization of the impinging droplets.
2. Minute fuel droplets, fuel vapor, induction air, and recycled exhaust gas swirl in a compound cyclonic motion

Table 1 - Results of California Standard "Driving Cycle" Test; Fuel Atomizing Carburetor - Type MV

| Hot 7-Mode Cycles | | As Measured | | | | C Ratio 15/ | Fact or | Corrected | | Wtg. Fact. | Weighted | |
| | | Cycle | | Average | | | | HC, ppm | CO, % | | HC, ppm | CO, % |
		6	7	HC	CO	CO_2							
Idle	HC	65	80	72								
	CO	0.1	0.2		0.2								
	CO_2	11.6	11.6			11.6							
							11.8	1.271	92	0.25	0.042	4	0.01
0-25	HC	65	80	72								
	CO	0.2	0.2		0.2								
	CO_2	12.2	12.5			12.4							
							12.6	1.190	86	0.24	0.244	21	0.06
30	HC	50	50	50								
	CO	0.2	0.2		0.2								
	CO_2	10.7	11.3			11.0							
							11.2	1.339	67	0.27	0.118	8	0.03
30-15	HC	65	50	58								
	CO	0.6	0.2		0.4								
	CO_2	12.0	11.8			11.9							
							12.3	1.219	71	0.49	0.062	4	0.03
15	HC	80	80	80								
	CO	0.2	0.2		0.2								
	CO_2	10.7	10.7			10.7							
							10.9	1.376	110	0.28	0.050	6	0.01
15-30	HC	50	80	65								
	CO	0.2	0.3		0.2								
	CO_2	12.5	12.5			12.5							
							12.7	1.181	77	0.24	0.455	35	0.11
50-20	HC	455	540	498								
	CO	1.2	1.1		1.2								
	CO_2	11.3	11.1			11.2							
							12.7	1.181	588	1.42	0.029	17	0.04
											Sum	95	0.29

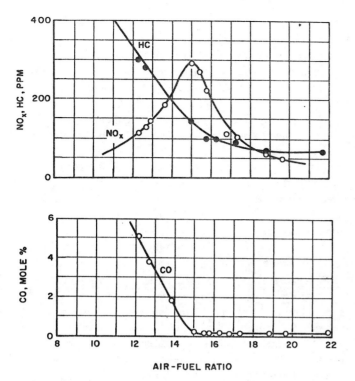

Fig. 13 - Exhaust emission as function of air-fuel ratio. Car speed 30 mph; 12% exhaust gas recycling; fuel atomizing carburetor - Type MV

effective for homogeneous intermixing of all components.

The atomizing carburetor operates in the following way: at idling or car cruising speeds up to 60 mph, as long as the manifold vacuum is at least 15 in. Hg, all the fuel supplied to the engine is aspirated from a float chamber through the atomizing nozzle. To ensure that the fuel flow remains constant even if the manifold vacuum rises above 15 in. Hg, the atomizing nozzle must be of the internal mixing type (16). When engine load is increased and the manifold vacuum drops below 14.2 in. Hg (critical pressure), the airflow is no more constant for a given position of the throttle but decreases according to a flow function, as presented in Fig. 10. Similarly, the fuel flow decreases according to a flow function determined by the design of the atomizing nozzle. For good acceleration response, the air-fuel mixture must be rich (about 12.5:1). Since the air-fuel ratio setting of the carburetor during car cruising is lean (about 18:1), a quick enrichment of the mixture can be conveniently accomplished by a booster venturi located upstream of the throttle. The booster venturi of the atomizing carburetor, as shown in Fig. 17, functions on the same principle as the booster venturi in a conventional carburetor; however, it is set in operation by a bleeder control

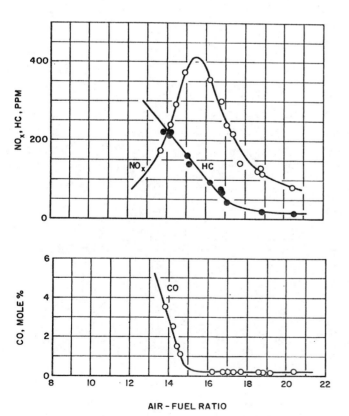

Fig. 14 - Exhaust emission as function of air-fuel ratio. Car speed 45 mph; 13% exhaust gas recycling; fuel atomizing carburetor - Type MV

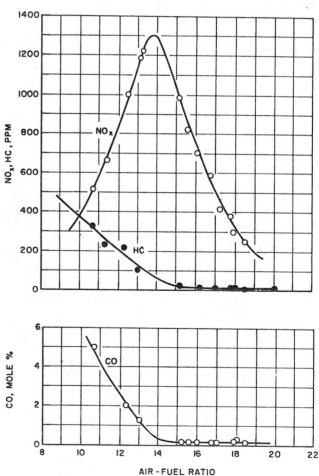

Fig. 15 - Exhaust emission as function of air-fuel ratio. Car speed 60 mph; 8% exhaust gas recycling; fuel atomizing carburetor - Type MV

valve responding to the manifold vacuum. As long as the manifold vacuum is above 15 in. Hg, the control valve is open, admitting air into the fuel passage between the booster venturi and the float chamber. Consequently, at engine operation resulting in a high manifold vacuum, the booster venturi is inoperative. During engine operation at high loads (or acceleration) and corresponding low manifold vacuum, the bleeder valve closes, setting the booster venturi in operation.

The fuel atomizing carburetor of Type AC, then, has two control circuits: fuel atomization and a conventional booster venturi. Evidently this carburetor does not require any fuel pressure regulator. Its operation could also be visualized as the same as that of the conventional carburetor, except that the functions of the idle jet, of the transfer slot, and of the low speed jet are all integrated into one, executed by the atomizing nozzle. For the control of unburned hydrocarbons during deceleration, the carburetor of Type AC was equipped with a throttle cracker operating on a similar principle as described by Wiese, et al (17).

Fig. 18 shows the atomizing carburetor, Type AC, installed on the V-8 engine of test car D. Since this was a rental car, no changes were made on the inlet and exhaust manifolds. All tests were performed with factory standard ignition, inlet, and exhaust. A standard "Driving Cycle" test was performed by the California Vehicle Pollution Laboratory in Los Angeles. The averaged emission from the sixth and seventh hot cycle was 198 ppm of unburned hydrocarbons and 0.15 mole per cent of carbon monoxide. Table 2 presents data obtained from the same test-run by collecting exhaust into plastic bags. The average value from both sampling bags was slightly lower for hydrocarbons and higher for carbon monoxide than that determined by calculation from the cycling procedure.

When comparing the test results obtained on test car A equipped with an atomizing carburetor Type MV and on test car D equipped with an atomizing carburetor Type AC, it may appear striking that for the latter the emission of unburned hydrocarbons is doubled while the emission of carbon monoxide is halved. The reason for this lies probably in the following facts:

1. Because of better homogenation of the air-fuel mixture, carburetor Type AC was operated at a leaner setting than carburetor Type MV. Therefore, lower emission of carbon monoxide resulted.

2. The exhaust manifold of test car D was not thermally insulated. Thus emission of unburned hydrocarbons was higher, as could be predicted from Fig. 12.

Other factors, such as a less favorable spark advance characteristic, could have contributed to this difference. In a like manner, it could be expected that the emission of nitrogen oxides, as determined by the sampling bag, would be

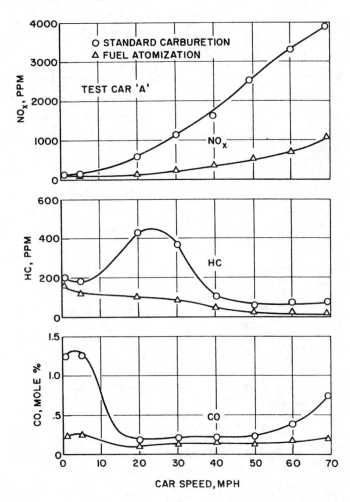

Fig. 16 - Comparison of emission resulting from standard carburetion and from fuel atomization carburetion - Type MV

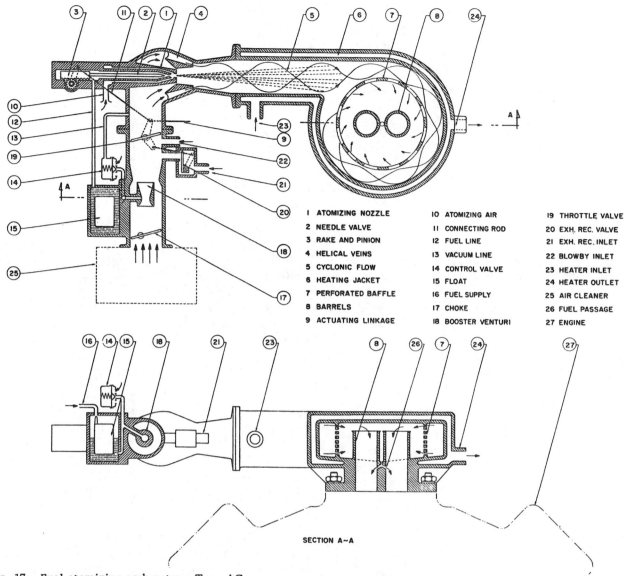

Fig. 17 - Fuel atomizing carburetor - Type AC

Fig. 18 - Fuel atomizing carburetor - Type AC, installed on test car D

Table 2 - Bag Sample Data from California Standard "Driving Cycle" Test
Fuel Atomizing Carburetor - Type AC

Hot 7-Mode Cycles: 6 and 7

Cycle No	Bag No.	Measured % CO	Measured % CO_2	Measured ppm HC	$1/2 CO + CO_2$ + 10 HC	Correction Factor	Corrected % CO	Corrected ppm HC	NO_x Data ppm
6	1	0.2	12.6	125	12.8	1.171	0.23	146	577
7	2	0.2	12.6	115	12.8		0.23	135	656

considerably lower with cold (not heated) inlet manifold, (Compare Fig. 7.)

Several road tests were performed to determine driving performance and fuel economy of atomizing carburetors Type MV and AC. Both driving performance and fuel economy were comparable to those of conventional carburetors. In this limited study, no attempt was made to investigate the effect of other engine variables in order to establish guidelines for optimal tuneup. According to the publication by Robison and Brehob (13), some gain in fuel economy should be obtainable by modification of the spark advance characteristic.

In this study, only two prototypes of the pneumatic fuel atomization carburetor were investigated. Several other carburetion systems for an effective emission control could be developed by various combinations of a few basic functional elements (18). Such systematic investigation might prove to be fruitful, especially if additional means of control, as for example the exhaust manifold gas reactor, were to be included in the study.

CONCLUSION

Simultaneous control of unburned hydrocarbons, carbon monoxide, and nitrogen oxides in automotive exhaust gas can effectively be approached from two directions:
1. Rich air-fuel mixture carburetion, exhaust gas recycling, and addition of an afterburner to, or air injection into, the exhaust system.
2. Lean air-fuel mixture carburetion and exhaust gas recycling, achievable by means of air-fuel mixture homogenation.

The first approach may prove to be more effective for control of nitrogen oxides than of hydrocarbons and least effective for control of carbon monoxide. Although it can be accomplished with a conventional carburetor, an exhaust gas recycling device and an afterburning system must be added. This control may be accompanied by a slight decrease in fuel economy.

The second approach may prove to be most effective for control of carbon monoxide, and more effective for control of unburned hydrocarbons than of nitrogen oxides. It can be accomplished by a carburetion system which produces a well homogenized air-fuel mixture and may be accompanied by a slight increase in fuel economy. The reported exploratory research and testing performed on two prototypes of the fuel atomizing carburetor indicate that automotive emission can be effectively controlled by this means to a level well below California Standards. Only large-scale research applied to this concept of automotive emission control could establish the furthest limits of emission control and fuel economy gain obtainable by this method.

ACKNOWLEDGMENTS

The author wishes to express his gratitude for the enthusiastic cooperation of all students and technical staff of the Dept. of Engineering, University of California, Los Angeles, who at various times were active on this project. He is especially grateful for the administrative support and encouragement given him by C. Martin Duke, Chairman, Dept. of Engineering, as well as for the continuous interest of L. M. K. Boelter, Coordinator of the Air Pollution Research projects.

He also wishes to convey his sincere thanks to Miles Brubacher of the Motor Vehicle Pollution Control Board, State of California, and to the staff of the California Vehicle Pollution Laboratory for their interest and assistance in evaluation of the experimental carburetion system according to California standards.

REFERENCES

1. "Vehicle Emissions." SAE Technical Progress Series, Vol. 6, 1964.
2. "Motor Vehicles, Air Pollution, and Health." A Report of the Surgeon General to the U. S. Congress, U.S. Dept. of Health, Education, and Welfare, June 1962.
3. R. W. Bishop and G. J. Nebel, "Control of Oxides of Nitrogen in Automobile Exhaust Gases - The Technical and Economical Aspects of One Approach." Paper presented at 22nd Annual Meeting of the Industrial Hygiene Foundation, Pittsburgh, October 1957.
4. J. A. Robison, "Exhaust Emissions Evaluation of a 144, 223, and 390 CID Production Engine." Technical Report No. PR 63-10, Product Research Office, Ford Motor Co., November 1963.
5. J. A. Robison, "Effect of Valve Timing on Performance and Exhaust Emissions of a Single Cylinder Engine."

Technical Report No. PR 64-16, Product Research Office, Ford Motor Co., Oct. 6, 1964.

6. R. D. Kopa and H. Kimura, "Exhaust Gas Recirculation as a Method of Nitrogen Oxides Control in an Internal Combustion Engine." Paper presented at APCA 53rd Annual Meeting, Cincinnati, May 1960.

7. R. D. Kopa, R. G. Jewell, and R. V. Spangler, "Effect of Exhaust Gas Recirculation on Automotive Ring Wear." Paper S321 presented at SAE Southern California Section, Los Angeles, March 12, 1962.

8. R. D. Kopa, B. R. Hollander, F. H. Hollander, and H. Kimura, "Combustion Temperature, Pressure, and Products at Chemical Equilibrium." Paper 663A presented at SAE Automotive Engineering Congress, Detroit, January 1963.

9. H. D. Daigh and W. F. Deeter, "Control of Nitrogen Oxides in Automotive Exhaust." Paper presented at American Petroleum Institute, 27th Mid-year Meeting, San Francisco, May 1962.

10. J. A. Robison, "Effect of Exhaust Gas Recycling on Performance and Exhaust Emissions of a Single Cylinder Engine." Technical Report No. PR 64-15, Product Research Office, Ford Motor Co. October 1964.

11. F. N. Scheubel, "On Atomization in Carburetors." NACA - Technical Memo No. 644.

12. E. A. Dörges, "Gemischverteilung an einem Vierzylinder-Vergasermotor." ATZ, Vol. 59, (January 1957), 194.

13. J. A. Robison and W. M. Brehob, "The Influence of Improved Mixture Quality on Engine Exhaust Emissions and Performance." Paper presented at Western States Combustion Institute Meeting, Santa Barbara, October 1965.

14. M. Gilbert, J. N. Howard, and B. L. Hicks, "An Analysis of the Factors Affecting the State of Fuel and Air Mixtures." NACA-Technical Note No. 1078, May 1946.

15. M. M. El Wakil, R. J. Priem, H. J. Brikowski, P. S. Myers, and O. A. Uyehara, "Experimental and Calculated Temperature and Mass Histories of Vaporizing Fuel Drops." NACA-Technical Note No. 3490, January 1956.

16. W. R. Marshall, Jr., "Atomization and Spray Drying." AICE, Chemical Engineering Progress Monograph Series, No. 2, Vol. 50, 1954.

17. W. M. Wiese, R. J. Templin, and P. C. Kline, "An Improved Device to Reduce Exhaust Hydrocarbons During Deceleration." Paper 486H presented at National Automobile Week, Detroit, March 1962.

18. R. D. Kopa, "Pneumatic Fuel Atomization as Applied to Automobile Air Pollution Control." Report No. 63-61, Dept. of Engineering, University of California, Los Angeles, December 1963.

This paper is subject to revision. Statements and opinions advanced in papers or discussion are the author's and are his responsibility, not the Society's; however, the paper has been edited by SAE for uniform styling and format. Discussion will be printed with the paper if it is published in Transactions, or in a Technical Progress or Advances in Engineering volume. For permission to publish this paper in full or in part, contact the SAE Publications Division and the authors.

16 page booklet - Printed in U.S.A.

ABSTRACT

The effects of air-fuel mixture quality and cylinder-to-cylinder air-fuel distribution on exhaust emissions have been determined on two engine-vehicle combinations. California Motor Vehicle Pollution Control Board (CMVPCB) test cycle emissions were measured on vehicles using a pre-mixed and pre-heated air-fuel charge supplied by a steam jacketed, nine cubic foot vaporization tank. The vaporization tank provided a near constant air-fuel mixture ratio for all operating modes of the 7-mode CMVPCB test cycle. The two vehicles were evaluated at nominal air-fuel ratios of 14:1, 16:1 and 18:1.

Cylinder-to-cylinder air-fuel distribution during the transient operation of the 7-mode CMVPCB test cycle was measured on a 200-CID six cylinder and a 289-CID eight cylinder engine. The procedure employed was to record the total carbon emissions ($CO + CO_2 + CH_4$ equivalent) for each cylinder during successive test cycles. Distribution measurements thus established were found to be repeatable within ± 0.40 air-fuel ratios. Comparing data with that of a conventional carbureted induction system showed that the vaporization tank significantly improved cylinder-to-cylinder air-fuel distribution for all modes of the cycle. This improvement permitted lean operation and resulted in low carbon monoxide concentrations. CMVPCB cycle hydrocarbon emission levels were similar for both the vaporization tank and the carbureted induction system when compared at equivalent air-fuel ratios. At the leaner air-fuel ratios obtainable with the vaporization tank, an actual increase in cycle hydrocarbon levels was noted.

Mixture quality was visually observed through a transparent section of the intake manifold installed on a six cylinder engine. Carbureted mixtures were found to be extremely heterogeneous while wet and globular in form. Manifold wetness with the vaporization tank induction system was nonexistent throughout the 7-mode test cycle.

The installation of a 4-inch high intake manifold riser extension, which provided an additional intake charge mixing volume, resulted in an improved cylinder-to-cylinder air-fuel distribution with the conventional lean carburetion. Application of exhaust heat to this mixing chamber accomplished little in improving the cylinder-to-cylinder air-fuel distribution.

Vehicle Exhaust Emission Experiments Using a Pre-Mixed and Pre-Heated Air Fuel Charge

by
J.H. Jones
and
J.C. Gagliardi
FORD MOTOR COMPANY

INTRODUCTION

Uniform cylinder-to-cylinder air-fuel distribution is an important criteria in multicylinder engine design. The effects of maldistribution on parameters such as power, fuel economy and octane requirements have been well documented. (1, 2, 3, 4)* An additional parameter, exhaust emissions, has become an important, if not the most important, criteria to be considered. Although the published literature on the effects of various engine variables on exhaust emissions is becoming more extensive, there still remains only limited information on the effects of cylinder-to-cylinder air-fuel distribution.

The techniques employed by previous investigators for the measurement of cylinder-to-cylinder air-fuel distribution have varied considerably; they include a direct measurement of fuel and air, Orsat analysis, radioactive tracers, spark plug temperatures, infra-red analysis, etc. (2, 3, 5, 6) Many of these investigations have been limited to engine operation of steady state and/or simulated acceleration conditions on an engine dynamometer. A new technique, covered in this paper, extends the limits of distribution measurements to include transient conditions. By means of a chassis dynamometer, cylinder-to-cylinder air-fuel distribution has been determined during all modes of operation — accelerations, decelerations and steady state cruises.

Since the advent of exhaust emission control a new emphasis has been placed on induction system design. (7) With minimum hydrocarbon and carbon monoxide emissions occuring at mixture ratios leaner than stoichiometric, the restrictions on lean operation which can result from poor cylinder-to-cylinder air-fuel distribution have become more critical. In addition, mixture quality, which is known to be an important factor in cylinder-to-cylinder air-fuel distribution, has also received added attention. In order to more fully determine the effects of these parameters on exhaust emissions, CMVPCB cycle data were measured on vehicles operating on both a pre-mixed and pre-heated air-fuel mixture and a conventional carbureted mixture.

---

* Numbers in parentheses designate references at end of paper.

DESCRIPTION

Test Vehicles

Two test vehicles with low mileage engines, a 200-CID six cylinder and a 289-CID eight cylinder, were evaluated in this test program. A complete description of the vehicles is given in Table 1.

TABLE 1

Test Vehicle Description

Year and Model	Engine Displ. (cu. in.)	No. of Cyl.	Carburetion Venturi	Transmission Type
1965 Comet	200	6	1V	Automatic
1966 Galaxie	289	8	2V	Automatic

The vehicles were equipped with conventional (non-thermactor) distributors and carburetors; however, carburetor calibration was varied for the rich and lean carburetion test conditions.

Vaporization Tank

The pre-mixed and pre-heated air-fuel mixture was obtained from a 9 cu-ft vaporization tank installed adjacent to a chassis dynamometer as shown in Figure 1. The tank was equipped with a steam jacket and external insulation and an internal baffle, centrally located, which directed the air-fuel mixture in a reverse flow arrangement as shown in Figure 2. Mounted on top of the tank was a conventional 1V carburetor, altered to deliver a constant air-fuel ratio in the 30 to 120 cfm flow range. The actual air-fuel ratio delivered was dependent on the size of the main metering jet, as shown in Figure 3. Since engine air flow during the CMVPCB test cycle could be as low as 6 to 10 cfm, it was necessary to employ auxiliary air pumps connected to the tank outlet in order to maintain the total flow above the minimum 30 cfm. The excess combustible air-fuel mixture was discharged to the atmosphere through a large capacity blower and ducting, as shown in Figure 1. A condensation trap at the bottom of the tank detected any liquid fuel escaping vaporization. The air-fuel mixture from the tank to the engine was carried in a 2" diameter insulated pipe. At the engine, the mixture flowed through a modified carburetor which was essentially only a "throttle body" that controlled engine speed and distributor vacuum. All external vents and air bleeds of the throttle body were plugged, as were the idle and main fuel systems. The crankcase ventilation valve was removed from the rocker cover and capped to prevent any dilution of the air fuel charge. Starting the vehicles with the vaporization tank induction system could usually be accomplished with the tank-supplied mixture, but in some instances it was necessary to use an auxiliary fuel system that could be removed after start-up.

Figure 1

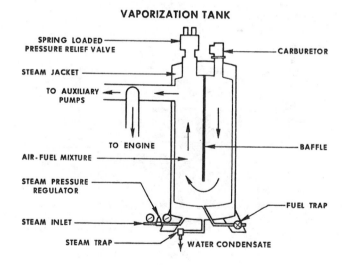

Figure 2

With vaporization tank operation at the leaner air-fuel ratios it was necessary to open the throttle of the throttle body in order to maintain idle speed. This throttle opening in effect resulted in an overadvanced condition at idle by allowing the distributor spark advance port to be exposed to manifold vacuum. To

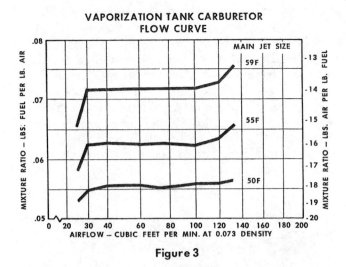

Figure 3

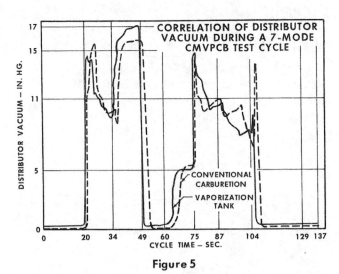

Figure 5

avoid this condition a throttle by-pass was installed on the throttle bodies; the installation for the 289-CID eight cylinder engine is shown in Figure 4. The by-pass allowed a fixed throttle position while permitting idle speed adjustments to be made with the flow adjustment valve in the by-pass line. With this arrangement, the spark advance characteristics could be held constant for both the carbureted and vaporization tank tests, as shown by the close agreement of the distributor vacuum traces shown in Figure 5.

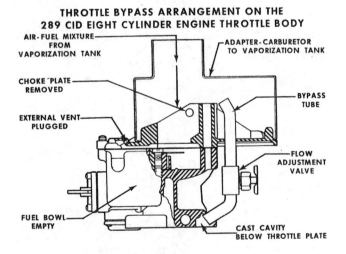

Figure 4

Windowed Intake Manifold

In order to observe and compare mixture quality of the two induction systems, the intake manifold of a 240-CID six cylinder engine was equipped with transparent windows in the outside wall (see Figure 6). In addition to these windows, the manifold was also modified by blocking the exhaust flow to the "heat box."

The manifold of the 240-CID six cylinder engine was selected because of the ease with which the windows could be installed.

Figure 6

Exhaust Heated Intake Mixing Chamber

Attempting to duplicate the vaporization tank concept in a practical induction system, a heated, high riser aluminum carburetor spacer was installed on the 289-CID eight cylinder engine-vehicle combination. It was felt that the riser incorporated the two salient features of the vaporization tank, namely, heat and mixing volume. The riser, four inches high, was heated by exhaust gases drawn from the exhaust crossover (see Figure 7).

PROCEDURE

All tests were conducted on a chassis dynamometer at road load conditions. The road load horsepower setting was 8 horsepower for a 50 mph cruise condition. Test conditions for

EXHAUST HEATED INTAKE MIXING CHAMBER

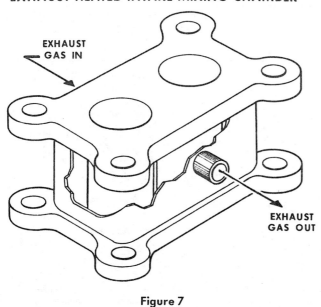

Figure 7

both the conventional carburetion and vaporization tank induction systems consisted of hot start 7-mode CMVPCB test cycles and steady state operation, including idle and cruises in the 15 to 50 mph range. Exhaust analysis included the continuous measurement of carbon monoxide (CO), carbon dioxide (CO_2) and hydrocarbon (HC) concentrations by infrared instrumentation.

For conventional carburetion, tests were conducted with both rich and lean mixture ratios, the mixtures being varied by adjustments in the idle mixture screws. This resulted in changes only in the idle and off-idle regions; however, the majority of the test conditions are in these areas and it was not deemed necessary to make major calibration changes in the carburetor. Vaporization tank tests were conducted at nominal mixture ratios of 14:1, 16:1 and 18:1.

Cylinder-to-cylinder air-fuel distribution for both types of induction systems under the transient operation of the 7-mode test cycle was determined for the 200-CID six cylinder and 289-CID eight cylinder engine-vehicle combinations. Exhaust sampling probes located near the valve in each exhaust port were used to obtain individual cylinder traces of CO, CO_2 and HC, which served as the basis of the distribution measurements. Since readings of the individual cylinder emissions were not made simultaneously, it was necessary to run successive test cycles — one for each cylinder.

The mixture quality and manifold wetness during the test cycle was observed through the windowed manifold for the conventional carburetor and vaporization tank induction systems. The effects of both heat and mixing volume on cylinder-to-cylinder air-fuel distribution were determined with the heated high riser carburetor spacer. Both transient and steady state conditions were examined while operating on lean, conventional carburetion.

All air-fuel ratio data reported herein were determined from a previously established empirical relationship between a flow measured air-fuel ratio and a total carbon balance (CO + CO_2 + CH_4 equivalent) of the exhaust. This relationship, as published by Robison and Brehob, (8) is shown in Figure 8. When reference is made to the air-fuel ratio of the mixture supplied to a cylinder during a transient mode of operation, what is actually meant is an average air-fuel ratio, integrated over the entire length of the mode.

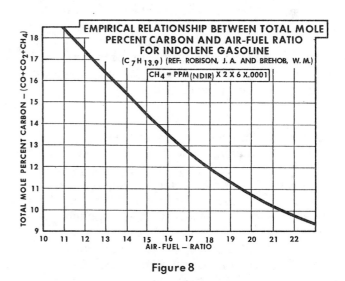

Figure 8

RESULTS AND DISCUSSION

Cylinder-to-Cylinder Air-Fuel Distribution

The method used in determining cylinder-to-cylinder air-fuel distribution, that is, the recording of CO, CO_2 and HC emissions for each cylinder during successive test cycles is wholly dependent upon the repeatability of the test cycles. In order to determine the repeatability of this procedure, the following check was conducted on the 289-CID eight cylinder engine-vehicle combination. On three consecutive days the vehicle was tested, each day performing a total of 15, 7-mode cycles. During the first seven warm-up cycles, total engine exhaust emissions were monitored; during the last 8 cycles, the distribution of the eight cylinders were individually sampled in random order. The three sets of CO, CO_2 and HC data thus obtained were then statistically analyzed to determine the mean and variance of the replicate data distributions. Random sampling and statistical treatment of the data were used to eliminate the effect of uncontrollable variation in test conditions on CO, CO_2 and HC emissions. Cylinders were not sampled in the same order each day, but rather were sampled in an order determined purely by chance. In this manner, each cylinder was exposed to both early and late sampling. Any changes in the exhaust concentrations due to variations in experimental parameters were reflected in the variance of the data.

As noted, the exhaust sampling probes were located near the valve in each of the exhaust ports. Previous investigators (9) have reported the disadvantages of such an arrangement due to the hydrocarbon "pulse" phenomenon. However, it must be remembered that for air-fuel ratio determinations, the total carbon (CO + CO_2 + CH_4 equivalent) concentration of the exhaust, which is independent of sample tap location, is of interest and not the absolute level of any one particular constituent.

The repeatability check was conducted for the two different levels of conventional carburetion—namely, rich and lean. The results shown in Figure 9 (rich) and Figure 10 (lean) are presented as the interval between plus or minus two standard deviations from the mean air-fuel ratio for the individual cylinders. Repeatability thus measured was found to be somewhat independent of mixture ratio, with only a slight improvement shown with the rich carburetion—0.74 average spread in plus or minus two standard deviations as compared to 0.79. With both the rich and the lean carburetion, the idle mode exhibited the poorest repeatability. The other modes established no particular trends, the average spread about the mean being approximately equal.

The repeatability of the cycle data has been shown graphically by superimposing the CO, CO_2 and HC traces of cylinder no. 1 at the lean setting over the three day test period, as shown in Figure 11. Considering that each trace was made on a different day and that on each day the cylinder was sampled during a different cycle of the eight cycle test pattern repeatability, in general, was considered acceptable.

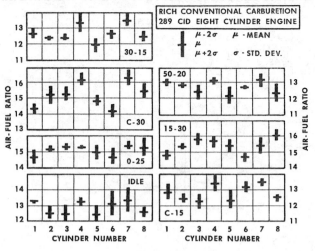

Figure 9

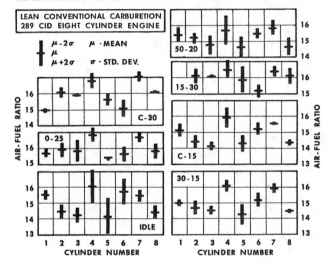

Figure 10

After the repeatability of the procedure was established, tests were conducted on the 200-CID six cylinder and the 289-CID eight cylinder engine-vehicle combinations. The results with lean carburetion, as shown in Figures 12 and 13, are presented as the superimposed traces of the CO, CO_2 and HC emissions and as the average air-fuel ratio of the individual cylinders, as determined from time averaged values over each of the seven modes of the test cycle. The tabular data indicate fair distribution for the 200-CID six cylinder engine, with the maximum spread of air-fuel ratios for the individual cylinders in the range of 0.2 to 1.2 ratios. The greatest spread in cylinder-to-cylinder air-fuel distribution occurred at idle and the two deceleration modes, 30 to 15 mph and 50 to 20 mph. The individual traces show a large degree of variability during the

during the 30 to 50 mph acceleration (a mode that is not included in the seven mode time averaging procedure); however, it is believed this condition is due to variability in the power enrichment valve rather than maldistribution.

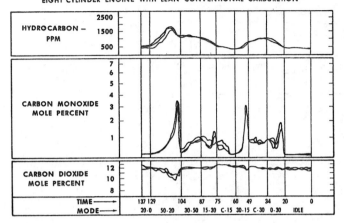

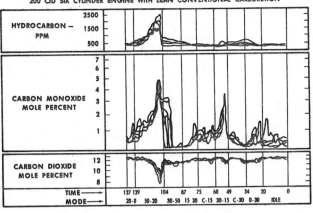

REPEATABILITY OF CMVPCB CYCLE MODAL EMISSION DATA
DAY-TO-DAY VARIATION OF CYLINDER NUMBER 1 OF A 289 CID EIGHT CYLINDER ENGINE WITH LEAN CONVENTIONAL CARBURETION

AVERAGE INDIVIDUAL CYLINDER AIR-FUEL RATIO
DAY-TO-DAY VARIATION OF CYLINDER NUMBER 1 OF A 289 CID EIGHT CYLINDER ENGINE WITH LEAN CONVENTIONAL CARBURETION

	TEST NUMBER				
MODE	1	2	3	AVG	MAX SPREAD
IDLE	16.7	16.4	16.4	16.50	0.3
0-25	15.5	15.7	15.7	16.63	0.2
C-30	14.7	14.9	14.9	14.83	0.2
30-15	15.1	14.9	15.1	15.03	0.2
C-15	15.2	14.9	15.2	15.10	0.3
15-30	15.2	15.1	15.2	15.16	0.1
50-20	15.8	15.2	15.4	15.46	0.6
AVG	15.46	15.30	15.41		
MAX SPREAD	2.0	1.5	1.5		

Figure 11

CYLINDER-TO-CYLINDER DISTRIBUTION DURING A 7-MODE CMVPCB TEST CYCLE
200 CID SIX CYLINDER ENGINE WITH LEAN CONVENTIONAL CARBURETION

AVERAGE INDIVIDUAL CYLINDER AIR-FUEL RATIO
7-MODE CMVPCB TEST CYCLE — 200 CID SIX CYLINDER ENGINE WITH LEAN CONVENTIONAL CARBURETION

	CYLINDER NUMBER							
MODE	1	2	3	4	5	6	AVG	MAX SPREAD
IDLE	15.3	15.2	15.1	15.0	16.0	14.8	15.23	1.2
0-25	15.6	15.4	15.2	15.8	15.4	15.5	15.48	.6
C-30	14.8	14.5	14.6	14.6	14.6	14.8	14.65	.3
30-15	14.1	14.2	14.1	13.9	14.7	13.8	14.13	.9
C-15	15.1	14.9	15.4	14.9	15.5	15.1	15.15	.6
15-30	15.7	15.5	15.6	15.5	15.6	15.5	15.57	.2
50-20	14.4	13.9	14.1	13.9	13.9	13.6	13.97	.8
AVG	15.00	14.80	14.87	14.80	15.10	14.73		
MAX SPREAD	1.3	1.6	1.5	1.9	2.1	1.9		

Figure 12

Similar data on the 289-CID eight cylinder engine shows a greater cylinder-to-cylinder maldistribution with a maximum spread of individual cylinders of 1.3 to 2.3 air-fuel ratios. No definite trends were established in regards to modal maldistribution as with the low air flow modes (idle and deceleration) of the 200-CID six cylinder engine.

With the pre-mixed and pre-heated mixtures supplied with the vaporization tank, cylinder-to-cylinder air-fuel distribution was significantly improved. Shown in Figures 14 and 15 are the vaporization tank results at a nominal air-fuel ratio of 16:1. The maximum spread of cylinder-to-cylinder air-fuel ratios was considerably reduced for both engines. For the 200-CID six cylinder engine the spread ranged from 0.2 to 0.5 air-fuel ratios, with generally equal magnitudes for all modes. For the 289-CID eight cylinder engine the maximum spread was approximately the same, ranging from 0.3 to 0.7 ratios. The cycle emission traces show a relatively constant level for each of the individual constituents. One exception is the 50 to 20 mph deceleration, where large changes are seen in the CO_2 and HC concentrations, the results of poor combustion under the conditions of low manifold pressure experienced during this mode. The remaining maldistribution exhibited by the tabular data in Figures 14 and 15 is believed due to inherent variation of the test procedure.

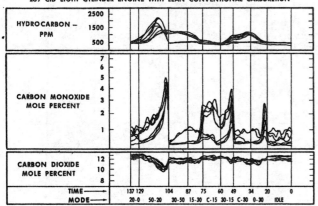
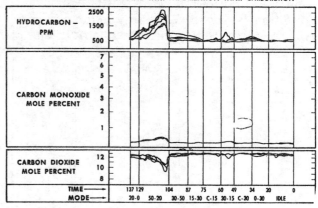

Figure 13

MODE	1	2	3	4	5	6	7	8	AVG	MAX SPREAD
IDLE	16.7	15.2	15.5	16.3	14.8	16.7	16.7	15.2	15.89	1.9
0-25	15.5	16.0	16.6	16.5	15.4	15.6	16.8	15.9	16.04	1.4
C-30	14.7	15.9	15.8	17.0	15.3	15.2	17.0	16.0	15.86	2.3
30-15	15.1	14.5	14.4	16.0	14.0	15.3	15.8	14.5	14.95	2.0
C-15	15.2	14.1	14.3	15.7	14.2	15.1	15.4	14.2	14.78	1.6
15-30	15.2	16.1	16.1	16.4	16.0	15.0	16.5	16.4	15.96	1.5
50-20	15.8	15.0	14.6	15.9	14.6	15.6	15.5	14.8	15.23	1.3
AVG	15.46	15.26	15.33	16.25	14.90	15.50	16.24	15.29		
MAX SPREAD	2.0	2.0	2.3	1.3	2.0	1.7	1.6	2.2		

Figure 14

MODE	1	2	3	4	5	6	AVG	MAX SPREAD
IDLE	15.8	15.9	15.6	15.9	15.8	16.1	15.85	0.5
0-25	16.0	15.9	15.6	15.9	15.8	15.9	15.85	0.4
C-30	15.7	15.6	15.7	15.9	15.6	15.8	15.72	0.3
30-15	15.8	15.7	16.0	15.9	15.8	15.9	15.85	0.3
C-15	15.8	15.9	15.7	15.9	15.8	15.9	15.83	0.2
15-30	15.7	15.8	15.6	15.9	15.8	15.7	15.75	0.3
50-20	15.9	16.2	16.0	16.1	16.1	15.9	16.03	0.3
AVG	15.81	15.86	15.74	15.93	15.82	15.89		
MAX SPREAD	0.3	0.6	0.4	0.2	0.5	0.4		

With the vaporization tank induction system at the 18:1 air-fuel ratio, some difficulty was encountered in achieving the correct vehicle speed within the allotted time of the test cycle. This condition occurred only with the smaller six cylinder engine. Besides this loss of horsepower, no other conditions—surge, acceleration stumble, misfire, etc.,—characteristic of lean operation, were observed.

In all instances of air-fuel ratio determination a total carbon balance was employed. The importance of this factor becomes readily apparent with the vaporization tank induction system in the 50 to 20 deceleration mode. Had either the CO or CO_2, or even both the CO and CO_2 concentrations been used, a mixture leaning would have been concluded, since the results show a large decrease in the CO_2 concentrations with only a minimal change in the CO. When the HC concentration is included in the carbon balance, the result is a near constant air-fuel ratio for all modes of the cycle. This is in agreement with the flow stand calibration of the vaporization tank carburetor, which resulted in a constant air-fuel ratio delivery in the 30 to 120 cfm range — the auxiliary pumps maintaining the minimum 30 cfm flow.

Mixture Quality

Mixture quality of both induction systems was observed through the windowed manifold during the CMVPCB test cycle. Conventional carburetion resulted in a wet manifold interior during the entire cycle with mixture quality ranging from a heavy fog at idle to a very heterogenous globular mixture during the 30 to 50 acceleration. At the count of 104 seconds (the start of the 50 to 20 deceleration) the puddles and large fuel globules developed during the preceding acceleration were quickly vaporized. Actual boiling of the fuel was observed through a major portion of the deceleration mode. The effects of this "flash off" are seen in the mixture enrichment observed during the deceleration modes of both carbureted test runs (Figures 12 and 13). With the vaporization tank induction system a dry homogeneous mixture was observed for all cycle modes. Only upon an extended idle was there any wetness observed, and then only a slight dampness.

Exhaust hydrocarbon concentrations have been shown to be a function of air-fuel ratio; therefore, in comparing the carbureted and vaporization tank data, considerations must be given to air-fuel ratio levels. Assuming that the calculated CMVPCB cycle CO values are indicative of the cycle air-fuel ratio, the hot start hydrocarbon results were compared at equivalent levels of cycle CO. On this comparative basis the pre-mixed and pre-heated vaporization tank mixture and the conventional carbureted mixture resulted in equal levels of hydrocarbon emissions, as shown in Figure 16. Similar trends were established with steady state operation, as shown in Figures 17 and 18. Here hydrocarbon concentrations are

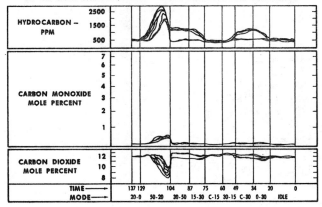

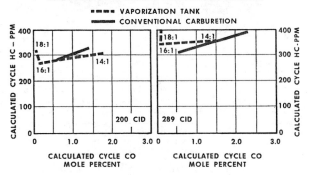

Figure 16

shown as a function of air-fuel ratio at various cruise conditions for the two engine-vehicle combinations. In general, equal air-fuel ratios resulted in equal hydrocarbon emissions, regardless of the induction system that was employed. However, the vaporization tank did allow an extension of the lean operating range of the engine. The hydrocarbon reduction resulting from this extension is small and may even result in an increase, as exhibited by the relatively flat curves of the cyclic and steady state data (Figures 16, 17 and 18). The real

AVERAGE INDIVIDUAL CYLINDER AIR-FUEL RATIO
7-MODE CMVPCB TEST CYCLE – 289 CID EIGHT CYLINDER ENGINE WITH VAPORIZATION TANK CARBURETION

MODE	1	2	3	4	5	6	7	8	AVG	MAX SPREAD
IDLE	15.7	16.0	16.2	15.9	16.1	15.9	15.9	15.9	15.95	0.5
0-25	16.1	15.9	16.1	15.8	16.2	15.9	15.6	15.9	15.94	0.6
C-30	15.7	15.6	15.6	15.5	15.5	15.5	15.6	15.3	15.54	0.4
30-15	15.4	15.5	15.2	15.2	15.2	15.2	15.4	15.2	15.29	0.3
C-15	15.8	15.8	15.9	15.7	15.9	15.7	15.5	15.6	15.74	0.4
15-30	16.1	15.7	15.9	15.8	15.9	15.9	15.7	15.9	15.86	0.4
50-20	15.8	16.1	15.6	15.9	15.6	16.1	16.2	15.5	15.85	0.7
AVG	15.80	15.80	15.79	15.69	15.77	15.74	15.70	15.61		
MAX SPREAD	0.7	0.6	1.0	0.7	1.0	0.9	0.8	0.7		

Figure 15

Exhaust Emissions – Carbureted vs Vaporization Tank

A list of the total engine exhaust emissions for hot start CMVPCB test cycles is shown in Table 2.

TABLE 2

Total Engine Exhaust Emissions – Hot Start CMVPCB Test Cycles

Vehicle Engine Size		Vaporization Tank			Carbureted	
		14:1	16:1	18:1	Rich	Lean
65 Comet	HC	300	282	310	321	294
200-CID	CO	1.81	0.12	0.08	1.39	0.66
66 Galaxie	HC	360	350	388	372	311
289-CID	CO	1.46	0.12	0.14	1.81	0.57

benefit of the pre-mixed and pre-heated mixture is the improved distribution which allows lean operation without approaching the misfire lean limit.

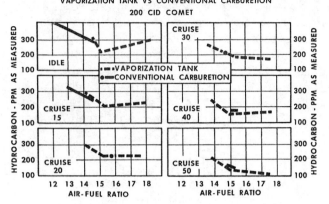

Figure 17

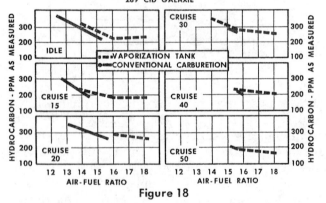

Figure 18

Exhaust Heated Intake Mixing Chamber

The effects of both heat and mixing volume were investigated with the exhaust heated intake mixing chamber. All tests were with conventional lean carburetion. Using the production spacer with its hot spot heat source as a base line (Figure 13), the 4-inch riser was first tested without the flow of exhaust gases through the heating chamber. These results, presented as the superimposed cycle traces, are shown in Figure 19. Comparison with the production base line shows that the added volume of the 4-inch riser significantly improved the cylinder-to-cylinder air-fuel distribution, the maximum spread of air-fuel ratios being reduced from a range of 1.3 to 2.3 to a range of 0.3 to 1.1. Adding heat to the 4-inch riser had little effect

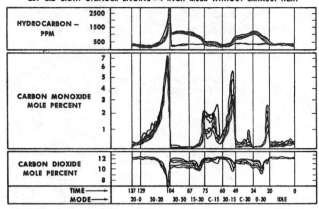

AVERAGE INDIVIDUAL CYLINDER AIR-FUEL RATIO
7-MODE CMVPCB TEST CYCLE – 289 CID EIGHT CYLINDER ENGINE – 4 INCH RISER WITHOUT EXHAUST HEAT

MODE	CYLINDER NUMBER								AVG	MAX SPREAD
	1	2	3	4	5	6	7	8		
IDLE	15.3	15.3	16.4	16.1	15.5	16.1	15.7	15.6	15.75	1.1
0-25	16.8	16.1	16.8	16.6	16.1	16.7	16.7	16.1	16.49	.7
C-30	16.4	15.8	15.6	16.4	15.6	16.2	16.5	15.5	16.00	1.0
30-15	14.8	14.6	14.6	14.6	14.0	14.9	15.1	14.3	14.61	1.1
C-15	15.1	14.5	14.9	14.8	14.4	14.6	15.0	14.5	14.73	.7
15-30	16.6	16.2	16.7	16.9	16.5	16.9	16.4	16.2	16.55	.7
50-20	14.5	14.8	14.8	14.5	14.6	14.7	14.6	14.6	14.64	.3
AVG	15.64	15.32	15.69	15.70	15.24	15.73	15.71	15.26		
MAX SPREAD	2.3	1.7	2.2	2.4	2.5	2.3	2.1	1.9		

Figure 19

on improving distribution, for with the flow of exhaust gases through the heating chamber of the riser, the maximum spread of air-fuel ratios remained essentially the same — 0.5 to 1.1 (see Figure 20). Mixture temperatures with the added exhaust heat were approximately 25 to 30° F higher than the no heat condition (see Figure 21). Also shown in Figure 21 are the mixture temperatures with the production spacer and with the vaporization tank at 16:1 air-fuel ratio.

As shown, the 4-inch riser with no exhaust heat results in essentially the same mixture temperatures as the production spacer with the same lean conventional carburetion. The differences that are present may be indicative of improved vaporization occurring with the added mixing volume of the 4-inch riser.

Table 3 shows a comparison of hot start CMVPCB test cycle results of the intake charge mixing chamber, with and without exhaust heat, and the production spacer. It is noted that the cycle HC and CO results follow the same trends that were established with the vaporization tank, as was shown in Figure 16.

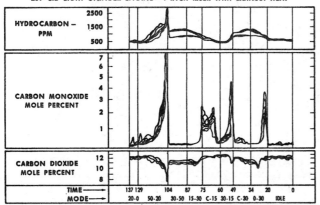

Figure 20

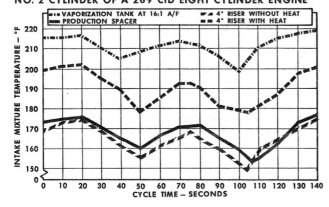

Figure 21

TABLE 3

Total Engine Exhaust Emissions With Intake Mixing Chamber
Hot Start CMVPCB Test Cycles
66 Galaxie – 289-CID

	Production Spacer	4-Inch Riser W/O Exhaust Heat	4-Inch Riser W/Exhaust Heat
HC	311	353	328
CO	0.57	0.94	0.68

SUMMARY

A comparison of the exhaust emission levels and cylinder-to-cylinder air-fuel distribution during the transient operation of the CMVPCB test cycle with a vaporization tank and with a conventional carbureted induction system has led to the following observations:

1. The vaporization tank induction system results in significantly improved cylinder-to-cylinder air-fuel distribution.

2. With a conventional carbureted induction system, mixture quality is wet and globular during the entire CMVPCB test; the vaporization tank provides a dry and homogeneous mixture.

3. At equivalent air-fuel ratios both the vaporization tank induction system and the conventional carbureted induction system result in equivalent hydrocarbon emissions.

4. Exhaust heating of the intake charge has little effect on improving distribution when compared to the improvements gained by increasing the intake charge mixing volume.

5. Recording total carbon emissions ($CO + CO_2 + CH_4$ equivalent) serves as a useful and repeatable basis for determined cylinder-to-cylinder air-fuel distribution during transient operation.

6. For air-fuel ratio determination from an exhaust analysis a total carbon balance (including carbon monoxide, carbon dioxide and hydrocarbons) must be employed. This is especially true for deceleration modes, where a significant portion of the exhaust carbon is in the form of hydrocarbons.

7. The vaporization tank can serve as a useful basis of comparison for new carburetor and induction system designs.

REFERENCES

1. Clark, J. S., "Initiation and Some Controlling Parameters of Combustion in the Piston Engine," Institution of Mechanical Engineers-Automotive Division Council, 1960-1961.

2. Donahue, R. W. and Kent, R. H., Jr., "A Study of Mixture Distribution," SAE Quarterly Transactions, October 1950.

3. Yu, T. C., "Fuel Distribution Studies—A New Look at an Old Problem," Paper No. 603B, presented at SAE National Fuels & Lubricants Meeting, Philadelphia, October 1962.

4. Cleveland, A. E. and Bishop, I. N., "Several Possible Paths to Improved Part-Load Economy of Spark-Ignition Engines," Paper No. 150A, presented at SAE National Automobile Week, Detroit, March 1960.

5. Cooper, D. E. et al, "Radioactive Tracers Cast New-Light on Fuel Distribution," Paper No. 59-6-T, presented at SAE Annual Meeting, Detroit, January 1959.

6. Rabezzena, H. and Kalmar, S., "Mixture Distribution in Cylinder Studies by Measuring Spark Plug Temperature," Automotive Industries, Vol. 66, March 1932, pp. 450-454, 486-491.

7. Bartholomew, E., "Potentialities of Emission Reduction by Design of Induction Systems," Paper No. 660109, presented at SAE Annual Meeting, Detroit, January 1966.

8. Robison, J. A. and Brehob, W. M., "The Influence of Improved Mixture Quality on Engine Exhaust Emissions and Performance," Western States Combustion Institute Meeting, October, 1965.

9. Chandler, J. M. et al, "Development of the Concept of Non-Flame Exhaust Gas Reactors," Paper No. 486M presented at SAE National Automobile and Product Meetings, Detroit, March 1962.

(THIS PAGE INTENTIONALLY LEFT BLANK)

ABSTRACT

Carburetor repeatability at idle was investigated both on the carburetor precision flow stand and on a chassis dynamometer equipped with exhaust emission test equipment. It was determined that some factors which significantly affect idle fuel flow repeatability on the carburetor test stand are:

1. Surface roughness of mating parts in the fuel bowl inlet system.
2. Inlet valve eccentricity.
3. Hydraulic shock in the fuel supply circuit.

In addition, vehicle testing had demonstrated that carbon monoxide changes at idle may occur with constant fuel-air ratio due to changes in fuel temperature. Cycle tests have also shown a relationship between fuel temperature and carbon monoxide levels.

"A Study of Factors Affecting Carburetor Performance at Low Air Flows"

K. C. Bier
J. J. Frankowski
D. K. Gonyou

Colt Industries, Holley Inc Carburetor Division

As exhaust emission requirements became a reality in the automotive industry, the performance of carburetors at low air flow assumed critical proportions. It became necessary to meet flow curve and idle set point requirements which were not within the capability of the existing production flow equipment. It also became common to find unexplained variations in the exhaust emission products of vehicles during idle which were frequently attributed to carburetion variables. This paper reports on a study of factors affecting these two facets of carburetor operation.

Even though the effort to obtain greater and greater precision in production flow measuring techniques is a continuing one, it is a matter of history that precision test stands are in common usage today. Unfortunately, far from eliminating all of the flow problems, the availability of precision measuring equipment served to demonstrate the need for refinements within the carburetor itself.

One of the more common limits, for example, allowed a tolerance of plus or minus 3% from the mean fuel flow at the curb idle set point. This requirement, at first glance, would seem to present little difficulty inasmuch as an adjustment was available to set the fuel flow which, in turn, was indicated on equipment having less than 1% error. The result of a production test stand survey of the idle set point of a large quantity of four barrel carburetors is shown in Figure 1. Figure 1 may be interpreted as a statement of one of the problems since it demonstrated an ability to set, and repeat, the idle fuel flow to a total spread of about 12% while the intended limit required half that spread.

One of the first questions which presented itself was whether the observed variations in fuel flow, shown in Figure 1, were the result of actual variations in carburetor metering or resulted from inconsistency in the fuel inlet system of the carburetor. Since no practical means of measuring fuel flow, or fuel-air ratio, downstream of the carburetor has been demonstrated to date, it has been the usual practice to measure fuel flow entering the carburetor and to assume that this flow is equal to that which the carburetor is metering into the airstream.

Over a relatively long period of time this relationship must hold true, of course, unless the fuel bowl floods or runs dry. Over a short period of time, however, the indicated fuel flow will be greater than the true flow from the carburetor when the fuel level is rising and, conversely, will be less than the true value when the fuel level is decreasing. It, therefore, appeared mandatory to separate the fuel inlet system effects, if any, from the carburetor metering system.

An experimental fuel bowl was constructed which contained a float with an integral hinge mounted in jeweled instrument bearings, a .060 inch diameter fuel inlet orifice, and a sharp, polished, stainless steel inlet needle. This bowl was installed on vibration isolating mounts adjacent to, but separate from, the remainder of the carburetor.

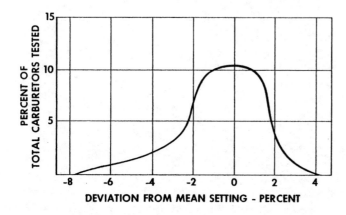

Figure 1. Production Carburetor Idle Fuel Setting Capability.

An amplifying liquid level readout was used to continuously monitor the level in the fuel bowl, as shown in Figure 2, while fuel flow measurements were made. The

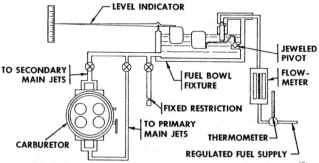

Figure 2. Schematic of Idle Flow Repeatability Test Installation Including Fuel Level Amplifying Indicator.

level amplifier consisted of a counter-weighted float and bracket assembly to which a 17 inch pointer was attached. This assembly was mounted on needle point pivots and was allowed to free float in the fuel reservoir. The resultant magnification factor was approximately 5½ times the actual level change, so a deviation of as little as .002 inch in level could be readily observed by the pointer.

The carburetor metering system was supplied fuel directly from the reservoir by means of plastic tubing to the main metering jets. The remainder of the installation consisted of the usual flow stand facilities. Fuel flow was measured by a flowmeter. (It must be remembered here that repeatability rather than absolute accuracy was sought.)

The carburetor was tested twice daily over a period of 78 working days. The data which were recorded during each test consisted of: 1) carburetor primary metering system fuel flow, 2) carburetor secondary metering system fuel flow, 3) scale reading on liquid level amplifier, 4) ambient temperature, 5) barometer, 6) fuel temperature. In addition a fixed restriction which was supplied fuel by a gravity head from the reservoir, was placed 13.5" below the fuel level. This dimension, which was the largest permitted by the physical limitations of the installation, was chosen such that a 1/16" change in the reservoir level resulted in less than a 1/2% change in head on the fixed restriction. The fuel flow through this restriction was measured after each test to serve as a check against any unexplained change in the metering system flow. Fuel viscosity and specific gravity were monitored periodically. The fuel flow measurements were recorded only when the fuel level indicated by the level amplifier had reached a stable, equilibrium value.

The results of this four month survey are shown in Figure 3. The total carburetor fuel flow which is obtained by adding the primary and secondary flows showed a total variation approaching that of the fixed restriction even though the average deviation is somewhat greater with the carburetor than with the fixed restriction. Since the change in fuel and air temperatures, as well as fuel viscosity and density was negligible over the period of the test, no correlation between these factors and fuel flow was obtained. Also, no measurable effect on fuel flow could be attributed to the barometric variation of 0.9 in. hg under these idle conditions. The most significant conclusion drawn from this extended test was that the variation in the carburetor metering system flow represented only a minor portion of the total variation displayed by the group of carburetors in Figure 1. Apparently the variation in the indicated fuel flow was due predominantly to the effects of the fuel inlet system.

Accordingly, the emphasis was directed to a study of the flow characteristics of the fuel inlet system. Using the level amplifier shown in Figure 2 on a quantity of production fuel bowl and inlet assemblies revealed that the inlet system required from 20 to 90 seconds to stabilize the fuel level to the point that fuel flow remained constant within .02 pounds per hour in the idle fuel flow range. Certain assemblies did not stabilize within 5 minutes. In addition it was found in several instances that a variation in liquid level of .020 to .040 inches existed at any given fuel flow depending upon whether this flow was arrived at by increasing or decreasing the flow. This phenomena was described as "overshoot and undershoot" due to the mechanical friction hysteresis of the fuel inlet system.

The addition of external vibration to the simple mass-force-spring system, which the fuel bowl assembly comprises, reduced the hysteresis effect significantly. Fuel level stabilization occurred within 20 seconds in almost all cases,

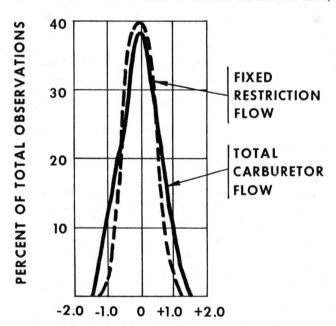

DEVIATION FROM MEAN READING - PERCENT
Figure 3. Carburetor Idle Flow Repeatability Compared to Fixed Jet Flow.

the level hysteresis at any fuel flow was practically eliminated, and a very repeatable level at zero flow was attained. The most effective frequency for the system studied was found to be approximately 47 cps at an acceleration level of 0.5 g peak. Excessive vibrations produced an erratic liquid level and promoted a hunting type of fuel flow indication.

In view of this data, one approach might be to subject the entire carburetor to the desired frequency and amplitude for acceptance testing. However, because of the difficulty in isolating unwanted vibrations from the production test equipment and replacing them with the desired frequencies, it was decided to investigate the frictional characteristics of the inlet system hardware to discover the specific cause of the hysteresis.

It was subsequently found that polishing the inlet needle contact surface to the float hinge tab to a surface finish of 6 to 10 microinches rms resulted in a decrease in time required for fuel level stabilization (without vibration) by decreasing the overshoot. Revisions to the radius of the burnished inlet orifice seat, together with a needle tip to orifice concentricity of .001 inch, centerline to centerline, eliminated the hysteresis almost entirely.

The reduction of friction forces from the fuel inlet components produced a very responsive and repeatable fuel level control system with correspondingly consistent fuel flow readings. As might be expected, however, the removal of the frictional damping forces resulted in a system that was extremely susceptible to excitation from external mechanical or hydraulic shock. These caused an instability which we termed "float bounce" that resulted in substantial instantaneous flow variations.

Efforts to eliminate the bounce, without reintroducing friction, involved modifications to all parts of the inlet system with little or no effect. These modifications included changes in the resiliency of the needle tip, decreased inlet system volume, various float spring rates, revised rigidity of the float bracket, modified clearance between the float hinge pin and float arm, and alternations to the float shape.

Changes that did eliminate the bounce but that proved impractical included 1) mounting a free standing spring and weight to the float arm to create a counter vibrating mass system to cancel the bounce, 2) adding a large baffle to the bottom of the float to cause viscous damping between the float and fuel, and 3) increasing the float lever ratio to increase the buoyant force.

One device which was successful in eliminating the bounce was a small hydraulic accumulator which was inserted in the fuel supply line just upstream of the carburetor fuel inlet fitting. This accumulator, or surge chamber, consisted of an unrestricted flow-through chamber containing a diaphragm with an adjustable calibration spring. It absorbed and dampened the system hydraulic shock as it was being caused and prevented flow instability. It should be noted that the hydraulic shock originating within the carburetor itself as a result of mechanical shock which would cause the float and inlet needle to move was also eliminated by the accumulator as a source that would cause bounce.

The output of an electronic transducer system which produced a linear voltage vs. flow relationship was supplied to a recording oscillograph to obtain a visual record of instantaneous fuel flow over a period of time. This technique was used to obtain the traces in Figure 4 which shows a comparison between a standard production fuel bowl, a bowl having the special construction previously described, and this same bowl used with the hydraulic accumulator.

It may be seen that the standard production bowl shows a cyclic variation of fuel flow with time which, incidentally, could also be seen as a slight cyclic variation in fuel level by the mechanical level amplifier. The trace obtained with the revised bowl was taken after float bounce was generated by a mechanical shock. It is seen that both the frequency and amplitude of the fuel flow variations are of a greater magnitude than those obtained with the standard bowl. It should be noted here that, although these instantaneous fluctuations in fuel flow are sufficiently large to present great difficulty with respect to setting an average fuel flow at idle, they do not, in fact, approach a magnitude sufficient to create operational problems. This bounce may be seen as a series of ripples when the fuel level is observed. This same bowl, when used with the accumulator, resulted in a variation of fuel flow with time of only one tenth of one percent.

It was felt at that time that a construction which included polished rubbing surfaces on the float tab, minimum eccentricity of needle and seat assemblies, and a burnished fuel inlet seat with a specific radius would yield consistent fuel flow measurement when used with the hydraulic accumulator. Therefore, it was decided to include the above construction in a production volume survey to establish its merits. One hundred production carburetors were built using the construction described above and were compared to an equal quantity of standard production units for idle setting repeatability.

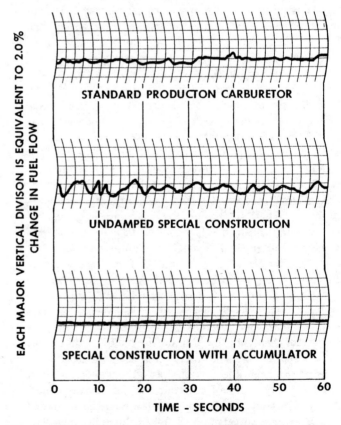

Figure 4. The Effect of Carburetor Fuel Inlet Construction on instantaneous Inlet Fuel Flow at Idle.

All of the carburetors were set to nearly identical flow conditions on one production flow stand and then inspected on an audit test stand where the idle mixture was measured and recorded after opening and closing the throttles several times. The carburetors having special inlet systems were set and audited with the accumulator mounted in the fuel circuit.

The repeatability of the auditing stand of the group of carburetors containing special inlet systems is compared to an equal size group of standard carburetors in Figure 5. It may be seen that, not only was the total spread reduced from 12% to less than 8% by the revision in the inlet system, but also the number of carburetors within the acceptable limit was increased greatly.

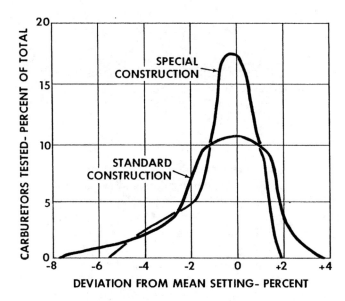

Figure 5. Effect of Inlet System Construction on Idle Fuel Setting Repeatability.

One of the more disappointing facets of this investigation is revealed by the non-symmetry of the curves in Figure 5 caused by the number of carburetors shown outside the −3% limit. Subsequent evaluation of this group of carburetors has shown this trend to be due to non-repeatability of the secondary airflow of these four barrel carburetors which had not been predicted by the small group of carburetors used in the earlier laboratory investigation.

The continuing investigation into the means of building more precise carburetors is currently concentrated on factors affecting airflow repeatability of secondary throttles.

In spite of the problems that remain, however, this investigation has indicated means by which the idle flow repeatability of carburetors may be significantly improved. These means may be summarized briefly as follows:

1. Reduced surface roughness of mating parts to achieve minimum friction in the inlet system.
2. Minimum eccentricity between the inlet valve and the valve seat together with a specific seat radius.

These factors work in conjunction with each other to eliminate the tendency for the valve to stick in the seat.

3. A fuel pressure pulse adsorption device to eliminate the flow variations which might otherwise result from the use of the above means.

The preceding description has enumerated some of the difficulties encountered in building and setting production carburetors to prescribed stringent flow requirements, but what about the performance of those carburetors on the vehicle?

Earlier publications (1, 2)* have shown the relationship between fuel-air ratio and carbon monoxide concentration as a constituent of the exhaust products. Some engines

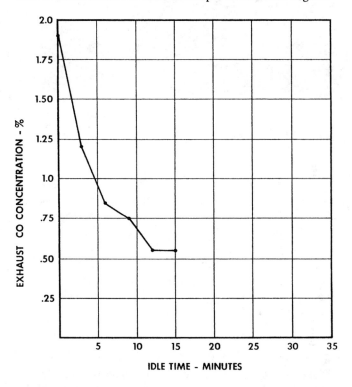

Figure 6. Extended Idle Test-Normal Carburetor Installation.

would idle well on a 1% CO mixture — others would not. This was attributed in general to variations in the usual engine parameters such as valve and ignition timing. Some engines would idle well enough on a 1% CO mixture with HC concentrations on the order of 130 ppm or less. Others that would idle well (at least in terms of smoothness) with this 1% CO setting yielded in excess of 300 ppm. We have usually been successful in reducing this HC concentration by an improvement in idle distribution such as by a relocation of the idle discharge port or a change in the bore-to-bore balance in the case of a multi-bore carburetor.

One phenomena that has been observed many times on many engines, however, for which no explanation was readily available, is the decrease in carbon monoxide at idle over a period of time which is shown in Figure 6. It is

---

*Numbers in parentheses designate References at end of paper.

Figure 7. Remote Carburetor Installation on Test Vehicle.

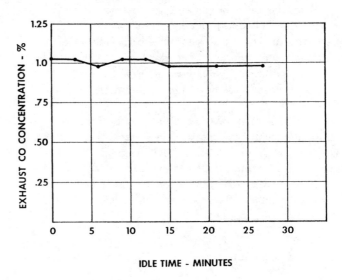

Figure 8. Extended Idle Test - Remote Carburetor and Remote Fuel Supply.

common to set the idle of 1% CO, for example, and then during the time required to check out other test variables find that the idle CO setting has deteriorated to perhaps 0.6% CO. Usually running the engine to a higher speed for a few minutes to "clean it out" will result in a return to the 1% setting. What is the mechanism by which this phenomena occurs? Is the carburetor metering changing during idle or is the engine converting a greater part of the same fuel flow to CO_2 and less to CO?

A high production volume vehicle, having a six cylinder engine with a single barrel carburetor, was selected for the investigation of this problem. The entire project was conducted on the chassis dynamometer and all of the data reported were obtained using the instrumentation prescribed by the Federal Test Procedure.

One theory which had been proposed to account for the CO reduction during idle suggested that, as idling continued, the temperature increase of the zinc or aluminum throttle body resulted in a greater expansion of the throttle bore than was experienced by the steel throttle plate which, of course, also had a less severe temperature rise. The increase in the size of the annular opening between the throttle plate and the throttle bore would result in an increased airflow and thus a leaner mixture.

The substitution of an aluminum throttle plate for the steel plate, which at least eliminated the difference in the coefficient of expansion between the plate and the bore, showed no significant effect on the CO trace over a period of time. It was recognized here that the temperature rise difference between the plate and the bore still existed.

In an effort to eliminate the temperature rise which was experienced by the carburetor during idle, a five foot length of pipe was installed between the engine and the carburetor to remove the carburetor from the underhood environment. This installation, which is shown in Figure 7, has the advantage of maintaining both air inlet and carburetor metal temperatures near the ambient. The results of this experiment were somewhat surprising; the CO trace was relatively unaffected in that CO still decreased with time.

The test installation was then modified to permit measurement of both air and fuel flow. We desired to use a flowmeter for the measurement of fuel flow for reasons of convenience and simplicity. This in turn required that the fuel supply reservoir, fuel pump and plumbing all be located external of the vehicle to eliminate the changes in fuel density and viscosity which would otherwise result. A laminar flow element and a draft gauge were used to indicate airflow.

The data obtained with this installation are shown in Figure 8. For the first time we failed to show the usual decrease in CO as idling continued. We reasoned that the only variable between this and prior tests was the removal of the complete fuel system from the underhood location and we should, therefore, investigate the effect of fuel temperature on the formation of CO. Accordingly, a heat exchanger was added to the installation which would allow heating or cooling the fuel. This heat exchanger was installed between the flowmeter and the carburetor such that the fuel flowing through the flowmeter remained at room temperature.

It was then possible to obtain the relationship between fuel temperature and CO emissions which is shown in Figure 9. It may be seen that as the fuel temperature increased the carbon monoxide concentration in the exhaust decreased until a fuel bowl temperature of about 116° was reached. Continued heating of the fuel beyond this temperature resulted in a rapid increase in CO emission. It should

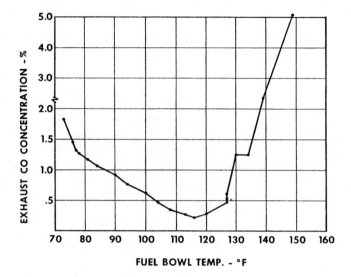

Figure 9. Extended Idle Test - Normal Fuel Bowl Venting.

be noted here that an increase in fuel flow was found when the fuel temperature exceeded 116°.

It was speculated that the fuel flow increase and the associated increase in CO concentration were the result of unmetered fuel vapors escaping from the carburetor vent system. All of the preceding data was obtained with a

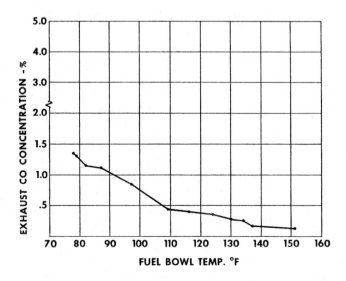

Figure 10. Extended Idle Test - Vent Tube Routed Externally.

fully balanced carburetor having a ¼ inch internal vent tube and no external vent. A piece of plastic tubing was fitted over the vent tube and then routed from the top of the air cleaner and also out from the engine compartment.

Figure 10 depicts the results obtained from this configuration. As before, the CO decreased with an increase in fuel temperature to 116°. Further heating of the fuel beyond this temperature, however, resulted in a continued slight decrease in the CO concentration in spite of the increase in fuel flow which was still observed. Fuel condensation was observed in the plastic vent line at these temperatures. This test confirmed the speculation that the rapid increase in CO beyond 116° was the result of unmetered fuel vapor escaping from the carburetor vent.

Each of the above experiments was conducted using Indolene as the fuel. It was felt that it should be possible to correlate the CO emission to fuel volatility inasmuch as it

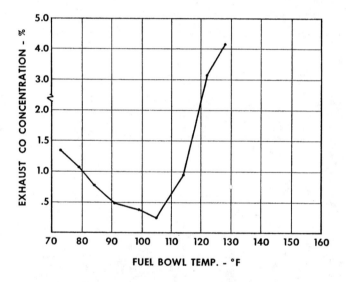

Figure 11. Extended Idle Test - Normal Fuel Bowl Venting - Test Fuel Commercial Gasoline.

could be related to fuel temperature. Therefore, a repeat of the test with conventional venting was made using commercial gasoline. Figure 11 shows results similar to those obtained with Indolene with the notable exception that in this case the increase in CO occurred at 105° as opposed to the 116° with Indolene. It may also be seen that the slope of the CO curve at temperatures above 105° is considerably steeper than was observed with Indolene.

Figure 12 shows partial distillation curves of both Indolene and regular gasoline. It is interesting to note that if vertical lines are drawn at the respective temperatures at which a CO increase was first observed, these lines intersect

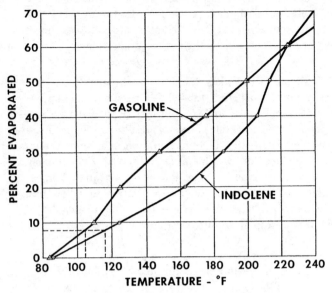

Figure 12. Fuel Distillation Curve

the respective distillation curves at about the 8% evaporated point in both cases. It is clear then that gasoline volatility is a factor in the emission products formed at low airflows.

The relationship between the CO formation at idle and both fuel temperature and volatility have been shown above as a variation which may occur independently of a variation in fuel-air ratio. This means that if the fuel-air ratio delivered to the engine has in fact not varied as the fuel temperature was increased, a carbon balance made at any exhaust CO concentration should give the same total as at any other concentration.

Figure 13 shows the effect of fuel temperature on the exhaust products at idle. Table 1 shows the tabulation of total carbon taken at two different fuel temperatures which resulted in the formation of significantly different carbon monoxide concentrations.

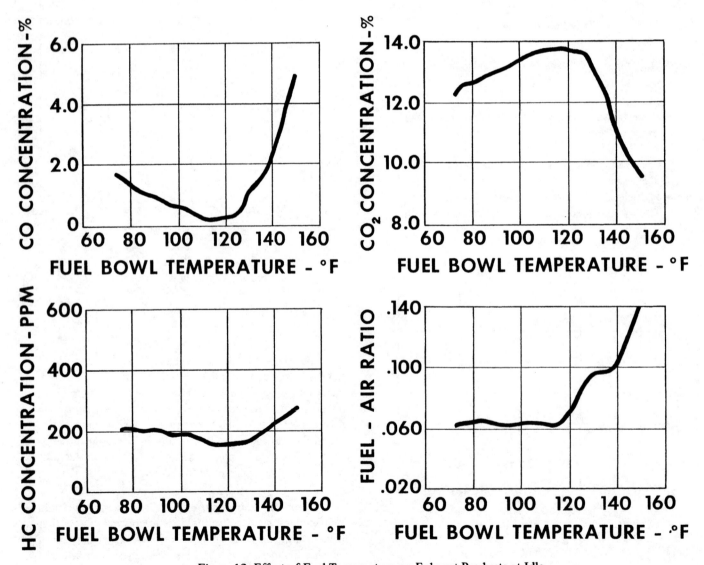

Figure 13. Effect of Fuel Temperature on Exhaust Products at Idle.

Note that the total carbon concentration in the exhaust gas is essentially the same at both conditions of fuel temperature although the proportions of CO, CO_2 and HC vary markedly with temperature. This lends credence to the hypothesis that exhaust products may vary with fuel temperature independently of fuel-air ratio. It is expected, of course, that this variation in exhaust products would be less marked under engine operating conditions which are more conductive to complete combustion than the idle mode.

An evaluation of the effects of variation in fuel temperature observed over the complete seven mode cycle, however, has shown that the carbon monoxide concentration is affected to a significant degree. Table 2 shows the total cycle results obtained over a range of fuel temperature from $0°$ to $125°F$. Of the most significance is the fact that even in the normal or practical range of fuel temperatures encountered in the Federal cycle, relatively modest variations in fuel temperature resulted in substantial variations in CO concentration.

Various thermal devices have been reported previously and indeed the 1970 model year will undoubtedly bring widespread use of the "hot and cold" air cleaner. Although this device is under consideration primarily as a driveability aid which may be required with the carburetor calibrations necessary to meet the tightened emission standards, some cycle gains may also be realized. Our limited data in this respect suggests, however, that the CO gains which might be expected from the improved vaporation are more than offset by the decrease in air density and the resulting increase in fuel-air ratio which resulted from heating the air.

The findings presented in this paper suggest that consideration be given to controlling fuel temperature directly as a means of improving both cycle results and driveability.

TABLE 1 – CARBON BALANCE FROM EXHAUST PRODUCTS SHOWN IN FIGURE 13.

Exhaust Products	Fuel Bowl Temperature - $°F$	
	$84°$	$113°$
CO Concentration - PPM	10,600	3,000
CO_2 Concentration - PPM	129,000	137,500
HC Concentration - PPM	207	174
Total Carbon Present - PPM	139,807	140,674

TABLE 2 – EFFECT OF FUEL TEMPERATURE ON 7 MODE CYCLE RESULTS.

Fuel Bowl Temperature - $°F$	HC Concentration PPM	CO Concentration Volume %
0	244	1.6
60	238	1.5
108	217	1.2
118	216	.75
125	377	2.8

REFERENCES

1. Fagley, Walter S., Jr. and Nunez, Richard R., "Exhaust Gas Analysis as a Tool for Measuring Fuel-Air Ratios", SAE Paper 670483, presented at the SAE Mid-Year Meeting, Chicago, Illinois, May 15-19, 1967.

2. Eltinge, Lamont, "Fuel-Air Ratio and Distribution from Exhaust Gas Compositions", SAE Paper 680114, presented at the SAE Automotive Engineering Congress, Detroit, Michigan, January 8-12, 1968.

690138

Process and Control for Producing Anti-Emission Carburetors

James E. Eberhardt and John A. Beck
Carter Carburetor Div., ACF Industries, Inc.

THE ADVENT OF EMISSION CONTROLS for automobiles in California in 1966 triggered a period of expansion in carburetion technology that is still in progress.

The industry's design and application engineers have been diligently developing carburetors that satisfy the early California and the 1968 nation-wide emission limits while maintaining performance characteristics. At Carter, a parallel and equally intensive continuous program has been underway to develop manufacturing and quality control processes which complement the design effort. The objective is not merely to provide the tools and techniques to convert the engineering innovations into hardware, but also to translate engineering specifications, old or new, into a product with a degree of functional uniformity heretofore thought unattainable.

This presentation will:
1. Explain the need for improved carburetor uniformity.
2. Describe Carter's program currently in operation to achieve uniformity.
3. Illustrate, by example, the process changes being generated as a result of the program.

NEED FOR UNIFORMITY

The need for uniformity is understood best when variations are related to the basic carburetor performance requirement, the "flow specification" (Fig. 1). The flow specification, or more simply the "flow curve," shows how the air-fuel ratio varies with an increase in air flow as the throttle is advanced from the curb idle position to the wide open position.

The flow curve is the result of the design engineer's calibration effort. It describes, under standardized environmental conditions, an optimum calibration that will perform satisfactorily over a wide range of temperature and altitude conditions. Deviations from the flow curve infringe upon this optimum calibration and compromise the carburetor's performance.

Of necessity, some tolerance from the optimum, called a "flow band," must be conceded to manufacturing since all pieces cannot be made exactly alike. Over the years prior to emission control, continued pressure to attain uniform carburetor performance reduced the flow band until it closely approximated total manufacturing capabilities.

To satisfy the emission control requirement, the previously established balance between flow performance and capability was upset. The flow band was reduced to the point where manufacturing could no longer meet the demands; therefore, faced with the conflict between "least possible variation" and "capability," manufacturing and quality control had to further raise the level of capability to manufacture, adjust, and test anti-emission carburetors

―――――――――――――――――ABSTRACT―――――――――――――――――

The advent of emission controls for automobiles in California in 1966 triggered a period of expansion in carburetion technology. The industry's design and application engineers have been diligently developing carburetors that satisfy the early California and the 1968 nation-wide emission limits while maintaining performance characteristics. The present paper explains the need for improved carburetor uniformity; describes a program currently in operation to achieve uniformity; and illustrates process changes being generated as a result of the program.

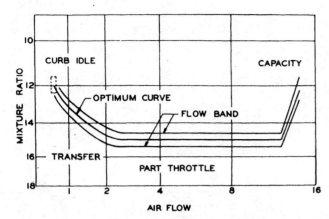

Fig. 1 - Carburetor optimum part throttle flow curve with flow band limits

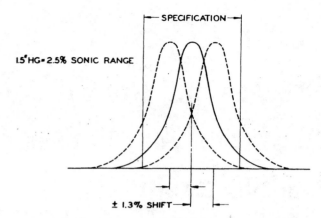

Fig. 2 - Barometric effect on mixture ratio

to the least possible variation from the optimum curve established by the engineering department.

In 1966, a team representing manufacturing, quality control, and engineering was formed to study the problem.

PROGRAM TO ACHIEVE UNIFORMITY

Increasing uniformity is a general problem. There are no quick solutions. For each carburetor model there are many specifications affecting functional performance. Manufacturing and quality control, working together, have found and continue to find ways to reduce part-to-part variation.

The program to accomplish this was as follows:

1. Existing operations were screened to determine their relationship to the functional performance of the carburetor. Follow-up priorities were then established.

2. According to priority, the operations were objectively examined to determine if functional consistency could be improved. This was done irrespective of the ease or difficulty in maintaining existing specifications. In this way, dormant opportunities were not passed by from force of habit.

3. Critical metering parts and sub-assemblies were examined to determine if functional tests could be added to supplement dimensional specifications, thereby increasing their reliability as components of the complete assembly.

By following these steps to reexamine existing operations, product variations have been reduced. The following examples illustrate some of the changes made as a result of the "fresh look" approach.

FLOW TESTING

Each carburetor is final-flow-tested and adjusted for conformance to the flow band. This operation is performed on flow stands provided with a large-capacity vacuum source and a constant and pressure-regulated source of stable liquid which simulates fuel. The carburetor is mounted so that suction is applied posterior to the throttle valve and a fuel line is connected to the carburetor fuel inlet. The throttle lever is attached to a positioning device which locates the throttle plate as required to simulate steady-state flow at specified points on the flow curve. Ideally, the stand's sensitive instrumentation measures the deviation from the optimum curve and nothing more. In practice, however, measurement error is introduced in the flow process. The importance of the measurement error grows as the flow band is reduced for anti-emission calibrations.

Variations and/or repeatability attributable to the gage, the part, or the operator can consume a large part of the allowable tolerances. The critical adjustable points are the major determinants of flow conformity.

Because of the direct relationship of flow testing error with carburetor flow conformance, this operation was given the top priority for reevaluation.

EFFECT OF ENVIRONMENTAL CONDITIONS

The team representing the manufacturing, product engineering, and quality control departments studied the problems and determined early the course of action for 1968 when all carburetors were to be inspected and adjusted to the closer flow limits. A major item in this undertaking was a study of the effect of environmental conditions upon flow test readout. The findings were disturbing.

In pre-anti-emission carburetor final testing, the flow targets were established on the flow stand by running a master carburetor. The validity of the target setting under varying environmental conditions was maintained by frequent remastering, which at best provided a "saw tooth" control. This was acceptable with a relatively wide flow band. However, with the reduction in band width, the height of the "saw tooth" could conceivably consume the allowable manufacturing limits if the remastering was out of time phase with the natural and normal changes in environmental conditions. This is illustrated in the following examples.

The solid line bell curve of Fig. 2 describes a typical carburetor flow inspection distribution at standard barometric pressure, and the dotted curve shows the shift of the distribution from the optimum flow curve when the barometric pressure rises or falls 3/4 in. Hg.

Fig. 3 shows the effect of air temperature on air-fuel ratio. Fig. 4 shows the effect on air-fuel ratio with a change

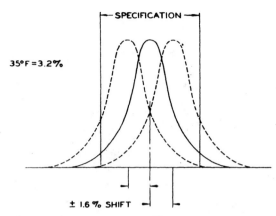

Fig. 3 - Air temperature effect on mixture ratio

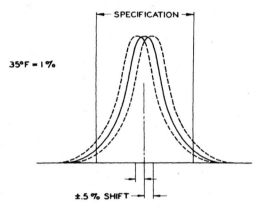

Fig. 4 - Fuel temperature effect on mixture ratio

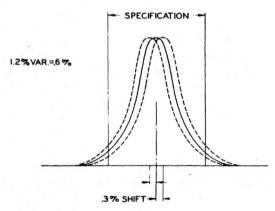

Fig. 5 - Fuel specific gravity effect on mixture ratio

of 35 deg in fuel temperature. Fig. 5 shows the effect on air-fuel ratio with variations in fuel specific gravity. This is based on the use of a fuel of low volatility with a stable distillation curve.

These values are not additive but the summation closely approximates the total shift shown in Fig. 6.

TEAM DECISION

As a solution to the flow band drift problem, the final choice of the team was either to rely on frequent test stand remastering which would allow the band to shift as environ-

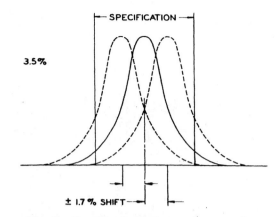

Fig. 6 - Combined effect on mixture ratio

Table 1 - Carter Standard Production Test Conditions and Limits

Air	
Barometric Pressure	29.50 in. ± 0.05 in. Hg
Temperature	75 ± 2 F
Relative Humidity	35 ± 5%
Test Fluid	
Specific Gravity	0.788 ± 0.004
Temperature	75 ± 3 F
Viscosity	0.095 ± 0.01 cp

mental conditions changed, or to control the environmental conditions.

The team concluded that in order to be sure the readings taken on each carburetor were meaningful, measurements against a fixed standard were necessary. The only satisfactory way to accomplish this was through environmental control.

THE TEST FACILITY

A direct consequence of this conclusion was the establishment of standard for the environmental conditions.(Table 1). Close tolerances were established for barometric pressure, relative humidity, air temperature, test fluid temperature, specific gravity, and viscosity.

To achieve control to these standards, a very special 48,000 sq ft facility was designed. A two-story building was constructed exclusively for this purpose. The first floor contains support equipment, such as vacuum pumps for the test stands, air conditioning chillers, boilers, a test fluid filtration system, and automatic test fluid blending equipment. The 24,000 sq ft second floor houses the constant environment test room. Designed to withstand a pressure differential of ± 1-1/2 in. Hg, it is constructed with 12 in. double reinforced concrete walls, floors, and ceiling. Temperature-and humidity-controlled air is supplied to the room at a constant rate. Barometric pressure is controlled automatically by modulating the speed of the exhaust blowers.

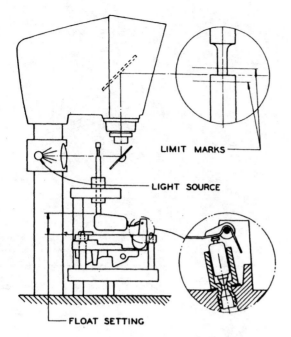

Fig. 7 - Optical float gaging

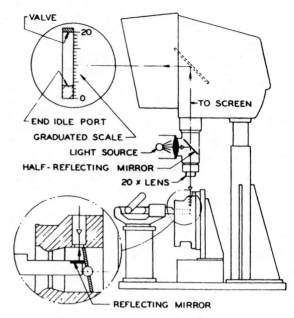

Fig. 8 - Optical gaging of idle port relation

Because of the differential pressure between the test room and normal atmosphere, personnel and material access to the room is through transfer chambers.

TEST EQUIPMENT

There are 52 new test stands in the environmental building which were specifically designed for flowing anti-emission carburetors. A portion of this equipment is presently controlled by a process computer. Carburetor flow technology is advancing rapidly, and Carter is already considering replacements and additions which will include greater and more advanced computer control.

The new test equipment under environmental control has produced significant results. However, the drive for uniformity extends beyond this flow test facility in revising old and creating new manufacturing processes to improve the basic uniformity of the product. Examples of this work follow.

FLOAT SETTING

Float setting (Fig. 7) is a standard operation in carburetor assembly. It consists of adjusting the floats so that the carburetor fuel level attains the proper height in the bowl. The float setting is one of the factors that determine the point at which the fuel nozzles begin to feed and the transition from the idle system to the part throttle system occurs. Variations in float settings cause flow variations in the transfer range. This setting had been made by mechanical gaging. Through the application of an improved version of an old technique -- optical gaging -- gage resolution has been amplified 10 times and the uniformity of the setting has been improved.

PORT RELATION GAGING

The relationship of the idle port to the throttle valve is a critical dimension (Fig. 8). Variations in the relationship affect the shape of the "off idle" range of the flow curve. Precision equipment has been developed for punching the port in relation to the valve but it must be set up and constantly monitored with a mechanical gaging operation.

Because of the involved geometric relationships, the gaging operation is a critical one. Again, an optical gaging method was developed for measuring the port relation. Although this measurement procedure is at present too complex for constant use by the port punch operator, it is used for setup purposes and quality control monitoring.

IDLE SCREW BALANCE ADJUSTMENT

Like most test equipment, flow test stands do not simulate all operating conditions. One parameter heretofore not checked is the distribution of flow between the bores on a two-barrel carburetor. The flow stand measures composite flow only. Formerly the normal flow test procedure was to back out each idle mixture screw an equal number of turns from the seat to obtain the specified fuel flow on the test stand. The screws were then fine tuned in small but equal increments to obtain the total idle mixture.

This procedure does not always produce best results, however, because minor dimensional variations within tolerance could cause an imbalance of idle fuel feed at equal turn settings. With the leaner idle calibration necessitated by anti-emission requirements, a poorly balanced carburetor can cause a rough idle.

To prevent this, an idle air flow balancing operation has been added to the carburetor assembly line prior to flow test (Fig. 9). Using pressure-regulated air as the test me-

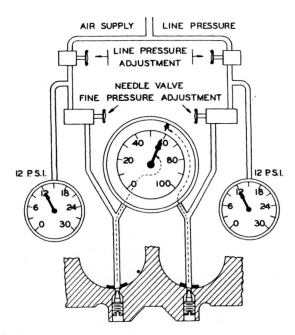

Fig. 9 - Idle adjustment screws bore-to-bore balancing

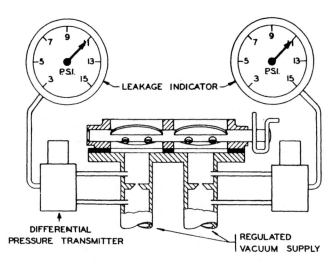

Fig. 10 - Bore-by bore leak test

dia, each idle mixture screw is set independently to obtain a specified pressure drop across the idle screw orifice of each bore.

By using this air flow method, the initial adjustment position is better balanced bore-to-bore. In the flow test area the idle screws are "fine" adjusted to give the desired total air-fuel ratio.

BALANCED BORE-TO-BORE AIR LEAKAGE

For the same reason described above in regard to idle fuel feed, the total air flow must also be divided bore-to-bore on anti-emission carburetors. Since leakage around the throttle valves and throttle shaft bearings could account for up to one-half of the curb idle air flow, leakage must be controlled, both in amount and in balance. To assure balance, the assembly line leak test, which formerly checked composite leakage only, has been revised so that each bore is individually tested against a standard (Fig. 10).

AIR TESTING OF METERING COMPONENTS

It is difficult to translate multiparameter, low flow control into dimensional specifications. Hole concentricity and surface finish, angular relationships of cross passages, and length of passages, are typical variables that can occur individually or in combination to influence performance. Control of these variables by dimensional gaging techniques alone is not totally reliable.

To supplement mechanical gaging, Carter has been developing functional tests for critical metering components and sub-assemblies. One application is the venturi cover assembly (Fig. 11) that is used on a dual carburetor. When properly machined and assembled, the restrictions in the tube ends are the only metering points in the sub-assembly,

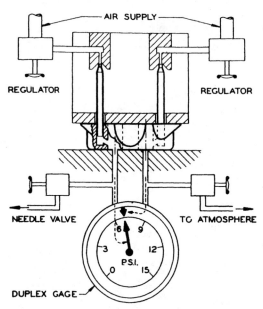

Fig. 11 - Venturi cover air flow schematic

but a burr shaved off the casting in the tube assembly, a tube pressed in too far, or a misalignment of the cross passages will create an uncontrolled restriction and cause a flow reject. This air flow test adjacent to the tube assembly operation detects and stops such defects at the source. Also, by air flowing each tube independently but simultaneously, the flow balance between the tubes is checked and conveniently displayed on a double indicating gage.

CONCLUSION

This paper has been presented to explain why manufacturing and quality control share with engineering the challenge of emission control and to illustrate by example the continuing program to improve functional performance through uniformity. This has required a positive concept of capability not limited by established methods, but expanded through the creative belief that everything can be done better.

750371

Closed Loop Carburetor Emission Control System

R. A. Spilski[1] and W. D. Creps[2]

ABSTRACT

Electronic control logic and system components are described that are used with a venturi carburetor to accomplish the fuel metering required in a closed loop emission control system. Special open loop functions are used because of the nature of the venturi carburetor and the warm-up characteristics of the three-way catalytic converter and exhaust gas sensors. Strip chart recordings illustrate the dynamic characteristics of the electronic control logic, and emission test results with a low mileage three-way converter are set forth.

RECENT PAPERS HAVE DESCRIBED the characteristics of a three-way catalytic converter and a zirconia dioxide exhaust gas sensor, and their use in a closed loop emission control system in an attempt to meet the 1978 emission levels of 0.41 g/mile HC, 3.4 g/mile CO and 0.40 g/mile NO_x (1-7).* The three-way converter is capable of reducing NO_x as well as oxidizing HC and CO provided the air-fuel ratio is controlled at the converter window, a range of air-fuel ratios approximately 0.05 ratio wide near stoichiometry, Fig. 1. The exhaust gas sensor signal provides a measure of whether the air-fuel ratio is at the converter window or whether a rich or lean correction in the fuel metering mechanism is required.

The fuel metering mechanism used to control the air-fuel ratio in closed loop systems reported in previous articles has been electronic fuel injection. However, an economic advantage could be realized if a venturi carburetor was used, since conventional induction system hardware could be employed with minimal modification.

This paper describes the electronic control logic and system components used to adapt a four barrel carburetor to closed loop air-fuel ratio control. The system is installed on a full-sized sedan with a V8 engine using the spark advance and exhaust gas recirculation calibrations employed in the oxidizer-only catalyst equipped vehicles designed to meet the 1975 California emission standards. An air injection reactor pump is not used with the closed loop system. A description of the system components and the characteristics of the carburetor mechanism is followed by discussion of the closed loop system block diagram. The functions of the electronic control logic are described using a block diagram and strip chart recordings of system control signals. Emission test results using low mileage catalyst material are used to gage the performance of the control logic and carburetor mechanism.

SYSTEM COMPONENTS

The basic components of the closed loop carburetor emission control system are illustrated in Fig. 2. The three-way catalyst material is located in the same underfloor converter used in the oxidizer-only catalyst systems designed to meet 1975 emission standards. Two exhaust gas sensors are used; one, designated as Z_2, which samples the tail pipe exhaust gas at the converter outlet, and one, designated as Z_1, which samples the engine-out exhaust gas at the exhaust manifold. The four barrel carburetor, Fig. 3, contains a special set of metering rods and a vacuum regulator to provide the mechanism to vary air-fuel ratio in response to command signals from the electronic controller. The closed throttle switch on the carburetor assembly in Fig. 3 provides for one of the open loop functions to be described and the cold start switch in Fig. 2 senses engine coolant temperature for another open loop function.

The solid state electronic controller, Fig. 4, provides the control logic necessary to process the signals from the various sensors and generate the command signal to modulate the air-fuel ratio. A development version of the electronic controller installed in a test vehicle is shown in Fig. 5. The potentiometer adjustments provide flexibility in changing the system calibration parameters for emission development work with test points located on the front panel to provide access to various control signals for strip chart recordings.

CLOSED LOOP CARBURETOR

The closed loop carburetor performs all of the functions of a conventional carburetor in metering fuel over a

---
[1] Chevrolet Engineering Center, Warren, Mich.
[2] General Motors Research Lab, Warren, Mich.
*Numbers in parentheses designate References at end of paper

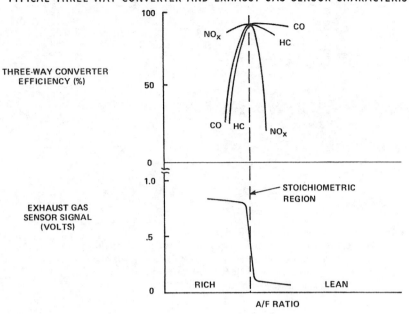

Fig. 1 - Typical three-way converter and exhaust gas sensor characteristics

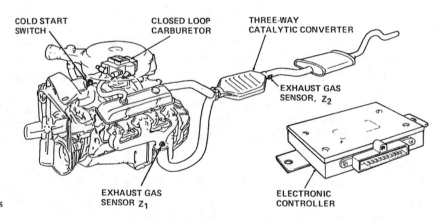

Fig. 2 - Closed loop carburetor system components

Fig. 3 - Closed loop carburetor

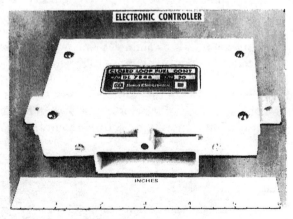

Fig. 4 - Electronic controller

wide variety of engine speeds, loads, and operating temperatures. The closed loop function is designed into the primary main metering circuit to provide closed loop control of air fuel ratio at the converter window for part throttle, warmed-up driving conditions. The other fuel metering circuits provide conventional pre-calibrated, or open loop, control of fuel for additional driving conditions, which include wide-open throttle, closed throttle, cold start and rapid throttle openings. In describing the carburetor mechanism, the closed loop function is followed by each of the open loop functions, including interaction between them during transient driving conditions.

CLOSED LOOP FUNCTION - In order to provide the mechanism for varying air-fuel ratio in response to an electrical command signal, a vacuum regulator and a special set of metering rods are used in the primary main circuit of a four barrel carburetor, Fig. 6. The power system, actuated by manifold vacuum, and the adjustable part throttle circuit, are retained.

The vacuum regulator is an electro-mechanical transducer that supplies a control vacuum signal ranging from zero to approximately 8 in Hg as a function of an electrical current as shown in Fig. 7, with manifold vacuum used as the source vacuum. Because of the internal characteristics of the vacuum regulator, the control vacuum varies inversely as the current and remains constant for a given value of current provided the source vacuum (manifold vacuum) is greater in magnitude than the control vacuum by approximately 1 in Hg. The control piston and spring are designed to move the closed loop metering rods between their two extreme positions, known as their rich and lean authority limits, for a control vacuum range of 1-6 in Hg. Therefore, the mid-range of the curve in Fig. 7 is used to position the closed loop metering rods to any point between their rich and lean limit as a function of a current command signal ranging from approximately 100-225 ma.

The primary main metering circuit of Fig. 6 is designed to provide closed loop control only at part throttle driving conditions, where manifold vacuum is greater than approximately 6 in of Hg. Under these conditions, the power system is in its full lean position and only the movements of the closed loop metering rods affect the quantity of fuel delivered for a given venturi signal. In addition, with the manifold vacuum above 6 in Hg, the source vacuum to the vacuum regulator is sufficient to allow positioning the closed loop metering rods to any point between their rich and lean limits.

The metering rods and jets of Fig. 6 are selected so that under part throttle conditions, movement of the closed loop metering rods from their full rich to full lean positions results in air-fuel ratios from approximately 14:1-17:1. A typical carburetor characteristic for a steady state part throttle condition is shown in Fig. 8 as observed on a chassis dynamometer with air-fuel ratio calculated from exhaust emissions. The stoichiometeric air-fuel

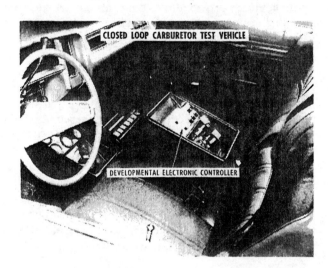

Fig. 5 - Closed loop carburetor test vehicle

CLOSED LOOP CARBURETOR MECHANISM

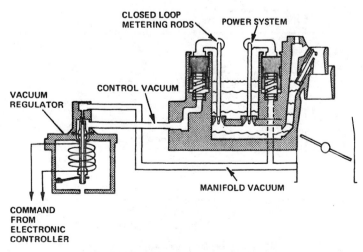

Fig. 6 - Closed loop carburetor mechanism

ratio is obtained at a control vacuum of around 3 in Hg with approximately two ratios of authority in the lean direction and one ratio in the rich direction.

WIDE OPEN THROTTLE - For wide open throttle and deep crowd conditions, the air-fuel ratio must be richer than stoichiometry to obtain the required detonation free performance and driveability. Under these driving conditions, manifold vacuum drops below 6 in Hg, and the power system metering rod begins to move upward in Fig. 6 to provide the necessary enrichment. Since manifold vacuum is used as the source to the vacuum regulator, the control vacuum drops to a maximum level equal to manifold vacuum which causes the closed loop metering rods to also provide enrichment and prevents them from responding to lean commands from the control logic. For very deep crowds, as well as wide-open throttle, the secondary system provides additional flow capacity just as in a conventional four barrel carburetor.

Although the wide-open throttle and deep crowd conditions constitute a special open loop driving mode, the loss of manifold vacuum is all that is required to provide open loop control, with no external sensor necessary.

CLOSED THROTTLE - In any venturi carburetor, the air flow drops to such a low level at closed throttle that the venturi signal becomes negligible and the primary main metering circuit does not provide the fuel required for idles and decelerations. For these conditions, a separate idle circuit is used to provide the necessary fuel with the quantity determined by the idle mixture screw adjustments. Because the closed loop metering rods have no ability to regulate fuel at these closed throttle conditions, a closed throttle switch is used to signal the electronic controller to switch to a special open loop, closed throttle logic mode. The control strategy used during the closed throttle mode has been shown to significantly affect the emission results and is discussed in detail in the section on the electronic control logic.

COLD START - The enrichment required for starting a cold engine is provided by a conventional choke mechanism. Another cold start consideration, that of converter warmup, is discussed in the section on the electronic control logic.

ACCELERATOR PUMP - The fuel required to overcome the momentary leanness during rapid throttle openings is supplied by an accelerator pump. Open loop metering of the fuel is accomplished by the same mechanical linkage and pump hardware used in a conventional carburetor.

SYSTEM BLOCK DIAGRAM

With the closed loop and open loop characteristics of the carburetor mechanism described above, and with the three-way converter and exhaust gas sensor character-

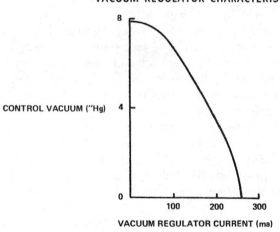

Fig. 7 - Vacuum regulator characteristic

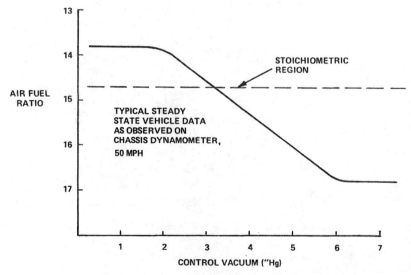

Fig. 8 - Closed loop carburetor characteristic

EMISSION CONTROL SYSTEM

istics shown in Fig. 1, a block diagram of the entire feedback control system is formulated in Fig. 9. The objective of the system is to minimize the production of HC, CO, and NO_x in the tail pipe exhaust gas by operating the engine at the air-fuel ratio which produces the highest mutual conversion efficiencies in the three-way catalytic converter. The two exhaust gas sensors analyze the gas composition as it leaves the engine, Z_1, and again as it leaves the three-way converter, Z_2. The electronic control logic utilizes the information from the exhaust gas sensor signals to generate commands to the vacuum regulator to either hold the air-fuel ratio constant or make a correction in the rich or lean direction to provide the correct mixture for operation at the three-way converter window. When the electronic controller is responding to the two exhaust gas sensors, the system is said to be in the closed loop mode since the tail pipe exhaust is being analyzed. When the control logic is responding to the closed throttle switch or cold start switch, the system is operating in an open loop mode.

The implementation of the control logic for both the closed and open loop modes is detailed in the next section on the electronic control logic.

ELECTRONIC CONTROL LOGIC

The block diagram of Fig. 10 illustrates the functional characteristics of the electronic control logic. The two exhaust gas sensors are the closed loop inputs and the cold start switch and closed throttle switch are the two open loop inputs to the logic circuitry. The only output is the current command signal I_0 which provides power to the

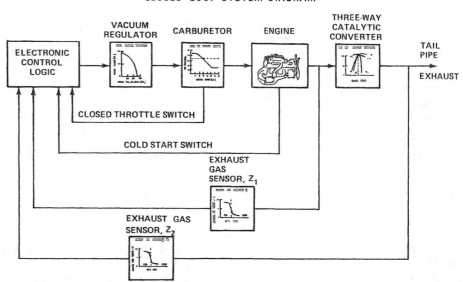

Fig. 9 - Closed loop system diagram

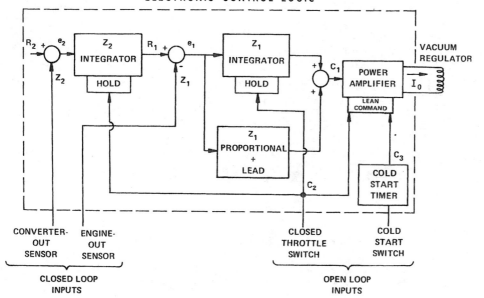

Fig. 10 - Electronic control logic

vacuum regulator. The major topics of the control logic discussion are the power amplifier, exhaust gas sensor signals, closed loop logic, closed loop stability, closed throttle logic, hot start circuit, a strip chart recording of several of the control signals for two cycles of the Federal Test Procedure, and the cold start timer.

POWER AMPLIFIER - The power amplifier is the common element that generates the current driving signal I_0 to power the vacuum regulator in response to the three command signals C_1, C_2, and C_3. The command C_1 is an analog signal produced by the closed loop portion of the logic circuitry. Excursions of C_1 over its range result in output currents of from 50-250 ma to position the closed loop metering rods to any point between their lean and rich limits. A current regulator within the power amplifier ensures that the output current I_0 is insensitive to coil resistance—which changes with temperature—and supply voltage. The two open loop commands, C_2 and C_3, drive the output current to zero and cause the C_1 command to be ignored.

EXHAUST GAS SENSOR SIGNALS - Since the closed loop control strategy is dependent on the nature of the two exhaust gas sensor signals, a description of the characteristics of these signals is given. A time based strip chart recording of the two exhaust gas sensor voltage signals is shown in Fig. 11. Four values of air-fuel ratio in the vicinity of the converter window during a steady state, part throttle test condition are presented with the observed conversion efficiencies listed for each of the data points. Data points 1 and 2 are slightly lean of the converter window, with data points 3 and 4 slightly rich of the converter window. The total range of air-fuel ratios between data points 1 and 4 equals 0.35 ratio.

While the performance characteristics of the two exhaust gas sensors themselves are similar, the signals obtained from them are quite different due to the differences in exhaust gas dynamics, as well as exhaust gas composition, at their two locations. The engine-out sensor Z_1 is shown both unfiltered and filtered in Fig. 11. The filtered portion is necessary to place a numerical value on its voltage level. The converter-out sensor signal Z_2 is unfiltered.

The Z_2 sensor exhibits a very sharp voltage characteristic at the converter window in Fig. 11 and is an accurate indicator of system operation at the point of highest mutual conversion efficiency. The action of the three-way converter in bringing the converter-out gases to chemical equilibrium results in the sharp transition characteristic of the Z_2 sensor at the converter window. This action has also been shown to result in Z_2 signals that are insensitive to driving conditions over a wide range of speeds and loads. The lower frequency oscillation of the Z_2 signal is due to the mixing action of the converter in averaging out individual cylinder firings as well as mal-distribution effects that are present in the engine-out exhaust gas. This homogeneous mixture at the tail pipe also makes the Z_2 sensor insensitive to car-to-car variations in mixture distribution.

The nature of the Z_1 signal in Fig. 11 indicates that it is not as accurate a measure of operation at the converter window as the Z_2 sensor. It does not exhibit an abrupt change in voltage level at the converter window. Other tests have shown that the Z_1 sensor is not as insensitive as Z_2 to driving conditions and mal-distribution differences between cars. The Z_1 sensor does, however, provide quicker response to a change in air-fuel ratio which contributes significantly to the dynamic performance of the system. The Z_2 sensor responds more slowly to changes in air-fuel ratio due to its location downstream in the exhaust system, but it contributes significantly to system accuracy over a wide range of operating conditions. Thus, the two sensors each provide useful information to the electronic control logic.

The two sensors are combined using a single fixed operating point for the Z_2 sensor and an operating point for the Z_1 sensor that is varied by the Z_2 sensor to correspond to the converter window for all driving and distribution conditions. In the discussion to follow, these two operating points are termed the reference voltages R_2 and R_1, respectively.

CLOSED LOOP LOGIC - The closed loop portion of

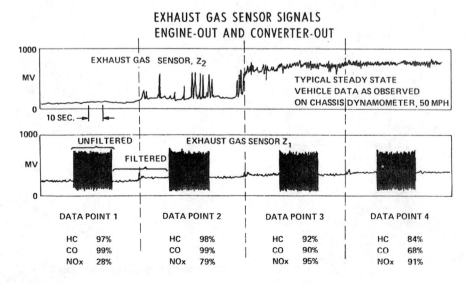

Fig. 11 - Exhaust gas sensor signals, engine-out and converter-out

EMISSION CONTROL SYSTEM

the control logic of Fig. 10 is the circuitry required to generate the command signal C_1 in response to the two exhaust gas sensor signals. A control strategy is used that attempts to drive the two error signals e_1 and e_2 to zero. The error signal e_2 is the difference between the desired operating point—fixed reference voltage R_2—and the Z_2 sensor signal. The error signal e_1 is the difference between the variable reference voltage R_1 and the engine-out sensor signal Z_1. The reference R_1 is the time integral of the error e_2. This arrangement allows the converter-out sensor Z_2 to dynamically establish the operating point, R_1, for the engine-out sensor Z_1 that provides for peak mutual conversion efficiency of the converter.

The closed loop command signal C_1 is the sum of the Z_1 integrator output and the Z_1 proportional plus lead circuit. The Z_1 integrator generates the time integral of the e_1 error signal which results in operation at the converter window since the R_1 reference is established by the Z_2 integrator.

CLOSED LOOP STABILITY - In order to obtain the quickest response in driving the error signals to zero, it is desirable to operate the system with the highest possible loop gain, which in this case is a function of the integrator gains. As in any closed loop system, however, the higher the loop gain, the greater the tendency for overshoots and oscillations. Based on experimental test results, an overshoot of approximately one-half inch Hg on the control vacuum signal for steady state driving conditions is permissible. Therefore, the integrator gains are adjusted as high as possible while limiting the maximum control vacuum overshoot to one-half inch Hg.

Experimental test results indicate that the gain and phase contributions of the proportional plus lead circuit permit larger integrator gains while maintaining the stability requirement. Fig. 12 illustrates the effects of four values of the proportional plus lead component on the Z_1 sensor, Z_1 integrator and control vacuum signals for a steady state, part throttle driving condition. Data point 1 contains a zero proportional plus lead component, data point 4 contains the full magnitude required in meeting the stability requirement, and data points 2 and 3 contain intermediate values of 30% and 60% of the maximum magnitude, respectively. The Z_1 reference (R_1) and the Z_1 integrator gain are held constant for all four data points so that the effects of the proportional plus lead circuit on the Z_1 integrator stability are isolated.

The control vacuum excursions for data point 1 in Fig. 12 exceed the stability requirement, producing excessive excursions in air-fuel ratio as evidenced by the slow cycling of the Z_1 sensor between rich and lean saturation. With the increased proportional plus lead component in data points 2 and 3, the frequency content of the control vacuum increases while the peak to peak magnitude decreases. Data point 4 has control vacuum excursions within the stability requirement which result in smaller air-fuel ratio excursions as indicated by the Z_1 sensor signal spending less time in rich and lean saturation.

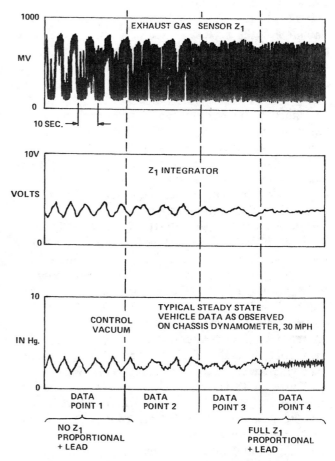

Fig. 12 - Stabilizing effect of proportional plus lead circuit

CLOSED THROTTLE LOGIC - An open loop mode is required for closed throttle driving conditions because the air-fuel ratio is determined by the idle circuit of the carburetor and not by the closed loop metering rods. At throttle closure, the closed loop metering rods are positioned to the lean limit and when the throttle is opened again they are returned to the same position they were in before the throttle was closed. Therefore, if the closed loop metering rods were providing the optimum quantity of fuel before the throttle closure, by being returned to that same position for the next part throttle maneuver, they will again be near the optimum position. The closed loop metering rods are placed in the full lean position during the throttle closure so that a consistent transition is made from the part throttle metering circuit to the idle circuit, and back again.

To accomplish the closed throttle logic, the command signal C_2 from the closed throttle switch is fed to both the power amplifier and an integrator hold circuit for each of the integrators in Fig. 10. The command signal C_2 to the power amplifier drives the output current to zero to position the closed loop metering rods to the lean limit. The integrator hold circuit switches the integrator to a hold mode during a throttle closure so that the value of integrator output voltage just prior to the throttle closure is stored and used as the initial condition on the integrator at the next throttle opening.

HOT START CIRCUIT - Another open loop function involves the exhaust gas sensor warm-up time. It is desirable to bias the system lean until the sensors have reached their operating temperature, such as a hot start condition when the engine and converter may be warmed up before the sensors.

Since the internal impedance of the exhaust gas sensor varies as a function of its temperature, an impedance sensing circuit is employed at the two summing junctions in Fig. 10 to generate artificially rich error signals e_1 and e_2 when the sensors are cool. These error signals drive the two integrators to lean saturation, resulting in lean operation until the sensors warm up.

STRIP CHART RECORDING, WARMED UP OPERATION - A strip chart recording of the control signals for warmed up operation illustrates the dynamic characteristics of both the closed loop and open loop control logic. Fig. 13 shows the 20th and part of the 21st cycles of the Federal Test Procedure with six signals plotted against a common time base: Z_2 sensor, Z_1 sensor, reference R_1, Z_1 integrator output, control vacuum, and car speed. The major events are marked along the time base with time zero at the beginning of the 19th cycle of the Federal test. The beginning of the 20th cycle acceleration mode is at the 163 second mark, the beginning of the 20th cycle cruise at the 205 second mark, the end of the 20th cycle cruise at the 300 second mark, and the beginning of the 21st cycle acceleration at the 346 second mark.

The dynamic characteristics of the closed loop logic are illustrated during the 20th cycle cruise (205-300 mark). Just as in the steady state driving mode discussed in Fig. 11, the engine-out sensor is a more responsive signal with high frequency components, while the converter-out sensor varies rather slowly. The reference voltage trace, R_1, shows how the converter-out sensor generates the optimum value of R_1 for two different driving conditions. The reference R_1 has a value of about 200 mv for the acceleration mode, prior to the 205 s mark, and increases in value to approximately 400 mv during the high speed cruise. Operation at the converter window during this time period was indicated by efficiencies of 97%, 97%, and 71% for HC, CO and NO_x, respectively, that were observed over the 20th cycle cruise mode.

The error signal e_1 is the difference between the reference R_1 and the engine-out sensor Z_1 and it is the time integral of this error signal that produces the Z_1 integrator trace of Fig. 13. The average value of control vacuum is a function of the Z_1 integrator signal, with the magnitude of control vacuum increasing for a decreasing integrator voltage. The high frequency component on the control vacuum signal is a result of the proportional plus lead circuit.

A closed throttle mode occurs between the end of the

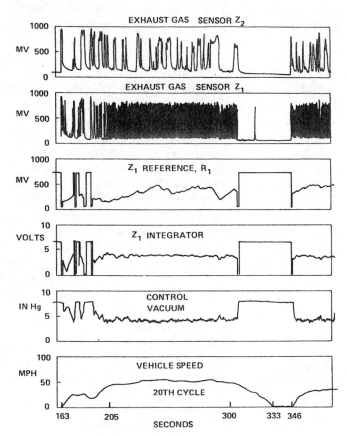

Fig. 13 - Emission cycle strip chart recording

20th cycle cruise and the beginning of the 21st cycle acceleration, approximately from the 300 to the 346 s marks. The control vacuum is driven to its maximum value of 8 in Hg by the closed throttle logic. The integrator hold circuits provide an initial condition on the two integrators at the throttle opening (346 s mark) that results in the same value that occurred just prior to the throttle closure. During the throttle closure mode, the integrators become comparators and their output voltages are either plus or minus saturation as shown in Fig. 13, depending on the sign of the error signals.

COLD START TIMER - Experimental test results have shown that the emission levels are improved if the system is run in a lean mode until the three-way converter is warmed up to its operating temperature. When the converter is cool, the conversion efficiencies are low enough that the tail pipe emissions are very close to the engine-out emissions. During this time period, lean operation significantly reduces the production of HC and CO with little effect on NO_x. To accomplish the lean cold start mode, the cold start switch senses that the engine is cold and is running, and actuates a timer circuit in the control logic, Fig. 10. The timer circuit generates a lean command C_3 to the power amplifier for a predetermined length of time that has been chosen to correspond to the warm-up time of the converter.

EMISSION CONTROL SYSTEM

Operation of the cold start timer is illustrated with a strip chart recording of the first cycle and part of the second cycle of the Federal Test Procedure, Fig. 14. The cold start timer operates for approximately 190 s, producing a full lean control vacuum of approximately 8 in Hg which results in a Z_1 sensor signal lean of stoichiometry. The two integrators function as comparators during this time period, similar to their operation during a throttle closure. The Z_2 sensor is below its operating temperature as indicated by the high output of approximately 800 mv produced by the hot start circuit.

Closed loop control begins after the cold start timer function elapses as indicated by the lower values of control vacuum and the shape of the Z_1 integrator and Z_1 sensor traces after the 190 s mark in Fig. 14. The Z_2 sensor reaches its operating temperature during the second cycle cruise (205-300 s mark) and begins to drive the Z_1 reference to its optimum value for operation at the converter window.

EMISSION TEST RESULTS

Emission test results obtained from the 1975 Federal Test Procedure were used to gage the performance of the electronic control logic and carburetor mechanism, Table 1. Subjective evaluation indicates that the vehicle performance and driveability on the emission test cycle are satisfactory. Fuel economy on this test is similar to vehicles equipped with an oxidizer-only catalyst system designed to meet the 1975 California emission standards. Because of the measurement variability experienced at the low emission levels shown in Table 1, three individual test results are listed to illustrate the range obtainable. Since questions regarding catalyst durability are unanswered, the catalyst material in the test vehicles was replaced periodically so that converter deterioration would not influence the emission test results.

The emission results in Table 1 indicate that the system has not demonstrated the ability to comply with the 1978 standards of 0.41 g/mile HC, 3.4 g/mile CO and 0.40 g/mile NO_x. In fact, the HC and CO levels are about the same as comparable production 1975 California cars equipped with oxidizer-only catalysts which certified to standards of 0.9 g/mile HC, 9.0 g/mile CO and 2.0 g/mile NO_x. The experimental closed loop test vehicles do not have emissions below the 1978 standard by a margin sufficient to insure certification or allow for production variability and test repeatability. In addition, although the subject of durability of the three-way catalyst and system components is not within the scope of this paper, durability experiments have indicated catalyst deterioration too great to be acceptable for certification purposes.

SUMMARY

This paper presents the electronic control logic and system components employed in adapting a four barrel carburetor for use in a closed loop emission control system. The closed loop and open loop fuel metering functions are discussed, including consideration of interactions during transient driving conditions. The description of the electronic control logic includes strip chart recordings to illustrate the exhaust gas sensor signals, system stability, and system operation. Emission test results using low mileage catalyst material are used to

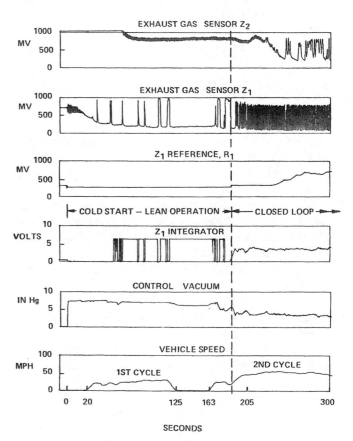

Fig. 14 - Cold start timer operation

Table 1 - Emission Test Results

	1975 Federal Test Procedure Results		
	HC G/Mile	CO G/Mile	NO_x G/Mile
1978 Federal emission standard	0.41	3.4	0.40
Typical closed loop carburetor emission results	0.50	4.76	0.31
	0.47	3.37	0.33
	0.31	3.72	0.41

Vehicle specifications.
- 4500 lb sedan
- 400 in³ V8 engine with 4 bbl. closed loop carburetor
- 260 in³ underfloor three-way converter (low mileage)

gage the performance of the control logic and carburetor mechanism. Although the emission levels do not meet the 1978 standard, they are sufficiently low to conclude that the venturi carburetor may now be considered a possible means of accomplishing the fuel metering in a closed loop emission control system.

ACKNOWLEDGMENTS

The system presented in this paper represents the efforts of many individuals. Members of the General Motors Closed Loop Task Force under L. J. Faix of Chevrolet have been responsible for component development and system design.

The specific contributors and contributions are: A. C. Spark Plug for exhaust gas sensors and catalyst material, Delco Electronics for electronic controls, Rochester Products for carburetors, Cadillac for system development, General Motors Research Laboratories for exhaust gas sensor development and system concepts, and Chevrolet for system development and project coordination. The support and resources of Chevrolet's engineering laboratory are appreciated.

The contributions of L. L. Cuttitta of Chevrolet in the continuing development and practical implementation of the closed loop carburetor have been invaluable.

The contributions of the Research Laboratories' Electronics Department under E. F. Weller for initial exhaust gas sensor and system development, and of G. Casey of Engineering Staff for early vehicle work are acknowledged.

REFERENCES

1. J. F. Cassidy, Jr., "Electronic Closed Loop Controls for the Automobile." Paper 740014 presented at SAE International Automotive Engineering Congress, Detroit, January 1974.
2. J. G. Rivard, "Closed-Loop Electronic Fuel Injection Control of the Internal Combustion Engine." Paper 730005 presented at SAE International Automotive Engineering Congress, Detroit, January 1973.
3. T. L. Rachel and R. Gunda, "Electronic Fuel Injection Utilizing Feedback Techniques." Paper presented at Session 36, IEEE International Convention and Exposition, New York, March 1974.
4. R. Zechnall, G. Baumann, and H. Eisele, "Closed-Loop Exhaust Emission Control System with Electronic Fuel Injection." Paper 730566 presented at SAE National Automotive Engineering Meeting, Detroit, May 1973.
5. W. J. Fleming, D. S. Howarth, and D. S. Eddy, "Sensor for On-Vehicle Detection of Engine Exhaust Gas Composition." Paper 730575 presented at SAE National Automotive Engineering Meeting, Detroit, May 1973.
6. D. S. Eddy, "Physical Principles of the Zirconia Exhaust Gas Sensor." Paper 73CH0718-7VT-C-3 presented at the IEEE Vehicular Technology Conference, Cleveland, December 1973.
7. J. Gyorki, "Fundamentals of Electronic Fuel Injection." Paper 740020 presented at SAE International Automotive Engineering Congress, Detroit, January 1974.

760226

Emission Control With Lean Mixtures

John F. Schweikert
James J. Gumbleton
Advance Product Engineering
Engineering Staff
General Motors Corporation

ABSTRACT

NOx emissions can be controlled through engine operation with lean homogeneous air/fuel mixtures. This emission control approach precludes the need for exhaust gas recirculation (EGR) and secondary air injection systems. The Lean Mixture concept results in similar emissions, fuel economy, and driveability when compared to EGR systems tailored to similar emission levels with similar aftertreatment systems. The Lean Mixture approach does offer the potential for less engine emission control hardware.

The minimum NOx level achieved experimentally at the lean driveability limit was about 1.2 g/mi but with significantly higher HC emissions. Lean Mixture systems are sensitive to variations in engine air/fuel ratio which produce a significant effect on their emissions and fuel economy. Due to this sensitivity, it appears that the Lean Mixture concept is limited to the current NOx emission standards (3.1 g/mi) unless technological advances in fuel/air metering occur that reduce engine and carburetor air/fuel ratio variations.

Vehicle programs have demonstrated the effect of lean engine air/fuel ratios on base emissions and fuel economy as well as in conjunction with various aftertreatment systems; oxidizing catalytic converters, and lean manifold reactors. Lean Mixture systems are not compatible with reducing catalysts.

A Lean Mixture–Manifold Reactor emission control system was used to evaluate the effect of compression ratio and leaded fuel on vehicle emissions and fuel economy. With the constraint of equal HC emission levels, the classical relationship between compression ratio and fuel economy appears to have been altered. The use of higher compression ratio did not result in improved fuel economy. In addition, the use of leaded fuels resulted in both a direct and a long term increase in HC emissions.

System durability is shown for a Lean Mixture–Catalytic Converter system with lead-free fuel and for a Lean Mixture system without aftertreatment with leaded fuel. An engine dynamometer program demonstrated the incentive for improved mixture distribution.

LEAN MIXTURE SYSTEM + CATALYTIC CONVERTER

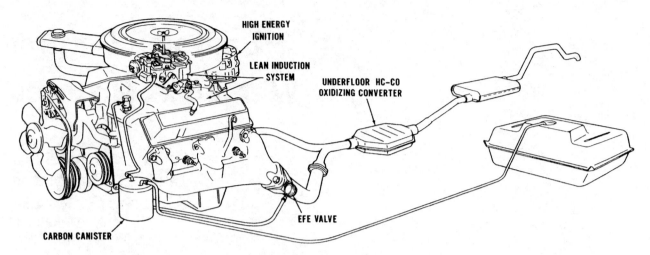

Fig. 1—The Lean Mixture emission control system utilizes less engine control hardware and is compatible with oxidizing exhaust aftertreatment systems.

THE LEAN MIXTURE EMISSION CONTROL SYSTEM is based upon achieving low engine-out emissions through homogeneous operation at very lean air/fuel ratios. These lean mixtures provide the incentive of simultaneous control of HC, CO, and NOx emissions with good fuel economy and a minimum of engine control hardware. Figure 1 schematically depicts the essential hardware requirements for the Lean Mixture concept. Included in this approach are an intake induction system which has good cylinder-to-cylinder mixture distribution, a High Energy Ignition (HEI) system which promotes burning of lean air/fuel mixtures and an Early Fuel Evaporation (EFE) system to assist on cold starts. An oxidizing catalytic converter is shown as the aftertreatment system.

DYNAMOMETER DATA—Steady-state dynamometer testing provides insight into the tradeoffs of emissions and fuel consumption with spark advance and air/fuel ratio. Testing of a Lean Mixture 350 CID (5.7 ℓ) engine shows the classical relationships of exhaust emissions and fuel consumption versus air/fuel ratio and illustrates the incentive for operating with lean mixtures (Figure 2). At the best economy spark advance (MBT), maximum NOx emissions and minimum HC emissions occur at about a 16:1 A/F ratio. Leaner operation to about a 20.5:1 A/F ratio results in the minimum fuel consumption, comparable CO emissions, and a reduction in NOx emissions. However the HC emissions have increased.

Reducing the HC emissions requires spark retard which results in a fuel economy penalty. Figure 3 shows the effect

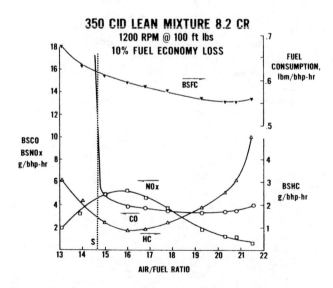

Fig. 2—Operation at lean air/fuel ratios can simultaneously reduce NOx emissions and fuel consumption but results in higher HC emissions.

Fig. 3—Spark retard reduces HC and NOx emissions but with significantly higher fuel consumption.

on exhaust emissions when the spark is retarded to produce a 10% increase in fuel consumption from the best economy (MBT) timing. At a nominal 19:1 A/F ratio, the HC emissions with the retarded spark are equivalent to the minimum HC emissions (16:1 A/F ratio) with MBT spark timing. The retarded spark also results in lower NOx emissions and higher CO emissions.

With Lean Mixture systems, the amount of spark retard required for HC control will depend on the emission standards and the exhaust aftertreatment system used. Additional data on the inter-relationship of air/fuel ratio and spark advance are presented in Appendix A.

MIXTURE DISTRIBUTION—The lean limit for stable engine operation is strongly influenced by the air/fuel ratio of the leanest cylinder. Therefore, reducing the variance of individual cylinders from the mean air/fuel ratio enables the engine to operate at a leaner combustion stability limit.

Figure 4 compares the mixture distribution of a 1975 production intake system to the modified system used in the Lean Mixture concept. Modifications to the 1975 production dual-plane intake manifold reduce the mixture distribution spread of about 1.0 - 1.5 A/F ratio to about 0.5 - 1.0 A/F ratio for speeds and loads normally encountered on the EPA emission test. These data were obtained on a dynamometer engine installation with individual cylinder exhaust manifolding. Accurate measurement of mixture distribution with conventional exhaust manifolding is difficult due to interference effects of adjacent cylinders.

Figures 5 through 8 show the effects of improved mixture distribution on exhaust emissions and octane requirements. Improving mixture distribution produced no effect on HC emissions except near the combustion stability limit

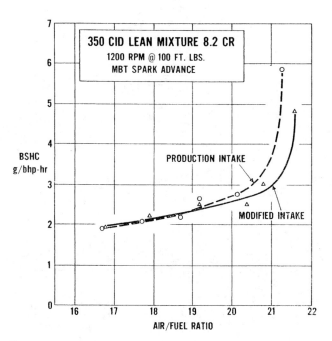

Fig. 5—Improved mixture distribution does not lower HC emissions but extends the combustion stability limit by about 0.5 A/F ratio

(Figure 5). The modified intake manifold permitted about 0.5 A/F ratio leaner operation before combustion instability occurred as evidenced by the sharp rise in HC emissions. CO emissions not shown, were unaffected by the reduced variance in mixture distribution. A slight reduction in NOx emissions was shown for the improved mixture distribution (Figure 6). Fuel consumption characteristics for both

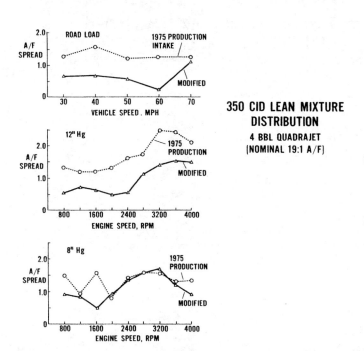

Fig. 4—The modified intake manifold improves mixture distribution over the operating conditions encountered on the EPA emission test.

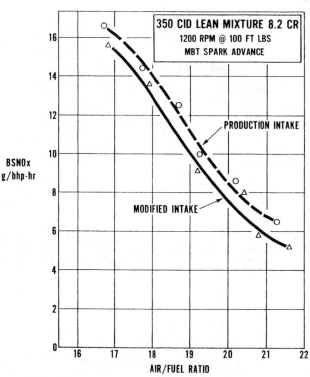

Fig. 6—BSNOx emissions were reduced slightly by the improved mixture distribution.

induction systems were similar except at the combustion stability limit where the modified intake allows slightly leaner operation (Figure 7). Vehicle tests also showed that the 1975 production manifold and the modified manifold resulted in equal emissions and fuel economy when calibrated at a nominal 19:1 cruise A/F ratio.

Mixture distribution also affects an engine's octane requirement. For Lean Mixture systems, the richest cylinder has the highest octane requirement which dictates the engine's overall part-throttle octane rating. Figure 8 details the part-throttle borderline spark and MBT spark requirements for the production and modified intake systems operated on commercial 91 RON unleaded fuel. Note that operation at the leaner air/fuel ratios tends to reduce the part-throttle octane requirement due to the diluent effect of excess air. Thus the use of lean air/fuel ratios and improved mixture distribution provides increased margin to borderline knock. However, the octane requirement at wide open throttle remains unaffected since normal power mixtures (approximately 13:1 A/F ratio) are employed.

LEAN MIXTURE VEHICLE SYSTEMS—The application of the Lean Mixture emission control concept to a vehicle system requires careful scheduling of both the engine air/fuel ratio and spark advance to achieve good emission and driveability performance. All Lean Mixture vehicles utilized the modified production intake manifold and a 4 bbl. Quadrajet carburetor. Leaner than stoichiometric choke schedules were possible due to the EFE system and a Cold

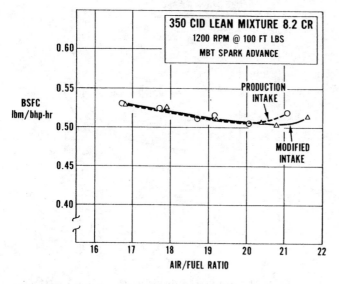

Fig. 7—Improved mixture distribution shows equal fuel consumption characteristics except near the combustion stability limit.

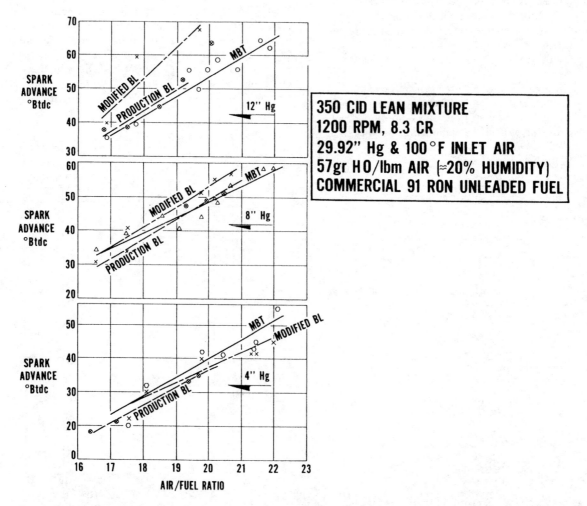

Fig. 8—Improving the mixture distribution offers an increased margin to part-throttle borderline knock.

Spark Advance (CSA) feature that provides additional spark advance during cold engine operation to maintain cold driveability. The production High Energy Ignition (HEI) system was tailored with vacuum advance schedules to provide the best fuel economy possible within the required emission control constraint. Prior to any emission testing all vehicles accumulated 4K miles.

Lean Mixture vehicle programs have evaluated the effect of lean engine air/fuel ratios on base emissions and fuel economy and in conjunction with various aftertreatment systems; oxidizing catalytic converters and manifold reactors.

WITHOUT EXHAUST AFTERTREATMENT—Table 1 details the emission and fuel economy data for a Lean Mixture vehicle without exhaust aftertreatment. These data indicate that the Lean Mixture system without aftertreatment at low mileage can achieve emission levels below the 1976 Federal standards of 1.5 HC, 15 CO, 3.1 NOx. HC emissions are substantially higher than the catalyst equipped 1976 Certification vehicles and do not provide sufficient margin for deterioration with miles, car variability and test variability. The 400 CID (6.5 ℓ) Lean Mixture data indicates composite (55 City/45 Hwy.) fuel economy losses of 6% (400 CID Baseline @ 5000 lb. I.W.) to 9% (350 CID Baseline @ 4500 lb. I.W.). This fuel economy loss is primarily due to the spark retard necessary to reduce the HC emission levels.

Figure 9 details the effect of durability mileage accumulation for the above 400 CID vehicle operated on leaded

TABLE 1

WITHOUT EXHAUST AFTERTREATMENT

NOMINAL 8.2 CR

SYSTEM	I.W.	ENGINE DISP.	# TESTS	1975 FTP EMISSIONS g/mi			FUEL ECONOMY		
				HC	CO	NOx	CITY	HIGHWAY	55/45
LEAN MIXTURE 18.3 A/F @ CRUISE	4500	400	AVG.(3)	1.3	7.8	2.0	11.9	18.5	14.1
'76 CERT. DATA CAR W/CATALYST CCS-EGR	4500	350	(1)	0.38	5.6	1.5	13.4	19.1	15.5
'76 CERT. DATA CAR W/CATALYST CCS-EGR	5000	400	(1)	0.32	3.4	2.3	12.7	19.9	15.2

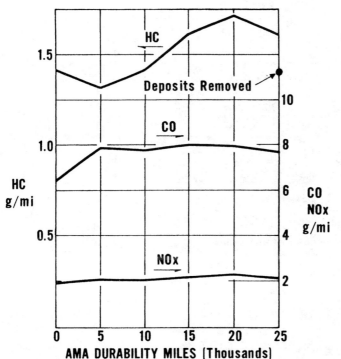

Fig. 9—The use of leaded fuel increases HC emissions due to the long term deposit effect.

fuel (0.5 g/gal TEL). The HC emissions exceeded the 1.5 g/mi standard at 15K miles. After 25K miles, the deposits were removed from the combustion chamber and the HC emission levels were nearly equal to the 0 mile level. This long term deposit accumulation is attributable to the lead deposits resulting from operation on 0.5 g/gal TEL leaded fuel. Additional spark retard would be required to compensate for the increase in HC emissions with miles and this would result in an additional loss in fuel economy.

MANIFOLD REACTOR AFTERTREATMENT—Incorporation of manifold reactor aftertreatment offers potential reductions in tailpipe HC emission levels. One experimental manifold reactor design is shown in Figure 10. It consists of an increased manifold volume, baffles to aid mixing and prevent short circuiting, as well as insulation and exhaust port liners to aid in sensible energy conservation. These manifold reactors operate in a 1200°F (650°C) – 1450°F (790°C) temperature range on the EPA test. At these temperature levels the reactors primarily provide HC control. Temperatures in excess of 1400°F (760°C) are required for CO control. Partial HC oxidation below this temperature yields the intermediate product of CO and thus reactor operation in this range can increase tailpipe CO emissions. Manifold reactors operating at lean air/fuel ratios do not produce any appreciable exothermic reactions and therefore depend on the conservation of the sensible energy of the exhaust gas. With a given system, higher exhaust gas temperatures can only be obtained through spark retard.

Table 2 presents emission levels for two separate 350 CID (5.7ℓ) Lean Mixture–Manifold Reactor vehicles targeted toward the 1977 Federal standards of 1.5 HC, 15 CO, 2.0 NOx. These vehicles, operated on Indolene Clear fuel, show a 7 to 10% loss in composite (55/45) fuel economy. As in the case without aftertreatment, the fuel economy loss results from the spark retard required to control HC emissions to these levels. Even with this spark retard, the Lean Mixture–Manifold Reactor system exhibits higher HC emissions than the production catalyst equipped vehicle. However, deterioration factors for Lean Mixture–Manifold Reactor systems would be expected to parallel engine-out emission deterioration and therefore can be targeted at higher initial HC emission levels than oxidizing converter systems.

Thus with some fuel economy penalty, these data indicate a potential to meet the 1977 Federal emission standards at low mileage. Durability and system deterioration on the emission certification procedure as well as structural durability under all types of driving conditions need to be established.

LEAN MIXTURE–MANIFOLD REACTOR A/F RATIO SENSITIVITY—With the Lean Mixture–Manifold Reactor control system, variations in engine air/fuel ratio significantly affect emissions and fuel economy. Figure 11 shows the emission interaction produced by varying the carburetor part-throttle metering calibration. All other engine control parameters such as choke schedule, spark schedule,

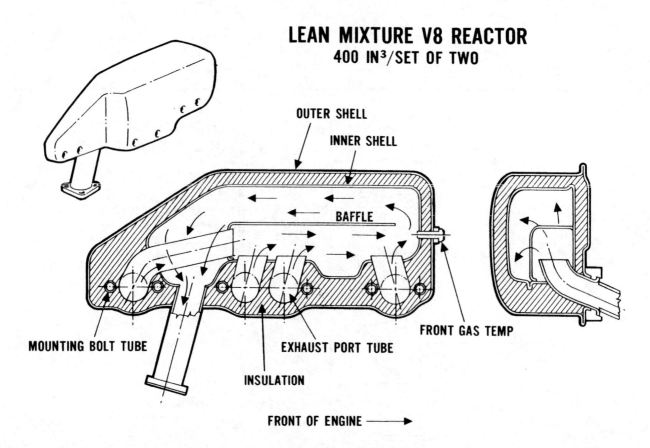

Fig. 10–Lean Mixture–Manifold Reactor systems require conservation of the exhaust gas sensible energy for best oxidation efficiency.

TABLE 2

4500 LB. INERTIA WEIGHT
350 CID ENGINE 8.2 CR
120 IN³/SIDE MANIFOLD REACTOR
INDOLENE CLEAR FUEL

SYSTEM	# TESTS	'75 FTP EMISSIONS g/mi HC	CO	NOx	FUEL ECONOMY CITY	HWY.	55/45
CAR A LEAN MIXTURE 19.4 A/F @ CRUISE	AVG.(3)	0.79	4.4	1.5	12.6	17.4	14.4
CAR B LEAN MIXTURE 19.8 A/F @ CRUISE	AVG.(3)	0.95	6.0	1.4	12.3	16.8	14.0
'76 CERT. DATA CAR W/CATALYST	(1)	0.38	5.6	1.5	13.4	19.1	15.5

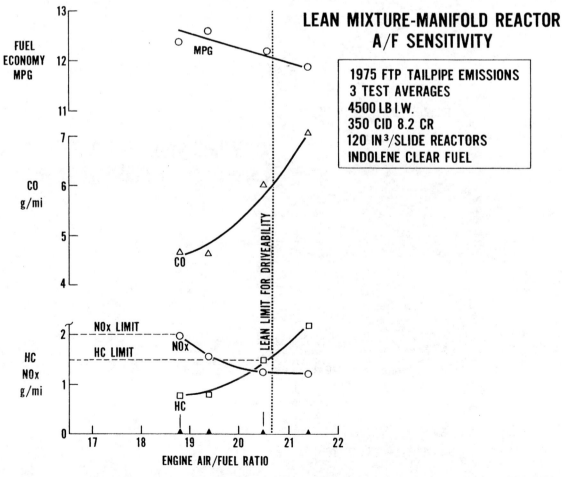

Fig. 11–NOx and HC emission sensitivity to air/fuel ratio variations limit the Lean Mixture–Manifold Reactor system calibration range to less than 2.0 A/F ratios.

and idle air/fuel ratio were maintained constant. These data were obtained on a 350 CID (5.7 ℓ) Lean Mixture–Manifold Reactor vehicle tailored toward the 1977 Federal emission standards of 1.5 HC, 15 CO, 2.0 NOx.

NOx emissions constrain the rich calibration limit, while HC emissions and driveability define the lean calibration limit. These data indicate that an air/fuel ratio calibration within a range of less than 2 A/F ratios must be maintained during initial vehicle build and throughout the service life of the car for emission compliance. With this inherent high sensitivity to air/fuel ratio variations, this system appears applicable only to the present NOx standard (3.1 g/mi). Technological improvements to reduce this A/F ratio sensitivity are required to allow application to lower NOx levels. The data show a minimum NOx of 1.2 g/mi and a significant HC increase at the lean A/F ratio limit.

COMPRESSION RATIO EFFECT–Non-catalytic forms of exhaust aftertreatment potentially permit the use of tetraethyl lead (TEL) additives in the fuel to raise the octain rating without extra refining. This offers the possibility of using higher compression ratios (CR). A 400 CID (6.5 ℓ) Lean Mixture–Manifold Reactor vehicle was operated at compression ratios between 7.3 and 9.3 CR to evaluate the interaction of compression ratio, exhaust emissions and fuel economy. Figure 12 details the combustion chamber variations required versus the base 8.3 CR configuration. The piston to head quench areas and volume were maintained constant. Total flame travel distance is similar for all three configurations. The basic spark timing setting was retarded 4° per CR increase to compensate for the burn rate effect relative to the MBT spark requirement.

Table 3 presents the effect of compression ratio on emissions and fuel economy for this Lean Mixture–Manifold Reactor vehicle. The baseline data were obtained with the 8.3 CR engine. With a 9.3 compression ratio, a gain in composite fuel economy of about 6% was realized when the spark was adjusted for the increase in burn rate with the higher compression ratio. However, the HC and NOx emissions were increased substantially and the reactor temperature was reduced. The large increase in tailpipe HC was attributed to two factors: (1) an increase in engine-out HC due to the increase in compression ratio, and (2) the reduced effectiveness of the reactor because of lower exhaust temperature. The increase in NOx was attributed to the higher compression ratio. Adjusting the vacuum spark advance to produce HC emissions similar to those obtained at the 8.3 compression ratio resulted in a loss in fuel economy on the city schedule, while retaining some of the fuel economy gained on the highway schedule and with no change in the composite (55/45) fuel economy.

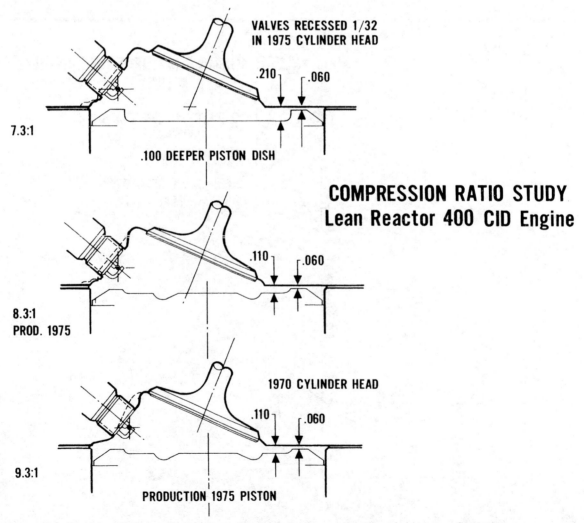

Fig. 12–All three compression ratios have similar combustion chamber configurations.

TABLE 3

4500 LB. INERTIA WEIGHT
400 CID ENGINE
200 IN³/SIDE MANIFOLD REACTOR
INDOLENE CLEAR FUEL

ENGINE	# TESTS	'75 FTP EMISSIONS g/mi			FUEL ECONOMY			REACTOR TEMP. (°F) AVG.
		HC	CO	NOx	CITY	HWY.	55/45	
8.3 C.R.	AVG. (3)	.62	4.8	1.7	12.9	16.2	14.2	1330
9.3 C.R.	AVG. (2)	1.20	5.4	2.6	13.4	17.5	15.0	1240
9.3 C.R. ADJUSTED SPARK	AVG. (3)	.50	4.9	2.1	12.4	17.1	14.1	1320
7.3 C.R.	AVG. (3)	.72	5.8	1.9	11.9	16.2	13.5	1330
7.3 C.R. ADJUSTED SPARK	AVG. (3)	.64	4.6	2.4	11.7	16.9	13.6	1310

With a 7.3 compression ratio, adjusting the basic spark to compensate for the slower burn rate still resulted in both a loss in fuel economy as well as an increase in HC emissions when compared to the 8.3 compression ratio baseline. Adjusting the vacuum spark advance for best fuel economy at equal HC emissions also showed a loss in fuel economy from the baseline 8.3 CR engine.

Thus, based on this preliminary data, it would appear that with a Lean Mixture–Manifold Reactor system, the 8.3 compression ratio is about optimum for emissions and fuel economy when tailored with the constraint of equal HC emission levels. Similar results have been reported for a single cylinder engine study (Reference 1). Thus, the emission constraints appear to have altered the classical relationship between compression ratio and fuel economy.

Alteration of the engine/vehicle size, different emission standards, or aftertreatment systems could affect this conclusion. A catalytic converter system may be less sensitive than the Lean Mixture–Manifold Reactor system to the lower exhaust gas temperature-associated with increased compression ratio. However, using a catalytic converter would require that the higher octane fuel for the increased compression ratio must be lead free.

LEADED FUEL EFFECTS—Lean Mixture–Manifold Reactor systems are sensitive to the long term HC emission increase resulting from lead deposit accumulation with miles as previously shown (Figure 9). In addition, leaded fuels produce a direct lead effect independent of deposit accumulation as shown in Figure 13. With the Lean Mixture–Manifold Reactor system, increasing the fuel lead level from 0 to 1.5 g/gal TEL results in approximately a 40% increase in HC emissions as measured on the EPA emission test. CO and NOx emissions were unaffected by the addition of TEL to the base fuel. Repeat tests with lead-free fuel after the maximum TEL level (as indicated by the 'x's within the circles) reduced the HC emissions to the original level. For this experiment, there was no mileage accumulation on leaded fuel other than that to perform the EPA emission test. As a precaution, the fuel system and engine were purged with lead-free Indolene Clear base fuel between emission tests. It appears that the lead inhibits

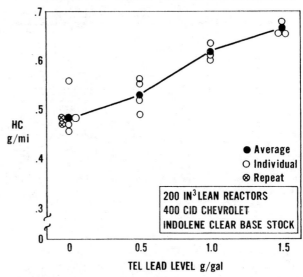

Fig. 13–HC emissions increase directly with the use of leaded fuel due to an oxidation inhibiting effect in the manifold reactors.

both the oxidation within the manifold reactor as well as in the engine. Previous studies (References 2 and 3) have shown a direct lead effect on HC emissions for engines without aftertreatment as well as with rich reactors.

Thus in meeting any emission standard with leaded fuel, these effects must be considered since a lower emission system target would be required than with lead-free fuel.

CATALYTIC CONVERTER AFTERTREATMENT—Catalytic converters with their higher oxidation efficiencies at lower exhaust temperatures provide significant control of both HC and CO emissions when used with the Lean Mixture emission control concept. Table 4 compares a 350 CID (5.7ℓ) Lean Mixture–Catalytic Converter vehicle tailored toward the 1977 Federal standards of 1.5 HC, 15 CO, 2.0 NOx. At low mileage, the Lean Mixture–Catalytic Converter system produced comparable HC and NOx emission levels and fuel economy. The use of lean carburetion and a lean choke schedule result in the lower CO emission level.

A direct comparison between a Lean Mixture–Catalytic Converter concept versus an AIR-EGR-Catalytic Converter concept was made on two 400 CID (6.5 ℓ) vehicles targeted toward emission standards of 0.41 HC, 3.4 CO, 2.0 NOx. Table 5 summarizes the low mileage emission results with fresh catalyst (2K miles). The Lean Mixture–Catalytic Converter system utilized a 19.5:1 A/F ratio at cruise whereas the AIR-EGR system was 16:1 A/F ratio with backpressure EGR to control NOx emissions to the 1.5 g/mi level. The Lean Mixture approach showed no advantage in emissions, fuel economy, or driveability when compared to the AIR-EGR system. The lower fuel economy for the AIR-EGR system is attributable to its lower engine-out HC calibration. As compared to the nearest 1976 Certification vehicles these vehicles show a 10-14% fuel economy loss even at their lighter inertia weight setting (4500 lb. versus 5000 lb.). However, the Lean Mixture–Catalytic Converter system does require less engine emission control hardware than the AIR-EGR system.

TABLE 4

**4500 LB. INERTIA WEIGHT
350 CID ENGINE 8.2 CR
260 IN^3 U/F CONVERTER (4K MILES)
INDOLENE CLEAR FUEL**

SYSTEM	# TESTS	'75 FTP EMISSIONS g/mi			FUEL ECONOMY		
		HC	CO	NOx	CITY	HWY.	55/45
LEAN MIXTURE 20.8 A/F @ CRUISE	AVG.(3)	0.48	0.7	1.4	13.4	19.0	15.4
'76 CERT. DATA CAR CCS–EGR	(1)	0.38	5.6	1.5	13.4	19.1	15.5

TABLE 5

**1975 FTP EMISSIONS
260 IN^3 U/F CONVERTER (2K MILES)**

SYSTEM	I.W.	CID	# TESTS	ENGINE OUT g/mi			TAILPIPE g/mi			MPG
				HC	CO	NOx	HC	CO	NOx	
LEAN MIXTURE 19.5 A/F	4500	400	AVG. (3)	1.3	5.4	1.4	0.20	1.0	1.5	11.5
AIR–Bp EGR 16.0 A/F	4500	400	AVG. (3)	1.0	4.7	1.3	0.13	1.0	1.5	10.9
'76 CERT. DATA CAR CCS–EGR	5000	400	(1)	–	–	–	0.32	3.4	2.3	12.7
'76 CERT. DATA CAR CCS–EGR	4500	350	(1)	–	–	–	0.38	5.6	1.5	13.4

LEAN MIXTURE–CATALYTIC CONVERTER A/F RATIO SENSITIVITY–With Lean Mixture–Catalytic Converter systems, variations in engine air/fuel ratios produce a significant effect on emissions and fuel economy. Figure 14 shows the tailpipe emission dependence on air/fuel ratio for a 400 CID (6.5 ℓ) Lean Mixture–Catalytic Converter vehicle tailored toward emission standards of 0.41 HC, 3.4 CO, 2.0 NOx. All other engine parameters such as choke schedule, spark scheduling, and idle air/fuel ratio were held constant. With a 2.0 g/mi NOx emission level constraint for the rich calibration limit and driveability as the constraint for the lean calibration limit, the calibration range for the system was less than 2 A/F ratios. Figure 15 shows that at the lean driveability limit, the engine-out HC emissions are rapidly increasing. This increase in engine-out HC emissions is masked at the tailpipe (Figure 14) by the fresh catalyst (2K miles) used; however at 50K miles, these high HC levels would be very significant. At the 3.1 g/mi NOx standard, the rich calibration limit provides about a 3.0 A/F ratio calibration range.

With this high dependence of NOx emissions on engine air/fuel ratio, Lean Mixture–Catalytic Converter systems appear to be limited to the present NOx standard (3.1 g/mi) unless technological advances in fuel/air metering occur that reduce engine and carburetor A/F ratio variations. The minimum NOx level of the Lean Mixture–Catalytic Converter system shown is about 1.2 g/mi at the lean A/F ratio limit but with significantly higher HC emissions. Application of NOx reducing catalysts is not feasible due to the high exhaust oxygen concentrations.

LEAN MIXTURE–CATALYTIC CONVERTER DURABILITY–Durability testing on the EPA certification procedure has been run on a 400 CID (6.5 ℓ) Lean Mixture–Catalytic Converter vehicle targeted toward an emission standard of 0.41 HC, 3.4 CO, 2.0 NOx. Figure 16 details the vehicle emission performance with miles. HC emissions exceeded the 0.41 g/mi limit at both the 25K and 30K mile tests. Post diagnostic testing revealed the increasing HC emissions were caused by a deterioration of catalytic converter efficiency and by a leaning of the engine air/fuel ratio (approximately 1 A/F ratio) due in part to a loss of exhaust valve sealing. The leaner engine air/fuel ratio resulted in significantly higher engine-out HC levels and a decreasing NOx trend with system durability miles.

This limited durability experience demonstrates the emission sensitivity of the Lean Mixture concept to variations in engine air/fuel ratio. Whether these variations are a result of initial vehicle build tolerances or a deterioration of the system calibration as a function of service life, the effect is a significant change in exhaust emissions.

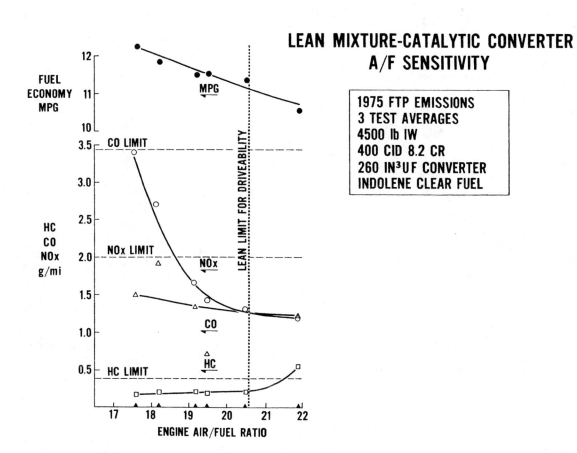

Fig. 14–Sensitivity of the NOx emissions to variations in engine air/fuel ratio limit the Lean Mixture–Catalytic Converter system calibration range to less than 2.0 A/F ratios.

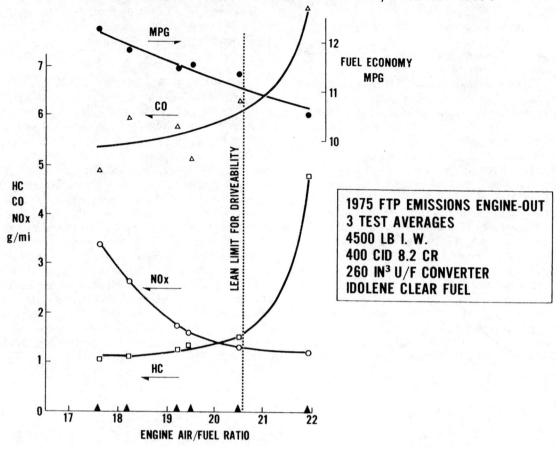

Fig. 15 – Engine-out HC emissions increase rapidly at the lean driveability limit.

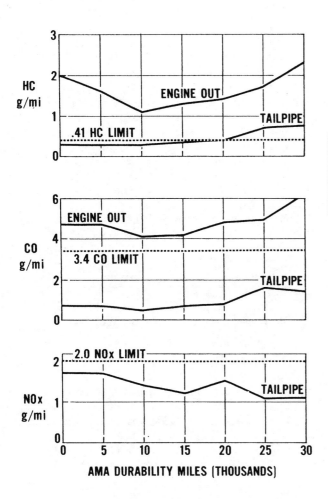

Fig. 16–The Lean Mixture–Catalytic Converter system exceeded the HC emission standard at 25K miles primarily due to the effect of leaner engine operation.

SUMMARY

The application of Lean Mixture engine technology has demonstrated another approach to the control of exhaust emissions. Driveability and fuel economy are comparable to EGR systems. Less engine hardware may be required since the EGR and AIR systems are eliminated. The major disadvantage to the Lean Mixture emission concept is the inherent high sensitivity of the NOx emissions to changes in engine air/fuel ratio. For a 2.0 g/mi NOx standard, a range of less than 2 A/F ratios must be maintained during initial calibration and throughout the service life of the emission control system. In practical terms, this limits the Lean Mixture approach to the current 3.1 g/mi NOx standard unless technological improvements occur to reduce the A/F ratio sensitivity. Experimentally, the minimum NOx level obtained was about 1.2 g/mi at the lean driveability limit, but with significantly higher HC emissions.

Lean Mixture systems without aftertreatment at low mileage show potential to meet current emission standards (1.5 HC, 15 CO, 3.1 NOx) with a fuel economy penalty. The spark retard required to control HC emissions results in a 6-9% composite (55/45) fuel economy loss.

Lean Mixture—Manifold Reactor systems reduce HC emissions and show a potential to meet the 1977 Federal standards (1.5 HC, 15 CO, 2.0 NOx) at low mileage but with a fuel economy loss of 7 to 10%. A/F ratio sensitivity on HC and NOx emissions limit the calibration range to less than 2.0 A/F ratios. System deterioration factors and durability are undetermined.

The addition of an oxidizing catalytic converter to the Lean Mixture concept produces lower HC and CO emission levels. At the 1977 Federal standards (1.5 HC, 15 CO, 2.0 NOx), the Lean Mixture—Catalytic Converter system shows emissions, driveability and fuel economy comparable to EGR systems. Retailoring for targets to meet a lower emission standard of 0.41 HC, 3.4 CO, 2.0 NOx results in a 10-14% fuel economy loss. A Lean Mixture — Catalytic Converter system targeted for these levels failed durability after 25K miles.

The use of higher compression ratios, normally associated with leaded fuel, did not improve fuel economy for a Lean Mixture — Manifold Reactor system at equal HC emission levels. In addition, leaded fuel has been shown to increase HC emissions due to a long term deposit effect and by directly inhibiting HC oxidation rates.

Engine dynamometer testing has shown that minimum HC emissions occur at the maximum NOx emission level (16:1 A/F ratio) and that leaner operation reduces NOx emissions but with higher HC emission levels. Spark retard reduces both HC and NOx emissions, but increases fuel consumption.

ACKNOWLEDGEMENT

The authors gratefully acknowledge the contribution of Mr. D. A. Singer in the acquisition of data.

REFERENCES

1. C. Morgan and S. Hetrick, "Tradeoffs Between Engine Emission Control Variables, Fuel Economy, and Octane," Paper 750415 presented at SAE Automotive Engineering Congress, Detroit, February, 1975.

2. A. Pahnke and W. Bettoney, "Role of Lead Antiknocks in Modern Gasolines", Paper 710842 presented at SAE National Combined Fuels and Lubricants, Powerplant and Truck Meeting, St. Louis, October, 1971.

3. R. Schwing, "The Effects of TEL Oxidation in an Exhaust Manifold Reactor — A Single-Cylinder Engine Study", Paper 710844 presented at SAE National Combined Fuels and Lubricants, Powerplant and Truck Meeting, St. Louis, October, 1971.

APPENDIX A

The Lean Mixture system exhaust emission control is accomplished by controlling the engine air/fuel ratio and spark advance. Figures 17-20 detail the interaction of air/fuel ratio and spark advance and their combined effect on exhaust emissions and fuel consumption.

NOx emissions can be reduced through operation at increasingly leaner air/fuel ratios or by spark retard. Leaner operation at fixed spark (relative to MBT) reduces NOx emissions but always with an increase in HC emissions (Figure 17). Spark retard is effective at reducing NOx and HC emissions but at a loss in fuel consumption (Figures 18 and 19). CO emissions are relatively unaffected by air/fuel ratio and spark advance (Figure 20).

These steady-state dynamometer emission results were obtained on a modified 350 CID (5.7 ℓ) V-8 engine with a measured 8.2 compression ratio. The engine hardware consisted of a modified dual-plane intake manifold with a 4 bbl. Quadrajet carburetor. No exhaust gas recirculation was used at any point. A production High Energy Ignition (HEI) system was used in conjunction with .060 in. (1.5 mm) spark plug gaps. Individual cylinder exhaust headers permitted checks of individual cylinder effects. The carburetor inlet conditions were controlled to a 29.92 in. Hg (760 mm Hg) barometer at a specific humidity of 57 grains H_2O/lbm air.

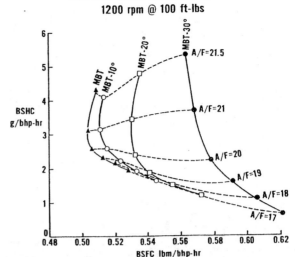

Fig. 18—HC emissions can be reduced by spark retard at the expense of fuel consumption.

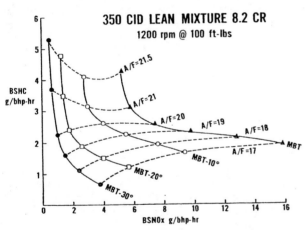

Fig. 17—Leaner air/fuel ratios reduce NOx emissions and increase HC emissions.

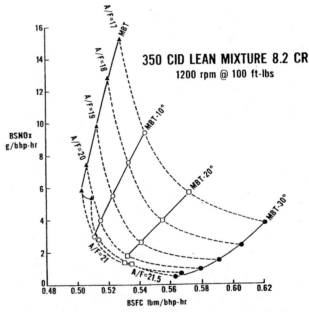

Fig. 19—NOx emissions can be reduced by spark retard which also increases fuel consumption.

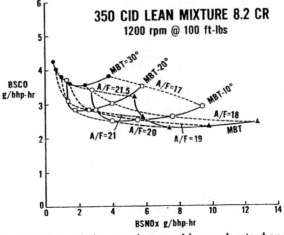

Fig. 20—CO emissions are increased by spark retard and operation near the combustion stability limit.

A Vaporized Gasoline Metering System for Internal Combustion Engines

D. R. Hamburg and J. E. Hyland
Engineering and Research Staff, Ford Motor Co.

ONE OF THE MOST widely accepted techniques for achieving the statutory NO_x standard of 0.4 gram/mile for automobiles is the use of a "reduction" catalytic converter. Unfortunately, such devices exhibit a relatively narrow range of air-fuel ratio over which useful conversion efficiency can be realized. This characteristic is illustrated in Figure 1 which shows the conversion efficiency versus air-fuel ratio for a typical noble metal reduction catalyst. It should be pointed out that the so-called three-way catalysts have an even narrower air-fuel ratio range over which efficient operation is possible. To effectively utilize catalytic converters to control NO_x, it is therefore necessary to employ a fuel metering system which provides very tight control of air-fuel ratio for both steady state and transient engine operation. A viable approach for obtaining the required tight control is to use feedback from a suitable engine exhaust gas sensor to "trim" an appropriate fuel metering system as

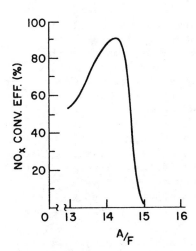

Fig. 1-Conversion efficiency versus air-fuel ratio for typical metal reduction catalyst

depicted in Figure 2. (1-3)*

---

*Numbers in parentheses designate references at end of paper.

---------ABSTRACT

A prototype vaporized gasoline metering system is described which utilizes engine exhaust heat to vaporize liquid gasoline prior to being combined with inlet air. It is shown that the system (1) exhibits minimal time-fluctuations in air-fuel ratio, (2) essentially eliminates the transient variations in air-fuel ratio due to load changes, and (3) provides a very uniform cylinder-to-cylinder distribution of air-fuel ratio. The use of the vapor system at very lean air-fuel ratios is considered, and a CVS cycle prediction of the lean-limit operation is presented.

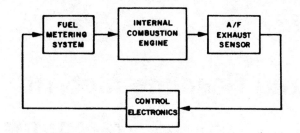

Fig. 2-Block diagram of basic A/F feedback system

BACKGROUND

Ford Motor Company became involved with such a system several years ago during the early development of the TiO_2 exhaust gas sensor. (4) At that time, the output of a prototype TiO_2 sensor was used successfully to control the air-fuel ratio produced by a Bendix electronic fuel injection system. As a result of this effort, the feasibility of the feedback concept was established. Because of the inherent complexity and attendant high production costs of fuel injection, however, it was decided to explore feedback using a much simpler fuel metering device. The particular device chosen for this exploration was a modified carburetor having an air-bypass adjustment which could be controlled electronically by the TiO_2 sensor. A simplified diagram of the basic carburetor showing the air-bypass section is presented in Figure 3. For clarity, actual carburetor details relating to such elements as the main metering system, the idle system, the power enrichment system, etc., are not shown in this diagram.

The air-bypass carburetor was installed on a 351 CID engine in a 1973 Ford Galaxie and evaluated on a chassis dynamometer. A typical recording of the open loop air-fuel ratio versus time as indicated by a TiO_2 exhaust sensor for this configuration operating at a 30 MPH steady-state cruise is shown in Figure 4. When the feedback loop which coupled the exhaust sensor to the air-bypass adjustment was closed and properly compensated to prevent instability, the recording of air-fuel ratio versus time shown in Figure 5 resulted. Examination of this recording reveals that although the long term drift has been eliminated, there is no appreciable reduction in the high-frequency fluctuations in the air-fuel ratio. The reason that feedback is incapable of reducing the high-frequency fluctuations is that the propagation delay through the engine imposes a fundamental limitation on the minimum response time of the closed loop system. To be more explicit, a change in air-fuel ratio occurring at the carburetor takes several engine revolutions before it can be detected in the engine exhaust.

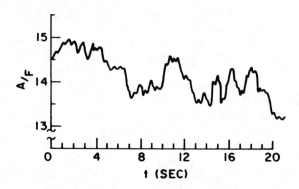

Fig. 4-Open loop air-fuel ratio versus time for air-bypass carburetor operating at 30 mph road load

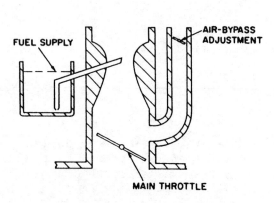

Fig. 3-Simple air-bypass carburetor

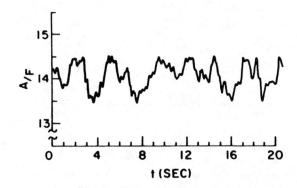

Fig. 5-Closed loop air-fuel ratio versus time for air-bypass carburetor operating at 30 mph road load

The ability to initiate any trimming of the air-bypass adjustment before several engine revolutions have occurred is therefore impossible, and any attempt to effect the necessary trim too rapidly after the change has been detected will result in an oscillatory condition.

It is thus apparent that since feedback cannot eliminate rapid fluctuations in air-fuel ratio, a fuel metering system should be employed which does not exhibit such fluctuations. Since it is generally believed that these fluctuations are caused to a great extent by random detachment of liquid gasoline from wet manifold and carburetor surfaces (5), it would appear that the difficulty could be circumvented by using a vaporized gasoline metering system such as described below.

GENERAL SYSTEM DESCRIPTION

The basic vaporized gasoline metering system utilizes engine exhaust heat to fully vaporize liquid gasoline entering an exhaust gas heat exchanger. The resulting gasoline vapors pass through a pressure regulating mechanism into the throat of a venturi through which engine intake air flows. The pressure regulating mechanism maintains a zero pressure differential between the gasoline vapors and the intake air at the entry ports to the venturi. This causes the fuel flow to be essentially proportional to airflow and thus produces a nearly constant air-fuel ratio independent of airflow as discussed in the following section. After passing through the venturi, the air and vaporized fuel are homogeneously mixed and subsequently enter the engine intake system through a suitable throttle. In order to compensate for variations in air-fuel ratio arising from changes in temperature, fuel composition, etc., feedback from an exhaust gas sensor is used to vary the area of the fuel metering orifice and thereby automatically maintain the desired air-fuel ratio. Since exhaust heat is generally not available prior to starting the engine, a supplementary heater is employed to vaporize the gasoline required to start and operate the engine until sufficient vapors are available from the exhaust heat exchanger. Provision is made to collect any gasoline condensate which is produced during the warm-up period and recirculate it back to the vaporizer without contaminating the main fuel supply.

BASIC METERING CONCEPT

The basic fuel metering element of the vaporized gasoline system is the venturi section shown in Figure 6. Engine intake air flows through this venturi and causes a pressure depression at the throat which draws in vaporized gasoline through the fuel nozzle located in the center of the venturi. When the vaporized gasoline and intake air are properly combined, the resulting homogeneous mixture will flow uniformly to all cylinders of the engine with negligible intake manifold wall-wetting and hence minimal time-fluctuations in air-fuel ratio. If properly implemented, the fuel metering venturi will produce an essentially constant air-fuel ratio independent of mass airflow through the venturi, and will thus result in the elimination of air-fuel ratio variations during transient engine operation. The necessary conditions required to produce the constant air-fuel ratio can be determined by examining the following expression which describes the air-fuel ratio for the metering venturi: (6)

$$\frac{A}{F} = K \left(\frac{A_A}{A_F}\right)\left(\frac{P_A}{P_F}\right)\left(\frac{T_A}{T_F}\right)^{-\frac{1}{2}} \left[\frac{\left(\frac{P_T}{P_A}\right)^{\frac{2}{\gamma_A}} - \left(\frac{P_T}{P_A}\right)^{\frac{\gamma_A+1}{\gamma_A}}}{\left(\frac{P_T}{P_F}\right)^{\frac{2}{\gamma_F}} - \left(\frac{P_T}{P_F}\right)^{\frac{\gamma_F+1}{\gamma_F}}} \right]^{\frac{1}{2}}$$

The derivation of this equation with definitions of the nomenclature used is given in Appendix A.

Referring to the above expression, if the fuel supply pressure P_F is made equal to the air supply pressure P_A, then

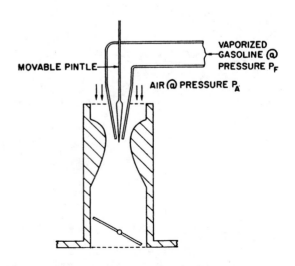

Fig. 6-Vapor system metering venturi

variations in the air-fuel ratio as a function of the venturi throat pressure P_T (and hence airflow) can be made quite small for the proper choice of the P_T range. This is illustrated in Figure 7 which shows air-fuel ratio as a function of airflow for an airflow range of 60 pph to 1200 pph. (This airflow range is typical for a 351 CID engine operating from idle to wide-open throttle.) The venturi cross-sectional area used to derive the plot of Figure 7 was chosen to provide values of P_T which were depressed from P_A by 0.1 inches of water at 60 pph and 45 inches of water at 1200 pph. If higher depression values for P_T were used, the variation in air-fuel ratio would be greater. Before discussing the implications of these small depression values, it should be pointed out that the actual air-fuel ratio established by the metering venturi is a function of the ratio of the air cross-sectional area A_A and the fuel cross-sectional area A_F. Either or both of these areas could thus be used to set the desired air-fuel ratio value as well as to provide a feedback trim mechanism to compensate for temperature variations, etc. In the basic metering venturi shown in Figure 6, adjustment of the fuel cross-sectional area is provided by movement of the tapered pintle rod within the fuel discharge nozzle.

As indicated above, in order for the metering venturi to yield an essentially constant air-fuel ratio independent of airflow, the fuel vapor supply pressure has to equal the air supply pressure, and the venturi throat depression has to be very small for low airflow values. To meet these requirements, a very accurate fuel pressure regulator is required which is capable of operating at the high temperatures necessary to vaporize gasoline. (A variable area venturi having a constant air-to-fuel area ratio could conceivably be used to relax these requirements, and such a device is being explored.) The pressure regulator selected for use in a laboratory evaluation of the vaporized gasoline metering system is a simple bladder-type regulator whose volume automatically changes to maintain its interior pressure equal to exterior pressure. In use, the bladder would have an input and an output port separated by an appropriate baffle structure, and vaporized gasoline would be supplied to the input port in a coarsely controlled manner so as to keep the bladder partially full. The output port would be connected to the fuel nozzle in the metering venturi and would deliver vaporized gasoline at a pressure equal to that exerted on the bladder. Since the air supply pressure for a conventional internal combustion engine is simply atmospheric pressure (neglecting the air cleaner), such a pressure regulating bladder with its exterior surface exposed to atmospheric pressure will make $P_F = P_A$. If an air cleaner is employed, a housing placed over the bladder and referenced to the actual inlet pressure of the metering venturi will insure this condition.

EXPERIMENTAL SYSTEM

The basic vaporized gasoline metering concept discussed above has been implemented on a 351W V-8 engine coupled to a laboratory dynamometer. A diagrammatic representation of the complete system is shown in Figure 8. Referring to this diagram, operation of the system can be described as follows: Fresh gasoline is pressure fed from a main fuel tank to a small holding tank through a conventional float-actuated valve. The liquid gasoline in the holding tank is pumped through an electronically controlled coarse metering valve into a heat exchanger located in the engine exhaust system. The metering valve employed is a conventional electronic fuel injector whose "on" time is automatically controlled to regulate the fuel flow through the heat exchanger and hence the amount of gasoline vapors which are generated. The heat exchanger used is a conically shaped stainless steel tube helix having a total surface area of approximately 70 square inches, and is located inside the normal exhaust pipe just downstream from the "Y".

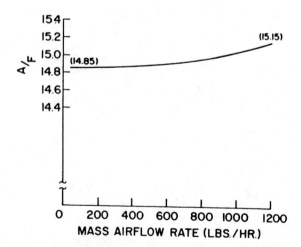

Fig. 7—Air-fuel ratio versus mass airflow rate for vapor metering system

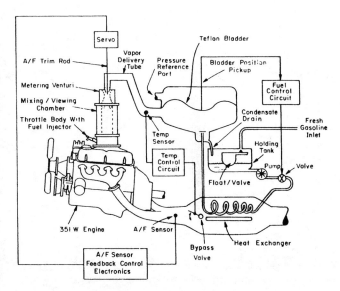

Fig. 8-Diagrammatic representation of vaporized gasoline delivery system

The gasoline vapors generated in the heat exchanger flow into the variable volume pressure regulator previously discussed and cause the bladder to billow up. The resulting displacement is sensed by a pickup whose output is fed back to the coarse fuel control and is used to automatically keep the bladder approximately half full of gasoline vapors. The particular bladder employed has a maximum volume of approximately 0.15 cubic foot and is constructed of 1 mil Teflon® PFA film which has a melting point of approximately 600°F. Since the gasoline currently being used in the laboratory is completely vaporized at approximately 400°F, the vapor temperature at the pressure regulator outlet is maintained at approximately 420°F by a simple closed-loop exhaust bypass control which regulates the amount of heat supplied to the heat exchanger. Any gasoline which condenses on interior surfaces of the pressure regulator and associated plumbing during warm-up is returned to the small holding tank and is subsequently re-vaporized. In this manner, the heavy gasoline fractions will not build up in the main fuel tank, but will be recirculated through the heat exchanger and finally consumed when the proper operating temperature is reached.

The outlet vapors from the pressure regulator pass through an insulated delivery tube and are discharged coaxially into the throat of the metering venturi. The venturi employed has a throat diameter of 1.3 inches while the fuel discharge nozzle has an orifice diameter of 0.31 inch. The fuel discharge nozzle is heated electrically to prevent cooling by the intake air which would otherwise cause condensation of fuel vapors on the nozzle. A tapered pintle capable of being positioned within the fuel nozzle is used to vary the orifice area and thus the air-fuel ratio. This pintle is connected to a servomechanism which can control the pintle position using feedback from an exhaust gas sensor located in the exhaust system.

The venturi is connected to the engine intake manifold through a mixing/viewing chamber mounted above a conventional butterfly-valve throttle body. The mixing/viewing chamber consists of a seven inch long cylindrical tube attached directly to the venturi exit port and mounted inside a somewhat larger air-tight chamber. The chamber itself, which is physically fastened to both the venturi and the throttle body, contains two viewing windows which make it possible to visually examine the outlet end of the venturi extension tube while the engine is running. At the point where the extension tube connects to the venturi, a circular swirling section having canted fins around its circumference and a hole in its center is located inside the tube in order to promote mixing of the air and fuel. This particular design allows the pure gasoline vapors to pass through the center hole and avoid condensation on the cool swirling fins*, but imparts sufficient turbulence to the air to encourage downstream mixing of the air and fuel.

In order to expedite the initial fabrication and evaluation of the vaporized gasoline metering system, an electronic fuel injector was installed in the throttle body and is used routinely for cold engine starts. Vapors can be used to start the engine when cold, however, by employing an auxiliary vaporizer such as a battery-powered heater. One such system which was implemented uses a 500 watt electric vaporizer during engine cranking to supply gasoline vapors directly to a metering valve in the throttle body. As soon as the engine starts, a 2 KW electric vaporizer is automatically energized which fills the pressure regulator with gasoline vapors and enables normal fuel metering through the venturi nozzle instead of the throttle body. After approximately 20 seconds of operation using the electric

---

*At atmospheric pressure, pure gasoline vapor has a dew point of ≈ 400°F while a mixture of air and gasoline vapor with an air-fuel ratio of 15:1 has a dew point of ≈ 125°F.

vaporizer, sufficient exhaust heat is available to permit operation of the normal exhaust system vaporizer in place of the electric unit.*

INITIAL EXPERIMENTAL RESULTS

The initial laboratory evaluation of the vaporized gasoline metering system was performed to verify the anticipated system advantages previously noted in this paper. To be specific, it was anticipated that the open loop vaporized gasoline system would (1) exhibit minimal steady state high-frequency** time-fluctuations in air-fuel ratio, (2) essentially eliminate the transient variations in air-fuel ratio due to airflow changes, and (3) provide a very uniform cylinder-to-cylinder distribution of air-fuel ratio. The evaluation, which was performed using a 351W V-8 engine coupled to an absorption dynamometer, did in fact substantiate the expected results. Specifically, the open loop vapor system exhibited steady state time-fluctuations in air-fuel ratio of less than $\pm 1\%$ for a wide range of engine operating loads and air-fuel ratios. Furthermore, the system displayed transient variations in air-fuel ratio of less than $\pm 1\%$ for step changes in airflow exceeding 400%. Finally, the system consistently provided cylinder-to-cylinder air-fuel ratio distributions of within $\pm 0.75\%$ for cylinders fed from each plane of the dual plane manifold used on the 351W engine.

The steady state and transient air-fuel ratio values reported above were measured with a TiO_2 exhaust gas sensor having a time constant of approximately 0.25 seconds. (7) A typical time recording of the air-fuel ratio along with the corresponding engine torque is shown in Figure 9. In an effort to corroborate these results, similar measurements were made using an NDIR CO analyzer to indicate air-fuel ratio variations. Since the response time of the CO analyzer was much slower than the TiO_2 sensor, the resulting recordings did not reveal the rapid high-frequency fluctuations in air-fuel ratio observed with the TiO_2 sensor, but did

---

*The electric vaporizer has been used for "chokeless" cold starts at 70°F ambient temperature and air-fuel ratios near stoichiometry.

**In this context, high-frequency refers to values which are too high to be eliminated by feedback from an exhaust gas sensor.

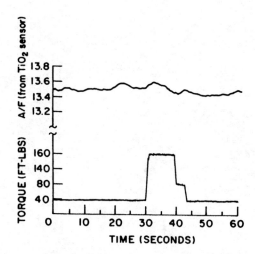

Fig. 9-Air-fuel ratio and engine torque versus time for 351W engine operating at 2000 rpm with open loop vapor system

show longer term fluctuations due to both temperature and airflow variations. A typical time recording of such an air-fuel characteristic together with the corresponding engine torque is shown in Figure 10. The use of feedback from an exhaust gas sensor to eliminate the low-frequency fluctuations in air-fuel ratio has been successfully demonstrated with the vapor system, and a detailed discussion of the feedback work will be included in a future paper.

The cylinder-to-cylinder air-fuel ratio distribution values reported were obtained using specially shaped sample probes located just downstream from each exhaust valve and connected through appropriate switching valves to conventional emission monitoring equipment. A typical cylinder-to-cylinder air-fuel ratio distribution achieved with the vapor system is shown in Figure 11. For comparison, a conventional liquid carburetor having the same venturi area and using the same mixing/viewing chamber as the vapor system was substituted for the vapor system, and a cylinder-to-cylinder distribution was obtained for the same engine operating condition. The resulting characteristic, shown in Figure 12, clearly illustrates the distribution advantage of a vapor system.

LEAN-LIMIT EXPERIMENTAL RESULTS

The vaporized gasoline metering system was originally devised as a scheme to provide very tight control of air-fuel ratio at values slightly rich of stoichiometry for use with NO_x catalysts. This is a very important application of

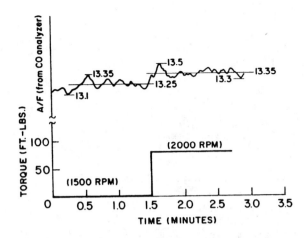

Fig. 10-Air-fuel ratio and engine torque versus time for 351W engine operating at 1500 rpm and 2000 rpm with open loop vapor system

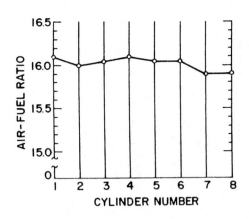

Fig. 11-Air-fuel ratio versus cylinder number for 351W engine operating at 2000 rpm, 40 ft-lb with vapor system

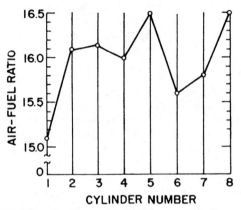

Fig. 12-Air-fuel ratio versus cylinder number for 351W engine operating at 2000 rpm, 40 ft-lb with conventional carburetor

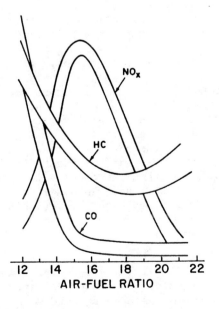

Fig. 13-Qualitative relationship of HC, NO_x, and CO emissions to air-fuel ratio

the vapor system and should be pursued further. However, the ability of the vapor system to provide a very uniform cylinder-to-cylinder distribution of air-fuel ratio with minimal time-fluctuations suggests that the system might also be useful in extending the lean misfire limit of a multi-cylinder engine. This is apparent since the engine operation would not be limited by a single "lean" cylinder as is the usual case. Instead, all cylinders would consistently receive the same air-fuel ratio and hence would be uniformly capable of operating at leaner air-fuel ratios. The use of extended lean-limit operation is an intriguing approach to the control of exhaust emissions, and is based on the relation of such emissions to air-fuel ratio shown qualitatively in Figure 13.

In order to evaluate the potential advantages of lean-limit vapor system operation, a CVS simulation method developed at Ford Motor Company was employed. (8) Basically, this technique utilizes emission and fuel consumption data obtained from steady state engine-dynamometer tests at specific speed-torque points to analytically predict the performance in a complete CVS cycle. The actual speed-torque points used are appropriately chosen to correspond to a particular powertrain-vehicle combination. The fundamental idea behind the simulation technique is that when an actual vehicle is operated over a CVS cycle, a unique trajectory or map is defined in the engine speed-torque-time space. The technique assumes that engine performance along this trajectory can be approximated by steady state operation at discrete speed-torque points for specific intervals of time. The particular speed-torque-time map

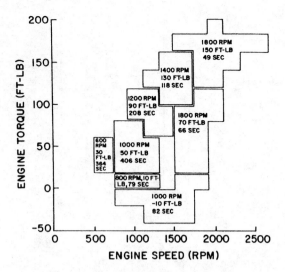

Fig. 14-Speed-torque-time map for CVS simulation of a 351W engine in a 5500 lb vehicle

employed for the lean-limit evaluation of the vapor system is shown in Figure 14. This map, with the particular segmentation shown, was developed for a 351W engine in a 5500 pound vehicle. Such an engine-vehicle combination was chosen for the simulated CVS cycle evaluation of the vaporized fuel metering system because comparable data using both conventional carburetion and electronic fuel injection were available from earlier work performed at Ford.

The previously described vapor system implemented on the 351W engine-dynamometer setup was operated at each of the speed-torque points specified in Figure 14 with the exception of the 1000 RPM, - 10 foot-pound point. This particular point could not be run because it required the use of a motoring dynamometer which was not available for the evaluation. At each speed-torque point explored, measurements of fuel consumption as well as CO, CO_2, HC, NO_x, and O_2 exhaust concentrations were obtained for various values of air-fuel ratio in the lean region. At each air-fuel ratio, the measurements were made for both MBT ignition timing as well as for a retard from MBT timing. The amount of retard used was arbitrarily chosen to give a torque loss of approximately 7% from the MBT value. The resulting torque loss was compensated for by increasing the throttle opening to give the correct torque value. In order to provide adequate combustion initiation at the lean air-fuel ratios, a high energy ignition system* with 0.100 in. gap spark

---

*The particular ignition system employed was a Ferroresonant Capacitive Discharge Ignition System developed at Ford Motor Company. (9)

plugs was used throughout the evaluation.

The experimental data obtained in the lean-limit evaluation of the vapor system were analytically processed and used to produce individual curves of HC and NO_x emissions (in grams) and fuel consumption (in pounds) versus air-fuel ratio for each speed-torque-time point explored. (Since the CO emissions were essentially invariant for the lean air-fuel ratios examined, CO emissions were not included in these plots.) A typical curve of HC, NO_x, and fuel consumption versus air-fuel ratio obtained in the lean-limit evaluation is shown in Figure 15. The complete set of curves for all the speed-torque-time points explored is presented in Appendix B. It should be emphasized that the emission and fuel consumption values given in each curve are calculated from steady state measurements to correspond to the time expended at each speed-torque point as dictated by the CVS simulation technique.

Each of the curves presented in Appendix B was examined to determine a "lean-limit" air-fuel ratio. In general, the particular air-fuel ratio chosen for each curve was a compromise between decreasing NO_x values and increasing HC and fuel consumption values. The results of this determination are tabulated in Figure 16 for the MBT situation and in Figure 17 for the retarded spark situation. The values used for the 1000 RPM, - 10 foot-pound point in these tables were estimates based on previous work with other systems.

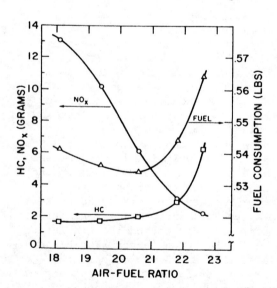

Fig. 15-HC, NO_x, and fuel consumption versus air-fuel ratio for 351W engine operating at 1400 rpm, 130 ft-lb for 118 s with vapor system and MBT timing

SPEED (RPM)	TORQUE (FT-LB)	TIME (SEC)	A/F	HC (GMS)	NOx (GMS)	CO (GMS)	FUEL (LBS)
600	30	364	18.0	6.60	.24	2.70	.333
800	10	79	15.7	2.34	.07	.56	.079
1000	-10	82	15.9	3.49	.015	5.61	.093
1000	50	406	20.0	8.10	1.42	5.94	.727
1200	90	208	22.0	6.60	1.90	5.71	.636
1400	130	118	21.8	2.95	3.13	4.84	.544
1800	70	66	21.5	1.60	.89	3.08	.270
1800	150	49	22.0	2.30	2.91	2.88	.321
		TOTALS		33.98	10.575	31.32	3.003

Fig. 16-Emission and fuel consumption values used for CVS cycle simulation of 351W engine with vapor system in 5500 lb vehicle-lean limit with MBT spark

SPEED (RPM)	TORQUE (FT-LB)	TIME (SEC)	A/F	HC (GMS)	NOx (GMS)	CO (GMS)	FUEL (LBS)
600	30	364	18.0	4.88	.11	2.70	.379
800	10	79	16.7	.92	.03	.56	.091
1000	-10	82	15.9	3.49	.015	5.61	.093
1000	50	406	19.6	7.71	1.02	5.94	.761
1200	90	208	21.0	3.88	1.60	5.71	.664
1400	130	118	21.5	2.50	2.00	4.84	.560
1800	70	66	21.3	1.30	.70	3.08	.275
1800	150	49	22.0	1.71	1.54	2.88	.331
		TOTALS		26.39	7.015	31.32	3.154

Fig. 17-Emission and fuel consumption values used for CVS cycle simulation of 351W engine with vapor system in 5500 lb vehicle-lean limit with retarded spark

The complete CVS cycle prediction for the lean-limit vapor system operation was found by dividing the total HC, NO_x, CO, and fuel consumption values shown in Figures 16 and 17 by the total distance covered in the CVS cycle. The results of this prediction, which apply to a 351W engine in a 5500 pound vehicle, are shown in Figure 18 for both the MBT timing condition and the retarded timing condition. Also included in this figure for comparison are CVS predictions for the same engine-vehicle combination with (1) a vapor system having a constant air-fuel ratio of 19:1, (2) a conventional carburetor having a 1974 production calibration, and (3) an electronic fuel injection system having air-fuel ratio and timing optimized at each speed-torque point to give best fuel economy consistent with reasonable emission levels. Comparison of the results presented in Figure 18 indicates that lean-limit vapor system operation potentially provides

CONFIGURATION	HC (GM/MI)	NOx (GM/MI)	CO (GM/MI)	FUEL (MPG)
VAPOR SYSTEM (Lean Limit-MBT)	4.6	1.4	4.2	15.3
VAPOR SYSTEM (Lean Limit-Retard)	3.5	.94	4.2	14.5
VAPOR SYSTEM (A/F≈19:1-MBT)	3.4	4.7	4.2	15.3
VAPOR SYSTEM (A/F≈19:1-Retard)	2.7	2.2	4.2	14.6
BASELINE CARB (Production Calib)	2.2	3.9	5.1	12.4
EFI (Best Economy)	4.3	3.4	4.4	13.9

Fig. 18-CVS cycle predictions for various configurations used with a 351W engine in a 5500 lb vehicle

appreciable improvements in fuel economy and NO_x emissions, but at the expense of higher HC levels.

SUMMARY

Evaluation of the vaporized gasoline metering system has shown that the system exhibits numerous beneficial characteristics which make it very appealing for use with conventional internal combustion engines. To be specific, it has been demonstrated that the vapor system (1) exhibits minimal steady state high-frequency fluctuations in air-fuel ratio, (2) displays negligible transient variations in air-fuel ratio for changes in engine load, (3) provides very uniform distribution of air-fuel ratio from cylinder to cylinder, and (4) enables cold engine starts at air-fuel ratios close to stoichiometry using vaporized gasoline supplied from an auxiliary electric vaporizer.

The first two characteristics listed above will permit very tight control of air-fuel ratio when coupled with feedback from an exhaust gas sensor. The third characteristic, in addition to the first two, will enable extended lean-limit operation which in turn will result in improvements in fuel economy and NO_x emissions as previously shown. In order for lean-limit operation to be viable, however, a practical method of lowering the HC levels as well as programming the air-fuel ratio as a function of engine load must be provided. The fourth characteristic listed above should result in significantly lower emission levels during the warm-up period following a cold engine start.

CONCLUSIONS

The favorable characteristics which have been demonstrated with the vaporized gasoline metering system justify its continued development as an alternative to more conventional fuel metering systems. It should be emphasized that the system described in this paper is an experimental one, however, and many unexplored areas must be investigated before production feasibility can be established. These unexplored areas include actual vehicle emission testing, low and high temperature starting and operation, practical component design and durability, and overall system safety.

ACKNOWLEDGEMENTS

The authors gratefully acknowledge the assistance and collaboration in this project of L. R. Foote, W. D. Plensdorf, and J. D. Zbrozek of the Ford Motor Company Engineering and Research Staff.

REFERENCES

1. J. Rivard, "Closed Loop Electronic Fuel Injection Control of the Internal Combustion Engine." Paper 73005 presented at the SAE International Automotive Engineering Congress, Detroit, January 1973.
2. R. Zechnall, G. Baumann, and H. Eisele, "Closed Loop Exhaust Emission Control System with Electronic Fuel Injection." Paper 730556 presented at the SAE Automobile Engineering Meeting, Detroit, May 1973.
3. M. Hubbard, Jr., and J. D. Powell, "Closed Loop Control of Internal Combustion Engine Exhaust Emissions." Stanford University Report, SUDAAR No. 473, February 1974.
4. T. Y. Tien, H. L. Stadler, E. F. Gibbons, and P. J. Zacmanidis, "TiO_2 as an Air to Fuel Ratio Sensor for Automobile Exhausts." The American Ceramic Society Bulletin, Vol. 54, No. 3, March 1975.
5. A. A. Zimmerman, L. E. Furlong, H. F. Shannon, "Improved Fuel Distribution - A New Role for Gasoline Additives." Paper 720082 presented at the SAE International Automotive Engineering Congress, Detroit, January 1972.
6. J. E. Hyland, "Venturi Metering Considerations for a Vapor Carburetor." Ford Scientific Research Staff Technical Report (to be published).
7. E. F. Gibbons, A. H. Meitzler, L. R. Foote, P. J. Zacmanidis, and G. L. Beaudoin, "Automotive Exhaust Sensors Using Titania Ceramics." Paper 750224 presented at the SAE International Automotive Engineering Congress, Detroit, February 1975.
8. P. N. Blumberg, "Powertrain Simulation: A Tool for the Design and Evaluation of Engine Control Strategies." Paper to be presented at the SAE International Automotive Engineering Congress, Detroit, February 1976.
9. J. R. Asik and B. Bates, "The Ferroresonant Capacitor Discharge Ignition (FCDI) System: A Multiple Firing CD Ignition with Spark Discharge Sustaining Between Firings." Paper to be presented at the SAE International Automotive Engineering Congress, Detroit, February 1976.

APPENDIX A

VENTURI METERING CONSIDERATIONS
FOR A VAPOR CARBURETOR

NOMENCLATURE

A/F = the air-fuel ratio
A = the cross-sectional area at the venturi throat
A_A = the air cross-sectional throat area
A_F = the fuel cross-sectional throat area
A_o = the cross-sectional area at the zero velocity state
C_P = the fluid specific heat at constant pressure
g_c = a proportionality constant
h_o = the enthalpy of the fluid at zero velocity
h = the enthalpy of the fluid at the venturi throat
K = a constant = .2231 for gasoline and air
m = the mass of the fluid
$\dot{M}$ = the fluid mass flow rate
$\dot{M}_A$ = the air mass flow rate
$\dot{M}_F$ = the fuel vapor mass flow rate
P = the pressure at the venturi throat
P_A = the air pressure at the zero velocity state (supply pressure)
P_F = the fuel pressure at the zero velocity state (supply pressure)
P_o = the pressure at the zero velocity state
P_T = the pressure at the venturi throat

R = the individual gas constant
R_A = the gas constant for air
R_F = the gas constant for fuel vapor
T = the absolute downstream or throat temperature
T_A = the absolute upstream air temperature
T_F = the absolute upstream fuel vapor temperature
T_o = the absolute upstream temperature
u = the internal energy at the venturi throat
u_o = the internal energy at the zero velocity state
v = the velocity at the venturi throat
v_o = the velocity at the stagnation point
V = the volume at the venturi throat
V_o = the volume at the zero velocity state
Z = the elevation at the venturi throat
Z_o = the elevation at the zero velocity state
γ = the specific heat ratio (C_p/C_v) for gas
γ_A = the specific heat ratio (C_p/C_v) for air
γ_F = the specific heat ratio (C_p/C_v) for fuel vapor
ρ = the density at the venturi throat
ρ_o = the density at the zero velocity state

SUBSONIC MASS FLOW THROUGH A VENTURI METER

The behaviour of the mass flow per unit time of a gas through a venturi meter can be predicted given the following assumptions: 1) the fluid in question is assumed to obey the perfect gas law and 2) the flow may be treated as isentropic one dimensional steady flow of a compressible fluid. Such a system is shown in Figure A-1 where the subscripted quantities refer to conditions in a large reservoir upstream of the venturi and the unsubscripted quantities refer to conditions at the throat of the venturi.

The first law of thermodynamics (conservation of energy) states that

$$u_o + P_o V_o + \frac{v_o^2}{2g_c} + mg_c Z_o =$$
$$u + PV + \frac{v^2}{2g_c} + mg_c Z \quad (A-1)$$

For the system being evaluated,

$$v_o \simeq 0$$

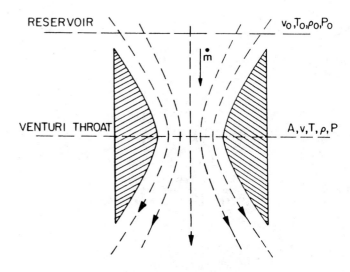

Fig. A-1-Basic venturi meter

$$Z_o \simeq Z$$

Using these conditions and the definition of enthalpy (h = u + PV), Equation A-1 reduces to

$$h_o = h + \frac{v^2}{2g_c} \quad (A-2)$$

For a perfect gas, the following conditions hold:

$$C_P = \frac{h_o - h}{T_o - T}$$

and

$$C_P = \frac{\gamma}{\gamma - 1} R$$

Furthermore, for a perfect gas during an isentropic process, it can be shown that

$$\frac{T}{T_o} = \left(\frac{P}{P_o}\right)^{\frac{\gamma-1}{\gamma}}$$

Substituting these relationships into Equation A-2 and solving for v yields

$$v = \left\{\frac{2g_c \gamma R T_o}{\gamma - 1}\left[1 - \left(\frac{P}{P_o}\right)^{\frac{\gamma-1}{\gamma}}\right]\right\}^{\frac{1}{2}} \quad (A-3)$$

The continuity equation (conservation of mass) states that

$$\rho_o v_o A_o = \rho v A = \dot{M} \qquad (A-4)$$

Substituting Equation A-3 into Equation A-4 yields

$$\dot{M} = \rho A \left\{ \frac{2g_c \gamma RT_o}{\gamma - 1} \left[1 - \left(\frac{P}{P_o}\right)^{\frac{\gamma-1}{\gamma}} \right] \right\}^{\frac{1}{2}} \qquad (A-5)$$

Again for a perfect gas during an isentropic process, it can be shown that

$$\rho = \frac{P_o}{RT_o} \left(\frac{P}{P_o}\right)^{\frac{1}{\gamma}}$$

Substituting this relationship into Equation A-5 and rearranging gives the desired form of the mass flow rate equation:

$$\dot{M} = \frac{P_o A}{T_o^{1/2}} \left[\frac{2g_c \gamma}{(\gamma - 1)R} \right]^{\frac{1}{2}}$$

$$\cdot \left[\left(\frac{P}{P_o}\right)^{\frac{2}{\gamma}} - \left(\frac{P}{P_o}\right)^{\frac{\gamma+1}{\gamma}} \right]^{\frac{1}{2}} \qquad (A-6)$$

Equation A-6 is only valid for subsonic flow; i.e., when the ratio of static to total pressure at the venturi throat (P/P_o) is greater than the critical pressure ratio.* When the critical pressure ratio is reached, the velocity of mass flow at the venturi throat becomes sonic and, by definition of sonic flow, the maximum mass flow rate for fixed area and upstream conditions is attained.

METERING PRINCIPLE APPLIED TO TWO GAS PHASE FLUIDS

From Equation A-6, the mass flow rate equation for air through a venturi meter is

---

*The critical pressure ratio is defined as

$$\left(\frac{P}{P_o}\right)_{CRIT} = \left(\frac{2}{\gamma + 1}\right)^{\frac{\gamma}{\gamma-1}}$$

$$\dot{M}_A = \frac{P_A A_A}{T_A^{1/2}} \left[\frac{2g_c \gamma_A}{(\gamma_A - 1)R_A} \right]^{\frac{1}{2}}$$

$$\cdot \left[\left(\frac{P_T}{P_A}\right)^{\frac{2}{\gamma_A}} - \left(\frac{P_T}{P_A}\right)^{\frac{\gamma_A+1}{\gamma_A}} \right]^{\frac{1}{2}} \qquad (A-7)$$

When air flows through a venturi with a constant upstream pressure P_A, Equation A-7 states that a pressure $P_T < P_A$ is experienced at the throat of the venturi. As the mass flow rate $\dot{M}_A$ increases, the pressure at the throat decreases; this is the basis of the metering principle of the venturi. Referring to Figure 6, if a fuel vapor nozzle is placed with its opening at the venturi throat, the throat pressure $P_T = f(\dot{M}_A)$ can be used to meter the mass flow rate of fuel vapor as a function of mass flow rate of air. Accordingly, from Equation A-6, the mass flow rate of fuel vapor is

$$\dot{M}_F = \frac{P_F A_F}{T_F^{1/2}} \left[\frac{2g_c \gamma_F}{(\gamma_F - 1)R_F} \right]^{\frac{1}{2}}$$

$$\cdot \left[\left(\frac{P_T}{P_F}\right)^{\frac{2}{\gamma_F}} - \left(\frac{P_T}{P_F}\right)^{\frac{\gamma_F+1}{\gamma_F}} \right]^{\frac{1}{2}} \qquad (A-8)$$

Since the air-fuel ratio at the venturi throat is equal to the ratio of the mass of air to the mass of fuel, it follows from Equations A-7 and A-8 that

$$\frac{A}{F} = \frac{\dot{M}_A}{\dot{M}_F} = \left(\frac{P_A}{P_F}\right)\left(\frac{A_A}{A_F}\right)\left(\frac{T_F}{T_A}\right)^{\frac{1}{2}} \left[\frac{\gamma_A(\gamma_F - 1)R_F}{\gamma_F(\gamma_A - 1)R_A} \right]^{\frac{1}{2}}$$

$$\cdot \left[\frac{\left(\frac{P_T}{P_A}\right)^{\frac{2}{\gamma_A}} - \left(\frac{P_T}{P_A}\right)^{\frac{\gamma_A+1}{\gamma_A}}}{\left(\frac{P_T}{P_F}\right)^{\frac{2}{\gamma_F}} - \left(\frac{P_T}{P_F}\right)^{\frac{\gamma_F+1}{\gamma_F}}} \right]^{\frac{1}{2}}$$

Assuming that γ_A and γ_F are constant over the temperature range of interest, the following constant is defined:

$$K = \left[\frac{\gamma_A(\gamma_F - 1)R_F}{\gamma_F(\gamma_A - 1)R_A} \right]^{\frac{1}{2}}$$

Therefore, the air-fuel ratio for a metering venturi is

$$\frac{A}{F} = K \left(\frac{A_A}{A_F}\right)\left(\frac{P_A}{P_F}\right)\left(\frac{T_A}{T_F}\right)^{-\frac{1}{2}} \left[\frac{\left(\frac{P_T}{P_A}\right)^{\frac{2}{\gamma_A}} - \left(\frac{P_T}{P_A}\right)^{\frac{\gamma_A+1}{\gamma_A}}}{\left(\frac{P_T}{P_F}\right)^{\frac{2}{\gamma_F}} - \left(\frac{P_T}{P_F}\right)^{\frac{\gamma_F+1}{\gamma_F}}}\right]^{\frac{1}{2}}$$

APPENDIX B

The curves of HC, NO_x, and fuel consumption versus air-fuel ratio obtained in the lean limit evaluation of the 351W engine equipped with the vaporized fuel metering system are shown in Figures B-1 through B-14. These figures are presented on the following pages.

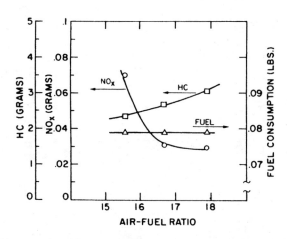

Fig. B-3-800 rpm, 10 ft-lb, 79 s operating point with MBT timing

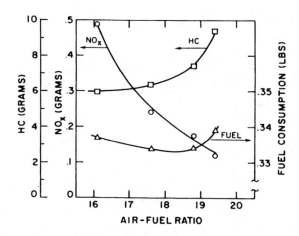

Fig. B-1-600 rpm, 30 ft-lb, 364 s operating point with MBT timing

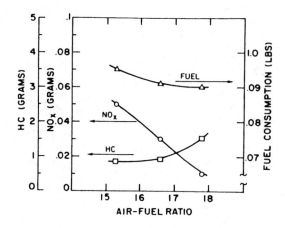

Fig. B-4-800 rpm, 10 ft-lb, 79 s operating point with retarded timing

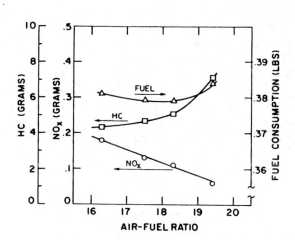

Fig. B-2-600 rpm, 30 ft-lb, 364 s operating point with retarded timing

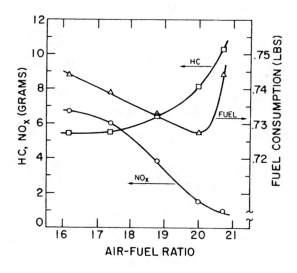

Fig. B-5-1000 rpm, 50 ft-lb, 406 s operating point with MBT timing

439

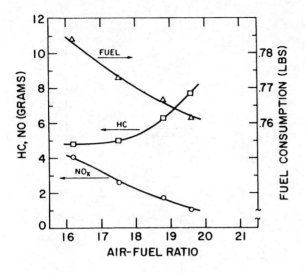

Fig. B-6-1000 rpm, 50 ft-lb, 406 s operating point with retarded timing

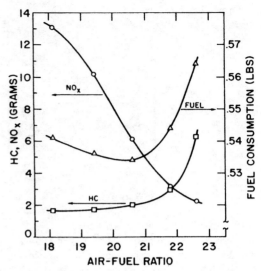

Fig. B-9-1400 rpm, 130 ft-lb, 118 s operating point with MBT timing

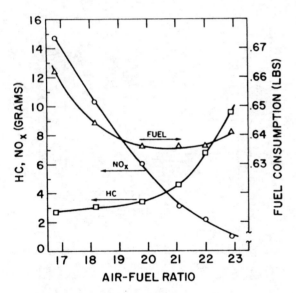

Fig. B-7- 1200 rpm, 90 ft-lb, 208 s operating point with retarded timing

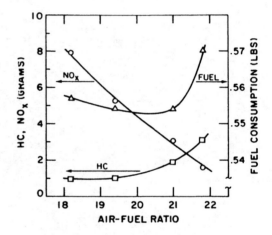

Fig. B-10-1400 rpm, 130 ft-lb, 118 s operating point with retarded timing

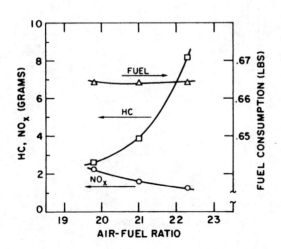

Fig. B-8-1200 rpm, 90 ft-lb, 208 s operating point with retarded timing

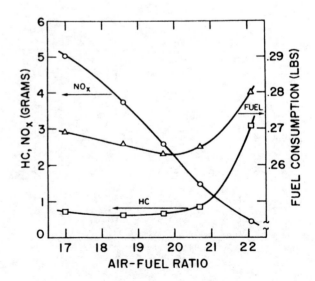

Fig. B-11-1800 rpm, 70 ft-lb, 66 s operating point with retarded timing

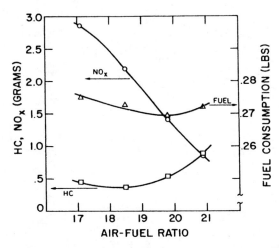

Fig. B-12-1800 rpm, 70 ft-lb, 66 s operating point with retarded timing

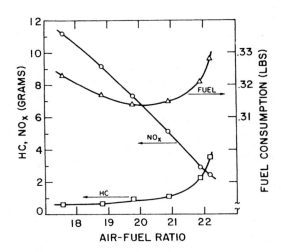

Fig. B-13-1800 rpm, 150 ft-lb, 49 s operating point with MBT timing

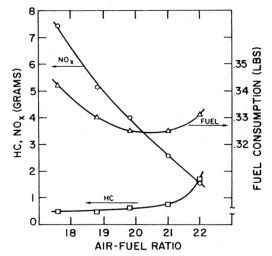

Fig. B-14-1800 rpm, 150 ft-lb, 49 s operating point with retarded timing

770352

A Feedback Controlled Carburetion System Using Air Bleeds

T. R. Gantzert, D. L. Hicks, and M. A. Jefferis
Carter Carburetor Div., ACF Industries, Inc.

IF FEDERAL EMISSION STANDARDS FOR the early 1980's are to be achieved, carburetors must be developed or modified to deliver a more precisely controlled air-fuel ratio so that exhaust gases will be of uniform composition. Three-way and dual-bed catalytic converters are highly efficient in reducing emissions only if the air-fuel ratio is held constant. Figure 1 shows ideal efficiencies. Excessively rich or lean ratios decrease catalytic converter efficiency.

When Carter began development of a feedback or closed loop carburetion system, the approach chosen was to control air bleeds in the low and high speed carburetor fuel circuits. Other approaches are available and were considered. (1)*

This paper explains the reason for choosing the air metering method and the basic concepts of the control system. The preliminary prototype phase of the program was successful based on engineering tests and on system evaluation by various automotive manufacturers in July and August of 1976. Figure 2 shows a schematic of components.

If a carburetor could be calibrated to deliver a precise air-fuel ratio under all conditions for the duration of its functional life, there would not be a need for a feedback system. However, there are a number of conditions which cause undesirable variations in the air-fuel ratio delivered by a carburetor and which cannot be compensated for by the vehicle operator or by preadjustment.

Two objectives had to be met in order to meet the requirements of the catalytic converter. First, the carburetor had to be calibrated to provide a relatively constant air-fuel mixture. Second, an electro-mechanical system had to be developed to automatically compensate for internal and external variations in air-fuel ratio caused by changes in engine speed and load, barometric pressure, carburetor and air temperature, humidity and fuel composition and viscosity. (2) Variations are also introduced by inherent manufacturing tolerances of the engine and carburetor as well as wear on individual components.

---

*Numbers in parentheses designate References at end of paper.

ABSTRACT

A feedback controlled carburetion system has been developed that maintains a flow of exhaust gases of uniform composition. This is a requirement if three-way or dual bed catalytic converters are to be used in meeting projected emission standards. Exhaust gas uniformity depends upon delivery of a constant air-fuel ratio by the carburetor. Instead of metering fuel directly, Carter Carburetor finds that precise and responsive control of the air-fuel ratio is obtained by using variable air bleeds in the carburetor fuel circuits.

DESIGN CONCEPTS

The major factor leading to the selection of the variable air bleed control method was the development at Carter of carburetor altitude compensation using variable air bleeds. (3) This provided experience in controlling the air-fuel ratio by bleeding air into the main and idle fuel circuits as well as providing hardware for development work.

The variable air bleeds consist of tapered metering pins positioned in orifices by a linear solenoid. See Figure 3. This drive mechanism moves the pins in defined steps in response to signals from the electronic control unit. A sensor in the exhaust stream provides the input signal to the electronic control unit. This signal varies with air-fuel ratio. See Figure 6. The solenoid moves the pins until the exhaust sensor indicates that the desired air-fuel ratio has been reached. Thus, the pin movement adjusts the air-fuel ratio to compensate for changes detected in the exhaust gases.

This approach allows Carter carburetors, with minor changes, to be adapted for feedback control and allows the air metering unit to be mounted either integrally with the carburetor or remotely.

In place of the linear solenoid, other electro-mechanical drives such as a stepper-motor or a vacuum solenoid can be used to position the air metering pins, depending on the electronics/carburetor interface requirements.

CARBURETOR OPERATION

As can be seen in Figure 4, the basic carburetor contains two fuel supply subsystems, the high-speed (main) circuit and the low-speed (idle) circuit. The high-speed circuit meters fuel with a tapered metering rod positioned in the jet by the throttle. Fuel is metered into the nozzle (main) well where air from the feedback controlled variable air bleed is introduced. Since this air is delivered above the fuel level, it reduces the vacuum signal on the fuel consequently reducing the amount of fuel delivered from the nozzle.

The idle circuit is needed at low air flows through the venturi because there is insufficient vacuum at the nozzle to draw fuel into the air stream. After leaving the main jet, fuel is supplied to the idle circuit by the low-speed jet. It is then mixed with air from the first idle bleed, accelerated through the channel restriction and mixed with additional air from the second idle bleed before being discharged from the idle ports below the throttle. Air from the variable air bleed is generally introduced between the first idle bleed and the channel restriction. This air reduces the vacuum signal on the low-speed jet and, consequently, the amount of fuel delivered to the idle circuit.

All power enrichment conditions override the feedback operation.

RANGE OF CONTROL

The basic purpose of feedback control is to provide a narrow air-fuel ratio band so that the catalytic converter will operate at maximum efficiency.

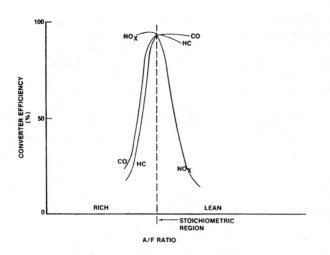

Fig. 1 - Ideal three-way catalytic converter efficiency

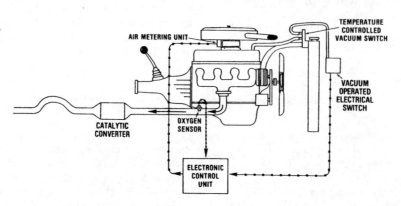

Fig. 2 - Feedback system

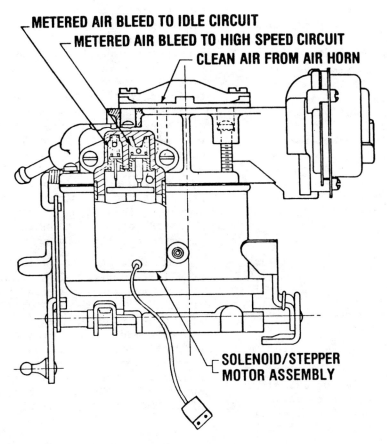

Fig. 3 - Air metering unit

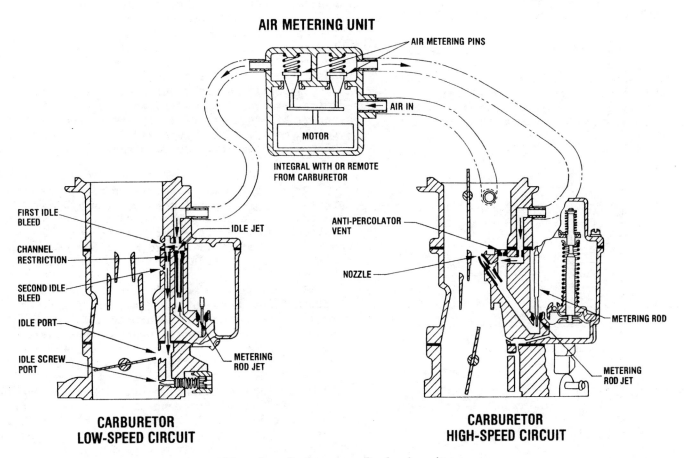

Fig. 4 - Carburetor fuel circuits

It was determined from experience and consultation with OEM customers that a feedback system might require a range of authority of approximately four air-fuel ratios. This range, in the present system, corresponds to control from two ratios leaner to two ratios richer than stoichiometry. This division is arbitrary and can be altered to any proportion desired.

Prior flow test work using carburetors with aneroid controlled altitude compensation shows that satisfactory control can be maintained to 7500 feet (2286m). Figure 5 (3). With feedback control it is expected that more precise compensation can be achieved at altitude.

COMPONENT DESCRIPTION

Oxygen Sensor - It is known that the partial pressure of oxygen in the exhaust reflects the air-fuel ratio with considerable accuracy. (4) Oxygen sensors are commercially available from several sources. Recent development work by manufacturers on the stabilized zirconia type has increased their usable life to a point where they are feasible for automotive use. (5) This type of sensor produces an output voltage that varies with the air-fuel ratio as shown in Figure 6. It is located in the exhaust pipe approximately six inches from the exhaust manifold flange. Care must be taken that maximum operating temperature does not exceed 900°C (1652°F) to prevent sensor damage.

Electronics - Figure 7 is a block diagram of the basic electronics. The sensor output signal is amplified by a buffer-amplifier and converted by logic circuitry into a digital count-up or count-down signal. Counting occurs at a rate determined by a clock oscillator and can be set to match vehicle and catalytic converter characteristics.

The counter/memory circuitry simultaneously stores the current count and decodes it to supply a control signal to the output amplifier(s). The configuration of the amplifier(s) is determined by the type of mechanism selected to drive the metering pins.

When the circuitry detects wide-open throttle, cold starts or other predetermined conditions, it stops the counter. This, in effect, switches the feedback system to the open loop mode until operating conditions permit return to automatic (closed loop) control. During periods of open loop operation the solenoid retains the position it held just prior to entering the open loop mode.

A variable rate switch permits the system to effect a change in air-fuel ratio more quickly under certain conditions. For example, if it is determined empirically that a faster rate of air-fuel change is desirable during light-to-medium accelerations, the switch can be configured to close at the beginning of the acceleration. The amount of increase in clock oscillator frequency that occurs when the switch closes is adjustable, as is the duration of the frequency increase.

Because the actual characteristics of catalytic converters vary slightly from the ideal characteristics shown in Figure 2, it is sometimes necessary or desirable to maintain the carburetor air-fuel ratio at some ratio slightly removed from stoichiometry. A variable bias control contained in the closed loop electronics allows such variation. This control causes the air-fuel ratio excursions which might normally be distributed equally about stoichiometry to be shifted so that the time-average air-fuel ratio is different from stoichiometry. The amount and direction of shift may be selected as required.

Drive Mechanism - The device used at present to move the air metering pins is a linear solenoid that was developed for this application. This actuator has two coils which position the armature according to the voltage being applied by the electronic control unit to each coil.

The armature is attached to a disk upon which the low and high speed air metering pins rest. At zero voltage the armature is in the center or neutral position which is the basic stoichiometric sea level calibration. The metering pins are capable of moving +0.100 inches (2.54mm) from this neutral position. The 0.200 inches (5.08mm) total travel represents a four air-fuel ratio change, e.g., 12.7 to 16.7 and is divided into 32 steps by the electronics. Thus, for each step correction of the electronics the solenoid moves 0.006 inches (0.15mm).

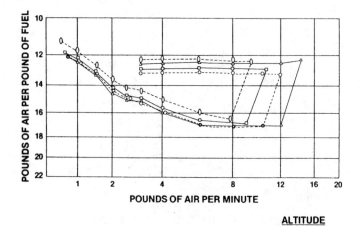

Fig. 5 - Flow curves of altitude compensated carburetor

OPERATING CONDITIONS

Means are provided for the feedback system to automatically disengage (open loop mode) during cold starts and heavy accelerations. Feedback operation under these conditions would counteract the desirable operation of the carburetor.

Two conditions must be met before closed loop operation can begin or resume.

First, the oxygen sensor must be at operating temperature. This is determined by sending a signal to the oxygen sensor to measure its impedance. As sensor temperature increases, the impedance decreases until operating temperature is reached. This signals the electronic control unit that the sensor is prepared for closed loop operation.

The second condition necessary for closed loop functioning is the termination of choke operation. Choke termination is indicated by a specific engine coolant temperature. When the desired coolant temperature is reached, a switch is activated to indicate that choke enrichment is no longer necessary.

With both oxygen sensor and engine coolant at proper operating temperature, the electronic control unit initiates closed loop operation.

In case of a hot start, the system operates as it does for a cold start. Should either sensor or engine coolant temperature have dropped below the required levels, the system

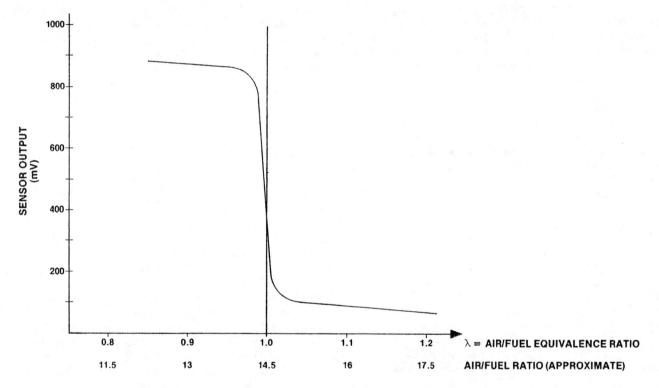

Fig. 6 - Oxygen sensor characteristics

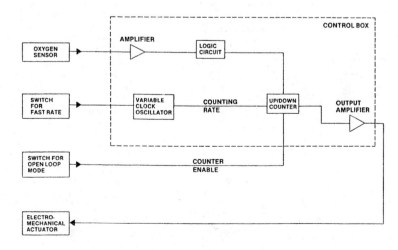

Fig. 7 - Control system block diagram

operates in the open loop mode until both reach their operating levels.

In the case of heavy acceleration or wide open throttle operation, which require power enrichment, the system switches into the open loop mode; this is accomplished by a diaphragm actuated electric switch connected to manifold vacuum. The switch is set to activate at the vacuum level required for power enrichment. When the vacuum rises above this level, the electronic control unit switches back into closed loop operation.

Idle, light-to-medium acceleration, cruise and deceleration modes are all under closed loop control at present.

The system is designed to ignore transient changes in the oxygen sensor output. An occasional misfire, for example, might produce a momentary indication from the oxygen sensor that the carburetor air-fuel ratio is incorrect. An attempt by the system to correct for this erroneous indication is not desirable. The system therefore ignores any indicated air-fuel ratio change which does not last longer than one clock period. See Figure 8 for graphic representation. The clock period is adjusted empirically for each type of vehicle on which the system is installed. Results obtained so far have indicated that a clock period of approximately 0.6 second is near optimum for the six and eight cylinder vehicles that were tested.

SYSTEM EVALUATION

By monitoring the movement of the air metering pins, it is possible to determine the frequency and magnitude of corrections required to maintain a constant air-fuel ratio. The fewer and smaller the corrections, the more precise is the basic carburetor calibration. The instrument used to determine the position of the linear solenoid and, thus, the movement of the metering pins, is a LVDT (Linear Variable Differential Transformer) which is attached to the bottom of the solenoid. Output of the LVDT can be monitored on a meter or recorded on a strip chart as shown in Figures 10 and 11.

At the time this paper was written, metering pin travel did not exceed 0.040 inches (1.02mm) during a 1975 EPA emission test. This represents approximately 0.7 air-fuel ratio.

Figure 9 shows a typical feedgas carbon monoxide trace produced by an engine equipped with a carburetor that was calibrated to stoichiometry. The feedback electronics were disabled during this test.

Figure 10 shows the carbon monoxide trace with the feedback electronics activated. As can be seen, the carbon monoxide is more controlled than in Figure 9.

Figure 11 shows the carbon monoxide trace after maladjustment of the carburetor. The feedback system is activated and the metering pin activity is considerably greater than in Figure 10. The carbon monoxide trace is not as controlled as in Figure 10, but is better than that in Figure 9. Traces were obtained by using a Beckman IR315A nondispersive infrared CO analyzer.

SUMMARY

A practical carburetor feedback system which meters air to control fuel flow has been developed to provide an essentially constant air-fuel ratio. This is a requirement for efficient uses of three-way and dual bed catalytic converters if automobiles are to meet projected emission standards. It has been found that: 1) Air bleed control can be integrated with present Carter carburetors with minor hardware modifications, 2) The system is insensitive to transient air-fuel ratio changes indicated by the oxygen sensor, 3) The system can be tailored to various types of catalytic converters and vehicle combinations, 4) The system provides a more constant air-fuel ratio than is presently obtainable

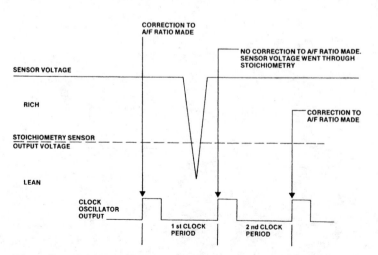

Fig. 8 - System response to O_2 sensor output

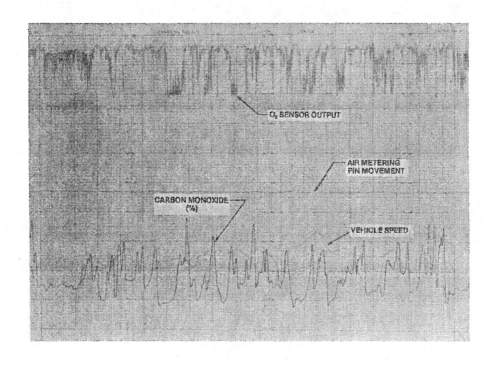

Fig. 9 - Base calibration

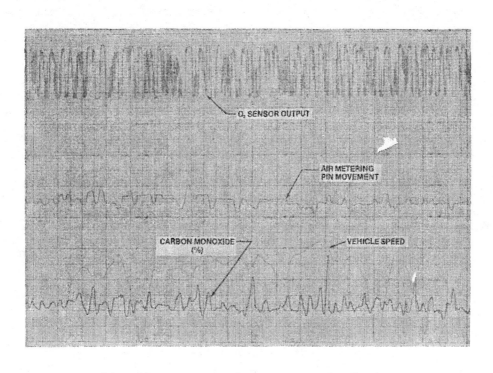

Fig. 10 - Base calibration with feedback

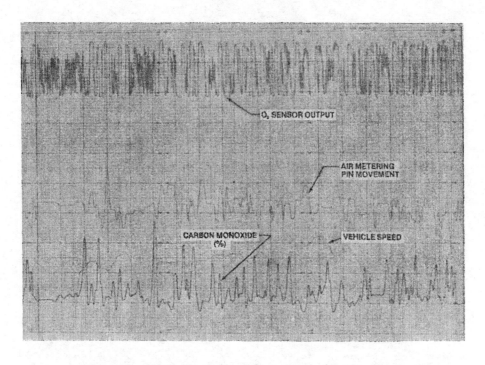

Fig. 11 - Maladjusted carburetor with feedback

with carburetors alone by compensating for internal and external effects.

ACKNOWLEDGEMENTS

The authors wish to thank the people at Carter Carburetor who contributed to this work, in particular, Thomas Byrnes, Jr., Pete Hoekstra, and Mike Sloboda. Also, that of Allen W. Lindberg, P.E., of St. Louis, Missouri.

REFERENCES

1. R. A. Spilski and W. D. Creps, "Closed Loop Carburetor Emission Control System". Paper presented at SAE Automotive Engineering Congress, Detroit, February, 1974.

2. Jay A. Bolt and Michael Boerma, "The Influence of Inlet Air Conditions on Carburetor Metering". Paper 660119 presented at SAE Automotive Engineering Congress, Detroit, January, 1966.

3. Richard C. Wrausmann and Robert J. Smith, "An Approach to Altitude Compensation of the Carburetor". Paper 760286 presented at SAE Automotive Engineering Congress, Detroit, February, 1976.

4. Heinrich Dueker, Karl-Hermann Friese and Wolf-Dieter Haecker, "Ceramic Aspects of the Bosch Lambda-Sensor". Paper 750223 presented at SAE Automotive Engineering Congress, Detroit, February, 1975.

5. J. David Powell, "Closed Loop Control of Internal Combustion Engine Exhaust Emissions" Paper PB-239850 prepared for Department of Transportation, February, 1974.

C244/85

Engine ignition and carburettor electronic control

G TAYLOR, BSc and **M FREELAND**, BSc
Ford Motor Company Limited, Basildon, Essex
R HUNNINGHAUS, MSc and **H STOCKER**, Dipl-Ing
Motorola Automotive and Industrial Electronics Group, Milton Keynes, Buckinghamshire
S BONFIGLIOLI, Dr-Ing
Weber SPA, Bologna, Italy

SYNOPSIS This paper describes a joint development which incorporates a Weber S.P.A. idle speed control concept in a unique electronic control module developed by Motorola AIEG for Ford Motor Company Limited.

Discussed are specific points relating to system concept, control strategy, hardware design, manufacturing and testing techniques.

The use of an industry standard microcomputer provided an easy introduction for engineering staff and the availability of development support equipment allowed extreme flexibility in the control functions performed.

The introduction of this system is believed to be a world first for Ford, there are no known similar packages which incorporate the control of the ignition and carburettor functions, together with a fully electronic pressure transducer and coil driver stage, in an under-hood installation.

1 INTRODUCTION

A single-chip microcomputer based engine management system has been developed as a joint activity between Ford Motor Co Ltd., Motorola Automotive and Industrial Electronics Group and Weber S.P.A. The system is named Electronic Spark Control II (ESC II). Two versions of the ESC II module are in production on the Ford Sierra and the new Granada. The 2.0L OHC carburetted engine features 3D spark control, manifold heater control and closed loop idle speed control (Fig 1). The 1.8L OHC carburetted engine uses a lower feature module which does not have idle speed control, but uses the same spark and manifold heater control strategy.

Other variants include a spark only version with boost control for the Escort Turbo. The boost control was implemented by re-programming the manifold heater control function.

The electronic control module is designed for underhood mounting. The manifold pressure transducer and the coil driver are integrated into the housing thus permitting a compact single package solution.

Four calibration versions exist for 2.0L engines; one each for manual and automatic transmission for the countries of the European community and two for Swedish and Swiss emissions requirements.

2 SYSTEM DESCRIPTION

Figure 2 shows the system components for the 2.0L variant with idle speed control. Engine speed and position information is supplied to the ESC II module through a distributor mounted Hall sensor. Engine load is detected by a pressure transducer located inside the module.

These two inputs together with the engine coolant temperature are used to determine spark and dwell angle. The maximum coil primary current is controlled through the on-board driver stage. A stepper motor driven actuator (Fig. 3) is mounted on the carburettor which provides a variable throttle stop. An idle switch on the actuator detects closed throttle conditions. Switch signals to adjust idle speed control are received from the air conditioning clutch and the automatic transmission. The modules also control two other functions, an intake manifold heater assembly and a power hold relay which maintains the system active for a short time after the ignition is turned off. Additional external electrical connections to the module provide adjustments of the ignition timing for low octane fuels and of the idle speed set point.

3 ELECTRONIC CONTROL MODULE

3.1 Module Block diagram

Fig. 4 shows a block diagram of the ESC II module together with its input and output signals for the 2.0L variant. These are :-

- Power supply and reset logic:

    This provides the required internal voltages and controls the reset and operating mode programming during power up

- Microcomputer:

    The single-chip microcomputer is the Motorola MC 6801U4, a 4 K byte memory version of the 6801 family. The 1.8L spark only and the turbo variants use the Motorola MC 6801 microcomputer with 2K byte memory. Both are widely used industry standard products and have been selected to meet the real time requirements in this application which include:

C244/85 © IMechE 1985

o   engine period measurement

o   pressure transducer signal measurement

o   spark and dwell timing

o   stepper motor control

The availability of EPROM versions provided high flexibility during the development cycle

- Pressure transducer:

   Engine load is determined with a Motorola produced Silicon Capacitive Absolute Pressure (SCAP) transducer which transforms the manifold pressure into a digital frequency. The transducer is mounted on the PC board as a hybrid subassembly.

- Input signal conditioning:

   This section contains the signal level adjustment and protection circuitry for all digital inputs.

- A/D Convertor:

   Microcomputers of the MC 6801 family do not provide on chip analog inputs. Therefore, for the only analog input signal, the engine coolant temperature, a discrete 6 bit A/D convertor was implemented which is controlled by the microcomputer to perform the conversion. The 6 bit A/D accuracy produces a 2°C resolution in the critical control region.

- Coil driver:

   The coil driver function provides current limiting and soft shut down of the current under engine stall condition. The current on/off timing is controlled by the microcomputer.

- Relay drivers:

   These are standard switch outputs to control the intake heater and power hold relays.

- Stepper motor driver:

   The stepper motor drive circuit provides a two level constant current drive to the stepper motor. Step sequence and step frequency are under software control.

All versions of the ESC II module are packaged in the same aluminium die cast housing. The spark only and turbo variants make use of standard leaded resistor and capacitor components. To maintain the same PC board size for the idle speed variant, with its higher functional requirement, it was necessary to introduce leadless chip resistors and capacitors which are mounted on the under side of the PC board. (Fig. 5). This enabled much higher component densities to be achieved. Chip components, or surface mounted devices (SMD's) are, with the exception of resistors, the same components as used in hybrid thick film circuits. The resistors are manufactured by thick film techniques, so all devices have proven reliability.

While chip components are relatively new in Europe they are viewed as standard practice in Motorola's US based automotive products.

3.2 Environmental specification and design considerations

The control module is mounted underhood on the fender wall. The temperature range is specified from -40°C to +100°C. The module being constantly exposed to a harsh environment in terms of vibration, humidity, water splash, salt spray and elevated temperatures required specific design attention to the application of the microcomputer.

The microcomputer used in production is a plastic packaged NMOS part specified for +85°C ambient air temperature operation. When the system design started in 1982 no CMOS parts with the functional capability of the MC 6801 or MC 6801 U4 microcomputers were available at cost effective prices. To enable the use of NMOS parts in this application a heat sinking arrangement for the microcomputer was necessary. This is shown in Fig. 6.

The microcomputer (C) is held against a heatsink (D) by a clamp (B). The heatsink together with the PC board (E) is fastened to mounting studs (A) which form part of the die cast aluminium housing. The heat flow is routed through the module mounting foot (G) to the vehicle metal work to gain benefit of extended heatsinking. With this arrangement the junction temperature is maintained below 150°C, the maximum allowable junction temperature for a plastic part.

This technique of heatsinking the microcomputer has been used by Motorola for several years in other engine controls. This example shows that in order to get the best and most cost effective design it is necessary to look at the complete packaging concept as well as electrical circuit design.

The underhood environment also required appropriate protection of the electronic components. A true hermetic package was determined not to be cost effective, so the PC board is conformally coated and mounted in a splash proof housing.

4 ACTUATOR

Initial consideration was given to either a DC motor driven actuator or one incorporating a stepper motor. If the system was to perform merely a closed loop idle speed control function, the DC motor approach would have been a satisfactory solution in terms of the control accuracy and would have allowed a simpler and cheaper driver configuration in the control module. However, since the system was intended to perform other functions, such as setting throttle position for crank and decel fuelling control, it was necessary to have an actuator position feedback loop. This led the investigation into a stepper motor driven actuator, the position of which could be determined by the controlling microcomputer. This solution was adopted because of the accuracy of its setting, despite the more complicated (and more expensive) drive circuitry required in the module.

The actuator (Fig. 3) itself is a carburettor mounted device developed by Weber for use on their DFTH carburettors, which directly controls the throttle lever. It contains a four

© IMechE 1985 C244/85

phase stepper motor with a permanent magnet rotor. This drives a cam via a toothed belt reduction system. The cam in turn lifts a spring loaded plunger which acts on the throttle lever. The cam profile is designed to linearize the engine response in the idle control region while producing large throttle plate movements in the carburettor venting and cranking positions.

The initial design requirement for the Stepper Motor Driver was to supply a relatively large current to the motor, whilst keeping the power dissipation within the module to a minimum This led to the decision to employ a chopping technique, with the chopping frequency being approximately ten times the highest step rate envisaged for the motor. One chopper drive is used for each pair of phases. At this stage it was also necessary to provide adequate de-coupling around the chopper drive components to ensure no radio frequency interference concerns.

'Misregistration' concerns, especially when driving the motor against the actuator end stop were solved by employing a dual current driving system, the lower current being used only when a low torque is required.

The control strategy defined a need for more than one step rate, so to increase torque a half step rate start up routine was introduced each time the motor was moved from rest. These drive requirements produced the need for a variety of step rates, in order to perform the desired control strategy functions, but meant that the motor was occasionally being run near its natural resonant frequency. To prevent 'misregistration' mechanical damping was incorporated within the stepper motor.

The system is now capable of being accurately run at a variety of step rates, regardless of the natural frequency of the motor.

5 CONTROL STRATEGY

5.1 Introduction

The strategy description relates to the 2.0L variant of the ESC II system.

The following inputs are used by the module to control the spark advance and the manifold heater:

O Engine speed (RPM) and crankshaft position, obtained from a Hall Effect sensor mounted in the distributor.

O Manifold Absolute Pressure (MAP), obtained from the integral SCAP sensor.

O Engine Coolant Temperature (ECT), obtained from a thermistor mounted in the water channel in the intake manifold.

O Two service octane adjustment inputs which are in the form of fly leads grounded to vehicle body to select.

The Idle Speed Control (ISC) system uses the following inputs in addition to those mentioned above:

O Neutral drive select for automatic transmission versions.

O Air conditioning on, as a pre-emptive load input.

O Service idle speed adjust input, similar to the octane adjust inputs.

O Idle tracking switch to detect closed throttle conditions.

5.2 3D Spark Control

The spark control system was developed for maximum flexibility and ease of calibration.

Both spark advance and dwell time were mapped during engine development and the required data stored in the following tables in the microcomputers Read Only Memory (ROM):

O Base Spark Advance (BSA), as a function of RPM and MAP.

O Spark Advance Offset Temperature (SAOT), as a function of RPM and MAP.

O Spark Advance Offset Temperature Multiplier (SAOTM), as a function of ECT.

O Dwell time, as a function of RPM and ECT.

During steady state running the calculated spark advance (CSA) is obtained by the background loop of the programme by looking up these tables with linear interpolation and applying the following formula:

$$CSA = BSA + SAOTM * SAOT$$

There are four exceptional cases where this calculation is modified.

1. If the engine is running in an area where Pre-Ignition (PI) is likely to occur, i.e. at high RPM, high load (MAP) and if the ECT is above a calibrated threshold then an additional retard is used. This allows the base calibration to be nearer the optimum required by the engine.

2. If one or both of the service octane inputs are selected then an additional offset is used.

3. If the RPM is above a calibrated maximum then the dwell time is set to zero, thus preventing the engine from over-revving.

4. If the ISC system has failed and the module detects that the engine should be at idle (see ISC Failsafe), then the spark advance is set to TDC. This considerably reduces the maximum idle speed with a failed actuator.

The interrupt driven foreground loop then uses this information and the dwell time from a look-up table to turn the ignition coil primary current on and off at the correct times.

The dynamic response of the spark system is controlled in two ways.

(A) By applying an exponential digital filter to the raw data obtained from the SCAP sensor to calculate the MAP used for the look up tables. The filter time constant used is calibrated as a function of RPM, ECT and current MAP. This filter is similar in effect to vacuum sustain and delay valves.

(B) The foreground loop is used to control the maximum rate of change in spark advance from one ignition cycle to the next. This is used to reduce the 'harsh' feel that the driver would otherwise experience due to the system's instant response to changes in RPM. In a conventional mechanical system friction and inertia in the bob weights perform the same function.

5.3 Manifold Heater Control

The electrical intake manifold heater element (Positive Temperature Coefficient Pill) was incorporated in the system to improve the drive flexibility during the first few miles after a cold start. It draws approximately 55A when it is first switched on, reducing to approximately 15A with the engine at idle when it is fully warmed up. Thus it is important to make sure that it is switched off while the engine is not running and also switched off as soon as the water in the intake manifold is hot enough to heat the hot spot. For these reasons the ignition module is used to control the manifold heater.

Providing the ECT is below 71°C the heater is switched on when the RPM reaches a calibrated threshold. If the ECT rises above 77°C, or the RPM falls below the threshold speed the heater is switched off.

5.4 Idle speed Control

The initial intention of the project was to provide a closed loop Idle Speed Control (ISC) system to operate during normal running and independently of the choke during engine warm up. During development several additional functions, were incorporated without increasing the hardware complexity.

The final system has the following main features, see figure 7:

(I)   Cranking position control.

(II)   Synchronous kickdown from highcam to idle.

(III) Closed loop idle speed control.

(IV)   Idle tracking switch self cleaning.

(V)   Deceleration damping control.

(VI)   Deceleration fuel reduction.

(VII) Anti-dieseling.

(VIII) Manifold venting.

(I) An area of concern to many car owners is inconsistent cold and or hot starting performance. One of the major causes of this is the inability of conventional mechanical systems to set the throttle to the optimum position for all engine starting conditions. The ISC actuator is used to perform this function.

The calibration data contains a table of the optimum throttle position for cranking as a function of ECT. On power up the ISC actuator is at a known position, the vent manifold position. After initialisation, the actuator is driven to the position obtained from the look-up table and is held there until the engine has started.

(II) Once the engine has achieved a satisfactory running speed, which is stored in the calibration as a function of ECT the system performs a controlled kickdown from the cranking or high cam position down to idle. Two factors are used to determine the speed at which the throttle closes, ECT and current RPM.

A synchronous stroke rate is looked up in the calibration data for the current ECT. Then every time the engine completes this number of strokes the throttle is closed by a fixed (small) amount. This gives a very smooth transition from high cam to idle.

(III) The ISC system is a two term Porportional (P) and Integral (I) closed loop feedback system, controlling to the desired RPM around a first guess open loop throttle position.

The positive and negative gains of both the P and the I terms are separately calibrated to best match the engines performance and current running conditions. This gives the system a very fast transient response to changes in the load demanded from the engine, the limiting factor being the time constant of the intake system.

The minimum and maximum throttle positions during ISC must be controlled.

(A) To avoid excessive undershoot or a stall occuring after a coasting deceleration when the driver dips the clutch a minimum throttle opening must be maintained. This minimum value is calibrated as a function of ECT and is used to override the closed loop system.

(B) To avoid excessive drive from the engine if the driver tries to brake without depressing the clutch, a maximum deviation from the open loop throttle position is used.

The desired idle speed and corresponding open loop throttle positions are calibrated in a series of tables as a function of ECT. There are eight pairs of tables, one for each permutation of the neutral/drive, air conditioning and service idle speed adjust inputs.

(IV) When the idle tracking switch is opened and the RPM is below the idle drive threshold speed the actuator is set into a rapidly oscillating mode. This is to

© IMechE 1985 C244/85

clean any dirt that may be preventing the idle tracking switch from closing.

(V) Above the idle drive threshold speed the actuator is first retracted to its mechanical end stop at which time the stepper motor's position register is zeroed, to ensure that the controller is in synchronisation with the actuator. It is then driven out to a position which is a function of the current RPM and ECT. This is known as the damper position.

The damper position is continually modified as the RPM changes, and is used in place of a mechanical dashpot to increase the air and fuel flow during the initial part of a deceleration, hence reducing Hydro-Carbon emissions.

(VI) The damper is only used for a fixed time after the driver takes his foot off the accelerator, typically two seconds in an automatic transmission calibration. After this time the actuator is retracted fully. This allows the throttle butterfly and the idle fuel progression system to close thus reducing the fuel flow.

If the deceleration continues down to idle then as the RPM falls below the drive idle threshold the actuator is driven to the last known idle position plus or minus a small amount to compensate for the rate of deceleration.

If the deceleration rate is low, i.e. if the car is driving the engine during a coasting deceleration, a negative adder will be used to reduce the engine torque change to a minimum thus improving the "drive feel".

If the deceleration rate is high, i.e. the driver dips the clutch while the system is in the deceleration fuel reduction mode, a positive adder is used to prevent excessive undershoot or a stall.

If the deceleration does not continue down to idle but the engine accelerates again then the damper mode is again entered.

(VII) When the driver turns the ignition off the actuator is retracted fully, and held for 4 seconds. This is used in place of an anti-diesel solenoid to prevent engine run on.

(VIII) The actuator is then driven out to the vent manifold position and the module turns the power hold relay off.

Driving the actuator to the vent manifold position has two purposes.

1. To allow any build up of fuel in the intake manifold to evaporate and escape. This improves hot restarts.

2. To pre-position the actuator for the next power up.

5.5 Failsafe Features

If the module detects an erroneous reading from either the ECT or the SCAP sensor it ignores the current reading and substitutes a failed sensor value. For the ECT sensor this is 99°C and for the SCAP sensor it is 760 mmHg. These values will provide the greatest possible safety against engine damage.

The most important failsafe mechanism is for the ISC actuator.

In order to provide protection from the above condition the module assumes that the actuator has failed and sets the spark advance to T.D.C. when the following conditions are satisfied:

O RPM is above the idle drive threshold.

O Idle tracking switch is closed.

O MAP is in the first idle range.

In failsafe the no load idle engine speed is limited to approximately 1900 RPM.

6 DEVELOPMENT TOOLS

The implementation of microcomputer based engine control systems require special support tools for software development and calibration. In addition to the standard laboratory development systems a mainframe computer based assembly and calibration support system was developed for the ESC II. With the proper equipment this system is accessible via the worldwide telephone network. It allows engineers to run programme assemblies and exchange programme code and calibration data between various locations. This is of specific importance in the case where the development and calibration activities are distributed throughout the world (e.g. cold climate tests in Finland, hot climate tests in Arizona).

To support engine and vehicle calibration development an in vehicle calibration console system was used.

It consists of two units, the emulator and the display. Both units are microcomputer based and exchange data via a serial link. The display unit can be mounted conveniently on the dashboard of a vehicle to be operated by the driver or calibration engineer.

The main features of the calibration systems are :

O operation transparent to the control software in the module

O display of 5 variables simultaneously in engineering units

O display of 8 logic variables

O analog output channels for 4 variables

O control module operation out of EPROM or 2 RAM areas

O switch-over between memory areas on-line

A typical installation in a vehicle is shown in Fig. 8.

7. MANUFACTURING AND TESTING

From the Electronics manufacturing standpoint the primary objective was the successful installation of a reliable and cost effective facility in Europe which would be capable of producing electronic engine controls of this complexity yet retaining flexibility to accommodate subsequent product developments.

Motorola has been producing engine controls in the USA for many years but ESC II was the first to be fully manufactured in Europe. To minimise risks in launching ESC II in the new engine electronics facility it was necessary to maintain commonality in manufacturing techniques and processes used by Motorola manufacturing operations throughout the world.

In the new facility state-of-the-art equipment was installed for automatic insertion of leaded components, automatic placement of chip components, and dual wave soldering to accommodate the chip components. By striving for a high degree of commonality a maximum amount of experience and know-how was transferred to the new operation.

This approach was also applied to the design and fabrication of production automatic test equipment. The design of the ESC II functional testers was derived from testers used on similar products in the USA which effectively minimised tester 'debug' time which can jeopardise an on time product launch.

In high volume manufacture optimization of the test cycle time is critical to maintain production line capacity. In this respect it is not practical to test each module throughout the full calibration nor is it desirable to have unique tests to accommodate each minor change in calibration. Since the calibration is contained in the microcomputer masked ROM code it is necessary only to ensure that the correct microcomputer is used and test only the module hardware.

This was achieved by implementing a special test routine in the microcomputer ROM activated by grounding a test pin in the module connector, which maintains the foreground real time programme active while changing the background loop in such a way that only 1 to 1 relationships between inputs and outputs are set up. For example, spark advance is controlled only by the pressure input, dwell time only by the coolant temperature input to check the pressure sensor and A/D convertor operation. The input switches from transmission and air conditioning create specific pulse patterns on the stepper motor outputs to check the phase drivers.

8 CONCLUSION

The use of standard microcomputer with powerful real-time capabilities proved to be an efficient way to implement the control of a carburettor and ignition system. The flexibility of the design approach is further demonstrated by the continuing development of similar engine management systems.

A high upgrading capability is ensured by the introduction of new additions to the MC 6800 series of microcomputers such as the MC68HC11. The increased functional capability of the MC68 HC11 will enable expansion of the ESC control to incorporate other functions, such as integral fuel handling.

ACKNOWLEDGEMENTS

A. Benstead, MOTOROLA AIEG for systems applications support

S. Bonfiglioli, WEBER, S.P.A. for idle speed control, system concept and actuator design and development

D. Cox, MOTOROLA AIEG for hardware design

P. Day, FORD MOTOR CO. for mainframe software support

B. Evrard, FORD MOTOR CO. for engine calibration

K.I. Forster, FORD MOTOR CO. for module and strategy design

R. Jaques, FORD MOTOR CO. for engine application

P. McBride, MOTOROLA AIEG for software support

J. Mercelat, MOTOROLA AIEG for software support

D. Newman, FORD MOTOR CO. for mainframe system support

To all members of the groups represented by the above and to all others who contributed to the success of the project.

© IMechE 1985 C244/85

Fig 1   2.0L OHC carburettor

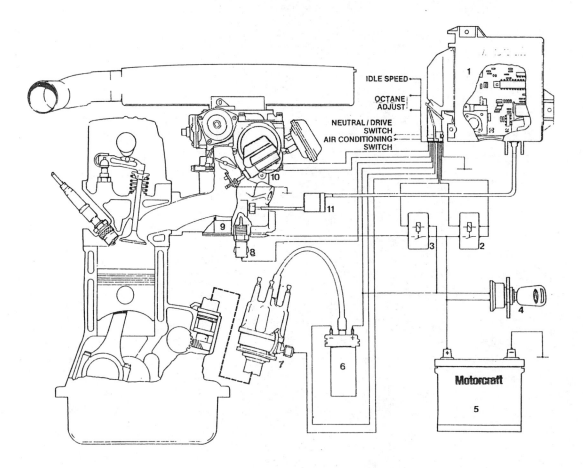

Fig 2   Electronic control system ESC II — system components
1 ESC II module
2 Power hold relay
3 Manifold heater relay
4 Ignition switch
5 Battery
6 Ignition coil
7 Hall effect distributor
8 Engine coolant temperature sensor
9 Intake manifold heater
10 Idle speed control actuator
11 Fuel trap

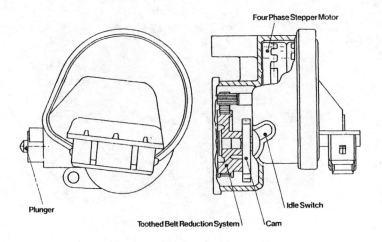

Fig 3 Stepper motor cross-section

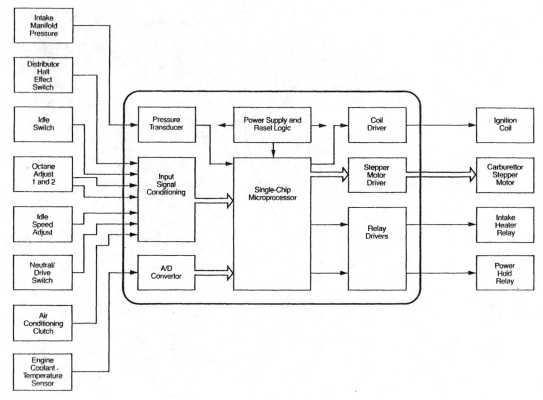

Fig 4 Block diagram of the ESC II module

Fig 5 Surface mounted devices

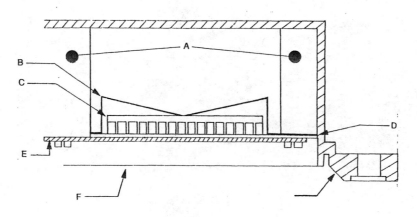

A. Micro Heatsink Mounting
B. Top Micro Hold-Down
C. 40 Pin Plastic Micro
D. Micro Heatsink
E. Printed Circuit Board
F. Module Cover
G. Module Mounting Foot

Fig 6 Heat sink arrangement for the microcomputer

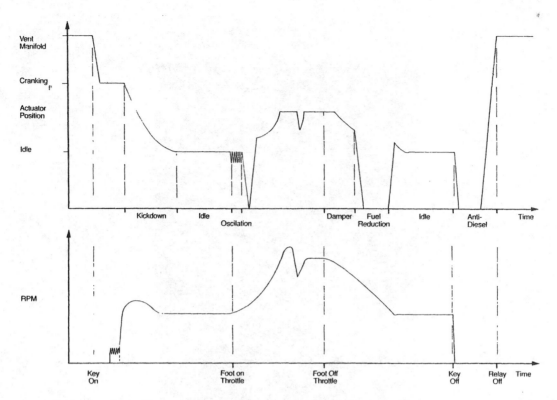

Fig 7 Idle speed control actuator position and engine speed versus time

Fig 8 Vehicle calibration console

871293

Electronic Gaseous Fuel Carburetion System

Colin G. Young

N.G.V. Fuel Technologies, Inc.

ABSTRACT

The design of the Dynex electronic carburetion system for gaseous fuels (propane and methane), by N.G.V. Fuel Technologies, Inc., is presented. The design and draw-backs of previous and existing systems are described, leading to the parameters dictating the design of an "ideal" system. Prototypes of three variants of the electronic system have been made and tested with promising results. Patents for the invention have been issued. The system does not require the use of a computer or fuel injectors, but senses the engine speed and manifold pressure in order to regulate the pressure and flow rate of the gaseous fuel being admitted to the engine. No carburetors (mixers) or adapters are required. Performance and emissions on (dual-fuel) gasoline operation are not impaired. The system can be used for gaseous fuel supplementing of two-stroke and four-stroke diesel engines, and also for controlling a liquid propane injection system.

* * * * * *

IN THE PAST, propane and natural gas (methane) carburetion systems have been mechanical/fluidic in design, using either the North American or European concepts. The North American approach has been to use a simple (non-adjustable) pressure regulator and a complex (moving-element, adjustable) carburetor, or mixer. The European approach has been to use a simple (fixed orifice) mixer and a complex (adjustable, variable) pressure regulator. The latter concept is somewhat analogous to having a truck with a single-speed transmission, in which case it is impossible to have the one ratio suit all driving conditions; it does offer some advantages in terms of ease of fitment and under-hood appearance, and in some cases does provide reasonable performance.

The North American design has certainly served the alternative fuels industry well and has provided acceptable performance, when compared with gasoline engine performance of the day; however, it also has some major limitations and short-comings, particularly in terms of mixer and adapter multiplicity, and installation and appearance considerations. There have been several attempts to improve the basic carburetion concepts, but in all cases a number of disadvantages have still existed. In light of smaller, more complex and computerized engines, a completely new approach to gaseous fuel carburetion is required, if these attractive alternative fuels are to find wide acceptance.

THE "IDEAL" CARBURETION SYSTEM

The ideal propane/methane carburetion system must:
* Accurately meter the fuel for all engine operating conditions,
* Be universal for all engine makes, models and sizes,
* Be universal for all gaseous fuels,
* Compensate for different heating values,
* Require no adapters, or range of sizes,
* Retain the under-hood aesthetics,
* Retain all O.E.M. engine components, especially the air induction system,
* Be compact and easily installed,
* Not impair gasoline performance and emissions (on dual-fuel systems),
* Be reliable and durable,
* Be easily serviced,
* Impose no hood-clearance or other fitment limitations,
* Be compatible with computerized engines,
* Be quiet in operation,
* Have good cold-weather performance,
* Provide instant starting,
* Have acceptance by owners,
* Have acceptance by mechanics,
* Be competitively priced.

DISADVANTAGES OF CURRENT DESIGNS

Most current designs only measure the engine air flow, and not the engine loading, hence they cannot vary the air:fuel metering as required to suit all operating modes. All mechanical

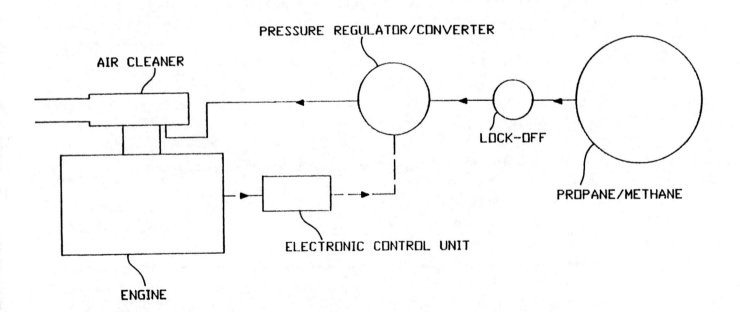

Figure 1. Schematic of the Dynex Electronic Carburetion System

carburetion systems require a range of several carburetor (mixer) sizes to suit the range of engine sizes, and also require an extensive range of adaptors to suit all engine-carburetion/injection combinations. There is a long lead-time for the designing, testing, tooling, manufacturing and distributing of new adapters to suit newly released vehicles and engines.

Fixed-venturi systems are seriously compromising in that a small venturi is required to accurately sense the air flow rate at cranking and idling conditions, while a large venturi is required at wide throttle openings so as not to severely choke the induction air. Due to the unacceptable option of choking the air flow - because of the large power loss and drastic increase in (dual-fuel) gasoline emissions - a large venturi must be used, with the resulting problem of unstable and unreliable starting and idling. With this design, a sensitive and delicate outlet pressure from the regulator is required; hysteresis, vehicle inertias, air flow disturbances (from the ram effect and cooling fans), and induction air pulsations can play havoc with the fuel metering.

Some designs use under-throttle adaptors which usually necessitate the use of longer manifold bolts (with the usual proportion of sheared and dropped bolts!), and often cause hood-clearance problems. Also, manifold vacuum (pressure) operates in the opposite manner to that desired; at idle, when very little fuel is required, there is a large vacuum; while, at wide-open-throttle, when a very large fuel flow rate is required, there is very little vacuum available.

Apart from the time required and the expense of having to replace or modify the standard air intake and gasoline carburetion/injection components, systems that affect in any way the operation of the engine computer can cause serious problems.

Injector systems for gaseous fuels, using timed pulsed injection, can have major fitment, noise, durability, cold-weather and flow-rate problems. With gaseous fuels, a much larger volumetric flow rate is required, and sub-zero temperatures with propane operation will not allow even moderate injection pressures to be used, due to the delicate liquid/vapor balance; even a small percentage of unvaporized fuel will cause excessive richness.

Carburetion systems that restrict the air flow in order to produce a signal or generate a partial vacuum to induce fuel flow, can seriously impair the performance, economy and emissions when operated on gasoline in a dual-fuel set-up. The expense of recertifying all affected vehicles for emissions regulations can be very high.

Designs employing close-fitting valves controlled by a low-force actuator can have sticking problems due to the formation of ice or heavy oil deposits. Delicate electronics that can be damaged by roadway corrosion, static electricity, engine heat or vibration - or have a non-failsafe mode - are unacceptable.

DEVISING AN OPTIMUM CARBURETION SYSTEM

The first step in attempting to design the perfect carburetion system is to list (a) all the desired features (without worrying about how they will be achieved!), and (b) all the limitations and disadvantages of all the past and present systems. A fascinating way to spend a long winter is to read all the U.S. patents on gaseous fuel carburetion; the paper pile will be quite high, but it will provide much inspiration and thought-starting potential.

Universality has to rate high in priority, so that the system will work with any gaseous fuel, and will suit all makes, models and sizes of engines. A major problem in the past has been the need to design, build and store all the adaptors and mixers required to cover all the combinations of engines, carburetors/injectors and vehicles. Along with lead-time hassles, adaptors usually look agricultural in nature and necessitate the deletion of the standard air-cleaner and intake ducting. Apart from the double expense of deleting and replacing, under-hood neatness and factory-appearance are becoming more important for vehicle owners. In many cases, deleting the standard air-cleaner will cause an objectionable raucous induction noise. The number one task then is to design a system that can be readily fitted to any vehicle, without creating any other problems.

An equally important requirement (and one that has been completely over-

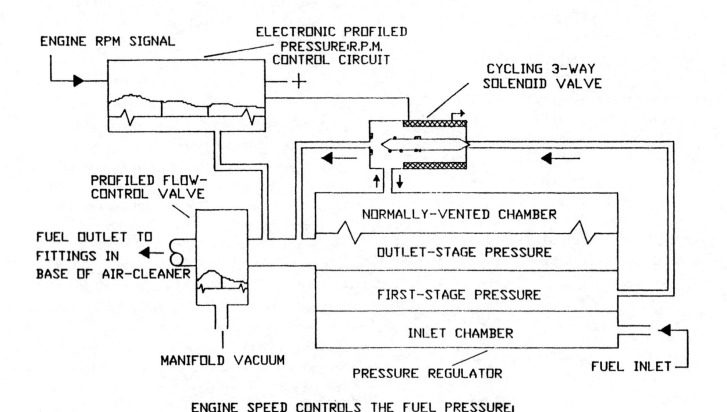

Figure 2. Fuel Pressure & Flow Rate Control Systems

looked in the past on more than one occasion!) is that the carburetion system must actually WORK, under all conditions, all the time. It is totally unacceptable to have a system that only works when new, on a car cruising on a smooth level highway, on a nice warm day. A professional design must work reliably under all realistic combinations of driving modes, atmospheric conditions, road conditions, fuel cleanliness and composition, and vehicle forces (accelerating, braking, cornering). The carburetion system must accurately supply the correct air:fuel ratio for all engine operating conditions, regardless of whether the vehicle is on a rough or smooth road, turning left or right, the (electric) cooling fan is on or off, or an inlet valve is leaking or not.

Propane and natural gas carburetion systems have always had the image of being an after-thought or stuck-on device. Vehicle owners have come to expect (gasoline) vehicles to function faultlessly, and they do - if the computer is still working! They will accept nothing less from alternative fuels, in terms of power, economy, maintenance and reliability.

An engine capable of 800 BHP[kW] and 200 mpg[km/l] is useless if it will not start. A major problem with fixed-venturi carburetion systems has been that of difficult (if not impossible) starting, and the hit-and-miss approach of flooding an engine with fuel and hoping there will be a combustible mixture available before the battery runs down, is completely unacceptable. It is vital that a carburetion system accurately measures the cranking air flow, or engine speed, and supplies the precise amount of fuel to ensure instant starting.

Installation time and complexity, along with compatibility with the engine computer control system, are prime factors in the total cost and acceptance of carburetion designs. Bulky components that have to be installed in a specified manner (typically vertically and longitudinally), or necessitate highly-skilled and/or tedious vehicle modifications soon lose their market appeal. Acceptance by mechanics has long been an over-looked parameter. Mechanics are mechanical in nature (not surprisingly!) and while electronics are being slowly accepted, still with some reluctance, computers-modules-blackboxes will be a sore point for a long time to come. Mechanics take pride in being able to diagnose and understand; they cannot do this with a 30 wire chip or module.

Tuning and trouble-shooting ease must be considered, and in the ideal carburetion system, mixture settings should be attainable without the need of a dynamometer and expensive analyzing equipment. Vehicles will not always be near a well-equipped propane/methane service facility, hence the system must be able to be readily understood by a competent mechanic in any geographical location.

NEW-GENERATION CARBURETION

The Dynex electronic carburetion system is shown diagrammatically in Figure 1. The methods of regulating the fuel pressure and flow rate, using ANY existing North American pressure regulator, are shown in Figure 2; while three variants of the system have been evaluated, only the preferred embodiment has been illustrated.

The most obvious advantage of this new-generation gaseous fuel carburetion technology is that it INDIRECTLY measures the air flow rate - and engine loading - by measuring the engine speed and inlet manifold pressure (vacuum). By doing this, it is not necessary to restrict the induction air flow, and it is not necessary to have restrictors/venturis to suit each and every vehicle-engine-carburetion combination.

The other major advantage is that NO propane/methane mixers or adaptors are needed. Rather than using the pressure drop created by an air restriction to induce fuel into the engine, gaseous fuel is supplied to the engine at a SLIGHT positive pressure. The fuel is induced through simple hose fittings in the base of the air cleaner, close to and symmetrical with the carburetor/throttle-body barrels.

Fuel flow is quiet, continuous and sufficient for even the largest of engines. The cost, and potential problems of fuel injectors, are eliminated. Fuel pressure does not exceed 1.5 psi [75 mm Hg], hence cold-weather propane operation is not impaired. No parts of the computer system or factory-fitted induction system are replaced.

CARBURETION CONCEPT

The carburetion principle involves "mapping" engines - recording air flow for all combinations of engine speed and inlet manifold pressure, with temperature compensations, and then, by employing engine speed and loading factors, determining the required fuel flow to suit every possible operating mode, thus providing the usual richening during idling and heavy engine loading conditions, and leaning during light loading conditions.

The second part of the carburetion concept is to design the fuel system to provide the correct amount of fuel for the various combinations of engine speed and inlet manifold pressure. The two ways of controlling a gas flow are by regulating (a) the gas pressure, and (b) the cross-sectional area of a valve through which the gas passes. It is possible to vary either the pressure or area (while the other remains constant), or to vary both the pressure and area.

Prototypes using all possible combinations have been tested. The first design combined (or computed) the engine signals, and then controlled, in the first case, the fuel pressure (keeping a fixed valve area), and in the second case, the valve area (keeping a fixed fuel pressure).

The second design used the engine signals to independently control the fuel pressure and valve area. The first variant had the engine speed regulating the valve area by means of a linear stepper motor, and the manifold pressure controlling the fuel pressure by means of a spring-biased diaphragm.

The second, and preferred, variant had the engine speed regulating the fuel

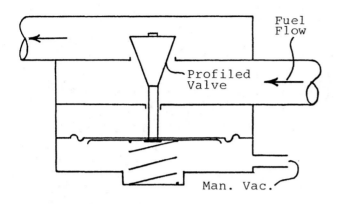

Figure 3. Flow Control Valve

pressure, and the manifold pressure controlling the flow valve area. The flow valve unit is shown in Figure 3, and employs a spring-loaded diaphragm moving a profiled valve in response to the manifold pressure. Various valve profiles can be used to suit particular engines if required, but this should only be the case if widely differing valve timings, bore:stroke ratios and combustion chamber shapes are encountered.

Several designs of speed:pressure control systems have been tested. The essential requirement of the control device is the ability to have a selectable profiled relationship between the fuel pressure and engine speed. All designs utilize the first stage pressure of the propane/methane pressure regulator to control the outlet pressure; in all cases, the static outlet pressure has been sub-atmospheric (usually minus 1-1/2" [38 mm] water column) so that the regulator acts as a fuel lock-off when the engine is stopped. The pressure is regulated to a maximum of 1.5 psi [75 mm Hg] above atmospheric, in a profiled manner from cranking to maximum rpm.

Mixtures to suit all engine sizes and maximum speeds - as well as all (gaseous) fuels - can be set electronically without the need of a dynamometer; in fact, the maximum power setting is made with the engine stopped! Alternatively, sized flow control valves in relation to engine capacity can be used. In addition, an engine speed governor is included in the system; the moving profiler in the pressure:speed comparator can readily be used to short-out the ignition system at a pre-set engine speed.

Variable resistance, inductance, capacitance and luminesence have all been tested in the pressure:speed control system, but the preferred design uses an L.E.D. (light emitting diode) tachometer with a mating coded photo-diode circuit to control a 3-way solenoid, as shown in Figure 4. A profiled blanking plate, controlled by the fuel outlet pressure acting on a spring-biased diaphragm, operates between the mating pairs of light emitting diodes and photo diodes, to permit or prevent light passage, as required. The always-open port of the 3-way valve connects with the normally vented side of the pressure regulator final stage. The normally-closed controlled port is connected to the first stage pressure chamber, and the normally-open controlled port is connected to the outlet pressure stage of the regulator.

When light passes across the sealed darkened chamber of the control device, first-stage pressure is applied to the normally-vented side of the final stage diaphragm, thus causing the outlet pressure to increase until the blanking plate moves sufficiently to block the

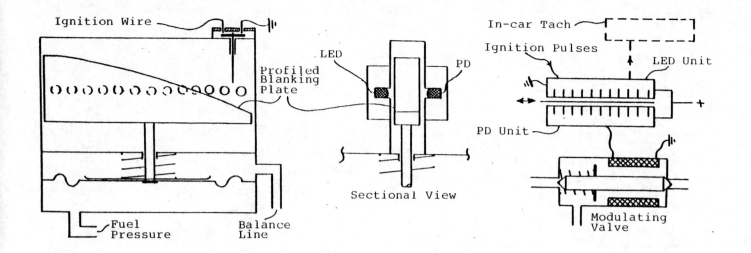

Figure 4. Fuel Pressure v. Engine Speed Control System

light passage, at which time the first-stage pressure passage will be closed and some of the control gas will be vented down-stream. As the engine is consuming fuel, the outlet pressure will decrease, so that the blanking plate will now move in the opposite direction until it again permits light passage. The cycle repeats a few times a second, so that a smooth pressure:speed relationship, in accordance with the blanking plate profile, is maintained.

This design allows an inexpensive in-car L.E.D. tachometer to be installed. While the basic design uses a linear progressive display, it is possible to use a coded display, in which case much smaller engine speed increments can be detected. While development has been conducted with moving-element comparators, tests with solid-state units are being made.

The system is compatible with computerized engines employing an EGO (exhaust gas oxygen) sensor. As shown in Figure 5, the EGO control can operate a solenoid valve so as to regulate the air:fuel ratio. The flow control valve manifold vacuum chamber contains an air bleed, so that the control believes the engine is loaded more than it actually is, and consequently supplies a mixture richer than that required. The EGO control will respond by activating the solenoid valve, thus closing the air bleed and leaning the mixture until the

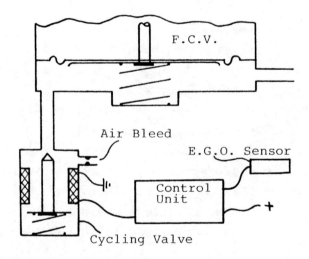

Figure 5. E.G.O. Control Option

cycle repeats. An optional method is to enrich the pressure:speed comparator, and employ the EGO control to interupt the comparator circuit in order to lean the mixture.

While most people may believe the only way to go is with a black-box computerized concept, it is believed that the Dynex electronic system can achieve substantially all of the benefits available with a micro-processor system, and because most of the applications will be after-market fitments, "straight" (understandable) electronics will have greater acceptance in the market.

However, due to the complementary nature of other M.T.A., Inc. products being developed, the computerized "call-up" system is still being pursued. Figure 6 shows the logic diagram of the computerized version of the carburetion system. Prime signals such as the engine speed, manifold pressure, throttle position and air flow rate, along with supplemental inputs such as the air temperature, the status of the air-conditioner, the road gradient, the transmission ratio engaged and the engine water temperature may be used to provide step or percentage variations in the gaseous fuel flow.

Full details of the carburetion system will be found in U.S. patents 4,449,509 and 4,505,249 and subsequent additions.

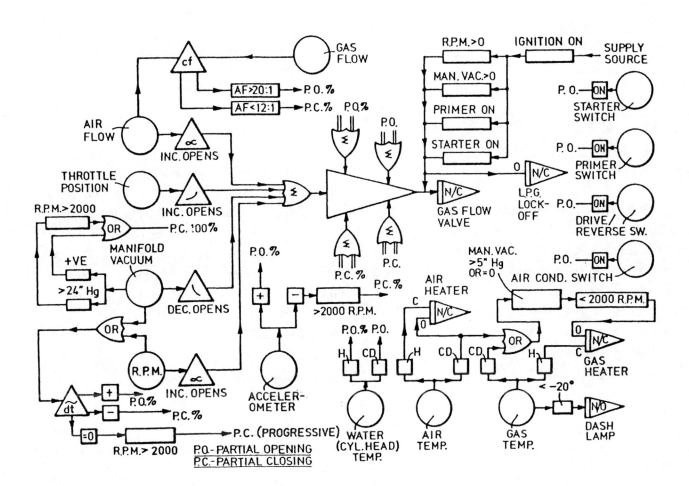

Figure 6. Logic Diagram of the Computerized Carburetion System

FUTURE APPLICATIONS

N.G.V. Fuel Technologies, Inc. are developing the Dynex electronic carburetion system for three applications, in addition to the immediate market for single- and dual-fuel gaseous fuel conversion of spark-ignition gasoline engines.

The system allows gaseous fuel supplementing of four-stroke diesel engines, and is particularly suited to providing timed pulsed induction on super-charged two-stroke diesel engines, where most of a continuous flow supplementing fuel would be wasted with the scavenging air. The manifold pressure control is replaced by an inter-related diesel rack (or equivalent) modulator, such that depending on engine parameters and the power demand as dictated by the driver, the diesel flow is gradually reduced while the gaseous fuel flow is increased, within safe pre-determined limits. Other features of the carburetion system include its ability to regulate the fuel flow to each cylinder in (inverse) response to variances between individual diesel injector flow rates, in order to smooth out the engine running.

The Dynex carburetion concept eliminates all mixers and adapters. The next phase is to eliminate the converter/regulator on propane systems; the control system lends itself ideally to controlling a liquid propane induction system, utilizing the induction air to vaporize the propane in an open heat-exchanger, with the benefit of cooling the air so as to increase power and drastically reduce pre-ignition potential.

The carburetion system is most compatible with the Dynex engine performance-optimizing control system that electronically "draws" the P-V (pressure-volume) diagram for each cylinder of an engine, and continually varies the ignition timing, valve timing and fuel flow so as to provide the most efficient and reliable engine performance.

SUMMARY

A discussion of past and present gaseous fuel carburetion systems has been made in order to evaluate all advantages and disadvantages. An electronic carburetion system that over-comes all the recorded problems has been designed; patents have been granted. Prototypes have performed extremely well. It is hoped that this discussion will be of assistance to all others involved with propane/methane carburetion, and that it may inspire them to come up with an even better "mouse-trap".

ACKNOWLEDGEMENTS

The assistance of the personnel who have contributed to the development of the carburetion system is gratefully acknowledged. Thanks to all in the propane and natural gas industries in the U.S.A., Canada and Australia who participated in the market research program conducted by Mech Tech Auto, Inc. Special thanks to Kevin Hunt, Bill Bird and Russell Clarke for their computer and technical assistance.

CHAPTER 3

IGNITION TIMING

FULL VACUUM CONTROLLED IGNITION SYSTEM

The purpose of this paper is to describe briefly, a full vacuum controlled ignition system and some of the factors governing its use in motor vehicles. Emphasis will be placed upon considerations arising in its application as well as some of the advantages and disadvantages of this type of system.

CONVENTIONAL IGNITION SYSTEM CONTROL

In order to compare the function of this system, it will be necessary to consider briefly, the predominant method of obtaining automatic spark advance used in motor vehicles during the past two decades.

Control of ignition timing to obtain automatic spark advance has, in the past, resolved itself into a rather conventionalized system. Centrifugal weights have been utilized to rotate the contact breaker cam in respect to the driving shaft in a direction opposite the shaft rotation. This produces advance spark timing as a function of the engine speed. The spark advance is controlled by spring tension reacting against the centrifugal weight movement and the resulting motion is linked to the breaker cam so as to rotate the cam and advance the timing. The springs and linkages are adjusted and designed to meet requirements of wide open throttle or full power demands of the engine. In order to obtain the additional spark requirements of the engine under partial power or part throttle operation, it is necessary that an additional mechanism be used. This usually comprises a vacuum actuated diaphragm linked to a movable breaker plate so as to rotate the breaker plate and contact points about the cam to produce the spark advance. This vacuum diaphragm is usually connected to the carburetor throttle plate so that the diaphragm will be actuated by manifold vacuum present in the throttle body except under idling or closed throttle plate position. The vacuum diaphragm reacts against a spring which is usually pre-loaded to prevent movement until the manifold vacuum has increased to some determined value and any additional manifold vacuum will cause the diaphragm and breaker plate to move, advancing the spark timing. This produces a spark advance as a function of engine load.

The conventional system comprises two independent means within the distributor for producing timing advance. The centrifugal weights satisfy the full power requirements of the engine and the manifold vacuum system satisfies the additional advance requirements of part load operation.

FULL VACUUM IGNITION CONTROL

The system to be described herein does not employ the use of centrifugal weights or manifold vacuum applied directly to the distributor. Considerable mechanical simplification has resulted in this distributor design. A single spark advance actuating force is used at the distributor in the form of a vacuum diaphragm controlling the breaker plate position. This diaphragm is actuated by a controlled differential vacuum derived in the carburetor through the use of specially drilled orifices which are interconnected in the carburetor body. The relationship of the vacuums produced for various speeds and loads are determined by the size and placement of these orifices in the carburetor so that the differential vacuums produced will be a function of engine speed and load conditions. It is, therefore, necessary that the carburetor be engineered for a specific type of engine so as to meet the requirements of that engine.

FULL VACUUM CONTROLLED DISTRIBUTOR

Figure (1) illustrates a cut-away section of a typical full vacuum controlled distributor showing its principal parts and assemblies. The design is such that the breaker cam is fixed to a solid distributor shaft and can at no time vary in position from the driven end of the shaft. This means that good concentricity tolerances may be maintained between the breaker cam lobes and shaft.

This was not always true in systems used in the past where additional tolerances must be considered due to the cam movement in respect to the shaft. The fixed cam to shaft relationship also eliminates wear considerations in service. This construction eliminates the possibility of torsional vibration appearing at the driven end of the distributor shaft from being amplified at the breaker cam as is the case where centrifugal weight governors are employed due to the coupling necessary through the breaker springs and connecting parts and bearings. Spark scatter or spread from this cause is held at a minimum.

The breaker plate assembly includes the breaker contacts and condenser and rotates about the cam so as to produce spark advance. The breaker plate is restrained in its movement by springs from a fixed or initial timing position. The springs are chosen as to rate and free length to permit a reasonable latitude in adjusting spark advance to meet the requirements of the engine. The distributor vacuum diaphragm is connected by a rod linkage to the breaker plate. This diaphragm is vented to atmosphere on the distributor breaker plate side and is actuated by the distributor differential carburetor vacuum on its closed side.

A specially designed sintered type bearing is employed and serves a dual purpose. The internal diameter of this bearing pilots the distributor shaft and its external diameter pilots the breaker plate assembly. By manufacturing this bearing to close concentricity limits, little variation results between the breaker plate and the cam. Lubrication for the bearing is provided from an oiler located in the base of the distributor assembly so as to furnish lubrication for both the shaft and the breaker plate. This bearing also provides a thrust face for the distributor cam and shaft. It is necessary that no excess lubricant be permitted to emerge from the bearing surface, otherwise dilution of the breaker cam lubricant and contamination of the contact breaker points would result. This is achieved through the use of a high density bearing impregnated with an SAE 40 viscosity oil. The O. D. of the bearing is ground as well as the end adjacent to the cam so as to maintain concentricity limits and obtain a partial sealing of its surface to prevent bleeding of oil. An additional sintered bearing is used at the bottom of the distributor base as well as a thrust collar to limit end-play to the desired amount.

CARBURETOR CONSIDERATIONS

As mentioned previously, the vacuum control for this ignition system is one of the major considerations in its application. Its design must be carefully integrated with the distributor design in order to meet the sparking requirements of the particular engine that it is required to serve. Figure (2) represents a cross-section of a typical carburetor showing the positioning of the special distributor vacuum passages. We notice at (A) that a drilling is made at the throat of the main venturi. Here we find that air passing through the venturi produces a depression as a function of carburetor air flow which is, in turn, related to engine speed. If this orifice were only used, a vacuum would be produced at the distributor diaphragm as a function of engine speed and this, in turn, could be applied to the distributor breaker plate movement to produce spark advance as a function of engine speed. The full power spark requirements of the engine could be met by applying the venturi vacuum to the distributor diaphragm with the proper spring loading. However, this would give us spark advance comparable to that obtained through the use of centrifugal weights only at full power and would not satisfy part load requirements of the engine. A second orifice is drilled at (B) into the throttle body slightly above the closed throttle plate position. This senses manifold vacuum present in the throttle body above the throttle plate and is blocked off by the throttle plate in its closed position where no vacuum appears in the system and spark advance is at full retarded due to the action of the breaker plate springs. The manifold vacuum orifice is interconnected to the venturi orifice through a vertical passageway designated by (C) in the diagram. A restriction is placed in the upper portion of the vertical passageway between the two orifices. The orifices are affected by each other under various speed and load conditions of the engine. During wide open throttle operation, the resultant manifold vacuum is less than that developed in the venturi and air flow occurs from (B) through the vertical restriction at (C) and out at (A). The restrictions presented by these three orifices determine the amount of bleeding one will produce on the other and their hole sizes must be carefully worked out. If (B) and (C) are large in respect to (A), a large amount of bleeding will occur and the resultant vacuum applied at the distributor is low and a weak spring adjustment must be used to provide a given degree of spark advance. Conversely, if orifice (A) is made large in respect to (B) and (C), we can expect an increase in distributor carburetor vacuum and a heavier spring adjustment must be made for the same degree of advance.

Under part throttle operation, particularly at the lower engine speed ranges, the manifold vacuum sensed at (B) is many times greater than that developed at the venturi and air flows into the venturi orifice down through the vertical passageway and out at (B). We find (A) is now bleeding the higher vacuum at (B) and again the relative sizes of their restrictions are important. If (A) and (C) are large in respect to (B), the carburetor distributor vacuums produced are low. For a given spring adjustment under wide open throttle conditions, the resultant part throttle spark advance would be lower than that produced if their relative restrictions were reversed. If restrictions (B) and (C) are made larger with respect to (A), higher part throttle distributor vacuums will result and a higher part throttle spark advance is obtained. To present a complete analysis of the function of the distributor vacuum passages located within the carburetor would be beyond the scope of this paper. However, certain fundamental considerations should be discussed.

In Figure (2) the carburetor was diagrammed showing a closed throttle plate position which is the case under engine idling conditions. This results in engine manifold vacuum being removed at the throttle body orifice and as no depression exists at the venturi, there is no actuating force applied at the distributor diaphragm and the spark timing is fully retarded.

WIDE OPEN THROTTLE OPERATION

Wide open throttle carburetor operation is diagrammed in Figure (3). Here, a relatively high venturi vacuum is developed over the speed range of the engine with attendant lower manifold vacuum being present in the carburetor throttle body and controlled bleeding of the venturi results. This differential vacuum is applied to the distributor advance diaphragm.

Figure (4) diagrams the resultant distributor differential vacuum referenced to air flow through the carburetor in cubic feet of air per minute.

Figure (5) illustrates the distributor differential vacuum in terms of engine R.P.M. when the carburetor is operated in conjunction with a specific engine at full power.

Figure (6) illustrates distributor spark advance determined by the differential vacuum described in Figure (5) reacting against the spark advance breaker plate springs within the distributor. The distributor spark advance springs have been adjusted to meet the sparking requirements of the engine for wide open throttle or full power operation. Once this relationship has been established, it is necessary in the carburetor design to produce differential carburetor vacuums which will further advance the spark timing to satify road load and part load engine spark advance requirements. This can only be accomplished through design of the orifice at (A), (B) and (C) so as to produce the correct relative vacuum relationships for full and part throttle operation.

PART THROTTLE OPERATION

The schematic diagram of Figure (7) illustrates the throttle position for part load operation. Here, the throttle has been advanced from its closed position to uncover the manifold vacuum orifice. The manifold vacuum in the throttle body increases due to lighter engine loading and greater vacuums will be sensed by this orifice. If the manifold vacuum present at (B) becomes greater than that present at the venturi, air flow will result from the venturi through the manifold vacuum orifice. We now have the venturi orifice bleeding the manifold vacuum orifice and controlling the resultant differential vacuum present at the distributor diaphragm. The only condition where no air flow is resulting in the vertical passageway (C) is when the developed venturi vacuum for any given speed equals the manifold vacuum present in the throttle body.

Examination of Figure (8) will show the resultant differential vacuums produced for carburetor air flow in terms of road load power requirements. The slope of this curve advances rapidly as the throttle plate uncovers the manifold vacuum orifice at a relatively low carburetor air flow. This cut-in characteristic is determined by the placement of the manifold vacuum orifice above the closed position of the throttle plate. As its drilling is made higher in relationship to the closed position of the throttle plate, the differential vacuum and resultant spark advance cuts in at a higher carburetor air flow. It is also possible to vary the slope at the cut-in point by changing the shape of the manifold vacuum orifice. If a horizontal type slot is used instead of a circular hole, the slope of the curve tends to be more abrupt and steep. If the orifice is made in the form of a vertical slot, the slope of the curve is decreased as the orifice becomes less sensitive to throttle plate position. The distributor differential vacuum is controlled in its maximum value by the relative bleeding that occurs under road load conditions.

Figure (9) illustrates the resultant distributor differential vacuums produced by the carburetor referenced against engine speed for vehicle road load operation.

Figure (10) describes the resultant spark advance in terms of engine speed produced by the applied distributor differential vacuums.

Figure (11) shows the resultant spark advance obtained of a typical L-head type engine over its normal speed range. The minimum spark for best torque requirements were determined through the use of a manual distributor. Rerunning this engine with full automatic spark control has produced very close approximations of the desired spark advance for both full power and road load engine operation. These conditions may be readily realized for a particular type of engine through proper carburetor and distributor design. At part loads and low engine speeds, the sparking requirements and automatic spark do not always find such complete agreement as at full and road load operation. There is a tendency for this distributor to be over-advanced at low speed partial loads approaching the full power requirements. This will be readily understood when it is remembered that increased manifold vacuums from wide open throttle rapidly approach the developed venturi vacuum and exceed it while the sparking requirements of the engine have not materially increased. The conventional type distributor, due to its preloaded vacuum diaphragm, is usually under-advanced for the same conditions. The spark advance remains at that produced by the centrifugal weights and no advance occurs until the manifold vacuum has built up to relatively high value. A similar type operation can be produced in the full vacuum ignition system by introducing a small poppet valve, actuated by manifold vacuum, in the vertical passageway. This prevents manifold vacuum from affecting the venturi orifice until the poppet valve spring is deflected due to higher manifold vacuums. At this time, the vertical passageway is cleared and the carburetor differential vacuums are produced in their normal manner. To date, experience on L-head engines has shown that this is not a necessity and, therefore, has not been used in current designs.

VACUUM CONTROL OF MAXIMUM ENGINE SPEED

Figure (12) illustrates a method of controlling maximum engine speed combined with the full vacuum controlled ignition system. Here we have combined within the distributor base, a speed sensitive shut-off valve designed to close due to a spring loaded centrifugal weight action at a pre-determined engine speed. Closing of this valve interrupts bleeding of manifold vacuum in the diaphragm at the base of the carburetor so that existing manifold vacuum is applied to the diaphragm causing it to move against its spring load. This movement acts to close the throttle plate until the engine speed is reduced and the governor valve again partially opens, balancing the throttle plate movement against distributor shaft speed.

Figure (13) illustrates a schematic of this system. Air enters the vacuum passageway at the carburetor air horn, travels to the distributor governor rotor body through a vacuum passage into the body of this assembly. When the governor valve is open, air flow occurs down through the distributor shaft and out through the distributor body into the governor diaphragm and manifold vacuum connection within the carburetor body. This air flow is interrupted at the governor rotor when the distributor shaft speed has increased to its governing point. When this occurs, the governor diaphragm is acted upon by manifold vacuum causing the throttle plate to be repositioned. This system produces positive engine governed speed control.

CONCLUSIONS

Full vacuum control of ignition timing represents several advantages as listed below:

(1) Simplification of assembly details of the distributor system.

(2) Elimination of centrifugal weight spark advance mechanisms.

(3) Longer life of wearing parts.

(4) Employment of a single spark advance mechanism within the distributor.

(5) Establishment of a definite relationship of full and part load differential vacuums within the carburetor.

(6) Simplification of service re-adjustment procedures.

(7) Fewer parts in assembly to be reworked if service overhaul is necessary of the distributor.

(8) Unaffected by torsional vibration causing spark timing spread through elimination of centrifugal weight assemblies.

There are several possible disadvantages in the application of this system which must be taken into account in its application.

(1) De-acceleration with closed throttle plate position fully retards the spark to the initial engine timing.

(2) The distributor rotor position is fixed in its relationship to the distributor driving shaft and rotor sparking to terminal housing post varies throughout the entire timing range.

(3) Additional vacuum passages must be introduced into the carburetor design to supply the differential distributor vacuums.

This design has been under development for the past five years and has been used on production vehicles for the past twenty months.

CAPTIONS FOR ILLUSTRATIONS

Fig. 1. Constructional details of full vacuum controlled distributor showing principal assemblies.

Fig. 2. Cross-sectional view of carburetor diagramming special passages for full vacuum control of distributor.

Fig. 3. Wide open throttle carburetor operation.

Fig. 4. Resultant carburetor differential vacuum for wide open throttle operation vs. cubic feet of air per minute through carburetor.

Fig. 5. Resultant carburetor differential vacuum for wide open throttle operation vs. engine speed.

Fig. 6. Spark advance vs. engine speed produced by carburetor differential vacuum for wide open throttle operation.

Fig. 7. Part throttle carburetor operation.

Fig. 8. Resultant carburetor differential vacuum for road operation vs. cubic feet of air per minute through carburetor.

Fig. 9. Resultant carburetor differential vacuum for road load operation vs. engine speed.

Fig. 10. Spark advance vs. engine speed produced by carburetor differential vacuum for vehicle road load operation.

Fig. 11. Differential vacuums and resultant spark advance for full load and road load engine operation obtained from dynamometer operation of a typical L-head engine. Also shown is minimum spark advance for best torque obtained by manual distributor operation as well as borderline detonation at full power operation.

Fig. 12. Full vacuum controlled ignition system combined with vacuum control of maximum engine speed employed in bus and heavy-duty truck engines.

Fig. 13. Schematic diagram showing vacuum connections and air flow of engine speed governor system.

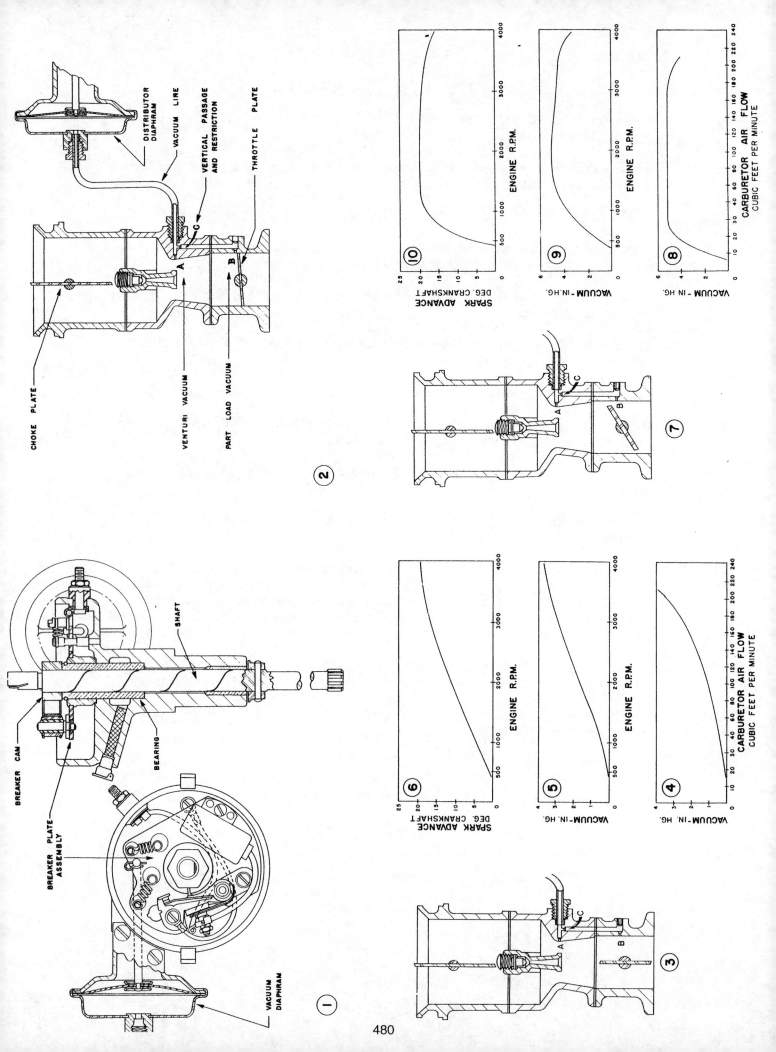

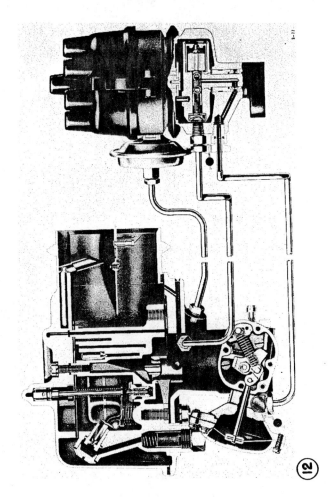

⑪

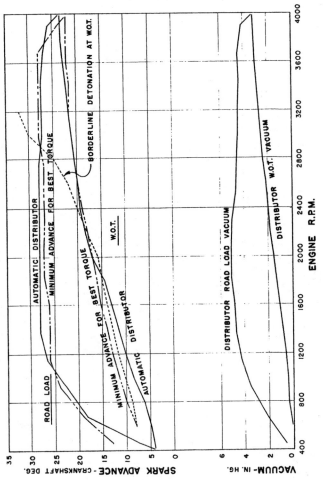

⑫

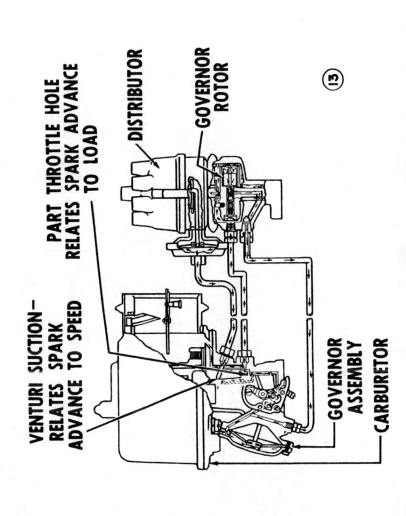

⑬

Capacitor Discharge Ignition - A Design Approach

Charles L. Shano and Arthur G. Hufton
Motorola, Inc.

SEVERAL FORMS OF CAPACITOR DISCHARGE (CD) ignition systems came into existance even before the emergence of transistor ignition systems.

Early designs used vacuum tubes in the d-c conversion circuit and a thryratron for discharging the high voltage capacitor into the ignition coil. These systems demonstrated that CD ignition could be superior to conventional in firing a fouled plug and extending spark plug life. However, because they did not provide the degree of freedom from maintenance that the auto industry was striving for, they saw very limited use. Other factors limiting use were high cost and incompatibility of components to shock, vibration, and temperature environments.

Transistors eventually replaced the tubes in the converters; however, it was the development of the solid state controlled rectifier which sparked the development of a practical ignition system.

The capacitor discharge approach to solving the ignition system problems became economically feasible with the development of low cost silicon controlled rectifiers capable of handling forward voltages in excess of 400 v.

What is there about this technique of producing ignition voltage that stimulates so much activity in our industry? To answer this question, we must briefly examine the shortcomings of the present induction type ignition system and identify the problems to be overcome. We must also examine the economic need for ignition system improvement to justify the development cost.

REVIEW OF CONVENTIONAL IGNITION SYSTEMS

DIRECT POINT OPERATED SYSTEMS - In the conventional system, seen in Fig. 1, the ignition transformer design is a compromise between low speed operation and spark duration time consistant with requirements for high speed operation. The compromise is particularly noticeable on

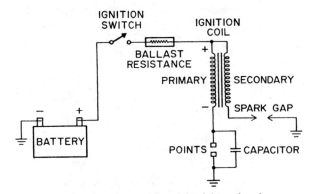

Fig. 1 - Conventional ignition circuit

ABSTRACT

A capacitor discharge ignition system tailored to modern automobile requirements has been developed which can be triggered direct from the regular distributor cam with breaker points or magnetic pick-up sensor. The system features improved fouled plug firing capability using the conventional ignition coil, draws current from the battery only when engine is rotating, and promises good performance in cold weather starting applications. Special attention has been given to keep ignition radiated radio interference to a minimum without sacrificing spark energy requirements for the many conditions of engine operation.

high rpm 8 cyl engines. The system has serious shortcomings at either end of the speed range and requires regular maintenance to prevent these from becoming a problem in the average car. During low temperature cranking requirements, the high voltage will fall off due to the slow rate of breaker point movement, as seen in Fig. 2.

The breaker point capacitor under these conditions is ineffective in aiding a fast fall off of current in the ignition coil primary winding, and arcing of the points results. The dwell time is also long and the system draws maximum average current at a time when battery reserves are already taxed. The ignition coil, however, requires this high energy storage to overcome the losses due to breaker point arcing. In most automobiles, the ballast resistor is shorted out during cranking to offset lower battery voltage and increase this energy to the coil. The need for this compromise also arises from the engine ignition requirements while being operated through the low speed range. During low speed acceleration, the demand for close to full available high voltage is realized because the engine cylinder is operating near its full volumetric efficiency and the combined centrifugal and vacuum advance requirements place the spark timing closer to top dead center. At high speed operation, the dwell time is too short to enable enough energy to be supplied to the coil, and again we see fall off of available high voltage generally resulting in a reduced high speed operating economy and capability. (See Fig. 3.)

It has been observed, under certain engine loading conditions with vacuum advance equipped distributors, that combustion may not necessarily take place until approximately 200 microsec have elapsed after the spark gap is ionized. Our observations are that in order to accommodate this error in timing, the ignition system should transfer and supply energy to keep the spark gas ionized long enough to insure combustion for all conditions of engine operation. The conventional ignition system has adequate spark duration to meet these requirements with the present methods of spark advancement.

Spark plug gaps tend to increase with engine life. This is apparently due to the combined effects of spark energy magnitude and electrode temperatures. The resulting gap growth reduces the ability of the ignition system to produce sufficient voltage to ionize the gap during acceleration and high speed operation. This results in the need for periodic maintenance or replacement of the spark plugs to prevent a loss in performance and operating economy.

Engine performance degradation begins to take place at the point where the secondary voltage is no longer high enough to fire the gap, and results in poor high speed performance and fuel economy.

Another and perhaps equally important shortcoming of the conventional system is its relative inability to fire a spark plug which has been fouled by lead deposits. The conductive fouling condition presents to the ignition system secondary circuit a terminating load resistance of variable value. (See Fig. 4.)

The effect of this fouling is shunting of the spark plug gap causing the spark gap voltage to be a function of the voltage divider network consisting of the coil secondary source impedance, harness resistance, and the resistance of the fouled plug. This can be shown as a reduction in high voltage and energy available, as seen in Fig. 5.

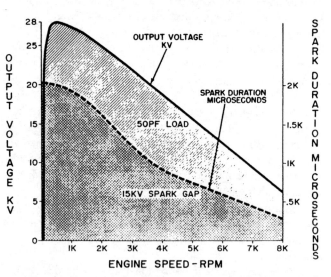

Fig. 3 - Conventional ignition voltage requirements and spark duration time as function of engine speed (8 cyl)

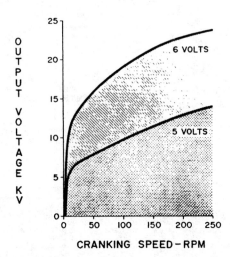

Fig. 2 - Voltage fall off at low speed cranking, representing cold weather conditions

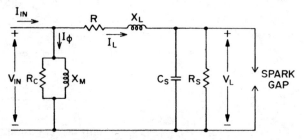

Fig. 4 - Ignition secondary equivalent circuit with resistor R_s in shunt with spark gap. (Coil secondary leakage inductance XL, lead wire resistance R, and shunt capacitance C_s)

The major mechanical constraint of the ignition system is in the use of the breaker points. Degradation of breaker points is caused by pitting and the build-up of an oxide layer between the breaker contacts due to the heat generated by the energy of the electric current which they continually interrupt. The value of the capacitor associated with the conventional system breaker points is designed nominally to minimize the effect of the inductive current on their contact interface. However, since the capacitor, ignition coil, and wiring harness have a considerable tolerance, the nominal design compatibility of these components working together is usually not realized. This results in breaker point replacement intervals of from 5000 to sometimes over 25,000 miles. If this contact deterioration were not present, the limiting factor on the life of breaker points would be the rubbing block wear, and the interim adjustments required for maintaining proper dwell necessary before complete wear out.

TRANSISTOR CONTROLLED SYSTEMS - The first major attempt at improving ignition system performance was the application of transistors to replace the current switching requirements of the breaker points. The transistors used in this application are capable of handling much higher currents than the breaker points, and this capability was utilized to obtain a performance advantage in the high speed operation of the ignition system. The high voltage available at cranking was also improved, because the transistor provides a more efficient method of switching. (See Figs. 6 and 7.)

Most ignition coils used with the transistor system required a turns ratio of 250:1 or higher to prevent transistor voltage breakdown failure. This resulted in coil primary currents having an average value of from 6 amps to over 10 amps, as shown in Fig. 8.

The high average current requirements of these systems resulted in early reliability problems. Transistorized ignition systems did improve breakerpoint life and high speed performance, however, they had little effect on fouled plug

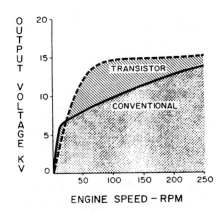

Fig. 6 - Spark plug voltage at cranking compared with conventional

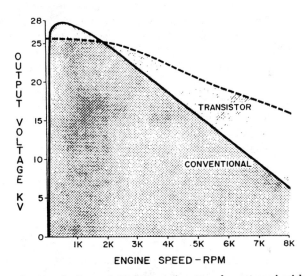

Fig. 7 - Spark plug voltage at running speed compared with conventional

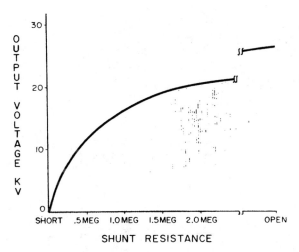

Fig. 5 - Secondary voltage as function of fouled plug load resistance

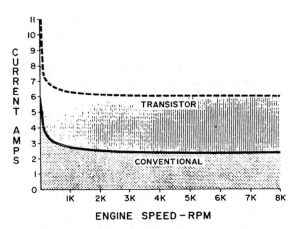

Fig. 8 - Typical transistor ignition system current drain versus conventional

firing ability. The promises of the transistor ignition system never materialized to the extent that it saw widespread acceptance.

CAPACITOR DISCHARGE SYSTEMS

DESIGN CRITERIA - The foregoing review of conventional and transistorized ignition systems points out some of their problems and shortcomings. The capacitor discharge ignition system is capable of solving these problems. However, it is intrinsically more costly at present because of its sophistication and complexity. In order for the system to have significant economic value, it must provide benefits at least equal in value to the extra cost. It is necessary, in the design planning stage, to estimate the savings the consumer can realize and target the cost of the system accordingly. Some of the factors considered were: consumer savings in service cost and inconvenience, alternative cost of the performance advantage, and reduction in operating cost over the longer maintenance period. After consideration of these and other factors, it was determined that the advantages could outweigh the cost and thus make the system economically practical for a sufficiently large number of applications. It was further assumed that component cost reductions would materialize rapidly, and that this would bring larger markets within reach.

Using an estimated end cost to the consumer as a guideline, a system design approach was formulated. It consisted of these elements:

1. Basic ignition system - breaker point operated and standard ignition coil.

2. Options - Breakerless adaptor to fit in place of points without distributor modifications; or pulse transformer ignition coil.

This breakdown permits the consumer to have cost and performance options at his discretion.

In order to minimize the cost of this apparently superior approach to providing the ignition requirements for gasoline engines and maintain the environmental capabilities of the system, a design for the basic ignition system evolved which is thought to fulfill the present market requirements for cost and performance.

BASIC SYSTEM FUNCTION - The basic capacitor discharge ignition, to be described, shows promise of overcoming the problems and shortcomings described in the previous discussion. The circuitry which has been developed for the basic system is unique in its simplicity of operation and its inherent regulation of output over a wide range of operating conditions. (See Figs. 9 and 10.)

The operation of the basic ignition system can be seen from the schematic diagram. When the ignition switch is closed and the cranking process started, breaker points or another suitable triggering mechanism produce a signal which is transformer coupled by T2 to the SCR gate. Battery current flows through the ignition coil primary, through the SCR gated on, through diode D4 to primary of transformer T1 and ground. The current increase in the primary of transformer C1 induces a current in winding FB which flows through diode D2, resistor R2 to the base of the power transistor Q1. (See Figs. 11 and 12.)

This voltage provides forward bias to transistor Q1 such that the transformer and transistor regenerate to increase the current through the primary winding of the transformer. At a particular current level through the transformer, saturation occurs. The voltage applied to the base circuit of the transistor decreases and the transistor turns itself off. During the turn off cycle, the energy stored in the transformer is coupled through the turns ratio and through the secondary winding and charges capacitor C4 through diode D3.

The following pulse from the timing mechanism fires the SCR through the driver transformer T2 and completes the circuit for discharging the capacitor through the ignition coil primary. Immediately after the secondary charging pulse decays, the transistor is again pulsed on repeating the capacitor charge cycle.

It is interesting to note that no current flows through the power transistor (Q1) in any quiescent state of the triggering mechanism. When breaker points are used for triggering, 80 milliamps will flow through the points when closed, be-

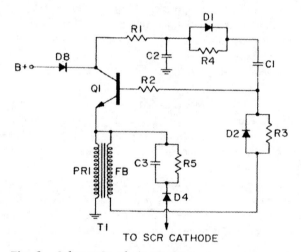

Fig. 9 - Schematic of circuit diagram converter unit

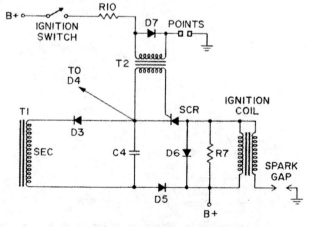

Fig. 10 - Schematic of circuit diagram converter unit

ing limited by a 150 ohm resistor in series with transformer T2. Cleaning of the points by the use of a higher current flow is not necessary and has been verified by the use of simulated contaminated points in excess of 1000 ohm contact resistance. This much resistance will not interfere with the system operation even at -40 F.

The transistor Q1 is designed to operate well within its load line requirements for safe area operation. (See Fig. 13.)

The transistor load line is controlled and limited by the saturation of transformer T1 and the circuits consisting of:

1. Diode D1, resistor R4, and capacitor C1 connected between collector and base.

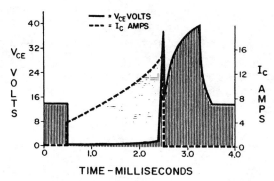

Fig. 11 - Transistor voltage and current waveforms

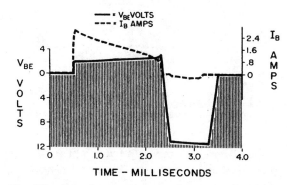

Fig. 12 - Transistor voltage and current waveforms

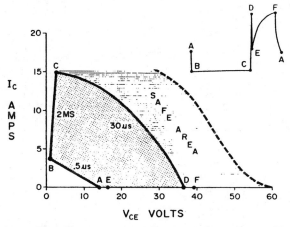

Fig. 13 - Transistor load line and safe area

2. The capacitor C4 and diode D3 connected in series with transformer T1 secondary winding.

The saturation of transformer T1 limits the current in the collector of transistor Q1 by causing reduced base drive and is represented by points A, B and C on the load line.

The circuit connected between the collector and base of transistor Q1 limits the high frequency component of the backswing voltage occurring when the transistor switches off, and is represented by points C, D and E on the load line. The capacitor C4 limits the low frequency component of the backswing voltage which also provides the stored energy for discharging into the ignition coil. This is represented by points E, F and A on the load line.

When transistor Q1 conducts, capacitor C1 is discharged through resistors R4, R1 and R2; capacitor C4 is discharged through the ignition coil. When transistor Q1 switches off, capacitor C1 absorbs the energy from the high frequency component of backswing voltage, which is approximately 70 millijoules.

Capacitor C2 is required to prevent false triggering by reducing the rate at which capacitor C1 is charged when battery voltage is abruptly applied.

The function which transistor Q1 performs is that of a d-c to d-c converter, enabling the operation of the capacitor discharge system from a capacitor stored energy level sufficient to provide the voltage and energy necessary to ignite the combustion mixture in the engine for all conditions of engine operation.

BASIC SYSTEM CAPABILITY - The electronic system is capable of producing a minimum of 26,000 v of normal battery voltages and sufficient output at low battery voltages during cranking to start a high compression engine. Cranking voltage available is approximately 18,000 v with 5 v input, and 20,000 v at 6 v input from the battery. It will operate over a temperature range of -40 F or lower to in excess of 225 F and over a speed range of well below 10 rpm to around 8000 rpm for an 8 cyl engine.

The requirement for spark duration coincides with the maximum discharge energy necessary for igniting the fuel mixture. The energy in the capacitive component of the discharge is approximately 4 millijoules for most automobile ignition systems using the regular high voltage cable and is of very short duration. The capacitive component of energy is usually sufficient to fire the fuel mixture; however, in cases where the gasoline is imperfectly vaporized under relatively low cylinder pressures (that is, as at full vacuum advance), some of the discharge energy will be used to vaporize a small part of the fuel mixture before combustion takes place.

If the energy in the high voltage capacitive component is insufficient to cause combustion, the additional energy must be provided by the inductive component where the capacitor discharge ignition system delivers 12 millijoules in a duration time of 250 microsec with resistive harness or 400 microsec with no resistance or Essex radio suppression wire of the magnetic induction type. (See Fig. 14 for comparison curves.)

SPARK PLUG LIFE - A great deal has been said about the improvement in spark plug electrode life with capacitor discharge ignition. The electrode erosion rate reduction can perhaps be understood from a comparison of the spark gap energy differences of the two systems. Assuming that the spark plug electrode temperatures would be the same for each system type under similar engine operating conditions, the erosion rate would be a function of the spark gap discharge energy. Referring to Fig. 14, comparing the spark gap energy for a conventional system and Motorola CD system, it can be observed that the delivered energy difference between the system is most pronounced at the lower speed ranges of the spark gap energy curve.

It is probable that the significantly smaller spark gap energy of the capacitor discharge system provides the understanding of why the capacitor discharge type of ignition will truly extend spark plug life. It remains to be seen, from experience, how significant the increase in plug life will be.

IGNITION COIL REQUIREMENTS - To effect a cost savings to the consumer, the basic ignition system has been designed to operate with conventional ignition coils having a primary inductance of 7 to 12 millihenry and a turns ratio of approximately 100 to 1. This has been well justified in that the foul plug firing capability has been improved by 2 to 1.

The curve of Fig. 15 shows a 2 to 1 improvement in fouled plug firing ability due to the application of a fast rise pulse to the primary of the ignition coil, which results in approximately a 2:1 rise time difference at the spark plug favoring the CD system.

A great deal has been said about the virtues of the capacitor discharge ignition when associated with fouled plug conditions, however, only a portion of this capability can be achieved when a conventional ignition coil is used.

If the user desires to operate the system in such a way that he can take advantage of the maximum capabilities for firing fouled plugs, it is necessary that he use an ignition coil which is designed to utilize the system capabilities. Such a coil is more truly a pulse transformer device in that it is not designed for energy storage.

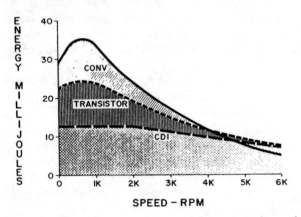

Fig. 14 - CDI spark energy versus engine rpm compared with conventional and transistor ignition

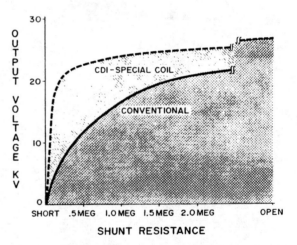

Fig. 16 - Available voltage versus fouled plug resistance with specially designed ignition coil compared to conventional

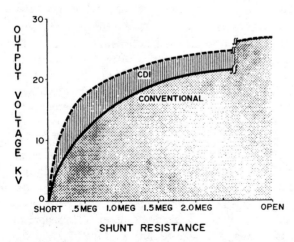

Fig. 15 - Conventional and CD ignition available voltage as function of simulated spark plug fouling resistance

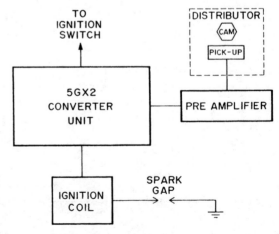

Fig. 17 - Pick-up, preamplifier, and converter block diagram

Fig. 18 - Power converter, preamplifier, and pick-up

As you can see from Fig. 16, the ability to fire a fouled plug has been improved 5 to 1 compared with conventional ignition by the use of a pulse transformer ignition coil, however, due to the high rate of energy release and low secondary inductance, the duration time is relatively short and can produce severe engine stumbling where vacuum advance is employed.

BREAKERLESS OPERATION - An electronic triggering package has been designed and will soon be in production to operate with the above discussed system. The package in Fig. 17 consists of a magnetic pick-up sensor and a preamplifier.

The magnetic pick-up will fit in place of the breaker points without modification to the existing distributor and is energized by the rotation of the distributor cam. The complete system will operate at engine cranking rates of lower than 20 rpm at a temperature of -40 F with appropriate reduction in battery voltages. The system is not affected by starter motor radiation or vibration of engine component parts. (See Fig. 18.)

RELIABILITY - Indications of reliability at the present time are favorable to the extent that the system has been completely encapsulated and is nonrepairable. This is accomplished in volume production by paying particular attention to component specification and testing of the individual components and subassemblies prior to making the final production encapsulation.

The use of silicon components for all semiconductors is mandatory in view of the system reliability and environmental temperatures encountered. The system packaging insures protection against shock and vibration to which it may be exposed in a wide variety of installations. For general applications, the system is compatible with either negative or positive ground electrical applications, and it is capable of being operated by either conventional breaker point configuration or electronic triggering, depending upon the users desire.

Its circuit is designed against improper or reverse polarity connections and is virtually electrically indestructable by laymen and service personnel. It is also relatively immune to shock, vibration, and humidity, which is another requirement necessary for a successful system.

This paper is subject to revision. Statements and opinions advanced in papers or discussion are the author's and are his responsibility, not the Society's; however, the paper has been edited by SAE for uniform styling and format. Discussion will be printed with the paper if it is published in SAE Transactions. For permission to publish this paper in full or in part, contact the SAE Publications Division and the authors.

8 page booklet - Printed in U.S.A.

Some Factors to Consider in the Design and Application of Automotive Ignition Systems

Paul C. Kline
Delco-Remy Div., General Motors Corp.

INTRODUCTION

The primary purpose of this paper is to provide the engineer who is relatively unfamiliar with automotive ignition with a general understanding of the subject, together with an appreciation of the importance of proper component design and application. Discussion centers on the standard or "Kettering" type system in general automotive use; however, the merits of semiconductor systems are also recognized. A number of potential ignition systems problems the engine designer must recognize are presented, together with approaches for avoiding them. The paper is divided into six basic sections:

(1) Electrical considerations
(2) Mechanical considerations
(3) Exhaust emissions considerations
(4) Environmental considerations
(5) Radio frequency suppression considerations
(6) Semiconductor ignition systems

THE AUTOMOTIVE IGNITION SYSTEM

Automotive engines are fitted with battery powered electrical ignition systems to provide the spark to ignite the fuel-air mixture compressed in the combustion chambers. This system must:

(1) Provide sufficient voltage to the spark plug to cause a spark of sufficient intensity to ignite the combustible mixture.
(2) Supply this spark at the exact instant in the compression cycle to provide, as far as is practical, the maximum power and best economy for the specific operating condition. This optimum spark timing varies as engine speed and manifold vacuum vary; therefore, the system must have means of automatically changing "spark timing" as speeds and loads vary.

ABSTRACT

Ignition is such a vital factor in the performance, reliability and service life of the internal combustion gasoline engine that its basic design and application deserve major attention from the engine designer. This paper reviews the electrical and mechanical functions of present-day standard, or "Kettering" type ignition systems, and discusses potential design and application problems, together with ways to avoid them. Distributor mounting and drive, environmental factors such as moisture and dirt, radio frequency suppression, and exhaust emission considerations are briefly discussed. Advantages and limitations of typical semiconductor systems are also presented because these comparatively new systems offer major advantages that, on many applications, more than justify their premium initial cost.

ELECTRICAL CONSIDERATIONS

A modern 12-volt ignition system, Figure 1, includes: Battery, Switch, Resistor, Coil, Distributor, Spark Plugs, and Necessary Wiring.

1. The *Battery* is the source of power for the system.
2. The *Switch* turns the system on or off. In Figure 1 it is also the starting switch—for this the key is turned beyond "On" into "Start" position, and is so constructed and wired that during starting the ignition resistor is bypassed, and full battery voltage is supplied the coil. After starting, the key returns to "on," and the resistor operates in series with the coil.
3. The *Resistor* provides part of the resistance necessary to limit current through the coil primary circuit. The resistor will be discussed further under coils.
4. The *Coil* is a "pulse" transformer that steps up the 12-volt battery voltage to the thousands of volts necessary to "jump" the spark plug gap.

Figure 2 shows construction of a typical coil. Its three essential components are: (1) The primary winding, normally consisting of 200 to 300 turns of comparatively heavy (20 to 23 gage) enamel or nylon coated copper wire. (2) The secondary winding, having thousands of turns (usually 20,000 to 30,000) of very fine (40 to 44 gage) enamel coated copper wire. The "secondary" is assembled inside of the "primary." (3) The "magnetic" or "iron" circuit, consisting of the "core" and "outside" iron. The "core" is made of many strips of soft silicon steel assembled into the center of the windings. The "outside" iron consists of laminated or sectioned sheets of similar steel wrapped around the outside of the windings.

These "essential" parts, with necessary insulating parts, are housed in a metal case and covered with a molded insulating cap. Connection to the windings is made through terminals in this cap. The space around the windings is filled with oil, or other insulating compounds.

Figure 3 shows schematically the primary ignition circuit (in gray) and the coil portion of the secondary "high voltage" circuit (in black). When the contacts or "points" in the distributor close, current flows from the battery, through the resistor, coil primary, contacts, and back to the battery through ground. This current through the primary winding creates a magnetic field within the coil (see dotted lines)—the majority of the flux lines in this field take the easiest magnetic path from the core iron, to the outside iron, and back to the core iron; thus, the iron concentrates the magnetic field. This current does not reach a maximum instantly, but requires a fraction of a second "build-up" time. At low speeds this time of contact closure is sufficient for the coil to reach the maximum permitted by circuit resistance; however, at higher speeds this time is less and current reached somewhat less. As this current builds up, it causes a comparable increase in the strength of the magnetic field "loading" the coil with magnetic energy.

Energy stored expressed mathematically:

Energy "Stored" in the Coil $= \frac{1}{2} Li^2$ Joules
where L is the primary inductance in henries and i is the primary current in amperes at the time the contacts open.

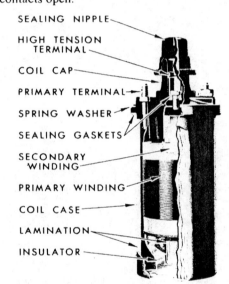

Fig. 2 Ignition Coil

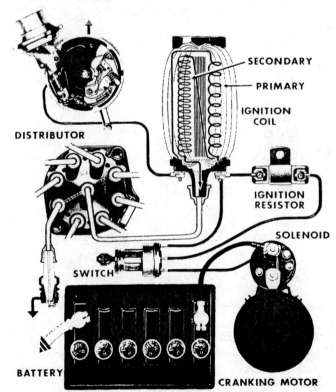

Fig. 1 Present-Day 12-V Ignition System

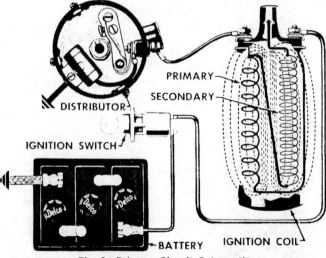

Fig. 3 Primary Circuit Schematic

This equation shows that energy stored does not involve resistance; however, resistance does affect current. Sufficient circuit resistance must be provided to limit the low speed current to a value the contacts can safely carry, and since it does not affect coil energy, part of the resistance can be provided by an external resistor if desirable from a design standpoint. The graph in Figure 4 illustrates energy input to a typical 12-volt coil.

When the contacts are opened by action of the distributor cam, current flow is stopped, and the magnetic field collapses very rapidly (much more rapidly than it was created). During this collapse, the magnetic flux lines cross or "cut" each of the primary and secondary turns, inducing a high voltage into the windings. Normally the voltage induced into the secondary winding increases as the field collapses until it reaches a value sufficient to "fire" the spark plug gap to which it is connected through the distributor cap, rotor, and wiring.

The peak energy transferred to the secondary system is stated as:

Peak energy, in joules $= \frac{1}{2} CE^2$

where $C =$ capacity of secondary system in farads
$E =$ peak voltage in volts

From the foregoing equation it can be seen that—

Secondary voltage, $E = \sqrt{\frac{2 \times \text{Energy in joules}}{C}}$

The value of the primary current when the contacts open decreases with increases in engine speed. The current for any given time of contact closure can be determined from the relation:

$i = E/R \, (1 - e^{-Rt/L})$

where $i =$ current in amperes
$E =$ supply voltage in volts
$R =$ total resistances of the primary circuit in ohms
$t =$ the length of time the contacts are closed in seconds
$L =$ primary inductance in henries
$e =$ Naperian log base, 2.718

At low speed the quantity $e^{-Rt/L}$ is insignificant and the primary current is approximately equal to E/R. With increased engine speed the quantity becomes greater, because t becomes smaller, and reaches a value of about 0.5 at top engine speed, thus reducing the current at break to about 50% of the low speed value. See Figure 5. Any increase in L will further decrease the current at break at high speeds, hence, will decrease the energy input.

The ignition system designer tailors the system to the application by varying coil primary inductance and resistance, and time the contacts are closed (distributor dwell time). He is restricted, however, by the limited current carrying capacity of the distributor contact points.

IGNITION PERFORMANCE

To make certain the plug is always "fired" an engine must have an adequate ignition reserve. This means that the high voltage the ignition system is capable of producing (called "available" voltage) must always be greater than the voltage necessary to provide a spark at the plug (called "required" voltage). A cathode-ray oscilloscope is used to determine reserve by measuring ignition voltages during engine operation. Figure 6 illustrates a typical performance curve. Available voltage data is obtained by making a connection from the secondary coil terminal to the input of the oscilloscope (through a voltage divider), then disconnecting the longest spark plug lead from its plug and observing the voltage developed on that lead. The longest lead is selected because the higher electrostatic capacity of that lead will give it the lowest available secondary voltage. A typical open circuit or "available voltage" wave is shown in Figure 7.

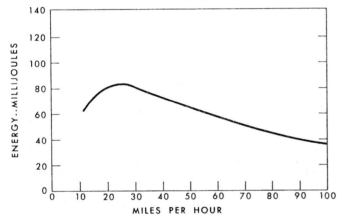

Fig. 4 Energy "Stored" in Typical 12-V Coil

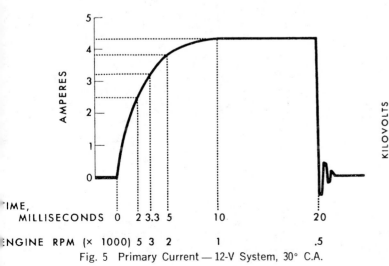

Fig. 5 Primary Current — 12-V System, 30° C.A.

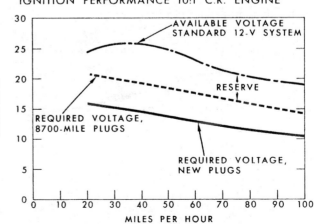

Fig. 6 Typical Ignition Performance Curves

The main factors influencing available voltage in a given ignition system are:

(1) *Engine Speed.* The length of time the distributor contacts are closed for each pulse is expressed by the equation:

$$t = \frac{6N}{D}$$

t = the time the contacts remain closed (in seconds)
D = the dwell of the contacts (in cam degrees)
N = the cam speed (in RPM)

It can be seen that closure time decreases as engine speed increases.

(2) *Contact or Dwell Angle.* From the equation expressed in the engine speed statement it can be seen that coil "build-up" time can also be increased by increasing the contact angle.

(3) *Distributor Point Opening.* This influences available secondary voltage because of its effect in determining contact angle. The greater the point opening on a given circuit breaker cam, the lower the contact angle.

(4) *Contact Arcing.* When an arc occurs between the contacts as they open, a part of the energy stored in the coil is lost in the arc. With less energy the output of the coil is less. Arcing is most noticeable at low speeds because the contacts are opened slowly. If contacts are in reasonably good condition, arcing nearly disappears at speeds above 1,000 rpm. As contacts deteriorate with extended use, arcing increases, causing available voltage to drop. Contacts must be replaced periodically to maintain adequate available voltage.

(5) *Battery Voltage.* Available voltages are affected directly by battery voltage.

(6) *Secondary Capacity.* Capacity in the secondary system has a lowering effect on available voltage. Peak energy transferred to the secondary = $\frac{CE^2}{2}$, where C is the capacity of the secondary system in farads, and E is peak voltage in volts; therefore, for a given energy transfer, a higher voltage can be obtained by decreasing secondary system capacity. Figure 8 illustrates how one car builder increased available voltage 15% by improving lead arrangement.

(7) *Losses.* Fouled spark plugs, "leaky" secondary leads, etc., all contribute to lower available secondary voltages.

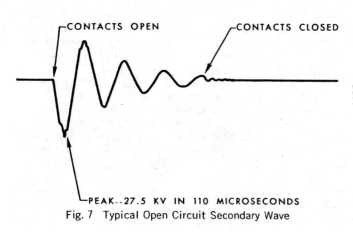
Fig. 7 Typical Open Circuit Secondary Wave

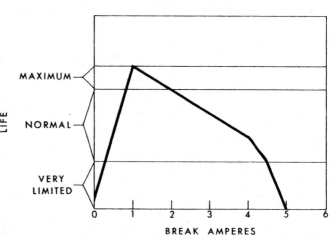
Fig. 9 Effect of Break Amperes on Contact Life

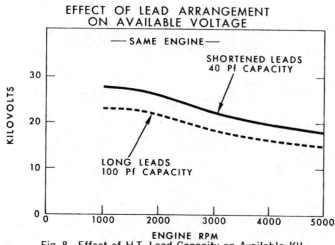
Fig. 8 Effect of H.T. Lead Capacity on Available KV

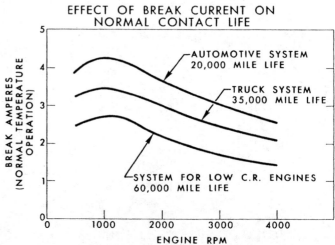
Fig. 10 Break Current vs. Contact Life

The two primary important available secondary voltage limiters on present standard ignition systems are:

(1) *Distributor Contacts*—Although tungsten contacts presently used are made of the best material commercially available for the purpose, they deteriorate rather rapidly if worked much above 4.0 amperes. See Figure 9. Conversely, contact life can be extended if a lower available voltage curve will provide an adequate reserve. Figure 10 illustrates typical break current curves, together with anticipated contact life. Figure 11 indicates how available voltage reduction must be accepted to gain extended contact life.

(2) *Resistive Shunt Loads*—Parallel leakage paths, particularly deposits that tend to accumulate on spark plug procelain near the spark gap "bleed off" secondary ignition energy, and reduce available voltage. Figure 12 illustrates how available voltage is affected by parallel secondary resistance. Moisture on the high tension system can also cause such losses. When plug deposits, moisture, or other deterioration lowers available into the required voltage area, servicing is necessary if "missing" is to be avoided.

Required voltage is that voltage required to fire the spark plug. Figure 13 illustrates a typical required voltage wave shape as observed on an oscilloscope.

The magnitude of secondary voltage requirements is primarily dependent upon the following items:

(1) *Spark Plug Electrode Shape*. New spark plugs have comparatively low requirements because electrodes have sharp corners. The rounding off of the center electrode during service causes the secondary voltage requirements to increase. The 5 KV requirement increase between new and used spark plugs shown in Figure 6 is largely due to this electrode shape change.

(2) *Spark Plug Gap*. The greater the gap the greater the required voltage. Gap growth due to normal erosion causes required voltage to increase as mileage on a plug accumulates.

(3) *Compression in the Cylinder*. Increasing compression ratios with the resulting increases in combustion chamber pressures increases secondary voltage requirements. See Figure 14 for the effect of increased compression ratios on required secondary voltages.

(4) *Mixture Ratio*. Lean fuel mixtures tend to cause higher secondary voltage requirements. It is also possible that extremely rich mixtures can have this same effect.

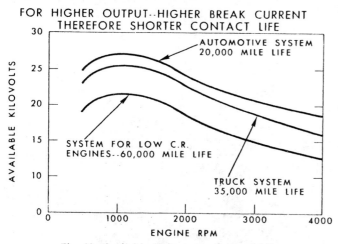

Fig. 11 Available Voltage vs. Contact Life

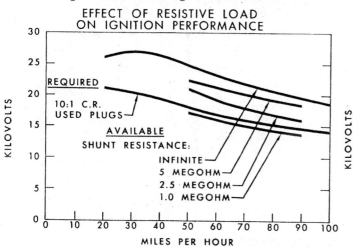

Fig. 12 Effect of Shunt Resistance Losses on Ignition

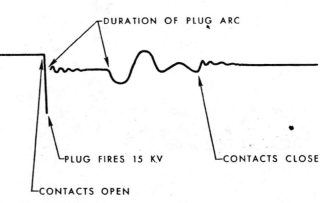

Fig. 13 Typical Secondary Voltage Requirement Wave

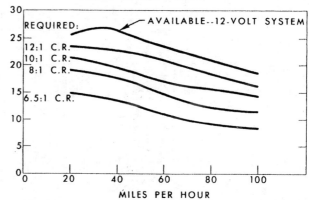

Fig. 14 Effect of Compression Ratio on Ignition

Cold weather cranking presents a special ignition performance challenge because battery voltage supply to the ignition system might be quite low and spark plug requirements are often quite high. Figure 15 illustrates a "resistor bypass" circuit commonly used to provide high available voltage during cranking. With this circuit, the resistor is bypassed during cranking, permitting normal or even higher than normal ignition primary break current even though the battery voltage has been pulled down severely by the cranking motor load. The -15° F cranking ignition voltages shown in Figure 16 illustrate that available voltages exceed required by an adequate margin even though the 12-volt battery has been pulled down to 5 volts by the severe cranking load.

MECHANICAL CONSIDERATIONS

The distributor, Figure 17, contains all of the mechanically moving parts in the ignition system. Normally it is fitted to the engine so as to be driven ½ engine speed by a gearing arrangement with the camshaft. Its three basic functions are:

(A) *Opening and closing the primary circuit* to activate the coil. Figure 18, the Circuit Breaker Assembly, consisting of the contacts, the breaker cam, and the condenser, perform this function. The cam has the same number of lobes the engine has cylinders, and is coupled to the distributor main shaft. Thus, each revolution it opens and closes this circuit once for each cylinder, causing the coil to produce one high voltage pulse per cylinder.

The condenser, connected across the contacts, prevents or minimizes arcing at the contacts as they open, and also contributes to the quick collapse of the coil's magnetic field.

(B) *Provides Spark Advance*, which may be defined as providing the spark earlier in the compression stroke as speed is increased, or under light loads when manifold vacuum is high. To understand the need for advance it should be noted that: 1) The "explosion" in the cylinder is not instantan-

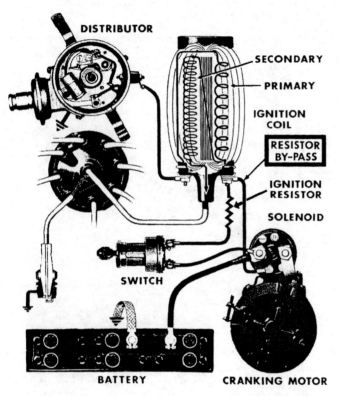

Fig. 15 Ignition Resistor Bypass During Cranking

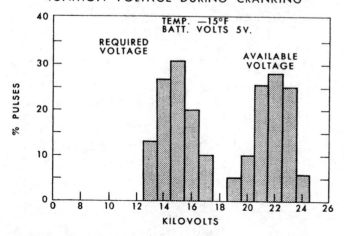

Fig. 16 Cranking Performance — Resistor Bypass System

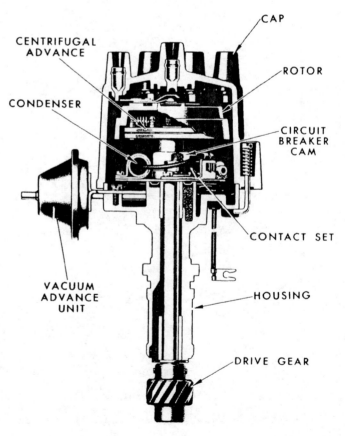

Fig. 17 Ignition Distributor Construction

eous, but a very rapid burning process; 2) The time from the instant the spark occurs until combustion is complete is generally .001 to .004 seconds, depending on fuel mixture, compression, combustion chamber shape, and other factors; 3) Most of the explosion power is given up during the latter part of this burning time; and 4) For the piston to obtain maximum power, it should be slightly past T.D.C. when this maximum power of explosion is delivered.

To illustrate "advance," assume that, at full throttle, conditions on an engine are such that .003 seconds "burning time" is required, and that maximum power results if burning is complete 10° past T.D.C. At 1,000 rpm, the crankshaft will rotate 18° in .003 seconds, while at 2,000 rpm it will rotate 36° in the same time. As Figure 19 shows, the spark must be initiated at 8° before T.D.C. at 1,000 rpm, and 26° before T.D.C. at 2,000 rpm if burning is to be complete by 10° after T.D.C. (The constant "burning time" assumed is not absolutely true; however, the idea presented shows why "advance" is necessary.)

Under heavy throttle conditions, when the fuel mixture is "rich" and highly compressed, a certain amount of advance will provide maximum power. Under part throttle "cruising" conditions, when manifold vacuum is high, the mixture "lean" and not so highly compressed, burning is slower and additional advance will increase economy.

Most engines are fitted with distributors that provide a combination of centrifugal and vacuum advance.

(1) *Centrifugal Advance*

Figure 20 shows a centrifugally actuated mechanism that moves or advances the breaker cam ahead of the distributor shaft, causing the lobes to activate the points earlier in the compression stroke, "advancing" the spark. Centrifugal force on the spring loaded weights varies with speed, and construction is such that advance depending on speed is accomplished. A typical centrifugal advance curve plot is shown in Figure 21. Centrifugal advance provides best full throttle timing.

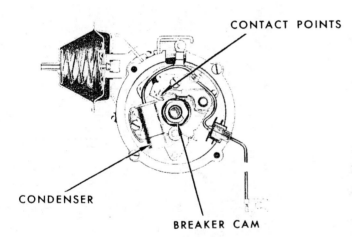

Fig. 18 Distributor Circuit Breaker Assembly

 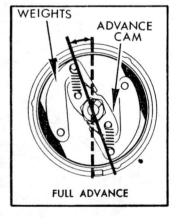

Fig. 20 Centrifugal Advance Mechanism

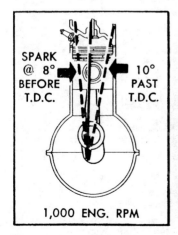 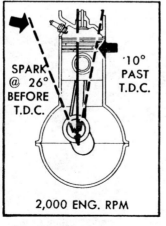

Fig. 19 Reason for Spark Advance

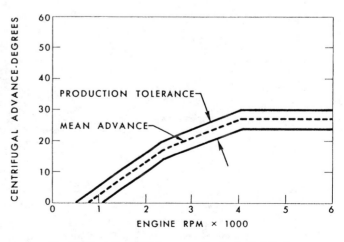

Fig. 21 Centrifugal Advance — Typical Distributor

(2) *Vacuum Advance*

Figure 22 shows a system whereby a vacuum unit is linked to a movable plate on which the contacts mount. Construction is such that under part throttle conditions, manifold vacuum can rotate this plate in a manner to cause advance. Figure 23 illustrates a typical vacuum advance curve. Figure 24 illustrates how the centrifugal curves in Figure 21 and vacuum curve in Figure 23 combine during engine operation. Centrifugal advance, depending on speed is present, and added vacuum advance may or may not be present, depending on intake manifold vacuum.

(C) *Distribution* of the high voltage occurs through the circuit shown in Figure 25. The pulse travels from the coil, its source, to the center tower of the cap, through the rotor, to an outer cap tower, and thence to the spark plug intended to "fire." The rotor is coupled to the main shaft, and distributor construction is such that the rotor tip aligns with an outer tower insert when the high voltage pulse is produced.

DISTRIBUTOR MOUNTING AND DRIVE

The exact timing of the spark provided the spark plug involves a number of mechanical tolerances and operating variables, each of which must be considered if the benefits of precise spark timing at all speeds are to be gained. Figure 26 illustrates a few of these factors:

(1) *Distributor Advance*—Centrifugal and vacuum curve tolerances.

(2) *Timing Gears Or Chain*—Tolerances, whip, backlash, and eccentricities.

(3) *Distributor and Camshaft Gears*—Tolerances, concentricity, backlash.

(4) *Engine Timing Marks And Pointer Location* with respect to the exact piston position are affected by a large stack-up of tolerances.

(5) *Reading Parallax*—Engine manufacturers find that "human elements," even with conscientious workmen, may cause variations of ± 2° in initial timing.

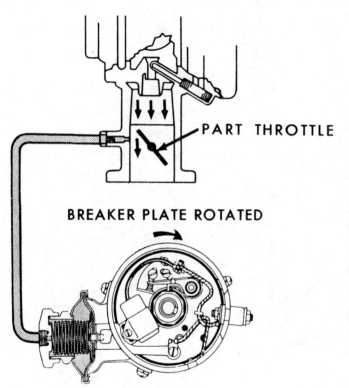

Fig. 22 Vacuum Advance System

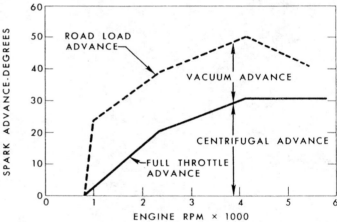

Fig. 24 Spark Advance on a Typical Engine

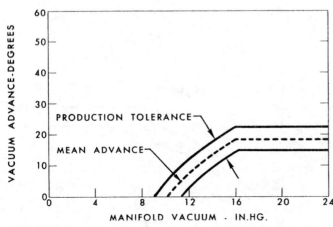

Fig. 23 Vacuum Advance — Typical Distributor

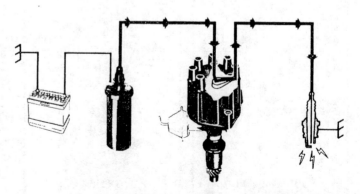

Fig. 25 High Voltage Distribution System

Driving the oil pump through the distributor, Figure 27, is a design practice proven very effective in stabilizing engine timing and reducing variations. The steady torque load which may reach 3 h.p. or more at higher speeds dampens torsionals in the total timing system, and keeps the distributor gear loaded against the drive side of the cam shaft gear so that it will not torsionally oscillate through the gear backlash. Without such a load, backlash may permit several degrees of "spark scatter." Unfortunately "scatter" caused by backlash is always in the direction to advance the spark, and may lead to detonation problems.

Damping "snubbers" may also be required to eliminate "whip" from timing chains. End play of the camshaft must be held to a minimum to prevent axial oscillations from working the camshaft-distributor gears through their helix angles, causing timing variations.

EXHAUST EMISSIONS CONSIDERATIONS

Until four years ago, spark advance determination for an engine was rather straightforward. The centrifugal curve was selected to provide maximum full throttle power, without detonation, while vacuum advance was based on part throttle tests to determine the amount of additional advance necessary to gain maximum fuel economy at "cruising" loads. In 1966 the new phrase *"Unburned Exhaust Emissions Control"* became a byword and a challenge throughout the gasoline engine industry. "Emissions" added a third, and major requirement to advance curve selection, *"provide minimum unburned exhaust emissions under all operating conditions."*

In addition to new carburetor designs and calibrations, new combustion chamber shapes, and a host of other engine changes, special spark advance characteristics have proven effective in lowering emissions levels. In some cases, power and economy objectives must be compromised slightly to meet these new requirements.

Nearly all late model cars have advance curves tailored specifically to meet emissions objectives. Some employ spark advance control devices outside the distributor. For example, the Transmission Controlled Spark (TCS) system used widely on '70 GM cars (Fig. 28) eliminates vacuum advance at

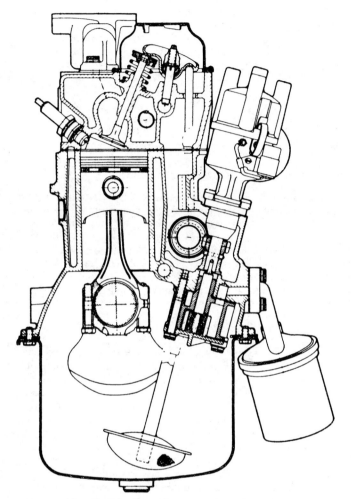

Fig. 27 Distributor Gear Drives Oil Pump

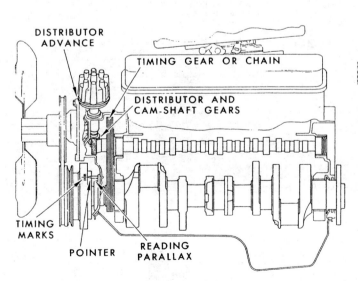

Fig. 26 Factors Affecting Spark Timing

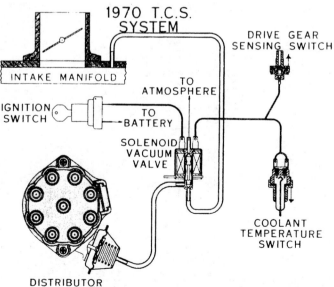

Fig. 28 Transmission Controlled Spark Schematic

idle, and during first and second gear operation, except when engine coolant temperature exceeds 220°F. "Timing" control components outside the distributor include: 1) Drive gear sensing switch on the transmission, 2) Solenoid vacuum valve, and 3) Coolant temperature switch. Other 1970 vehicles use a double diaphragm vacuum unit on the distributor. Figure 29 shows a system designed to provide normal vacuum advance anytime the throttle is opened past the carburetor "spark port," but retard timing appreciably during idle and coast-down.

Spark timing to meet emissions needs will continue to challenge the ignition designer. Unique new systems will be required, and present approaches will demand improvement if the reduced emissions levels and solid vehicle driveability objectives of vehicle manufacturers are to be met.

ENVIRONMENTAL CONSIDERATIONS

Water, in various concentrations from precipitation, road splash, and condensation has long been a nemesis of the ignition system. Dirt, dust, road salt, heat, oil, and inadequate distributor ventilation may also present problems. Techniques for overcoming such environmental problems are well established and application proven. The engine designer who takes the following precautionary measures will generally gain an ignition system that is free from normal environmental problems.

I. *Consider The Ignition System Environment Early In New Engine Programs*

Ignition system environment problems can be avoided by due concern for this area early in a new engine design program. If flexibility in the distributor location exists, a mounting should be selected where other engine or vehicle components will provide natural splash protection. Spark plug locations too close to hot exhaust manifolds may cause leads and boots to deteriorate rapidly, and must be avoided. Spark plugs should not be located in depressions that might trap water, and high tension leads should be routed to avoid hot exhaust manifolds. Plans for new engine designs should include periodic "Failure Prevention Analysis" reviews to identify potential problem areas.

By keeping ignition component design and environment in mind from the start of the design, and by reviewing this area at regular intervals, the designer can avoid future field problems and save the cost of embarrassing field fixes that are often painfully expensive.

II. *High Tension Cables*

High tension cables must be selected that are made of a durable, high quality material, capable of withstanding millions of high voltage pulses, corona, ozone, moisture, dirt, salt, oil, heat, and more often than not, mechanic abuse at spark plug replacement time. Figure 30 shows cable construction used as original equipment on most U.S. cars. The non-metallic conductor core was developed to minimize radio and TV interference created by the ignition system. Temperature is a major consideration in selecting the proper insulation material. Neoprene may be used for ambient temperatures to 260° F, while hypalon is used up to 300° F, however, for higher temperatures, in the 300° to 425° F range, silicone rubber insulation should be used. Metallic conductor cores are used for applications over 300° F because the non-metallic suppression cores deteriorate at these temperatures. In this case, RFI suppression, if

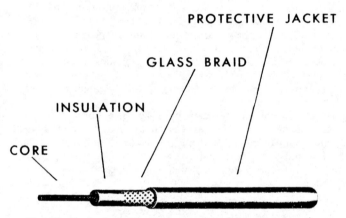

Fig. 30 Typical O.E.M. High Tension Cable Construction

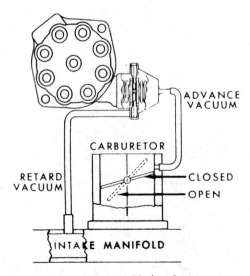

Fig. 29 Advance-Retard Timing System

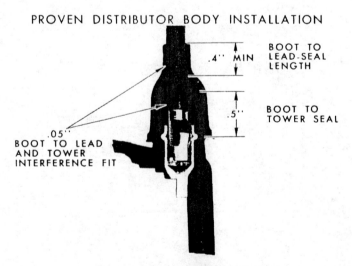

Fig. 31 Distributor High Tension Connection

required, must be accomplished with devices other than suppression cable. Cable with vinyl or other plastic insulations are marketed, however, temperature limitations and heat aging tendencies discourage their use.

III. *Distributor, Coil Tower, and Spark Plug Boots*
High tension boots are intended to seal; if they do not seal, and continue to maintain that seal, they will not assure trouble free ignition.

Figure 31 shows a good distributor boot installation. The boot must grip the cable tightly (.05" interference fit) for .4" of the cable length to assure a good moisture seal, and grip the tower for at least .50". Neoprene is satisfactory for applications up to 260° F, while hypalon may be used up to 300° F. Vinyl hardens and shrinks with heat aging, and is not dependable for long-life protection.

A proven spark plug boot installation is shown in Figure 32. Hypalon is used where temperatures below 300° F are anticipated, while silicone rubber is necessary for use in the 300° F to 425° F range.

IV. *Distributor Cap Design*
In humid weather moisture films often condense inside distributor caps during engine shut-down periods. This moisture, when added to dust and other surface contamination can cause significant ignition energy losses. The engine may fail to start in damp weather if these losses reduce available ignition voltage below spark plug firing potential. The engine designer must, therefore, allow space for, and specify distributor caps that are sized and designed to minimize the effects of moisture.

Figure 33 shows a cap designed to handle 25 KV with minimum losses even when wet. The cap is large enough to provide generous flashover spacing between inserts, with sharp cornered ribs across each of the possible leakage paths to break up moisture films and minimize losses.

V. *Distributor Ventilation*
High voltage corona and the rotor gap sparking breaks down the air inside the distributor into nitrous oxide compounds. Inadequate ventilation may permit acid concentrations sufficient to corrode and roughen the breaker cam, causing rapid rubbing block wear, plus corrosion of other internal parts. Cap and rotor surfaces may also deteriorate, permitting conducting carbon tracks to form across insulating surfaces. Distributors having vent openings to outside moving air with a total area equivalent to a ¼" diameter hole are generally free from such trouble.

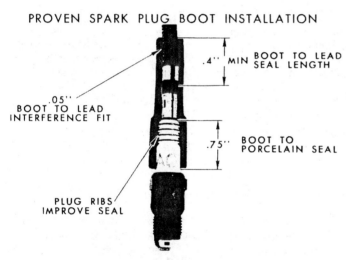

Fig. 32 Spark Plug Termination

Fig. 33 Distributor Cap

Fig. 34 Dust Shielded Distributor

Some years ago the maintenance department of a large bus fleet with undercoach engines installed large rubber boots to enclose the distributor and adjacent leads. These boots prevented adequate distributor internal ventilation, did not keep all splash out, trapped part of the water that gained entrance and prevented trapped moisture from evaporating by restricting overall ventilation—results were disastrous. Moral of this story—"be cautious in design, and thorough in testing any splash protection scheme that encloses the distributor and its leads."

VI. *Special Features To Tailor Systems To Application Needs*

Typical special features include: 1) Dust shield distributors, Figure 34, for dusty environments. 2) Coil and distributor caps molded of alkyd, a premium material that will not carbon track as a result of temporary arcing across its surface. 3) Cam lubricators, (Fig. 35) to extend periods between maintenance by assuring cam lubrication even under higher temperature and somewhat dusty conditions. 4) External-ignition-proof features for mine or marine installations where flammable gasses might surround the engine at times. These four features are representative of a large variety that most ignition manufacturers offer to meet special needs.

VII. *Waterproof "Military Type" Systems For Submersion Or Heavy Splash Applications*

Figure 36 shows an ignition system designed to operate even when submerged in salt water. The coil and distributor are built into an integral housing, and the entire unit sealed. Ventilation is provided by tubing which draws air from the air cleaner, through the distributor, to the intake manifold. High tension leads are enclosed in water-proof conduits which are gasket sealed at the distributor and spark plug ends. Such systems are often used for cement hauling trucks and vehicles in chemical plants where engines are washed or steam cleaned regularly.

RADIO FREQUENCY INTERFERENCE (RFI) SUPPRESSION CONSIDERATIONS

Since ignition is a major source of electro-magnetic radiation in a vehicle, Radio Frequency Interference Suppression is another major application consideration. Briefly stated, the ignition system must be suppressed to the extent that it will not seriously interfere with RF communication or other electronic services within the vehicle or in the environment outside the vehicle, without reducing ignition performance or durability below acceptable levels.

Engineers face ever increasing suppression challenges arising from:

(1) Increased use and complexity of vehicle entertainment centers.
(2) Increasing use of land mobile communications equipment.
(3) Changes in vehicle antenna designs and locations.
(4) Increasing F.M. and TV broadcast services across the country.

The voltage in an ignition system rises very rapidly until the spark plug gap ionizes. The capacitive component of the spark then discharges through the gap, followed by the discharge of the inductive component. Interference is generated by transients in these two components which are very complex and unstable. The spark may also be struck and restruck innumerable times as the combined energies of the inductive and capacitive components are irregularly dissipated through the gap. These rapid breakdowns produce steep voltage peaks which are short in duration and rich in radio frequency interference. A similar condition exists at the ignition contacts when arcing takes place. Spark plug leads, although not a source of interference, act as an antenna to transmit interference generated at the plug gap into the atmosphere and to nearby wiring and sheet metal which may then reradiate into the atmosphere.

In general, there are three basic methods of suppressing ignition RF interference:

(1) Capacitors
(2) Resistors
(3) Shields

Fig. 35 Distributor Cam Lubricator

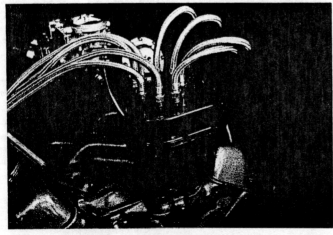

Fig. 36 Waterproof Ignition System

Capacitors: Capacitors are used to suppress RF interference because of their low impedance to the flow of high frequency currents. It is common practice to place a capacitor at the battery terminal of the ignition coil (Fig. 37) to suppress any RF current that might be reflected into the ignition switch to coil wiring.

Resistors: Resistor suppressors are often used in spark producing devices such as ignition systems. The resistor acts as a series impedance and tends to eliminate the snuffing out and restriking of the spark. It also serves to isolate interference generated by the capacitive discharge of the spark. Resistor suppressors may be found in the form of lumped resistors, resistor spark plugs or distributed resistance spark plug leads.

Lumped resistors, although not in broad usage in the United States, are commonly used in Europe. These resistors are placed at either end of the spark plug lead or built into the distributor rotor.

Resistor spark plugs (Fig. 38) are in broad use today. Used alone they provide some degree of suppression; however, when used to supplement resistive ignition cable, they prove quite beneficial.

Distributed resistance cable, (Fig. 39), is the most widely used ignition suppression device in the U.S. Its construction consists of a resistive core sheathed with insulation, glass reinforcing, and a protective jacket. Suppression resistors must be evaluated as regards suppression quality, durability, reliability, and effect on engine and ignition performance. Since excessive resistance in the ignition circuit could compromise spark energy, and affect engine startability and performance, it is desirable to keep the total suppression resistance to the minimum value consistent with effective suppression.

Shielding: Shielding can be quite effective in the suppression of radiated interference from an ignition system. Its purpose is to contain RF energy within a given area, in the case of an automotive vehicle, within the engine compartment. Shielding of the ignition system may take any one or combination of:

(1) Braided conduit over the ignition cables.
(2) Sheet metal shields adjacent to the ignition components.
(3) Wire mesh screening.
(4) Natural shielding of the vehicle body and structure.

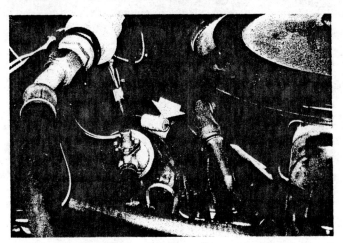

Fig. 37 RFI Suppression Capacitor on Ignition Coil

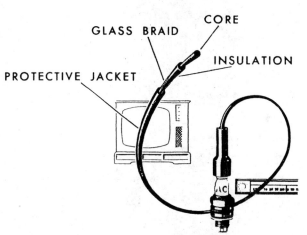

Fig. 39 RFI Suppression Cable

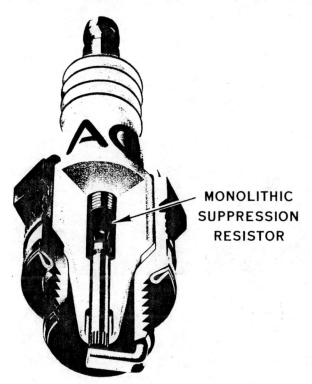

Fig. 38 Resistor Spark Plug

Fig. 40 Shielded Ignition for Sports Car

Braided conduit over high tension cables is commonly used for RFI suppression on military vehicles and in some extreme cases on commercial vehicles. For maximum suppression quality, it is necessary to completely shield the ignition distributor as well as the spark plugs. This method of shielding imposes a severe penalty on the performance of the ignition system due to the capacitance loading as discussed in a previous section. The natural shielding of automotive body and structural members is very effective. On vehicles such as trucks and tractors where such natural shielding may be inadequate, added sheet metal or mesh shielding along the ignition system might be necessary. Vehicles having fiber glass bodies (Fig. 40), plastic hoods, etc., lack this natural shielding and impose quite a challenge to the engineer responsible for RFI suppression.

Figure 41 shows an RFI shield in a distributor to prevent interference generated by normal contact arcing from interfering with a vehicle's own radio reception.

The alert ignition engineer will constantly monitor styling and other sheet metal design changes, substitution of plastics for metal in the area surrounding the engine, and changes in radio or antenna design, location or sensitivity to avoid RFI problems.

ELECTRONIC IGNITION SYSTEMS

Electronic ignition systems have been under development for over 35 years, first through the use of vacuum tubes, and later through the use of semiconductors. Developments in the semiconductor industry have made these type systems commercially available in recent years. Several fine technical papers have been presented on this subject; therefore, this paper will not cover this area in detail, but merely review four systems which are commercially available. These systems are:

(1) A contact-triggered inductive system.
(2) A breakerless inductive system.
(3) A contact-triggered capacitor-discharge system.
(4) A breakerless capacitor-discharge system.

In order to evaluate these systems in comparison to the standard ignition system, the engineer must consider the cost of the system relative to its ability to:

(1) Eliminate periodic maintenance of the distributor.
(2) Increase spark plug life.
(3) Provide increased ignition output.
(4) Improve engine startability at low temperatures.

(1) CONTACT-TRIGGERED INDUCTIVE SYSTEM

A diagram of the contact-triggered system is shown in Figure 42. This system uses a transistor to control the ignition coil current, while the distributor contacts control the transistor base current. Approximately 1.0 ampere of base current through the contacts can control 7.0 to 10.0 amperes of coil primary current. This is the least costly of the electronic systems and will provide the following:

A. *Reduced periodic maintenance of the distributor* because the current switched by the contacts is reduced substantially. This greatly extends contact life, however, distributor maintenance is not eliminated because rubbing block wear continues to make breaker cam lubrication and contact adjustment a necessity.

B. *Increased ignition output voltage in the higher speed range* is gained as a result of the increased coil currents. However, high speed operation may still be limited by contact bounce.

(2) BREAKERLESS INDUCTIVE IGNITION SYSTEM

To obtain further improvements over the contact-triggered inductive system in the area of distributor maintenance and high speed operation it is necessary to consider the breakerless inductive system. A diagram of this system is shown in Figure 43. This system also uses a transistor to control the

Fig. 41 RFI Point Shield

Fig. 42 Contact Triggered Inductive System

Fig. 43 Breakerless Inductive System

ignition coil primary current, however, in this case the transistor base current is controlled by a triggering circuit and a breakerless pick-up in place of the distributor contacts. This system is the least costly of the breakerless systems and will:

A. *Eliminate the need for periodic maintenance* of the distributor by eliminating the contact set and breaker cam.
B. *Provide increased output voltage* in the higher speed range as well as eliminate contact bounce as a factor in high speed operation.

Since both of the preceding electronic systems are of the inductive type, their spark characteristics do not differ appreciably from a standard system. Therefore, large improvements in the areas of spark plug life or cold weather startability are not gained. Small improvements may be realized in cases where contact deterioration has taken place, however, to obtain consistent improvement in these areas one should consider the Capacitor-Discharge type system.

CONTACT-TRIGGERED CAPACITOR DISCHARGE SYSTEM

The contact-triggered capacitor discharge type system as shown in Figure 44 contains:

(1) a storage capacitor for storage of the primary energy
(2) a power supply to step the battery voltage up to approximately 400 volts to charge the capacitor
(3) a triggering circuit to allow the capacitor to discharge through the output transformer
(4) an output transformer to step the 400 VDC up to spark plug firing voltage.

The contact-triggered version of this system uses the distributor contacts to control the triggering circuit, and thus requires no changes in the standard distributor. This system will provide:

(1) *Reduced periodic maintenance of the distributor* as a result of the low level currents switched by the contacts.
(2) *Increased spark plug life* as a result of the faster output voltage rise time afforded by the capacitor discharge type system.
(3) *Increased ignition output voltage* in the higher speed range as a result of flexibility in transformer design.
(4) *Improved engine startability* at low temperatures as a result of the faster discharge of energy from the capacitor discharge type system.

Like the contact-triggered inductive system, the contact triggered C-D system is inhibited by the contacts in the areas of distributor periodic maintenance and high speed contact bounce. To eliminate these problems, the breakerless version of the C-D system must be considered.

BREAKERLESS CAPACITOR DISCHARGE SYSTEM

The breakerless C-D system (Fig. 45) contains the same basic elements as the contact-triggered version except the triggering circuit is controlled by a breakerless pick-up in place of the distributor contacts. This system offers the same advantages as the contact-triggered version plus the breakerless feature to eliminate periodic maintenance of the distributor and contact bounce at high speeds.

Although semiconductor systems have not been used widely in automobiles, they do deserve serious consideration for two types of applications:

(1) Long-life commercial applications where savings on maintenance costs and down time can pay a sound return on the original cost premium.
(2) Special applications where engine performance or durability might be limited by the ignition system.

SEMICONDUCTOR SYSTEMS RECOMMENDATION

The alert engine engineer will constantly monitor vehicle needs in the area of ignition performance, maintenance objectives, and serviceability; and match these needs against the advantages, availability, and economics of premium semiconductor systems.

REFERENCES

(1) Hartzell, H. L. "Ignition Problems In Damp Weather," Paper No. 566 presented at SAE Annual Meeting, Detroit, January 1951
(2) Norris, J. C., "Delcotronic Transistor-Controlled Magnetic Pulse-Type Ignition System," Paper No. 617B presented at S.A.E. Automotive Engineering Congress, Detroit, January 1963
(3) Hetzler, L. R. and Kline, P. C., "Engineering C-D Ignition for Modern Engines," Paper No. 670116 presented at SAE Automotive Engineering Congress, Detroit, January 1967
(4) "Bibliography on Ignition and Spark Ignition Systems," U.S. Department of Commerce, National Bureau of Standards, Miscellaneous Publication No. 251

Fig. 44 Contact Triggered C-D System

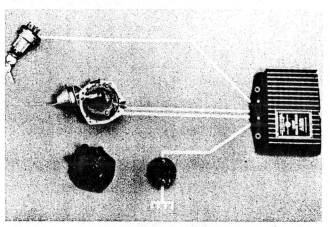

Fig. 45 Breakerless C-D System

750346

HEI – A New Ignition System Through New Technology

Gerald O. Huntzinger and Gerald E. Rigsby
Delco-Remy Div. General Motors Corp.

FOR MANY YEARS, THE IGNITION SYSTEM for automotive engines has attracted almost continuous attention from individuals and organizations intent on developing system improvements. This flow of new ideas, some good, some not, has involved a great deal of engineering evaluation to determine real value to a vehicle owner. Evaluation of new ignition system developments is a difficult process, and it is possible to be greatly misled by the evidence of a single test or series of tests. It is understandable, then, that new ignition developments are not readily accepted. The sheer persistence of this effort toward new ignition developments is real evidence that a better ignition system is desired and needed. This paper describes the basis on which the new high-energy ignition (HEI) system was developed and explains the technical barriers that had to be overcome to obtain its high-energy performance and associated design characteristics.

BACKGROUND

Past experience with various types of production and experimental ignition systems provided an excellent background for setting design objectives for the new HEI system.

Typical previous production systems shown in Figs. 1–3 each represented a step forward at the time; however, as described, each also had major limitations.

The standard breaker system, Fig. 1, has been subject to continuous improvement from the time of its conception by

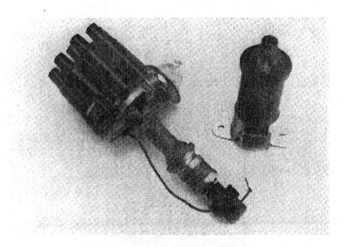

Fig. 1 – Standard "flat top" breaker ignition system

Charles Kettering to the present. The distributor shown represented major serviceability and environmental protection improvements when introduced; however, its maintenance requirements and performance are not compatible with the needs of the modern automobile.

Fig. 2 shows the Delcotronic breakerless transistor system introduced as an option on 1963 General Motors vehicles (1).*
It was the first production automotive system to eliminate

---

*Numbers in parentheses designate References at end of paper.

ABSTRACT

A new high-energy, electronic ignition (HEI) system with substantially improved performance and durability has been developed for automotive gasoline engines. All components of the system are integrated within the distributor, and no scheduled parts replacement or adjustment of system components is required. The requirements of system performance and durability, which guided the development of the new system, are outlined. New concepts, materials, and technology used in the design and development are illustrated and explained. A discussion of the operation and limitations of both breaker-operated and transistor systems explains the need for dwell time control and coil primary current regulation. These are new control functions incorporated in the HEI system which result in reduced component size and power dissipation which is essential for the integral design.

Fig. 2 — Delcotronic "breakerless" transistor system

Fig. 3 — "Unitized" transistor ignition system

breaker contacts and as such represented a very significant technological advance. This system added complexity to the vehicle by requiring mounting of additional components, and it added electrical connections and an increase in generator capacity to provide 4 A of additional current. Environmental protection improvements were not gained, since it retained the standard type of ignition coil, distributor cap, and high-tension leads and boots. This is a typical transistor system similar to others now used in the industry.

The unitized system, Fig. 3, introduced by Delco-Remy and Pontiac in 1972, was the first to use custom integrated-circuit electronics, and also the first to combine the distributor, coil, and electronics in a single integrated package. It removed the major limitations of previous electronic systems and provided greatly improved environmental protection for the high-voltage terminations. Several of its design innovations were carried over to the HEI system. Its performance, however, was limited because it retained the same basic size and therefore had no better high-voltage distribution than previous systems.

HEI PROGRAM OBJECTIVES AND DESIGN SPECIFICATIONS

This previous work in the development of new ignition systems has identified the elements of ignition system performance that are important for proper engine operation and true customer value. Similarly, development and application of new materials provided the basis for improvements which prolong the useful life of the system. Combining this technical background with an analysis of ignition system needs of the future resulted in the following development objectives for the new ignition system:

1. Maintenance-free operation of distributor, coil, and electronics
2. Extended spark plug life
3. Improved ignition of fuel mixtures
4. Improved reliability and life

The following paragraphs describe the need for each objective and define the design specification required to accomplish it.

MAINTENANCE-FREE OPERATION—The goal of a maintenance-free system is to eliminate scheduled parts replacement and adjustment, thereby retaining new-car performance. The benefits of such a system become apparent when typical maintenance recommendations for United States cars of prior years are studied. Distributor points and cam lubricators required attention at 12,000 mile intervals. Spark plugs required replacement at either 6000 or 12,000 miles, depending on whether the fuel used was leaded or nonleaded. Engine timing needed to be checked and adjusted at 12,000 mile intervals.

Since it is common practice to replace breaker points and retime the engine when spark plugs are replaced, proper maintenance becomes an expensive procedure for the car owner. If car owners do not follow the owner's manual recommendations, fuel economy and exhaust emissions suffer. The only sure remedy is a maintenance-free ignition system. Therefore, the design specification requires an electronic ignition system triggered by an ignition pulse generator which maintains original timing without wear or adjustment.

EXTENDED SPARK PLUG LIFE—It is important to extend the replacement interval for spark plugs so that the engine will maintain proper operation with a minimum of attention from the owner. This requires that the ignition system develop a higher output voltage for two reasons:

1. New spark plugs are initially set to wider gaps. GM has increased gap settings from 0.030 in to as much as 0.080 in on

some cars because this extends the ability to ignite the fuel mixture under a wider range of engine operation.

2. The extended mileage causes the spark plug electrodes to erode to wider gaps, requiring a higher breakdown voltage.

The voltage level required was determined by conducting a large-scale field test on GM cars equipped with high energy ignition systems and wide-gap spark plugs, and operated on nonleaded fuel. Results showed that to meet this objective, the design specification for the new system must require an available ignition voltage of 35 kV.

IMPROVED IGNITION OF FUEL MIXTURES—Engine tests have shown that ignition of the fuel mixture under some conditions of operation is improved by extending the duration of the spark, for example, the length of time the spark burns. The typical breaker ignition system can produce a spark duration of about 1200 ms under a given set of conditions for spark plug requirements and engine speed. For these same conditions, it appears that about 1800 ms is closer to optimum value. Accordingly, the design specification set for HEI requires 50% additional spark duration.

IMPROVED RELIABILITY AND LIFE—Conventional ignition systems are susceptible to the effects of temperature, moisture, and salt deposits. A substantial upgrading in both mechanical design and materials is needed if the ignition system is to continue to function at top effectiveness for the normal life of the vehicle. Evaluation of various mechanical design approaches made it apparent that integrating the system into a single package, which brings all of the elements into close association, would greatly improve system reliability. Improved insulating materials in the coil, cap, rotor, and distribution system are also essential for long life under increased electrical stress. Therefore, the design specification for HEI calls for an integrated design with improved insulation materials.

SUMMARY OF MAJOR DESIGN SPECIFICATIONS—The major design specifications for the HEI system can thus be summarized as follows:

1. Electronic ignition with maintenance-free pulse generator
2. 35 kV available voltage
3. 50% additional spark duration
4. Integrated design with improved insulation materials

The following sections describe the HEI system that resulted from these design specifications.

HEI SYSTEM DESCRIPTION

From the beginning, the HEI system was developed as a total system in which the distributor, coil, and electronic module were combined into one integrated package. Fig. 4 is a photograph of the system, and Fig. 5 is a section view in which the arrangement of the major elements of the system is shown.

INTEGRAL DESIGN—The integrated design is important to the integrity and reliability of this new ignition system. The magnetic pulse generator, electronic module, and ignition coil are all contained within the distributor, where they are well protected from physical and environmental abuses. The conventional high-voltage coil-to-distributor lead is completely

Fig. 4 — Photograph of HEI system

Fig. 5 — Section view of HEI system

eliminated, and all other interconnections are contained within the unit. This permits a complete test of the system as manufactured to assure that it functions properly. Vehicle installation requires only mounting the unit, connecting the eight spark plug leads and one battery wire, and adjusting initial timing.

HIGH-VOLTAGE DISTRIBUTION SYSTEM—Since the HEI system must produce 40% more output voltage than previous systems, a new distribution system design was necessary. The size of the distributor cap was increased so that the higher voltages could be distributed to the correct spark plug without risk of cross fire to the wrong cyl.

Insulation Materials—New insulation materials were required for the coil, rotor, and cap to withstand the increased electrical stress and to prevent ignition failure caused by carbon tracking. A great deal of development and testing was required to find a material and its optimum molding process to provide the dielectric and insulation properties needed to eliminate this failure mode, and which also had the physical properties needed to permit mounting the coil in the cap. The material selected for HEI is outstanding in both respects. It is a thermoplastic, injection-molded, glass-reinforced polyester.

High-Voltage Connections – New high-voltage terminals (Fig. 6) were developed for the HEI system. They are similar to spark plug terminals, permitting easier attachment, and providing improved dielectric and environmental sealing of the connections. When desired, the spark plug wires can be assembled into a harness and held together by means of a plastic hard shell. This permits very rapid and positive installation of the spark plug wires to the HEI system. A latching device (Fig. 7) helps provide for proper connection of plug wires and helps to prevent any loosening or movement, which might defeat the moisture protection and dielectric seal of the connection.

IGNITION COIL – A major breakthrough in ignition coil design was required to permit integrating the coil into the HEI distributor. Conventional coils have an inefficient flux path and winding arrangement which are not suitable for this purpose. A smaller-size coil with low-resistance primary winding is essential to the new system and required the development of a more efficient, closed-core type of ignition coil. The significance of this feature will be discussed in a later section of the paper.

ELECTRONIC MODULE AND MAGNETIC PULSE GENERATOR – The unique performance of the HEI system is the result of special features designed into the ignition coil, electronic circuits, and the magnetic pulse generator. The electronic module and magnetic pulse generator work together to provide new control functions for the critical parameters in the ignition system. This permits the use of smaller components and reduced power dissipation while reducing the risk of undue overstress on system components during operation in extreme conditions. The following sections of this paper will develop the important elements of this concept and explain their significance in terms of system performance and component stress.

Fig. 6 – HEI high-voltage terminals

Fig. 7 – Photograph of HEI secondary terminal latch

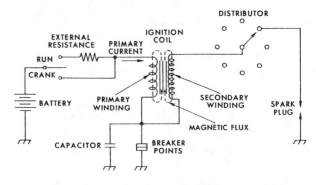

Fig. 8 – Circuit of typical breaker ignition system

FUNDAMENTALS OF IGNITION SYSTEM OPERATION

Before discussing the new high-energy ignition system, it will be helpful to review typical breaker and transistor ignition systems. The basic operation of these systems will be described, and the limitations which prevent them from achieving the increased performance levels required of the HEI system will be explained.

BREAKER-OPERATED INDUCTIVE IGNITION SYSTEMS – The circuit of a typical breaker type ignition system is shown in Fig. 8. This circuit functions as follows: when the distributor breaker points close, current flows from the battery through the primary winding of the ignition coil. This current produces and maintains a magnetic flux within the iron core of the coil. When ignition is required, a cam will open the breaker points, interrupting the current. The resulting decay of flux will induce a voltage in both the primary and secondary windings. The voltage induced in the secondary winding is routed by the distributor to the correct spark plug to produce the ignition spark.

The wave forms of the current and voltage are shown in Fig. 9. The primary current starts at zero when the points close and rises slowly to a maximum value. When the points open and the current falls to zero, a voltage of several thousand volts is induced in the secondary winding.

If the coil is not connected to a spark plug, this induced voltage will have a damped sinusoidal wave form, as shown by the center trace. The peak value of this voltage is the maximum that can be produced by the system and is called the available voltage of the system.

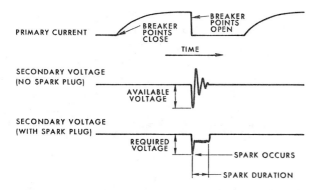

Fig. 9 – Breaker ignition voltage and current wave forms

When the coil is connected to a spark plug, as it is in normal operation, the secondary voltage will rise to the breakdown potential of the spark plug, and a spark will occur. This alters the wave form, as shown in the bottom trace. After the spark occurs, the voltage is reduced to a low value by the characteristics of the arc and remains at this level until all ignition energy is dissipated and the arc goes out. The value of voltage which caused the spark to occur is called the required voltage of the spark plug. The interval during which the spark burns is called the period of spark duration.

These characteristics represent two requirements for proper ignition performance:

1. The available voltage of the ignition system must always exceed the required voltage of the spark plug to assure a spark.
2. The spark must possess sufficient energy and duration to initiate combustion under all conditions of operation.

The characteristics of the new HEI system are needed because present ignition systems are not able to satisfy these two requirements adequately for modern automobile engines.

Typical Performance – The performance of an ignition system is generally expressed as the value of available secondary voltage as a function of engine speed. This is illustrated in Fig. 10 for a typical system with a maximum performance of approximately 25 kV. The performance shown is about the best that can be obtained reliably from a breaker-operated ignition system, and it is far short of the 35 kV required from the new HEI system. Unfortunately, the breaker system is subject to a limitation in current switching capability which prevents it from achieving additional performance improvements.

Primary Current Limitation – Although the tungsten material used for the breaker points is the best available for the purpose, the life of the points is dependent upon the amount of current that they are required to switch. This is shown in Fig. 11 (2). The typical breaker point system provides acceptable life characteristics when operated in the range of 3.5–4.0 A (depending on temperature). However, increasing the current to obtain additional performance causes a rapid reduction in reliability and breaker point life. Because of this current limitation, a significant improvement in performance cannot be obtained with the breaker system, and a more sophisticated system must be developed to provide the needed improvement.

TRANSISTOR INDUCTIVE IGNITION SYSTEMS – When the transistor first came on the scene, it was immediately expected to permit much-improved ignition system performance because transistors could switch greater current than breaker points. During the several years which ensued, however, the major improvement was to eliminate the breaker points and thus to remove this requirement for periodic maintenance. As far as actual spark and voltage characteristics were concerned, little significant improvement occurred. In fact, it was not uncommon to find that some transistor ignition systems placed on the market were actually less effective and resulted in shorter spark plug life than was previously obtained with the breaker point system. It is important to understand why this occurred before the new approach used in the HEI ignition system is discussed.

The Effects of Operational Extremes – Even though transistors are available which can switch more current than breaker points, it is not possible to take full advantage of their ratings. The electrical system of the vehicle is subject to wide operational extremes which place severe stress on all components connected to it. For many electromechanical devices, switches, motors, contacts, etc., the extremes of voltage and current are transient in nature and can be endured without permanent damage. For electronic devices, however, even a few ms of excessive stress may cause degradation or failure. Thus, to provide reliable operation, the system is designed to operate at something less than its maximum capability.

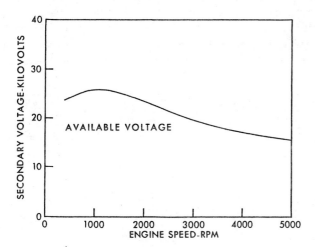

Fig. 10 – Typical breaker ignition performance

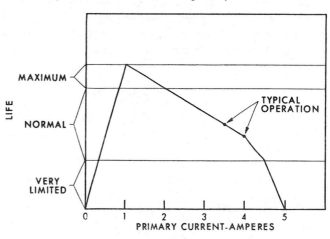

Fig. 11 – Effect of break amperes on contact life

The effects of these operational extremes are illustrated in Fig. 12, where the normal run and cold-starting modes of ignition operation are compared. In the normal run mode, a typical, regulated battery voltage would be in the 14.3 V range. The ignition coil primary winding will have a warm, stabilized resistance of about 2.65 Ω and, in series with it, an additional external resistance of 1.35 Ω. This provides a total resistance of 4.0 Ω which will limit the maximum current through the breaker points to a safe value of 3.6 A. Under these conditions, the typical system will produce an output voltage of 25 kV, and about 360 V will be developed across the opening breaker points. A useful figure of merit for the electrical stress sustained by the breaker points is the product of the voltage and current. In this normal mode of operation, this product is about 1300 V·A. This is very normal operation, and breaker points will give long and satisfactory life at this stress level. Power transistors are readily available which can meet this figure of merit requirement, for example, 130 V–10 A, 400 V–3.5 A, etc. Thus, transistors can easily replace the points in this normal mode of operation, with no loss in performance.

Starting the car in very cold weather may impose severe momentary overloads on the ignition system components. Under these circumstances, the cranking motor may draw so much current that the battery can only produce about 4.5 V at the time ignition is required. To insure an adequate coil current at this very low battery voltage, the external resistance is bypassed, and about 2.9 A is supplied to the ignition coil. This will start the engine with no undue stress on the points (only 875 V·A); however, immediately after start, a cold overrun condition occurs which creates severe electrical stress. At this point, the engine speeds up and cranking motor current decreases, permitting the battery voltage to rise abruptly to as much as 11 V. This will cause coil current to rise momentarily above 7 A, until the cranking motor is disengaged and the external resistance is again connected into the circuit. During this overrun period, with coil current above 7 A, the primary voltage may exceed 500 V. This represents a momentary overload stress level of 3500 V·A, which can be handled by contact points for short periods without permanent damage. Transistors, however, cannot tolerate these momentary excursions above their capability without failure, and the system design must be compromised to reduce these extreme stresses to acceptable levels. Consequently, most transistor systems provide performance about equivalent to a standard breaker system. Their merits are elimination of contact points, improved high-speed operation, and consistent cranking performance.

If a conventional transistor system were to be designed to produce HEI performance levels, the stresses that would result from these operating extremes would require a semiconductor of such high capability as to be economically impractical. The solution to this problem is discussed in the following section.

The foregoing analysis has shown that neither the breaker-operated nor the conventional transistor ignition system is capable of producing the performance levels specified for HEI.

NEW TECHNOLOGY FOR HIGH-ENERGY IGNITION

New concepts of circuit operation are needed to remove the previously described limitations of conventional ignition systems so that the required performance improvements can be obtained. In the high-energy ignition system, one new concept in circuit operation is to provide automatic current regulation. By this means, the circuit protects itself from momentary overloads caused by operational extremes and therefore does not need to be derated from its maximum performance capability. A second new concept is use of a low-resistance coil design with automatic dwell control. This improves high-speed performance and minimizes power dissipation.

These additional features represent a considerable increase in electronics circuit complexity which, fortunately, can now be provided by integrated circuit electronics in a reliable and practical manner.

IMPROVEMENT BY LOW-RESISTANCE DESIGN—A substantial part of the performance loss at high speed is due to the resistance of the primary circuit. This is normally adjusted to limit the maximum current, which occurs at low speed, to the desired value. Unfortunately, it also restricts the current available at high speed to much lower values. If this resistance is reduced to a very low value (0.5 Ω is a minimum practical value), the current will rise to the desired level more rapidly, and the high-speed performance will be substantially improved. Electronic current limiting must be provided to limit the current after it has reached the desired value.

The reasons that high-speed performance is improved by this low resistance mode of operation are illustrated in Fig. 13, which compares the current buildup of a conventional resistance-limited system (2.6 Ω) to that of the low-resistance, HEI system (0.5 Ω). At the 3.4 ms point (about 3100 rpm), the HEI system will have reached full current and thus will deliver its full output voltage of 35 kV. At the same speed, the

	NORMAL RUN	COLD CRANK	COLD OVERRUN
BATTERY VOLTAGE	14.3	4.5	11.0
COIL RESISTANCE, OHMS	2.65	1.54	1.54
EXTERNAL RESISTANCE, OHMS	1.35	0	0
TOTAL RESISTANCE, OHMS	4.00	1.54	1.54
MAX BREAK AMPERES	3.6	2.9	7.14
PRIMARY VOLTAGE	360	300	500
SECONDARY VOLTAGE	25 KV	14-22	34-40
VOLT-AMP PRODUCT	1300	875	3575

Fig. 12 – Typical operational extremes

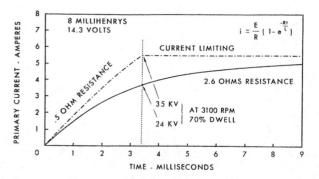

Fig. 13 – Resistance effect on current buildup

conventional resistance-limited system will have reached only 70% of full current, resulting in an output voltage of only 24 kV. Fig. 14 shows the performance of these two systems over the entire speed range, and the superiority of the low-resistance HEI system is evident.

SYSTEM REQUIREMENTS FOR LOW-RESISTANCE DESIGN—To obtain the performance improvements made available by the low-resistance mode of operation, the HEI system must provide the following unique control functions:

1. Electronic current limiting at 5.5 A to prevent excessive current stress
2. Automatic dwell time control to minimize power dissipation in the electronics circuit

Current limiting is accomplished by absorbing the excess battery voltage in the switching transistor. This causes a substantial power dissipation during the time the current is being limited. The automatic dwell control delays the turn-on of coil current until there is just enough time for the current to reach full value before turn-off to produce the spark. As a result, the time that current limiting is required is reduced to a minimum. This function, while somewhat complex, eliminates excessive power dissipation in the system and increases its temperature tolerance and reliability.

SUMMARY OF BENEFITS OF NEW TECHNOLOGY—These new concepts of low-resistance design, electronic current limiting, and automatic dwell control provide a number of important benefits for the HEI system. They are:

1. 35 kV output voltage to 3100 rpm
2. Reduced primary current—5.5 A do work of 11. (A conventional resistance-limited design requires a 2 mH coil with 11 A of primary current to achieve 35 kV with acceptable high-speed performance.)
3. 30 kV output voltage during cranking at battery voltages as low as 6 V
4. No cranking bypass switch required
5. Substantially increased coil efficiency

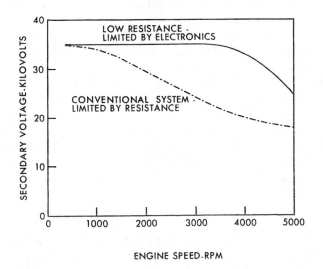

Fig. 14 – Ignition performance comparison, resistance-limited versus low-resistance design

COMPONENT DESIGN

Because of the unique features and requirements of the HEI system, a special design was necessary for each component part. In many aspects, the designs differ appreciably from conventional practice. The following sections outline the major design features of each component part and include some additional discussion of the component function.

IGNITION COIL—The specifications and considerations of the previous sections have defined some of the electrical characteristics of the coil. These characteristics are:

Secondary voltage—35,000 V
Primary current—5.5 A
Primary resistance—0.5 Ω

Primary inductance, secondary inductance, turns ratio, primary voltage, and physical characteristics follow.

Primary Inductance—The primary inductance influences both high-speed performance and the level of stored energy from which the output voltage is developed. A value of 8 mH is used in the HEI system to produce the desired output voltage throughout the engine speed range.

Secondary Inductance—The secondary inductance is one of several factors that influence spark duration. In the HEI system, a value of 80 H provides the correct spark duration to meet the requirements for the system.

Turns Ratio and Primary Voltage—The turns ratio of the coil is fixed when the primary and secondary inductances are established. In the HEI coil, the turns ratio is 100:1.

The primary voltage developed by the coil is a function of turns ratio and secondary voltage. By design, the HEI system develops 35 kV and, with a turns ratio of 100:1, a peak primary voltage of 35,000/100 = 350 V will be developed. This voltage appears across the transistor junctions after the current has been interrupted and is therefore the maximum voltage stress applied to the transistor. The stress level which is imposed on the transistor by the ignition coil, expressed as the volt-ampere product, is therefore 350 × 5.5 = 1925 V·A.

Physical Design—The major influences on the physical design of the coil are the desire to integrate it into the distributor cap and the requirement for low primary resistance. Conventional coil design, shown in Fig. 15, does not work well from either of these standpoints because the outside position of the primary winding causes high resistance, and the tall physical arrangement is not suitable for mounting in the distributor cap.

Instead, a new closed-core design is used, as shown in Fig. 16. This design is more like a conventional transformer

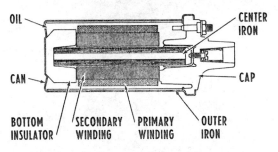

Fig. 15 – Conventional oil-filled ignition coil design

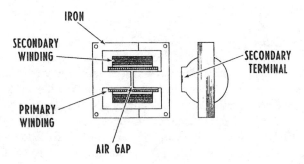

Fig. 16 – HEI coil design

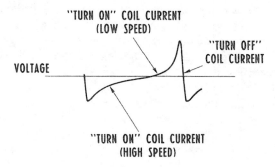

Fig. 18 – Pulse generator wave forms

with the primary winding placed next to the magnetic lamination and the secondary winding on the outside. This allows low primary resistance with a reasonable size of wire. In addition, the high-voltage terminal of the secondary winding is exposed on the flat side where it is directly accessible to the rotor when mounted in the top of the distributor cap. The winding assembly is potted in an epoxy compound for permanent protection against moisture and other degrading effects.

The iron assembly for the coil is made from E-shaped laminations stacked to the desired thickness and assembled through the windings end to end. One of the center legs is shortened just enough to provide a precisely calibrated air gap which establishes the primary and secondary inductance values.

MAGNETIC PULSE GENERATOR—There are many types of pulse generators which could trigger the electronic circuit of an ignition system. The HEI system uses a concentric variable reluctance design, shown in Fig. 17. It consists of a stationary magnet and pickup coil assembly which surrounds the distributor shaft and a rotary pole piece fitted to the advance mechanism on the shaft.

Pulse Generator Wave Form—The voltage wave form of the pulse generator is shown in Fig. 18. As the distributor shaft rotates and the pickup teeth approach alignment, the output voltage increases in positive polarity. As the teeth pass through alignment, the output voltage abruptly reverses and passes through zero to a negative polarity. The electronic module switches off the coil current to produce the spark on this rapid negative-going portion of the voltage wave form. This gives a precisely located switching point for accurate spark timing.

Automatic Dwell Control—The relatively slow positive-going portion of the wave form is used by the electronic module to establish the point at which coil current is turned on in preparation for the next ignition pulse. From this portion of the wave form, the module anticipates the next ignition pulse and adjusts its turn-on point so there is just enough time for the coil current to reach its design value. Thus, turn-on occurs just prior to ignition at low speed and advances toward the previous ignition pulse at high speed. This automatic function greatly minimizes power dissipation while maintaining maximum performance throughout the speed range.

Additional Pulse Generator Characteristics—Several other characteristics of this pulse generator are important to overall ignition system performance and reliability. They are:

1. Simple, reliable construction. The materials are rugged and stable and there is no need for adjustment. The signal is self-generating without external power, and only two connections are required.

2. High-output, speed-related signal. The signal is produced by alignment of all teeth for each pulse, yielding a relatively high output voltage and a low source impedance. Further, the amplitude of the signal increases with speed, thus providing information for the automatic dwell control function.

3. High synchronism accuracy. Each pulse is produced by the average position of all teeth, thus producing a very constant angular interval between pulses.

4. Minimum vibration-induced signals. The concentric nature of the pickup inherently cancels flux changes and resulting output signals which are due to undesired radial motions of the rotor which might result from vibration.

5. Continuous indication of angular position. The wave form, although of complex shape, is a very stable function of angular position, and this information is also used by the electronic module for timing the turn-on point of coil current to accomplish automatic dwell control.

ELECTRONIC MODULE—The electronic module was developed jointly by Delco Electronics and Delco-Remy specifically for use in the HEI system. It is designed to be mounted inside the distributor. Fig. 19 is a photograph of the module with the cover removed so that the interior parts can be seen.

Physical Description—A custom integrated circuit is used to provide all of the signal processing and control functions. This chip is mounted on a ceramic substrate with the additional circuit components necessary for its operation. The output signal from this chip is the control signal for the output transistor. This device is separately mounted adjacent to the ceramic substrate and controls the current through the ignition coil.

These electronic components are contained in the sealed plastic housing which protects them from atmospheric contamination and physical abuse. The bottom of the housing is aluminum and is designed to conduct the heat generated within

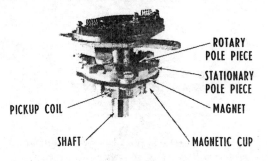

Fig. 17 – Magnetic pulse generator

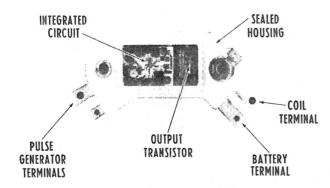

Fig. 19 – Photograph of electronic module

the module into the distributor base, where it is safely dissipated.

Electrical Functions – The details of electrical operation of the module will not be described since it is more properly the subject of another paper. The fundamental functions performed by the electronic system are as follows:

1. Limit and regulate ignition coil current at 5.5 A
2. Interrupt this current at time of ignition and withstand the resulting induced voltage of 350 V
3. Automatically control the dwell time of coil current so that full output is developed with minimum power dissipation
4. Provide the above functions at battery voltages ranging from 4.5–18 V
5. Operate through a temperature range of -20–257°F (-30–125°C)

HEI TEST PROGRAM

Since the HEI system is almost totally new in terms of design, materials, and components, the testing program it entailed was one of the most comprehensive ever conducted at Delco-Remy. The total program included component, material, and system testing in the laboratory, and extensive field testing on all types of vehicles under various operating conditions all over the United States and at one test site in Canada. Some of the most significant and interesting portions of this work will be described briefly.

LABORATORY TESTING – In the laboratory test program, thousands of components and total systems were tested for performance and durability. In this program, as with any new product development, the test equipment required was special and had to be designed and built within the time frame of the testing program. A great deal of laboratory trouble-shooting was involved as problems were revealed in both field and laboratory testing. Some problems are quite obvious and quickly solved, but frequently problems discovered during a test have obscure causes, and proof of correction requires many hours of testing on many units.

Component Testing – Both performance and durability testing were required for all of the major components, such as the electronic module, ignition coil, magnetic pulse generator, parts molded from new insulation materials, and vacuum advance units. The HEI objectives set some pretty tough goals. Performance tests had to prove that the objectives had been met before moving to the next step in the development program. Durability tests had to include mechanical and vibration durability, as well as electrical.

The electronic module development is a good example of the component testing undertaken. Since the electronic module is the very heart of the HEI system, the test program to qualify it was exhaustive.

Fig. 20 shows a special test stand which was developed to test the electronic module for conformance to performance specifications. This stand will measure and record 30 operating parameters of a functioning electronic module in one minute. This information, stored on magnetic tape, is then statistically analyzed by an engineering computer and stored in memory for future comparison and evaluation. This permits data on all modules prior to, during, or subsequent to durability or field testing to be compared and evaluated.

System Testing – Laboratory testing of the total HEI system was similarly extensive. Over 10,000 systems have been evaluated in the laboratory for performance, which includes electrical functions and mechanical advance functions. Fig. 21

Fig. 20 – Photograph of module testing facility

Fig. 21 – Photograph of precision distributor test stand

shows a computer-controlled test stand in which the HEI system can be evaluated in terms of spark advance to within 0.025 deg accuracy.

Durability testing is done under the following four conditions:

1. Room ambient—speed cycled continuously between 500 and 5000 erpm
2. High temperature—ambient maintained at 240°F
3. Thermal cycling—ambient cycled between -20°F and 225°F
4. High humidity—test run at 140°F and 70% RH

Vehicle tests are similarly an important part of laboratory testing. Hot and cold cranking must be thoroughly evaluated, temperature studies must be made, and electrical system compatibility on the vehicle must be thoroughly investigated to avoid unexpected system malfunction under certain conditions.

FIELD TESTING—The final proof of any new product design is in a successful field test program. The field test program in HEI was probably the most extensive ever conducted on any automotive component. The test program began in February 1972. Since that time, vehicles have been operated at 38 test sites all over the continental United States and one location in Canada (see Fig. 22). Most tests have been in severe operating service, such as taxis and police cars. When 1975 car production started, a total of 1651 test units had been operated over more than 50,000,000 miles of test. Many of the field test sites used lead-free fuel. This allowed us to measure spark plug requirement voltages when the systems were installed and then use follow-up measurements to determine a spark plug requirement profile versus mileage. All HEI systems field tested were given a serial number so that a performance comparison of the total system and all components could be made at the conclusion of the test.

	BREAKER IGNITION	DELCOTRONIC (TRANSISTOR)	UNITIZED	HEI
MAINTENANCE-FREE DISTRIBUTOR		✓	✓	✓
NON-TRACKING INSULATION MATERIALS				✓
EXTENDED SPARK PLUG LIFE WITH WIDE GAP PLUGS				✓
INCREASED ENERGY FOR LONG BURN				✓
IMPROVED STARTING CAPABILITY		✓	✓	✓
MAJOR ENVIRONMENTAL PROTECTION (Moisture, Splash, Dirt)			✓	✓
AVERAGE ELECTRICAL CURRENT DRAW (Amps)	2	5	2	2
CAR PLANT CONSIDERATIONS PREPACKAGED, PRETESTED SYSTEM			✓	✓
MINIMUM CAR PLANT HANDLING			✓	✓
NUMBER COMPONENTS TO MOUNT	2	5	1	1
NUMBER ON CAR PRIMARY ELECTRICAL CONNECTIONS	6	14	2	2
NUMBER ON CAR SECONDARY ELECTRICAL CONNECTIONS (V-8 ENGINE)	18	18	8	16

Fig. 23 – Ignition system design comparison

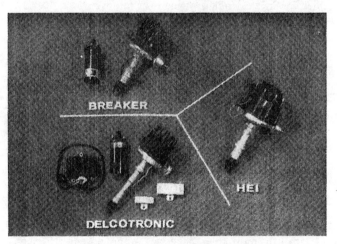

Fig. 24 – Delco-Remy ignition systems

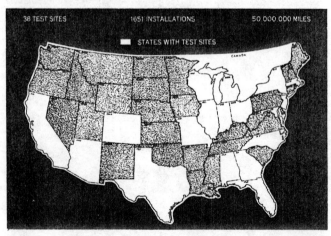

Fig. 22 – HEI field test locations

Fig. 25 – Components eliminated with HEI

SUMMARY

HEI has already demonstrated that it can meet or exceed the initial design objectives set. Fig. 23 shows a design comparison between breaker, Delcotronic, unitized and HEI systems. The comparison in Fig. 24 shows how HEI reduces the complexity of the ignition system. Fig. 25 illustrates a major benefit for the car plant with HEI—all the tagged components in the illustration are eliminated.

System performance, as shown in Fig. 26, has actually come out slightly higher than anticipated with the design, and has already shown that spark plug life is substantially extended. For 1975, GM driver manual recommendations on spark plug replacement have been extended from 12,000 miles to 22,500 miles with HEI, when the engine is operated on lead-free fuel.

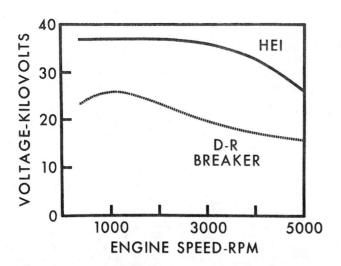

Fig. 26 — Ignition performance, HEI and D-R breaker systems

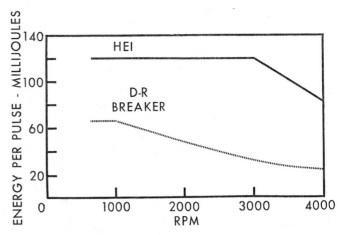

Fig. 27 — Ignition system energy, HEI versus D-R breaker

Cold-starting performance is greatly improved because of the inherent primary current regulation provided by HEI. The comparison in Table 1 shows that during cold starting, when system voltage drops to 6 V, HEI produces more spark plug voltage than a typical transistor system will produce during normal running operation.

Table 1 — Cold-Starting Performance

System	Voltage to Spark Plugs
HEI	30 kv
Delcotronic	16 kv
Breaker	13-22 kv

The increased energy provided by HEI will improve the probability of ignition of hard-to-fire mixtures when they occur by producing extended spark duration and permitting the use of wider spark plug gaps (0.060–0.080 in). Fig. 27 compares spark energy provided by HEI and conventional Delco-Remy breaker type systems.

HEI, then, is unique in that it combines new concepts, new materials, and new technology, in a system which is geared to future engine requirements. For the car owner, it reduces maintenance costs, provides an ignition system that retains new-car performance, and improves engine starting. As a benefit to the vehicle emissions control system, HEI provides maintenance-free, consistent ignition, eliminates field service adjustments that affect emissions, and improves spark characteristics to ignite mixtures that may have marginal combustibility. Finally, HEI offers the car assembly plant a prepackaged, pretested system, fewer components to schedule, handle, and install, fewer electrical connections, and reduced installation complexity.

REFERENCES

1. SAE paper 617B–Jan. 1963.
2. Paul C. Kline, "Some Factors To Consider In The Design and Application of Automotive Ignition Systems." SAE Transactions, Vol. 79 (1970), paper 700083.

770854

A Production Computerized Engine Timing Control System

William P. Winstead
Delco Remy Division General Motors Corp.

The purpose of this paper is to describe the digital electronic spark timing system used on the 1978 Oldsmobile Toronado. The system will be described with regard to the Design Requirements, the System Hardware, the System Design Capability and the Application Requirements.

The spark timing control system as applied to the 1978 Oldsmobile Toronado was developed to meet the increasingly complex requirements of engine spark timing control, in an effort to meet the future needs to optimize the vehicle fuel economy, exhaust emissions and driveability. The 1978 system was developed primarily to control engine timing, but with the basic capability of being expanded to control additional engine functions such as exhaust gas recirculation, idle speed, etc. The control of these engine functions requires multiple inputs and the ability to optimize complex relationships. Therefore, a system was required which would accept a variety of input signals, perform calculations using this input information and provide the necessary outputs to control electrical and electro-mechanical actuators. This was accomplished using microprocessor technology, and applying that technology to the automotive environment and production methods.

The basic design concept of this system centers around the method used by the Engine Engineer to define the relationships of the input signals to the desired output control. Since these relationships are normally defined in terms of vehicle data, the system was designed to accept these input/output relationships in data tables rather than equation form.

Typical data used for the design of mechanical control of spark advance is illustrated in Figures 1 and 2. By combining these two curves, a data table is generated which defines the spark advance as a function of engine RPM and manifold vacuum. A typical example of such a speed/vacuum advance map is shown in Figure 3. The data in this advance map or data table is determined by the Engine Engineer through various engine and vehicle test sequences and is utilized directly by the microprocessor to determine the spark timing for all engine operation conditions.

As a result of the above consideration, the system design evolved around a digital microprocessor, custom designed to

1) accept a variety of inputs and generate outputs determined by preprogrammed calculations.
2) utilize data tables as the fundamental definition of the input/output relationships.

ABSTRACT

A digital electronic spark timing system has been developed and placed in production for use on the 1978 Oldsmobile Toronado. This system consists of an ignition distributor which contains a magnetic pickup and ignition module, a coolant temperature sensor and a controller package containing electronics and a pressure transducer. Engine RPM, manifold vacuum and coolant temperature inputs are used to derive and control the engine spark timing.

The controller electronics design is based on a digital microprocessor, custom designed to perform engine control functions such as spark timing, exhaust gas recirculation, idle speed, etc. The system design criteria includes data storage and calculation capability, automotive sensor and actuator interface, flexibility and accuracy.

DESIGN REQUIREMENTS

The principle design requirements for the 1978 Toronado system may be defined as the inputs, outputs, control functions, operating modes and environmental considerations.

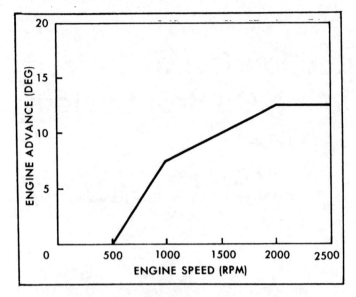

Fig. 1—Spark Advance vs. Engine RPM for a conventional ignition system

Inputs. In order to provide the necessary spark timing control it was necessary to provide for the following input signals:

1) Manifold Vacuum from 0 to 60 kilopascals
2) Engine speed and position over the range of 0 to 6000 RPM
3) Engine coolant temperature from −40°F to 300°F

In addition to the above engine parameter inputs it was also necessary to provide a "Reference Timing" input line to facilitate setting the initial engine timing.

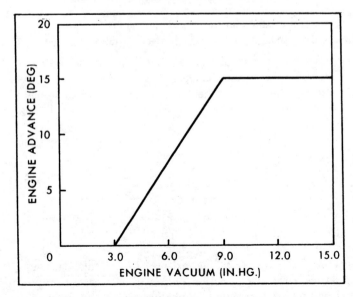

Fig. 2—Vacuum spark advance for a conventional ignition system

Outputs. A total of 3 output lines are necessary to provide control for the following output functions.

1) An Ignition Output Line is required to provide for switching of the ignition coil current, thus controlling the dwell (or coil current on-time) and the spark timing.
2) Two Indicator Light Lines are required to provide for control of a coolant over-heat indicator and an ignition function indicator.

Function Controls. To provide the above outputs as functions of the engine inputs, the following basic control functions are necessary.

1) Timing Advance vs. Engine RPM and Manifold Vacuum is the basic timing function and is defined by a data table similar to that shown in Figure 3. A graphic representation of this data table is shown in Figure 4 and illustrates the detail and flexibility with which engine data may be defined to obtain the desired engine timing.
2) A Timing Advance vs. Coolant Temperature function provides the ability to add a timing adjustment to the above basic timing as a function of the engine coolant temperature. The data table for this function is shown in table form and graphically in Figure 5.
3) The Dwell (On-Time) vs. Engine RPM function provides the ability to control the coil current on-time as a function of engine speed. Data for this function is also contained in a data table.
4) A Coolant Hot-Light Control is based on the coolant temperature being above or below a given temperature.
5) A Check Ignition Light Control provides driver notification of certain system malfunctions.

Operating Modes. In order for the system to function in a predictable manner under all vehicle operating conditions, it is necessary to provide for alternate operating modes. Under normal vehicle operating conditions, the system is designed to compute timing advance and dwell as previously described. However, under some extreme operating conditions, it is necessary to switch control from the microprocessor to a "Reference Timing" mode. This mode is defined as controlling the spark timing and dwell directly from the magnetic pickup input signal. The conditions for this reference timing mode are:

1) All conditions where the engine speed is less than 200 RPM.
2) All conditions where the vehicle system voltage is less than 9.5 volts.
3) Certain detectable system malfunctions.
4) All conditions where the "Reference Timing" select line is grounded.

Environmental Considerations. In order to insure the reliability of the system under all anticipated environmental conditions that might be expected on the vehicle, and since the controller was to be mounted in the passenger compartment, it was designed to operate over a temperature, vibration and humidity range compatible with under dash mounting.

ENGINE (RPM) \ ENGINE VACUUM (IN. HG.)	0	1.5	3.0	4.5	6.0	7.5	9.0	10.5	12.0	13.5	15.0	16.5	18.0
600	-2	-2	10	15	•	•	•	•	•	•	•	•	•
800	-2	-2	10	15	•	•	•	•	•	•	•	•	•
1000	5	5	•	•	•	•	•	•	•	•	•	•	•
1200	7	7	•	•	•	•	•	•	•	•	•	•	•
1400	9	•	•	•	•	•	•	•	•	•	•	•	•
1600	10	•	•	•	•	•	•	•	•	•	•	•	•
1800	9	•	•	•	•	•	•	•	•	•	•	•	•
2000	12	•	•	•	•	•	•	•	•	•	•	•	•
2200	13	•	•	•	•	•	•	•	•	•	•	•	•
2400	14	•	•	•	•	•	•	•	•	•	•	•	•
2600	•	•	•	•	•	•	•	•	•	•	•	•	•
2800	•	•	•	•	•	•	•	•	•	•	•	•	•
3000	•	•	•	•	•	•	•	•	•	•	•	•	•
3200	•	•	•	•	•	•	•	•	•	•	•	•	•
3400	•	•	•	•	•	•	•	•	•	•	•	•	•
3600	•	•	•	•	•	•	•	•	•	•	•	•	•
3800	•	•	•	•	•	•	•	•	•	•	•	•	•
4000	•	•	•	•	•	•	•	•	•	•	•	•	•

ADVANCE (ENGINE DEGREES) 234 DATA POINTS
• INDICATES ADDITIONAL DATA LOCATIONS

Fig. 3—Spark advance data table as a function of both vacuum and engine speed

Fig. 4—Three dimensional spark advance surface

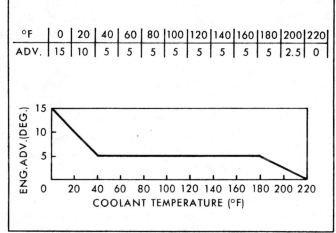

°F	0	20	40	60	80	100	120	140	160	180	200	220
ADV.	15	10	5	5	5	5	5	5	5	5	2.5	0

Fig. 5—Example of spark advance correction as a function of engine coolant temperature

SYSTEM HARDWARE

The hardware required for the 1978 system includes a temperature sensor, ignition distributor and a controller package. The components are shown in Figure 6 and the functions of each component and subcomponent are as follows.

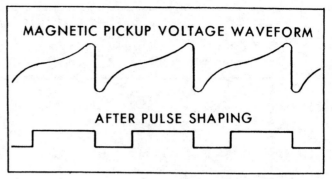

Fig. 7—Reference signal generated by the magnetic pickup before and after pulse shaping

2) An electronic ignition module, as used in the HEI System, accepts the ignition output signal from the controller and switches the ignition coil current as determined by the controller output. The ignition module also provides for ignition coil current limiting.

Controller Package. The controller package is located in the vehicle passenger compartment and contains the following major sections, as shown in Figure 8.

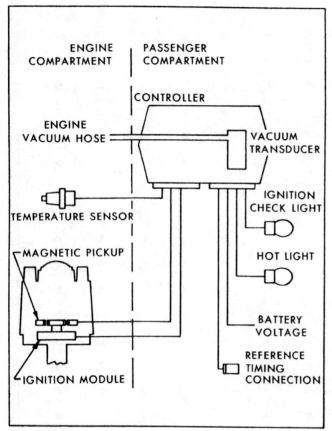

Fig. 6—Major components of the 1978 spark timing control system

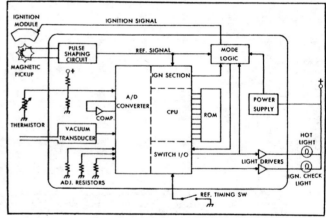

Fig. 8—Functional block diagram of the 1978 spark timing control system

Temperature Sensor. The coolant temperature sensor consists of a thermistor packaged to mount in the engine water jacket. The resistance of the thermistor varies as a function of the coolant temperature; and in conjunction with a resistor network, a voltage is developed which is proportional to the coolant temperature and ratiometric to the power supply voltage.

Ignition Distributor. The ignition distributor is similar to a standard HEI distributor except the mechanical advance mechanisms have been omitted. In addition to the ignition coil and rotor, the distributor contains the following.

1) A Magnetic Pickup is mounted on the distributor shaft and generates a voltage signal consisting of 4 pulses per revolution of an 8 cylinder engine. The controller package contains a signal conditioning circuit which converts the varying voltage generated by the pickup into a pulsed signal as shown in Figure 7. This square wave signal is then used by the controller to determine the engine speed and crankshaft position.

1) The power supply section provides an 8 volt DC source for the controller package and is designed to provide the necessary filtering and protection from charging system transients.
2) The pressure transducer senses engine manifold vacuum and provides an analog voltage proportioned to the sensed differential pressure and ratiometric with the power supply voltage.
3) The microprocessor is the computer portion of the controller and consists of a central processing unit (CPU), a read-only memory (ROM), a comparator and a crystal. The temperature and vacuum signals, both of which are presented to the CPU as voltages, are converted to internal digital signals via an analog-to-digital converter which is contained almost entirely within the CPU. The CPU processes the reference signal from the distributor pickup to obtain RPM data. It then performs the necessary

calculations as required by the instruction sequence. These calculations are based on the digital temperature, vacuum, RPM and dwell data which is stored in the ROM. The results of these calculations are then used to control the outputs of the system. The ignition output section of the microprocessor receives the desired dwell and timing advance angles as calculated by the microprocessor section. Using the reference signal from the magnetic pickup to derive a crankshaft position indicator, the desired dwell and timing advance angles are then compared with the position indicator and the ignition output signal is generated (see Figure 10) such that;

I) The ignition coil current is switched on at a predetermined crankshaft position to provide the desired dwell, and

II) it is switched off again at a predetermined crankshaft position to provide the desired timing advance.

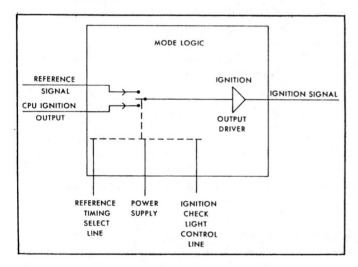

Fig. 9—Functional diagram of the Mode Logic

4) The mode logic section controls the ignition output signal which drives the ignition module. This section uses either the CPU ignition output signal or the reference signal from the magnetic pickup as determined by the various inputs as shown in Figure 9. The mode logic section will allow the CPU to control the timing advance and dwell, based on the various engine inputs, if the mode logic inputs are as follows.

I) The Reference Timing Select Line is open (not grounded).
II) The 8 volt power supply is in regulation.
III) The control signal to the Check Ignition Light indicates the light is off.

Under any other set of mode logic input conditions, the ignition module will be controlled by the reference signal. This generates a fixed timing advance based on the fall (or negative slope) of the distributor pickup signal.

5) The Hot Light indicator is controlled by a switch type output from the CPU, which controls a power driver to the

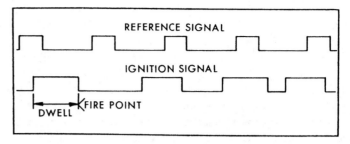

Fig. 10—Ignition output signal with respect to the shaped reference signal

actual light indicator. The microprocessor makes a periodic comparison between the coolant temperature input signal and the reference data stored in the ROM.

6) The Check Ignition indicator is controlled by a switch output from the CPU. The microprocessor combines RPM information, self check data, inputs from the mode logic section and a reference pulse check which determines that they are functioning properly, and CPU then provides a signal to the power driver, thus providing a light indicator for any recognized malfunction.

SYSTEM DESIGN & CAPABILITY

In developing the basic system design criteria, it was necessary to consider the system requirements, the hardware design and the desired capability. The following describes those items considered in determining the basic system capability.

The data and algorithm configuration is the primary system specification as determined by the Engine Engineer. This information completely describes the interaction of the inputs and outputs. To accept large quantities of data and instructions in the memory was a basic requirement. It was also necessary to retain maximum flexibility in allowing for reasonable turn-around time for changes in both data and instructions. These considerations lead to the CPU and ROM configuration used in the final design.

Read Only Memory (ROM). The ROM contains two types of information, instructions and data. Each piece of information is coded into a ten place binary number called a word. Data can then be expressed to 1 part in 1024. Table 1 describes the present utilization and the capability of the ROM.

Central Processing Unit (CPU). The CPU has the capability of implementing instruction codes effecting the following operations.

1) Addition of two variables (A+B).
2) Subtraction of two variables (A−B).
3) Compare two variables (A≥B or A≤B).
4) Multiplication of a variable by a fraction expressed to the nearest $1/32$nd. (A:K/32).
5) Implement in one instruction the equation X=[(A−B):K/32]+B. This is used for linear interpolation between data points.

Any number of the above type calculations may be used to accomplish a required computation. In addition to its calculating capability, the CPU also contains temporary storage registers for up to 18 different variables and a number of custom designed input and output functions. The variables used in any of the above operations may be data values stored permanently in the ROM, data from temporary storage registers or digitized input data. The results of a given sequence of operations may be used to control any given output of the CPU.

The input capability of the CPU was designed to meet the need of interface with sensor outputs of the voltage, switch or frequency type. Table 2 shows the type of input capability designed into the CPU.

1) The switch type inputs are capable of detecting either a low (grounded) or high (regulated voltage) state of the input signal. One such input is used in the 1978 system to detect if the mode logic is passing the reference pulse directly to the ignition module.
2) The voltage inputs can recognize any voltage signal whose level is between 15 and 75 percent of the regulated voltage. These inputs are parts of an analog-to-digital converter, which samples all input voltages once per program cycle, converting then to a digital number with a resolution of one part in 511. The analog inputs used for the 1978 system are shown in Table 3.
3) The frequency inputs are of 2 types:

   I) a low frequency input (1 to 500 Hz), and
   II) a high frequency input (100 Hz to 100 KHz).

The 1978 system uses the low frequency input as an RPM detector. This input is also part of the Ignition Output Section to provide crankshaft reference position information. The high frequency input is not used in the 1978 system.

The output capability of the CPU was designed to provide various output lines as shown in Table 4.

The switch outputs are—

1) designed to provide either a high or low signal as determined by the program instructions. Six of these switch outputs are the same six switch inputs listed in Table 3. The program instructions determine whether they are inputs or outputs, while the seventh switch output functions only as an output. The 1978 system uses one of these outputs to control the Hot Light and another to control the Check Ignition Light.
2) The ignition output is a custom designed section of the CPU designed specifically to control the timing advance and dwell signal to the ignition module. The resolution of this output is variable; at 2000 RPM and 45° timing advance it is approximately 0.6 engine degrees.
3) The frequency output may vary in frequency and duty cycle by program control. The maximum duty cycle resolution is approximately 1 part in 1024 for a minimum frequency of 27 Hz. This output, while not used for the 1978 system, might be used to control the current to a solenoid type actuator.

PRODUCTION APPLICATION

The application of a microprocessor based timing control system involved several considerations to insure performance and reliability. The following considerations were given in the areas of System Specifications, Environmental Considerations and Serviceability.

Performance Specifications. In order to derive the performance specification and limits for a system with large and varied data inputs, it was necessary to develop an extensive computerized error analysis program. These programs were developed to simulate all the significant error sources and compute the accuracy of the various outputs with respect to the variable inputs. It was necessary for this program to include such items as part repeatability, part-to-part variation and variations in component interaction. It was then necessary to confirm these system error estimates through a sample test program.

Power Supply Specifications. Since the system is required to operate over the total range of vehicle battery voltages, it was necessary to specify the system regulated power supply for operation in both the running and the cranking mode. The system power supply is regulated at $8.0 \pm .5$ volts, and the system operates in a normal running mode while the power supply is in regulation. However, under cranking condition, where the vehicle battery voltage is too low to support the regulated 8 volts, some minimum dwell and timing advance must be maintained to insure startability. To meet this requirement, the system mode logic senses an out of regulation state and will provide for the ignition module to be driven directly by the reference signal.

In addition to the voltage considerations, it was also necessary to design the regulated power supply to withstand, without damage, all vehicle charging system transients.

Self Diagnosis. It was desirable to provide for detection of certain system malfunctions, provide driver notification and revert control of the ignition module to the pickup reference signal. It was also necessary to revert to this reference signal operation during extremely low engine RPM to avoid calculation and output errors due to the high acceleration rates experienced under starting conditions. To meet this requirement, the ROM was coded such that the CPU detects either a low ROM condition or a failure of the instruction to sequence. Under either of these conditions, the CPU uses a switch output to turn on the "Check Ignition" Light, and to signal the mode logic to revert to reference pulse operation.

Environmental Specification. Because the controller package was designed to mount in the passenger compartment, all components and assemblies were designed to operate over a $-40°$ to $70°C$ ambient temperature range. A combination of thermal cycle testing over this temperature range and continuous high temperature durability testing was used to qualify to this specification.

The system was also designed and qualified to operate over the total humidity range including actual condensation on the assembly. Special tests were conducted to verify system operation under conditions of vibration, shock and electromagnetic interference.

Field Experience. To evaluate both the system performance and reliability, extensive field testing was conducted using police and taxi test fleets. These fleets experienced high operating hours and mileage over a short period of time and provided rapid feedback of any design or durability problems.

The test fleets were also useful in helping to develop field service diagnostic procedures. These procedures were written to enable the mechanic to diagnose system malfunction to the major component level with the use of tools and equipment readily available at the dealership level.

CONCLUSIONS

The 1978 Toronado spark timing system is a digital microprocessor control system designed to meet the requirements of an automotive application and to provide the engine designer with the ability to control spark timing in the most versatile manner possible. Although spark timing is not the only item affecting exhaust emissions, fuel economy and driveability, it is recognized as a major contributor and a more versatile spark timing control system will certainly contribute to improvements in these areas.

Although spark timing was the only engine parameter controlled in the 1978 system, future expansions and capability utilization can add further benefits as the need for more complex and versatile engine controls are defined.

TABLE 1—READ ONLY MEMORY CHARACTERISTICS

Characteristic	1978 System	Capability
Total Words	1024 (1 Chip)	4096 (2 to 4 Chips)
Words Used As Data	350	Any Combination
Words Used As Instruction	350	Possible

Data Type	Storage Resolution
Advance vs. RPM & Vac.	An advance data value is stored for every 200 rpm increment (600 to 4000 rpm) and 1.5 in. of hg. increment (0 to 18 in. of hg.) for a total of 234 data points. At each point the advance is stored to the nearest .25 degrees (40° = 160).
Advance vs. Coolant Temp.	An advance data value is stored for every 20°F increment (0 to 260°F) for a total of 13 points. The advance is stored to the nearest .25 degree.

TABLE 2—CPU INPUT CAPABILITY

Type Input	Used in 1978	Capability
Switch	1	6
Voltage	6	7
Frequency	1	2

TABLE 3—1978 ANALOG INPUTS

Input	Resolution
1) Vacuum transducer	2 mmHg
2) Temperature thermistor network	2°C
3) Adjustments—four voltage inputs are used to allow in production adjustments by using selected resistors in voltage divider networks	N/A

TABLE 4—CPU OUTPUT CAPABILITY

Type Output	Used in 1978	Capability
Switch	2	7
Ignition	1	1
Output Frequency	1	1

NOTES

780118

Microprocessor Engine Controller

T. W. Hartford
Bendix Electronics and Engine Control Systems Group
Troy, MI

THE NEED FOR SOPHISTICATED ELECTRONIC CONTROL of Otto-cycle internal-combustion engines has been discussed by many researchers (1,2,3).* Some have established the superiority of digital over analog control and the need for a high degree of flexibility in the selected architecture. Others have projected the expected improvement in air/fuel-ratio accuracy with the implementation of fuel injection, and the necessity for simultaneous control of spark advance and exhaust-gas recirculation (4,5).

This paper describes a microprocessor engine controller developed with a view to improving the emissions levels, economy, and driveability of fuel-injection-equipped vehicles presently outfitted with analog electronic controllers. Cost reduction was also a goal, despite expansion of the control function to include spark-ignition advance as well as more complex fuel-injection and exhaust-gas-recirculation control.

The development program addressed four major technical issues:

• The ability of standard microprocessor designs to provide adequate dynamic response for the complex multiple control functions required for Otto-cycle engines.
• The feasibility of adequate ignition control based on crankshaft-position pulses which in an eight-cylinder engine occur once for each crankshaft rotation of 90 degrees.
• The feasibility of providing the digital electronics required to interface the microprocessor to the engine in a single custom input/output (I/O) circuit.
• The feasibility of so configuring the system as to allow control-law changes or application of the controller to other engines without redesign of the custom input/output electronics.

The sections that follow describe the microprocessor-engine-controller hardware and software subsystems, and summarize some of the experimental results obtained. A companion paper (6) provides a more detailed discussion of the input/output hardware design.

HARDWARE

A microprocessor-based architecture was selected for the engine controller, with a view to reducing cost and achieving greater flexibility. An industry standard microprocessor

---
*Numbers in parentheses designate References at end of paper.

ABSTRACT

This paper describes a microprocessor engine controller designed for onboard control of closed-loop fuel injection, spark advance, and exhaust-gas recirculation. Developed through the preproduction prototype stage, the controller utilizes software subroutines to accomplish such operations as multiplication and interpolation. Ignition is controlled by means of time delays from relatively widely spaced crankshaft-position pulses. Emissions, driveability, fuel-economy, and hardware-cost comparisons are made between the microprocessor engine controller and a 1977 production analog electronic-fuel-injection controller with a mechanical-advance-controlled distributor. Directions for future development efforts are also described.

unit (MPU)--the MC 6800--was chosen over several
custom and standard designs, by reason of its
superior performance in engine control. Use of
a standard microprocessor unit can be expected
to result in lower production costs because of
the economies of scale obtained through shared
consumer markets. It can also be expected to
mean that product-improvement developments--
including ongoing process and mask improvements
to enhance reliability--will be financed by the
semiconductor supplier through reinvestment of
his profit dollars; custom-product improvements,
by contrast, generally stop the day the product
is frozen for production unless the procurer
himself provides development funds. Finally, the
selection of a standard microprocessor reduced
the development time and cost, making it possible to utilize hardware and software development
tools readily available from the microprocessor
manufacturer.

Figure 1 is a simplified block diagram of
the microprocessor engine controller, showing
both production-model and engineering-model
configurations. Engineering-model units are
equipped with added electronics to implement
display and calibration, thus increasing engineering productivity during control-law and
control-schedule development. Otherwise, the
two configurations do not differ functionally.

The microprocessor unit communicates with
both memory and input/output electronics via two
data buses: a 16-bit parallel address bus,
which selects one of 65,536 possible word locations in memory or input/output, and an 8-bit
parallel bidirectional data bus, by means of
which the microprocessor unit reads a word to
memory or input/output. Arithmetic operations
within the microprocessor unit are performed in
8-bit parallel binary arithmetic.

Three types of memory are used, depending
on system configuration. The production configuration uses a mask-programmed Read-Only Memory
(ROM) for storage of control laws and control
schedules. A read/write Random-Access Memory
(RAM) is used as a scratch pad for the program
memory and contains such data as sensor values,
output control words, and intermediate computation results. To facilitate rapid system development, the engineering configurations contain
Erasable Programmable Read-Only Memory (EPROM)
in place of Read-Only Memory.

Figure 2 illustrates memory partitioning in
the microprocessor engine controller. The
control-law program defines the equations and
solution procedures for engine control. The
control-law schedules represent the actual data
surfaces for a specific engine application.
Partitioned in this manner, the memory permits
substitution of different calibration schedules
when the microprocessor engine controller is
applied to different engines. In addition, all
production controls can be built to the same
specifications, with insertion of the appro-

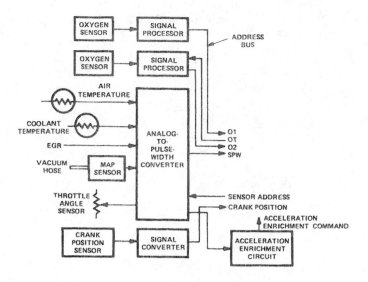

Fig. 1 - Microprocessor engine controller
simplified block diagram

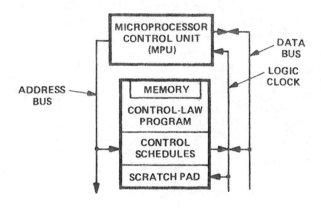

Fig. 2 - Memory partitioning in the
microprocessor engine controller

priate calibration module occurring at final
assembly or at the automobile dealership.

Figure 3 details the input-signal-
conditioning circuitry, which converts sensor
input signals to pulse widths and/or standard-
izes their levels to those acceptable to the
digital input/output electronics. The digital
input/output electronics convert pulse-width
inputs to binary numbers and binary-number
output words to output pulse widths. These
circuits also synchronize the information flow
between the microprocessor unit and the
microprocessor-unit clock cycles.

The input-signal-conditioning circuitry
implements two-channel oxygen-sensor signal
processing to convert zirconia oxygen sensor
voltage levels to standard levels acceptable to
digital logic. It also converts crankshaft-
position-sensor output to standard logic levels.
The acceleration-enrichment (AE) circuit indi-

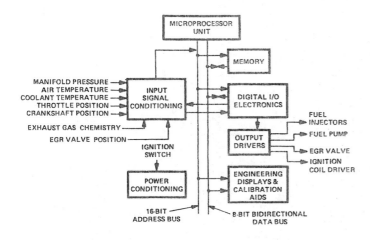

Fig. 3 - Input-signal conditioning

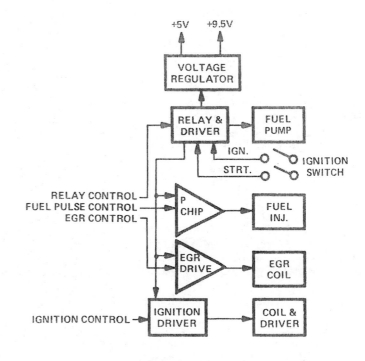

Fig. 4 - Output drivers

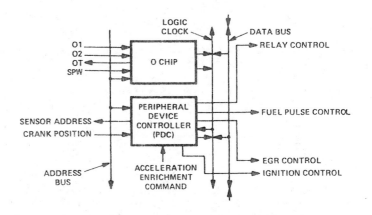

Fig. 5 - Digital input/output electronics

cates to the computer when acceleration has been demanded. The analog-to-pulse-width converter changes analog sensor inputs into pulse widths which are proportional to the sensed physical variable for coolant temperature, inlet-air temperature, manifold absolute pressure, throttle position, and exhaust-gas-recirculation valve position.

The output drivers, detailed in Figure 4, provide such impedance matching, power amplification, and current regulation as may be required to enable the output control signals to operate the engine actuators. Each driver serves as a necessary interface between the low-power metal-oxide-silicon (MOS) logic circuits and the individual actuator. A relay and relay driver control the electric fuel pump, a current-regulator circuit drives the fuel injectors, a power-amplification circuit drives the exhaust-gas-recirculation solenoid, and an impedance-matching driver delivers the trigger signal to a high energy ignition driver assembly. The power-conditioning circuitry controls and regulates battery power to the electronics by means of a voltage regulator. Suppressed and protected from undesirable electrical-system transients, this circuitry supplies +5-volt and +9.5-volt power for the digital and the analog circuits, respectively.

Figure 5 details the large-scale-integration (LSI) digital input/output electronics. The 0-chip block in this figure represents a custom large-scale-integration circuit which contains the logic-clock oscillator and drivers, a digital input-signal multiplexer, and a mechanism for preprocessing the exhaust-gas-chemistry sensor inputs. The use of custom large-scale integration for digital circuits and custom large-scale integration combined with thick-film hybrid construction for analog interface circuits results in reduced costs and improved reliability.

The balance of the digital signal processing is performed by a custom large-scale-integration circuit called the peripheral-device controller (PDC). Here input pulse widths are converted to binary words, which are transmitted to the microprocessor unit in synchronism with its data-processing rate. Binary words received under program command from the microprocessor unit are converted into pulse-width outputs to control the fuel injectors, the exhaust-gas-recirculation valve, and ignition firing. Engine period is measured by a counter, and this information is synchronized and transmitted in binary form to the microprocessor unit. Detection circuitry for engine stall, logic-clock failure, and schizophrenic computer operation is also located in the pheripheral-device controller, which processes into the microprocessor unit all computer interrupts. These are distinguished from one another by flag signals derived within the peripheral-device-controller circuit. A 3-bit output port is provided for control of on/off functions.

The microprocessor unit controls all information flow to the microprocessor engine controller via address-bus states; that is, all binary-word transfers occur on the data bus in response to specific address-bus states. This architecture provides for considerable flexibility since the system can be expanded by adding additional devices between the two buses. Figure 6 illustrates this bus-expansion capability with respect to the calibration aids. Note that the production-model buses are first buffered to avoid overload of the microprocessor-unit circuit. Additional memory, an external-device control assembly, and an instrumentation assembly are then added between the buses, under control of the production-assembly microprocessor unit. The separate mechanical assembly which contains all these items--called the calibration control box (CCB)--is wired to the electronic control unit (ECU) by means of a plug-in cable. This calibration control box is useful for problem isolation during the development stage. In addition to providing much of the required automobile instrumentation, it permits modification of the data schedules which determine the control calibration for specific engine or engine/car combinations while the control is still installed on the automobile.

Figure 7 shows the programmable microprocessor-engine-controller assembly typically used with the calibration control box, seen in Figure 8. Figure 9 is a preproduction prototype microprocessor engine controller.

SOFTWARE

The software architecture of the MC 6800 microprocessor unit features 8-bit parallel processing with a 16-bit parallel address bus and 72 variable-length instructions with seven addressing modes. It is also characterized by interrupt vectoring, two accumulators, an index register, a stack pointer and variable-length stack, a condition-code register, and memory-mapped input/output. The unit we selected has a standard logic-clock rate of 1 megahertz, although rates of 1.5 and 2 megahertz are available on more recent models. Instructions require from two to twelve clock periods to execute. The register-to-register add is two periods and the memory-to-register add is three or four periods, depending on address mode. Only simple instructions--add, substract, and shift, for example--are contained in the instruction set; complex operations--such as multiplication, division, and surface and table interpolations--are accomplished via subroutines.

The microprocessor-engine-controller software is functionally partitioned in memory into a control-schedules section and a control-

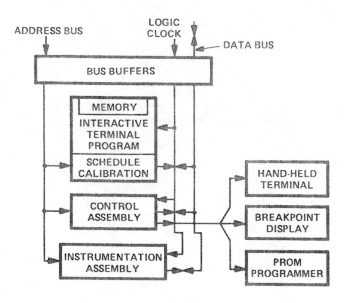

Fig. 6 - Engineering displays and calibration aids

program section (see Figure 2). The control-schedules section of the Read-Only Memory contains data points which represent the six surfaces and ten tables stored in the electronic control unit. The control-law algorithms, which are solved by the computer in order to control the engine, are contained in the control-program section of the Read-Only Memory. The control program is modularized into subroutines. The main program, which first organizes data into registers, calls these subroutines in the proper order to solve the control algorithms.

The software block diagrams seen in Figures 10 and 11 illustrate the computation scheme which is used for control of multipoint closed-loop fuel injection, ignition spark advance, ignition dwell, and exhaust-gas recirculation. Control-program execution is started by an interrupt from the peripheral-device-controller circuit. Interrupt-driven software developed in subroutine modules permits control laws to be reconfigured without major software redevelopment.

Five interrupts may occur: an engine-crankshaft-position pulse, a fuel pulse No. 1 complete, a fuel pulse No. 2 complete, an acceleration request, or a system reset. The crankshaft-position interrupt occurs once for every fixed increment of crankshaft rotation; the increment in degrees between each interrupt can be calculated by dividing 720 by the number of engine cylinders. The fuel-pulse-complete and acceleration-request interrupts are used to provide acceleration enrichment by injection of extra fuel. The system-reset interrupt, which occurs whenever power is first applied to the electronic control unit, causes the computer to execute a program sequence that initializes known values in specific scratch-pad registers

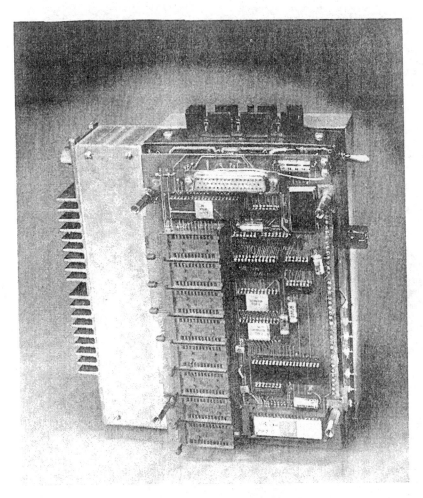

Fig. 7 - Programmable microprocessor engine controller assembly

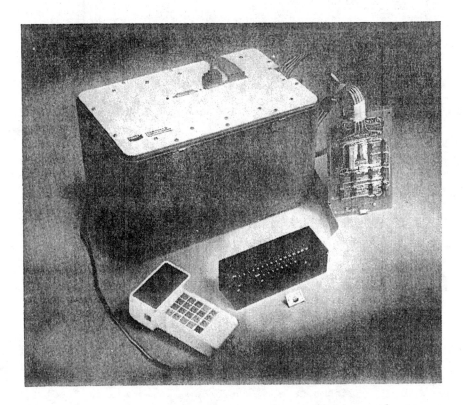

Fig. 8 - Engineering calibration control box and associated displays

Fig. 9 - Preproduction prototype microprocessor engine controller

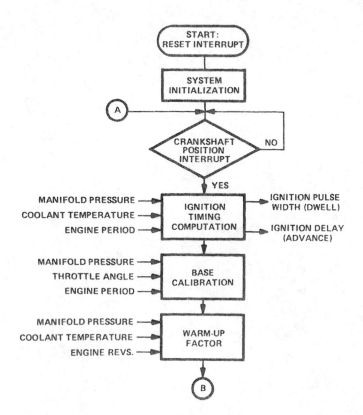

Fig. 10 - Software block diagram -- Part I

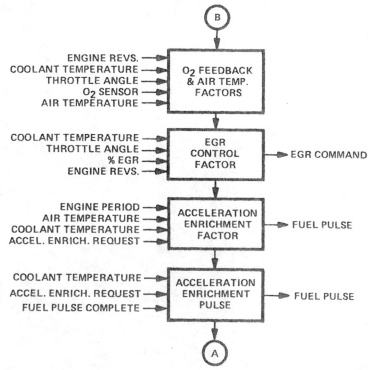

Fig. 11 - Software block diagram -- Part II

and thereby prevents undesirable system transients.

The computations of the control program are interleaved in real time with analog-to-digital conversions. For example, the control program may command the analog-to-digital converter to convert the manifold-absolute-pressure (MAP) reading. During this conversion, the fuel-injection air-temperature factor may be computed, after which the software retrieves the manifold-absolute-pressure value from the peripheral-device controller. Ignition advance and ignition dwell may be computed while a second conversion is in progress. This interleaving process continues and is carefully timed so that the software does not wait any significant time for analog-to-digital conversion. When these computations are complete, all sensor values will have been converted for use in other computations. The software program then stops execution until another interrupt is received.

Ignition control by means of ignition-delay and pulse-width data is illustrated in Figure 12. A single ground-going pulse turns off the power to the ignition coil in the ground state, thus initiating the spark firing. The ignition-delay quantity controls the time between the last crankshaft-position pulse and coil-power-off, and the ignition-pulse-width quantity controls the coil-power-off duration. Figure 12 also shows how these quantities control ignition advance and ignition dwell.

The use of interrupt-driven software which computes synchronously with engine operation has resulted in simplification of some engine-dynamics control algorithms. Since the engine is itself a sampled-data system, engine-response expressions are generally complex when expressed in real time. Transport lag varies with engine speed when expressed in real time, for example, but may be nearly constant when expressed in revolutions (7).

A second advantage of interrupt-driven software is the availability of additional real time at low engine speeds, which allows slowly varying tasks to be accomplished in addition to the engine-control functions. In developmental systems, the available real time is used to display to the operator on a hand-held terminal selected engine operating parameters via a special program residing in the calibration control box. Man/machine communication is enhanced by the use of a display program which converts binary values in computer storage to decimal numbers in such common engineering units as revolutions per minute, torr, and degrees Fahrenheit. Interactive prompter symbols are also generated to clue the operator in the use of the display.

SYSTEM EVALUATION

Emission and fuel-economy comparison tests have been conducted on a microprocessor engine controller of the type herein described and on a 1977 production analog controller, each controlling open-loop fuel injection. With both control units programmed according to the same calibration schedule, emissions, fuel-economy, and air/fuel-ratio results were obtained which correlated within 1 percent--the magnitude of the experimental error. It was therefore concluded that analog-technology capabilities can be duplicated in a digital system which utilizes a microprocessor-unit-based architecture. Moreover, since fast hardware multiply and divide operations are not required, standard microprocessor-unit designs can provide adequate dynamic response for engine control.

The emissions and fuel-economy tests were conducted with a fuel-management control law requiring the solution of four surfaces and ten tables to complete one fuel computation. The computation time was approximately 8 milliseconds using a 1-megahertz microprocessor-unit clock frequency. Computations were made once per engine revolution.

An ignition-control algorithm requiring the solution of two surfaces and one table was then added to the fuel-management control laws. The mechanical-distributor advance curves were programmed into the microprocessor-engine-controller schedules, and the computer spark-advance control was compared to the mechanical control by means of an oscilloscope. In this test, the mechanical-distributor advance was used to operate the engine. Some jitter was evident in the electronic spark-advance control at constant engine speeds, but this was to be expected because of cyclic speed variations and the use of time delays to control the ignition. Under transient conditions, however, the required spark-advance changes occurred faster with the electronic advance system. Emissions and fuel-economy results for the vehicles equipped with either electronic fuel injection and electronic advance control or with mechanical-distributor advance and analog electronic control still correlated within 1 percent, but the microprocessor engine controller resulted in improved driveability. It was therefore concluded that electronic spark advance and dwell control can be accomplished with time delays which are initiated by relatively widely spaced crankshaft-position pulses. The solution of the fuel-control algorithm once per revolution and the ignition-control algorithm once per engine firing resulted in a 14-millisecond solution time for one eight-cylinder-engine revolution. At higher engine speeds, the frequency of ignition-control computation is reduced to ensure one fuel computation per revolution.

Intensive microprocessor-engine-controller control-law and control-schedule development, along with the addition of closed-loop fuel management to both the analog control and the microprocessor engine controller, led to the preliminary conclusion that any emissions breakthrough attained with either the microprocessor engine controller or the analog control can also be duplicated in the other architecture. However, not all areas of the microprocessor-engine-controller control surfaces have been exploited to date. The microprocessor engine controller, programmed for ignition and exhaust-gas-recirculation control, did attain better driveability and fuel economy than the analog control while meeting 1980 emissions standards.

The flexibility of the large-scale-integration architecture was demonstrated through its application not only to conventional

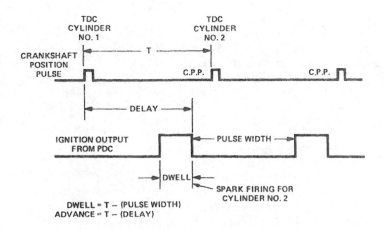

Fig. 12 - Ignition-control timing

electronic-fuel-injection systems but also to single-point injection systems and closed-loop carburetion systems. Control-law changes were facilitated because no architectural redesign was necessary; indeed, in one six-month period, 50 control-law changes were implemented without any digital-hardware modification. In general, then, the degree of control the computer software has over the input/output hardware modules enhances the flexibility of the microprocessor engine controller.

The goal of providing all needed digital input/output electronics in a single custom large-scale-integration circuit was not achieved in the microprocessor-engine-controller implementation described here. The necessary location of the logic-clock oscillator and special clock drivers within the input/output circuit--50 percent of the 0-chip area is devoted to the system clock--proved to be the primary barrier. However, a microprocessor unit with the system clock oscillator and drivers included internally has recently become available. Use of this unit--the MC 6802--in the next microprocessor-engine-controller design iteration will permit all digital input/output functions to be incorporated in a single input/output chip.

Based on identical volume projections, production-cost estimates for a microprocessor engine controller configured for closed-loop multipoint fuel injection, exhaust-gas recirculation, and spark-advance control are slightly lower than those for the Model Year 1977 production analog open-loop fuel-injection and exhaust-gas-recirculation controller. Future development efforts are likely to result in still lower costs for the microprocessor engine controller.

FUTURE DIRECTIONS

Further microprocessor-engine-controller development is ongoing and proceeding in several directions. Control laws are being refined and simplified. Improvements in semiconductor technology are reducing solution time; some versions of the microprocessor engine controller currently operate at a 2-megahertz clock frequency, for example, reducing solution time by 50 percent. Higher levels of integration are also being pursued through normal improvement of standard semiconductor parts and through architectural improvements in the custom large-scale-integration design.

CONCLUSION

A microprocessor-based engine electronic-fuel-injection controller has been developed which can achieve emissions levels comparable to those of current analog controls. By including programmable spark-advance control, the microprocessor engine controller can surpass the fuel-economy and driveability performance of the analog electronic-fuel-injection controller. Moreover, it can provide closed-loop fuel-injection control, exhaust-gas-recirculation control, and spark-advance control at about the same cost as a Model Year 1977 production analog open-loop fuel-injection and exhaust-gas-recirculation controller.

The microprocessor-engine-controller development program has successfully resolved the four major technical issues previously cited:

- Standard microprocessors--without fast hardware multiply and divide features--can provide adequate dynamic response to control fuel injection, exhaust-gas recirculation, and spark advance in Otto-cycle internal-combustion engines.
- Ignition spark advance can be adequately controlled through the use of time delays in association with one crankshaft-position pulse per cylinder firing, there being in an eight-cylinder Otto-cycle engine four such pulses per revolution.
- The digital electronics required to interface the microprocessor to the engine can be provided in a single custom input/output circuit if the system logic clock is supplied by the microprocessor circuit.
- The custom large-scale-integration digital input/output circuitry can be so configured that control-law changes do not require circuitry redesign.

REFERENCES

1. H. G. Bruijning, W. J. Kleuters, and P. J. Poolman, "Ignition and Electronic Injection Control for the Future," Institution of Mechanical Engineers Paper No. C 342/73 (May 1973).

2. J. Camp and T. L. Rachel, "Closed-Loop Electronic Fuel and Air Control of Internal-Combustion Engines," SAE Paper No. 750369 (February 1975).

3. P. H. Schweitzer and C. Volz, "Electronic Optimizer Control for Internal-Combustion

Engine: Most Miles per Gallon for Any Miles Per Hour," SAE Paper No. 750370 (February 1975).

4. R. S. Oswald, N. L. Laurance, and S. S. Devlin, "Design Consideration for an Onboard Computer System," SAE Paper No. 750434 (February 1975).

5. J. B. Russell and R. G. Nedbal, "Air/Fuel-Ratio Control Using A Single Microprocessor," SAE Paper No. 770006 (February 1977).

6. A. W. Barman and R. S. Henrich, "A Microprocessor Input/Output for Automotive Engine Control," presented at the SAE International Automotive Engineering Congress and Exposition, Detroit, Michigan, 27 February - 3 March 1978.

7. R. Gunda and T. L. Rachel, "Electronic Fuel Injection Utilizing Feedback Techniques," *IEEE Intercon Conference Record*, Seminar 36 (March 1974).

780655

Electronic Spark Timing Control for Motor Vehicles

Paul H. Schweitzer and Thomas W. Collins
Optimizer Control Corp.

INTRODUCTION

WE ARE WITNESSING a burst of activity in electronic engine controls. Electronic fuel injection has been available for several years. Electronic skid control and other controls followed. Last year electronic spark timing controls have been introduced by Chrysler and General Motors. It is safe to predict that before long they will be widely used.

Electronic control design for automobiles has regularly been in the hands of electronic specialists and automotive aspects were often ignored. Automotive electronics must operate under extreme environmental conditions. A multiplicity of delicate sensors is undesirable in an automobile. Simplicity and ruggedness have top priorities.

A general outline of electronic engine controls were presented in Ref. 1. This paper deals specifically with spark timing control. This will be preceded, however, by a brief overview of control theory fundamentals.

TYPES OF CONTROLS

The function of a control or control system is to set the setting of a machine in such a manner that its performance is improved in the broad sense that it becomes more satisfactory to the operator. In an internal combustion engine this may mean increased power, reduced specific fuel consumption, reduced exhaust emissions, stable operation and responsiveness (drivability) or other performance characteristics.

A control can be manual (Steuerung in German) or automatic (Regelung in German). This paper treats only automatic controls, and control will mean automatic control unless otherwise specified.

Control systems can be mechanical, hydraulic, pneumatic, fluidic, electric, electronic, or a combination of these. Irrespective of the mechanism used, every control system consists basically of three components; Sensor, Logic, and Actuator.

---ABSTRACT

The spark-Optimizer is a closed-loop type electronic control device that continuously corrects the ignition timing; in effect it re-tunes the engine some ten times every second. In contrast to the better known pre-programmed controls, the Optimizer is an adaptive type system, in which the output influences the input. By providing the correct spark timing all the time, the Optimizer reduces fuel consumption considerably.

This paper describes the spark-Optimizer in its advanced version including recent improvements, like biasing. A comparison with other electronic spark control devices show its superiority in accuracy, flexibility, simplicity, and production costs. Samples of test results show considerable fuel savings with the Optimizer.

0148-7191/78/0605-0655$02.50
Copyright © 1978 Society of Automotive Engineers, Inc.

The sensor (one or more) provides the input to the control. The logic is the decision maker. The actuator receives the output of the logic which consists of instruction to set the setting. The scheme is shown on Fig. 1.

Two types of control systems are in use. In a pre-programmed control system (programmed control) the logic receives its input from the sensor which responds to external influences, external in the sense that they do not originate from inside the control mechanism or its target setting. An example where engine speed is used to set the spark timing is shown in Fig. 2.

In a programmed control the logic has built-in instructions from the designer how to respond to a sensor signal in formulating its instructions to the actuator. In the case of spark timing, the logic can order, by a suitable linkage, or otherwise, an advance of the spark timing with engine speed in a linear or nonlinear fashion. The control designer determines in advance the functional relationships between engine speed and spark advance and programs that relationship into the control system.

Another type of control system is designated as adaptive control. In an adaptive control system the control output influences its input. A classic example of the adaptive control is the engine speed governor, invented by James Watts in 1788. A better known example is the room thermostat, shown in Fig. 3.

Adaptive controls use feedbacks for input. Popularly they are called closed-loop controls.

In the case of the thermostat, the target of the control is the room temperature. A sensor senses the room temperature and feeds that information to the logic. The logic compares that with a reference temperature that has been selected by the user. The control accomplishes its objective by equalizing these two temperatures, by sending instructions to the heat supply for increase or decrease. In this example the reference is set by the user.

In the case of spark timing control the reference must vary with engine speed, load, and perhaps other parameters. This means that the reference itself must be programmed, based on information supplied by sensors that sense the influencing parameters.

In another type of closed-loop control a single output characteristic of an engine provides the reference for the logic. In the three-way catalytic converter for emissions control an exhaust sensor senses the air-fuel ratio and the logic selects the stoichiometric ratio for matching.

A spark timing control of this category could sense engine knock and cease advancing the timing when engine knock appears. A schematic of such a control is shown in Fig. 4.

A most advanced type of adaptive control is the optimizing control, invented by Charles Draper and Y. T. Li in 1951 (Ref. 2).

The optimizing control also uses the closed-loop feedback principle, but instead of aiming at a fixed target it seeks an optimum. The basic concept of an optimizing control, applied to spark timing, is shown in Fig. 5.

The sensor monitors the engine power output and sends this information to the logic. Simultaneously the logic also receives information on the spark setting. The logic compares these sets of information and figures out what change of spark setting would favor the power output. Accordingly it sends orders to the actuator to advance or retard the spark timing as the circumstance requires.

DITHERING TECHNIQUE

In order to facilitate monitoring engine performance, or rather the change of the performance with a change in spark timing, it is very helpful to deliberately disturb the spark setting. One way of doing this is starting from zero or a low value, the control advances the spark timing continuously until a drop in power is sensed. Then the control reverses and retards the timing until again a drop in power is sensed. This play is repeated continuously. The spark timing, therefore, can never get far off the optimum. In the dithering technique the spark timing is periodically changed up and down a small amount from a middle point. Therefore the spark timing is alternately advanced and retarded a number of times per second.

The oscillator (dither) keeps the logic continuously informed of the dithering phase. The logic is able to determine whether the engine responds favorably or unfavorably after the spark was advanced (or retarded). Thus the logic can form correct decisions, when to advance and when to retard the spark timing. It then sends appropriate commands to the actuator as shown in Fig. 6.

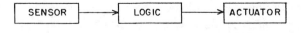

Fig. 1 - Basic scheme of a control system

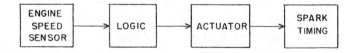

Fig. 2 - A programmed control of the spark timing

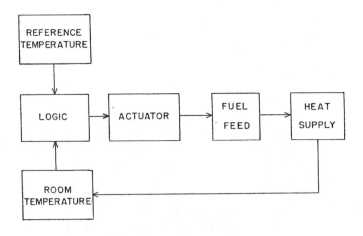

Fig. 3 - Room thermostat, a closed-loop control

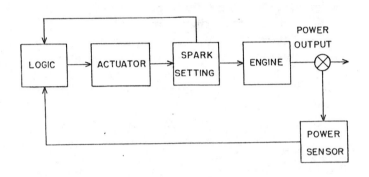

Fig. 5 - Optimizing control of spark timing

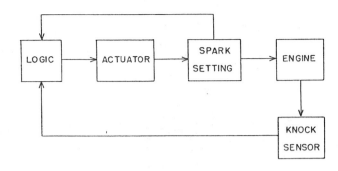

Fig. 4 - Closed-loop control of spark timing with knock sensor

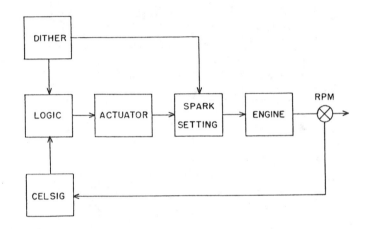

Fig. 6 - Optimizing control, using the dithering technique for spark timing

If MBT (minimum for best torque) timing is sought, the sensor must sense torque or rather a change of torque. This could be accomplished by sensing torque directly, but a more practical way is to sense the change in rotational speed (rpm), which can be sensed conveniently by an acceleration sensor, named Celsig.

If a slotted disc is fastened to the crankshaft, or another driven shaft, and the teeth of the disc pass a magnetic sensor, the rpm is proportional to the number of pulses generated in a fixed period of time. If during the advance dither period the pulse frequency increases and during the retard period it decreases, the engine favors advance and the control is to provide advance, and vice versa.

Fig. 7 shows schematically a slotted disc type celsig with its facing magnet as used in an optimizing spark timing control.

The adaptive closed-loop system depicted in Fig. 3, is superior to the straight programmed system inasmuch as it continuously checks and corrects its own performance. Nevertheless, the error correcting systems, some of which are now under active development for automotive use, share one important source of inaccuracy with the straight programmed system. In selecting the setting sought, both rely on the designer's judgment. However, the designer's information and intelligence is limited, and the setting he decides is right for the engine may be very different from what is really best for the engine and its operator.

In an optimizing control system the incentive for correcting a setting, and the direction it is to go, derives from the engine itself. It is the engine, not the designer that decides. This eliminates the most fundamental inaccuracy automatic controls are burdened with. Other sources of errors in various control systems will be discussed later in the paper.

The above outlined fundamentals apply to all control systems, from mechanical to electronic, and for any operating system subject to control. In the case of internal combustion engines the controlled setting may pertain to the spark timing, the air-fuel ratio, the valve timing (if variable), the injection timing (in diesel engines), or other influencing variables. The target of the control may be maximum power output, minimum fuel consumption, minimum exhaust emissions, or

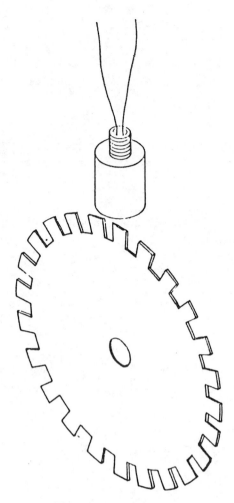

Fig. 7 - Slotted disc type celsig in a spark timing controller

other performance characteristics. The control theory, and to a large extent the technology applied are the same.

The rest of this paper is limited to the problem of spark timing in spark ignition engines. The problem of controlling the air-fuel ratio was discussed in a companion paper, (Ref. 3).

SPARK TIMING IN S. I. ENGINES

The timing of the ignition or spark plug discharge is one of the most important parameters that influence engine performance. It is usually expressed by spark advance, in degrees crank angle before top dead center. For power output the optimum timing is MBT. If the spark timing differs from MBT, combustion efficiency suffers, reducing power and increasing specific fuel consumption. If the spark setting is far from MBT the engine is unable to operate.

Spark timing has also decisive influence on exhaust emissions and engine knock. Late timing always reduces NOx emissions and to a lesser degree also the unburned hydrocarbons. Advanced timing may cause engine knock. MBT timing which is most favorable for performance may be unacceptable for other reasons.

The range of possible spark timing is very wide, from about 10 deg ATC to about 70 deg BTC, depending on the circumstances. The range of MBT timing is almost as wide. The function of the spark timing control is to find the best timing under actual circumstances.

MBT timing depends on at least ten variables, engine-related and environmental, as listed, where the numbers in parentheses indicate a normal range (in deg c. a.) of their influence, not considering extreme conditions. Engine speed (35), Load (15), Metal temperature (10), Ambient air temperature (10), Ambient air pressure (10), Ambient air humidity (10), Fuel octance number (10), Air-fuel ratio (10), Residual gas from preceeding cycle (5), Exhaust gas recirculation (10).

Combustion efficiency depends on how much the actual timing is off MBT. When the actual timing is 5 deg off MBT, plus or minus, the degradation is seldom more than 1%, at 10 deg it may amount to 5%, at 20 deg 16%. These figures are approximate and vary from engine-to-engine. But even in freshly tuned engines actual spark timing is often over 20 deg later than MBT and in out-of-tune engines even more. There is no justification for such late timing but there is an explanation for it.

Under certain operating conditions, some engines knock at MBT timing. Factory timing is so chosen that engine knock be avoided under almost any circumstances encountered in practice. The safety margin used is often more than 20 deg, not because engine knock is so unpredictable but because the car makers know no way to predict it, given the wide variations in operating conditions and in individual engines of the same model. Furthermore, even knowing the requirements, the available control technology is inadequate to cope with them.

ENGINE KNOCK

Engine knock, of the type frequently identified with detonation, is the result of abnormal combustion. It varies in intensity from incipient or trace knock which only instruments or a trained ear can detect, to medium knock which is annoying to the operator, to heavy knock which is damaging to the engine.

Knock is commonly controlled by retarding the spark timing but this also reduces thermal efficiency. When spark timing is advanced, the cycle efficiency is increased. When the advance is carried into the knock region, the increased turbulence associated with detonation causes an increased heat loss which reduces the combustion efficiency.

Indicated thermal efficiency is a product of the cycle and the combustion efficiency. This product reaches maximum at a certain spark timing which is either more or less advanced than the point of incipient knock (IK).

Some engines knock at MBT, others do not. It greatly depends on the rate of heat transfer from the combustion chamber to the cylinder head. Engines with air cooling or with high-temperature pressure cooling are more likely to knock at MBT, especially if they also have a compact combustion chamber. But knock at MBT is never heavy and ordinarily only occurs at high load operation.

Whether the IK point is more advanced or less advanced than the MBT point, the two invariably move together when operating conditions change. The IK point is ordinarily more advanced; when not, seldom more than 5 deg later.

EMISSIONS

Spark timing, however, is often retarded a substantial amount because of emission requirements. NOx is particularly sensitive to spark timing. Fifteen degrees or more retard from MBT has been used in the past to suppress NOx emissions. This practice has been largely discontinued lately because of the heavy fuel penalty involved. Five percent economy loss "buys" almost three times as much NOx reduction through EGR than through spark retard (Ref. 4).

A moderate spark retard has also been used for the sake of reducing HC emissions. In this case the effect is indirect. By delaying combustion the engine actually ends the power stroke with more HC. However, the increased exhaust temperature enables the exhaust system to burn more of the unburned hydrocarbons in the exhaust pipe and muffler. A thermal or catalytic reactor accomplishes this objective more economically.

CONTROL TYPES FOR SPARK TIMING

The control system for controlling spark timing used in the last half century is mechanical-pneumatic. Its principle is depicted, in terms of our methodology, in Fig. 8.

The engine speed is sensed by the flyweights mounted on the distributor shaft. The flyweights, through linkage, raise a sleeve with helical splines which change the phase between the distributor shaft and the crankshaft. The intake vacuum is transmitted through a tube to a membrane, mounted on the distributor housing and through a rod, rotates the timing disc of the distributor. Without being readily distinguishable, all elements of a pre-programmed control system are present.

Even though it has been successfully used for a long time, the conventional spark control has numerous shortcomings. An incomplete list of them follows:

    1. Out of not less than ten influencing variables it corrects only for two. This alone easily can cause an error of 30 deg c.a. in the correct spark timing.

    2. Manufacturing tolerances in the control mechanism and engine-to-engine variations in the same model increase the control inaccuracy.

    3. Drift in use, wear in the mechanism, points and cam wear in the distributor cause unintended spark retardation.

    4. For mechanical reasons the vacuum advance is generally linear, and the centrifugal advance is segmentally linear. The two advances add up irrespective of the requirements.

    5. Correct spark timing is a function of the engine load. Light load requires more spark advance than heavy load. Conventional spark control senses intake vacuum rather than torque. The two are not equivalent.

    6. Because of engine-to-engine, cylinder-to-cylinder and device-to-device variations, the control designer has to provide a safety margin to surely avoid engine knock. The necessary safety margin may amount to 20 deg c.a. retard and even 30 deg c.a. if knock in very hot weather under heavy load operation is to be prevented. Ignition beginning that late reduces combustion efficiency and fuel economy intolerably.

ELECTRONIC SPARK TIMING CONTROLS

In 1976 electronic spark timing controls made their appearances as original equipment on automobiles. Chrysler's ESA (Electronic Spark Advance) and General Motor's MISAR (Microprocessed Sensing and Automatic Regulator) replaced mechanical-pneumatic mechanism with electronic circuitry. This is regarded as a significant step in the ever widening penetration of electronic controls in automotive engineering.

As all controls, electronic controls can be classified as pre-programmed and adaptive controls even though sometimes they are so mixed

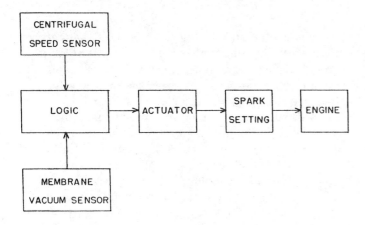

Fig. 8 - Principle of the conventional centrifugal-vacuum spark timing control

that it is hard to tell them apart. Electronic implementation lends greater flexibility to the controls but leaves the basic principles intact.

The two electronic spark timing controls presently in production are of the straight pre-programmed type. Both use speed sensors and vacuum sensors similar to the conventional control but in addition MISAR senses coolant temperature and, ESA in addition intake air temperature.

Being programmed, these controls share some of the shortcomings of the conventional controls. Going over the six items discussed:

1. Even with the increased number of sensors, the number of influencing variables the control corrects for is small. Theoretically the number could be increased to include most, if not all of those listed, but the translation of the theory into practice runs into exponentially increasing difficulties. The capacity of the logic's memory is not the only problem. The sensors themselves are often delicate analog devices and their accuracy is limited by the production cost and calibration. For some variables, like fuel octane rating, no practical sensor can be visualized. So even with sophisticated electronic control, in exteme cases a deviation of 30 deg c. a. from the correct spark timing is entirely possible.

2. Manufacturing tolerances can, with electronics be greatly reduced but not cheaply. The capital and operating costs of quality controls is high.

3. Drift in use and wear in the mechanism can possibly be almost eliminated if used with breakerless ignition systems.

4. Non-linearities continue to be a problem. For avoiding interpolations an over-sized memory bank may be used. With n independent sensors, each supplying five data levels, the logic would have to choose the right one from 5^n values. Interpolations add to the complexity of the circuitry.

5. Substituting intake vacuum for load causes a falsification. An inexpensive torque sensor is not available now.

6. Although his choices are greatly widened, the control designer still has to use his own limited experience in choosing the proper safety margin for all speed/torque points to be covered by the control.

From this analysis it appears that electronic spark timing control devices that are used in present day automobiles offer some advantages over the conventional spark control. It brought, at the same time, increased complexity and cost. On their durability and serviceability, experience is still lacking. This, however, should not discourage automakers. Innovations, even if costly, often pay off in the long run.

Progress should be expected from going to adaptive type controls using closed loop and feedback for error correcting.

In controlling spark timing the number of needed sensors would remain the same but transfer errors could be largely eliminated. Since no such system is presently in use, its evaluation would be speculative. But more than mere theory is available on the optimizing control of spark timing, which is described below.

SPARK-OPTIMIZER

For general evaluation an optimizing control with dithering technique will be used that seeks maximum power (MBT) timing. Going over the six items:

1. It is readily apparent that for homing in on maximum power, only one sensor is needed, which senses power. This can be either a torque sensor or a speed sensor. All the logic needs is information on the direction the power is changing at a particular moment. This is sufficient for the logic to decide whether an advance or a retard is required in that particular moment, to give the right instruction to the actuator.

Thus the number of sensors is reduced to one irrespective of how many influencing variables the control corrects for. By correcting for all influencing variables, rather than for some of them only, the greatest error source of the programmed spark control is eliminated.

2. The inaccuracies caused by the limitations imposed by manufacturing tolerances are also largely eliminated. Those involve mostly the sensors and their interfaces. Having only one sensor which needs not sense magnitude only plus signals, close calibration is no longer required.

3. Drift in use is as good as eliminated because the optimizing control re-tunes the engine continuously. In view of the fact that the average privately owned car is 5 - 10 deg out of tune, the fuel saving achieved by proper spark tuning is of economic consequence.

4. Whether mechanical/pneumatic or electronic, a programmed control system has difficulties handling nonlinearities. The correction desired is hardly ever proportional to the sensed parameter. The optimizing control is completely free from such restraints.

5. The falsification caused by substituting intake vacuum for engine load does not arise when the vacuum sensor is completely eliminated.

6. Since the control designer does not select the settings, he does not select a safety margin for the settings either. Any deviation from the normal is taken care of automatically. The setting selected by the optimizing control is the correct one for that individual engine running under the particular operating conditions.

FLEXIBILITY

There is one aspect however, where the programmed system appears superior. It is its flexibility.

Spark timing must satisfy several requirements. The engine must be powerful, economical, drivable, must not knock, and comply with existing exhaust emission regulations. These requirements are often in conflict with each other.

True, a control system can neither create nor eliminate conflicts between requirements. If no set of spark timing exists which satisfies all these requirements, a control system cannot make it exist. But not all requirements are equally rigid. Power and fuel economy must occasionally be sacrificed to some extent. But to decide where and how much, involves a value judgment. The designer exercises it by means of a program.

A smart designer gives in on performance where it is necessary but only where and not more than is necessary. An optimizing control is viewed as being incapable of permitting needed flexibility. This was true until recently but is not any longer.

The biasing technique resolved this problem and made the optimizing control not less flexible than any pre-programmed control.

The biasing technique enables an optimizing control to home in not on the optimum but at a setting off optimum. In case of spark timing the biased Optimizer can find a setting a certain number of degrees retarded from MBT. The bias can be a fixed or a variable amount. It can even be programmed. The biased optimizing control has all the flexibility of a straight programmed control but is incomparably simpler and still preserve the benefit of the optimizing control and corrects for many more variables than it has sensors for.

The biasing technique is so intricately tied in with the electronic circuitry employed that its description must be left to a later part of this paper.

Next a brief account is presented of the evolution of the spark-Optimizer to its present stage.

HARDWARE DEVELOPMENT

A spark timing Optimizer was first installed in an automobile in 1967. It followed closely the patent issued on "Selftuning Engine System" in 1964. (Ref. 5) with an arrangement such as seen in Fig. 9.

Dithering of the spark setting was done by an electric motor through a crank mechanism that turned the timing plate of the distributor back and forth about ±4 deg c.a.

The celsig used in the beginning was of the inertial (seismic) type in which a weight is loosely mounted by ball bearings on an engine shaft. It moved a few thousandths of an inch forward when the engine decelerated and backward when accelerated. Electric contacts sensed the presence of either acceleration or deceleration.

Seismic celsigs were accurate when new but lost sensitivity when the balls brinnelled in use. They were, therefore, replaced early with an electric tachometer (as in Fig. 9) whose voltage output was electrically differentiated by the logic circuit.

The mechanical dither was subsequently replaced by a double point distributor with points 3 deg (actually 180 + 3 deg) apart and a clock-controlled electronic oscillator that switched the circuit alternately to advance and retard eight times per second. An account of early road trials with such a prototype is given in Ref. 6.

Further advance in hardware development was achieved in a prototype by Harmon Electronics in 1976. As shown in Fig. 10.

The celsig consists of a slotted disc with a facing magnetic sensor. The disc was mounted on the alternator shaft. The number of electrical pulses per second was proportional to the engine speed.

Dithering was created by an electronic oscillator which alternated the triggering pulse to the distributor coil and thus advanced and retarded the spark discharge with a cycle frequency of 8 Hz.

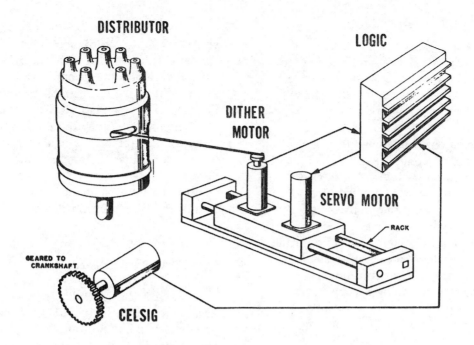

Fig. 9 - Spark timing Optimizer assembly, electromechanical

The mechanical actuator was similarly eliminated; or rather replaced by electronics. The logic controlled the timing of the triggering pulse.

These were significant advances, less in the performance of the device than in its simplicity and adaptability. Production cost was reduced and a single model could be retrofitted to many makes and models of engines without replacing or altering any part of the engine. The distributor was not modified, in fact it was left strictly alone.

A number of fleet vehicles, taxis, and trucks were fitted with the Harmon prototype and except for occasional electronic interference (noise) and trace knock on some engines at heavy load, the road trials were favorable. The fuel savings exceeded the expectations.

Test results with eleven such vehicles over more than 12 000 miles are shown in Table 1.

Tests conducted by automobile and parts manufacturers largely confirmed the fuel economy improvements. The NOx emissions naturally increased over the timing set deliberately late by the manufacturer.

In the early Optimizer prototypes electronics was used at most for reliability. Gradually electronic dithering, electronic celsig, and electronic actuators were developed to add to the logic which was electronic from almost the beginning.

ELECTRONIC CIRCUITRY

More than by hardware development, the Optimizer was helped by the dramatic advance in

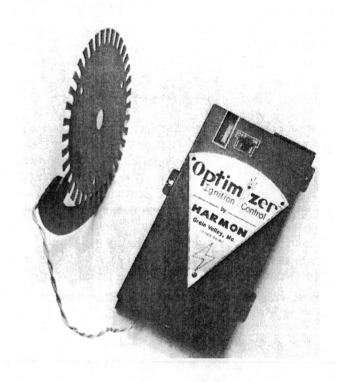

Fig. 10 - Spark-Optimizer by Harmon electronics

electronics in recent years. Microelectronics with digital circuitry had substantially reduced cost and improved reliability of such and similar devices.

Digital circuitry is particularly adaptable to a spark-Optimizer which has no analog sensors. The actuator also gets by without mechanical components.

Table 1 - Fuel Economy Improvement with Optimizer

Car	Engine CID	Tuning	Driving	Distance	MPG Improvement
Chevrolet '73	454	Factory	Highway	170m	6.5%
same			City		20.1%
Ford '71	400	Calif. std.	Highway	170m	29.6%
Pontiac '69	428	MBT			0
Comet '71	200	no	Highway	300m	11%
same		MBT	Highway	600m	4%
Ford '73 Taxi	302	no	City	1700m	57% (suspect)
Ford '74 Taxi	351	no	City	1000m	8%
Ford '73 Taxi	400	no	City	700m	25%
same		MBT	City		less than 5%
Ford '73 Taxi	429	no	City	5000m	18-20%
Truck 25,000 lb	330			1200m	5%
Oldsmobile '73	350	no	City		31%
same			Highway		14%
Ford '73 pickup	361	no		1000m	32%

The celsig, the only sensor, operates by sensing engine speed through pulse counting. Pulse counts in consecutive time intervals shows the presence of acceleration or deceleration.

Through segmentation and dithering techniques, a processing of the pulse counts enables the logic to make correct decisions and send proper orders to the actuator. The block diagram in Fig. 11 shows the principle of the digital optimizing circuit.

The inputs to the logic originate from two electronic pulse generators. The dither pulser is slow, not over ten pulses per second but the pulses are long. In using square waves, the duration of the positive voltage is one half dither cycle. The other half has zero voltage. The dither frequency is constant, independent of engine rpm.

The celsig pulser is fast, some two hundred pulses per engine revolution. Their duration is of no consequence, but their frequency must be proportional to the engine rpm.

These two pulse generators furnish the information the logic needs to institute the proper correction. The dither pulse frequency is constant, that of the celsig pulses changes with the engine speed. While the engine accelerates the celsig pulse frequency increases, during deceleration it decreases. The logic perceives an acceleration or a deceleration from pulse counts in successive time intervals.

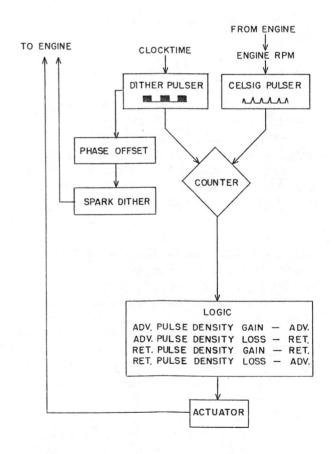

Fig. 11 - Block diagram of digital Optimizer control

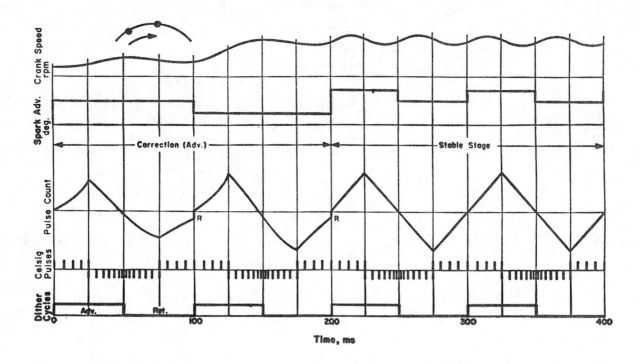

Fig. 12 - Optimizer correction time history

SIGNAL PROCESSING IN A DIGITAL SPARK-OPTIMIZER

The counter continuously counts the dither pulses, up or down, depending on the phase of the dither cycle they coincide with. The complete dither cycle is divided into four segments: first half-advance, second half-advance, first half-retard, second half-retard. In our example each segment lasts $1/40$ s = 25 ms.

At the beginning of a dither cycle the counting begins at zero and the count is up for 25 ms. At that point the counting is reversed and, unless the engine changed speed, it gets down to zero again after 50 ms. During the retard period the count is first down and then up, but the net is again zero.

When the engine speed varies during a time interval, the net counts after the two segments are other than zero. A negative net signifies acceleration, a positive net deceleration.

When the spark timing is dithered, alternating advance and retard periods, the engine speed fluctuates. If the spark timing is at its optimum (MBT) setting, the fluctuations are mild and the net counts are zero. If the spark timing is later than MBT, the engine will speed up during the advance period and slow down during the retard period. The net count is negative after the advance and negative after the retard period. If the spark timing is early the opposite is the case. This permits the logic to determine whether the spark needs advance or retard to improve its performance.

By choice the counter regularly is counting up during the first half-advance segment, then down during the second half-advance and first half-retard segment, finally up again during the second half-retard segment. If advancing the spark timing helps, the net count after the completion of the entire dither cycle will be a negative number. The counter reports this to the logic and resets to zero. The logic advances the timing an incremental amount and the game starts all over.

This sequence is shown in Fig. 12 from 0 - 100 ms. The procedure repeats during the next dither cycle 100 - 200 ms.

After 200 ms in Fig. 12 the spark was sufficiently advanced to reach optimum. At MBT the engine speed becomes stabilized and the up-counts and down-counts no longer differ from each other. The Optimizer has successfully completed the correction process.

In Fig. 12 the bottom line shows the alternating advance and retard segments of the dither cycle. The one above the dispersion of the celsig pulses is shown. The sawtooth line above that represents the cumulative celsig pulses and its slopes correspond to the fluctuating engine speed. The square wave above that represents the change of spark advance and the top line the sinusoidal speed fluctuations during the correction and the stable stages.

It will be noticed that at stable MBT stage the frequency of the spark timing and speed fluctuations is double of the dither frequency. No surge or jerkiness can be perceived with the Optimizer, and ordinarily a person is unable to tell whether the engine is dithered or not.

PHASE OFFSET

Unless the dither frequency is very low, which is impractical in a motor vehicle, a phase offset between the dither pulser and the spark dither is necessary. As seen in Fig. 11, the dither pulser sends out two signals. One goes to the counter and gets there almost instantaneously. The other dithers the engine speed which is being monitored by the celsig. In order to produce any change in engine speed, the altered spark must either improve or spoil the combustion at least in one cylinder. This takes time; in a 6-cylinder engine, it may take 2/6 revolution. The improved (spoiled) combustion must accelerate (decelerate) the crank rotation sufficiently for the counter to detect a difference from the preceding pulse count. The counter must receive these counts, that correspond to each other, at the same time. In order to make up for the time delay, the dither pulser must send each signal to the spark dither earlier than the corresponding signal to the counter. This is accomplished by a phase offset covering a fraction of a dither cycle. The phase offset is a function of the dither frequency and need not be changed as long as the dither frequency is left unchanged.

BIASING

Biasing is effected by making the first and second "half" of the advance period unequal. For example, instead of each lasting 1/40 s = 25 ms, we make the first segment be 26 ms and the second segment 24 ms. When the logic counts fewer pulses in the second "half" of the advance period than in the first "half", it will conclude that the engine has lost speed during the advance period. Actually the engine may have maintained or even gained speed during the period. The logic gets fooled and will react the "wrong" way, the way we want it to react. The same miscounting is forced during the retard half of the dither cycle. This time, however, the first half is 24 ms and the second half is 26 ms. This longer time causes the logic to believe the engine has speeded up even though it may have maintained speed or even slowed down. Again the desired bias results. This method will give a retard bias. The opposite time durations would give an advanced bias.

Fig. 13 is similar to Fig. 12 only the two "halves" of the counting periods are unequal.

The biased Optimizer can operate in a stable manner at 5, 10 or more deg off MBT. Naturally, a fuel penalty must be paid for off MBT operation. If the bias is less than 5 deg (which ordinarily

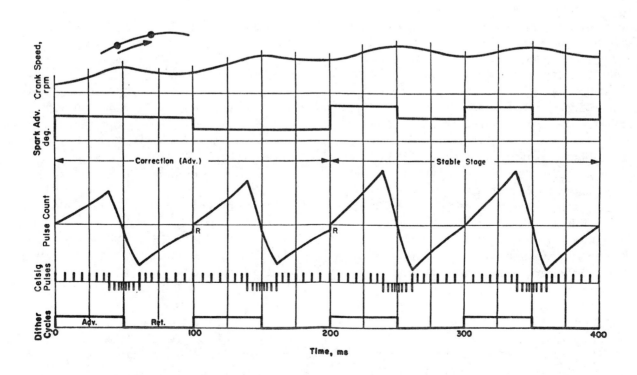

Fig. 13 - Optimizer correction time history, biased

suffices to eliminate knock) the loss in fuel economy is seldom more than 1%.

When the Optimizer is biased in favor of retard, the advance correction process is terminated before peak performance is reached. The engine attains stable operation on the ascending side of the rpm versus spark advance curve.

The Optimizer control box may include an adjustment for setting the amount of the bias according to the need of the engine. Some engines need no bias and probably none as much as 10 deg for the sake of suppressing knock.

PROGRAMMED BIAS

The bias designed into an optimizing circuit can be a fixed bias; it can be adjustable and it also can be programmed.

For the sake of knock elimination a programmed variable bias is hardly justified, because the bias necessary will seldom exceed 5 deg c.a. Many engines need no bias at all, for the rest 5 deg is ample. The fuel penalty for 5 deg bias is around 1%.

Retarded timing is often used for the control of nitrogen oxides emissions. This practice has become more limited recently because it involves heavy fuel penalty. EGR was found more acceptable for the control of NOx because it is more cost effective. Nevertheless, for short intervals when the NOx emission is particularly heavy, spark retardation deserves attention. Such is the case in the limited periods of hard accelerations.

For sensing accelerations, the present Optimizer control circuit can be used with only a slight modification. The slotted disc celsig generates the pulses, the frequency of which is proportionate to the rpm. An up/down-count of two consecutive (say 1/2 s) time periods would show the "secular" change in engine speed, distinct from the dithering. The counter report would activate the pre-selected bias.

A possible program could consist of MBT timing (zero bias) during very small or no acceleration and an adequate bias during accelerations that exceed a selected amount, for instance 100 rpm per second. Since the sum of the acceleration periods is small in time or miles travelled, we may sin with relative impunity. Accuracy in sensors is, therefore, of no consequence.

With bias programming the control is still completely digital and, therefore, less sensitive to maladjustments and electrical noise than an analog or partially analog control.

Naturally, programming the bias of an optimizing control is very much easier than programming all of the spark advance. The range is much smaller (like 15 against 60 deg c.a.) and the programmed part is small. Variations in non-engine related variables can be ignored and the control still responds to, and corrects for, all pertinent variables, instead of the number limited by the number of sensors used do.

Even the biased Optimizer is incomparably simpler and sturdier than pre-programmed electronic controls are.

ELECTRONIC SPARK TIMING CONTROLS

In Table 2 six electronic control systems for spark timing are evaluated on the basis of presently available information. With complete information, some statements and conclusions may have to be changed.

Two, or perhaps three, of these systems are in production now, and others are being readied for production.

Information on Chrysler's ESA was obtained from Refs. 7 and 8. On General Motors' MISAR from Ref. 9 and from private information from Delco Remy. Ford Motor Co. supplied the information on its EEC. For the Robert Bosch column information derived from Ref. 10 has been used with the oral consent of Bosch representatives.

Optimizer Control Corporation's (OCC) spark-Optimizer has been evaluated in its two versions on the basis of available prototypes. On the unbiased version information has been published in Ref. 6. On the biased version with a programmed bias this paper is the sole information publicly available.

Without claiming to be exhaustive, the evaluation was done under these headings:

1. Accuracy. The term here means how close the setting by the control approaches the setting the designer desires. This is not necessarily - in fact very rarely is - identical with what is best for the engine and its operator. The inaccuracies here considered result mainly from restraints imposed by the control mechanism.

2. Simplicity. If the control mechanism and its operation can easily be understood by an automobile mechanic and the device does not make him uncomfortable we shall consider it simple.

3. Reliability. This includes sturdiness, durability, maintainability, and low rate of deterioration in use.

4. Flexibility. This term covers the inherent ability of the control system to accommodate trade-offs between requirements, allowing choices based on value judgment by the control designer. It also includes the ability of one device to be adapted to many engines with little or no modification.

Table 2 - Spark Timing Control Features

Type of Control \ Feature	Conventional	Chrysler ESA	GM MISAR	Ford EEC	Robert Bosch	OCC Optimizer	OCC Biased Optimizer
Accuracy	±10 deg	±5 deg	±3 deg	±3 deg	±3 deg	±3 deg	±3 deg
Simplicity	Good	Poor	Poor	Poor	?	Good	Good
Reliability	Fair	Fair	Fair	?	?	Good	Good
Flexibility Deg. adv.	Fair 10 - 45	Fair 10 - 65	Good 0 - 70	Good 0 - 70	?	Poor Unlimited	Good Unlimited
Digitalization	Does not apply	None all analog	Limited	Limited	Limited	Complete	Complete
Production Cost	Low	High	High	High	High	Low	Low

5. Digitalization. This is a term applied to electronic circuitry. The more of its components, sensors, interfaces, logic, actuators, etc. are purely digital, on/off in nature, the more completely digital the circuitry is. This was given an independent heading because digitalization crosses over several characteristics. It helps simplicity, reliability, and production cost. It ordinarily also makes the device less sensitive to false signals and external electrical noise.

6. Production Cost. This feature surely is not the last in importance as no device does the user any good if he cannot afford it.

In Table 2 the control systems are given a rating in general terms.

For the sake of comparison the conventional (non-electronic) spark control is listed as a sort of base line.

Some explanations to the table are in order. The estimates on control accuracy are rough and on the optimistic side.

Ratings on simplicity reflect the belief that no pre-programmed control with several non-linear analog sensors and a memory bank to store thousands of instructions can be made simple.

A dominant criterion in flexibility is the control's ability of accommodating trade-offs between conflicting requirements.

Regarding digitalization, we shall consider a control system completely digital if all its sensors are digital and the componentry of the logic and actuator are essentially digital. If all or some of the sensors are analog with an analog-to-digital convertor in the interface, and the circuitry is substantially digital, the qualifying adjective "limited" shall be used. All other electronic control systems are classified as analog.

In production cost, the cost of development was ignored but the cost of quality control considered. Unless the device is produced in very large numbers, where it can be automated, quality control and calibration are a substantial part of the production cost.

IGNORED VARIABLES

A dominant defect of the conventional and other pre-programmed controls is less the way it handles the data it receives than the information it does not get at all. Out of the not less than ten influencing variables the conventional spark timing control gets two, the others three or four. A pre-programmed control that would process the information supplied by many more sensors would be prohibitively complicated. For some pertinent parameters, such as fuel octane rating, no practical sensor can be thought of.

Even the information that is sensed is often not the one that really counts. Instead of supplying the truly needed information some sensors substitute data that are only similar but not identical or equivalent. The falsification resulting from using intake vacuum for engine load has been mentioned already. Another substitution is that of the coolant temperature, when it is the metal temperature that affects the combustion and so influences the correct spark timing.

Sometimes throttle position or its rate of change, or even clock time are used for control parameters, while it is obvious that they, by themselves, do not affect the correct spark timing. They only anticipate changes in influencing variables such as engine speed and metal temperature with some success under normal environmental conditions and driving habits. In Table 3 the ten mentioned relevant variables are listed with indication how the various control systems correct for them.

Ignored are such variables as throttle plate position and clock time although some control systems make use of them. They are obviously not influencing parameters, only anticipations of changes in influencing parameters.

Instead of providing continuous corrections, some control systems only make a one-step correction when a particular parameter exceeds a threshold. The control's response to starting, stopping, idling, and coasting may be included in this category.

The shortcoming of all pre-programmed systems in response is readily apparent.

HYBRID CONTROLS

Mixed or hybrid controls have both pre-programmed and closed loop components. They switch from programmed to adaptive circuit at certain operating points or when a particular variable reaches a threshold value. In some hybrid controls of spark timing the programmed and the adaptive parts work side-by-side continuously. This is the case with an optimizing spark timing control with a programmed bias.

The adaptive part of the control automatically homes in on a spark setting that is a certain amount off optimum. Only the amount that represents the difference is programmed. The program, selected by the control designer, may be a fixed or variable percentage of the spark advance. It may increase with engine load or with other influencing variables.

Usually a load sensing sensor is sufficient to enable the bias control to produce the desirable trade-offs.

CIRCUITRY

Fig. 14 shows a typical circuit used to perform the biased optimizing control process. In this figure down-counters DC1 and DC2 along with set/reset flip-flop FF1 establish the dither period. The advance time established by the binary number N1 may be made longer or shorter than 1/2 of the dither period (established by N2). This non-symmetrical dither cycle coupled with the fact that DC4 is constant for both dither half cycles provides the biasing technique required.

DC3 provides a phase offset or delay in up/down-counting after dither to allow for the inertia of the machine that is being optimized.

The up/down-counting of the celsig pulses is under control of Exclusive or gate X01, Inverter II, and AND gates A7, A8, and A9. The counter will count up for DC4 time, down for the remainder of the ADVANCE time, continue down for a DC4 time and up for the remainder of the retard time. If the answer in the up/down counter is positive at the end of the dither period, a retard command will be issued thru A5 and if the answer is negative an advance command will be issued thru A6.

As a final note on this technique, N1, the binary number used to establish bias may be a constant which is always present, a constant to be switched in by monitoring some engine parameter or a device within the Optimizer (such as the up/down counter), or more than one number selected as required by advanced programming by the designer, based on either external selectors or parameters internal to the Optimizer.

OPTIMIZATION USING A MICROPROCESSOR INSTEAD OF HARD WIRED INTEGRATED CIRCUITS

The previously described circuits, if reduced to a single chip and produced in large quantities, would at present cost less than the cost of using a microprocessor. It would be wrong, however, not to mention the fact that the Optimizer can be programmed into a microprocessor. The word "program" as used in this section is not a predetermined setting based upon input parameters as mentioned earlier in this paper; but rather a set of instructions that will cause the microprocessor (Ref. 11) to perform the optimizing function.

Fig. 15 shows a simplified flow chart for the program required. It performs all of the functions provided in the previously mentioned electronic circuits. The functions are self explanatory on the flow chart. The system requires a celsig input and provides advance and retard commands to the firing time circuit. The firing time circuit can also be programmed into the microprocessor or hard wired as desired. The bias in this system could be programmed so that either an outside sensor or a parameter within the microprocessor could change the bias for knock, NOx, etc. as required.

Fig. 16 shows a typical firing time circuit program flow chart for a microprocessor.

The normal time between successive firings is determined by the time set in Delay counter No. 2 and the decrementing counter. If advance or retard are required, the time between the next two firings is adjusted, thereby resetting the firing time from that point on. If retard is required the added time of delay counter No. 1 will retard the firing. If advance is required, bypassing both delay counters will advance the firing. The

Table 3 - Spark Timing Control Corrections

Type of Control	Conventional		Chrysler ESA		GM MISAR		Ford EEC		Robert Bosch		OCC Optimizer		OCC Biased Optimizer	
Parameter	Sensor	Correction	Sensor	Correction	Sensor	Correction	Sensor	Correction	Sensor	Correction	Sensor	Correction	Sensor	Correction
Engine Speed	Analog (3)	Yes	Analog (3)	Yes	Digital (7)	Yes	Digital	Yes	Digital (7)	Yes	Digital (7)	Yes	Digital (7)	Yes
Engine Load	Analog (4)	Yes	Analog (4)	Yes	Analog (4)(8)	Yes	Analog (8)	Yes	Analog (4)(8)	Yes	No	Yes	Digital (7)	Yes
Metal Temperature (1)	No	No	Analog (5)	Yes	Analog (5)	Yes	Analog (5)(8)	Yes	Analog (5)	Yes	No	Yes	No	Yes
Amb. Air Temperature	No	No	Analog	Thrhld. (6)	No	No	Analog (8)	Yes	No	No	No	Yes	No	Yes
Amb. Air Humidity	No	No	No	No	No	No	No	No	No	No	No	Yes	No	Yes
Amb. Pressure (Alt.)	No	No	No	No	No	No	Analog (8)	Yes	No	No	No	Yes	No	Yes
Fuel Octane Rating	No	No	No	No	No	No	No	No*	No	No	No	Yes	No	Yes
A/F Ratio	No	No	No	No	No	No	No*	No*	No	No	No	Yes	No	Yes
Residual (2)	No	No	No	No	No	No	No	No	No	No	No	Yes	No	Yes
EGR	No	No	No	No	No	No	Analog (8)	Yes	No	No	No	Yes	No	Yes

* can be added

NOTES:
(1) The temperature of the metal surfaces in contact with the cylinder charge.
(2) The left over gas from the previous cycle.
(3) Centrifugal advance.
(4) Vacuum advance.
(5) Substitute: Coolant temperature
(6) Threshold switch, step correction.
(7) Slotted disc.
(8) Analog sensor with analog-digital converter.

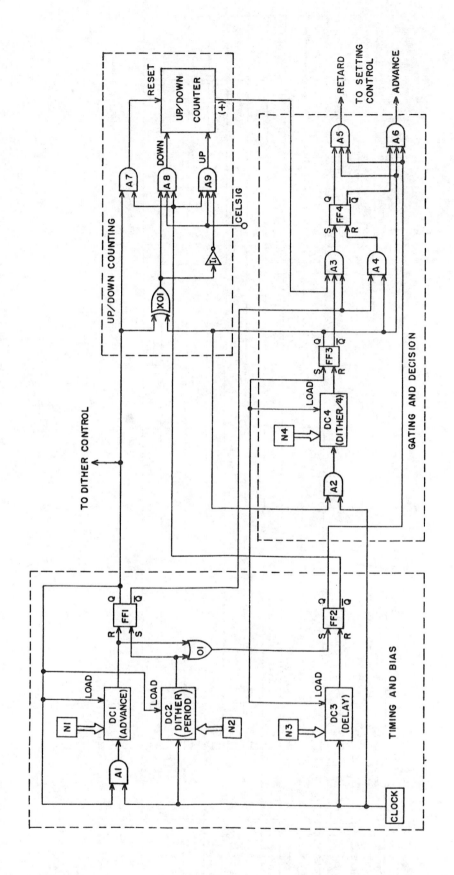

Fig. 14 — Typical digital circuit to provide biased optimizing control

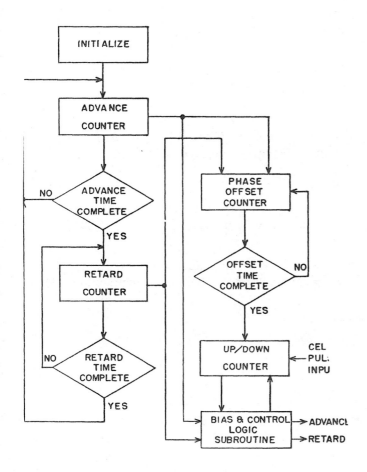

Fig. 15 - Microprocessor program flow chart for performing Optimizer control

counters are clocked by the celsig input. The number of degrees between each celsig pulse determines the smallest angle that the firing can be adjusted to.

Also a reference pulse is required so that the first firing when the engine is started will be at the proper time. If it is desired to reference every firing to an outside signal instead of letting the system free run, as shown in the flow chart, the reference pulse should drive into point (A) and the line from (B) to (A) should be removed.

The output of this program goes through the dwell counter section to drive the coil transistor. The dwell can either be crank angle dependent by counting celsig pulses, or time dependent by counting the system clock. In either event the dwell may also be programmed to respond to a special angle or time requirement at various RPM (celsig pulse rate).

In review, it should be evident, by way of the small parts count that the optimizing system could be fabricated on a single chip at a low cost or incorporated into a microprocessor. Another point that should be covered is that since the optimizing operation is relatively slow (milliseconds) as compared to the speed of most microprocessors (sub-microseconds) the system could be time

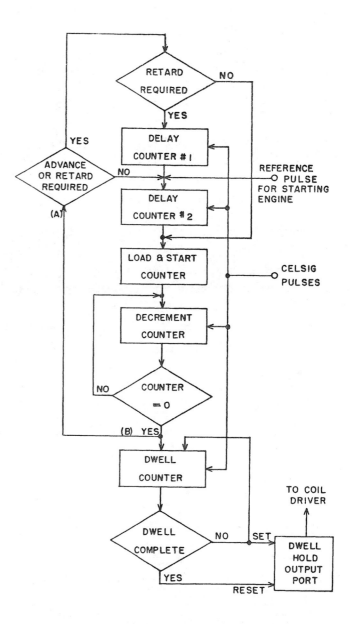

Fig. 16 - Firing time flow chart

shared to perform other functions, even optimize other parameters of the engine such as air/fuel ratio. Even if the microprocessor were optimizing two functions, there would still be enough time to monitor other engine functions and command corrective action such as drive actuators for EGR, carburetor choke, or inform the vehicle operator about the engine behavior.

REFERENCES

1. P. H. Schweitzer, "Electronic Engine Controls: Pre-Programmed, Error Correcting or Optimizing." 1976 Joint Automatic Control Conference, Purdue University, West Lafayette, IN. July 27-30, 1976.

2. C. S. Draper and Y. T. Li, "The Principles of Optimalizing Control Systems and an

Application to the Internal Combustion Engine." ASME, Sept. 1951.

3. P. H. Schweitzer, "Engine Controls: Programmed and Adaptive." Automobile Engine Control Workshop, U.S. Department of Transportation Systems Center, Cambridge, MA. July 8-9, 1975.

4. C. R. Morgan and S. S. Hetrick, "Trade-Offs Between Engine Emission Control Variables, Fuel Economy and Octane." SAE Paper 750415.

5. P. H. Schweitzer, "Maximum Power Seeking Automatic Control System For Power-Producing Machines." U.S. Patent No. 3,142,967. August 4, 1964.

6. P. H. Schweitzer and Carl Volz, "Electronic Optimizer Control For I. C. Engines: Most MPG For Any MPH." SAE Paper 750370.

7. Chrysler Corporation, "Inside The Electronic Lean Burn System." '76 May Reference Book.

8. Chrysler Corporation, "Diagnosing Electronic Lean Burn System Performance." '76 June Reference Book.

9. Dennis J. Simanaitis, "MISAR: An Electronic Advance." Automotive Engineering, Jan. 1977.

10. Hansjoerg Manger, "Digital Electronic Spark Advance Systems." SAE Paper 760265.

11. M. Schwartz and B. Kute, "Anatomy Of A Microcomputer." Machine Design, March 11, 1976, Volume 48, No. 6.

ELECTRONIC SPARK TIMING CONTROL FOR MOTOR VEHICLES

Paul H. Schweitzer — Thomas W. Collins

Introduction

We are witnessing a burst of activity in electronic engine controls. Electronic fuel injection has been available for several years. Electronic skid control and other controls followed. Last year electronic spark timing controls have been introduced by Chrysler and General Motors. It is safe to predict that before long they will be widely used.

Electronic control design for automobiles has regularly been in the hands of electronic specialists and automotive aspects were often ignored. Automotive electronics must operate under extreme environmental conditions. A multiplicity of delicate sensors is undesirable in an automobile. Simplicity and ruggedness have top priorities.

A general outline of electronic engine controls were presented in Ref. 1. This paper deals specifically with spark timing control. This will be preceded, however, by a brief overview of control theory fundamentals.

Types of Controls

The function of a control or control system is to set the setting of a machine in such a manner that its performance is improved in the broad sense that it becomes more satisfactory to the operator. In an internal combustion engine this may mean increased power, reduced specific fuel consumption, reduced exhaust emissions, stable operation and responsiveness (drivability) or other performance characteristics.

A control can be manual (Steuerung in German) or automatic (Regelung in German). This paper treats only automatic controls, and control will mean automatic control unless otherwise specified.

Control systems can be mechanical, hydraulic, pneumatic, fluidic, electric, electronic or a combination of these. Irrespective of the mechanism used, every control system consists basically of three components; Sensor, Logic and Actuator.

The sensor (one or more) provides the input to the control. The logic is the decision maker. The actuator receives the output of the logic which consists of instruction to set the setting. The scheme is shown on Fig. 1.

Fig. 1 Basic Scheme of a Control System

Two types of control systems are in use. In a pre-programmed control system (programmed control) the logic receives its input from the sensor which responds to external influences, external in the sense that they don't originate from inside the control mechanism or its target setting. An example where engine speed is used to set the spark timing is shown in Fig. 2.

Fig. 2 A Programmed Control of the Spark Timing

In a programmed control the logic has built-in instructions from the designer how to respond to a sensor signal in formulating its instructions to the actuator. In the case of spark timing, the logic can order, by a suitable linkage, or otherwise, an advance of the spark timing with engine speed in a linear or nonlinear fashion. The control designer determines in advance the functional relationships between engine speed and spark advance and programs that relationship into the control system.

Another type of control system is designated as adaptive control. In an adaptive control system the control output influences its input. A classic example of the adaptive control is the engine speed governor, invented by James Watts in 1788. A better known example is the room thermostat, shown in Fig. 3.

Fig. 3 Room Thermostat, A Closed Loop Control

Adaptive controls use feedbacks for input. Popularly they are called closed-loop controls.

In the case of the thermostat, the target of the control is the room temperature. A sensor senses the room temperature and feeds that information to the logic. The logic compares that with a reference temperature that has been selected by the user. The control accomplishes its objective by equalizing these two temperatures, by sending instructions to the heat supply for increase or decrease. In this example the reference is set by the user.

In the case of spark timing control the reference must vary with engine speed, load and perhaps other parameters. This means that the reference itself must be programmed, based on information supplied by sensors that sense the influencing parameters.

In another type of closed-loop control a single output characteristic of an engine provides the reference for the logic. In the three-way catalytic converter for emissions control an exhaust sensor senses the air-fuel ratio and the logic selects the stoichiometric ratio for matching.

A spark timing control of this category could sense engine knock and cease advancing the timing when engine knock appears. A schematic of such a control is shown in Fig. 4.

Fig. 4 Closed-Loop Control of Spark Timing
with Knock Sensor

A most advanced type of adaptive control is the optimizing control, invented by Charles Draper and Y. T. Li in 1951 (Ref. 2).

The optimizing control also uses the closed-loop feedback principle, but instead of aiming at a fixed target it seeks an optimum. The basic concept of an optimizing control, applied to spark timing, is shown in Fig. 5.

Fig. 5 Optimizing Control of Spark Timing

The sensor monitors the engine power output and sends this information to the logic. Simultaneously the logic also receives information on the spark setting. The logic compares these sets of information and figures out what change of spark setting would favor the power output. Accordingly it sends orders to the actuator to advance or retard the spark timing as the circumstance requires.

Dithering Technique

In order to facilitate monitoring engine performance, or rather the change of the performance with a change in spark timing, it is very helpful to deliberately disturb the spark setting. One way of doing this is starting from zero or a low value, the control advances the spark timing continuously until a drop in power is sensed. Then the control reverses and retards the timing until again a drop in power is sensed. This play is repeated continuously. The spark timing, therefore, can never get far off the optimum. In the dithering technique the spark timing is periodically changed up and down a small amount from a middle point. Therefore the spark timing is alternately advanced and retarded a number of times per second.

The oscillator (dither) keeps the logic continuously informed of the dithering phase. The logic is able to determine whether the engine responds favorably or unfavorably after the spark was advanced (or retarded). Thus the logic can form correct decisions, when to advance and when to retard the spark timing. It then sends appropriate commands to the actuator as shown in Fig. 6.

Fig. 6 Optimizing Control; Using the Dithering Technique for Spark Timing

If MBT (minimum for best torque) timing is sought, the sensor must sense torque or rather a change of torque. This could be accomplished by sensing torque directly, but a more practical way is to sense the change in rotational speed (rpm), which can be sensed conveniently by an acceleration sensor, named Celsig.

If a slotted disc is fastened to the crankshaft, or another driven shaft, and the teeth of the disc pass a magnetic sensor, the rpm is proportional to the number of pulses generated in a fixed period of time. If during the advance dither period the pulse frequency increases and during the retard period it decreases, the engine favors advance and the control is to provide advance, and vice versa.

Fig. 7 shows schematically a slotted disc type celsig with its facing magnet as used in an optimizing spark timing control.

Fig. 7 Slotted Disc Type Celsig in a Spark Timing Controller

The adaptive closed-loop system depicted in Fig. 3, is superior to the straight programmed system inasmuch as it continuously checks and corrects its own performance. Nevertheless, the error correcting systems, some of which are now under active development for automotive use, share one important source of inaccuracy with the straight programmed system. In selecting the setting sought, both rely on the designer's judgment. However, the designer's information and intelligence is limited, and the setting he decides is right for the engine may be very different from what is really best for the engine and its operator.

In an optimizing control system the incentive for correcting a setting, and the direction it is to go, derives from the engine itself. It is the engine, not the designer that decides. This eliminates the most fundamental inaccuracy automatic controls are burdened with. Other sources of errors in various control systems will be discussed later in the paper.

The above outlined fundamentals apply to all control systems, from mechanical to electronic, and for any operating system subject to control. In the case of internal combustion engines the controlled setting may pertain to the spark timing, the air-fuel ratio, the valve timing (if variable), the injection timing (in diesel engines), or other influencing variables. The target of the control may be maximum power output, minimum fuel consumption, minimum exhaust emissions, or other performance characteristics. The control theory, and to a large extent the technology applied are the same.

The rest of this paper is limited to the problem of spark timing in spark ignition engines. The problem of controlling the air-fuel ratio was discussed in a companion paper, (Ref. 3).

Spark Timing in S. I. Engines

The timing of the ignition or spark plug discharge is one of the most important parameters that influence engine performance. It is usually expressed by spark advance, in degrees crank angle before top dead center. For power output the optimum timing is MBT. If the spark timing differs from MBT, combustion efficiency suffers, reducing power and increasing specific fuel consumption. If the spark setting is far from MBT the engine is unable to operate.

Spark timing has also decisive influence on exhaust emissions and engine knock. Late timing always reduces NOx emissions and to a lesser degree also the unburned hydrocarbons. Advanced timing may cause engine knock. MBT timing which is most favorable for performance may be unacceptable for other reasons.

The range of possible spark timing is very wide, from about 10 deg ATC to about 70 deg BTC, depending on the circumstances. The range of MBT timing is almost as wide. The function of the spark timing control is to find the best timing under actual circumstances.

MBT timing depends on at least ten variables, engine-related and environmental, as listed, where the numbers in parenthesis indicate a normal range (in deg c.a.) of their influence, not considering extreme conditions. Engine speed (35), Load (15), Metal temperature (10), Ambient air temperature (10), Ambient air pressure (10), Ambient air humidity (10), Fuel octane number (10), Air-fuel ratio (10), Residual gas from preceeding cycle (5), Exhaust gas recirculation (10).

Combustion efficiency depends on how much the actual timing is off MBT. When the actual timing is 5 deg off MBT, plus or minus, the degradation is seldom more than 1%, at 10 deg it may amount to 5%, at 20 deg 16%. These figures are approximate and vary from engine-to-engine. But even in freshly tuned engines actual spark timing is often over 20 deg later than MBT and in out-of-tune engines even more. There is no justification for such late timing but there is an explanation for it.

Under certain operating conditions, some engines knock at MBT timing. Factory timing is so chosen that engine knock be avoided under almost any circumstances encountered in practice. The safty margin used is often more than 20 deg, not because engine knock is so unpredictable but because the car makers know no way to predict it, given the wide variations in operating conditions and in individual engines of the same model. Furthermore, even knowing the requirements, the available control technology is inadequate to cope with them.

Engine Knock

Engine knock, of the type frequently identified with detonation, is the result of abnormal combustion. It varies in intensity from incipient or trace knock which only instruments or a trained ear can detect, to medium knock which is annoying to the operator, to heavy knock which is damaging to the engine.

Knock is commonly controlled by retarding the spark timing but this also reduces thermal efficiency. When spark timing is advanced, the cycle efficiency is increased. When the advance is carried into the knock region, the increased turbulence associated with detonation causes an increased heat loss which reduces the combustion efficiency.

Indicated thermal efficiency is a product of the cycle and the combustion efficiency. This product reaches maximum at a certain spark timing which is either more or less advanced than the point of incipient knock (IK).

Some engines knock at MBT, others don't. It greatly depends on the rate of heat transfer from the combustion chamber to the cylinder head. Engines with air cooling or with high-temperature pressure cooling are more likely to knock at MBT, especially if they also have a compact combustion chamber. But knock at MBT is never heavy and ordinarily only occurs at high load operation.

Whether the IK point is more advanced or less advanced than the MBT point, the two invariably move together when operating conditions change. The IK point is ordinarily more advanced; when not, seldom more than five deg later.

Emissions

Spark timing, however, is often retarded a substantial amount because of emission requirements. NOx is particularly sensitive to spark timing. Fifteen degrees or more retard from MBT has been used in the past to suppress NOx emissions. This practice has been largely discontinued lately because of the heavy fuel penalty involved. Five percent economy loss "buys" almost three times as much NOx reduction through EGR than through spark retard (Ref. 4).

A moderate spark retard has also been used for the sake of reducing HC emissions. In this case the effect is indirect. By delaying combustion the engine actually ends the power stroke with more HC. However, the increased exhaust temperature enables the exhaust system to burn more of the unburned hydrocarbons in the exhaust pipe and muffler. A thermal or catalytic reactor accomplishes this objective more economically.

Control Types for Spark Timing

The control system for controlling spark timing used in the last half century is mechanical-pneumatic. Its principle is depicted, in terms of our methodology, in Fig. 8.

Fig. 8 Principle of the Conventional Centrifugal-Vacuum Spark Timing Control

The engine speed is sensed by the flyweights mounted on the distributor shaft. The flyweights, through linkage, raise a sleeve with helical splines which change the phase between the distributor shaft and the crankshaft. The intake vacuum is transmitted through a tube to a membrane, mounted on the distributor housing and through a rod, rotates the timing disc of the distributor. Without being readily distinguishable, all elements of a pre-programmed control system are present.

Even though it has been successfully used for a long time, the conventional spark control has numerous shortcomings. An incomplete list of them follows:

1. Out of not less than ten influencing variables it corrects only for two. This alone easily can cause an error of 30 deg c.a. in the correct spark timing.

2. Manufacturing tolerances in the control mechanism and engine-to-engine variations in the same model increase the control inaccuracy.

3. Drift in use, wear in the mechanism, points and cam wear in the distributor cause unintended spark retardation.

4. For mechanical reasons the vacuum advance is generally linear, and the centrifugal advance is segmentally linear. The two advances add up irrespective of the requirements.

5. Correct spark timing is a function of the engine load. Light load requires more spark advance than heavy load. Conventional spark control senses intake vacuum rather than torque. The two are not equivalent.

6. Because of engine-to-engine, cylinder-to-cylinder and device-to-device variations, the control designer has to provide a safety margin to surely avoid engine knock. The necessary safety margin may amount to 20 deg c.a. retard and even 30 deg c.a. if knock in very hot weather under heavy load operation is to be prevented. Ignition beginning that late reduces combustion efficiency and fuel economy intolerably.

Electronic Spark Timing Controls

In 1976 electronic spark timing controls made their appearances as original equipment on automobiles. Chrysler's ESA (Electronic Spark Advance) and General Motor's MISAR (Microprocessed Sensing and Automatic Regulator) replaced mechanical-pneumatic mechanism with electronic circuitry. This is regarded as a significant step in the ever widening penetration of electronic controls in automotive engineering.

As all controls, electronic controls can be classified as pre-programmed and adaptive controls even though sometimes they are so mixed that it is hard to tell them apart. Electronic implementation lends greater flexibility to the controls but leaves the basic principles intact.

The two electronic spark timing controls presently in production are of the straight pre-programmed type. Both use speed sensors and vacuum sensors similar to the conventional control but in addition MISAR senses coolant temperature and, ESA in addition intake air temperature.

Being programmed, these controls share some of the short-comings of the conventional controls. Going over the six items discussed:

1. Even with the increased number of sensors, the number of influencing variables the control corrects for is small. Theoretically the number could be increased to include most, if not all

of those listed, but the translation of the theory into practice runs into exponentially increasing difficulties. The capacity of the logic's memory is not the only problem. The sensors themselves are often delicate analog devices and their accuracy is limited by the production cost and calibration. For some variables, like fuel octane rating, no practical sensor can be visualized. So even with sophisticated electronic control, in extreme cases a deviation of 30 deg c.a. from the correct spark timing is entirely possible.

2. Manufacturing tolerances can, with electronics be greatly reduced but not cheaply. The capital and operating costs of quality controls is high.

3. Drift in use and wear in the mechanism can possibly be almost eliminated if used with breakerless ignition systems.

4. Non-linearities continue to be a problem. For avoiding interpolations an oversized memory bank may be used. With n independent sensors, each supplying five data levels, the logic would have to choose the right one from 5^n instructions. Interpolations add to the complexity of the circuitry.

5. Substituting intake vacuum for load causes a falsification. An inexpensive torque sensor is not available now.

6. Although his choices are greatly widened, the control designer still has to use his own limited experience in choosing the proper safety margin for all speed/torque points to be covered by the control.

From this analysis it appears that electronic spark timing control devices that are used in present day automobiles offer some advantages over the conventional spark control. It brought, at the same time, increased complexity and cost. On their durability and serviceability experience is still lacking. This, however, should not discourage automakers. Innovations, even if costly, often pay off in the long run.

Progress should be expected from going to adaptive type controls using closed loop and feedback for error correcting.

In controlling spark timing the number of needed sensors would remain the same but transfer errors could be largely eliminated. Since no such system is presently in use, its evaluation would be speculative. But more than mere theory is available on the optimizing control of spark timing, which is described below.

Spark-Optimizer

For general evaluation an optimizing control with dithering technique will be used that seeks maximum power (MBT) timing. Going over the six items:

1. It is readily apparent that for homing in on maximum power, only one sensor is needed, which senses power. This can be either a torque sensor or a speed sensor. All the logic needs is information on the direction the power is changing at a particular moment. This is sufficient for the logic to decide whether an advance or a retard is required in that particular moment, to give the right instruction to the actuator.

Thus the number of sensors is reduced to one irrespective of how many influencing variables the control corrects for. By correcting for all influencing variables, rather than for some of them only, the greatest error source of the programmed spark control is eliminated.

2. The inaccuracies caused by the limitations imposed by manufacturing tolerances are also largely eliminated. Those involve mostly the sensors and their interfaces. Having only one sensor which needs not sense magnitude only plus signals, close calibration is no longer required.

3. Drift in use is as good as eliminated because the optimizing control retunes the engine continuously. In view of the fact that the average privately owned car is five to ten degrees out of tune, the fuel saving achieved by proper spark tuning is of economic consequence.

4. Whether mechanical/pneumatic or electronic, a programmed control system has difficulties handling nonlinearities. The correction desired is hardly ever proportional to the sensed parameter. The optimizing control is completely free from such restrains.

5. The falsification caused by substituting intake vacuum for engine load does not arise when the vacuum sensor is completely eliminated.

6. Since the control designer does not select the settings, he does not select a safety margin for the settings either. Any deviation from the normal is taken care of automatically. The setting selected by the optimizing control is the correct one for that individual engine running under the particular operating conditions.

Flexibility

There is one aspect however, where the programmed system appears superior. It is its flexibility.

Spark timing must satisfy several requirements. The engine must be powerful, economical, drivable, must not knock, and comply with existing exhaust emission regulations. These requirements are often in conflict with each other.

True, a control system can neither create nor eliminate conflicts between requirements. If no set of spark timing exists which satisfies all these requirements, a control system cannot make it exist. But not all requirements are equally rigid. Power and fuel economy must occasionally be sacrificed to some extent. But to decide where and how much, involves a value judgment. The designer exercises it by means of a program.

A smart designer gives in on performance where it is necessary but only _where_ and not more than _is_ necessary. An optimizing control is viewed as being incapable of permitting needed flexibility. This was true until recently but is not any longer.

The biasing technique resolved this problem and made the optimizing control not less flexible than any pre-programmed control.

The biasing technique enables an optimizing control to home in not on the optimum but at a setting off optimum. In case of spark timing the biased optimizer can find a setting a certain number of degrees retarded from MBT. The bias can be a fixed or a variable amount. It can even be programmed. The biased optimizing control has all the flexibility of a straight programmed control but is incomparably simpler and still preserve the benefit of the optimizing control and corrects for many more variables than it has sensors for.

The biasing technique is so intricately tied in with the electronic circuitry employed that its description must be left to a later part of this paper.

Next a brief account is presented of the evolution of the spark-Optimizer to its present stage.

Hardware Development

A spark timing Optimizer was first installed in an automobile in 1967. It followed closely the patent issued on "Selftuning Engine System" in 1964 (Ref. 5) with an arrangement such as seen in Fig. 9.

Fig. 9 Spark Timing Optimizer Assembly, Electro-Mechanical

Dithering of the spark setting was done by an electric motor through a crank mechanism that turned the timing plate of the distributor back and forth about ± 4 deg c.a.

The celsig used in the beginning was of the inertial (seismic) type in which a weight is loosely mounted by ball bearings on an engine shaft. It moved a few thousands of an inch forward when the engine decelerated and backward when accelerated. Electric contacts sensed the presence of either acceleration or deceleration.

Seismic celsigs were accurate when new but lost sensitivity when the balls brinnelled in use. They were, therefore, replaced early with an electric tachometer (as in Fig. 9) whose voltage output was electrically differentiated by the logic circuit.

The mechanical dither was subsequently replaced by a double point distributor with points 3 degrees (actually 180 + 3 deg) apart and a clock-controlled electronic oscillator that switched the circuit alternately to advance and retard eight times per second. An account of early road trials with such a prototype is given in Ref. 6.

Further advance in hardware development was achieved in a prototype by Harmon Electronics in 1976. As shown in Fig. 10.

Fig. 10 Spark-Optimizer by Harmon Electronics

The celsig consists of a slotted disc with a facing magnetic sensor. The disc was mounted on the alternator shaft. The number of electrical pulses per second was proportional to the engine speed.

Dithering was created by an electronic oscillator which alternated the triggering pulse to the distributor coil and thus advanced and retarded the spark discharge with a cycle frequency of 8 Hz.

The mechanical actuator was similarly eliminated; or rather replaced by electronics. The logic controlled the timing of the triggering pulse.

These were significant advances, less in the performance of the device than in its simplicity and adaptability. Production cost was reduced and a single model could be retrofitted to many makes and models of engines without replacing or altering any part of the engine. The distributor was not modified, in fact it was left strictly alone.

A number of fleet vehicles, taxis and trucks were fitted with the Harmon prototype and except for occasional electronic interference (noise) and trace knock on some engines at heavy load, the road trials were favorable. The fuel savings exceeded the expectations.

Test results with eleven such vehicles over more than 12,000 miles are shown in Table 1.

Table 1 Fuel Economy Improvement with Optimizer

Tests conducted by automobile and parts manufacturers largely confirmed the fuel economy improvements. The NOx emissions naturally increased over the timing set deliberately late by the manufacturer.

In the early Optimizer prototypes electronics was used at most for reliability. Gradually electronic dithering, electronic celsig, and electronic actuators were developed to add to the logic which was electronic from almost the beginning.

Electronic Circuitry

More than by hardware development, the Optimizer was helped by the dramatic advance in electronics in recent years. Microelectronics with digital circuitry had substantially reduced cost and improved reliability of such and similar devices.

Digital circuitry is particularly adaptable to a spark-Optimizer which has no analog sensors. The actuator also gets by without mechanical components.

The celsig, the only sensor, operates by sensing engine speed through pulse counting. Pulse counts in consecutive time intervals shows the presence of acceleration or deceleration.

Through segmentation and dithering techniques, a processing of the pulse counts enables the logic to make correct decisions and send proper orders to the actuator. The block diagram in Fig. 11 shows the principle of the digital optimizing circuit.

Fig. 11 Block Diagram of Digital Optimizer Control

The inputs to the logic originate from two electronic pulse generators. The dither pulser is slow, not over ten pulses per second but the pulses are long. In using square waves, the duration of the positive voltage is one half dither cycle. The other half has zero voltage. The dither frequency is constant, independent of engine rpm.

The celsig pulser is fast, some two hundred pulses per engine revolution. Their duration is of no consequence, but their frequency must be proportional to the engine rpm.

These two pulse generators furnish the information the logic needs to institute the proper correction. The dither pulse frequency is constant, that of the celsig pulses changes with the engine speed. While the engine accelerates the celsig pulse frequency increases, during deceleration it decreases. The logic perceives an acceleration or a deceleration from pulse counts in successive time intervals.

Signal Processing in a Digital Spark-Optimizer

The counter continuously counts the dither pulses, up or down, depending on the phase of the dither cycle they coincide with. The complete dither cycle is divided into four segments: first half-advance, second half-advance, first half-retard, second half-retard. In our example each segment lasts 1/40 sec = 25 ms.

At the beginning of a dither cycle the counting begins at zero and the count is up for 25 ms. At that point the counting is reversed and, unless the engine changed speed, it gets down to zero again after 50 ms. During the retard period the count is first down and then up, but the net is again zero.

When the engine speed varies during a time interval, the net counts after the two segments are other than zero. A negative net signifies acceleration, a positive net deceleration.

When the spark timing is dithered, alternating advance and retard periods, the engine speed fluctuates. If the spark timing is at its optimum (MBT) setting, the fluctuations are mild and the net counts are zero. If the spark timing is later than MBT, the engine will speed up during the advance period and slow down during the retard period. The net count is negative after the advance and negative after the retard period. If the spark timing is early the opposite is the case. This permits the logic to determine whether the spark needs advance or retard to improve its performance.

By choice the counter regularly is counting up during the first half-advance segment, then down during the second half-advance and first half-retard segment, finally up again during the second half-retard segment. If advancing the spark timing helps, the net count after the completion of the entire dither cycle will be a negative number. The counter reports this to the logic and resets to zero. The logic advances the timing an incremental amount and the game starts all over.

This sequence is shown in Fig. 12 from 0 to 100 ms. The procedure repeats during the next dither cycle 100-200 ms.

Fig. 12 Optimizer Correction Time History

After 200 ms in Fig. 12 the spark was sufficiently advanced to reach optimum. At MBT the engine speed becomes stabilized and the up-counts and down-counts no longer differ from each other. The Optimizer has successfully completed the correction process.

In Fig. 12 the bottom line shows the alternating advance and retard segments of the dither cycle. The one above the dispersion of the celsig pulses is shown. The sawtooth line above that represents the cumulative celsig pulses and its slopes correspond to the fluctuating engine speed. The square wave above that represents the change of spark advance and the top line the sinusoidal speed fluctuations during the correction and the stable stages.

It will be noticed that at stable MBT stage the frequency of the spark timing and speed fluctuations is double of the dither frequency. No surge or jerkiness can be perceived with the Optimizer, and ordinarily a person is unable to tell whether the engine is dithered or not.

Phase Offset

Unless the dither frequency is very low, which is impractical in a motor vehicle, a phase offset between the dither pulser and the spark dither is necessary. As seen in Fig. 11, the dither pulser sends out two signals. One goes to the counter and gets there almost instantaneously. The other dithers the engine speed which is being monitored by the celsig. In order to produce any change in engine speed, the altered spark must either improve or spoil the combustion at least in one cylinder. This takes time; in a 6-cylinder engine, it may take 2/6 revolution. The improved (spoiled) combustion must accelerate (decelerate) the crank rotation sufficiently for the counter to detect a difference from the preceding pulse count. The counter must receive these counts, that correspond to each other, at the same time. In order to make up for the time delay, the dither pulser must send each signal to the spark dither earlier than the corresponding signal to the counter. This is accomplished by a phase offset covering a fraction of a dither cycle. The phase offset is a function of the dither frequency and need not be changed as long as the dither frequency is left unchanged.

Biasing

Biasing is effected by making the first and second "half" of the advance period unequal. For example, instead of each lasting 1/40 second equal 25 ms, we make the first segment be 26 ms and the second segment 24 ms. When the logic counts fewer pulses in the second "half" of the advance period than in the first "half," it will conclude that the engine has lost speed during the advance period. Actually the engine may have maintained or even gained speed during the period. The logic gets fooled and will react the "wrong" way, the way we want it to react. The same miscounting is forced during the retard half of the dither cycle. This time however, the first half is 24 ms and the second half is 26 ms. This longer time causes the logic to believe the engine has speeded up even though it may have maintained speed or even slowed down. Again the desired bias results. This method will give a retard bias. The opposite time durations would give an advanced bias.

Fig. 13 is similar to Fig. 12 only the two "halves" of the counting periods are unequal.

Fig. 13 Optimizer Correction Time History, Biased

The biased Optimizer can operate in a stable manner at 5, 10 or more degrees off MBT. Naturally a fuel penalty must be paid for off MBT operation. If the bias is less than 5 deg (which ordinarily suffices to eliminate knock) the loss in fuel economy is seldom more than one percent.

When the Optimizer is biased in favor of retard, the advance correction process is terminated before peak performance is reached. The engine attains stable operation on the ascending side of the rpm vs spark advance curve.

The Optimizer control box may include an adjustment for setting the amount of the bias according to the need of the engine. Some engines need no bias and probably none as much as 10 deg for the sake of suppressing knock.

Programmed Bias

The bias designed into an optimizing circuit can be a fixed bias; it can be adjustable and it also can be programmed.

For the sake of knock elimination a programmed variable bias is hardly justified, because the bias necessary will seldom exceed 5 deg c.a. Many engines need no bias at all, for the rest 5 deg is ample. The fuel penalty for 5 deg bias is around one percent.

Retarded timing is often used for the control of nitrogen oxides emissions. This practice has become more limited recently because it involves heavy fuel penalty. EGR was found more acceptable for the control of NOx because it is more cost effective. Nevertheless, for short intervals when the NOx emission is particularly heavy, spark retardation deserves attention. Such is the case in the limited periods of hard accelerations.

For sensing accelerations, the present Optimizer control circuit can be used with only a slight modification. The slotted disc celsig generates the pulses, the frequency of which is proportionate to the rpm. An up-down-count of two consecutive (say 1/2 sec) time periods would show the "secular" change in engine speed, distinct from the dithering. The counter report would activate the pre-selected bias.

A possible program could consist of MBT timing (zero bias) during very small or no acceleration and an adequate bias during accelerations that exceed a selected amount, for instance 100 rpm per second. Since the sum of the acceleration periods is small in time or miles travelled, we may sin with relative impunity. Accuracy in sensors is, therefore, of no consequence.

With bias programming the control is still completely digital and, therefore, less sensitive to maladjustments and electrical noise than an analog or partially analog control.

Naturally, programming the bias of an optimizing control is very much easier than programming all of the spark advance. The range is much smaller (like 15 against 60 deg c.a.) and the programmed part is small. Variations in non-engine related variables can be ignored and the control still responds to, and corrects for, all pertinent variables, instead of the number limited by the number of sensors used do.

Even the biased Optimizer is incomparably simpler and sturdier than pre-programmed electronic controls are.

Electronic Spark Timing Controls

In Table 2 six electronic control systems for spark timing are avaluated on the basis of presently available information. With complete information some statements and conclusions may have to be changed.

Two, or perhaps three, of these systems are in production now, and others are being readied for production.

Information on Chrysler's ESA was obtained from Refs. 7 and 8. On General Motors' MISAR from Ref. 9 and from private information from Delco Remy. Ford Motor Co. supplied the information on its EEC. For the Robert Bosch column information derived from Ref. 10 has been used with the oral consent of Bosch representatives.

Optimizer Control Corporation's (OCC) spark-Optimizer has been evaluated in its two versions on the basis of available prototypes. On the unbiased version information has been published in Ref. 6. On the biased version with a programmed bias this paper is the sole information publicly available.

Without claiming to be exhaustive, the evaluation was done under these headings:

1. **Accuracy.** The term here means how close the setting by the control approaches the setting the designer desires. This is not necessarily — in fact very rarely is — identical with what is best for the engine and its operator. The inaccuracies here considered result mainly from restraints imposed by the control mechanism.

2. Simplicity. If the control mechanism and its operation can easily be understood by an automobile mechanic and the device does not make him uncomfortable we shall consider it simple.

3. Reliability. This includes sturdiness, durability, maintainability, and low rate of deterioration in use.

4. Flexibility. This term covers the inherent ability of the control system to accomodate trade-offs between requirements, allowing choices based on value judgment by the control designer. It also includes the ability of one device to be adapted to many engines with little or no modification.

5. Digitalization. This is a term applied to electronic circuitry. The more of its components, sensors, interfaces, logic, actuators, etc. are purely digital, on/off in nature, the more completely digital the circuitry is. This was given an independent heading because digitalization crosses over several characteristics. It helps simplicity, reliability and production cost. It ordinarily also makes the device less sensitive to false signals and external electrical noise.

6. Production Cost. This feature surely is not the last in importance as no device does the user any good if he cannot afford it.

In the Table that follows (Table 2) the control systems are given a rating in general terms.

For the sake of comparison the conventional (non-electronic) spark control is listed as a sort of base line.

Some explanations to the Table are in order. The estimates on control accuracy are rough and on the optimistic side.

Ratings on simplicity reflect the belief that no pre-programmed control with several non-linear analog sensors and a memory bank to store thousands of instructions can be made simple.

Table 2 Spark Timing Control Features

A dominant criterion in flexibility is the control's ability of accommodating trade-offs between conflicting requirements.

Regarding digitalization, we shall consider a control system completely digital if all its sensors are digital and the componentry of the logic and actuator are essentially digital. If all or some of the sensors are analog with an analog-to-digital convertor in the interface, and the circuitry is substantially digital, the qualifying adjective "limited" shall be used. All other electronic control systems are classified as analog.

In production cost, the cost of development was ignored but the cost of quality control considered. Unless the device is produced in very large numbers, where it can be automated, quality control and calibration are a substantial part of the production cost.

Ignored Variables

A dominant defect of the conventional and other pre-programmed controls is less the way it handles the data it receives than the information it does not get at all. Out of the not less than ten influencing variables the conventional spark timing control gets two, the others three or four. A pre-programmed control that would process the information supplied by many more sensors would be prohibitively complicated. For some pertinent parameters, such as fuel octane rating, no practical sensor can be thought of.

Even the information that is sensed is often not the one that really counts. Instead of supplying the truly needed information some sensors substitute data that are only similar but not identical or equivalent. The falsification resulting from using intake vacuum for engine load has been mentioned already. Another substitution is that of the coolant temperature, when it is the metal temperature that affects the combustion and so influences the correct spark timing.

Sometimes throttle position or its rate of change, or even clock-time are used for control parameters, while it is obvious that they, by themselves, do not affect the correct spark timing. They only anticipate changes in influencing variables such as engine speed and metal temperature with some success under normal environmental conditions and driving habits. In Table 3 the ten mentioned relevant variables are listed with indication how the various control systems correct for them.

Table 3 Spark Timing Control Corrections

Ignored are such variables as throttle plate position and clock time although some control systems make use of them. They are obviously not influencing parameters, only anticipations of changes in influencing parameters.

Instead of providing continous corrections, some control systems only make a one-step correction when a particular parameter exceeds a threshold. The control's response to starting, stopping, idling and coasting may be included in this category.

The shortcoming of all pre-programmed systems in response is readily apparent.

Hybrid Controls

Mixed or hybrid controls have both pre-programmed and closed loop components. They switch from programmed to adaptive circuit at certain operating points or when a particular variable reaches a threshold value. In some hybrid controls of spark timing the programmed and the adaptive parts work side-by-side continuously. This is the case with an optimizing spark timing control with a programmed bias.

The adaptive part of the control automatically homes in on a spark setting that is a certain amount off optimum. Only the amount that represents the difference is programmed. The program, selected by the control designer, may be a fixed or variable percentage of the spark advance. It may increase with engine load or with other influencing variables.

Usually a load sensing sensor is sufficient to enable the bias control to produce the desirable trade-offs.

Circuitry

Fig. 14 shows a typical circuit used to perform the biased optimizing control process.

Fig. 14 Typical Digital Circuit to Provide
Biased Optimizing Control

In this figure down-counters DC1 and DC2 along with set/reset flip-flop FF1 establish the dither period. The advance time established by the binary number N1 may be made longer or shorter than 1/2 of the dither period (established by N2). This non-symetrical dither cycle coupled with the fact that DC4 is constant for both dither half cycles provides the biasing technique required.

DC3 provides a phase offset or delay in up/down-counting after dither to allow for the inertia of the machine that is being optimized.

The up/down-counting of the celsig pulses is under control of Exclusive or gate XO1, Inverter I1, and AND gates A7, A8, and A9. The counter will count up for DC4 time, down for the remainder of the ADVANCE time, continue down for a DC4 time and up for the remainder of the retard time. If the answer in the up/down counter is positive at the end of the dither period, a retard command will be issued thru A5 and if the answer is negative an advance command will be issued thru A6.

As a final note on this technique, N1, the binary number used to establish bias may be a constant which is always present, a constant to be switched in by monitoring some engine parameter or a device within the Optimizer (such as the up/down counter), or more than one number selected as required by advanced programming by the designer, based on either external selectors or parameters internal to the Optimizer.

Optimization Using a Microprocessor Instead of Hard Wired Integrated Circuits

The previously described circuits, if reduced to a single chip and produced in large quantities, would at present cost less than the cost of using a microprocessor. It would be wrong however, not to mention the fact that the Optimizer can be programmed into a microprocessor. The word "program" as used in this section is not a predetermined setting based upon input parameters as mentioned earlier in this paper; but rather a set of instructions that will cause the microprocessor (Ref. 11) to perform the optimizing function.

Fig. 15 shows a simplified flow chart for the program required.

Fig. 15 Microprocessor Program Flow Chart for Performing Optimizer Control

It performs all of the functions provided in the previously mentioned electronic circuits. The functions are self explanatory on the flow chart. The system requires a celsig input and provides advance and retard commands to the firing time circuit. The firing time circuit can also be programmed into the microprocessor or hard wired as desired. The bias in this system could be programmed so that either an outside sensor or a parameter within the microprocessor could change the bias for knock, NOx, etc. as required.

Fig. 16 shows a typical firing time circuit program flow chart for a microprocessor.

Fig. 16 Firing Time Flow Chart

The normal time between successive firings is determined by the time set in Delay counter No. 2 and the decrementing counter. If advance or retard are required, the time between the next two firings is adjusted, thereby resetting the firing time from that point on. If retard is required the added time of delay counter No. 1 will retard the firing. If advance is required, bypassing both delay counters will advance the firing. The counters are clocked by the celsig input. The number of degrees between each celsig pulse determines the smallest angle that the firing can be adjusted to.

Also a reference pulse is required so that the first firing when the engine is started will be at the proper time. If it is desired to reference every firing to an outside signal instead of letting the system free run, as shown in the flow chart, the reference pulse should drive into point (A) and the line from (B) to (A) should be removed.

The output of this program goes through the dwell counter section to drive the coil transistor. The dwell can either be crank angle dependent by counting celsig pulses, or time dependent by counting the system clock. In either event the dwell may also be programmed to respond to a special angle or time requirement at various RPM (celsig pulse rate).

In review, it should be evident, by way of the small parts count that the optimizing system could be fabricated on a single chip at a low cost or incorporated into a microprocessor. Another point that should be covered is that since the optimizing operation is relatively slow (milliseconds) as compared to the speed of most microprocessors (sub-microseconds) the system could be time shared to perform other functions, even optimize other parameters of the engine such as air/fuel ratio. Even if the microprocessor were optimizing two functions, there would still be enough time to monitor other engine functions and command corrective action such as drive actuators for EGR, carburetor choke, or inform the vehicle operator about the engine behavior.

REFERENCES

1. P. H. Schweitzer, "Electronic Engine Controls: Pre-Programmed, Error Correcting or Optimizing." 1976 Joint Automatic Control Conference, Purdue University, West Lafayette, Ind. July 27-30, 1976.

2. C. S. Draper and Y. T. Li, "The Principles of Optimizing Control Systems and an Application to the Internal Combustion Engine." ASME, Sept. 1951.

3. P. H. Schweitzer, "Engine Controls: Programmed and Adaptive." Automobile Engine Control Workshop, U.S. Department of Transportation Systems Center, Cambridge, Mass. July 8-9, 1975.

4. C. R. Morgan and S. S. Hetrick, "Trade-Offs Between Engine Emission Control Variables, Fuel Economy and Octane." SAE paper 750415.

5. P. H. Schweitzer, "Maximum Power Seeking Automatic Control System For Power-Producing Machines." U.S. Patent No. 3,142,967. August 4, 1964.

6. P. H. Schweitzer and Carl Volz, "Electronic Optimizer Control For I. C. Engines: Most MPG For Any MPH." SAE paper 750370.

7. Chrysler Corporation, "Inside The Electronic Lean Burn System." '76 May Reference Book.

8. Chrysler Corporation, "Diagnosing Electronic Lean Burn System Performance." '76 June Reference Book.

9. Dennis J. Simanaitis, "MISAR: An Electronic Advance." Automotive Engineering, Jan. 1977.

10. Hansjoerg Manger, "Digital Electronic Spark Advance Systems." SAE paper 720265.

11. M. Schwartz and B. Kute, "Anatomy Of A Microcomputer." Machine Design, March 11, 1976, Volume 48, No. 6.

Table 1.

FUEL ECONOMY IMPROVEMENT WITH OPTIMIZER

Car	Engine CID	Tuning	Driving	Distance	MPG Improvement
Chevrolet '73 same	454	Factory	Highway City	170m	6.5% 20.1%
Ford '71	400	Calif. std.	Highway	170m	29.6%
Pontiac '69	428	MBT			0
Comet '71 same	200	no MBT	Highway Highway	300m 600m	11% 4%
Ford '73 Taxi	302	no	City	1700m	57% (suspect)
Ford '74 Taxi	351	no	City	1000m	8%
Ford '73 Taxi same	400	no MBT	City City	700m	25% less than 5%
Ford '73 Taxi	429	no	City	5000m	18-20%
Truck 25,000 lb	330	no	City	1200m	5%
Oldsmobile '73 same	350	no	City Highway		31% 14%
Ford '73 pickup	361	no		1000m	32%

P. Schweitzer - Th. Collins

Table 2

SPARK TIMING CONTROL FEATURES

Type of Control / Feature	Conventional	Chrysler ESA	GM MISAR	Ford EEC	Robert Bosch	OCC Optimizer	OCC Biased Optimizer
Accuracy	±10 deg	±5 deg	±3 deg	±3 deg	±3 deg	±3 deg	±3 deg
Simplicity	Good	Poor	Poor	Poor	?	Good	Good
Reliability	Fair	Fair	Fair	?	?	Good	Good
Flexibility Deg. adv.	Fair 10 - 45	Fair 10 - 65	Good 0 - 70	Good 0 - 70	?	Poor Unlimited	Good Unlimited
Digitalization	Does not apply	None all analog	Limited	Limited	Limited	Complete	Complete
Production Cost	Low	High	High	High	High	Low	Low

P. Schweitzer – Th. Collins

Table 3

SPARK TIMING CONTROL CORRECTIONS

Type of Control / Parameter	Conventional Sensor	Conventional Correction	Chrysler ESA Sensor	Chrysler ESA Correction	GM MISAR Sensor	GM MISAR Correction	Ford EEC Sensor	Ford EEC Correction	Robert Bosch Sensor	Robert Bosch Correction	OCC Optimizer Sensor	OCC Optimizer Correction	OCC Biased Optimizer Sensor	OCC Biased Optimizer Correction
Engine Speed	Analog (3)	Yes	Analog (3)	Yes	Digital (7)	Yes	Digital	Yes	Digital (7)	Yes	Digital (7)	Yes	Digital (7)	Yes
Engine Load	Analog (4)	Yes	Analog (4)	Yes	Analog (4)(8)	Yes	Analog (8)	Yes	Analog (4)(8)	Yes	No	Yes	Digital (7)	Yes
Metal Temperature (1)	No	No	Analog (5)	Yes	Analog (5)	Yes	Analog (5)(8)	Yes	Analog (5)	Yes	No	Yes	No	Yes
Amb. Air Temperature	No	No	Analog	Thrhld. (6)	No	No	Analog (8)	No	No	No	No	Yes	No	Yes
Amb. Air Humidity	No	No	No	No	No	No	No	No	No	No	No	Yes	No	Yes
Amb. Pressure (Alt.)	No	No	No	No	No	No	Analog (8)	Yes	No	No	No	Yes	No	Yes
Fuel Octane Rating	No	No	No	No	No	No	No*	No*	No	No	No	Yes	No	Yes
A/F Ratio	No	No	No	No	No	No	No	No	No	No	No	Yes	No	Yes
EGR	No	No	No	No	No	No	Analog (8)	Yes	No	No	No	Yes	No	Yes
Residual (2)														

NOTES:
(1) The temperature of the metal surfaces in contact with the cylinder charge.
(2) The left over gas from the previous cycle.
(3) Centrifugal advance.
(4) Vacuum advance.
(5) Substitute: Coolant temperature
(6) Threshold switch, step correction.
(7) Slotted disc.
(8) Analog sensor with analog-digital converter.

* can be added

P. Schweitzer – Th. Collins

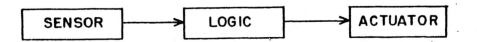

Fig. 1.
Basic Scheme of a Control System

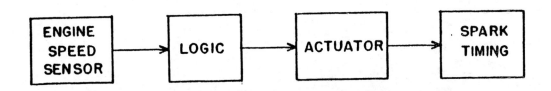

Fig. 2.
A Programmed Control of the Spark Timing

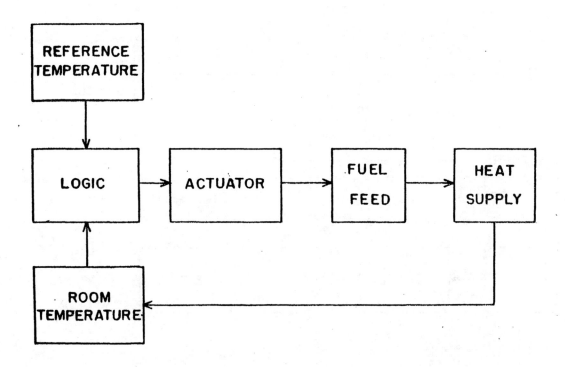

Fig. 3.
Room Thermostat, A Closed-Loop Control

P. Schweitzer . Th. Collins

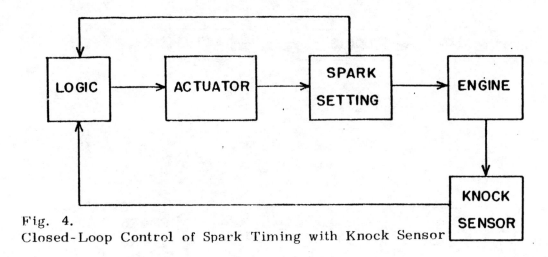

Fig. 4.
Closed-Loop Control of Spark Timing with Knock Sensor

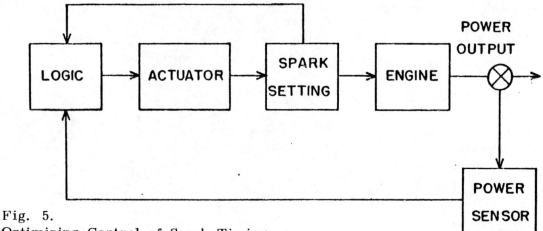

Fig. 5.
Optimizing Control of Spark Timing

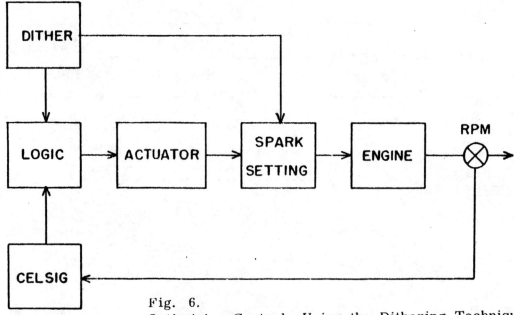

Fig. 6.
Optimizing Control, Using the Dithering Technique for Spark Timing

P. Schweitzer - Th. Collins

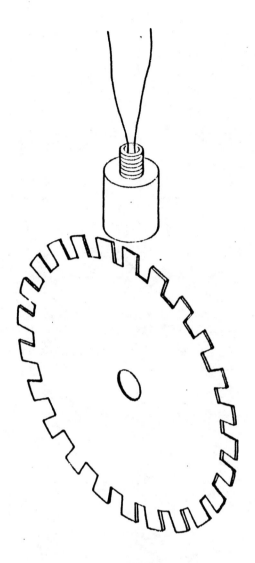

Fig. 7.
Slotted Disc Type Celsig in a Spark Timing Controller

P. Schweitzer - Th. Collins

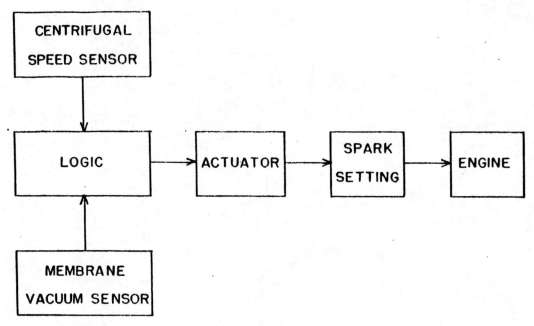

Fig. 8.
Principle of the Conventional Centrifugal-Vacuum Spark Timing Control

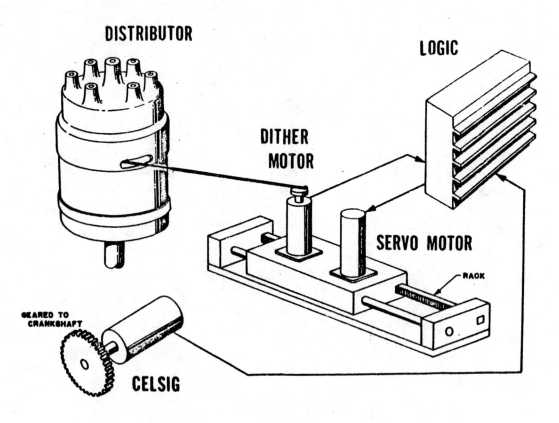

Fig. 9.
Spark Timing Optimizer Assembly, Electro-Mechanical

P. Schweitzer - Th. Collins

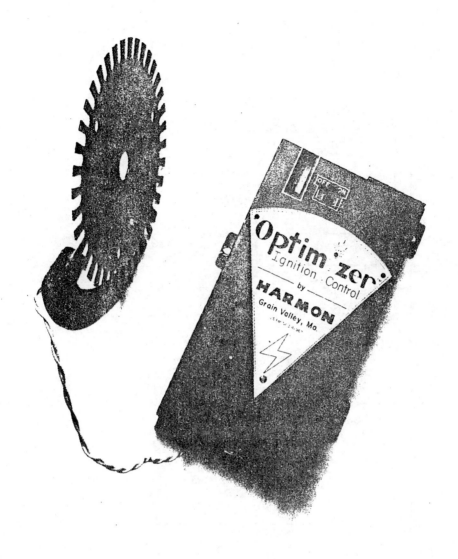

Fig. 10.
Spark-Optimizer by Harmon Electronics

P. Schweitzer - Th. Collins

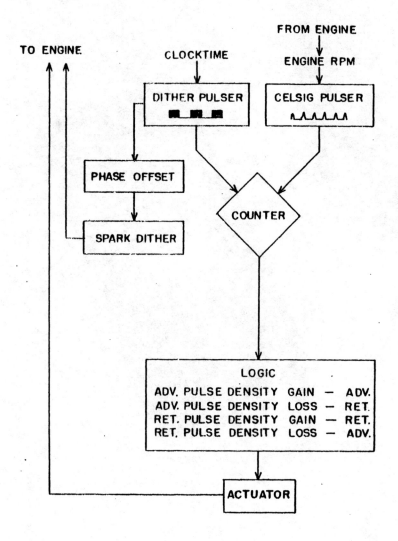

Fig. 11.
Block Diagram of Digital Optimizer Control

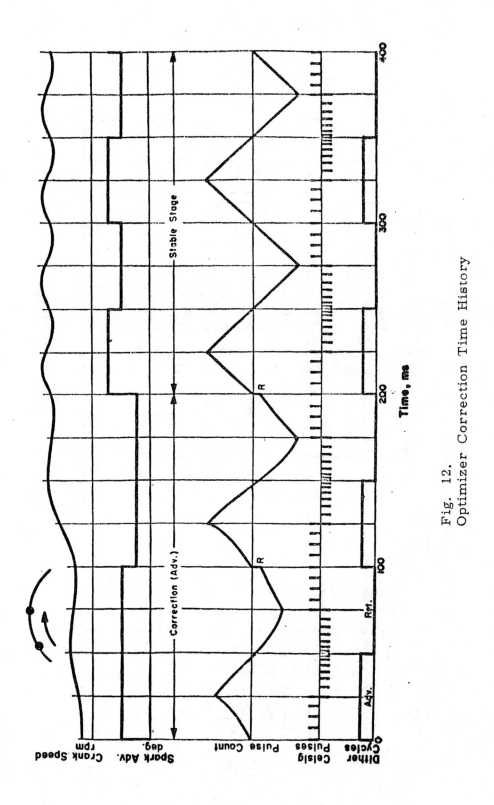

Fig. 12.
Optimizer Correction Time History

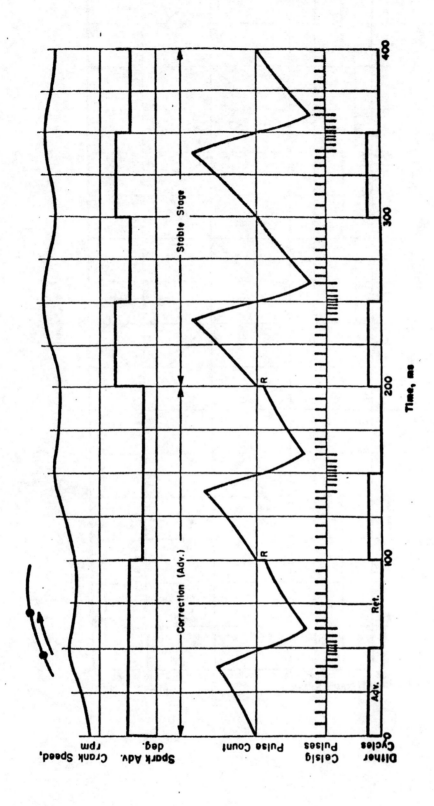

Fig. 13.
Optimizer Correction Time History, Biased

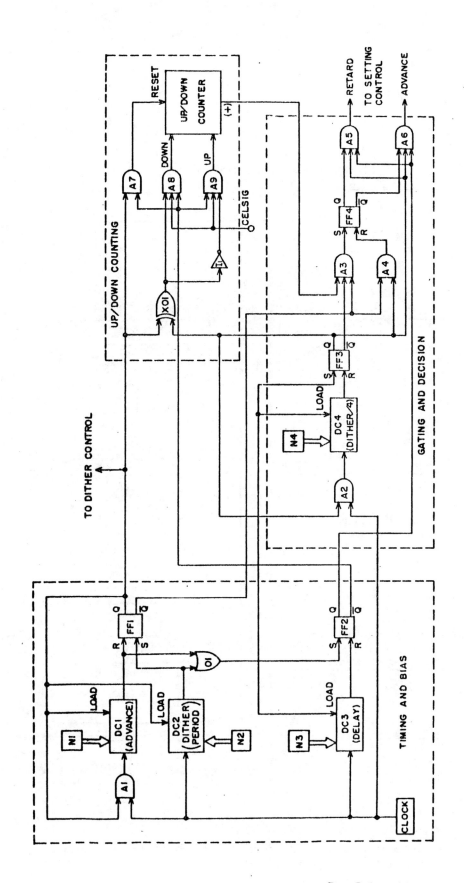

Fig. 14.
Typical Digital Circuit to Provide Biased Optimizing Control

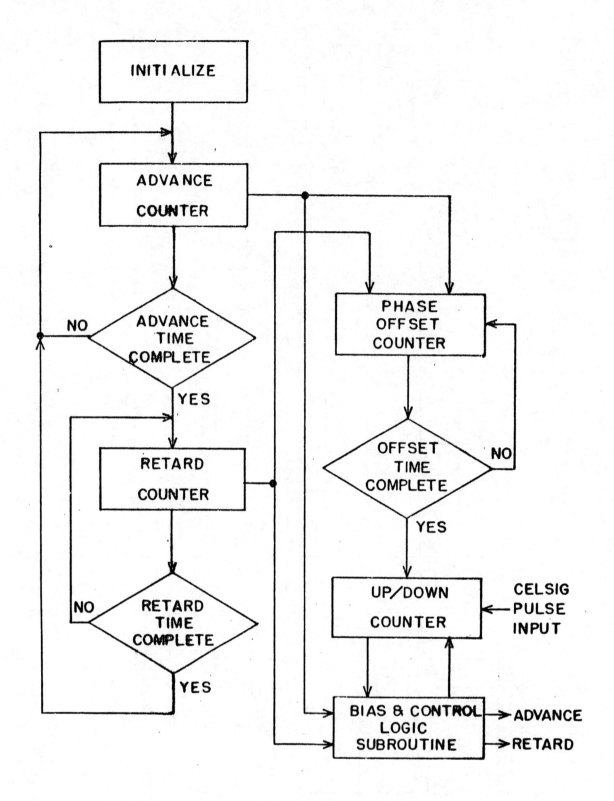

Fig. 15.
Microprocessor Program Flow Chart for Performing Optimizer Control

P. Schweitzer - Th. Collins

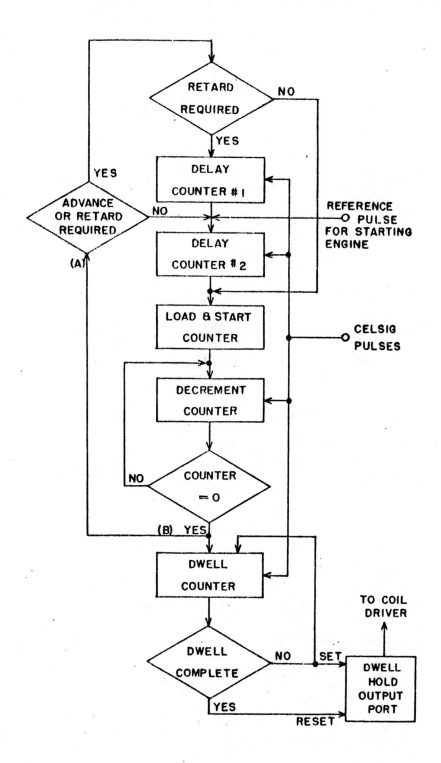

Fig. 16.
Firing Time Flow Chart

P. Schweitzer - Th. Collins

820004

Electronic Control of Ignition Timing

M. Welch, J. A. Cooke, F. Kelly,
and W. G. Lloyd
Shell Research Limited

ELECTRONIC CONTROL OF IGNITION TIMING

AS A MAJOR SUPPLIER OF GASOLINE, the Shell Group must ensure that the performance of gasoline sold in a particular market meets the needs of the vehicles that make up that market.
One of the most important aspects of gasoline quality is its road anti-knock performance. One technique by which this is measured is called knock-limited spark advance (k.l.s.a.), whereby the ignition timing in an engine using the gasoline is advanced until trace knock is detected by a trained operator. By doing this for a series of gasolines, the k.l.s.a. values obtained provide a relative assessment of their respective road performances.

In the past, results were obtained at Thornton Research Centre by a mechanical system which physically rotated the distributor to control the ignition timing. This was done using a control cable linking the distributor body to a calibrated quadrant located in the driving compartment. Such a system had several disadvantages, such as the different linkages needed for each type of car and the length of time (typically 1 or 2 days) needed to set up and calibrate the system. Also, mechanical hysteresis is unavoidable, causing timing errors, and lack of space in the engine compartment can make it difficult to establish a true 1:1 relationship between quadrant and distributor degrees.

Electronic control of the ignition timing from within the driving compartment should not only remove these disadvantages, but also provide extra facilities.

An instrument to fulfil the following requirements was therefore specified:
 (i) Installation should be simple and quick
 (ii) Control of two types of k.l.s.a. test in which
    (a) the ignition timing is varied while the engine speed is constant, over the range 1000 to 5000 rev/min and
    (b) the ignition timing is preset while accelerations are performed from 1000 to 5000 rev/min over a period of 10 seconds or longer
  should be possible.
 (iii) Static ignition timing should be variable over the range $10°$ a.t.d.c. to $25°$ b.t.d.c. in increments no larger than $0.25°$, without altering the the centrifugal or vacuum advance characteristics of the distributor.

ABSTRACT

An electronic system for the control of engine ignition timing has been developed for use in studies of the road anti-knock performance of gasolines. It does not require a crankshaft encoder, and operates using existing signals from the distributor. Static timing is controlled without alteration to the centrifugal or vacuum timing mechanisms. An additional feature identifies the cylinder(s) most prone to knock.

0148-7191/82/0222-0004$02.50
Copyright 1982 Society of Automotive Engineers, Inc.

(iv) The operation of the car's ignition system must be as designed, i.e. spark energy and duration must be unchanged.
(v) The system must operate on four or six cylinder engines and be compatible with not only the standard Kettering but also transistorised ignition systems.
(vi) It should be possible to retard the spark timing on one or more cylinders independently to identify those that are knock critical.
(vii) Aggregate ignition timing should be displayed to indicate the characteristics of the centrifugal and the vacuum advance timing mechanisms.

THE ELECTRONIC TIMING CONTROLLER

The Electronic Timing Controller (E.T.C.) has been developed at Thornton Research Centre to meet these requirements. The principle by which the E.T.C. operates is to electronically delay the signal from the contact breaker points before using it to operate the ignition coil. Control of the ignition timing is effected by varying this delay as shown in Figure 1. The required value of ignition timing can then be selected anywhere up to 99.9 degrees after the points have opened. For some tests it may be necessary to set the contact breaker to open in advance of its normal timing.

The delay must be in units of crank-angle degrees, and fitting an encoder to derive these would be a lengthy procedure. To avoid this, the instrument generates its own quasi crank-angle degrees by multiplication of the contact breaker frequency as indicated in Figure 2. The input circuitry is designed to handle signals from either contact breaker points or a magnetic pick-up and make them suitable for manipulation by the frequency multiplier circuit (shown schematically in Figure 3). The distributor signal frequency is converted to a voltage from which the ripple is filtered before being applied to a voltage-to-frequency converter. By setting the overall frequency gain to 1800 for 4 cylinders and 1200 for 6 cylinders, the output generates clock pulses every 0.1 crank-angle degree.

These clock pulses are then used to delay the contact breaker signal by the use of down counters (see Figure 4). The counters are preset to the value of the delay required by thumbwheel switches located on the front panel and then made to count down to zero by the clock pulses, on receipt of the contact breaker signal. The dwell angle in the coil switching cycle is increased by the same amount as the delay specified.

If an engine is not fitted with an electronic ignition system, one must be provided between the E.T.C. and the ignition coil. The MAX, a commercially available transistor-assisted system where coil-switching action of the contact breaker points is replaced by a transistor switch has been found to be suitable. The operation of the coil thus remains that of inductive charge and discharge as in the Kettering system. The use of a capacitor discharge type of ignition

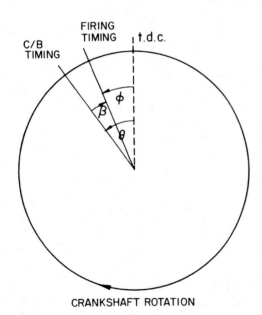

θ = angle at which contact breaker points open
β = angle by which contact breaker signal is delayed
ϕ = $\theta - \beta$ = angle by which timing is advanced before t.d.c.

Fig. 1 - Principle of electronic system

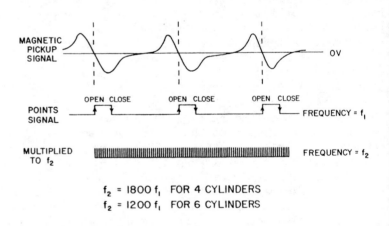

f_2 = 1800 f_1 FOR 4 CYLINDERS
f_2 = 1200 f_1 FOR 6 CYLINDERS

Fig. 2 - Frequency multiplication

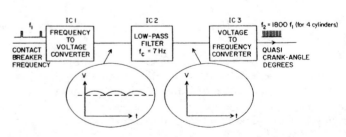

IC1 = TELEDYNE PHILBRICK 4708 IC2 = BARR AND STROUD EF14 IC3 = TELEDYNE PHILBRICK 4705

Fig. 3 - Schematic of frequency multiplier

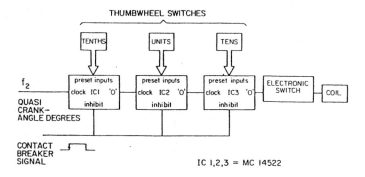

Fig. 4 - Schematic of electronic delay using down counters

was not considered since its virtues stem from its ability to produce higher ignition voltages at higher frequencies than the Kettering system and this would not comply with item (iv) of the instrument specification.

The resulting instrument requires connections only to the car battery, the contact breaker points, and either the low tension side of the coil, or the car's own electronic ignition system.

Once trace knock has been detected, it may be desirable to identify the knocking cylinder(s)[1]*. The E.T.C. has the facility to delay the ignition timing to individual cylinders by an extra five degrees (see Figure 5). When this is done to the cylinder(s) causing knock, trace knock disappears. To synchronise the instrument with the engine firing order this facility requires a small pulse transformer to be fitted onto the spark-plug lead of the last cylinder in the firing order. Cylinder selection is by means of push-buttons mounted on the instrument's front panel.

A further feature of the instrument is a timing indicator which provides the operator with a display of the aggregate of the static ignition timing and the centrifugal and vacuum advance. This is intended for use in long-term programmes, particularly in bench tests. The E.T.C. has a moving coil meter scaled from 0 to 60 crank-angle degrees of advance and this provides an indication of centrifugal and vacuum advance, in addition to the static timing. The indicator is

---

*Numbers in parentheses designate Reference at end of paper.

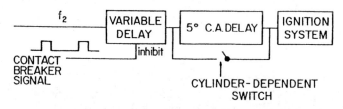

Fig. 5 - Identification of knock-prone cylinder(s)

driven by a flip-flop which is set by the coil-firing pulse and reset at top dead centre as shown in Figure 6. The reference point for t.d.c. is obtained by a proximity detector mounted close to the crankshaft pulley.

The instrument is shown in Plate I. It has a wide, low profile to allow positioning over a car's instrument panel. In addition to the controls previously mentioned, the front panel provides switches for selecting 4- or 6-cylinder operation, for magnetic pick-up or contact breaker signals from the distributor, and adjustment controls for the trigger points of the magnetic pick-up and t.d.c. proximity detectors.

EXPERIMENTAL WORK

Preliminary tests were carried out on a number of vehicles chosen to establish the compatibility of the E.T.C. with a wide range of ignition systems, and its ability to function in vehicles known to have electrical systems prone to interference. This work showed the need, when using vehicles with magnetic pickups, to be able to adjust the firing point on the pick-up waveform. Different systems not only fire on either positive or negative going edges of the pick-up waveform but also fire at voltages other than the zero crossing point.

Tests to measure k.l.s.a. values of the primary reference fuels were then carried out: first with the mechanical system, and then with the electronic system. Extensive tests have been carried out with a variety of cars on a chassis dynamometer; the acceleration tests were carried out over the range 1500 to 4000 rev/min, and the constant speed test at 3400 rev/min with wide-open throttle. By way of example, typical results obtained for two cars are given. Figures 7 and 8 show the results for a vehicle with a standard Kettering ignition system (Car A), and Figures 9 and 10 for a vehicle with a transistorised ignition system (Car B). The small differences between results obtained with each method are within the normal experimental scatter associated with use of a mechanical system.

Having established trace knock, different operators used the cylinder identification facility, and good agreement was obtained between them.

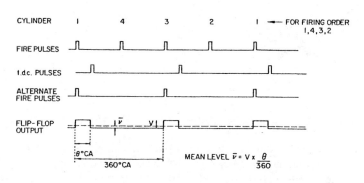

Fig. 6 - Principle of timing indicator

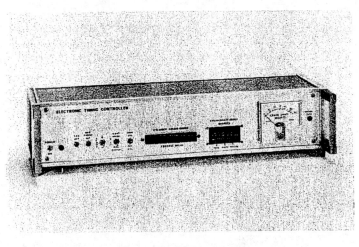

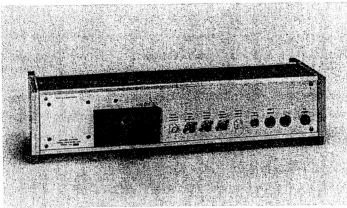

Plate I - The electronic timing controller

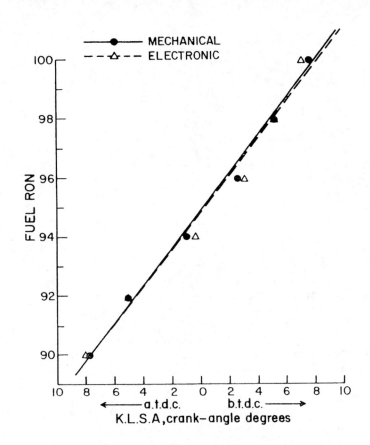

Fig. 8 - Car A: knock-limited spark advance measured mechanically and electronically - wide-open-throttle constant speed of 3400 rev/min

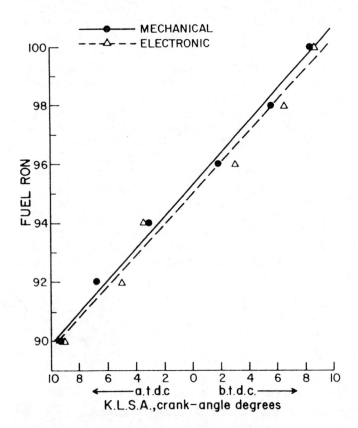

Fig. 7 - Car A: knock-limited spark advance measured mechanically and electronically - acceleration from 1500 to 4000 rev/min

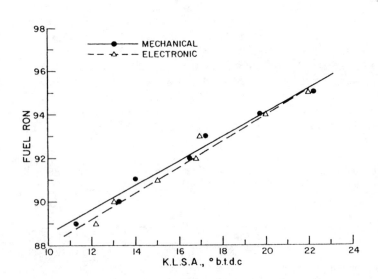

Fig. 9 - Car B: knock-limited spark advance measured mechanically and electronically - acceleration from 1500 to 4000 rev/min

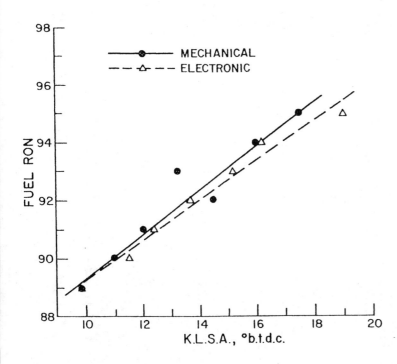

Fig. 10 - Car B: knock-limited spark advance measured mechanically and electronically - wide-open-throttle constant speed of 3400 rev/min

CONCLUSION

The E.T.C. has been successfully used in knock-rating tests for which it was designed, and will be used for all such future work at Thornton. Other possible uses for the instrument are:
- (i) to study the effects of ignition timing on vehicle fuel economy, and
- (ii) to construct engine power curves as a function of timing.
- (iii) to study distributor centrifugal and vacuum advance characteristics in long-term tests.

REFERENCE

1. L.B. Graiff, W.M. Ehrhardt and E.J. Haury, "A device and technique for determining the octane requirements of individual cylinders of an engine", SAE 801353.

C217/85

Evolution of ignition systems

H DECKER, Dipl-Ing
Robert Bosch GMBH, Stuttgart, West Germany

SYNOPSIS The evolution of ignition systems is discussed in terms of market shares, functions, technologies, the different components as well as the state of the art of static high tension distribution.

1 Influence of ignition systems

The current discussion about antipollution legislation in Europe exerts an important influence on the development of spark ignited engines as well as on the application of electronic engine management systems. The United States and Japan have been in this situation a long time before it started in Europe. The general approach to suit stringent antipollution laws is by using fuel mixture preparation systems such as injection. These solutions are already described in publicised literature. However besides the exact control of air/fuel mixture, ignition always plays an important part in overcoming the pollution problems. Regarding the situation of the time schedule of the European antipollution laws as shown in Fig. 1, it appears to be the appropriate moment to discuss how ignition can help to solve the problems. By doing this it should be discussed how the ignition systems will look like in the future in terms of function and technology.

The influence of the ignition system on the engine can be separated into two functional groups:
- influence of ignition timing
- influence of ignition characteristics.

Looking at the goals, which must be achieved, leads us immediately to the influence of ignition timing on the exhaust gas pollutants. The most important impact of ignition timing can be seen in Fig. 2. The emission of nitrous oxide depends very heavily on the ignition angle. As also does the emission of hydrocarbons. The later the ignition angle is situated, the lower is then the efficiency of the combustion, and therefore the exhaust gases become hotter; this reduces hydrocarbons by thermal reactions which occur outside the combustion chamber. The third pollutant which has to be minimized i. e. carbon monoxide is almost not influenced by ignition timing.

The ignition angle not only influences the antipollution characteristics but is also important for the consumption of the engine, the torque and therefore for the driveability of the car.

Ignition characteristics, such as available secondary voltage, spark current, spark duration influence the combustion and therefore the exhaust gas contents as well. Misfiring due to inadequate ignition systems may even damage the catalyst, and by doing so, inactivate the whole antipollution system. Complete results are treated in [7] .

2 Trends of the market

Depending on the type of engine and on the specific objectives of the car and the market where it is sold, different ignition systems will also be seen in the future, as shown in Fig. 3. There is a definite trend to replace the breaker point triggered coil ignition by transistorized coil ignitions due to the fact that the secondary output of the coil must be increased and a reliable spark must be generated. The transistorized coil ignition will see a comparatively constant market in the years to come since this system suits the very small engines which are common in Europe.

Electronic ignition systems, with sophisticated ignition maps and additional functions, such as temperature dependent ignition angle will gradually penetrate the market, either as stand alone systems or in combination with for example a fuel mixture system such as Motronic. Another important combination will be the knock control, since the action on the ignition angle in case of knocking combustions is the quickest and simplest way to handle the problem.

The tendencies shown depend very much on how exactly the antipollution legislation will proceed because any increase in sophistication is directly related to on-costs of the car.

3 Objectives for transistorized coil ignition systems

The transistorized coil is an interesting example of how a product has been adapted to objectives from the market. It is only necessary to look at the differences in development of such systems which have been optimized to meet the differing requirements in Western Europe on one hand and in the United States on the other hand.

European objectives were
- increase in compression ratio
- use of leaded fuel
- high engine speed
- comparatively lean mixtures
- splitting of the market to Hall sensor and inductive pick up triggered systems.

The increase in compression ratio leads to higher voltage demand and therefore to a higher stored energy. Table 1 gives the correlation between stored energy and the ignition characteristics. Today up to 200 mJ of stored energy is used which corresponds to a primary current in the range of 10 Amps. Fig. 4 shows how the stored energy of transistorized coil ignition increased over time in comparison to the breaker point ignition.

The use of leaded fuel adds decisively to the problem of fouled plugs which is even amplified by the European habit of short distance driving. To overcome this problem the internal resistance of the ignition systems decreased gradually and reached values of as low as 100 kOhm, i. e. almost the values of condensor discharge ignition systems.

High engine speed demands a fast coil and to maintain the level of stored energy, a high primary current.

The trend to leaner mixtures, not only at low, but even at high speeds increased the secondary voltage demand and the necessity of a long spark duration. In both cases as shown in Table 1, stored energy is involved. Fig. 4 shows how the stored

energy increased in roughly one decade. In parallel to the change in the characteristic data of the systems a major step in technology took place. The introduction of hybrids led to reduced cost, improved reliability and a remarkable reduction in weight and size. Today an ignition module's weight is less than 25 % compared to the previous generation. The relations in size are shown in Fig. 5. However it has to be pointed out, that the smallest modules differ from each other in their capability. The intelligence which has to be designed into an IC needs space and peripheral components whilst the high current requires space in the module as well as a sufficiently designed heat sink.

The splitting of the market into inductive pick up sensors and Hall vane switch sensors in the distributor caused the entire redesign of the electronic features in the control module.

Table 1:

stored energy
$$E_{stored} = \frac{1}{2} L_1 \cdot I_1^2$$

available secondary voltage
$$V_{sec} \sim \sqrt{E_{stored}}$$

internal resistance
$$R_i \sim \sqrt{\frac{L_2}{C_2}}$$

spark duration
$$t_{spark} \sim \sqrt{E_{stored} \cdot L_2}$$

spark current
$$i_{spark} \sim \sqrt{\frac{E_{stored}}{L_2}}$$

where:
L_1 = primary inductance,
I_1 = primary current at ignition timing
L_2 = secondary inductance,
C_2 = secondary capacitance.

Inductive triggered and Hall triggered modules have basically the same functions, current limitation and dwell control. Both tasks are achieved by means of closed loop control circuits, and in the case of a Hall triggered module where a dwell control function has to be developed, the electronic circuit has to generate a ramp as a function of time. This ramp can be made by analogue voltage or digital counting. The slope of this ramp serves to change the dwell angle in cases where adaption is needed. Those cases are mainly changes in battery voltage, temperature, speed. Since the output signal of the Hall sensor is purely digital with one information in the moment of the ignition time and the other information sometimes beforehand, the application of such systems encounters a problem especially when the engine is accelerating under no load conditions as happens during cranking.

From one spark to the next the speed might increase or decrease by a factor of two or more. Thus the beginning of the dwell angle is calculated on the basis of obsolete digital information. Therefore additional precautions have to be taken to avoid misfiring and to avoid prolonged dwell angle which would mean additional power dissipation. The solution to overcome this problem is an acceleration monitor circuit which observes the primary current and looks after the maximum dwell angle in the case of decreasing current.

The typical time delay of inductive pick up systems, which will be discussed below, can be reduced by employing a similar electronic slope function as with Hall sensor systems.

4 Discussion of the range of functions of current transistorized ignition systems

Fig. 6 shows two typical systems representing the low and high end of the range of functions. Both are built in hybrid technology. The combination of an inductive pick up and a coil with a stored energy of almost 100 mJ is a well chosen compromise between cost and function and especially suitable for small cars where a maintenance free system is required or in cases where catalysts must be protected from misfiring. A minimized low cost system can even be implemented without a current limit or dwell control by optimizing the resistance of the coil and using the duty cycle of the inductive pick up signal as dwell angle. In order to avoid the additional advance at cranking speed a special feature to accomplish this may be incorporated into this system.

This additional advance is caused by the position of the trigger signal on the output wave form of the inductive pick up as shown in Fig. 7. Two wave forms with different amplitudes, one for very low and one for high speed are shown, but both are normalized to the same time period. If the ignition point trigger level is based on a voltage level ① it can be seen that with decreasing speed there is an additional advance because of the distance between the crossing point and the geometrical (i. e. mechanical) ignition point. This additional advance may cause cranking problems at low engine speed, for example at very low temperature. In this case another solution is chosen. The trigger level for the ignition point is situated on a voltage level ②, which is in the negative part of the wave form. Due to this, the crossing point between the inductive pick up signal and the trigger level at lower speeds shows an increasing retard when the speed decreases. This behaviour simplifies the cranking of the engine.

This retard feature requires an additional function in the ignition module, a stall current switch off, which prevents a current flow when the engine is not running.

The high end system contains features like a high current level in order to achieve a stored energy close to 200 mJ so that it can ignite very lean mixtures. It can be used for engines with up to 8 cylinders together with very fast coils and can be adapted either to inductive or to Hall pick ups. A very accurate closed loop current and dwell control combined with a spark duration control at high speed guarantees a constant and elevated high tension voltage versus speed. In order to obtain ideal delay conditions at cranking speed, a stall current cut off is implemented as well. Further options are a powerful engine speed output capable of driving different other electronics and an exact speed limitation, if desired. This accumulation of functions and the required accuracy is achieved by employing digital and analogue technology combined in the integrated control circuit of the control module.

5 Electronic ignition

As shown above in Fig. 2, the ignition angle is the key parameter in an ignition system. Therefore the electronic ignition, where the ignition point is defined electronically by means of an ignition map, is penetrating the market gradually as discussed in chapter 2. The advantage of electronic ignition systems shall not be treated here in detail since it is well known from literature [2], [3], [7], [5].

Nevertheless one important fact which is connected to the advent of catalyst systems in Europe should be discussed. Fig. 8 shows the operational area of an engine as a function of the air/fuel ratio. This area is defined by the knock limit, the lean misfire limit and the rich operation limit [4]. In the region of the stoichiometric mixture ($\lambda = 1$) the engine can be operated at a comparatively low compression ratio only. In order to maintain a high engine performance combined with low specific consumption as desired in Europe, it is necessary to operate the engine very close to or even at the knock limit. Furthermore there is an obvious tendency to poorer quality of the fuel which will be even more important when unleaded fuel will be come into wide spread use. All these facts increase the advantage of using a knock control system. Since the easiest way to avoid knock is to retard the ignition angle the knock control is often combined with an electronic ignition system. An electronic ignition system can minimize the safety margin to a fixed and defined knock limit. Combined with knock control the ignition map can be set to optimal values at ideal conditions because an appropriate knock control protects the engine in all points of operating conditions.

These arguments are partly the reason, why electronic ignition as a stand alone system will be limited in application, whereas combinations with knock control, with injection etc. will see an extended use.

6 Evolution of the technology

The most important part of an engine management system is up to now built in a pc-board technology where separately packaged components are used. Integration took place by means of integrated circuits. Since a short time integration of the whole control module is a topic of discussion started by the following arguments:
- reduction of space for mounting
- weight reduction
- trend towards engine compartment installation with elevated temperature level.

Already a decade ago these requirements led to the application of hybrid technology for small electronic modules which are produced at high volumes, especially for voltage regulators and transistorized ignition modules. But now integration is in dicussion for complex modules as well. Fig. 9 shows which technologies are employed currently:
- hybrid technology
- surface mounted device technology.

The hybrid technology is an integration method and is defined by mixing printing technique for resistors and connections with separate components as capacitors or integrated circuits on aluminium oxide substrate. Semiconductors are preferably mounted as chip and bonded directly to the substrate in order to minimize the number of contact points. Therefore the reliability of this technology is very high. Furthermore the substrate can be mounted directly on a heat sink which leads to low thermal resistance and increased reliability or to higher operating temperatures.

Another integration method is the use of packaged leadless surface mounted devices on a pc-board. By doing this the small size of hybrid technology can almost be reached. Pc-boards can even be fitted with components from both sides. The availability of components for this technology is improving considerably each year. With regard to the operating temperature, the same values are applicable as for the current pc-board technology.

Fig. 10 shows the difference in size of two electronic ignition systems with similar features, one in conventional pc-board technology and the other in hybrid technology. This should prove, that these integration methods of modules are feasible for complex ignition systems too.

7 Evolution of other ignition components
Since the voltage requirement depends heavily on the mixture, the spark gap, the compression ratio etc. it was inevitable, that the voltage demand increased. As an example Fig. 11 shows the influence of the spark gap. The resulting increase of all factors in the last decade was roughly 10 kV. This had a major influence on the design of distributors, coils, high tension cables and connectors and even on the spark plug. There is also a definite trend to a kind of integration, where the entire high tension cabling is mounted into a spinster-like one piece assembly. The objective of this development was a well defined state of all the connecting points. The connections where defects might occur are reduced by 50 %. The ignition coil encountered new impulses for evolution too. The bottle like and the transformer like coil are well known since a long time. But nevertheless this component was exposed to the following influences:
- elevated high tension voltage
- weight reduction
- minimized primary resistance
- double ended coils for the generation of two sparks

The requirement of weight reduction and minimized primary resistance leads to the solution to add a magnet, which is placed in the air gap of a transformer type coil. This magnet causes a bias in magnetic induction. Therefore a higher current can be driven through the coil until magnetic saturation of the core is reached, or, on the other hand, if the current is already given, less iron is needed for the magnetic circuit. By this means a weight reduction, as obtained in a real application, of almost 50 % can be achieved. Fig. 12 shows this example. A transformer type coil can also easily be adapted to two high tension outputs for use as a double ended or twin spark coil.

8 State of the art and trends of the static high tension distribution
This topic has to be taken into consideration when the evolution of ignition systems is discussed. The main objectives for the development of static distribution systems are:
- reduction of electric noise
- avoidance of the mechanical distributor drive
- shorter high tension leads
- reduction of high tension contacts
- weight reduction.

Suitable solutions are:
- separate coil per cylinder
- double ended coils.

A system which is in an advanced stage of development is
- quadruple ended coil with 4 high tension diodes.

In the case of double or quadruple ended coils only two power stages are needed for a four cylinder engine. A somewhat different version of the separate coil per cylinder solution is a condensor-type ignition where each cylinder has a transformer coil for itself, as recently published by Saab.

Fig. 13 shows that only engines with an even number of cylinders are suitable for use with these systems and that a phase sensor for the definition of the cylinder or cylinder group has to be incorporated. The static distribution must be combined with electronic ignition.

With increasing numbers of cylinders static distribution shows no advantages in terms of weight and cost because these two factors are dominated by the number of coils and power stages. At higher cylinder numbers it is even more advisable to place the power stages separately from the control module and combine them with the coil.

The application of static distribution confronts the engineer with a couple of problems which he has to look at very closely. One example which should be discussed is the case of double ended coils. Theo coil generates a twin spark. One spark ignites fresh mixture in the working cycle cylinder.

The second spark is directed to a cylinder which is 360° more advanced in the exhaust cycle. Normally the intake valves are opening prior to the end of the exhaust cycle. Therefore special consideration must be taken that in this case no fresh mixture in the intake manifold is ignited. Similar problems exist with the other systems [1].

Further development of static high tension distribution has to focus on these priorities:
- safe function
- availability of the technology
- low cost.

Therefore, although the development is advancing, the rotating distribution will be the safest and least expensive solution for many engines in the near future.

9 Conclusion

Ignition systems are in a decisive phase of development in terms of function, application and technology which will continue in the years to come. Not only control modules are affected but each single component of the whole system. In addition to this there exists a definite trend towards integration of the ignition into other systems and vice versa as for example fuel injection and knock control. These efforts will be accelerated by more stringent antipollution laws which will be seen in Europe soon.

REFERENCES

[1] H. Decker, R. Kaufmann, R. Schleupen: L'allumage électronique avec considération particulière des distributions haute tension statique et rotative, SIA, Mai 1984, Toulouse

[2] H. Decker, H. U. Gruber: Knock Control of Gasoline Engines - A Comparison of Solutions and Tendencies, with Special Reference to Future European Emission Legislation, SAE-Paper 1985, No. 850298

[3] H. Schwarz, H. Bertling: Einfluß der Zündung auf Motor und Abgas, Bosch Technische Berichte 5 (1977), Heft 5/6, S. 220 - 225

[4] R. Fritz: Elektronik im Auto, Umschau 83 (18), p. 535-539

[5] H. Spies, M. Knoke: Dignition - ein intelligentes Zündsystem, Impulse, Fahrzeug-Elektrik/-Elektronik von Volkswagen, Heft 2/1982, S. 22 - 28

[6] H. Decker, I. Gorille, M. Zechnall, S. Rohde: New Digital Engine Control Systems, IEE, London, November 1984

[7] H. Schwarz: Ignition Systems for Lean Burn Engines, I Mech E (1979) C 95/79, S. 87 - 96.

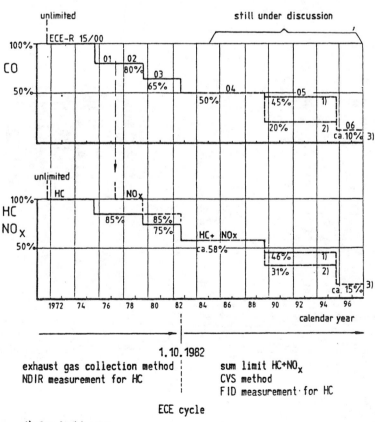

Fig 1 Pollution limits for passenger cars, ECE

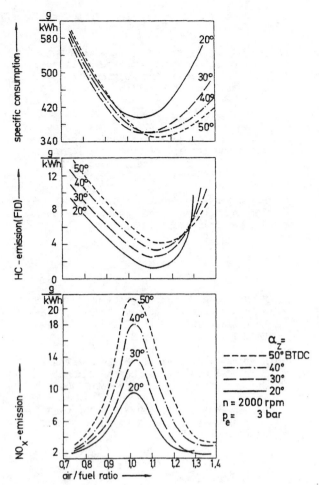

Fig 2 Influence of ignition timing on specific consumption and exhaust emissions

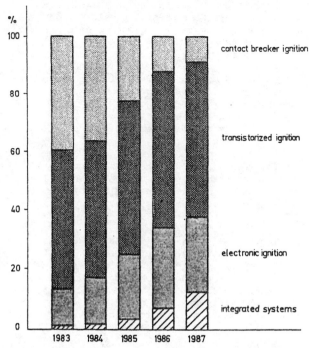

Fig 3 Market share of ignition systems in Western Europe

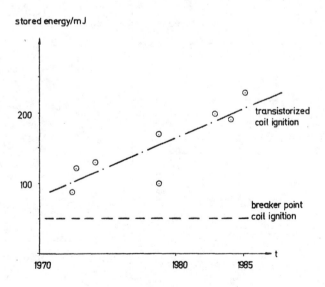

Fig 4 Evolution of stored energy

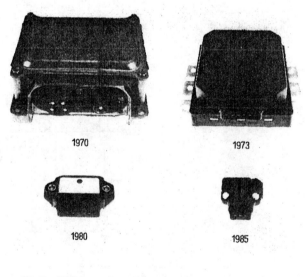

Fig 5 TCI control modules, comparison of size

function	low end system	high end system
stored energy	90 mJ	180 mJ
cylinders served	4	up to 8
trigger system	inductive	inductive, Hall
closed loop dwell control	no	yes
spark duration control	no	yes
acceleration detection	no	yes
current limitation	no	yes
stall current cut-off	unnecessary	yes
engine speed output	no	yes
speed limitation	no	yes

Fig 6 Transistorized coil ignition, functions

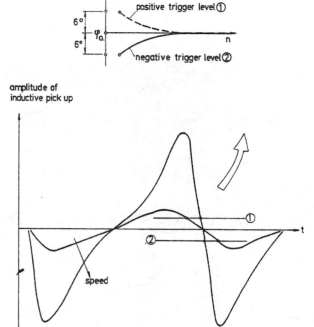

Fig 7 Inductive pick-up, phase shift at low speed

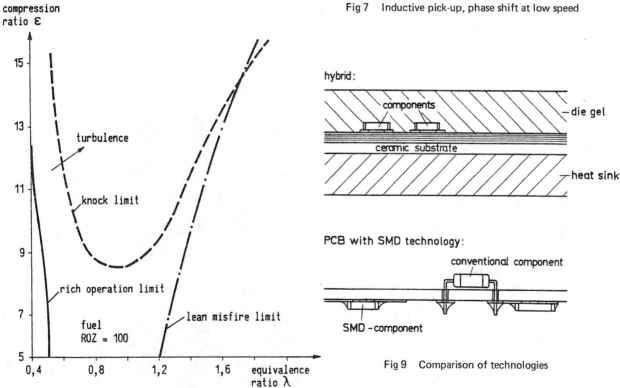

Fig 8 Operational area of the engine

Fig 9 Comparison of technologies

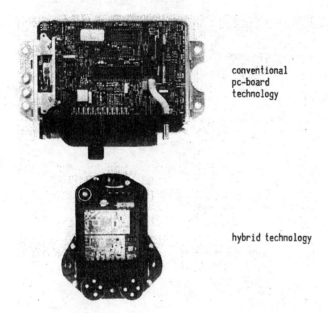

Fig 10 Electronic ignition, comparison of size and technology

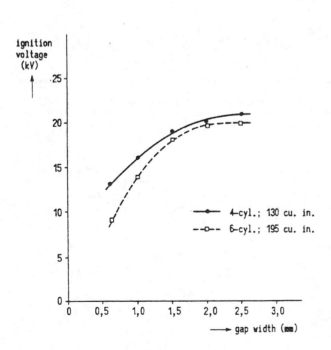

Fig 11 Influence of spark plug gap width on ignition voltage

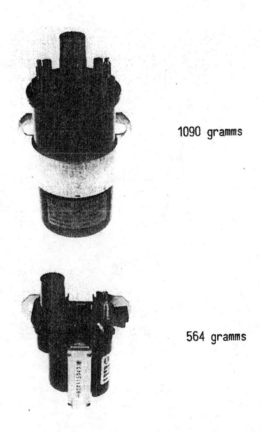

Fig 12 Weight reduction of ignition coils

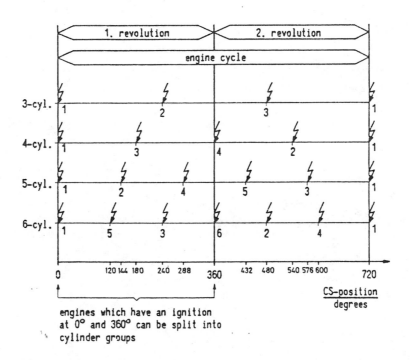

Fig 13 Engine ignition points

860249

General Motors Computer Controlled Coil Ignition

R. F. Gardner
Powertrain Electronics Group BOC Flint
GMC

R. J. Waring
Industrial Products Div.
Magnavox Electronic Systems, Co.
Auburn, IN

T. C. Marrs
Marshall Electric, Co.
Rochester, IN

ABSTRACT

The design of an ignition system for a modern internal combustion engine had to take into account all the functions currently being performed by the present system as well as the requirements of the new system. The need to keep the ignition transparent to the engine computer was paramount because of not wanting to redesign hardware and software requiring long leadtimes. The need to drive the engine oil pump on the first application was still necessary but the desire was to provide "Net build timing" from the crankshaft which negated the need for service and engine position timing adjust. The electronics strategy had to be consistent with the sensing and manufacturing techniques that were selected for the application. The ignition coil and its' development were planned for the size, performance and appearance desired.

The design and development of a new ignition system was necessitated when packaging requirements were considered for Flint built BOC V-6 engines for the 1980's and beyond. With the advent of front wheel drive vehicles that have bottom loaded powertrains, the length of the engine became of paramount importance. The need for a new method of firing the spark plugs was apparent when down-sizing of the engine resulted in the elimination of the mounting position and the camshaft drive gear for the distributor. The parameters considered for the new ignition system are listed below:

1. Performance - Greater than or equal to current production energy levels, as well as equal or greater EMI immunity levels.
2. Interchangeability - transparent to the engine computer and other control strategies.
3. Sensor Strategy - zero speed sensing and improved electrical noise immunity desired. Crankshaft sensing required for greater accuracy.
4. Powertrain Manufacture - net build timing required to eliminate timing set in engine plants and the elimination of service required timing adjust, due to wear of engine components (timing chain stretch and distributor gear wear).
5. Component Manufacture - within the realm of existing technologies and consistent with the manufacturing techniques required for the projected volumes.
6. Reliability - consistent with the goals of the overall powertrain reliability.
7. Appearance - styled to meet the overall powertrain theme.

The electrical pinout illustrated in figure #2 was selected. The use of a 14 pin integral connector and putting the coils into a single package were reliability considerations. The use of a waste spark strategy, firing cylinder pairs with a single coil, was dictated by economics and the use of fuel injection allowed its use without encountering harmful backfires during hot restart conditions.
Hall sensors for camshaft and crankshaft position sensing were selected because of their packaging, noise immunity and operating characteristics.
The inductive ignition coil strategy was selected because of known technology and manufacturing techniques and evolved into section bobbin coil for the efficiency,

0148-7191/86/0224-0249$02.50
Copyright 1986 Society of Automotive Engineers, Inc.

performance, manufacturing and size considerations.

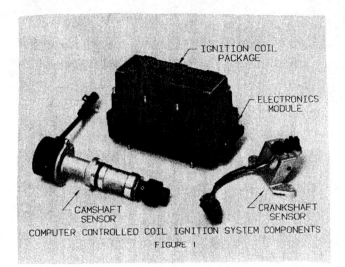

FIGURE #1

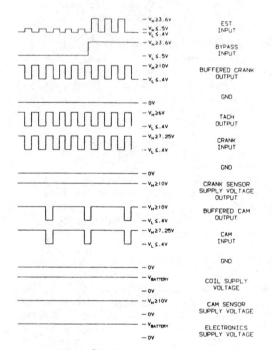

FIGURE #2

SYSTEM DESCRIPTION

The components of the computer controlled coil ignition system that were developed to meet the needs of this 6 cylinder, sequentially fuel injected and turbo charged engine are shown in figure #1.

These components are interconnected in the system as shown in figure #3. This block diagram shows the direct connection of the high voltage ignition coil secondary windings to the spark plug pairs in the "waste spark" configuration. Figure #5 shows this concept in more detail with one of the spark plugs in a cylinder under compression and the other plug in a cylinder in its exhaust stroke. The delivery of the high voltage to the spark plugs in this system is controlled by the interruption of the primary current in the appropriate ignition coil.

As shown in figure #3, the major components of this system are:
1. Crankshaft sensor
2. Camshaft sensor
3. Electronics module
4. 3 coil ignition package

These components are described in greater detail in the following section. Also included with the component description are details relating to the important technical considerations in the design of the component and its functions as they relate to the system.

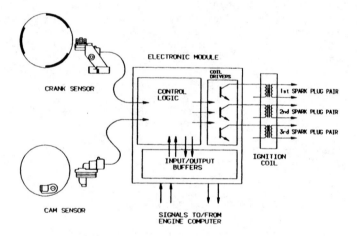

FIGURE #3

SENSORS

In developing the sensors for the computer controlled coil ignition system, a number of factors were considered important.

It was decided that a crankshaft sensor would be used to improve the timing accuracy of the spark event. this sensor would also provide the basic timing of the ignition system if engine computer timing was lost.

A camshaft sensor would also be required to provide the cylinder identification required by a sequential fuel injection system. This signal would also be used to synchronize the cylinder pair electronics ring counter and to control the

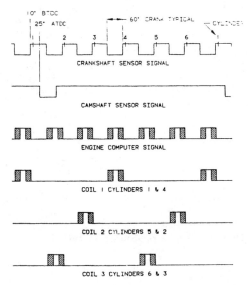

SENSOR SIGNALS - TIMING - CYLINDER PAIR DIAGRAM
FIGURE 4

▨ INDICATES COMPUTER CONTROL OF DWELL AND ADVANCE

FIGURE #4

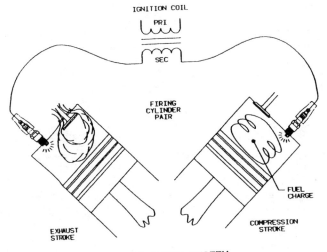

WASTE SPARK SYSTEM

FIGURE #5

fuel injection and ignition relationship under start-up conditions.

Because of the critical nature of these sensor signals to the successful operation of the system, a number of characteristics were considered very important:
1. Immunity to electromagnetic interference.
2. Wide operating voltage ranges.
3. Low or zero speed operation.
4. Sensitivity to mechanical noise and vibration.
5. Compatibility to electronics interfacing.
6. Clean fast switching signals.
7. High reliability over wide temperature range.
8. Small size for ease of application.

Based upon comparative data and considerable experience with hall cell sensors in rough vehicular applications, the hall cell type sensor was selected for both applications. Techniques have been developed to handle the hall cell disadvantages of stress sensitivity, wide distribution of parameters and parameter shifts with temperature.

The crankshaft sensor is mounted in the area of the harmonic balancer/pulley assembly and consists of the hall cell sensor and a sumarium cobalt magnet in an integrated package. The slot between the hall device and the magnet is interrupted by a ferrous metal ring. The ring has 3 teeth and is mounted on the harmonic balancer/pulley assembly. This sensor generates 3 symmetrical pulses per crankshaft revolution as shown in figure #4. Note that the rising edge of this signal occurs at 10 degrees before top dead center of each cylinder.

The camshaft sensor is mounted in place of the distributor and also consists of a hall cell sensor and a magnet in an integrated assembly. The space between the sensor and the magnet is interrupted by a ferrous metal cup with a single slot that is connected to a distributor-type shaft. This cup is rotating at camshaft speed and generates a single pulse every 2 crankshaft revolutions as shown in figure #4. Note that the falling edge of the camshaft sensor signal identifies cylinder number 1.

It should be noted that later design for engines with no distributor shaft have had this sensor incorporated in the timing chain cover. The signal produced is basically the same as shown in figure #4.

ELECTRONICS MODULE

The electronics module performs the following electronic functions (See figure #3).
1. Engine starting ignition control
2. Cam- and crankshaft sensor signal buffering for engine control computer.
3. Controls coil firing sequence.
4. Ignition "back up" strategy if computer signal is lost.
5. Provides coil power driver.
6. Timed power shut down circuits if ignition is on and engine not running.
7. Tachometer signal.

8. Ignition coil primary current limiting (2nd generation).

In addition the electronics module also performs these functions:
1. Provides a strong secure mounting for the ignition coil package.
2. Provides heat sinking for the ignition coil driver transistors.
3. Provides a mechanical mounting to engine bracket.
4. Provides mechanical and environmental protection for the electronic circuitry and ignition coil connections.

The electronics module consists of an aluminum die-cast housing incorporating a 14 pin waterproof connector for all input and output connections. Internally, this 14 pin connector is soldered to the single printed circuit board on which is mounted all other electronic components. The ignition coil driver transistors over-hang the printed circuit board on which is mounted all other electronic components. The ignition coil driver transistors over-hang the printed circuit board so that they can be fastened to the die-cast housing for maximum heat-sing efficiency. Studs in the bottom of the die-cast housing for mounting to the engine mounting bracket are provided.

After the printed circuit board with its power transistors and 14 pin connector has been mounted in the die cast housing, the housing is filled with an epoxy potting compound. This potting stabilizes all components against shock and vibration, and provides protection from any environmental contaminants including moisture and other automotive solvents.

When the module potting has cured and the final tests have verified an acceptable device, the ignition coil package is wired to the appropriate leads that protrude from the module potting.

The ignition coil package is then attached to the electronics module with a gasket secured in the interface. The sealed surface provides additional protection for the ignition coil connections and the potted surfaces.

The electronic circuitry of the module consists of bipolar transistors, CMOS digital devices, and other discrete components appropriate to performing the functions.

Upon cranking the engine, the electronic module receives crankshaft and camshaft sensor signals as shown in figure #4. During this cranking period, the input signal from the engine computer is grounded by the electronic module . After recognition of the appropriate sequence of crankshaft and camshaft signals, fuel injection and ignition events begin.

When the engine starts, the ignition events are controlled by the crankshaft sensor and the electronic module only. The start of dwell is initiated by the falling edge of the crankshaft sensor signal (reference figure #4) and the spark plug firing is controlled by the rising edge of this same signal.

This mode of operation continues until a minimum RPM is achieved for a required period of time that permits the engine control computer to verify that it is capable of providing ignition data to the electronic module. At this point, the engine computer generates a signal that causes the electronic module to remove the ground on the engine computer signal line and relinquishes ignition control to the computer.

At this time, normal ignition control is established with the computer initiating start of dwell with the rising edge of its digital signal and firing time control with the falling edge of this same signal. This mode of operation is shown in figure #4. Also shown is the effect on dwell initiation and advance of the firing event as a result of computer calculations based on other engine sensor inputs.

It should be noted that the ignition "back-up" strategy is essentially the same as the engine start mode of operation. When the engine computer recognizes that it cannot, for any reason, provide valid ignition data, it signals the electronic module which in turn grounds the engine computer signal and restores control directly from the crankshaft sensor. During the "back-up" mode of operation, dwell begins at 70 degrees before top dead center and firing at 10 degrees before top dead center as shown in figure #4. It should be noted that "back-up" operation is intended for driving to a service center only, not continous operation.

Other circuitry within the electronic module recognizes the condition of ignition turned on without engine operation. If this condition exists for an extended period of time, the output transistors are turned off to minimize output circuitry over-heating. Re-initiation of cranking restores normal operation.

In addition to the analog and digital circuitry involved in these functional modes of operation, circuits are incorporated to:
1. Minimize the effect of electrical noise on power lines as well as data lines.
2. Operate with a 24 volt battery.
3. Tolerate reverse battery conditions.
4. Withstand intermittent battery conditions.

5. Provide power to the cam and crank sensor.
6. Provide a 3 pulse per crank revolution tachometer signal.
7. Provide required impedance and voltage level sensor signals for the engine computer.

IGNITION COIL PACKAGE

A 2 spark coil was used to provide minimum size, weight and cost to the package. As previously noted, the engine was able to accept the fact that 2 spark plugs will fire at the same time. It's exhaust stroke produces a waste spark of approximately 2000 volts and produces the ground for one end of the secondary winding, allowing the opposite end of the secondary to produce the required voltage to fire the compression stroke spark plug. This voltage is expected to be anywhere from 10,000 volts to 28,000 volts depending on the plug condition and the cylinder pressure.

The 2 spark coil was designed to give balanced output voltages. This was accomplished by using 2 secondaries over one primary with the secondary start wires connected and the winding direction of each secondary being opposite to the other. The secondary finish wires become the high voltage terminations. Since neither end of the secondary is grounded, the ground is obtained through the exhaust stroke spark plug. The start wires of the secondary are at one-half the potential of the secondary output and must be insulated from the primary to withstand this voltage. The secondary traverse is necessarily short to facilitate placing 2 secondaries over 1 primary winding. The short traverse gives a low secondary capacitance and thus a higher peak output voltage under light capacitive loading conditions. This coil construction is illustrated in figure #6.

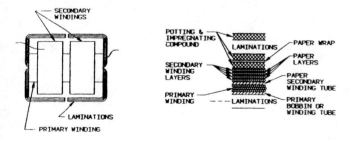

COIL CONSTRUCTION   COIL ASSEMBLY SECTION

TWO SECONDARY COILS
FIGURE #6

The magnetics are obtained by utilizing one-half inch wide M-36 steel shunts formed to contain the magnetic field in a low profile. The magnet wire, case material, insulating materials, and the potting compound were chosen for high dielectric strength and high operating temperature capabilities. The coil package is impregnated under vacuum with 3 coils needed for a 6 cylinder engine, all potted in one case. Putting 3 coils in 1 package forces high quality since 1 defective coil would cause 3 to be discarded. However, this concept also gave a better appearance and simplified mounting to the electronic module with improved sealing characteristics.

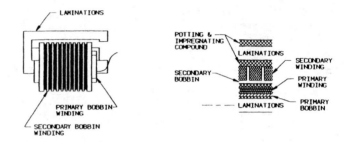

COIL CONSTRUCTION   COIL ASSEMBLY SECTION

BOBBIN COIL

FIGURE #7

A second generation coil package was designed to preserve the attributes of the original design while reducing size, weight, and cost. This design was developed around a sectioned bobbin secondary rather than the conventional paper section windings utilized in the original design. The section bobbin gives balanced output with just one secondary winding, gives an even lower secondary capacitance, and allows a much higher degree of automation. The magnetics were redesigned to allow a lower profile by using a stamped "U" lamination with the return path portion of the "U" two-thirds the width of the core portion without significant reduction of coil performance.

SUMMARY

An electronic ignition system that requires no high voltage distributor has been described. This system has been put into production successfully and has met the original design goals.

1. Elimination of the high voltage distributor.
2. Reduction of vehicle radiated radio frequency interference.
3. Reduction of operating spark scatter.
4. Minimized factory and service adjustments.
5. More flexible engine packaging.

The design of the components described in this paper have demonstrated the ability to perform their described functions reliably in an automotive environment.

A Compact Microcomputer Based Distributorless Ignition System

Giuseppe Cotignoli and Enrico Ferrati
Marelli Autronica S.P.A.
Torino, Italy

The digital electronic ignitions are suitable in Europe for applications on medium or medium/high class cars. Generally, the high voltage mechanical distributor is present in such system, but now, due to the continuous improvement of cost/performance ratio of the ignition components, a completely distributorless system becomes actractive. In particular, the decreasing cost of microprocessors and ignition coils, reduces or cancels the cost gap between the distributorless and mechanical solution. In addition, the high voltage performances improve since the high voltage cable length is reduced, especially when the coils are mounted directly on the engine. Also, the reliability of this system increases due to non-moving parts. This paper describes a distributorless digital electronic system with only one pick-up, a low cost microprocessor and a cost effective solution for driving the ignition power stages.

INTRODUCTION

The electronic ignition systems - Breakerless or stand alone ignitions with digital electronic control of the spark timing - at present are widely used in the European market.
It's well known, Breakerless ignition allows a constant spark energy in any working conditions and a reliability higher than usual contact-breaker ignitions. The majority of medium/low cars are equipped with such systems. The electronic digital ignitions, in addition to the Breakerless performances, permits the controlling of the variable spark advance versus engine RPM, pressure at the intake manifold and some other parameters. In this case, the condition for optimal operation of the engine is defined by the engine designer. He can give priority to car performance, or gasoline consumption or pollution or a proper compromise of these quantities. (1)*
In fact, the applications of these ignitions gave priority first to fuel consumption at the end of the 70's and at the beginning of the 80's, and then, starting from 82/83 the performances, especially for turbo charged engine cars.
The spark advance electronic digital control systems in production now, use a mechanical distributor for high voltage. This involves the necessity to have a rotary part in the ignition system with the relevant inconvenients due to the wear and tear of moving mechanical parts. To eliminate the distributor, two different solutions have been proposed: one of these with a coil for each cylinder, the other based on double output coil. The present paper describes a system with the latter configuration. In addition to better performances and higher reliability typical of distributorless system, this one has a cost comparable with the mechanical distribution version, due to the configuration of the engine references

and the electronic circuits for the power stages driving.

SYSTEM DESCRIPTION

The distributorless electronic digital ignition system (trade name "Digiplex II"), object of the present paper, is composed by the following components (Fig. 1):

. Wheel for RPM and TDC detection
. Wheel pick-up
. Electronic control unit with pressure sensor
. 2 coils

The block diagram of the system is shown in Fig. 2.

WHEEL FOR RPM AND TDC DETECTION - For a 4 cylinders engine, a simple 5 teeth wheel has been designed: two teeth relating to cylinders TDC, two teeth at 90° from the first one, and one at 6° from a TDC tooth of a couple of cylinders (Fig. 3). This configuration allows the identification of the couple of cylinders which the ignition spark has been supplied to without using a complex hardware or software strategies.

As well, the simple construction of such a wheel allows the use of the pulley, properly modified, already present for driving the water pump and other mechanical parts. Consequently, the additional installation cost is very low. On the other hand, in order to calculate and carry out the spark advance angle, few angular references require to manage time intervals instead of angular counting. Using suitable calculation algorithms, the tolerance of spark advance in dynamic conditions has been maintained within acceptable limits.

WHEEL PICK-UP - The wheel pick-up supplies the electrical pulses corresponding to the mechanical references on the wheel.
The pick-up is a variable reluctance type.

CONTROL UNIT WITH PRESSURE SENSOR - The control unit is composed by an elaboration circuit and a pressure sensor.
This latter is a sensor based on piezoresistive effect realized with properly made

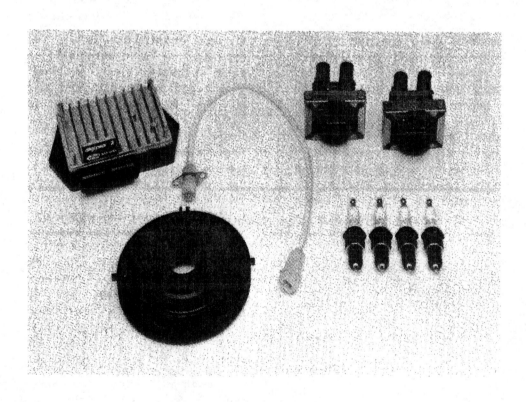

FIG. 1: system components

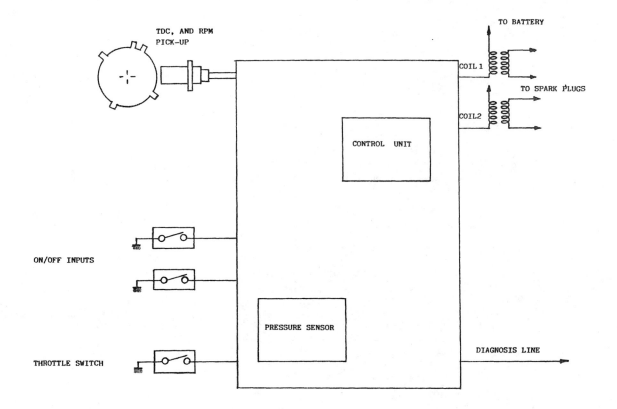

FIG. 2: system block diagram

thick film resistors; the electronic circuit for the signal, amplification and thermal adjustment is integrated on the sensitive element and it's made with thick film technology (2).

The single technics used to manufacture it, the resulting ruggdness, its performance and reliability make it a valid competitor to the SI pressure sensor. The main features of the sensor are shown in table 1.

The block diagram of the elaboration unit is shown in Fig. 4.

The ECU is based on the Motorola microcomputer MC6805S2 whose main characteristics are:

- n. 2 timers
- n. 4 A/D channels
- watch dog
- ~ 1,4 Kbytes ROM
- 64 bytes RAM

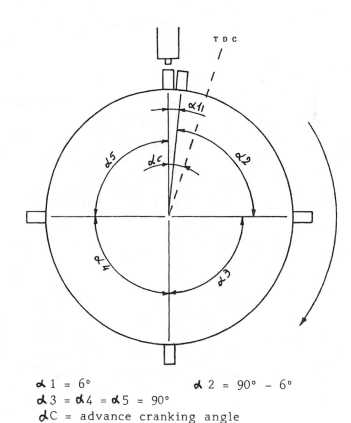

$\alpha 1 = 6°$ $\alpha 2 = 90° - 6°$
$\alpha 3 = \alpha 4 = \alpha 5 = 90°$
αC = advance cranking angle

FIG. 3: input configuration

TABLE 1: ECU CHARACTERISTICS

. Tridimensional map : = f (RPM, pressure)

- pressure : 10 values
- RPM : 16 values
- Linear interpolation along the RPM axis with 25 RPM resolution

. Spark advance

- steady state: +/- 1°
- dynamic conditions: +/- 5° (with constant acceleration of +/- 5000 rpm/sec from 1000 to 6000 rpm and with advance curve slope of 3°/100 rpm)

. Integral sensitive type trigger : 10 V x 150 μ sec

. ON/OFF inputs : grounded type

. Manifold pressure sensor : relative type

- range : 0/-600 mmHg (aspirated engine)
- accuracy: 3% full scale

. Power stages : coil current typical 6,5A +/- 0,2A

primary clamp voltage 350V +/- 25V

desaturation time: 7% of the engine phase time

secondary voltage:
35 KV with 50 pF
29 KV with 50 pF // 1 MΩ
21 KV with 50 pF // 0,5 MΩ
14 KV with 50 pF // 0,25 MΩ

. Self diagnosis output : open collector
. Diagnosis patterns : ISO standard identification

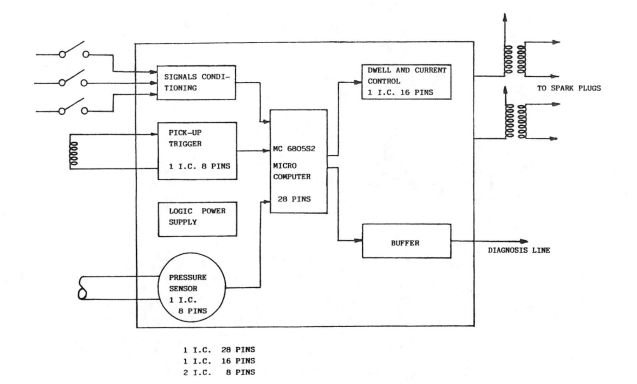

FIG. 4: ECU block diagram

The threshold of the electromagnetic pick-up interface is proportional to the input signal integral.
In particular, the circuit supplies the pulses to the microcomputer only if the input signal energy is higher than a determined level.
Therefore, a high immunity is guaranteed because the noise generally is characterized by high voltage and short time.
As well, the ECU has an ON/OFF inputs interface to shift the advance spark map, to decrease the spark advance during the overrun or for self diagnosis functions.

The dwell angle and the ignition coil current control is performed by a custom integrated circuit which is driven by the microcomputer. Since the current is never at the same time flowing into the two coils, the integrated circuit output can drive alternatively the power Darlingtons and consequently the ignition coils.
The Darlington selection is performed by the microcomputer processing the information of input signals (Fig. 5).

The solution allows it to reduce, regarding the control unit, the cost between distributorless and mechanical distributor versions; in fact for the distributorless system it's necessary to add a power Darlington and a transistor only.
For self diagnosis purpose, the information relating to the system working - RPM, manifold pressure, advance angle, input signals status - is available using a serial line.
The detailed characteristics of the ECU are shown in table 1.
The ECU is composed by few integrated circuits and made with thick film technology.
Therefore, a compact device with small dimensions and high reliability, is obtained.
As well, the ECU characteristics are suitable for under hood application.

COILS - The coils are of double output type, with a closed magnetic circuit.
With this configuration, on each spark plug there are two sparks per engine cycle: one during the compression phase

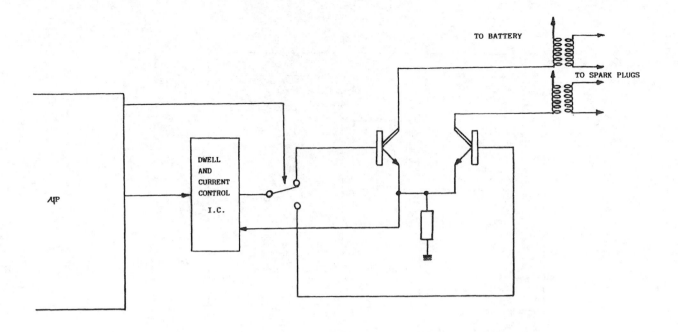

FIG. 5: multiplexing of the ignition power stages

and one during the exhaust phase. This latter occurs with a very limited energy due to the ambient characteristics in that engine phase.
The coils characteristics permit the installation on the engine as shown in Fig. 7.
Such configuration represents an optimization of the ignition system: reducing the length of high voltage cables,

FIG. 6: E C U

FIG. 7: ignition system installation

the voltage at spark plugs is raised and the capacitive effects of the cables are eliminated.

CONCLUSIONS

A distributorless ignition system has been described.
This system is able to:

- utilize only one pick-up for RPM and TDC detection
- determine the advance versus RPM, manifold pressure and other eventual parameters (coolant temperature, throttle position)
- drive two double output coils for applications with 4 cylinders engine
- perform the diagnosis of ECU inputs and sensors

The total elimination of moving mechanical parts now present in the ignition systems, brings to a considerable increase of the system reliability.

The system described results as compact, with small dimensions and limited cost, due to the technical solutions implemented.

A peculiarity of the system is the flexibility: few changes of hardware and software permit other additional functions (i.e.: cut-off, knock control, EGR control). In this way, said system could be utilized even when the strict regulations on the exhaust gas pollution will be adopted in Europe.

REFERENCES

(1) R. Dell'Acqua - F. Forlani - Technological Approaches to electronic ignition Systems in the Euopean Market - SAE paper 840449 Detroit, 1984

(2) R. Dell'Acqua - G. Dell'Orto - F. Forlani - A. Punzone - C. Canali - Characteristics and performances of thick film pressure sensors for automotive applications - SAE paper 820319, Detroit, 1982

871914

Crankshaft Position Measurement with Applications to Ignition Timing, Diagnostics and Performance Measurement

Yibing Dong
Department of Electrical Engineering and Computer Science
Univ. of Michigan

Giorgio Rizzoni and William B. Ribbens
Vehicular Electronics Laboratory
The University of Michigan
Ann Arbor, MI

Abstract

This paper introduces a high accuracy method of measuring crankshaft angular position of an I-C engine. The method uses a sensor which couples magnetically to the starter ring gear. There are many automotive applications of this measurement of crankshaft angular position including ignition timing reference, engine performance measurement and certain diagnostic functions. The present paper disusses only the ignition timing application. Engine performance measurements are reported in refs. (1-3). The diagnostic application is discussed in refs. (4-5).

The passage of a starter ring gear tooth past the sensor axis causes a pulse to be generated in the sensor output. The waveform of this sensor voltage is independent of engine angular speed (including zero speed). However, this waveform is a function of gear tooth profile and is consequently influenced by gear wear.

The present method uses a finite state machine to process the sensor output signal. This method is highly versatile and is readily adaptable to provide a signal having suitable format for timing reference for an arbitrary electronic ignition system. The method is particularly well suited for a distributorless ignition system.

Another benefit of the finite state machine signal processing is its ability to correct for sensor errors. Sensor errors can occur randomly due for example to electrical noise and, in extreme cases, to broken ring gear teeth. The error correcting feature makes this method highly fault tolerant and essentially insensitive to the condition of the ring gear.

System Description

The structure of the instrument for the fault-tolerant detection of crankshaft angular position is shown in block diagram form in figure 1. The system is comprised of three major blocks: sensor/interface, signal conditioning, and state table.

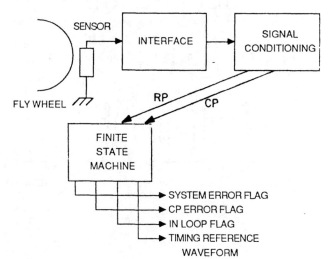

FIGURE 1

Sensor/interface

The sensor employed in the study is of the magnetoresistive type. In addition to the advantages generally afforded by magnetic sensors, e.g.: immunity to environmental factors and ruggedness, the magnetoresistive sensor also provides an output which is independent of speed, unlike transducers of the inductive type. The sensor is magnetically coupled to the starter ring gear on a modified flywheel. The output

0148-7191/87/1019-1914$02.50
Copyright 1987 Society of Automotive Engineers, Inc.

waveform is quasi-sinusoidal and provides a pair of zero crossings for each tooth on the flywheel. A small number of teeth are modified to provide a reference signal at known angles of rotation of the crankshaft. In this particular study, the modification consisted of removing part of the metal from selected reference teeth, so as to obtain a lower signal level from the sensor in correspondence and the reference locations. The teeth were modified on the side opposite that of the starter motor, so as not to weaken the structure.

In general, it is not necessary to employ this particular technique, so long as a predetermined number of reference signal are available. Figure 2 shows the raw output of the sensor, and the corresponding logic waveform, including on of the reference signals.

Signal Conditioning

The signal conditioning circuit converts the raw sensor output into TTL-level pulse waveforms. These are shown in figure 3, and consist of a clock waveform at a frequency of M cycles per revolution, where m is the number of ring gear teeth, and of a reference pulse waveform consisting of M unevenly spaced pulses per revolution. The reason for the uneven spacing is that a unique sequence of clock pulses can be generated in this fashion for each engine revolution. In our case we experimented on a 6-cylinder engine, and selected m = 144 and M = 6.

Note that the first waveform provides a high resolution measurement of _relative_ angular position, while the second yields six references of _absolute_ engine position.

State Table

The third block performs the computations and error-checking functions of the system.

It consists of logic and memory circuits which process the information provided by the signal conditioning circuit. The relative and absolute crankshaft angular position signals are processed to obtain accurate engine position information, as can for example be employed in implementing a spark ignition strategy.

The inputs to the finite state machine are the clock waveform (CP) and the reference pulses (RP); the outputs can be tailored according to need to provide the required engine position information as well as a number of error-checking and self-diagnostic signals.

A RESET input is connected to the

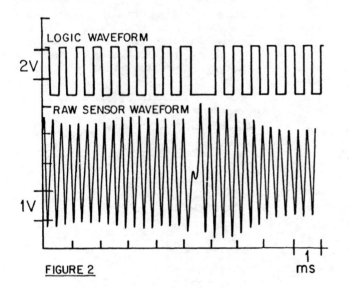

FIGURE 2

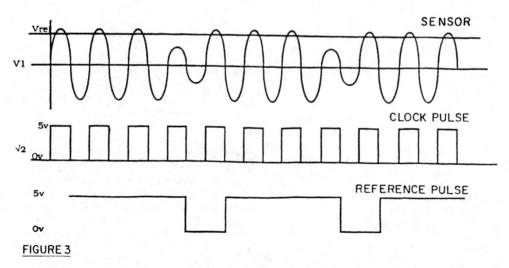

FIGURE 3

ignition circuit to ensure correct initialization of the system on start-up.

The finite state machine can access the unique sequence generated by the CP and RP waveforms in a state table, and uses the clock signal (CP) to cycle through the states as the crankshaft rotates.

Thus, any desired output related to crankshaft angular position can be generated by suitable programming of the finite state machine. For example, a typical output might consist of a reference pulse to mark a fixed baseline value of spark advance angle for each cylinder, to be used in computing the actual ignition timing based on operating conditions.

Operation

A state table stored in the finite state machine represents each increment of engine revolution by one fixed state, and generates a one-to-one correspondence between states and degrees of engine rotation relative to a fixed reference, e.g. TDC position for cylinder number 1.

The duration of each state is determined by the number of zero crossing in the sensor waveform, i.e. by the number of teeth in the ring gear. In this experimental system, we selected to have each state represent one tooth (two zero crossings), i.e. 2.5 degrees (exactly) of engine rotation. Different applications may require varying levels of resolution.

To be emphasized is that the mapping between engine angular position and states in the table is unique, and repeats exactly for each revolution of the crankshaft. For ignition timing applications, it is essential that after initialization (i.e. ignition) the system start up as quickly as possible. Out technique ensures that, on the average, the system is fully operational after at most 1/3 crankshaft revolution.

Figure 4 is a flowchart indicating the sequence of events following the ignition.

1. The system is reset.
   State table "address" is set at zero position. State table awaits first RP. "IN-LOOP" output reads logic 0.
2. First RP, arrives.
   State table starts cycling according to state clock, CP.
3. Second RP is detected
   Based on the (unique) number of clock cycles between RP_1 and RP_2 the state table can compute the exact angular position of the crankshaft.
4. The State Table enters the loop corresponding to normal operation. From this point on, barring any sensor errors, the state table cycles through its states without the need to receive any absolute position information. The only input required is CP.
   At this point the "IN-LOOP" output registers a logic high.
5. The RP signal is continuously monitored to detect any discrepancies between the state table and sensor inputs. This feature enables the implementation of self-diagnostics and error correcting strategies.

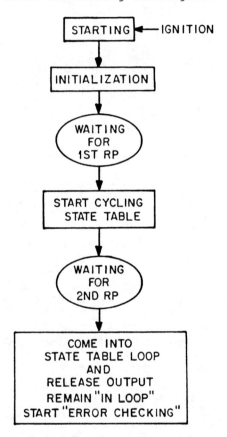

FIGURE 4

Error Correcting Capabilities

The status of the system is at any time defined by one of three conditions: no error, CP error or RP error.

When no error takes place in the inputs provided by the sensor, the output of the state table provides at all times the correct engine position information, thanks to the repeatability of the unique pulse sequence generated by the sensor. If the sensor signal were to fade momentarily and one or more clock pulses were missed (CP error), then an instantaneous discrepancy would be detected between the sensor output and the engine position state table. This

INPUT 1	INPUT 2	DESCRIPTION	RESPONSE
0	0	No modified tooth comes neither expected one	Continue timing table
0	1	No modified tooth comes but we expect one	Miscounting due to mistakes in the tooth detect circuit. System warning !
1	0	Modified tooth comes but we do not expect it	Miscounting . switching forward to the right state
1	1	Modified tooth comes as we expected	Continue timing table

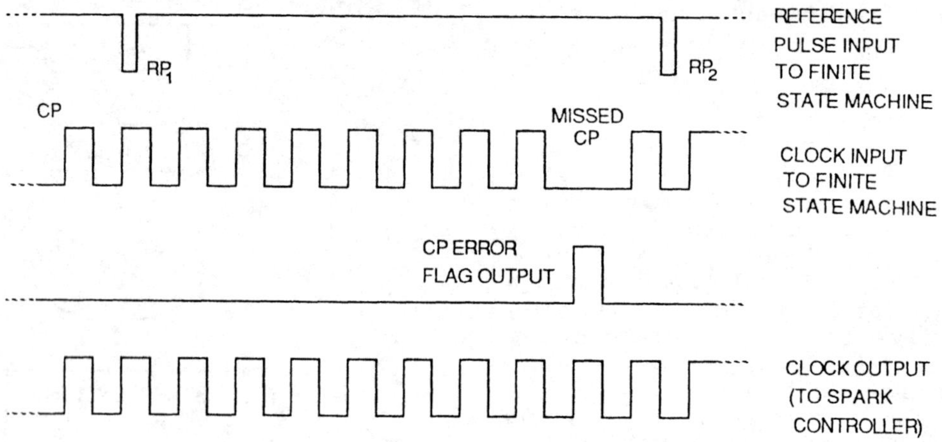

FIGURE 5

discrepancy would be detected between the sensor output and the engine position state table. This discrepancy would be detected at the occurrence of the first RP following the missing CP. The state table would then automatically correct for the missed CP by advancing to the state corresponding to the engine position. Figure 5 depicts the situation which would be encountered if two consecutive CP's were missed during one revolution.

It is important that the signal conditioning circuit be designed for glitch free operation, to prevent the occurrence of an excessive CP count.

If a missed detection occurred in the case of a RP, the engine position output generated by the state table would not be affected since that is determined exclusively by the CP signal once the system is "IN LOOP". However, this could be used as self diagnostic feature, since it might identify a failure in the sensor subsystem.

A false alarm in RP could not at this point be detected in a fail safe manner. In the experimental system we designed the signal conditioning/sensor interface circuit in order to minimize the possibility of a false alarm in the RP waveform. Note however that an occasional error in this signal would cause only one or, at most, two misfires before the finite state machine entered the IN LOOP state again and started cycling correctly.

Experimental Results

The concepts described in the first part of this paper were implemented in practice by duplicating the information provided by an existing commercial crankshaft position sensor system on a production vehicle, equipped with a 3.8 L 6 cylinder fuel injected engine and a distributorless ignition system. The crankshaft position information obtained by means of the state-table based system was used as an input to the ignition timing control module, in place of the existing system.

The vehicle successfully ran under our experimental system. Figure 6 displays oscilloscope photographs of timing waveforms simultaneously obtained form the commercial and the experimental systems. The choice of waveforms was dictated exclusively by the availability of a commercial system which could be compared to our experimental device.

Amongst the potential sources of error limiting the accuracy of such a

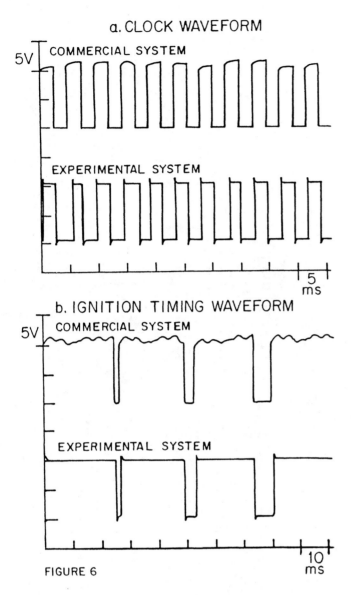

FIGURE 6

position measurement system, we considered primarily those due to the initial positioning of the ring gear on the flywheel and to the variation in the relation between zero crossing and crankshaft position due to tooth-by-tooth variations in ring gear geometry. This last error may be due to aging as well as to imperfections in the ring gear manufacturing process.

The tooth aperiodicity error may be viewed as a pseudo-random process and analyzed as such. It is clearly bounded:

$$|\varepsilon| \le \frac{\delta\theta}{2}$$

where $\delta\theta$ is the mean distance in degrees between zero crossings. In our case, ε_{max} = .625 deg., although in practice the actual error is much smaller.

With current manufacturing technologies we believe it is feasible to obtain an accuracy in the measurement of engine position of the order of a quarter of a degree.

Conclusion

This paper has introduced a high accuracy method of measuring crankshaft angular position in the IC engine.

The method is based on a noncontacting measurement of angular position taken by means of a magnetic sensor placed near the started ring gear on the flywheel.

The sensor output is processed by a finite state machine, which generates digital output signals indicating engine position (referenced to TDC for cylinder number one) in increments of 2.5 degrees.

By suitable programming of the finite state machine, it is possible to generate any desired timing waveform. The system was tested successfully on a production vehicle equipped with distributorless ignition, by replacing the existing crankshaft position sensor system with the experimental system.

In addition to ignition timing application demonstrated in this study, the same measurement of crankshaft angular position can be utilized for measurement of engine performance and for diagnostic applications.

REFERENCES

(1) G. Rizzoni, "A Passenger Vehicle Onboard Computer System for IC Engine Fault Diagnosis, Performance Measurement and Control," in Proc. 37th IEEE Vehicular Technology Conference, Tampa, Fl., June 1-3 1987, pp. 450-457.

(2) W. B. Ribbens "A New Metric for Torque Nonuniformity," SAE Intl. Congress & Exposition, Cobo Hall, Detroit, Mich., Feb. 28 - Mar. 4, 1983, SAE 830425.

(3) N. Henein, "A Diagnostic Technique for the Identification of Misfiring Cylinders," SAE Intl. Congress and Exposition, Cobo Hall, Detroit, Mich.,1987, SAE 870546.

(4) G. Rizzoni, "A Dynamic Model for the Internal Combustion Engine," Ph.D. Dissertation, Department of Electrical Engineering and Computer Science, The University of Michigan, Ann Arbor, February 1986.

(5) W. B. Ribbens and D. Gross, "Torque Nonuniformity Measurements in Gasoline Fueled Passenger Cars Equipped with automatic Transmission - Theory and Experimental Results," SAE Intl. Congress and Exposition, Cobo Hall, Detroit, Mich., Feb. 24-28, 1986, SAE 860414.

CHAPTER 4

VALVE TIMING

Design and Development of a Variable Valve Timing (VVT) Camshaft

Carl A. Schiele
Environmental Activities Staff, General Motors Corp.

Stephen F. DeNagel and James E. Bennethum
Research Laboratories, General Motors Corp.

WHEN THE NECESSITY FOR controlling the exhaust emissions from the internal combustion engine became evident General Motors, like all the others in the industry, investigated the influence of various engine design parameters such as combustion chamber geometry, ignition timing, air-fuel ratio, and valve timing on emissions.

Early efforts in this area were directed primarily at reductions of hydrocarbon (HC) emissions. As engine modifications and exhaust treatment methods achieved significant reductions of this pollutant, the oxides of nitrogen (NO_x) began receiving more attention. Detailed studies of this pollutant made it apparent that NO_x control might be accomplished within the cylinder of the engine since formation took place during the period of high gas temperatures and concentrations were essentially frozen at that time. The published results of several studies indicated that factors which reduced peak gas temperatures in the cylinder also reduced NO_x formation (1-5).* Dilution of the fresh charge was shown to be one of the most effective ways of limiting peak cylinder temperatures. In conventional automotive engines, the products of combustion from earlier engine cycles present a very convenient diluent mixture for this purpose. Dilution with exhaust products has come to be known as exhaust gas recirculation (EGR).

EGR can be accomplished either externally or internally. The external method involves diverting the exhaust gases from some point in the exhaust system back to the intake system. For naturally aspirated engines, the exhaust system operates at a mean pressure above ambient and the intake system has a mean pressure below ambient. As a result, a natural flow of the exhaust gases will take place through a recirculation tube and back into the cylinder along with the fresh charge.

All naturally aspirated engines have some inherent internal

---

*Numbers in parentheses designate References at end of paper.

ABSTRACT

The development of a variable valve timing (VVT) camshaft was initiated as a potential means of controlling exhaust emissions from a spark ignition piston engine. This approach was based on the fact that valve overlap influences internal exhaust gas recirculation which in turn affects spark ignition engine emissions and performance. The design, fabrication, bench tests and engine durability tests of a unit incorporating splines to allow the intake cams to move relative to the exhaust cams is discussed. Preliminary test data from a 350 CID (5700 cm^3) engine fitted with the VVT camshaft are discussed with regard to durability and emissions.

EGR; that is, some portion of the products of combustion remain in the cylinder (residual gas) which mix with the incoming fresh charge before combustion is initiated. The degree to which this dilution occurs is a function of the timing of the inlet and exhaust valve event as well as other engine design and operating parameters. Equal changes in the timing of each valve event, such as would be achieved by displacing the angular position of the camshaft relative to the crankshaft, or increasing the valve overlap period during which both inlet and exhaust valves are open, influences the amount of residual gas. The amount of charge dilution, and consequently some degree of NO_x control, can be accomplished by internal EGR through valve timing changes.

As well as having the potential for reducing NO_x, the variation of valve overlap has a direct effect on engine combustion and power output. Large valve overlap enhances operation at high-speed wide-open throttle (WOT), while low-speed part-throttle operation requires low overlap for smooth running. These interrelationships indicate that a variable valve timing (VVT) device capable of adjusting valve overlap while running may be desirable for optimizing engine operation for both emissions and performance.

Recent automotive publications show that several firms throughout the world are now actively involved in developing hardware to accomplish this. This paper describes GMR's successful design and development of such a device, the VVT camshaft, and presents some preliminary test results obtained from a 350 CID (5700 cm^3) engine equipped with this hardware.

DESIGN

DESIGN CRITERIA - The approach adopted to achieve variable valve timing was to design a multipiece camshaft having the capability of rotating the intake cams relative to the exhaust cams. The decision to advance the intake cams was based on information (1, 4, 5) showing that this was the most efficient means of reducing NO_x by valve overlap. However, the most important point at the time this work was started was to develop a practical means of affecting cam timing changes in an operating engine.

It became apparent that the camshaft would require fewer parts if it were built for use in an engine which had siamesed inlet and exhaust ports. Camshafts for these engines have the cams arranged in the following sequence: two exhaust, four intake, four exhaust, four intake, and two exhaust. This allowed the camshaft to be divided into five segments, with only two intake segments carrying four intake cams each. Therefore, a high production 350 CID (5700 cm^3) engine with siamesed ports was chosen for the initial design studies.

The general design concept was developed to have a central actuating member translatable along the axis of rotation of the camshaft. Sliding this central member in and out would have no effect on the exhaust cams, but would rotate the intake cams and shift their angular position relative to the exhaust cams. This concept is shown schematically in Fig. 1.

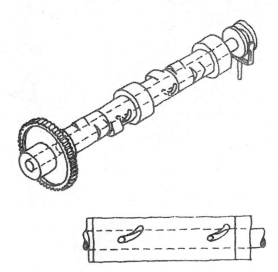

Fig. 1 - General design concept

There were several basic considerations governing the evaluation of the potential designs for such a camshaft:

1. It must be capable of withstanding the maximum loads.
2. It should be capable of the full range of timing variations as determined from piston-valve interference.
3. It should be adaptable to current production engines and fit into the standard camshaft space without major alteration.
4. It should be durable and inexpensive.

POTENTIAL DESIGNS - Several potential design concepts were reviewed.

Pins and Grooves - Pins protruding from the internal member and mating with internal grooves in the cam segments were considered. The exhaust segments would have straight grooves and the inlet segments helical grooves. This idea was rejected because the pins would be highly stressed and subject to fatigue failure.

Ball Splines - A similar arrangement with the pins replaced with circulating balls was considered but this system presented size, cost, and assembly problems.

Splines - By far the most promising design idea involved full splines, a straight spline driving the exhaust cam segments, and a helical involute spline driving the intake cam segments. This design, shown conceptually in Fig. 2, was ultimately chosen on the basis of compactness, strength, and ease of assembly.

LOADING ANALYSIS - An existing valve train analysis was modified to handle the VVT camshaft and used to check the stresses expected in the various components of the system. Given the valve train geometry and inertias, spring data, cam lift or acceleration versus cam angle, and cylinder gas pressure versus cam angle, the program computed the cam lift, velocity, acceleration and curvature, instantaneous rocker arm ratio, pressure angle, and contact stresses. Additional calculations were included to determine the valve-piston clearance, minimum permissible tappet diameter, system forces, and the instantaneous camshaft torque due to the forces created by the springs, inertia, and cylinder gas pressure.

Initially, there was some thought that system friction would be substantial and should be included when determining the maximum camshaft drive torque. No definitive friction data

EXHAUST LOBE SEGMENT   CENTRAL DRIVING MEMBER   INTAKE LOBE SEGMENT   Fig. 2 - Full spline design concept

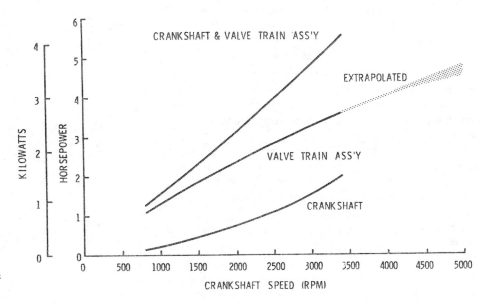

Fig. 3 - Experimental power absorption curves

were available for this system, so special engine motoring tests were run to determine this information. An engine minus pistons, connecting rods, and accessories was motored and the power recorded. The test was then repeated after removing the valve train so that only the crankshaft remained. The difference in absorbed power fixed the average power absorbed by the valve train. A separate bearing analysis was used to determine the proportion of this power that could be attributed to the sliding friction of the valve train components. By neglecting frictional losses in camshaft drive chain and rocker arm pivots, valve train frictional torque loss was determined and assigned to the cam-lifter interface. Using this loss in conjunction with the load analysis, the theoretical dynamic coefficient of friction was determined and used to introduce a conservative influence on the design analysis.

Using the load analysis, modified by the inclusion of friction, the average computed torque to drive the complete valve train was determined to be 59.7 in-lb (6.74 N · m). By comparison, the extrapolated experimental drive torque for the entire valve train ranged 55.4-60.5 in-lb (6.25-6.83 N · m) which corresponds to 4.5-4.8 hp (3.36-3.58 kW) at 5000 rpm as shown in Fig. 3. The contribution due to friction was 30.4 in-lb (3.43 N · m) or 51% of the average drive torque.

Drive torque, as a function of crankshaft angle, was derived for each of the five proposed cam segments and for the total camshaft at the drive sprocket by summing the instantaneous torques according to the timing of each particular cam at various engine speeds. Peak, or maximum, driving torques were found to be 180 in-lb (20.3 N · m) for an exhaust segment and 145 in-lb (16.4 N · m) for an intake segment at 5000 rpm. The peak torque at the drive sprocket for the same engine speed is 187 in-lb (21.1 N · m). Simulated advancement of the intake cams did not influence peak or average driving torque significantly. Therefore, all subsequent analyses, with the exception of the torsional analyses, treated loading for the standard camshaft timing at an engine speed of 5000 rpm.

Two torsional vibration analyses of the engine were then performed to ascertain what resonant conditions might exist in the segmented camshaft. One analysis simulated a source of excitation due to the harmonics of the valve train torque loads and the other simulated the cylinder gas forces on the crankshaft as providing the excitation force. The results indicated that no oscillatory conditions would exist between the camshaft segments. However, they did indicate that a resonant vibration of the timing chain can result in sprocket loading double the normal drive torque, or 187 ± 187 in-lb (21.1 ± 21.1 N · m). The production camshaft drive sprocket is designed to handle this additional load but it was necessary to consider the effect of this loading on the components of the VVT camshaft. The stresses in the splines of the central drive member were, therefore, examined for a maximum torque load of 374 in-lb (42.2 N · m). The stresses in the splines of the intake and exhaust cam segments were calculated for total torque loads of 159 and 196 in-lb (18.0 and 22.1 N · m), respectively. These loads reflect the maximum torsional vibration torque superimposed on the nominal drive torque.

In diagnosing potential spline failure problems there are four types of stresses that must be considered: shear stress in the spline, shear stress in the shaft, compressive stress in the splines, and circumferential tensile stress in the internal spline part.

Using these stresses, the splines were evaluated on the basis of five modes of failure: fracture of the drive shaft beneath the spline, shear at the spline pitch line, fracture at the spline root in a cantilever failure similar to that of a gear tooth, fretting corrosion, and rupture of the segment shell.

This analysis showed that the VVT camshaft could be built to fit into the normal camshaft space in the engine and meet all the loading requirements but one. It appeared that failure would occur in the case of fretting corrosion as compressive stresses on the faces of the splines exceeded the allowable limit, even with the best materials. However, because of the ambiguities involved in using fatigue factors for establishing these allowable stresses, it was decided to proceed with the design and evaluate these loading effects on physical specimens in separate bench tests.

MATERIAL SELECTION - The results of the loading analysis made it necessary to use steel rather than the normal camshaft material for the VVT components. AISI-SAE 4140 was selected for the driveshaft on the basis of its nitriding capability. This permitted case hardening of the external helical splines while maintaining maximum dimensional stability over the entire shaft length. Similarly, AISI-SAE 4620 was specified for the five camshaft segments on the basis of compatibility with interfacing components and machinability of the internal splines. This steel provided necessary strength in the unhardened internal spline core, yet facilitated carburization and hardening of the cam lobes to Rc 52-55.

Chilled iron valve lifters were chosen based upon previous Corporate experience with steel camshafts. The modulus of elasticity of this iron is nearly that of the 4620 steel, a highly desirable quality from the standpoint of cam-lifter face wear and spalling. A cam-lifter materials compatability test was run to substantiate optimum wear resistance. Results showed scuffing of the cam and spalling of the lifter could be minimized in this application by phosphate coating the cam lobes and by addition of 0.5-1.0% zinc dialkydithiophosphate to the engine oil. This additive is common in commercial high quality SE oils.

MANUFACTURE AND TEST

PROCESSING CONSIDERATIONS - The anticipated process control of this multipiece assembly was considerably more complex than that of a single piece production camshaft. Fabrication was to progress in five stages: cam lobes, intake segments, exhaust segments, driveshaft, and complete assembly. Initially, intake and exhaust cam segments would be processed identically. Intermediate steps included normalizing the steel and rough grinding bearing journals and cam lobes to a master. It was during this phase that the reference was to be established for the critical cam and spline timing relationship of each segment. Carburization and hardening of the segments to a depth of 0.020 in (0.05 cm) was next, followed by removal of core hardness to prepare each segment for the inter-

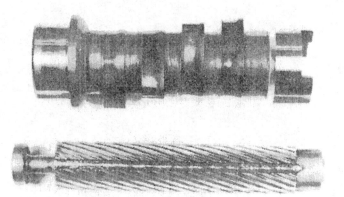

Fig. 4 - Cam test samples

nal splining. While intake cam segments progressed through the helical broaching operation, the internal straight splines and and pilot diameters would be machined into the exhaust cam segments. Simultaneously, the external helical splines and straight spline keyways would be machined into the driveshaft prior to nitriding. At this point, the five segments would be assembled onto the driveshaft according to predetermined timing references and clamped as an assembly for final cam lobe grinding. The final step was to lap and fit internal and external splines. This expensive and tedious operation was necessary so that this articulated assembly would have maximum slip fit with minimum spline clearance.

TEST SAMPLES - Because of the concern for the fretting of the splines, test samples were made and tested prior to fabrication of a complete camshaft. These samples duplicated a single intake segment and driveshaft section. The assemblies followed the same processing path as outlined for the complete camshaft assembly with one exception. The broach tooling required a long lead time and, therefore, a more expeditious means was sought to machine the internal helical splines in the cam segment. An investigation of electro-discharge machining (EDM) processing revealed that the test samples could be produced by this technique although it was expensive and had never been used before to machine helical splines over 4 in (10.16 cm) in length. Some loss in machining accuracy was anticipated due to anode errosion, but it was found that this could be corrected by lapping. Four test samples similar to those shown in Fig. 4 were produced by the EDM process and tested.

BENCH TESTS - A special apparatus was designed and built to bench test the sample segments. Provisions were made to duplicate loading, speed, and axial driveshaft motion that would be encountered in actual engine operation. This equipment (Fig. 5) consists of a camshaft housing and drive system. The housing not only retains the cam segment sample, but also provides equivalent valve train loading through hydraulic valve lifters. The drive system is comprised of a 5 hp (3.73 kW) electric drive motor and hydraulic actuator. The motor drives the test segment through its splined driveshaft while the actuator provides an axial force to the driveshaft to control cam segment relative rotation. This relative rotation, caused by the

Fig. 5 - Apparatus for sample bench tests

helix angle of the splines, simulates the engine valve timing change. A single lubrication system provides continuous oil flow to cam and hydraulic lifters and regulates oil flow to the splines through the driveshaft itself.

The purpose of these bench tests was to ascertain helical spline wear characteristics and to insure that the splines were capable of sustaining typical loads. Durability tests were run at constant speeds up to 2500 rpm, with relative rotation cycling and with various degrees of spline lubrication. The final results of testing two EDM splined segments and eventually four broached spline segments indicated that fretting corrosion was indeed a problem. The evidence of reddish brown iron oxide formation on the splines, lubricated only at assembly, was the result of metal-to-metal contact following intermittent breakdown of the lubricant boundary layer (6). The damaged splines are shown in Fig. 6. However, engine oil, continuously supplied to the splines during operation, was sufficient to support the squeeze film necessary to virtually eliminate effects of fretting corrosion.

Additional testing was conducted using a broached spline test segment and driveshaft with all but three equiangular external splines removed. This modification was made so that the full drive load would be carried by some combination of these three splines alone. After 30 h of testing at 2500 rpm with continuous spline lubrication, none of the splines showed evidence of distress. These results indicated that the steel cam segments might be overdesigned and that reversion to cast iron segments may be feasible from the standpoint of spline strength in the future.

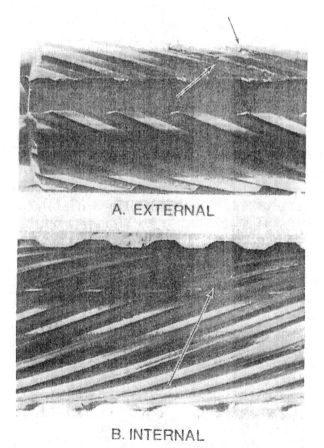

Fig. 6 - Internal and external spline failure

ENGINE-CAMSHAFT ASSEMBLY - With the reassurance that the design was inherently sound, several complete camshaft assemblies were readied for durability testing in engines. Five such assemblies were manufactured, all having the original helical spline specifications of 21 teeth, 7.5 in (120 cm) lead, and 21° 30′ 4″ right-hand helix angle. Each assembly had five specific cam segments fit to a particular driveshaft which precluded interchangeability.

A coil spring provides the force to load the drive sprocket and segments axially in compression against the rear support. Thrust bearings are used both fore and aft to minimize friction. Axial control of the central driveshaft is transmitted through the slide and ball bearing. Minor machining of the rear cam bearing journal was required to facilitate use of the rear thrust bearing.

The actuator used to control cam timing change was hydraulically operated in this test installation. A strain gage load cell (Fig. 7) was inserted between the hydraulic ram and cam actuator shaft for the purpose of sensing axial loading on the cam driveshaft. The gage output signal was amplified and continuously monitored using a digital voltmeter during all subsequent tests.

ENGINE TESTS - A series of engine tests was run to evaluate the VVT camshaft engine mechanically. The primary objectives were to:

1. Compare engine performance with the modified valve train to that of a stock engine.
2. Check the durability of the VVT camshaft components.
3. Determine the forces necessary to rotate the intake cam segments in the running engine.

Initial testing was done with the camshaft set at 0 deg advance which corresponds with the standard or stock timing for this engine. WOT runs at several speeds 1000-4500 rpm resulted in brake torque values that differed from stock engine data by less than 2%. The engine was then run at WOT and part throttle over the entire range of intake cam advance and retard positions. No mechanical difficulties were encountered, and it was therefore concluded that the modified valve train was functioning properly.

The WOT performance data for four intake cam positions are plotted in Fig. 8 as a function of engine speed. Brake torque and power output are generally reduced with either advance or retard from the standard cam timing. However, 20 deg advance produces slightly higher brake torque values at engine speeds below 2250 rpm. At 40 deg advance and low speed the brake torque values approach the standard timing values, but the engine runs rough and there is an occasional misfire.

During the above tests, the intake cam was set and locked in position with the hydraulic actuator inoperative. For the durability tests the actuator continuously rotated the intake cam segments at various frequencies between the 0 and 40 deg advance positions. All tests were run at WOT and MBT spark timing determined for the standard cam position. The engine was tested for a minimum of 5 h at each of five speeds from 2500-4500 rpm. A total of 90 h of operation was accumulated, 7 of which were at 4500 rpm, before the engine was dis-

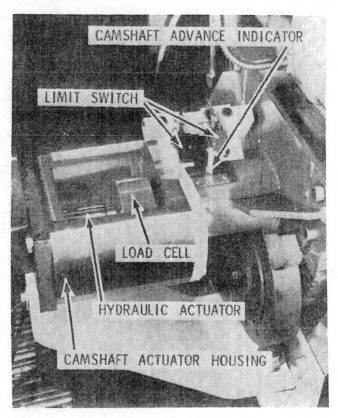

Fig. 7 - VVT camshaft actuator and control

assembled. A detailed examination of all the valve train components showed no evidence of distress and, in particular, the splines were in excellent condition.

The force required to hold or move the central actuating member of the VVT camshaft was continuously monitored with the load cell during all the tests. The purpose of these measurements was to provide information necessary for the design of a control system which could be programmed to adjust the intake cam advance in accordance with other engine operating parameters. The maximum forces measured are plotted in Fig. 9 as a function of the rate at which the intake cam segments were rotated. These data are of interest because they did not match expectations. It was anticipated that there would be a small increase in force as the actuation rate went up since the relative velocity between the parts would increase. However, it was felt that the large forces should be balanced in the multicylinder arrangement by almost equal and opposite effects on other cams. For example, the valve spring compression forces on one cam should be offset by valve spring expansion forces on another. Obviously, this cursory analysis was in error and it appears that large actuation forces would be required if high rates of actuation must be developed by the controller.

With the exception of actuation force behavior, these test results completely substantiated the design analysis and the conclusions drawn from the special bench tests relative to spline fatigue.

PRELIMINARY EMISSION TESTS - Since the VVT design had been shown to be mechanically sound, it was decided to

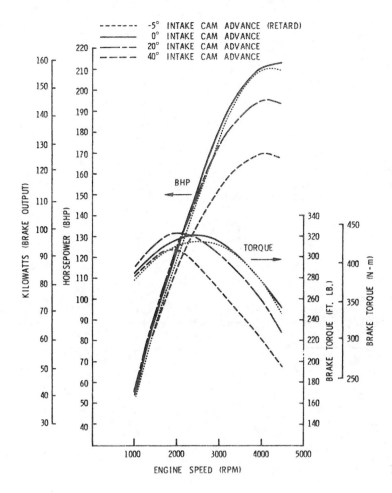

Fig. 8 - VVT engine performance curves

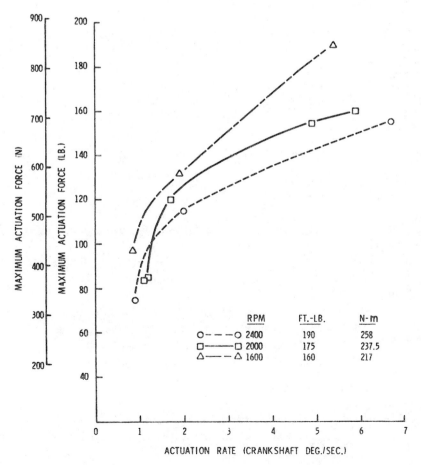

Fig. 9 - Intake cam positioning forces

Table 1 - Cam Advance Influence at Three Load/Speed Conditions

bmep						
psi	42.2		35.5		28.2	
kPa	291		244.5		194.5	
Speed, rpm	1469		1600		1469	
Spark timing, deg btdc*	50		40		50	
Cam advance, deg	0	45	0	45	0	45
bsfc						
lb/hp·h	0.633	0.608	0.656	0.660	0.740	0.746
kg/kW·h	0.384	0.369	0.398	0.401	0.449	0.453
$bsNO_2$						
g/hp·h	9.14	0.81	7.61	0.69	8.16	0.96
g/kW·h	12.23	1.085	10.20	0.93	10.92	12.87
bsHC						
g/hp·h	5.12	4.10	4.93	4.90	6.05	9.57
g/kW·h	6.86	5.50	6.60	6.56	8.11	12.82
bsCO						
g/hp·h	109.0	102.2	103.2	114.9	121.1	121.9
g/kW·h	146.0	137.0	138.3	154.0	162.3	163.2

Table 2 - Back Pressure Effect

bmep						
psi			35.5			
kPa			244.5			
Speed, rpm			1600			
Exhaust pressure						
in Hg	1.0		2.0		4.0	
kPa	3.38		6.76		13.52	
Spark timing, deg btdc	40	50	40	50	40	50
Cam advance, deg	0	45	0	45	0	45
bsfc						
lb/hp·h	0.645	0.636	0.656	0.644	0.656	0.659
kg/kW·h	0.392	0.386	0.398	0.391	0.398	0.400
$bsNO_2$						
g/hp·h	7.36	0.99	7.41	0.95	7.64	0.77
g/kW·h	9.86	1.33	9.93	1.27	10.23	1.03
bsHC						
g/hp·h	4.74	4.42	4.78	4.49	4.93	4.63
g/kW·h	6.35	5.93	6.40	6.01	6.60	6.20
bsCO						
g/hp·h	105.5	103.2	106.3	103.5	103.2	109.2
g/kW·h	141.3	138.3	142.7	138.7	138.3	146.3

proceed with some preliminary emission tests. The same engine was fitted with a special manually adjustable carburetor in order to maintain the air-fuel ratio constant at 13:1. This richer than normal mixture was used to maintain good combustion at all test conditions with the large amounts of charge dilution anticipated. Several load-speed combinations were chosen to represent operation in the range of the 1972 light-duty vehicle federal exhaust emission test and were similar to test conditions for other experimental low emission engines being run at the time. Preliminary results are presented in Tables 1 and 2 which show the influence of load and exhaust back pressure, respectively, at standard and advanced intake cam timing. The 0 deg cam advance represents the standard timing condition.

The following general conclusions can be drawn from these preliminary emission test results:

1. Mass emissions of NO_x were substantially reduced from standard valve timing values with 45 deg advance of the intake cam.
2. Mass emissions of HC and CO and the bsfc values were essentially unaffected by intake valve timing changes.

These conclusions are all in agreement with the results of the single cylinder engine studies of Siewert (1).

STATUS

This work has demonstrated that a VVT camshaft can be designed to fit into the space now occupied by a standard camshaft in a high production V-8 engine. Engine tests indicate that the GMR camshaft has acceptable durability and that controlling internal EGR with intake valve timing may be an effective means of reducing NO_x emissions. However, the rich 13:1 air-fuel ratio, partially responsible for these low NO_x values, also accounts for the relatively high HC and CO levels.

When examining the possibilities of a future engine control package including the VVT concept, the following factors must be given further consideration.

MATERIALS AND PROCESSING - The original design analysis resulted in the choice of steel for the camshaft components which made it necessary to use complicated and expensive processing. Bench tests have shown that the unit may be overdesigned but whether normal camshaft materials can be used and economical processing techniques developed remains to be shown.

CONTROLS - More attention is needed in the design and development of VVT controls. In the ultimate arrangement, it may be desirable to adjust the exhaust valve timing relative to the crankshaft as well as the intake timing relative to the exhaust. These ideas must be given further consideration; in any case, it is apparent that the controls required will be sophisticated and therefore could be expensive and large.

ADDITIONAL TESTING - Finally, it must be emphasized again that the emission results presented here are only preliminary. Further extensive testing to substantiate these conclusions as well as evaluating the performance and fuel economy effects of VVT must be carried out.

For these reasons the VVT camshaft is best described, at its present state of development, as a tool to be used for further engine research and development.

ACKNOWLEDGMENTS

During a program of this magnitude a considerable number of people contribute to the ultimate achievement of the project goals. The following individuals deserve special mention for their assistance in making the VVT camshaft a reality: J. R. Burns and W. J. Norkus for their contributions to the design and fabrication; P. Vernia of the Metallurgy Department for his materials and processing recommendations and for performing the materials compatibility tests; C. Goohs for his thorough torsional analysis; A. O. DeHart for his lubrication suggestions and assistance with the bearing friction analysis; J. B. Feiten for running the emissions tests; R. M. Siewert for his editing help; and E. Scheibe of the Diesel Equipment Division for providing the special valve lifters required. To all of these and the many others who were associated with this effort, we express our sincere appreciation.

REFERENCES

1. R. M. Siewert, "How Individual Valve Timing Events Affect Exhaust Emissions." SAE Transactions, Vol. 80 (1971), paper 710609.

2. A. A. Quader, "Why Intake Charge Dilution Decreases Nitric Oxide Emission From Spark Ignition Engines." SAE Transactions, Vol. 80 (1971), paper 710009.

3. R. M. Campau, "Low Emission Concept Vehicles." SAE Transactions, Vol. 80 (1971), paper 710294.

4. M. A. Freeman and R. C. Nicholson, "Valve Timing for Control of Oxides of Nitrogen (NO_x)." Paper 720121 presented at SAE Automotive Engineering Congress, Detroit, January 1972.

5. G. B. K. Meacham, "Variable Cam Timing as an Emission Control Tool." SAE Transactions, Vol. 79 (1970), paper 700673.

6. C. Lipson, "Wear Considerations in Design." Englewood Cliffs, New Jersey: Prentice-Hall, 1967, pp. 99-109.

Variable Valve Timing as a Means to Control Engine Load

P. Nuccio

SUMMARY

This work is the prosecution of a theoretical research which proposed a method to control the engine load by means of a variable valve timing. As can be predicted, the early closing of the inlet valve allowes to lower the engine load with reduced air-throttling, so that a fuel economy can be obtained.
A hydraulic system was designed to perform the variable valve timing as follows. A rotating distributor, controlled by the camshaft, empties a capacity in the timing system, so the free closing stroke of the valve is determined by its spring. A particular device virtually eliminates the valve impact against its seat.
The hydraulic system was applied to a single cylinder engine and bench tests were carried out so as to allow a comparison between the engine performances obtained by air-throttling and those obtained by variable valve timing. Preliminary experimental results are reported and discussed.

INTRODUCTION

The possibility of improving the Otto engine fuel consumption at part loads has been studied since the beginning of this century (1)[*] in order to get near the control characteristic of the Diesel engine.
For this purpose either stratified charge engines with reduced air-throttling, even without throttle valve, (1,2,3) or particular devices in the place of the throttle valve have been developed in order to lower the losses of the engine charge replacement.
At the moment the first method (i. e. the charge stratification) is more common than the second, but satisfactory results have not yet achieved. With regard to the second method, the most important obstacle for its practical application is the complexity of these devices, which change the usual timing system: in fact this part of the engine has often been modified. Really there are many mechanisms: with a variable clearance between the cam and the tappet, with the variable valve lift by means of hydraulic tappet (4), with tridimensional cams (5), with variable cam track, with oscillating cams (6), etc. . In this category all mechanical solutions, which change the engine timing, can be included, even if their purpose is not to control the engine load, but, e.g. it is the reduction of the exhaust pollutants (7,8, 9), in particular nitric oxides and unburnt hydrocarbons, and the optimisation of the volumetric efficency. Therefore the prosecution of research work on this subject could be useful, because there is the possibility of improving the engine timing system, which in the present configuration seems to be at the maximum efficiency limit.

OBJECT OF RESEARCH INVESTIGATION

In a previous paper of E. Antonelli and G. Colasurdo (10) a method of controlling the Otto engine load through the early closing of intake valve together with the arrangement for the corresponding hydraulic system has been proposed. Furthermore the obtainable improvement in fuel economy has been calculated: it could be of the order of some percent. Figure 1 shows a typical diagram of the four-stroke inlet and exhaust process. The throttled cycle is compared with that obtained by the early closing of intake valve, so that it is possible to appraise qualitatively the reduction of the work during the inlet stroke.
This method is preferable to the retarded closing of the valve, because the intake air is premixed with the fuel. In fact, after the bottom dead center, the charge would be expelled with further enrichment in the carburation zone. Therefore the research work has been continued in order to terminate the design of this hydraulic system, to build it and finally to determine by some experimental tests the real fuel economy, when the air-throttling is reduced or entirely avoided. This design has been developed according to the least complexity as far as possible, because in this way a greater reliability is often obtained.

[*] Number in parentheses designate references at end of paper.

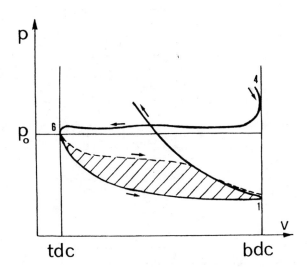

Fig. 1 Four-stroke inlet and exhaust process.
Solid line: cycle obtained by throttle valve.
Dashed line: cycle obtained by early closing of intake valve
Dashed area: pumping work reduction (point 1 being equal).

HYDRAULIC DEVICE

Figure 2 shows the hydraulic device to achieve a variable valve timing for an engine with valve push rods and rocker arms; the working principle is as follows:

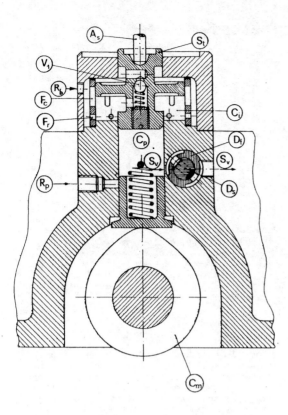

Fig. 2 Hydraulic device for the variable valve timing.

the capacity C_p is filled with oil before the intake stroke by means of a reinstatement duct R_p provided with an unidirectional valve. This capacity can empty quickly during the period of virtual valve event, through the emptying duct S_v connected with the crankcase (the oil is the same used for the engine lubrication, SAE 30). In this case the free closing stroke of the intake valve is determined by its spring. The emptying is achieved by means of the rotating distributor D_s controlled by the camshaft C_m with an unitary gear ratio. The distributor is furnished with an oblique hole, which means that the capacity C_p is emptied only every 720° of crankshaft rotation. Whenever the rotating distributor lowers the engine load, it is necessary to refill the capacity C_p, in order to restore the oil which has been lost during the engine control and due to undesired leakages.

Valve Braking

When the capacity C_p empties, the cam will not control the closing stroke of the inlet valve, therefore it is necessary to eliminate the impact against its seat. For this purpose there are some holes, F_c and F_r, on the cylinder C_i, which the little piston S_t slides into; moreover the push rod A_s of the intake valve leans on this piston (the cylinder C_i is also refilled by a proper duct R_b). The largest holes F_c are covered progressively in order to give a corresponding braking effect and at last moment the small holes F_r follow. In fact the braking effect should be neither too weak, otherwise the valve damages itself and the seat, nor too strong because in this case there is an undesirable air-throttling between the valve and the seat. Therefore the valve closing should be very quick up until the seat, then the braking effect becomes intense to avoid the valve impact. Figure 3 shows this free motion of inlet-valve at three engine speeds and for two different crank angles.

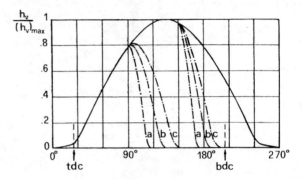

Fig. 3 Typical inlet-valve lift curve (solid line) and free closing curves (dashed lines).
a) 1000 RPM; b) 2000 RPM; c) 3000 RPM

There is also a ball check valve V_1 in the piston S_t, in order to avoid the braking effect during the valve lift. Obviously for this hydraulic tappet is not necessary a clearance adjuster, nevertheless the usual system in the rocker arm remain to put the piston S_t in the best position for the final braking.

Variable valve timing

Of course the moment when the capacity C_p empties can be changed. Figure 4 shows the particular device, a "fixed distributor" D_f, which allowes the timing control between the emptying of the capacity C_p and the camshaft. In fact the fixed distributor has two splines S_c on the outer part, besides the

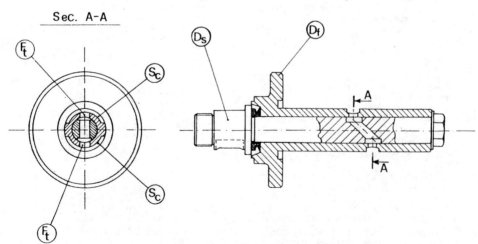

Fig. 4 Rotating (D_s) and "fixed" (D_f) distributors.

necessary radial holes F_t. These splines are always in comunication with the emptying duct S_v; therefore with a rotation of the fixed distributor by the proper drive the holes F_t are angularly shifted and consequently the angular position of the rotating distributor which empties the capacity C_p. The fixed distributor can operate in 200° of crankshaft rotation, whereas the total valve lift occurs in 240°; this fact allowes a wide range for the engine control.

ENGINE CHARACTERISTICS

The hydraulic system has been applied to a single cylinder four stroke engine (Lombardini La 490), which is fitting for fixed users with restricted power requirements; the engine data are reported in table I. In this engine the tappet's case has a suitable position, so that it was easy enough to carry out the necessary modifications in order to install the hydraulic device. Figure 5 shows the engine after the modifications: these are mainly the camshaft replacement, in order to control the rotating distributor with a gear pair and to change the cam ramp lenght; the suddivision of the crankcase-cover to install in the new upper part the hydraulic tap-

TABLE I

ENGINE DATA

Piston Stroke	80 mm
Cylinder Bore	88 mm
Displacement	487 cm^3
Compressio Ratio*	6 : 1
Max. Ang. Speed	3000 RPM
Max. Power	8.8 kW
Max. Torque	32 Nm
Cooling System	by air

pet and the two distributors, which must be near the capacity C_p for a rapid emptying. As a the longitudinal section of fig. 5 shows, the engine is not provided with the lubricating gear pump, but has a centrifugal device only; therefore an external oil circuit, see fig. 6 for the hydraulic tappet and valve braking has been introduced. The lack of the oil pump together with the constant spark advance (24° of crankshaft rotation) reduces the

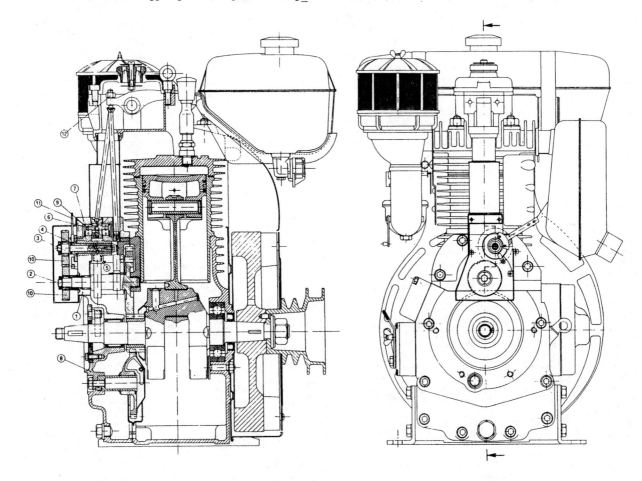

Fig. 5 Engine after the modifications to install the hydraulic device
1 - Tappet's case
2 - Camshaft
3 - Rotating distributor
4 - Fixed distributor
5 - Inlet valve lifter
6 - Braking piston of the inlet valve
7 - Tappet's case-cover
8 - Crankshaft case-cover
9 - Braking cylinder
10 - Gear pair
11 - Drive of the fixed distributor
12 - Valve train clearance adjuster

* At the present this engine has a low compression ratio, but it will increase after these preliminary tests.

permitted range of angular speed of this engine considerably.

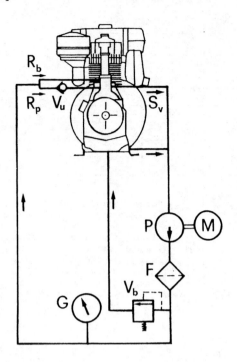

Fig. 6 External oil circuit for the hydraulic tappet.
P - Gear pump
M - Electric motor
F - Oil filter
G - Pressure gauge
V_b - By-pass valve
V_u - Ball check valve
R_p - Reinstatement duct of the tappet
R_b - Reinstatement duct of the valve brake
S_v - Emptying duct

PROCEDURE OF EXPERIMENTAL TESTS

At first bench tests were carried out to tune up particularly the valve braking. In fact, while reducing engine load, a weak braking effect is noted from the valve knocking against its seat and consequently adjusted by the taking up system; on the other and a strong braking effect increase the inlet valve closing delay (the valve spring can not exceed the braking and inertial forces due to the cam contour) and modifies also the maximum torque characteristic or, at least, causes an undesirable air-throttling near the inlet valve; so the valve braking has been located in the precise position, in which starts to disappear the valve knocking. In any case the two engine performance characteristics at full load (i.e. brake mean effective pressure versus engine speed) for the unmodified engine and after installation of the hydraulic tappet have been carried out and compared (see fig. 7); actually the two characteristics are very close. However some problems arise : the speed range has been more restricted, because exceeding 2500 RPM strong vibrations appeared, due to low balance of this engine; the fuel consumption was very high and there was a high scattering in the tests data. Therefore the engine has been fitted with a new more suitable carburetor, so the fuel consumption decreased from 400 to 300 g/kW h (see fig. 8) and an excellent repeatability of the experimental results was obtained.
The direct comparison of the two control methods has been performed by means of control characteristics (at wide open throttle with the hydraulic tappet). In order to avoid the undesirable effects on the fuel consumption of the air-fuel ratio change, it was maintained nearly constant during the engine control ($\pm$ 1% of the full load point, approx. A/R=14).

This was made possible by changing the idling and the main carburetor jets; otherwise the air-fuel ratio diminished with both the throttle valve and with the hydraulic tappet (less with the latter).

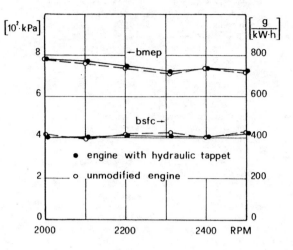

Fig. 7 Engine performance characteristics at full load versus engine speed (current carburetor).

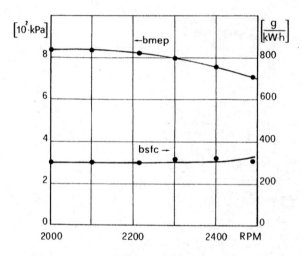

Fig. 8 Engine performance characteristic at full load versus engine speed; engine equipped with hydraulic tappet (new carburetor).

EXPERIMENTAL RESULTS

In the figure 9, 10 and 11 the control characteristics at three different angular speeds are compared. The brake mean effective pressure (bmep) and the brake fuel consumption (bsfc) are related to the maximum bmep and to the respective fuel consumption. In figure 9, at 2000 RPM, the two methods are equivalent up to 70% of the maximum bmep, then the characteristic obtained with the hydraulic tappet improves as regards that obtained with the throttle valve. This improvement in the fuel economy is of the order of 5% at medium loads and of 11% at low ones. Sometimes engine instability appeared at very low loads (20% of the max. bmep), when the engine was controlled by the hydraulic tappet. Probably the air speed through the carburetor was too slow for a sufficient homogeneous carburation; in fact there was no turbulence in the air flow, the throttle valve being wide open. Really with a small part-throttle (10°÷15°) the engine instability was reduced or disappeared completely; obviously when the engine instability appeared, the test result has been eliminated.

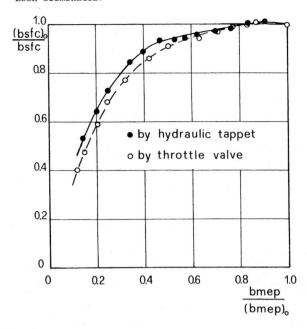

Fig. 9 Engine control characteristics at 2000 RPM

Figure 10 shows the same characteristics at 2200 RPM; the fuel economy is a little better, of order of 6 ÷ 12% at the same loads; in this case the engine instability appeared again at low loads.

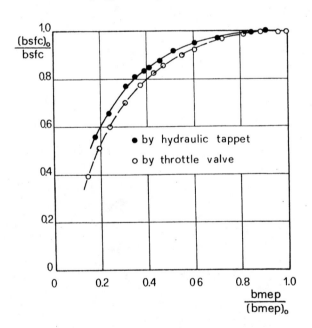

Fig. 10 Engine control characteristics at 2200 RPM

The situation at 2400 RPM (see fig.11) is quite different, because the two characteristics are not too far apart, on the average of 2% at medium loads and 6% for low ones. Evidently air-throttling occurs near the valve, because at this angular speed the spring can not close the inlet valve quickly enough; however in this case there was no engine instability.
At the end of the experimental tests the engine has been dismounted and checked, in order to verify certain parts, in particular the conditions of the intake valve and of its seat. no particular wear was noted after these preliminary tests, even if the total number of working hours was considerable: about one hundred (including the first cheking tests).

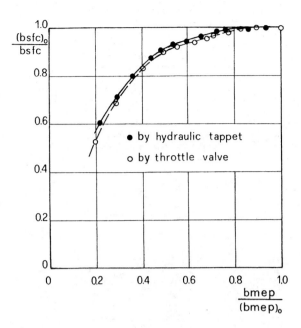

Fig. 11 Engine control characteristics at 2400 RPM

CONCLUSIONS

The hydraulic device made it possible to test the engine performance when the throttle valve is avoided and replaced with the early closing of the inlet valve. At the moment the reduction in fuel consumption is of the order of 5 ÷ 6% at medium loads and of 11 ÷ 12% at low ones; in fact from the maximum bmep to 70% of it, the two methods are equivalent. Two incoveniences have been found with early closing of the intake valve: engine instability at low loads, resulting from difficult carburation owing to the slow air speed through the carburetor, and air-throttling near the inlet valve, when the engine speed increases, this fact reduces fuel economy. These first results encourage to continue the research work, however during the future systematic tests some modifications will be necessary: in particular an injection system replacing the carburetor, in order to eliminate the first inconvenience, and a different type of valve spring which allows a rapid valve closing. A further improvement could be obtained by modifying the valve brake, in order to eliminate its effect (even if moderate), while the valve is opening.

REFERENCES

1. J. H. Weaving Stratified Charge Engines. Status Report C. E. C., February 1975.
2. W. U. Roessler, A. Muraszew. Evaluation of Prechamber Spark Ignition Engine Concepts. EPA Report N. 650/2-75-023, February 1975.
3. E. Antonelli. Un nuovo motore a carica stratificata in camera aperta: Definizione e scelta dei principali parametri geometrici della camera di combustione e degli apparati di alimentazione e di distribuzione. Pubblicazione N. 209, Ottobre 1978, Istituto di Macchine e Motori per Aeromobili, Politecnico di Torino.
4. H. J. Bruins. Variable Valve Timing for Cruise Efficiency. Automotive Design Engineering, January 1968.
5. A. Titolo. Vantaggi di una camma a fasatura variabile e soluzione meccanica per realizarla. ATA, Ottobre 1971.
6. G. Torazza. Distribuzione per motori alternati

vi ad alzata ed angolo di apertura delle valvole variabili. ATA, Novembre 1972.
7. G. B. Kirby Meacham. Variable Cam Timing as an Emission Control Tool. SAE, Paper 700673.
8. R. M. Siewert. How Individual Valve Timing Events Affects Exhaust Emissions. SAE, Paper 710609.
9. M. C. Paulmier. Distribution Variable pour Moteur a Combustion Interne a Distribution par Soupape. Effets sur les Performances et les Emissions de Gaz Polluants à l' Echappement. Ingenieurs de l'Automobile, Jiun 1974.
10. E. Antonelli, G. Colasurdo. Contributo al miglioramento della caratteristica di regolazione dei motori ad accensione comandata. Pubblicazione N. 165, Luglio 1974, Istituto di Macchine e Motori per Aeromobili, Politecnico di Torino.

850074

Effects of Intake-Valve Closing Timing on Spark-Ignition Engine Combustion

Seinosuke Hara,
Yasuo Nakajima
and Shin-ichi Nagumo
Nissan Motor Co., Ltd.

ABSTRACT

In spark-ignition engine pumping loss increases and fuel economy decreases during partial load operation. Methods to reduce this pumping loss by controlling the intake-valve closing timing are currently under study. The authors, also, have confirmed that pumping loss can be reduced by controlling the amount of intake air-fuel mixture through making changes in the intake-valve closing timing.

However, when pumping loss was reduced by controlling intake-valve closing timing, an improvement in fuel economy equivalent to the reduction in pumping loss was not obtained.

In this study, it was found that the major contributing factor to this phenomenon was the deterioration of the combustion, namely, increase in combustion duration and in combustion fluctuation. Therefore, an analysis was conducted on how the various factors which influence the combustion, such as the gas temperature, pressure and residual gas fraction in the cylinder during the compression stroke, change when the intake-valve closing timing is modified. As a result of this analysis, which was carried out through experiments and simulation based on computations, it was found that the principal cause of the combustion deterioration was the drop in cylinder gas temperature and pressure which was traced to a decrease in the effective compression ratio.

INTRODUCTION

The load of a spark-ignition engine is regulated by controlling the intake pressure through throttling, which regulates the amount of the air-fuel mixture drawn into the cylinders. This leads to an increase in pumping losses during partial-load operation, however, which causes fuel economy to drop. Pumping losses tend to increase with lighter operating loads, and when cruising at 40 km/hr under a road-load condition, fuel economy decreases by about 12%.

Various studies have been done on schemes for reducing pumping losses and improving fuel economy. For instance, instead of regulating the intake pressure through throttling, methods have been researched for controlling the length of the intake stroke (1, 2)* and for regulating the intake-valve closing timing (3, 4, 5). However, in previous investigations where such methods were applied for reducing pumping losses, a thorough analysis was not done of other factors that also have a significant bearing on fuel economy.

In this study, attention has been focused on combustion as one of main factors influencing fuel economy. An attempt has been made to clarify how combustion phenomena would change when the intake-valve closing timing was utilized to reduce pumping losses. The factors producing changes in combustion phenomena have been analyzed both experimentally and by simulation.

EXPERIMENTAL APPARATUS AND PROCEDURE

The test engine used in this study was a water-cooled, 4-in-line, carbureted engine, the main specifications of which are given in Table 1. As illustrated in Fig. 1, a splined camshaft and cams were employed, and the intake-valve closing timing alone was varied in a step-by-step fashion, rather than varying it continuously. The exhaust valve timing and the opening timing of the intake valves were not varied. The numbers noted in this figure under each cam represent the crank angle between the intake-valve closing timing and the bottom dead center of the intake stroke. The intake valve lift obtained with each cam is shown in Fig. 2 together with the respective valve timing.

The cams were divided into two groups depending on whether the intake valves closed before or after the bottom dead center of the intake stroke.

---

* Numbers in parentheses designate references at end of paper.

0148-7191/85/0225-0074$02.50
Copyright 1985 Society of Automotive Engineers, Inc.

An "early intake-valve closing," abbreviated as early IVC, was obtained with the cams in the former case and a late IVC was obtained with those in the latter case. For example, "early IVC at 60°" means the closing timing of the intake valves coincided with a 60° crank-angle position before the bottom dead center of the intake stroke.

Table 1 Test Engine Specifications

TYPE	WATER COOLED 4-STROKE
NUMBER OF CYLINDERS	4 (IN-LINE)
BORE × STROKE	85 × 78 mm
DISPLACEMENT	1,770 cc
COMPRESSION RATIO	8.5 TO 1
NUMBER OF SPARK PLUGS	2/CYLINDER

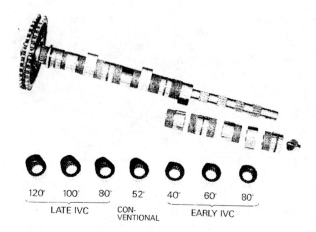

Fig. 1 Splined Camshaft and Cams

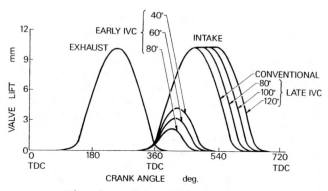

Fig. 2 Valve Lift Diagrams

The experimental apparatus is illustrated schematically in Fig. 3. Cylinder pressure measurements were taken with a piezoelectric pressure transducer attached to the No. 4 cylinder and the recordings were analyzed statistically.

The EGR rate, residual gas fraction (RGF) and air motion in the No. 4 cylinder were also measured. The following equations were used in calculating the EGR rate and RGF.

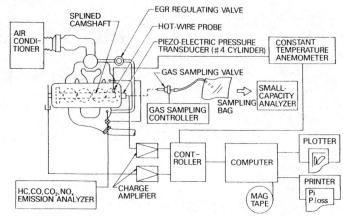

Fig. 3 Schematic Diagram of Experimental Apparatus

$$\text{EGR rate} = \frac{\text{EGR gas volume}}{\text{Intake air volume}} \times 100$$

$$= \frac{[CO_2]i - [CO_2]o}{[CO_2]e - [CO_2]i} \times 100 \; (\%) \ldots\ldots (1)$$

$$\text{RGF} = \frac{\text{Residual gas volume}}{\text{Intake air volume}} \times 100$$

$$= \frac{[CO_2]c - [CO_2]o}{[CO_2]e - [CO_2]c} \times 100 \; (\%) \ldots\ldots (2)$$

where

$[CO_2]i$: CO_2 concentration in the intake manifold (with EGR)

$[CO_2]o$: CO_2 concentration in the intake manifold (without EGR)

$[CO_2]c$: CO_2 concentration in the cylinder gas before ignition

$[CO_2]e$: CO_2 concentration in the exhaust manifold

Air motion in the cylinder was measured with a hot-wire probe at the spark gap on the intake side. The probe output was calculated using a stationary time-averaged method (6).

Furthermore, a simulation model was employed to compute the gas temperature in the cylinder and the residual gas fraction. In this model, a quasi-steady state was adopted for the intake system disregarding the non-steady effect and the intake manifold was treated as an infinite plenum (3, 7).

CHANGES IN COMBUSTION PHENOMENA ACCOMPANYING A PUMPING LOSS REDUCTION

PUMPING LOSS REDUCTION - Figure 4 shows a pressure-volume diagram of the cylinder pressure that was measured when the early-IVC-60° cam was used and throttling was performed to obtain a brake torque of 29.4 Nm at an engine speed of 1,400 rpm. The solid line in the figure shows the result for the early IVC at 60° and the broken line represents a conventional IVC under the same operating conditions.

For the early IVC at 60°, the cylinder pressure was higher during the first half of the intake stroke than for the conventional IVC. Subsequently, the pressure dropped as the closing timing of the intake valve approached. Following valve closing the drop in cylinder pressure was nearly consistent with adiabatic expansion and at the bottom dead center of the intake stroke it was lower for the early IVC at 60° than for the conventional IVC. As the screen-tone indicates, pumping losses produced by cylinder pressure variation during the intake stroke have been reduced with the early IVC at 60° in comparison with the conventional IVC.

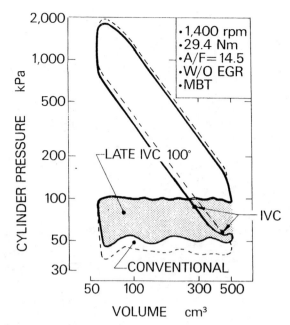

Fig. 5 Pressure-Volume Logarithmic Diagram for Late Intake Valve Closing

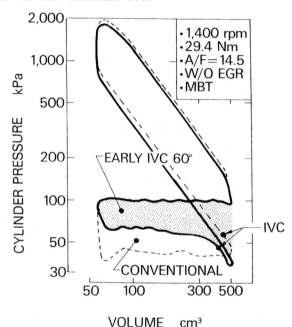

Fig. 4 Pressure-Volume Logarithmic Diagram for Early Intake Valve Closing

Figure 5 shows a comparison between a late IVC at 100° and the conventional IVC under the same operating conditions as for the early IVC at 60°. Throughout the entire intake stroke the cylinder pressure was higher for the late IVC at 100° than for the conventional IVC. As the screen-tone shows, pumping losses have been reduced with the late IVC at 100°.

The pumping losses for the other cams besides the two mentioned above were also investigated and Fig. 6 presents a comparison of the variation in pumping losses for all the cams used in this study. As this figure shows, a greater reduction in pumping losses was achieved as the closing timing of the intake valves moved farther away from the bottom dead center of the intake stroke. In comparison with the conventional IVC, pumping losses have been reduced by 30% with the early IVC at 60° and by 24% with the late IVC at 100°.

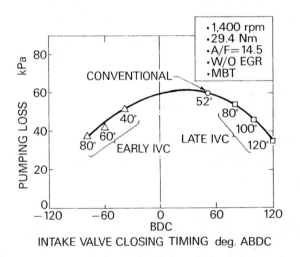

Fig. 6 Relationship Between Intake-Valve Closing Timing and Pumping Loss

As the next step, the indicated specific fuel consumption (ISFC) was investigated for various intake-valve closing timings and the results are presented in Fig. 7. In Figs. 7, 8 and 9, estimated ISFC values are also shown. These values were estimated by assuming that the pumping loss reduction achieved by varying the intake-valve closing timing (Fig. 6) would translate into an improvement in power output of an equal amount. In making these estimations it was presupposed that the combustion duration, cooling loss and other factors besides pumping losses which affect the ISFC would not vary.

From Fig. 7 it can be seen that there was a slight difference (screen-tone portion) with an early or a late IVC, but for the most part the improvement in the ISFC coincided with the estimated values. With an early IVC at 80°, a 7%

improvement has been achieved as compared with the conventional IVC.

Figure 8 shows the results of an investigation on pumping losses and the ISFC for the early IVC at 60° when EGR was used to reduce NOx emissions. Although there was no change in the pumping loss reduction achieved with the IVC at 60°, the difference (screen-tone portion) between the measured and the estimated ISFC values became larger when NOx emissions were lowered through the application of EGR.

Investigation results for the late IVC at 100° are presented in Fig. 9. Similar to the results for the early IVC at 60°, there was no change in the pumping loss reduction when EGR was used to reduce NOx emissions, but the difference (screen-tone portion) between the measured and the estimated ISFC values became larger.

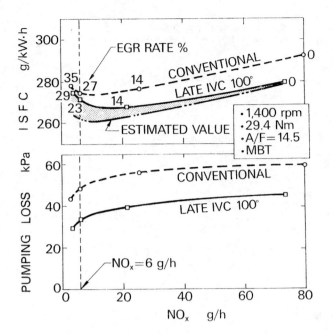

Fig. 9 Pumping Loss and ISFC for Late IVC at 100° When Nox Was Reduced

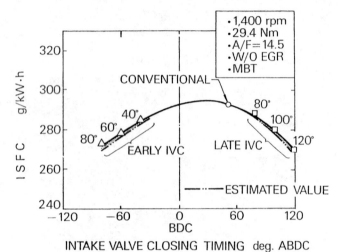

Fig. 7 Relationship Between Intake-Valve Closing Timing and Indicated Specific Fuel Consumption (ISFC)

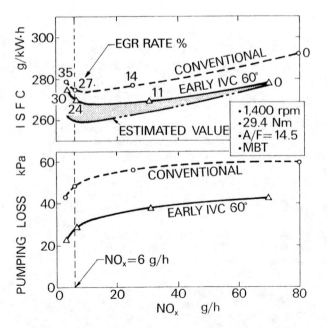

Fig. 8 Pumping Loss and ISFC for Early IVC at 60° When Nox Was Reduced

CHANGES IN COMBUSTION DURATION AND PI FLUCTUATION RATE - As explained earlier, the estimated ISFC values were calculated on the assumption that the factors exerting influence on ISFC besides the pumping losses would not vary. In actuality, though, varying the intake-valve closing timing produced various changes including variations in the effective compression ratio and air motion in the cylinder. The effects of those changes on combustion were presumed to have caused the differences noted in the foregoing section.

The delay period and the main combustion duration (MCD) were investigated for the early IVC at 60°, the late IVC at 100° and the conventional IVC when NOx emissions were reduced using EGR, and the results obtained are presented in Fig. 10. The delay period was defined as the interval from ignition to a burnt mass fraction of 10% and the MCD as the period from 10% to 90%.

From Fig. 10 it can be seen that both the delay period and the MCD were longer for the early IVC at 60° and the late IVC at 100° than for the conventional IVC. A comparison made at a NOx level of 6g/hr revealed that the delay period and the MCD were 23% and 33% longer, respectively, for the early IVC at 60° than for the conventional IVC. Moreover, in comparison with the delay period and the MCD for the late IVC at 100°, the delay period of the early IVC at 60° occurred shorter and the MCD was longer.

As the EGR rate was increased to lower the NOx level further, the difference between the delay period and the MCD for both the early IVC at 60° and the late IVC at 100° and those for the conventional IVC grew larger. This trend was virtually the same for both the delay period and the MCD and it corresponded well with the tendency shown in Figs. 8 and 9 for an increasing difference between the estimated ISFC values and the measured ones.

In order to investigate the engine stability, the rate of fluctuation Cpi in the indicated mean effective pressure pi was calculated using the following expression:

$$C_{pi} = \frac{\sigma_{pi}}{\overline{p_i}} \times 100 \ (\%) \quad \ldots \ldots \ldots \ldots \ldots (3)$$

where
σ_{pi} : standard deviation of pi
$\overline{p_i}$: mean pi

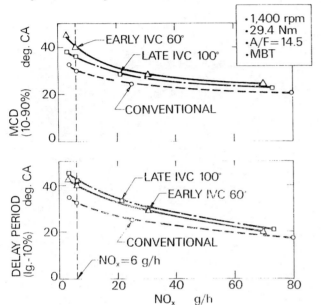

Fig. 10 Comparison of Delay Period and Main Combustion Duration (MCD)

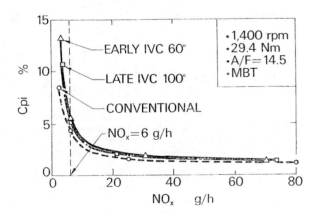

Fig. 11 Comparison of Pi Fluctuation Rates

From Fig. 11 it can be seen that for both the early IVC at 60° and the late IVC at 100° Cpi was larger than for the conventional IVC. Furthermore, this difference increased as the NOx level was further reduced. This trend matched the one mentioned earlier for the combustion duration and it agreed well with the tendency shown in Figs. 8 and 9 for an increasing difference between the estimated and the measured ISFC values.

This increase in combustion duration and in the pi fluctuation rate was concluded to be one reason why an ISFC improvement corresponding to the pumping loss reduction was not obtained even though pumping losses were reduced by varying the intake-valve closing timing.

FACTORS AFFECTING COMBUSTION

CYLINDER GAS TEMPERATURE AND PRESSURE – Figure 12 shows the results of an investigation of the cylinder gas temperature and pressure when NOx emissions were reduced and there was a large difference between the estimated and measured ISFC values, as was illustrated in Figs. 8 and 9. For the cylinder gas temperature, a mean value was calculated based on the measured values of the cylinder pressure, intake air volume and residual gas fraction. From Fig. 12 it can be seen that in the compression stroke the cylinder gas temperature and pressure were lower for the IVC at 60° than for the conventional IVC. At the point indicated in this figure for ignition timing the cylinder gas temperature and pressure were 70 K and 75 kPa lower, respectively.

This drop in cylinder gas temperature and pressure was presumed to be one reason for the longer combustion duration with the early IVC at 60° than with the conventional IVC. The same thing was true for the late IVC at 100°. The main factors causing the drop in cylinder gas temperature and pressure were concluded to be the decrease in the effective compression ratio and the reduction of the EGR rate and residual gas fraction. The influence of these factors will be taken up in the following sections.

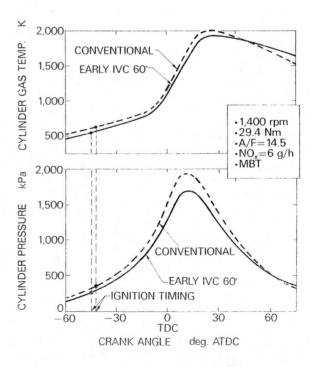

Fig. 12 Comparison of Cylinder Pressure and Gas Temperature

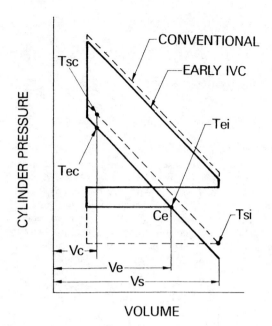

Fig. 13 Pressure-Volume Logarithmic Schematic Diagram for Early and Conventional IVC

$$T_{ec} = T_{ei} (V_e/V_c)^{n-1} \quad \quad (4)$$

$$T_{sc} = T_{si} (V_s/V_c)^{n-1} \quad \quad (5)$$

n : polytropic index

From the above equations, the cylinder gas temperatures T_{ec} and T_{sc} for the early IVC at 60° and the conventional IVC were found to be 406 K and 429 K, respectively, at a 50° crank-angle position before the top dead center of the compression stroke. The difference in temperatures was 23 K. In this case, cylinder gas temperatures T_{ei} and T_{si} before compression were both assumed to be 300 K and polytropic index n was 1.3.

The temperature difference of 23 K was the result of a simple calculation based on the idea that a decrease in the effective compression ratio would cause the cylinder gas temperature to drop. The difference, however, corresponded well with the temperature difference calculated from the measured values for the cylinder pressure, intake air volume and residual gas fraction when no EGR was used (Fig. 14). From these findings, it is possible to conclude that a decrease in the effective compression ratio was one cause of the drop in cylinder gas temperature that occurred when an early IVC was used.

Decrease in Effective Compression Ratio - Figure 13 shows a pressure-volume schematic diagram which compares the pressure volume data for the early IVC and the conventional IVC. It should be noted that in this case the air-fuel mixture was regarded as a perfect gas. Thus it was assumed that there would be expansion due to throttling with the throttle valve in the carburetor and that no exchange of heat would occur between the mixture and the intake manifold, cylinder wall, combustion chamber and other parts. Furthermore, it was also assumed that the intake-valve closing timing for the conventional IVC would coincide with the bottom dead center of the intake stroke.

Proceeding on the basis of the foregoing assumption, it was presumed that the temperature T_{si} for the conventional IVC at the bottom dead center of the intake stroke and the temperature T_{ei} for the early IVC at the moment the valves closed would both be equal to the intake air temperature upstream from the throttle valve (8). Furthermore, it was assumed that the cylinder gas would expand adiabatically from the time the intake valves closed to the bottom dead center and subsequently contract adiabatically until the valve closing point was reached again. Therefore, at point C_e in the compression stroke, the temperature and pressure should return to the same levels that were present when the intake valves closed. Thus it was concluded that if a comparison were made of the cylinder gas temperature at any arbitrarily chosen points during the compression stroke, T_{ec} for the early IVC would be lower than T_{sc} for the conventional IVC. The rationale behind this conclusion was that the effective compression ratio V_e/V_c for the early IVC in following Eq. (4) would be lower than V_s/V_c for the conventional IVC in Eq. (5).

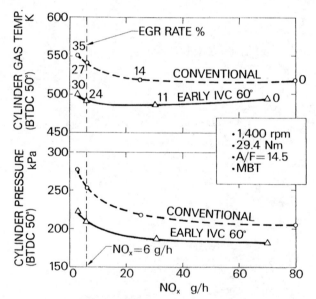

Fig. 14 Cylinder Pressure and Gas Temperature When Nox Was Reduced

The same conclusion can also be drawn regarding the late IVC. Figure 15 presents a pressure-volume schematic diagram that compares the pressure-volume data for the late IVC and the conventional IVC. In the case of the late IVC, air was drawn into the cylinder at a uniform cylinder pressure until the bottom dead center was reached. During the subsequent compression stroke the air that had initially been drawn into the cylinder was returned to the intake manifold at the same cylinder pressure as in the intake stroke. Compression began at the point C_1, where the intake valve closed.

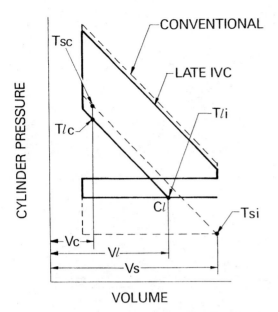

Fig. 15 Pressure-Volume Logarithmic Schematic Diagram for Late and Conventional IVC

The air-fuel mixture drawn into the cylinder was regarded as a perfect gas and the same assumptions were made as for the early IVC. Thus the temperature T_{li} for the late IVC was presumed to be equal to the temperature T_{si} for the conventional IVC at the bottom dead center of the intake stroke. Moreover, the temperature T_{li} corresponded to the temperature T_{ei} in Eq. (4). Consequently, it was concluded that the temperature T_{lc} at any arbitrarily selected point during the compression stroke would drop due to a decrease in the effective compression ratio, in the same manner as for the early IVC.

The decrease in cylinder pressure due to the early or late IVC can be explained from the fact that the cylinder gas temperature dropped. Another explanation is that since the brake torque was kept constant, the amount of the air-fuel mixture drawn into the cylinder was reduced because of the pumping loss reduction, and, as a result, there was less amount of gas in the cylinder.

In the foregoing explanation of the decrease in the effective compression ratio, it has been presupposed that the cylinder gas temperatures T_{si} and T_{ei} (or T_{li}) were equal. In actuality, however, these temperatures are believed to be influenced by the EGR rate and the residual gas fraction.

Influence of EGR - As was shown in Fig. 14, the EGR rate needed to achieve the same level of NOx emissions became progressively smaller for the early IVC at 60°, in comparison with the conventional IVC, as the level of NOx emissions was lowered. Similar to this tendency, the difference in both the cylinder gas temperature and pressure between the conventional IVC and the early IVC at 60° grew larger as the level of NOx emissions was lowered.

The required EGR rate for the early IVC at 60° decreased because of the drop in the cylinder gas temperature due to the lower effective compression ratio, as was explained in the preceding section. It has been concluded that the reduction in the EGR rate caused the increased difference in temperature and pressure as noted above. The reason for this was that the reduction of EGR gas, which was hotter than the newly induced air in the intake manifold, meant the latter was heated less, and, as a result, the cylinder gas temperature was lowered.

Therefore, it is concluded that the reduced EGR rate was one cause of the drop in cylinder gas temperature with the early IVC in the region of low NOx emissions, as compared with the conventional IVC. The same conclusion can be drawn for the late IVC.

Influence of Residual Gas - As can be seen in Fig. 2, there was an overlapping area in valve lift between the intake and exhaust valves of the test engine. Under partial-load operation, it was possible that a portion of the burnt gas might have flowed from the exhaust manifold into the intake manifold during the overlap interval, since the intake manifold pressure was lower than that of the exhaust manifold. Such an influx of burnt gas would have caused an increase in the residual gas fraction in the cylinders.

In the case of the early IVC, it was inferred from the pressure-volume data (Fig. 4) on cylinder pressure during the first half of the intake stroke that the intake manifold pressure rose. Thus, the reverse flow phenomenon should have been abated, i.e., the residual gas fraction should have been reduced.

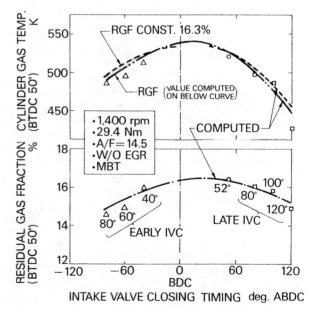

Fig. 16 Relationship Between Intake Valve Closing Timing and Residual Gas Fraction/Cylinder Gas Temperature

In order to investigate the residual gas fraction in the cylinder, gas samples were taken and analyzed when no EGR was applied. As the results shown in Fig. 16 confirmed, the residual gas fraction was smaller for the early IVC than for the conventional IVC. The same finding was verified for the late IVC. The dot-dash line in this figure represents the computed result obtained with the aforementioned simulation model. It can be seen that there was good correspondence between the computed and measured results.

Because the reduction in the residual gas fraction was thought to have lowered the cylinder gas temperature, the same model was employed to compute the cylinder gas temperature at a 50° crank-angle position before the top dead center of the compression stroke. The computed result is shown with the two-dot-dash line in the upper half of Fig. 16. This computed value agreed well with the temperatures indicated with circle, triangle and squares calculated from the measured values of the cylinder pressure, intake air volume and residual gas fraction.

The broken line in this figure represents the computed value of the cylinder gas temperature when the exhaust-valve closing timing was varied to make the residual gas fraction for the early IVC and late IVC the same as that for the conventional IVC. The broken line indicates a higher value than the two-dot-dash line, and the difference (screen-tone portion) shows the drop in the cylinder gas temperature due to the reduction of the residual gas fraction. It can be seen that this difference accounted for about 15% of the drop in the cylinder gas temperature that occurred as a result of changing the conventional IVC to the early IVC at 60°.

Other Factors - In explaining the decrease in the effective compression ratio in an earlier section, the intake air-fuel mixture was regarded as a perfect gas. In actuality, however, it is a real gas and therefore other factors could be considered as possible causes of an increase in the cylinder gas temperature with the early and late IVC. For instance, the rate of expansion around the throttle valve in the carburetor is smaller for both the early and late IVC than for the conventional IVC, thus the drop in temeprature due to throttling would be smaller. Moreover, the higher pressure following expansion should raise the heat transfer rate, thereby increasing the amount of heat received.

However, all the data confirmed that the cylinder gas temperature dropped. Therefore, it is concluded that the decrease in the effective compression ratio, the reduction of the EGR rate and the reduction of the residual gas fraction exerted a governing influence over any other factors.

The reduction of the EGR gas and residual gas fraction is thought to have raised the burned gas temperature and thus shortened the combustion duration (9, 10). Therefore, even if such reductions lowered the cylinder gas temperature before combustion, they did not contribute to any lengthening of combustion duration.

AIR MOTION IN THE CYLINDER - As was shown in Fig. 4, the shortening of the intake-valve opening period also reduced the amount of valve lift.

Consequently, for the early IVC, the air-fuel mixture had to be induced into the cylinder through a narrower valve opening in a shorter period of time, as compared with the conventional IVC. Thus, the maximum flow velocity of the air-fuel mixture flowing through the valve opening increased with the early IVC, as is shown in Fig. 17. This figure also shows that along with the faster flow velocity, there was an increase in the kinetic energy of the mixture as it passed through the valve opening. The flow velocity and kinetic energy values presented here were computed using the model mentioned earlier.

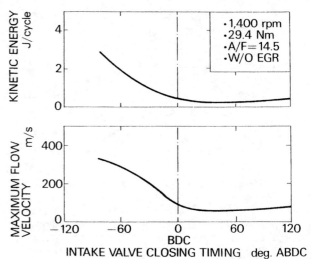

Fig. 17 Maximum Flow Velocity and Flow Kinetic Energy at Intake-Valve Opening

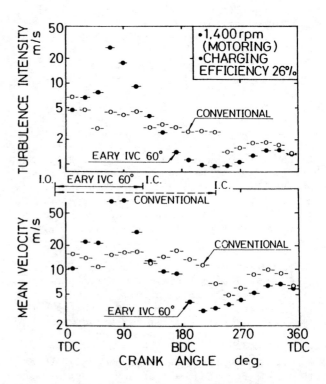

Fig. 18 Mean Velocity and Turbulence Intensity at Intake-Side Spark Gap

Measurements of the air motion in the cylinder were made with a hot-wire anemometer because it was thought that the higher kinetic energy of the intake air with the early IVC would improve the air motion. Fig. 18 shows the mean velocities and turbulence intensities that were measured at the intake-side spark gap for the early IVC at 60° and the conventional IVC. From the figure it can be seen that the mean velocity and turbulence intensity of the mixture as it passed through the intake-valve opening were larger for the early IVC at 60° than for the conventional IVC.

In the compression stroke, however, the mean velocity and turbulence intensity were larger for the conventional IVC, and a large decrease took place with the early IVC at 60° following the closing of the intake valve. This decrease is thought to be attributable to a change in the flow direction of the air-fuel mixture into the cylinder due to the reduction in valve lift.

It is concluded that this deterioration in air motion with the early IVC was one reason for the increase in the combustion duration (11). For the late IVC, on the other hand, no major difference in air motion was seen in comparison with the conventional IVC.

For both the early and late IVC, the fuel did not vaporize as readily because of the higher intake manifold pressure. There was a tendency for a larger percentage of droplets to be drawn into the cylinder. Such a situation would normally have an adverse impact on combustion. However, the increased flow velocity at the valve opening in the case of the early IVC tended to promote fuel vaporization (12), and this is thought to have cancelled out the inimical influence of the rise in intake manifold pressure. For the late IVC, on the other hand, there was no flow velocity effect, and the deterioration in fuel vaporization due to the rise in intake manifold pressure is thought to have had a direct, adverse effect on combustion.

CONCLUSION

This study has examined the effects of intake-valve closing timing on spark-ignition engine combustion when the closing timing was varied for the purpose of reducing pumping losses. The principal results of this investigation are summarized below.

(1) Accompanying the reduction of pumping losses, there was an increase in combustion duration. As a result, an ISFC improvement corresponding to the pumping loss reduction could not be obtained.

(2) An even larger increase in combustion duration occurred when the EGR rate was increased in order to reduce NOx emissions. In this case the ISFC improvement achieved by reducing pumping losses was diminished. At an EGR rate of around 25% (NOx emissions level of 6 g/hr), the improvement effect was only about one-third as large as when no EGR was applied.

(3) One of the principal causes of the increase in combustion duration was the drop in cylinder gas temperature and pressure which was traced to a decrease in the effective compression ratio.

(4) When the intake-valve closing timing was advanced to a point before the bottom dead center of the intake stroke, the air motion in cylinder deteriorated and this became one cause of the increase in combustion duration.

ACKNOWLEDGEMENT

The authors wish to thank Mr. Y. Yoshikawa for his help in programming the simulation model, and Mr. T. Itoh and Mr. S. Kamegaya for their help in measuring the air motion in the cylinder.

We also thank the personnel of Engine & Powertrain Research Laboratory of Nissan Motor Company for their help in conducting the experiments and processing the data.

REFERENCES

(1) H. N. Pouliot, W. E. Delameter and C. W. Robinson, "A Variable-Displacement Spark-Ignition Engine" SAE Paper 770114

(2) Donald C. Siegla and Robert M. Siewert, "The Variable Stroke Engine-Problems and Promises" SAE Paper 780700

(3) R. H. Sherman and P. N. Blumberg, "The Influence of Induction and Exhaust Processes on Emissions and Fuel Consumption in the Spark Ignited Engine" SAE Paper 770880

(4) J. H. Tuttle, "Controlling Engine Load by Means of Late Intake-Valve Closing" SAE Paper 800794

(5) J. H. Tuttle, "Controlling Engine Load by Means of Early Intake-Valve Closing" SAE Paper 820408

(6) M. Horvatin and A. W. Hussmann, "Measurements of Air Movement in Internal Combustion Engine Cylinders" DISA Information, No. 8 July 1969

(7) H. Tasaka and S. Matsuoka, "Analysis of Gas Exchange Process of S. I. Engines Under City-Driving Conditions" SAE Paper 800535

(8) J. Kestin, "A Course in Thermodynamics" Xerox College Publishing 1966

(9) F. Nagao, "A Lecture on Internal-Combustion Engines (In Japanese)" First Volume Yokendo Co., Ltd. Tokyo 1974

(10) S. D. Hires, R. J. Tabaczynski and J. M. Novak, "The Prediction of Ignition Delay and Combustion Intervals for a Homogeneous Charge, Spark Ignition Engine" SAE Paper 780232

(11) D. R. Lancaster and R. B. Krüger, S. C. Sorenson and W. L. Hull "Effects of Turbulence on Spark-Ignition Engine Combustion" SAE Paper 760160

(12) C. F. Aquino, "The Design and Development of the Upper-Pivoted Sonic Carburetor" SAE Paper 780078

C241/85

Electronics – the key to engine management

E W MEYER, BSEE, MSAE
Austin Rover, Birmingham

In early 1984, the Federal Republic of Germany proposed through their legislative process that all automobiles sold in Germany would have to meet the 1983 U.S.A. Federal emission levels by 1988/89. The European Commission under Count Davignon opposed this and reacted in mid 1984 by proposing two new emission regulations. E.C.E. 15.05 would have reduced the combined HC and NOx levels to 15 grams/test and the CO level to 45 grams/test for all cars dependent of their inertia weight, and would have become effective in 1989. E.C.E. 15.06 would have introduced 1983 U.S.A. federal emission control levels in 1995 for all cars. Considerable discussion and political activity ensued during the winter of 1984/5 and the spring of 1985 trying to resolve German desire for more rigorous emission control sooner and the rest of the European communities desire for less severe control levels at a pace which would allow all automobile companies and the regulatory groups in each country to be prepared. The resulting compromise announced by the entire European Community in March and June 1985 was to separate cars into categories dependent on the displacement of their engines.

ENGINE DISPLACEMENT	2.0L or larger	1.4-1.999L	1.399 or smaller
TYP. VEHICLE INERTIA, WGT. KG	>1.590	1130-1360	<1130
EMISSION CO	25	30	45
HC + NOx	6.5	8	15
NOx	3.5	-	6
DATE	10/88	10/91	10/90

As can be seen above the requirement for the intermediate cars is virtually as severe as the transposed 1983 American standard which is shown in E.C.E. cycle terms for 2.0L and larger cars.

The combined HC + NOx value would have had to be at least 11 to allow emission control using lean mixture engines and the 8-10% better fuel economy they produce. Perhaps engines can be designed and developed with very low emission outputs at lean mixtures but this certainly would be on the leading edge of engine technology.

C241/85 © IMechE 1985

The technology to meet the 25/6.5/3.5 requirement for cars with engines over 2.0L is based on the prior work done to meet the American and Japanese regulations. Most of these cars would be in the executive or luxury market segments and as indicated would have inertia weights above 1590 Kg.

The technology for the intermediate engines will require a threeway catalyst, electronic controlled single point fuel systems with or without exhaust sensor feedback. These cars will largely fall into the medium car sector and have inertia weights of 1130 and 1360 Kg.

The third sector will contain some lower medium marketing sector cars plus all of the small and basic sector vehicles with inertia weights below 1130 and mostly below 910 Kg. It was this last category that was initially chosen to be investigated by ARG, as the technology to meet it with outstanding vehicle driveability and minimum fuel economy loss was understood the least. A typical ARG engine and vehicle were chosen for this investigation work.

The investigation determined that these emission values could be achieved on cars of this size with careful adjustment of the ignition calibration and an accurate fuel system. The final emissions results achieved were:

   30.0 grams/test CO

   12.8 grams/test NOx

   2.8 grams/test NOx

There was 2% loss in fuel economy with the work accomplished to date but it is believed that this can be recovered with newer lean burn engines producing lower output of engines hydrocarbons.

The techniques to achieve these results involved manipulating spark advance to retard it from the conventional pattern to reduce HC at specific speeds and loads and NOx at specific intermediate speeds and loads. Concurrently spark advance at other specific lower speeds and loads could be advanced to recover as much of the fuel economy as possible lost by the emission changes.

The resulting ignition map now had slopes and steps in it that a conventional, distributor-type spark advance mechanism could not achieve. These are illustrated in Fig 1.

It clearly indicated the necessity for electronically controlled programmed ignition with its intrinsic capability to respond to these conditions.

This same investigation showed that conventional carburettors had to be improved as well in both the consistency and uniformity of metering and in the increased capability of adjustment of calibrations between idle, off idle and road load. Consistency is defined as the ability of a single fuel device to reproduce the same calibration when tested repeatedly under the same conditions whereas uniformity is defined as the ability of a number of like fuel devices to reproduce the same calibration when compared with one another under the same conditions. Obviously consistency will affect and be part of uniformity.

In the air-fuel-ratio range where these tests operated, the effect of varying AFR was most dramatic with respect to carbon monoxide content in the exhaust gas. These results are illustrated in Fig 2.

However, the combined HC plus NOx results illustrated in Fig 3 also shows a rich mixture limit.

The clear conclusion of these tests was that the fuel system had to be able to meter consistently and uniformity within +3% both in the cold enrichment areas and in normal idle - road load areas. Typical present carburettors used on European cars cannot meet the requirement without improvements. Tests of several fixed choke units and a variable choke unit in populations large enough to allow meaningful statistical calculations to be applied to the results were made. A simplified table of the results is shown below.

	Idle	Road Load		FULL LOAD
Goal	$\pm$.87 AFR*	$\pm$1.09 AFR*		$\pm$.72 AFR*
Carburettor A‡	$\pm$2.33 "	$\pm$1.14 "	"	$\pm$1.06 "
Carburettor B‡	$\pm$3.01 "	$\pm$1.38 "	"	$\pm$1.99 "
Carburettor C‡	$\pm$5.63 "	$\pm$1.14 "	"	$\pm$1.27 "

* A $\pm$ 3% mixture variation translated into appropriate air-fuel ratios at the three different operating conditions.

‡ Indicates $\pm$ two standard deviations or 95% of the population.

To allow a whole product population to achieve a $\pm$ 3% consistency plus uniformity metering condition, a carburettor needs additional metering techniques achieved with electronic controls particularly in the cold enrichment areas where crude bimetal or wax capsule devices are normally used.

With the emission regulations for vehicles with engines below 1.4L identified and the minimum necessary hardware to meet these emissions established it becomes clear to us that both the fuel and ignition systems requires microprocessor control for the most cost effective solution. Because there are several new powerful single chip micro-processors with sufficient memory available which can handle both jobs, it is our judgment that a combined micro-processor control would be the optimum design. The microprocessor candidates we would suggest as being most viable for this job are Motorola's MC68HC 05B4 and NEC's PD 78C11. Both have CMOS construction with its low power dissipation and noise immunity which is very useful in automobile controls. We believe that this combined system of fuel and programmed ignition will cost very little more than the existing carburettor plus breakerless electronic ignition. This situation is important as it illustrates the potential of electronic engine management controls to achieve a more complex task and yet hold the original cost levels.

It is believed that two other conditions can be achieved with a design of this kind. First, full diagnostics of sensors, actuators and the control itself can be built into this unit, within our cost goal described above, to help manufacturing and field service. Secondly, we believe, from the analytical analysis, we can achieve and validate a reliability level of 99+% at an 80% confidence level with this system which is very important. Every new product introduced on an automobile is immediately the source of all problems whether they are related or not. The field service people always start way down the learning curve and the less intrinsic problems generated by a new system, the better the chance of controlling and eliminating false diagnosis and warranty. A block diagram for the system we would propose is shown in Fig 4.

Because single point fuelinjection puts all fuel through one or two injectors, it can meter more accurately. It does this by avoiding the lower flow rates where multipoint injector metering accuracy is poor. Also injecting into a plenum and entraining and preparing the fuel in the intake manifold produces a more homogenous mixture which can be ignited and burnt more effectively in the cylinder allowing the leanest possible mixtures to be used. The transient problem of moving from road load mixtures to higher load mixtures during accelerations requires careful matching of fuel input to engine dynamic characteristics to prevent NOx spikes with single point injection but the difference between it and gang-injected multipoint injection is much overplayed in our experience and will not be a problem at these emission levels. And last but not least, it costs about 35% less than multipoint fuel injection.

Various marketing studies of European car manufacturers, which have talked to engineering and purchasing departments, have shown a distinct projected growth of this new form of fuel injection in Europe. This is shown in Fig 5.

With the new emission regulations now specifically in place and becoming effective in 1991 it is certain to be used on more cars than were anticipated earlier and to become the predominant fuel system on engines with displacements of 1.4L and above. Renault has already announced the use of single point fuel injection on its 1.4 and 1.7L engines in its 5,9, and 11 models for Switzerland.

© IMechE 1985 C241/85

With the need for single point fuel injection and, almost certainly, programmed ignition in a combined low cost electronic control, the selection of a micro-processor for this control becomes very important.

Again the semiconductor industry has several good ones from which to choose. We would suggest that the Motorola 68HC11, the Intel 8906 the NEC 7811, the National Semiconductor HPC and the Texas Instruments' RR/6 micro-processor are all excellent products. The extra memory in the Motorola and Intel units is very attractive however, as is the CMOS construction of the Motorola, National and T.I. units. All the processor have adequate computational performance, timers and A/D convertors. Using surface mounted components and chip carrier type semiconductors along with these micro-processors should produce a compact, low power, low cost combined electronic control. The lower dissipation of the CMOS processors should significantly help reliability and will also allow location of the control in the engine bay to further improve reliability by reducing wiring harness connector interfaces. Full diagnostics could be included in this design as well to help servicing.

A block diagram of the combined ignition - fuel electronic control we suggest, and believe is required, for this level of emission control is shown in Fig 6.

In summary, the new European Community emission regulations along with normal, competitive, fuel economy programmes will continue to accelerate the use of electronic engine management systems on European cars. It is gratifying that the semiconductor components to achieve these controls are available in the excellent form that they are. The solutions to the emission control problems are thus in the area of product design rather than in the year-by-year inventions required in the earlier programme in the United States. Properly recognising the requirements of each emission category early enough and responding with the correct product design will meet society's needs for clean air and yet achieve those solutions with reliable, cost effective devices. Also the electronic controls described in this paper are evolutionary. From non-catalyst to threeway catalyst engine management systems, much of the control techniques are similar. Each step means only adding to the simpler design rather than creating a whole new product.

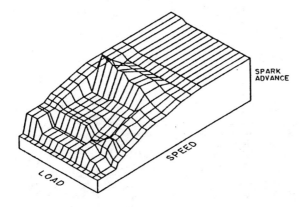

Fig 1

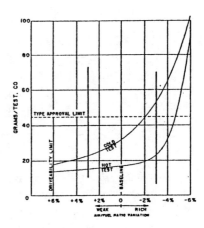

Fig 2 Variation of carbon monoxide mass emissions with mixture

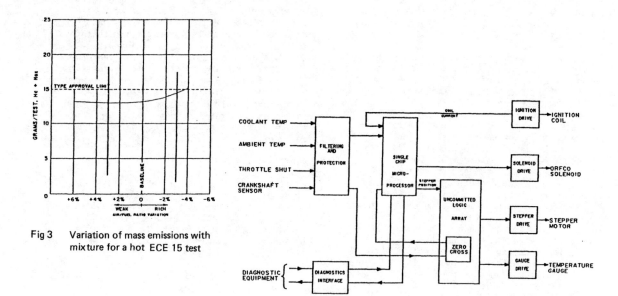

Fig 3 Variation of mass emissions with mixture for a hot ECE 15 test

Fig 5

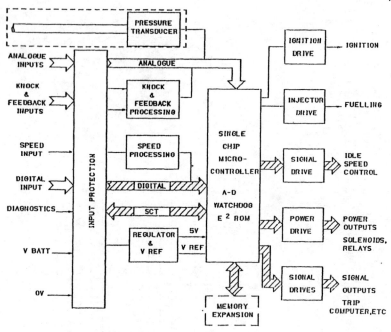

Fig 4 Combined EFC and programmed ignition

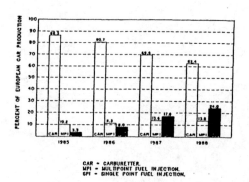

CAR = CARBURETTER.
MPI = MULTIPOINT FUEL INJECTION.
SPI = SINGLE POINT FUEL INJECTION.

Fig 6 Lucas advanced engine management system

© IMechE 1985 C241/85

860537

Development of a Variable Valve Timed Engine to Eliminate the Pumping Losses Associated with Throttled Operation

Alvon C. Elrod
Department of Mechanical Engineering
Clemson Univ.

Michael T. Nelson
Lucas CAV
Div. of Lucas Industries and Co.

ABSTRACT

The primary objective of this research program was to reduce pumping losses associated with the standard air intake system for spark ignition engines. This was accomplished by controlling the induction of the air/fuel charge with variable valve timing applied to the intake valves.

The intake manifold pressure was maintained at atmospheric pressure and the timing of the intake valves was electronically adjusted to allow the induction of an air/fuel mixture sufficient to support the required engine load. This variable valve timing allowed flexibility in scheduling the intake valve timing events, simulating throttled operation with improved efficiency. Brake-specific-fuel-consumption (BSFC) was improved significantly, while air/fuel ratios remained stabilized, with no sacrifice in load control.

ALTHOUGH SOME EFFORTS have been made for many years to improve combustion engine efficiencies, it is only recently that the economic considerations have warranted more serious attention. The references following this report, though not all inclusive, indicate the nature of work that has been done. Until the mid-sixties, performance took precedence over economy of operation. However, now with the greatly increased cost of fuel, there exists a strong economic incentive for improved operational efficiency. On a broad scale, the work that can be done to improve gasoline engine efficiencies can be directed toward decreased friction, higher compression ratios, improved combustion, and reduction of an engine's pumping losses. The pumping losses contribute significantly to the lowering of an engine's operational efficiency and are the subject of this report. These losses are negative work required by an engine to pump air through it. They are due primarily to the resistance associated with the fresh air charge as it flows past the throttling valve and into the individual cylinders.

With a standard spark-ignition engine, the most efficient mode of operation is at wide-open-throttle where the pumping losses are minimal. However, an automobile engine spends the majority of its time at part-throttle, so the efficiency for this mode is of primary importance. This is illustrated by the fact that modern intermediate sized automobiles require about 15 hp to maintain 60 mph speed on a level road, though these vehicles have engines ranging up to about 150 hp capacity. Blackmore and Thomas (Ref. 1) indicate that the penalty in part-load performance of the conventional spark-ignition engine, attributed primarily to the throttling process, varies from 3.5% of the indicated mean-effective-pressure at wide-open-throttle to nearly 100% for a fully throttled idling engine. They conclude that elimination of the throttling process (i.e. running the engine at wide-open-throttle throughout its load/speed range) would improve the average overall efficiency of the engine by 20%.

Since the purpose of throttled operation is to control the engine power output by varying the amount of air/fuel charge available for combustion, elimination of the throttle valve requires the development of an alternative means of controlling the amount of charge inducted to support the required engine load. The purpose of this report is to describe substantial improvements in the fuel economy of a gasoline engine equipped with an alternative means of controlling breathing. Research efforts have been directed toward the following two objectives:

1) eliminating the pumping losses of a spark-ignition engine while maintaining the same useful output, and

2) accomplishing the elimination of the throttling process by controlling the induction of air/fuel charge into the engine, and thus the engine load, with variable valve timing applied to the intake valves.

To accomplish these objectives, a variable valve timing device has been designed and used to con-

0148-7191/86/0224-0537$02.50
Copyright 1986 Society of Automotive Engineers, Inc.

trol the intake valves of a four cylinder Fiat engine with double overhead camshafts. This engine was chosen because of the accessibility of the intake camshaft.

AN IMPROVED OPERATING CYCLE

The air-standard Otto cycle approximates a spark-ignition engine's behavior and is a useful tool in explaining how the objectives of this research were accomplished (Fig. 1). The idealized cycle consists of:
- 1-2 isentropic compression,
- 2-3 constant volume energy addition,
- 3-4 isentropic expansion,
- 4-5 constant volume energy rejection,
- 5-6 the exhaust stroke, and
- 6-1 the intake stroke.

The area 1-2-3-4-5 is proportional to the useful output of the engine.

Both the intake and exhaust processes occur at atmospheric pressure for an engine operating at wide-open-throttle (full load - Fig. 1a). During the intake stroke, the atmospheric air being inducted does a positive amount of work on the engine as the piston moves toward bottom dead center. During the exhaust stroke, a negative quantity of work is done by the engine in expelling the exhaust gases into the atmosphere. Consequently, since both processes occur at the same pressure and displace the same volume, the positive work associated with induction and the negative work associated with the exhaust process cancel. This is represented graphically by the straight line 6-5-6-1 on Fig. 1a.

During throttled operation (part load - Fig. 1b), the basic cycle remains essentially the same, and the exhaust process with its related work is approximately the same as for full-load operation. However, during the intake stroke, cylinder pressure falls below atmospheric by an amount determined by the throttle setting. Therefore, the amount of positive work associated with the intake stroke is less than the negative exhaust work. Thus, there is a net negative effect which helps cause the throttled engine to be less efficient than one which is operated at full throttle. The negative net work, represented in Fig. 1b by the shaded region, is termed the pumping loss attributed to throttle operation and is eliminated with an improved operating cycle.

Fig. 1c represents the improved operating cycle. This figure is the same as Fig. 1b except that the negative area representing pumping work for throttled operation does not appear. In this improved cycle, the pumping work for part-load (throttled) operation is eliminated by replacing the conventional throttle with variable intake valve timing. The engine load is controlled by variable intake valve timing, and both the intake and exhaust processes occur at atmospheric pressure. It is noted that for this idealized analysis, energy rejection is accomplished first at constant volume (4-5), and then by the constant pressure process (5-1). The intake and exhaust processes are graphically represented by the lines (6-1) and (1-6), respectively.

As the above discussion shows, it is possible to achieve a part-load cycle that eliminates the pumping losses associated with throttled operation. The intake camshaft of a commercial spark-ignition engine has been redesigned to allow the engine to approximate this improved operating cycle. The test results presented in a later section of this report show that substantial improvements in the economy of part-load operation are achievable with the application of this method of controlling the engine breathing.

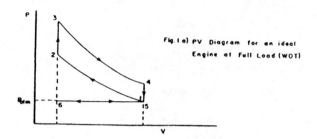

Fig. 1 a) PV Diagram for an ideal Engine at Full Load (WOT)

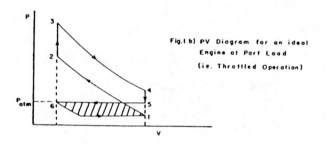

Fig. 1 b) PV Diagram for an ideal Engine at Part Load (i.e. Throttled Operation)

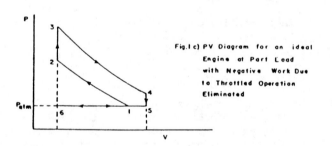

Fig. 1 c) PV Diagram for an ideal Engine at Part Load with Negative Work Due to Throttled Operation Eliminated

FIG. 1 - PRESSURE-VOLUME DIAGRAMS

ENGINE MODIFICATIONS

THE ELECTRONICALLY CONTROLLED VARIABLE CAMSHAFT - Variable valve timing is, as the name implies, flexibility in the scheduling of valve timing events throughout a combustion engine's load/speed range. Several inherent advantages exist with variable valve timing. However, the primary one is that the cam profile does not have to be a compromise of the shapes best suited for operation at various combinations of load and speed. Thus with variable valve timing, various engine parameters related to emissions, economy, and performance can be optimized at an infinite number of load/speed points.

Since pumping losses result from operating the intake process at subatmospheric pressure, it is desirable for an engine to operate with the intake process at atmospheric pressure. Such an engine, using variable valve timing, has been developed at Clemson University. Utilizing an electronically-operated variable camshaft, this design allows the camshaft to regulate the amount of air/fuel mixture available for the combustion process. Thus, the engine maintains load control without a throttle valve. The actual process is represented by the idealized air standard cycle shown in Fig. 1c. During the intake stroke (6-5), a full charge of air/fuel is inducted into the engine. The intake valve is then left open during a portion of the compression stroke (5-1) so that a controlled amount of the air/fuel charge is pushed back into the intake manifold prior to the valve being allowed to close (1) and compression (1-2) to begin. With flexibility in control of the amount of air effectively inducted into the engine (6-1), load control is maintained while the losses associated with throttled operation are eliminated. It is noted that since the intake valve remains open during a portion of the compression stroke (5-1), the effective length of the compression stroke is reduced. Similarly, during throttled operation of a conventional engine, a portion of the compression stroke is used in pumping air at sub-atmospheric conditions back to atmospheric pressure. The part of the conventional engine's compression process that is above atmospheric pressure is comparable with the complete compression process of the improved cycle (see Figs. 1b and 1c). Considering only these comparable compression processes, it can be said that the conventional engine and one operating on the improved cycle have nearly the same effective compression ratio. These compression ratios would be the same for the same engine loads except that the improved cycle is more efficient than the conventional one. Initial testing of this improved cycle consistently showed that a fuel-rich mixture was being supplied to the engine. Inspection revealed that, during the portion of the compression period in which the intake valve remained open, the air/fuel mixture was pushed back through the carburetor and reenriched. Therefore, a reed-valve assembly (i.e. essentially a check valve) was designed to alleviate this problem. This reed system was attached to the baseplate of the carburetor to restrict the air/fuel mixture from being pushed back through it. Hence, the air/fuel mixture pushed back into the intake manifold is simply available to another cylinder when its induction stroke begins. This check valve may not be required with all engine applications.

The test engine was a four cylinder Fiat engine with double over-head camshafts. Fig. 2 shows a top view of the engine's camshaft arrangement. The intake camshaft is shown in the upper portion of this photograph while the exhaust camshaft is in the lower region. A close-up view of the redesigned intake camshaft is shown in Fig. 3. Each original intake cam lobe has been replaced by a paired set of half-width, adjacent cam-lobes. This paired set of cam-lobes is different in that one part (No. 1) is attached rigidly to the timed camshaft and the other part (No. 2) is movable. When these split cam-lobe pairs are aligned, the manufacturer's original profile is maintained. Fig. 4 shows the adjustable cam-lobes rotated to a fully retarded intake valve closing position. This setting can be adjusted between 0 and 44 cam angle degrees. Engine rotation takes the cam-lobes up and into the picture. Thus the intake valve opening time is retained with the fixed cam-lobe while the closing time is retarded with the movable cam-lobe (No. 1). The amount the valve closing time is delayed is a function of the angular distance between the fixed and adjustable cam-lobes. Since the closing time of the intake valve can be extended well into the compression stroke, the amount of air/fuel charge available for the combustion process can be regulated and load control can be maintained without conventional throttle operation.

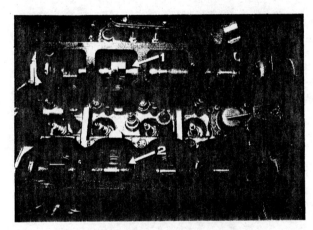

1) Redesigned Intake Camshaft
2) Standard Exhaust Camshaft

Fig. 2 - Top View of Engine's Camshaft Arrangement

1) Fixed Cam Lobe
2) Adjustable Cam Lobe

Fig. 3 - Magnified View of New Intake Camshaft

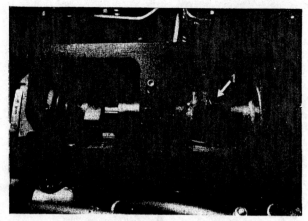

1) Fully Retarded Adjustable Cam Lobe

Fig. 4 - View of the Intake Camshaft with Valve Timing Adjusted

The intake camshaft includes a hollow tubular shaft (Fig. 5) upon which the fixed cam-lobes are attached. A hexagonal inner shaft runs concentric with this shaft and controls the movable cam-lobes. Concentricity of the two shafts is assured with bushings. The movable cam-lobes are attached to the hexagonal inner shaft (Fig. 6) and ride in slots machined in the tubular shaft, in a position adjacent to the fixed cam-lobes.

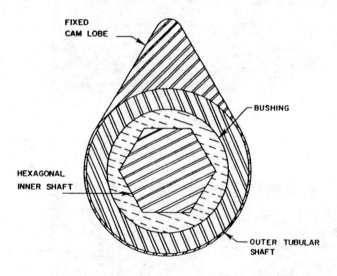

SECTION C-C

FIG. 5 - CAMSHAFT SECTION

THE ELECTRONICALLY CONTROLLED INTAKE CAMSHAFT ACTUATOR - Fig. 7 shows the engine with the camshaft's control actuator attached. The control mechanism's primary function is to provide the desired angular positioning of the intake camshaft's adjustable cam-lobes under dynamic operating conditions. This feature permits the intake valve closing time to be adjusted to support various engine load requirements. The heart of this control actuator is a harmonic drive, which is a mechanical power transmission unit developed by the Harmonic Drive Division of the Emhart Machinery Group. The specific power transmission unit has a gear reduction ratio of 160:1, and is connected directly to the hexagonal camshaft which drives the movable cam-lobes. This unit is well suited for this purpose. It has the accuracy and repeatability required for locating the adjustable cam-lobes at the various positions needed to map engine parameters over a wide range of simulated road-load conditions.

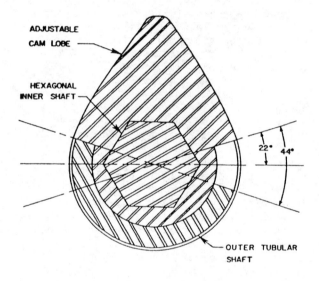

SECTION B-B

FIG. 6 - CAMSHAFT SECTION

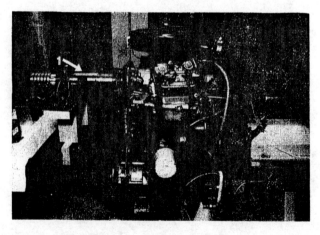

1) Camshaft Control Actuator

Fig. 7 - Engine with Camshaft Control Actuator Attached

An electric motor is used to give the harmonic drive unit its required input. The harmonic drive and motor configuration rotate at camshaft speed as a solid body except when valve timing adjustments are being made. Electrical connections are made via a slip-ring and brush-block assembly. When signalled by an electronic control unit, the electric motor activates the harmonic drive's wave generator. The wave generator causes the power transmission unit to act as a differential gear box, thus rotating the hexagonal camshaft relative to the tubular camshaft and changing the intake valve closing time. A valve timing indicator is used to provide feedback signals, from the electronic control unit, which indicate the amount of adjustment of the camshaft.

It is noted that careful selection of components can reduce the size of the control actuator to about a cylinder 4 inches in diameter and 3 inches in length. Fig. 8 shows the entire electronic control system. The control actuator is shown in the foreground; and from left to right in the background are a power supply, the actual electronic control unit, and a valve timing indicator. The valve timing indicator permits the relative angular position of the adjustable lobe to be known at all times and is useful mainly for research purposes.

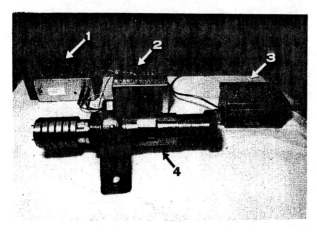

1) Power Supply
2) Electronic Control Unit
3) Valve Timing Indicator
4) Camshaft Control Actuator

Fig. 8 - Electronic Control System

EXPERIMENTAL WORK

POWER-CURVE TESTING (VARIABLE VALVE TIMED OPERATION) - Using a dynamometer, a set of power curves (Fig. 9) was generated for an engine running at wide-open-throttle. Each power curve resulted from operation at a specific intake valve closing time with the engine speed being varied from 1000 rpm to 4000 rpm. These power curves were run in order to determine the load control capabilities of the variable intake camshaft. The amount of intake valve closing delay ranged from 0 to 40 camshaft degrees (0 to 80 crankshaft degrees). Various engine parameters were monitored. However, the ones of primary importance in this work were power and brake-specific-fuel-consumption (bsfc). This "power-curve" nomenclature can be misinterpreted since part-load operation was achieved with the variable camshaft as the intake valve closing time was delayed. Thus, an explanation of this conventional term used in an unconventional manner is required. With this variable valve timed operation, the engine is always run at wide-open-throttle. Varying intake valve closure allows the effective size of the engine to be changed without any reduction of efficiency due to pumping losses. Hence, for each camshaft setting, the engine is producing its maximum possible output. Because the engine's output can be varied in this manner, the term "power-curve" is properly applied to part-load operation.

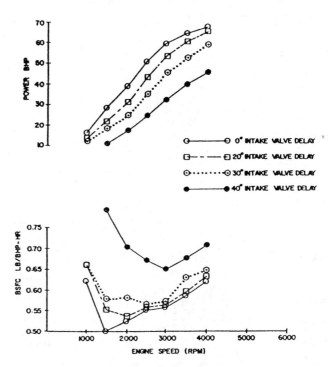

FIG. 9 - POWER-CURVE RESULTS

Driving applications require that the effective size of the engine be changed continually in order to meet ever-changing road-load requirements. Preliminary test results show that the control actuator, when attached to the variable camshaft, provides the engine with this capability.

ROAD-LOAD TESTING (CONVENTIONAL THROTTLED OPERATION) - The power-curve results, generated for the improved operating cycle, could now be examined to determine the suitability of the variable valve timed engine for satisfying the normal road-load requirements of an engine. Since the purpose of the research was to improve the economy of operation of automotive engines as they are applied in everyday use, testing under simulated road-load conditions (Fig. 10) was necessary. Baseline data were obtained using the conventionally throttled engine with its original intake camshaft. Data were generated for the engine run through a speed range corresponding to vehicle speeds of 20 mph to 90 mph. The power required to move a suitable automobile at various highway speeds was calculated using the equation (Ref. 2):

$$BHP = \frac{V}{375} \left[0.0165\{1 + 0.01(V-30)\}W + 0.0013 \, AV^2 \right]$$

A list of symbols and a table of values calculated using this equation are provided in the appendix.

During testing, the engine's throttle position and the dynamometer's load setting were adjusted to match these calculated values and fuel consumption data were recorded.

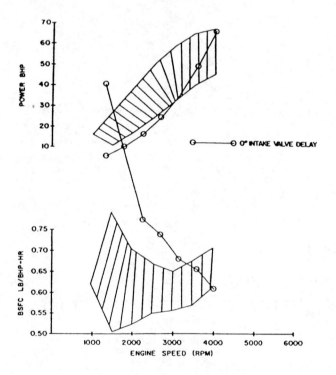

FIG. 10-ROAD-LOAD RESULTS

DISCUSSION OF RESULTS

Fig. 9 shows power curve results with power and bsfc plotted against engine speed. The power curve obtained using a fixed intake valve closing delay of 0° (the standard camshaft arrangement) represents full-load operation. When an engine is operating at full-load, it is generating its maximum possible power output at each speed point.

Delaying the closing time of the intake valves by 20°, 30°, and 40° reduces the amount of air/fuel mixture available for the combustion process and consequently the amount of power produced by the engine. Thus, the delayed intake valve closing times of 20°, 30°, and 40° each respectively represent part-load operation at smaller percentages of full-load. Fig. 9 shows that the horsepower results follow these predicted trends, and throttled operation is simulated with the variable camshaft since load control of the engine is maintained. The bsfc results support this same conclusion. Theoretically, as the engine load tends to zero the bsfc tends to infinity. The test results show this trend since an appreciable increase in bsfc is seen as the intake valve closing time is delayed. This is because as engine power requirements decrease for a constant speed, the amount of horsepower needed to overcome frictional effects is a greater percentage of the engine's net power output. Thus the engine becomes somewhat less efficient and the amount of fuel required per unit horsepower increases.

A curve of horsepower and one of bsfc for road-load testing are plotted on Fig. 10. The crosshatched areas on Fig. 10 show the range of test data for the modified engine with intake valve closing time delay ranging from 0° to 40° of camshaft rotation. These areas represent the data shown on Fig. 9. It is evident that a little more than 40° delay of the intake valve closing time will enlarge the power curve area for the modified engine enough to allow the road-load test curve of the original engine to be matched. This increased valve timing delay will result in the upper boundary of the bsfc area being raised a small amount, but it can be seen that the modified engine can produce the power required for part-load operation at all speeds while consuming less fuel than the conventional engine. The reason for this is that at part-load conditions of a conventional engine the bsfc suffers not only because of mechanical friction but also because of pumping losses. Both of these losses combine to form a very large percentage of the engine's power output and the bsfc suffers greatly. At full-load, wide-open-throttle, there is no difference between the standard cycle and the improved operating cycle since 0° valve delay represents the standard camshaft configuration. The primary purpose of conducting the road-load testing is to assess improvements in efficiency due to elimination of the pumping losses associated with throttled operation. Since the pumping losses for throttled operation are greatest when the engine is lightly loaded, the most significant improvements are realized in this zone. Because bsfc measures the amount of fuel required for a specific power output, lower values of it indicate a more efficient engine. It can be seen in Fig. 10 that bsfc values for the engine equipped with the variable camshaft are much improved. Thus, these combined results show that controlling engine breathing with variable valve timing permits the engine to maintain the same power output exhibited during the entire range of throttled operation while significantly reducing the fuel required for that output. Therefore, quantifying the economy improvements of the new operating cycle at each road-load operating condition is the next step in the overall evaluation procedure.

CONCLUSIONS

The research objective of eliminating the pumping losses associated with throttled operation in a spark ignition engine while maintaining the same useful output is successfully achieved. This goal is accomplished by controlling the induction of air/fuel charge into the engine, and thus the engine load, with variable valve timing applied to the intake valves. Furthermore, a variable valve timing control mechanism which eliminates normal throttle operation is successfully demonstrated for a commercial engine. Improvements in bsfc of the modified engine as compared with a conventional one are realized. These improvements are greatest at low speed, light load conditions where the pumping losses associated with throttling are large. In general, allowing the engine to operate at wide-open-throttle conditions (i.e. without a throttle plate) throughout its load/speed range significantly reduces the amount of fuel required for a specific power output for all part-load power settings.

REFERENCES

1. Blackmore, D. R. and Thomas, A., "The Scope for Improving the Fuel Economy of the Gasoline Engine," in the Passenger Car Power Plant of the Future, I. Mech. E. Conference Publications, 1979.
2. Obert, E. F., Internal Combustion Engines and Air Pollution, Harper and Row, 1973.
3. Anon., "Variable Valve Timing Has Electronic Control," Automotive Engineering, V. 92, No. 5, May 1984.
4. Herrin, R. J., "A Lost-Motion, Variable-Valve-Timing System for Automotive Piston Engines," SAE paper no. 840335.
5. Richman, R. M. and Reynolds, W. C., "A Computer Controlled Poppet-Valve Actuation System for Application of Research Engines," SAE paper no. 840340.
6. Stojek, D. and Stwiorok, A., "Valve Timing with Variable Overlap Control," FISITA paper no. 845026.
7. Torazza, G., "A Variable Lift and Event Control Device for Piston Valve Operation," paper 2/10, pp. 59-67, 14th FISITA Congress, 1972.
8. Parker, D. A. and Kendrick, M., "A Camshaft with Variable Lift-rotation Characteristics, Theoretical Properties and Application to the Valve Gear of a Multicylinder Piston Engine," paper B-1-11, pp. 224-232, 15th FISITA Congress, 1974.
9. Paulmier, C., "Variable Valve Timing for Poppet Valve Internal Combustion Engines," Ingens de L'Auto, p. 442, June/July 1974. (Translation from the BL)
10. Kerr, J., "Variable Valve Timing to Boost Any Engine," The Engineer, Vol. 54, No. 3, pp. 28-29, July 1980.
11. Beresford, N. and Ruggles, K., "An Investigation of Induction Ramming," Dept. of Mech. Eng., Brunel University Project Report 284, 1984.
12. Tuttle, J. H., "Controlling Engine Load by Means of Early Intake-valve Closing," SAE paper no. 820408, 1982.
13. Meacham, G. B. K., "Variable Cam Timing as an Emission Control Tool," SAE paper no. 700673, 1973.
14. Smith, P. H., Valve Mechanisms for High Speed Engines, pp. 108-109, Foulis, 1967.
15. Zappa, G. and Franca, T., "A 4-stroke High Speed Diesel Engine with Two-stage Supercharging and Variable Compression Ratio," session B3 paper D19, 13th CIMAC Congress, Vienna, 1979.
16. Roe, G. E., "Variable Valve-timing Unit Suitable for Internal Combustion Engines," Proc. Int. Mech. Engrs., Vol. 186, 23/72, 1972.
17. Scott, D., "Eccentric Cam Drive Varies Valve Timing," Automotive Engineering, pp. 120-124, October 1980.
18. Anon., "Variable Inlet Valve Timing Aids Fuel Economy," CME, Vol. 31, No. 4, p. 19, 1984.
19. Mansfield, W. P., "Development of the Turbocharged Diesel Engine to High Mean Effective Pressures Without High Mechanical or Thermal Loading," paper A6, 6th CIMAC Congress, 1965.
20. Lilly, L. R. C. (ed.), Diesel Engine Reference Book, Butterworths, 1984.

APPENDIX

Table 1 - Table of Values Calculated Using the Road-load BHP Equation:

RPM	MPH	BHP
1000	22.2	3.3
1500	33.3	6.5
2000	44.4	11.9
2500	55.6	19.9
3000	66.7	31.0
3500	77.8	45.8
4000	88.9	64.9

Equation for calculation of road-load BHPs:

$$BHP = \frac{V}{375}\left[0.0165\{1 + 0.01(V-30)\}W + 0.0013\,AV^2\right]$$

where

BHP = Brake Horse Power produced by the engine
V = Road Speed (mph)
W = Vehicle Weight (lbs.)
A = Vehicle Frontal Area (ft^2)

(Table 1, continued)

Note: if vehicle speed is less than 30 mph the (V-30) term is given a value of zero (i.e. it can never be negative).

Assumed values for constants and conversion factors:

1 mph = 45 rpm
W = 2600 lb.
A = 20 ft^2

885065

Adaptive Ignition and Knock Control

M. Holmes, D. A. R. W. Willcocks, B. J. Bridgens

ABSTRACT

This paper will explore the knock and emissions control capabilities of the Lucas Adaptive Ignition System.

A detailed description is given of the operation of the adaptive knock control together with its integration with efficiency seeking control. Comparisons are made between the performance of adaptive and conventional knock control strategies.

Results are presented on the ability of the adaptive ignition system to control the trade-off between NOx and fuel consumption on a number of vehicles. Operation on lean burn engines will be studied with particular reference to the influence of adaptive ignition on the lean and rich limits of operation.

Suggestions are made on the simplified application strategy for the adaptive ignition system.

OPERATION

THE LUCAS ADAPTIVE IGNITION SYSTEM controls spark advance in order to optimise the performance of individual engines in terms of fuel economy, emissions and knock.

Two feedback parameters can be used to control spark timing: the slope of the torque versus spark advance curve and an accelerometer knock sensor. The technique for measuring the slope of the torque curve involves using small perturbations in spark advance to infer the slope of the curve from the engine speed response to the perturbations.

A block diagram of the system is shown in figure 1. The knock sensor feedback is optional depending on requirements.

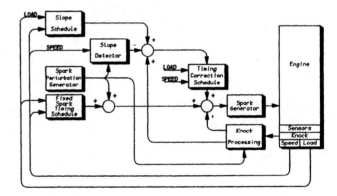

Fig 1. Block diagram.

PERTURBATION PATTERN - For a four-cylinder engine the perturbation pattern is as in figure 2.

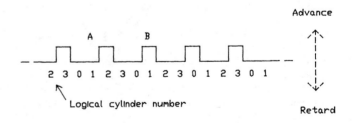

Fig 2. Ignition angle perturbation waveform (4 cyl. engine).

The gradient of the torque curve is inferred from the change in engine speed between an advanced and a retarded fire of the same cylinder. The waveform is asynchronous to the cylinder repetition cycle and therefore each cylinder contributes an equal amount to the signal.

If the speed change between cylinders 0 and 1 at points A and B are subtracted, a signal is obtained which rejects linear engine accelerations. The perturbation is of a higher frequency than the resonant frequency of the drivetrain and therefore as engine speed, and hence perturbation frequency, increase, the gain of the system will fall and the signal level will drop but so also will the noise level, so that a similar signal to noise ratio will result. A similar arrangement is used for a six-cylinder engine with a perturbation pattern as below (fig. 3), which is again asynchronous to the cylinder repetition cycle. In this case a torque-gradient signal can be calculated every other fire compared with every three fires in the four cylinder case.

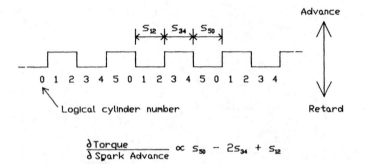

$$\frac{\delta \text{Torque}}{\delta \text{Spark Advance}} \propto S_{50} - 2S_{34} + S_{12}$$

Fig 3. Ignition angle perturbation waveform (6 cyl. engine).

Once a correction to spark timing has been calculated, this is stored in a non-volatile correction map.

MAPPED CONTROL - In real driving situations the operating conditions of the engine are continually changing with the driver's response to varying road conditions. It is difficult for a control system with a single integrator element to react fast enough to these changes whilst rejecting unwanted noise under steady-state conditions. Changes in load can be particularly rapid. One way of improving the transient performance is to use a mapped ignition system. Digital mapped ignition is widely used in modern vehicles but in a fixed schedule manner. The operating conditions of the engine are represented by a point on a square two-dimensional array of, typically, 256 sites, indexed by speed and load. In our adaptive system each site in the correction map is an integrator and is read from and updated when it is one of the four nearest neighbours of the current operating point. Linear interpolation is used when reading from the map and also for allocating the correction signal between the four sites when writing to the map. This technique is described in more detail in (1).*

* Numbers in parentheses designate references at end of paper

KNOCK CONTROL

In petrol engines it is often not possible to operate at the spark advance which would correspond to the peak of the torque curve owing to the onset of knock, particularly at high loads.

If knock were allowed to occur unchecked, progressive damage would be sustained by the piston and valves within the cylinder on account of the rapid rises in in-cylinder pressure associated with knock.

An adaptive ignition system using only torque-gradient feedback would cause the engine to knock at any condition where the sought spark advance setting was advanced of the knock limit. Some strategy must therefore be used to prevent this from happening. One way of preventing knock is to program the controller to seek a certain gradient of the torque curve rather than the maximum. This gradient can be chosen such that knock does not occur over the spread of production engines. If however the engine type is substantially knock limited, or if varying fuel quality is anticipated, it is better to use a sensor to detect the occurrence of knock and adjust the ignition timing accordingly.

Traditional methods of knock control, widely used in non-adaptive systems, rely on information obtained from a piezo-electric accelerometer mounted on the engine block. This can be either broadband or tuned, but the frequency component measured in either case is that characteristic of the knock signal which is typically around 6kHz. The output is rectified and integrated every fire but the signal-to-noise ratio available from this method is poor and so the signal from the sensor is usually gated to give an output only during the window of crank angle around TDC where knock is likely to occur. When the signal rises significantly above the background average a knock is deemed to have occurred. The ignition is then retarded and subsequently ramped back at a fixed rate until knock is detected again. However the signal quality is such that, especially at higher engine speeds where mechanical vibrations are greater, it is difficult to distinguish a true knock from the background noise level. The ignition timing is therefore programmed away from knock and it is also common to allow a safety margin between the spark advance used and that at which knock is likely to occur for that speed and load in order to allow for variability in fuel quality and in the characteristics of individual production samples. Unfortunately this results in a reduction in the fuel economy of the engine.

Lucas has improved the performance of knock control systems in two ways: firstly by using a novel signal processing technique which correlates knock with spark advance, and secondly by using a map to retain information about knock at different engine conditions.

CORRELATED KNOCK - We have developed a means of improving the signal to noise ratio of a conventional knock sensor so that closer control of spark advance can be used, with associated benefits in fuel economy. The process used relies upon the increase in average knock intensity with spark advance. The output of the knock sensor is measured for successive fires on each cylinder with one spark slightly advanced of the other. The retarded measurement is then subtracted from the advanced measurement, giving the gradient of knock with spark advance at that condition, ("correlated knock"). Not only does this process remove most of the background noise, (since this will be approximately the same for each fire), but the derived signal will have a mean level of zero during conditions of zero knock allowing the signal to noise ratio to be improved still further simply by integration of successive signals. The mean level of knock can then be controlled by setting the ignition timing to give any desired position on the knock gradient characteristic.

Figure 4 shows measurements obtained from an engine running at a medium load and high speed without knock control while the ignition timing is steadily advanced. The correlated knock signal can be compared with the raw knock signal and with the knock signal obtained from in-cylinder measurements. The apparent onset of knock is well correlated between in-cylinder measurements and the knock-gradient measurement, and both are more useful than the raw knock signal. The knock-gradient technique also has a great advantage over in-cylinder techniques in terms of the number of sensors required per engine and sensor unit-cost.

The small changes, or perturbations, in spark advance applied to successive fires follow the same set pattern which is used also to detect the gradient of the torque curve at the operating point.

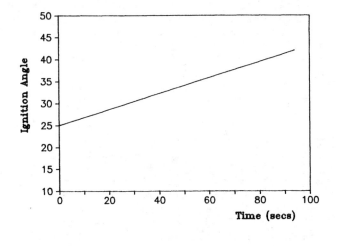
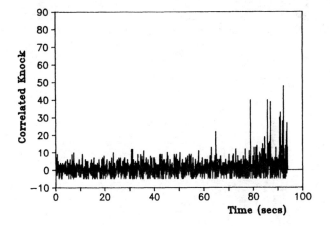
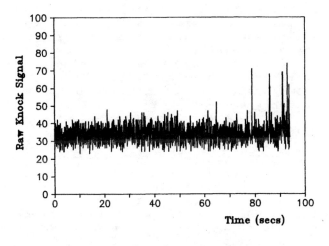
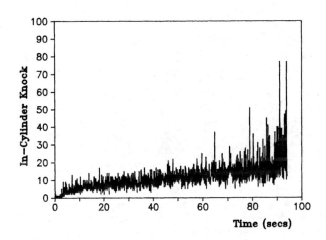

Fig 4. Knock signal comparisons.

INTEGRATED CONTROL - We now have two feedback signals from the engine which can be used to control the spark advance setting using the simple control strategy of figure 5.

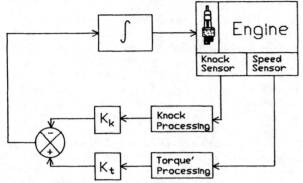

Fig 5. Schematic of combined control loop

At a knock limited condition the signals will oppose one another; the torque-gradient signal will be trying to advance the spark while the knock-gradient signal will try to retard it. Equilibrium will be attained when these two are equal.

This is the simplest form of control but it is found to work well. The gains K_k and K_t can be chosen to give the best compromise between steady-state and transient performance, as illustrated by figure 6.

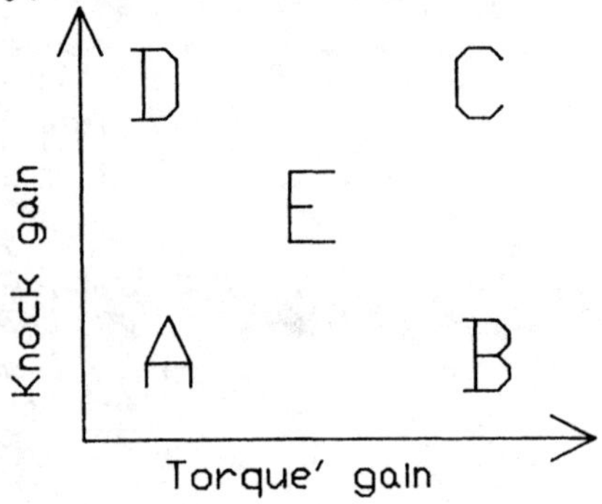

Fig 6. Effect of different settings of controller gains K_k and K_t.

In region A the control system is overdamped and responds very slowly to changes in engine characteristics. The opposite is true in region C where the spark timing will vary greatly as noise affects the system. Region B would give good torque output but would have a tendency to knock while region D would allow very little knock but the overcompensation after knock would lead to unnecessarily reduced torque output. Obviously the best choice of gains lies somewhere within region E, with K_k and K_t finely tuned to give the most acceptable trade off between the average engine efficiency and the level of knock. This point can be chosen from empirical results.

The use of combined torque gradient and knock sensing means that the rate of advance after knock is fast where a large torque loss would otherwise be incurred. If a more rapid transient response to knock is required, an additional fast-acting correction may be made to the spark advance without substantially affecting the optimised setting of spark advance. It should be noted however that, since knock is a somewhat random process, the system performance is rather more difficult to control than would be a deterministic system.

RETENTION OF KNOCK LIMIT - With a mapped ignition system, as the operating conditions change so the spark timing changes to the value which is the expected optimum without knock for those conditions. A conventional non-adaptive mapped system might cause knock then retard in response, and slowly ramp back at a fixed rate towards the original spark timing until further knock occurs. This is wasteful in terms of fuel economy and performance since the retard will often overcompensate and torque output will be unnecessarily reduced. The adaptive ignition system will provide the most suitable ignition timing, learned from previous experience, immediately the speed or load changes. Moreover after knock at conditions where the torque maximum is significantly advanced of the knock limit the ignition will tend to advance more rapidly, minimising any torque loss.

Performance with different Fuel Qualities - In the presence of varying fuel quality an adaptive ignition system will behave better than a fixed map with knock control. A higher octane rating will cause the ignition timing to advance to the new knock limits while a lower octane rating will retard the map at knock limited sites. The map will gradually converge to suit the characteristics of the new fuel. A vehicle fitted with adaptive ignition will therefore gain maximum benefit from any quality of fuel, without risk of damage to the engine, although power output at high loads would suffer as a result of retarded ignition when using low octane fuels.

EMISSIONS CONTROL

SLOPE CONTROL - In order to evaluate the performance of the Adaptive Ignition System when operating to slope values on the torque against spark advance curve to reduce NOx, a multi-vehicle test program was undertaken. During the tests systems which optimised to the top and to a retarded slope of the torque curve were evaluated. These are designed to give maximum fuel economy and reduced NOx from the engine respectively. A series of ten medium sized vehicles fitted with 1.6 litre four cylinder carburetted engines were tested.

Test Procedure - All results were obtained by using the ECE 15.04 type test from a hot start.

The target slope values were mapped in a 16 by 16 map indexed by engine load using manifold pressure and engine speed. Slopes for conditions between breakpoints were looked up by linear interpolation.

Three different 'slope maps' were tested, together with the standard mapped ignition system fitted to the vehicle. The standard system was used as a base line for comparisons between vehicles. The three slope maps which were used are described below. The first map, designated 'Zero Slope' contained all zeros and so would cause the system to seek the top of the torque curve. The other two maps contained retarded slope targets in the part of the map where NOx emissions are important. The values of these slopes were 15 and 30 slope measurement units and the maps were designated 'Half Slope' and 'Full Slope' respectively.

The value of full slope used was selected to give a significant reduction in NOx emissions. The value was chosen by performing a test on a single vehicle to determine the slope that gave the required reduction in NOx emissions for that vehicle.

It was found that a single slope value over the whole NOx sensitive area gave good results. The probable reason for this is that the fuel consumption penalty increases with the square of the retard from optimum whereas NOx is approximately linear with spark advance. Therefore if the amount of retard is minimised at any one map site by distributing it equally around the map the best NOx to fuel economy trade off should be obtained.

The Half Slope value was tested to give more understanding of the variation on NOx emissions with slope. These values were then used for all of the ten vehicles in the main test.

The adaptive system was always driven for 8 drivecycles (which is equivalent to two ECE 15.04 tests) before testing began. These 'learning cycles' allowed the adaptive maps to be built up to maturity. The adaptive map was cleared prior to each set of learning cycles.

For each of the four systems tested, two ECE 15.04 tests were conducted back-to-back. This allowed a check to be made on any experimental errors present in the results and ensured, for the adaptive systems, that all learning was complete. The full test program undertaken for each vehicle is given below.

1 ECE Test - Standard Ignition.
8 Zero Slope learning cycles.
2 ECE Tests - Adaptive Zero Slope.
8 Half Slope learning cycles.
2 ECE Tests - Adaptive Half Slope.
8 Full Slope Learning cycles.
2 ECE Tests - Adaptive Full Slope.
1 ECE Test - Standard Ignition repeat.

The above is only an example schedule. The order of the adaptive tests was switched around from vehicle to vehicle to remove any effect of test ordering from the results. The standard ignition test was always performed at the start of testing and repeated at the end of the day. The results from these two tests were compared to verify that no major changes in the vehicle or in test conditions took place in the day as these would cause the two results to differ. As long as these two results were the same to within the expected experimental error it was assumed that the other results from that days testing were valid.

Results - Figures 7 and 8 show the major results obtained from the tests.

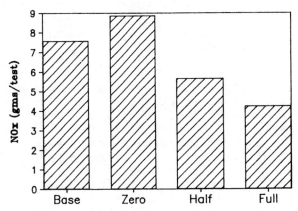
Fig 7. Average NOx emissions.

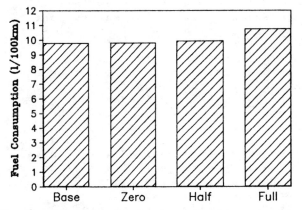
Fig 8. Average fuel consumption.

Figure 7 shows the values of NOx emissions for each of the systems averaged over all of the vehicles. Figure 8 shows the average fuel consumption figures.
The fuel consumption was not measured directly but calculated from the emissions carbon balance.

These results allow the effect of the Adaptive Ignition system in reducing NOx emissions and the fuel consumption penalties imposed by this reduction to be assessed. The results show that half slope operation gave a 17% reduction in NOx compared with fixed timing and a 36% reduction compared with zero slope. This was at the expense of 1.5% in fuel economy. At full slope the trade off worsens with a 40% reduction compared with fixed timing and a 52% reduction compared with zero slope at the expense of 10% in fuel economy.

Analysis of the car to car variability of NOx shows improved control of the spread of NOx at full and half slope compared with zero slope. This may be due to the fact that the top of the torque curves are relatively flat allowing a small amount of wander of the spark advance.

Conclusion - It has been shown that by using the Adaptive Ignition system to control to a slope value is a particularly effective way of controlling the NOx - fuel economy trade off. The fuel consumption increase in the Full Slope test was larger than expected, this was caused by the initial choice of slope value being too large and causing too great a degree of retard on certain vehicles. This effect can be explained because the specifications of the vehicles concerned had been changed and changes in engine or transmission dynamics can cause a given slope target to give different degrees of retard. This problem can be overcome by creating a new slope map when engine or transmission dynamics are changed.

LEAN BURN EMISSIONS CONTROL - As the adaptive system responds to variability between engines and in the environment, the use of adaptive control will give better control of emissions (NOx and HC), particularly on lean burn engines.

To demonstrate this we can look at two major forms of variability:

AFR Variability - Operation on lean burn engines normally requires tight control of AFR (2). If the AFR is too rich then NOx will become too high and if the AFR is too lean then HC will increase dramatically and driveability will become unacceptable. It is also important to use the correct spark advance but this depends on AFR. Adaptive ignition control will produce different spark timings at different AFR's and it can be shown that these will compensate for tolerances in fuelling components.

Richer engines will suffer from too much NOx but the richer AFR will have a faster burn rate resulting in a more retarded spark timing using adaptive control. This retard will reduce the higher NOx caused by the richer AFR.

Leaner engines will suffer from high HC and poor driveability because of partial burn effects caused by insufficient time for the mixture to burn. Adaptive control will advance the spark timing because of the slow burning mixture and reduce the likelihood of partial burn occurring.

Figure 9 shows NOx + HC for a high swirl lean burn engine as a function of AFR. Curves are shown for fixed timing corresponding to MBT at the nominal AFR of 20:1 and for timing corresponding to MBT at each AFR to predict adaptive performance. For the reasons outlined above the minimum of NOx + HC is flatter using adaptive control allowing in most cases an extension of the AFR window of operation by 1 AFR in both the rich and lean directions for the same level of emissions.

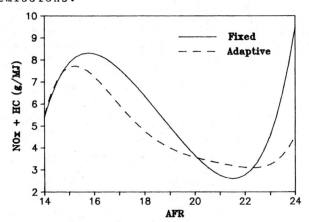

Fig 9. Fixed and adaptive NOx + HC against AFR.

Compression Ratio Variability - Higher compression ratio engines also produce higher NOx because of faster burn rates. We would expect adaptive control to compensate for this by retarding the spark timing and reducing the rise in NOx.

APPLICATIONS ENGINEERING

Applications engineering is of course easier for adaptive systems but there is still the need to map areas where maximum torque or fuel economy is not desired. These areas are mapped in terms of a fixed spark advance for idle and overrun and in terms of a target slope value for the other areas.

IDLE AND OVERRUN - There is no advantage in controlling to a position on the torque curve under these conditions. Spark advance is calibrated in the normal way and adaptive control can either be disabled or the fixed timing can act as a constraint in the advanced direction to the adaptive optimisation.

EMISSIONS CONSTRAINED AREAS - As previously mentioned, under these conditions there is a distinct advantage in allowing adaptive control to operate; particularly on lean burn engines. It may be necessary to operate to a target slope value corresponding to a retarded spark timing. This target slope would need to be mapped. This can take place by the normal method of adjusting spark timing until the desired emissions levels are reached at each operating condition. To record the slope measurement a special piece of equipment is supplied which can make the spark advance adjustment and measure the target slope value. This value is then fed into a slope map for each particular operating condition.

However we have found a simpler method which takes advantage of the fact that we are operating relative to the optimum timing at each operating condition even though the spark advances may be quite different. We have found that if a single slope value is used over the whole emissions limited area this will give a precise control of the economy against emissions trade-off at each operating point.

Using this method mapping can be accomplished in a matter of days simply by driving the car over the emissions cycle using first one slope value and then another slope value and interpolating between them if neither gives the desired emissions values.

When a large slope value is used it may be necessary to blend this in gradually with the zero slope values surrounding the emissions sensitive areas to improve driveability.

KNOCK LIMITED AREAS - The applications strategy will depend on whether closed loop control with a knock sensor is used or open loop control without a sensor is used.

Open Loop Control - Applications work is carried out in the normal manner by adjusting the spark advance at knock limited conditions to maintain a safety margin away from the knock limit. Target slope values are measured at these conditions using the special purpose equipment. These slope values are then used to produce a slope map over the knock limited areas.

Closed Loop Control - In this case the slope map is simply set to zero and the knock control will prevent the system adapting into the knock region. Some application work will be necessary to choose the gain of the knock control loop in order to produce the required knock sensitivity.

Obviously all the normal applications work to decide the optimum location of the knock sensor and the pass-band of the knock filter will need to take place. However these factors will not be so critical because of the improved signal to noise ratio using the knock correlation technique.

CONCLUSIONS

We have shown that the performance of a spark ignition engine can be optimised in terms of fuel economy and improvements to emissions control can be made using torque slope feedback of ignition timing.

The improved emissions control performance will compensate for errors in fuelling equipment allowing cheaper, less sophisticated fuelling systems to be used. Accurate control of the NOx against fuel economy trade-off can be made by optimisation to non-zero slope values on the torque curve.

Knock control using conventional accelerometer sensors can be improved using an improved signal processing technique which allows reliable knock control to take place even at high engine speeds. Further improvements result from exploiting the learning ability of the correction map already present in the adaptive system.

The use of the engine speed signal for torque slope feedback and the improved technique for processing accelerometer knock sensor signals gives improved ignition timing control without the need for costly in-cylinder sensors.

ACKNOWLEDGMENT

This work has been undertaken for the Gasoline Engine Systems Division of Lucas Automotive from whom support and permission to give this paper is gratefully acknowledged.

REFERENCES

1. Wakeman AC, et al, "Adaptive Engine Controls for Fuel Consumption and Emissions Reduction", SAE 870083.

2. Cucchi C, Cavallino F, "2.0 Litre CHT Engine - Development of a Lean Burn System", EAC International Conference on New Developments in Powertrain and Chassis Engineering, Strasbourg June 1987.

890673

Variable Valve Actuation Mechanisms and the Potential for their Application

Richard Stone and Eric Kwan

Brunel University
London

ABSTRACT

The numerous variable valve actuation mechanisms for poppet valves need to be classified, if sensible comparisons are to be made, and one possible taxonomy is presented here. Not all the mechanisms proposed have been tested, but where they have it is usually with gasoline engines. It is well established that controlling the valve events can raise and flatten the torque curve. However, it is difficult to quantify and compare the gains in torque and consequential reduction in fuel consumption, as the results depend very much on the starting point. This is also the case when variable valve actuation is used to reduce engine emissions. Fortunately it is quite easy to realise suitable variable valve timing systems for controlling the valve overlap, and the point of inlet valve closure.

The other main application to gasoline engines, is in obtaining load control without throttling. The thermodynamic background is discussed here along with some experimental results; the section ends by assessing the suitability of some of the mechanisms for this purpose.

Finally, the application of variable valve timing to diesel engines is discussed. It is argued that the control of valve overlap can flatten the torque curve of some diesel engines, and that the control of the inlet-valve-closing angle should lead to improved starting. However, the most promising application on diesel engines, is the control of the valve overlap on highly turbocharged engines.

Since an earlier review of variable valve timing (1)[*] there has been a proliferation of both mechanisms and and papers on variable valve actuation (VVA). However, efforts are still almost exclusively directed towards spark ignition engines, in order to increase the torque curve, and to reduce the part load throttling losses. None the less, attention will be drawn here to other uses of variable valve actuation, and some possible diesel engine applications. Work is still also being undertaken on the potential for reducing emssions by using VVA.

It is difficult to make a paper such as this both compact and comprehensive; inevitably there are omissons. The next section categorises the mechanisms and illustrates each main type by a typical example. Where possible, purely theoretical studies are identified as such. The various reasons for applying variable valve timing are discussed in the subsequent sections, and the different references are segregated, as to whether they contain experimental results or theoretical/computational predictions.

In the section on gasoline engines, particular attention is paid to the reduction of the throttling losses at part load, and it is argued that few of the proposed mechanisms are suited to this task. There is also a discussion of the role of inlet valve closure and valve overlap on gasoline engine performance, and the mechanisms that might achieve these variations.

In the following discussion of variable valve actuation systems mechanisms that have been produced and tested are identified as such.

VARIABLE VALVE ACTUATION SYSTEMS

TAXONOMY - With so many VVA systems being developed, it is useful to classify the systems, and one possible taxonomy has been developed in Figure 1. Not all the mechanisms have been realised, and some that have been developed are best suited to experimental investigations of the effects of valve timing. Only a few of the simplest mechanisms have been used in production applications. Some of the mechanisms that have

---
[*] Numbers in parentheses designate references at the end of the paper

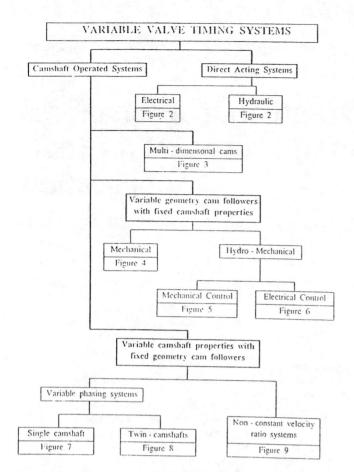

Figure 1 Taxonomy of Variable Valve Actuation (VVA).

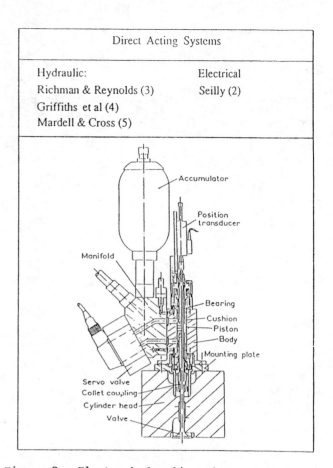

Figure 2 Electro-hydraulic valve actuator (4).

been proposed, and indeed some of the mechanisms that have been produced, are not suitable for use because of discontinuities introduced into the valve motion. Figure 1 should be self explanatory: examples of each type of system with an illustration are given in the following Figures.

DIRECT ACTING SYSTEMS – The direct acting valve actuation systems (Figure 2), may appear attractive because of their flexibility for controlling the valve lift and period. However, the disadvantages are the power requirements and size, which make these devices most suited to development work. The development of the Helenoid, a fast acting solenoid is described by Seilly (2), and the limit on operating speed corresponds to 6000 rpm, but with peak currents of 150 amps when operating with a 48V supply.

The hydraulic system described by Richman and Reynolds (3), only achieved the desired response at speeds up to 1000 rpm, and the discrepancies increased continuously up to the maximum operation speed of 3000 rpm. A more recent system described by Griffiths, Mystry and Phillips (4) maintained stable and accurate control up to the equivalent engine speed of 3000 rpm. The valve lift data are stored in a memory with 1024 8 bit locations; the actuator system is shown in Figure 2. However, no experimental results from engine performance appear to be available yet from any of these systems. Mardell and Cross (5) describe a hydraulic system with solenoid control valves, that achieve fast operation by adopting the design philosophy originating with the Helenoid (2). Mardell and Cross present computer simulations of valve operation, and the effect of the valves on engine performance; experimental results are not presented.

CAMSHAFT OPERATED SYSTEMS – The remaining systems described here all make use of one or more camshafts. Multi-dimensional cams are discussed first, and this is followed by a description of variable geometry cam followers with fixed camshaft properties, and systems and variable camshaft properties and fixed geometry cam followers.

Multi-dimensional cams – The simplest VVA system would appear to be a multi-dimensional cam, with axial movement to select the desired cam profile. The multi-dimensional cam system proposed by Simpson - Dalzell (6) leads to a theoretical point contact. This is clearly unacceptable in one of the most severely loaded elements in the engine, and not surprisingly there is no evidence of its use. The mechanism described by Azouz et al (7), would appear to suffer the same shortcoming, despite employing a roller follower. However, this mechanism has been used on a test engine to vary the inlet

Variable camshaft properties, with fixed geometry cam followers:
Variable phasing systems with multiple camshafts
Elrod & Nelson (36)
Lenz et al (37)
Saunders & Rabia (38)

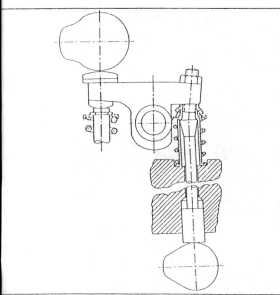

Figure 8   VVA mechanism for late inlet valve closure (38).

able position idler wheels to control the
th of the timing belt between the crankshaft
ey and the camshaft pulley on twin overhead
haft engines. Sapienza et al (34) present
showing the effect of the inlet valve
sing angle on the torque curve, whilst the
system (35) has only been used on a bench
.

The VVA systems that employ two camshafts in
bination, use a variable phasing mechanism
een the camshaft drives, to control their
bined action. The system developed by Elrod
Nelson (36) employs concentric shafts
ing separate cam lobes, both of which act on
same follower. Such a system introduces
continuities into the valve motion, but it
used in an experimental engine to reduce the
r output without throttling. The approaches
d by Lenz et al (37) and Saunders and Rabia
) are similar, and are illustrated by
ure 8. However, Lenz et al (37) developed
ir mechanism to obtain early inlet valve
sing, whilst Saunders and Rabia (38) were
king late inlet valve closing: both
eriments were investigations of the benefits
m using inlet valve closure control, to
uce the throttling losses.

The most complex VVA mechanisms are those
t have non-constant velocity drives to the
shaft. Fuenmayor et al (39) analyse some

Variable camshaft properties, with fixed geometry cam followers:
Non constant velocity ratio systems
Roe (17)
Payri et al (41)
Scott (42)
Parker & Kendrick (43)
Ma (44)

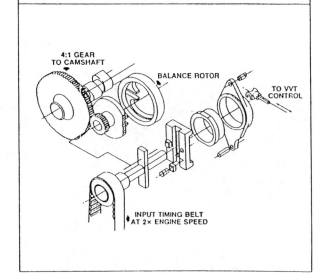

Figure 9   The Ford 4:1 harmonic drive VVA, Ma (44).

four-bar mechanisms that can be used for this type of VVA systems, whilst Freudenstein et al (40) present a fuller and broader treatment, that describes the synthesis and analysis of VVA mechanisms. The mechanism described by Roe (17) uses an epicyclic gear train in a differential mode that can take a modulating input. Payri et al (41), and the Mitchell system described by Scott (42), use a four bar linkage to obtain a non-constant velocity system. The remaining systems (43, 44) utilise sliding elements to produce non-constant velocity variable valve timing systems. In general, a harmonic motion is superposed on the constant angular velocity input. The systems of Payri et al (41), Mitchell (42), A E Developments (38) and one of the Ford systems described by Ma (44) have one harmonic cycle per camshaft revolution, and this necessitates multiple mechanisms for multi-cylinder engines. However, Ma (44) also describes a system that gives 4 harmonic cycles per camshaft revolution, it is shown here in Figure 9. When this mechanism is appropriately phased with the valve events, there can be simultaneous reduction or increase, of the valve periods (and overlap) on a four cylinder single camshaft engine from a single mechanism. The papers describing these mechanisms (41, 42, 43, 44) all have engine test data.

Multi - dimensional cams
Simpson - Dalzell (5)
Azouz et al (7)
Titolo (8)

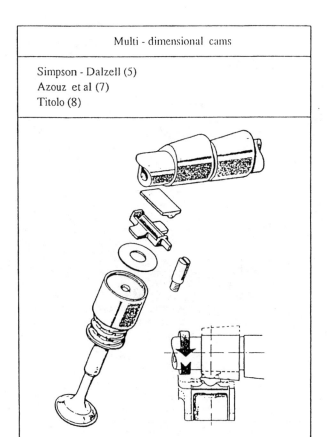

Figure 3   The Fiat Auto multi-dimensional cam system (8).

valve closing angle from 30 to $55°$ abdc, with a fixed inlet valve opening, in order to confirm predictions from a computer simulation.

By employing a tilting pad cam follower, Titolo (8) ensures line contact between the cam and follower, Figure 3. The tilting pad has a cylindrical seating that allows the follower to rotate about an axis perpendicular to the camshaft axis, timing variations are obtained by axial movement of the camshaft. This system has been used on a 2 litre Lancia engine, and a 4 litre V8 Ferrari engine with 4 valves per cylinder. Gray (9) provides a summary of the system performance; at high speeds the valve period increases by $42°$ to $270°$, and the lift increases by 2.97mm to 9.3mm. The phase angle of the cam has to remain constant, otherwise the cam surface would become hollow and destroy the line contact with the flat follower. However, the camshaft drive can incorporate a variable phase drive of the type shown in Figure 7. The line contact is also important, as it implies that the cam can be manufactured without recourse to a profiled grinding wheel.

Variable Geometry Followers, Fixed Camshaft Properties - VVA systems with fixed camshaft properties can employ either purely mechanical variable geometry cam followers, or systems with hydraulic elements. Figure 4 shows the system described by Torazza (10), in which a rocking

Variable geometry cam followers with fixed camshaft properties: Mechanical
Torazza (10)
Zappa & Franca (14)
Stivender (12)
Paulmier (13)

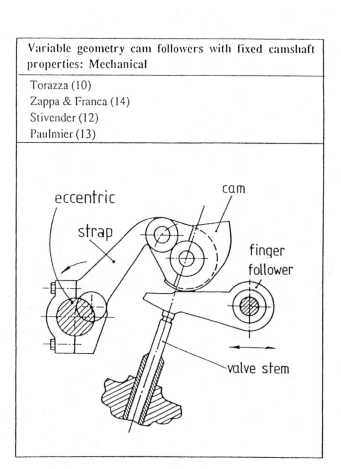

Figure 4   The Torazza (10) variable geometry cam follower system.

cam is driven by an eccentric and strap, with a variable position finger type follower between the cam and follower. One such system had scope for a $70°$ increase in the valve period, through a $40°$ delay in valve closing with a $30°$ delay in valve opening. Torazza presents data from tests on a gasoline engine at full throttle that show a gain in torque and reduction in fuel consumption at all speeds. The system described by Zappa and Franca (11) is similar, but employs a conventional cam with a variable position roller follower, in order to control the inlet valve closing angle. Stivender (12) used a system in an overhead valve engine that had control of the rocker pivot position, thereby controlling the valve lift and the power output of a gasoline engine. Any changes to the valve timing, would thus be a side effect from the way the valve train clearances were taken up. Paulmier (13) employed a system with a lightly loaded valve spring, in which the cam and valve train would lose contact at high speeds, thereby increasing the valve period and lift. The Paulmier system has been used on both a rig and an engine, with the aim of increasing the low speed torque and reducing emissions.

Variable geometry cam followers with hydraulic elements can have either mechanical control (14, 15, 19) or electrical control (20, 21). Herrin and Pozniak (14) used a flow rate sensitive valve incorporated into a hydraulic tappet. When the cam follower exceeded a certain speed the tappet would transmit the cam motion. This leads to a degree of lost motion, such that the valve period and lift are reduced at low speeds, as illustrated by Table 1.

Table 1. Measured performance of the hydraulic tappet system described by Herrin and Pozniak (14).

Engine speed rpm	Inlet valve opening btdc°	closing abdc°	lift mm	Exhaust valve opening bbdc°	closing atdc°	lift mm
450	5	53	9.76	50	9	9.76
800	9	64	9.97	55	20	9.97
2000	13	74	10.12	58	30	10.12
4000	15	78	10.15	59	34	10.15

The system devised by Charlton and Shafie-Pour (15, 16) is shown in Figure 5. The motion from the main cam is transmitted through a hydraulic chamber which has the facility for an additional motion input. This is the hydraulic analogue of the system devised by Roe (17), who used an epicyclic gearbox in a differential mode. However, as well as having a potential for control of the opening and closing events, the Charlton and Shafie-Pour system can also control the valve lift. The magnitude of the input from the secondary cam to the modulation piston, is controlled by a rocker arm with a variable pivot position. The problem of defining the motion from the secondary cam is solved very elegantly in this case, as the primary cam is a polydyne (18) design. As the primary cam is defined as a polynomial function, it can be shown that the motion from the secondary cam is also a polynomial function whose coefficients are all the same function of speed. The system performance has been computer modelled, and the predictions compared with experimental data from a test rig that simulates the exhaust blow-down. The Roe (17) system does not seem to have been applied to an engine.

Nuccio (19) describes a hydraulic tappet in which the pressure can be relieved by a rotating spill valve. The phasing between the spill valve and the camshaft, determines the point at which the valve starts to close. The impact of the valve on its seat is limited by a hydraulic damper. Nuccio used this system to control the power output of a gasoline engine without recourse to throttling.

There are several examples of variable geometry cam follower systems in which hydraulic elements are controlled electrically (20, 21). Payri et al (20) use a hydraulic tappet similar to that of Nuccio (19); the pressure is relieved here by an electrically controlled valve, again allowing the valve lift to be reduced and the valve to close earlier. Payri et al (20) have

Variable geometry cam followers with fixed camshaft properties:
Hydro - mechanical with mechanical control

Herrin and Pozniak (14)
Charlton and Shafie - Pour (15, 16)
Nuccio (19)

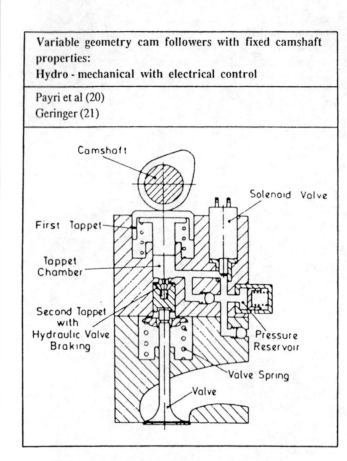

Figure 5 The hydraulic VVA system developed by Charlton and Shafie-Pour (15,16).

produced their VVA system but not applied it to an engine; their engine performance data are derived from simulations. The system described by Geringer (21) appears to be similar, and is shown here as Figure 6.

Fixed Geometry Camshaft Systems - Most VVA systems have fixed geometry cam followers, but variable phasing between the crankshaft and the camshaft(s). When the phasing is varied continuously within a cycle, a non-constant velocity ratio system is obtained, and the valve period and timing can both be affected. Only variable phasing systems using more than one camshaft can obtain variable lift. Variable phasing systems are by far the most common VVA system, and the phasing variation is either achieved in the coupling between the camshaft and the cam wheel (6, 22, 23, 24, 25, 26, 27, 28), or in the drive to the cam wheel (29, 30).

Meacham (22) tested a variable phasing system applied to a single camshaft, in which the camshaft was advanced by up to 36° ca, in response to the manifold vacuum. The control strategy was developed for tests with varied cam timings, and the aim was to reduce emissions by controlled EGR. Emissions of NOx and HC are presented for varying loads.

The Alfa-Romeo variable phasing system (23) has entered production, with a phasing variation of 32° ca, the control strategy is not entirely clear from the published sources (23, 24, 25, 26). The Renold variable phasing system (27) has a phasing range of 30° ca, but results from its application to a 6 cylinder, twin camshaft, 24 valve engine have not been published.

Schiele et al (28) had a 45° ca phasing variation applied to a single camshaft engine, and presented full load performance data, with some emissions and fuel consumption data for several low speed, low load operating points (below 1600 rpm, 2.91 bar bmep). Stojek and Stwiorok (29) had a variable phasing system applied to both the inlet and exhaust camshafts, and their results are discussed later.

The Toyota variable phasing system (307) has a 20° ca range, but no performance data are presented. Freeman and Nicholson (31) conducted a similar investigation to Meacham (22), on the effect of phasing variation of a single camshaft on the emissions of HC and NOx. The phasing control system was tested in a vehicle by Freeman and Nicholson (31), but like the Alfa-Romeo system (23), the unit did not have a progressive phasing action.

Mercedes-Benz (32) have announced the use of a variable phasing system on the inlet camshaft, for two high performance engines scheduled for 1989. At low engine speed, as well as during part load, an electronically controlled unit

Variable geometry cam followers with fixed camshaft properties:
Hydro - mechanical with electrical control

Payri et al (20)
Geringer (21)

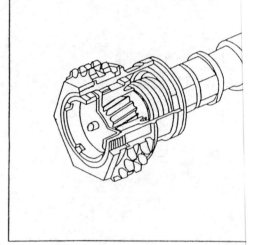

Figure 6 The Geringer (20) VVA hydro-mechanical system with electrical control.

VVA In Practice - Not surprisingly, it is the simplest mechanisms that have entered production, notably variable phasing devices. Other systems that would appear suited to production are those based on hydraulic tappets, and the Titolo (8) multi-dimensional cam and follower. The combination of two simple systems could offer a very versatile control of valve timing, more so than a complex system such as the non-constant velocity ratio system, where there are complex (and sometimes undesirable) interactions.

An alternative approach is to have an additional valve in each inlet port, this is usually of a rotary form as described by Walzer (45), and is used as an alternative to early inlet valve closure for reducing the throttling losses. Other elements in the inlet system that can affect the gas exchange process are: swirl control valves, and variable geometry induction systems. Neither of these will be considered further here, but in practice, they should form an integral part of the induction system design.

The requirements for VVA systems are demanding, and not surprisingly they are not always met. Firstly, it must be remembered that the valve train includes some of the most heavily stressed components in an engine. Secondly, if the desired valve motion is to be achieved, then the valve train has to be very stiff, as otherwise small deflections can have a significant effect on the valve acceleration. The valve motion must be continuous, such that the maximum accelerations are within the limits imposed by the valve spring(s) and cam contact forces, and nor should there be discontinuities in the acceleration, as these cause impulsive loads that can induce vibrations in the valve train.

In selecting a VVA system there will be many considerations, including: the ease and cost of manufacture and assembly, durability, reliability, the speed of response of the system, the stiffness of the system, the patent/legal position, and the frictional losses. Frictional losses have a particularly severe effect on the part load efficiency, and might thus negate any benefits attributable to the better control of the gas exchange processes. Clearly the non-constant velocity camshaft drive systems will usually have the highest frictional losses of the purely mechanical systems.

VVA APPLIED TO GASOLINE ENGINES

The reasons for applying VVA to gasoline engines include:

1. Improving the fuel economy at part load. By using early or late inlet valve closure, the throttling or pumping losses can be reduced - this is widely quoted, but not always correctly shown.
2. Increasing the low speed power output (by flattening the torque curve) and

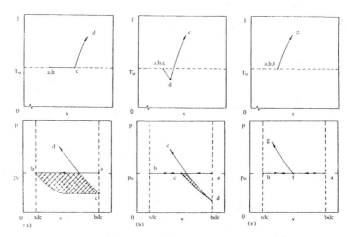

Figure 10 State diagrams for load control by throttling (a), early inlet valve closure (b), and late inlet valve closure (c). In reality all processes are irreversible.
▨▨▨ dissipated work

the full throttle fuel economy. This is usually through the control of the valve overlap and or the inlet valve closure angle. The part throttle performance can be improved by reducing the valve overlap period, to minimise the interference between the induction and exhaust processes.

3. The reduction of emissions.

The discussion developed here is in the context of naturally aspirated engines. In turbocharged engines, the turbocharger characteristics are such that at high speeds the boost pressure would be too high in the absence of a wastegate system. As pointed out by Watson and Janota (46), a valve timing can be used which is optimised for low speed performance (small valve overlap, with early inlet valve closing), and the turbocharger boost pressure can be used to compensate for the lower volumetric efficiency at high speeds. An additional reason for a low valve overlap, is to prevent unburnt mixture flowing directly to the exhaust system.

Results for each of these benefits will be identified here from the literature, including the results obtained from studies where a series of fixed camshaft timings have been used. Particular attention will be paid to the results from reducing throttling losses, and some of the associated underlying thermodynamics will be reviewed. This section ends by defining the requirements of VVA systems for reducing the throttling losses, and examining the suitability of the mechanisms.

LOAD CONTROL WITHOUT THROTTLING - Both early and late inlet valve closing can reduce the mass of charge trapped in the cylinder, without recourse to throttling, and the consequential dissipation of work that is often termed the 'pumping loss'. Figure 10 shows the effect of throttling, early inlet valve closure and late inlet valve closure on the pumping losses.

With early inlet valve closure, the expansion of the gas to below atmospheric pressure, and the subsequent compression back to ambient pressure, will have associated irreversibilities. The irreversibility is such that for a given trapped mass, the specific volume and temperature will be slightly higher than with late inlet valve closure at the commencement of the compression from ambient pressure.

If throttling is considered to be isothermal, then it is evident that the compression temperatures will be significantly higher than with late or early inlet valve closure. It should also be noted that for the same trapped mass, the pressure at bottom dead centre with throttling is higher than with early inlet valve closure, since the throttling process is isenthalpic, whilst the expansion is tending towards an isentropic process with a consequentially lower temperature at bottom dead centre.

Figure 10 has been constructed on the basis of the trapped charge being half the maximum. The properties used have been those of air, whilst in practice the presence of fuel will lower the ratio of gas specific heat capacities, thus moderating the changes in temperature and pressure in isentropic processes. In reality the processes will be irreversible, and should be represented with broken lines, for example, during the expansion from c to d (Figure 10b), heat transfer from the cylinder increases the entropy, reducing the fall in temperature and pressure.

In practice vaporisation of the fuel continues during the induction and compression processes. The assumed isothermal fall in pressure during induction with throttling, b to c (Figure 10a), will aid vaporisation; this does not occur with either early or late inlet valve closure. Furthermore, with early inlet valve closure, the fall in pressure c to d (Figure 10b) is accompanied by a fall in temperature. However, there will be heat transfer from the cylinder to its contents, that should aid the formation of a homogenous mixture. With late inlet valve closure, mixture will be displaced from the cylinder a to f (Figure 10c), and this may be drawn into another cylinder. If backflow occurs through the flow sensing element, be it the venturi of a carburettor or an air flow transducer, then care is needed in the mixture control system design if the correct air fuel ratio is to be maintained.

The Effect of Compression Ratio on the Cycle Efficiency - In all three cases shown in Figure 10, the expansion ratio is the same, but the effective compression ratio is reduced with either early or late inlet valve closure. Figure 11 shows that the expansion ratio has a more significant effect on cycle efficiency than the compression ratio. However, if the compression ratio is reduced too much, the fall in efficiency will not be offset by the reduction in the pumping losses.

The cycle efficiencies in Figure 11 have

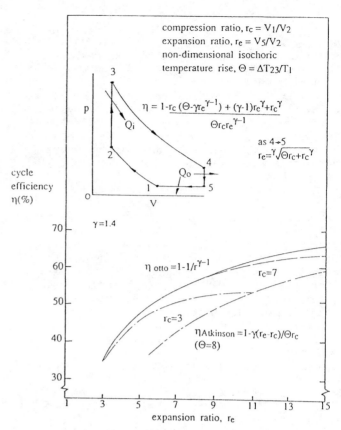

Figure 11 The influence of Compression Ratio and Expansion Ratio on Cycle Efficiencies.

been calculated using the normal air standard cycle assumptions. For different expansion ratios, the efficiency has been plotted for two different compression ratios. These efficiencies, and the limiting case when full expansion occurs (the Atkinson cycle) are influenced by the isochoric temperature rise. As the isochoric temperature rise increases, the divergence from the Otto cycle reduces. Thus with early or late inlet valve closing, and a fixed expansion ratio, there is a slight fall in the ideal air standard cycle efficiency. For example, with an expansion ratio of 9, when the compression ratio is reduced from 9 to 7 the cycle efficiency falls from 58.5 to 58.25%, and even with a compression ratio of 3, the cycle efficiency is still 52.9%. As throttling changes neither the compression nor expansion ratio there should be no change to the cycle efficiency. Full expansion to ambient pressure (the Atkinson cycle) is unlikely to occur as very high expansion ratios are required.

Figure 11 also illustrates the benefits if the clearance volume could be reduced, such that the compression ratio is kept constant. Suppose the compression ratio could be maintained at 7, as the expansion ratio increases from 7 to 11, the cycle efficiency increases from 54.1 to 60.9, but a further increase of expansion ratio to 15 only gives a 2.8 percentage point increase in the cycle efficiency. Luria et al (47) present a similar analysis, and also report on some experiments in which three combinations of the inlet valve closing angle and the clearance volume were tested. The greatest recorded improvement in efficiency was 19% at a load of 2.24 bar, after the expansion ratio had been increased from 7.2 to 22.2, and the inlet valve closure was delayed from 45° to 136° abdc.

Experimental Results From Early and Late Valve Closure - Many data are available from experiments with late (36, 38, 48, 49) and early (19, 20, 37, 41, 49, 50) inlet valve closure. The earliest work by Tuttle (48, 50) used special camshafts with fixed timing, and this method was also used by Hara et al (49).

Tuttle used special camshafts in which the inlet valve closure was delayed by 60°, 80° and 96° from the standard inlet valve closing point of 82° abdc. The gross indicated efficiencies with late inlet valve closure were below those of the conventional engine, but there was a reduction in bsfc that increased to about 14g/kWh at 3 bar bmep - the lowest output without recourse to throttling. The only significant variations in the emissions was lower NOx, and that was attributed to the lower cycle temperatures.

The fall in indicated efficiency was attributable to the lower compression ratio, and a reduction in the burning rate. Hara et al (49) present results for a bmep of 2.1 bar, obtained with the inlet valve closing 60° bbdc and 100° abdc. In both cases the pmep was reduced by about 0.2 bar from 0.6 bar, and this led to a net isfc reduction of about 10g/kWh. When EGR was used to reduce the NOx emissions, the pmep was reduced almost equally in all cases. However, the effect on combustion duration and variability was such that with about 25% EGR, the net isfc was almost the same with conventional throttling, early and late inlet valve closure.

When Tuttle (50) used early inlet valve closure the bsfc was reduced by about 20g/kWh at 3 bar bmep (a 7% reduction). This was a greater improvement than Tuttle had with late inlet valve closure (48). With early inlet valve closure Tuttle (50) found an improvement in the gross indicated efficiency, (the gross indicated efficiency had reduced with late closure). Tuttle attributed the higher gross indicated efficiency with early inlet valve closure, to there being more time for heat transfer to the cylinder contents prior to combustion, thus leading to a higher cycle efficiency.

Elrod and Nelson (36) present limited data, in part because they could only delay inlet valve closure by 40°. At about 3000 rpm this enabled the full throttle bmep to be reduced by 44%; when compared with the same output obtained by throttling, this gave an improvement of about 3.5% in efficiency. Saunders and Rabia (38) could delay the inlet valve closure by 135° abdc from the standard timing of 45° abdc. The greatest gains in efficiency were around 10% in the 4.5 bar bmep range, at lower loads the gains reduced. These results were in broad agreement with their computer modelling predictions, that had included the wave action effects in the induction system.

Nuccio (19) obtained reductions in the fuel consumptions of up to 5.11% at 2000rpm, by using a system that could close the inlet valve early by up to 80° bbdc. Payri et al (20) had a similar system to Nuccio. Ma (51, 52) presents results from a computer simulation of late inlet valve closing, which assumed twin inlet valves per cylinder. By having two inlet camshafts Ma suggest that phase control can be used to obtain late inlet valve closure. When such a system is used in conjunction with a variable compression ratio, the benefits are said to be even greater (52).

It would thus seem, that late or early inlet valve closure, can be used to reduce the throttling losses at part load operation. The greatest gains of 10-15% occur in the 3 to 4.5 bar bmep range. At lower outputs, the reduced compression ratio and lower combustion quality negate the benefits of the reduced throttling losses.

FULL AND PART THROTTLE PERFORMANCE - Much data have been published for gasoline engine performance with results at full and part throttle. Most data are for full throttle, from engines with VVA systems (9, 10, 25, 29, 31, 34, 51, 53). Siewert (54) presents data for performance and emissions obtained from tests in which each valve event was varied independently. Siewert (54) conducted tests at a fixed speed of 1200 rpm and an imep of 3.45 bar, and his results are mostly concerned with the effect of the valve events on emissions.

Effect of Overlap Control on Part Throttle Performance - The part throttle performance can be improved by reducing the valve overlap period, to minimise the interference between the induction and exhaust processes. The gains are greatest under idle conditions (a very important operating point the European Urban Test Cycle) as this is when the inlet manifold pressure is lowest.

Herrin and Pozniak (14) identified that reducing the valve overlap period in a throttled engine, led to an improvement in idle quality as a result of better scavenging. However, they only present results for the engine speed fluctuations and the inlet manifold pressure fluctuations as a function of the valve overlap area.

Ma (51) quantified the effect of valve overlap, as enabling a 200 rpm reduction in idle speed and 12% reduction in fuel flow, and predicted that this could give a 6.1% reduction in the ECE-15 urban cycle fuel consumption.

Alfa Romeo obtained a 28% reduction in the urban fuel consumption when using their variable phasing system (23), which varied the valve overlap period from 32° to 0°. However, it is not clear how much of this reduction in fuel consumption is attributable to: the change from a carburettor to fuel injection, and a possible reduction in the pumping loss through later inlet valve closure.

Gray (19) shows the combined effect of inlet valve closure and valve overlap, on two unspecified part load conditions. At the lower load with the inlet valve closing before 50° abdc, reducing the valve overlap period reduced the fuel consumption. The better control of the scavenging is also usually associated with a reduction in the emissions.

The emissions data from Stojek and Stwiorok (29) presented here in Table 2, show that increasing the valve overlap leads to significant NOx reductions at higher loads, but at low loads the emissions are inherently lower and not reduced much. However, a high valve overlap at low load gives high hydrocarbon emissions.

Table 2 Emissions data from Stojek and Stwiorok (29) at 1500 rpm with 268° period cams and symmetrical overlap.

Emissions	HC (ppm)			NOx (ppm)		
bmep (bar)	1	2.1	4.2	1	2.1	4.2
Overlap (°)						
20	120	100	80	1180	1890	4800
50	250	160	100	820	1210	2860
80	700	350	160	610	680	1040

Full Throttle Performance - In many tests both the inlet valve closure angle and the valve overlap have been varied simultaneously, and this makes it difficult to establish the source of the benefits. The results from Torazza shown in Figure 12 are typical, and the corresponding valve timings are listed in Table 3. As there was no apparent control over the valve phasing, these results may not be the optimum.

Table 3 The Valve Timings used by Torazza (10) for the Results Shown in Figure 10.

Engine speed rpm	INLET			EXHAUST		
	opening °btdc	closing °abdc	period °	opening °bbdc	closing °atdc	period °
1500	10	34	224	47	11	238
4000	24	53	252	59	27	266
6500	40	74	294	71	43	294

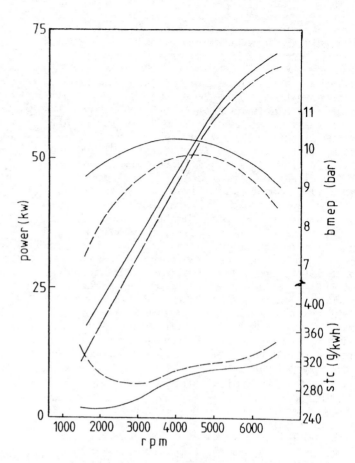

Figure 12 The performance of a VVA system on the full load performance, adapted from Torazza (10).

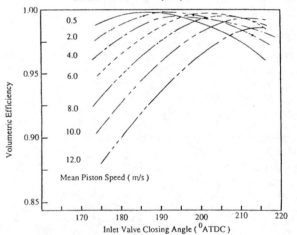

Figure 13 The variation of the optimum inlet valve closure angle with the mean piston speed.

A simple cylinder filling model that neglects pulsation effects, shows that as the engine speed increases, the volumetric efficiency is optimised by having later inlet valve closing angles. Such a result can be made largely independent of engine speed, by plotting the results in terms of the mean piston speed, as shown in Figure 13. The duration of the valve overlap period also has an effect on the volumetric efficiency, but to predict this, it is necessary to model the wave action effects in both the induction and exhaust systems.

Similar results are derived by Nagao et al (55) who present a more rigorous analysis. They show that the volumetric efficiency is a function of the inlet valve closing angle, and the nominal Mach number of the flow through the mean effective valve area into a cylinder, in which the piston is travelling at its mean speed. This confirms earlier work reported by Taylor (56) in terms of the 'Inlet Valve Mach Index'. Asmus (57) considers the influence of all valve events, and suggests that the inlet valve closure angle has the dominant influence in determining automotive engine performance.

In a simulation of a motored low heat rejection diesel engine, Morel et al (58) considered both the timing and the rate of the valve opening and closing on the volumetric efficiency. As before, they found that as the engine speed was increased, then the optimum inlet valve closure angle became later. However, as the opening/closing rate was increased the optimum timing came closer to bdc, and was less advanced as engine speed was increased.

Gray (9) presents results for a 2 litre gasoline engine with a fixed 20° valve overlap, that show the maximum torque occurring: at 3100 rpm with the inlet valve closing 35° abdc, and at 3800 rpm with the inlet valve closing 65° abdc. These timings are somewhat different from those shown by Torazza (10) in Table 2 for a similar sized engine. In many cases the gain in low speed torque will be limited by the onset of knock.

Schiele et al (28) and Stojek and Stwiorok (29) present experimental results of the emissions and fuel consumption from variable phasing tests. In the tests reported by Schiele et al (28) only the phasing of the inlet camshaft was varied, and only results at full throttle are presented. Stojek and Stwiorok (29) varied the phasing of both the inlet and exhaust camshafts equally to give a variation in the valve overlap period. Some different period camshafts were used, but the change in period was only slight. The results are plotted in terms of the valve overlap period, but some of the effects may be more dependent on the consequential change in the inlet valve closure angle. Stojek and Stwiorok (29) conclude that a low overlap (10°) is required at all low load conditions, that 45° overlap is required for the highest rated output, and that a 90° overlap gives the highest low speed torque (a result that is more likely to be a consequence of closing the inlet valve at 31° abdc). As mentioned before, the additional advantage of low overlap with a fixed period cam event, is the consequential late inlet valve closing angle (71° abdc), that can lead to a reduction in the throttling losses at low loads.

Freeman and Nicholson (31) were primarily interested in the use of VVA for reducing NOx emissions, and they present full load data for the trade offs between NOx and HC emissions at different speeds and mixture strengths, plotted against the valve overlap area. The initial experiments were with four fixed timing camshafts, but data are also presented from tests on their variable phasing system.

Sapienza et al (34) present torque curves from a Toyota engine fitted with a system for varying the phase of the inlet camshaft by 33° ca; four different cam profiles were investigated, and the greatest torque increase from the stock cam was 19% at 3000 rpm using a different (but unspecified) camshaft. Additional results are also presented for emissions from the EPA 1975 FTP, and idle quality.

The significance of the valve overlap period on full throttle performance is difficult to establish, in the absence of wave action modelling or definitive experimental results. In high speed and output racing engine applications, it is probable that wave dynamics can be used for improved scavenging with large valve overlap periods. However, good driveability and fuel economy are less signficant in this application. Usually associated with a larger valve overlap area is a greater valve period, that permits higher valve lifts. The effective flow area at the maximum valve lift is not likely to be increased much, so any gains are a consequence of the increased duration at which the valve lift is greater than (for example) a quarter of its diameter. In conclusion, it is argued here that inlet valve closing angle is probably a more important determinant of full throttle engine performance than the valve overlap period.

Performance of Non-Constant Velocity Systems
- The complex interactions associated with non-constant velocity VVA systems, makes it very difficult to obtain generalised conclusions fromthe published performance data. Consequently, in this section it will only be possible to discuss each set of results in turn.

Ma (51) presents a torque curve and fuel consumption data from an engine fitted with a non-constant velocity ratio VVA system, but there are few experimental details. Griffiths and Mistry (53) present torque curves and fuel consumption maps obtained from an engine fitted with the Mitchell non-constant velocity ratio VVA system (42). Since this application was with a single camshaft engine, there is an interdependence between variations of: the inlet valve opening, the inlet valve closing, and the exhaust valve opening; the exhaust valve closing was contrived to remain constant at 21° atdc. As the inlet valve closing point was advanced the opening point was delayed, thereby reducing the inlet valve period and the valve overlap period. Associated with these changes was a reduction in the exhaust valve period as the exhaust valve opening was retarded. The gains in fuel consumption were mostly below 2500 rpm and increased with reducing speed, there was also an increase in the low speed torque.

Parker and Kendrick (43) experimented with a non-constant velocity ratio VVA system with separate inlet and exhaust cams. The torque curve is presented with its corresponding specific fuel consumption, and the associated valve timings are tabulated; some emissions data are also presented. An increase in torque of 20.9% was obtained at 1500 rpm with a 19%

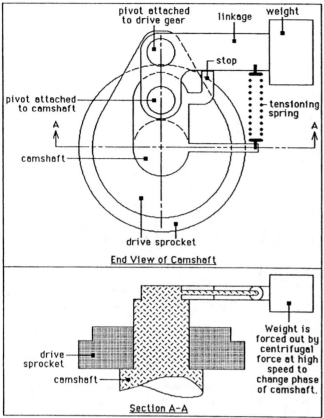

Figure 10E: Linkage (With Centrifugal Weight) for Variable Camshaft Advance

CONCEPT 11: Valves actuated by multiple, discrete cams, one at a time

General description: Usually, a fairly standard camshaft is used which has two or more cams located adjacent to each other, only one of which actuates the valve at any time (Fig. 11). The active cam, and thus the timing, is changed by either translating the camshaft axially relative to the tappet (76), or by moving the tappet relative to the camshaft (77). Soeters (78) used this concept with two cams and two rockers per valve, using latches to change the active rocker.

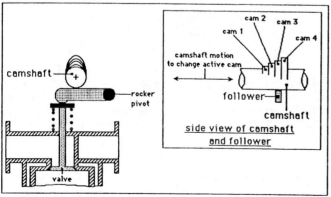

Figure 11: VVT Using Multiple Discrete Cams One at a Time

Advantages: For cases in which only a limited range of timing is desired, simple implementations can be devised.

Disadvantages:
1. The number of timing variations is severely limited by the space available for the cams and for the means for changing the active cam.
2. Since timing is usually changed step-wise, timing cannot always be optimal. Step changes can also lead to control problems, resulting in a loss in the quality of the driving characteristics.
3. When the active cam is changed, impacts will be experienced at some point in the linkage train unless great care is taken in making the transition. Many patents and papers do not address this problem at all, and none addresses it adequate.

CONCEPT 12: Adjust the valve train tappets transversely relative to the cams

General description: The tappet is moved either transversely across or rotationally around the cam to change the phase angle of the lift event, normally without much effect on the lift and duration (Fig. 12 and (41, 79)).

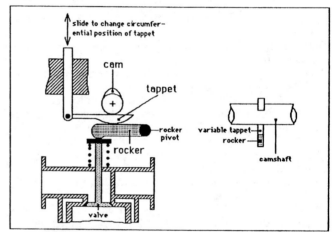

Figure 12: VVT Using Circumferentially Adjustable Tappet

Advantages: This technique allows individual control of each valve's advance, while the otherwise comparable Concept 10 cannot. This added flexibility is especially useful for independently controlling the intake and exhaust valves when they are actuated by the same camshaft ((41, 80)).

Disadvantages: This technique is a complex way to make a simple phase change. The total change in phase obtainable is usually limited by cam pressure angle and the desire to minimize follower complexity, to a range of only about 30 degrees.

CONCEPT 13: Suppressed valve movement

General description: The valves are actuated in a normal manner except that valve movement can be completely suppressed when desired. Valve motion is disabled by releasing two parts of the linkage train so that they move relative to each other without moving the valve (Fig. 13). Some examples of this concept are:

A. Many systems suppress valve movement by disabling the hydraulic lifter (31).
B. Some 1981 Cadillacs employed suppressed valve movement using a solenoid to rotate a perforated plate, which released the fulcrum of the valve rockers (9, 10, 81, 82). This technique was originated by Eaton Corporation (83).

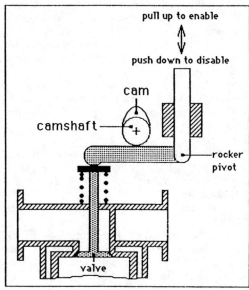

Figure 13: VVT Using Suppressed Valve Movement

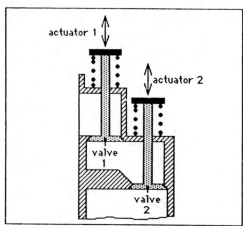

Figure 14: VVT Using Two Valves Operating in Series

C. Bates et al. (5) suppressed valve motion by moving the rocker fulcrum to directly above the valve stem.

D. A unique variation of this VVT concept was once used on an American production car, the 1917 Enger Twin-Unit Twelve (83). On driver demand, the system halved the engine displacement by blocking open half the exhaust valves and shutting off their matching intake valves.

Another unique variation of this technique, in which variable displacement was effectively produced without changing valve timing at all, was used by Stivender (84). Stivender simply eliminated fuel input to those cylinders which were to be inoperative.

Advantages: This is one of the simplest VVT techniques. This is one of the few VVT techniques that reduces the average mechanical losses, since the moving parts are standard when all cylinders operate and some of these parts do not move when any valves are disabled. The inherently higher thermodynamic efficiency combined with lower mechanical losses allow these systems to achieve high average fuel economy improvements (up to 15% has been claimed (5)).

Disadvantages:
1. Engine displacement can only be varied in steps of one cylinder, limiting how well valve timing can be optimized and leading to engine roughness at some settings.
2. The act of making the discontinuous displacement changes can cause problems in control, throttle response, jerks in the power output, and spikes in emissions (5).
3. Valve timing is not optimized on the active cylinders.

CONCEPT 14: Two valves operate in series on each intake and/or exhaust port

General description: With two valves in series operating the port, it is only effectively open when both valves are open (Fig. 14). A separate, relatively simple VVT mechanism (such as change in camshaft phase angle) is applied to each valve to vary the overall port timing (28, 53, 85). Alfa Romeo (86) has considered using a simple, passive check valve in series with each inlet valve to prevent reverse flow at low engine speeds.

Advantages: The individual valve trains are essentially standard, and thus maintain standard lift curve properties at all adjustments. The valve operators usually consist of standard camshafts with simple phase changers.

Disadvantages:
1. The increased number of valves and valve operators increases cost and complexity and makes packaging more difficult.
2. The two valves inherently have an additional volume between them, adding an accumulator effect that washes out some of the effect of the valve farther from the combustion chamber.
3. Since the flow must pass through two valves, flow losses are greater than for a single valve.

CONCEPT 15: Two valves operate in parallel on each port

General description: The valves-in-parallel concept provides intake or exhaust flow whenever either of the two valves is open (Fig. 15). A simple VVT concept (such as

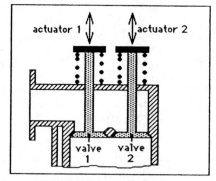

Figure 15: VVT Using Two Valves Operating in Parallel

phase angle change) is applied to each to vary either the overall valve timing (28, 54, 86, 87) or the effective flow area (88).

Advantages: The individual valve trains are essentially standard, and thus maintain standard lift curve properties at all adjustments. The valve operators usually consist of standard camshafts with simple phase changers.

Disadvantages: Doubling the number of valves and valve operators and independently operating them all adds much mechanical complexity. The system sometimes operates with the flow area of only one valve, and thus with greatly increased flow restriction.

Table 1: Summary of VVT Concepts

Concept Number and Title	Major Limitations of Most Concept Members	Approximate Ranges of Duration and Advance that Can be Achieved	Best Features	Overlapping and Related Concepts
1: Electrically actuated valves	seating and backseating impacts; high energy consumption; duration range and engine speed range limited by actuator system response and impacts	unlimited, except that minimum duration is limited by the actuator response	potentially unlimited range of variability	none
2: Hydraulically actuated valves	seating and backseating impacts; high system cost, complexity, and energy consumption; actuator response and impacts limit maximum engine speed	unlimited, except that minimum duration is limited by the actuator response	potentially unlimited range of variability	none
3: Valves actuated through variable characteristic hydraulic lifters	seating impacts; often, liftoff impacts; impacts limit practically obtainable range of duration and maximum engine speed	unlimited duration range, except as limited by impacts; small changes in advance result from duration changes	one of the simplest methods for obtaining large ranges of duration	usually a subset of Concept 5; sometimes a subset of Concept 13
4: Axially varying cam profiles	point loading causes high contact stress between cam and follower	theoretically unlimited, but practically limited by the need for long cams to get high ranges of duration and advance	simultaneous variations in duration and advance can be obtained at each valve in a relatively simple manner	similar to some forms of Concept 11
5: Valves actuated through lost-motion mechanisms	lift-off and seating impacts; engine speeds limited by impacts	unlimited duration range, except as limited by impacts; small changes in advance result from duration changes	one of the simplest techniques for obtaining large ranges of duration	most members of Concept 3 and some concepts 6 and 7 are also members of this concept; Concept 13 is basically 100% lost motion
6: Valves actuated through a variable lever ratio linkage	most have little effect on duration and advance; added valve train weight and flexibility can limit engine speed	most vary lift only	simplest method for varying lift scale	sometimes a member of Concept 5; those that are members of Concept 5 have the basic properties of Concept 5
7: Valves actuated by two independent cams in series	some designs add mass and compliance, limiting maximum engine speed	unlimited range of advance; some can vary duration up to 1.5:1; the "Adjusticam" variation allows 3:1 or more	large ranges of duration and advance can be obtained without impacts, particularly with the "Adjusticam" variation of this technique	sometimes a member of Concept 5; those that are members of Concept 5 have the basic properties of Concept 5
8: Valve-actuating cam is driven through a variable-motion drive	high cost and complexity; usually somewhat bulky, making packaging difficult	theoretically unlimited range of advance and duration; most systems limit these ranges to less than 25 camshaft degrees	large ranges of duration and advance without impacts; some use a single input to vary all valves simultaneously to nearly maximize torque	none
9: Valves actuated by two cams in parallel	valve train impacts at high durations; complicates camshaft design the two independent cams for each follower are provided on a common camshaft	duration ranges up to 1.5:1, but ranges longer than the duration of the maximum lift cam dwells cause impacts	simple, compact method for obtaining small ranges of duration up to high engine speeds	none
10: Valves actuated through a standard camshaft with variable advance	provides changes in advance alone, except when combined with another concept (see last column)	changes advance only	avoids impacts and engine speed limitations; used with other VVT concepts; this technique alone can provide significant VVT advantages	often used with the other concepts (except concepts 1 and 2) to change cam advance; Concept 12 is basically a subset of this class
Variation 10-1: Camshaft driven by timing belt with movable idlers	space limitations make the system more difficult to apply to two or more camshafts run off a single belt	the range of advance is unlimited	compact and mechanically simple when applied to a single camshaft	see Concept 10
Variation 10-2: Camshaft driven through a helical spline	adds to the length of the engine; many have high friction, increasing wear and the force and energy requirements for phase changes	theoretically any range of advance, but spline length and pressure angle limitations usually limit the range	low wear rates and energy losses, since relative motion occurs only when adjusting	see Concept 10
Variation 10-3: Camshaft driven through a hydraulic motor	use of engine oil usually necessitates multi-stage actuators, usually limiting the range of advance	most designs limit the range of advance to less than 30 camshaft degrees	low wear rates and energy losses, since relative motion motion occurs only when adjusting	see Concept 10
Variation 10-4: Camshaft driven through a differential gear mechanism	many moving parts; in most versions, phase changer gears are always active, causing mechanical losses and wear	the range of advance is unlimited	unlimited range of advance; those versions that avoid relative motion when not adjusting minimize energy losses and wear	see Concept 10
Variation 10-5: Camshaft driven through a linkage	range of advance is very limited; advance usually cannot be actively controlled, limiting the applications	the range of advance is usually limited to less than 30 camshaft degrees	no active control is used, eliminating sensors, control systems, and control mechanisms	see Concept 10
11: Valves actuated by multiple, discrete cams, one at a time	requires careful system design and operation to avoid impacts when changing cams; space limitations limit the number of cams per valve	range of advance and duration only limited by the number of cams used	valve train has an effectively standard configuration at each setting, minimizing engine speed limitations and eliminating impacts	those forms that have all the cams for each valve in sequence on a single camshaft are similar to Concept 4
12: Adjust the valve train tappets transversely relative to the cams	for most applications, this is complex for the amount of variation available	usually no duration effect; the range of advance is usually limited to less than 30 camshaft degrees	good method for making small, independent changes in advance at each valve	similar to Concept 10, except that this concept works independently on each valve
13: Suppressed valve movement	no advance variation, only two durations, 0 or standard; impacts occur if the cam is in its lift phase when motion is suppressed or unsuppressed	duration is either zero or that of the actuating cam; does not change advance	one of the simplest methods; has proven to provide significant VVT benefits through changes in the effective engine displacement	this is Concept 5 extended to the limit of lost motion
14: Two valves operate in series on each intake and/or exhaust port	very complex when VVT is provided on both cams (as required to obtain the full range of duration and advance)	any, depending on the VVT technique used on each valve	can provide large ranges of effective duration and advance using relatively simple VVT techniques on each valve	one or both valves is controlled by one or more of VVT concepts 1 through 12
15: Two valves operate in parallel on each port	very complex when VVT is provided on both cams (as required to obtain the full range of duration and advance)	any, depending on the VVT technique used on each valve	large duration/advance ranges from simple, reliable VVT concepts on each valve; especially good when two values of inlet flow area are desired	one or both valves is controlled by one or more of VVT concepts 1 through 13

Combinations of VVT Concepts - Some VVT systems use combinations of two or more of the concepts mentioned above. For example, (89) combined Concepts 1 and 2 and (90) discussed combining Concepts 14 and 15.

SUMMARY

After an extensive search, we have categorized all VVT mechanisms that we found into 15 basic concepts. None of these 15 concepts is ideal, all have inherent basic limitations. The highlights of our categorization system, including the relationships between the categories, are given in Table 1.

As a consequence of the limitations we have noted, only a very few concepts have ever been translated into functional units for experimentation, and we found only three very simple versions of variable valve timing that have ever been put into production use.

CONCLUSIONS AND RECOMMENDATIONS

A wide variety of VVT mechanisms have been proposed, each with its own advantages and limitations. Because of the tradeoffs in the selection of a VVT mechanism that these advantages and limitations imply, a thorough combination of theoretical and experimental studies are needed to find the true potential of VVT, and for each specific VVT mechanism. Once the optimum modes (for a variety of criteria) of VVT operation and the size and nature of the benefits available from VVT are accurately known, the cost-effectiveness of particular VVT mechanisms can be better evaluated.

We believe that some forms of Concept 8 and the "Adjusticam" form of Concept 7 (as listed above) have the most VVT potential, because of the wide range of duration and advance they can provide without impacts or maximum engine speed limitations. Thus, we think these concepts should be emphasized in future efforts to apply VVT.

Although there are relative few patents on VVT mechanisms of the form of Concept 8, this concept has a wide range of truly original concepts, and we believe that it has the most potential for more truly original concepts. (In fact, we could have split Concept 8 into two or more separate concepts.) Thus, Concept 8 has the most potential for a breakthrough in VVT mechanisms.

REFERENCES

1. Dresner, T.L. and Barkan, P., (1989), "A Two-Input Cam-Actuated Mechanism and its Application to Variable Valve Timing", SAE Paper 890676

2. Novak, J.M. and Blumberg, P.N. (1978), "Parametric Simulation of Significant Design and Operating Alternatives Affecting the Fuel Economy and Emissions of Spark-Ignited Engines", SAE Paper 780943

3. Freeman, M.A. and Nicholson, R.C., (1972), "Valve Timing for Control of Oxides of Nitrogen", SAE Paper 720121

4. Siewart, R.M., (1971), "How Individual Valve Timing Events Affect Exhaust Emissions", SAE Paper 710609

5. Bates, B., Dosdall, J.M., and Smith, D. H., (1978), "Variable Displacement by Engine Valve Control", SAE Paper 780145

6. Stecklein, G.L., Southwest Research Institute, (1987), Personal Communication

7. Elrod, A.C. and Nelson, M.T., (1986), "Development of a Variable Valve Timed Engine to Eliminate the Pumping Losses Associated with Throttled Operation", SAE Paper 860537

8. Clerk, D., (July 27, 1880), United States patent number 230,470

9. Dunne, J., (October, 1980), "Cadillac's Revolutionary 3-in-1 V8", *Popular Science*, 121-122

10. Givens, L., (October, 1980), "Cadillac Introduces V-8-6-4 Engine", *Automotive Engineering*, 52-54

11. Simanaitis, D., (December, 1985), "Nissan Aims for the Top", *Road & Track*, 42-48

12. Inoue, K., Nagahiro, K., Ajiki, Y., Katoh, M., (December 6, 1988), U.S. patent 4,788,946

13. Longstaff, K., Holmes, S., (May 13, 1975), U.S. patent 3,882,833

14. Seilly, A.H., (1979), "Helenoid Actuators - A New Concept in Extremely Fast Acting Solenoids", SAE Paper 790119

15. Scott, D., (May, 1980), "Variable Valve Timing to boost engine efficiency", *Popular Science*, 96-97

16. Stefanides, E.J., (April 23, 1979), "Super-Fast Solenoids to Operate Auto Engine Valves?", *Design News*, 9-10

17. Engler, W.B. and Kryzanowsky, C.J., (April 21, 1908), U.S. patent 885,459

18. Petsche, P.E., (September 28, 1909), U.S. patent 935,323

19. Gavrun, M.T. and Loyd, R.W. Jr., (September 7, 1976), U.S. patent 3,978,826

20. Mardell, J.E. and Cross, R.K., (1988), "An Integrated Full Authority Electrohydraulic Engine Valve and Diesel Fuel Injection System", SAE Paper 880602

21. Michelson, G.P. and Ule, L.A., (December 16, 1975), U.S. patent 3,926,159

22. Richman, R.M. and Reynolds, W.C., (1984), "A Computer-Controlled Poppet- Valve Actuation System for Application on Research Engines", SAE Paper 840340

23. Ule, L.A., (March 1, 1977), U.S. patent 4,009,695

24. Eaton, A., Stanford University, (1988), Personal Communication

25. Nagao, F., Nishiwaki, K., Otsubo, K., Yokoyama, F., and Okada, T., (1968), "Relation between Valve Angle-Area and Low Speed Performance of a Uniflow-Scavenged Engine", *Bulletin of JSME*, **11**, #48, 1181-1187

26. Herrin, R.J. and Pozniak, D.J., (1984), "A Lost-Motion, Variable-Valve-Timing System for Automotive Piston Engines", SAE Paper 840335

27. Geringer, B., (1988), "Variable Intake Valve Control and Optimized Intake Manifolds for Torque Curve Improvement and Fuel Consumption Reduction", (Translation by Stephen Carlson, Bendix Electronics), *Internationales Wiener Motorensymposium*, **12-99**, 54-74

28. Gray, C., (1988), "A Review of Variable Valve Timing", SAE Paper 880386

29. Payri, F., Desantes, J.M., and Corberan, J. M., (1988), "A Study of the Performance of an SI Internal Combustion Engine Incorporating a Hydraulically Controlled Variable Valve Timing System", SAE Paper 880604

30. Hausknecht, L.A., (May 8, 1979), U.S. patent 4,153,016

31. Benson, C.F., Humphrey, H.W. Jr., (January 9, 1979), U.S. patent 4,133,332

32. Nagao, F., Nishiwaki, K., and Kajiya, S., (1968), "Valve Timing Control by a Hydraulic Tappet with a Leak Hole", *Bulletin of JSME*, **11**, #48, 1175-1180

33. Nagao, F., Nishiwaki, K., Otsubo, K., and Yokoyama, F., (1969), "Relation between Inlet Valve Closing Angle and Volumetric Efficiency of a Four-Stroke Engine", *Bulletin of JSME*, **12**, #52, 894-901

34. Akiba, K. and Kakiuchi, T., (1988), "A Dynamic Study of Valve Mechanisms: Determination of the Impulse Forces Acting on the Valve", SAE Paper 880389

35. Dyer, G.L., (October 19, 1976), U.S. patent 3,986,484

36. Rust, R.H., Fulghum, L.O., (October 28, 1975), U.S. patent 3,915,129

37. Norbye, J.P. (April, 1977),"Variable Valve Timing & Rotary Valves", *Popular Science*, 121-125

38. Allen, C.H., (November 9, 1971), U.S. patent 3,618,573

39. Codner, S.J. Jr., (May 1, 1973), U.S. patent 3,730,150

40. Miller, M.J., (November 9, 1971), U.S. patent 3,618,574

41. Stone, C.R. and Kwan, E.K.M., (August/September, 1985), "Variable Valve Timing for IC Engines", *Automotive Engineer*, **10**, 54-58

42. Paxton, D.R., (August 29, 1972), U.S. patent 3,687,010

43. Stivender, D.L., (1968), "Intake Valve Throttling (IVT) - A Sonic Throttling Intake Valve Engine", *SAE Transactions*, **77/2**, 1293-1302

44. Entzminger, W.W., (1988), "Variable Valve Action (VVA) Through Variable Ratio Rocker Arms", SAE Paper 880730

45. Charlton, S. and Shafie-Pour, M., (1988), "A Hydraulic Valve Control System and Its Application to Turbocharged Diesel Engines", SAE Paper 880603

46. Aoyama, S., (November 9, 1982), U.S. patent 4,357,917

47. Dresner, T.L., (1988), "Multi-Input Cam-Actuated Mechanisms and Their Application to IC Engine Variable Valve Timing", PhD Thesis, Stanford University

48. Stecklein, G.L., (June, 1987), "Vehicle Systems for the Future", *Technology Today*, Southwest Research Institute Publ., San Antonio, Texas, 11-12

49. Semple, H.F., (July 23, 1985), U.S. patent 4,530,318

50. Whiting, H.W., (November 7, 1950), U.S. patent 2,528,627

51. Lenz, H.P., Wichart, K., and Gruden, D., (1988), "Variable Valve Timing: A Possibility to Control the Engine Load Without Throttle", SAE Paper 880388

52. Freudenstein, F., Maki, E.R., Tsai, L-W., (1988), "The Synthesis and Analysis of Variable Valve Timing Mechanisms for Internal Combustion Engines", SAE Paper 880387

53. Ma, T.H., (1987), "Recent Advances in Variable Valve Timing", *Alternative and Advanced Automotive Engines*, New York, Plenum Press, 235-252

54. Ma, T.H., (1988), "The Effect of Variable Engine Valve Timing for Fuel Economy", SAE Paper 880390

55. Kerr, J., (July 30, 1980),"Variable Valve Timing to Boost any Engine", *The Engineer*, 28-29 & 54

56. Scott, D. and Yamaguchi, J., (October, 1980), "Eccentric Cam Drive Varies Valve Timing", *Automotive Engineering*, 120-124

57. Scott, D., (February, 1981), "Trick Camshaft Boosts Engine Power, Economy", *Popular Science*, 8-12

58. Griffiths, P. and Mistry, K.N., (1988), "Variable Valve Timing for Fuel Economy Improvement - The Mitchell System", SAE Paper 880392

59. Parker, D.A. and Kendrick, M. (1974), "B-1-11 - A Camshaft with Variable Lift-Rotation Characteristics; Theoretical Properties and Application to the Valve Gear of a Multicylinder Piston Engine", *Fisita Congress*, Paris, 224-232

60. Anonymous, (1973), "Variable Camshaft Boosts Engine Efficiency", *Automotive Engineering*, **81**, #12, 11-13

61. Anonymous, (November, 1978), "Four-stroke Variable Valve Timing and Lift", *Motorcycle Sport*, 480-485, England

62. Raggi, L., (January 11, 1972), U.S. patent 3,633,555

63. Roe, G.E., (1972), "Variable Valve-Timing Unit Suitable for Internal Combustion Engines", *Proc. Instn. Mech. Engrs.*, **186 23/72**, 301-306

64. Torazza, G. and Dante, G., (February 15, 1972), U.S. patent 3,641,988

65. Luria, D., (April 18, 1978), U.S. patent 4,084,557

66. Meacham, G.B.K., (1970), "Variable Cam Timing as an Emission Control Tool", *SAE Journal*, 2127-2144, Paper 700673

67. Ball, G.A., Elliott, C.M., (1982), "Passenger Car Spark Ignition Data Base Variable Valve Timing", prepared by Chrysler Corporation Engineering Office for U.S. Department of Transportation

68. Yamashita, R., Hiromitsu, M., (October 3, 1978), U.S. patent 4,117,813

69. Sapienza, S.J. IV and van Vuuren, W.N., (1988), "An Electronically Controlled Cam Phasing System", SAE Paper 880391

70. Miokovic, S., (June 10, 1975), U.S. patent 3,888,216

71. Schiele, C A., DeNagel, S.F., and Bennethum, J.E., (1974), "Design and Development of a Variable Valve Timing (VVT) Camshaft", SAE Paper 740102

72. Clemens, J.D. and Williams, E.A., (May 3, 1978), U.S. patent 4,091,776

73. Pouliot, H.N., Delameter, W.R., and Robinson, C. W., (1978), "A Variable-Displacement Spark-Ignition Engine", *SAE Transactions*, **86/1**, 446-464

74. Isakson, W.R., (June 14, 1938), U.S. patent 2,120,612

75. Nichols, R.G., (May 23, 1970), U.S. patent 3,516,394

76. Mueller, R.S., (May 1, 1979), U.S. patent 4,151,817

77. Beal, R.G., (April 22, 1975), U.S. patent 3,878,822

78. Soeters, R.A., (November 7, 1977), "2.3 L Ford Engine Camshaft Investigation Interim Report", Ford Motor Company, Project #1776-41, Library # 77170

79. Scherenberg, H.O., Gassman, J., (June 26, 1962), U.S. patent 3,040,723

80. Roan, H.A., (December 16, 1941), U.S. patent 2,266,077

81. Mueller, R.S., Uitvlugt, M.W., (1978), "Valve Selector Hardware", SAE Paper 780146

82. Anonymous, (October 21, 1976), "Dual-Displacement Engine Boosts Fuel Economy", *Machine Design*, 10-12

83. Givens, L., (May, 1977), "A New Approach to Variable Displacement", *Automotive Engineering*, 30-34

84. Stivender, D.L., (1978), "Engine Air Control -- Basis of a Vehicular Systems Control Hierarchy", SAE Paper 780346

85. Dave, S.M., (October 2, 1973), U.S. patent 3,762,381

86. Birch, S., (August, 1988), "Engine Technology from Alfa", *Automotive Engineering*, 87-88

87. Ma, T.H. and Rajabu, H., (1988), "Computer Simulation of an Otto-Atkinson Cycle Engine with Variable Timing Multi-Intake Valves and Variable Compression Ratio", *Proc Instn Mech Engrs*, 273-277

88. Mayersohn, N. and Zino, K., (March, 1988),"Power for Tomorrow", *Popular Mechanics*, 53-57

89. Woods, R.L. and Katz, S., (October 19, 1976), U.S. patent 3,986,351

90. Trenne, M.U., (April 29, 1980), U.S. patent 4,200,067

910445

Perspectives on Applications of Variable Valve Timing

T.W. Asmus
Chrysler Corp.

Abstract

While Variable Valve Timing (VVT) holds the potential to alleviate some of the compromises normally associated with engine operation over broad speed and load ranges, generalizing its benefits in a quantitative manner is difficult. Details of base engine design and the means of VVT execution and engine duty cycle play major roles in determining the outcome of such an application.

With the profusion of VVT means available, the matter of selecting an appropriate system for a given engine and duty cycle is difficult particularly when there is a host of competing features and technologies.

This treatment of VVT is based on assessments of compromises associated with static valve timing and provides a framework for considering VVT as an engine performance option. Extensive use is made of pressure and geometric data, and several generalizations are made relative to strategies for incorporating common VVT systems.

INTRODUCTION

Automotive I.C. engines are designed to accommodate a highly diverse duty cycle, and therefore incur well-known compromises. Certain of these result from the broad ranges of speed and load under which they operate. The flow processes involved in gas exchange are time dependent and don't precisely scale with speed and load. Valve timing along with many other design parameters, therefore, are subject to compromise over this diverse duty cycle. Valve events, in their broadest sense, can impact volumetric, thermal and mechanical efficiencies. The primary role of VVT in today's environment is to enable improved management of volumetric efficiency compromises over the broad range of operating conditions. Applying VVT to this end may impact engine processes other than those directly related to volumetric efficiency, and these are the following: burn rate, expansion ratio, pumping losses and mechanical efficiency.

The number of reported methods for the implementation of VVT are too numerous to mention here but are well summarized by others [1, 2], and the myriad of candidate engines is well known. The means by which engine processes are affected by valve events is the major focus of some reports [3-13], and others deal mainly with mechanisms. It is difficult to obtain generalized conclusions as to the relative merits of the various categories of VVT systems, and specific claims often reflect more on base engine characteristics than on the specific VVT system in question. This can leave the engine planner and designer in a difficult position relative to selecting a sensible course of action particularly when a broad range of performance enhancement technologies is available.

It is helpful to view engine processes as they relate to valve timing at the extremes of the engine operating envelope, and these are the following:

- <u>Idle</u> - minimum load and speed
- <u>Torque</u> - maximum load at minimum usable speed, i.e., generally below the speed where intake tuning plays a role
- <u>Power</u> - maximum load at maximum desired speed

Hereafter these will be referred to simply as <u>Idle</u>, <u>Torque</u>, and <u>Power</u>. All other operating conditions are represented by combinations of the above or are dominated by factors other than valve timing.

Figure 1 is included for orientation and shows cylinder pressure trends over the gas exchange process for <u>Idle</u>, <u>Torque</u> and <u>Power</u>, time-area "windows" created by valve lift in a conventional, static valve-timing engine along with normalized cylinder volume and its derivative. Some of the valve-timing compromises are apparent from this figure, and these will be highlighted in a later section.

TYPES OF LOSSES ASSOCIATED WITH VALVE TIMING

A cam-actuated poppet valve may contribute to engine performance compromises either because it affords too much or too little flow capacity at a particular point in the engine cycle, i.e., the "shape" of the time-area window does not precisely match the requirement. This is not surprising since a valve lift profile is designed to satisfy, first, mechanical requirements and, second, flow requirements. This mismatch may manifest itself as a charging or cylinder filling loss, a pumping loss or an expansion loss.

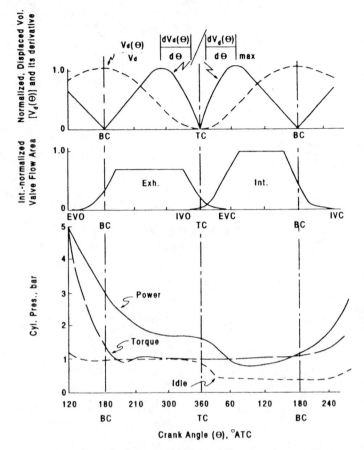

Figure 1: Generalized features of the gas exchange process

Frequent reference is made to optimized static valve timing, and this is intended to imply that a design and development process has been conducted which yields a valve timing scheme which gives equal priority to the attainment of <u>Idle</u>, <u>Torque</u> and <u>Power</u>. Base engines having similar types of valve mechanisms were used in comparisons of the effects of bore-stroke ratio to the number of valves per cylinder. Each of the four valve events will be discussed qualitatively here in terms of aforementioned losses at the three operating conditions and will be treated more quantitatively in a subsequent section. This discussion will treat blowdown, the beginning of exhaust valve opening (EVO), as the start of the gas exchange process and will proceed through the cycle. Figure 2 shows ln P vs.

ln V for the three operating conditions of concern, and these are helpful in supporting the discussion which follows:

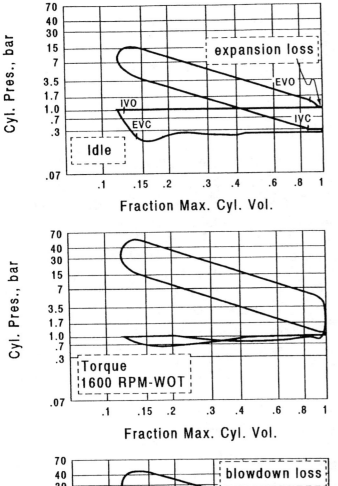

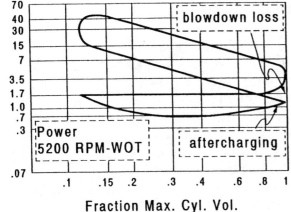

Figure 2: ln P vs. ln V for <u>Idle</u>, <u>Torque</u> and <u>Power</u>

<u>EVO</u>: From Figures 1 and 2 it is clear that with optimized, static timing at <u>Idle</u> there is a small expansion loss and at <u>Power</u> there is incremental pumping work associated with this event. At <u>Torque</u> there is some of each. The expansion loss at <u>Idle</u> will manifest itself as a small thermal efficiency compromise, and the incremental pumping work at <u>Power</u> (hereafter referred to as $PMEP_{bd}$) will compromise performance via an increase in PMEP. Ideally, isentropic expansion proceeds to BC whereupon instantaneous cylinder depressurization occurs such that cylinder and exhaust manifold pressures are essentially equal throughout the displacement phase of the exhaust event.

<u>EVC</u>: The goal of an optimum static valve timing configuration is to insure that, at <u>Power</u>, exhaust lockup is <u>just</u> avoided. This then insures that the exhaust valve closing (EVC) part of valve overlap contributes minimally to the overlap time-area window. If this condition had not been met, there would be an upturn in cylinder pressure near the end of the exhaust stroke especially if the design includes low overlap [3]; and this would have been apparent in Figure 2, at <u>Power</u>. When the above condition is satisfied, there is a small exhaust time-area window after TC which allows some exhaust re-induction when only intake flow is desired. This yields a small <u>Torque</u> compromise to satisfy a <u>Power</u> requirement. (This has consequences at part load as well but as a practical matter is not limiting the engine in any significant way.)

<u>IVO</u>: The major consequence of this is its contribution to overlap and its indirect effect on lift later in the intake valve time-area window. If this is biased early, and overlap is high, then there is a readily identifiable feature present at the TC end of the <u>Idle</u> and <u>Power</u> pumping loops shown in Figure 2, i.e., the sharp apex followed by the negative sloping portion becomes rounded. <u>Power</u> enhancement may or may not be the result of early intake valve opening (IVO) depending on intake tuning factors, i.e., this event initiates an intake system expansion wave which affects the phase angle of a recompression which

drives the cylinder charging process near intake valve closing (IVC).

IVC: This event, or more precisely the time-area window between BC and intake valve seating, strongly influences volumetric efficiency over the speed range, therefore is involved in major tradeoffs between Torque and Power. Figure 2 illustrates, in the P-V domain, some of the consequences of a compromised IVC. Note in the Idle and Torque modes that the isentropic compression does not begin until roughly 15% and 10% swept volume after BC respectively and that this is a direct consequence of intake valve reverse flow. Figure 2 also illustrates evidence of aftercharging in the Power mode, i.e., just before and just after BC cylinder charging is apparent by the non-zero slope of ln P vs. ln V. The magnitude of this slope just before BC is indicative of the aftercharging potential. (The slope just after BC is difficult to ascertain as this is close to the start of compression.) IVC biased for Power will tend to degrade idle combustion stability relative to a Torque-biased IVC owing to the effects of reduced compression temperatures and pressures on ignition delay.

VALVE EVENTS AS TIME-AREA WINDOWS

Cam-actuated poppet valves yield characteristic time-area windows whose shape is largely governed by mechanical constraints resulting from a focus on high-speed performance, durability and functional robustness. It is interesting to envision idealized time-area windows specified solely on the basis of flow requirements [14]. Bore-stroke ratio and valve mechanism stiffness play roles in determining the sizes and shapes of the optimized time-area windows and, thus, valve timing compromises. Figure 3 illustrates idealized time-area windows along with those of a 2-valve, undersquare (B/2a < 1.0) and a 4-valve, oversquare (B/2a > 1.0) engine designed to operate over similar speed ranges.

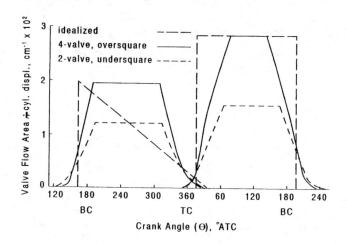

Figure 3: Displacement (cyl.)-normalized time-area windows for a 2-valve, undersquare and a 4-valve, oversquare engine along with idealized time-area windows [14]

The flat portion on top of these time-area windows is a consequence of the flow area at high lift being limited by the cross-sectional area of throat minus that of the valve stem [15]. This tends to occur when the lift to inner valve seat diameter ratio $L_v/D_{vis} \gtrsim 0.25$.

It is interesting to note the departure of the real, time-area windows from the idealized ones shown in Figure 3. Other things being equal (especially speed range, valve mechanism type and tuning), it is clear that this degree of departure is greater for the 2-valve, undersquare than the 4-valve, oversquare engine.

METHODS

POPPET VALVE FLOW AREA: The minimum valve flow area was computed as the surface area of frustums of cones as a function of valve lift and all other geometric details until this area exceeds the throat-minus-stem area [15]. The following depictions show the upper limit of each phase of valve lift (L_v), and Eqs (1)-(4) were used to compute minimum flow areas (A_{vi}) for the respective phases designated (I)-(IV).

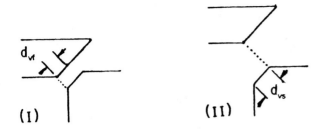

$$A_{vf}(I) = \pi(D_{vs} - \tfrac{1}{2}L_v \cdot \sin 2\beta)L_v \cdot \cos \beta \qquad (1)$$

for $0 \leq L_v \leq d_{vl}/\sin \beta$

$$A_{vf}(II) = \pi L_v \cdot \cos \beta (D_{vs} - 2d_{vl} \cdot \cos \beta + L_v \cdot \cos^2 \beta) \qquad (2)$$

for $d_{vl}/\sin \beta \leq L_v \leq (d_{vs} + d_{vl})/\sin \beta$

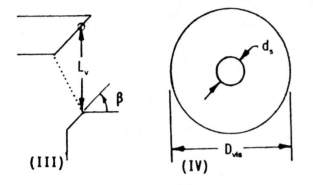

$$A_{vf}(III) \approx \pi[D_{vs} + 2d_{vs} \cdot \cos \beta - (d_{vs} + d_{vl})\cos \beta]$$
$$[L_v^2 + (d_{vs} + d_{vl})^2 - 2L_v(d_{vs} + d_{vl})\sin \beta]^{\tfrac{1}{2}} \qquad (3)$$

for $(d_{vs} + d_{vl})/\sin \beta \leq L_v$

$$A_{vf}(IV) \approx \tfrac{1}{4}\pi(D_{vs}^2 - d_s^2) \qquad (4)$$

whenever the frustum area is greater than the throat area

DISPLACED VOLUME AND ITS DERIVATIVE: The following crank-slider equation and its crank angle derivative were used:

Displaced Volume and its Derivative:

$$V_d(\theta) = \tfrac{1}{4}\pi B^2 \left[L_R + a - L_R \cdot \left[1 - \left(\tfrac{a \sin \theta + D_{os}}{L_R}\right)^2 \right]^{\tfrac{1}{2}} - a \cdot \cos \theta \right] \qquad (5)$$

$$\frac{dV_d(\theta)}{d\theta} = \frac{\pi^2 B^2}{720} \left[a \cdot \sin \theta + L_R \cdot \left(\frac{\left(\tfrac{a \sin \theta + D_{os}}{L_R}\right)\tfrac{a}{L_R} \cdot \cos \theta}{\left[1 - \left(\tfrac{a \cdot \sin \theta + D_{os}}{L_R}\right)^2\right]^{\tfrac{1}{2}}} \right) \right] \qquad (6)$$

VALVE PSEUDO-FLOW VELOCITY (v_{ps}): This is obtained by dividing the crank angle derivative of displaced volume by the crank angle-dependent valve flow area [3] as follows:

$$v_{ps} = \frac{dV_d(\theta)}{d\theta} \cdot \frac{1}{A_{vf}(\theta)} \qquad (7)$$

While v_{ps} is based solely on geometry, it has questionable significance in terms of real flows; however, it is useful in terms of establishing perspectives on certain aspects of the gas exchange process. This is particularly true when valve flows are principally driven by piston motion and where only small pressure gradients are involved, i.e., when compressibility incurs only small errors. This is illustrated in Figure 6.

THERMAL CONVERSION EFFICIENCY (η_t): Eq (8) was used to estimate the consequences of varying EVO under various engine operating conditions.

$$\eta_t = \frac{\int_{\theta_i}^{\theta_i + \theta_b} \left[1 - \cos\tfrac{\theta - \theta_i}{\theta_b} \cdot \pi \right] \left\{ 1 - \left[\tfrac{V(\theta)}{V(\theta, EVO)}\right]^{k-1} \right\} d\theta}{\int_{\theta_i}^{\theta_i + \theta_b} \left[1 - \cos\tfrac{\theta - \theta_i}{\theta_b} \cdot \pi \right] d\theta} \qquad (8)$$

where $V(\theta, EVO)$ is data-based cylinder volume at virtual EVO

This equation expresses thermal conversion efficiency where the finite burn interval is simulated *via* a cosine function representation of mass fraction burned, the expansion loss associated with compromised EVO is factored in *via* $V(\theta, EVO)$ and the polytropic exponent, k, is experimentally based.

PRESSURE MEASUREMENTS: Kistler model 6121 pressure transducers were used to establish the data base upon which this work is based, and extreme caution was exercised in establishing low-pressure calibration.

QUANTIFICATION OF VALVE TIMING COMPROMISES

Valve timing issues will be treated in the same order here as in the previous section for a 2-valve and a 4-valve engine assuming equivalent duty cycles.

EVO: This involves expansion and pumping loss trade-offs which are apparent in Figure 2. Figure 4 is a focused P-V analogue of Figure 2 and enables direct evaluation of losses as MEPs by physical area measurement. Figure 5 shows the component losses and their sum for the two engines plotted against EVO timing expressed as the cylinder-displacement-normalized, time-area window from EVO to BC. Note that the optimum EVO timing is where the component losses sum to a minimum [3] and that this point changes with engine operating condition. However, since the component losses countervail, the compromise attributable to optimized, static valve timing is small, i.e., less than 1%. The level of total MEP loss associated with EVO largely reflects the non-ideal shape of the leading edge of the exhaust time-area window, and furthermore EVO compromises are smaller for 4-valve than 2-valve engines because this time-area window is more concentrated near BC as can be noted in Figure 3.

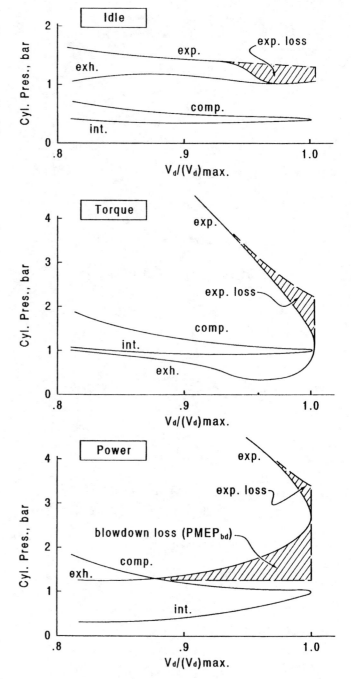

Figure 4: P-V traces focused on BC region

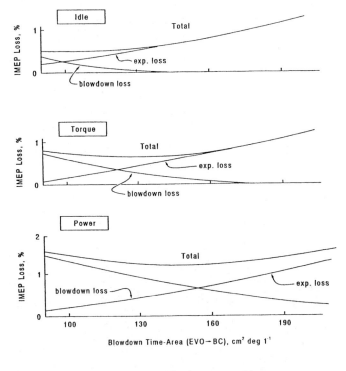

Figure 5: Expansion and blowdown component losses associated with EVO.

EVC: The final portion of the displacement phase of an exhaust event involves a flow that is largely piston driven, and if the EVC timing is reasonable the pressure gradient across the valve is small. Therefore, exhaust valve pseudo-flow velocity (v_{ps}) Eq (7) is useful for assessing and/or establishing EVC timing. Figure 6 is a plot of v_{ps} over the entire gas exchange process for the two types of engines mentioned earlier. Note the maximum exhaust v_{ps} occurring at roughly 20° and 10°BTC respectively for the 2 and 4-valve engines. This represents a potential flow constriction that is a geometric consequence of piston and valve kinematics. The magnitude of this can be manipulated via valve timing, or more precisely, the valve flow area in this crank angle regime. It has been established that there is a quantitative relationship between peak exhaust v_{ps} and cylinder pressure at TC [3]. If exhaust lockup is just avoided at the desired peak power speed, then further delays in EVC are of little or no Power benefit and, in fact, tend to be detrimental to Torque and Idle due to exhaust induction and valve overlap effects respectively. Having established EVC in this manner, it can be concluded that Torque and Idle sustain minimum detriment.

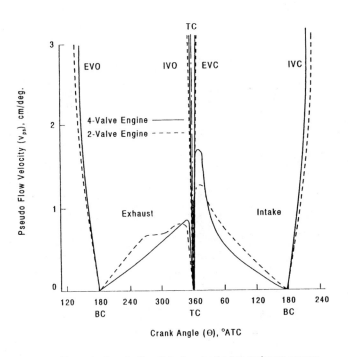

Figure 6: Pseudo Flow Velocity over the gas exchange process for engines optimized for Idle, Torque & Power

Torque compromises associated with EVC are a result of exhaust induction which is mainly driven by piston motion. Figure 7 illustrates significant features of this valve event including the critical lift region if Power losses are to be avoided and the exhaust induction time-area window from TC to EVC. The magnitude of the volumetric efficiency loss due to this event was calculated to be less than 1% for both 2 and 4-valve versions of 2.5L 4-cylinder base engine having EVC optimized as described earlier. This calculation was based on displaced volume when a particular v_{ps} criterion has been met, an assumed burned-to-unburned gas temperature ratio of 2.7 and some reference volumetric efficiency in the Torque range.

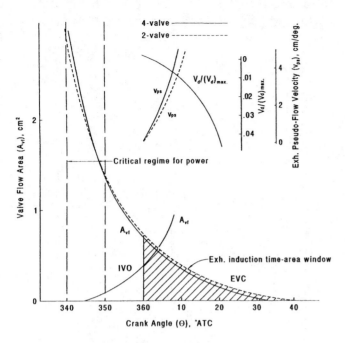

Figure 7: Significant features associated with EVC for a 2.5L short engine with 2 and 4-valve heads.

Significant exhaust system gas dynamic effects [17] complicate matters by imposing a time-dependent pressure gradient across the exhaust valves. Perhaps the most significant consequence of this occurs in I-4 engines having log-type exhaust manifolds where blowdown from one cylinder sends a pressure wave to an adjacent cylinder during its EVC period, and this has the potential to significantly reduce scavenging efficiency in the Torque range. This condition is attenuated by long-runner, exhaust manifold designs.

IVO: This event, as a determinant of valve overlap, has a significant effect on idle combustion stability. The overlap mass flow contribution to the total burned gas fraction is directly proportional to the magnitude of the effective time-area window resulting from overlap as backflow is choked over all reasonable levels of overlap and idle operating conditions. Where EVC is reasonably set for performance, there is no evidence suggesting that the phase angle of the overlap time-area window has any direct relevance in terms of Idle.

As mentioned earlier, IVO has an indirect effect on the intake time-area window mainly after the overlap period. Figure 8 illustrates and places in perspective the effect of doubling the overlap time-area window on the overall intake time-area window where an attempt was made to maintain a constant time-area window from BC to IVC. The initial overlap level is acceptable for idle and the latter is far outside of the range of acceptability.

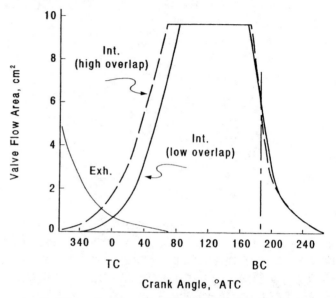

Figure 8: The effect of doubling the overlap time-area window on the overall intake time-area window

Interactions between IVO timing and unsteady intake pressures will be discussed with respect to performance in the next section.

IVC: Retardation of this event from the torque-maximized condition toward a power-maximized one yields predictable trends in Torque and Idle. At low speed conditions piston-driven, reverse flows mainly act to reduce volumetric efficiency and compression temperature respectively. The crank angle of reverse-flow cessation can be estimated using a v_{ps} calibrated via P-V data using effective start of compression as a v_{ps} criterion. The fractional volume of charge displaced and corresponding Torque compromise can therefore be estimated as shown in Figure 9.

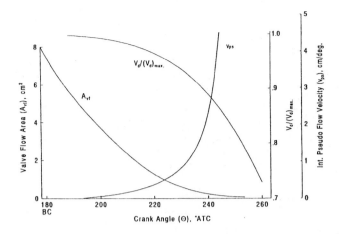

Figure 9: Assessment of <u>Torque</u> compromise associated with static valve timing.

Continued cylinder filling after BC becomes increasingly important at higher engine speeds. The details of this are highly engine specific and reflect the nature of intake tuning. During the mid-portion of the intake stroke the local intake pressure decreases owing to high rates of air withdrawal by the rapidly expanding cylinder volume. As the piston approaches BC a recompression follows the depression and aids the aftercharging process. The degree to which aftercharging is effective depends on the timing and magnitude of the intake recompression and on the availability of a time-area window at the appropriate phase angle; this is where valve timing assumes a role. It is noted that IVO timing can affect the aftercharging process by establishing the phased relationship between the intake recompression and the IVC time-area window as this is acoustically coupled to the timing of the onset of the intake depression. Figure 10 illustrates the unsteady intake pressures for a tuned and an untuned engine along with corresponding ln P vs. ln V plots. Note the close relationship between intake and cylinder pressures during the intake process particularly near BC.

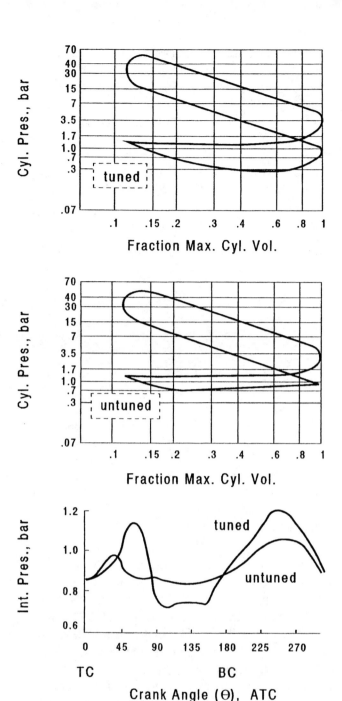

Figure 10: Unsteady int. pres. at <u>Power</u> for a tuned and an untuned engine along with corresponding ln P vs. ln V plots

Figure 11 is a generic illustration of a means of quantifying the aftercharging process based on cylinder and inlet pressures and on the instantaneous intake valve flow area. It is emphasized that both timed, positive pressure gradient and flow area are essential to the aftercharging process. Using a standard flow equation the mass flow rate and the cumulative mass flow can be computed. This is illustrated in reference [18].

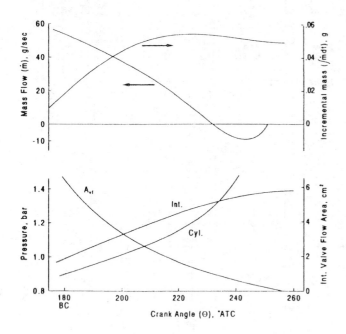

Figure 11: General illustration of aftercharging.

DISCUSSION

Means are available to manipulate poppet valve time-area windows in a wide variety of ways ranging from those that alter the phased relationship between pistons and fixed valve lift profiles to those that alter both the size and phase angle of the time-area windows. The former are generally associated with modest valve event changes and performance enhancement while the latter may additionally provide some measure of load control. Generally those means which offer the greatest valve event flexibility are also the most complex and technologically risky.

Many factors contribute to the desirability of VVT, and some of these areas follows:
- The intended engine speed range
- Packaging the VVT mechanism
- Ease of integrating a VVT mechanism with an existing cam drive
- The accrual of net benefits, e.g., performance, fuel economy and emissions control.

Some base engine characteristics on which performance generalization can be made are listed below.

<u>Speed Range</u>: Any engine can be made to be highly compromised by expansion of its speed range, e.g., <u>Torque</u> will become increasingly compromised to achieve <u>Power</u> and vice versa.

<u>Number of Valves/Cylinder</u>: Other things being similar, more valves/cylinder give larger time-area windows over shorter durations; therefore, are less compromised for a given speed range or are equally compromised for a broader speed range.

<u>Bore/Stroke</u>: This is an indicator of the size of the time-area windows normalized for displacement, i.e., for a given speed range undersquare is more performance-compromised than oversquare.

<u>Rate of Valve Lift Based on Mechanism Stiffness</u>: Mechanisms capable of providing high lift rates generally can provide larger time-area windows over shorter durations than those limited in this respect, e.g., a pushrod design will likely be more compromised than a direct-acting, OHC design.

VVT STRATEGIES

Applying principles outlined earlier, several common approaches to VVT are discussed below.

GENERAL CAM PHASER: This type of VVT has been evaluated over many years when most automotive engines had a single camshaft. In addition to changing the desired events this scheme changes all events, and this can yield compromised benefits. For example, retarding such a cam for <u>Power</u> imparts directionally proper changes in EVC and IVC however moves EVO in a countervailing direction. Caution must be exercised to avoid exhaust lock-up due to early EVC at low speeds with the cam advanced for <u>Torque</u>. This factor may limit the range of authority of this system. Pseudo-flow velocity (v_{ps}) in time domain can be used to determine the degree to which the cam can be advanced without causing this condition. Optimizing such a system for Power requires that EVO be sufficiently early to avoid excessive blowdown losses ($PMEP_{bd}$) in the retarded mode, and this will incur an excessive expansion loss when the cam is advanced for Torque and possibly Idle. This will have the effect of reducing thermal efficiency whenever the cam is advanced. Applying Eq (8) and assuming a 20 crank angle degree range of authority an incremental expansion loss as high as 3.5% may be expected. Alternatively, this expansion loss (η_t) can be avoided by selecting EVO timings that will incur incremental $PMEP_{bd}$ and compromise the Power increase associated with this type of VVT.

INTAKE CAM PHASER: This system obviates the problem of the general cam phaser but has a limitation of its own. As discussed earlier both low overlap and early IVC are desireable for Idle, and they can only co-exist with a shortened intake duration, i.e., shorter than what would likely be used for its static valve timing counterpart. This corresponds to a reduced intake time-area window which may compromise Power, i.e., the level of Power benefit may be reduced to satisfy Idle. Like the general cam phaser, this system with fixed phase angle between IVO and IVC has an additional limitation in intake-tuned engines owing to the fact that retarding for Power delays the intake recompression which the delayed IVC seeks to capture. Because of this interaction, it is desireable to consider intake tuning variables while optimizing such VVT systems.

VARIABLE DURATION INTAKE CAM: This includes a low lift, short duration cam for Torque and Idle and high lift, long duration cam for Power. The compromises associated with both types of cam phasers are avoided, but a more complex mechanism is required. Given the timing variations in both IVO and IVC the tuning interactions are expected to be an important factor, i.e., at long durations (Power mode) the intake recompression is advanced as IVC is retarded. Seemingly, this would be an attribute for very high speed operation. If such a system were optimized for Idle, Torque and Power it is doubtful that it could play any major role in load or emissions control.

FLEXIBLE INTAKE VALVE SYSTEM: This might be a variable fulcrum, a lost motion, a hydraulic or a pneumatic device which can impart a significant measure of load control. Controlling load in this way alters in-cylinder flows in ways that may cause burn rates to change. Reports on this yield diverse conclusions ranging from burn rate increases [4, 5] to burn rate decreases [6, 7]. It is probable that the effect of intake-flow generated turbulence on burn rates depends on the large scale flow characteristics designed into the base engine. In modern, fast burn engines it is likely that load control via intake valve event manipulation will have a tendency to reduce thermal conversion efficiency and thus negate some of the fuel economy benefits of reduced PMEP.

VVT FOR EMISSIONS CONTROL: It has been reported that the last portion of exhaust to leave the cylinder at the end of the displacement phase of the exhaust event contains disproportionally high concentration of HC emissions [19]. Valve timing strategies have been studied which effectively retain some of these emissions in the cylinder by either early EVC or by reinduction of them by using late EVC.

The regulation of residual fraction via either early EVC or by variable valve overlap have been investigated as a means to reduce engine out NO_x emissions. While this is directionally similar to the use of an external EGR system there may be some net benefits resulting from the higher temperature of the internally recycled exhaust versus that which is recycled via an external, EGR system. Furthermore, if the internally recycled exhaust is higher in HC, then overall emissions picture relative to both HC and NO_x may be different. The generality of these findings is unclear at this point.

CONCLUSIONS

1. It is clear that the net benefits of VVT are highly base-engine and duty-cycle specific.

2. In engines whose valve timing is optimized, variations in IVC (only) holds the potential to improve performance. Whether this improvement occurs mainly at high speeds or low, strongly depends on the datum, e.g., intake tuning and speed range and valve timing.

3. For engines designed to satisfy the automotive duty cycle as we know it today, the exhaust valve event compromises are very small; therefore, exhaust VVT is not independently justified.

4. Interactions between VVT and intake tuning are sufficiently important to warrant consideration in any application.

5. VVT has an established role as a performance feature. Its role as a fuel economy and emission control feature is unclear.

6. The rationalization of VVT can be enhanced by understanding the static valve timing compromises which VVT tries to overcome.

NOMENCLATURE

Where possible those used by Heywood [16] are applied here and are as follows:
- a Crank radius
- A_{vf} Minimum, geometric valve flow area
- B Bore
- d_s Valve stem diameter
- d_{vf} Valve face inner extension
- d_{vs} Valve seat width
- D_{os} Piston pin or crank offset
- D_{vis} Valve inner seat diameter
- k Experiementally-based polytropic exponent
- L_R Connecting rod length
- L_v Valve lift
- v_{ps} Valve pseudo-flow velocity
- V Cylinder volume
- V_c Clearance volume
- V_d Displaced volume
- β Valve seat angle
- θ_b Rapid burning angle
- η_t Thermal conversion efficiency
- θ Crank angle
- θ_l Start of rapid burning angle

- BC, ABC, BBC   Bottom-center crank position, after BC, before BC
- EVC, EVO   Exhaust valve closing, opening
- IVC, IVO   Intake valve closing, opening
- PMEP, $PMEP_{bd}$ Pumping mean effective pressure, blowdown
- TC, ATC, BTC   Top-center crank position, after TC, before TC

ACKNOWLEDGEMENTS

Chrysler Engineering colleagues L. A. LaPointe and R. G. Mosier are recognized, respectively, for contributing a formidable cylinder and manifold pressure data base and mathematical consultation.

REFERENCES

1. Ahmad, T. and M. A. Theobald, "A Survey of Variable-Valve-Actuation Technology," SAE Paper 891674, SAE Trans., sect. 3, 1989.
2. Gray, C., "A Review of Variable Engine Valve Timing," SAE Paper 880386, SAE Trans., sect. 6, 1988.
3. Asmus, T. W., "Effects of Valve Events on Engine Operation," <u>Fuel Economy in Road Vehicles Powered by Spark Ignition Engines</u>, J. C. Hilliard and G. S. Springer, Eds., Plenum Press, New York, 1984.
4. Stivender, D. L., "Intake Valve Throttling (IVT) - A Sonic Throttling Intake Valve Engine," SAE Paper 680399, SAE Trans., vol. 77, 1968.
5. Davis, G. C., R. S. Tabaczynski and R. C. Belaire, "The Effect of Intake Valve Lift on Turbulence Intensity and Burnrate in S.I. Engines - Model Versus Experiment," SAE Paper 840030, SAE Trans., 1984.

6. Newman, C. E., R. A. Stein, C. C. Warren and G. C. Davis, "The Effects of Load Control with Port Throttling at Idle-Measurements and Analyses," SAE Paper 890679, SAE Trans., sect. 3, 1989.
7. Tuttle, J. H., "Controlling Engine Load by Means of Late Intake-Valve Closing," SAE Paper 800794, SAE Trans., vol. 89, 1980.
8. Siewert, R. M., "How Individual Valve Timing Events Affect Exhaust Emissions," SAE Paper 710609, SAE Trans., vol. 80, 1971.
9. Tuttle, J. H., "Controlling Engine Load by Means of Early Intake-Valve Closing," SAE Paper 820408, SAE Trans., vol. 91, 1982.
10. Tasaka, H. and S. Matsuoka, "Predictions of Combustion, Fuel Economy, and Emissions Characteristics Influenced by The Gas Exchange Process of S.I. Engines," SAE Paper 810821, SAE Trans., vol. 90, 1981.
11. Tasaka, H., and S. Matsuoka, "Analysis of Gas Exchange Process of S.I. Engines Under City-Driving Conditions," SAE Paper 800535, SAE Trans., vol. 89, 1980.
12. Saunders, R. J. and E. A. Abdul-Wahab, "Variable Valve Closure Timing for Load Control and The Otto Atkinson Cycle Engine," SAE Paper 890677, SAE Trans., sect. 3, 1989.
13. Ma, T. H., "Effects of Variable Engine Valve Timing on Fuel Economy," SAE Paper 880390, SAE Trans., sect. 6, 1988.
14. Taylor, C. F., The Internal-Combustion Engine in Theory and Practice, Vol. 1, pp. 202, M.I.T. Press, Cambridge, MA, 1985.
15. Kastner, L. J., T. J. Williams and J. B. White, "Poppet Inlet Valve Characteristics and Their Influence on the Induction Process," Proc. Instn. Mech. Engrs., Vol. 178, p. 1, No. 36, 1963-64.
16. Heywood, J. B., Internal Combustion Engine Fundamentals, McGraw-Hill Book Company, New York, 1988.
17. Blair, G. P., J. R. Goulburn, "An Unsteady Flow Analysis of Exhaust Systems for Multicylinder Automobile Engines," SAE Paper 690469, SAE Trans., vol. 78, 1969.
18. Grohn, M., "The New Camshaft Adjustment System by Mercedes-Benz-Design and Application in 4-Valve Engine," SAE Paper 901727, 1990.
19. Tabaczynski, R. J., J. B. Heywood, and J. C. Keck, "Time-Resolved Measurements of Hydrocarbon Mass Flow Rate in the Exhaust of a Spark-Ignition Engine," SAE Paper 720112, SAE Trans., vol. 81, 1972.

C427/1/154

A High Speed Variable Valve Timing Mechanism for Engines

by R D NORRIS, MEng, Ricardo Consulting Engineers Ltd

SYNOPSIS. Variable Valve Timing (VVT) is becoming an increasingly common feature in internal combustion engines due to demands for increased performance fuel economy and lower exhaust emissions. This paper reviews the principles and benefits of VVT and describes the prism mechanism, a modified form of a device used in steam engines, which is suitable as a VVT mechanism in modern high speed engines. The mechanism is of compact design, light weight, high stiffness and has a dynamic performance suitable for high speed operation over a wide range of valve periods.

1. INTRODUCTION

It is well known that VVT can be used to increase the performance, particularly of gasoline engines, and indeed a number of systems have entered mass production during the last decade. The main application of VVT has been to enhance the performance of high specification vehicles, giving increased market appeal. VVT has also been used as a development tool to study the complex optimisation of fixed timings on conventional valve trains.

In general VVT is an attractive goal, but one which can be difficult to realise in practice due to the complexity of some of the valve actuation systems. Nevertheless, some of the simpler devices have reached production; namely phase change mechanisms on the inlet camshaft of DOHC engines and a mechanism for switching between two different cam profiles (Ref. 1). This paper brings attention to a relatively little known VVT mechanism which is potentially very suitable for engine use. The basic principles of the device are derived from a mechanism previously used in steam engines. However, the mechanism is presented in a modified form which is considerably more compact and allows continuously variable timing with high speed operation. To date most work on VVT has been concerned with gasoline engines and this paper concentrates on these.

2. VVT PRINCIPLES

Typical valve timings for a 4-stroke engine are shown in figure 1. In theory there are 4 different events that can be changed independently:

EVO	Exhaust valve opening.
EVC	Exhaust valve closing.
IVO	Inlet valve opening.
IVC	Inlet valve closing.

However, in practice the variables with most effect are the overlap period and the IVC timing. Both of these variables can be changed by a VVT device on just the inlet valves of an engine. Inlet camshaft phase mechanisms change both of these variables but not independently, and experimental data shows that variable period devices provide greater benefits than variable phase devices (Ref. 2). The EVO timing also has some effect on the maximum power and torque produced by the engine, but experience has shown that this feature can be optimised for a wide range of engine conditions without requiring significant timing variation.

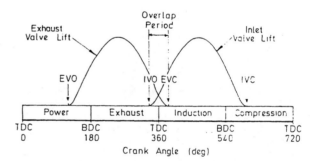

Fig 1 4-stroke cycle valve timing

The benefits of changing the overlap period and IVC at different engine conditions are shown in Table 1. These benefits are derived from the effects of valve timing on the processes within the engine, as described in the following sections.

2.1 Gas Dynamics

The gas dynamics or ram effects in the inlet and exhaust systems cause the gases to continue to flow, into or out of the cylinders, for longer than the theoretical 180° periods of the 4 stroke cycle. Because these gas dynamics effects increase with speed, the optimum valve timings generally result in longer valve periods as engine speed increases. The increased trapping efficiency through the use of optimum timings naturally improves performance but it also improves cycle efficiency and hence fuel economy.

VARIABLE	BENEFITS	
IVC	EARLY (35° ABDC)	LATE (65° ABDC)
	Low speed torque Part load fuel economy	High speed torque
OVERLAP	SMALL (20°)	LARGE (50°)
	Low load stability Part load fuel economy HC Emissions	Full load torque NOx Emissions

Table 1   Main variable valve timing effects

2.2 Internal EGR

During the overlap period when inlet and exhaust valves are both open, it is possible for exhaust gases to enter the inlet manifold and create exhaust gas recirculation (EGR). This process is particularly significant when large overlap periods are used and at light loads due to the high manifold depression. The recirculated exhaust gases within the combustion mixture act to stifle the combustion process. This reduces combustion pressures and temperatures and hence reduces NOx exhaust emissions. This is beneficial in view of increasingly stringent exhaust emission legislation, however, excessive EGR also reduces combustion stability causing cycle-to-cycle variation which results in poor idle quality. Hence under idle conditions a small overlap period is required which conflicts with the requirements under full load conditions. On fixed timing engines a compromise overlap is therefore required, and much of modern combustion system development is aimed at achieving better full load performance whilst maintaining acceptable part load combustion stability and idle quality (Ref.3). VVT can therefore be seen as adjunct to multivalve combustion systems for wide speed range engines.

2.3 Short Circuiting

During the overlap period under high load conditions, incoming mixture may exit through the exhaust valve and short circuit the combustion cycle. This unburnt mixture leads to high hydrocarbon levels in the exhaust. Longer overlap periods tend to promote this effect.

2.4 Summary

The above are the main processes which are affected by changes to valve timing but there are other effects dependant on the combustion system being used. For instance, tests at Ricardo show that HC and NOx emission levels can also be affected by the IVC timing. The actual valve timings used on a particular engine are usually a result of extensive development work in combination with computer simulation.

3. TYPICAL ENGINE VALVE TIMINGS

For a typical two valve per cylinder gasoline engine the valve timings might be as follows:-

    Valve period   245°
    Overlap   30°
    IVC   50° ABDC
    EVO   50° BBDC

For a four valve per cylinder engine the increased curtain area of the valves provides greater rates of gas exchange so that a reduced overlap period ($\simeq$ 15°) is necessary in order to maintain idle stability. This is, however, in conflict with the performance objectives for this type of engine.

For racing engines the valve timing strategy is aimed purely at increasing the maximum power to its limit. Idle stability is therefore ignored and very high engine speeds are used with high inertia ram effects. This leads to considerably wider valve periods and higher overlap as follows:-

    Valve period   330°
    Overlap   120°
    IVC   90° ABDC
    EVO   90° BBDC

The principal benefit of VVT in automotive use is that the engine design is not constrained by this compromise between idle stability and maximum power. The designer is therefore able to use a small overlap period at idle and to approach the racing engine timings at maximum power conditions. To achieve this, a typical VVT device might change the valve timings by 13° in phase and 60° in period. To achieve maximum benefit the valve timings should be continuously variable within these extremes.

4. VVT BENEFITS

Typical benefits from adding VVT to the inlet valves of an automotive engine are shown in table 2. These benefits are based on engine testbed data, with a VVT system capable of achieving optimum valve timing points at all speed and load conditions. The actual benefits realised on a particular engine are, however, highly dependant on the characteristics of the base engine and also on the VVT system and its control strategy. For, example, the system can be configured to provide balanced gains throughout the speed range, or to give a substantial increase in maximum power with little improvement in low speed operation.

The 25 to 30% emission reductions are considered to be very significant even if they appear small compared with a typical catalyst efficiency. Since VVT reduces the engine out emissions, the combined use of VVT and catalyst technologies will result in significant emission reductions and this approach will become increasingly attractive as emission regulations become more restrictive around the world. For further emission reductions it may also be possible to use VVT on the EVO timing to reduce the warm up time of the catalyst. At the present time, the exhaust emissions during the cold start period of the test cycle form a large proportion of the total emissions emitted through the whole cycle. For the future American ULEV (ultra low emission vehicle) limits, VVT combined with a catalyst must be considered as a candidate technology.

Another benefit of VVT is that it enables the working speed range of the engine to be extended by raising the maximum power speed. The wider valve periods at high speeds allow higher engine speeds to be achieved whilst maintaining the same maximum valve lift and acceleration levels. This wider speed range gives more flexibility to the choice of transmission ratios and enables further driveability and fuel consumption benefits to be derived.

TYPICAL OPTIMUM IMPROVEMENTS	
HIGH SPEED TORQUE	10%
LOW SPEED TORQUE	7%
PART LOAD FUEL ECONOMY	5%
HC EMISSIONS	30%
NOx EMISSIONS	25%

Table 2 Typical variable valve timing benefits

5. VVT MECHANISMS

Since the early stages of the commercialisation of the internal combustion engine, many proposals have been made for VVT systems. However, relatively few of these mechanisms are of practical use due to their complexity, speed limitations and durability constraints. Specific reviews of these mechanisms are presented in references 2, 4 and 5 and will not be repeated here.

A potentially viable VVT mechanism must be simple, lightweight, high in stiffness and have kinematics suitable for durable operation at high speeds. It should also ideally be capable of continuously varying not only the phase of the valve event, but also the lift and the period. The rest of this paper is concerned with the description of one such mechanism. The operating principle of the mechanism is derived from that of a device formerly used in the valve operating gear of a steam engine. This device is first described followed by a description of the modified form suitable for internal combustion engines.

6. CAPROTTI MECHANISM

The Caprotti valve gear (Ref. 6), and the similar Cossant valve gear were used on some steam locomotives and ships in the 1920's. The actual mechanisms are quite complex, but a simplified drawing illustrating their basic operating principle is shown in figure 2. The mechanism consists of two cams, both rotating at half engine speed, which act on either end of a beam. The valve motion is taken from the centre of the beam via a series of levers. The lift of the centre of the beam is equal to the sum of the lifts from the two cams with an associated rocker ratio of 0.5.

The variable timing is achieved by varying the relative phasing of the two cams, the phase of one cam controls the valve opening point and the phase of the second controls the closing point. If continuously variable phase change devices are used on both camshafts then both the valve opening and closing points may be independently adjusted giving variable period as well as variable phase of the valve event.

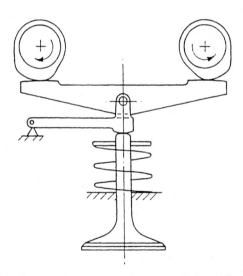

Fig 2   Caprotti valve gear (simplified)

6.1 Cam Profiles

Although, in principle, standard automotive cams can be used with this mechanism, the amount of timing variation is limited by the kinematics of the resultant valve motion. Instead, the Caprotti mechanism uses a special profile which dates from even earlier than the valve gear. These cam profiles are shown in figure 3, together with the resulting valve motion. Each of the cams has a very wide period with a long dwell at maximum lift.

The valve gear operation starts with a certain amount of backlash at the valve, such that the

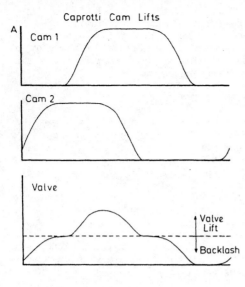

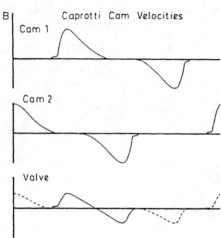

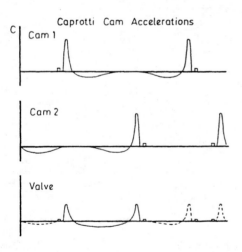

Fig 3   Caprotti valve gear cam profiles

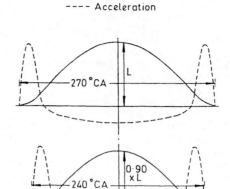

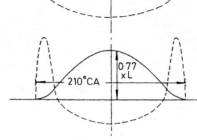

Fig 4   Valve lift and acceleration diagrams for different periods

rising curve of cam 2 takes up all the backlash when it reaches maximum lift and the beginning of its dwell period. Then, while cam 2 is on its dwell period, the rising curve of cam 1 contributes purely to valve lift. Hence the opening valve lift, velocity and acceleration are purely defined by cam 1. Similarly the valve closing profile is purely defined by the falling curve of cam 2 while cam 1 is on its dwell. The backlash then reappears with the falling curve of cam 1. The rising profile of cam 1 and falling profile of cam 2 can be designed to give a smooth transition between the cams near to maximum valve lift. This type of cam profile gives acceptable valve kinematics for a wide range of valve periods as shown in figure 4. The valve acceleration diagram is smooth and continuous which is ideal for high speed operation.

6.2   Application to high speed engines

When applying Caprotti type mechanisms to engines (Ref. 7) two basic difficulties are encountered.

i)   The mechanism is inevitably large due to the 0.5 rocker ratio associated with the cam lifts. This also leads to a mechanism which exhibits relatively high inertia and low stiffness.

ii)   The backlash that appears when the valve is closed prevents the use of conventional hydraulic lash adjustment. This is unfortunate because hydraulic lash adjustment is highly desirable on modern engines to minimise maintenance.

The mechanism presented next is a modified form of the Caprotti valve gear which does not suffer from problems.

7.   PRISM MECHANISM

The prism mechanism is shown in figure 5, and its operation through the valve event cycle is

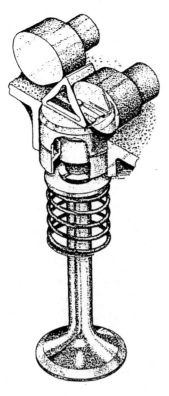

Fig 5   Prism VVT mechanism

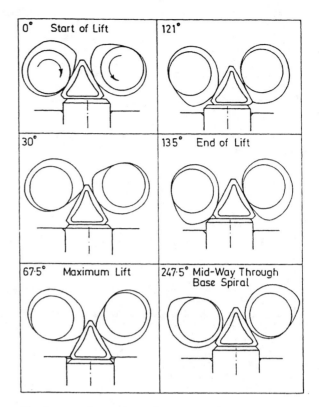

Fig 6   Operation of Prism mechanism through cycle

shown in figure 6. The beam part of the Caprotti valve gear is replaced with a triangular prism shaped follower. This is acted on by the two cams in a similar way to the Caprotti gear, but instead of pivoting, the follower is allowed to slide from side to side. This sliding action is controlled by a conventional bucket tappet, which maintains the horizontal attitude of the prism and prevents a side force acting on the valve stem. The mechanism is shown applied to a direct attack valve train but the prism may also be added to pushrod or rocker follower designs.

7.1 Operation

As was described for the Caprotti valve gear, the valve lift results from the addition of the individual cam lifts with an associated rocker ratio. However, because of the wedging action of the prism acting between the cams, the rocker ratio is higher than that of the Caprotti mechanism. The rocker ratio is dependant upon the base angle of the prism. With the 60° base angle shown in figure 6, the rocker ratio is 1.0, so that the cam lifts directly add to give the valve lift. This gives a very compact mechanism and is of a similar size to a conventional direct attack valve train.

The compactness and relative simplicity of the mechanism can be seen in figure 7 which shows a working model of the Prism mechanism applied to a 4-valve, cylinder head in a triple overhead camshaft arrangement. In theory the mechanism can be made smaller by using a larger prism base angle and increasing the rocker ratio further. However, this is undesirable since the increased wedging action causes high friction levels. The 60° base angle used in the model leads to cam forces similar to the equivalent direct attack valve train.

The addition of the prism follower to what is essentially a direct attack valve train will inevitably increase the valve train mass. However, the increase in effective mass is only approximately 10% if the mechanism is used in its optimum geometry form, and the system has inherently high stiffness.

Fig 7   Working model of prism VVT mechanism

7.2 Spiral Cam Profiles

The prism mechanism described above is a compact version of the Caprotti valve gear. The cam profiles used in the prism mechanism could be exactly the same as the special form used in the Caprotti system. However, this would mean that conventional hydraulic lash adjustment could not be used. To overcome the second problem of the Caprotti valve gear therefore, a new form of cam profile can be used. This cam profile is shown in Figure 8.

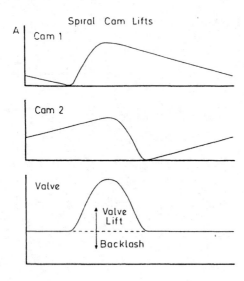

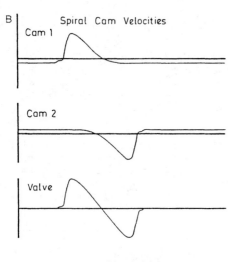

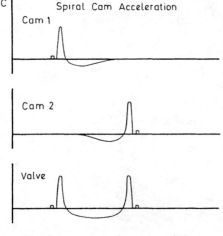

Fig 8   Spiral cam profiles

The new profile consists of two parts:-

i) An active part which either opens or closes the valve.
ii) A constant velocity ramp which forms a spiral taking the active part of the cam back to its starting point

The cam profiles are carefully designed, such that there is a smooth transition between the two parts, in lift velocity and acceleration.

The active part of each cam works in exactly the same way as the standard Caprotti profile cams and gives the same beneficial kinematic properties. The difference is in the constant velocity base spiral during the base event (valve closed period). The constant velocity ramp on cam 2 is of equal magnitude and opposite sign to the ramp on cam 1. This means that when the two cam lifts are added together, the ramp velocities cancel out giving a constant value for the base event. Consequently, because the resultant lift is constant there is no backlash during the base event so that a conventional hydraulic lash adjuster can be used.

Unfortunately, this type of cam profile cannot be used for a mechanical lash adjusted valve train because, as the valve period is changed, the height of the follower during the base event also changes. This amounts to a change in valve clearance which would cause a deterioration in the valve kinematics because the valve would not open at the start of the active part of the cam.

This problem does not occur with hydraulic adjustment because the change in follower base event height (approximately 0.5 - 0.7mm) is compensated by a change in the lash adjuster height, so that the valve still opens at the correct time.

If the cam/follower contact conditions are compared with a conventional valve train, it is found that the spiral cam forms have large radii of curvature throughout their event. The contact forces are similar to the conventional design so that for a given material combination and contact stress limits, cams having a smaller base circle radius can be used with the prism system. This makes the design more compact and also reduces friction. The elastohydrodynamic lubrication conditions between the cam and prism are very similar to those between a conventional cam and flat follower.

7.3 Camshaft Drives

The valve opening characteristics of the prism mechanism are adjusted in a similar manner to the Caprotti mechanism, that is by altering the phasing of the two cams. For complete variation of the valve event a variable phase mechanism must be included in the drive to both of the camshafts acting on the prism. This allows independent variation of the valve opening and closing points, and with suitable phase change devices, this variation can be continuous.

Figure 7 shows the prism mechanism operating the inlet valves of the 4 valve per cylinder design, using three camshafts. In this case, the exhaust camshaft has fixed timing but phase change devices are included on both inlet camshafts. The design of the camshaft drive is constrained by the need to install these phase change devices, and also by the close centre distances of the three camshafts. There are a number of solutions available, using gear or chain transfer drives between the camshafts, but the optimum solution will depend upon the application.

The camshaft drive can be simplified if only one phase change device is included. In this case only one valve timing point, either opening or closing, can be varied and the other timing point remains fixed. If, as would normally be the case, the VVT is applied to the inlet valves then two inlet camshafts are still needed, the first with variable timing, and the second with fixed timing. The exhaust valves can be actuated from this second camshaft via rockers so that the need for three camshafts is avoided.

8. CONCLUSIONS

VVT is an attractive technical solution to the challenge of providing tomorrows refined, highly rated gasoline engine with improved fuel economy and exhaust emissions. The greatest gains are achieved with VVT systems that provide continuous, and independent, variation of both of the valve opening and closing points.

The prism VVT mechanism provides this flexibility of operation in a compact and relatively simple manner. The mass, stiffness and kinematics of the systems are suitable for high speed operation over a wide range of valve periods.

9. ACKNOWLEDGEMENTS

I would like to thank my colleagues at Ricardo Consulting Engineers for their contributions and support for this work. I would also like to thank the directors of Ricardo for permission to publish this paper.

REFERENCES

1. Inoue,K., Nagahiro,K, Ajiki,Y. and Kishi,N. 'A high power, wide torque range, efficient engine with a newly developed variable - valve-lift and timing mechanism'. SAE paper 890675

2. Gray,C. 'A review of variable engine timing'. SAE paper 880386.

3. de Boer,C.D., Johns,R.J.R., Grigg,D.W., Train,B.M., Denbratt, I. and Linna J.R.- 'Refinement with performance and economy for four-valve automotive engines'. IMechE paper C394/053.

4. Dresner, T. & Barkan, P. 'A review and classification of variable valve timing mechanisms'. SAE paper 890674.

5. Demmelbauer-Ebner,W., Dachs,A. & Lenz,H.P. 'Variable valve actuation systems for the optimisation of engine torque'. SAE paper 910447.

6. Caprotti,A. 'Valve gear for reversing steam engines'. UK patent GB170,855. Accepted July 1922.

7. Dresner, T. & Barkan, P. 'The application of a two-input cam mechanism to variable valve timing'. SAE paper 890676.

960496

Spark-Ignition Engine Knock Control and Threshold Value Determination

Yun Young Ham and Kwang Min Chun
Yonsei Univ.

Jae Hyung Lee and Kwang Soo Chang
Hongik Univ.

ABSTRACT

Knock control algorithms were developed for a spark-ignition engine. Spark timing was controlled using cylinder block vibration signal. The vibration signal of a 1.5 L four cylinder spark-ignition engine was measured using an accelerometer which was attached to the cylinder block. The maximum amplitude of the bandpass-filtered accelerometer signals was used as the knock intensity.

Three different spark-ignition engine knock control algorithms were tested experimentally. Two algorithms were conventional algorithms in which knock threshold values were predetermined for each engine condition. Spark timing was retarded and advanced depending on the knock intensity in one algorithm and the knock occurrence interval in the other algorithm.

The third algorithm was a new algorithm in which knock threshold values were automatically corrected by monitoring knock condition. Knock condition was determined by the nondimensional variance of the 10 cycles' cylinder block vibration data which were sampled continuousely for each cylinder. It was confirmed that the value of nondimensional variance could be used as a tool of knock threshold value determination.

INTRODUCTION

It is well known that by increasing compression ratio we can improve both thermal efficiency and power output of a spark-ignition engine. But increasing compression ratio is usually accompanied by engine knock[1]. The occurrence of knock in an engine not only decreases the operator comfort and worsens the fuel economy of a vehicle[2], but it can also damage an engine[3]. Spark timing control is an useful way of eliminating knock by retarding spark timing on knock occurrence.

Knock can be characterized by many different quantities such as cylinder pressure[4-7], cylinder block vibration[8], exhaust gas temperature[2], sound[9], etc. Generally, knock sensors which measure the block vibration are used in knock control systems due to their advantages in cost and ease of maintenance[10-15].

In this study, we measured vibration signals using an accelerometer attached to the cylinder block and tested three different knock control algorithms. Two algorithms are conventional algorithms in which knock threshold values are predetermined for each engine condition. Spark timing is retarded and advanced depending on the knock intensity in one algorithm and the knock occurrence interval in the other algorithm. The third algorithm is a new algorithm in which knock threshold values are automatically corrected by monitoring knock condition.

EXPERIMENTAL APPARATUS

The schematic diagram of the experimental setup is shown in Fig.1. A 1.5L four cylinder engine was used. Cylinder pressure was measured using a Kistler 6053 piezoelectric pressure transducer and block vibration was measured using a B&K 4371 accelerometer. Engine torque was measured using a Kistler 9031 piezoelectric load washer. Data were taken and digitized using a DT 2839 board and stored

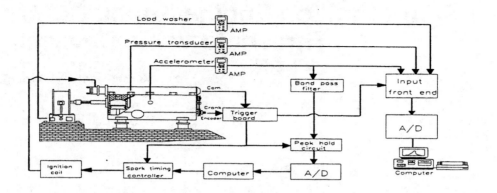

Fig.1 Schematic diagram of an experimental apparatus

in the data acquisition computer. The cylinder block vibration signal was amplified and filtered by 5-10 kHz band-pass filter. The output signal of the filter was passed through the peak-hold circuit which held the maximum value of the signal. The peak level was saved in a control PC using an A/D converter. Knock occurrence was determined by comparing the digitized peak value and the knock threshold value. The amount of retard/advance of spark timing was calculated using the knock control algorithm. Data of spark timings and peak values were stored in the control PC.

To measure the torque gain from knock control, crankangle based engine torque is required. Kistler 9031 load washer is used which is a piezoelectric transducer with a fast torque response. The load washer signal was acquired at every 0.2 crank degrees for one cycle. As the load washer signal represents relative value, it is compensated by the signal of the load cell. Engine torque of a cylinder for each firing cycle is defined as the average of torque data during 180 crank angles.

RELATION OF PRESSURE AND VIBRATION SIGNALS

Knock characteristics were determined using pressure signals of cylinder 4 and block vibration signals. Typical vibration signals of 4 cylinders for a knocking combustion cycle is shown in Fig.2 and compared with the raw and band-pass filtered pressure signals of cylinder 4. Knock intensities of the pressure and vibration signals were defined as the maximum amplitude of the filtered pressure and acceleration. The relation of knock intensity based on pressure and vibration signals is shown in Fig.3 which were acquired for 100 consecutive combustion cycles at each spark

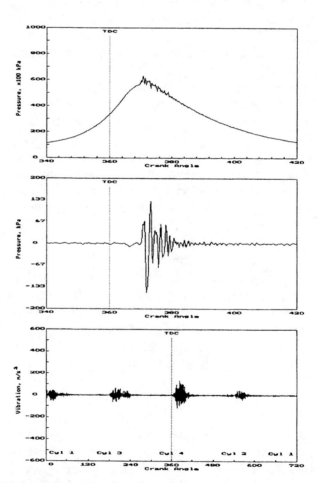

Fig.2 Comparison of pressure and vibration signals taken simultaneously at knocking condition (2000 rpm, WOT)

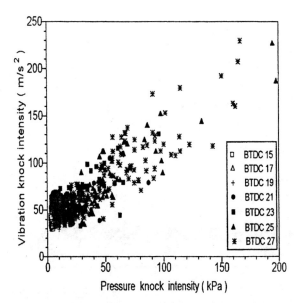

Fig.3 Comparison of knock intensities of pressure and vibration data for 100 consecutive cycles (2000rpm, WOT)

timing. It is shown that pressure and vibration signals have a good correlation.

Fig.4 illustrates the cumulative frequency distribution of knock intensity of 100 consecutive combustion cycles calculated from the block vibration. At a spark timing of 14 crankangle degrees before TDC, it is shown that there is no knock occurrence for any cylinder. Vibration signals of cylinders 2 and 3 are higher than those of cylinder 1 and 4 due to a difference in signal transmission path. At MBT spark timing, which is the upper limit of knock control algorithm, it is shown that there are large differences of distribution characteristics for each cylinder. Knock intensities of cylinder 3 are larger than those of any other cylinders. This trend can be found in the case of spark timing of 24 crankangle degrees before TDC. This cylinder-to-cylinder variation is due to the variations in signal path difference, compression ratio, mixture composition, mixture distribution and combustion chamber cooling.

KNOCK CONTROL ALGORITHMS WITH PREDETERMINED THRESHOLD VALUE

Determination of the knock threshold value is difficult and time consuming but very important for spark timing determination during engine development and knock control. First, we determined knock occurrence percentage using cylinder pressure signals. And then, the knock threshold value for vibration signals were determined which gave same knock occurrence percentage as those determined by cylinder pressure signals. The knock threshold value for vibration signals were about twice those of average non-knocking vibration signals. If the maximum filtered acceleration value is greater than knock threshold value, it is decided that knock occurred. In this study, when knock occurs, spark timing is controlled by using two parameters. One is knock intensity and the other is the knock occurrence interval. For each case, the upper limit of spark advance is the MBT timing and the lower limit of spark advance is the spark advance mapped in the commercial ECU at the same operating condition.

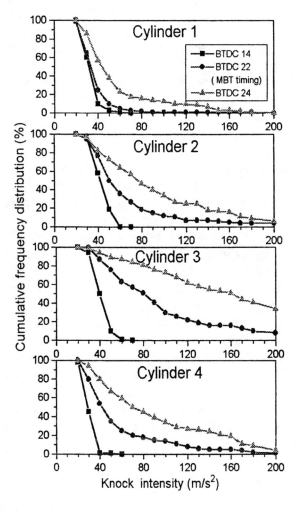

Fig.4 Cumulative frequency distribution of knock intensity for each cylinder (2000rpm, WOT)

SPARK TIMING CONTROL USING KNOCK INTENSITY - When knock intensity is the parameter of spark timing control, retard angle is calculated as follows:

Retard angle = (KI - TV)/TV × CV + MR

where, KI : current cycle's knock intensity
TV : knock threshold value
CV : retard correction value (2 degrees)
MR : minimum retard angle (1 degree)

Fig.5 shows the result of the knock control in which the spark timing of all cylinders is retarded simultaneously according to knock intensity. If knock doesn't occur, the spark timing is advanced 1 degree for every 20 cycles. Data on the torque gain curve are the relative values of the averaged 100 cycles' torque compared to the values at the reference spark timing. It is shown that there is about 3% torque gain due to knock control. After heavy knock occurrence spark timing is retarded in large amount and knock is prevented with sudden torque reduction.

Fig.6 shows the result of the knock control in which spark timing of each cylinder is retarded individually according to knock intensity. It is shown that spark timing of each cylinder approaches the upper limit of spark advance more often than that with the simultaneous knock control and knock occurs more often in cylinder 3 than other cylinders. In spite of heavy knock occurrence in one cylinder, there is no serious torque loss because the spark timing of other cylinders is not retarded. Total torque gain is a little greater than the case of the simultaneous knock control. The knock frequency of all cycles is about twice the frequency with the simultaneous knock control.

Fig.7 shows the comparison of the knock frequency distributions of the individual knock control and the simultaneous knock control for each cylinder. Also it shows comparison of cases when retard value on knock occurrence is constant and when retard value varies according to knock intensity. The knock occurrence percentage decreases as the retard angle on knock occurrence becomes larger. The knock occurrence percentage for each cylinder is higher for individual control cases than for simultaneous control cases except for cylinder 3

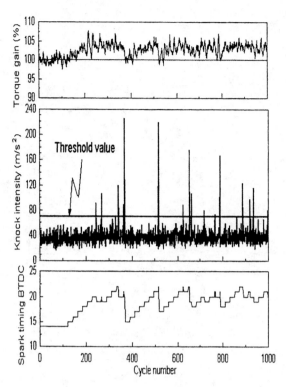

Fig.5 Result of the simultaneous knock control whose parameter is knock intensity (2000rpm, WOT)

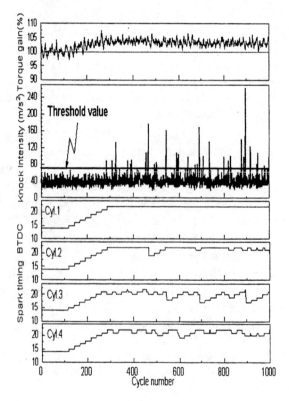

Fig.6 Result of individual knock control whose parameter is knock intensity (2000rpm, WOT)

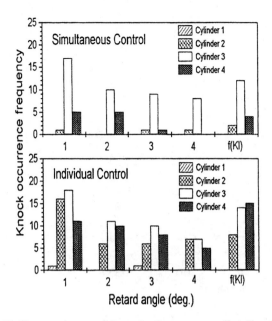

Fig.7 Comparison of knock frequency distribution for each cylinder of simultaneous and individual knock control (2000rpm, WOT)

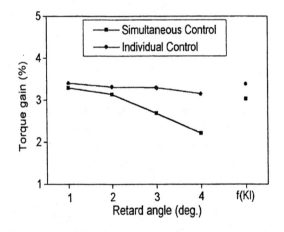

Fig.8 Comparison of torque gain for each cylinder of simultaneous and individual knock control (2000rpm, WOT)

which has the highest knock occurrence percentage. The spark timing of the simultaneous knock control is determined mainly by the cylinder in which knock occurs often. It is shown that the knock frequency distributions resulted from the individual knock control are more uniform than those resulted from the simultaneous knock control. Also, we can see that the knock frequency with knock intensity based control (shown in the x-axis as f(KI)) is higher than that in the case of 2° retard and lower than that in the case of 1° retard.

As shown in Fig.8, the individual knock control results in higher torque gain than the simultaneous knock control. The torque gain with the individual knock control decreases slightly as the retard angle on knock occurrence becomes larger and it decreases more with the simultaneous knock control. The torque gain with knock intensity based control (shown in the x-axis as f(KI)) is almost same as those of retard angle 1° or 2°.

SPARK TIMING CONTROL USING KNOCK OCCURRENCE INTERVAL - The knock occurrence interval is the time between one knock cycle and the next knock cycle. The knock occurrence interval algorithm is designed to determine the retard angle according to the knock occurrence interval so as to make the knock occurrence interval close to the disired knock interval. The retard angle is determined by using the gain function. Fig.9 shows the gain function which we use in the knock control algorithm. The gain function decides the retard angle according to the nondimensional knock occurrence interval ε.

$$\varepsilon = \frac{T_{int} - T_d}{T_d}$$

where, T_{int} : kncock occurrence interval
T_d : desired knock interval

Fig.10 shows the result of knock occurrence interval control when the desired knock occurrence interval is 20 cycles. It is shown that there is about 3 % torque gain and the knock occurrence frequency of total cycles is higher than the case of simultaneous knock intensity based control. The knock occurrence frequency according to the desired knock interval is shown in Fig.11. The knock occurrence frequency decreases as the desired knock interval increases. Here again cylinder 3 has higher knock occurrence compared to other cylinders. We can see that the knock occurrence frequency can be controlled by changing the desired knock interval. In Fig.12, the torque gain decreases as the desired knock interval increases.

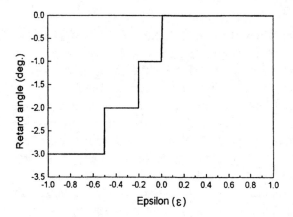

Fig.9 Gain function of knock control algorithm whose parameter is knock occurrence interval

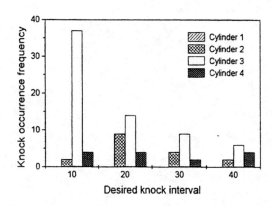

Fig.11 Knock occurrence frequency against desired knock interval with the knock control algorithm whose parameter is knock occurrence interval (2000rpm, WOT)

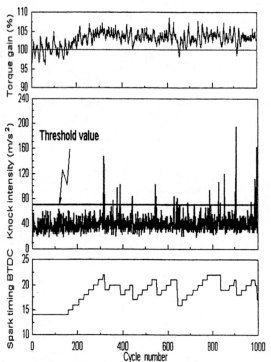

Fig.10 Result of knock control algorithm whose parameter is knock occurrence interval (2000rpm, WOT)

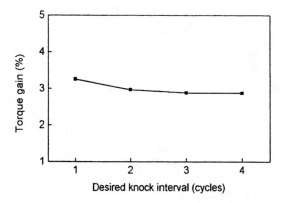

Fig.12 Torque gain against desired knock interval with the knock control algorithm whose parameter is knock occurrence interval (2000rpm, WOT)

AUTOMATIC CORRECTION OF THRESHOLD VALUE

Generally, the knock threshold value of an engine is determined by multiplying a mean non-knock vibration level by a constant. In such a knock control system, there is a problem of not considering manufacturing dispersion and the possible change of an engine with time. To solve the above mentioned problem, the present study developed an algorithm which enables automatic correction of the knock threshold value by monitoring the knock condition of the last 10 cycles for each cylinder. Knock condition of last 10 cycles is determined by the nondimensonal variance value normalized by the reference variance value at the non-knocking state for each cylinder. Fig.13 shows the comparison of the maximum filtered acceleration data and the nondimensional variance value at every 10 cycle steps which are obtained by changing the spark timing from BTDC 14° to BTDC 20° with crank angle 2° step. The reference variance value is the variance value of the first 100 cycles.

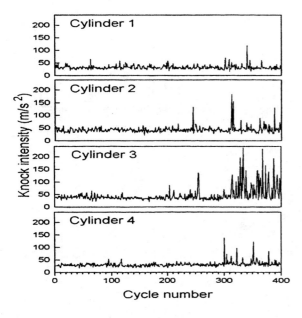

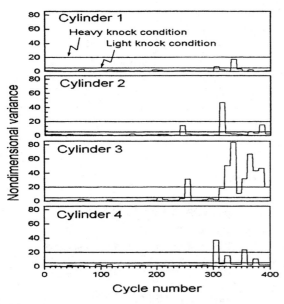

Fig.13 Comparison of knock intensity and nondimensional variance of last 10 cycles' knock intensity for each cylinder (2000rpm, WOT)

The nondimensional variance level for light knock condition is 5 and heavy knock condition is 20 for all cylinders which are determined from the comparison with knock intensity data.

Fig.14 shows the block diagram of the threshold value automatic correction algorithm. The knock threshold value is corrected when the decisions made by using nondimensional

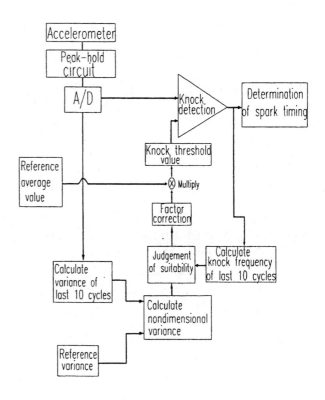

Fig.14 Block diagram of the threshold value automatic correction algorithm

variance and knock threshold are different. If there are vibration signals which are over the threshold value when the last 10 cycles were determined as a non-knocking by using nondimensional variance, it means that last knock next 10 cycles' threshold value increases. On threshold value was set too low. Accordingly, the contrary, if there are no vibration signals which are over threshold value when past 10 cycles were determined as knocking, it means that last 10 cycles' threshold value was set too high. So, next 10 cycles' threshold value decreases.

In Fig.15, results of the automatic correction of the knock threshold value are shown. When the initial threshold value is too high as shown in Fig.15 (a), the threshold value decreases. When the initial threshold value is too low as shown in Fig.15 (b), the threshold value increases. In all three cases including the case shown in Fig.15 (c) where the threshold value does not change much, the final threshold value becomes about 70 ~ 80 m/s^2.

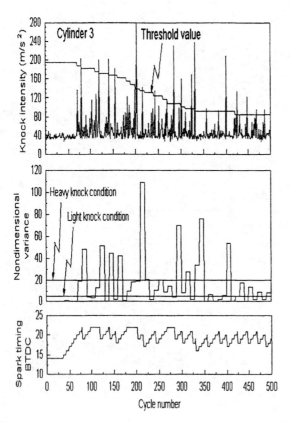

(a) When the initial threshold value is too high

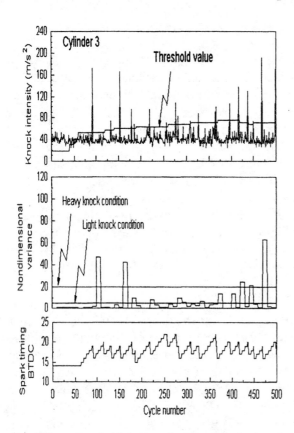

(b) When the initial threshold value is too low

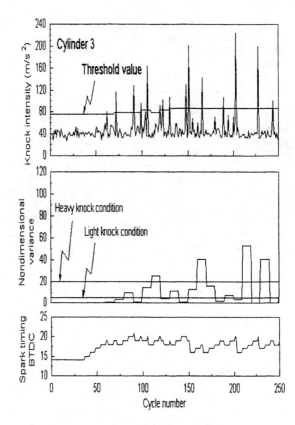

(c) When the initial threshold value is moderate

Fig.15 Results of the threshold value automatic correction

CONCLUSION

In this research, three different spark-ignition engine knock control algorithms were tested experimentally using the cylinder block vibration signal. The result can be summarized as follows.

1) The results of studies done on the knock control algorithms which use knock intensity and knock occurrence interval as a parameter of the retard/advance value control confirmed the improvement of WOT torque with acceptable knock occurrence. Both control algorithms give similar torque gain of about 3 % when they are optimized.

2) It is shown that the value of vibration signal's nondimensional variance can be used as a tool of knock threshold value determination. The best knock threshold value can be used for each engine during its life time by using the knock threshold value automatic correction algorithm.

ACKNOWLEDGMENT

This project is carried with the support of G7 project. The authors also thank to Daewoo Motors Company for their hardware support.

REFERENCES

1. John B. Heywood, "Internal Combustion Engine Fundamental", Mc-Graw Hill, 1988
2. Matthew L. Franklin, Thomas E. Murphy, "A Study of Knock and Power Loss in the Automotive Spark Ignition Engine", SAE Paper 890161, 1989
3. G. König, R.R. Maly, D. Bradley, A.K.C. Lau and C.G.W. Sheppard, " Role of Exothermic Centres on Knock Initiation and Knock Damage", SAE Paper 902136, 1990
4. K.M. Chun, J.B. Heywood, "Characterization of Knock in a Spark-Ignition Engine", SAE Paper 890156, 1989
5. K. M. Chun, K. W. Kim, "Measurement and Analysis of Knock in a SI Engine Using the Cylinder Pressure and Block Vibration Signals", SAE Paper 940146, 1994
6. Tassos H. Valtadoros, Victor W.Wong and John B. Heywood, "Engine Knock Characteristics at the Audible Level", SAE Paper 910567, 1991
7. K. Sawamoto, Y. Kawamura, T. Kita and K. Matsushita, "Individual Cylinder Knock Control by Detecting Cylinder Pressure", SAE Paper 871911, 1987
8. O. Hirako, N. Murakami and K. Akishino, "Influence of Valve Noise on Knock Detection in Spark Ignition Engine", SAE Paper 880084, 1988
9. T. Priede, R. K. Dutklewicz, "The Effect of Normal Combustion and Knock on Gasoline Engine Noise", SAE Paper 891126, 1989
10. Y. Boccadoro, T. Kizer, "Adaptive Spark Control with Knock Detection", SAE Paper 840447,1984.
11. Heinz Decker, Hans-Ulrich Gruber, "Knock Control of Engines - A Comparison of Solutions and Techniques, with Special Reference to Future European Emission Legislation", SAE Paper 850298,1985.
12. T.Iwata, K.Sakakibara and H.Haraguchi, "A New Method to Automatically Optimize the Knock Detection Level in the Knock Control System", SAE Paper 891964,1989.
13. S. Rohde, M. Philipp, "Combined Boost Pressure and Knock Control System for S.I. Engines Including 3-D Maps for Control Parameters", SAE Paper 890459, 1989.
14. Karl P.Schmillen, Manfred Rechs, "Different Methods of Knock Detection and Knock Control", SAE Paper 910858, 1991.
15. T. A. Fauzi Soelaiman, D. B. Kittelson, "Detecting Knock in Noisy Spark Ignition Engines", SAE Paper 931900, 1993.

960584
Comparison of Variable Camshaft Timing Strategies at Part Load

T. G. Leone, E. J. Christenson, and R. A. Stein
Ford Motor Co.

Copyright 1996 Society of Automotive Engineers, Inc.

ABSTRACT

In this paper, four Variable Camshaft Timing (VCT) strategies are described: Intake Only, Exhaust Only, Dual Equal, and Dual Independent. The strategies utilize internal residual at part load for NOx reduction and fuel consumption improvement. The emphasis of the paper is a detailed comparison of part load data from steady-state engine dynamometer testing. Projections of EPA cycle fuel economy and emissions benefits relative to external EGR are also shown. Only limited data was acquired at idle and WOT.

Implications of the strategies on the engine control system are briefly addressed.

INTRODUCTION

Variable camshaft timing (VCT) involves phase-shifting the camshaft(s) relative to the crankshaft as a function of engine operating conditions. With DOHC engines, there are 4 possible types of VCT: phasing only the intake cam (Intake Only), phasing only the exhaust cam (Exhaust Only), phasing the intake and exhaust cams equally (Dual Equal), and phasing the intake and exhaust cams independently (Dual Independent). The Dual Equal strategy (1) is also applicable to SOHC and pushrod engines.

The subject of this paper is a comparison of the four types of VCT, with emphasis on part load fuel consumption and emissions. Effects at idle and WOT are only briefly addressed.

All four strategies as described in this paper utilize continuously variable cam phase-shifting to increase internal residual dilution at part load for NOx reduction and fuel consumption improvement. A primary objective is to allow elimination of the external exhaust gas recirculation (EGR) system, which would provide a significant offset in variable cost, complexity, and warranty cost.

In the following sections, concept descriptions are given for the four strategies at part load; idle and WOT are described later in the paper.

DUAL EQUAL CONCEPT DESCRIPTION

The Dual Equal VCT strategy is described in detail in a previous paper (1), and a substantial portion of that description is repeated in this paper for completeness. In the Dual Equal strategy, the camshaft events are significantly retarded at part load. The primary objectives of the cam retard at part load are the following: delayed valve overlap for increased residual dilution, late intake valve closing (IVC) for pumping work reduction, and late exhaust valve opening (EVO) for increased expansion work. Figure 1 illustrates the concept of Dual Equal cam retard for a 2.0L 4-valve I4 engine with the valve events retarded 30 crank angle (CA) degrees relative to standard cam timing. The effects of the retarded valve events are described below.

DELAYED OVERLAP - With the cam events significantly retarded, overlap is delayed into the intake stroke. At the beginning of the intake stroke, the intake valve is closed while the exhaust valve is still open. As the piston moves down, exhaust gas is drawn back into the cylinder from the exhaust port. This has three benefits relative to standard cam timing:

1) NOx is reduced because of the increased internal residual.
2) Unburned HC are reduced because the last part of the exhaust, which is relatively high in unburned HC (2,3), is drawn back into the cylinder to be 're-burned' during the next combustion event.
3) Intake stroke pumping work is reduced for two reasons:
- Initially the pressure of the exhaust gas drawn back into the cylinder is at exhaust backpressure. As shown in Figure 1, this results in higher cylinder pressure during the first part of the intake stroke, which reduces pumping work.
- With the increased internal residual, a higher manifold absolute pressure (MAP) is required to maintain a given load. The higher MAP results in reduced intake stroke pumping work, also as shown in Figure 1.

Figure 1: Dual Equal Camshaft Phase Shifting

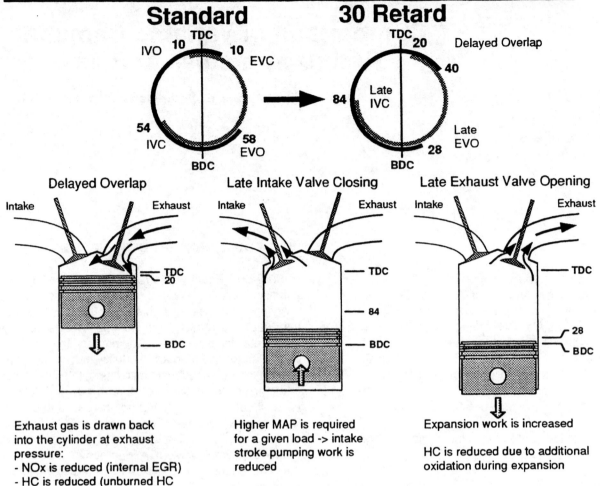

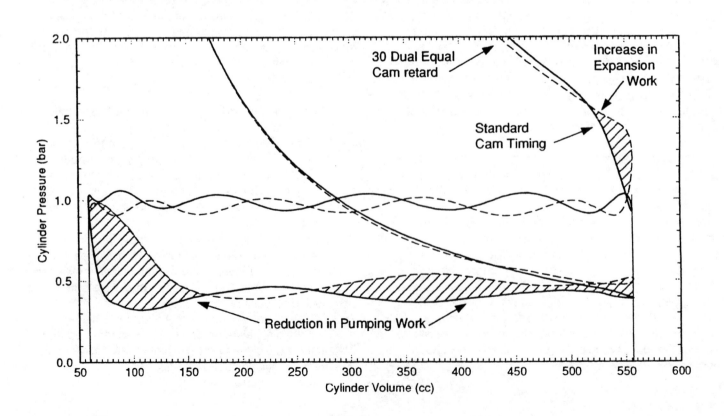

LATE INTAKE VALVE CLOSING - At low engine speeds, later IVC causes more of the fresh charge to be pushed back into the intake port during the first part of the compression stroke. To maintain a given load, a higher MAP is required, and intake stroke pumping work is reduced.

Previous studies (4,5) have been conducted on the use of late IVC for pumping work reduction. These studies showed that although pumping work is significantly reduced with late IVC, the consequent reduction of effective compression ratio and temperature near the end of the compression stroke results in degradation of burn rate and dilution capability, limiting the fuel consumption benefit of using late IVC.

In the Dual Equal strategy, this reduction of temperature rise during compression is offset by the higher initial temperature of the residual / fresh charge mixture. Higher initial temperature results from the use of internal instead of external exhaust gas recirculation (EGR).

Thus, the combination of late IVC with increased internal residual in the Dual Equal VCT strategy results in offsetting effects on unburned gas temperature, such that the burn rate degradation usually associated with late IVC is avoided.

LATE EXHAUST VALVE OPENING - With fixed cam timing, the EVO timing is a compromise between expansion work at low speed (later EVO results in increased expansion work) and exhaust stroke pumping work at high speed (earlier EVO allows more time for blowdown before BDC). The expansion work loss associated with EVO can be calculated from cylinder pressure vs. volume diagrams (1), and is shown as the shaded area in Figure 2. This expansion work loss is termed the EVO loss, and can be expressed as a percentage of the IMEP to determine the fuel consumption effect.

With Dual Equal cam retard, EVO occurs later in the expansion stroke, resulting in increased expansion work and improved ISFC. As illustrated in the pumping loop at the bottom of Figure 1, with 30 degrees of cam retard, the expansion work loss associated with the standard EVO timing is nearly fully recovered.

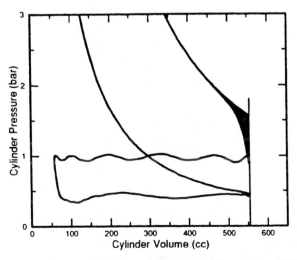

Figure 2: EVO Loss at 1500PRM / 2.62 bar BMEP

EVO timing also has an effect on unburned HC emissions at low speeds and light loads. Later EVO results in additional crevice outgassing and time for in-cylinder oxidation of HC before blowdown. For example, at 1500 RPM, 36 degrees of retard allows 4 msec additional time for oxidation.

DUAL INDEPENDENT CONCEPT DESCRIPTION

Because the intake and exhaust camshafts can be phased independently, the Dual Independent strategy has the flexibility to optimize cam phasings for operation at idle and WOT, in addition to part load.

The optimum strategy at part load is simply a refinement of Dual Equal retard, whereby both cams are significantly retarded, but with variable instead of fixed overlap. This will be discussed in more detail later in the paper.

EXHAUST ONLY CONCEPT DESCRIPTION

In the Exhaust Only VCT strategy, the exhaust camshaft is significantly retarded at part load. The primary objectives of the cam retard at part load are increased valve overlap for increased residual dilution, and late EVO for increased expansion work. Figure 3 illustrates the concept of Exhaust Only cam retard for the 2.0L I4 engine with the exhaust valve events retarded 30 crank angle (CA) degrees relative to standard cam timing. The effects of the retarded exhaust valve events are described below.

INCREASED OVERLAP - With the exhaust events significantly retarded, overlap is extended into the intake stroke. At the beginning of the intake stroke, the exhaust valve is still open. As the piston moves down, exhaust gas is drawn back into the cylinder from the exhaust port, and also flows back into the intake port (due to intake vacuum). As with Dual Equal cam retard, this has three benefits relative to standard cam timing:

    1) NOx is reduced because of the increased internal residual.
    2) Unburned HC are reduced because the last part of the exhaust is drawn back into the cylinder.
    3) Intake stroke pumping work is reduced for two reasons:
    - The cylinder pressure is higher during the first part of the intake stroke, due to backflow from the exhaust port (as shown in Figure 3).
    - With the increased internal residual, a higher MAP is required to maintain a given load, resulting in reduced intake stroke pumping work, also as shown in Figure 3.

LATE EXHAUST VALVE OPENING - With Exhaust Only cam retard, EVO occurs later in the expansion stroke. As discussed for Dual Equal, this results in increased expansion work and improved unburned HC emissions.

Figure 3: Exhaust Camshaft Phase Shifting

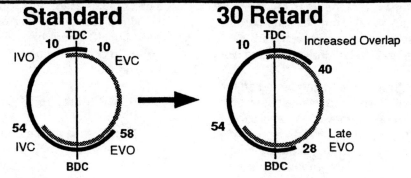

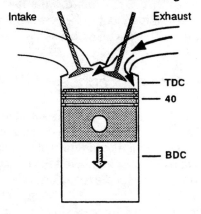

Late Exhaust Valve Closing

Exhaust gas is drawn back into the cylinder:
- NOx is reduced (internal EGR)
- HC is reduced (unburned HC 're-burned')
- Intake pumping work is reduced

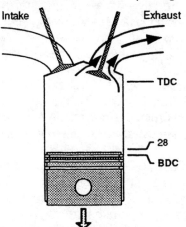

Late Exhaust Valve Opening

Expansion work is increased

HC is reduced due to additional oxidation during expansion

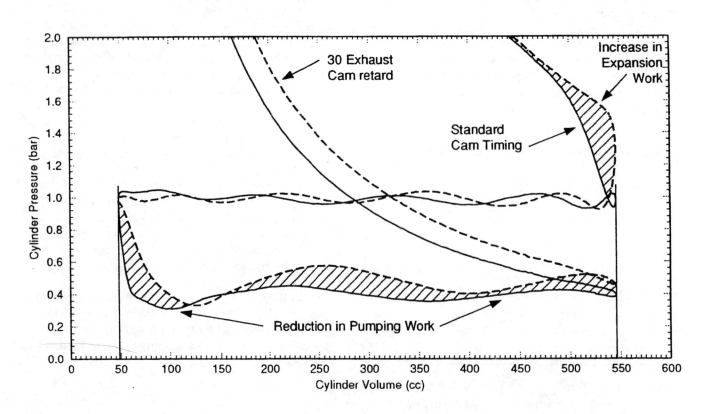

INTAKE ONLY CONCEPT DESCRIPTION

In the Intake Only VCT strategy, the intake cam is significantly advanced at part load. The primary objective of the cam advance at part load is increased valve overlap for increased residual dilution. Figure 4 illustrates the concept of Intake Only cam advance for the 2.0L I4 engine with the intake valve events advanced 30 crank angle (CA) degrees relative to standard cam timing.

With the intake events significantly advanced, overlap is extended into the exhaust stroke. Near the end of the exhaust stroke, the intake valve starts to open. The intake manifold is at low pressure, so exhaust gas backflows from the exhaust port and the cylinder into the intake port. This exhaust gas is then drawn into the cylinder on the subsequent intake stroke. As with Dual Equal and Exhaust Only, this has three benefits relative to standard cam timing:

1) NOx is reduced because of the increased internal residual.
2) Unburned HC are reduced because the last part of the exhaust is recycled.
3) Pumping work is reduced. With the increased internal residual, a higher MAP is required to maintain a given load, which results in reduced intake stroke pumping work, as shown in Figure 4.

Note that advancing the intake cam results in earlier IVC. When the intake valve closes earlier in the compression stroke, less of the fresh charge is pushed back into the intake port and a lower MAP is required to maintain the load. This tends to counteract the effect on MAP of the increased residual. However, because the change in cylinder volume with crank angle is non-linear, advancing IVC has a relatively small effect on MAP, which is more than offset by the effect of the internal residual.

TRANSIENT CONTROL OF RESIDUAL DILUTION

The transient control of residual dilution was discussed in (1), and that discussion is repeated in this paper for completeness.

Previous work on fast burn engines has demonstrated significant improvement in fuel consumption and NOx emissions with high rates of EGR at steady state. However, these benefits are not fully realized in vehicles because the transient control of external EGR is difficult due to the large volume of air/EGR mixture in the intake system. In addition, HC emissions increase rapidly at high rates of external EGR. As a result of these two factors, vehicles typically are calibrated to use EGR rates much lower than allowed by the steady state stability limit. For example, an engine which can tolerate 25% EGR at steady state may only be calibrated to run 10 - 15% EGR in a vehicle.

With VCT, internal residual is increased via backflow from the exhaust, as described earlier. Because the residual is directly controlled at the poppet valves, the manifold filling and emptying effects associated with external EGR are eliminated. In principle, the transient control of dilution is improved, and therefore VCT should allow the use of increased dilution for greater NOx and fuel consumption benefits. Also, as shown later, the HC emissions with VCT at increased rates of dilution (lower NOx) are comparable to or less than those for 10% external EGR.

Additionally, if introduced near the main throttle for good mixing with the fresh air, the use of external EGR causes throttle body icing problems in cold weather. External EGR also aggravates deposit build-up problems in the intake port. Both of these concerns are eliminated with Dual Equal, Dual Independent, or Exhaust Only VCT, because the residual is drawn back from the exhaust port directly into the cylinder. Intake Only VCT eliminates the throttle body icing concern, but not the intake port deposit concern (because large amounts of residual flows back into the intake port).

PART LOAD DATA AT 1500 RPM / 2.62 BAR BMEP

Cam phasing sweeps were run back-to-back with external EGR sweeps at a number of speed/load test points. The part load data shown were evaluated at the spark timing which resulted in a 1% loss in BMEP relative to MBT (from regressed constant throttle spark sweeps). All data was acquired at stoichiometry. The test engine was a 2.0L DOHC 4-valve I4 with intake ports designed to enhance tumble in-cylinder motion with no intake valve masking. The engine has the capability to independently phase-shift intake and exhaust camshaft timing during engine operation. Figure 6 shows results for Exhaust Only, Intake Only, and Dual Equal cam phase sweeps without external EGR, as well as an EGR sweep at standard cam timing, at 1500 RPM, 2.62 bar BMEP, a test point which is typically heavily weighted in the EPA City Cycle. Dual Independent will be discussed in a later section, as a refinement of Dual Equal.

FUEL CONSUMPTION AND EMISSIONS - Figure 6 includes plots of BSFC, PMEP, EVO Loss, BSHC, and BSCO vs. BSNO. As indicated in Figure 6, cam phasing is evaluated at lower NOx when compared to EGR because improved transient control of dilution is assumed.

Compared to 10% EGR, 40 degrees Dual Equal retard improves BSFC by 4.4%, 35 degrees Exhaust Only retard improves BSFC by 3.8%, and 35 degrees Intake Only advance improves BSFC by 3.4%. Cam phasing was limited by combustion stability for Exhaust Only and Intake Only. A maximum phasing of 40 degrees was taken as representative of a typical production limit for valve-to-piston clearance.

The largest portion of the fuel consumption improvement comes from reduced pumping work. PMEP is reduced by 0.21 bar for Dual Equal, by 0.11 bar for Exhaust Only, and by 0.09 bar for Intake Only at this speed/load. As expected, Dual Equal shows the greatest benefit in pumping work. Exhaust Only does not have the pumping work reduction of late IVC, and has slightly less pumping work reduction with increased overlap compared to delayed overlap (see pumping loop diagram in Figure 5). The pumping work benefit of Intake Only is

Figure 4: Intake Camshaft Phase Shifting

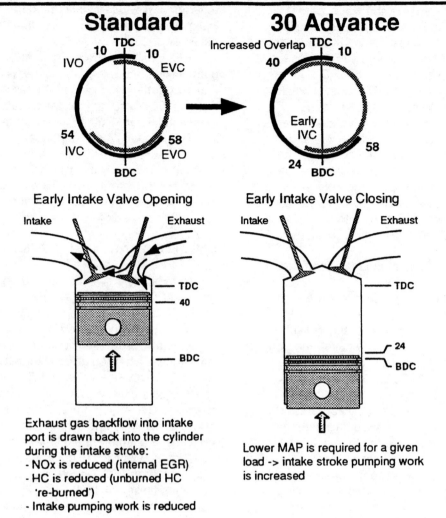

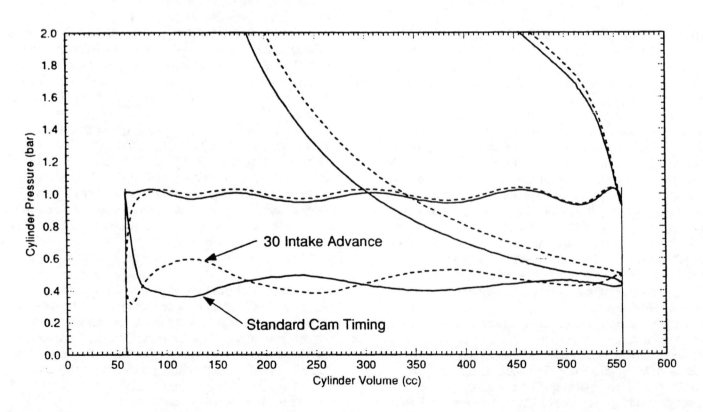

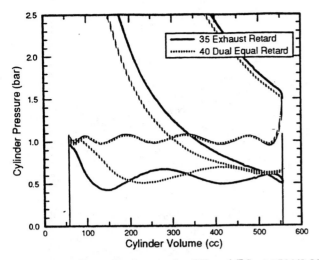

Figure 5: Pumping Loops for DE and EO at 1500/2.62

smaller still, due to early IVC.

The fuel consumption effect of a PMEP improvement can be calculated by expressing the difference as a percentage of the IMEP. (Note that "Area C" effects are important for Dual Equal and have been taken into account, see Appendix.) The reduced pumping work gives an expected fuel consumption benefit of 4.5% for Dual Equal, 3.1% for Exhaust Only, and 2.5% for Intake Only.

Similarly, an increase in expansion work due to later EVO improves fuel consumption for Dual Equal and Exhaust Only. The fuel consumption effect is calculated by expressing the difference in EVO loss as a percentage of the IMEP. The increased expansion work gives a fuel consumption benefit of 0.6% for Dual Equal and Exhaust Only (EVO is fixed for Intake Only). The amount of this benefit depends on the EVO timing of the base engine; for example, this benefit was 1.0-1.2% for the engine in reference (1), which had an earlier base EVO timing.

BSHC are reduced about 0.8 g/kW-hr for Dual Equal. Since unburned HC represent lost heating value of the fuel, a reduction in HC results in improved fuel consumption. This fuel consumption effect can be estimated by expressing the BSHC difference as an Emissions Index (mass flow of emission/mass flow of fuel). A difference of 0.8 g/kW-hr corresponds to a fuel consumption improvement of about 0.2% for Dual Equal. Because they are evaluated at significantly lower NOx than that for 10% EGR, Exhaust Only and Intake Only showed slightly higher BSHC than 10% EGR at this speed/load, for a fuel consumption penalty of about 0.1%. However, all three VCT strategies have lower BSHC than EGR at equal BSNO.

BSCO are increased about 15 g/kW-hr for Dual Equal at this speed/load. CO emissions also represent lost fuel energy. The fuel consumption effect can be estimated as 0.3 times the difference in CO Emissions Index (1). A difference of 15 g/kW-hr corresponds to a fuel consumption penalty of about 1.3% for Dual Equal. Exhaust Only and Intake Only had slight benefits due to lower BSCO at this speed/load (0.3 and 0.2% respectively). At stoichiometry, CO is affected by air-fuel mixing, and this aspect will be discussed in detail later in this paper.

In the above comparison, the BSNO for VCT is lower than the BSNO for standard cam timing with 10% EGR. Operating at lower BSNO results in slightly better ISFC (in addition to the ISFC benefits due to lower EVO loss and HC), because of lower burned gas temperatures, which result in reduced heat transfer and dissociation losses. The magnitude of this effect is estimated by referring to a plot of ISFC vs. BSNO for external EGR. This gives an estimated fuel consumption benefit of 0.9% for Dual Equal and Intake Only, and 1.1% for Exhaust Only.

As shown in Table 1, the sum of the effects due to reduced PMEP, EVO loss, BSHC, and BSCO, and to operating at lower BSNO, total 4.9% for Dual Equal, 4.9% for Exhaust Only, and 3.5% for Intake Only, compared to the measured benefits of 4.4%, 3.8%, and 3.4% respectively. Although the sums of the calculated components do not agree exactly with the measured benefits, they illustrate the relative importance of each component.

Table 1 also shows results with air-fuel mixing effects eliminated (fuel consumption corrected for CO). This is more representative of engines with valve masking, as discussed later.

BURN RATE - Figure 7 is a plot of 0-10% Burn Time, 10-90% Burn Time, Spark Advance, and COV IMEP vs. BSNO at 1500 RPM/2.62 bar BMEP. The burn rates and stability are fairly comparable for the three VCT strategies relative to EGR at a given BSNO level at this speed/load. Previous results for Dual Equal VCT on an engine with valve masking (1) showed comparable burn times and stability for VCT and EGR at a given BSNO at all speed/loads tested. However, with this tumble-port engine with no valve masking, VCT shows significant effects on burn rate as engine speed is increased. This is discussed in the next section.

Table 1: Components of Fuel Benefit at 1500/2.62

	Dual Equal Retard	Exhaust Retard	Intake Advance
Pumping	4.5%	3.1%	2.5%
Dilution	0.9	1.1	0.9
EVO	0.6	0.6	0.0
HC	0.2	-0.1	-0.1
CO	-1.3	0.3	0.2
Total Calc.	4.9	4.9	3.5
Meas.	4.4	3.8	3.4
Corr. for CO	5.7	3.5	3.3

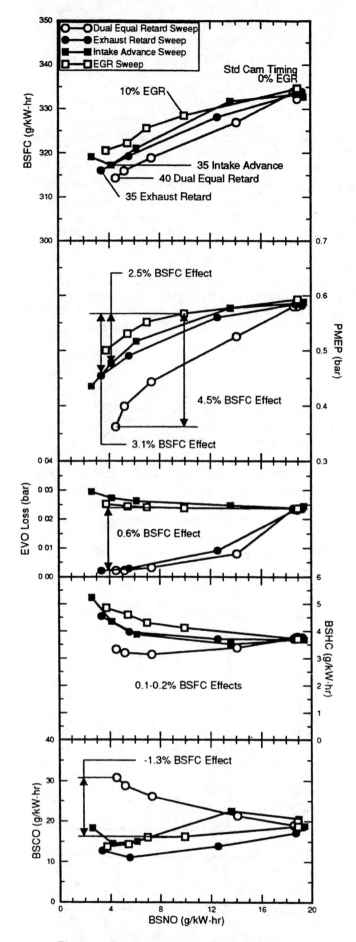

Figure 6: Fuel Consumption and Emissions for Cam Phasing vs EGR at 1500/2.62

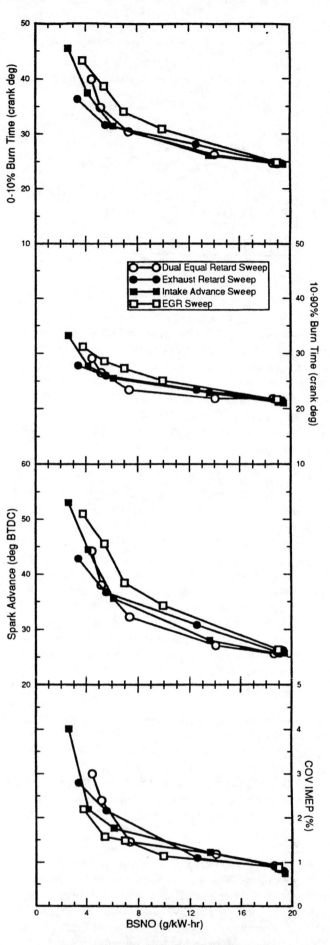

Figure 7: Burn Rates and Stability for Cam Retard vs EGR at 1500/2.62

EFFECTS ON AIR-FUEL MIXING AND BURN RATE

AIR-FUEL MIXING - The effects of VCT on air-fuel mixing were discussed in reference (1), and a substantial portion of that discussion is repeated in this paper for completeness. At stoichiometry, the in-cylinder air-fuel mixing can affect the level of CO emissions, and consequently the fuel consumption. As shown in the previous section, Dual Equal VCT significantly affected the BSCO emissions at 1500 RPM / 2.62 bar BMEP on the 2.0L I4 engine. Phasing the cam events can affect the in-cylinder air-fuel mixing due to a number of factors:

- Because the piston is moving downward before IVO for Dual Equal, the amount of backflow into the intake port is reduced. This is expected to adversely affect mixing, since the backflow of hot exhaust into the intake port assists fuel vaporization and the subsequent mixing of the fuel with the fresh air. The opposite case applies for Intake Only cam advance.
- Once forward flow into the cylinder begins, the flow velocity through the intake valves is increased for Dual Equal, because retarding the intake valve timing results in lower valve lift at a given piston velocity for about the first 2/3 of the intake stroke. This increases inlet turbulence and would be expected to improve in-cylinder air-fuel mixing. The opposite case applies for Intake Only cam advance.
- A change in valve lift at a given piston velocity can also alter the in-cylinder flow field. For example, if valve masking is used to generate either tumble or swirl, then a reduction in valve lift with Dual Equal would be expected to increase the resulting angular momentum of the incoming flow, and also the degree of in-cylinder mixing.
- The initial temperature of the in-cylinder gases is higher for all four VCT strategies, because of the increased internal residual. This should assist fuel vaporization and mixing.

Thus, the net effect on the degree of in-cylinder air-fuel homogeneity is a complex trade-off between these effects, and is a function of engine speed and cam phasing. These effects, especially differences in the level of in-cylinder motion, could also affect the burn rate.

At stoichiometry, the level of CO emissions is a direct function of the degree of in-cylinder air-fuel homogeneity. Both CO and H_2 are formed in higher concentrations during combustion of regions which are rich of stoichiometry. Because CO and H_2 have significant heating value, the CO level has a significant effect on ISFC. The fuel consumption effect of a change in CO can be estimated as 0.3 times the change in CO Emissions Index (1).

Figure 8 is a plot of percent improvement in BSFC and change in BSCO vs. Dual Equal cam retard at 2.62 bar BMEP for 4 different engine configurations (1). Three of the engines had valve masking to generate either tumble or swirl. The fourth engine is the 2.0L 4-valve I4 with a port designed to enhance tumble, but without valve

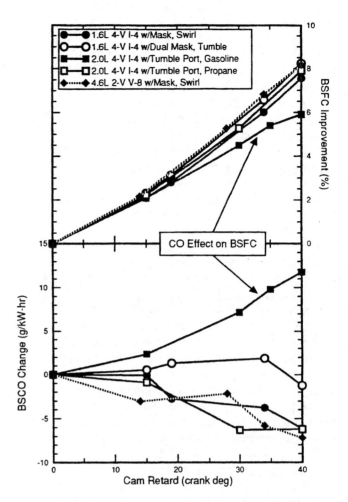

Figure 8: Dual Equal Cam Retard for Four Engines At 2.62 bar BMEP, 1500 RPM (1200 RPM for 2-valve)

masking. This engine was also run with pre-mixed propane as the fuel to eliminate air-fuel mixing effects.

As shown, BSCO for the engines with valve masking is nearly unaffected or decreases slightly with increasing Dual Equal cam retard. This indicates that the effect of reduced backflow into the intake port on in-cylinder mixing is offset by an increase in forward flow velocity through the valves, increased angular momentum of the in-cylinder flow field, and higher initial temperature. The net effect is constant or slightly improved air-fuel mixing with increasing cam retard for engines with valve masking.

For the 2.0L engine with the tumble port, however, the CO level increases significantly with increasing cam retard at this speed/load, and the improvement in BSFC is correspondingly reduced. The in-cylinder mixing is adversely affected with cam retard, possibly because a port designed to enhance tumble typically is not very effective in generating tumbling motion at low valve lifts. When the same engine is run with pre-mixed propane as the fuel, the BSCO level decreases slightly with increasing cam retard, and the BSFC improvement is in good agreement with the other engines.

Additional data acquired with gasoline on the 2.0L engine shows that the net effect on in-cylinder mixing and CO levels is a function of engine speed and cam phasing. Results at 2500 RPM/2.62 bar BMEP are shown in Figure 9. BSCO decreases at this speed/load with increasing EGR, Dual Equal retard, or Exhaust Only retard. However, BSCO first increases and then decreases with Intake Only advance. This illustrates the wide range of effects that cam phasing can have on mixing at a given speed/load.

In order to make the fuel consumption results of this study more generic and applicable to engines with valve masking, subsequent results presented in this paper have been corrected to the BSCO level of 10% EGR at each speed/load. This was done for two reasons: (1) The CO effects of VCT are dependent on the amount of cam phasing and the engine speed, and therefore the CO effects on vehicle fuel economy projections will be dependent on the specific speed/load weighting factors for a particular vehicle; and (2) Engines with valve masking showed no effect of Dual Equal cam retard on CO at all speed/loads tested (no Intake Only or Exhaust Only data was acquired on an engine with valve masking).

The fuel consumption results were corrected by using a factor of 0.3 times the difference in CO Emissions Index for VCT relative to 10% EGR. This correction gave good agreement with propane results at four speed/load points (pre-mixed propane was assumed to eliminate air-fuel mixing effects). The absolute value of the difference between propane and corrected gasoline fuel economy benefits, averaged for the four speed/loads and three VCT strategies, was less than 0.5%.

The actual values of the CO correction factors for the three VCT strategies at the eight speed/loads tested ranged from 0.982 to 1.010, with an average of 0.999. In other words, the fuel consumption impact of CO for this engine (for VCT relative to 10% EGR) ranged from a 1.8% penalty to a 1.0% benefit, with an average effect of 0.1%. As stated previously, subsequent fuel consumption results presented in this paper include these corrections for CO.

BURN RATE - Figure 9 shows BSCO, Spark Advance, and 0-90% Burn Time vs. BSNO for Dual Equal, Exhaust Only, and Intake Only at 2500 RPM, 2.62 bar BMEP. The BSCO results, discussed earlier, showed that mixing is significantly affected by cam phasing at this speed/load. Burn rate is also significantly affected, with Intake Only having slower burn rate than EGR, Dual Equal comparable burn rate to EGR, and Exhaust Only significantly faster burn rate than EGR.

Testing with pre-mixed propane at this speed/load (Figure 10) showed that the burn rate effects are still present, indicating that the effects on burn rate are due to altered in-cylinder motion and turbulence levels. A comparison of pumping loops is shown in Figure 11 for Intake Only and Exhaust Only at equal NOx. This comparison illustrates that the inlet wave dynamics are very significantly altered by the cam phasing, resulting in differing mass flow rate vs. crank angle histories for the two VCT strategies.

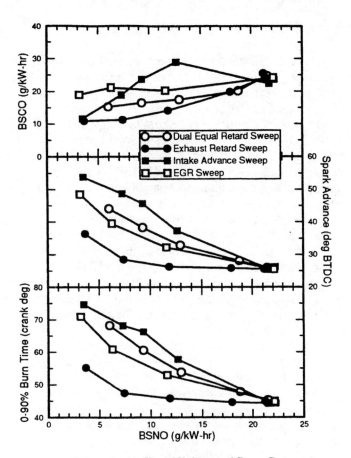

Figure 9: Air-Fuel Mixing and Burn Rates at 2500/2.62, Gasoline

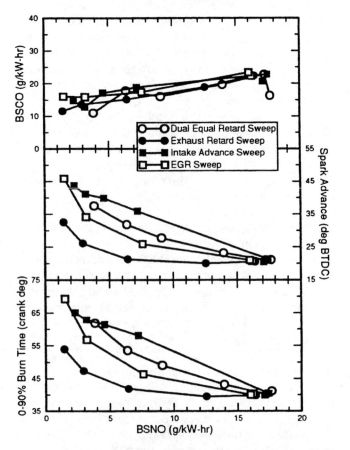

Figure 10: Air-Fuel Mixing and Burn Rates at 2500/2.62, Propane

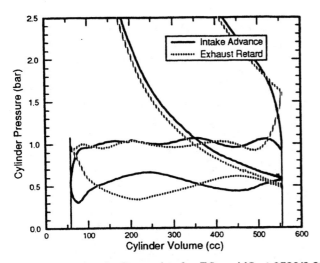

Figure 11: Intake Dynamics for EO and IO at 2500/2.62

EFFECT ON KNOCK AT MEDIUM-HIGH LOAD

The alternative VCT strategies affect the safety margin between MBT and borderline knock spark timing. All the strategies utilize internal residual dilution for NOx reduction and fuel consumption improvement at part load. The use of internal residual instead of external exhaust gas recirculation (EGR) results in higher initial temperature of the residual / fresh charge mixture. This results in a higher unburned gas temperature at the end of the compression stroke, and increases the tendency to knock.

As mentioned earlier, late IVC in the Dual Equal strategy results in lower effective compression ratio and temperature rise during the compression stroke. This offsets the effect of increased internal residual on unburned gas temperature.

In contrast, Exhaust Only retard does not have late IVC to offset the higher temperature of internal residual, and is therefore expected to be more knock-limited than EGR. Intake Only advance has early IVC, which slightly increases effective compression ratio, and is therefore expected to be even more knock-limited than Exhaust Only.

These factors are shown in Table 2. Advancing IVC changes effective compression ratio much less than retarding IVC, due to the non-linear relationship between crank angle and piston position.

Tests were conducted with 91 RON fuel on the 2.0L I4 at part load test points to determine the safety margin between MBT and borderline knock spark timing for the three VCT strategies compared to external EGR. Results for 1500 RPM/ 5.0 bar BMEP are shown in Figure 12. As shown, the safety margin with external EGR decreases slightly with increasing EGR. The safety margin with Dual Equal improves with increasing cam retard and exceeds 20 degrees at 15 degrees cam retard. The safety margin with Exhaust Only decreases rapidly with increasing cam phasing, and that of Intake Only decreases slightly more rapidly than for Exhaust Only. Consequently knock limits the amount of cam phasing which can be applied at this speed/load for these latter two strategies.

All subsequent results presented in this paper have included the knock constraint, whereby Exhaust Only and Intake Only cam phasing were limited to maintain a steady-state borderline safety margin of 2 degrees from MBT. It is important to note that the amount of cam phasing which can be applied will depend on the knock safety margin of the base engine, and the impact of this constraint on the overall vehicle results will depend on the vehicle speed/load weighting times.

Table 2: Effect of VCT on Part-Load Knock

	Internal Residual	Effective Compression Ratio (IVC)
Dual Equal	--	+++
Exhaust Only	--	0
Intake Only	--	-

Key:

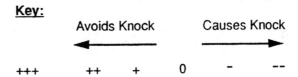

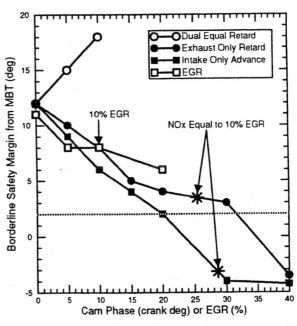

Figure 12: Borderline Knock Safety Margin from MBT

EFFECTS OF ENGINE LOAD

Figure 13 is a plot of percent improvements in fuel consumption and NOx emissions vs. engine load for the three VCT strategies compared to 10% EGR. To make the results generic, all fuel consumption results discussed in this section were corrected to equal BSCO as discussed previously.

As shown in Figure 13, the largest fuel consumption benefit at part load is provided by Dual Equal, followed by Exhaust Only, and then Intake Only. Relative to Dual Equal, Exhaust Only does not have the pumping work reduction of late IVC, and has slightly less pumping work reduction with increased overlap compared to delayed overlap (Figure 5). In addition, the amount of retard for Exhaust Only is limited by knock at 5 bar BMEP. Relative to Exhaust Only, Intake Only does not have the expansion work increase of late EVO, and is knock-limited at both 4 and 5 bar BMEP.

At light load, the three strategies give comparable NOx emissions benefits. At higher loads, however, knock limits the NOx benefits of Exhaust Only and Intake Only. For this engine, Exhaust Only gives little NOx emissions benefit compared to 10% EGR at 5 bar BMEP, and Intake Only shows a NOx penalty at 5 bar BMEP.

Because the improvements in fuel and NOx emissions decrease with increasing load, VCT will provide less benefit in the vehicle when applied to engines which are heavily loaded (small ratio of engine displacement to vehicle weight).

Cam phasings for the three strategies are shown in Table 3 for these and additional speed/load points. At low speed/light load, the amount of cam phasing is limited by combustion stability, whereas at heavier loads (4 and 5 bar), the cam phasing is limited by knock for Intake Only and Exhaust Only. The maximum cam phasing was constrained to 40 degrees, which is approximately the production limit for some engines to avoid valve-to-piston interference. If the maximum cam phasing constraint were eliminated, Dual Equal would be limited by the engine becoming unthrottled (see later discussion on Dual Independent).

Benefits in fuel consumption and emissions for VCT relative to standard cam timing with 10% EGR for these speed/load points are shown in Tables 4, 5, and 6.

Table 3: Cam Phasing (*Indicates Knock-Limited)

RPM	BMEP	Dual Equal Retard	Exhaust Retard	Intake Advance
1500	2.62	40	35	35
2000	2.62	40	40	40
2500	2.62	40	40	40
1500	4.0	40	40	33*
2000	4.0	40	40	37*
2500	4.0	40	38*	35*
1500	5.0	40	27*	21*
2000	5.0	40	32*	30*

Table 4: Fuel Economy Benefits of VCT vs 10% EGR

RPM	BMEP	Dual Equal Retard	Exhaust Retard	Intake Advance
1500	2.62	5.7%	3.5%	3.3%
2000	2.62	2.8	2.4	2.0
2500	2.62	3.0	2.0	1.1
1500	4.0	3.4	2.0	0.9
2000	4.0	1.4	1.0	1.2
2500	4.0	1.2	0.6	0.8
1500	5.0	2.1	0.0	-1.6
2000	5.0	0.9	0.1	-0.4

Table 5: NOx Emissions Benefits of VCT vs 10% EGR

RPM	BMEP	Dual Equal Retard	Exhaust Retard	Intake Advance
1500	2.62	55%	66%	58%
2000	2.62	28	56	37
2500	2.62	19	36	19
1500	4.0	36	48	19
2000	4.0	9	35	17
2500	4.0	5	20	9
1500	5.0	24	5	-24
2000	5.0	6	14	-12

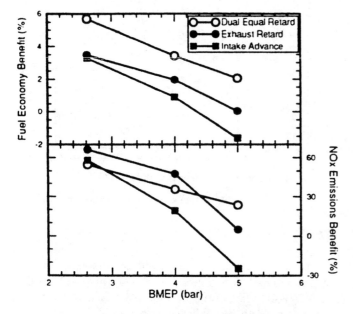

Figure 13: VCT Benefits vs Load At 1500 RPM, Compared to 10% EGR

Table 6: HC Emissions Benefits of VCT vs 10% EGR

RPM	BMEP	Dual Equal Retard	Exhaust Retard	Intake Advance
1500	2.62	19%	-10%	-6%
2000	2.62	8	-9	-5
2500	2.62	24	3	15
1500	4.0	20	10	7
2000	4.0	21	16	16
2500	4.0	15	8	17
1500	5.0	16	8	7
2000	5.0	18	14	10

EFFECTS ON IDLE

In contrast to part load, where VCT is used for increased residual dilution and improved fuel economy and emissions, VCT can be used at idle to decrease residual dilution and improve idle stability.

For an engine with fixed valve timing, the amount and phasing of valve overlap is a trade-off between idle quality and high speed power. Intake Only, Exhaust Only, and Dual Independent VCT allow decreased valve overlap for better idle stability, without compromising high speed power.

Dual Equal does not allow changes in the amount of valve overlap. However, the phasing of overlap can be optimized at idle, especially on engines with asymmetric overlap (1). The 2.0L I4 already has symmetric overlap, but slight Dual Equal advance still gives small benefits at idle. This is due to earlier IVC, which increases effective compression ratio and consequently the temperature rise during compression, which increases laminar flame speed and improves idle stability.

Cam phase sweeps are shown in Figure 14 for the 2.0L I4 engine at an idle-in-neutral condition of 800 RPM/ 0.5 bar BMEP and a fixed spark timing of 25 degrees BTDC. Idle-in-neutral is the worst case for idle stability, although idle-in-drive is more important for fuel consumption during the EPA cycle. Test cell availability did not allow testing at idle-in-drive.

As shown, the best fuel consumption is achieved with Dual Independent phase-shifting. The Dual Independent point in Figure 14 is 10 degrees exhaust advance and 10 degrees intake retard, for zero nominal valve overlap. Intake Only and Exhaust Only with 5 degrees overlap attain nearly the same benefit as Dual Independent, while 5 degrees Dual Equal advance gives only a slight benefit at idle for this engine.

The idle sweeps were run at a constant spark timing of 25 degrees BTDC (typical of production vehicle calibrations), which is retarded from MBT. Therefore, a decrease in 0-90% Burn Time causes the fixed spark timing to be closer to MBT, which is directly reflected as an improvement in fuel consumption (Figure 14).

The 0-90% Burn Time and Standard Deviation of IMEP (a measure of combustion stability) are improved due to reduced valve overlap and residual dilution for Exhaust Only, Intake Only, and Dual Independent. The improvement for Dual Equal is due to early IVC, as described above. The measured fuel consumption benefits are 6.3% for Dual Independent, 5.4% for Intake Only, 4.9% for Exhaust Only, and 1.2% for Dual Equal. The full amount of these benefits may not be attained in a vehicle if spark retard is used to maintain constant torque reserve for idle speed control. However, improved combustion stability may allow for lower idle speed in some applications.

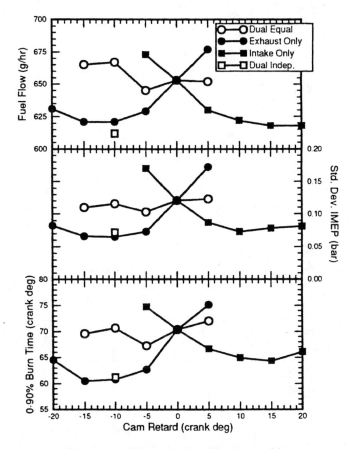

Figure 14: Effects of Cam Phasing at Idle

EPA CYCLE PROJECTIONS

As shown in Tables 3 to 6 earlier, data were acquired at a number of part load test points for the 2.0L I4 engine. Using a vehicle simulation program, weighting factors for the EPA cycle were assigned to these test points to estimate fuel economy and emission improvements with cam phasing for a 3125 lb. vehicle during the EPA cycle.

Test cell availability did not permit testing at idle-in-drive (650 RPM, 1.17 bar BMEP), which is heavily weighted in the EPA cycle. The only idle condition tested was idle-in-neutral (800 RPM, 0.5 bar BMEP), which is the worst case for idle stability.

The benefits of VCT are expected to be less at idle-in-drive than at idle-in-neutral. Therefore EPA cycle fuel economy projections are presented as a range. The lower value assumes zero benefit for VCT at idle, and the higher value assumes that the measured benefit at idle-in-neutral also applies at idle-in-drive. The actual EPA cycle results should fall within this range.

The results of these projections are shown in Table 7. As mentioned earlier, VCT will provide more benefit when applied to engines which are less heavily loaded in the EPA cycle. For example, reference (1) shows a Metro-Highway cycle (M-H) fuel economy improvement of 2.8% for Dual Equal on a 4.6L engine in a 4000 lb. vehicle (assuming no benefit at idle). The benefits of Dual Equal and Exhaust Only would also be greater for an engine with earlier base EVO timing.

Table 7: Projected Benefits of VCT vs 10% EGR

	Dual Equal	Exhaust Only	Intake Only
M-H Fuel	2.0 - 2.2%	1.1 - 1.7%	0.7 - 1.4%
HC	13	4	4
NOx	20	31	9

Projections were also made to compare VCT to a more aggressive EGR calibration. Table 8 shows cycle-average VCT benefits compared to 15% EGR at 2.62 bar BMEP, and 10% EGR at higher loads:

Table 8: Projected Benefits of VCT vs 10-15% EGR

	Dual Equal	Exhaust Only	Intake Only
M-H Fuel (corr for CO)	1.8 - 2.0%	0.9 - 1.5%	0.5 - 1.2%
M-H Fuel (uncorrected)	1.5 - 1.6%	0.8 - 1.4%	0.7 - 1.4%
HC	14	5	5
NOx	14	26	3

The fuel consumption projections in the first row are corrected to equal BSCO, as discussed previously. For reference, projections based on the uncorrected data are shown in the second row.

These projections are based on steady state fully warmed-up engine dynamometer data. A large fraction of the total tailpipe HC emissions are due to transients and to the initial portion of the EPA test before the catalyst has reached high conversion efficiency.

Because the transient control of residual dilution is improved with VCT compared to external EGR, VCT has the potential to achieve reduced HC emissions during transients.

During the start and initial idle of the test, cam phasing for minimum dilution and best stability would be used, and Dual Independent, Exhaust Only, and Intake Only could be used to reduce overlap. The use of phasing to increase dilution during the initial cruises and accelerations of the test would be limited, and benefits relative to an external EGR calibration would differ from those determined for steady-state. For these reasons, the actual EPA cycle HC benefits are likely to be different than those shown in Tables 7 and 8.

NOx CONTROL AT MEDIUM-HIGH LOAD - Because a significant fraction of the total NOx mass flow during the EPA test cycle results from operation at medium-high load, the capability to eliminate external EGR is dependent upon achieving adequately low NOx at these loads with VCT. For the 2.0L 4-valve I4 engine in a 3125 lb. test weight vehicle, about 62% of the NOx is produced at loads of 4.0 bar and above, and about 50% at 5.0 bar and above. For a 4.6L engine in a 4000 lb. test weight vehicle, the corresponding percentages are 34% and 15%.

As shown in Table 5, NOx equivalent to that obtained with 10% EGR can be achieved with all three VCT strategies at 4.0 bar BMEP, and with Dual Equal and Exhaust Only at 5.0 bar BMEP. Table 8 indicates that Dual Equal and Exhaust Only VCT both give better NOx control over the EPA cycle than 10-15% EGR, even in this relatively heavily-loaded vehicle. Intake Only gives about the same NOx control as 10-15% EGR in this application. This indicates that the external EGR system could be deleted for this vehicle with any of the three VCT strategies.

However, it is important to note that the amount of NOx control possible with Intake Only and Exhaust Only is strongly dependent on the knock tendency of the base engine at medium-high loads (see previous section on knock). On a slightly more knock-limited engine, elimination of the external EGR system would not be feasible with Intake Only for this vehicle.

DUAL INDEPENDENT RESULTS

The Dual Independent strategy incorporates the most flexibility to optimize valve events for operation at part load, idle, and WOT. The Dual Independent strategy at part load is simply a refinement of Dual Equal retard, whereby both cams are significantly retarded, but with variable instead of fixed overlap. The primary advantages of Dual Independent compared to Dual Equal are improved idle and WOT.

Although previous results presented in this paper have been limited to a maximum retard of 40 degrees, Dual Independent results presented in this section are at more aggressive levels of cam retard. This is because the main part load advantage of Dual Independent relative to Dual Equal is at medium-high loads at high cam retards.

Figure 15 shows intake cam phase sweeps for exhaust cam retards of 44 and 50 degrees at 2500 RPM /

4 bar BMEP. The zero of the x-axis represents Dual Equal phasing. The overlap is varied as the intake cam is phased. With large Dual Equal cam retard, residual dilution (and NOx) are primarily controlled by EVC timing because valve overlap is delayed well into the intake stroke. Exhaust is drawn back into the cylinder during the first part of the intake stroke, until the exhaust valve closes. As shown in Figure 15, when the cams are significantly retarded, the NOx and HC are determined by EVC timing, and are independent of overlap.

At medium-high load, the amount of cam retard with Dual Equal is limited by MAP approaching atmospheric pressure (the engine becomes unthrottled) due to late IVC. By advancing the intake cam at constant EVC, the torque reserve can be increased (MAP decreased) without an effect on emissions or pumping work. MAP is decreased due to earlier IVC. Pumping work is not increased (even though MAP is lower) because the cylinder pressure does not dip as low during the first part of the intake stroke. This occurs because IVO is earlier and the intake valve is at higher lift at the time of maximum piston velocity. This is illustrated by the pumping loops in Figure 16 comparing 44 Dual Equal retard with 44 Exhaust/28 Intake retard.

Alternatively, at equal torque reserve (MAP), the exhaust retard can be increased to result in lower NOx emissions at high loads (arrow in Figure 15).

EFFECT ON WOT PERFORMANCE

Optimized cam phasing at WOT is expected to result in improved low speed torque and high speed power (6, 7).

At low speed, volumetric efficiency can be improved by advancing IVC with Intake Only advance or Dual Equal advance. However, with Dual Equal advance EVO also occurs earlier, which reduces expansion work. Also, EVC may occur too early to allow exhaust out of the cylinder at the end of the exhaust stroke. This limits the benefit which can be obtained with Dual Equal advance at low speed.

At high speed, volumetric efficiency can be improved by retarding IVC with Intake Only retard or Dual Equal retard. With Dual Equal retard later EVC also assists scavenging at the end of the exhaust stroke, however, later EVO increases pumping work during the first part of the exhaust stroke.

Cam phase sweeps were run at speeds from 1500 RPM to 3500 RPM on the 2.0L 4-valve I4 engine at borderline knock spark timing with 91 RON fuel. These data are shown in Figure 17. For this speed range, the optimum Dual Equal phasing was 8 degrees advance. The optimum Intake Only phasing was 8 advance at 1500 RPM, and 20 advance from 2000 to 3500 RPM. The benefits in torque, shown vs. RPM in Figure 18, ranged from 0.9 to 5.3% for Intake Only, and from 1.2 to 2.5% for Dual Equal. Exhaust Only phase sweeps were also run, but benefits were less than 0.5%.

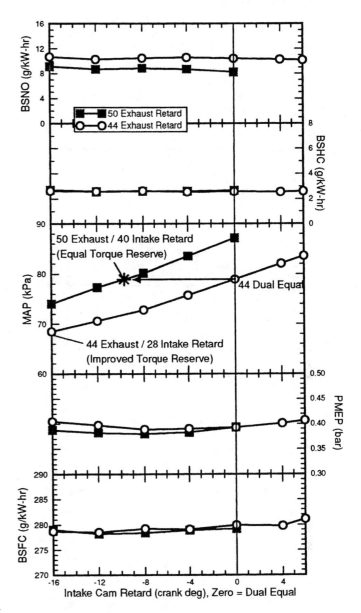

Figure 15: Dual Independent VCT at 2500/4.0

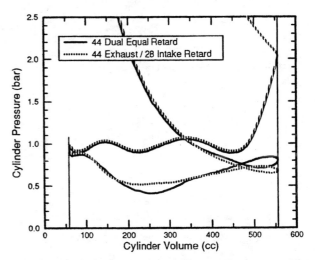

Figure 16: Dual Independent vs Dual Equal at 2500/4.0

Overall, the gains in low-RPM torque were significant with Intake Only, modest with Dual Equal, and negligible with Exhaust Only. Similar trends in high-RPM power are expected, but the experimental phase-shifter hardware used for this study (where the primary emphasis was on comparison of the strategies at part load) was not suitable for operation above 3500 RPM.

It should be noted that the gains in WOT performance which can be achieved with VCT are dependent upon the degree of exhaust tuning (8), and on the optimization of the length of the camshaft events for use with VCT. The data in this study was acquired on an I4 engine with close-mounted catalyst (minimal exhaust tuning), and event lengths were not optimized for VCT.

Although no WOT data was acquired for Dual Independent, this strategy should give even better low speed torque and high speed power than Intake Only. Optimum IVC could be maintained, while independently optimizing exhaust phasing for the best compromise between expansion work, exhaust stroke pumping work, and scavenging at each RPM.

Improvements at WOT could also allow a reduction in final drive ratio, for a fuel economy benefit at equal vehicle performance.

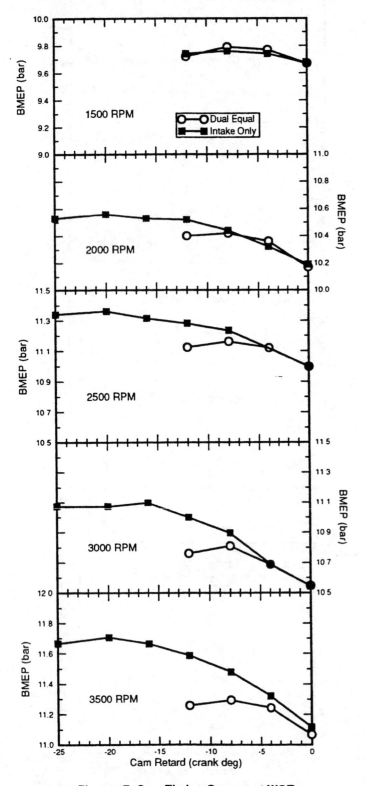

Figure 17: Cam Timing Sweeps at WOT

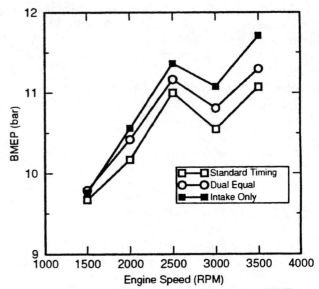

Figure 18: Effects of Cam Phasing at WOT

CONTROL SYSTEM ASPECTS

Figures 19 to 21 show BMEP vs. throttle angle at various camshaft phasings for the 2.0L I4 engine (Exhaust Only, Intake Only, and Dual Equal phasing respectively). Exhaust Only and Intake Only have relatively small effects on BMEP at a given throttle angle. Therefore Exhaust Only or Intake Only cam phasing could be scheduled as a function of speed and airflow in a manner similar to EGR without major impacts on throttle feel.

However, maximum torque is significantly reduced with Dual Equal cam retard, due to the late IVC (increased pushback during the first part of the compression stroke). Therefore, the cams must be phased back towards standard cam timing as higher torque is required. The dotted line in Figure 21 shows a potential trajectory for cam phasing scheduled as a function of throttle angle to achieve smooth torque response.

A possible control system strategy for Dual Equal is to determine desired cam phasing as a function of speed and airflow in the light to medium load region (up to approximately 4 bar BMEP), and as a function of throttle angle at higher loads. This strategy permits precise control at light to medium loads where airflow is very sensitive to throttle angle, plus tailoring of 'drive feel' with throttle angle control in the high load region.

The optimum Dual Independent cam phasing at part load is simply a refinement of Dual Equal, so in principle the control strategy would be similar to Dual Equal. In practice, this would be complicated by coordinating control of two individual phasers.

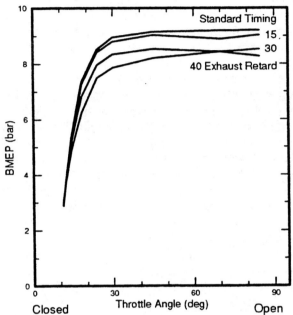

Figure 19: Effect of Exhaust Only on BMEP at 1500 RPM

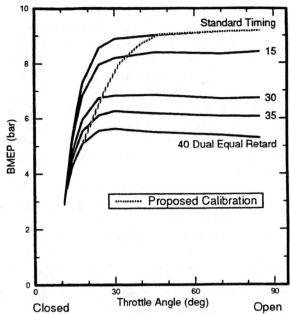

Figure 21: Effect of Dual Equal on BMEP at 1500 RPM

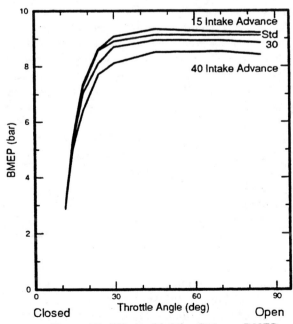

Figure 20: Effect of Intake Only on BMEP at 1500 RPM

SUMMARY

PART LOAD FUEL ECONOMY AND EMISSIONS - Dual Equal, Exhaust Only, and Intake Only VCT (Variable Camshaft Timing) strategies were compared at part load. All three strategies utilize continuously variable camshaft phase-shifting to delay or increase valve overlap for increased residual dilution and reduced pumping work at part load. Dual Equal and Exhaust Only also have late exhaust valve opening (EVO) for increased expansion work, and Dual Equal also has late intake valve closing (IVC) for reduced pumping work.

Projections based on engine dynamometer data for a 2.0L 4-valve I4 engine with a tumble port and no valve masking in a 3125 lb. vehicle during the EPA Metro-Highway (M-H) test cycle are shown below:

Projected Benefits of VCT vs 10-15% EGR

	Dual Equal	Exhaust Only	Intake Only
M-H Fuel (corr. for CO)	1.8 - 2.0%	0.9 - 1.5%	0.5 - 1.2%
M-H Fuel (and longer exh.)	2.1 - 2.3%	1.2 - 1.8%	0.5 - 1.2%
HC	14	5	5
NOx	14	26	3

The fuel economy ranges indicated depend on the benefits assumed at idle. The lower values assume no benefit at idle, while the higher values assume that improvements determined at constant spark timing for idle-in-neutral also apply to idle-in-drive.

Benefits will be greater for Dual Equal and Exhaust Only when applied to engines with earlier base EVO timing, or when the exhaust event length is increased for use with VCT. The values in the second row above are based on test results with 8 degree longer exhaust event.

The benefits of VCT would be greater when applied to engines which are less heavily loaded in the EPA cycle. Previous work (1) showed 2.8% M-H fuel economy benefit for Dual Equal VCT (with 38 degree cam retard constraint) applied to a 4.6L engine in a 4000 lb. vehicle.

Greater benefits would also be attained for Intake Only and Dual Equal if final drive ratio were adjusted to obtain equal vehicle performance (see 'WOT' below).

The NOx results above indicate elimination of external EGR should be feasible for all three VCT strategies, even in this relatively heavily-loaded vehicle. However, on a more knock-limited engine, deletion of the external EGR system would not be feasible with Intake Only for this vehicle (see 'Knock at Medium Load' below).

Unburned HC are reduced (even with more dilution and at lower NOx levels) because the last part of the exhaust, which is high in unburned HC, is recycled. Dual Equal and Exhaust Only also have late EVO, which allows more time for in-cylinder oxidation. Actual HC benefits in the vehicle are likely to differ from those shown above due to transients and the cold start portion of the test.

KNOCK AT MEDIUM LOAD - Knock limits the amount of cam phasing which can be applied at medium-high load for Intake Only and Exhaust Only. The use of internal residual dilution increases unburned gas temperature and increases the tendency to knock. For Intake Only, early IVC also slightly increases effective compression ratio. Knock at medium-high load limits the NOx control capability and fuel economy improvement of Intake Only and, to a somewhat lesser extent, Exhaust Only. In contrast, Dual Equal has late IVC, which decreases effective compression ratio and decreases the tendency to knock.

AIR-FUEL MIXING - VCT affected the air-fuel mixing and consequently the CO levels and the fuel consumption on this tumble port engine without valve masking. These effects varied with engine speed and the amount of cam phasing. Previous work (1) showed that air-fuel mixing was not affected by Dual Equal VCT on engines with valve masking. To make the results of this study more generic and applicable to engines with valve masking, the fuel consumption at each speed/load point was corrected for the difference in CO Emissions Index relative to that with 10% external EGR.

IDLE - Exhaust Only and Intake Only allow reduction of valve overlap and residual dilution at idle for improved idle stability. Dual Equal allows only the phasing of fixed overlap. Fuel consumption benefits measured at idle-in-neutral are reflected in the higher numbers in the table above. Improved combustion stability may also allow lower idle speed in some applications.

WOT - Limited data acquired at WOT (up to 3500 RPM) showed up to 5.3% improvement in torque for Intake Only and up to 2.5% for Dual Equal. Benefits with Exhaust Only were negligible.

Improvements in WOT performance would be greater with optimization of camshaft event lengths for use with VCT and with a tuned exhaust system. Adjustment of final drive ratio for equal vehicle performance would result in further improvements in fuel economy.

CONTROL SYSTEM - Exhaust Only or Intake Only cam phasing could be scheduled as a function of engine speed and airflow, in a manner similar to EGR. Scheduling of Dual Equal cam phasing would require throttle angle as an additional input, to ensure smooth torque response at medium to high loads.

DUAL INDEPENDENT VCT - Dual Independent cam phase-shifting has the flexibility to reduce overlap at idle and optimize intake and exhaust phasings at WOT. Thus, benefits at idle and WOT should be equal to or greater than those for Intake Only.

At part load, the Dual Independent strategy is a refinement of Dual Equal, whereby both cams are significantly retarded, but with variable instead of fixed overlap. Benefits at part load are equal to or somewhat greater than those for Dual Equal.

ACKNOWLEDGEMENTS

The authors wish to acknowledge the contributions of the following: J. Biundo for conceiving and designing dual independent phase-shift hardware; R. Bergquist, J. Aucott, and M. Tomas for data acquisition; C. Weaver and C. Warren for data analysis; H. Fader and C. Warren for software for data acquisition, analysis, and plotting; P. Gentile for hardware preparation; personnel in Ford's production organizations for hardware support, especially B. Keeble; and A. O. Simko for inspiration.

REFERENCES

1. Stein, R. A., Galietti, K. M., and Leone, T. G., "Dual Equal VCT - A Variable Camshaft Timing Strategy for Improved Fuel Economy and Emissions", SAE 950975, 1995.
2. Daniel, W. A., "Engine Variable Effects on Exhaust Hydrocarbon Formations (Single Cylinder Engine Study with Propane as Fuel)", SAE 670124, 1967.
3. Tabaczynski, R., et al., "Time-Resolved Measurements of Hydrocarbon Mass Flow Rate in the Exhaust of a Spark Ignition Engine", SAE 720112, 1972.
4. Tuttle, J. H., "Controlling Engine Load by Means of Late Intake-Valve Closing", SAE 800794, 1980.
5. Hara, S., Nakajima, Y., and Nagumo, S.,"Effects of Intake-Valve Closing Timing on Spark-Ignition Engine Combustion", SAE 850074, 1985.
6. Asmus, T. W., "Valve Events and Engine Operation", SAE 820749, 1982.
7. Asmus, T. W., "Perspectives on Applications of Variable Valve Timing", SAE 910445, 1991.
8. Fraidl, G. K., Quissek, F., and Winklhofer, E., "Improvement of LEV/ULEV Potential of Fuel Efficient High Performance Engines", SAE 920416, 1992.
9. Heywood, J. B., Internal Combustion Engine Fundamentals, McGraw-Hill, 1988, p. 47.

NOMENCLATURE

BMEP	Brake Mean Effective Pressure
BSCO	Brake Specific Carbon Monoxide
BSFC	Brake Specific Fuel Consumption
BSHC	Brake Specific Hydrocarbons
BSNO	Brake Specific Nitric Oxide
COV IMEP	Coefficient of Variation of IMEP
EVC	Exhaust Valve Closing
EVO	Exhaust Valve Opening
IMEP	Indicated Mean Effective Pressure
ISFC	Indicated Specific Fuel Consumption
IVC	Intake Valve Closing
IVO	Intake Valve Opening
MAP	Manifold Absolute Pressure
MBT	Minimum spark advance for Best Torque
PMEP	Pumping Mean Effective Pressure
VCT	Variable Cam Timing
WOT	Wide Open Throttle

APPENDIX - AREA C EFFECTS

Heywood (9) defines areas A, B, and C of a cylinder pressure vs. volume (P-V) diagram, as shown below. Indicated work per cycle is area A + area C, while pumping work per cycle is area B + area C.

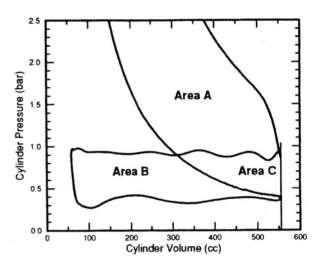

Definition of Areas A, B, and C

The use of late IVC (e.g. with Dual Equal cam retard) reduces both area B and area C, as shown below. The reduction in area B is a useful reduction in pumping work. The reduction in area C, however, reduces both pumping work and indicated work, so it has no net effect on cycle efficiency. The reduction in area C is shaded in the P-V plot below.

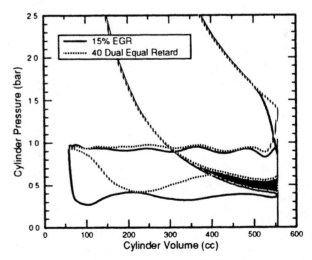

Difference in Area C with Late IVC

Therefore, when late IVC is used, a simple comparison of PMEP will overstate the fuel economy benefit attributable to reduced pumping work. The reduction in area C does not contribute to improved fuel economy, so it should be subtracted from the difference in PMEP. This "Area C effect" is accounted for in the calculation of fuel consumption effects in the PART LOAD DATA section of this paper.

980769

VAST: A New Variable Valve Timing System for Vehicle Engines

W. Hannibal
Märkische Fachhochschule Iserlohn

A. Bertsch
Ingenieurbüro KORO, Neckarsulm

Copyright © 1998 Society of Automotive Engineers, Inc.

ABSTRACT

VAST is a variable timing system developed for spark-ignition and compression-ignition engines.
It makes variable control of the intake and exhaust valves possible for SC and OHC engines.
The system is based on the principle of the purely mechanical effect of the cam's irregular angular speed. It makes it possible to vary the valve opening angle (up to approx. 50° crankshaft angle) and the spread (up to approx. 30° crankshaft angle) continuously for each valve lift movement.

When the valve opening period is prescribed, longer time profiles can be realized with this system than with any mechanical VVT system known to the authors.

The valve lift is influenced directly at the cam, which means this system can be used with all types of known valve gear– both for engines/ cylinder heads with just one camshaft and for multi-valve cylinder heads with two camshafts.

The system is being tested on functioning spark-ignition and compression-ignition engines. The results show that the components of the VVT system meet the high requirements for series-production engines. The measured friction is on the same level as with conventional valve gear.

An adaptation of VAST to an American passenger car engine is planned.

INTRODUCTION

Variable valve timing systems are being used more and more frequently in modern passenger car engines [1]. Basically, they function according to two different principles.

1. Camshaft control systems:

These are systems that turn the entire camshaft relative to the crankshaft during engine operation. This requires DOHC cylinder heads.

The first system in this VVT group was employed in an production passenger-car engine from Alfa Romeo [2]. With this system, the intake camshaft spread was adjusted to two positions.

Other systems operate in a way similar to the Alfa Romeo system, which varies the angle through straight/ helical gearing.

BMW has several engines in production that operate on a similar principle. In some of their versions, however, the camshaft can be adjusted steplessly [3].

2. Shifting arm systems:

These are systems in which the valve timing is changed during engine operation through the use of coupling elements in the cam follower or rocker assembly that have different cam contours.

Honda has used systems in this VVT group for some years and in large numbers in series production [4]. Depending on the design, the system can be shifted between two or three different valve lifts. Mitsubishi uses this functional principle in various engines and is also able to use it to realize cylinder shut-off [5].

Numerous analyses and own experiments on the technical realization of variable valve timing systems on series engines were performed within the framework of the development of the system presented here. They led to the following goals for the development of a VVT system that would make the technology more advanced and could be realized in series-production engines:

- The primary development goal was the improvement of the target variables for the engine, such as the lowering of fuel consumption and emissions.
- Despite improved exhaust gas quality, a high level of power and torque is required.
- Stepless variation of the opening period and spread of the intake and the exhaust valve lifting movements are desirable [6].
- Achieving the longest possible time profiles is desired.
- The design should be inexpensive and suitable for use with all kinds of cylinder head and valve gear concepts, particularly for engines with just one camshaft.

In the opinion of the authors, fully variable VVT systems that permit free selection of the progress of valve lift are still in the pre-development stage. It is not to be expected that these systems will be ready for series production use in the near future.

For this reason, only VVT systems that operate according to mechanical principles alone can be recommended for series production use [6].

Based on these specifications and goals, the VAST VVT system was developed for passenger car engines by the KORO consulting engineers' office.

MECHANICAL PRINCIPLE

The fundamental principle upon which VAST is based is the effect of the irregular angle speed of the cam, which is already known from the Mitchell system [7].

The operating principle is shown in **Fig. 1**. The rotational axis of the camshaft and cams is at point M. By shifting the rotational axis (M*) of the coupling element that connects the camshaft with the cams, the cam angle is also shifted. The value for this angle shift changes constantly as the camshaft rotates. Thus, a periodic change arises in the differential angle (Δ) between the camshaft and cams (see **Fig. 2**).

The position of the rotational axis (M*) is described by the parameters of eccentricity (e) and setting angle (phi). The position of the rotational axis (M*) of the coupling element does not change during rotation of the camshaft. By shifting the rotational angle (M*) to a position above the center axis (Z), the cam moves over the follower more quickly in the entry zone and more slowly in the exit zone. The maximum cam lift, however, does not change. As shown in the example in **Fig. 2**, based on the initial lift, this cam movement causes a change in the lift with reference to the opening period and spread. Thus, the opening angle and the spread can be varied based on the position of the rotational angle (M*).

The eccentricity (e) and setting angle (phi) parameters can be selected freely during operation, over wide ranges.

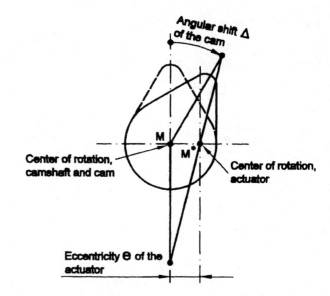

Fig. 1: Diagram of the operational principle of non-uniform cam drive

The maximum cam lift for a specific opening angle is derived for a particular cam design based on different peripheral conditions as necessitated by the laws of physics. Therefore, a cam designed for conventional valve gear must be a compromise between maximum power output at a full load and good part-load/ idle-speed performance with good emissions behavior. For this reason, the maximum cam lift – and with it the maximum time profile of the valve lift – is limited.

Since with VAST it is possible to reduce the opening angle and the spread in the lower and middle speed range, the cam contour can be designed in such a way that the maximum opening angle and cam lift can be reached.

The valve lift that results from the reduction of the opening angle has a significantly larger time profile than the same opening angle for a conventional cam contour (see **Fig. 3**).

When the follower has the maximum diameter – by this we mean tappet diameter or the contact surface of the cam on the follower or the rocker arm – a significantly larger valve lift can be achieved with VAST. As an alternative, when cam lift is at its maximum, the diameter of the follower can be reduced significantly. This is because maximum excursion of the cam contact line on the follower is derived solely from the cam contour, and not from the angular speed.

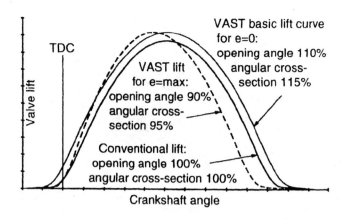

Fig. 3: Comparison between conventional and typical VAST valve lift

A valve lift created with a conventional cam corresponding to the valve lift with a reduced opening angle shown in **Fig. 3** would require a tappet diameter enlarged from 35 mm to 43 mm.

Furthermore, one of the system's inherent advantages is that, due to the enlarged opening angle, the radius of the cam tip becomes larger without any change in the basic radius. This reduces the maximum surface pressure between the cam and the follower. Furthermore, the larger tip radius makes it possible to reduce the maximum force of the valve spring. This has a positive effect on the friction value at the cams.

SYSTEM DESCRIPTION

An example of the way VAST is constructed is shown in **Fig. 4**.

The camshaft, which is driven by the crankshaft, consists of a smooth tube. A radial pin, which connects the camshaft with the driver, is fastened inside the camshaft. The driver is in turn connected with the cam, which can be rotated on the camshaft. The driver is mounted on an eccentric pack, which consists of an inner and an outer eccentric. By rotating the two eccentrics in opposite directions, the driver's rotational axis can be shifted eccentrically in relation to the rotational axis of the camshaft or the cam. The eccentrics are only turned in the event of a change in the setting for the position of the driver's axis. This means that they do not rotate along with the camshaft.

The cam supports itself radially directly on the camshaft. The power flow (cam-driver-eccentric pack-camshaft bearing) is very short and stiff.

Conventional cam contours and cam/follower pairings are used, which means that no additional work or costs arise for development.

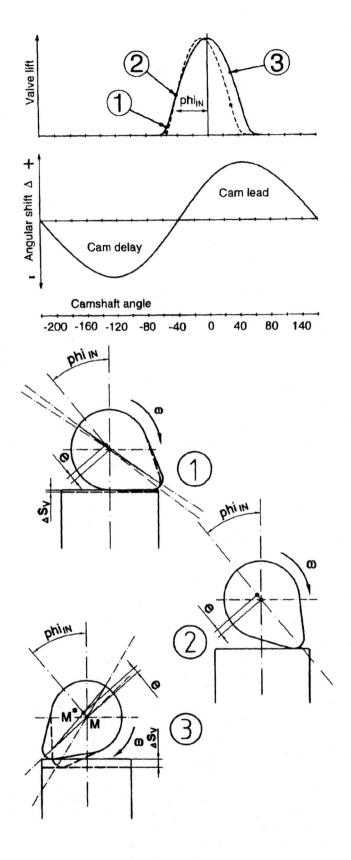

Fig. 2: Relationship between phase angle and cam motion or cam position [8]

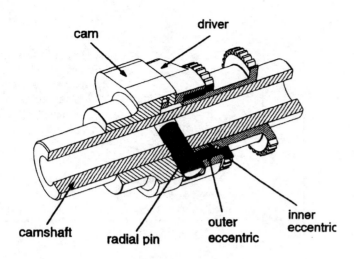

Fig 4: Sectional diagram of a cam drive for a two-valve per cylinder engine

EFFECTS ON THE VALVE CURVES

As **Fig. 5** shows by way of an example, the opening period (up to approx. 50° crankshaft angle) and spread (to approx. 30° crankshaft angle per valve stroke) are varied. Besides the fact that the intake is closed significantly earlier and the exhaust is opened significantly later, the valve overlap at top dead center is reduced.

The position of the setting angle (phi) corresponds to the distance in degrees of crankshaft angle from the point of the maximum valve opening of the initial contour to the intersection of each valve lift. Depending on the design, this angular setting can be freely selected within a wide range. The values for the setting angle (phi) of the intake and exhaust lifts can also be selected freely and independently of one another.

For kinematic reasons, it makes sense to have the VAST initial exhaust valve lift (i.e. without adjustment) represent the largest possible opening angle. When the opening angle is reduced, the positive valve acceleration rate rises. This is desirable, since it causes an increase in the fullness of the lift. The increase in acceleration is not critical, since it only causes control only to take place at engine speeds lower than the maximum engine speed. This means that the absolute values for maximum positive valve acceleration are not exceeded (see **Fig. 12**). In principle, however, a VAST design in which the opening angle is enlarged is also possible.

With diesel engines, and also with sports engines and special engines, it is desirable to pass as much air as possible through the engine in all engine speed and load ranges. It is precisely here that long time profiles are assential, despite reduced opening angles. For such applications, it is particularly important that the smallest distance between the top of the piston and the valve head can be kept constant (see **Fig. 6**). **Fig. 6** shows piston lift, as related to valve lift, against the degree of the crank angle for the top dead center area. The smallest distance for the two curves in the vertical direction represents the reserve to the piston crown.

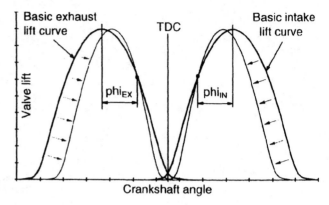

Fig. 5: Example of the variation in the valve lift curves

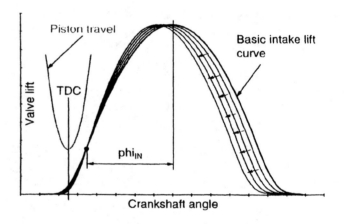

Fig. 6: Reserve distance between the top edge of the piston and the inlet valve

To improve mixture preparation when this VVT system is used in multi-valve engines, as shown in **Fig. 7**, it is possible to achieve specific intake swirl in the charge.

By shifting the two inlet valve lift curves assigned to one cylinder in opposite directions, a significant difference in valve lift can be achieved. Depending on the design, at the point of maximum piston speed, this difference can be as much as 50 % of the valve lift. To do this in an engine with two inlet valves, for example, the spread is enlarged for the first cam and reduced for the second. Depending on design and construction, there is a wide field of application here.

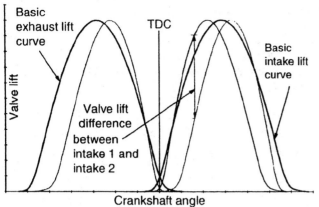

Fig. 7: An example of VAST valve lifts on a multi-valve engine for creating swirl during the intake phase

VAST - APPLICATION TO SEVERAL CYLINDER HEADS

Due to cost and production requirements, engine developers are faced with the task of finding variable valve timing systems on which the added costs and complexity for the application a variable valve timing system remains at a reasonable level. With the camshaft control systems such as those described in the introduction, this is possible. For the development of the VAST system that is covered in this paper, this challenge proved to be one of the most important.

During development work, the KORO engineering office equipped several cylinder heads for both gasoline and diesel engines with the new variable valve timing system. This showed that the effort required to make the changes to the cylinder heads that were available is relatively low in each case. This is due primarily to the fact that the valve gear already being used could be retained. It was only necessary to make modifications in the area of the camshaft bearing and the corresponding actuation.

Fig. 8 shows the complete camshaft of a four-cylinder OHC engine equipped with VAST. To vary the timing, the control shaft – not shown here – intervenes at the gear wheels of the elements on the camshaft.

The scope of the overall system required for the application shown in **Fig. 8** is represented in **Fig. 9** as a 3D model. In the version shown here, as well as the VAST camshaft a control shaft and a corresponding turning device, such as a positioning motor with an angle sensor, are required.

An example for the design of the components for a multi-valve cylinder head is shown in **Fig. 10**.

Fig. 8: Example of VAST camshaft design for a four-cylinder OHC engine

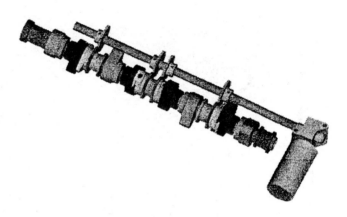

Fig. 9: 3D-model of the overall system for **Fig. 8**

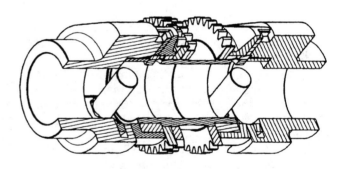

Fig. 10: Sectional diagram of the components for a multi-valve cylinder head

DISCUSSION OF THE RESULTS

In consideration of the interests of customers with whom this system was developed, we will only discuss the basic results at this point. The results of the charge flow calculations show the potential for increasing power and torque. **Fig. 11** shows an example of results that show the increase in torque compared to the basic version of an 2.0-liter, four-cylinder engine.

An example of the effects of the VAST VVT system on the valve acceleration is shown in **Fig. 12**. Compared to the basic version, there was no change in the maximum positive accelerations that arise. The value limits indicated here were taken into consideration. What can be seen clearly, however, is the reduction in the maximum negative accelerations that arise. This is due to the enlarged opening angle (+4.5 % in this case) of the VAST cam contour compared to the basic version. In this way, the final valve spring force (-12% in this case) can be reduced. From an engine speed of 4000 rpm down to idle speed, the opening angle and the spread are reduced continually. In terms of the valve accelerations, this causes the values to be higher than the basic values.

mechanical components stand up to the tough requirements of series use.

In deference to our customers, the thermodynamic results for the individual engines cannot be shown in detail now. The potential for lowering fuel consumption and emissions, however, is already known [8].

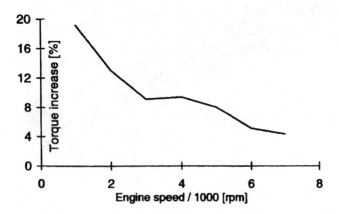

Fig. 11: Results of a charge flow calculation for a 2.0-liter, four-cylinder engine

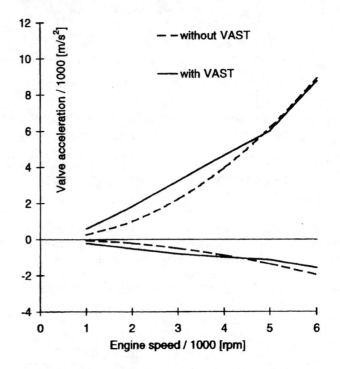

Fig. 12: Valve acceleration comparison

Fig. 13 shows a summary of the friction power measurements of the 2.0-liter OHC, four-cylinder head.

It is notable that as eccentricity increases, friction decreases in the idle speed range. This is due to the high sliding speeds of the cam on the follower when the opening angle is reduced. This improves the formation of the lubricating film. As the speed increases, the influence of the system friction increases at larger eccentricities and thus larger reductions in the opening angle. It can be clearly seen that the curve for the system's moment of friction with a continuous adjusting curve and reduced valve spring force does not differ significantly from the curve for the basic valve gear. When the adjusting curve is continuous, beginning at an engine speed of 4,000 rpm, the opening angle is reduced continuously as the engine speed drops.

The acoustic requirements for a VVT system are met. This is due in part to the fact that the standard hydraulic valve clearance adjustment can be used without changes. In addition, all components are located inside the valve housing.

Results from long-term running show that the

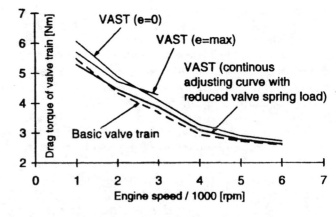

Fig 13: Comparison of friction moments for the OHC cylinder head

PERSPECTIVES FOR VAST

Further development steps will consist of optimization of the components from the perspective of production costs. An alternative form of control is being tested which would eliminate the need for the control shaft (see **Fig. 9**).

A solution is currently being worked on that will also make it possible to install the system on engines that have a central, lower camshaft.

An application on an American V-engine is planned in order to show the advantages of the VAST VVT system, particularly with regard to lowering emissions while maintaining the same high level of performance.

Technical evaluation of the system in relation to other system show that VAST has comparable chances for being realized in series engines [1]. The authors also see good possibilities for realizing the system, particularly with two-valve engines that use a camshaft to control engine breathing. Since it is possible to integrate the system into a cylinder head concept already in existence, development and investment costs remain at a reasonable level.

During the course of the development, more than 20 patents were applied for, five of which have already been granted. The issue of license options is possible.

CONCLUSIONS

The VAST variable valve timing system presented here is a system developed under series production conditions that can be applied to all kinds of engines. It features variable control capability for the inlet and exhaust valves of SC and OHC engines. With continuous control of the opening period and the spread, it is possible to optimize target variables for the engine. The system's friction energy levels do not vary significantly from those for the basic valve gear. The operating principle of non-uniform cam drive makes it possible to apply the system to a wide variety of modern engine concepts.

REFERENCES

[1] **Hannibal, W.:** Anforderungen, Funktionsprinzipien und technische Bewertung variabler Ventilsteuerungen für Serien-Ottomotoren. Vortrag Haus der Technik, 18.06.1997

[2] **Bassi, A.; Arcari, F.; Perrone, F.:** C.E.M.- The Alfa Romeo Engine Management System-Design Concepts-Trends for the Future. SAE-Paper 85 0290, 1985

[3] **von Versen, O.:** Technik der neuen VANOS-Motoren. Mot-Sonderteil 03/93, S. 82 ff

[4] **Inoue, K.; Nagahiro, R.; Ajiki, Y.:** A High Power, Wide Torque Range, Efficient Engine with a Newly Developed Variable Valve-Lift and -Timing Mechanism. SAE-Paper 89 0675, 1989

[5] **Hatono, K.; Iida, K.; Higashi, H.; Murata, S.:** Ein neuer Mehrphasen-Motor mit variabler Ventilabschaltung. MTZ 54 (1993) 9, S. 412-418

[6] **Hannibal, W.:** Vergleichende Untersuchung verschiedener variabler Ventilsteuerungen für Serien-Ottomotoren. Dissertation Universität Stuttgart, 1993

[7] **Mitchell, S. W.:** Valve Timing Mechanism. US Patent 4,131,096, Mar. 21, 1977

[8] **Lancefield, T. M.; Gayler R. J.; Chattopadhay, A.:** The Practical Application and Effects of a Variable Valve Timing System. SAE-Paper 93 0825, 1993

Application of a Valve Lift and Timing Control System to an Automotive Engine

Seinosuke Hara and Kenji Kumagai
Atsugi Motor Parts Co., Ltd.

Yasuo Matsumoto
Nissan Motor Co., Ltd

ABSTRACT

This paper describes a new variable valve lift and timing control system, which varies the lift and timing by changing the fulcrum of the rocker arm. The fulcrum is varied according to the inclination of a lever that engages with a control cam. The control cam is driven by an actuator mechanism so as to provide multistage control over valve lift and timing. As a result, this new system reduces valve train noise and delivers stable valve operation through the high-speed range. The compact actuator mechanism provides excellent response for controlling valve lift and timing. When applied to a 1.8-liter 4-cylinder gasoline engine, the system contributed to higher power output, lower fuel consumption and reduced noise.

INTRODUCTION

One approach to raising the charging efficiency of a high-speed automotive engine so as to boost its maximum power output is to increase the valve lift and to lengthen the interval the valves are open, i.e. controlling the valve opening timing and valve closing timing (Fig. 1). This approach has its drawbacks, however, including larger intake-exhaust valve overlap, reduced charging efficiency in the low speed range and deterioration of combustion conditions under low-speed, low-load operation.[1][2]

Various attempts have been made over the years to resolve this trade-off between low-speed and high-speed performance by varying the valve lift and valve timing to match the engine operating conditions.[3] Listed below are examples of the many different types of methods that have been tried.

(1) Interchanging of multiple cams[4]
(2) Use of a 3-dimensional cam[5]
(3) Varying the fulcrum of the rocker arm[6][7]
(4) Varying the rotational phase of the camshaft[8][9][10]
(5) Varying the angular velocity of cam rotation[11]
(6) Operating valves by hydraulic[12] or electromagnetic[13] force

However, for a variety of reasons nearly all of these methods have yet to find application in production engines. Some of the problems preventing practical application include the lack of sufficient capacity for high-speed operation, increased control losses and difficulties in mounting the mechanism to existing engines. The only examples that have been seen recently of applications in production engines are for systems that vary the rotational phase of the camshaft.[8][10]

These phase control systems do not vary the valve lift, and their control over valve timing is limited because they do not change valve opening duration.

The aim of this research was to achieve high engine performance throughout the entire speed range by controlling both the valve lift and timing. First, an investigation was carried out to examine the effects on torque and fuel consumption of the differences seen in valve lift and timing between high-speed and low-speed engines. The results of that

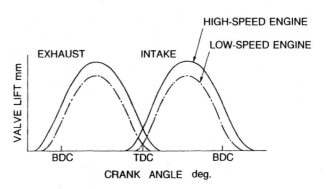

Fig. 1 Comparison of valve lift and timing for high-speed and low-speed engines

investigation made clear the performance benefits that could be obtained by controlling these two variables.

With an eye toward achieving those performance benefits, a valve lift and timing control system was developed, which was designed for application to a SOHC engine. This system employs a variable mechanism for changing the fulcrum of the rocker arm (method (3) mentioned above). The system was subsequently installed and tested in a 4-cylinder gasoline engine. The results confirmed that it was a practical and effective approach to obtaining higher performance from an automotive engine. The following sections present an outline of the system and its effects.

RELATIONSHIP BETWEEN VALVE LIFT AND TORQUE AND FUEL CONSUMPTION

In the present work, an investigation was first carried out to determine what effect various intake and exhaust valve lift settings and the accompanying changes in valve opening duration would have on torque and fuel consumption. The test engine used was a water-cooled 4-cylinder SOHC engine, having the specifications indicated in Table 1. Many different types of conventional camshafts were used to induce a wide range of variation in the valve lift and timing as indicated in Fig. 2. The resulting changes in torque that occurred are shown in Fig. 3.

Table 1 Test engine specifications

NUMBER OF CYLINDERS	4 (IN-LINE)
BORE × STROKE	83 × 83.6
DISPLACEMENT	1809 cc
COMPRESSION RATIO	8.8 to 1
COMBUSTION CHAMBER	HEMISPHERICAL

At a low speed of 1600 rpm, reducing the intake valve lift produced a large improvement in torque. An intake valve lift of 7 mm increased torque by 5% as compared with the conventional camshaft. Reducing the exhaust valve lift also produced an improvement, though not as large as that seen for a smaller intake valve lift. With an intake valve lift of 7 mm, an exhaust valve lift of 9 mm improved torque by 2% over the level obtained with the conventional camshaft, for a total improvement of 7%.

In the high speed range at 6000 rpm, making the intake valve lift larger than the conventional setting of 9 mm resulted in improved torque. An intake valve lift of 11 mm improved torque by 14%. However, changing the exhaust valve lift was found to have little effect. Increasing the exhaust valve lift to 11 mm from the conventional setting of 10 mm did not produce any improvement in torque.

An investigation was then conducted to determine the effects of valve lift on engine stability and fuel consumption under idling operation. The results obtained are shown in Fig. 4.

The Cpi values shown in the figure are given by the following equation, and a smaller Cpi value indicates better engine stability.

$$Cpi = \frac{\sigma pi}{\overline{pi}} \times 100 \ (\%)$$

where, σpi is the standard deviation of mean effective pressure pi and $\overline{pi}$ is the mean of pi.

When the exhaust valve lift was set at 10 mm, reducing the intake valve lift resulted in a smaller Cpi, i.e., engine stability improved. This is attributed to the fact that valve overlap was reduced, which decreased the amount of residual exhaust gas in the combustion chamber, thereby improving

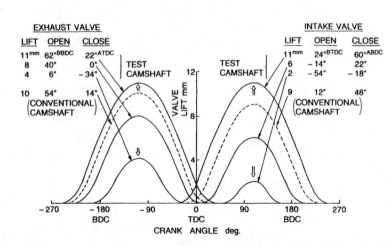

Fig. 2 Valve lift and timing changes in a conventional valve train

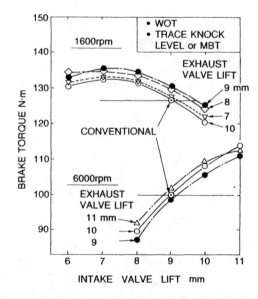

Fig. 3 Effects of valve lift on brake torque

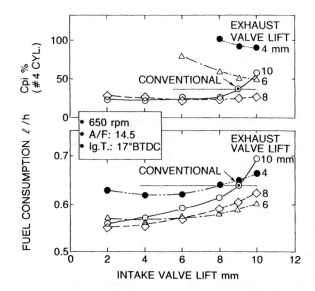

Fig. 4 Effects of valve lift on engine stability and fuel consumption under idling operation

combustion. The figure also shows that the fuel consumption decreased with a smaller intake valve lift. This was due, in addition to the improved combustion, to the reduction in camshaft driving loss, thanks to the reduced valve lift. When the intake valve lift was reduced from 9 mm to 2 mm, fuel consumption decreased by 13%.

Reducing the exhaust valve lift was also effective in lowering the Cpi and the fuel consumption. An exhaust valve lift of approximately 8 mm produced the largest reduction in both Cpi and the fuel consumption. With a smaller exhaust valve lift, Cpi showed a large increase and fuel consumption also increased. The reason for this is that combustion deteriorated due to the fact that the advanced closing timing of the exhaust valves increased the residual exhaust gas in the combustion chamber.

As indicated by the foregoing results, varying the intake and exhaust valve lift and timing makes it possible to achieve the desired torque level in both the low and high speed ranges. In addition, fuel consumption and engine stability under idling operation can also be improved. These results thus indicate that valve lift control makes it possible to obtain high engine performance over a wide range of operating speeds from low to high.

The results presented above indicated that controlling the intake valve lift was particularly effective. Therefore, considering the cost of the mechanism, it was decided that it would be better to control only the intake valve lift. The following section will describe a valve lift control system for the intake valves only, which was developed for a 1.8-liter 4-cylinder gasoline engine having the same specifications as the above-mentioned test engine.

VALVE LIFT AND TIMING CONTROL MECHANISM

The general performance requirements for an automotive valve train can be summarized as follows.
(1) It should be capable of high-speed operation.
(2) It should operate quietly.
(3) It should have a low friction level.
(4) It should provide high wear resistance.

In adopting a variable valve mechanism, the following points should be taken into consideration.
(5) Control losses should be low.
(6) The mechanism should be compact, maintaining the cylinder head height.
(7) It should be easy to mount to existing cylinder heads.

A number of mechanisms have been tried in the past to vary the valve lift.[14][15] As the example in Fig. 5 illustrates, one of the major issues is the need to provide a buffer mechanism to suppress shock at the moments when the valves lift and seat. The valve lift and timing control (VLTC) system presented here successfully resolves this issue and it also satisfies the conditions noted above.

VLTC MECHANISM - The construction of the VLTC system is illustrated in Fig. 6. It consists of a rocker arm with a gently curved back surface, a lever that supports and runs along the back surface of the rocker arm, a hydraulic lash adjuster that supports the end of the lever and a control cam that varies the slant of the lever. A rotatable rocker shaft is provided in approximately the center of the rocker arm. The shaft is able to move vertically thanks to the fork that extends downward from lever.

The operation of the mechanism can be understood as follows. First, assume that the control cam pushes the lever down to increase the valve lift as illustrated on the left side of Fig. 6. Under this condition, the back surface of the rocker arm moves along the lever to allow large valve lift when the rocker arm is lifted by the drive cam. During this time the point of contact between the lever and the rocker arm switches from point A to point B in Fig. 6. The rocker ratio

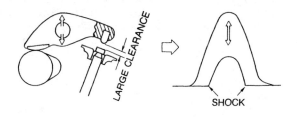

Fig. 5 Problem accompanying variable valve lift and timing

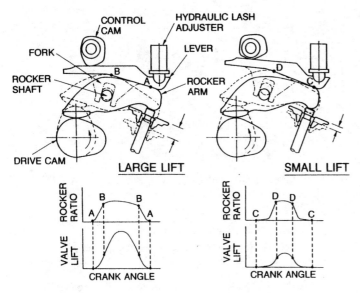

Fig. 6 Working principle of VLTC mechanism

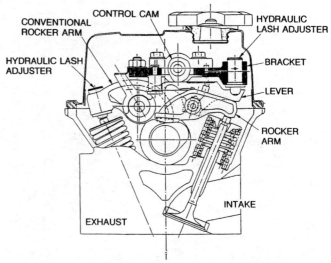

Fig. 7 Cross section of cylinder head when equipped with VLTC system

gradually increases from nearly zero at the onset of valve opening and there is a resulting smooth increase in the valve lift, as the lift curve in the figure indicates. This design works to prevent the impact noise usually heard with a conventional mechanism during valve lift (Fig. 5). The same effect is achieved when the valve closes.

Now assume that the lever rotates upward to reduce the valve lift by creating a larger gap between it and the rocker arm, as shown on the right side of Fig. 6. Under this condition, the rocker arm moves to fill in the gap, when the drive cam turns and the rocker arm begins to lift, so that no valve lift is initially obtained. Subsequently, the back surface of the rocker arm approaches the lever and the point of contact shifts from C to D. As a result, the rocker ratio increases and the valve begins to lift. In this case, the remaining cam lift determines the valve opening, since most of the cam lift is used up to fill in the gap between the rocker arm and the lever. This means that only a small amount of valve lift is obtained. In addition, the valve timing is also changed substantially as the valve opening timing is retarded and the closing timing is advanced.

One important feature of the VLTC system is that it is designed to allow the fulcrum of the rocker arm to shift which contributes to quiet operation regardless of the amount of valve lift used. Another important feature is that the moving parts have been limited to the rocker arm and rocker shaft so as to assure high-speed operation; thus the equivalent mass of the rocker arm is smaller than that of a conventional product. The smaller equivalent mass makes it possible to use conventional valve springs, which helps to prevent any increase in camshaft drive torque.

Fig. 7 shows the VLTC system installed in a water-cooled 1.8-liter 4-cylinder gasoline engine. Fueling is provided by a single point injection system. The VLTC system is installed only on the intake valve side, and a conventional valve train is employed for the exhaust valves. It should be noted that, in terms of the layout, it is possible to install the VLTC system on both the intake and exhaust valve sides.

Fig. 8 Cutaway model of cylinder head when equipped with VLTC system

Fig. 9 Component parts of VLTC system

With the engine shown in Fig. 7, brackets are employed to support the control cam and the hydraulic lash adjuster. The use of these brackets has made it possible to achieve a more compact VLTC system, thereby making it easier to mount it to the cylinder head. In addition, the increase in the height of the rocker cover was kept to 20 mm as compared with a conventional valve train. A cutaway model of the cylinder head is shown in Fig. 8 and the component parts are shown in Fig. 9.

ANALYSIS OF VALVE MOTION - The rocker arm in the VLTC system undergoes complex motion owing to the fact that its shaft moves along a straight line. Since the rocker arm motion differs from that of conventional arms, it cannot be calculated with traditional valve motion computation procedures. Thus a new program was developed to calculate valve motion in the VLTC system.

A flow chart of the design procedure is given in Fig. 10. The motion of the rocker arm is found geometrically. First, the rocker arm and lever layout and maximum and minimum valve lift are determined. In addition, the back surface profile of the rocker arm is formed from a number of arcs. Next, the drive cam profile is approximated from the required valve lift and back surface profile of the rocker arm. The valve lift and acceleration are then calculated from the two profiles, and both profiles are optimized through repeated modifications so as to match the required valve lift and acceleration. If necessary, modifications are also made to the rocker arm and lever layout.

The calculated results for the valve lift and valve acceleration in the VLTC system are given in Fig. 11. The maximum valve lift used during high-speed engine operation was set larger than the conventional valve lift, but the equivalent mass of the rocker arm was made smaller. Moreover, the use of brackets increased the support stiffness of the rocker arm. These measures enabled the VLTC system to display high-speed operating performance equal to that of a conventional valve train (Fig. 12).

The intermediate level valve lift showed a large positive valve acceleration value (Fig. 11). Since this level of valve lift is used in the low speed range, it does not present any problems for practical application.

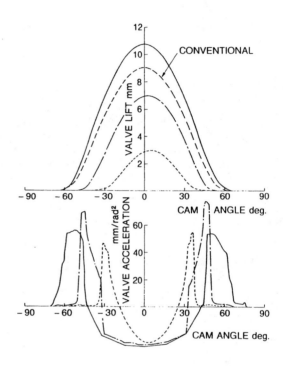

Fig. 11 Calculated results for valve lift and acceleration with VLTC system

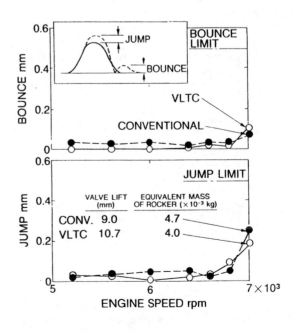

Fig. 12 Comparison of valve jump and bounce under high speed operation

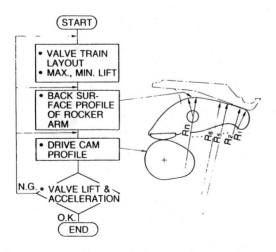

Fig. 10 Flow chart of VLTC design procedure

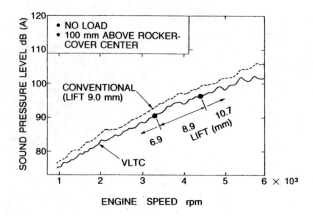

Fig. 13 Sound pressure comparison

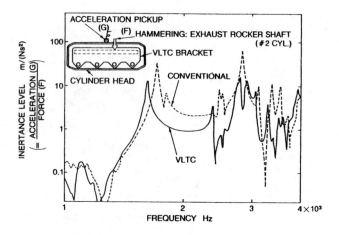

Fig. 14 Inertance level comparison

The noise level of the VLTC system has also been reduced by 2 to 3 dB from that of a conventional valve train (Fig. 13). This noise reduction is largely due to the increased cylinder head stiffness resulting from the use of brackets, which gives the VLTC system an integrated construction with the four cylinders (Fig. 14).

VLTC CONTROL SYSTEM - As mentioned earlier, various approaches to valve lift control have been tried over the years. One of the major issues has been to develop a control system that would provide a low level of control loss. Achieving such a control system for a multi-cylinder engine with four or more cylinders is especially difficult.

The reason for this difficulty is explained here in reference to an in-line 4-cylinder engine. As shown in Fig. 15, the reactive force of the valve spring is constantly acting on the control cam of one of the cylinders via the lever. Because of the integrated construction for the four cylinders, a large driving torque is required in order for the control camshaft to drive the lever. This means high control losses and it also necessitates the use of a large-size actuator.

These problems have been overcome in the VLTC system by using a torsion spring to link the control cam to the control shaft (Fig. 16). With this approach, rotational force is initially stored in the torsion spring and then used to turn the control cam during the interval when the valves are closed. This makes it possible to control valve lift with a lower torque level (about 0.2 to 0.3 N·m) as shown in Fig. 17. Since the valve lift can be switched in roughly 0.2 sec., sufficiently rapid response is assured for practical application.

The profile of the control cam consists of many gently curved surfaces, as illustrated in Fig. 18. This type of profile works to reduce impact noise when the valve lift is switched.

In selecting the best actuator to drive the control cam, a comparison was made of a pneumatic system, a stepping motor and a DC motor. The DC motor system was chosen because it offers the advantages of compact size, less power consumption and lower cost. The DC motor is actuated by a control unit according

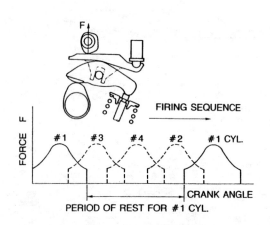

Fig. 15 Force acting on control cam

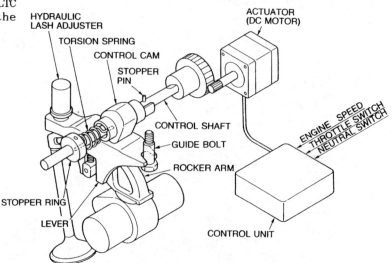

Fig. 16 VLTC control mechanism

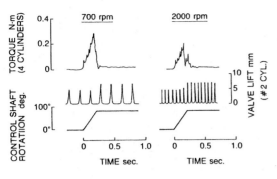

Fig. 17 Torque acting on control shaft during valve lift change

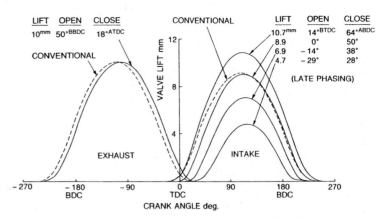

Fig. 20 Valve lift and timing with VLTC system

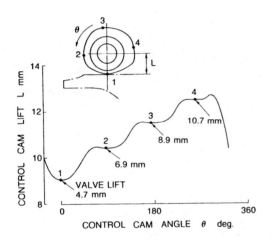

Fig. 18 Lift and profile of control cam

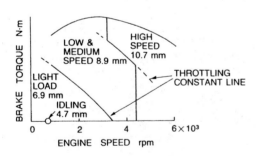

Fig. 19 Valve lift control map

the valve operation, the VLTC can deliver two types of performance patterns: improved maximum power output or improved torque at low and medium speeds.

The torque and power output obtained with the VLTC system for the late phasing valve timing shown in Fig. 20 is indicated by the solid line in Fig. 21. In the low and medium speed ranges, the VLTC system provided only a small improvement in torque over a conventional system, but in the high speed range, it improved the torque level substantially. The maximum power output was improved by 7%.

With the early phasing, the phase of both the intake and exhaust valves was advanced by a crank angle of 6 degrees. The results obtained shown by the dot-dash line in Fig. 21. A large improvement in torque was obtained in the low and medium speed ranges. At an engine speed of 1600 rpm, an improvement of 7% was achieved.

to various inputs that include the engine speed, throttle opening and a neutral switch signal. A typical valve lift control pattern provided by the motor is shown in Fig. 19.

BENEFITS OF VLTC SYSTEM

The valve lift and timing variation provided by the VLTC system is illustrated in Fig. 20. The full load performance obtained with the VLTC system is shown in Fig. 21. Depending on the phase control selected for

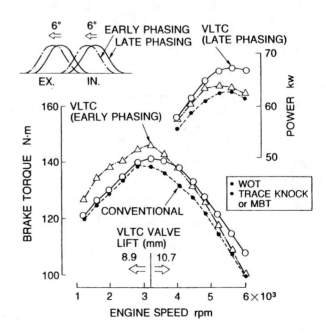

Fig. 21 Full load performance obtained with VLTC system

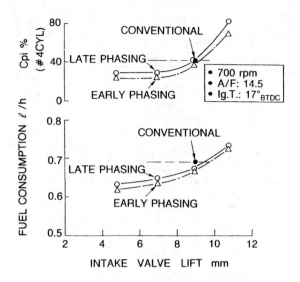

Fig. 22 Idling performance obtained with VLTC system

These results confirmed that the VLTC system was very effective in both the high speed range and in the low and medium speed ranges. If phase control over the cam rotation were added to the VLTC system, it is thought that the system would be effective in improving torque over the entire speed range from low to high.

The effects of the VLTC system on fuel economy will be discussed next. Idling performance is shown in Fig. 22 relative to the intake valve lift. The results show that idling performance varied somewhat depending on the phase of the valve operation. Early phasing valve timing resulted in better fuel consumption and engine stability than late phasing timing. For both early and late phasing valve timing, engine stability was improved by reducing the valve lift. However, no effect was seen with a valve lift of less than approximately 7 mm. By contrast, fuel consumption continued to decrease as the valve lift was reduced. This result is attributed to the reduction in valve train friction. With early phasing valve timing, fuel consumption was improved by 11% when the intake valve lift was set at 4.7 mm.

Vehicle tests were carried out to determine the effects of the VLTC system when the valve lift control pattern shown in Fig. 19 was employed. The results of a Japanese 10 mode test are given in Fig. 23. A 4% improvement in fuel economy was obtained. In addition, improved combustion reduced CO and HC emissions by 22% and 17%, respectively.

CONCLUSION

The results of this research into a system that simultaneously controls valve lift and timing in an automotive gasoline engine are summarized below.

(1) Valve lift and timing control, which reduced both the valve lift and the valve opening duration, was found to have a large effect on improving the performance of a high-speed engine in both the low and high speed ranges. The effect of intake valve control was particularly large.

(2) A valve lift and timing control (VLTC) system, which provides the type of control mentioned above, has been developed and refined to the point where it is applicable to production engines.

(3) The VLTC system features a buffer mechanism to reduce valve shock, reduced mass of its moving parts and a multiple stage control mechanism employing a control cam. These features enable it to deliver low noise levels, high speed operation, excellent durability and good control response, thereby resolving any problems with regard to practical application.

(4) The noise level of an engine equipped with the VLTC system was found to be 2 to 3 dB lower than that of a conventional engine. This noise reduction is largely due to the increased stiffness provided by the VLTC system brackets, in addition to the valve shock buffer mechanism.

(5) The VLTC mechanism has sufficient potential to be further refined into an optimum valve lift and timing control system. This could be done by combining it with a cam phase control system that is already in practical use.

(6) Issues that will be addressed in future work include cost reductions so as to make mass production feasible, further refinement of the system as mentioned in (5) above and expansion of its range to application to include engines with multi-valve cylinders.

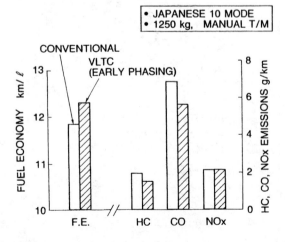

Fig. 23 Fuel economy and HC, CO and NOx emission levels

ACKNOWLEDGMENTS

The authors would like to thank K. Katayama, H. Ofuji, K. Miisho, H. Satoh and many others at Nissan Motor Co. for their cooperation in the development of the VLTC engine system. Thanks are also due to U. Nagai, H. Watanabe and many others at Atsugi Motor Parts Co. for their assistance in connection with the development of the VLTC mechanism.

REFERENCES

(1) D. Stojek and A. Stwiorok, "Valve Timing with Variable Overlap Control," XX FISITA Congress, 845026, 1984.
(2) S. Hara, Y. Nakajima and S. Nagumo, "Effects of Intake-Valve Closing Timing on Spark-Ignition Engine Combustion," SAE Paper 850074.
(3) C. Gray, "A Review of Variable Engine Valve Timing," SAE Paper 880386.
(4) C. Henault and J. Josas, "Variable Gas Distribution Device for Internal Combustion Motors," US Patent Specification 4448156, Published May 1984.
(5) A. Titolo, "Variable Valve Control from Fiat," MTZ, May 1986, pp. 185-188.
(6) G. Torazza, "A Variable Lift and Event Control Device for Piston Engine Valve Operation," 14th FISITA Congress 1972, Paper 2/10, pp. 59-67.
(7) H.P. Lenz, K. Wichart and D. Gruden, "Variable Valve Timing—A Possibility to Control Engine Load Without Throttle," SAE Paper 880388.
(8) D. Scott, "Variable Valve Timing has Electronic Control," Automotive Engineering, May 1984, pp. 86-87.
(9) S. Sapienza, B. Shirey and N.V. Vuuren, "An Electronically Controlled Cam Phasing System," SAE Paper 880391.
(10) K. Maekawa, N. Osawa and A. Akasaka, "Development of Valve-Timing Control System," SAE Paper 890680.
(11) D. Scott and J. Yamaguchi, "Eccentric Cam Drive varies Valve Timing," Automotive Engineering, October 1980, pp. 120-124.
(12) R.H. Richman and W.C. Reynolds, "A Computer-Controlled poppet-Valve Actuation System for Application on Research Engines," SAE Paper 840340.
(13) D. Scott, "Variable Valve Timing to Boost Engine Efficiency," Popular Science, May 1980, pp. 96-97.
(14) L.A. Hausknecht, "Valve Control System," US Patent Specification 4134371, published January 1979.
(15) C.A. Hisserich, "Valve Timing Overlap Control For Internal Combustion Engines," US Patent Specification 3897760, published August 1975.

2004-01-1869

Study on Variable Valve Timing System Using Electromagnetic Mechanism

Chihaya Sugimoto, Hisao Sakai, Atsushi Umemoto, Yasuo Shimizu and Hidetaka Ozawa

Honda R&D Co., Ltd.

Copyright © 2004 SAE International

ABSTRACT

In recent years, increasing attention has been paid to a non-throttling technology that is expected to contribute to a reduction in fuel consumption. This paper describes a study on the technology behind the electromagnetic variable valve timing mechanism (electromagnetic valve mechanism). The electromagnetic valve mechanism ensures highly efficient and stable valve opening/closing control. The detailed information and findings will be described in the main body. In addition, the advantages of the mechanism's application to a homogeneous charge compression ignition engine (HCCI engine) will also be described.

1. INTRODUCTION

There have been many attempts to reduce pumping losses of the four-stroke engine by applying a non-throttling mechanism that would lead to improved fuel consumption. The variety of the mechanisms ranges from utilizing variable intake valve timing to adjust the intake air volume[1], [2], [3], to varying the intake valve timing by releasing hydraulic pressure acting on the cam and valves, which is known as the hydraulically driven valve system[4]. In addition to these, there is a mechanism that uses intermediate levers and an offset camshaft to vary the contact angle between the base of the intermediate lever and the roller follower. This technology is fully mechanical and was invented by BMW and applied to mass-produced vehicles. There are, however, technical issues for each of these mechanisms. More specifically, a hydraulic mechanism is vulnerable to environmental issues when the ambient and engine temperatures are low, and concerning any kind of mechanical system, a typical unsolved technical problem is the increased number of parts and the increased complexity of the system. To cope with these issues, an engineering investigation was carried out on an electromagnetic valve mechanism that would ensure simplicity and greater freedom in valve timing control. This paper describes the details of the configuration, the control logic of the mechanism and the engine performance achieved.

2. BASIC PRINCIPLE OF THE ELECTROMAGNETIC VALVE MECHANISM AND OPERATIONAL MODE

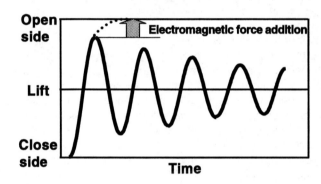

Fig.1 Basic Principle of the Electromagnetic Valve

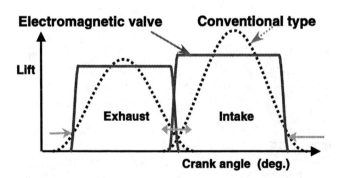

Fig.2 Comparison of Electromagnetic Valve Profile and Conventional Valvetrain Profile

Figs.1 and 2 show the basic principle of the electromagnetic valve mechanism and a comparison of the profiles between an electromagnetic valve and conventional system, respectively. The valve operation is triggered by viscous damped oscillation of the spring and other active components while an electromagnet attracts the valve and supports it to make up for the shortfall in valve lift. The operation of the valve is regulated by charging and discharging the electromagnet, resulting in a constant lift height independent of engine speed. As a result, the profile against the crankshaft angle forms a rectangular wave for improved volumetric efficiency at low and mid engine speeds. Infinitely variable valve timing can be provided for both opening and closing, and the target is to utilize the intake/exhaust air pulsation and to reduce pumping losses.

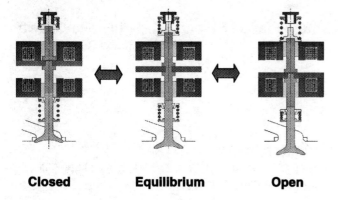

Fig.3 Motion of Electromagnetic Valve

The operational mode is shown in Fig.3. At the initial state (center figure), the forces from the upper and lower coil spring are equilibrated, leaving the valve and armature in the middle of the valve lift stroke. When the valve operation is triggered, the upper and lower coils are alternatively charged according to a predetermined electric cycle, which results in the armature being vibrated and attracted so that it closes the valves (left figure). After this, the mechanism starts to perform a continuous operation (right figure). When the mechanism receives the command to stop, all the valves simultaneously stop their operations and return to the equilibrium (center figure).

3. STRUCTURE AND SYSTEM CONFIGURATION

The major structure and specifications of the electromagnetic valve mechanism are shown in Fig.4 and Table 1. The system configuration is depicted in Fig.5. A square armature disk, featuring springs above and beneath it, supports two shafts. A lift sensor is attached to the upper shaft whereas a small hydraulic tapet used for a valve damping mechanism is located between the lower shaft and valve. Considering wear and thermal expansion of the parts, the mechanism features a separated operation where the valve comes into contact with the seat metal first and then the armature is consequently attracted.

Table1 Valvetrain Specification

Valve diameter (IN/EX)	Φ30/24mm
Valve lift (IN/EX)	7/6mm
Valve pitch (IN/EX)	35.5/37.5mm
Spring force (IN)	100/480N
Spring force (EX)	100/550N

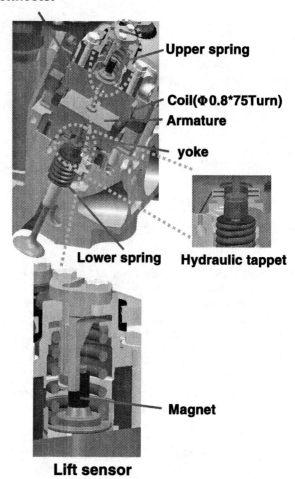

Fig.4 Structure and Hydraulic Tappet and Lift Sensor

This mechanism corresponds to the tapet clearance in the conventional cam-driven valvetrain. In order to minimize the negative effect caused by in-cylinder pressure, the exhaust mechanism has a larger spring load than the intake system.

The yoke features a laminated structure consisting of silicon and steel plates for reduced eddy current loss and costs. Inside the yoke is a coil surrounding a resin bobbin. The electromagnetic valve system shares the timing sensor with other systems while it has an exclusive ECU and driver in addition to a conventional ECU. Also, taking future installation to a mass-production engine into consideration, the mechanism is opereted by a low voltage of 42V. Each valve is regulated independently in accordance with the input from the lift sensor.

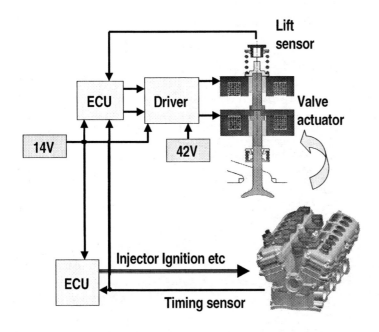

Fig.5 Electromagnetic Valve Management System

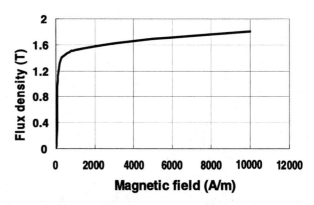

Fig.6 Magnetic Flux and Magnetic Field Characteristics

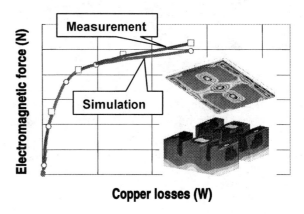

Fig.7 Static Magnetic Field Simulation

4. PURSUANCE OF HIGHER EFFICIENCY THROUGH SIMULATIONS

Another important factor affecting fuel consumption in an engine that controls intake airflow through infinitely variable valve timing control and non-throttle operation is the amount of electricity required. General measures to reduce the amount of electricity consumed include; improved attraction by increasing the magnetic path area, a reduced leakage flux, upgraded material properties, reduced eddy current loss by using a laminated structure and cooling of the coil.

Fig.6 shows a B-H curve of the silicon steel used during the study. B is the magnetic flux density and H is the magnetic field. As shown in the graph, in strong magnetic fields, the flux density is saturated. Based upon this observation, the attraction capability was investigated through a static magnetic field simulation. Furthermore, during the electricity/mechanism compound simulations, the actual behavior of the armature observed during attraction was investigated. Specifically, a series of operations caused initially by electric voltage application, which allows electric current flow as well as magnetic flux generation, were simulated. Also, a mechanism of sudden flux variation that causes eddy current loss was observed. By applying several patterns of calculations to these simulations, it was possible to determine a structure and driving method that would ensure low electricity consumption. Fig.7 shows a comparative study between an attraction capacity simulation and actual measurement under an infinitesimal distance. Although there is a difference under a saturated condition, both values agree closely with each other.

Fig.8 shows a transformer model used for the electric/mechanical compound simulation as well as the result of the simulation. Eqs.(1), (3) and (4) represent the left circuit whereas Eq.(2) corresponds to the right circuit. Both self and mutual inductances are proportional to time. This is attributable to the fact that the B-H curve features non-linear characteristics and thus the electric permeability tends to vary. Addtionally, the phenomenon is also caused by variations in magnetic resistance along with displacement of the armature. Therefore, the generation of inductive electromotive force due to electric current I_1 is affected not only by the product of L_1 and dI_1/dt but also of that between I_1 and dL_2/dt. During the valve lift, the attraction force is not immediately available even if an electric current is thanks to a constant voltage control. A sudden increase, however, can be observed under an infinitesimal distance condition. Eddy current loss occurs when the megnetic flux abruptly changes. No eddy current loss occurs with the armature firmly held.

$$V = R_1 I_1 + L_1 \frac{dI_1}{dt} + \frac{dL_1}{dt} I_1 + M \frac{dI_2}{dt} + \frac{dM}{dt} I_2 \quad ----(1)$$

$$0 = R_2 I_2 + L_2 \frac{dI_2}{dt} + \frac{dL_2}{dt} I_2 + M \frac{dI_1}{dt} + \frac{dM}{dt} I_1 \quad ---(2)$$

$$L_1 = \frac{N_1^2 A}{\frac{1}{\mu} + \frac{2x}{\mu_0}} \quad ---(3)$$

$$\frac{dL_1}{dt} = -\frac{N_1^2 A}{\left\{\frac{l}{\mu} - \frac{2x}{\mu_0}\right\}^2}\left\{\frac{1}{\mu_0}*\frac{d\mu}{dB}*\frac{dB}{dt} + \frac{2}{\mu_0}*\frac{dx}{dt}\right\} \quad --(4)$$

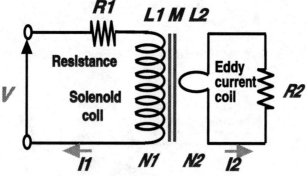

Where, M represents mutual inductance, μ is the magnetic permeability of a magnetic substance, and μ_0 is the atmospheric magnetic permeability. i is the effective magnetic path length. A, N and χ correspond to the effective cross section of the magnetic path, the number of coil twists and armature lift, respectively.

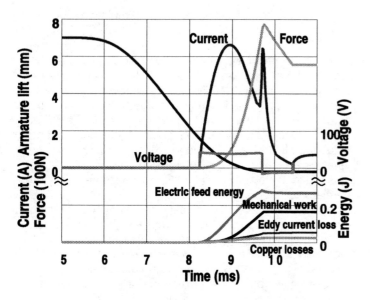

Fig.8 Electric/Mechanical Compound Simulation

5. ELECTRICITY CONSUMPTION PERFORMANCE

Based upon the simulation results, an in-line four-cylinder engine equipped with a highly efficient electromagnetic valve mechanism featuring a laminated structure was assembled. The measurement result of electricity comsuption between the coil terminals are shown in Fig.9. Electricity consumption was measured at approximately 150W at an engine speed of 1,500rpm and with deactivation of one intake and exhaust valve. It almost doubled to 300W when all 16 valves were activated.

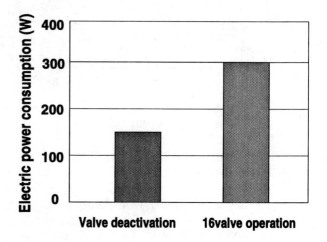

Fig.9 Comparison of Electric Power Consumption

6. CONTROL SYSTEM

Fig.10 explains the basic principle of the control logic. During valve lift, the system provides a constant voltage regulation while it supplies a constant current control when the valves are held in place. A typical constant current control, however, is easily affected by disturbances such as friction occurring during valve lift, which results in a shortened reach of free vibration. This makes the variation in the actuator inductance slower, allowing an easier current flow. Since this type of control logic ensures the electricity current can be kept constant, the electric voltage can be lessened and this acts on reducing the attraction power. The control logic, on the other hand, ensures greater stability in behavior whereby the electric current becomes larger and acts towards increasing the attraction power by ensuring the electric voltage remains constant. The method of the constant voltage control is to regulate the electricity charge timing in accordance with the results of the comparison between the actual valve displacement detected with the valve lift sensor and the target value, which comprises of the controls for the beginning and end of the charge. The charge beginning control is aimed at a stable operation and reduced valve speed for when the valve comes into contact with the seat. The speed of contact is determined mainly by when the charge begins, which forms the basis for the measurement of the valve movement time. The value is then compared with the target to determine the timing of the charge beginning for the following cycle. The electricity charge ending control is a method for calibrating the retardation time for the following cycle by compensating for variations in the retardation time, which are caused due to residual flux and by measuring the time required to achieve a 1mm lift after the charge is complete. This is based upon the idea that the valve timing is determined by the timing of the charge completion.

The mechanism features a system to judge whether or not the valve appropriately sits on the seat metal. In case of a failure, which keeps the valve at an equilibrium state, the mechanism attempts to let the valve open or close in accordance with the valve position when the failure occurs.

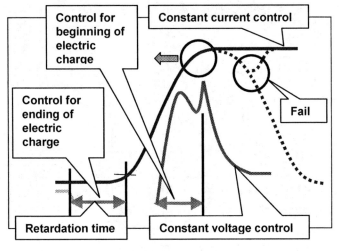

Fig.10 Control Principle

7. NOISE AND VIBRATION PERFORMANCE

The conventional valvetrain is characterized by a damping curve that copes with the impulse sound generated when the valve comes into contact with the seat metal, which thereby leads to improved durability. The electromagnetic valve mechanism, on the other hand, does not feature such a characteristic and thus the attraction characteristics curve becomes extremely sharp especially when the distance is infinitesimal, leading to the serious problem of the impulse sound being generated when the valve and armature come into contact with the seat metal. The contribution ratio of the valvetrain noise to the overall mechanical noise generation is highest at idle speed. Fig.11 shows the result of the noise evaluation at an area in close proximity to and directly above the cylinder head. The test was conducted inside a simplified unechoic chamber at idle engine speed. The measurement with the electomagnetic valve mechanism was approximately 19dB worse in terms of sound pressure level (SPL) than the conventional cam-driven valvetrain.

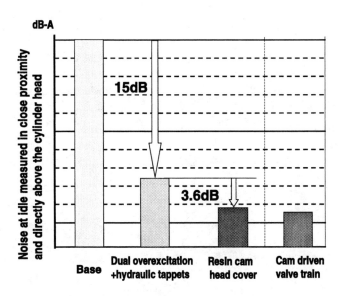

Fig.11 Controlling Methods, Structural Difference and Effects

As shown in Figs.12 and 13, it was possible to inhibit the abrupt variation in attraction characteristics observed in the neighborhood of the valve contact with its seat. The method incorporated was to provide a dual overexcitation control under a constant voltage condition, which in turn reduced the contact speed of the armature to 0.1m/sec. from 0.45m/sec. As a result, the impact acceleration at the yoke reduced to one fourth or less than the original setup. A further improvement was achieved when this was used in conjunction with an integrated compact-type hydraulic tappet, a resin cam head cover and a reduced idle speed setting, as shown in Fig.14. This combination allowed noise generation characteristics similar to those of the conventional cam-driven valvetrain.

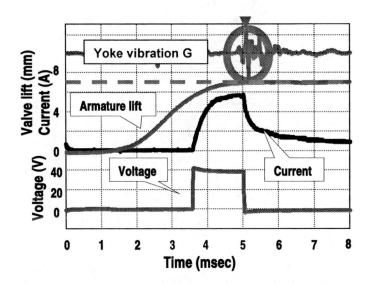

Fig.12 Overexcitation Control under Constant Voltage Condition

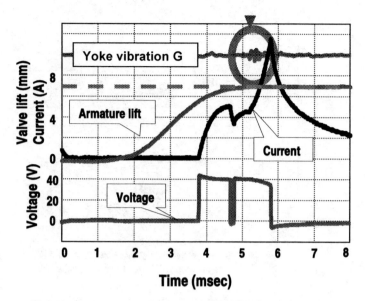

Fig.13 Dual Overexcitation Control under Constant Voltage Condition

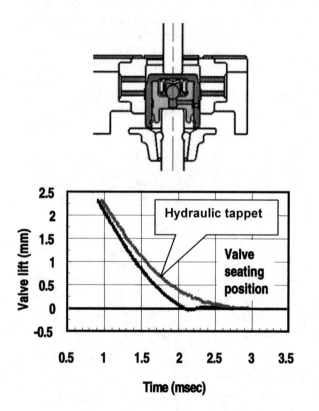

Fig.14 Hydraulic Tappet Effect

One significant characteristic of the noise pattern for the electromagnetic valve as a unit is, as shown in Fig.15, that the increasing ratio of SPL along with that in the engine speed is less than in the conventional cam-driven system. This can be explained by the fact that the contact speed of the electromagnetic valve does not depend on the revolution of the engine. Therefore, if the noise generation at idle remains similar to the conventional system, the new mechanism would ensure better SPL at higher revolution ranges.

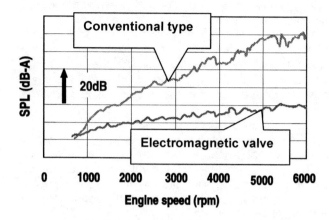

Fig.15 Relationship between Engine Speed and SPL

8. ENGINE PERFORMANCE

The major specifications of the engine used for the experiment are listed in Table 2.

Table 2 Test Engine Specifications

Bore*stroke	Φ75 × 90
Displacement	1590cc
Bore Pitch	84mm
Valve Lift (IN/EX)	7/6mm
Valve Train Type	Electromagnetic Valve

Fig.17 shows the confirmed result of the engine's actual fuel consumption. Thanks to the incorporation of the electromagnetic variable valve timing mechanism, the engine was able to feature a non-throttling operation. It was also confirmed that an intentional retardation of the intake valve close timing led to a 10% improvement in brake specific fuel consumption (BSFC) including electricity consumption over the conventional cam-driven valvetrain. The PV curve observed during the test is shown in Fig.16. The figure shows that the intentional close timing retardation reduces pumping losses (Test conditions: 2.9kW at 1,500rpm with deactivation of one intake and exhaust valve). A 10-15 mode fuel consumption simulation suggested a 7% improvement in fuel economy.

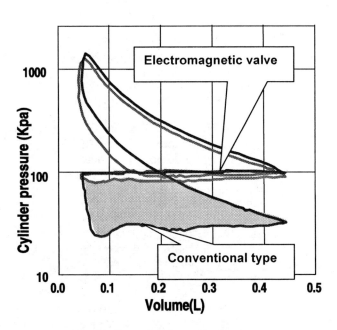

Fig.16 Pumping Loss Reduction

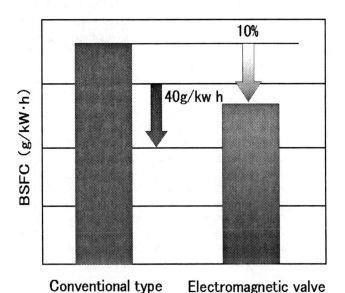

Fig.17 Comparison of Fuel Consumption Values

Concerning exhaust gas emissions, the engine with the new mechanism achieved a similar level to the conventional cam-driven valvetrain while operating under a stoichiometric air-fuel ratio condition.

Fig.18 shows the performance under wide-open-throttle (WOT) operation. The torque generation under low and mid engine speeds was improved by approximately 20% by setting the close timing of the intake valve at bottom dead center (BDC), which increases the actual stroke volume and by utilizing the scavenging effect caused by intake/exhaust pulsation, which in turn improves volumetric efficiency.

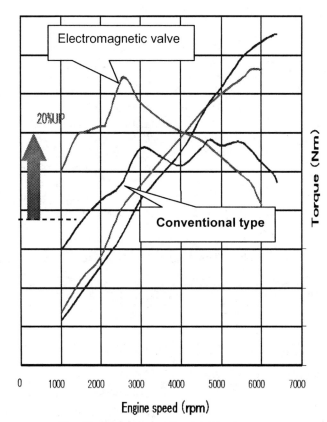

Fig 18 Wide-Open Throttle Performance

9. HCCI ENGINE WITH ELECTROMAGNETIC VALVE MECHANISM

As the next logical step in the research, attention was turned to HCCI operation using an in-line four-cylinder engine incorporating the mechanism explained here and a direct-injection system.

Fig.19 Test Engine Equipped with Electromagnetic VVT Mechanism

Table 3 provides the major specifications of the HCCI engine. With a premixed homogeneously compressed ignition using negative overlap, as shown in Fig.21, no satisfactory results could be achieved either in low or high load conditions. Along with the increase in fuel mixture concentration by port injection under the homogeneous premixed condition, the self-ignition timing advances while the in-cylinder pressure increases, resulting in a restriction in engine power generation due to the occurrence of knocking. To cope with this, increasing the compressed ignition range through direct injection was attempted. The injection during negative overlap allowed an increase in the HCCI operational range (Fig.20).

Table 3 Test Engine Specifications

Bore*stroke	Φ75×90
Displacement	1590cc
Bore pitch	84mm
Compression ratio	15(11.5)
Combustion chamber	Pent roof
Valve train type	Electromagnetic valve mechanism
Fuel system	Direct&port injection

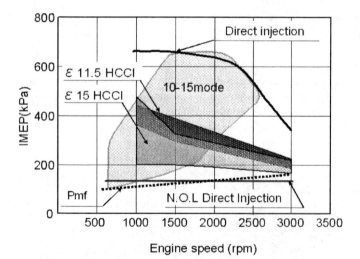

Fig.20 Possible Region for HCCI Operation with Negative Valve Overlap and Direct Injection

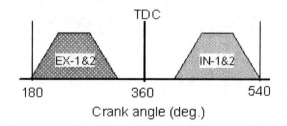

Fig.21 Negative Valve Overlap

10. CONCLUSION

The following results were achieved through the development of constant-voltage valve operation control, which incorporated an electromagnetic valve mechanism with high efficiency and stable, low-speed seating while taking into consideration installation on an actual engine:

1. The adoption of non-throttling technology permitted a 10% improvement in BSFC (2.9kW, 1,500rpm) compared to the conventional cam-driven valvetrain. Also, a 7% improvement of the 10-15 mode fuel economy was confirmed through simulations.

2. The use of the best possible valve timing and scavenging effect improved torque generation at low and medium engine speeds by approximately 20%.

3. The use of compact-type integrated hydraulic tappets, dual overexcitation voltage control and a resin cam head cover ensured a noise level at idle engine speed as low as that observed with the conventional cam-driven valvetrain.

4. The combination of timing control for charge commencement and charge completion allowed stable valve operation and seating speed.

5. A method for expanding the HCCI operations was investigated using the engine with electromagnetic valves and direct injection system. As a result, almost all the loads and engine speeds used during the 10-15 mode operation were found to be fully covered.

6. The electric/mechanical compound simulation that combined the static magnetic field and transformer model helped to reveal the mechanism of eddy current losses, copper losses, mechanical work and energy required during the actual engine operation.

REFERENCES

[1] Controlling Engine Load by Means of Late Intake-Valve Closing: SAE Paper 800794
[2] Controlling Engine Load by Means of Early Intake-Valve Closing: SAE Paper 820408
[3] A Review of Variable Engine Valve Timing: SAE Paper 880386
[4] Study of Vehicle Equipped with Non-Throttling S.I. Engine with Early Intake Valve Closing Mechanism: SAE Paper 930820

CHAPTER 5

FUEL INJECTION

680043

A New Distributing Injection System and Its Potential for Improving Exhaust Gas Emission

Hans May and Harry Schulz
Institut für Wärmetechnik und Verbrennungsmotoren
Technical University Aachen, West-Germany

THE REQUIREMENTS on otto engines for automobiles are today different than those of 10-20 years ago. The modern development is mainly characterized by the aim to reach not only high specific power, a favorable torque characteristic, low fuel consumption, and good operating reliability, but also to obtain a reduction in the exhaust gas components injurious to health.

In principle, there are two approachs to cleaner engine exhaust gas composition: afterburning systems or improved combustion. To regulate the combustion process to obtain good exhaust gas composition, special requirements for the control precision are necessary. At today's state of development, these can be realized more easily by fuel injection than by carburetion. Especially, when operating with a carburetor, auxiliafy equipment is necessary to reach good transient stages. The control precision, however, obtainable in this way is usually unsatisfactory for unsteady operating conditions.

Though the advantages of fuel injection are well known and long since proved, (1 - 4),[*] it is used on a only small scale because of the higher manufacturing costs in comparison to a carburetor system. Presently in order to obtain favorable exhaust gas composition with a carburetor, auxiliary equipment (such as afterburning systems) and resulting additional costs are necessary. Therefore, fuel injection is of more interest today.

Until now, a number of fuel injection systems have been proposed and some have been realized. These systems operate with pneumatic, mechanical, or electronic governors. The required fuel quantity is delivered by displacement injection pumps or distributing systems (5-7, 12).

In this paper, the possibilities and experiences with an injection system proposed and developed by F.A.F. Schmidt is described. It is a system which works on the principle of distributing discs. In this injection system, the requirements for mixture control and formation within the cylinder, with regard to good atomization, are solved separately in order to permit optimal control of both processes.

For this purpose, two independent controls, injection time and injection pressure, are used for metering the required fuel quantity. In addition to manifold injection, injection into the cylinder, eventually in connection with a stratified charge, is possible.

Some examples will show that an exact adaptation of

---

[*]Numbers in parentheses designate References at end of paper.

ABSTRACT

The control principles and the design of a fuel injection system, developed by F.A.F Schmidt, are described. In this system, injection time and injection pressure are controlled independent of each other. The injection time is controlled by two rotating discs having slots, which are turnable to each other and which are turned by the influence of a centrifugal governor in connection with a three-dimensional cam. With the three-dimensional cam, a punctiform scanning of engine characteristics can be realized.

Some results obtained with this injection system are shown, for example, fuel quantity characteristic, CO and n-hexane characteristic of a 4-cyl 4-stroke engine, injection pressure distribution dependent on crank angle, and consumption loops for injection and carburetor operation.

the injected fuel quantity to the characteristics of the engine can be realized.

GENERAL REQUIREMENTS ON CONTROLLING

In the otto engine, mixture control is necessary to obtain good operating conditions, which means a variation of the fuel-air ratio. The coefficient used for this is the relative air-fuel ratio λ, expressing the ratio of the air quantity really working in the engine to the quantity of air needed for stoichiometric combustion of the supplied fuel.

An economical operation, for example, low fuel consumption, will be determined by the average value of λ. For the behavior of the engine primarily, the mixture ratio near the spark plug is important. Even if the average lean mixture is adjusted, there should be a rich mixture near the spark plug. For good idling, generally, an excess of fuel near the spark plugs is necessary. The same is demanded for cold starting, during which eventually 40-50% excess of fuel is required.

Unwanted CO appears with the excess of fuel and increases with it. The principal dependence of the CO concentration from λ can be calculated considering the ideal otto process.**

** The authors thank Dipl. Ing. H. Cremer for computing the ideal otto process.

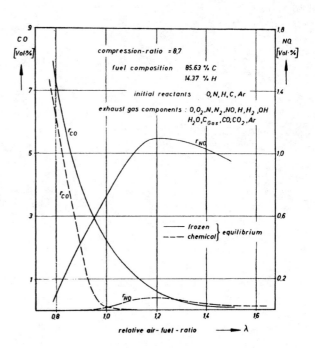

Fig. 1 - Effect of relative air-fuel ratio on exhaust gas CO and NO concentrations considering the ideal otto process (r_{CO} = Volume % of CO, r_{NO} = Volume % of NO in the exhaust gas)

Table 1 - Dissociation and Equilibrium Equations Used for Calculating Exhaust Gas Components in Ideal Otto Process

$$O + O \rightleftarrows O_2 \qquad \frac{\nu_{O_2}}{\nu_O^2} = K_{P_{O_2}} \frac{P_0}{\sum \nu_i}$$

$$O + H \rightleftarrows OH \qquad \frac{\nu_{OH}}{\nu_O \cdot \nu_H} = K_{P_{OH}} \frac{P_0}{\sum \nu_i}$$

$$N + N \rightleftarrows N_2 \qquad \frac{\nu_{N_2}}{\nu_N^2} = K_{P_{N_2}} \frac{P_0}{\sum \nu_i}$$

$$H + H + O \rightleftarrows H_2O \qquad \frac{\nu_{H_2O}}{\nu_H^2 \cdot \nu_O} = K_{P_{H_2O}} \left(\frac{P_0}{\sum \nu_i}\right)^2$$

$$N + O \rightleftarrows NO \qquad \frac{\nu_{NO}}{\nu_N \cdot \nu_O} = K_{P_{NO}} \frac{P_0}{\sum \nu_i}$$

$$C + O \rightleftarrows CO \qquad \frac{\nu_{CO}}{\nu_C \cdot \nu_O} = K_{P_{CO}} \frac{P_0}{\sum \nu_i}$$

$$H + H \rightleftarrows H_2 \qquad \frac{\nu_{H_2}}{\nu_H^2} = K_{P_{H_2}} \frac{P_0}{\sum \nu_i}$$

$$C + O + O \rightleftarrows CO_2 \qquad \frac{\nu_{CO_2}}{\nu_C \cdot \nu_O^2} = K_{P_{CO_2}} \left(\frac{P_0}{\sum \nu_i}\right)^2$$

where:

ν = Mole number
P_0 = Total pressure
K_P = Equilibrium constant

Fig. 1 shows the CO and NO concentration plotted against the air-ratio λ for the ideal otto process. The calculation was made for a compression ratio of 8.7 and a fuel C/H ratio of 5.96 for chemical and frozen equilibrium. The equilibrium reactions taken into account are given in Table 1. The different published experimental results regarding CO and NO concentration agree satisfactorily with the given theoretical results. They lie within the range of chemical and frozen equilibrium as shown in Fig. 1 (4, 17, 22).

As is known, in engine operation unburned hydrocarbons are in the exhaust gas, the concentration of which depends primarily on the design of the combustion chamber. The appearance of hydrocarbons depends on rapid cooling effects at the combustion chamber wall; therefore, the ratio of surface to volume of the combustion chamber is important. Also, an influence of valve overlapping on the hydrocarbon concentration is assumed (4, 16, 17).

As the calculation of the ideal otto engine process was made under the assumption of equilibrium conditions, theoretically no unburned hydrocarbons are obtained in the exhaust gas. To obtain further information about this, investigations regarding the temporal and local course of combustion should be made.

For a given engine, each operating point is exactly defined by speed and throttle valve position, if the influence of the atmospherical and thermal engine conditions are neglected at first. The function of the fuel governor is to supply the correct fuel quantity to the air quantity within the combustion chamber. This is realized practically by the carburetor; however, correction terms are needed for an additional mixture control for the different operating conditions (10). In general, this control works imperfectly (11, 12).

A more exact metering of the fuel quantity for each operating point is possible with fuel injection, because speed and throttle valve position can be used as two independent standard sizes and correction terms, such as mixture enrichment at cold starting, pressure and temperature of the atmosphere, cooling water temperature, fuel cutoff at deceleration, etc., can be considered easily by the controller.

To decrease the exhaust gas components injurious to health, a superfattening of the mixture, an operation in the region of excess of fuel, should be avoided. Especially at acceleration, this means that the fuel quantity must be metered most exactly, as demanded by the characteristic of the engine. Because with the same engine a higher output can be obtained with fuel injection than with carburetor, it is possible to shift the operating range even at full load in directing of greater air excess without power loss. In this way, lower fuel consumption and generally smaller toxious exhaust gas pollution are obtainable.

OPERATING PRINCIPLE OF
DISC-DISTRIBUTOR SYSTEM

The injection system consists of an unregulated high-pressure pump with a subsequently added control distributor.* Two independent regulating members are used for metering the fuel quantity required by the engine at any operating point, that is, for any desired combination of speed and load.

Discs with holes or slots are used as the first metering control member, consisting of the following parts, as shown in Fig. 2:

1. A stationary disc (3) with as many holes as engine cylinders.
2. A rotating disc (1) with a slot running at cam shaft speed on 4-stroke engines or at the crankshaft speed on 2-stroke engines.
3. A disc (2) with a slot running at the same speed as disc (1) and turnable to disc (1) by the controller.
4. A stationary disc (0) with the same number of holes as disc (3).

The arrangement of the slots in the discs 1 and 2 represents the effective control area, which determines the injection time. By turning both rotating discs against each other, the injection time can be changed at fuel cutoff up to zero. Fig. 3 shows the different control possibilities, which are given by adequate selection of rotation and control direction and by displacing the holes in the stationary

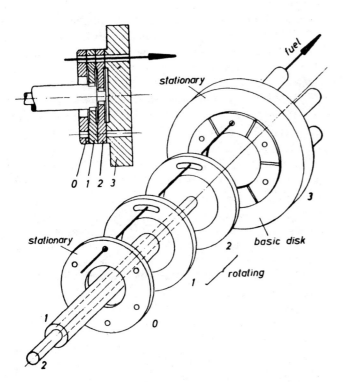

Fig. 2 - Disc arrangement of distributor injection system

---

*For more than 10 years, several 2- and 4-stroke car engines with manifold injection and with cylinder injection have been equipped with this injection system and have been proved partly successful by driving tests. Thus, the clearness can be considered sufficiently proved.

discs. The set of discs shown here is developed along the radius of the holes and slots.

If, for example, the rotation direction of the discs is in agreement with the control direction, a constant injection end (IE) and a regulated injection begin (IB) is obtained. By changing the control direction, inversed conditions are obtained, for example, a constant injection begin and a regulated injection end. In this way at injection into the cylinder, the time history of the mixture formation can be regulated within a certain range depending on the position of the regulating slots. This can be used to improve the course of combustion

An essential point to note is that in this fuel injection system the necessary effective area of the slots for metering the fuel quantity has to be adjusted only once, as this one effective area supplies all engine cylinders in the same way via the corresponding holes. This is advantageous regarding homogeneous mixture distribution to the single cylinders in comparison to the carburetor and to displacement injection pumps. Due to the different length of the suction pipes and the different flow conditions, a homogeneous mixture distribution cannot be realized exactly by the carburetor. Also with displacement injection pumps, an exact uniformity of fuel supply is difficult to obtain, because each engine cylinder is supplied by a separate plunger, which meters fuel quantity by changing its stroke and which has to be adjusted separately.

To study the uniformity of the single injection sprays, the complete injection aggregate has to be considered, that is, nozzles, injection pipes, and metering system.

On displacement injection systems, the injected fuel quantity depends primarily on the volume of the fuel, which will be displaced by the plungers with different velocity. On the disc-distributor system, the fuel quantity injected per time unit will be influenced by the minimum opening cross-section of the injection nozzle. Using the same nozzles and injection pipes, the average deviation of fuel quantity in the whole operating range was found to be only a third on the described distributor system in comparison to the average deviation of fuel quantity measured on a displacement injection system (7).

Commercial atomizing nozzles, as developed for displacement injection pumps, are used for this injection system. The accuracy of these nozzles as to their minimum opening cross-section and the nozzle spring was found to be so high that it is not necessary to use a set of matched nozzles. Single nozzels can be exchanged without a noticeable deterioration of the uniformity.

At constant fuel pressure and without turning the two rotating discs relative to each other, and thus at constant effective slot area (for example, without controlling), the fuel quantity injected per working cycle and cylinder changes inversely to the speed as a hyperbolic function.

The second control member is a pressure valve working as overflow valve, which controls the injected fuel quantity in relation to engine load, or which is used only for correction control independent of load. In this case, the influence of load is controlled by a three-dimensional cam, as shown later. If no great precision is demanded, the three-dimensional cam is not necessary.

Usually, a fuel quantity characteristic is used to describe the operating conditions of an engine. For any operating point, the required injection quantities are plotted against engine speed with throttle valve position as parameter for maximum power and minimum specific fuel consumption. As is known, the required quantity of an engine decreases within a wide operating range as a hyperbolic function of speed, especially in the partial load range, which is important for the usual driving conditions.

Due to the control principle of the distributor injection system, a good adaptation to the engine characteristic is reached automatically, that is, only small control ways are needed at partial load.

In higher load and speed ranges, the fuel quantities differ essentially from the hyperbolic characteristic. In this range the governor has to do the work primarily. As the working capacity of the speed governor (a centrifugal governor, in general) is better at higher speeds and higher control device power is also available, exact fuel metering can be realized more easily. Injection systems, however, working on the displacement principle, have to be controlled more in the partial load range (11).

LOAD CONTROL BY FUEL PRESSURE - In this design example of the injection system, the fuel supply is regulated, depending on the speed as described above, by changing the effective metering area of the rotating discs. As second standard size, independent from the first, fuel pressure is chosen for considering the engine load. This simple design can be applied in such cases where no high precision of control is demanded. The influence of speed is considered by a centrifugal governor. With increasing speed, the flyweights of the centrifugal governor move outward against a spring tension. This motion causes, by means of a crank gear, a relative turning of both rotating discs 1 and 2.

Thus, the effective area of the metering slots increases with speed. The adaptation of the injected fuel quantity to

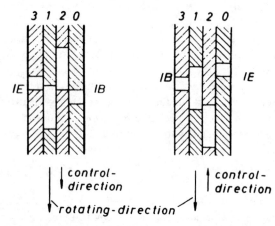

Fig. 3 - Arrangement of slots and holes for controlling injection time

a given characteristic line of the engine at a given pressure is made by fixing the initial overlap of the metering slots (Fig. 2) and by suitable design of the governor spring.

Load is controlled by the fuel pressure in connection with an overflow valve. Throttle valve position and fuel pressure are coupled by a cam plate, the contour of which will be determined by engine tests.

The injection pressure lies within a range of 20-45 atm. The lower limit is used for idling, while pressure increases with increasing load. Examples for values of injection duration and injection pressure in dependance on crank angle are given in the next section.

On the high pressure fuel pump, an uncontrolled 4-cyl wobble plate pump is used as shown in Fig. 4. The fuel flows through the hollow plungers into the cylinders of the pump. The wobble plate consists of a metallic ring, coated with a synthetic having good sliding characteristics. The plungers are sealed by O-rings to reduce leakage losses. Thus, at low speeds a high delivery rate is obtained.

The quantity, delivered by the pump, is about four times the injected quantity, in order to have sufficient fuel for cold start and for controlling and cooling purposes.

LOAD CONTROL BY THREE-DIMENSIONAL CAM - With this design, the injected fuel quantity can be adapted punctiformly to any engine characteristic by means of a three-dimensional cam. Such high requirements for control precision are generally demanded for modern engines, especially regarding the exhaust gas composition (15). Generally, the adaptation to the characteristic is chosen in such a way that a rich mixture is adjusted at full load to obtain maximum power and good acceleration, whereas in the other regions of the characteristic, especially at partial load, a relatively lean mixture is provided to get a low fuel consumption in the normal operating range.

In this design example, the load influence is also considered by changing the effective area of the metering slots. By means of the three-dimensional cam, the rotating discs are turned relative to each other, depending on the engine load corresponding to the correlating speed given by the engine characteristic.

Fig. 5 shows a schematic of the design with the three-dimensional cam. In contrast to the design with the simple governor, the motion of the centrifugal weights is transmitted to the inner shaft by the three-dimensional cam in connection with a feeler. The cam is connected with the centrifugal weights of the governor and moves in a radial direction together with the centrifugal weights when the speed is changing. The construction design can also be solved in a different manner. The feeler can be moved in axial direction dependent on the throttle valve position, and, therefore, dependent on the load.

Thus, the assigned curves of the three-dimensional cam are picked up depending on the different throttle valve positions. The feeler is ducted in a lateral slit of the hollow

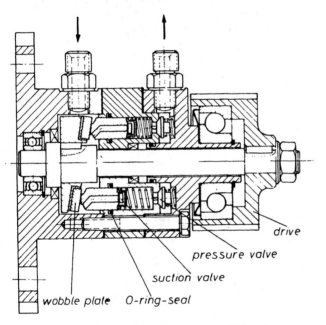

Fig. 4 - Cross-section of high pressure wobble plate pump

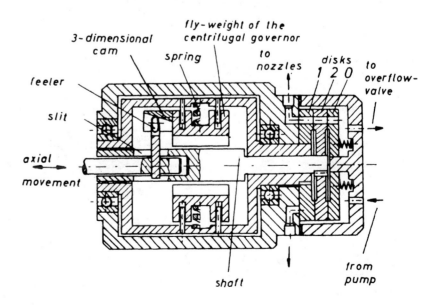

Fig. 5 - Schematic view of the distributor system with three-dimensional cam shaft. Thereby, the hollow shaft is turned when scanning the three-dimensional cam and both rotating discs 1 and 2 are turned against each other. In this way, any required effective area of the regulating slots can be realized by an adequate profiling of the three-dimensional cam for any given coordination of throttle valve position and speed. The spring is made with a variable characteristic (for example, a conical spiral spring). Therefore, the elevations of the cam can be chosen in such a way that at low revolutions the controlling way will be relatively large and high control precision will be obtained.

Load control can also be obtained partly by the fuel pressure and partly by the three-dimensional cam. This method has the advantage that the basic control is made by the fuel pressure respective the overflow valve, and the cam has only small additional control functions. So only a small control way and small controlling device power are necessary. Therefore, the dimensions of the governor can be kept small and a higher control accuracy is reached. This is important for the transition regions, such as acceleration to obtain a decrease of the undesired exhaust gas components.

If the fuel pressure is used for load control as well, it is possible to influence the formation of the spray, its range, and the distribution of the droplets in order to obtain improved combustion. This is important, especially for direct injection into the cylinder.

If load is controlled only by the three-dimensional cam, the overflow valve is designed for constant fuel pressure. Then it is possible, in general, to use the same aggregate without greater alterations for different engines, only the cam has to be changed and adapted to the given characteristics. An example of the governor design and the three-dimensional cam is given in Fig. 6.

Fig. 7 shows a distributor design with a changed arrangement of the discs and a changed centrifugal governor.

The application of the different specifications will be determined primarily by the control requirements and the engine characteristic.

CORRECTIVE CONTROL - At different atmospheric conditions, for cold starting and acceleration, the required fuel quantities differ from the basic adjustment. These corrections are made by the fuel pressure through overflow valve in connection with corresponding sensing elements, that is, by changing the initial tension in the spring of the overflow valve.

The cold starting enrichment is made with a thermometer

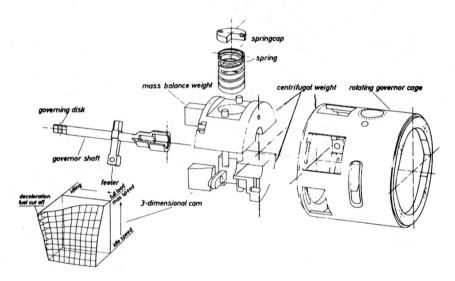

Fig. 6 - Exploded view of governor

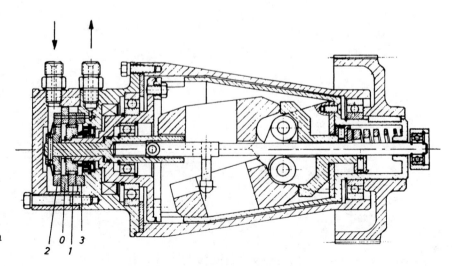

Fig. 7 - Distributor design example with three-dimensional cam

probe. This probe regulates the fuel pressure level dependent on the cooling water temperature or oil temperature (on air-cooled engines). The influence of the pressure and the temperature of the atmosphere is taken into account by means of an evacuated box (11).

EXPERIMENTAL RESULTS

Several investigations have been made to study the control characteristics, the uniformity of injection, spray formation at different pressures, acceleration behavior, friction characteristic of the discs for different materials, erosion, and operating temperature of the discs, etc.

Results of these experiments are described in Refs. 7 and 24. It was found, for example, that the operating temperatures of the discs are almost the same as the fuel temperature and that erosion has no disadvantagous effect on the closeness of the discs, because these regrind themselves.

Further investigations have been made on friction torque at the discs for several material combinations and rubbing speeds. As known, static friction has a higher friction coefficient than sliding friction. Arranging the rotating discs as shown in Figs. 2 and 5, static friction has to be overcome when turning the discs relative to each other. At the configuration shown in Fig. 7, both the rotating discs run between stationary discs. Thus, only sliding friction occurs, which leads to a better hysteresis behavior.

The power needed for the fuel injection system is very small. Even at small 4-stroke engines, about 10% higher output is gained compared with the carburetor.

The different designs of the distributor injection system have been tested on several 2- and 4-stroke engines, as well as on the test stand, under normal driving conditions. On the tested 2-stroke engines, no high control precision was demanded, so the simple governor described above could be used. Due to the considerable scavenging losses in 2-stroke engines operating with a carburetor, an essential consumption decrease along with power increase and better acceleration was obtained with fuel injection (14, 24).

Fig. 8 shows the fuel quantity characteristic of a 2.2 liter 6-cyl 4-stroke engine. The upper limit of the single load ranges corresponds to the quantity for maximum power and the lower limit corresponds to lowest specific fuel consumption. The governor was adapted to the characteristic to provide maximum power at high load and low fuel consumption at partial load. This adaptation is valid for hot running conditions. For cold starting, a greater fuel quantity is supplied by the mentioned corrective control. As shown in Fig. 8, a punctiform adaptation to the requirements of the engine over the whole operating range can be achieved by the distributor system with the mechanically controlled three-dimensional cam.

In the range of partial load, the regulation curves join closely to the engine curves for minimum specific fuel consumption without great control work, as mentioned already. The injected quantities have been measured on an injection test stand and are drawn in Fig. 8. In important operating ranges, the requirement for a low fuel consumption corresponds generally to the requirement for a good exhaust gas composition (19). The engine operating with the adjusted injection equipment has given good performance on the test stand and in driving tests with reference to fuel consumption, output, and acceleration.

The loops of consumption of a German 4-cyl 4-stroke engine with a stroke volume of about 1.3 liter (Fig. 9) demonstrate the advantage of injection operation with this system concerning the CO emission. Abandoning the possible power gain in comparison to the carburetor, at full load the engine can run within a CO range of about 0.5% compared to 2-4% with a carburetor. The CO characteristic shown in Fig. 10 is on the same engine type as in Fig. 9. The test results described here are valid for the adjustment of minimum specific fuel consumption.* The tests were made with the distribution injection system shown in Fig. 7 with mani-

---

* Fig. 10 is from a paper, which will be published in near future, by Dipl. Ing. U. Hattingen and Dipl. Ing. H. Waldeyer. The authors thank both gentlemen for making these test results available.

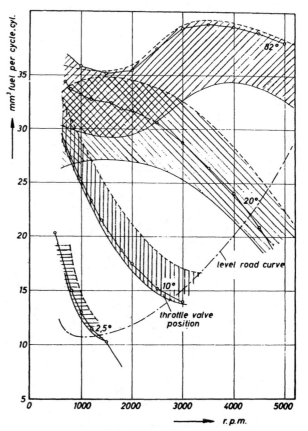

Fig. 8 - Fuel quantity characteristic of a 6-cyl 4-stroke automobile engine

fold injection at a series engine designed for carburetor operation. The fuel was injected into the suction pipe immediately in front of the inlet valve in direction of the air flow, so that a part of the fuel could be injected directly into the cylinder. Constructive alterations on this engine (for example, at the cylinder head) were not made. Thereby, an improvement of the mixture preparation would have been possible within certain limits during injection operation.

As can be seen from Fig. 10, above the road level curve, that is, when accelerating, CO concentrations of less than 1% are obtained. Below the road level curve, that is, when decelerating, the CO values are a little higher. However, in all operating ranges, the CO concentration is less than 2.3%, the value prescribed as average in the California test for this engine. In practice, for the range below the road level curve, the engine can be driven with a leaner mixture corresponding to the adjustment for minimum fuel consumption, as in the case when only engine operation free of jerks is important.

Fig. 11 shows a CO characteristic, which was measured on another engine of the same type as mentioned in Fig. 10, having a distributor system with a three-dimensional cam. The three-dimensional cam was designed in such a way that favorable driving behavior and good acceleration, as well as low toxic exhaust components, were obtained.

The corresponding unburned hydrocarbon characteristic (n-hexane) measured simultaneously with the CO concentrations given in Fig. 11, is shown in Fig. 12. The nondispersive infrared analyzer was used for these measurements. As can be seen from the figure, the unburned hydrocarbon concentrations are between 60 and 120 ppm in all ranges. Both the CO and n-hexane characteristics were measured on the test stand under steady-state conditions. As the governor of the mechanically controlled injection system scans the engine characteristic, also under transient operating conditions, it can be expected that in a California test procedure the CO and n-hexane concentrations will not exceed the values of 2.3% CO and 410 ppm n-hexane, which are prescribed as average for the investigated engine.

The injection duration and injection pressures chosen for the above experiments with manifold injection, are evident from Fig. 13 for idling and for full and half load at an engine speed of 3000 rpm. The pressure distribution was

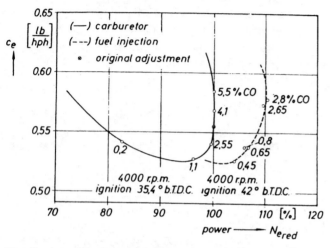

Fig. 9 - Specific fuel consumption and CO concentration of a 4-cyl 4-stroke automobile engine at full load (injection begins 40 deg after top dead center, injection duration 40 deg). For comparison: specific fuel consumption = 0.46 lb/hph at 3/4 load and 2500 rpm

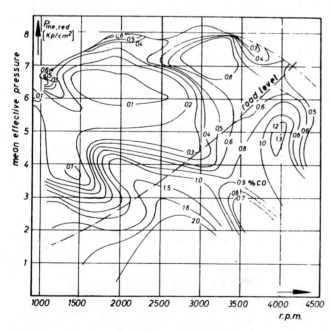

Fig. 10 - CO characteristic of a 4-cyl 4-stroke otto engine for adjustment minimum specific fuel consumption

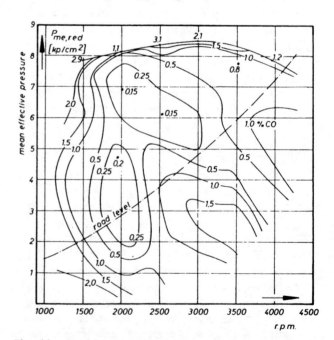

Fig. 11 - CO characteristic of a 4-cyl 4-stroke engine with actual distributor injection system

measured by means of a piezo pressure pickup and on oscillograph in the injection pipe near the nozzle.

The origin of the ordinate axis corresponds to the opening pressure of the injection nozzle (15.2 kp/cm^2), and the origin of the abscissa axis corresponds to the static injection beginning at 40 deg crank angle after top dead center in the intake stroke. As is known, the injection delay increases with increasing speed, for instance from 40 deg CA at idling speed up to 62 deg CA at 3000 rpm. The duration of injection is 35 deg CA at full load. The end of injection is controlled by load, corresponding to the conditions given by the control principle shown in Fig. 3.

The optimum time and duration of injection depends on engine design, among other things. For example, when testing an American 6-cyl 2-stroke boat engine, the best values for power and fuel consumption at full load (6000 rpm) were obtained when injection began at 54 degree crank angle before bottom dead center. The fuel was injected into the transfer port from the crankcase through partly opened inlet slots. Thus, a part of the fuel was injected immediately into the cylinder.

Applying manifold injection, the time and duration of injection are of subordinate importance.

For direct injection into the cylinder, the optimum values have to be determined by experiments, as does stratified charging.

FURTHER POSSIBILITIES FOR REGULATING MIXTURE PREPARATION AND COMBUSTION

For further development, a stratified charging seems to be of great interest (1, 4, 16, 21, 23). This is possible to realize by a suitable combustion chamber design in connection with a divided injection.

With the described injection system, a divided injection is possible without great additional expenditure. Only an additional opening in the rotating discs is necessary to obtain a stratified charge in the combustion chamber. Thereby, a nearly stoichimetric mixture is provided near the spark plug to provide a good ignition. In the remaining combustion chamber, however, a greater excess of air is prepared simultaneously. As known, this measure improves the knocking limits and economy. The engine becomes, however, more sensitive to control, for example, mistakes in mixture preparation lead to an impairment of running efficiency. Therefore, a stratified charge is possible only by an injection system which can be adapted exactly to the engine characteristic. For practical application, a number of further questions, such as the influence of the air motion in the combustion chamber, ignition point, control of the injection point, etc., are important.

In principle, control of injection start and injection end dependent on load and speed exists with the disc-distribution system. An optimal adaptation to an engine has to be found experimentally. Furthermore, with special constructive measure, it is possible to control the injection start and end simultaneously.

CONCLUSION

The working principle and the possibilities of application of a distributing disc fuel injection system are described in this paper. It is shown that all technical requirements, which will be demanded for an injection system, can be realized. Also a stratified charge is possible in a relatively simple manner.

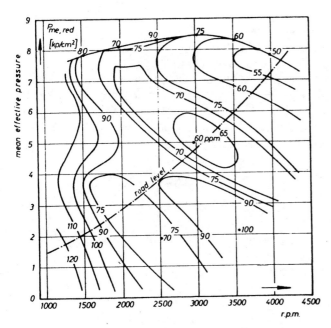

Fig. 12 - n-hexane characteristic of a 4-cyl 4-stroke engine with actual distributor system

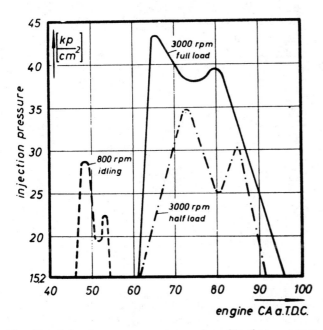

Fig. 13 - Injection pressure measured on distributor system for 4-cyl 4-stroke engine

The tests carried out with this system have shown that a punctiform scanning of the engine characteristic can be made with high control precision.

We have modified several German made 4-stroke automobile engines to use this injection system and find the CO and hydrocarbon emission reduced from the carbureted version. With further development of the injection system and matching with the engine, additional gains could be expected in this area.

REFERENCES

1. F. A. F. Schmidt, Verbrennungskraftmaschinen, 3. Auflage 1951, Verlag lag Oldenbourg, München.
2. F. A. F. Schmidt, "The Internal Combustion Engine." London: Chapman & Hall, 1965.
3. F. A. F. Schmidt, Benzinmotoren mit Kraftstoffeinspritzung und Funkenzündung, MTZ 11 (1950) Nr. 6.
4. H. Knapp, E. U. Joachim, and G. Baumann, "Beeinflussung der Kraftfahrzeugabgase durch Benzineinspritzung." MTZ 26 (1965), Nr. 9.
5. H. Scherenberg, "Der Erfolg der Benzineinspritzung bei Daimler-Benz." MTZ 22 (1961), Nr. 7,
6. H. Grözinger, "Die Benzineinspritzung des 230 SL - Motors von Daimler-Benz." ATZ 65 (1963), Nr. 6.
7. H. Heitland and N. Jeschke, "Möglichkeiten zur Erzielung optimaler Betriebsweise von Fahrzeug - Ottomotoren durch Benzineinspritzung." MTZ 21 (1960), Nr. 8.
8. F. A. F. Schmidt, Gegenseitige Beeinflussung von Gemischbildung und Zündvorgängen in Verbrennungsmotoren, Heft 9 der Schriften der Deutschen Akademie der Luftfahrtforschung.
9. H. Kühl, Dissoziation von Verbrennungsgasen und ihr Einfluss auf den Wirkungsgrad von Vergasermaschinen, VDI - Forschungsheft 373 (1935).
10. W. E. Meyer, Grundsätzliche Gedanken bezüglich der Regelung von Einspritz-Viertakt - Ottomotoren. MTZ 20 (1959), Nr. 12.
11. U. Anders, "Entwicklungsprobleme der Benzineinspritzung von Personenwagen-Motoren." ATZ 63 (1961), Nr. 10.
12. H. de Lavenne, "Die Benzineinspritzung am Motor Peugeot 404." MTZ 24 (1963), Nr. 1.
13. H. Heitland, "Berechnung der Gasgewichte von Saugmotoren mit nicht zu grosser Ventilüberschneidung unter Benutzung der Schmidt - Taylorschen Füllungskonstante." MTZ 24 (1963), Nr. 6.
14. A. Beckers and K. Restin, "Einige Probleme der Regelung von Einspritz-Zweitakt-Ottomotoren." MTZ 20 (1959), Nr. 12.
15. P. H Schweitzer, "Amerikanische Gesetzgebung gegen Luftverunreinigung durch Verbrennungsmotoren des Kraftfahrzeuges." MTZ 26 (1965), Nr. 3.
16. P. H. Schweitzer and L. J. Grunder, "Hybrid Engines." SAE Transactions, Vol. 71 (1963) p. 541.
17. Eberan - Eberhorst, "Die motorische Verbrennung und ihre Abgasprodukte als Häufigkeitsproblem." ATZ 67 (1966), Nr. 8.
18. E. S. Starkman, "Reciprocating Engine Combustion Research - A Status Report." Ninth Symposium on Combustion. New York: Academic Press, 1963.
19. E. S. Starkman, "Engine Generated Air Pollution - A Study of Source and Severity." 11. Internationaler automobiltechnischer Kongress, Juni 1966 München, Bericht Nr. A 12.
20. K. Lohner, H. Müller, and W. Zander, "Entwicklung der Verfahrenstechnik zur Nachverbrennung der Abgase von Ottomotoren." MTZ 27, (1966), Nr. 7.
21. A. W. Hussman, F. Kahoun, and R. A. Taylor, "Charge Stratification by Fuel Injection into Swirling Air." SAE Transactions, Vol. 71 (1963), p. 421.
22. H. Luther, "Ergebnisse und Probleme bei Untersuchungen der Abgase von Verbrennungskraftmaschinen." MTZ Jahrg. 20, Heft 12, Dez. 1959, S. 460 bis 463.
23. J. E. Witzky and J. M. Clark, "A Study of the Swirl Stratified Charge Combustion Principle." SAE Transactions, Vol. 75 (1967), paper 660092.
24. H. May and H. Schulz, "Benzineinspritzung bei Kraftwagenmotoren." MTZ 28 (1967) Nr. 5.

This paper is subject to revision. Statements and opinions advanced in papers or discussion are the author's and are his responsibility, not the Society's; however, the paper has been edited by SAE for uniform styling and format. Discussion will be printed with the paper if it is published in SAE Transactions. For permission to publish this paper in full or in part, contact the SAE Publications Division and the authors.

Society of Automotive Engineers, Inc.

741224

Electronic Fuel Injection in the U.S.A.

Jerome G. Rivard
The Bendix Corp.

FUEL INJECTION OF THE INTERNAL COMBUSTION ENGINE has been viable almost since the origin of this engine. However, the high cost of fuel injection has led to wide application of the less expensive carburetor. On the other hand, mechanical fuel injection systems have been, and still are, routinely applied to performance engines.

Electronic fuel injection (EFI) had its origins during the 1950's, when The Bendix Corporation secured worldwide basic patent coverage (1)*. Successfully operating systems were developed and installed on vehicles; however, the field of electronics, and especially semiconductors, was not advanced to the point where a durable electronic control unit could be designed. The vacuum tube was still in vogue, the automobile environment was much too harsh, and costs were prohibitively high. The significant achievement at this time was the limited production of approximately 300 systems for Chrysler Corporation during model year 1958.

The development remained essentially dormant at Bendix for a period of several years until 1967 when pending exhaust emission controls indicated that a market might be evolving for EFI. Robert Bosch of West Germany was also developing EFI and succeeded in marketing in Europe the first high-volume, production EFI system to Volkswagon.

---

*Numbers in parentheses designate References at end of paper.

By this time, electronic semiconductor technology and manufacturing techniques had evolved to make the application of EFI practical.

The Bendix Corporation has for several years been aggressively developing EFI for the American automotive industry; the first system will be introduced during 1975-model-year production.

CURRENT EFI SYSTEM CONCEPTS

BASIC SYSTEM - The basic function of EFI is to provide to the engine cylinders precise quantities of fuel in the correct proportion with air to achieve the desired vehicle performance, the legal emissions levels, good fuel economy, and pleasing driveability. A typical system installation is depicted in Fig. 1. The Bendix EFI system can be categorized as a low pressure, two group, pulse-timed, intake manifold injection system that meters fuel to individual cylinders by injecting it in the vicinity of the cylinder intake valve. The fuel is controlled by electrically actuated, solenoid-injection valves operating with a supply pressure differential of 39 lb/in^2 (268.7 k pa). The injection phasing and duration is controlled by an electronic control unit (ECU) which computes the needed quantity of fuel from measurements of intake-manifold pressure, engine speed, and air temperature combined with a knowledge of engine

---------- ABSTRACT

A brief evolutionary history is followed by a technical description of the current Bendix EFI system concepts. Application requirements are reviewed in relation to vehicle emissions, fuel economy and driveability. The advantages of feedback control are discussed with emphasis on the need for low-cost durable sensors. EFI is compared to the carburetor and other competitive systems in terms of cost, fuel control accuracy, and fuel economy. The current status of EFI electronic circuit technology and a projection of future generation designs are reviewed. System manufacturing considerations, including costs, are covered. Finally, the necessary application developments are reviewed, including the future potential of integrated electronic controls.

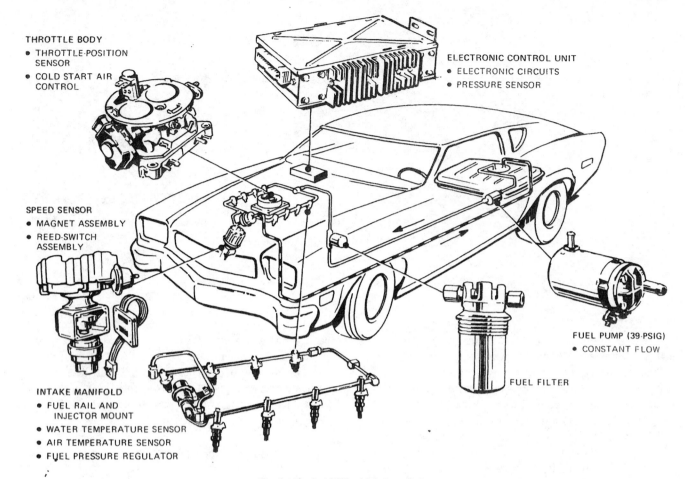

Fig. 1 - Typical EFI vehicle installation

physical and operating parameters, engine phasing, and engine temperatures. A detailed description of the fundamental operation of EFI is contained in Appendix A.

This system approach is commonly referred to as the speed-density concept because an intake-manifold pressure sensor and an engine-speed sensor are used to compute air flow. Sensors for directly measuring air flow (2) have recently appeared, with the choice between the two sensors ultimately resting on system cost. Indeed, both approaches may continue to be viable, so that the selection will depend on application.

In the basic EFI system, the ECU is calibrated to accept sensed engine parameters and compute a fuel-delivery schedule based on engine demands. This does present limitations in reasonable tolerances relative to engine variations and engine changes with operating life.

CLOSED-LOOP CONTROL - The fixed-calibration ECU described earlier will satisfy the requirements of many EFI applications. For other applications, a second-generation EFI system with classical feedback control is now evolving.

The fundamental advantage of feedback control is illustrated in Fig. 2. In a closed-loop engine fuel control system, the air/fuel ratio is actively controlled so as to continuously produce a desired engine output characteristic. One such characteristic is exhaust-gas chemistry, a parameter which correlates to engine exhaust emissions and vehicle performance. It has further significance for exhaust-gas-treatment devices such as catalysts.

Electronics facilitates measuring the engine output characteristics and providing feedback control to such inputs as air/fuel ratio, ignition timing, and perhaps the flow rate of exhaust gas recirculation. The significant element in such a system is the device for sensing the desired output characteristic. One promising system, illustrated in Fig. 3, uses a

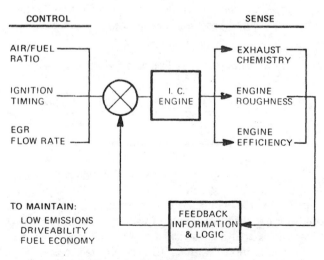

Fig. 2 - Concepts for closed-loop control

ELECTRONIC FUEL INJECTION IN THE U.S.A.

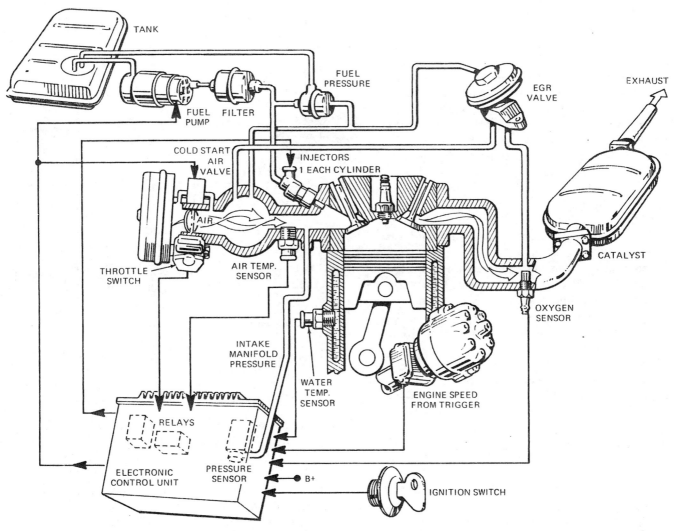

Fig. 3 - Bendix speed-density closed-loop EFI system

sensor which measures the free oxygen in the exhaust. Further detail on this system can be found in the literature (3), (4), (5).

APPLICATION REQUIREMENTS

The application of EFI to the internal combustion engine has been primarily motivated by the requirement for more precise fuel control under all engine operating conditions. The federal exhaust emission standards have probably influenced engine design more than any other single requirement in recent years. The exhaust-gas treatment systems evolved to meet these standards require the engine to maintain exhaust-gas chemistry to exact levels. In the process, the fuel economy and driveability of automobiles have degraded severely.

The carburetor has been the classic fuel control. In the quest for more precise control over the air/flow ratio, the carburetor has been improved by closer manufacturing tolerance, improved calibration and testing, and functional design. In the process, it has evolved from the rather rudimentary 1968 model to the very complex design for 1975.

During this evolution, function has been provided by auxiliary control devices.

By contrast, EFI provides a total systems approach to fuel management. Since it provides a fuel-delivery calibration addressing all engine requirements, function compromise has been avoided. The primary consideration then becomes one of providing sensed engine parameters which accurately measure the desired engine operating characteristic.

The question always arises as to the fuel economy of EFI-equipped vehicles relative to that of carbureted systems. Such a question is difficult to answer since exact one-to-one comparison has not yet been validated. However, a comparison of engineering data for EFI-equipped vehicles being developed for 1975 with similar data for carbureted vehicles of the same model year shows, in some cases, as much as a 10% advantage in fuel economy. The credibility of this figure should be provided by comparisons of actual production vehicles during model year 1975.

Besides fuel economy, EFI-equipped vehicles consistently show driveability advantages. Cold starts and driveaway over a wide range in ambient temperatures have been very impressive, again because more exact calibration is provided compared to the carburetor.

Reliability of the EFI system in United States production has not yet been established. Projections have been made using test experience with similar components and classical prediction techniques. The ultimate goal is less than 1.5% warranty returns for the first year, or 12,000 miles. We are not optimistic enough to think that this level can be achieved in the first year of production because of the usual production-startup intangibles. Also, the first-generation ECU has undergone a function evolution up to production release, and ultimate production-design optimization cannot be achieved the first year. Second-generation ECUs are already under test; these units will be inherently more reliable by virtue of a reduced number of components and interconnections.

System malfunctions will require new diagnostic test equipment — a likely benefit to the owner and servicing agency because diagnosis will be fast and accurate. Servicing will be accomplished by parts replacement. The ECU is not designed to be repaired at the dealer service level. For this reason, qualified repair depots must be established.

Environmental tolerance of the system has been thoroughly addressed during design and development by documenting the system environment, designing with margin, laboratory testing in a simulated environment, and finally field testing on vehicles.

STATUS OF EFI — ECU TECHNOLOGY

ECU FUNCTIONAL DESCRIPTION - The electronic control unit is the heart of the EFI system. By use of sophisticated electronic circuit design, it is possible to deliver fuel to the engine at a rate that is a function of continuously measured engine input and output parameters. In addition, the ECU provides power control to the fuel pump and can easily compute and actively control other engine functions, such as EGR flow, ignition timing, and secondary air pump flow. Any other engine or power train component which is a function of the same sensed parameters as those used to control fuel delivery can likewise be regulated by the ECU.

Fig. 4 shows a functional block diagram of the electronic control unit. This controller can be operated either open or closed loop. For open-loop operation, a calibration is programmed into the ECU as a function of measured engine parameters and the selected control logic. This basic calibration can be easily modified in closed-loop control. For example, a desired air/fuel ratio can be commanded by sensing the engine exhaust chemistry and then modifying, by appropriate circuit logic addition, the basic air/fuel calibration to provide the desired output. This process is shown conceptually in Fig. 3.

The ECU functional characteristic is shown in Fig. 5. Two primary measured parameters are used for air-flow computation: intake-manifold absolute pressure, and engine speed. Air temperature is also measured to correct for air density.

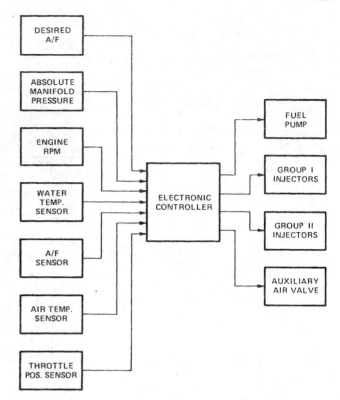

Fig. 4 - Fuel injection electronic control system

The engine-speed sensor also provides injector phasing to engine rotation. The desired injection valve, or injection-valve group, is commanded to start fuel injection by means of the timing pickup. The duration of injection-valve opening is then established by the computed air flow and any other modifying parameter. Modifying parameters include enrichment for cold start, provided by engine temperature measurement; wide-open-throttle enrichment for acceleration performance; closed-throttle idle mixture; and closed-loop feedback. Any other desired parameter influencing engine operation can be measured and introduced into the ECU to modify the basic fuel calibration.

CURRENT ECU DESIGN - The EFI electronic control unit is a hybrid and is configured as a mix of technologies. For control functions which are standard for EFI, custom bipolar integrated circuits are employed utilizing emitter-coupled logic (ECL). For functions which are currently developing, off-the-shelf bipolar integrated circuits and discrete elements are employed. In addition, thick-film technology is being utilized both as active subcircuits and passive resistor networks. This is shown conceptually in Fig. 6.

This packaging, shown in Fig. 7, consists of all circuit elements and networks mounted on standard printed circuit boards housed in a metal case for EMI protection and equipped with a custom hard-mounted connector.

The ECU is designed to operate at a maximum temperature limit of 185°F (85°C). Current models are being installed in the vehicle passenger compartment behind the dashboard, where temperatures rarely exceed 150°F (66°C). While engine-compartment installations have been successful, the reliability of the ECU is considerably enhanced by locating

ELECTRONIC FUEL INJECTION IN THE U.S.A.

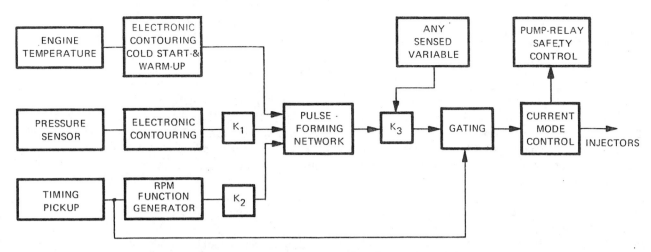

Fig. 5 - Electronic control unit of the functional block diagram

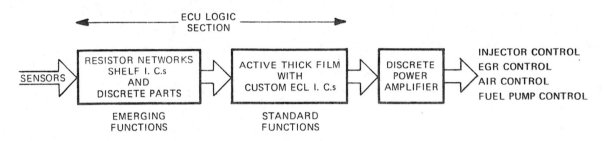

Fig. 6 - Current ECU concept

it in the passenger compartment. Moreover, such installation reduces unit cost.

Extreme package consolidation has not been attempted. In fact, flexibility for functional and calibration changes has been deliberately designed into the unit in order to accommodate changes dictated almost up to the day of the start of production. At this stage of application, this design is probably the most cost-effective.

The current ECU design will see changes in 1976 and 1977. Cost and reliability aspects of the present ECU will be significantly improved by custom integrating the "emerging" functions into one new integrated circuit. Such consolidation is expected to eliminate approximately 60% of the existing ECU space presently consumed by the discrete parts. This stage of customization will be combined with many other value improvements which potentially can reduce the ECU cost by an estimated 65%.

We are continually asked why we aren't using digital circuits and LSI. There is no reluctance to use these promising, cost-reducing manufacturing techniques; and indeed, our advanced ECU developments are directly aimed at exploiting digital LSI circuit designs. The major reason for the described design is to maintain the flexibility needed for addressing major and minor functional and calibration changes. The optimization of the EFI system to vehicle requirements for emissions, performance, fuel economy, and driveability is at an extreme level of continuous change. However, control logic is expected to begin settling down when the emission laws stabilize and desired fuel economy is achieved concurrent with an acceptable emissions level. In our opinion, this will probably not occur until model year 1978, at the earliest.

FUTURE ECU GENERATIONS - Future designs which utilize the MOS technology are expected to replace the EFI bipolar technology ECU. The state of the MOS technology relative to ECU cost indicates that this development should take place in two phases:

    Phase I - MOS and ROM's as look-up tables
    Phase II - MOS and the microprocessor

The design for Phase I has already begun. Preliminary

Fig. 7 - Electronic control unit – production type

cost figures show this approach to be competitive with the bipolar approach. Through innovative design practices, it appears to be possible at essentially the same ECU cost, to include one more function (e.g., ignition advance) into the ECU, thus providing even more incentive and benefit to the consumer. A side benefit of such an inclusion is field experience with a full-up MOS control system preliminary to a fully integrated on-board control system. Using this approach, Phase I is expected to reveal design preferences and provide guidance for Phase II; at the same time, it should add new customer incentives.

Phase II will treat the development of an on-board computer utilizing a microprocessor which time-shares the sensors and electronics for a multiplicity of tasks even beyond the Phase I functions. Such a computer will operate from algorithms and from non-volatile MOS memories. This approach can be justified only on the basis that it can provide many control functions at an installed cost which is much lower than that of providing each control function individually. In 1975, 200 mil microprocessors are expected to cost in the neighborhood of $75.00 (to the OEM manufacturer). By the time all of the supporting electronic circuits are added, the selling price would be entirely prohibitive. However, the cost of the microprocessor is projected to decrease over the next three or four years to the neighborhood of $10 - $15. Such a low price can very well permit its use in the automobile.

Integrated controls are already being designed. The most complex portion of an integrated controller is that part dealing with the EFI fuel control and the related control functions in the Phase I ECU design.

SYSTEM MANUFACTURING CONSIDERATIONS

MANUFACTURING APPROACH - EFI requires a systems approach to manufacturing. Several manufacturing technologies must be applied, ranging from highly specialized electronics and precision electrohydraulic injectors to conventional mechanical air-flow control valves. Assembling this capability at one location for high-volume production is not practical.

The OEM supplier can better address the production of the EFI system by establishing a central system engineering group operating along with a central manufacturing group to do final assembly and checkout. The central manufacturing group then acts to control the many specialty suppliers for the various system components. In this manner, it will control overall system integration to specifications and will develop and control the component suppliers.

A strong reliability and quality assurance operation is required to monitor quality at the supplier and system-assembly levels. As a result, well-defined performance and procurement specifications are mandatory for both systems and components.

PRODUCTION VOLUME - Initial system installations will be costly because of relatively low volume. Major cost reductions are expected to occur in the range of 200,000 to 500,000 systems production per year. At this level, capital investments to realize high-volume cost reductions are justified.

The ECU tooling and capital investment is modular, and the major cost reduction is achieved by the time production reaches 250,000 units per year.

SYSTEM COST - The question of system cost is, of course, of primary significance. The reference for comparison is the carburetor fuel-control system, since it is produced in high volume by techniques which have been refined to the point of least cost.

A fair cost assessment must be based on a net installed cost impact using an add-and-delete cost makeup, and a production quantity comparable to high production, namely a minimum of one million systems per year.

Independent cost estimates have been made on this basis by the OEM supplier and the automotive manufacturers. Assuming current 1977 emission laws (HC = 0.41 gm/mi; CO = 3.4 gm/mi; and NO_X = 0.4 gm/mi. - 1975 FTP), the net installed cost of EFI is essentially equal to that of the carburetor.

With EFI the desired emissions, driveability, and fuel economy can generally be achieved with less complexity in exhaust-gas treatment devices. For example, in some cases, the emissions levels can be met using only a catalyst without the addition of a secondary air pump. This contributes significantly to a reduction in overall system cost.

REQUIRED APPLICATION DEVELOPMENTS

ENGINE: OPTIMIZED INTEGRATION - To date, the application of EFI to U.S. engines has not been design integrated. This is a natural part of system evolution, since existing engines are tooled for several production years and currently intake manifolds are designed for carburetor installation. Design constraints have been induced to provide as uniform a mixture as possible to each cylinder.

EFI meters the fuel at the intake valve of each individual cylinder. The intake manifold can thus be designed for uniform air distribution to each cylinder, thereby giving the designer more latitude. The fuel manifolding and injector installation can be optimized for cost reduction by designing the intake manifold concurrently. Also, the air control valving can be designed integral with the manifold, eliminating both interfaces and certain components. A further advantage is provided by the fact that while carburetor systems are increasing in complexity and cost EFI is on a reducing cost curve. Optimized integrated engine design is yet to be realized.

SENSORS - The sensors in the EFI system measure the engine input and output parameter to be controlled. The adaptation of aerospace technology to the automobile environment has resulted in a major evolution in the sensors used to measure intake-manifold absolute pressure. Currently being tested are miniature, semi-conductor, strain-gage pressure sensors which show a potential for the lowest high-volume cost.

ELECTRONIC FUEL INJECTION IN THE U.S.A.

Other sensors being examined for replacement are those for engine speed and throttle position. Engine speed is now measured by reed switches actuated by a magnet located on an engine rotating member. In the future, the solid-state ignition system will probably provide this signal—a special advantage if the electronic controls for the fuel system and the ignition system, are integrated. The costly position sensors hopefully will be superceded by simplified devices now being developed.

Closed-loop controls require measuring engine output parameters. One very effective, inexpensive device for this purpose is the exhaust oxygen sensor, while other sensors are being developed to sense engine-power output characteristics, new concepts for exhaust sensing are still needed.

The sensors, along with the ECU, will always form the heart of the EFI system, and considerable development for function, cost, and durability is still needed. High confidence exists that desired results can be achieved.

INTEGRATED ELECTRONICS

Integrated electronic controls for the automobile have been considered imminent for about ten years. Many prophets have predicted that integrated electronics would by now be controlling everything from the engine and steering and braking systems, to systems monitoring and information displays. As is usual with technology application to the automobile, we are seeing the application of electronics based on need and constrained by economics. There are not, as yet, any really "integrated electronics."

Based on current applications, there is a logical evolution starting with engine fuel control as a base. It is then simple to add electronic ignition control and controls for emissions systems. EGR is a very logical starting point, especially with the application of EFI. The sensed engine parameters (manifold pressure, inlet air temperature, and engine speed) provide the intelligence for control of all three subsystems.

With little added cost, this first generation of integrated electronic control could also provide the driver with an indication of fuel consumption (either rate or integrated total) and a display of engine conditions, either to indicate significant change or to advise of necessary maintenance.

Based on current system evaluation, a first-generation integrated electronic module could be introduced by model year 1978. The cost trade-offs to meet this goal should realistically be made within the next year. As shown in Fig. 8, such a module has broad potential for adding functions (6).

Multiplexing and level logic have shown promise in simplifying vehicle wiring harnesses, thereby reducing the amount of copper in automotive electrical systems. A significant reduction in the number of wires and interconnections is currently hindered by the cost of the transducers. In most cases, the transducers are simple on-off switches which, to be effective, must be located at the device being controlled. The cost of these remote control devices is still prohibitive.

SUMMARY AND CONCLUSIONS

EFI is being recognized in the United States as a necessary fuel management control system, and the American production is starting for model year 1975. Compared to the traditional carburetor, the EFI system provides more accurate fuel delivery for all engine operating conditions. EFI provides the best tool to achieve legal exhaust emission levels while preserving good performance and driveability; it also appears to improve economy. We predict that the high-volume production cost of an EFI-equipped vehicle will eventually be equal to that of the carburetor-equipped vehicle.

In conclusion, the application of electronics to engine and vehicle control has only just started. The automotive industry moves very deliberately without year-to-year major changes. Evolution with application based on functional need and competitive cost will be the prime motivators for the growth in application of electronics to the automobile.

REFERENCES

1. R. W. Sutton, S. G. Woodward, and C. A. Hartman, "Fuel Injection System," U.S. Patent No. 2,980,090, April 18, 1961.

2. H. Scholl, "Electronic Fuel Injection," Institution of Mechanical Engineers, March 1972.

3. J. G. Rivard, "Closed-Loop Electronic Fuel Injection Control of the Internal-Combustion Engine," Paper No. 73005 presented at SAE International Congress, Detroit, January 1973.

4. R. Zechnall, G. Bonmann, and H. Eisele, "Closed-Loop Exhaust Emission Control System with Electronic Fuel Injection," Paper No. 730566 presented at SAE Automobile Engineering Meeting, Detroit, May 1973.

5. T. L. Rachel and R. Gunda, "Electronic Fuel Injection Utilizing Feedback Techniques," Paper presented at IEEE Intercon Conference Record, Seminar 36, New York, March 1974.

6. L. Hogan "Semiconductor Technology Advances in the 1970's," paper presented at IEEE Vehicular Technology Group, Detroit, May 1970.

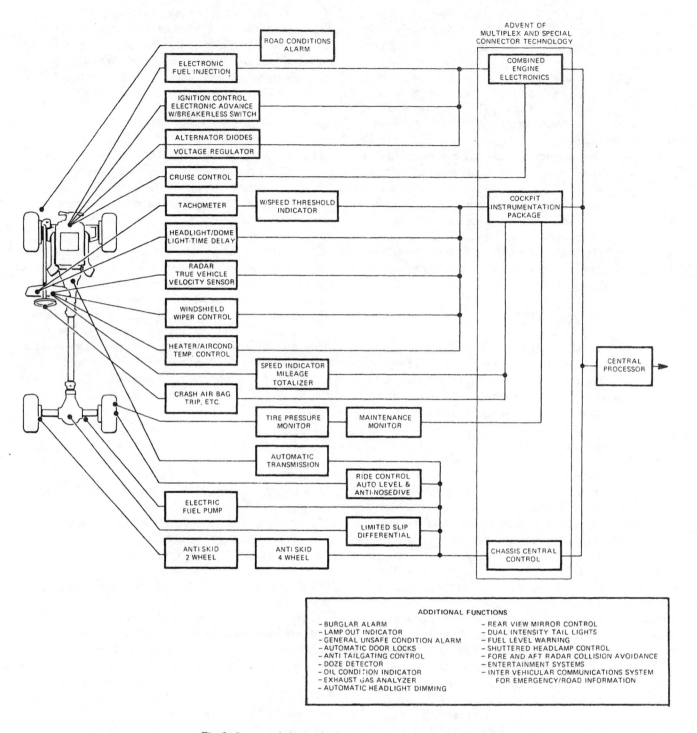

Fig. 8 - Integrated electronics for automotive control applications

APPENDIX
BASIC ELECTRONIC-FUEL-INJECTION SYSTEM

SYSTEM CONFIGURATION AND OPERATION

The EFI fuel-control system, shown schematically in Fig. A-1, is made up of four basic building blocks: the fuel-delivery subsystem, the air-induction subsystem, the primary sensors, and the electronic control unit.

FUEL-DELIVERY SUBSYSTEM - The fuel-delivery subsystem includes the fuel-tank pickup, the electrically driven constant-displacement fuel pump, a filter, the fuel manifold, injectors for each cylinder, a fuel-pressure regulator, and supply and return lines. The fuel is delivered to the injectors at a nominal pressure of 39 lb/in^2 gage (268.7 kPa). The maximum flow-rate capability in a typical eight-cylinder engine system is 32 gal. (121 l) per hour; excess fuel is returned to the tank. The pressure regulator is referenced to the intake-manifold pressure to maintain a constant differential pressure across the injectors, independent of atmospheric or intake-manifold pressure variations, thus ensuring the delivery of a precise quantity of fuel for each commanded increment of injector-open time.

AIR-INDUCTION SUBSYSTEM - The air-induction subsystem includes the integrated intake-manifold and throttle-body assembly for primary air-flow control, as well as an auxiliary fast-idle air valve controlled by either engine water temperature or a heating coil. The fast-idle valve supplies increased cold-starting air independent of the primary throttles. The throttle body incorporates a throttle-position sensor. An air-temperature sensor provides air-density correction data to the electronic logic.

PRIMARY SENSORS - There are three primary engine

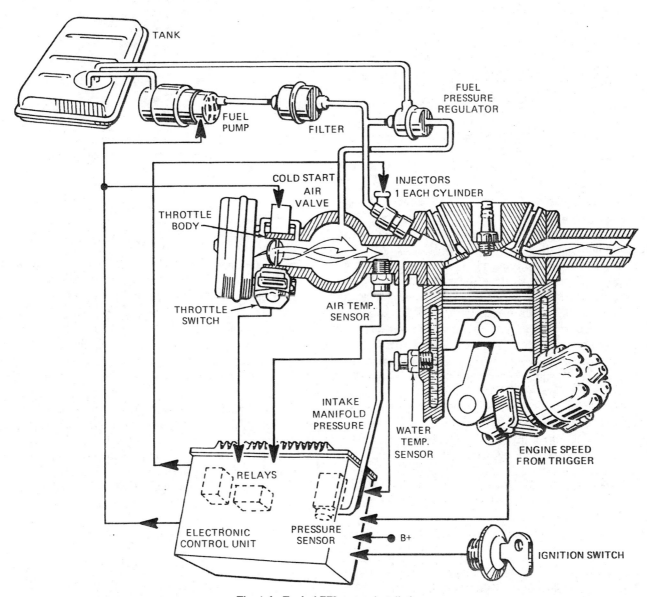

Fig. A-1 - Typical EFI system installation

sensors. An intake-manifold pressure sensor measures absolute pressure in the intake manifold to provide a parameter for continuously computing air flow to the engine. A speed sensor, usually mounted integral with the distributor, provides engine-speed data for computing air flow, as well as engine-phasing data for synchronizing injector-open timing with engine requirements. A temperature sensor, installed in the engine water jacket, provides a signal to indicate fuel-enrichment requirements for cold start and engine warm-up.

ELECTRONIC CONTROL UNIT - The electronic control unit is the heart of the system. It receives information from the sensors that monitor key engine parameters and, using a selected control logic, computes the exact fuel requirement—relative to air flow—for each cylinder on each engine cycle. These computation results are translated into injector-open time signals, which energize the injector solenoid valves to deliver a specific quantity of fuel to each cylinder. The electronic-control-unit calibration is flexible enough to determine precise fuel requirements for any unique operating point over a wide range of engine-performance requirements.

SYSTEM CONTROL LOGIC

The control logic employed in electronic fuel injection to determine the fuel-delivery schedule that will exactly meet the requirements of an engine are summarized in Fig. A-2. Engine speed and air density are used to determine engine-air flow rates since, for any given engine configuration, air flow is proportional to the product of cylinder air density and engine speed. Cylinder air density is equal to the product of manifold air density and engine volumetric efficiency, the latter of which must be determined experimentally for each engine configuration. It follows that air flow is proportional to the product of manifold air density,* engine speed, and engine volumetric efficiency. Fuel flow, then, can be scheduled as any desired function of air flow—or the parameters used to determine air flow—and engine speed.

Carburetor operation is based on less exact logic. Fuel delivery as scheduled by the carburetor is a function of a venturi pressure drop, this drop being a measure of air flow for a particular air density. In a typical application, there can be a 40-to-1 difference between minimum and maximum air flow, with the result that precise fuel metering is often compromised at low flow rates.

The speed/density approach enjoys a number of inherent advantages over the venturi-flow approach. In the implementation of the speed/density approach, fuel for each cylinder is injected once per engine cycle (every second engine revolution), based on a signal from the EFI trigger. This injector-trigger signal is developed exactly once per engine cycle, with the result that the injection repetition rate is exact with speed. The quantity of fuel delivered per injection is regulated in accord with the air density in the manifold, and this

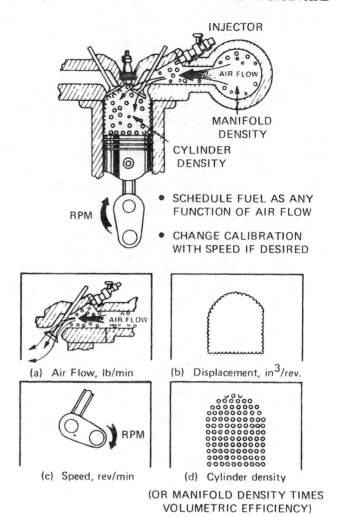

Fig. A-2 - Speed-density control logic

density varies only over a range of 4 or 5 to 1 in a warm engine. The accuracy of fuel delivery is therefore significantly greater than that of which the carburetor is capable.

The EFI system enjoys other inherent advantages. Since no venturi is required at the inlet of the intake manifold, engine air flow is no longer restricted at this point, and the throttle valves can be sized for minimum air restriction and good driver "feel." Moreover, the EFI sensing logic compensates for air-temperature variations, eliminating the tendency to operate "too lean" at low ambient temperatures and "too rich" at high ambient temperatures. Finally, closed-loop control has added a new degree of flexibility to electronic fuel injection. The basic fuel calibration can now be easily modified in response to such sensed engine-output parameters as exhaust-emissions level, engine torque, and power.*

In the majority of EFI system applications to date, it has been common to utilize two-group injection, each group containing half the cylinders. On an eight-cylinder engine,

---

*Manifold air pressure and temperature are used to define density.

*A partial embodiment of this flexibility is described in the early sections of this presentation.

then, four individual injectors are powered from a single output stage of the electronic control unit, and all four are energized simultaneously during an engine cycle; the other four, as a group, are energized half an engine cycle later. This configuration reduces system cost and has thus far provided satisfactory performance.

Sequential or individually phased injection has been considered and may become a necessity in future systems where better fuel-delivery accuracy on transients is required. This configuration would also provide greater flexibility in that it would make possible individual injector control in response to sensed engine parameters.

1975 SYSTEM COMPONENTS

The following paragraphs describe each of the components scheduled to be used in EFI systems designed for installation on 1975-model-year vehicles.

ELECTRONIC CONTROL UNIT - The electronic control unit, shown in Fig. A-3, utilizes integrated circuits to the greatest possible extent. Four custom-developed integrated circuits form its basic module; other integrated circuits provide calibrations unique to a particular vehicle. Information is supplied to the electronic control unit by the various sensors that monitor prevailing engine conditions. The unit processes these signals and, on the basis of previously selected criteria, computes and communicates appropriate pulse-width commands to the fuel injectors, which deliver the fuel to the individual cylinders. The unit can also control exhaust-gas recirculation and other special operations, such as ignition-advance mechanism changes, which may be required under particular conditions.

PRESSURE SENSOR - The pressure sensor, an analog device with associated electronic processing circuitry, outputs a voltage that is proportional to the intake-manifold pressure. The manifold-pressure signal is a measure of engine load and is therefore an important parameter in speed-density fuel metering. The pressure measurement is made with an accuracy of ±1 percent and takes into account hysteresis, resolution, wear, and aging. The sensor will withstand temperatures of 300°F (149°C) and substantial levels of vibration, shock, and overpressure. The unit shown in Fig. A-4 is 1.125 x 1.125 x 1.75 inches (2.86 by 2.86 by 4.45 centimeters) in size and is one of several units being evaluated for production systems.

Fig. A-4

SPEED SENSOR - The speed sensor or "trigger" provides the electronic control unit with data on engine speed and phase. These data are used to relate the frequency of injector operation and the time at which fuel injection takes place with actual engine operation. The sensor shown in Fig. A-5 incorporates magnetic-reed switches, which were selected because of their high reliability, their freedom from contamination, and the absence of any rubbing or sliding contacts. These stationary switches are activated by and situated in close proximity to a rotating magnet driven from the engine. The total number of switches used is determined by the number of injectors that are to be fired in sequence. If two groups of injectors are to be activated in one complete engine cycle, two switches are required; four-group and eight-group injection, in turn, require four and eight switches, respectively.

Fig. A-3

Fig. A-5

FUEL-PRESSURE REGULATOR - The pressure regulator, shown in Fig. A-6, maintains the pressure of the fuel-transfer system at a preset level 39 lb/in^2 above the intake-manifold pressure. Regulation is accomplished by bleeding excess flow not needed for engine operation from the high pressure portion of the fuel-transfer system. Used in this operation is a flat-plate valve, which varies the outlet or bleed-flow orifice area in response to a force balance between the fuel pressure and the sum of a reference-spring force and a force representing the intake-manifold absolute pressure. The regulator typically operates at a natural frequency of 400 Hz with a gain of 0.043 lb/in^2 (296.3 Pa) per gallon-per-hour flow over a flow-rate range of 5 to 40 gal. (18.9 to 151.4 l) per hour.

INJECTOR - The injector, shown in Fig. A-7, is a solenoid-actuated on/off valve, with a poppet pintle design that breaks up the fuel into small particles. Since a constant pressure differential is maintained across the injectors by the fuel-distribution system, the amount of fuel delivered by a given injector depends on the length of time the injector is open. Typical injection times for most engine applications range from 2.5 to 9 ms. Injector design assures fast opening and closing: at 12 V direct current, the opening time is 1.7 ms. and the closing time 1.2 ms. Injectors of various sizes are available, the steady-state capabilities of which range from 285 to 440 cm^3/min. at a fuel-supply pressure of 39 lb/in^2 gage (268.7 kPa). The relatively narrow spray-cone angle of the injector minimizes intake-manifold wall wetting.

FUEL PUMP - The fuel pump, shown in Fig. A-8, is the basic component of the fuel-delivery system. It provides sufficient flow to maintain a nominal regulated system pressure of 39 lb/in^2 gage. The pump has a positive-displacement rollor-cam design and is driven by an integral wet-brush 12-V direct-current motor. It incorporates two internal check valves, one for over-pressure protection and the other to maintain system pressure after pump shutoff. Typical pump-flow characteristics are 33 gal. (125 l) per hour at a pressure

Fig. A-7

of 39 lb/in^2 gage and 42 gal (159 l) per hour at a pressure of 0 lb/in^2 gage. The nominal current draw at 39 lb/in^2 is 3.5 A.

THROTTLE-POSITION SENSOR - Throttle position and rate of change of throttle position are two bits of sensed information constantly needed for fuel-injection control, since they are used to modulate, by small percentages, the basic fuel calibration of the engine. The throttle-position sensor senses closed-throttle, full-throttle, or part-throttle position and rate of change of throttle motion, and conveys this information precisely and without delay to the electronic control unit for electronic processing. As shown in Fig. A-9, the sensor is so designed that sliding mechanical contacts carrying electric current divide the throttle angle into multiple but discrete voltage levels, and these are processed by the control unit to yield the required data.

TEMPERATURE SENSORS - Intake-air temperature is used by most EFI control systems in combination with manifold pressure to very precisely determine the density of the inducted air. Engine-temperature data, normally supplied by a sensor mounted to read engine-coolant tempera-

Fig. A-6

Fig. A-8

ELECTRONIC FUEL INJECTION IN THE U.S.A.

Fig. A-9

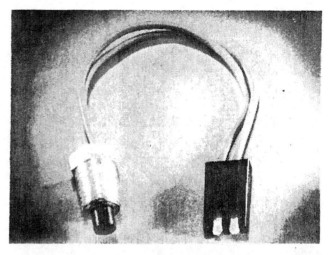

Fig. A-10

ture, are needed to provide the fuel enrichment during cold operation. Both the water-temperature sensor and the air-temperature sensor have the physical design shown in Fig. A-10. They are two terminal devices comprised of a coil of high-temperature-coefficient nickel wire sealed into an epoxy case and molded into a brass housing. The resistance of the wire changes as a function of temperature. The voltage drop across the sensor is monitored by the ECU. Sensor output is linear over the temperature range.

THROTTLE BODY - The primary function of the throttle body is to control engine air flow in response to driver commands. Although its specific configuration may vary from one application to the next, the general requirements imposed on it remain essentially the same. A typical throttle body is shown in Fig. A-11.

Main air flow is controlled by means of conventional butterfly valves, which are actuated by the throttle pedal. The number of throttle blades may vary, but two are generally preferred. The throttle body also provides the mounting for a throttle-position sensor, which is actuated by rotation of the throttle shaft.

Idle air is controlled by an idle stop on the main throttle blades, in combination with a bypass system having an adjustable orifice. The throttle-blade opening typically provides 60% of the idle air and the bypass control provides 40%.

Start-up air control is also incorporated into the throttle

Fig. A-11

body in the form of a fast-idle valve. This control provides increased air flow for starting, with automatic shutoff after start, and thus makes possible automatic foot-off-the-throttle engine starts.

Full-vacuum and/or ported-vacuum signals can be incorporated as needed for individual applications.

C345/87

The Ford central fuel injection system

M A HODGSON, BEng, AMIMechE
Ford Motor Company Limited, Basildon, Essex

SYNOPSIS

This paper presents an overview of the Ford central fuel injection (CFI) system, developed to ensure that vehicles meet standards similar to the 83 US Federal emission regulations, which are now being legislated in some European countries. The electronic hardware and aspects of the control strategy are described and certain peripheral advantages (in addition to emission control) are also discussed.

1. INTRODUCTION

Traditionally the fuel delivery system used in most European vehicles has been the carburettor, fuel injection being reserved for the up market, high performance models. For 1987 however Ford Motor Company will be marketing an injection system on 1.4 litre Escort, Orion and Fiesta vehicles for sale in Switzerland and Germany.(1)

The move away from the carburettor, which for years has proved a cheap and reliable method of controlling the fuel and air supply to the internal combustion engine, has arisen primarily because of new and more stringent exhaust emission regulations now being mandated in certain European countries. These new regulations are similar to those currently applied in the USA and have necessitated the development of fuel systems capable of delivering fuel with greater precision than hitherto achieved with the carburettor. The concept of throttle body injection (TBI) or central fuel injection (CFI) as it is sometimes referred to, is a system developed to meet the more exacting fuel metering requirements at a reasonable cost.

2. SYSTEM HARDWARE

The Ford 1.4 litre CFI system is based around the standard 1.4 `compound valve hemispherical head` (CVH) engine with the only change being a reduction in compression ratio from 9.5:1 to 8.5:1. The principal system components are the CFI unit and a microprocessor based electronic engine control unit (EECIV) with associated sensors and actuators. Development work demonstrated that the emissions and drivability objectives could be achieved without additional emissions control hardware, such as exhaust gas recirculation or secondary air systems. This is due to the favorable power to weight ratio of the vehicles and the fuel control that can be attained with the EECIV CFI system. The fuel control is based on a `speed density` method of air measurement (section 3) which does not require the use of a costly `vane air meter`.

2.1 The central fuel injection unit

The CFI unit is mounted on the inlet manifold in the same manner as a conventional carburettor and is of similar size and general appearance. The injector is mounted centrally in the single venturi upstream of the throttle plate. Integral with the unit is a fuel pressure regulator which maintains fuel pressure to the injector at 100 kN/m^2, thus this is a low pressure system compared to the 300 kN/m^2 or more pressure used in most multipoint injection systems. All the fuel supplied by the high flow electric pump is fed to the injector, the fraction not required for injection flowing back to the tank via the pressure regulator. This continuous fuel flow helps to prevent vapour formation and injector heating. The fuel required for combustion is injected by the EECIV management system applying a current controlled pulse to the injector, causing the pintle to open and emit a finely atomised fuel spray. The quantity of fuel injected is controlled by the time the injector is held open. Also integral to the CFI unit are an air charge temperature (ACT) sensor which provides air temperature data to the EECIV and a d.c. motor throttle plate actuator. The d.c. motor is controlled by the EECIV to modulate the throttle and hence to provide idle speed control (ISC) and other functions.

2.2 Electronic engine control

The EECIV is a commonly packaged unit fitted to several Ford vehicles. It consists of a custom designed 16 bit microprocessor, (2) a read only memory (ROM) based control strategy, with random access memory (RAM) for variable storage and power circuitry tailored to the application. The package, which is approximately 160 by 180 by 40 millimetres is mounted under the vehicle dash panel and is connected to the sensors and actuators via a sixty way plug and socket. On the CFI package the EECIV processes eleven input signals, (Fig 1) although it is capable of handling many more, these signals are:

C345/87 © IMechE 1987

Air charge temperature
Battery voltage
Self test demand
Engine coolant temperature
Engine speed and crankshaft position
Exhaust gas oxygen level
Idle tracking
Knock level
Manifold absolute pressure
Throttle position
Vehicle speed

The EECIV controls five outputs (Fig 1) these are to the:
D.C. motor throttle actuator
Fuel pump relay
Injector
Ignition coil (via an ignition module)
Diagnostics connector

3 CONTROL STRATEGY

3.1 Fuel control

As stated previously the need for sophisticated fuel delivery control arose because of new emission requirements. It is difficult to make a direct comparison between the 15.04 European and the current 83 US Federal standards, because of the difference in drive cycles. However, the severity of the US standards are such that different technology is necessary to comply. One way such an improvement could be accomplished in the time allowed was by the adoption of the main approach used for vehicles marketed in the USA, the use of a three way catalyst (TWC). This is the method of control which has been adopted for the 1.4 CFI system. The TWC is so called because under the correct conditions it can simultaneously assist the conversion of the three major pollutants, carbon monoxide (CO), unburnt hydrocarbons (HC) and oxides of nitrogen (NOX), by reduction to nitrogen and oxidation to water and carbon dioxide. However one of the necessary conditions is that the air to fuel ratio (AFR) of the combustible mixture should be stoichiometric, which is about 14.7:1 for air/gasoline. It is also necessary for adequate conversion of all three pollutants that the mixture be maintained to within 0.1 of an AFR from stoichiometric, (Fig 2) since rich combustion will only favour reduction reactions (NOX conversion) and lean combustion oxidation reactions (conversion of HC and CO). Maintaining a stoichiometric AFR within a tenth of an AFR is an impossible task for a conventional carburettor, hence the need for an injection system. Even though an injection system has the required precision to deliver the correct quantity of fuel, some system must exist to ascertain accurately what the correct quantity of fuel is. Practically this requires a means of measuring the mass of air entering the engine so the required fuel can be calculated and secondly, since the former can never be accomplished accurately enough, a means of measuring the deviation from stoichiometric to provide feedback to the fuel calculation. Moreover the system must have flexibility because it would not be satisfactory to always run at a stoichiometric AFR, for example a rich mixture will be required for cold engine and probably full load operation.

In the Ford CFI system the requirement of measuring the mass of air is met indirectly by the use of a `speed density calculation`. This is accomplished by processing data on the manifold pressure, air charge temperature, engine speed, engine volumetric efficiency and swept volume. In a simplified form this gives:

$$AMF = K*Eff*MAP*N/ACT \qquad (1)$$

where AMF is the air mass flow.
K is a constant depending on the capacity of the engine.
Eff is the volumetric efficiency, taken from a look up table of engine speed versus load. The table being a series of predetermined values derived by `mapping` engines on a dynamometer.
N is the engine speed.
ACT is the air temperature relative to absolute zero.

From the inferred air mass flow the EECIV calculates the required injector on time, based on programmed data on the injector flow characteristics, adjusting this time to compensate for injector nonlinearities, principally the on time required before fuel starts to flow, which varies with battery voltage. Normally this time will be the injection time to give a stoichiometric AFR, however if the throttle position sensor indicates that the throttle opening is large or the MAP value is high, then the injection time will be adjusted to give the AFR for best power. Similarly the injection time will be adjusted if the engine is cold or has just started to ensure the vehicle exhibits good drive characteristics. In a simplified form the fuel on time or pulse width is then given as:

$$PW = C*AMF*M/(14.7*N) \qquad (2)$$

where C is a constant depending on the injector characteristics.
M is a multiplier for enrichment.

Although the fueling control from the speed density calculation is generally better than a carburettor and can provide excellent full throttle and cold engine operation, it does not provide accurate enough AFR control for the TWC to function efficiently. Since the mass air flow is only calculated indirectly anything which affects the calculation, for example engine to engine variability of capacity and volumetric efficiency, sensor and injector inaccuracies, will affect the AFR obtained. To overcome this whenever the strategy requires operation at a stoichiometric AFR it uses feedback from a heated oxygen sensor mounted in the exhaust to adjust the fueling until a stoichiometric AFR is obtained. Effectively the exhaust gas oxygen sensor (HEGO) gives a high signal (0.8 volts) when the exhaust gas is resulting from rich combustion and a low signal (0 volts) when the combustion is lean, the system only resolving whether the combustion is rich or lean. Depending on the initial HEGO state the strategy makes incremental increasing or decreasing adjustments to the injector time (ramps) until the HEGO output switches, then alternately increases and decreases the pulse time so that the HEGO continuously switches states. Now however instead of ramping the fuel between switches the system attempts to predict the fueling required for midway between HEGO switches. On detecting a switch the fueling is changed in a single step to this midpoint and then ramped the rest of the way to the next HEGO switch, this `jumpback´ (Fig 3) providing a faster response than simply ramping between switches. Although at any one instant the fuel injected may not be providing a stoichiometric AFR (in fact it may vary by half an AFR), on the average it is exactly correct and the properties of the catalyst enable it to process the exhaust as if it were a homogeneous mixture resulting from stoichiometric combustion. (3)

It will be appreciated that any deviation between the AFR resulting from the speed density calculation and the stoichiometric AFR, will take a finite length of time to correct. Such a correction may be required every time the engine speed or load changes and may result in a momentary deviation from the stoichiometric AFR. To reduce the error between the calculated fuel and actual fuel required, an adaptive fuel strategy (4) is used. The adaptive system consists of a multiplier to the fuel pulse width equation (equation 2) and is taken from a speed versus load table contained in battery backed RAM. Initially, or if the vehicle battery is disconnected, all the multiplier values are one. Whenever the system is operating closed loop the correction required to achieve stoichiometric AFR is written to the corresponding part of the adaptive table. The next time the engine is operating at this particular speed and load point the speed density calculation is modified by the corresponding adaptive value giving a more accurate result. Moreover since the values in the adaptive table equate to the differences between that engine and the nominal engines on which the calibration was developed, they can be used to correct the fueling during open loop operation.

Even with the closed loop and adaptive fuel control the AFR ratio would fluctuate during transient operation, unless a system existed to compensate the fuel injected. The AFR fluctuations which occur during accelerations and decelerations are mainly due to manifold `wetting´ effects. Expressed simply at low manifold pressures high temperatures the manifold walls support a smaller equilibrium fuel film than at high manifold pressures and low temperatures. The observation is that on accelerations (low to high manifold pressures) some of the injected fuel goes into increasing the manifold fuel film to the higher equilibrium quantity, instead of into the combustion chamber, thus the engine momentarily runs lean. During decelerations the reverse effect occurs and the engine momentarily runs rich. The transient fuel strategy compensates for this by increasing or decreasing the calculated injection time after a change in manifold pressure has occurred. The amount of compensation and the rate at which it reduces after the start of the transient event, is calculated as a function of change in manifold pressure, air charge and coolant temperature and engine speed. Ideally if the system was fully compensated no fluctuation would occur during any transient though this proves difficult to achieve in practise across the complete load, speed and temperature operating range of the engine.

In addition to the transient fuel the strategy provides acceleration enrichment fuel as a function of rate of change and magnitude of throttle position. This is required because the transient fuel cannot provide compensation any faster than about fifty milliseconds, which is the time required to detect the transient and begin compensation. While this is fast enough to provide the catalyst with an on average stoichiometric exhaust gas it is not fast enough to prevent possible torque and power loss caused by the engine running momentarily lean. The drive problems that would result are prevented since the acceleration enrichment can be delivered within ten milliseconds of the transient occurring.

3.2 Spark advance control

Unlike a mechanical system the EECIV system can provide the optimum spark advance under all speed and load conditions. The EECIV receives data on the engine speed and crank position in the form of a square wave signal from a hall effect device mounted in the distributor, this is known as the profile ignition pickup or PIP signal. Each rising edge of this signal corresponds to 10 degrees BTDC for each cylinder, the strategy calculates the spark advance required and sends a similar square wave signal to the ignition module. The timing of the rising edge of this signal now corresponding to the advance required and the ignition module fires the coil on receiving this edge. Essentially the spark advance calculated is the minimum spark advance for best torque (MBT), retarded where necessary at full load to ensure that the advance is clear of detonation. Moreover since the MBT requirement is higher at lower coolant temperatures the spark advance is increased at colder engine temperatures to improve the drive response of the vehicle. If fuel quality problems are encountered the spark control system includes a `knock´ detector, this is a broadband vibration detector mounted on the engine block. The output of the sensor is conditioned by hardware circuitry within the EECIV to detect the magnitude of characteristic detonation frequencies occurring around TDC. If detonation is detected then the spark advance is retarded in steps until it discontinues, and then readvanced in smaller increments to the point which is just clear of detonation, or when the programmed advance is again reached.

3.3 Throttle actuation control

To perform an idle speed control function, assist in starting and improve the emission control the CFI unit is fitted with an idle speed control system (ISC). This consists of a d.c. motor which via a system of gearing moves a plunger to modulate the throttle to a limited extent (a maximum of about 3500 r/min no load). Integral to the plunger is an idle tracking switch so the system can detect whether the driver or d.c. motor is controlling the throttle. The system performs a variety of functions these include:

'Dashpot control'. On a deceleration the d.c. motor controls the throttle closing rate, preventing HC 'spikes' being generated. If the throttle were allowed to snap shut, then at high speed there might be insufficient intake mixture causing partial misfires.

Idle speed control. When the vehicle is stationary the d.c. motor is used to control the idle to the desired speed irrespective of engine loading. The desired idle speed is calculated as a function of the engine coolant temperature, so that for example the engine has a faster idle speed when cold.

Engine start control. While the engine is being cranked for a start the speed density equation is not used, rather the fuel is injected on each intake event, the quantity of fuel injected varying with coolant temperature. The d.c. motor is used to position the throttle during starts so that the air flow is optimal for the fuel being injected, this ensures that the engine can normally be started under all ambient conditions with no driver intervention.

3.4 Self test strategy

An important part of the control strategy is its self test capabilities. On the 1.4 litre CFI system the self test consists of dynamic and static 'on demand' tests and a continuous monitoring ability. For the on demand tests the self test input to the EECIV is grounded which causes the system to examine all sensor inputs for validity, reporting in coded form via the diagnostic output on any faults. During the dynamic part of the test the idle speed and fueling are cycled in a preprogrammed fashion and the operator is required to 'blip' the throttle, checks are then made that the system and sensors respond as predetermined. Since faults can be transitory in nature the system continually monitors all sensory inputs during normal operation, if a fault is detected a corresponding code is stored in the battery backed RAM for later retrieval. Some sensor failures might prevent the engine functioning, for example failure of the MAP sensor, hence the system is designed so that where possible replacement values are used that will keep the vehicle in operation. In the case of the MAP sensor failure the strategy estimates the manifold pressure from the throttle position. Substitutions are made in a similar manner for all sensor inputs, except for the PIP signal without which the system cannot operate. Even in the unlikely event that the microprocessor or strategy programme within the EECIV fails then the engine can still be operated provided the PIP signal is present. In this situation hardware circuits in the EECIV energise the fuel pump and pulse the injector for a predetermined time on the rising edge of each PIP signal from the distributor, similarly the ignition module fires the coil on the rising edge of the PIP signal giving a fixed ten degrees of spark advance. This hardware, limited operation strategy (HLOS) gives sufficient operation to drive the vehicle to, for example the nearest garage.

4 CFI SYSTEM ADVANTAGES

As stated previously, the CFI system has been developed to meet more stringent emission standards, however several additional advantages do result, particularly when compared to the carburettor.

4.1 Cold start and drive

The precise control of fueling, ignition timing, idle speed and cranking throttle position versus engine temperature, means that first time 'no touch' starts can be accomplished down to minus thirty degrees centigrade, and ensures excellent 'drivability' from this temperature. In this respect the CFI system performance is comparable to a multipoint injection systems and is generally considered to be superior to the quality of driveability achievable with the choke systems of carburettors. The only disadvantage compared to multipoint injection is that care must be taken with the manifold design, since liquid and vapour distribution effects can adversely affect performance.

4.2 High ambient operation

Unlike the traditional carburettor the CFI system does not suffer from fuel vaporisation problems leading to poor starts and drives after 'soaking' at high temperatures. The high flow fuel system design, where fuel is forced to circulate around the injector before returning to the tank, ensures that no vapour locks can form at the injector or in the fuel lines. In tests the CFI system often performed better than multipoint systems mainly because being manifold mounted the injector was further away from the engine mass, the main source of heat generation.

4.3 Cost and complexity

Compared to the modern carburettor the CFI unit is relatively simple containing only about ten per cent of the number of components. The major cost component of the unit is the injector and as a rough approximation the CFI system is cheaper than a multipoint by at least the cost of the extra injectors. Additionally the lower pressure fuel system means that the pump and other fuel system components are cheaper. A further advantage of the CFI is that the same basic unit can be fitted to a wide range of engines, thus eventually economies of scale can be realised.

5 CONCLUSION

The Ford central fuel injection system has been designed to meet more exacting emission requirements. It should also bring customer benefits in improved vehicle operation and performance.

REFERENCES

(1) SULLIVAN, J.A. European applications of Ford central fuel injection engine control system. SAE 840546, 1984.

(2) BREITZMAN, R.O., KERINS, J.H. Development of an optimal control custom microprocessor. SAE 820250, 1982.

(3) KANEKO, Y., KOBAYASHI, H., KOMAZONE, R. Effect of air-fuel modulation on conversion efficiency of three-way catalysts. SAE 780607, 1978.

(4) GAMBERG, E.M., HULS, T.A. Adaptive air/fuel control applied to a single point injection system for SI engines. SAE 841297, IMechE C444/84, 1984.

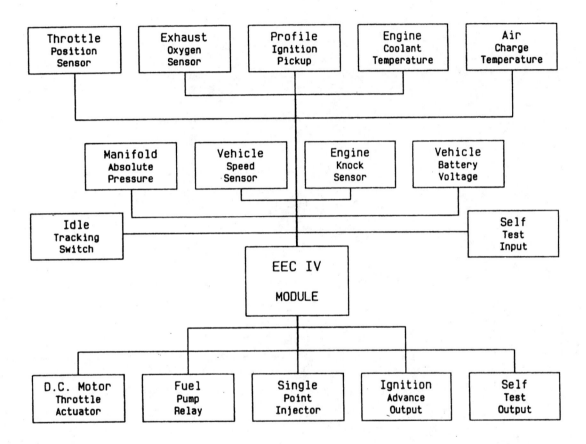

Fig 1

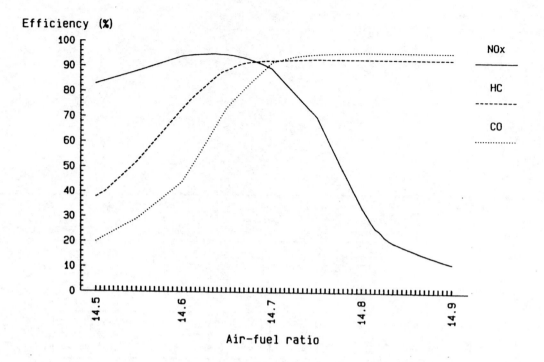

Fig 2

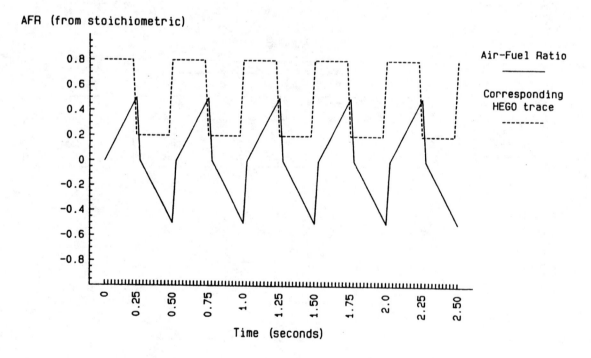

Fig 3

891330

Current Status of New Development in Electronic Fuel Injection System

Jai Chun Chang
Technical Center Bosch, Korea Branch, Korea

ABSTRACT

The engine management systems are using more and more electronic fuel injection system.
The main reasons for this are the more stringent requirements concerning lower exhaust gas emission, lower fuel consumption, better driveability, and improved performance.

The technical advantages of the electronic fuel injection compared with the conventional carburettor system are very well known.

- Exact measuring of the fuel under all operating conditions of the engine.

- Direct supply of the exactly measured fuel quantity to the inlet valve of the engine.

The advantages of the electronic fuel injection have led to a rapid increase in the production of the injection system. The two types of fuel injection systems being used are the multipoint injection (MPI) and the singlepoint injection (SPI). Besides of the Motronic system which will be introduced in this paper there are four other main systems (K-Jetronic, L-Jetronic, Motronic and Mono-Jetronic) which are currently being used in different projects.

THROUGH modern digital electronics, it is possible to optimize fuel metering together with ignition control and with an exact and fast registration of the engine condition and it is also possible to influence the mixture together with the ignition angle of the engine.

The injection duration and the ignition timing are registered in different characteristic maps and characteristic lines and are chosen and calculated from different operation parameters like engine-speed, intake air quantity and temperature, engine temperature, throttle-valve position, etc.

The best combustion in every speed load point is reached, when the injection duration and the ignition angle are optimum controlled.
Through the use of this new system the following advantages can be attained:
- Optimal combustion of precise measured fuel quantity.
- Fast reaction to sudden changes of the motor operation.
- Stable and fast motor specific adjustment.

CONSTRUCTION AND OPERATING PRINCIPLE BY MOTRONIC-SYSTEMS

A digital engine control system that combines injection and ignition, is called Motronic, which uses microcomputers to process engine-specific data on injected fuel quantity and spark advance.

After sensors deliver data to the microcomputer concerning the engine speed, the engine temperature, the crankshaft position, the intake air quantity, and air temperature, the computer calculates the optimum ignition timing and the optimum quantity of fuel to be injected.

Modern digital technology makes it possible to precisely match the quantity of the fuel injected and the ignition timing to the different operating conditions, such as idle, part load, full load, warm-up overrun, and transient modes.
In addition, a number of auxiliary functions can be realized, for instance overrun fuel cutoff, cold-start control and camshaft control, to name only a few.

Therefore, with Motronic it is possible to achieve reduction in fuel consumption up to 20% depending on peripheral conditions, driving cycle, and reference basis.

Diagram 1 shows the scheme of a Motronic-system.

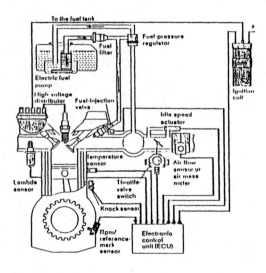

Diagram 1
Schematic of the Motronic

SCOPE OF FUNCTIONS

The scope of functions of a Motronic-system can be varied by using different steps of construction (Diagram 2).

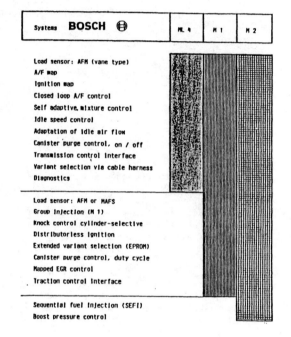

Diagram 2
Motronic-System
Scope of Functions

The selected scope of functions depends upon the request of the vehicle manufacturer.
To reach an optimum effect, one should consider the possibilities and requests of an engine control already in the development of new engines and vehicles.

IGNITION SYSTEMS

The Motronic uses a microcomputer to control the electronic ignition system, of which there are three types (Diagram 3).

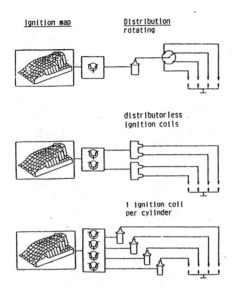

Diagram 3
Ignition Distribution

Ignition timing and ignition characteristics are influencing the operational behaviour of the engine.
The ignition point should be selected such that the following can be achieved:
- Maximum engine power
- Lower emissions
- Economical fuel consumption
- No engine knock

Therefore, it has to be optimized for all engine operating points.
The optimum ignition point depends upon a number of factors, in particular engine speed, load and design, fuel and operating conditions (e.g. starting), idling, and overrun (Diagram 4).

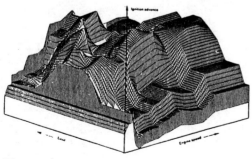

Diagram 4
Motronic Characteristic Map (1)

FUEL INJECTION SYSTEMS

Today's and future requirements to fuel metering systems call for complex system strategies, which cannot be realized without electronic control.

Fuel injection started as multipoint injection, where fuel is injected by one injector per cylinder. The timing of the injections is controlled as a function of the operation parameters of the engine.

One important feature is the sequential injection. The fuel is injected once per two revolutions for each cylinder individually, so that each cylinder gets the appropriate amount of fuel at the same time with reference to its cycle (Diagram 5).

Diagram 5
Sequential Injection

SIGNAL PROCESSING

The microcomputer calculates fuel-injection duration on the basis of the air-flow and engine speed signals and corrects the end stage with this calculated signal (Diagram 6).

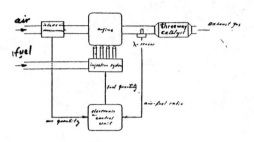

Diagram 7
Emission Control with Threeway Catalyst

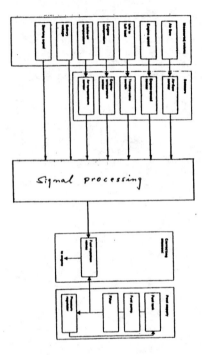

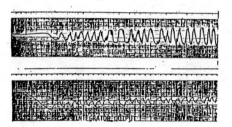

Diagram 8
Typical Initialization

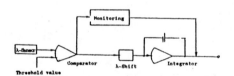

Diagram 6
Schematic of Fuel Metering

LAMBDA CLOSED-LOOP CONTROL

The lambda closed-loop control operates between $\lambda = 0.8 \ldots 1.2$, in which normal disturbances are compensated for by controlling λ to $1.00 \pm 1\%$ (Diagram 7).

The closed-loop control scans the signal of the lambda sensor and corrects the injection signal by a factor (Diagram 8).

As long as the signal "too lean" appears, the control permanently increases its correction factor (mixture gets richer) until the sensor signal reverses and indicates too rich. When the signal "too rich" appears then the factor is decreased permanently (mixture gets leaner) until the sensor signals "too lean" (Diagram 9, 10).

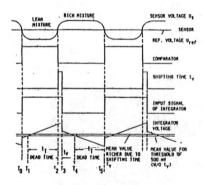

Diagram 9
λ-Feedback Control, Wave Forms

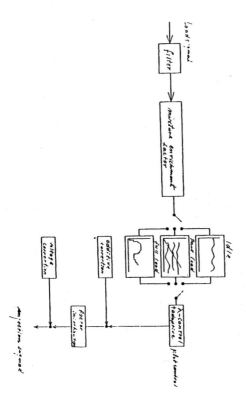

Diagram 10
Motronic with Lambda-Control Principle
Duration of Injection Calculation

CLOSED-LOOP IDLE-SPEED CONTROL

Sensor detects engine speed,
λ-temperature and throttle-valve
position. The electronic controller
compares the instantaneous engine
speed with the desired idle speed.
The idle-speed actuator is located in
an air passage which functions as a
throttle-valve bypass. The controller
sends a signal to the idle-speed
actuator, which supplies less air if
the engine speed is too high, and
more air if the engine speed is too
low (Diagram 11).

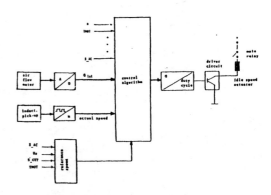

Diagram 11
Idle Speed Control System

FUTURE DEVELOPMENT TRENDS

- Camshaft control
- Swirl control
- Variable manifolds
- System communication
- Learning control
- Computer aided application
- Use of in-cylinder signals

CONCLUSION

The development status in electronic
fuel systems has been described.
These Motronic systems have reached a
high technical standard.
There is a continuous need for improvement
of the existing systems and components.
Limited through the still increasing
exhaust gas limit and the striving for
a minimum fuel consumption, the shares
of the fuel injection and the electronic
ignition in new engine developments are
still increasing in the coming years.

REFERENCES

(1) Bosch Technical Instruction Combined
    Ignition and Fuel-Injection System
    1987 72 2011

(2) Otto Gloecker, Heinrich Knap, and
    Hansjoerg Manger, "Present Status
    and Future Development of Gasoline
    Fuel Injection.Systems for Passenger
    Cars."
    SAE paper 800467

(3) Jai Chun Chang,
    Electronic Control of Fuel Injection
    and Ignition
    Korea Society of Automotive
    Engineers
    SAE Korea Science Lecture
    April 1988

From the Simple Carburetor to the Electronic Fuel Injection S.P.I. and M.P.F.I.

Silvério Bonfiglioli

Weber do Brasil Ind. Com. Ltda.

SUMMARY

The requirement to reduce the fuel consumption and the continuous pressure of legal standards upon the emission of pollutants produced by engines combustion, such as: HC (unburned hydrocarbons), CO (carbon monoxide) and NOx (nitrogen oxide), mainly in areas of heavy urban traffic, requests a constant technical and technological evolution of the carburetor.

The progression of legal limits upon the emission of pollutants, from the Phase I to Phase II in 1992, and then to Phase III in 1997 (U.S.A. 83 limits) is putting pressure upon automobile manufacturers to define, with their components manufacturers, increasingly more technologically advanced fuel systems, with higher quality and reliability standards.

Car performance, drivability and comfort are also continually improving, requiring increasingly more sophisticated technical solutions at competitive prices.

All this emphasizes the growing requirement for the use of electronics with a microprocessor to assist the carburetor or even as the heart of more complex systems with higher performance, such as S.P.I. and M.P.F.I. electronic fuel injection.

In order to meet these problems, Weber do Brasil, a company belonging to the Magneti Marelli Group, has for many years been engaged in the research and development of electronic systems, either with carburetor or fuel injection.

* The Twin Barrel carburetor with Fully Automatic Choke and simple electronic functions such as: fuel cut-offt during deceleration, idle speed control and mapped ignition, as an immediate low-cost solution, to comply with the limits imposed by Emissions Phase II, with a T.W.C. catalyzer open loop.

* A "Speed Density" closed loop digital fuel injection system with a Lambda Sensor and no catalyzer, integrated ignition, idle speed control and self-diagnosis as a medium- to long-term solution, as an alternative to the Phase II carburetor and replacing it at the beginning of Phase III.

Cheaper solutions, such as S.P.I. G7 Single Point Injection or ones with a higher performance such as the M.P.F.I , G7 Full Group or Sequential and Phased 0A, with or without knocking control, will eventually be chosen by the car manufacturer, optimizing the cost/performance ratio for each application.

Based on European production experience, started in 1982, Magneti Marelli/Weber do Brasil have developed modern technical solutions, with advanced technology, both for Gasohol (E22), and for Alcohol (E98), making the maximum use of the flexibility of digital electronics to obtain the required quality levels.

INTRODUCTION
Ladies and Gentlemen,

On-board electronics has been, for America as for Europe and Japan, a solution able to help the automobile to comply with the continuous requirements of legislation on the emission of polluting exhaust gases. It started with the feed-back carburetor via a Lambda Sensor and through central fuel injection (S.P.I.), it arrived at digital systems such as Sequential and Phased Multi Point (M.P.F.I) for normal or turbocharged engines.

More and more sofhisticated on board diagnosis and continuous reliability targets (OBD II) , have accelerated, in Europe, the integration of digital vehicle controls. The same will happen here in Brazil: from the present Phase II we shall rapidly pass to Phase III (U.S.A. '83 Emissions), with electronic vehicle control systems ever closer to worldwide standards in terms of technology, performance, reliability and price.

Considering the above circunstance, I will present how Weber do Brasil, a Company of the Magneti Marelli Group and Latin-American leader in carburetor production, is preparing its products for the Brazilian market.

1) THE ELECTRONIC CARBURETTOR

1.1 Idle Speed Control - ISC

By enhancing the simple Twin Barrel carburetor with more sophisticated electro-mechanical auxiliary devices, such as the fully automatic water heated choke, gasoline cutoff upon deceleration, automatic control of idle speed, and with the help of simple digital electronics, it is possible to comply with the limits imposed by Phase II with a T.W.C. catalyzer or Open Loop Oxidant. Figure 1 shows the 30/32 TLDE Weber electronic carburetor developed for engines of 1000 cc to 2000 cc.

It is possible to obtain a vehicle performance comparable to a Single Point Fuel Injection System at lower cost and improving the traditional carburetor on:

- Totally automatic starting, with no foot action;
- Reduced emission of hydrocarbons;
- Improved drivability when cold and at low speed;
- Reduced service frequency;
- Catalyzer protection;
- Reduced maintenance;
- Avoids wrong setting of a manual choke during cold start.

Figure 2 illustrates the block diagram of the ISC Digital Electronic Control Unit - CML 001. Besides controlling idling RPM, fuel cut-off and the throttle position during deceleration, it also performs an auto-diagnosis of the input and output signals, and is protected from short circuits to ground or to the battery positive terminal.

Figure 3 shows the experimental results obtained with a TLDE carburetor plus T.W.C. Catalyzer on 1800 cc and 2000 cc engines, in a complete USA 75 emissions cycle.

1.2 Mapped Digital Ignition with Knock Control - MICROPLEX

The use of mapped 3D digital ignition with the possibility of modifying the spark angle on the table as a function of the engine's knocking allows the engine to work at its maximum limit of efficiency, with an increased compression rate.

Advanced spark angles are calculate within the map, linearly interpolating (16 x 8) points and measuring: manifold pressure, engine RPM, battery voltage, engine temperature and idle switch

Microplex digital electronics sends the high voltage to provide the spark energy, recognizing the correct pair of cylinders, through two small-sized coils with no conventional mechanical distributor, (each coil attends to two cylinders, as in S.P.I. G7 digital fuel injection).

By combining the use of the 28/32 TLDF twin barrel carburetor with the strategies of ignition control, it is possible to obtain the best performance out of the engine, at the same time attaining the limits of emission of Phase II with a possible small C.O.C. Catalyzer. The Microplex system uses a diagnosis similar to digital electronic injection. In this manner it is possible to have, for small engines, an advanced technological system at a lower cost as compared to electronic injection, and capable of complying with the requirements of our market.

Figure 4 shows the Magneti Marelli / Weber MICROPLEX digital ignition distributorless system.

2) S.P.I. INJECTION DIGITAL SPEED - DENSITY - G7

Weber's G7 Single Point injection is a digital system "Speed Density" type where the quantity of air admitted by the engine is calculated as a function of its angular velocity, or its RPM, called "SPEED", and the air density, calculated by measuring the absolute pressure corrected by the air temperature, called "DENSITY".

The main function of this final injection system is to process, defining the correct air-fuel ratio under various engine requirement conditions, using a Lambda oxygen sensor in a closed loop, mounted on the exhaust system. In this manner, it complies with the requirements of Performance, Fuel Savings, and to the Emission Laws, without T.W.C. or C.O.C.

Fuel injection is done by a completely stainless steel "Bottom Feed" injector, mounted in the center of an aluminum throttle body, injecting in a manner that is phased and synchronous or linear asynchronous with top dead center. This throttle body is installed on the engine intake manifold in the place of the traditional carburetor. See the system layout G7 in Figure 5.

Another important function of the system is the Static Type Ignition Control (distributorless), where the spark is produced by two coils controlled by Power Modules inside the Electronic Control Unit and modified in accordance with the engine timing.

By means of an air passage by-passing the main throttle of the body, a stepper motor regulates the air flow required to maintain a constant idle speed, compensating variations produced by water temperature, air conditioning, lights, etc., permitting variations of, at most, 40 RPM in relation to the nominal speed.

An Electronic Control Unit has the function of processing the information on the state of the engine received by the Sensors and Potentiometers and actuate the Fuel Control Injector, the Idling Air Actuator and the Ignition Coils. See block diagram in Figure 6.

It is produced on S.M.D. technology and designed to be used inside the engine compartment.

The G7 injection system is supported by a Self-Diagnosis Strategy which analyzes input and output signals, identifying any possible hardware problem.

The software in turn enables a Recovery Strategy to use specific, constant parameters, memorized in the E2PROM to handle fault conditions.

At the same time, a visual signal via diagnosis lamp mounted on the car panel is given to the user, indicating a problem to be corrected.

Complete maintenance diagnosis is done through an electrical diagnosis connector cable and digital equipment outside the system, available at car manufacturer dealer and at Magneti Marelli / Weber Technical Assistance, since the beginning of 1992.

The G7 fuel injection system uses an Autoadaptive Strategy enabling the automatic correction of the main adjustable parameters such as Injection Time, Ignition Timing and Idle Speed whenever variations are identified such as: differences between production engines, aging of engines, fuel quality, etc.

A knocking sensor mounted on the engine permits a strategy of retarding the Ignition Timing whenever knocking is identified, avoiding damage to the engine.

At the same time the system also controls a Trip Computer, the Canister Valve and the Air Conditioning.

The Throttle Body used in the S.P.I. is shown in Figure 7.
This is a two-part structure in Aluminum with a surface protection of Chemical Nickel for use with gasohol and alcohol.

It is interchangeable with the carburetor and applicable on suction intake or supercharged engines of 40 to 110 HP DIN. Possible throttle bore are from 28 to 42 mm diameter.

The principal components, such as the stainless steel "Bottom Feed" injector, Viton Pressure Regulator, the Idle Speed Actuator and sensors are all from European technology, tropicalized and resistant to gasohol (E22) as well as to alcohol (E100).

All have been tested in accordance with the car manufacturer's requirements and also comply with our test standards of at least one year long in contact with alcohol or gasohol.

The G7 system, currently in production in the models for 1500 cc engines, uses a small fuel pump (38 x 115 mm) mounted on the fuel tank. The pump was developed with technology for alcohol (E100) and complies with the specifications of the systems, either S.P.I. or M.P.F.I.

3) M.P.F.I. DIGITAL SPEED DENSITY INJECTION - G7

The M.P.F.I. G7 system is an integral digital assembly developed for suction intake or turbocharged engines, with high performance and driving comfort, where it is possible to use an electronic fuel injection system at higher cost. As can be seen in Figure 8 it is a Speed Density type, uniting all the engine control functions such as: sequential or simultaneous injection in a Closed Loop via a heated Lambda Sensor, Static Ignition with a Knocking Sensor, Idle Speed Control, intelligent Air Conditioning Control, Canister, EGR, Active Self Diagnosis, and Trip Computer.

All the functions are self adapting; this is to obtain a better definition of the air-fuel ratio, better vehicle performance and exhaust control, maintaining these under control regardless of the life of the engine.

The G7 system uses only components compatible with alcohol : a Top Feed Injector in stainless steel with a fuel atomizer or Multi-Hole Plate Director for better corrosion resistance and elimination of gum deposit; a Fuel Pressure Regulator in Viton; a Fuel Gallery in nickel plated steel and a Fuel Pump mounted on the tank.

It is controlled by a Digital Electronic Unit with 64 KBytes of ROM memory, 4 KBytes of RAM memory and 1 KBytes of E2PROM, also used for self diagnosis, as Limp Home strategies, and during servicing as a connection to the Diagnosis Tester.

The G7 Electronic Unit is manufactured with S.M.D. technology and may be mounted in the engine compartment. When on board it has been designed to guarantee immunity from Electro Magnetic Interference (E.M.I.) better than 100 v/m, in the frequency range of 10 KHz to 1 GHz.

As in Figure 9, the M.P.F.I. Throttle Body is in Aluminum with a single barrel. It is manufactured under a technology permitting a large addition of components, such as a throttle potentiometer, an idling speed control actuator and a sensor for the temperature of air drawn into the engine. It is important to emphasize the particular law of progressive throttle opening, and its high resistance to engine oil, dust and corrosion.

FIGURE 1

FUEL AUTOMATIC CHOKE CARBURETOR WITH I.S.C. SYSTEM

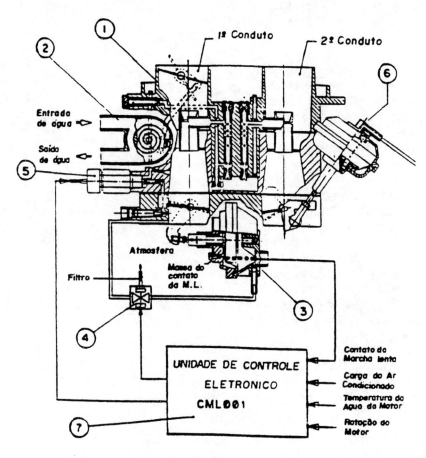

1. TLDE carburetor
2. Automatic choke
3. Idle speed actuator
4. Electro-pneumatic valve
5. Cut-off solenoid
6. 2nd barrel vacuum
7. Digital E.C.U.

FIGURE 2

BLOCK DIAGRAM

ELECTRONIC CONTROL UNIT C.M.L. 001

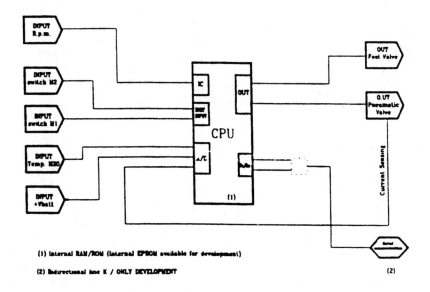

(1) Internal RAM/ROM (Internal EPROM available for development)

(2) Bidirectional line K / ONLY DEVELOPMENT

. Technology : thick film.
. Microprocessor : 68705 S 3
    3712 Bytes ROM
    104 Bytes RAM
. Inputs / OUtputs : Total nr. 11.
. R.F.I. imunity : 100 v / m.
. Engine compartment mounted.

FIGURE 3

EMISSIONS CONTROL - EPA 75

VEHICLES WITH T.W.C. AND 30/32 TLDE I.S.C. CARBURETOR

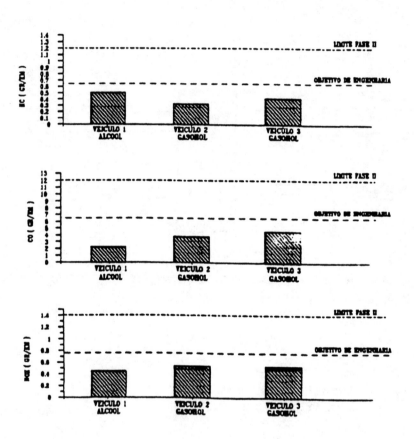

FIGURE 4

DISTRIBUTORLESS IGNITION SYSTEM MICROPLEX

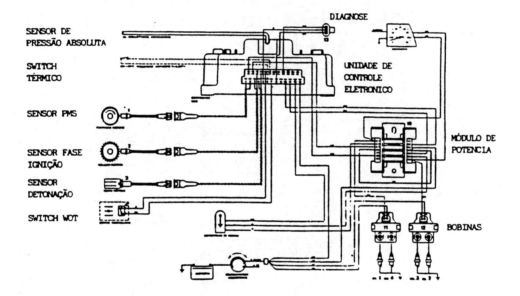

S.M.D. technology.
Engine compartment mounted.

FIGURE 5

SINGLE POINT INJECTION SYSTEM G7

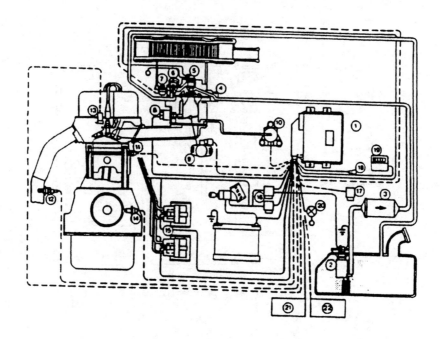

COMPONENT LIST

01. Microprocessor control unit
02. Fuel pump
03. Fuel filter
04. Throttle body
05. Bottom feed injector
06. Pressure regulator
07. Air temperature sensor
08. Idle speed actuator
09. Throttle position sensor
10. Absolute pressure sensor
11. Water temperature sensor
12. Oxygen sensor (lambda sensor)
13. Knock sensor
14. Engine RPM/Timing sensor
15. Ignition coils dual spark output
16. Relays
17. Trimmer
18. Diagnostic connector
19. On-board computer
20. Diagnostic lamp
21. Purge canister valve
22. Air conditioning A/C

FIGURE 6

BLOCK DIAGRAM

ELECTRONIC CONTROL UNIT G7

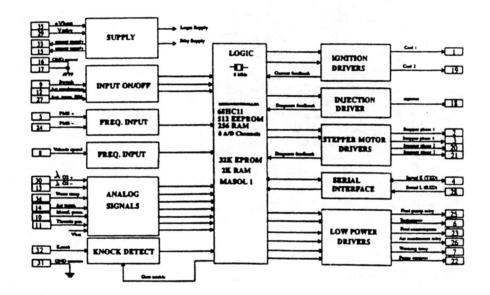

INPUTS

Analogic : 6
Frequency : 2
Digital : 2

OUTPUT

Digital : 8
P W M : 6

Custom circuits.
ROM memory size: 32 K or 48 K expanded.
Connector : 35 pins.

FIGURE 7

38 MB8 SINGLE POINT INJECTION THROTTLE BODY ASSEMBLY

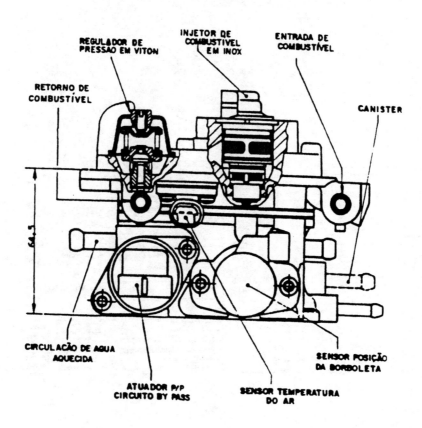

Alluminium Nickel Coated Body and Top Cover.
Full Stainless Steel injectors.
Linear stepper motor actuator.
Linear T.P.S.

FIGURE 8

MULTI POINT INJECTION SYSTEM G7

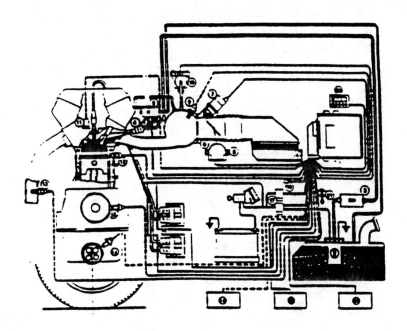

COMPONENT LIST

01. Microprocessor control unit
02. Fuel pump
03. Fuel filter
04. Top feed Injectors
05. Pressure regulator
06. Throttle body
07. Air valve for idle speed control
    (Stepper motor / V.A.E.)
08. Throttle position sensor
09. Air temperature sensor
10. Absolute pressure sensor
11. Knock sensor
12. Water temperature sensor
13. Oxygen sensor (lambda sensor)
14. Vehicle speed sensor
15. Engine RPM sensor
16. Nr. 2 ignition coils
17. Relays
18. Air conditioning input (A/C)
19. Diagnostic connector
20. Fuel consumption device
21. Diagnostic lamp
22. E.G.R. valve
23. Purge canister valve

FIGURE 9

MULTI POINT EVOLUTIVE THROTTLE BODY

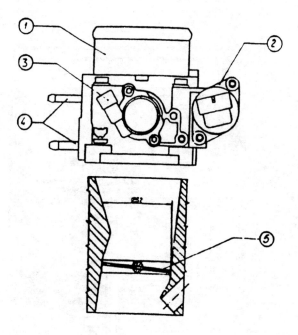

1. Body casting in Alluminium
2. Stepper motor by pass.
3. Throttle body sensor (T.P.S.)
4. Vacuum tube
5. Evolutive single barrel shape

950228
Development of The First Volume Production Thermoplastic Throttle Body

Simon Hughes
SU Automotive

Michael Welch
DuPont (U.K.) Ltd.

Gurradish Sanghera
Rover Group

ABSTRACT

With Automotive emission legislation become ever more stringent, fuel injection systems are rapidly replacing the carburettor. The demand for high volume fuel injection systems at a competitive cost has led to the development of a low cost thermoplastic throttle body.

This paper describes the intensive development programme undertaken to achieve the first high volume automotive thermoplastic throttle body.

INTRODUCTION

In 1989 Rover introduced the first derivatives of its new K series engine, family, which encompassed several features that were innovative for a high production volume engine.

The K16 1396cc engine has been subject to various phases of programmes that emphasise continuous product improvement allied with the reduction of cost, the introductions of plastic components have been a key part of these programmes with the largest contribution made by use of plastics for the throttle body.

From the outset of the project a multi-discipline core team was created, including Rover, SUA, DuPont and all component suppliers. This enabled vital contribution from all suppliers to ensure that all key project objectives were achieved.

OBJECTIVES - The team was given the following fundamental design objectives:
i. Provide a significant cost reduction over existing aluminium designs
ii. Be significantly lighter than the existing design
iii. Match or exceed the product quality / reliability of the existing aluminium designs
iv. Be compatible with current and future manifold technology
v. Low investment cost
vi. Be suitable for a wide potential range of engine capacities

Various design routes were considered to achieve these objectives with the final choice being to manufacture several components of the throttle body assembly by plastics injection moulding. These components are:
    Throttle Body Housing
    Throttle Disc
    Throttle Lever
    Throttle Potentiometer Mounting Adapter
    Stepper Motor Housing

This has enabled the targets to be met due to:

i. The reduction in the number of components in the design, together with the elimination of many matching, finishing and assembly operations, has reduced the product cost to approximately 65% of the current aluminium design.

A fundamental tool in this is the use by all parties of open book costing. This enables all involved to understand the cost courses and implications for every component and to optimise the design accordingly.

ii. The use of a Thermoplastic for many of the components has resulted in a weight reduction of 40% on the current aluminium design.

iii. The functional specifications of the plastic throttle body and idle speed control device to which they have been designed and validated, are identical to the current aluminium design.

However, due to the reduction in weight the throttle body and idle speed control device have proven to be much less susceptible to the effects of vibration. Also many of the design features which originally required complex assemblies can be simply produced as an integral part of the moulding eliminating many potential failure modes.

As a consequence the product quality and reliability are enhanced.

iv. Analysis of the investment cost of tooling the plastic throttle body in terms of moulding tooling versus die casting tooling, machine tooling and finishing equipment shows a 65% reduction in cost for plastic versus aluminium.

v. A 48mm diameter throttle body has proven successful for a range of engines from 75ps up to 150+ps without compromising performance or driveability characteristics.

DESIGN ROUTES

Having decided through the analysis described earlier to produce the major components in plastic, the team then had to make 3 fundamental design route choices;
i. Basic design and operating principle of throttle body and idle speed control device.
ii. Production technique of plastic components
iii. Choice of material for plastic components

DESIGN OF THROTTLE BODY / IDLE SPEED CONTROL DEVICE

The team was provided with detailed airflow specifications in terms of closed throttle leakage, idle airflow, throttle opening progression and idle speed airflow gain by the customer.

Particular attention was given to the closed throttle leakage and idle airflow setting to ensure that the device is suitable for use on small engines, or for low idle speeds or where other gas / air flow routes are required, e.g. exhaust gas recirculation or air assisted injection.

Several throttle body design concepts were investigated
i. An oval section throttle bore and butterfly disc
ii. A stepped oval section throttle bore and butterfly disc with one sealing surface of the throttle bore / disc being effected as a face edge seal.
iii. A conventional circular bore and butterfly disc at 0°
iv. A conventional circular bore and butterfly disc at 5°

The objectives / requirements against which they were appraised were;
i. To provide minimal air leakage past the disc
ii. To provide a smooth progressive increase of airflow, and hence driveability as the butterfly disc opened
iii. To be suitable for mass production moulding, assembly, setting and calibration techniques.

iv. To be suitable for use on MPI petrol engines.

The design solution which clearly provided the optimum in low air leakage and smooth airflow progression was the conventional circular bore with the butterfly disc at 5°.

A decision was made at an early stage to use an idle air bypass system to provide idle speed control, overrun manifold depression control and cold start airflow.

To simplify the overall design, save weight in a critical area and resultant effect on vibration it was decided to separate the idle speed control device from the throttle body and to mould it on the inlet manifold plenum connected by a hose.

This was done for the following reasons:

i. It simplified the overall thottle body design
ii. It creates a smaller, lighter, more compact and flexible unit than the system it replaces solving problems created by the tight packaging restraints imposed on modern engines.
iii. Locating inboard on the manifold reduces the mass located at the end of the manifold providing significant reductions in vibration levels and effect.

PRODUCTION TECHNIQUE FOR PLASTIC COMPONENTS

Due to the required level of component precision and the high volume production rates (initially planned at 200,000 per annum) decision analysis clearly identified the method of production to be injection moulding.

MATERIAL SELECTION

BODY HOUSING AND FLAP - Initial material selection was based on the following criteria:-

i. Engine bay temperature range of -40°C to 130°C
ii. Resistance to all fluids encountered in an engine bay i.e. fuel, oil, grease, anti-freeze and water etc.
iii. Ease of injection moulding with very low distortion resulting from crystallisation of material during injection moulding process.

Although amorphous thermoplastics give lowest mould shrinkage and hence distortion, they were rejected because of their poor solvent / fuel resistance.

From the semi crystalline thermoplastic available Polyamide 66 was selected because of its proven performance in underhood applications such as Inlet Manifolds, Fuel Rails, Camshaft Covers and Radiator Header Tanks.

From the range of Polyamide 66's commercially available, a 40% spheroidical mineral filled grade was chosen. This material offers a relatively low shrinkage rate of 1% which is uniform in all directions. Unfilled polyamide 66 has a shrinkage rate of 1.4%.

The bypass valve housing is also moulded in PA66 40%M because the same properties are needed for the housing and flap.

The lever arm of the throttle body is a mechanically loaded part having the throttle cable running in tension in a narrow groove. The higher mechanical properties of a 35% glass reinforced Polyamide 66 were selected for this application.

MOULDING PARAMETERS

The moulding parameters were reviewed at the outset of the development to assess their effect upon the dimensional variation of the part. The areas of concern were:-
- Resin flow pattern within the tool cavity
- The effect of the location of the gate/injection point had upon the resin flow
- The temperature profile within the tool
- The effect of injection and hold pressure

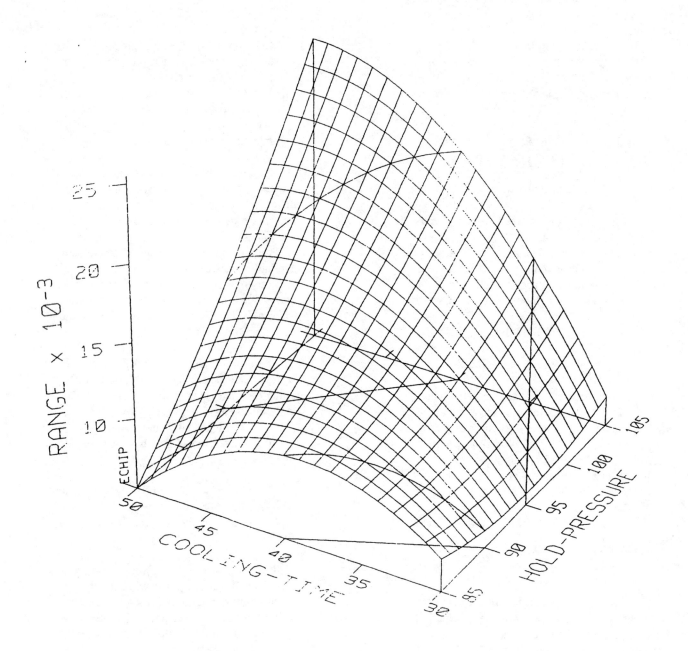

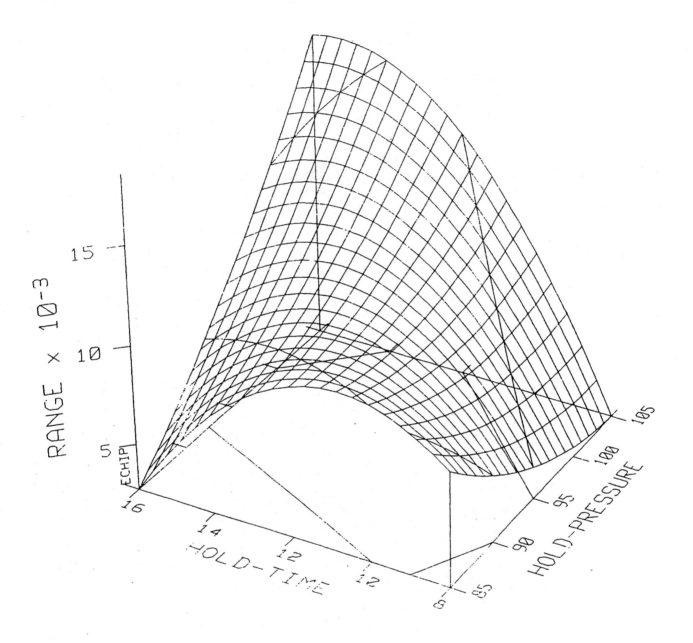

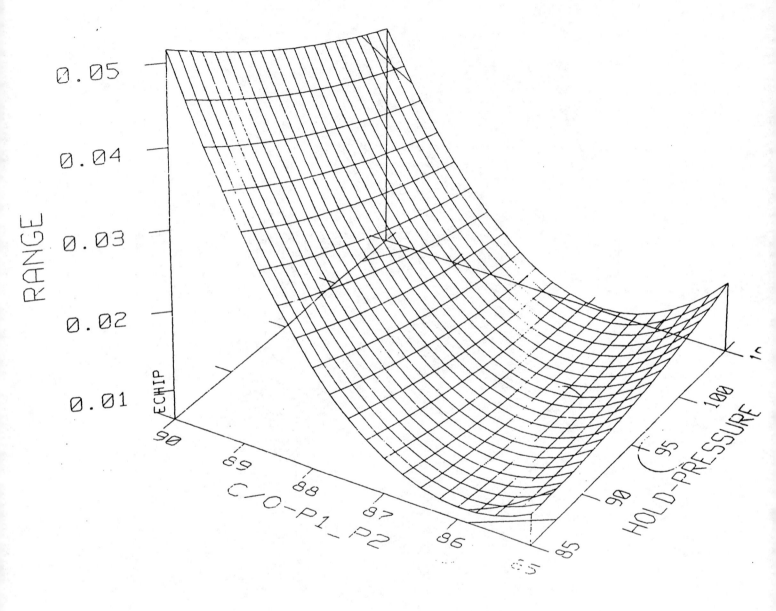

- The effect of the time of hold pressure
- The effect of the tool temperature

The first problem tackled was the type and number of gates/injection points into the tool cavity.

The ideal solution to give lowest distortion would have been a diaphragm gate. However, this would have required a secondary machine operation to remove it, which was rejected as too costly.

The compromise solution to avoid a diaphragm gate and give low distortion was a tool with 8 gates, this trim resulted in a low "out of round" condition.

The resulting resin flow and temperature profile of the development tool were then judged to be fixed parameters.

The next step in the tool development was a program of tool "sculpturing" to correct for the inherent "out of round" caused by melt flow and temperature differentials within the tool.

When this programme was complete the round parts were measured, the resulting bore dimension was then used as the reference point for the throttle body housing.

The next step in developing the commercial part was to address the dimensional variations caused by the moulding parameters, i.e. melt pressure, melt temperature, tool temperatures, pressure hold time etc.

Using Design of Experiments almost every combination of variables were examined versus the dimensional range seen on a large number of moulded parts. See typical charts, pages 4, 5, and 6.

Because of the required very low variation in dimension demanded by this application it was discovered that every moulding parameter had a significant effect upon dimensional range of this particular part.

The Design of Experiment results allowed the most robust processing window to be identified for the injection moulding process.

PRODUCT DEVELOPMENT

At each of the design / tool / process development stages the resulting assemblies were subjected to product validation and performance testing as detailed in that section of the paper.

The following areas are of particular note: Effect of temperature

The very nature of the operation of a throttle body creates significant temperature differentials.

Similar effects are seen on all throttle bodies although the extent of these effects will vary dependant upon the design and materials used in their construction and the engine installation.

Due to relatively high coefficients of expansion of plastics these effects were of particular significance.

Under conditions of thermal soak the engine radiates, convects and conducts heat with the result that under bonnet temperatures, including the throttle body can reach 130°C.

When the engine is started relatively cold air flows through the centre of the throttle body and over the butterfly disc causing rapid and uneven cooling.

Consequently, under these circumstances, for a short period of time the throttle bore shrinks less than the butterfly disc thus increasing the air leakage past the butterfly disc when it is closed.

Conversely, in cold climate conditions the system is required to function down to temperatures of -40°C. When the engine is started a flow of initially cold but then increasingly warm air flows through the throttle bore and over the butterfly disc causing it to expand more quickly than the bore. Under these conditions it is paramount that clearance between the throttle body / butterfly disc and hence freedom of throttle movement / closure is retained.

A considerable amount of work was undertaken by SUA to understand and quantify the effects of thermal growth and shrinkage and thus to be able to dimension, tolerance and specify the design so that freedom of throttle movement was permanent, whilst achieving low airflow leakage and minimal gain with high temperature.

These levels of gain were assessed against the range of control permissible from the idle speed control system and the effects upon the engine.

The net result was a design with a practicable minimum of airflow gain which falls well within the engine / idle speed control requirements and a minimal leakage specification of 0.3 - 1.0 cfm (0.6-2.0 kg 1hr)

Research and development on actual components and assemblies had been done by SUA to establish that where even lower leakage's / rates of gain are required for an application they are possible, through the use of other plastic materials.

EFFECTS OF HUMIDITY / MOISTURE - Due to the choice of PA66 as the thermoplastic material for this product a considerable amount of testing and product specification development was undertaken to understand, quantify and minimise the effect of humidity and moisture on the performance of the components and the assembly.

An initial problem was to find a realistic test technique as the conditions created in and the results obtained from artificial conditioning, i.e. with the use of humidity chambers, tended to be misleading.

The most accurate and effective method of test was found to be locating calibrated parts and assemblies over long periods in tropical and sub tropical environments and regularly checking the dimensional / flow calibration for shift. Parts and environments were similarly assessed on engines and under bonnet.

Once realistic and reliable data was available it was possible with the final product specification to detect a slight increase in leakage, especially under prolonged soak, e.g. storage, in a tropical environment. This is due to the butterfly disc absorbing moisture slightly faster that the body, because it has a higher ratio of surface area`` to mass than the body. With time the body absorbs the same amount of moisture and both parts then have the same size differential.

When fitted to an engine and operated after exposure to a tropical environment there was again a very small change in leakage as both parts gave up moisture as they adjusted to the lower levels of RH found under bonnet.

In sub-tropical environments the leakage changes due to storage or operation are minimal. The effects of water splash or immersion are in practice indiscernible.

The effects of moisture and humidity upon the final product specification fall well within control and operational requirements.

RESISTANCE TO THE EFFECTS OF VIBRATION - Due to the non-resonant nature of the material, prolonged exposure to high vibration levels such as those experienced on engines results in almost no wear or damage whatsoever.

RESISTANCE TO SALT SPRAY AND CORROSIVE FLUIDS - Again, due to the nature of the material the normally damaging effects caused by salt and corrosive fluids are virtually undetectable.

To provide idle speed control, overrun manifold depression control and cold start airflow the device employs a manifold mounted, idle air bypass system, connected to the throttle body via a hose.

Control is provided by a Sonceboz stepper motor / linear actuator with an airflow characterised pintle. This controls airflow through an orifice which is integrally moulded in the idle speed control housing. The pintle profile can be designed to meet the requirements of individual engines.

There is a Crankcase ventilation tapping upstream of the throttle disc to provide part and full throttle crankcase ventilation. (The closed throttle crankcase vent being located on the Inlet manifold.)

CONCLUSIONS

Through a combined team working approach the plastic throttle body had delivered the following for Rover:-

- large reduction in weight
- the ability to cater for a wide range of engine capacities from a single unit
- Functional benefits including vibration characteristics quality and reliability
- Providing circa 80% improvement in lifetime energy consumption and CO_2 production

1999-01-0171

A Comparison of Gasoline Direct Injection Systems and Discussion of Development Techniques

Mike Fry, Jason King and Carl White
Cosworth Technology Ltd.

Copyright © 1999 Society of Automotive Engineers, Inc.

ABSTRACT

An overview of work relating to the investigation of direct injection gasoline engines is presented.

Comparisons between conventional port-injected and two separate gasoline direct injection combustion systems are drawn. The comparison is between two of the main alternative GDI systems currently under consideration, reverse tumble charge motion with high-pressure swirl-type fuel injector (RTGDI) and low-pressure air-assist direct injection (AAGDI). This comparison was carried out using a 444cc single cylinder research engine of Cosworth Technology design. Included in the discussion are the influence of GDI systems on volumetric efficiency and performance at full load and fuel efficiency and emissions at part load.

Techniques to aid the development and calibration of GDI systems are discussed, including Moving Geometry CFD analysis and Design of Experiments (DOE). The development and application of these techniques is considered key to the successful development of GDI combustion systems.

Some design considerations for the practical implementation of direct injection gasoline concepts for inclusion in production programs are addressed.

1. INTRODUCTION

The motor industry is under a great deal of pressure to reduce fuel consumption and emissions from motor vehicles. For the near term, at least, there will be demand for both gasoline and diesel powered passenger cars. Gasoline direct injection offers a significant potential for improving the fuel economy and CO_2 emissions from spark ignited engines, whilst maintaining or improving the high specific power output of current MPI engines. In this respect the GDI engine promises a mixture of the benefits of both current gasoline and Diesel engines. Several types of GDI technology are reaching the stage of production feasibility and indeed some have entered production. The concepts so far presented differ significantly, and a clear leader has yet to emerge.

Most groups working on GDI systems are concentrating on a single concept and while comparisons are drawn between individual GDI systems and MPI, few direct comparisons between differing GDI concepts have been made in the literature.

The work presented here contains a comparison between two of the main alternative GDI systems currently under consideration, reverse tumble charge motion with high-pressure swirl-type fuel injector (RTGDI) and low-pressure air-assist direct injection (AAGDI).

In order to provide directly comparable results, as many features as possible were kept unchanged with each system considered. An MPI version of the same engine was also tested to provide a baseline.

2. COMPARISON OF MPI, RTGDI AND AAGDI USING A SINGLE-CYLINDER ENGINE.

A single cylinder engine was designed that was suitable for MPI and both types of GDI operation through the use of separate cylinder heads. Three configurations were tested.

1. Port Injection (MPI).
2. Reverse Tumble high-pressure GDI (RTGDI).
3. Air Assist low pressure GDI (AAGDI).

A description of the specification of each engine variant and the test facilities used is included as Appendix.1. The MPI and RTGDI engines were identical in most respects apart from fuel system and piston. The AAGDI engine required a different cylinder head that lead to differences in friction and VE that are addressed in the text. Previous published work, [including Noma et. al.], suggests that GDI engines can have compression ratios of 10% to 20% greater than comparable MPI engines due to beneficial knock characteristics. In this study the CR of all variants was the same at 10.5:1. This was selected in order to enable more direct comparison of results. Theory based on the Otto cycle suggests that increasing CR from 10.5:1 to 12:1 will increase efficiency by 3.9 % under non-knock limited operation. A practical benefit of nearer 3% might be expected if the CR of these GDI variants could be raised to 12:1. It may be possible to gain a benefit of this type with both GDI systems, and this should be borne in mind when drawing conclusions.

The comparison work included full and part load performance. The part load comparison took place under both homogenous and stratified operation.

The comparison is restricted to the engine speed range of 1000 to 3000 rpm.

3.0. RESULTS CORRECTION

In testing the single cylinder engine two areas were unrepresentative, relative to a production 4-cylinder engine:

1. During testing, high-pressure fuel supply for the RTGDI and compressed air supply for the AAGDI were externally provided. In practice these pump requirements would be driven from the crankshaft and so a small reduction in available torque and increase in brake specific emissions and fuel consumption would be experienced.
2. The single cylinder engine used, in common with many such devices, had higher specific friction levels than a multi-cylinder unit. The effect of this was to reduce maximum BMEP and to cause the specific emissions and fuel consumption figures to be worse than that which could be expected from a multi-cylinder engine.

The appropriate additional fuel pump/compressor load is different for each system as detailed below.

1. MPI - no correction for fuel pump load is appropriate. The standard low-pressure fuel pump has a modest electrical power (approx. 100W) requirement and a similar low-pressure fuel supply is required on both the RTGDI and AAGDI systems, hence this load has negligible effect in comparisons between systems.
2. RTGDI - The RTGDI system uses a high-pressure pump of cam-driven plunger design and this was driven by mains electric motor during the tests. The pump drive load was measured as equivalent to 2 kPa BMEP at 1000 rpm through to 3 kPa BMEP at 3000 rpm engine speed.
3. AAGDI - A 6.5 bar regulated air supply is required. This was provided by shop compressed air during the testing. It was assumed that a fixed volume positive displacement compressor, driven from the crankshaft would be used in practice. An estimate of the compressor load required was made as equivalent to -10.6 kPa BMEP. The derivation of this estimate is included as Appendix.2. With a more sophisticated air supply system on an engine, it would be possible to control supply to meet demand. This may allow some reduction in pump work, when air demand is low.

Over the 1000 to 3000 rpm range the engine friction measured was 20 kPa BMEP greater than that typical of a multi-cylinder production engine in the case of the MPI and RTGDI and 35 kPa greater in the case of the AAGDI, attributed to necessary differences in cam drive design.

The magnitude of the additional friction increase in the AAGDI was comparable to the estimated compressor load and so these affects tend to cancel one another and so were ignored. The fuel pump work compensation required in the case of the RTGDI was very small and so this was also ignored for simplicity. The overall effect of these factors being that the un-adjusted test bed results, as presented, closely indicate comparative affects that would be seen in practical vehicle engines, with the parasitic losses applied. The absolute performance, emissions and fuel economy levels are worse than those that might be expected from a practical multi-cylinder engine.

4. FULL LOAD PERFORMANCE

In comparisons between systems at full load, spark advance and injection timing were optimised for best torque (unless stated otherwise). An AFR of 12.5:1 was maintained at wide-open throttle, this being typical for a SI engine operating under full-power enrichment.

The full load performance is influenced by direct injection in two ways:

1. Volumetric efficiency (VE).
2. Brake specific air consumption (BSAC). – An expression of mass of air consumed per unit brake-energy generated.

The primary mechanism by which GDI systems can influence VE is through charge cooling. Latent heat of vaporisation of the fuel is removed from the air, increasing the charge density and hence the VE.

Under these full load conditions the theoretical maximum benefit to volumetric efficiency from the fuel charge cooling was estimated to be 6.5%. Other sources have quoted a similar (7%) value [Ref Andriesse & Ferrari]. The theoretical maximum benefit would be achieved when all fuel is fully vaporised, drawing its latent heat of vaporisation from the charge air, compared to a baseline where all the latent heat of vaporisation is drawn from the engine metal surfaces. Clearly in practice, the baseline MPI case will draw some latent heat from the charge, and some surface impaction of liquid fuel is inevitable in a practical GDI system. So the achievable benefit would be expected to be less than the theoretical maximum, and this was observed to be the case during testing, with benefits between 2% and 5% recorded. Under peak power operation, VE may be limited by the flow capability of the inlet ports and valves. Under these conditions a small additional improvement to VE may be achieved through GDI, in that the fuel is not required to pass through the inlet ports and so the ports are required to flow less mass for a given VE. This affect is not considered in the work presented.

The air assist injection system can have a further influence on VE, as it is possible to pump some air through the injector after Inlet Valve Closure (IVC) and so provide a small degree of supercharging. At a nominal 90% VE, the amount of air injected with the fuel represents 2.5% of the total air in the combustion system. This small but significant effect explains why the AAGDI system provides peak VE with the injection event straddling IVC, while the RTGDI engine achieves peak VE with earlier injection timing (see Fig.2.). In previous literature [Ref. Yang & Anderson] it has been suggested that, with a high-pressure injection system such as the RTGDI, peak VE occurs with injection late in the induction stroke due to heat transfer from the cylinder walls to charge. The argument being that a cooled charge will draw more heat from the walls, and so if the charge is cooled by the injection of fuel early in the stroke subsequent charge heating will increase and the VE benefit will be lost. This may well be an influence, but also early injection, with a high penetration rate, will lead to piston impingement and subsequent loss of charge cooling. This is thought to be the larger effect.

The AAGDI performance may be improved further using a strategy of early fuel injection, for homogenous charge and a second air-only injection after IVC to boost VE. This hypothesis was not confirmed at the time of writing, due to EMS limitations. It may also be necessary to increase compressor capacity in order to exploit any benefits that may be gained in this way.

The full load performance achieved by the three systems investigated is shown in Fig.1.

The RTGDI system improved VE by 2 to 4% over the MPI, due to charge cooling.

For full load operation, The AAGDI system produces finer droplets (8μm SMD) [Ref. Houston & Newmann] than the RTGDI system (around 25μm SMD) and has a lower penetration rate. These characteristics are likely to reduce wall and piston wetting and hence more latent heat of vaporisation is drawn from the charge. For this reason, and the supercharging affect previously described, the AAGDI was expected to give a higher VE improvement than the RTGDI. However, no additional benefit was seen at 1000rpm and the VE at 3000rpm was actually lower than the MPI case. A cycle simulation model (see section 11) was utilised in order to explain these results. The inlet and exhaust system of the AAGDI case was different to that of the RTGDI / MPI configuration. Differences were kept to a minimum but practical considerations meant that some differences were inevitable. The cycle simulation model indicated that exhaust tuning was the primary reason for the low volumetric efficiencies seen. This was due to an increased exhaust length between the exhaust port and the first reflective volume. Changes to the induction tuning and cylinder head cooling had secondary effects.

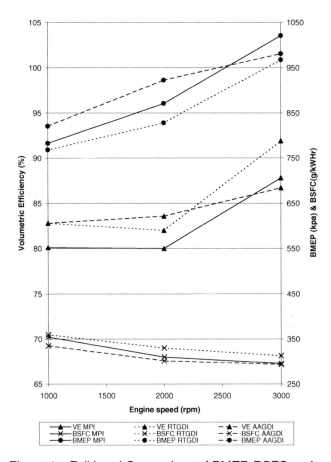

Figure 1. Full Load Comparison of BMEP, BSFC and VE.

Whilst both GDI systems demonstrated some improvements in VE, the same was not true of torque output. The AAGDI system produced improvements in torque more or less equal to the VE benefits. The RTGDI system produced reductions in torque output, reflecting an increased BSAC. Increased O_2 levels in the exhaust gas, and increased exhaust temperatures suggest that the RTGDI system was exhibiting a degree of charge stratification and associated incomplete/slow combustion that was reducing torque output.

The reason for this stratification affect is not clear but it is likely that injector/charge interaction was somewhat compromised in a system primarily optimised for part-load stratified operation. The injector location was selected analytically prior to the start of engine testing and it was not practical to carry out full optimisation within the scope of this project. Predictions of wall wetting when employing a side injector in the RTGDI may explain the difficulty of achieving good BSAC at full load. A more upright (relative to cylinder axis) injector spray would probably improve the full load performance. Other work presented [Ref Ando et al] states that VE benefits can be translated into torque with RTGDI systems and so it is likely that this could have been achieved in this case with further optimisation of injector location.

At 2000 rpm, optimum End of Injection (EOI) timing for best torque was achieved at 225° BTDC EOI for the RTGDI and at 130° BTDC EOI in the case of the AAGDI system. With both systems, as EOI was reduced from the optimum, torque reduced while VE was at least maintained or rose further. This is due to significant increases in charge stratification associated with the reduced charge formation time available at later EOI's. This influence caused the optimum EOI for best torque to increase with engine speed.

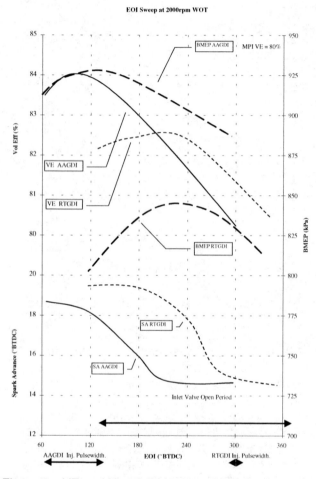

Figure 2. VE and Knock limited spark advance as a function of End of Injection (EOI) timing.

Knock characteristics are also effected by the charge cooling mechanism of the direct injection fuel spray (see Fig.2.). With both variants of GDI, the knock limited spark advance increased as the VE increased with variation in injection timing, from a value comparable to the MPI system at early EOI timings. This is explained as the same reduction in charge temperature that provides the VE effect will also reduce final compression temperature and so improve resistance to knock. Theory suggests that the final compression temperature will be reduced by about 2°C for each °C reduction in charge temperature, assuming an adiabatic and polytropic compression process. As the VE is being influenced through changes in charge temperature and not pressure, the final compression pressure will be largely unaffected.

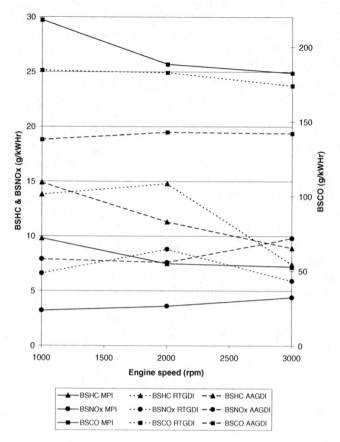

Figure 3. Full Load Comparison of Emissions

The full load emissions performance of the systems is included as Fig.3. Both GDI systems provide similar and significant increases in NOx and HC emissions in comparison with the MPI system. The increased NOx is likely to be due in part to some degree of stratification even in the nominally homogenous charge, in the case of the AAGDI system the higher achieved engine load will also be an influence. The increase in HC emissions is most likely the result of a combination of stratification and wall/piston impaction of fuel, leaving more fuel in regions where flame extinction will prevent burning. WOT CO brake specific emissions were reduced from the baseline by both GDI systems. These reductions in brake specific emissions were due to reductions in exhaust gas concentration and in the case of the AAGDI, to increases in engine torque. At first, these reductions seem strange as any degree of stratification under $\lambda<1$ conditions would lead to increased CO production at the richest combustion sites. The reduction seen could be the result of selective post flame oxidation of CO in preference to hydrocarbons. CO produced in richer regions will mix with O_2-rich gas from leaner regions and areas of flame extinction during expansion, these oxidise in preference to the less reactive hydrocarbons. By this mechanism the same degree of stratification that can increase HC and NOx production could cause the reductions in CO concentration recorded.

Smoke levels were below 0.2 BSU for all of the full load points presented.

5. PART LOAD PERFORMANCE

In comparisons between systems at part load, unless otherwise stated, spark advance and injection timing were optimised for best BSFC. The following three steady state part load speed/load sites were selected with the 20 kPa BMEP correction for the higher engine friction applied, in order to achieve indicated loads equivalent to a representative multi-cylinder unit:-

Table 1. Part Load Test Points

Nominal multi-cylinder Operating point	Equivalent single-cylinder test point
1000 rpm, 70 kPa BMEP	1000 rpm, 50 kPa BMEP
2000 rpm, 200 kPa BMEP	2000 rpm, 180 kPa BMEP
3000 rpm, 500 kPa BMEP	3000 rpm, 480 kPa BMEP

Table 2. shows direct comparisons between the three tested systems at the part load sites. In the comparisons drawn, the greatest emphasis was placed on the 2000 rpm test site. This is an industry standard reference point, typical of those encountered under cruise conditions in emissions tests and so was of primary interest. The comparative MPI points were taken at stoichiometric AFR as it was assumed that this system would operate in conjunction with a conventional three-way catalyst. However, the authors accept that optimum BSFC at part load would probably be achieved with a lean homogenous EGR tolerant combustion system and, as with GDI engines, reliance on the development of an affective lean NOx catalyst for stringent emissions legislation compliance.

Fig.4. shows specific fuel consumption of the three systems at the three tested part load sites. Comparison at the 2000rpm part-load site shows significant Brake Specific Fuel consumption improvements of approx. 9% and 15% for the RTGDI and AAGDI systems respectively over the conventional MPI system with EGR. This is primarily due to the increase in manifold pressure and subsequent reduction in pumping losses, achieved by lean stratified operation with late injection timings and high levels of both EGR and AFR. A reduction in combustion chamber heat losses due to the smaller temperature differential between the outer burned gases and wall also contributes to improved BSFC. A further small contribution to cycle efficiency is made through a raised γ due to lean operation.

Table 2. Comparison of systems at 2000 RPM, 180 kPa Part Load Test Point (optimised for BSFC)

	MPI	RTGDI	AAGDI
Torque (Nm)	6.3	6.3	6.4
BMEP (kPa)	177	179	183
SMOKE BSU	-	-	0.1
AFR	14.5	29.0	28.6
EGR %	10	21.7	21.3
EOI °BTDC F	-	77	20
SOI °BTDC F	-	84	97
EGT °C	562	341	344
BSFC g/kWh (without pumps)	459	420	397
SPARK ADV °BTDC	34	60	25
HC ppmC	3657	6372	3504
NOx ppm	282	745	668
CO %	0.4	0.3	0.13
CO2 %	14.4	6.4	6.9
O2 %	0.9	11.7	11.1
BSNOx g/kWh	2.4	13.9	10.2
BSHC g/kWh	11.0	36.6	18.8
BSCO g/kWh	24.7	31.0	14.5

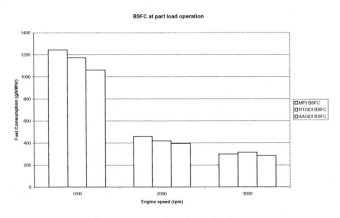

Figure 4. BSFC at the part load operating points

Both GDI combustion systems achieved best BSFC at the 2000rpm site with 29:1 AFR and 21.5% EGR added. The MPI combustion system was limited to 10% external EGR by rapidly degrading combustion stability beyond this value. At 1000 rpm the optimum BSFC point for the MPI system was at 0% EGR. In this configuration the engine had poor EGR tolerance, becoming unstable with more than 5% external EGR added. Equal sized inlet and exhaust valves have been shown by engine cycle simulation to result in higher levels of internal EGR at low engine speed, and thus a low external EGR tolerance despite the minimal valve overlap. These larger than desirable exhaust valves were inherited along with the cylinder head from a previous direct injection project. The change of predicted residuals and measured external EGR tolerance limit for the test points is included as Table 3.

Table 3. Residuals and external EGR at part-load points (MPI).

Engine Part-load Test Point (rpm)	Residuals (%) From cycle simulation	External EGR Tolerance (%) Experimental
1000	22.7	6.6
2000	13.3	13.0
3000	7.5	22.0

In terms of combustion characteristics, peak cylinder pressure is influenced by the mass in the cylinder, and the timing and nature of heat release during combustion. Differences between measured combustion characteristic of the three systems are shown in table 4.

Table 4. Comparison of combustion characteristics @ 2000 rpm, 180 kPa.

	MPI	RTGDI	AAGDI
Pmax (bar)	13.6	23.9	32.1
APmax (°ATDC)	12.6	10.0	7.2
Rmax (bar/°)	0.28	1.04	1.59
ARmax (°ATDC)	-6.1	-2.6	-2.1
Ign Delay SA to 10%	39.5	58.8	17.0
Burn Dur 10 – 90%	46.0	40.3	18.1
Spark Adv °BTDC	34	60	25

The higher maximum cylinder pressure of the RTGDI system over the MPI system is largely a function of the increased cylinder mass through less-throttled stratified operation. However, for the same AFR and EGR levels, the AAGDI system has an even higher peak cylinder pressure. It is obvious from looking at the other combustion characteristics, particularly ignition delay and 10 – 90% burn time, that the heat release characteristics of the two GDI systems are significantly different, and this results in the difference in maximum cylinder pressures.

Differences in charge motion and chamber geometry between the MPI, RTGDI and AAGDI systems will have a significant affect on burn duration through influences on fuel preparation and scheduling of turbulence generation. For conventional MPI engines, the bulk tumbling motion is broken down to small-scale turbulence near TDC by deformation of the flow field in high aspect ratio flat combustion chambers. High levels of turbulence intensity and a reduced bulk flow velocity is desirable at the spark plug gap just before initiation of combustion, and will assist in the tolerance of EGR in the combustion area.

The RTGDI combustion system utilises reverse tumble flow which, through conservation of angular momentum in the low aspect ratio chamber, maintains bulk motion throughout the compression stroke. For the single fluid GDI engines with late injection for stratified mixture generation; mixing rate control via high levels of bulk motion, and lower levels of small-scale turbulence are required in order to maintain stable stratification. A compromise between generating sufficient levels of bulk motion and the scheduling of turbulence generation to assist in increasing the flame speed of lean stratified mixtures must be achieved. The significant squish flow from the exhaust side into the chamber as the piston approaches TDC will assist in turbulence generation.

At the optimum fuel injection timing of 77°EOI the RTGDI engine is subject to a very long ignition delay and thus requires a very advanced 60°BTDC MBT spark timing in order to ensure sufficient levels of engine stability and optimum heat release. This ignition delay is a function of ignition timing and is largely independent of EGR level (see Fig.5). The delay characteristics are such that the 10% burn time remains around the TDC to 20°BTDC range, over wide variations in EGR and spark timing. Any ignition reactions initiated in the period before 20°BTDC progress at a very slow rate, and increase in speed rapidly around 20°BTDC as burn commences. This characteristic indicates that the conditions in-cylinder are not conducive to flame initiation and propagation until this crank angle. At around this crank position, squish flow from the exhaust side is entering the bowl and it is likely that the interaction of squish flow and bulk motion plays a part in allowing flame propagation to commence. Prior to this the charge into which the spark energy has been released has been seeded with pre-combustion reactions and the rich local AFR and low levels of small-scale turbulence prevent significant flame development. The amount of time available after spark and before flame development affects the extent of the pre-burn reactions and so affects the flame development that takes place after 20°BTDC. In this fashion varying spark timing influences burn characteristics and still provides a degree of spark sensitivity, with earlier timings required for best BSFC at higher levels of EGR (see Fig.6).

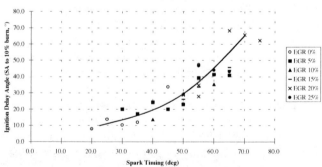

Figure 5. Ignition delay as a function of Spark timing RTGDI @ 2000 rpm 180 kPa BMEP

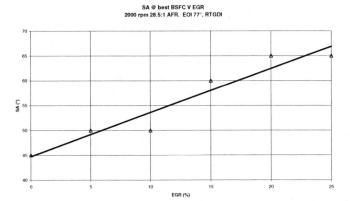

Figure 6. Spark advance for best BSFC V EGR. RTGDI @ 2000 rpm 180 kPa BMEP

The ignition delay for the MPI system is almost 20° shorter than the RTGDI system, but still quite long by conventional MPI standards. Unlike the RTGDI engine, the ignition delay is largely a function of the high levels of EGR concentration within the combustible charge, resulting in a significant reduction in flame speed. Slow flame propagation throughout the chamber may also be compounded by lower levels of small-scale turbulence due to early break down of bulk charge motion within this combustion system.

Once combustion has been initiated, it can be seen from table 4 that the 10 – 90% burn duration times of the MPI and RTGDI systems are similar, but are significantly longer than the AAGDI system. Factors affecting rate of heat release in the MPI system remain the influence of high levels of EGR and lower levels of micro turbulence on flame velocity. For the RTGDI system the global AFR and EGR levels are much higher than for the MPI engine. However, entrained levels of EGR within the stratified charge will be better tolerated due to richer local AFR in the reaction region. Whilst this rich local AFR will result in a slow flame speed propagation, the flame front has to travel a smaller distance than in a conventional MPI chamber. Higher levels of small scale turbulence than the MPI system may also exist later in the combustion process, as the combustion chamber geometry is able to maintain increased levels of bulk motion later in the compression stroke, which interact with the squish flow close to TDC. The RTGDI system may also have slightly accelerated burn characteristics compared to the MPI system due to reduced heat losses associated with stratified combustion. This series of influences result in comparable 10% - 90% burn durations for the RTGDI and MPI.

Previously published work [Ref. Fraidl et al] has identified a mechanism by which reverse squish flow during early combustion can destroy stratification in a RTGDI type system. This was addressed in the previous work by initiating the heat release early to allow stable combustion, at the expense of some cycle efficiency. This mechanism appears not to significantly affect this system as the burn duration occurs mainly after TDC. Combustion stability was also comparable to the MPI. One reason why the RTGDI engine used here did not exhibit the same characteristics as those earlier presented, may be due to a significantly greater 10–90% burn duration. This was probably caused by lower overall tumble motion.

The AAGDI combustion chamber utilises a conventional forward tumble port arrangement with a re-entrant bowl-in-piston. The bowl-in-piston chamber geometry is designed to contain the axially delivered fuel/air charge; this is important with this system as the nature of the air assist fuel injection enables good charge preparation to be maintained through to injection timings as late as 20°EOI BTDC. Squish flow from around the centrally located bowl further enhances charge containment and turbulence generation as the piston approaches TDC Measurements showed that cylinder pressure exceeds injector air pressure at 31°BTDC for the 2000 rpm part load case. Some back-flow into the injector must occur that will result in a slight reduction in charge mass. This reverse flow should not degrade mixture preparation unduly as the bulk of the fuel will be delivered earlier in the pulse width before the reverse flow occurs. The 6.5 ms pulse width utilised in testing equates to 78° crank rotation at 2000 rpm. In practice a shorter air pulse-width could be used at an appropriate earlier EOI timing to prevent the reverse injector flow whilst maintaining the timing of the main fuel delivery. This was thought to be a minor effect and was not investigated further.

The ignition delay and burn duration of the AAGDI engine are very short when compared with the other two systems. This is due to the late injection enabling good stratified mixture preparation, with the air/fuel ratio around the plug being slightly rich of stoichiometric just prior to ignition. In addition to this, entrained levels of EGR within the ignitable mixture are lower than the overall chamber mean as a result of providing air with the fuel during injection. This has been demonstrated by the fact that varying EGR levels has only a minimal affect on burn duration with the AAGDI system. As with the RTGDI engine, the flame front also has to travel a smaller distance than in a conventional MPI chamber, resulting in rapid heat release. The 10% mass fraction burn occurs at 8° BTDC with 90% reached at only 10°ATDC, and is thus too advanced for optimum thermodynamic heat release. Retarding the spark timing succeeds in improving the thermodynamic heat release characteristics, however, engine instability also increases dramatically. In this case it is likely that reverse squish flow as previously discussed in relation to the RTGDI is having an influence on the AAGDI performance. The AAGDI system is more likely to suffer this effect due to its fast heat release rate and comparatively large effective squish areas into which charge can be driven in advance of the flame front. This effect reduces stratification and increases cycle-to-cycle variability. The result is that the spark is required to be advanced beyond the optimum from a cycle efficiency viewpoint, in order to achieve stable combustion. This behavior is likely to be associated with the combustion chamber geometry, rather than a specific air-assisted fuel system characteristic.

6. PART LOAD EMISSIONS

Details of engine out emissions for the three systems are included in table 2.

In terms of unburned hydrocarbon emissions concentration; the AAGDI and MPI engines generate similar levels, and almost half the level of the RTGDI system. The uHC formation mechanisms for homogenous combustion of flame quenching at the chamber walls, crevice volumes and oil layer sources, and partial burn of some engine cycles, apply for the MPI system. Levels of uHC are also typical of values seen in conventional engines at part load with high levels of EGR applied.

The uHC formation mechanisms for the AAGDI system differ slightly from the MPI system. Flame quenching in this case will occur at the boundary of the stratified region, where the AFR exceeds the lean flammability limit of gasoline, preventing further propagation of the flame into excessively lean areas. Loss of unburned charge through reverse squish flow, which is assisted by the increased chamber pressure caused by the progressing flame front, will also reduce the definition of the stratified mixture leading to an increase in uHC. It is also likely that a degree of fuel spray adherence to the bottom of the bowl will result in some diffusion combustion from the piston surface, and any effect of partial burning of some engine cycles will still remain.

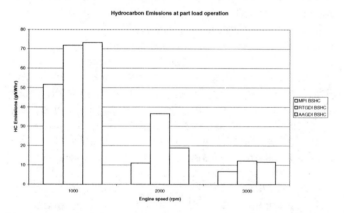

Figure 7. Specific uHC emissions at part load test sites

For the RTGDI engine, uHC formation mechanisms will include those that effect the AAGDI engine. However, due to the increased injection pressures, and the associated higher fuel penetration levels, increased wall wetting and piston top impingement will occur. Liquid droplet combustion of larger fuel particles resulting from reduced mixture preparation time and the fuel spray characteristics at the end of injection will also increase engine out uHC levels. It has been reported that in stratified combustion events with large combustion delays, as in the case of the RTGDI, marked increases in uHC emissions may be expected. This effect being due to a loss of stratification during the long ignition delay period. [Ref. Frank & Heywood]. The inherently lower exhaust gas temperature of the GDI engines operating in a stratified mode may also inhibit post flame oxidation of unburned hydrocarbons. A comparison between specific hydrocarbon emissions of the three systems is shown in Fig 7. At the 2000 rpm point it can be seen that although the AAGDI engine produces the lowest concentrations of engine out uHC, the higher exhaust mass flow due to un-throttled part load operation results in higher specific uHC emissions in comparison with the MPI. This effect applies to any system operating at a very lean/stratified condition.

In relation to smoke generation, which has been reported as a concern in GDI engines, AFR sweeps at part load have been performed. With EOI timing of 77° and 20°BTDC for the RTGDI and AAGDI systems respectively, it was demonstrated that negligible smoke emissions would be produced from a correctly calibrated engine of either fuel system. The results of the AFR sweep can be seen in Fig.8. Obviously, with a highly stratified charge, the smoke increase at a global AFR of less than 15:1 for the RTGDI system, and 13:1 for the AAGDI engine means that the AFR in the rich area around the spark plug will be significantly lower than this. This would be an area of the engine operating map that should never be utilised. In terms of absolute particulate mass it is likely that levels at part load will exceed those of a conventional MPI engine, but should still be lower than a common rail equipped HSDI Diesel.

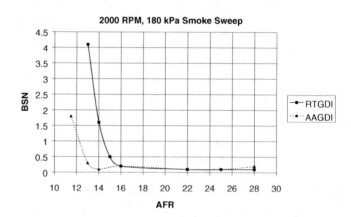

Figure 8. Part Load Smoke Behavior

A comparison of NOx emissions shows that both GDI engines produce significantly higher engine out concentrations than the MPI engine. As with the specific uHC emissions, the differences between MPI and GDI are further increased for the specific NOx measurement, as can be seen in Fig.9, by the higher exhaust mass flow resulting from lean operation. The higher local temperatures and pressures in the reaction zone of the stratified charge results in high NOx production in these areas even though mean chamber temperature is reduced by lean operation.

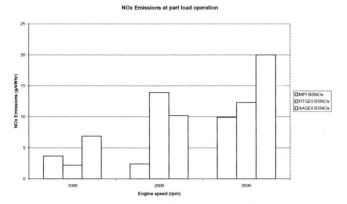

Figure 9. Specific NOx emissions at the part load test sites

Previous work has suggested that the effectiveness of different NOx reduction mechanisms varies between homogenous and stratified combustion systems. With a conventional MPI engine operating at stoichiometric AFR, EGR reduces NOx formation by increasing the level of burned diluent within the combustion chamber. This lowers peak combustion temperatures by providing an increased thermal mass and is therefore known as the "thermal effect". In essentially stratified combustion systems, such as HSDI diesel, it has been shown that the availability of oxygen for the reversible NOx formation reaction in the peak temperature zones is most significant. Replacing air with EGR reduces oxygen concentration and thus succeeds in minimising NOx formation. The oxygen availability effect also exists in conventional MPI engines were peak NOx occurs at $\lambda = 1.1$, despite a reduction in peak temperatures and pressures in comparison to $\lambda = 1$. The affect of charge temperature on final compression temperature has already been explained in section 4. As NOx formation is largely a function of temperature it follows that utilising cooled EGR to minimise charge temperature will therefore reduce NOx levels. Although this theory has not been tested, it is likely that due to the chemical nature of gasoline with its low cetane/high octane number, cooling EGR will result in significant increases in HC emissions. Finally, the chemical affect of reduced dissociation through addition of EGR has been investigated in MPI and HSDI engines and shown to have only a minimal influence. At the time of writing it is unclear which NOx reduction mechanism dominates.

Obviously, BSFC, NOx, uHC and engine stability have to be traded-off against each other for an optimum solution, with an added complication being compatibility of the exhaust gas with the after treatment system. When making this trade-off it should be borne in mind that the system must be optimised for minimum specific emissions, and this optima may not coincide with that required for minimum concentrations due to exhaust gas mass flow variations. This may be particularly significant when balancing air versus EGR as the charge diluent.

7. ANALYTICAL TOOLS TO AID GDI DEVELOPMENT

The GDI charge formation process can be complex in nature and laborious to optimise by traditional test and development processes. The application of advanced analytical techniques is greatly beneficial in allowing the development of these systems to reach some optimum in feasible project time frames. In the following sections (8 through 11) the key techniques used within this project are discussed.

8. MOVING GEOMETRY CFD

In order to enable this project and future projects an improved process of producing moving geometry CFD results was required, traditionally this has been the most labour intensive analysis tool available and hence not widely used in the industry.

The CFD code used was STAR-CD, which was supplemented with a suite of macros, known as RAMM-ICE (Rapid Meshing Methodology for Internal Combustion Engines). RAMM-ICE is a semi-automatic meshing tool, which generates the code, required to represent valve and piston motion in the 3D modeling of the in-cylinder volumes. The meshing of the system was divided into several areas - port, back of valve, underneath the head of the valve, combustion chamber and piston. In this way several simpler meshing and boundary interfaces were created and controlled by use of the macros. This approach enabled the assessment of different port / chamber / piston iterations each in a matter of days rather than weeks, as had previously been the case.

The direct injection fuel spray was included in the CFD simulation through the use of a Lagrangian two-phase model. The Lagrangian model allows for heat, mass and momentum transfers between the phases. The inclusion of a droplet break-up treatment allows for good qualitative predictions for charge stratification at part load operating conditions. The initial size, position and velocity distributions of the fuel sprays are entered as user defined routines. This approach makes simulating different injector and injection timing cases comparatively simple.

The first models were run without the added complexity of the fuel spray. Subsequent models included the fuel spray as described above. The combined air and fuel model provided valuable insight in to the interactions between the cylinder motion and the fuel spray, the fuel spray and the piston and cylinder walls. Figure 10. Shows an example of fuel distribution predictions for stratified charge operation as the piston approaches TDC. The sequence shown is for the RTGDI cylinder head with an early piston design. Contours indicate fuel concentration and the small circle indicates spark location. Prediction of temporal AFR distribution at the spark site is of key interest during combustion system development.

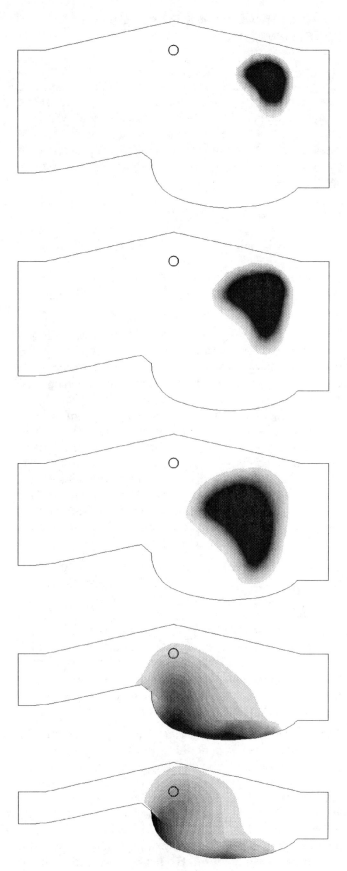

Figure 10. Moving Geometry CFD prediction of fuel spray behavior as piston approaches TDC (Contours indicate fuel concentration).

9. GEOMETRIC SPRAY PENETRATION ANALYSIS

Whilst moving geometry CFD is a very powerful technique for aiding the understanding of fuel-spray / charge interaction, the computer resource and the effort required to construct the model can be significant. In order to generate a first approximation of fuel spray behavior and piston interaction, 3D moving geometry CAD models were constructed to show nominal spray impingement on piston and other combustion system surfaces. This approach is very quick to produce results, but obviously ignores the affect of charge motion on spray behavior. This simplification gives a reasonable approximation of wall wetting regions under low speed stratified operating conditions with a high-pressure fuel injector. These models were animated to allow an initial appreciation of the influence of piston bowl shape and provide a first iteration design that could be further analysed and optimised through moving geometry CFD and engine testing.

Using this simple initial step in the combustion system design process, a reduction in the hardware iterations to be more fully analysed was achieved. In practice this enabled earlier test-hardware procurement.

10. DESIGN OF EXPERIMENTS

Compared to a port injection gasoline engine, a number of additional control variables must be optimised for every operating condition of a GDI engine. In addition to spark timing and EGR, the variables of fuel injection pressure, phasing and duration, air blast timing and duration, and AFR must be optimised. The number of, and strength of interactions among, these variables, renders the conventional "change one factor at a time" approach ineffective and inefficient. In making use of Design Of Experiments (DOE) techniques, many variables can be optimised simultaneously. Moreover, the models of BSFC, stability and emissions that are generated in the process greatly enhance the engineer's understanding of engine behavior.

Two DOE methods have been evaluated. The first involved a classical three-level design [Ref. Box] performed twice: once with broad variable ranges, then again with narrower ranges determined from analysis of the first experiment. In effect, "zooming in" on the region of optimal fuel consumption and emission.

The second method [Ref. Bisgaard, Myers] is a sequential process and relies to some extent on prior knowledge of the behavior of the system under investigation. In this case the knowledge that optimal conditions will exist at several points, namely, stratified charge operation and homogeneous charge operation. Application of this prior knowledge ensures that the appropriate local optimum is identified and achieved. A minimal two-level experiment is used to determine the direction in which the desired response, BSFC say, is improving at the highest rate.

Having ascertained the orientation of steepest descent (in k dimensions), new test points are defined in this direction and testing continues until the response again starts to worsen. At this point, the two level experiment is repeated and a new direction found and followed. The process continues until further improvement is impossible. At this time, further points may be added to the two level design to enable the construction of a full mathematical model of response close to the optimum.

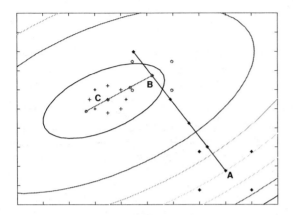

Figure 11. Sequential DOE experiment, in 2D. (Contour lines are BSFC).

Figure 11. is an illustration of a sequential approach in two dimensions only. Starting at A, the direction of greatest improvement in BSFC is calculated from data at five points. A new series of test points along this indicated direction is computed and data is recorded at each until no further improvement is observed (B). The process is repeated. At C, BSFC cannot be reduced in any direction. Additional test points (marked "+") can be included around C to give a full second order model around the optimum.

11. ENGINE CYCLE SIMULATION

An engine cycle simulation package with one-dimensional, transient, compressible fluid dynamics capability was used to model the port injected single cylinder engine. The model was constructed using a highly versatile, object-based program within a Windows based graphical environment, as shown in Fig.12.

A model of this type is suitable for investigating a wide range of engine issues, for example: -

1. Intake and exhaust system design and optimisation.
2. Acoustics.
3. Valve profile and timing development.
4. Thermal analysis.
5. EGR system development.

In addition, a model provides a great deal of information that is difficult to obtain experimentally. For example, residual fraction, trapping ratio and the variation of flow parameters with crank angle at any point in the induction or exhaust system.

Good correlation with experimental results was achieved. Volumetric efficiency was predicted to within 5% for all calculation points in the analysis range. In addition, the pressure fluctuations in the inlet and exhaust manifolds were predicted with good accuracy, see Fig.13.

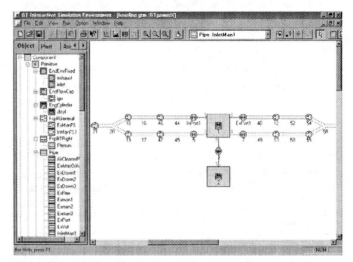

Figure 12. Engine Cycle Simulation Model.

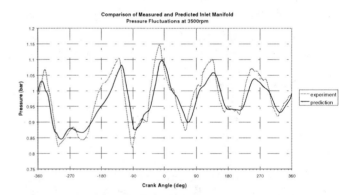

Figure 13. Inlet manifold pressure fluctuations Measured v Modeled

The engine cycle simulation code used was GTPower, running on a PC.

12. DESIGN CONSIDERATIONS

Direct injection gasoline engines present some new challenges to the engine design team, as part of this project all aspects of GDI engines were considered. Additionally, a concept scheme of a V6 engine was developed.

Perhaps the most obvious area of change relative to a conventional port injection engine is the cylinder head and intake system.

The intake system requires additional complexity to ensure seamless switching between throttle and fuelling control. Variable geometry may also be required to allow optimum charge motion for both stratified and homogeneous operation. Additional demands may also be placed due to the requirement for flowing and distributing large volumes of external EGR.

Positioning of the fuel injector is the key consideration for GDI engine design. The cylinder head has the added complexity of packaging the fuel injector. Central mounting of the injector along with the spark plug limits the valve sizes possible within a given bore diameter and hence the maximum power output, as well as requiring careful design to ensure adequate cooling around the spark plug and injector. The side-mounted injector needs to be suitably upright in order to minimise wall wetting, which has implications for the intake system routing. Piston bowl requirements can lead to an increase in piston mass and compression height. The effect of this upon engine balance, con-rod length and cylinder-block deck-height may be significant.

Other areas requiring some consideration are tribology of intake valve and seat as well as that between piston / rings and bore. Fuel pump drive requirements and, in the case of air injection systems, the air pump drive requirements must also be provided for.

As well as numerous other detail changes to the engine, there are knock on effects for the recipient vehicle such as fuel tank feed and return, vapor recovery, air box and induction system and mechanical noise suppression to combat high frequency noise from the injection systems.

In summary, the design of a direct injection gasoline engine presents new challenges to the design engineer.

13. CONCLUSIONS

Both the RTGDI and the AAGDI systems were successfully implemented into an engine and both ran in stratified and homogeneous modes.

Of the two systems considered the AAGDI was more flexible in its operation at part load, tolerating greater ranges of injection timing and AFR/EGR, whilst maintaining stable combustion.

Both GDI systems gave improvements to VE at full-load. The AAGDI gave improved full load performance while the RTGDI did not deliver the expected improvement in engine torque. This would probably be corrected by further optimisation of the combustion system.

Moving geometry CFD is a key analytical tool in the development of GDI combustion systems. The semi-automation of much of the model set-up process allowed CFD to be employed usefully within tight time-scales.

Cycle simulation proved useful in aiding the understanding of engine behavior in areas not readily amenable to experimental investigation, an example being in-cylinder residual prediction. Cycle simulation is particularly useful in conjunction with moving geometry CFD, as a source of realistic boundary conditions.

Design of Experiments techniques offer a reliable and robust method of finding optimum operating conditions in GDI systems. Two techniques were used either of which are appropriate depending on the level of knowledge available before testing, and the number of variables to be optimised.

ACKNOWLEDGMENTS

Thanks are due to Paul LeCornu, Andrew Rabbitt, Justin Seabrook, John Thornton, Ruaraidh Walker, and Phil Wilding for their contributions to this work.

REFERENCES

1. Ando, H et al. "Mitsubishi GDI Engine – Strategies to meet the European Requirements". Proceedings of AVL Engine and Environment Conference, Vol.2, Graz, Austria. 1997.
2. Andriesse, D. Ferrari, A. "Assessment of Stoichiometric GDI Engine Technology" Proceedings of AVL Engine and Environment Conference, Vol.2, Graz, Austria. 1997.
3. Bisgaard, S, "Quality Quandaries - Why Three-Level Designs are not so Useful for Technological Experiments", Report No. 148, Center for Quality and Productivity Improvement, University of Wisconsin-Madison, 1996
4. Box, G E P . Draper N R, "Empirical Model Building and Response Surfaces", John Wiley & Sons, New York, 1987
5. Fraidl GK, Piock WF, Wirth M. "Gasoline Direct Injection: Actual Trends and Future Stratagies for Injection and Combustion Systems" SAE960465.
6. Frank R, Heywood JB. "Combustion Characterisation in a Direct-Injection Stratified-Charge Engine and Implications on Hydrocarbon Emissions". SAE 89058
7. Heywood JB. "Internal Combustion Engine Fundamentals" International Edition, 1989. McGraw-Hill Book Co, New York. pp 252 eq 6.42.
8. Houston R, Cathcart G. "Combustion and Emissions Characteristics of Orbital's combustion Process Applied to Multi-Cylinder Automotive Direct Injection 4-Stroke Engines". SAE 980153
9. Myers R H, "Response Surface Methodology", Allyn and Bacon, 1971
10. Noma K, et.al. "Optimized Gasoline Direct Injection Engine for the European Market" SAE980150.
11. Stoker H, Archer M, Houston R, Alsobrooks D, Kilgore D. Application of Air-Assisted Direct injection to Automotive 4-stroke Engines – the "Total System Approach". Proceedings of Aachener Kolloquium Fahrzeug- und Motorentechnik, 1998.
12. Yang J, Anderson RW."Fuel Injection Stratagies to Increase Full-Load Torque Output of a Direct-Injection SI Engine" SAE 980495.

NOMENCLATURE

γ: Ratio of Specific heats (approx. 1.4 for air)
λ: Equivalence ratio (AFR/Stoichiometric AFR)
AAGDI: Air Assist low pressure GDI
ABDC: After Bottom Dead Centre
AFR: Air Fuel Ratio
Apmax: Crank angle at Max. cylinder Pressure °ATDC
Armax: Crank angle at Rmax °ATDC
ATDC: After Top Dead Centre
BBDC: Before Bottom Dead Centre
BMEP: Brake Mean Effective Pressure kPa

BSAC: Brake Specific Air Consumption kg/kWh
BSFC: Brake Specific Fuel Consumption g/kWh
BSU: Bosch Smoke Unit BSU
BTDC: Before Top Dead Centre
CAD: Computer Aided Design
CFD: Computational Fluid Dynamics
CP_{air}: Specific heat capacity of air at const. press.J/kg/K
CR: Compression Ratio
DOE: Design Of Experiments (using statistical methods)
EGR: Exhaust Gas Recirculation
EMS: Engine Management System
EOI: End Of Injection
EVC: Exhaust Valve closure
EVO: Exhaust Valve Opening
GDI: Gasoline Direct Injection
HSDI: High Speed Direct Injection
IMEP: Indicated Mean Effective Pressure kPa
IVC: Inlet Valve closure
IVO: Inlet Valve Opening
MBT: Minimum spark advance for Best Torque
MPI: Multi Point Injection (conventional port injection)
Pmax: Maximum cylinder Pressure Bar
Rmax: Maximum rate of rise in cylinder pressure Bar/°
RTGDI: Reverse Tumble high pressure GDI
SA: Spark Advance °BTDC
SI: Spark Ignition
SDIMEP: Standard deviation of IMEP kPa
TDC: Top Dead Centre
uHC: Unburned hydrocarbons
VE: Volumetric efficiency %
WOT: Wide Open Throttle

APPENDIX 1. TEST ENGINE AND FACILITY:

Features common to all variants:
Bore:	83mm
Stroke:	82mm
Capacity:	444cc
Compression Ratio:	10.5:1
Number of valves:	4
Inlet valve diameter:	29mm
Exhaust valve diameter:	29mm
Cam Period:	236
Valve lift:	9mm
Inlet valve opening:	6°BTDC
Inlet valve closing:	50°ABDC
Exhaust valve opening:	50°BBDC
Exhaust valve closing:	6°ATDC

Features specific to MPI:
Cylinder head:	Vertical entry, reverse tumble ports.
Angle between valves:	25°
Fuel injector:	Conventional MPI 30° hollow cone type.
Injector position:	Induction manifold.
Fuel Pump:	Low-pressure (3Bar) electric fuel pump.
Piston:	Flat piston crown.

Features specific to RTGDI:
Cylinder head:	As MPI
Angle between valves:	25°
Fuel injector:	Swirl type 60° cone angle, hollow cone type.
Injector position:	Cylinder head, inlet side, and 45° to vertical.
Fuel Pump:	50Bar nominal high-pressure plunger type pump. Driven from mains powered electric motor, with conventional low pressure Electric pump.
Piston:	Heart shaped bowl on inlet valve side, squish area on exhaust side.

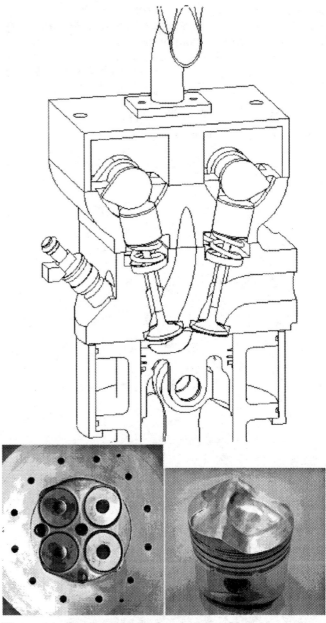

Figure 14. Sectional view, Combustion Chamber and Piston for RTGDI engine

Features specific to AAGDI:

Cylinder head:	"Conventional", forward tumble ports.
Angle between valves:	40°
Fuel injector:	Air Assist low-pressure type.
Injector position:	Cylinder head, near central location, 5° to vertical.
Fuel Pump:	Low-pressure (7Bar) electric fuel pump.
Air supply:	Shop compressed air.
Piston:	Central, shallow re-entrant bowl.

The fuel system used for the AAGDI testing was supplied by the Orbital Engine Company.

Engine Test Bed:

Motoring dynamometer	0-6000 rpm, 22 kW max power produce/absorb.
Torque-shaft torque measurement.	
Emissions measurement:	Horiba Mexa 7000 analyser HC, CO, CO2,NOx, O2.
Smoke measurement:	Bosch smoke meter.
Cylinder pressure recording:	AVL Indimaster with Kistler 6125 transducer.
Fuel Flow Measurement:	Gravimetric fuel weigher

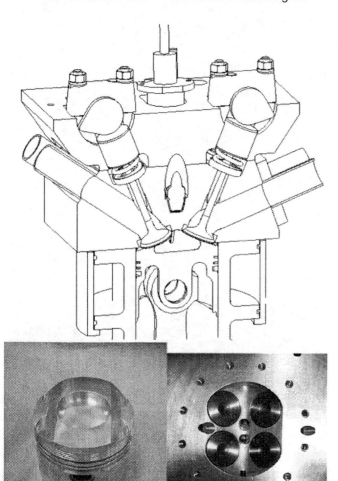

Figure 15. Sectional view, Piston and Combustion Chamber for AAGDI engine.

APPENDIX II. AAGDI COMPRESSOR WORK CALCULATION

The AAGDI system requires a compressed air supply, which will add a parasitic loss to a practical engine. This parasitic loss was allowed for in comparisons between systems. The additional compressor drive load required from the engine crankshaft was estimated from the following relationship:

$$-W_{comp} = \frac{\dot{m}_c CP_{air} T_0}{\eta_{ctt} \eta_{cm}} \left[(P_{cr})^{\left(\frac{\gamma-1}{\gamma}\right)} - 1 \right]$$

[Heywood]

This provides an estimated Compressor load of 10.6 kPa BMEP for a nominal 1.8l engine, with the following assumptions:

Compressor of reciprocating type, 30 cc swept volume.
Drive gear ratio from engine crankshaft 1:1.
Compressor VE = 75%
Compressor entry Density = 1.22 kg/m3
Engine capacity = 1.8l
Injector airflow requirement = 10 mg per shot.
Compressor cover (compressor flow capability/injector flow requirement) = 1.37
$\gamma = 1.4$

Where:

Specific heat capacity of air at constant pressure = CP_{air} = 1003 J/kg/K
Compressor and drive mechanical efficiency = η_{cm} = 81%
Total to total isentropic compressor efficiency = η_{ctt} = 80%
Compressor entry temp = T_0 = 288 K
Compressor pressure ratio = P_{cr} = 7.5 (to provide 6.5 Bar Gauge air supply)
Compressor drive Power (W) = W_{comp}

The assumption of 10mg per shot is appropriate for homogenous operation (Early injection). At stratified operating conditions (Late injection) the mass flow rate tends to be lower as the cylinder pressure is higher and so the injector is operating with a lower pressure drop. With a more sophisticated air supply system it would be possible to control air supply to meet demand and so reduce the parasitic losses to some degree. One method of achieving this would be to throttle the air entering the compressor, so reducing air supply at times of low demand [Ref: Stoker H, et al].

1999-01-0174

Influence of Fuel Injection Timing Over the Performances of a Direct Injection Spark Ignition Engine

Rakosi Edward, Rosca Radu and Gaiginschi Radu
Technical University Iasi

Copyright © 1999 Society of Automotive Engineers, Inc.

ABSTRACT

The paper presents an analysis of the influence of the most important factors that affect the working process of a direct injection spark ignition engine, especially the injection timing.

In the first part of the paper we study the engine's constructive solution and the changes in the engine's performances induced by the alteration of the compression ratio. This phase allowed us to choose the most appropriate variant.

In the second part we study the influence of the injection timing over the engine's performances, for two different compression ratios and two injection rates.

The most significant results are emphasized using the engine's speed characteristics, determined on the test bed.

INTRODUCTION

For the same engine constructive solution, the use of gasoline injection leads, obviously, to better results (lower fuel consumption, higher output power and torque). Nowadays, most of the automobiles are equipped with different fuel injection systems, the fuel being injected into the inlet manifold or inlet valve port. Less studied and developed (especially for automobiles) is the direct fuel injection (for the spark ignition engines).

An analysis of the processes inside the engine shows the major advantages of this type of injection: cycle heat is more reasonably used (the mixture is partly heated both by the walls of the combustion chamber and by the piston; thus, this part of the cycle heat is used instead of being transferred to the coolant) and cylinder filling is better. Moreover, mixture formation can be conducted according to the most appropriate solutions.

CHOSEN CONSTRUCTIVE SOLUTIONS

During the experiments a four stroke spark ignition engine was used (bore=97 mm, stroke=84.4mm). The engine has four in line cylinders and is water-cooled. At full load, the engine equipped with a carburetor has the following performances:

- rated power : 61 kW at 4500 rev/min;
- rated torque: 16 daN.m at 3000 rev/min;
- minimum BSFC: 308 g/kWh.

Figure 1 presents a schematic diagram of the engine.

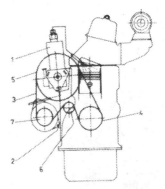

Figure 1. Schematic diagram of the engine

In comparison with the well-known advantages of fuel injection, in our case we could use higher compression ratios, due to the supplementary cooling effect of the charge and cylinder walls of the vaporizing fuel. In the meantime, the same fact led to an increase of the self-ignition delay and to a decrease of the knocking trend. Higher compression ratios lead a higher thermal efficiency of the engine ($\eta_t = 1 - \dfrac{1}{\varepsilon^{n_c}}$).

Tests were also developed for compression ratios higher then the ones imposed by the knocking limit and which are due to combustion chamber architecture and the fuel's octane number (COR=97, determined octane number). This high over-compression, although a modern procedure, is yet rarely used in engine design. This solution allows the preservation of the initial octane number (due to the reduction of the ignition advance under the knocking limit), even though power and specific fuel consumption at full load are altered. At part load, when ignition advance increases, the tendency towards knocking decreases.

The dihedral combustion chamber, placed in cylinder head, has been completed with a second volume, placed in the piston's head. Due to its geometry, this second volume creates a swirl of the charge, with favorable effects upon mixture formation and knocking.

As the basic design criteria was to minimize the changes of the basic engine, four compression ratios were established; the matrix of the compression ratio is as follows:

$$C\Re = \begin{vmatrix} \varepsilon_0 = 8.0 & \varepsilon_1 = 8.6 \\ \varepsilon_{oc-1} = 9.8 & \varepsilon_{oc-2} = 11.0 \end{vmatrix}$$ (**oc** stands for over-compression).

The engine was equipped with the original camshaft (marked D_0); the injection pump was equipped with eccentric cam profiles. As the fuel injection may occur anytime between the beginning of intake and the end of compression, we made some preliminary tests in order to determine the injection timing that guarantees the engine start and stability. Figure 2 presents the engine on the test bed.

Figure 2. Engine mounted on the test bed

Figure 3 shows the variation of the engine's power and torque for the aforementioned compression ratios. As these values were obtained for different engine speeds, we did not consider specific fuel consumption to be a significant factor, at least in this stage of our investigation.

Besides this, over-compression at full load has required a decrease of the injection delay (due to knocking), leading thus to higher specific fuel consumption.

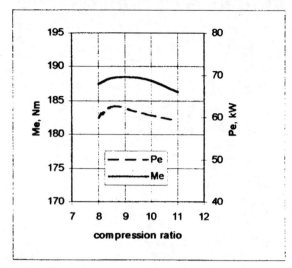

Figure 3. Output power and torque variation

On the basis provided by these preliminary tests, we selected two compression ratios, ε_1 and ε_{oc-2}, which led to close performances. They define, in fact, two different adjustment solutions, as the first compression ratio is specific to regular spark ignition engines, while the second one is specific to over-compressed engines.

INFLUENCE OF THE INJECTION TIMING

To develop a complete study of the performances of the direct fuel injection engine, we have taken into account some aspects concerning distribution phases, in order to improve cylinder filling. Thus, two new camshafts were achieved; they had different raising rates and heights and were marked D_1 and D_2 (the engine's original one was marked D_0). To test the influence of distribution over the engine's performances we have used the engine with the ε_1 compression ratio. In order to get more accurate results we have used different angular positions (θ) of the new camshafts compared with the standard camshaft. Table 1 presents the tested variants.

Using these variants and defining γ as the start of injection angle relative to TDC (fig. 4), we obtained the following results:

- D_{11} - insignificant results;
- D_{12} - the variation of the engine' power for different γ angles is shown in fig. 5. Rated power is obtained at 3800...4000 rev/min, depending on the injection start. The best results ere obtained at 4000 rev/min and γ = 135° C.A, that is when the injection takes place during intake.
- D_{13} - figure 6 shows the output power variation as a function depending of the angle of the start of injection.

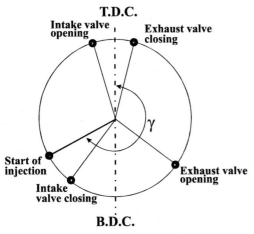

Figure 4. Definition of γ

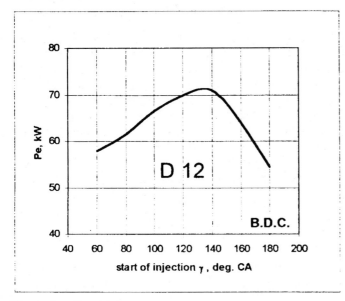

Figure 5. Output power

Table 1.

Camshaft	Tested variants	
D_0	D_{00} - D_0 camshaft, $\varepsilon = 8.6$	
D_1	D_{11} - D_1 camshaft, $\theta = 0°$ CAR	
	D_{12} - D_1 camshaft, $\theta = -8°$ CAR	
	D_{03} - D_1 camshaft, $\theta = +8°$ CAR	$\varepsilon_1 = 8.6$
D_2	D_{21} - D_2 camshaft, $\theta = 0°$ CAR	
	D_{22} - D_2 camshaft, $\theta = -8°$ CAR	
	D_{23} - D_2 camshaft, $\theta = +8°$ CAR	

The maximum power (aprox. 65 kW) is lower then the one obtained within the D_{12} variant (71 kW); maximum power is achieved when $\gamma = 190°$ C.A. The corresponding speeds are also between 3800 and 4000 rev/min. When displacing the start of injection towards the compression stroke, for $\gamma \geq 200°$ C.A., the engine becomes unstable and misfire occurs.

- D_2 - insignificant results concerning the engine's performances.

As a result of these tests, the original camshaft was preserved in order to equip the engine.

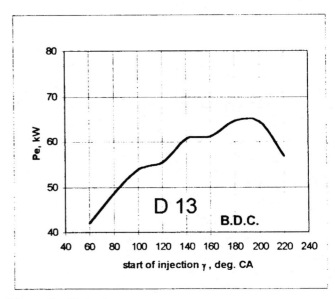

Figure 6. Output power

Researches concerning injection period and rate began with the use of two cam profiles for the injection pump.

First we have used an eccentric profile camshaft (8 mm eccentricity) - marked **e**. This type of profile ensures a good filling process for the pumping element (taking into account the high speeds of the spark ignition engine), but the injection period is too long. This is the reason why we looked for other variants, that would combine the advantages of the **e** variant with a shorter injection period and which would allow the injection of the most of the fuel quantity during intake. In order to achieve this purpose, we have used spiral profiles, marked s_1 and s_2. Tests were developed with an engine equipped with these types of camshafts. The results for the two profile types and for the ε_1 and ε_{oc-2} compression ratio are presented in fig. 7.

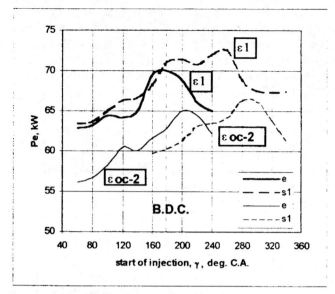

Figure 7. Output power for different camshafts

We noticed that profile e and compression ratio ε_1 led to a maximum power of 70 kW when the injection started at 172° C.A., that is towards the end of intake. Rated power was achieved at 4000 rev/min. When $\gamma \geq 180°$ C.A. the engine's working process became unstable, even though its indices did not suddenly alter.

For the same compression ratio ε_1 but an injection pump equipped with the s_1 type camshaft (which has been recommended by the previous tests of the injection equipment), we were able to use a much wider variation range for the start of injection. As fig. 5 shows, while displacing the start of injection towards the end of intake, the output power increases. In the B.D.C. area, the engine's power is about 70.6 kW and is then increasing significantly after the intake valve closes and compression begins ($\gamma > 270°$ C.A.). The optimum moment for the start of injection is for $\gamma \approx 270°$ C.A., when the engine's output power reaches 72.1 kW at 4000 rev/min. Any further displacement of the injection start ($\gamma > 280°$ C.A.) reduces the output power to an average of 56 kW. As we mentioned before, all rated powers were obtained for the same engine speed, 4000 rev/min. When γ is comprised between 200° C.A. and 240° C.A., the engine trends to knock.

Based on the aforementioned results, the optimum timing for the start of injection is:

- for the camshaft with **e** cam profile: $\gamma = 172°$ C.A.;
- for the camshaft with the s_1 cam profile: $\gamma = 270°$ C.A.

The use of the s_1 cam profile allows a later start of the injection, spraying, vaporizing and mixture formation needing a lower period. This fact is partly confirmed by the higher performances, as shown in table 2.

Tests made with an engine using the ε_{oc-2} compression ratio that was equipped with the e and s_1 type camshafts have partly confirmed the previous mentioned results. Thus, when the **e** profile is used, the output power has the same increasing tendency, while the start of injection is moved towards the end of the intake. The optimum injection timing is $\gamma \approx 195°$ C.A. The highest output power is 65.2 kW, at 4000 rev/min (rated speed). The explored domain of variation is basically the same as for ε_1, which leads us to the conclusion that, for the e type cam profile, the influence of the increased compression ratio is not very important. This might be due to a higher injection period and to a poor spraying quality achieved when **e** type cams are used in order to drive the injection elements.

Table 2.

Cam profile	P_e/n, [kw]/[rev/min]	c_e/n [g/kW.h]/[rev/min]	γ °C.A.
e	69.92/4000	322/4000	172
s_1	72.12/4000	307/4000	270

Note: compression ratio $\varepsilon_1 = 8.6$

For the same compression ratio (ε_{oc-2}), the s_1 type cam profile leads to lower variations of the performances due to the change of the injection start. Thus, the engine's brake power was comprised between 59 kW and 66,4 kW (at 4000 rev/min), when g was modified from 160° C.A. and 340° C.A. The optimum timing for the injection start was $\gamma = 290°$ C.A., while for the ε_1 compression ratio the optimum value was $\gamma = 270°$ C.A. We may presume that the 20° C.A. decrease of injection advance is due to the influence of the higher compression ratio.

The optimum timing for the start of injection, for the ε_{oc-2} compression ratio, is as follows:

- for the camshaft with **e** cam profile: $\gamma = 195°$ C.A.;
- for the camshaft with the s_1 cam profile: $\gamma = 290°$ C.A.

The best performances that resulted for this constructive variant are presented in table 3.

Table 3.

Cam profile	P_e/n, [kw]/[rev/min]	c_e/n [g/kWh]/[rev/min]	γ °C.A.
e	65.18/4000	328/4000	195
s_1	66.38/4000	312/4000	290

Note: compression ratio $\varepsilon_{oc-2} = 11.0$

The table shows that the s_1 type profile leads to higher performances then the **e** profile. As the ignition advance had to be restrained because of the knocking tendency, the output power at full load has decreased compared to the previous results (table 2). This variant has achieved the lowest fuel consumption's (at 72% load), for an excess air coefficient of 1.22...1.23.

As a result of the presented tests, we finally concluded that the engine should be equipped with its original camshaft, while the injection pump should be equipped with the s_1 type camshaft. Figures 8, 9, 10 and 11 show the engine's performances, at full load, for the aforementioned equipping variant and for the four compression ratios taken into account.

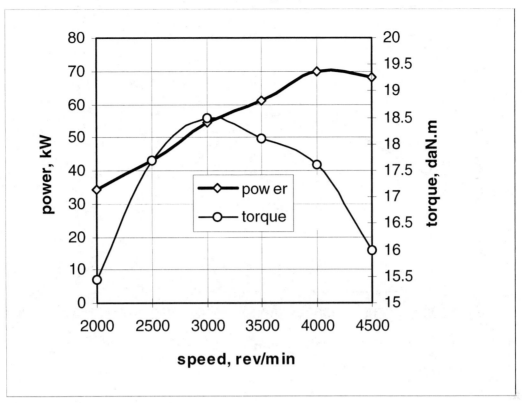

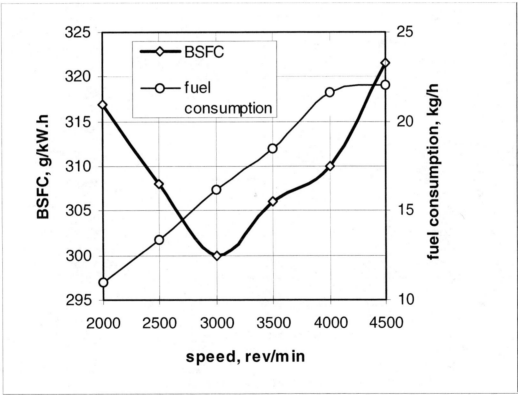

Figure 8. Performances at full load - ε_0

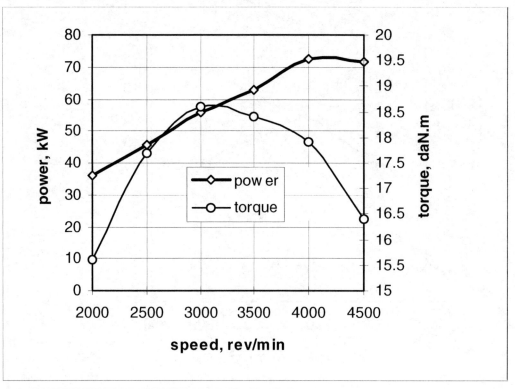
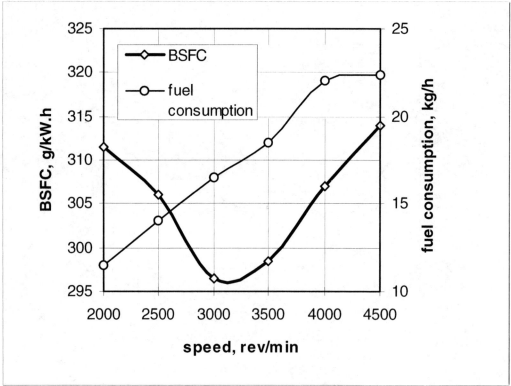

Figure 9. Performances at full load - ε_1

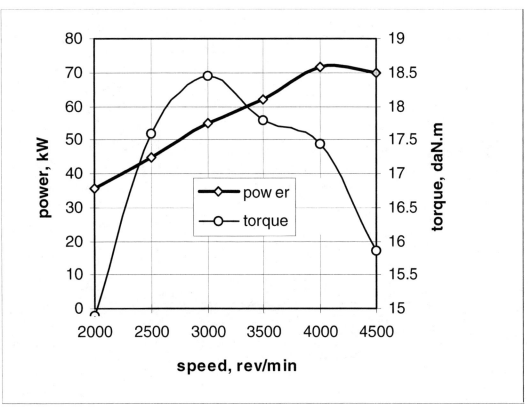

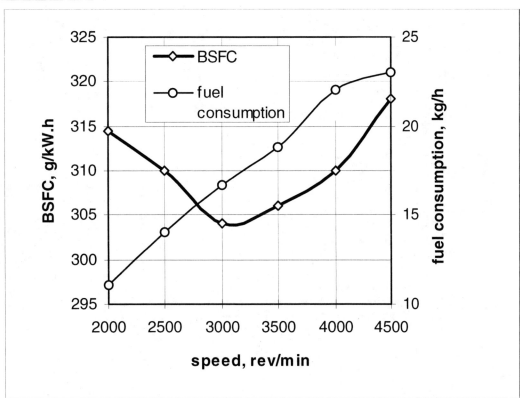

Figure 10. Performances at full load - ε_{oc-1}

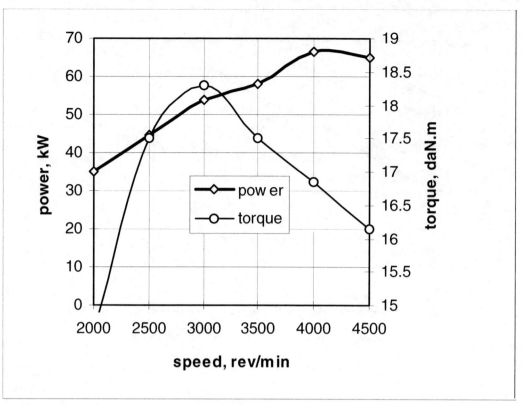

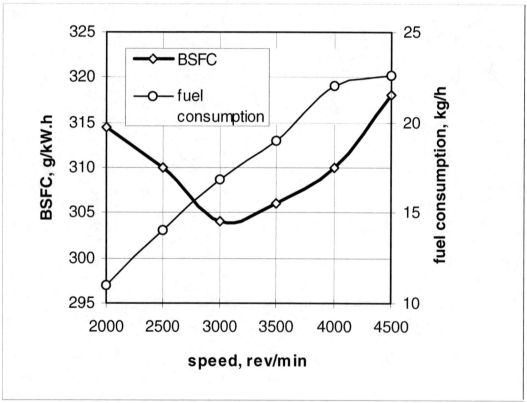

Figure 11. Performances at full load - ε_{oc-2}

CONCLUSIONS

1. The use of the $\varepsilon_0 = 8$ compression ratio led to higher performances compared to carburation fueling; the maximum output power has increased from 61 kW to 69.92 kW (at 4000 rev/min). Meanwhile, maximum torque has increased from 16 daN.m to 18,5 daN.m (at 3000 rev/min).
2. The best results concerning the output power and torque were achieved for the $\varepsilon_1 = 8.6$ compression ratio. For this compression ratio, the maximum power was 71.12 kW at 4000 rev/min and the maximum torque was 18.62 daN.m at 3000 rev/min; the specific fuel consumption was 296 g/kW.h. We noticed a favorable variation of the fuel consumption per hour in the range 4000 - 4500 rev/min (fuel consumption remains almost constant).
3. When compression ratio was raised to $\varepsilon_{oc-1} = 9.8$, the engine's performances slightly diminished due to the reduction of the ignition advance in order to avoid knocking. The specific fuel consumption has increased to 304 g/kWh.
4. Over-compressing the engine to $\varepsilon_{oc-2} = 11$ led to a more severe decrease of the performances at full load. The high compression ratios did partly compensate this decrease. The engine's power has reached 66.38 kW/4000 rev/min, the maximum torque was 18.26 daN.m/3000 rev/min and the lowest specific fuel consumption was 307 g/kWh at 3000 rev/min.
5. For all the four compression ratios the fuel consumption increases slowly between 4000 and 4500 rev/min. Moreover, the maximum power is achieved at 4000 rev/min instead of 4500 rev/min for the case of carburation. Thus, the engine is more suited for traction.

REFERENCES

1. Campbell, A.S. - Thermodynamic Analysis of Combustion Engines, Willey, New York, 1979.
2. Chang, Yon, Chen., Veshagh, A. - A Refinement of Flame Propagation Combustion Model for Spark-Ignition Engines, SAE Papers 920679, 1992.
3. Chmella, F. - High compression stratified charge engines and their suitability for conventional and alternative fuels, I. Mech. E. Conference Publications, 1979
4. May, M.G. - The high compression lean burn spark ignited 4-stroke engine, I. Mech. E. Conference Publications, 1979.
5. Mikulic, F., Quissek, F., Fraidl, G.K. - Development of a low emission high performance four valve engine, 1990, SAE International Congress and Exposition, Detroit, Michigan, Paper 900227.
6. Nakajima, Y. - Analysis of Combustion Patterns effective in improving anti-knock performance of a spark ignition engine, J.S.A.E., March 1984.
7. Rao, A.C., Bechar, L.B. - Theory of internal combustion engines, Belvedere Printing Work, 1981.

CONTACT

- **Rakosi Edward** - lect., Technical University "Gh. Asachi" Iasi, Romania; Buna Vestire 18, Iasi 6600 Romania.
- **Rosca Radu** - lect., Technical University "Gh. Asachi" Iasi, Romania; Dumbrava Rosie 2, Iasi 6600, Romania; e-mail: rrosca@sb.tuiasi.ro
- **Gaiginschi Radu** - professor, Technical University "Gh. Asachi" Iasi, Romania.

DEFINITIONS, ACRONYMS, ABBREVIATIONS

η_t: thermal efficiency;
ε: compression ratio;
c_e: break specific fuel consumption;
n_c: mean coefficient of the polytropic compresssion;
C_h: fuel consumption per hour;
M_{ec}: break corrected torque;
P_{ec}: break corrected power.

CHAPTER 6

EMISSIONS

THE GENERAL MOTORS CATALYTIC CONVERTER

by

M. F. Homfeld

R. S. Johnson

W. H. Kolbe

Engineering Staff

General Motors Corporation

Warren, Michigan

for presentation at the

1962 March SAE

Meeting

Detroit, Michigan

THE GENERAL MOTORS CATALYTIC CONVERTER

INTRODUCTION

The role of the automobile in air pollution has received a great deal of attention in recent years, both from organizations in California and from the automobile producers. The Automobile Manufacturers Association (AMA), representing all of the major automobile manufacturers in this matter, has been working closely with the California authorities in an effort to bring about a better understanding of the part the automobile plays in polluting the atmosphere, and to find ways of reducing its contribution to a minimum.

One device which has shown promise of reducing the emission of hydrocarbons and carbon monoxide from automobiles is the catalytic converter. However, a suitable catalyst was required before a practical converter could be developed. This catalyst must be one whose activity would not deteriorate rapidly when exposed to lead compounds in the exhaust gas.

In January of 1959, Bishop and Nebel of General Motors Research Laboratories presented a paper[1] before the SAE on the evaluation of a catalyst developed by Dr. Eugene J. Houdry of Oxy-Catalyst Inc. Although the device used to contain this catalyst was a laboratory test fixture, they concluded that the catalyst:

1. Substantially reduced hydrocarbons and CO.

2. Appeared to be resistant to lead poisoning.

3. Was expected to be low cost, since Oxy-Catalyst had said that it did not contain any precious metals.

4. Made a catalytic converter technically feasible; however, many problems remained which would require a major engineering effort to solve.

As a result of this work, General Motors decided to develop a practical converter using the Houdry catalyst.

The design and development of a practical system was assigned to the Power Development Group of the General Motors Engineering Staff. The principal objectives were:

1. Minimum size and weight consistant with maximum efficiency and installation considerations.

2. Durable under all operating conditions.

3. Low noise level.

4. Minimum cost.

To present this subject, we have divided the paper into a general description of the system, a discussion of the development of some of the components, a section on performance and durability, and a summary.

There is one factor which we would like to emphasize; this is the choice of a satisfactory test method. To some readers of this paper, the subject of test methods* will be as important as the design of the actual catalytic converter system.

DESCRIPTION

To concentrate our efforts and eliminate as many variables as possible, the system was designed and developed for cars of one particular make and model. These were Chevrolet sedans equipped with 283 in^3 V-8 engines and Turboglide transmissions.

A schematic diagram of the converter system is shown in Figure 1.

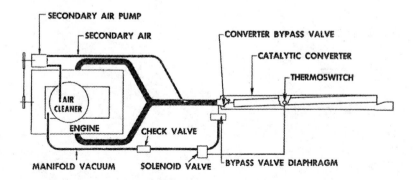

Figure 1 - Schematic diagram of the catalytic converter system.

* See Page 6.

An air pump, belt driven from the engine, takes in air through the carburetor air cleaner and discharges it into the exhaust pipe, far enough upstream of the entrance to the catalytic converter to provide sufficient time for mixing. This air must be added since exhaust gas seldom contains enough oxygen to burn all the combustibles. The diluted exhaust gas then flows to the catalytic converter.

The converter replaces the conventional muffler in the exhaust system. It is positioned as close to the engine as possible for maximum heat transfer to the catalyst during warm-up and low, constant speed driving. The unit is flexibly mounted to the frame using conventional mounting materials and procedures. The exhaust pipe and tailpipe have the same cross-section and material as production Chevrolet cars. The exhaust pipe is insulated.

At the entrance to the converter is a by-pass valve. This valve is used for over-temperature protection and can direct the exhaust gas either under or through the bed. A thermoswitch with its probe located in the center of the bed senses catalyst temperature and by means of a small solenoid valve, controls the application of manifold vacuum to a spring-loaded vacuum diaphragm. This vacuum motor actuates the by-pass valve.

The by-pass system is designed to "fail safe." Should either a vacuum hose or an electrical wire break, the system will bypass. In addition, the valve goes to the bypass position each time the ignition switch is turned off. This helps to keep the bypass valve bearings free. A check valve is provided between the intake manifold and the solenoid valve to maintain actuator vacuum, even at full throttle.

The bypass valve, shown in Figure 2, is fabricated primarily from stampings. In the normal position, the valve must be gas-tight. Therefore, a gas tight seal is provided in the form of a conical disk seat-

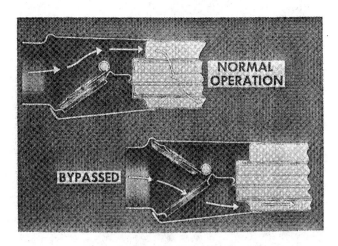

Figure 2 - Design and operation of the bypass valve.

ing on a circular, stamped seat. The disk is retained by a loosely fitting shoulder rivet, so it is self-centering on its seat.

The internal construction of the converter is shown in Figure 3. The converter shell is formed of two similar stampings, welded together around the sides and ends. Each shell stamping has a strengthening channel at the center, and the remaining surfaces are curved to control noise. A central rib ties the shell stampings together.

The catalyst bed is inclined with respect to the converter shell. This provides better gas velocity distribution throughout the bed and minimizes height. Two grid plates retain the catalyst and are supported so that they are free to expand in any lateral plane. These grids contain many slotted openings which run normal to the axis of the converter.

The converter is insulated with aluminum silicate fiber and the insulation is protected from water splash and tearing by a metal covering. The system

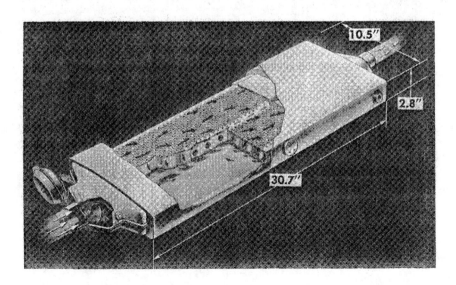

Figure 3 - Cut-away view of the catalytic converter.

is shown installed on a car in Figures 4 and 5.*

Figure 4 - Catalytic converter installation.

Figure 5 - Installation of the secondary air pump.

DEVELOPMENT

Background Experience

The early Oxy-Catalyst converter described by Nebel and Bishop is shown in Figure 6. It was 40 inches long by 12 inches wide by 8 inches high. Weight was 80 pounds including 25 pounds of catalyst. Secondary air was supplied by a venturi. Road clearance was limited and weight and noise were excessive.

Oxy-Catalyst submitted a more compact converter

* Converter system specifications may be found in Figure 1 of the Appendix.

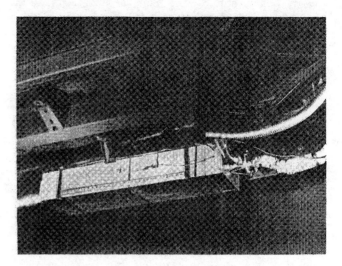

Figure 6 - Early Oxy-Catalyst converter.

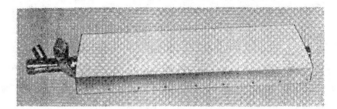

Figure 7 - Improved converter submitted by Oxy-Catalyst.

shown in Figure 7. This unit was 40 inches long by 11.5 inches wide by 4 inches high and weighed 57 pounds including catalyst. Secondary air was furnished by a venturi in this design also. Six of these units were evaluated, primarily by road testing. The results further substantiate the findings of Nebel and Bishop; however specific problems became apparent:

1. Durability was poor, primarily because of fatigue failure.

2. Shell noise was severe because of unsupported flat areas.

3. Warping was caused by excessive temperature.

4. The unit was difficult to mount because of size and weight, and the tailpipe size required for venturi operation was excessive.

5. Secondary air induction noise was excessive.

6. The quantity of secondary air provided was not satisfactory.

Following this evaluation, Engineering Staff decided to design an improved converter system. The objectives were:

1. Continue to flow the exhaust gas down through the bed. With up-flow or cross-flow, the bed

would be by-passed if not completely full of catalyst.

2. Decrease length and height to suit the installation space.

3. Adequately provide for grid expansion.

4. Devise a satisfactory punched grid.

5. Stiffen surfaces to avoid shell noise.

6. Provide over-temperature protection.

7. Provide a positive and quiet method of introducing secondary air.

The new design which evolved has already been described. However, some of the more important development items will be discussed in detail.

Secondary Air Supply

The Oxy-Catalyst converter system tested employed a venturi in the exhaust pipe to supply secondary air. This air was taken from the engine compartment and flowed through a hose into a plenum chamber which surrounded the inlet opening into the venturi throat. For adequate secondary air flow into the throat, a delicate pressure balance at the venturi inlet, outlet, and throat was required. Any restriction on the downstream side of the venturi would reduce the amount of air inducted and excessive restriction could force exhaust gas out of the air intake. The venturi system therefore had several disadvantages.

1. A tailpipe larger than normal was needed to maintain a low venturi back pressure.

2. A silencer was required on the air inlet.

3. The secondary air hose and silencer were necessarily large for minimum restriction.

4. Flow restrictions through the converter caused by catalyst packing, grid plugging, lead accumulation, etc. reduced the quantity of secondary air inducted.

5. It produced excessive engine back pressure at high speeds.

6. A whistle existed during accelerations which could not be eliminated with a resonator.

7. There was not enough secondary air inducted at idle and very low speeds.

Engineering Staff tried a more refined venturi; however, only the whistle problem was eliminated.

Therefore, it was decided to replace the venturi with an air pump, belt driven from the engine. The displacement of the pump was chosen to supply enough air for idle, road load, and normal acceleration. More air than necessary will cool the exhaust gas by dilution and decrease bed temperature and thus the activity of the catalyst. For this reason, the pump was intentionally sized to supply less air than required at full throttle.

Grid Design

In order to adequately retain the catalyst, the grid slots must be relatively narrow. To punch such narrow slots in sheet metal would normally require a thin, fragile punch. This problem was solved by the technique shown in Figure 8 which used a sturdy punch.

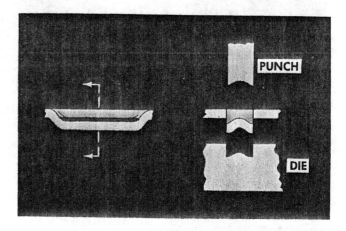

Figure 8 - Grid Slot punching method.

In our early grid the slots were arranged as shown in Figure 9. The slotted areas of the grid are certain to be at a higher temperature than the areas which are in contact with the shell. This caused high, lengthwise thermal stresses and fatigue cracks where indicated. Therefore, the slots were tried in the other direction in a subsequent grid design. It was reasoned that the full width of the grid would be available to resist the thermal expansion stresses. Unfortunately, this plan decreased the transverse beam strength, and the lower grid sagged under the load of the catalyst weight and gas pressure. Distortion was severe enough to open some slots so that catalyst was lost.

In the final, successful grid design, the slots were shortened 60 percent and were staggered. The opening of the shorter slots is more stable. With the staggered pattern, the slot connecting webs form hundreds of beams which can flex under non-uniform temperature conditions.

Some of the early failures were no doubt aggravated by the lack of good over-temperature protection. If so, this was a fortunate thing. A better grid design resulted.

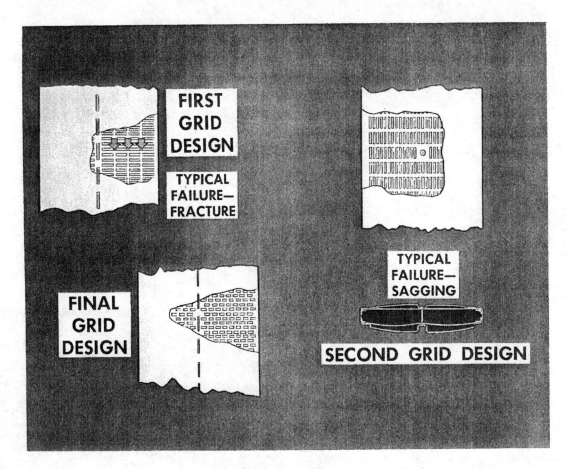

Figure 9 – Grid designs.

Over-Temperature Protection

Under severe driving conditions, a malfunctioning carburetor, or a fouled spark plug, increased quantities of fuel and air combine in the converter and burn, releasing an excessive amount of heat to the bed. These high temperatures can cause warpage of the grid and shell, sagging of the lower grid due to loss of strength, and even some loss of catalyst activity if the high temperature is sustained. An early attempt at over-temperature protection was to stop the flow of secondary air. This was not effective for a fouled spark plug. When a cylinder was not firing, a combustible mixture of fuel and air entered the bed, whether or not secondary air was supplied. From this it was concluded that the only safe protection was a by-pass valve.

The setting of the thermoswitch must be a compromise. The converter must never be damaged, yet a high setting is desirable so that bypassing does not seriously decrease efficiency. Another factor is that the thermoswitch senses temperature at only one location, and bed temperature may be considerably higher at some other location. A setting of 1300 - 1360°F. is the compromise used by Engineering Staff.

Backfire Resistance

The Engineering Staff converter originally had five round columns to connect the central ribs of the upper and lower shells, Figure 10. Even one backfire would shear some of the welds at the ends of the columns. The design was changed to use a continuous central strut, and is shown as the final design. The redesign provided increased length of weld. After this change, the converter was backfired 50 times at operating temperature, with no damage whatever.

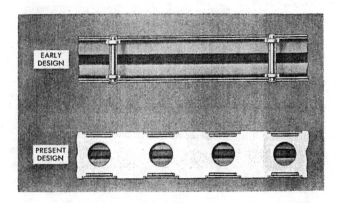

Figure 10 – Vertical reinforcement for converter shell.

PERFORMANCE AND DURABILITY

Test Methods

At the beginning of this paper, we stressed the fact that a good test method is extremely important for the proper evaluation of a catalytic converter. The basic reason for this is that the performance of a catalytic converter is dependent upon the temperature of the catalyst bed. This is illustrated in Figure 11 which shows the percent reduction in hydrocarbons plotted against bed temperature under several driving conditions. With the Houdry catalyst, temperature must be above 550°F. to achieve any reduction and should be above 1000°F. for good efficiency.

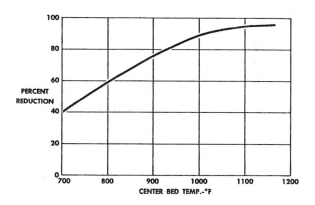

Figure 11 - Percent reduction in hydrocarbons or carbon monoxide vs bed temperature.

The temperature of the catalyst bed varies with driving conditions. At 30 mph road load, for example, the engine is operating on a lean, part throttle mixture and the concentration of combustibles and the temperature of the exhaust gas are low. Therefore, the bed will not stay hot and active for prolonged periods at constant, low speeds. Alternate accelerations and decelerations, or transient type driving, produces the opposite effect; that of increasing bed temperature. During accelerations gas temperatures are high, and the engine is operating on rich power mixtures; during decelerations, combustibles are very high. Because of its mass the catalyst bed does not react instantly to these changes in driving conditions. As a result the operating temperature of the bed at a given time is a function of many driving conditions and thus is very dependent upon the driving cycle it is exposed to.

Since driving conditions have such a large effect on test results, the selection of a satisfactory driving schedule is difficult. It is also extremely important if a good evaluation of a catalytic converter is to be obtained.

Three different performance tests have been used by General Motors to evaluate this converter. These are the California Driving Cycle, the AMA Test Route, and the California Dynamometer Cycle.

California Driving Cycle

The 1959 California Legislature enacted laws requiring the California Department of Public Health to develop standards for motor vehicle exhaust emissions. Such a standard was adopted by the Department on December 4, 1959. It stipulated that the exhaust emissions from an automobile could not exceed 275 parts per million total hydrocarbons and 1.5% carbon monoxide when tested according to the California Driving Cycle.*[2] In these standards, however, the Department of Public Health did not specify the driving conditions prior to any of the eleven modes of the cycle. Since these prior driving conditions are of paramount importance in the evaluation of a catalytic device, the standard could not be directly applied to catalytic converters. There were good reasons however, why these driving conditions were not specified - the standard was for all cars, not specifically for those with exhaust devices, and application of the standards was to be left to the Motor Vehicle Pollution Control Board.

Later in this paper, test results will be presented which show good performance on this driving cycle, even with very old catalyst, providing bed temperature is maintained by transient-type driving prior to the measurements.

AMA Test Route

The Exhaust System Task Group of the Automobile Manufacturers Association felt a need for a uniform road test procedure which they could use for comparison and development. They agreed on a 59 mile test route.** The route included suburban-type roads, some heavy city traffic, and some expressway driving.

The test procedure was very simple. An infra-red gas analyzer was used to alternately sample gas entering the converter and leaving the converter, once each minute. The percent reduction in hydrocarbons or carbon monoxide was thus based on an average of about 60 reading from each sampling point. This test procedure might be criticised on three items:

1. The data were not weighted in proportion to gas flow at the time of the reading.

2. The route included only one warm-up in 59 miles. An average trip is probably not this long.

* Appendix Figure II.
** Appendix Figure III.

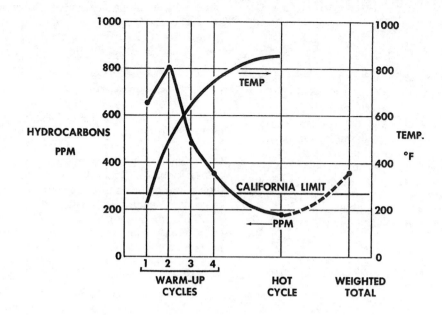

Figure 12 - Typical hydrocarbon emissions during the California dynamometer cycle.

3. The route may not represent average traffic and driving conditions.

However, the route was very useful for development work. It was used extensively by Engineering Staff, both for efficiency tests and for accumulating miles on the catalyst.

California Chassis Dynamometer Cycle[3]

This test procedure was approved by the California Motor Vehicle Pollution Control Board in May, 1961. It is based on the eleven mode driving cycle and takes warm-up into account. The test is run on a chassis dynamometer and the car is started at room temperature. The car is started and driven through six 7-mode warm-up cycles, each of which lasts 137 seconds. This is followed by one 11-mode hot cycle. The entire test is run in 20 minutes or 7.5 miles of driving. The results of the first four warm-up cycles are weighted and combined with the hot cycle results. Limits are unchanged at 275 parts per million and 1.5%.

Figure 12 illustrates the importance this cycle places on warm-up characteristics. Note that this particular test showed 355 parts per million although emissions were down to 180 ppm during the hot cycle. Bed temperatures are also shown. This is the procedure now being used by the Board to evaluate exhaust devices.

Catalyst Life

Figure 13 shows hydrocarbon emissions for the California Driving Cycle with bed temperature maintained in the 1000 - 1300°F. range. Note that the Standards are met, even with 12,000 mile catalyst, if bed temperature is maintained. These tests were run on the road.

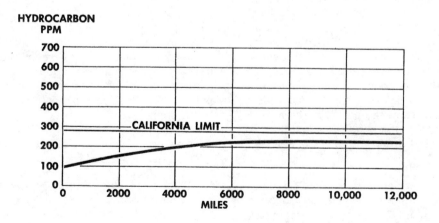

Figure 13 - California Driving Cycle - Bed temperature maintained in the 1000 - 1300°F. range.

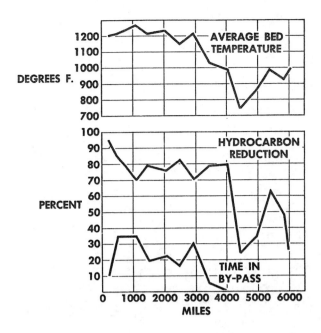

Figure 14 - Percent hydrocarbon reductions, percent of time in bypass, and bed temperature on the AMA Test Route.

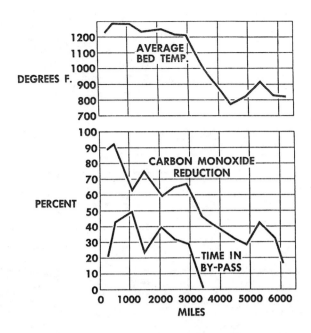

Figure 15 - Percent carbon monoxide reduction, percent of time in bypass, and bed temperature on the AMA Test Route.

Figure 14 shows percent hydrocarbon reduction vs miles based on the AMA Test Route procedure. Figure 15 shows similar data for carbon monoxide. For both gases, reduction dropped to 40% in about 5000 miles. We felt that this was a minimum acceptable percent reduction.

Figure 16 shows emissions vs miles as measured on the California Chassis Dynamometer Cycle. This car failed to pass the hydrocarbon requirement after 3300 miles.

The results presented for the three different test methods were not run on the same car or converter; however, we feel they are representative. All data were obtained with infra-red analyzers. Hydrocarbons are expressed as normal hexane.

Development work has been continued in an effort to improve catalyst life. For example, one way to improve life is to inject the secondary air into the exhaust ports rather than into the exhaust pipe. The burning of hydrocarbons and carbon monoxide by means of air injection into the ports is discussed in a paper[4] by Messrs. Brownson, Johnson and Candelise. They show that although the reduction in emissions by means of air injection is not as great as with a warmed-up catalytic converter,

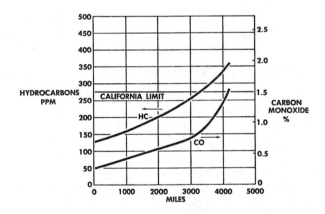

Figure 16 - Emissions as measured on the California Dynamometer Cycle vs miles on the catalyst.

there is a reduction immediately with no warm-up period required. This means that injecting the air into the exhaust ports can reduce emissions during the first warm-up cycles of the California Dynamometer Cycle, when the catalytic converter is ineffective. The combustion which takes place in the ports also increases the sensible heat in the exhaust gas, increasing the warm-up rate of the catalyst bed. A catalyst life test with this system is in progress, though it is too early to state how much catalyst life is increased.

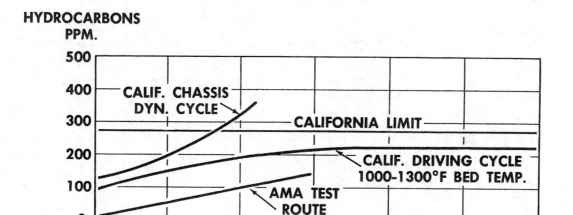

Figure 17 - Hydrocarbon emissions vs miles on the catalyst, as measured by three different test methods.

Comparison of Test Methods

The results from the California Driving Cycle, the AMA Test Route, and the California Dynamometer Cycle are compared in Figure 17. Two of the curves are identical to those in Figures 13 and 16. The emissions shown for the AMA Test Route is a smoothed curve from the same data that were used for Figure 14. The 275 ppm California limit does not, of course, apply to the AMA Test Route. Tabulated on Figure 17 are typical hydrocarbon concentrations entering the converter.

The California Driving Cycle and the California Dynamometer Cycle are two extremes in test procedures. The former seems to show that catalyst life is over 12,000 miles. The latter indicates a life of only 3300 miles. The reason for this discrepancy is that the California Driving Cycle does not consider warm-up at all, while the California Dynamometer Cycle places considerable emphasis on warm-up time.

On the AMA Test Route, the emissions both before and after the converter are much lower than with the other two test methods. There are two reasons for this: no weighting factors are applied to consider the higher gas flow during acceleration, and there is only one warm-up event in each 59 mile cycle.

These wide variations in results clearly show the importance of choosing a good test method. By a good test method, we mean one which:

1. Truly represents driving conditions in the area where pollution is a problem.

2. Has a length of trip and a cool-down period which represent average conditions.

3. Properly considers exhaust gas flow rate and dilution effects of secondary air.

Structure Durability

The accelerated durability test method used by General Motors is the Proving Ground 25,000 Mile Durability Test. This is run on a route which includes Belgian block roads, hills, and high speeds. Two converter systems successfully completed this test.

The total miles on the catalytic converter program to date is 375,000 miles.

Catalyst Attrition

Early converters had a reservoir to feed catalyst to the bed to replace attrition losses. With the latest design, attrition is only 0.1 pound per 1000 miles on the AMA Route and .38 on the 25,000 Mile Durability Test. A reservoir is no longer used.

Noise

The catalytic converter is not quite as effective in silencing as a conventional muffler, although most observers consider it commercial. If improved silencing is required, a resonator could be added to the tailpipe. No difference in noise is detectable when the bypass valve is actuated.

Odor

Early in the development program, the exhaust gas from the catalytic converter often had a strong, sweet odor. At that time, the odor was considered a problem. However, as the efficiency of the system was improved, the odor became less and less severe. With the present system, the exhaust gas has an odor which is different from that of untreated exhaust gas. The sweet characteristic is most noticeable when the converter is not completely warmed up.

SUMMARY

The General Motors Catalytic Converter system is functionally and structurally satisfactory from a mechanical standpoint.

The California Motor Vehicle Pollution Control Board requires a catalyst life of 12,000 miles. Our tests indicate that catalyst life is considerably less than this. However the level of exhaust emissions, as specified in the California Standards, can be met by periodic catalyst replacement.

A catalytic converter requires time to warm-up and become efficient. The current California test procedure places considerable emphasis on warm-up time.

Because of the stainless steel construction, the over-temperature protection, and the engine-driven air pump, the catalytic converter described will cost considerably more than a conventional muffler.

Development work is continuing in an effort to improve catalyst life.

REFERENCES

1. Bishop, R. W., and Nebel, G. J., "Catalytic Oxidation of Automobile Exhaust Gases - an Evaluation of the Houdry Catalyst," paper presented at the Annual Meeting, Society of Automotive Engineers, January, 1959.

2. "Technical Report of California Standards for Ambient Air Quality and Motor Vehicle Exhaust," State of California, Department of Public Health.

3. "Test Procedure for Vehicle Exhaust Emissions," California Motor Vehicle Pollution Control Board.

4. Brownson, D. A., Johnson, R. S., and Candelise, A., "A Progress Report on MANAIROX - Manifold Air Oxidation of Exhaust Gas", paper presented at the March, 1962 meeting, Society of Automotive Engineers.

APPENDIX

CONVERTER SYSTEM SPECIFICATIONS

Converter Dimensions (over-all)

Length	38.00
Width	10.56
Height	2.89

Converter Weight, Lbs.

Structure		26.0
Catalyst		13.5
	Total	39.5

Shell and Grid Material . . . . Stainless Steel

Thermoswitch

Actuation temp., °F.	1300 – 1360
Differential temp.	Minimum

Air Pump Displacement, in^3 . . . . 23.8

Figure I

CALIFORNIA DRIVING CYCLE

Mode	Rate of Speed Change MPH/SEC	Percent of Total Time	Percent of total sample volume
Idle		15.0	4.2
Cruise			
20 mph		6.9	5.0
30 mph		5.7	6.1
40 mph		2.7	4.2
50 mph		0.7	1.5
Acceleration			
0 – 60 mph	3.0	1.1	5.9
0 – 25 mph	2.2	10.6	18.5
15 – 30 mph	1.2	25.0	45.5
Deceleration			
50 – 20 mph	1.2	10.2	2.9
30 – 15 mph	1.4	11.8	3.3
30 – 0 mph	2.5	10.3	2.9
		100.0	100.0

Figure II

APPENDIX

AMA TEST ROUTE

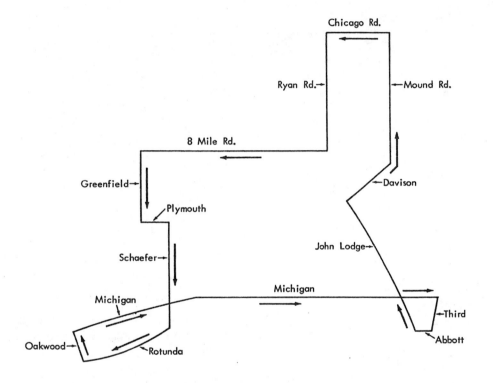

Detroit Area

Distance - Approx. 59 Miles

Travel Time - Approx. 2 Hours

Figure III

DEVELOPMENT AND EVAULATION OF AUTOMOBILE EXHAUST CATALYTIC CONVERTER SYSTEMS

ABSTRACT

For the past seven years, the Ford Motor Company has been working on the development of catalytic exhaust treating systems designed to minimize the emission of certain vehicle exhaust gas constituents. In 1959, the development of a low-temperature, catalytic-converter system for the oxidation of exhaust gas hydrocarbons was described in a paper presented to the SAE. That system, which used vanadium pentoxide as the catalyst, has since been extensively developed in a program that included 250,000 miles of converter evaluation on vehicles. Many of the basic system requirements and problems covered in those tests are relevant in vehicle applications of a catalytic converter system with any type of catalyst.

With the insertion of a carbon monoxide limit in the California Exhaust Standard, work on the low-temperature, catalytic converter system was discontinued since this system did not, and was not designed to, oxidize carbon monoxide. An engine-dynamometer catalyst screening procedure was developed and used to select catalysts capable of oxidizing both hydrocarbons and carbon monoxide for vehicle evaluation. Tests were made of catalysts submitted by outside suppliers, catalysts made by outside suppliers to Ford specifications, and catalysts made at the Ford Research Laboratories. From the results of the screening tests, three of the catalysts were accepted for vehicle tests.

Tests of vehicle systems employing catalysts selected from the screening tests showed the systems to be inadequate from the standpoint of the catalysts maintaining oxidation performance for extended mileage.

If a satisfactory catalyst can be found to improve the performance of the systems to an acceptable level, much work still remains to be done to develop an acceptable catalytic converter system for vehicles.

DEVELOPMENT AND EVALUATION OF AUTOMOBILE EXHAUST CATALYTIC CONVERTER SYSTEMS

H. C. Schaldenbrand and J. H. Struck

INTRODUCTION

Photochemical air pollution, or "smog" is a form of air pollution that occurs in Los Angeles County, California, and is reported to exist in other metropolitan areas of the United States.[1] The formation of photochemical smog has been attributed, in part, to emissions from automobiles. A photochemical reaction is believed to result from hydrocarbons and oxides of nitrogen, both present in automobile exhaust gases, combining in the presence of sunlight to form compounds causing eye irritation, reduced visibility and crop damage.[1,2]

Carbon monoxide has long been known to be toxic in high concentrations such as occur when a motor vehicle is operated in a closed garage. Recently, public health authorities have become concerned with the increasing atmospheric concentration levels of carbon monoxide. The main source of carbon monoxide in atmospheric air has been attributed to exhaust gases from automobiles.[3]

Treatment of automotive exhaust gases to alleviate the air pollution problems stated above can be approached from several directions. A method that has received considerable attention is the oxidation, by catalysis, of the hydrocarbons and carbon monoxide in the exhaust stream. In this manner, the quantity of one of the basic reactants in the formation of photochemical smog, hydrocarbons from automobile exhaust, would be lowered. By lowering the quantity of carbon monoxide emitted by automotive exhaust systems, the ambient-air carbon monoxide concentration would be correspondingly lowered.

Catalytic oxidation of exhaust gas hydrocarbons and carbon monoxide is accomplished by placing a device called a catalytic converter in the vehicle exhaust system. This catalytic converter contains a compartment, the catalyst bed, which is filled with catalytic material. Exhaust gas hydrocarbons and carbon monoxide are oxidized while passing through this catalyst bed. The catalytic material itself does not enter into the reaction, but only promotes the oxidation process. Usually, an air compressor is used to supply additional oxygen necessary for complete oxidation to the exhaust gas stream. A schematic drawing of a typical catalytic converter system is presented in Figure 1.

In August of 1957, the Ford Motor Company presented a paper to the SAE describing single-cylinder engine tests of oxidation catalysts for the removal of unburned hydrocarbons from automotive exhaust gases.[4] This work, which involved the screening of many catalysts, established the possibility of using vanadium oxide catalysts for this purpose. The vanadia catalysts were relatively efficient in oxidizing hydrocarbons without appreciable oxidation of carbon monoxide. The exclusion of carbon monoxide oxidation was advantageous from two standpoints. First, less additional air was required in the exhaust stream, reducing the cost of the secondary air pump; and secondly, lower-cost steels could be used in the converter because the high heat release associated with carbon monoxide oxidation is absent. At the time of this work, carbon monoxide was not considered an important air pollutant.[7] From the continuation of this work in 1959,

CATALYTIC CONVERTER SYSTEM

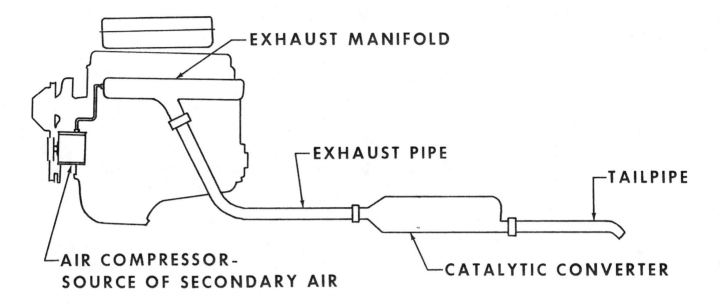

Figure 1

the Company presented two additional papers describing the development of a catalyst, vanadium pentoxide, and the application of this catalyst to a "low-temperature, catalytic-converter system", designed to oxidize exhaust gas hydrocarbons only.[5,6]

A program was conducted to determine the performance and catalyst life of the low-temperature, catalytic-converter system. This program was also aimed at defining and resolving problems associated with the operation of this system on vehicles. Much of this work involved principles common to those encountered with any catalytic converter system. These aspects of the program will be discussed in the first part of this paper.

An engine-dynamometer screening test was later devised and used to evaluate catalysts capable of oxidizing both hydrocarbons and carbon monoxide. Catalysts were selected from these tests for evaluation on vehicles.

The second part of this paper will present the development and results of the engine-dynamometer screening tests and the conclusions drawn from vehicle tests with selected catalysts.

PART I

LOW TEMPERATURE CATALYTIC CONVERTER SYSTEM DEVELOPMENT PROGRAM

Any catalytic converter system must meet three basic requirements if it is to perform its desired function of treating automobile exhaust emissions. These requirements are:

1. Performance

    The system must have a catalyst capable of oxidizing the desired exhaust constituents.

2. Catalyst Life

    The catalyst used must be able to maintain performance for extended mileage.

3. Vehicle Compatibility

    The system must be adaptable to a vehicle, including fitting within available space without affecting the performance of the vehicle or any of its components.

A catalytic converter system was designed to meet these requirements and was installed on a test vehicle. This system was evaluated as a base-line system for comparison in subsequent tests in which a particular system variable was studied to determine its effect on performance and catalyst life. Vanadium pentoxide catalyst was used in this base-line system. This catalyst had proven capable of oxidizing exhaust gas hydrocarbons for extended mileage.[6] An engine-driven air compressor with delivery characteristics as shown in Figure 2 was used to supply secondary air. Catalytic converter A-1 was incorporated as the base line converter. This converter, Figure 3, has a catalyst bed volume of 760 cubic inches with an 8 inch bed thickness. The system was adapted to a 1959 Ford sedan equipped with a V-8 engine and 2-speed automatic transmission. Except for a reduction in interior dimensions of the passenger compartment caused by altering the floor pan to fit the converter, compatibility of this converter system to the vehicle was considered adequate for test purposes.

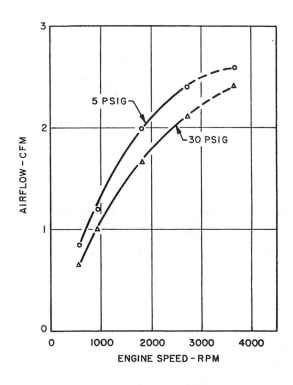

Figure 2

CONVERTER A-1 (CROSSFLOW DESIGN)

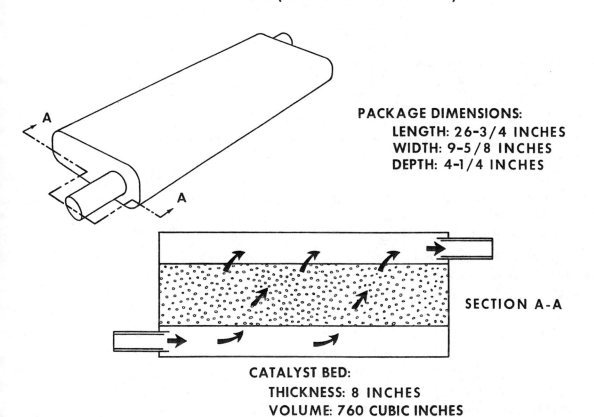

PACKAGE DIMENSIONS:
   LENGTH: 26-3/4 INCHES
   WIDTH: 9-5/8 INCHES
   DEPTH: 4-1/4 INCHES

SECTION A-A

CATALYST BED:
   THICKNESS: 8 INCHES
   VOLUME: 760 CUBIC INCHES

Figure 3

Evaluation of the Base-Line Converter System

The base-line converter system with converter A-1 was operated for 24,000 miles to evaluate oxidation performance and catalyst life. A premium commercial grade fuel was used throughout the test (see Appendix A for fuel analysis), and mileage was accumulated at the Ford Motor Company Dearborn Proving Ground according to a simulated city-suburban driving pattern. The exhaust pipe from the standard exhaust manifold to the converter inlet was wrapped with asbestos tape insulating material to conserve the sensible heat in the exhaust gases to improve performance. Similar converters were distributed to member companies of the Automobile Manufacturers Association and to California agencies for evaluation and for comparison of results.

The system was evaluated at periodic intervals according to an industry-recognized, stabilized-cruise efficiency test to determine oxidation performance of the catalyst with mileage. A warm-up procedure also conducted on a chassis dynamometer was incorporated in the evaluation test. Complete test details are given in Appendix B.

After a soak period of 12 hours, at ambient (indoor ambient during winter months) temperature, the vehicle was pushed into place on the chassis dynamometer and started. It was accelerated at a normal rate to 30 mph, road-load, and operated at constant throttle position for 30 minutes at this cruise condition. At the end of the "cruise 30 warm-up test", the car was operated at 60, 50 (or 45) and 30 mph, road-load conditions. Finally, the vehicle was operated at idle for six minutes. The cruise conditions were maintained at constant throttle until no change in temperatures or concentrations were observed. Exhaust gas concentrations from "before" and "after" the converter and system temperatures were monitored continuously during the entire test. From these measurements, oxidation efficiencies were calculated for the warm-up period and at the stabilized road-load conditions. The test was particularly advantageous from the standpoint of being reproducible in the determination of slight changes in oxidation performance with mileage.

The results of the stabilized cruise tests using the base-line converter A-1 are given in Figure 4. The oxidation performance as measured by this test showed relatively slow decrease in efficiency for 19,000 miles at each of the measured road load conditions. A more severe decrease in oxidation performance was noted at 24,000 miles. Some of the differences between the shapes of the curves for each condition can be attributed to differences inherent in the tests themselves such as measurement techniques, car tuning (ignition points and spark plugs were changed during the test as needed), and mechanical repairs to the converter. These test results are in general agreement with trends reported by other testing laboratories using the same converter system.[8]

Figure 5 shows the results of the cruise 30 warm-up test as measured initially and after 20,000 miles. In some of the tests, positive oxidation efficiency was noted for the first moments of operation during the warm-up test. This initial efficiency was probably due to adsorption of hydrocarbons in vapor form by the catalyst bed. The release of these adsorbed hydrocarbons would then account for the negative efficiencies always measured in the first few minutes of the test, as shown in Figure 5.

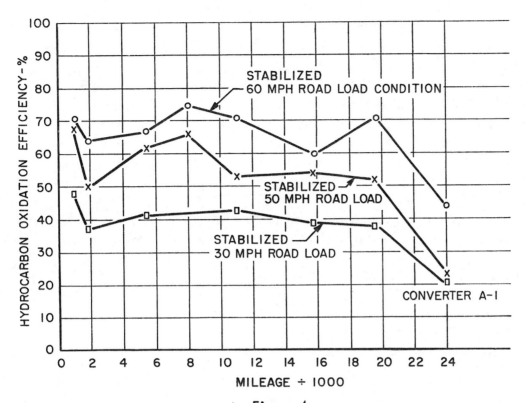

Figure 4

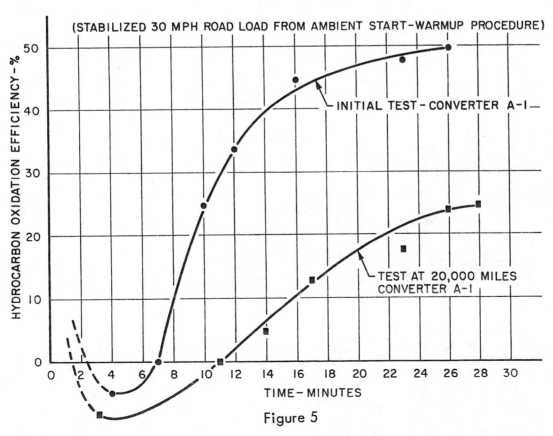

Figure 5

Measurements of hydrocarbon oxidation efficiency and catalyst bed temperatures (Figure 6) were taken at 30-mph operation either during the warm-up period, at stability, or during the cooling-off period following 50-mph operation. The results show that from 400°F to approximately 700°F catalyst bed temperature, hydrocarbon oxidation efficiency increased from 0 to 30%. In the 700°F range the change in efficiency is quite large with small temperature change, while from 700°F to 1280°F efficiencies of 60% to 95% were attained. While this particular curve is valid for only this catalyst and converter, other data were obtained showing the same general relationship of efficiency and temperature when using different converters and catalysts and at other operating conditions.

Development Program - Effect on System Performance and Catalyst Life of System Variables

A development program was conducted in which the following system variables were studied in tests to determine their effect on oxidation performance and catalyst life.

1. Converter Design - shape of catalyst bed and flow path
2. Duty Cycle - type of driving used to accumulate mileage
3. Amount of Catalyst Used
4. Use of Non-Leaded Fuel
5. Operation without Secondary Air

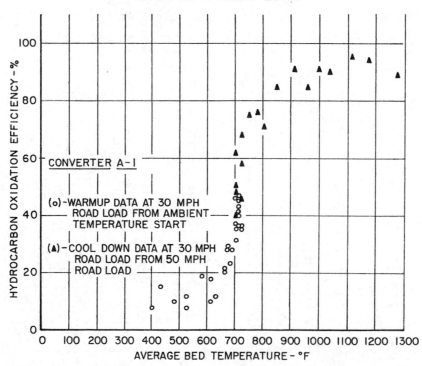

Figure 6

In each of these tests, with the exception of the duty-cycle test, the results were compared with those obtained using the base-line converter A-1. For the duty-cycle test, the results obtained from evaluating two other converters, B-1 and B-2, were used for comparison. The type of fuel used, test procedure, method of mileage accumulation and catalyst used (vanadium pentoxide) were identical in each case (except where noted) to those used in the A-1 tests.

Converter Design

The effects of converter design on oxidation performance and catalyst life were studied in tests of the converters shown in Figures 7 and 8. The converter

CONVERTER B-1 (DOWNFLOW DESIGN)

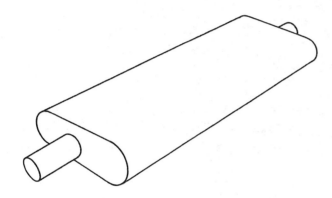

PACKAGE DIMENSIONS:
LENGTH: 26-3/4 INCHES
WIDTH: 11-3/4 INCHES
DEPTH: 4-1/4 INCHES

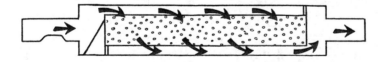

CATALYST BED DIMENSIONS:
THICKNESS: 3 INCHES
VOLUME: 800 CUBIC INCHES

Figure 7

shown in Figure 7 (B-1) was similar in exterior size and shape to the A-1 converter, but it differed in that the flow of exhaust gases was directed down through a bed of 3-inch thickness, in contrast to the cross flow of gases through the 8-inch thick bed in converter A-1. The amount of catalyst used was approximately the same in each converter (800 cubic inches - about 27 pounds of vanadia). Converter C-1, Figure 8, was also a downflow converter with catalyst capacity of approximately 800 cubic inches, but it differed radically from the other two models in shape. This round, pillbox-shaped converter incorporated an inclined, center-ramp inlet plenum from which exhaust gases were allowed to fan out through an upper chamber, drop downward through a 2-3/4 inch thick catalyst bed, and exit through an outlet plenum which was symmetrically opposite to the inlet plenum as shown in Figure 8.

CONVERTER C-1 (CIRCULAR DOWNFLOW DESIGN)

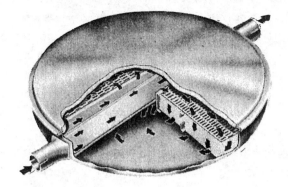

PACKAGE DIMENSIONS:
DIAMETER: 21-1/4 INCHES
THICKNESS: 4-7/8 INCHES

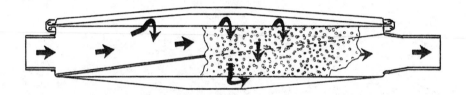

CATALYST BED DIMENSIONS
THICKNESS: 2-3/4 INCHES
VOLUME: 800 CUBIC INCHES

Figure 8

Hydrocarbon oxidation efficiency versus mileage for converters A-1, B-1, and C-1 is plotted in Figure 9 using an arithmetic average of the 60 and 50 mph road-load conditions of the stabilized efficiency test described above. The best overall performance throughout the accumulation of mileage was obtained with converter B-1, probably because the more uniform distribution of flow through the catalyst bed resulted in optimum utilization of available catalytic material. Flow tests performed on an airflow test stand with ambient air demonstrated that for the pillbox-shaped converter C-1, most of the flow was through the outer edges of the cylindrical catalyst bed, minimizing the effectiveness of a large portion of the bed. Overuse of the outer portion of the bed is the probable reason for the more rapid deterioration of this converter as compared to converters A-1 and B-1.

Duty-Cycle, High-Temperature Operation

The effect of the duty-cycle or driving pattern used in accumulating mileage on the converter system was investigated using converters B-1 and B-2. Converter B-2 is a downflow converter of the same size and shape as converter B-1. The only difference between these two converters is that the inlet and outlet pipes of converter B-2 were reshaped to improve flow distribution and reduce system back pressure. This point will be discussed in detail later.

Converter B-2 was tested for 6200 miles. Mileage was accumulated according to a high-duty test-track procedure which had an overall average speed of 42 miles per hour and a maximum speed of 80 miles per hour. The procedure used for accumulating mileage with converter B-1 and with all of the other converters tested (the city-suburban procedure) has an overall average speed of 25 miles per hour and a maximum speed of 60 miles per hour. The object of testing converter B-2 with the high duty cycle was to determine the effect of high-temperature operation on catalyst life.

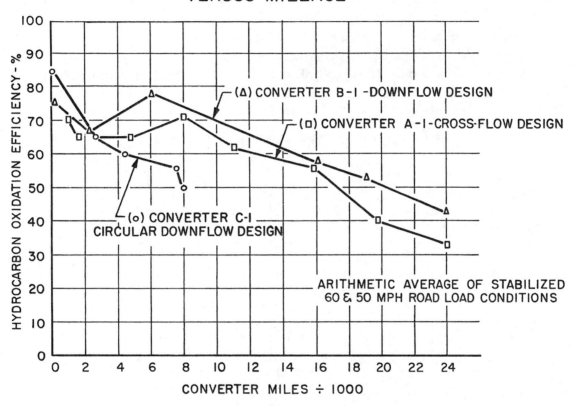

Figure 9

The results of this test are presented on Figure 10 along with those of converter B-1 for comparison, and show that catalyst life was severly affected by the high-temperature duty cycle. The upper operating temperature limit for vanadium pentoxide catalyst on gamma alumina support was known to be approximately $1200°F$ from previous work.[5] Converter B-2 catalyst bed temperatures were above $1200°F$ for 26 hours, or approximately 18 percent of its total operating time, and this was likely the cause of its shortened effective life.

Amount of Catalyst Used

The effect of the amount of catalyst on oxidation performance and catalyst life was explored with crossflow converters of the A-1 type. Converter A-2 has a catalyst capacity of 18 pounds versus the 27 pounds for converters A-1. The lower capacity was obtained by cutting off the corners of the catalyst bed, inlet plenum and outlet plenum, resulting in a rhombus-shaped converter. It was not expected that this alteration would affect flow distribution significantly since previous tests with crossflow converters had indicated that most of the flow was going through the center of the converter.

Converter A-2 was tested for 14,000 miles. Results in Figure 11 show that oxidation performance at 50 and 60 mph, as measured by the stabilized tests, depreciated faster with this converter than with converter A-1, which utilized the greater quantity of catalyst. It is believed that the longer catalyst life of converter A-1 was due to more exposed catalyst surfaces being available, and consequently a longer time was required to coat these surfaces with deposits that choke off the internal catalyst-covered surfaces of the support material. When the catalyst in the middle of the bed of converter A-1 did become coated with deposits, increased restriction in that area probably caused flow to increase in other portions of the bed (the corners) which were less used previously. Converter A-3 which contained only 10 pounds of vanadium pentoxide was evaluated only once to show that, when new, as little as 10 pounds of catalyst is sufficient for effective oxidation at cruise conditions as shown in Figure 11.

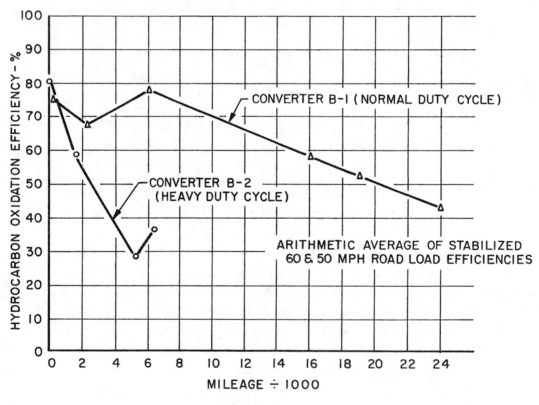

Figure 10

Use of Non-Leaded Fuel

Not all of the factors which affect catalyst life stem from the lead in the gasoline. The oxidation efficiencies of converter A-1 were compared with those of converter A-4, which was evaluated for 10,000 miles of vehicle operation with non-leaded fuel (Indolene Clear, Appendix A). As mentioned previously, a premium grade com-

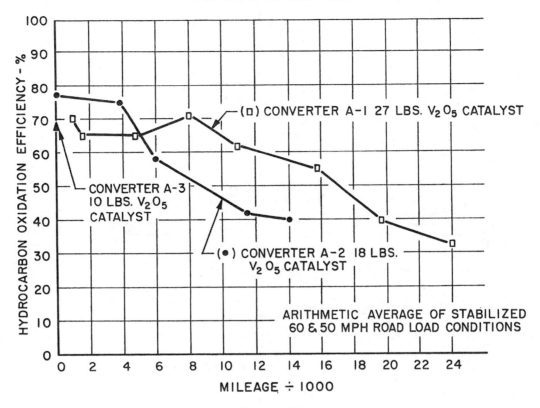

Figure 11

mercial fuel (with lead) was used throughout the test with converter A-1. It was recognized that the concentration levels measured with the Model 15A Liston Becker infrared analyzer vary somewhat with the types of hydrocarbons present in the exhaust stream, which in turn depend to some degree on fuel composition. However, the fuels used in this test were sufficiently similar that no serious change in Liston Becker measurements between the two systems could be expected. Furthermore, this effect was minimized by using oxidation efficiencies for comparison.

A comparison of the 60-mph, road-load condition oxidation efficiencies for converters A-1 and A-4 is presented in Figure 12 and shows no significant difference in the oxidation performance of these two converters. Similar results were shown at the other operating conditions measured (Appendix D). The results of this test are not considered conclusive evidence that lead has no effect upon catalyst life, since a full investigation of this subject would require a multitude of tests beyond the scope of this program.

Operation Without Secondary Air

A pump to supply additional air to the exhaust stream is needed to insure the availability of sufficient oxygen at all engine operating conditions. Figure 13 is a typical carburetor flow curve showing the air-fuel ratio of the mixture supplied to the engine at idle, road-load, and wide-open-throttle operating conditions versus airflow through the carburetor. Also shown are the limits of a band within which the air-fuel

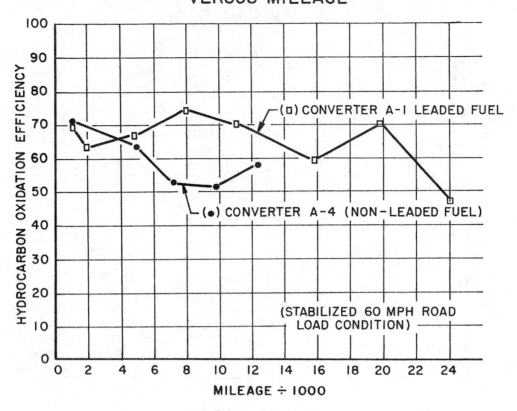

Figure 12

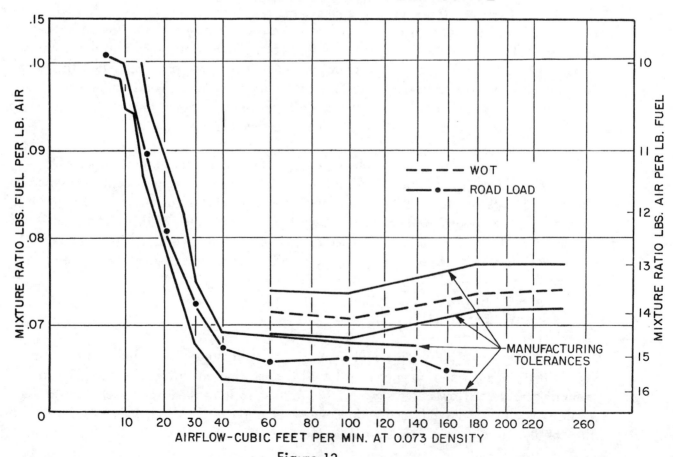

Figure 13

(or fuel-air shown by the left hand ordinate) ratio for a particular carburetor corresponding to a particular engine must lie in order to obtain satisfactory vehicle performance and economy. The point on the curve farthest to the left (at 6 cfm airflow) represents idle operation. All other points on the descending curve (solid line) to the right of this idle point represent road-load operating conditions. The horizontal dotted upper line denotes wide-open-throttle operation. For a fuel with a particular hydrogen-to-carbon ratio, there is one chemically correct air-fuel ratio, called the stoichiometric ratio, which represents the minimum air-to-fuel ratio or mixture with which it is theoretically possible to oxidize completely all of the carbon and hydrogen present in the fuel. On Figure 14, this ratio is shown on the same typical carburetor flow curve for a fuel with a hydrogen-to-carbon ratio of 1.86. The stoichiometric air-fuel ratio for this fuel is 14.6:1 (calculations are shown in Appendix C). The crosshatched region shown beneath this line represents those conditions when theoretically enough oxygen is present in the air-fuel mixture for complete oxidation. This area falls generally within the normal road-load cruise operating conditions of the engine. At idle, at road-load conditions below 38 cfm airflow, and at all wide-open-throttle conditions, insufficient air is supplied to the engine for complete oxidation of all of the fuel. Figure 15 is a plot, on this same carburetor flow curve, showing the added air needed for complete oxidation at the operating conditions where insufficient oxygen is available. Up to 18 cubic feet of air are required. Sample calculations of the stoichiometric air-fuel ratio and secondary air requirements are given in Appendix C.

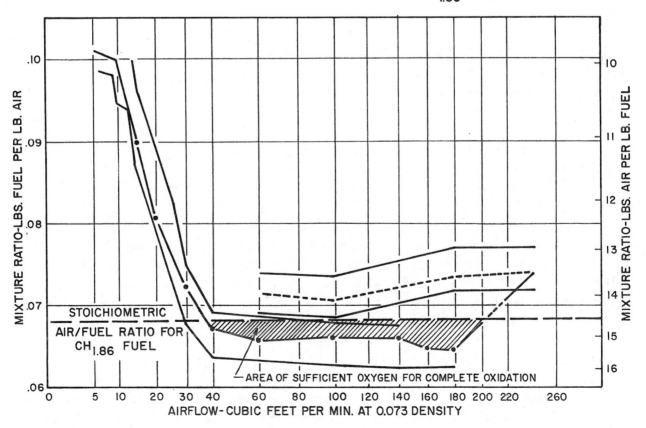

Figure 14

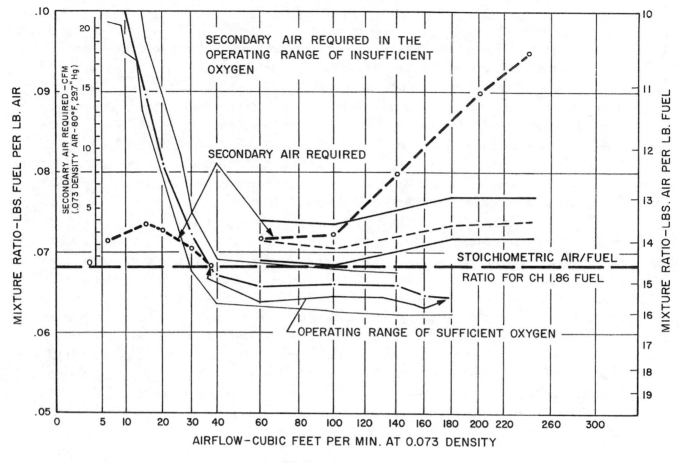

Figure 15

In the systems tested using vanadium pentoxide as the catalyst, additional air was supplied by one or two single-cylinder, positive-displacement pumps with flow characteristics shown in Figure 2. The air was introduced into the exhaust manifold. The capacity of these pumps was considered sufficient to cover the majority of engine operating conditions encountered in normal driving.

A test was run with converter A-5, a crossflow converter similar to the base line converter A-1, in which no secondary air was added. Vanadium pentoxide is known to have several oxidation states, and it is readily possible to change its oxidation state.

This test was conducted to determine if sufficient oxygen for acceptable oxidation performance could be supplied by the catalyst itself during periods of insufficient oxygen and be recovered during periods when an excess of oxygen is present without detrimental effect to catalyst life. Results of this test are shown on Figure 16 where hydrocarbon oxidation efficiency versus mileage is plotted for converters A-5 and A-1. No significant difference in either oxidation performance or catalyst life was shown in operating converter A-5 without secondary air. It was noted, however, that the odor emitted by converter A-5 was more intense than that emitted by other converters operated with secondary air. Tabulated data for this test with converter A-5 and for all of the tests in the controlled variables development program are presented in Appendix D. Stabilized efficiencies and catalyst bed temperatures for each condition of the stabilized cruise test are given.

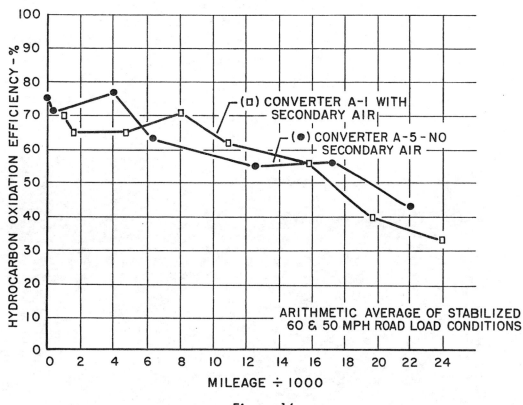

Figure 16

Problems Encountered in the Application and Use of Converter Systems on Vehicles

The development program described in the preceding section investigated the effects of certain controlled variables on catalyst performance and life. Secondary air requirements were also investigated. These tests were performed to evaluate the low-temperature catalytic converter system from the standpoint of the first two basic requirements of a catalytic converter system, performance, and catalyst life. The third basic requirement, vehicle compatibility, applies to the application and operation of a catalytic converter system on a vehicle. Particular problem areas concerning this requirement which were investigated include:

1. Effect on Vehicle Components
2. System Backpressure
3. Overtemperature Protection
4. Odor
5. Sound

Effect on Vehicle Components

This term defines the capability of the system to fit within the available space on the vehicle without affecting the operation or performance of the vehicle or any of its components. Compatability will be discussed here in terms of converter C-1, the pillbox-shaped converter which was designed to fit beneath the passenger compartment of certain 1960 vehicles with a minimum of alteration to the vehicle. To satisfy this design objective, it was necessary to alter the floorboard of the passenger compartment as shown in Figure 17. For passenger comfort and vehicle protection the floorboard above the converter was insulated. Use of the converter on this car also necessitated removal of the power seat motor and rerouting of the emergency brake cables. On this particular vehicle, it was not necessary to alter the fuel and brake lines, and no adverse conditions pertaining to their operation were encountered.

Figure 17

System Back Pressure

The use of an exhaust-system device will increase exhaust-system back pressure and, thus decrease engine performance, due either to the restriction imposed by the device itself or to thermal effects increasing the volume of flow through the system. This problem can be minimized with proper converter design, and this was attempted with the vanadia systems.

Pressure drop versus air-flow curves for some of the typical converters evaluated in this program are given in Figure 18. These measurements were made on a test stand with ambient air to determine the restriction imposed by the converters for comparison with each other and with a standard muffler. Converter B-2 (downflow design) was initially of the same design as converter B-1. After redesign of the inlet and outlet pipes of the converter to smooth the transition from the round inlet pipe to the thin, flat, plenum shape and from the outlet plenum to the outlet pipe, the restriction imposed by this converter was significantly reduced as shown.

PRESSURE DROP VERSUS AMBIENT AIRFLOW

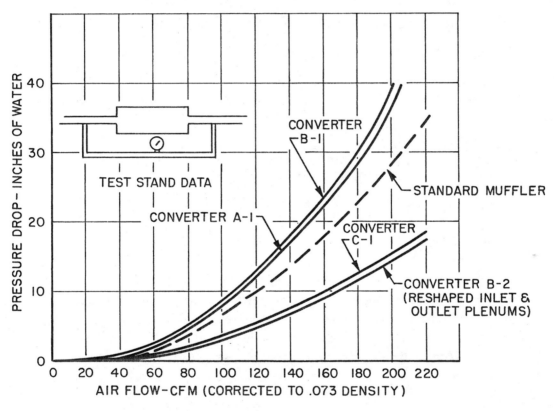

Figure 18

Exhaust system back pressures of a vehicle equipped with converter B-2 after 5300 miles of operation were then measured up to 60 mph road-load speeds using a chassis dynamometer, and the results were compared with those of a standard vehicle. Measurements were taken between points one pipe diameter downstream from the exhaust manifold flange and twelve inches upstream from the tailpipe exit. Results in terms of pressure drop versus road-load speed for the two tests are shown in Figure 19. The data show that the catalytic converter system had higher back pressures than the standard system. This was probably due to a combination of increased heat in the exhaust system due to inlet pipe insulation and operation of the converter which resulted in higher flow volume, plus a plugging of the retaining screens in the catalyst bed by the pellets, which resulted in increased flow restriction by the converter.

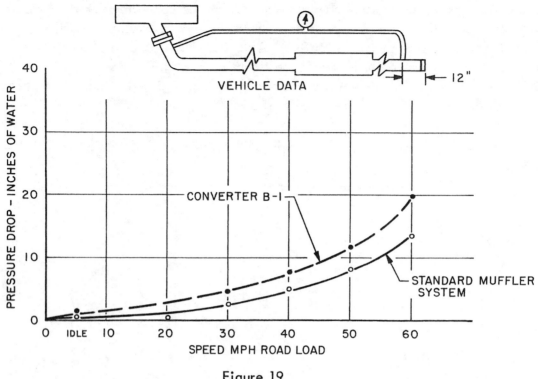

Figure 19

Over-Temperature Protection

With vanadia catalyst it is imperative to keep catalyst temperatures below 1200 °F. Higher temperatures change the structure of the gamma alumina support and reduce catalyst effectiveness. Continuous operation above 1200 °F changes gamma alumina to alpha-type alumina, or corundum, with loss of surface area and, consequently, activity. Many catalysts have an upper temperature limit within the probable operating range. This limit is dependent on the support material used or the thermal stability of the catalyst material itself.

The vanadia systems evaluated in the development program were not provided with a valve that would enable the exhaust gases to bypass the converter bed. Instead, an alarm system was used which warned the driver to reduce speed when the catalyst bed approached 1200 °F. It seems likely, however, that a positive acting bypass system would be necessary in the practical application of a catalytic converter system employing a catalyst which has an upper temperature limit.

Odor

In all of the tests with the low-temperature catalytic converter system, a characteristic odor was emitted from the tailpipe, which is unlike the odor of untreated exhaust gases. The odor varied in character and intensity between converter systems and in the same system at various oxidation efficiencies. The hedonic character of the odor was judged, at best, mildly pleasant, although most reaction tended toward strong objection. The typical odor was judged to consist of two components: an almond odor not unpleasant to some, and an acrid odor generally irritating.

Preliminary work on defining the intensity of the odor was attempted by collecting exhaust condensates from standard cars and converter-equipped vehicles at various engine operating conditions. These condensates diluted with distilled water, in known quantities, were subjected randomly to a panel of "sniffers," who were asked whether or not they could detect an odor in the sample. No definitive pattern of odor intensity was obtained from panel answers in this test.

Another series of tests was conducted with converter systems B-2 and A-5, the former system being operated with secondary air and the latter without secondary air. In these tests, exhaust gas was brought directly into the passenger compartment of the vehicle through a half-inch copper tube inserted into the exhaust system directly behind the catalytic converter. Various fuels were used for the engine, and the relative intensities of both the almond and acrid portions of the odor judged on a 10-point-maximum basis by two observers. Test results obtained at 60 mph road-load-cruise, which produced typical odors, are shown in Table I. A significant difference in relative intensity of the odors was observed on both converter systems with and without air when using different fuels. The limited scope of this experiment precludes any definite conclusions. The results also indicate that operation with secondary air decreased the intensity of the acrid portion of the odor. Similarly, odors were observed in testing catalytic converter systems utilizing catalysts other than vanadium pentoxide.

Sound

The systems tested were considered, in the opinion of automotive engineers active in muffler design and evaluation, to be adequate from the standpoint of sound although below the overall silencing quality of a standard muffler. It would be necessary on some cars to add a resonator to the system for filtering out low frequencies. A separate muffler system would be required if a bypass system were used for over-temperature control.

Summary of the Results of the Development Program for the Low-Temperature Catalytic Converter System

The results of this development program for the low-temperature catalytic converter system showed that this system, as tested, is not acceptable for use as a practical exhaust treating device for the following reasons:

- The large amount of catalyst necessary to sustain oxidation performance for extended mileages.
- The longer-than-desirable warmup time.
- The quick deterioration of warmup characteristics with mileage.
- The inherent difficulties in adapting the system to a vehicle: over-temperature protection, odor, back pressure, and associated problems.

TABLE I

RELATIVE INTENSITIES OF TYPICAL CATALYTIC CONVERTER EXHAUST ODORS

(Vehicle Operating Condition: 60 mph Road Load Cruise)

Converter System	Fuel	Odor Intensity (judged on basis of 10 points maximum by two observers	
		Almond	Acrid
B-2 (With secondary air)	full boiling range premium blend	10	7
	Iso-octane	1	1
	Toluene	3	3
A-5 (Without secondary air)	full boiling range premium blend	10	10
	Iso-octane	1	1
	Indolene Clear	5	7

The results achieved and conclusions drawn from the development of this system are significant, since the problems encountered are likely to appear in development of any catalytic converter system.

PART II

ENGINE DYNAMOMETER SCREENING TESTS AND VEHICLE EVALUATIONS OF SCREENING TEST CATALYSTS

Development of the low-temperature catalytic converter system was discontinued because of the inability of the catalyst, vanadium pentoxide, to oxidize carbon monoxide. The requirement for carbon monoxide oxidation was published in the California Exhaust Standard of December 4, 1959.[9] A program was then initiated to develop a catalytic converter system for the oxidation of both hydrocarbons and carbon monoxide. This work was carried out in two stages:

1. Development and application of a catalyst screening test procedure to find acceptable catalysts.
2. Evaluation in vehicle converter systems of the catalysts judged acceptable by the screening tests.

Evaluated in this program were catalysts submitted by outside suppliers, catalysts made by outside suppliers to Ford specifications, and catalysts made at the Ford Research Laboratories.

Of the 17 catalysts tested according to the screening test procedure, three were made by Ford. The remaining catalysts were selected from among those submitted by more than ten outside companies. Catalysts were selected for screening on the basis of manufacturers' data on the performance of the catalyst when subjected to vehicle exhaust streams. Some of these catalysts were resubmitted two or three times following changes by the supplier to correct deficiencies indicated in the screening tests.

Catalyst Screening Tests

An engine dynamometer screening test was devised to gain a perspective of the qualifications of various catalysts for potential application to a vehicle system. The engine dynamometer was accepted for use in screening catalysts since it is economically feasible from the standpoint of time and equipment required to complete a test and because repeatability of engine operating conditions can be maintained throughout the test for valid comparisons of catalyst performance. Basically, the screening test consists of:

1. Adaptation of each catalyst to a screening test converter system.
2. Evaluation of each catalyst according to a 100-hour test procedure.
3. Analysis of results to determine acceptability for vehicle evaluation.

Screening Test Converter Systems

The test converter used to evaluate catalysts suited for muffler type locations is shown in Figure 20. This converter (D-1) is designed to treat the exhaust gases from one-half of an engine (4 cylinders of a V-8 engine) and has a 3-3/4 inch-thick catalyst bed of 260-cubic-inch volume. The converter is fabricated from SAE 1020 carbon steel and is designed for easy access to the catalyst bed. To insure that the

CONVERTER D-1 - CATALYST SCREENING TEST CONVERTER

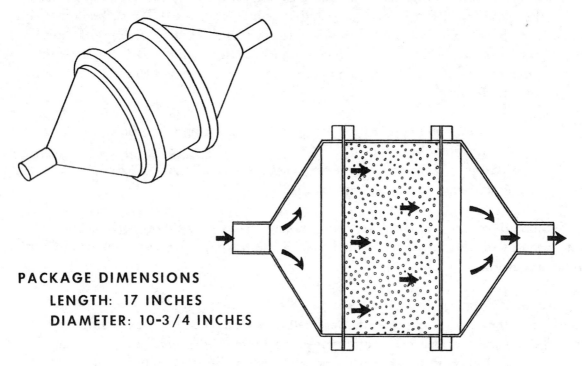

PACKAGE DIMENSIONS
LENGTH: 17 INCHES
DIAMETER: 10-3/4 INCHES

CATALYST BED DIMENSIONS
DEPTH: 4 INCHES
VOLUME: 260 CUBIC INCHES

Figure 20

converter accepted all of the exhaust from the four cylinders, the exhaust gas crossover (for carburetor heat) was blocked at its entrance into the left-hand exhaust manifold. The converter was then installed on the left-hand side of the engine as shown in Figure 21, which is an illustration of a typical screening test installation.

Catalyst operating temperatures were controlled, according to the particular manufacturer's recommendations, by placing the test converters closer to or farther from the exhaust manifold flange. Any other manufacturer's restriction or recommendations were also followed.

Exhaust-gas-sample taps were located in the inlet and outlet plenum cones of the cylindrical test converter. Thermocouple stations were placed adjacent to each sample tap and in the geometric center of the catalyst bed (shown on Figure 21). Secondary

Figure 21

air was supplied by either one or two engine-mounted compressors as required. The secondary air was introduced into the front of the exhaust manifold. The compressors were identical to those used in evaluating the low-temperature catalytic converter system.

Evaluations of several catalysts which were capable of high temperature operation were conducted using the manifold converter E-1 shown in Figure 22. With this converter, exhaust gas was sampled only from the outlet pipe of the converter, since it was not possible to obtain a representative inlet gas sample. Secondary air was added ahead of the manifold into the cylinder head. Temperatures were monitored in the inlet plenum, catalyst bed, and at the exhaust-gas-sample tap. A typical installation using this converter appears in Figure 23.

Engine Dynamometer Test Procedures and Techniques

Two dynamometer test cycles were used for evaluation of catalysts on the screening program. A one-hour evaluation cycle was used consisting of five modes typical of the engine operating conditions encountered in a vehicle. The cycle permits the evaluation of stabilized low- and high-speed cruises, idle, and an

CONVERTER E-1 - SCREENING TEST
MANIFOLD CATALYTIC CONVERTER

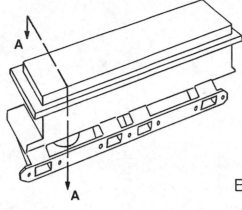

PACKAGE DIMENSIONS:
 LENGTH: 18-3/4 INCHES
 WIDTH: 5 INCHES
 DEPTH: 3-5/8 INCHES

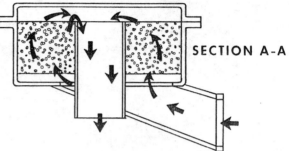

SECTION A-A

CATALYST BED DIMENSIONS:
 THICKNESS: 2-3/4 INCHES
 VOLUME: 180 CUBIC INCHES

Figure 22

Figure 23

indication of acceleration and deceleration performance of the catalyst. A warmup procedure similar to that used in testing the low-temperature catalytic system was also incorporated in the test (constant-throttle, 30-mph road-load from an ambient-temperature start). Engine speeds and loads are detailed in Appendix E.

A second one-hour cycle (the "durability" cycle), also given in Appendix E, was used to accumulate time or dynamometer "mileage." This cycle consists of 54 minutes of various simulated cruises and 6 minutes of alternate wide-open and part-throttle operation.

Exhaust gas analyses were conducted only during the one-hour evaluation cycles. Temperatures were monitored continuously during all cycles. The evaluation cycle was usually run at the beginning of the work-day from a cold start, after which the catalyst was continuously run using the durability cycle. Usually, at least ten hours were accumulated with the durability cycle before another evaluation cycle was conducted.

The standard test was terminated at 100 hours which, if the operating conditions were reproduced in actual driving, would amount to approximately 5,000 miles.

Screening Test Results

Catalysts were screened using the cylindrical converter D-1 and test procedures given above. Each test was made in compliance with the supplier's specifications as to basic catalyst requirements, such as secondary air and maximum operating temperature. Results of these tests will be presented in terms of stabilized, 30-mph road-load oxidation efficiencies versus dynamometer test hours. Tabulated data of a typical screening test containing the results for each mode of the evaluation cycle are given in Appendix F.

A test using vanadium pentoxide was run for 100 hours and used as a base line for comparison of the hydrocarbon oxidation performance of the other catalysts. This standard was desirable in the light of the experience accumulated on vehicles with this catalyst during development of the low-temperature catalytic converter system.

Figure 24 shows the results of 12 screening tests on the basis of 30-mph road-load hydrocarbon and carbon monoxide oxidation efficiencies. All of these catalysts were rejected for further testing on the basis of poor oxidation performance, deterioration of oxidation performance with time, or debilitation of the catalytic material.

Results from the tests of those catalysts accepted for vehicle application are shown in Figure 25. These three catalysts (M, N and O) oxidized both hydrocarbons and carbon monoxide and exhibited capability of maintaining oxidation performance for a significant time.

Four catalysts were evaluated using the manifold converter E-1. However, only catalyst "Y" had sufficient physical strength to remain in the converter for the

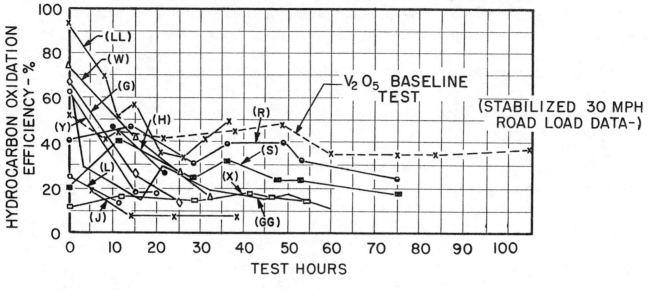
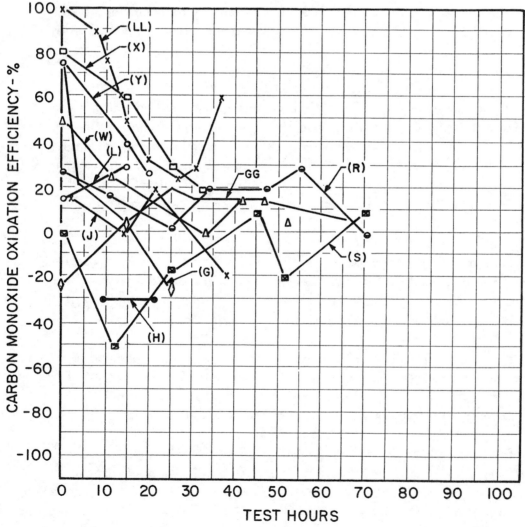

Figure 24

PERCENT HYDROCARBON AND CARBON MONOXIDE OXIDATION EFFICIENCY VERSUS TEST HOURS - CATALYSTS SELECTED FOR VEHICLE TESTS

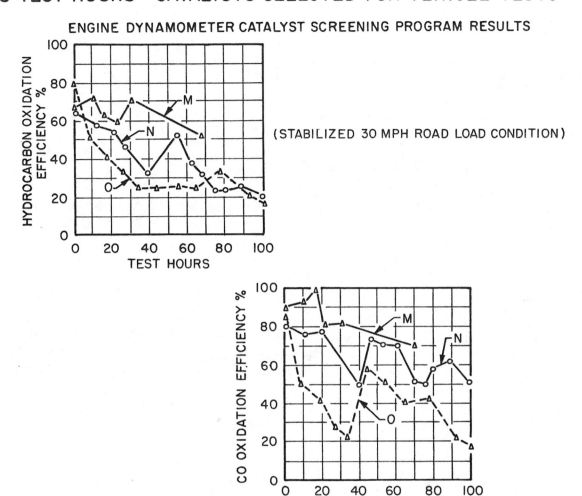

Figure 25

entire 100-hour test. The results from this test are shown as outlet concentrations versus dynamometer hours since it was not possible with this converter to obtain a representative inlet sample for calculation of oxidation efficiencies. The results of this test for the 60 and 30-mph road-load conditions of the evaluation cycle are presented in Figure 26. The application of this catalyst in a manifold converter was attractive in view of the low overall concentration levels obtained in all modes of the cycle and the physical durability of the catalyst after high temperature exposure for 100 hours.

Upon analysis of all of the screening tests, it was decided to evaluate catalysts "O" and "N" on a vehicle in a muffler type converter system. Catalyst "Y" was selected for further evaluation in a manifold catalytic converter system.

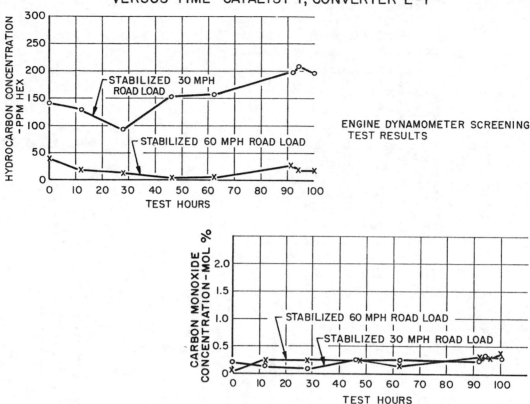

Figure 26

Vehicle Evaluations of Screening Test Catalysts

Two converters were used in vehicle tests for undercar placement, both of 400-cubic-inch capacity. The quantity of catalyst was reduced from the 800-cubic-inch volume used in the low-temperature converter systems, since that size was considered uneconomical and too large for practical vehicle installation. The smaller size also benefits warmup, since less time is required to heat the catalytic material to oxidation temperatures. The converter shown in Figure 27 (converter F-1) incorporated an inclined bed designed to improve flow distribution. The catalyst bed had a thickness of two inches and a volume of 420 cubic inches. Converter G-1 shown in Figure 28 was a half-length version of converter B-2 used for the vanadia catalyst and had a bed thickness of 3 inches and volume of 400 cubic inches. Converter F-1 was fabricated of SAE 1020 steel, with the exception of the retaining screens which were stainless steel. Converter G-1 was constructed entirely of stainless steel.

Secondary air was supplied to these systems in the same manner as with the vanadia systems using two of the same engine-mounted compressors. The systems were adapted to vehicles equipped with 352-cubic-inch displacement engines and three-speed automatic transmissions.

CONVERTER F-1 (DOWNFLOW DESIGN)

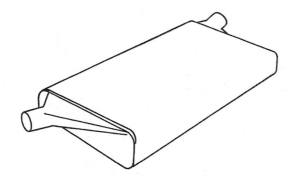

PACKAGE DIMENSIONS:
　　LENGTH: 18-7/8 INCHES
　　WIDTH: 10-7/8 INCHES
　　HEIGHT: 3-3/8 INCHES

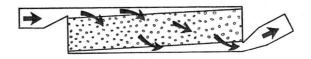

CATALYST BED DIMENSIONS:
　　THICKNESS: 2 INCHES
　　VOLUME: 420 CUBIC INCHES

Figure 27

CONVERTER G-1 (DOWNFLOW DESIGN)

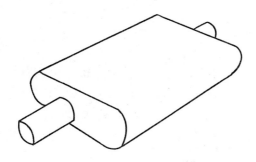

PACKAGE DIMENSIONS:
　　LENGTH: 12-5/8 INCHES
　　WIDTH: 11-5/8 INCHES
　　HEIGHT: 4-1/4 INCHES

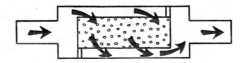

CATALYST BED:
　　THICKNESS: 3 INCHES
　　VOLUME: 400 CUBIC INCHES

Figure 28

Figure 29 is a schematic of the manifold converter (H-1) used in the vehicle tests. This converter was designed for use with a six-cylinder engine and was installed on a vehicle equipped with a 144-cubic-inch displacement engine and two-speed automatic transmission. Secondary air was added into the cylinder-head exhaust ports using a pump with delivery characteristics as shown in Figure 30. The material used in construction of this converter was high-temperature alloy steel, including the screen above and below the catalyst bed. Catalyst bed volume was 105 cubic inches with a thickness of 1 5/8 inches.

MANIFOLD CONVERTER H-1 (6 CYLINDER MODEL)

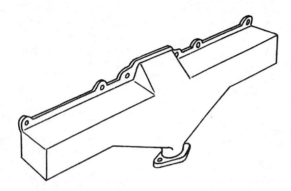

OVERALL LENGTH: 23-5/8 INCHES

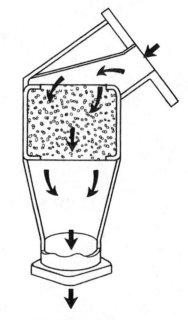

CATALYST BED:
VOLUME: 105 CUBIC INCHES
THICKNESS: 1-5/8 INCHES

Figure 29

Test Procedures

Vehicle tests were conducted according to the following test procedures:

- Stabilized cruise and cruise 30 warmup
- Detroit Traffic Route - operation in normal traffic.
- Individual 11 mode - selected conditions typical of urban driving.
- Bag technique - total exhaust from chassis dynamometer cycle.
- California Motor Vehicle Pollution Control Board chassis dynamometer test procedure

AIRFLOW VERSUS ENGINE SPEED AT 15 PSIG DISCHARGE PRESSURE - DOUBLE PISTON, POSITIVE DISPLACEMENT PUMP

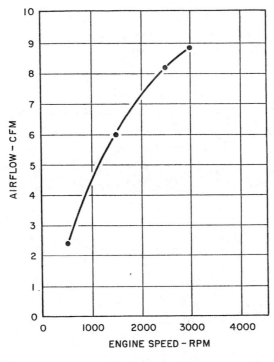

Figure 30

The stabilized-cruise and cruise-30 warmup procedures are the same as those used to evaluate the low-temperature catalytic converter system and are detailed in Appendix A. These procedures were reused because they are the only tests which give repeatable data on the oxidation state of a catalyst with mileage.

The Detroit traffic route procedure is a test developed by member companies of the Automobile Manufacturers Association and used in cooperative air-pollution work. It consists of sixty miles of city-suburban driving in the Detroit area during which exhaust gas concentrations are continuously monitored. Concentrations of hydrocarbons, carbon dioxide, and carbon monoxide are alternately taken from before and after the catalytic converter at one minute intervals. The one-minute concentration readings are averaged for the entire test, and the oxidation efficiency is determined. This procedure was used to gain knowledge of the oxidation performance of the device as the vehicle is driven in normal city traffic. Since the concentrations are averaged for a time duration of two hours, warmup characteristics are not shown by the final numbers.

The individual 11 mode tests are evaluated using a warmed-up converter. Time-average concentrations are obtained for each of 11 modes derived from an AMA traffic survey made in the Los Angeles area.[14] The time-average concentrations are then multiplied by volume-weighting factors as specified in the California Exhaust Standard. The final average concentrations obtained in this test do not show warmup characteristics.

The bag sample technique is a test developed in the Ford Research Laboratories in which the total exhaust gas from the vehicle is collected in an evacuated polyethylene bag.[13] Average concentration of hydrocarbons, carbon monoxide, and carbon dioxide in the bag sample are measured at the conclusion of the test. During collection of the exhaust sample, the vehicle is operated on a chassis dynamometer according to a representative driving cycle adapted from the one used by the Coordinating Research Council in the 1956 Field Survey.

The California Motor Vehicle Pollution Control Board chassis dynamometer test procedure (hereafter referred to as MVPCB test) was conducted in a manner similar to the specified requirements in the second draft of that procedure dated May 18, 1961.[10] Differences in instrument-sample system and method of analysis are included in Appendix G where the procedure, analysis technique, and sample systems used for all of the above techniques are described.

Results of Vehicle Tests with the Screening Program Catalysts

Catalyst "O" was tested using converter F-1 located beneath the passenger compartment of a 1960 station wagon. Results of the test to 8300 miles are shown on Figure 31 in terms of stabilized efficiencies at 30-mph road-load. Initial tests were

HYDROCARBON & CARBON MONOXIDE OXIDATION
EFFICIENCY VERSUS MILEAGE

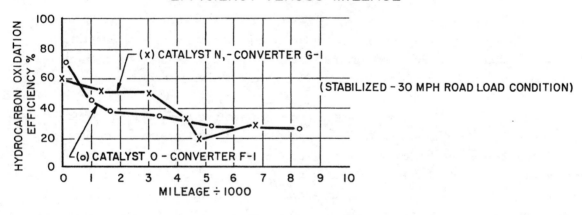

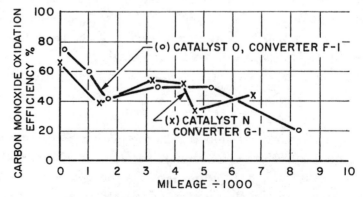

Figure 31

excellent, with conversion efficiencies of nearly 80% for both hydrocarbons and carbon monoxide. Oxidation efficiencies after 8300 miles, however, had declined to 20%. Cruise-30 warmup characteristics declined as shown by Figure 32. Initially, maximum oxidation was achieved in only 5 minutes from a cold start. After 8000 miles, it took approximately 10 minutes to achieve a maximum efficiency that was approximately only

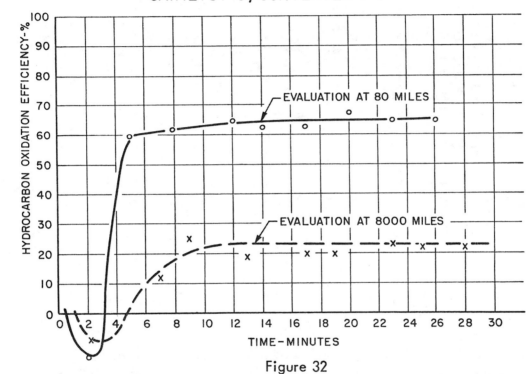

Figure 32

30% of the initial level. Evaluation of the system was discontinued at this point, and the converter was opened for visual inspection of the pellets. The material was heavily coated with a white deposit. The pellets themselves had not deteriorated physically other than a rounding off of sharp corners, due probably to a grinding down of the edges of the cylindrical pellets from physical contact with each other.

Converter G-1 was used to evaluate catalyst "N". This converter was also installed in a 1960 station wagon, which accumulated 9300 miles with the converter. Stabilized efficiencies obtained with this system at 30-mph road-load are shown plotted versus mileage on Figure 31. The system was also evaluated according to the Detroit traffic route procedure, with hydrocarbon and carbon monoxide oxidation efficiencies versus mileage shown on Figure 33. Upon inspection of this converter after 9300 miles, it was found that only 1/8 of the catalyst remained. The pellets showed signs of severe attrition, and the converter was filled with catalytic dust much different in color from the original material. The converter screens were checked for holes that might have opened up from heat to allow the catalyst to escape, but none were found.

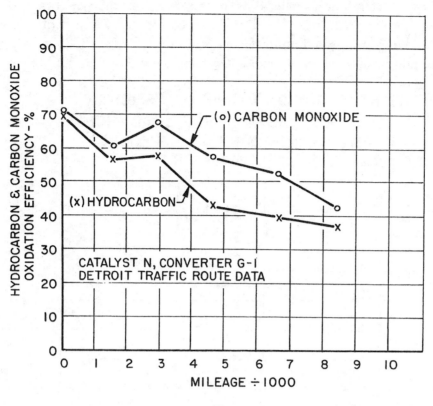

Figure 33

The loss of catalyst must have occurred from chemical decomposition or physical breakdown of the catalyst itself. Tabulated data for these tests with catalysts "N" and "O" appear in Appendix H.

Catalysts "O" and "Y" were evaluated in combination with the manifold catalytic converter H-1. Catalyst "Y" was selected on the basis of the 100 hours screening test described previously. It was later decided to evaluate catalyst "O" since its physical durability was good for 8000 miles on the vehicle test. It was theorized that at the high temperatures expected to be attained in the manifold converter, many of the possible exhaust gas compounds such as the lead oxides, which could be harmful to the catalyst, would pass through the converter in vapor form and not deposit on the catalyst surface. If this were true, a catalyst such as catalyst "O" might retain its oxidation performance in this environment. It was also desirable to test catalyst "O" since comparative tests with catalyst "Y" in the cylindrical screening test converter indicated it to be a more active catalyst.

The results of five tests with this converter are given in Table II. Catalyst "O" was evaluated first. Three MVPCB chassis dynamometer tests were conducted using this catalyst with hydrocarbon and carbon monoxide outlet concentrations ranging from 316 to 247 ppm and 0.91 to 0.69 mole % respectively. At the end of the third test, which covered a time period of about one week, the converter was found only half full. The remaining pellets were found still hard in composition, but had been worn down to a much smaller size permitting escape through the retaining screens.

TABLE II

TEST RESULTS USING MANIFOLD CATALYTIC CONVERTER H-1

(Vehicle equipped with 6 cylinder, 144 cubic inch displacement engine, 2 speed automatic transmission)

Test No.	Catalyst	Motor Vehicle Pollution Control Board Chassis Dynamometer Test Procedure	
		hydrocarbon concentration (ppm hex)	carbon monoxide concentrations (mole %)
1	O	316	0.69
2	O	271	0.69
*3	O	247	0.91
4	Y	261	1.21
5	Y	344	0.72
Test Results Using Standard Exhaust Manifold on Same Vehicle			
6	--	606	2.69

* This test run with one piston of double piston secondary air pump disconnected.

The converter was then filled with catalyst "Y." Two tests were conducted (MVPCB test) with resulting outlet concentrations of hydrocarbon and carbon monoxide attained being 261 and 344 ppm, and 1.21 and 0.72 mole % respectively. For a relative determination of the oxidation efficiency of this system, a test with this vehicle was conducted using a standard exhaust manifold. Results of this test (shown on Table II) reveal that the converter, filled with catalyst "Y," is approximately 50% effective in oxidizing hydrocarbons and approximately 65% effective in carbon monoxide oxidation. Temperatures of 1500 to 1700° F were measured in the catalyst bed during these tests. The converter was removed from the vehicle after the second test with catalyst "Y," since the converter inlet flange had severely warped from the high-temperature operation, permitting leakage of exhaust gases at the point where the flange is bolted to the cylinder head. The high-alloy-steel retaining screens were also found to have warped, although not severely enough to allow catalyst leakage. The catalyst itself did not appear to be structurally damaged.

At best, it appears that with catalyst "Y," this system will likely never be better than marginally acceptable from the standpoint of performance. If a catalyst with improved performance can be found for use in this converter, many problems remain to be resolved, not the least of which are catalyst life, vehicle compatibility of an engine compartment device and an effective over-temperature protection system.

Effect of Test Procedure on Oxidation Performance of a Catalytic Converter

A complete catalytic converter system was submitted by an outside source for evaluation. This system was evaluated at periodic intervals according to the Detroit traffic route and the bag-test procedures for determination of performance changes with mileage, as well as the stabilized-cruise efficiency test. The data shown in Table III are the results of the first two tests mentioned above. They are presented to show that resulting concentrations are dependent upon test procedure, and that different test procedures result in different performance changes with increasing mileage. As an example, the carbon monoxide concentrations measured initially were 0.15 mole % with the Detroit traffic route procedure and 1.0 mole % with the bag technique. The large (1.0 mole %) carbon monoxide concentration measured in the bag was probably due to a 0-60 mph acceleration mode in the cycle used to

TABLE III

CONVERTER OUTPUT CONCENTRATIONS VS. MILEAGE

AS MEASURED BY TWO TEST PROCEDURES

MILEAGE	CONVERTER OUTLET CONCENTRATIONS * (ppm hex and mol % CO)			
	City Traffic Route Test		Bag Test	
	HC	CO	HC	CO
0	110	0.15	115	1.0
1200	180	0.90	150	1.5
3000	200	0.80	250	1.9
4500	295	1.25	-	-

* Concentrations are "as read" without correction for air dilution.

fill the bag. Since this mode contributes more than one-half of the exhaust gas content in the bag, it is a controlling factor in the average concentration sampled from the bag contents. In other words, each test will yield results depending on the makeup of the test itself.

Using the tests at 0 and 1200 miles for comparison, the Detroit traffic route test showed an increase in hydrocarbon concentration at 1200 miles that was twice as large as that shown by the bag test. Carbon monoxide concentrations showed a gain at 1200 miles ten times as great with the traffic route test as was shown by the bag procedure.

Additional tests using fresh, unused catalyst (type "N") in converter G-1 were conducted to compare the results of tests using the individual 11-mode and MVPCB test procedures with those obtained using the traffic route and bag procedures. Results of these tests are shown on Figure 34 in terms of average concentrations of hydrocarbons and carbon monoxide measured from the outlet of the converter. The data shows that the MVPCB procedure, which takes warm-up into account, yields outlet concentrations approximately 50 percent higher than the average of the other three tests.

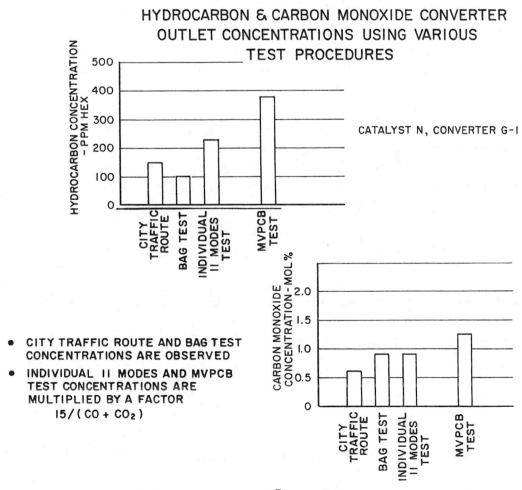

- CITY TRAFFIC ROUTE AND BAG TEST CONCENTRATIONS ARE OBSERVED
- INDIVIDUAL 11 MODES AND MVPCB TEST CONCENTRATIONS ARE MULTIPLIED BY A FACTOR $15/(CO + CO_2)$

Figure 34

Summary and Conclusions

The second part of this paper has discussed the development of an engine dynamometer screening test and the results of evaluating seventeen different catalysts with this procedure. Three catalysts selected from the screening tests were evaluated on vehicles using conventional muffler type systems and a manifold catalytic converter system. The results of these tests indicate that none of these catalysts meet the minimum requirements of adequate hydrocarbon and carbon monoxide oxidation and maintenance of oxidation performance for extended mileage.

These minimum requirements have been specified by the Motor Vehicle Pollution Control Board of the state of California as converter outlet concentrations below 275 ppm hex and 1.5 mole % CO for 12,000 miles as measured using their test procedure dated May 18, 1961.[10] The only system investigated which shows any liklihood of promise is the manifold converter system which inherently (by location) improves warm-up time. Catalyst life, materials and vehicle compatability problems of this system have not yet been investigated, since no catalyst has yet been found for use in this system which can meet the performance requirement.

If a catalytic material can be found that displays the capability of maintaining oxidation performance for extended mileage, much work remains to be done in development of a complete system. This fact becomes evident in the discussion on vehicle compatability, over-temperature protection, system back pressure, odor, and sound covered in Part I of this paper. It should also be pointed out that additional requirements in this area are specified in the Approved Step II Procedures of the California Motor Vehicle Pollution Control Board dated October 13, 1961. These include such criteria as device malfunction, effect of exhaust system backfire, effect of converter heat on pedestrians, operation of the system in heavy rains and during operation in mountainous and desert regions, and the possible creation of compounds which would have a general toxic effect on the population. None of the systems evaluated in this paper were considered developed to a point where investigations of this nature were justified.

REFERENCES:

1. The Occurrence, Distribution, and Significance of Photochemical Air Pollution in the United States and Canada - John T. Middleton and Arie J. Haagen-Smit - Air Pollution Control Association 53rd meeting, Cincinnati Ohio, May 1960 (60-13)

2. "Chemistry and Physiology of Los Angeles Smog," A. J. Haagen-Smit - Industrial and Engineering Chemistry vol. 44 - no. 6, pages 1342-1346, June 1952

3. Technical Report of California Standards for Ambient Air Quality and Motor Vehicle Exhaust, State of California, Department of Public Health, 1960

4. "Single Cylinder Engine Tests of Oxidation Catalysts" – W. A. Cannon, E. F. Hill and C. E. Welling, Presented at SAE National West Coast meeting, Seattle Washington, August 12-16, 1957

5. The Application of Vanadia-Alumina Catalysts for the Oxidation of Exhaust Hydrocarbons, W. A. Cannon and C. E. Welling, Presented at SAE Annual Meeting, Detroit Michigan, January 16, 1959

6. "The Development of a Catalytic Converter for the Oxidation of Exhaust Hydrocarbons, R. T. VanDeveer and J. M. Chandler, Presented at the SAE Annual Meeting, Detroit Michigan, January 16, 1959

7. Conference on Catalytic Decomposition of Vehicle Combustion Products, co-sponsored by the Air Pollution Control District – county of Los Angeles and the Air Pollution Foundation, March 21 and 22, 1957, Pasadena, California

8. Evaluation of a Ford Motor Company's Catalytic Converter for the Oxidation of Exhaust Hydrocarbons, John C. Chipman, Air Pollution Control District – County of Los Angeles, July 1959

9. Standards for Ambient Air and Motor Vehicle Exhaust adapted December 4, 1959, California Department of Public Health

10. California Motor Vehicle Pollution Control Board Second Draft May 18, 1961, Appendix F, Test Procedure for Vehicle Exhaust Emmissions

11. Vehicle Emissions Measurement Panel of the AMA, Vehicle Combustion Products Committee – Hydrocarbon Content Measurement Using the Beckman L/B Infra-Red Analyzer – Model 15A, February 24, 1961

12. Vehicle Emission Measurements Panel of the AMA, Vehicle Combustion Products Committee – Carbon Dioxide and Carbon Monoxide Content Measurement Using the Beckman L/B Infra-Red Analyzer – Model 15A, May 22, 1961

13. "A Simplified Method for Characterizing a Motor Vehicle's Exhaust Emissions," A. M. Smith and J. H. Struck, Presented at Annual Meeting of the Air Pollution Control Association, Cincinnati Ohio, May 23-26, 1960.

14. "Los Angeles Traffic Pattern Survey," Traffic Survey Panel of the Automobile Manufacturers Association, Presented at the SAE West Coast Meeting, August 12-16, 1957

Appendix A

Fuel Analysis

(1) A premium grade, commercial brand fuel (with TEL added) was used for all tests reported in Part I of this paper except the non-leaded test with converter A-4. Hydrocarbon breakdown of this fuel during the period of use (1959 - 1960) was:

Olefins	-	5%
Aromatics	-	36%
Saturates	-	59%

(2) The non-leaded fuel utilized in the test with converter A-4 had the following hydrocarbon breakdown:

Olefins	-	3%
Aromatics	-	46%
Saturates	-	51%

(3) In all of the tests included in Part II of this paper, a fuel similar to (1) above was used except that it included the addition of:

Sulfur	-	.085%
Phosphorous	-	0.3 theories

Appendix B

Stabilized Cruise Efficiency and Warmup Test

Stabilized cruise efficiencies of a catalytic converter are obtained from measurements of the concentrations of hydrocarbons and carbon monoxide in the exhaust stream before and after the converter at stabilized road load operating conditions. The procedure is particularly useful in determining differences in oxidation efficiency with mileage.

The warmup test consists of 30 minutes of operation at 30 mph road load from an ambient temperature start. During this time, exhaust gas concentrations before and after the converter are measured for calculation of oxidation efficiency. Oxidation efficiencies versus time are plotted for analysis. With this test, it is possible to determine differences in warmup time for a particular oxidation efficiency with increasing mileage.

Operating Procedure:

1. After an ambient soak of at least 8 hours, accelerate the vehicle to 30 mph using normal acceleration rates. Hold this speed for 30 minutes. (This part of the procedure is used for the warmup test.)

2. Using normal acceleration rates proceed to 60 mph and hold till stability has been reached. Stability is defined as no change in concentration or temperature measurements for a minimum period of 5 minutes.

3. Decelerate to 50 mph (or 45 mph) and hold until stability has been reached.

4. Decelerate to 30 mph. Maintain for stability.

5. Decelerate to idle. Hold for six minutes.

On the tests reported in this paper, the above procedure was utilized with the vehicle being operated on a chassis dynamometer. For these tests, a load was not applied to the dynamometer since this particular dynamometer has windage and bearing friction losses approximately equal to cruise road loads of a 3500 pound vehicle and an inertial mass of approximately 6000 pounds.

Data Accumulation:

Data is accumulated at ten minute intervals throughout the test for the following items:

(1) speedometer reading (mph)

(2) true vehicle speed (mph)

(3) engine speed (rpm)

(4) intake manifold vacuum (in. hg.)

(5) exhaust system backpressure (in. H_2O)

Temperatures are recorded continuously throughout the test at the following points:

(1) converter inlet, outlet and catalyst bed

(2) inlet carburetor air and ambient air

(3) any other point of interest

Exhaust gas concentrations of desired constituents are also continuously monitored with alternate samples being taken from before and after the converter in one minute intervals. An air purge is taken after each pair of exhaust gas readings.

The odometer is read at the start and conclusion of each test.

Sample System   - See Figure 1-B

Analysis:

(1) Stabilized Cruise - Efficiencies are calculated for the 60, 50, or 45 and 30 mph road load conditions from the stabilized before and after measurements. For the idle condition, efficiency is calculated from the last pair of readings obtained during the 6 minutes of operation. Calculations from the initial test with converter A-1 are given below:

	Hydrocarbon Concentrations (ppm hex)		
Cruise	Before	After	Efficiency (%)
60	140	100	71
50	160	50	69
30	250	130	48
Idle	390	100	74

(2) Warmup - Efficiencies are calculated in the manner as shown above, for each pair of readings obtained during the 30 minute procedure. Efficiency versus time is then plotted for analysis.

TYPICAL SAMPLING SYSTEM

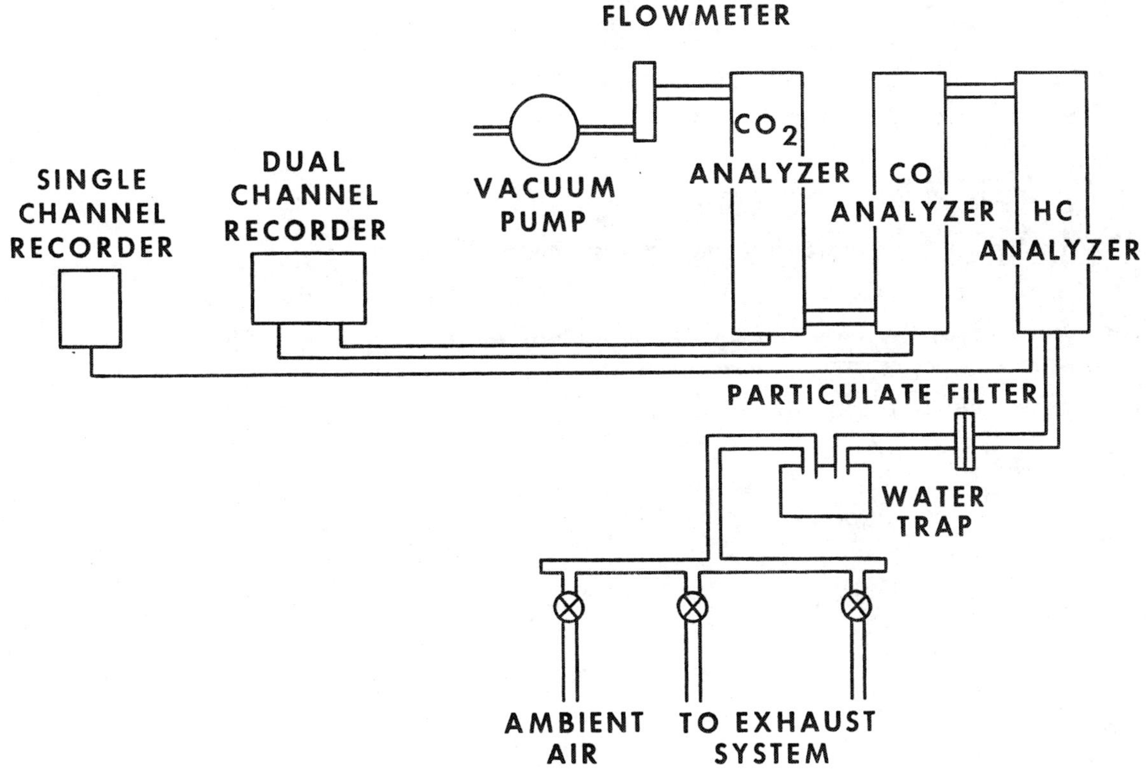

Figure 1-B

Appendix C

Determination of Stoichiometric Air-Fuel Ratio and Secondary Air Requirements

I   The stoichiometric air-fuel ratio is the chemically correct ratio, by weight, between a fuel of known hydrogen-to-carbon ratio and air such that complete oxidation of the fuel can be theoretically obtained. The equation for <u>complete</u> <u>oxidation</u> of a mixture of fuel with hydrogen to carbon ratio of 1.86 and <u>air is:</u>

$$CH_{1.86} + (1.47)(3.76) N_2 + 1.47 O_2 = CO_2 + 0.93 H_2O + (1.47)(3.76) N_2$$

(Assumes 3.76 moles of N_2, including inert gases, per mole of O_2 in air)

The small amounts of argon, carbon dioxide, and hydrogen present in air are included in the nitrogen constant (28.161) in the calculations below. From the equation above, the stoichiometric air-fuel ratio is calculated thusly:

Weight of fuel:

$$\begin{array}{rcl} \text{carbon} &=& 12 \\ \text{hydrogen} &=& \underline{1.86} \\ && 13.86 \end{array}$$

Weight of air:

$$\begin{array}{rclcl} \text{Nitrogen} &=& (1.47)(3.76)(28.161) &=& 155.65 \\ \text{Oxygen} &=& (1.47)(32) &=& \underline{47.04} \\ &&&& 202.69 \end{array}$$

Stoichiometric air-fuel ratio (by weight) = $\dfrac{202.69}{13.86}$ = 14.62:1

II   To calculate the amount of secondary air necessary for complete oxidation at a particular operating point on the carburetor flow curve (Figure 14, Part I in the text), it is necessary to perform the following calculation:
    (1) For the point at 240 cfm on the wide open throttle portion of the curve, the air-fuel ratio of the mixture is 13.5:1.
    (2) The amount of secondary air necessary for complete oxidation of the air-fuel mixture at this point would be:

$$\frac{14.6 - 13.5}{14.6} \times 240 \text{ cfm} = 18.1 \text{ cfm}$$

    (3) Therefore, 18.1 cfm of secondary air of .073 density (80° F and 29.7 in. Hg. same as carburetor air specified) must be supplied.

Appendix D

Stabilized Cruise Efficiency Test Results of Low Temperature Catalytic Converter Systems

The tabulated data presented in this appendix are the stabilized cruise efficiency test results of the various converter systems described in Part I of this paper. The data give inlet hydrocarbon concentrations (ppm hex) and the corresponding efficiencies for each operating condition of the test procedure. Oxidation efficiency is calculated in the following manner:

$$\text{oxidation efficiency (\%)} = \frac{\text{inlet concentration} - \text{outlet concentration}}{\text{inlet concentration}} \times 100$$

Temperatures are reported from the inlet plenum, catalyst bed, and outlet plenum of the converter. Hydrocarbon concentrations are "as read" without correction for air dilution.

STABILIZED CRUISE EFFICIENCY TEST RESULTS

CONVERTER: A-1

Mileage	60 MPH ROAD LOAD CRUISE					50 MPH ROAD LOAD CRUISE					30 MPH ROAD LOAD CRUISE					IDLE	
	HC "In"* (ppm hex)	Eff. (%)	Temp. "In" (°F)	Cat. Bed Temp. (°F)	Temp. "Out" (°F)	HC "In" (ppm hex)	Eff. (%)	Temp. "In" (°F)	Cat. Bed Temp. (°F)	Temp. "Out" (°F)	HC "In" (ppm hex)	Eff. (%)	Temp. "In" (°F)	Cat. Bed Temp. (°F)	Temp. out (°F)	HC "In" (ppm hex)	Eff. (%)
1000	140	71	1025	1150	1040	160	68	940	1060	980	250	48	700	810	715	390	74
1600	210	71	960	1060	1060	270	59	840	825	850	350	37	610	680	610	510	60
4700	120	67	1000	1000	950	160	62	860	860	820	290	41	610	610	610	410	52
8000	100	75	990	1030	1020	130	66	920	925	875	-	-	-	-	-	-	-
10,948	130	71	1120	1130	1085	150	53	1010	1030	990	170	43	780	750	700	310	61
15,818	100	60	1065	1060	1040	140	54	940	940	910	190	39	740	720	685	330	61
19,788	90	50	1090	1070	1010	110	31	980	990	870	170	20	700	640	570	340	37
24,228	80	44	1035	1040	990	80	22	920	-	860	120	21	690	670	610	290	37

* Converter Inlet Concentration

STABILIZED CRUISE EFFICIENCY TEST RESULTS

CONVERTER: A-2

Mileage	60 MPH ROAD LOAD CRUISE					50 MPH ROAD LOAD CRUISE					30 MPH ROAD LOAD CRUISE					IDLE	
	HC "In"* (ppm hex)	Eff. (%)	Temp. "In" (°F)	Cat. Bed Temp. (°F)	Temp. "Out" (°F)	HC "In" (ppm hex)	Eff. (%)	Temp. "In" (°F)	Cat. Bed Temp. (°F)	Temp. "Out" (°F)	HC "In" (ppm hex)	Eff. (%)	Temp. "In" (°F)	Cat. Bed Temp. (°F)	Temp. "Out" (°F)	HC "In" (ppm hex)	Eff. (%)
100	110	82	1040	1020	-	150	73	930	910	-	280	62	550	660	-	440	71
3865	110	77	970	1085	1080	150	74	870	985	970	220	61	660	760	755	370	78
6010	190	65	1085	1105	1115	230	53	910	940	950	330	42	690	705	685	550	62
11,700	140	50	1010	1060	1040	160	33	900	945	925	-	-	-	-	-	-	-
14,100	60	39	1030	1030	1040	80	42	910	970	930	140	36	700	730	700	260	50

CONVERTER: A-3

Mileage	HC "In"*	Eff. (%)	Temp. "In" (°F)	Cat. Bed Temp. (°F)	Temp. "Out" (°F)	HC "In"	Eff. (%)	Temp. "In" (°F)	Cat. Bed Temp. (°F)	Temp. "Out" (°F)	HC "In"	Eff. (%)	Temp. "In" (°F)	Cat. Bed Temp. (°F)	Temp. "Out" (°F)	HC "In"	Eff. (%)
0	140	71	1030	1030	900	170	65	940	950	850	370	37	700	700	600	560	44

CONVERTER: A-4

Mileage	HC "In"*	Eff. (%)	Temp. "In" (°F)	Cat. Bed Temp. (°F)	Temp. "Out" (°F)	HC "In"	Eff. (%)	Temp. "In" (°F)	Cat. Bed Temp. (°F)	Temp. "Out" (°F)	HC "In"	Eff. (%)	Temp. "In" (°F)	Cat. Bed Temp. (°F)	Temp. "Out" (°F)	HC "In"	Eff. (%)
1691	135	70	1120	1160	1130	170	62	900	920	865	230	44	715	715	660	-	-
4900	125	64	1090	1170	1115	140	50	925	975	925	225	42	690	730	710	370	50
7100	150	53	1070	1200	1160	195	54	880	960	920	280	32	640	690	660	480	46
9759	135	52	910	1190	1130	185	43	705	940	870	270	33	530	730	655	450	45
12,310	135	59	970	1250	1140	175	52	760	980	910	210	29	545	710	600	410	38

* Converter Inlet Concentration

STABILIZED CRUISE EFFICIENCY TEST RESULTS

CONVERTER: B-1

Mileage	60 MPH ROAD LOAD CRUISE					50 MPH ROAD LOAD CRUISE					30 MPH ROAD LOAD CRUISE					IDLE	
	HC "In"* (ppm hex)	Eff. (%)	Temp. "In" (°F)	Cat. Bed Temp. (°F)	Temp. "Out" (°F)	HC "In" (ppm hex)	Eff. (%)	Temp. "In" (°F)	Cat. Bed Temp. (°F)	Temp. "Out" (°F)	HC "In" (ppm hex)	Eff. (%)	Temp. "In" (°F)	Cat. Bed Temp. (°F)	Temp. "Out" (°F)	HC "In" (ppm hex)	Eff. (%)
70	130	76	1045	1060	915	120	74	940	960	840	240	62	725	750	650	500	80
2200	90	67	1070	1050	930	120	67	950	920	830	230	48	680	645	575	310	58
4000	180	55	1160	1185	1060	210	52	1030	1040	930	360	50	740	740	690	660	65
6026	150	83	1085	1190	1070	140	75	960	985	870	290	53	690	695	600	550	60
8756	150	60	1060	1080	940	180	42	955	950	855	310	33	750	735	650	550	43
16,125	120	64	965	980	880	140	54	910	910	800	250	39	680	660	560	410	64
19,478	110	62	1080	1100	970	140	44	950	940	790	260	31	690	670	540	440	55
24,260	110	46	1110	1100	1000	130	40	990	950	860	250	34	760	710	610	460	46

CONVERTER: B-2

Mileage	HC "In"* (ppm hex)	Eff. (%)	Temp. "In" (°F)	Cat. Bed Temp. (°F)	Temp. "Out" (°F)	HC "In" (ppm hex)	Eff. (%)	Temp. "In" (°F)	Cat. Bed Temp. (°F)	Temp. "Out" (°F)	HC "In" (ppm hex)	Eff. (%)	Temp. "In" (°F)	Cat. Bed Temp. (°F)	Temp. "Out" (°F)	HC "In" (ppm hex)	Eff. (%)
0	90	78	1000	1010	1010	120	83	890	900	900	200	50	650	650	680	290	69
1631	150	59	1030	1030	-	110	56	930	950	-	100	33	740	710	-	310	66
5351	120	33	975	950	870	170	29	870	820	720	240	21	660	590	495	300	23

* Converter Inlet Concentration

STABILIZED CRUISE EFFICIENCY TEST RESULTS

CONVERTER: C-1

Mileage	60 MPH ROAD LOAD CRUISE					50 MPH ROAD LOAD CRUISE					30 MPH ROAD LOAD CRUISE					IDLE	
	HC "In"* (ppm hex)	Eff. (%)	Temp. "In" (°F)	Cat. Bed Temp. (°F)	Temp. "Out" (°F)	HC "In" (ppm hex)	Eff. (%)	Temp. "In" (°F)	Cat. Bed Temp. (°F)	Temp. "Out" (°F)	HC "In" (ppm hex)	Eff. (%)	Temp. "In" (°F)	Cat. Bed Temp. (°F)	Temp. "Out" (°F)	HC "In" (ppm hex)	Eff. (%)
0	220	82	1150	1230	1060	145	66	1080	1090	940	140	64	790	770	640	–	–
200	60	83	1130	1130	96	90	77	980	890	810	170	58	710	710	560	150	68
2632	120	66	–	–	–	150	66	1080	1050	960	210	52	810	750	670	200	65
4300	100	60	1020	1150	1160	140	53	1030	1030	880	250	40	760	740	600	430	63
7540	120	54	1130	1140	985	155	48	1020	1010	850	220	41	760	725	590	410	62
7700	45	56	1250	1250	1080	75	60	1130	1125	950	135	52	855	840	655	335	69
8000	90	50	1220	1220	1070	120	46	1060	1040	925	175	40	790	750	625	320	63

* Converter Inlet Concentration

Appendix E

Engine Dynamometer Catalyst Screening Procedure Test Cycles

Evaluation Cycle	Speed (rpm)	Load (Ft.-Lbs.)	Time (Minutes)
Cruise 30	1375	54	15
Cruise 60	2475	92	15
Idle	475	18-20	10
Accel	1500	185	10
Decel	1000	4-6	10

Durability Cycle	Speed (rpm)	Load (Ft.-Lbs.)	Time (Minutes)
Step 1	1700	52	15
" 2	2200	58	15
" 3	2700*	65	12
" 4	3400*	70	12
" 5	1900	33	1
" 6	1900	WOT	1
" 7	1900	33	1
" 8	1900	WOT	1
" 9	1900	33	1
" 10	1900	WOT	1

*These two steps are optional depending on the ability of the catalyst to withstand high temperatures.

Appendix F

Engine Dynamometer Screening Test Results

The results of a typical engine dynamometer screening test are presented on the following table. The test presented is that of catalyst "N" which was later tested on a vehicle. The data is given to show the ranges of exhaust gas concentrations and the converter inlet, outlet, and catalyst bed temperatures encountered in the various modes of the screening test evaluation cycle. Hydrocarbon and carbon monoxide inlet concentrations are shown with the corresponding oxidation efficiencies. Oxidation efficiency was calculated as:

$$\text{efficiency (\%)} = \frac{\text{inlet concentration} - \text{outlet concentration}}{\text{inlet concentration}} \times 100$$

The concentrations shown are "as read" without correction for secondary air dilution.

CELL 5 CAT. EVALUATION
CATALYST "N"

Date	Conv. Hrs.	CRUISE 60							CRUISE 30									IDLE							ACCEL.							DECEL.		
		HC "In"* ppm	Eff (%)	CO "In" Mol.%	Eff (%)	Temp. °F "In"	Temp. °F Bed	Temp. °F "Out"	HC "In" ppm	Eff (%)	CO "In" Mol.%	Eff (%)	Temp. °F "In"	Temp. °F Bed	Temp. °F "Out"	HC "In" ppm	Eff (%)	CO "In" Mol.%	Eff (%)	Temp. °F "In"	Temp. °F Bed	Temp. °F "Out"	HC "In" ppm	Eff (%)	CO "In" Mol.%	Eff (%)	Temp. °F "In"	Temp. °F Bed	Temp. °F "Out"	CO "In" Mol.%	Eff (%)	Temp. °F "In"	Temp. °F Bed	Temp. °F "Out"
6-15	2	55	45	0.3	67	1240	1330	1150	190	63	0.5	80	920	1200	870	360	72	3.9	90	520	1390	750	150	7	3.25	17	1270	1510	1300	230	78	2.30	98	640	1500	950
6-16	12	55	55	0.40	63	1170	1250	1190	190	58	0.69	75	900	1070	815	250	88	2.4	96	625	1230	800	110	18	2.70	39	1290	1520	1280	310	93	1.05	91	610	1570	970
6-17	21	65	46	0.35	71	1210	1280	1090	180	56	0.65	77	870	1040	795	270	81	2.25	96	510	1290	765	150	20	2.70	43	1265	1510	1270	240	88	0.85	94	740	1560	985
6-20	28	90	44			1200	1265	1070	320	47			No Temp. Data			410	71			460	1280	720	210	19	3.30	45	1245	1510	1270	270	74			650	1600	935
6-21	-38	100	40	0.30	67	1140	1235	1050	310	32	0.80	50	770	910	730	330	70	2.20	86	500	1275	720	230	22	3.65	27	1190	1470	1190	410	83	1.4	93	580	1530	880
6-22	46	95	47	0.5	80	1190	1310	1125	300	47	0.9	78	810	1000	790	440	70	3.10	90	580	1340	800	215	12	3.75	36	1250	1490	1240	305	75	2.05	95	720	1470	930
6-23	54	95	47.5	0.40	87.5	1170	1260	1090	340	54	1.1	72.5	800	940	800	370	72	2.75	91	540	1220	750	230	11	3.75	45	1220	1460	1290	290	73	1.45	86	680	1400	900
6-24	62	55	41	0.40	63	1190	1270	1120	320	38	0.95	74	No Data		770	325	65	2.10	88	570	1260	770	210	14	3.35	45	1240	1470	1250	305	72	1.10	86	670	1300	900
6-27	69	100	40	0.45	44	1180	1240	1085	330	33	0.70	57	820	900	750	410	55	2.95	71	510	1180	750	235	13	3.55	32	1240	1460	1480	400	68	1.85	68	660	1300	880
6-28	74	120	38	0.40	50	1140	1200	1110	350	26	0.80	50	815	880	875	425	60	2.90	72	455	1150	1140	255	8	3.60	28	1190	1385	1220	420	64	1.80	64	600	1150	1050
6-29	80	110	36	0.40	62	1190	1235	1105	350	26	0.85	59	820	880	740	420	55	2.35	72	580	1070	740	255	6	3.65	21	1220	1375	1200	375	56	1.10	73	660	1100	830
7-5	89	100	20	0.40	38	1150	1170	990	290	27.6	0.70	64	820	820	710	390	38.5	3.20	58	520	910	630	210	2	3.75	21	1230	1340	1150	270	37	2.15	56	630	960	720
7-12	100	110	41	0.40	50	1150	1230	1040	340	21	0.75	53	790	840	680	355	28	2.20	43	590	1150	650	165	15	2.10	38	1270	1460	1210	305	25	1.30	50	670	1220	740

* Converter Inlet Concentration

Appendix G

Test Procedures Used in Vehicle Evaluations of Catalysts "O" and "N"

I. Detroit Traffic Route Test

    The Detroit traffic route test is used to measure the average exhaust gas concentrations emitted from a vehicle during normal city driving. The route utilizes 59 miles of paved roads consisting of 44.7 miles of city traffic, 7.4 miles of expressway and 6.9 miles of suburban streets. (see Figure 1G) It is a route proposed and recognized by member companies of the Exhaust System Task Group of the Automobile Manufacturers Association. Driving time required to complete the route is approximately two hours. During this time, exhaust gas concentration of hydrocarbons, carbon monoxide and carbon dioxide are sampled continuously, alternating for one minute intervals from taps before and after the catalytic converter. From these measurements the concentrations at one minute intervals from both the "after" and "before" sample taps are averaged to determine the average concentrations measured over the entire route.

General Test Requirements and Sample System

    The test utilizes two operators: one driver and one instrument technician. Morning tests did not begin before 9:00 a.m. and afternoon tests after 2:15 p.m. to avoid excessive traffic conditions. For the tests reported using catalyst "N", a Liston Becker Model 15A infrared analyzer with 13-1/2 inch sample cell was used to measure hydrocarbon concentrations. Two Liston Becker Model 28 analyzers were used to measure carbon monoxide and carbon dioxide concentrations. The hydrocarbon analyzer was calibrated to read 1000 parts per million hexane full scale while the carbon monoxide and carbon dioxide instruments were calibrated for 15 mol% full scale deflection. These instruments were placed in the car with the detector cell diaphragm vertical and longitudinal of the car to minimize the effects of vehicle road shock transmitted to the diaphragm.

    The sample system consisted of 1/4 inch copper tubing from the two locations before and after the converter to a panel board equipped with electric valves. Another valve on this board sampled outside air. From the board, the sample path was to an ice bath condenser, through a particulate filter and to the individual analyzer in series. From the instruments, the flow is through a 16 CFH flowmeter and out through a vacuum pump.

    Use and calibration of the instruments generally followed those procedures outlined in reports of the Vehicle Emissions Measurements Panel of the AMA dated 2/24 and 5/22/61.[11,12] A schematic of the sample system used appears in Figure 1 of Appendix B.

Data Analysis

(1) Record hydrocarbon, carbon monoxide, and carbon dioxide meter readings at one-minute intervals for both the "before" and "after" sample taps and average for the entire test.

(2) Record the average air purge readings and subtract from meter readings recorded in (1) above.

(3) Correct resulting average readings according to the following formula:

$$\text{Avg. meter readings} \times \left(1 + \frac{\text{gain at start} - \text{gain at end}}{\text{Gain at start}} \times 1/2\right)$$

$$\text{Equals } (=) \text{ Corrected avg. meter readings}$$

If the correction factor in the parenthetical expression above is less than 1.025, neglect this step.

(4) Determine average concentrations of the corrected average meter readings from the proper calibration curve corresponding to the instrument used for measurement of that particular exhaust constituent.

(5) Determine per cent oxidation efficiencies from the average concentrations.

II. **Individual 11-Mode Test**

The individual 11-mode test consists of measuring exhaust gas concentrations while the vehicle is being driven according to eleven separate driving modes, multiplying the observed concentrations by weighting factors given in the California Exhaust Standard published by the State of California Department of Public Health,[3] and totaling the resulting products to obtain final numbers for determination of compliance with the California Exhaust Standard (275 ppm hex and 1.5 mol% CO). The test was conducted on Northline Road in Wayne, Michigan. This road is relatively smooth, reasonably straight, and has normally light traffic conditions that permit the test to be carried out in a safe manner.

Driving Procedure

The test consists of the following driving modes (1) through (11):

(1) Idle

(2) Cruise - 20

(3) Cruise - 30

(4) Cruise - 40

(5) Cruise - 50

These conditions are stabilized for a minimum of three minutes. Cruise conditions are run in highest vehicle road gear for both standard and automatic transmissions.

(6) 0-60 at 3.0 mph/sec

(7) 15-30 at 2.2 mph/sec

(8) 0-25 at 2.2 mph/sec

Acceleration modes are preceded by stabilized cruise or idle operation for at least one minute. An accelerometer is used to maintain the rates given above so that each mode can be completed in the time specified by the corresponding acceleration rates. The vehicle is operated at wide open throttle if it cannot complete the mode in the specified time. Acceleration rates are maintained for a minimum of two seconds beyond the specified time. Automatic transmission vehicles are driven with selector in the normal drive position. Standard transmission vehicles are driven thusly:

0-60 shift to second gear at 15 mph, to high gear at 30 mph

15-30 operate in high gear

0-25 shift to second gear at 15 mph

(9) 30-15 at -1.4 mph/sec

(10) 50-20 at -1.2 mph/sec

(11) 30-0 at -2.5 mph/sec

Deceleration modes are preceded by cruise conditions stabilized for a minimum of one minute. All decelerations are driven at closed throttle, with braking, such that the deceleration rates given above are maintained and the mode is completed in the time specified by the corresponding deceleration rates. Vehicles which exceed the deceleration rate are also operated at closed throttle. Deceleration rates for conditions (9) and (10) are maintained for at least two seconds beyond the specified times given by the acceleration rates. All decelerations are run in the highest vehicle road gear for both standard and automatic transmissions, except that standard transmissions are declutched at 15 mph on the 30-0 mode.

General Test Requirements and Sample System

The test utilized two operators, one for driving and the other for instrument operation. The test is run with the converter fully warmed up. All conditions are run twice to compensate for wind effects. Efficiencies were not evaluated on this test and exhaust gas was sampled only from "after" the converter.

Vehicle installation of instruments and the sample system used were the same as shown in Figure 1 of Appendix B with the following exceptions:

(1) carbon monoxide concentrations were measured with a Liston Becker model 15A infrared analyzer calibrated for full scale deflection of 10 mol%. A Liston Becker Model 28 analyzer calibrated to 15 mol% full scale deflection was used to measure CO_2.

(2) hydrocarbon concentrations were obtained with a Liston Becker model 15A infrared analyzer equipped with a 5 inch sample cell (in place of the 13-1/2 inch cell). The smaller sample cell is used so that during the deceleration modes, the amplifier gain can be reduced to 1/4 so that recorder deflections always remain on scale. A 1/4 gain calibration curve is used to determine the deceleration concentrations.

Data Analysis

Stabilized cruise and idle hydrocarbon, carbon monoxide, and carbon dioxide concentrations are determined from the average meter readings during the last three seconds of these modes. Time-average concentrations are determined for the specified time of each transient condition by using either one-second interval averages, planimeter approximation or an observed estimate. The observed hydrocarbon and carbon monoxide concentrations of each mode are multiplied by a correction factor of $15/(CO+CO_2)$ for cruise and acceleration modes and $15/(.0006\ HC + CO + CO_2)$ for deceleration modes. Each hydrocarbon and carbon monoxide concentration was then multiplied by the corresponding weighting factors given below:

Condition	Weighting Factor
idle	.042
cruise 20	.050
cruise 30	.061
cruise 40	.042
cruise 50	.015
Acceleration	
0-60	.059
15-30	.455
0-25	.185
Deceleration	
30-15	.033
50-20	.029
30-0	.029
	1.000

The products obtained are then totaled giving the final hydrocarbon and carbon monoxide concentrations as measured by the test.

III. Bag Test

The Bag Test employs the technique of capturing in a large plastic bag the total exhaust gas from a vehicle being operated on a chassis dynamometer according to a 4-minute driving cycle. The contents of the bag are then sampled for hydrocarbon, carbon monoxide, and carbon dioxide concentration levels. This test was utilized in conjunction with the stabilized cruise efficiency test (Appendix A) and was conducted just prior to the 30 mph road load condition of that test. Therefore, the test was always conducted with a warmed-up catalytic converter. The instrumentation and sample system utilized was identical to that used for the stabilized cruise test.

Procedure

(1) Evacuate 10' x 14' polyethylene bag.

(2) Insert exhaust pipe extension into the bag, start car and proceed with the following driving cycle:

Mode	Elapsed Time (seconds)	Accum. Time (min. & sec.)
Acceleration 0-60	20	20
Cruise 60	15	35
Deceleration 60-50	5	40
Cruise 50	15	55
Deceleration 50-20	15	1-10
Cruise 20	15	1-25
Acceleration 20-30	10	1-35
Cruise 30	15	1-50
Deceleration 30-0	10	2-00
Idle	30	2-30
Acceleration 0-30	20	2-50
Deceleration 30-15	10	3-00
Cruise 15	15.	3-15
Acceleration 15-40	20	3-35
Cruise 40	15	3-50
Deceleration 40-0	10	4-00

(3) Remove exhaust pipe extension, insert sampling probe (1/4 inch copper tubing) and seal bag.

(4) Measure hydrocarbon, carbon monoxide and carbon dioxide level of bag content. Recheck after one and two hours have elapsed.

IV. <u>MVPCB Chassis Dynamometer Test</u>

The Motor Vehicle Pollution Control Board test was conducted using a Clayton

Model CN150 chassis dynamometer equipped with a 4000-pound inertia wheel. The dynamometer was adjusted to absorb 9 horsepower at 50 miles per hour. The vehicle was driven according to the procedure specified in the second draft of the Test Procedure for Vehicle Exhaust Emissions written by the Motor Vehicle Pollution Control Board.[10] The following interpretations of that procedure were made:

1. Vehicle accelerations were driven at a constant acceleration rate equal to the average rates specified in the procedure.

2. Air was directed to the air cleaner intake through a tube which collected ambient temperature air from the side of the vehicle.

Hydrocarbon concentrations were measured, in the 0-1000 ppm hex range, with a Liston Becker Model 15A Infrared Analyzer equipped with a 5 inch sample cell and 1/2 inch filter cell. Concentrations greater than 1000 ppm were measured using a Model 28 Liston Becker analyzer with a 2-1/2 inch sample cell. This instrument was calibrated to 11,500 ppm n-hexane full scale, and carbon dioxide response with this instrument was approximately 200 ppm.

Carbon monoxide and carbon dioxide concentrations were measured with model 15A Liston Becker analyzers utilizing 1/8 inch sample cells. The carbon monoxide instrument was calibrated for a full scale deflection of 10 mol%, while the carbon dioxide instrument was calibrated for 15 mol% full scale deflection.

A sample system schematic is shown in Figure 2-G.

DETROIT TRAFFIC ROUTE

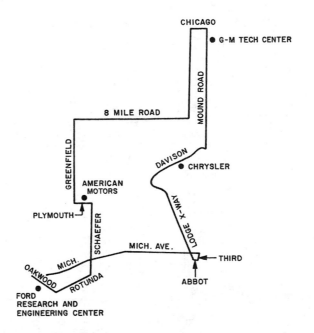

Figure 1-G

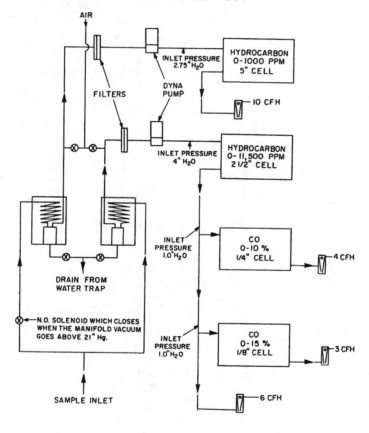

Figure 2-G

Appendix H

Data from Vehicle Evaluations of Screening Test Catalysts

The following tabulated data are from the vehicle tests of catalysts "O" and "N". Results are presented for stabilized cruise efficiency tests with both catalysts and for Detroit traffic route tests with catalyst "N". The stabilized tests report inlet hydrocarbon and carbon monoxide concentrations for each operating condition of the procedure plus the corresponding oxidation efficiency calculated from the outlet concentrations. Oxidation efficiency is calculated as:

$$\text{efficiency (\%)} = \frac{\text{inlet concentration} - \text{outlet concentration}}{\text{inlet concentration}} \times 100$$

Temperatures are also given from locations in the inlet plenum, catalyst bed and outlet plenum of the converter. The Detroit traffic route data gives inlet and outlet concentrations of hydrocarbons and carbon monoxide together with the corresponding oxidation efficiencies calculated as shown above. The concentrations reported from both tests are "as read" without correction for air dilution.

CONVERTER G-1 CATALYST "N"

	CRUISE 60							CRUISE 45							CRUISE 30							IDLE		
	HC "In"* (ppm hex)	HC Eff. (%)	CO "In" Mol (%)	CO Eff. (%)	Temperature–°F			HC "In" (ppm hex)	HC Eff. (%)	CO "In" Mol (%)	CO Eff. (%)	Temperature–°F			HC "In" (ppm hex)	HC Eff. (%)	CO "In" Mol (%)	CO Eff. (%)	Temperature–°F			HC "In"* (ppm hex)	HC Eff. (%)	CO "In" Mol (%)	CO Eff. (%)
Mileage					"In"	Cat. Bed	"Out"					"In"	Cat. Bed	"Out"					"In"	Cat. Bed	"Out"				
0	65	54	0.30	50	1035	1230	1030	115	40	0.45	67	880	1080	950	165	60	0.15	67	720	880	760	225	82	1.85	92
1356	80	50	0.80	38	1050	1240	1100	130	54	1.10	59	880	1110	975	190	50	0.40	38	705	855	750	240	83	0.95	90
3100	95	42	0.65	31	1140	1320	1180	175	43	0.85	41	940	1145	1040	240	37	0.30	50	740	850	760	305	75	1.75	88
4360	95	32	1.15	30	1160	1250	1195	160	41	1.35	48	960	1140	1045	215	33	0.60	50	810	920	800	685	71	2.55	84
4750	95	37	0.35	43	1110	1240	1140	155	32	0.35	29	995	1090	1005	190	18	0.15	33	800	885	770	260	95	1.40	86
6715	120	33	1.25	52	1120	1250	1150	185	38	1.10	50	910	1060	950	225	29	0.35	43	740	830	740	275	58	0.80	75

CONVERTER F-1 CATALYST "O"

Mileage	HC "In"* (ppm hex)	HC Eff. (%)	CO "In" Mol (%)	CO Eff. (%)	Temperature–°F			HC "In" (ppm hex)	HC Eff. (%)	CO "In" Mol (%)	CO Eff. (%)	Temperature–°F			HC "In" (ppm hex)	HC Eff. (%)	CO "In" Mol (%)	CO Eff. (%)	Temperature–°F			HC "In"* (ppm hex)	HC Eff. (%)	CO "In" Mol (%)	CO Eff. (%)
					"In"	Cat. Bed	"Out"					"In"	Cat. Bed	"Out"					"In"	Cat. Bed	"Out"				
82	175	74	0.75	87	900	1120	1045	255	73	0.55	82	720	920	855	285	63	0.20	50	570	685	625	440	79	3.75	97
163	190	74	1.10	86	1035	1260	1095	240	75	1.10	91	860	1090	920	280	71	0.20	75	720	790	610	400	83	2.05	98
1000	180	55	1.05	71	1030	1110	1030	190	47	0.95	63	860	950	870	200	45	0.25	60	690	720	550	270	70	1.40	90
1670	200	50	1.10	45	1030	1110	1010	200	45	0.95	53	835	920	810	210	38	0.25	40	670	735	635	330	33	1.90	71
3400	140	28	1.10	32	1030	–	960	210	29	1.30	39	860	960	770	230	35	0.30	50	680	700	550	310	42	2.40	77
5230	125	32	0.80	44	1030	1090	940	170	41	0.90	44	850	910	760	180	28	0.20	50	660	580	520	300	23	2.60	58
8344	155	26	1.10	36	980	1100	915	185	24	1.10	46	810	895	745	210	26	0.25	20	675	695	545	330	16	3.50	21

* Converter Inlet Concentration

ALL CONCENTRATIONS ARE AS READ – NO CORRECTION FOR DILUTION

STABILIZED CRUISE EFFICIENCY TEST PROCEDURE DETAILED IN APPENDIX B

CONVERTER G-1, CATALYST "N"

DETROIT TRAFFIC ROUTE TEST RESULTS

MILEAGE	CONVERTER INLET CONCENTRATIONS		CONVERTER OUTLET CONCENTRATIONS		OXIDATION EFFICIENCY - %	
	Hydrocarbons (PPM HEX)	Carbon Monoxide MOL %	Hydrocarbons (PPM HEX)	Carbon Monoxide MOL %	Hydrocarbons	Carbon Monoxide
105	265	1.35	80	0.40	70	70
160	270	1.65	70	0.40	74	76
1436	255	1.35	110	0.50	57	63
1494	210	1.50	65	0.40	69	73
2900	455	1.65	185	0.50	59	70
2960	430	1.60	180	0.50	58	69
3150	355	2.00	115	0.60	68	70
3200	310	1.90	160	0.50	48	74
4518	285	1.80	165	0.75	42	58
4677	245	1.45	140	0.60	43	50
4703	285	1.95	135	0.70	53	64
6574	200	0.85	120	0.40	40	53
6621	215	0.90	105	0.25	51	72

ALL CONCENTRATIONS ARE "AS READ" WITHOUT CORRECTION FOR AIR DILUTION.

Catalytic Converter Development Problems

D. L. Davis and G. E. Onishi
Studebaker-Packard Corp.

THE PRESENCE OF UNBURNED hydrocarbon compounds and carbon monoxide gas in the automobile exhaust stream is related to the combustion efficiency of the engine and is a function of engine design, maintenance, and specific operating conditions. A limited reduction of the unburned gases can be obtained by control of these factors; however, the need to provide substantial reductions has become pressing, particularly in areas of large vehicle populations. Methods of accomplishing this reduction include:

1. Direct flame type afterburner.
2. Induction system devices (to control the flow of gasoline to the combustion chamber during deceleration periods when combustion efficiency is low).
3. Flameless low temperature oxidation of the gases in the exhaust stream on active catalytic surfaces.

BACKGROUND

The technology of oxidizing the hydrocarbon compounds and carbon monoxide into carbon dioxide and water is neither new nor complex. A chamber must be provided in which these compounds can be mixed with oxygen in the presence of a catalytic material which will cause the reaction to occur. Designing such a reactor to accomplish only this conversion is relatively simple when the concentration, volume, and temperature of the gases are relatively constant, and space does not limit the size of the reactor. However, designing such a reactor for use on an automobile presents a unique and far more complicated problem.

During the past several years, a vast amount of research has been devoted to automobile exhaust emissions by the automobile industry, the chemical industry, municipalities, and public health services. Based upon these studies, the state of California has established certain specifications and performance requirements for exhaust emission control devices to be used in that state. Since these are the only standards that have been proposed to date, they have been accepted as the present design and performance criteria.

Cost - The device should not impose an undue cost burden upon the owner. One interpretation of a reasonable cost is one-half cent per mile of total vehicle operation, including amortization of original equipment costs and maintenance costs.

Maintenance - Frequent maintenance should not be required. The device should function properly without replacement of major components for at least 12,000 miles.

Size - The device must be of such size that it can be adapted to the space available in the conventional automobile with a minimum of car body design changes.

Operation - The device must be designed to have no serious effect on gasoline consumption and performance of the car. Exhaust system back pressure should not be increased more than 25% and, should it replace the conventional muffler, the device must provide equivalent silencing. Safety must be considered in that it should not apply excessive heat to any part of the vehicle and if a malfunction should occur, it should not create a danger or hazard.

Performance - The exhaust control device must maintain exhaust emission levels at not over 275 ppm (0.0275% by volume) of hydrocarbon and 1.5% by volume of carbon monoxide for a minimum period of 12,000 miles or approximately 1 yr of average use. A standardized test driving cycle has been established to include representative events that would occur during an average trip of 18 minutes duration throughout which the device must effect the required reduction. The cycle includes periods of idle, steady state driving at various speeds, accelerations, and decelerations.

ANALYSIS OF SPECIFICATIONS

The first four requirements represent sound engineering and design practices and present no serious obstacle. The

---

ABSTRACT

A research and development program by Studebaker-Packard on catalytic converter system designs is discussed, along with details of specific problem areas encountered, methods used in their solution, and an outline of remaining unresolved problems. Data are presented which show the effect on conversion efficiency of catalyst bed configuration, catalyst temperature, auxiliary air, and gasoline additives. Catalyst warmup rate and more effective utilization of available exhaust heat is discussed. Performance of several catalysts under various automobile operating conditions are illustrated.

major problems arise in attempting to meet the performance standards and become far more complicated than just oxidizing the unburned compounds.

The reactor and all associated components must be carried in the automobile where space to accommodate any accessory is limited. The temperature, volume, and concentration of hydrocarbon compounds and carbon monoxide in the exhaust gases have been found to vary considerably over a wide range of operating conditions. Exhaust gas temperatures at the exhaust manifold outlet can vary from 500 F at idle to over 1500 F at 70 mph wide-open-throttle operation. Exhaust gas volumes can vary from 5-250 cfm under the same two conditions. Hydrocarbon concentrations can vary from 100 ppm under some steady state cruise conditions to 4000 ppm during decelerations. Carbon monoxide can vary from a trace to 10% (by volume) under different engine operating conditions.

CONVERTER SYSTEM DESIGN

A typical catalytic converter system as applied to an automobile is illustrated schematically in Fig. 1. The basic configuration includes a converter, oxygen supply, and a temperature limiting device. The converter is a closed vessel containing a bed of catalyst material through which the exhaust gases are passed. Secondary air, from an air compressor operated off the engine, is mixed with the exhaust gas to provide the necessary oxygen for combustion. When the mixture of air and exhaust gas enters the catalytic bed, the hydrocarbon compounds and carbon monoxide are converted to carbon dioxide and water. This reaction generates heat which, under certain operating conditions, could exceed the safe operating temperature of the catalyst. To protect the converter system from overheating, a thermoswitch is inserted into the center of the catalyst bed to sense catalyst temperature and actuate a valve which allows the exhaust gases to bypass the bed.

An alternate method of controlling the conversion rate may be accomplished by regulating the amount of secondary air supply. A solenoid valve in the air compressor inlet line, actuated by the thermoswitch, provides this control. However, in the case of engine misfiring, conditions, the unburned gas is already mixed with adequate air for combustion and destructive temperatures would soon be reached in the converter. For this reason, the bypass method of control was chosen.

TEST OPERATIONS AND RESULTS

The heart of the system is the catalyst and its container. Based on preliminary laboratory tests by chemical manufacturers, catalysts were chosen which appeared to meet the necessary requirements. The problem then became one of evaluating the catalysts in converters on automobiles under actual operating conditions to explore the effects of the many variables on the conversion efficiency. These variables include bed configuration (the relation of the catalyst bed depth to the face area), space velocity (the relation of the volume of gas per unit of time to the volume of catalyst bed), and the effects of temperature and lead contamination.

Preliminary testing began on an engine dynamometer using quarter size converters. In addition to expediting the work with the use of a minimum of catalyst, this method provided accurate control of such variables as exhaust gas volume, temperature, and emission concentrations. The installation, as diagramed in Fig. 2, consisted of a 289 cu in. overhead valve V-8 engine operated at a speed and load that provided a volume of exhaust gas in excess of the test requirements. Secondary air, supplied from a high pressure source, was metered through a rotometer into the exhaust

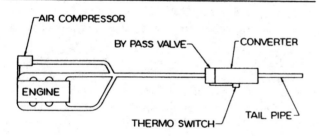

Fig. 1 - Catalytic converter system

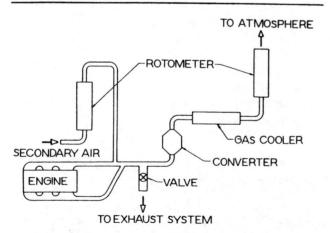

Fig. 2 - Quarter size converter test equipment

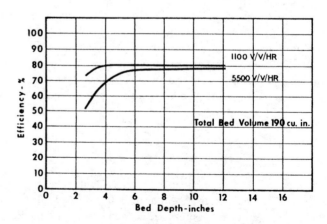

Fig. 3 - Effect of bed depth on catalyst performance

stream. The mixture of exhaust gas and secondary air was piped to the converter with provisions for a relief valve to bleed the excess gas mixture to the dynamometer exhaust system. The amount of mixed gas and secondary air entering the converter was metered accurately with this valve since the dynamometer exhaust system was at a negative pressure and the exhaust gas and air were at a positive pressure.

After leaving the converter, the gas temperature was reduced through a cooler and its volume measured with another rotometer. Thermocouples were inserted in the catalyst bed to record temperatures at several points and gas samples for analysis were removed from either end of the converter.

Catalyst Bed Configuration - Initial dynamometer testing was accomplished to establish the bed configuration of the converter necessary to house the catalyst. It was found that the bed configuration, as well as the bed volume (as it affects space velocity), has a decided effect on conversion efficiency, as shown in Fig. 3. For deeper beds there is very little change in efficiency with a change in space velocity. For more shallow beds there is a substantial change. Therefore, the most desirable bed thickness would be something more than 6 in.; however, the actual configuration must be a compromise, considering volume (as it affects warmup rate), maximum space velocity, available space in the car, and exhaust system back pressure. Thus, most of our present work is with full size converters with bed depths of 4 in. and volumes of 450-600 cu in., resulting in space velocities of 21,000-16,000 cu ft of exhaust gas per hour per cubic foot of catalyst bed at 70 mph cruise conditions.

Basic Converter Designs - Fig. 4 illustrates two basic converter designs; the horizontal or cross-flow type and the vertical-flow type where the flow is either up or down through the catalyst bed.

In the cross-flow design, the two "Vee" sections are pieces of angle iron welded to, and extending the full length of, the cover and bottom of the bed. They serve three-fold purpose:

1. To compress the catalyst when the cover is installed to assure a firmly packed bed.

2. To reinforce the converter to reduce the possibility of warpage.

3. To act as a dam to prevent the exhaust gases bypassing the catalyst (top "Vee" only).

The lower cross-section shows the normal gas flow, while the upper shows the gas bypassing the bed when the bed volume is reduced.

The cross-flow design lends itself to the desirable bed thickness; however, it has certain disadvantages. Much of the bed is exposed to radiant heat loss, thus requiring the entire converter to be insulated to conserve heat and assure a more uniform bed temperature. This design is more susceptible to an early decline in conversion efficiency with a slight decrease in bed volume due to either attrition of the catalyst or loss due to rupture or warpage of the container.

The down-flow converter has several advantages:

1. A 4 in. thick bed of sufficient volume can be built into a converter only slightly larger than the conventional oval tri-flow muffler, and only slight modification of the present floor pan is required to provide sufficient floor pan and ground clearance.

2. Only the top half of the converter need be insulated to protect the car floor pan and reduce radiant heat loss of the incoming exhaust gas and most of the bed--the outlet chamber is not insulated to allow for more rapid dissipation of the heat in the converter outlet gases to the atmosphere.

3. A loss in catalyst merely decreases the bed volume with a resultant increase in space velocities, and since the converter is designed to handle space velocities present under high speed operation, a decrease in bed volume will not affect conversion seriously under normal operating conditions.

One of the objections to the cross-flow converter design is dramatically illustrated in Fig. 5, showing the loss in conversion efficiency that occurred in one of the early cross-flow converters tested where 8.7% of the original catalyst was lost due to a warpage of the bed retaining screen. Work with down-flow converters of the same approximate external dimensions and slightly smaller catalyst beds indicates that this 8.7% change in bed volume has very little effect on the overall conversion efficiency in the down-flow design.

Catalyst Considerations - A catalyst must be sufficiently resistant to abrasion to withstand the constant vibration it is

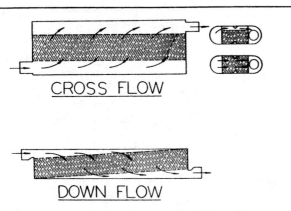

Fig. 4 - Catalytic converter designs

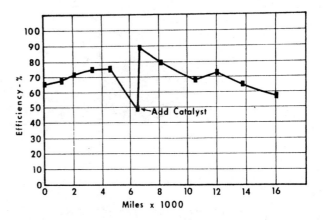

Fig. 5 - Effect of attrition loss on conversion efficiency

subjected to in an automobile. A catalyst that is too soft would, in a few thousand miles, be worn sufficiently to reduce the bed volume to a point where its conversion efficiency would be reduced seriously. To evaluate the resistance of a catalyst to attrition, the test equipment illustrated in Fig. 6 was designed. A weighed quantity of the catalyst to be tested was placed in a container of 6 sq in. horizontal cross-section with a bottom of 16 mesh screen. The container was mounted on a shake table operating at 1700-1800 cycles per minute at 1/16 inch horizontal travel. The per cent weight loss of the catalyst versus time in hours up to 100 hr was reported. The volume of catalyst sample was not important since the loss was reported as percentage of the original weight; however, we start with a bed depth of approximately 3 in.

The attrition rate of several catalysts investigated is shown on Fig. 7. Based on these tests, the following observations were made:

1. The variation in the rates of loss is due to the differences in the size, shape, and composition of the catalyst pellets. Some catalysts were formed in hard rice-size pellets, some were pure catalyst, and some were a metallic carrier impregnated with catalyst.

2. An exact correlation between the shake table loss and the loss in a converter mounted on a car has not been established. However, the attrition loss of catalyst B would seem to indicate that 100 hr on the shake table is equivalent to approximately 12,000 miles of operation on the road. Two converters of different design filled with catalyst B have been operated over 12,000 miles. The average catalyst loss rate was 0.101 lb/1000 miles or a 4.5% reduction in bed volume in one converter and approximately double this in the second.

Secondary Air Pump - Having once established a bed volume and configuration and verified the catalyst's resistance to abrasion, it was necessary to establish the capacity requirement of the secondary air pump.

In determining the amount of secondary air to be supplied, two parameters must be considered.

1. The volume should be held to a minimum because of the cooling effect on the exhaust gas with resultant loss in effectiveness of the catalyst.

2. Enough air must be supplied to assure the maximum conversion of which the particular catalyst is capable.

This conversion efficiency will vary depending upon the activity of the catalyst at the gas temperatures available.

The theoretical amount of secondary air necessary for complete conversion of hydrocarbons and carbon monoxide to water and carbon dioxide can be calculated by using 15 lb of air per pound of hydrocarbon and 2.5 lb of air per pound of carbon monoxide. This calculation assumes perfect mixture of the oxygen in the air with the carbon monoxide and hydrocarbon in the exhaust stream. Actually, this condition cannot be obtained. To assure a sufficient quantity of air for maximum conversion by the catalyst being evaluated, the amount supplied to a full size converter was adjusted to produce maximum conversion at the various speeds and loads. Fig. 8 shows typical curves of the air requirements of two different catalysts at three road-load car speeds. Plotting the air deliveries that result in the maximum conversion efficiencies at the various speeds gives the desired compressor air delivery curve. Under sustained operation at loads higher than normal cruise, more air would be required, but under normal conditions where the various driving modes are interspersed, this quantity is sufficient to assure a maximum utilization of the catalyst.

The theoretical air requirement is compared to Fig. 9

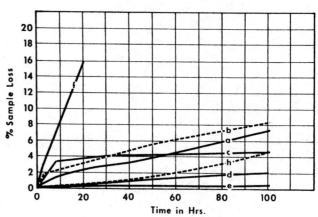

Fig. 7 - Bench test results of catalysts attrition

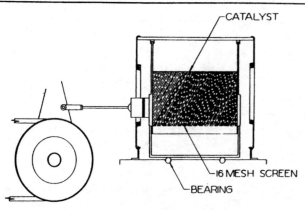

Fig. 6 - Bench test for rate of attrition

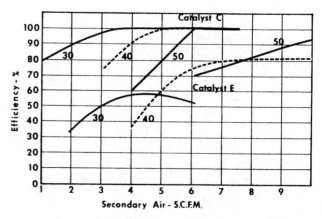

Fig. 8 - Secondary air requirements

with the actual requirement necessary to assure maximum conversion using catalysts C and E. As an expedient, commercially available air compressors have been used on cars for converter road testing. Although they perform satisfactorily, these would not be suitable for production automotive use because of their size, weight, and cost. However, developments are in progress to provide an air pump designed specifically for this application.

Catalyst and Converter Temperature Considerations - Any catalyst will induce a high degree of conversion if its temperature is maintained at a sufficiently high level. This optimum temperature varies with different catalysts and increases as the catalyst becomes contaminated and loses its activity. Fig. 10 shows the effect of bed temperature on the efficiency of a catalyst at two different mileages at a vehicle speed of 30 mph. The high bed temperature was obtained by cruising at 50 mph until the bed temperature stabilized, then dropping to 30 mph cruise. The volume and hydrocarbon concentration of the exhaust gas dropped immediately to that of 30 mph cruise; however, the efficiency decreased only as the bed temperature decreased.

Fig. 11 shows the efficiency that was obtained with an apparently deactivated catalyst when its temperature was maintained at a higher level. The solid line represents the conversion efficiency at 30 mph cruise over a period of 13,000 miles of vehicle operation. The activity of the catalyst after less than 3000 miles had decreased to a point of negligible conversion. The dotted curve shows the efficiency at the same mileages with an induced increase in bed temperatures, with the average bed temperatures noted on the applicable curves. These figures show that a 200 deg increase in bed temperature resulted in an approximate 40% increase in efficiency.

To confirm these data and evaluate the effect of inlet gas temperature on converter bed temperature and efficiency, a car was equipped with a water cooled exhaust pipe to permit operation with varying temperatures of the exhaust gas at the converter inlet. An appreciable gain in efficiency can be attained with an increase in bed temperature, as shown on Fig. 12. Increasing the exhaust temperature from 650 to 900 deg resulted in an efficiency increase from 32 to 81%.

With the converter located as close to the exhaust manifold as possible and the exhaust pipes insulated, the only other way to increase the temperature of the exhaust gas at the converter with the heat available is by the utilization of the heat of conversion produced by the converter. One design concept that will accomplish this is illustrated in Fig. 13. An extra valve and a preheater have been added to a bypass over-heat controlled converter.

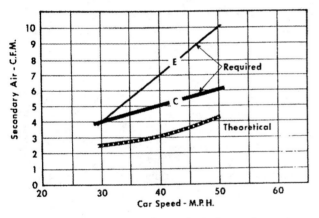

Fig. 9 - Actual versus theoretical air requirements

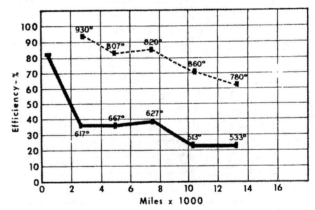

Fig. 11 - Effect of catalyst temperature on conversion efficiency

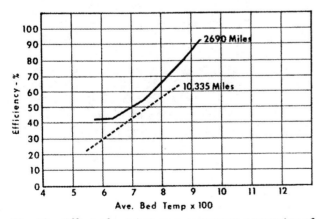

Fig. 10 - Effect of catalyst temperature on conversion efficiency

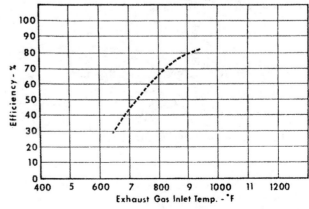

Fig. 12 - Effect of exhaust gas temperature on conversion efficiency

In the upper view, valve A is in a position to bypass the catalyst bed when the bed temperature exceeds 1100 F. Valve B is open, allowing the exhaust to escape through the tail pipe which offers less restriction to gas flow than the preheater. During warmup and low speed driving when catalytic bed temperatures are below 800 F, the valve positions shown in the middle view force the gas through the bed, where its temperature is increased due to the heat of conversion, then through the preheater where the heat of conversion is used to preheat the exhaust stream. Under normal bed operating temperatures of 800 - 1100 F the exhaust gas follows the flow path shown in the lower view.

Such a device operated satisfactorily under test, but the increase in inlet gas temperatures was below that anticipated. Further work with a more efficient heat exchanger is currently in progress.

Early in the test program it was recognized that to accomplish a contribution to the reduction of air pollution the converter should reach a reasonable conversion efficiency in a very short period of operating time. Most driving in the larger cities where the automobile's contribution to air pollution is most serious is in the form of short trips; however, longer trips at relatively low speeds also make serious contributions. The importance of a rapid warmup rate of the catalyst was emphasized further by the new California Motor Vehicle Pollution Control Board's performance specification previously noted.

temperature level required for satisfactory oxidation of the hydrocarbon and carbon monoxide is dependent upon at least five factors:

1. The sensible heat available in the exhaust stream.
2. The temperature of the exhaust stream.
3. The specific heat of the catalyst material.
4. The amount of catalyst used.
5. The activation temperature of the catalyst.

The specific heat and the activation temperature of the catalyst are fixed, as are the temperature and sensible heat of the exhaust gas as it leaves the exhaust manifold. However, as the exhaust gas travels the 7 ft through the exhaust pipe to the front of the muffler on a conventional car, its temperature drops 270 deg (from 840 to 570 F at 30 mph cruise). This loss of 270 deg can be reduced some 50% by insulating the exhaust pipes from the engine to the converter. Baffling the front vertical section of the pipe from the fan blast will also help to reduce this loss. Locating the converter as close to the manifold as possible will converse some of this heat although approximately 50% of the loss is in the 12 in. vertical section immediately below the manifold.

The quantity of catalyst can be decreased considerably to maintain a satisfactory space velocity for low speed operation and, when used in conjunction with a second larger converter, the quantity necessary for higher exhaust flows is available.

The converter shown on Fig. 14 was built and tested to evaluate this theory. It was fastened directly to the exhaust manifold flange of the engine. The catalyst bed volume was 175 cu in. compared to 800 cu in. in the full size converter which was then being tested. During warmup at low exhaust flows, and temperatures below 1100 F, the valve was in the position shown by the solid line. The exhaust gas followed the path of the solid line arrows through the catalyst bed and then to the exhaust pipe. When the catalyst bed temperature reached 1100 F the valve shifted to the position shown by the dotted line and the gas followed the flow indicated by the dotted arrows, bypassing the small converter completely, and travelling to the full size converter located under the car.

Fig. 15 compares the warmup rates of the small and full size converters. The small converter was operating at 45% conversion efficiency before the larger unit had started to convert. With the small converter, the time to reach 50% conversion efficiency was reduced from 25 to 8-1/2 minutes.

In addition to the rapid warmup rate experienced with

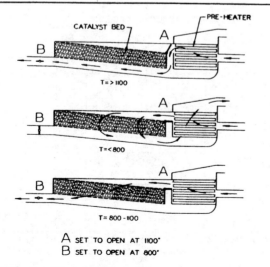

Fig. 13 - Catalytic converter preheater operation

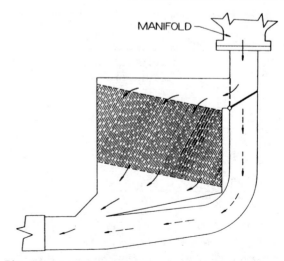

Fig. 14 - Auxiliary catalytic converter for rapid warmup

this converter, its heat of conversion resulted in an increase in exhaust gas temperature of approximately 100 F.

The importance of a rapid warmup rate in meeting the recently established California performance standards is illustrated in Fig. 16. These curves are from data obtained on dynamometer rolls with the converter equipped car operated in accordance with California's specified test procedure and the data calculated in accordance with California's recommendation. The temperatures are those at the center of the catalyst bed at the end of each cycle. The conversion efficiencies are based upon the weighted hydrocarbon concentrations in the exhaust stream before and after the converter during each cycle.

Even though the catalyst was new and operating at its maximum efficiency, no conversion occurred during the first cycle and very little during the second. It was not until the fourth cycle, or after approximately 8 minutes of vehicle operation, that the catalyst really became active. It reached its maximum activity during the last cycle.

Since the total emission level is based upon 45% of the average weighted emissions during the first four cycles plus 55% of the average during the seventh cycle, the importance of attaining a high rate of conversion during the first cycles becomes obvious.

The rapid loss in conversion efficiency at relatively low mileages is due primarily to contamination of the catalyst by the lead in commercial gasolines. There is some initial thermal destruction of the catalyst, but this affects efficiencies by only a few per cent when the catalyst temperature is maintained within safe limits. Fig. 17 shows the rate of efficiency loss during two different tests, one with a 2.6 cc/gal leaded gasoline and the second with a nonleaded gasoline. These runs were made on a dynamometer engine under identical operating conditions. The loss with the nonleaded gas is slightly greater than would normally be expected and might reflect some slight lead contamination although in preparation for this run new cylinder heads, intake and exhaust manifolds, and exhaust pipes were installed. The interior of the converter was sand blasted and the engine was flushed several times with clean oil.

Several attempts were made to filter the lead out of the exhaust stream by means of beds of inactive material in front of the converter with very limited success.

The effect of lead contamination on the effective life of several different catalysts is illustrated on Fig. 18. These data were obtained on a car operating at 50 mph cruise. The loss in activity is much more rapid at lower car speeds where the exhaust gas temperatures are not sufficiently high to spark

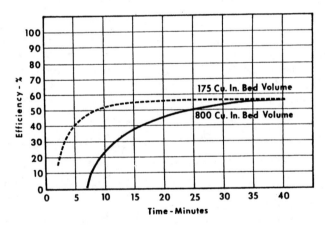

Fig. 15 - Effect of catalyst bed volume on warmup rate

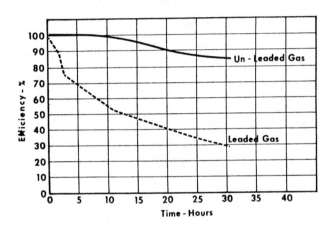

Fig. 17 - Catalyst life: leaded versus unleaded gasolines

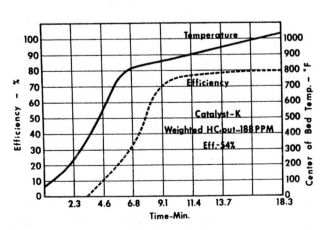

Fig. 16 - Catalyst efficiencies during warmup - California test cycle

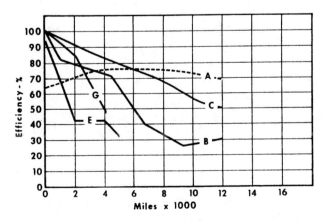

Fig. 18 - Catalyst performance at 50 mph

the reaction. The curves illustrate the vast difference in catalyst susceptibility to lead contamination.

Catalyst A, although its conversion efficiency when new is lower than that of the others shown, actually improved in activity for several thousand miles before gradually losing its effectiveness, and appeared to be relatively immune to contamination by lead or other gasoline additives. Unfortunately, it had no effect on carbon monoxide conversion.

Fig. 19 shows the warmup rate of a new catalyst now under evaluation. These data were obtained during 30 mph cruise operation. The vehicle was allowed to cool to ambient air temperature for a period of 16 hr. The engine was started and the car immediately accelerated to 30 mph where it was held for 30 minutes. Emission concentrations into and out of the converter were recorded every two minutes.

Succeeding curves at later mileages and higher degrees of catalyst contamination will move further to the right and show a lower final efficiency. This deactivation may occur in a few hundred miles or it may require several thousand miles. This rate of contamination is an indication of the effectiveness of the catalyst in meeting the emissions reduction performance specifications.

During these two tests, the temperature of the exhaust gas into the converter was the same. Had the exhaust gas temperature been increased as the catalyst lost its activity, the original 45 mile curve could have been duplicated fairly well regardless of mileage and catalyst contamination.

CONCLUSION

The main problem still unresolved in the catalytic converter development program is that of maintaining adequate catalyst conversion efficiency throughout the specified 12,000 miles. This might be accomplished in any one of three ways.

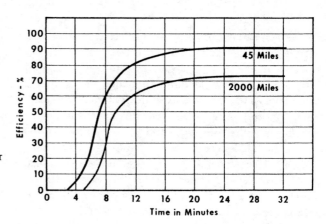

Fig. 19 - Catalyst efficiencies during warmup 30 mph cruise

1. By omitting the catalyst contaminating additives from commercial gasoline. This solution is not economically feasible at the present time.

2. By developing a catalyst that will not be affected adversely by the additives in gasoline. Much progress along this line has been made in the past two years and the problem is continuing to receive the attention of numerous catalyst development laboratories.

3. By providing sufficient heat in the catalyst bed to promote a high degree of conversion even after contamination. This method is also being thoroughly investigated.

This, then, is the problem that is being attacked by literally thousands of scientists and engineers in a spirit of cooperation that makes a final solution inevitable. The pool of knowledge that has been accumulated over the past several years should eventually provide the means of overcoming the last barriers.

Paper subject to revision. SAE is not responsible for statements or opinions advanced in papers or discussions at its meetings. Discussion will be printed if paper is published in Technical Progress Series, Advances in Engineering, or Transactions. For permission to publish this paper, in full or in part, contact the SAE Publications Division and the authors.

660106

DESIGN AND DEVELOPMENT OF THE GENERAL MOTORS AIR INJECTION REACTOR SYSTEM

WILLIAM K. STEINHAGEN
GEORGE W. NIEPOTH
STANLEY H. MICK

Power Development Group
Engineering Staff
General Motors Corporation

ABSTRACT

The General Motors Air Injection Reactor System meets the California standards for hydrocarbon and carbon monoxide emission and will be installed on most 1966 GM cars and light trucks sold in California. The various components of the system are described along with their calibration for optimum emission control, and the special techniques used to analyze the system performance on the California Cycle emission test.

INTRODUCTION

The General Motors Air Injection Reactor System meets the California standards for control of exhaust hydrocarbons and carbon monoxide. This system is used on most General Motors cars and light trucks sold in California in 1966. The system is basically an air injection system combined with engine modifications to increase its effectiveness. A photograph of an Air Injection Reactor System installation is shown in Figure 1.

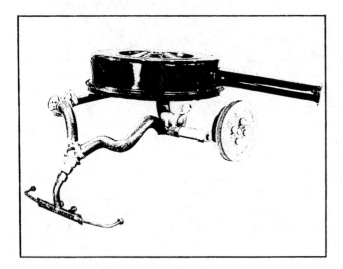

Figure 1 - A 1966 V-8 engine equipped with the GM AIR system shows the belt driven air pump and air distribution system.

Air injection, which is the heart of the control system, is used on all vehicles. Engine modifications have been made as required on specific engine-carburetor-transmission combinations to offer suitable control.

This paper will first describe the various components of the system and the durability testing procedure. Calibration of the components for optimum emission control will then be presented along with techniques used to analyze the performance of the AIR system on the California Cycle emissions test.

DESCRIPTION OF SYSTEM COMPONENTS

AIR INJECTION INTO EACH EXHAUST PORT

Injection of air into the exhaust port of an internal combustion engine for treatment of the exhaust gases is an old concept. References 1 and 2 in the Bibliography list two patents issued in 1940 and 1941 to inventors who claimed a reduction in partially burned and unburned exhaust constituents.

In 1962, the SAE was told of GM's activities with air injection (Reference 3). Since then there has been much effort to combine air injection with engine modifications to meet the California requirements and provide driveability characteristics which would be acceptable to our customers.

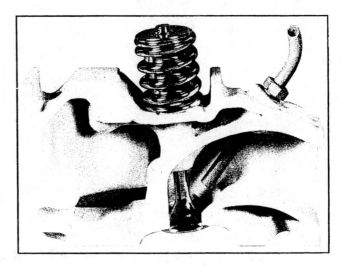

Figure 2 - Air is delivered to the exhaust port of each cylinder in the vicinity of the exhaust valve.

The mechanics of air injection are quite simple. Pressurized air is transmitted through hoses and distribution manifolds to individual tubes or passages in each cylinder exhaust port, close to the exhaust valve, as shown in Figure 2. A positive displacement, non-lubricated pump supplies the air. It is a semi-articulated vane pump, belt driven from the engine crankshaft. The development of this pump was one of the significant steps in the AIR system development (Reference 4).

Figure 3 shows the components of the Air Injection Reactor System. Some of these components are new and some are modifications of existing parts.

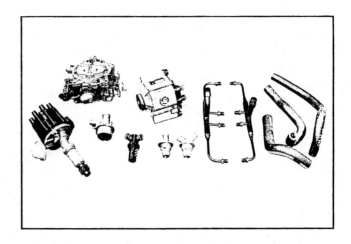

Figure 3 - The components of the AIR system make it adaptable to new cars only.

ENGINE MODIFICATIONS

Various engine modifications have proven helpful in controlling exhaust emissions and improving air injection system effectiveness. All modifications which are used in the General Motors exhaust emission control system are described even though the items and their degree of use may vary between engines.

Carburetor

Special carburetors are used on each of the equipped vehicles. These carburetors are calibrated for optimum air injection effectiveness as well as vehicle performance, economy and driving feel. The flow characteristics may, therefore, be different from carburetors on unequipped cars. In addition, the fuel flow tolerance band has been reduced in critical areas to assure optimum carburetion for emission control.

Ignition Distributor

All systems are equipped with ported vacuum spark advance units. At engine idle, this port is above the throttle plate and no vacuum is supplied to the distributor. The ignition timing is, therefore, retarded. As the throttle is opened, the port is uncovered providing vacuum spark advance for normal part throttle operation.

Some ignition distributors are provided with additional retard of the centrifugal advance unit at idle. This additional retard provides a 5° to 15° retarded idle ignition timing. Normal ignition timing is restored at an engine speed of approximately 1200 rpm and above.

Idle Speed

Some engines use a higher idle speed. These settings vary among engines. Typical specifications are 550 rpm for automatic transmission equipped vehicles and 700 rpm for vehicles with synchromesh transmissions.

Carburetor Idle Mixture

There is no special idle mixture adjustment required for this system. It has been shown that effective emission control is provided when the usual range of idle mixture adjustment is used.

Engine Cooling System

A modified engine cooling system is used on some cars due to the increased heat rejection at idle caused by the retarded ignition timing. Changes such as redesigned fans, higher fan speeds, fan shrouds, larger radiators, or increased radiator fin area are incorporated.

Thermal Modulation for Spark Control

A thermal modulation valve for the ported spark control is used on some cars. This valve provides full vacuum advance at idle when coolant temperatures become excessive. The advanced timing reduces the cooling requirements and also increases engine speed at idle which improves fan effectiveness. This valve operates only under extreme temperature conditions.

Anti-Backfire Valve

The purpose of this valve is to prevent explosions in the exhaust system during certain throttle maneuvers. The anti-backfire valve allows a gulp of air to enter the intake manifold following rapid throttle closure. The anti-backfire valve outlet is connected to the intake manifold or to a connection at the carburetor throttle body. The inlet to the valve is connected to the air pump discharge line or to the engine air cleaner for clean air and silencing. The control vacuum line is connected to a separate point in the intake manifold.

Check Valves

Check valves are located in the air pump discharge lines to prevent backflow of exhaust gases into the air injection lines or into the air pump. The valves prevent backflow when the air pump bypasses at high speed and load or in case the air pump drive belt fails.

Air Pump Relief Valve

A relief valve is located in the discharge cavity of the air pump. The valve allows pump outlet air to bypass the air injection system at high engine speeds and loads. This relieving is necessary to prevent excessively high exhaust system temperatures under certain extreme operating conditions.

DURABILITY TESTING

As of July 1, 1965, a total of 250 GM vehicles had been operated over 2-1/2 million miles with the AIR system installed. Basic durability testing, both vehicle and bench, has proven the durability of the AIR system components.

Long range emission control testing of the air injection reactor system was accomplished using eleven cars driven a total of 1 million miles at the Milford Proving Ground on a modified city driving schedule. Nine of these cars were driven 100,000 miles each, while two of them were driven 50,000 miles each. Discussion in this paper will be limited to the eleven cars that were operated solely on the modified AMA driving schedule.

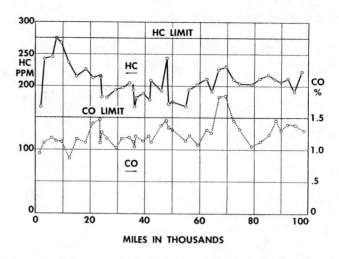

Figure 5 - A 1965 vehicle equipped with the GM AIR system operated with exhaust emissions below the California limits for 50,000 miles with regular maintenance followed by 50,000 miles with minimum maintenance. Eight other cars gave similar performance in proving the system for use in California.

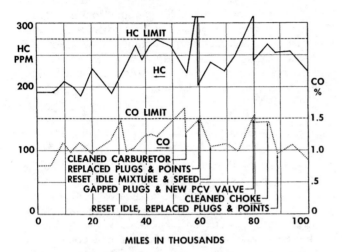

Figure 4 - A 1964 test car equipped with air injection was driven 100,000 miles on a simulated city driving schedule with minimum maintenance. Emissions exceeded the California limits only when the engine malfunctioned.

Two of the 100,000 mile cars were driven with a minimum of maintenance. Engine maintenance was performed only when the drivers reported substandard driveability. Figure 4 shows the emission results of a car that was driven 100,000 miles with minimum maintenance. Exhaust emissions exceeded the California limits only when the engine malfunctioned due to misfiring spark plugs, stuck choke, and dirt in the carburetor.

Nine cars were driven on the durability test with regular maintenance as part of our 1966 California certification program. A tune-up was performed every 12,000 miles. Figure 5 shows the California Cycle exhaust emission results for a 1965 car equipped with the AIR package as measured for 100,000 test miles. The engine received regular maintenance during the first 50,000 miles and minimum maintenance during the second 50,000 miles. Carbon monoxide emission exceeded the limit at about 66,000 miles because of a plugged fuel filter and burned ignition points.

The very satisfactory emission control history of this car was typical of the other eight durability cars. The results of these nine cars were used to describe the long term emission control characteristics of the AIR system and the maintenance requirements.

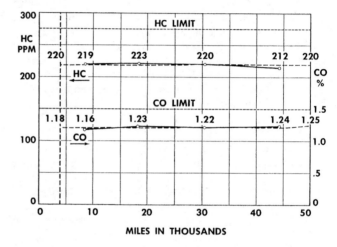

Figure 6 - Average exhaust emissions for nine AIR test cars run 50,000 miles with regular maintenance showed no increase in hydrocarbons once initial stabilization occurred. There was a small increase in carbon monoxide emission.

LONG TERM EMISSION CONTROL

The exhaust emission results for each of the nine cars were first averaged over the 12,000 mile tune-up intervals. The nine cars were then averaged together to arrive at a single plot to determine the aging characteristics of the emission control, Figure 6. On an average basis, with regular maintenance there was no increase in the hydrocarbon emissions. Some cars increased slightly, while some decreased slightly in hydrocarbon emissions. However, the average showed essentially no change once initial stabilization of deposits had occurred which took about 4,000 miles.

There was a small increase in carbon monoxide over the 50,000 mile period. On an average, this showed that the CO deterioration factor for that period would be about 1.06. These results indicate that there is very little deterioration with the GM AIR system, provided the engine receives maintenance sufficient to keep it operating properly.

MAINTENANCE

Analysis of exhaust hydrocarbon and carbon monoxide emissions measured before and after each tune-up produced the average results shown in the following table:

Effect of Tune-up on Exhaust Emissions

(29 tune-ups at 12,000 mile intervals)

Average Exhaust Emissions

	Before	After
Hydrocarbons	215	211
Carbon Monoxide	1.21	1.19

Change for Each Car

	Increased	Decreased	Same
Hydrocarbons	10	18	1
Carbon Monoxide	12	16	1

These results show that the average exhaust emissions did not change appreciably when the cars were tuned-up. Individual cars did change, but not in a uniform direction. Thus, an AIR car that is performing satisfactorily is not helped or hurt by a tune-up.

The results of the durability tests have shown that there is a minimum maintenance required in addition to normal vehicle servicing. The recommended annual maintenance is:

1. Replacement of the crankcase ventilation valve which is now presently required in California.

2. Check the air pump belt and tighten if required.

3. Service the air pump air filter element on those vehicles which employ a separate unit.

This maintenance is a minimum required to keep the GM AIR system in proper operating condition. It is necessary for emission control that the vehicles and engine receive proper maintenance to provide good driveability.

SYSTEM DEVELOPMENT

The calibration of components for the Air Injection Reactor System required a departure from previously established procedures. New techniques have been developed and will be discussed as they apply. In addition, new performance limits, different from optimum for other considerations, are required in some cars. All exhaust emission results shown have been corrected for exhaust dilution using the standard California correction factor. Exhaust emission results are shown to indicate the effect of variables and to show trends. The absolute emission values will, of course, vary between engines and vehicles.

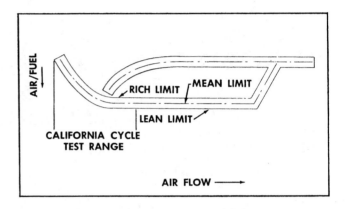

Figure 7 - A carburetor flow curve is established to describe the air-fuel mixture requirements for an engine. Such a flow curve is used to establish the allowable limits for carburetor production.

CARBURETOR CALIBRATION

Carburetor Flow Band

When standard carburetors are calibrated for production, a flow curve is established. Figure 7 shows such a flow curve established on a carburetor flow box. Air-fuel mixture ratio is plotted against air flow from idle to maximum carburetor capacity. Shown are the part throttle and full throttle flow bands. The lean limit curve shows the leanest carburetor calibration for good economy with acceptable performance and driveability. The general air flow range required for most American engines to perform the California Cycle test is indicated on Figure 7.

The flow band establishes the allowable limits for production. It is bounded by the lean limit and rich limit curves. The rich limit is chosen to provide a practical tolerance for manufacture -- usually 6% to 10% richer than the lean limit curve. Carburetor manufacturers frequently develop a mean limit curve which follows the center of the flow band.

Exhaust emission control has placed some new criteria which must be used in establishing the carburetor flow band. The carbon monoxide (CO) content of the exhaust is a function of the air-fuel mixture being delivered to the engine (Figure 8). Note that one air-fuel ratio represents about 2.5% carbon monoxide when the mixture is richer than stoichiometric. Thus, it is apparent that in critical portions of the carburetor flow curve special control of the mixture is required.

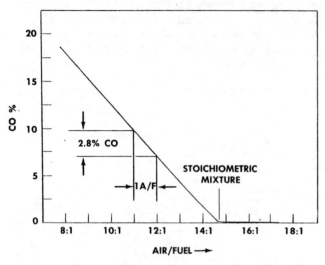

Figure 8 - Carbon monoxide in the engine exhaust is a function of the air-fuel mixture delivered to the engine. A small change in air-fuel ratio can cause a large change in carbon monoxide emission.

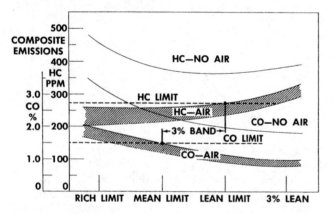

Figure 9 - Exhaust emission test results describe the air-fuel mixture limits for an engine to comply with the California limits.

Exhaust emissions for an air injection equipped car measured on the California Cycle test are shown for various carburetor calibrations in Figure 9. Shown also are the California limits of 275 ppm hydrocarbons (HC) and 1.5% carbon monoxide (CO). Emission curves for the same car with the air injection disconnected are also shown.

These curves illustrate two things. One, it is necessary to control the carburetor flow curve to assure compliance with the standards. Two, air injection becomes more effective in burning hydrocarbons as the mixture is richened. In this particular case, the carburetor calibration must be held between the mean and lean limit curves to comply with both the hydrocarbon and carbon monoxide standards.

Carburetor Manufacturing

These previous results show that close control of the carburetor mixture ratio is required for effective exhaust emission control and primarily in the low flow region. There are many factors that affect the carburetor calibration in the low air flow range. Extensive programs were initiated, and are continuing, to find ways to reduce the production flow band. Also, production carburetor flow stands, which are used to test all carburetors before shipment, had to be improved before they could be used to qualify carburetors for the closer limits. In addition to these significant improvements, the manufacturers have provided a factory adjustment of the flow calibration after final assembly to control the mixture ratios in the critical low flow range.

The adjustable off-idle carburetor shown in Figure 10 provides an additional air bleed into the idle system. The adjustment screw is set on the final carburetor assembly line, along with the idle mixture and air flow, and then sealed.

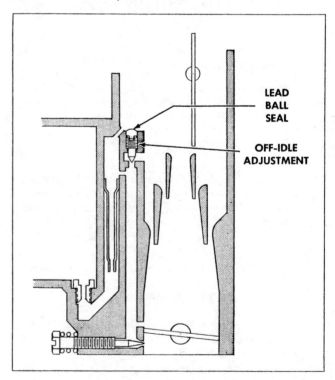

Figure 10 - Closer manufacturing tolerances have been achieved by adjusting the air-fuel mixture on the carburetor final assembly line.

With this modification, the carburetor manufacturers are able to hold the flow band to a 3% tolerance at the idle and off-idle set points as shown in Figure 11. The band widens on either side of the set point. Further improvements are limited by present measurement and manufacturing capabilities. This represents a subtantial improvement in the low air flow range encountered in the city driving and on the California Cycle test.

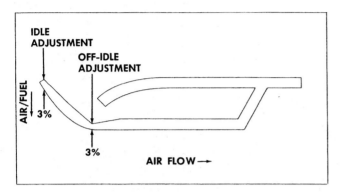

Figure 11 - A 3% tolerance band can be maintained at two points in the metering curve with assembly line adjustments.

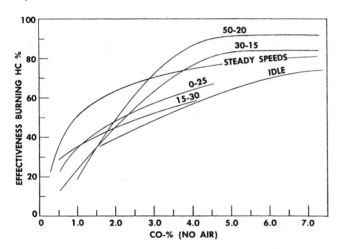

Figure 12 - Air injection burns hydrocarbons and carbon monoxide more effectively with richer carburetion. Thus rich carburetion is used at idle and on decelerations where the hydrocarbon concentrations are highest.

Air Injection Effectiveness

The GM AIR system adds a new criterion for establishing a lean limit flow curve. Air injection becomes ineffective at burning hydrocarbons when the mixture is too lean. Figure 9 shows that hydrocarbon emissions with air injection are usually lower when rich carburetion is used. On the basis of composite emissions from the California Cycle, air injection may burn two or three times more hydrocarbons with rich carburetion than with lean. This improved effectiveness with richer calibration can be seen in every mode on Figure 12. This plot is of air injection effectiveness in burning hydrocarbons at various carburetor calibrations.

The carbon monoxide reading without air injection is frequently used as a measure of carburetor calibration. Maximum effectiveness at burning hydrocarbons for any mode usually occurs at carbon monoxide concentrations of between 4% and 6%.

Even though richer mixtures would be desirable, it is necessary to use lean calibrations on all modes except the idle and decelerations. This must be done to keep the carbon monoxide concentration below the California limit of 1.5%. Using richer mixture at idle provides best idle quality (about 6% CO) and most effective burning of hydrocarbons at idle and on decelerations. (Since decelerations are made at closed throttle, the mixture is determined by the idle setting.) This provides the greatest reduction in hydrocarbons on decelerations where the concentrations are highest. The mixture for other modes is kept lean for low CO, but not so lean as to cause driveability problems.

Analysis Techniques

Carburetor calibration has such a large effect on emissions that we have had to develop different procedures for measuring the calibration. One of the best ways found is to record the carbon monoxide concentration in the exhaust. But when air injection is used, some of the CO is burned and the concentration is dependent on other factors. To provide calibration data, we have established a procedure for running an additional hot cycle of the California Cycle test with the air injection disconnected. This eighth cycle data is compared to the seventh cycle data of the standard California Emissions Test Schedule to determine the effectiveness of the AIR system. The Appendix shows the detailed procedure for this analysis to obtain the results shown in Figure 12.

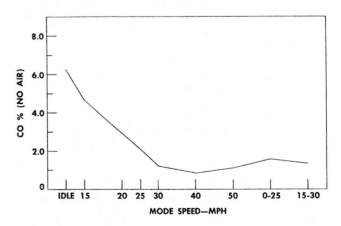

Figure 13 - Carbon monoxide in the exhaust is a more sensitive measure of carburetor calibration. A plot of the CO for the various modes of the California Cycle arranged by increasing air flow is an effective tool for evaluating the carburetor.

A ninth cycle is also run with air injection disconnected, but this cycle is modified to give steady state data at idle, 15, 20, 25, 30, 40, and 50 mph. The CO data for this ninth cycle plus the CO readings from the eighth cycle accelerations are plotted to provide an improved measure of carburetor calibration (Figure 13). The CO plotted on the ordinate indicates richer mixtures at the top. The car speed is arranged on the abcissa to approximate increasing air flow to the right. The shape of this curve follows closely the carburetor flow curve, but the CO curve can be used to detect smaller changes.

Idle Mixture

Historically, idle mixture adjustments have been made to obtain the best idle quality. This usually occurs at an air-fuel ratio of about 12.5:1. A non-air injected engine will produce an exhaust CO concentration of about 6% at this air-fuel ratio.

Figure 14 shows the results of California Cycle tests run on an air injected car with various idle settings. These settings range from 2% to 8% exhaust CO concentration and were measured with the air injection disconnected. With air injection connected, the exhaust concentrations are, of course, much lower. As shown in Figure 14, idle mixture affects emissions with air injection. Lean idle mixtures yield lower CO emissions while richer mixtures are best for reducing hydrocarbons.

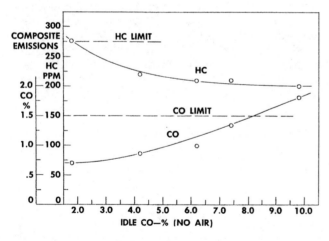

Figure 14 - Best AIR performance is obtained with an idle set at 6% CO. Richer idle results in reduced hydrocarbons while leaner idle yields lower CO emissions.

Best idle quality usually occurs at about 6% CO. This setting also provides the best compromise between CO and HC reductions with the GM Air Injection Reactor System. A study in California (Reference 5) has shown that the cars in the existing population have a mean setting of 5.75% CO at idle, with a normal distribution of lean and rich settings as shown in Figure 15. Thus, the AIR system requires no special idle mixture setting. Best mixture for idle quality is best for emission control.

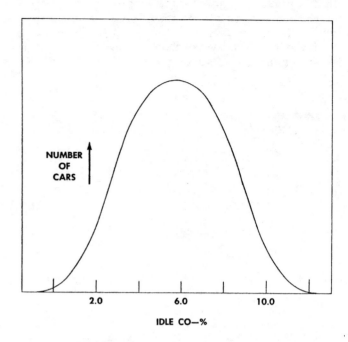

Figure 15 - A survey of about 1,000 cars in California showed them to have a mean idle CO setting of 5.75%. Thus no special idle mixture is required with the AIR system.

Idle Speed

Increasing idle speed reduces hydrocarbon emissions. Higher idle speed has the greatest effect on reducing emissions during decelerations. In effect, increased idle speed decreases the intake manifold vacuum on deceleration and thus reduces dilution and promotes combustion. Increased idle speeds are being used on most General Motors AIR equipped vehicles for 1966. Typical settings are 550 rpm for automatic transmission vehicles and 700 rpm for manual transmission vehicles. Several side benefits are derived from increased idle speeds such as improved engine cooling and improved idle quality.

Choke Calibration

Figure 16 shows exhaust emission results plotted for each cycle of the California Cycle test. Hydrocarbon and carbon monoxide emissions with and without air injection are shown. These results show that air injection is effective during the first cycle as well as the seventh. The richer mixtures (higher CO) associated with choke operation during the first cycle tend to make the air injection system more effective.

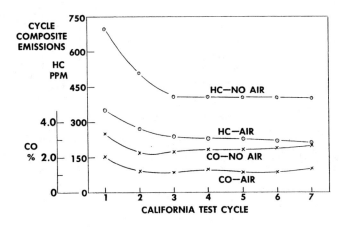

Figure 16 - Air injection is effective in reducing exhaust emissions with a cold engine as illustrated in these repeat California test cycles run from a 12-hour cold soak.

Thus, the Air Injection Reactor System will partially compensate for an over-rich choke. However, it is necessary to calibrate the choke for more precise operation with mild ambient temperatures since too rich a calibration will result in CO values above the California limit.

IGNITION TIMING

Retarding ignition timing has been shown to effectively reduce exhaust hydrocarbon emissions, but it also reduces fuel economy and performance. However, a considerable benefit can be derived by retarding the spark at idle with no significant change in fuel economy. The retarded spark at idle requires more throttle opening to maintain idle speed and thus improves combustion during decelerations.

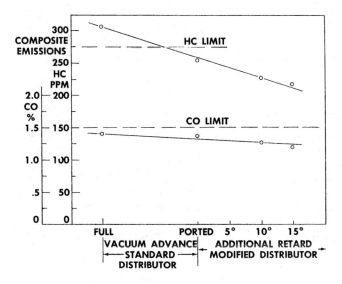

Figure 17 - Retarded ignition timing at idle reduces hydrocarbon emissions with no significant loss in economy or performance.

Figure 17 shows the reduction in composite hydrocarbons obtained on the California Cycle test with various amounts of idle spark retard. Porting the vacuum advance mechanism provided 15° retard for this particular vehicle at idle and during decelerations. Modifying the centrifugal advance mechanism provided additional retard of 5°, 10° and 15°. Each step was effective in reducing the hydrocarbon emissions. There was no effect on the CO emissions and previous studies have concurred with this finding.

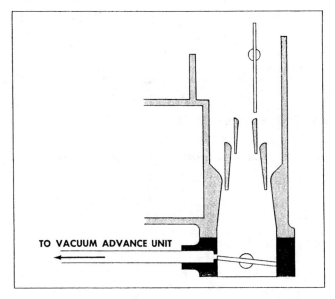

Figure 18 - Connecting the ignition distributor vacuum advance unit to a small port in the carburetor throttle body makes it inoperative at idle and during decelerations.

The spark retard at idle was obtained by two methods: ported spark, and centrifugal advance. Ported spark removes all vacuum advance at idle. This is done by connecting the vacuum advance diaphragm to a small port in the carburetor throttle body, just above the throttle blade in the idle position (Figure 18). Thus, at idle or during decelerations, the vacuum advance mechanism is inoperative. As the throttle begins to open, vacuum is applied to advance the timing. All GM AIR equipped cars use ported spark at idle.

When additional retard at idle is required, the centrifugal advance curve is modified (Figure 19). This figure shows a standard centrifugal spark advance vs. engine speed curve. Modification at low speed is accomplished by retarding the basic setting by the appropriate amount. Initial advance is very rapid until the normal spark curve is met at about 1200 engine rpm. This type of modification has a minimum effect on fuel economy and full throttle power. Some GM AIR-equipped cars use 5°, 10°, or 15° retard from standard at idle by using modified distributors. Most installations, however, use standard ignition distributors.

973

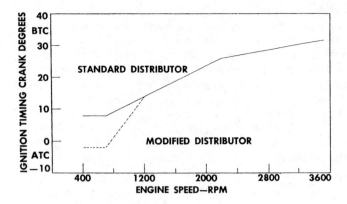

Figure 19 - Additional ignition timing retard is provided by modifying the ignition distributor centrifugal advance mechanism at idle with normal timing at speeds above 1200 rpm.

Spark retard at idle causes some problems, mainly from the standpoint of increased heat rejection into the cooling water. For example, ported spark plus an additional 10° of retard by using a modified ignition distributor could be expected to double the heat rejection to the coolant at idle. Thus, cooling capacity had to be increased on nearly all GM AIR-equipped cars. A few installations use a temperature-operated valve which reactivates the vacuum spark advance at idle in the event coolant temperatures become too high. This reduces heat rejection and increases idle speed to prevent overheating under extreme conditions.

AIR REQUIREMENTS

The amount of air required for optimum emissions control depends on many factors. Fortunately, it is not critical and some variation in the amount of air at any one operating condition has little effect on emissions.

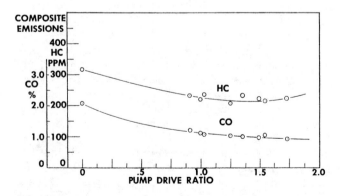

Figure 20 - The air pump was sized so that optimum air for exhaust emission control could be obtained with pump drive ratios from 1:1 to 1.5:1.

A convenient way of varying the amount of air to find the optimum is by changing the pump drive ratio as shown in Figure 20. This figure shows the composite California Cycle hydrocarbon and carbon monoxide emissions as a function of pump drive ratio. This is the ratio of air pump speed to engine speed. The amount of air required for minimum CO in the exhaust was always greater than for minimum HC, but the CO was usually near a minimum at the air rate for minimum HC.

Most GM AIR-equipped cars have the pump drive ratio chosen for optimum HC reduction. The size of the pump was chosen at 19.3 cu. in. per revolution to permit nominal drive ratios of from 1:1 to 1.5:1 which are convenient for installation.

A positive displacement pump works well as the air supply. When driven at some proportion to engine speed, it supplies approximately the right amount of air for all modes of operation without special controls as shown in Figure 21. The hydrocarbon emissions are plotted for each mode of the California Cycle. Optimum air flow for each mode is provided by a single pump drive ratio, in this case 1.25:1.

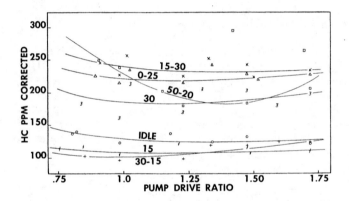

Figure 21 - The air pump provides the proper amount of air for each mode of the California cycle test without special controls.

During the GM AIR system development, it was frequently desirable to have some way of determining how much air the experimental air pumps were delivering to the exhaust system. Reference to the correction factor specified by California to correct for dilution of the exhaust gave an indication but was only approximate. Because of the nature of this correction factor, rich mixtures tend to reduce the factor and lean mixtures increase it. Measuring pump air flow during an emission test is difficult and may affect emission results.

To provide a check on pump performance on every test, a comparison is made of the measured emissions in the seventh and eighth cycles. The differences in the exhaust concentrations between these two cycles is due to the lack of air injection in the eighth cycle. The air injected during the seventh cycle reacts with unburned exhaust products and dilutes the exhaust. Using the chemical equations for these reactions, the dilution can be calculated as shown in the Appendix. The dilution factor is the air added expressed as a per cent of the original dry exhaust without air. Thus, the per cent air is on the same basis as the per cent CO or CO_2 in the eighth cycle.

These dilution factors are not affected by carburetor calibration. They have been plotted against pump drive ratio for each mode in Figure 22. Once this relationship has been established, it can be used to determine equivalent air flow of another type of air pump or continued performance of an air pump on durability test.

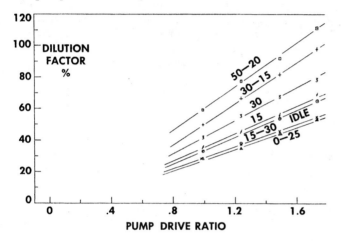

Figure 22 - Calculation of the exhaust dilution associated with air injection has been an effective way of monitoring pump performance.

ANTI-BACKFIRE VALVE

Early in the development of the AIR system, it became apparent that exhaust explosions would be a problem. These explosions occurred when the throttle was suddenly closed after an acceleration. While these backfires occurred on all engines, some were more susceptible than others. Use of a throttle return check, or dashpot, to retard throttle closure helped but did not completely prevent backfires on all cars. The backfires in the exhaust could be eliminated completely by cutting off the air injection for a few seconds after the throttle closing.

The explosions indicated the presence of a considerable amount of unburned fuel in the exhaust on these maneuvers. This unburned mixture in the exhaust after sudden throttle closing is a basic characteristic of automotive engines. During an acceleration, high intake manifold pressure results in the presence of liquid fuel in the intake system. When the throttle is suddenly closed, the accompanying decrease in intake system pressure causes this liquid fuel to evaporate and be carried into the cylinders. The result is an air-fuel mixture in the cylinders which is too rich to burn during several cycles. The unburned mixture is pumped through the engine into the exhaust. Normally this mixture does not ignite in the exhaust because it remains too rich. The only evidence of the problem on a standard car is that the engine may stall if the car speed is low. The hydrocarbon emissions are also extremely high for a brief period. With air injection, this rich mixture is mixed with additional air in the exhaust and becomes combustible. As soon as the engine begins firing again, a flame in the exhaust port ignites the combustible mixture in the exhaust system.

The solution to this problem has been to keep the engine firing at all times, regardless of how severe the throttle maneuver. This is done by means of a valve which bleeds air into the intake system in proportion to the amount of fuel on the sudden throttle closing. The valve senses increases in intake manifold vacuum. The greater the increase, or the more suddenly it increases, the more air is bled into the intake system. This air mixes with the excess fuel and maintains a combustible mixture to the engine.

The valve, variously called the intake air bleed or gulp valve (Figure 23), is connected to a supply of clean air and a silencing system such as the engine air cleaner or the air pump discharge line. The downstream side of the valve is connected to the intake system in a central location so that the gulp of air can be distributed evenly to all cylinders.

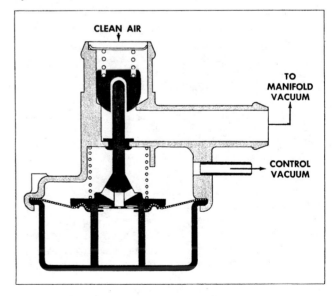

Figure 23 - The anti-backfire valve permits a controlled amount of air to enter the intake manifold when the engine throttle is closed preventing exhaust system backfires.

The diaphragm supplies the force to open the valve. The control line to the diaphragm connects to the intake system separate from the air flow path. This is necessary to prevent chatter caused by feedback when the two lines are not separated.

When vacuum increases rapidly, the check valve in the center of the diaphragm assembly closes. The high vacuum above the diaphragm causes an unbalance which compresses the spring and lifts the valve off its seat. A small bleed hole in the check valve allows the vacuum below the diaphragm to gradually equalize and the spring returns the valve to the closed position. The check valve allows the vacuum below the diaphragm to equalize quickly when manifold vacuum decreases rapidly on throttle opening.

The various engines equipped with the GM AIR system require different amounts of air to keep the engine firing. The length of time which the valve is open under the most severe throttle maneuvers varies from about one second on some engines to as long as four and a half seconds on others. This gulp duration is controlled by the size of the orifice in the check valve. The spring is set so that a minimum of about five inches of mercury increase is required before the valve is actuated.

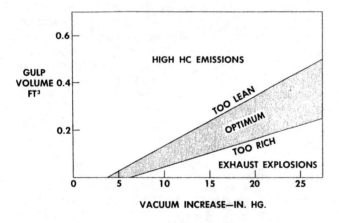

Figure 24 - There is an optimum amount of air required for each degree of throttle closure. Too little air results in backfires, too much air can cause higher emissions.

The gulp volume is controlled by the duration of the valve and the size of the connection to the intake system. Most gulp valves are made one size and intake connection size is chosen to provide the optimum air flow rate. Too little air results in backfires, too much air can cause engine miss and high emissions as shown on Figure 24. When the gulp volume is kept in the optimum range and with the correct duration, the exhaust emissions are lower with the valve operating. The gulp valve also eliminates the need for anti-stall devices such as throttle return checks or dashpots.

PRESSURE RELIEF VALVE

Two problems occur with air injection-equipped cars at high speeds and loads: (1) the exhaust system temperatures can become too high with the richer mixtures used for best power operation, and (2) the increased volume of flow through the exhaust system causes high back pressure and power loss.

Figure 25 shows exhaust gas temperatures measured at the muffler inlet during road load operation. Exhaust system temperatures are generally lower with air injection than without during normal operation. These cooler temperatures result from the relatively large volume of cool air supplied to the exhaust. With the lean mixtures used for part throttle, there is little heat generated in the exhaust because the percentage of combustibles is small. Also the exhaust system back pressure is low and the additional air flow has little effect on power output.

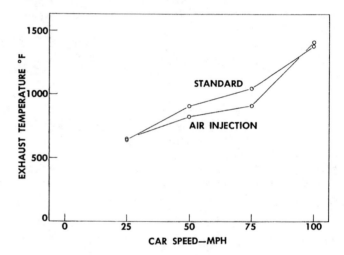

Figure 25 - Exhaust gas temperature measured at the muffler inlet are generally lower with air injection.

However, at heavy load, the amount of air injected is a smaller per cent of the total exhaust flow. Also, richer mixtures are used to obtain maximum power. This results in more combustibles in the exhaust and a chance for higher exhaust system temperatures as shown in Figure 26. These exhaust gas temperatures were measured at the exhaust manifold outlet. With a normal full throttle carburetor calibration, exhaust temperature increases at WOT are moderate. But when richer mixtures occur, any amount of air injection causes higher than standard temperatures. An increase in the amount of air injected with rich mixtures increases the exhaust system temperature. However, as shown in Figure 26, bypassing a portion of the air at full throttle can reduce the exhaust system temperatures.

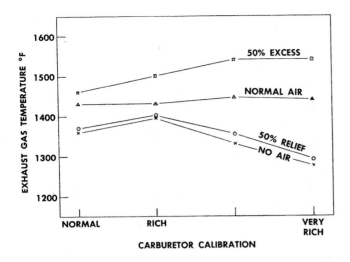

Figure 26 - At high vehicle speeds and heavy loads, exhaust system temperatures are higher with air injection. Exhaust gas temperatures can be controlled by bypassing a portion of the air.

A pressure relief valve has been included to insure against these high exhaust temperatures by bypassing some of the pump flow at high speeds and loads. Figure 27 shows the valve located in the air pump outlet cavity. It consists of a spring-loaded metal disc. On cars with varying engine size and exhaust system restrictions, the pressure settings are different. This is achieved by the use of color coded spacers which adjust the preload of the spring.

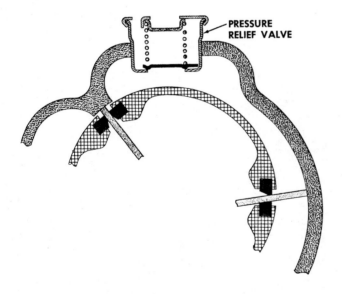

Figure 27 - A spring loaded disc valve located in the air pump discharge cavity allows a portion of the pump air to bypass the exhaust system at high speeds and loads.

CHECK VALVES

To prevent damage to the air pump in the event of failure of the belt that drives the pump, it is necessary to have check valves included in the air injection system. If the belt failed, the exhaust pressure would cause backflow of the exhaust into the pump and to the air cleaner. The temperatures would increase rapidly to well above the allowable limits for the air hoses and pump. Figure 28 shows the temperatures that resulted when a test car was operated at various speeds with the air pump drive belt disconnected to simulate a belt failure. For the installation shown, the back flow at low speed was from the right air manifold, through the pump discharge cavity and on through the left air manifold. The exhaust system pressure was higher on the right side. At about 55 mph, the pump began motoring backwards. The flow was then from both exhaust manifolds, through the pump, and into the air cleaner.

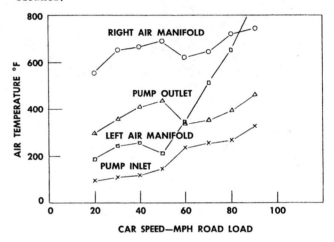

Figure 28 - Disconnecting the air pump drive belt to simulate belt failure results in backflow of exhaust gas into the air injection system. These high temperatures can fail the system components if check valves are not used.

To prevent damage in the event of a belt failure, it was necessary to install a check valve in each pump discharge line. A single check valve is enough on a six-cylinder in-line engine where a single air manifold is used. But on V-8s with an air manifold on each side, a check valve is needed in each line to prevent cross-flow which could overheat the air distribution hoses.

The check valve required a special development. Early designs with a spring-loaded metal disc failed to stop reverse flow at idle and low speeds and were noisy. The valve had to be made very light so that it responded to the rapid pressure fluctuation in the exhaust at idle and light loads. Each exhaust valve opening event produces a positive pressure wave followed by a considerable period below atmospheric pressure. The check valve designs with a heavy disc opened during the period of below atmospheric pressure, and did not close fast enough during the next pressure pulse. This resulted in back flow at idle along with excess noise. Back flow could be prevented by using a heavy spring in the valve, but this produced excessive pump back pressure during normal pump operation.

Figure 29 shows a schematic of the check valve being used on the GM AIR-equipped cars. It has a Viton disc valve attached to the valve plate at the center. Under steady flow conditions, the valve adds little to the overall system operating pressure as shown in Figure 30. At 50 mph, the pressure drop across the check valve is normally less than two inches of mercury.

These check valves work well to seal the exhaust from back flow into the pump at all speeds and loads. On most installations, it has been necessary to install the valves about two inches upstream of the air manifolds. This avoids excessive temperatures on the Viton valve.

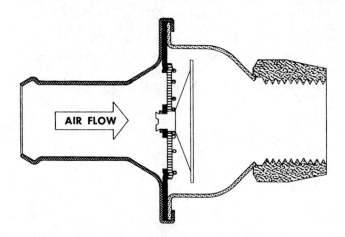

Figure 29 - Check valves are provided in the air distribution lines to prevent exhaust gas backflow in case of air pump belt failure.

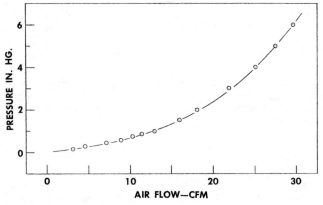

Figure 30 - The pressure drop across the check valves is low and does not interfere with air injection performance.

SUMMARY

The development of the 1966 GM Air Injection Reactor System is based on engineering information that has been gathered over the years pointing the way to reduce exhaust emissions. In addition, it is designed to meet the specific needs of the State of California as outlined in their test procedures.

The development of instrumentation to measure exhaust pollutants, the establishment of air quality standards and procedures for determining compliance with them, and finally the development of means for controlling exhaust emissions have been carried on concurrently. This cooperative effort between government and industry has resulted in the systems available today.

Our objective, of course, is cleaner air. Another objective is satisfied customers and we must be careful to provide a system that will do the required job with minimum required maintenance and maximum reliability.

APPENDIX

AIR INJECTION EFFECTIVENESS

The air injection effectiveness calculation of this program computes:

a. A Dilution Factor - percent of volume ratio of injected air to original dry exhaust.

b. Percent of CO in exhaust burned by the injected air. (Expressed as percent of original dry exhaust volume.)

c. Percent of HC in exhaust burned by the injected air. (Expressed as percent of original dry exhaust volume.)

d. Percent of injected O_2 which burns CO.

e. Percent of injected O_2 which burns HC.

f. Percent of injected O_2 which remains free.

g. Effectiveness in burning CO (% CO burned by injected O_2).

h. Effectiveness in burning HC (% HC burned by injected O_2).

Problem Statement and Solution

Given:

HC, CO and CO_2 measured values with air injection.

$\overline{HC}$, $\overline{CO}$ and $\overline{CO_2}$ measured values without air injection.

Chemical reaction equations -

$$2\,CO + O_2 \longrightarrow 2\,CO_2$$

$$2\,H_2 + O_2 \longrightarrow 2\,H_2O$$

$$2\,C_8H_{18} + 25\,O_2 \longrightarrow 16\,CO_2 + 18\,H_2O$$

Assume that only the above reactions take place and that the percent H_2 in original exhaust = 1/2 percent CO in original exhaust and percent H_2 that burns = 1/2 percent CO_2 that burns. These assumptions are based on approximations of exhaust analysis curves. They introduce only slight error in the end results and greatly simplify the computation.

Calculate:

a. % Dilution - (Volume air added/initial exhaust volume)

b. % CO in exhaust that burns

c. % HC in exhaust that burns

d. % effectiveness in burning CO (% of original CO in exhaust that burns)

e. % effectiveness in burning HC (% of original HC in exhaust that burns)

Let

V = Original volume of dry exhaust gas

$V_{O/CO}$ = Volume of O_2 burning CO

$V_{O/HC}$ = Volume of O_2 burning HC

$V_{O/F}$ = Volume of O_2 remaining free

$V_{O/H}$ = Volume of O_2 burning H_2

V_{O_2} = Volume of O_2 injected

V_A = Volume of injected air

V_{CO} = Original volume of CO before air injection

V_{HC} = Original volume of HC

Then $V_{O_2} = V_{O/CO} + V_{O/HC} + V_{O/F} + V_{O/H}$

and $V_A = 4.76\, V_{O_2}$

We may write an expression for the concentration of CO in percent after air injection in terms of the above volumes.

$$CO = \frac{\text{original volume of CO} - \text{Volume of CO burned by injected } O_2}{\text{final volume after air injection and reaction}} \times 100$$

The denominator (final volume after air injection and reaction) would be:

$$V + V_A - \begin{Bmatrix} \text{Vol } O_2 \text{ burning CO} \\ \text{Vol } O_2 \text{ burning } H_2 \\ \text{Vol } O_2 \text{ burning HC} \\ \text{Vol CO burned} \\ \text{Vol } H_2 \text{ burned} \\ \text{Vol HC burned} \end{Bmatrix} + \begin{Bmatrix} \text{Volume of} \\ \text{Product of } O_2 + CO \\ \text{Product of } O_2 + H_2 \\ \text{Product of } O_2 + HC \end{Bmatrix}$$

From the reaction equations

Vol CO burned = $2\, V_{O/CO}$

Vol H_2 burned = $2\, V_{O/H}$

Vol HC burned = $(1/12.5)\, V_{O/HC}$

Vol of Product, $O_2 + CO = 2\, V_{O/CO}$

Vol of Product, $O_2 + H_2$ = Zero (water is removed from exhaust before measurement)

Vol of Product, $O_2 + HC = (8/12.5)\, V_{O/HC}$

Then

$$(1) \quad CO\% = \frac{100\,(V_{CO} - 2\, V_{O/CO})}{V + V_A - \begin{Bmatrix} V_{O/CO} \\ + V_{O/H} \\ + V_{O/HC} \\ + 2V_{O/CO} \\ + 2V_{O/H} \\ + (\frac{1}{12.5})\, V_{O/HC} \end{Bmatrix} + \begin{Bmatrix} 2\, V_{O/CO} \\ + (8/12.5)\, V_{O/HC} \end{Bmatrix}}$$

if we define

$$X = \frac{V_{O/CO}}{V} \times 100 \text{ (percent } O_2 \text{ burning CO)}$$

$$Y = \frac{V_{O/F}}{V} \times 100 \text{ (percent } O_2 \text{ remaining free)}$$

$$Z = \frac{V_{O/HC}}{V} \times 100 \text{ (percent } O_2 \text{ burning HC)}$$

$$W = \frac{V_{O/H}}{V} \times 100 \text{ (percent } O_2 \text{ burning } H_2)$$

then $X + Y + Z + W = \%\ O_2$ injected

and $4.76(X + Y + Z + W) = \%$ air injected

Dividing numerator and denominator of Equation (1) by V we get:

$$(2)\ CO\% = \frac{(\frac{V_{CO}}{V} - 2\frac{V_{O/CO}}{V})\,100}{1 + \frac{V_A}{V} - \frac{V_{O/CO}}{V} - 3\frac{V_{O/H}}{V} - (1 - \frac{7}{12.5})\frac{V_{O/HC}}{V}}$$

$$= \frac{\overline{CO} - 2X}{1 + \frac{4.76(X+Y+Z+W)}{100} - \frac{X}{100} - 3\frac{W}{100} - \frac{.44Z}{100}}$$

but $W = \frac{1}{2}X$ by assumption.

$$(3)\ CO\% = \frac{100(\overline{CO} - 2X)}{100 + 4.64X + 4.76Y + 4.32Z}$$

Similarly

$$(4)\ CO_2\% = \frac{100(\overline{CO_2} + 2X)}{100 + 4.64X + 4.76Y + 4.32Z}$$

and

$$(5)\ HC = \frac{100(\overline{HC} - .08Z)}{100 + 4.64X + 4.76Y + 4.32Z}$$

Solving Equations (3) and (4) simultaneously,

$$(6)\ \text{From Equation (3)}\quad Y = \frac{100(\overline{CO} - 2X)}{4.76\,CO} - (\frac{100 + 4.64X + 4.32Z}{4.76})$$

$$(7)\ \text{From Equation (4)}\quad Y = \frac{100(\overline{CO_2} + 2X)}{4.76\,CO_2} - (\frac{100 + 4.64X + 4.32Z}{4.76})$$

Solving Equations (6) and (7) for X,

$$(8)\ X = \frac{\overline{CO}\,CO_2 - CO\,\overline{CO_2}}{2(CO_2 + CO)}$$

Putting the value of Y from Equation (7) into Equation (5) and solving for Z,

$$Z = \frac{100}{8} HC\left(\frac{\overline{HC}}{HC} - \frac{\overline{CO_2} + 2X}{CO_2}\right)$$

Percent Dilution = $4.76(X + Y + Z + W) = (1.5X + Y + Z) \times 4.76$

Percent CO in exhaust that burns = $2X$

Percent HC in exhaust that burns = $.08Z$

Percent effectiveness in burning CO = $\frac{2X}{CO}(100) = \frac{200X}{CO}$

Percent effectiveness in burning HC = $\frac{.08Z}{HC}(100) = \frac{8Z}{HC}$

REFERENCES

1. "Exhaust Manifold for Internal Combustion Engines," United States Patent Office, No. 2,217,241, Patented October 8, 1940, by Max Tendler.

2. "Internal Combustion Engine," United States Patent Office, No. 2,263,318, Patented November 18, 1941, by Harold C. Tifft.

3. Brownson, D. A., R. S. Johnson, and A. Candelise, General Motors Corporation, "A Progress Report on ManAirOx - Manifold Air Oxidation of Exhaust Gas," Paper No. 286N presented at SAE National Automobile Week Detroit, Michigan, March 12-16, 1962.

4. Thompson, W. B., Saginaw Steering Gear Division, General Motors Corporation, "The General Motors Air Injection Reactor Air Pump," Paper No. 660108 presented at SAE Automotive Engineering Congress and Exposition, Detroit, Michigan, January 11, 1966.

5. Tuesday, Charles S., General Motors Research Laboratories, et al., "Los Angeles Test Station Project, 1962," (1013 car idle survey, to be published).

680110

Emission Control by Engine Design and Development

D. L. Hittler and L. R. Hamkins
American Motors Corp.

THE INTENSE STUDY by state and federal agencies in combination with the automobile industry over the past decade or more, on the subject of air pollution, has brought to the forefront a problem affecting all of us. This study has shown that a significant part of air pollutants are contributed by the automobile. These facts coupled with the legislative activities of state and federal agencies in the field of vehicle emission control regulations have provided more than adequate motivation to invent, design and produce a practical and inexpensive system of emission control (1, 2)*
The effect of this motivation has been widespread within the automotive industry to include the supplier firms, the research groups, the production and development groups, and, finally, the manufacturing group. Considering the objectives of practicality and low cost, a maximum effort to reduce the emission of selected exhaust products from within the basic engine was established. This paper is devoted to a discussion and a description of the steps taken to reduce the carbon monoxide and hydrocarbon emission from the two engine displacements involved in the American Motors' 6-cylinder engine family. This engine family to include displacements of 199 and 232 cu in. was introduced in 1964 for production. (3)

THEORY

The desire to minimize the emissions of the unburned hydrocarbons and carbon monoxide, without the aid of a secondary system external to the basic engine, quite naturally focuses the interest of the engine designer on the physics and chemistry of the engine combustion process. Understanding of the mechanisms involving that process, which have been shown to influence specifically the relative proportions of these two products in the engine exhaust, has become extremely important. There are, of course, other compounds produced by the engine, such as nitric oxide, of which the engine designer must also have cognizance, since it is believed to be a necessary key to the chain of events constituting the formulation of photochemical pollutants. At the outset of the program which this paper describes, the reduction of nitric oxide concentrations was considered of secondary importance.

A portion of the exhaust emission research, which is not associated with the study of external control systems, points

---

*Numbers in parentheses designate References at end of paper.

ABSTRACT

Requirements of federal and state agencies to control the exhaust emissions of motor vehicles defined performance standards for 1968. Consideration of economics and practicality motivated a full scale design and development program to meet this required performance without the use of a secondary control system. An analysis of new low quench type combustion chambers compared with the standard chambers used on two displacements of a 6-cylinder engine family in prior model years is presented in terms of parameters related to exhaust emission and to conventional engine design.

The engine induction system, with special emphasis on carburetor characteristics, is contrasted for application to non-emission controlled engines with application to emission controlled engines.

out two areas or characteristics of most existing engine designs which hold the majority of potential in reducing their exhaust emissions. These areas involve mixture preparation and quality (the induction system) and combustion chamber configuration.

Induction System - The induction system has been the target of much development effort aimed at reducing exhaust emission levels. Many researchers have demonstrated the potential gains. (4-6) In the program described in this paper, the induction system was the object of more development effort than any other single item.

It has become well known that a more frugal proportioning of the inducted air-fuel charge, so that fuel quantities in excess of those strictly required by the engine to meet a particular work requirement are held to a minimum, could possibly result in substantial reductions of unburned hydrocarbons and primarily carbon monoxide. To accomplish this meant leaner carburetion for most part throttle operation, as achieved by more precise carburetor control of mixture ratio. Complete vaporization of the fuel in the air-fuel mixture and complete homogenization of this mixture were also shown to be beneficial when attempting to improve distribution, permitting even leaner mixture ratios and further reductions in exhaust emissions without severely depreciating normal vehicle performance. (6, 7)

Combustion Chamber - Combustion chamber shape is now widely understood to be influential in determining the concentrations of unburned hydrocarbons present in the exhaust of an engine. (8) Flame-quenching in the air-fuel boundary layer (0.005-0.010 in. thick) adjacent to the chamber walls, or in any other volume where sufficient heat can be transferred away from the burning gases to the material comprising the physical confines of the chamber, is the primary mechanism which accounts for unburned hydrocarbons in exhaust gases. Through the elimination or reduction of these quench volumes, which in part have evolved out of the desire to improve octane control and combustion smoothness, sizable reductions in the emission of unburned hydrocarbons may be realized. The reductions to be anticipated are proportional to the amount of quench volume eliminated. Since the internal surface area of the chamber cannot be entirely eliminated, it may become the dominant factor in determining an engine's relative emission level of unburned hydrocarbons. In terms of reducing exhaust concentrations, the amount of surface area in relation to the chamber volume it encloses is an important parameter for design consideration.

DESIGN GOALS, LIMITATIONS, AND
RESULTING DESIGNS

Considering the potential reductions in exhaust emissions, which recently published literature and the authors' own investigations have shown to be a possibility, along with the "clean engine" emission levels of the 199 CID and 232 CID engines, it was concluded that the existing Federal standards of 275 ppm for unburned hydrocarbons and 1.5 mole per cent carbon monoxide could be met without the aid of an external system. (1) This required maximum application of the principles expressed in the two areas previously discussed and moderate engine redesign. Enthusiasm spawned by this possibility launched design and development programs aimed at the 1968 model series.

Induction System - With regard to the induction systems for these same two engines the desired goals are outlined below:

1. Develop leaner carburetor calibrations with primary emphasis on leaner curb idle, part throttle and choke calibrations.
2. Control of production carburetor flow curves to a tolerance range closer than prior practice.
3. Improve mixture preparation within the intake manifold through increased velocities. NOTE: Increased heat transfer to the intake manifold was desired but considered impractical due to limitations in material characteristics and "coking" deposits on the floor of the intake manifold.
4. Remove the water heated distribution tube in the 199 CID intake manifold.

Being able to operate these engines at leaner air-fuel ratios for all conditions of vehicle operation except full load was the primary objective for redesign of the carburetors and intake manifolds. There was also a desire to eliminate the cost burden of the water heated distribution tube in the main runner of the 199 CID intake manifold.

Intake Manifold - The gas velocities of the intake manifold were re-examined from the predecessor engine. In the case of the 232 engine no change was made. Table 1 illustrates the conventional mean gas velocities for each manifold. In the case of the 199 engine a new manifold with a smaller cross section was designed. The mixture velocity of this new manifold is 15,000 fpm compared with a mixture velocity of the previous manifold of 13,100 fpm for the main runner and 14,400 fpm for the branches. A study of full load and part throttle fuel distribution for both the 232 and 199 engines comparing two combustion chamber types (detailed later), indicated no significant change in the cylinder-to-cylinder fuel distribution quality. Under steady state operation (engine dynamometer) the full load mixture distribution among the 6 cylinders was generally within one air-fuel ratio for each respective engine speed. Under part throttle conditions the cylinder-to-cylinder mixture distribution was slightly improved compared to full load operation.

Carburetion - The size of the carburetor to be used was scrutinized for both engine displacements. Prior to emission control the 199 and 232 engines used common sized carburetors. For the 199 application the carburetor venturi size was reduced from 1.5 in. diameter to 1.375 in. This was done to improve the fuel mixing and increase the velocity at low flow conditions.

Because of their ability to improve exhaust emission levels, an increased curb idle speed and a leaner curb idle air-fuel ratio (approximately 14:1) were applied. For manual transmission applications the curb idle speed was in-

Table 1 - Design Parameters for Quench and Low Quench Engines

Parameter	199 Engine Quench	199 Engine Low Quench	232 Engine Quench	232 Engine Low Quench
Bore, in.	3.75	3.75	3.75	3.75
Stroke, in.	3.00	3.50	3.00	3.50
Bore-Stroke Ratio	1.25	1.07	1.25	1.07
Compression Ratio	8.5:1	8.5:1	8.5:1	8.5:1
Valve Timing				
Intake opens, deg btdc	12.5	12.5	12.5	12.5
Intake closes, deg abc	51.5	51.5	51.5	51.5
Exhaust opens, deg bbc	53.5	53.5	53.5	53.5
Exhaust closes, deg atc	10.5	10.5	10.5	10.5
Intake valve overlap	23	23	23	23
Intake Valve Spring Load, lb[a]				
Valve closed	100	100	100	100
Valve open	180	180	180	180
Exhaust Valve Spring Load, lb[a]				
Valve closed	100	100	100	100
Valve open	180	180	180	180
Intake Manifold Velocity, fpm[b]				
Runner	13,100	15,000 with H_2O Tube	13,300	13,300
Branches	14,400	15,000	13,950	13,950
Combustion Chamber				
Quench height, in.	0.077	0.159 at cyl. wall	0.077	0.159 to cyl wall
Quench area, sq. in.	2.43	None	2.43	None
Quench volume, cu in.	0.187	None	0.187	None
Surface area at tdc, sq in.	37.11	35.35	37.52	35.60
Volume at tdc, cu in.	4.48	4.47	5.24	5.20
Surface to volume ratio, in.$^{-1}$ (Reference: Appendix)	8.28	7.90	7.16	6.84
Volume of head gasket void, cu in.	0.017	0.0083	0.017	0.0083

[a] Prior to 1966 models, the valve spring load was that listed below. This performance was changed for exhaust emission purposes relative to California system performance.
   Intake and exhaust valve closed, lb - 88, 88
   Intake and exhaust valve open, lb 155, 155

[b] Calculated at 4000 rpm at WOT at standard temperature and pressure for 100% volumetric efficiency.

creased from 550 rpm to 600 rpm. For automatic transmission applications, the curb idle speed was increased from 550 rpm in neutral to 525 rpm in gear (net increase was approximately 50 rpm). Curb idle mixture richness was leaned from a previous "best idle" to a condition leaner than "least fuel best torque" (LBT). The degree of this leanness compared to LBT in terms of speed drop is 50 rpm; however, the speed drop indicated was the result of the air-fuel ratio deemed necessary for emission compliance -- 14:1 air-fuel ratio. Consistent with the preceding discussion on idle air-fuel ratio, redesign of the carburetor idle system was performed to provide a carburetor which was intentionally limited in its idle fuel adjustability in consumer service. (9) The performance requirement for such a system was set at a rich limit of 3.0 mole percent carbon monoxide in the exhaust stream at the specified curb idle speed (equivalent to 13.5:1 air-fuel ratio for fuel $CH_{(1.86)_x}$). The basic carburetor calibration was modified to optimize the carbon monoxide and hydrocarbon emission during vehicle operation. A simply stated objective was to develop a calibration which metered leaner than stoichiometric during a maximum portion of the operating conditions. Merely leaning the metering level of the carburetor used on the previous version of this engine was not possible since severe driveability problems resulted.

Fig. 1 describes the carburetor flowbench characteristics of a typical carburetor used on the previous engines. The curve shown is a conglomerate of three basically different designs; it was drawn in this manner to point out the basic problem areas of pre-emission controlled carburetors. Referring to the "full curve", the idle air-fuel ratio is relatively rich, the idle air flow is relatively low, the off-idle and part-throttle metering level is richer than desired, for exhaust emission control, and the power mixture timing is earlier than desired. The full-throttle metering level is that desired since no reduction in full-throttle power is warranted. It is recognized that the most commonly used flowbench curve falls short of describing the carburetor flow characteristics as they would be described on an actual engine. However, a more representative characteristic of low speed operation can be described using the flowbench by generating a "cross section" curve, also shown in Fig. 1. In this instance, the flowbench is balanced at the same idle depression but at a much lower WOT depression (0.1 in. Hg compared to 3.0 in. Hg.). With the curb idle set at the air and fuel required for emission compliance, as described above, the cross section curve for the typical carburetor used on the nonemission controlled engine demonstrates a deep off-idle hole (extreme leanness) and a power mixture operation far richer than desired.

By contrast, Fig. 2 describes the flowbench characteristics for a typical carburetor used on the emission controlled engine. At the desired metering level for idle, the off-idle tendency for extreme leanness has been controlled in much closer relation to the basic curve, the power enrichment timing has been delayed and the tendency to "overshoot" at low air flow has been reduced to the level of the WOT curve. Each of these areas required substantial development and resulted in basic design changes within each carburetor in order to eliminate the problem.

The choke performance of a carburetor used on an emission controlled engine is extremely critical. Since the rating of exhaust emission performance is made at room temperature, little or no compromise in choke performance can be accepted at this temperature condition. Automatic chokes used on the nonemission controlled engines frequently allowed air-fuel mixture ratios richer than 9:1 during some portion of the first cycle of the current exhaust emission test procedure. (10) The automatic choke systems developed for the emission controlled engines employ more precisely articulated mechanisms, controlled maximum mixture ratios and precise "choke-off" time. Typical mixture levels for the first cycle of the emission test (complete cycle) result in a carbon monoxide level in the exhaust stream of approximately 3 mole percent carbon monoxide. Choke-off time is designed to provide nearly no enrichment during the 15-

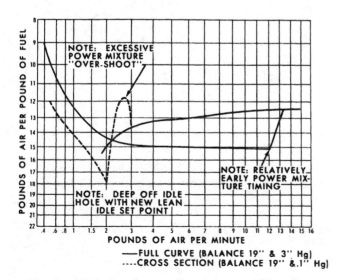

Fig. 1 - Typical carburetor flow curve-quench engine

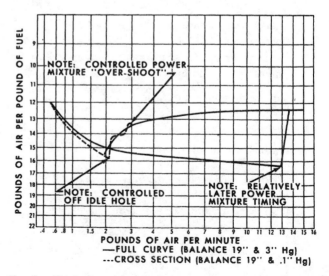

Fig. 2 - Typical carburetor flow curve - low quench engine

50 mph acceleration mode of the first cycle, but to allow a minimum of 100 rpm fast idle (articulated by the choke system) during the 50-20 mph closed throttle deceleration. (10)

The performance requirements for the accelerator pump becomes more critical when considered for the emission controlled engine. Prior to the days of emission control, accelerator pump capacity and timing were judged by the limit of rich "slugging" in warm ambient conditions and the elimination of transient "sag" and/or induction system backfire during low ambient temperature conditions. For exhaust emission control, a relatively low accelerator pump output is desired at room temperatures during the exhaust emission test procedure. (10) Several combinations of accelerator pumps were examined with this objective at hand including thermostatic control systems. The resultant system is a pump of greater displacement than used previously, but vented so that at slow throttle opening rates, a significant portion of the fuel is bled from the discharged volume. Accelerator pump performance played an important role in the early development of the low quench engine, primarily because the low quench engine appeared to require a greater

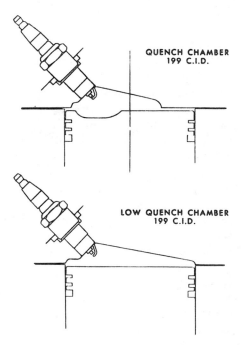

Fig. 3 - Cross section of 199 CID combustion chamber - quench and low quench

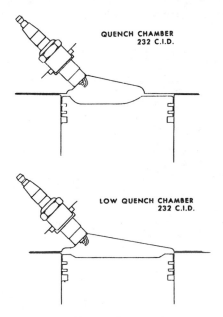

Fig. 4 - Cross section of 232 CID combustion chamber - quench and low quench

Fig. 5 - 199 CID combustion chamber comparison: left, quench chamber; right, low quench chamber

Fig. 6 - 232 CID combustion chamber comparison: left, quench chamber; right, low quench chamber

pump discharge volume for acceptable driveability. This observation was confirmed on several vehicles with common carburetors. The explanation offered is that the low quench engine employs less combustion chamber turbulence than the quench engine and, as such, requires more fuel during transient low air flow conditions to provide a continuous flame travel in the chamber. (11) This condition is magnified by a semi-warm engine.

Combustion Chamber - The pre-1968, non-California versions of the two engines in the 6-cylinder family both had normal quench type combustion chambers as shown in Figs. 3-6. At first glance it is quite evident that there are substantial quench volumes in the chambers of both engines; the quench volumes intentionally designed into these chambers represent 4.2% and 3.6% of the total chamber volumes at tdc for the 199 CID engine and the 232 CID engine respectively. Table 1 lists this information in more detail. Using the taping method (see Appendix) for determining the remaining surface area for the two chambers and assuming a boundary layer thickness of 0.010 in. (8) and adding the product of this area and thickness to the intentional quench volumes just mentioned, the total percent of quench volume for each of the engines under these highly qualified circumstances becomes: 11.4% for the 199 CID engine and 9.8% for the 232 CID engine. Thus, for these chambers, the basic wall surface area compared to the intentional quench volume is responsible for the greater part of the emission levels of unburned hydrocarbons.

Close scrutinization of the chamber designs for both engines also revealed another quench volume, although much smaller by comparison (0.017 in.^{3}), existing as the result of the cylinder head gasket bore diameter exceeding the cylinder bore diameter by 0.130 in. (Fig. 7). This results in an additional 0.4% quench volume for the 199 CID engine and 0.3% for the 232 CID engine.

To summarize, approximately 38% of the total quench volume for either engine could conceivably be eliminated, assuming little or no change in the surface-to-volume ratio, by simply removing the intentional quench volumes through cylinder head casting changes, a gasket bore diameter change and a redesign of the piston tops. Quench volumes existing because of the piston to cylinder wall clearance above the ring grooves, as well as the volume involving ring land clearance, were not considered large enough to be practically eliminated within the scope of the program, and were thus ignored.

Layout studies of new low quench type combustion chambers proceeded from this point for both engines. Economic considerations and a maximum 18 month leadtime made it mandatory that spark plug locations, valve positions and sizes, crankshaft dimensions, and connecting rod lengths, remain as they were in the 1967 quench type version. With the aforementioned goals and design limitations, a new combustion chamber shape evolved for each engine (Figs. 5 and 6) which could be credited with a 28% reduction in the existing quench volume for either engine. The pistons for both displacements had to be redesigned in order to retain an 8.5:1 compression ratio.

The 3.50 cc dish in the top of the piston for the 199 CID quench engine was eliminated with the low quench version having a flat top. The depression in the piston for the larger engine was reduced from 13.95 cc to 11.96 cc and redistributed symetrically over the top of the piston. These details are also shown in Figs. 5 and 6.

By these designs nearly three-quarters of the original objective was achieved. Total elimination of the intentionally designed quench volumes as they existed in the original chambers could not be achieved since the boundary layer associated with the chamber wall surface within this quench zone was included as part of the intentional quench volume. Only that portion which was centrally located (presumably more than 0.010 in. from a wall surface) was considered removed. As a point of academic interest, the resulting reduction in S/V ratio for either the 199 CID chamber or the 232 CID chamber was less than 10%. See Table 1.

In the initial design stage of these new chambers, serious emphasis was expended to avoid upsetting the basic burning characteristics of the chambers. However, the desire to eliminate quench volume required substantial revision. In the final design relatively large changes were made in the shape of these chambers. Experience has shown that the timing of the peak pressure point and its magnitude can be greatly changed by seemingly unimportant changes in chamber shape. (11, 12) Since the combustion chamber pressure is closely related to the mass of charge burned up to any given instant during the combustion event, it became obvious that the removal of the intentional quench volume, without being afforded the opportunity of moving the spark plug or making any major redesigns of the entire chamber, would result in substantial pressure rises late in the combustion process or at later crank angles. (11, 12) A closer look at the spark plug boss and its location within the cham-

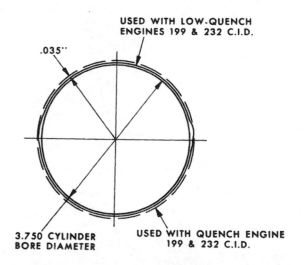

Fig. 7 - Head gasket bore configuration

ber revealed that there was a substantial amount of quench area (and cast iron) immediately behind it. See Figs. 5 and 6. By removing the cast iron surrounding the spark plug boss from within the cylinder head cavity, sizeable increases in chamber volume near the spark plug were achieved, which had the important effect of placing the ignition point more near the center of volume for the new chambers, thus improving the ultimate volume distribution of the low quench chambers. This fact made the ultimate designs less of a compromise in terms of knock control and engine performance.

Plotting the rate of change of volume with respect to flame travel (assuming constant flame speed) for the new low quench chamber designs and then comparing them with those for the two quench type designs (Figs. 8 and 9), shows very clearly that the new chambers may have markedly different pressure characteristics, especially in the later phases of flame travel. Also, comparing these new characteristic rates with the limits for the volume rates that Taub (13) as-

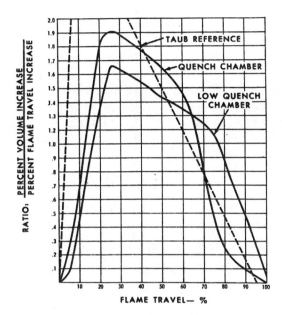

Fig. 8 - Theoretical flame travel and rate of volume burned for the 199 CID engine with quench type and low quench type combustion chambers

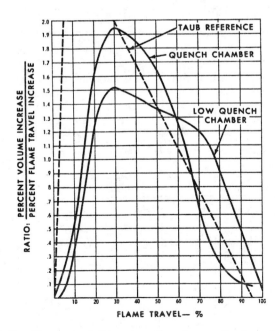

Fig. 9 - Theoretical flame travel and rate of volume burned for the 232 CID engine with quench type and low quench type combustion chambers

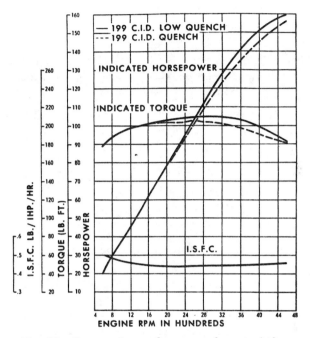

Fig. 10 - Bare engine performance characteristics

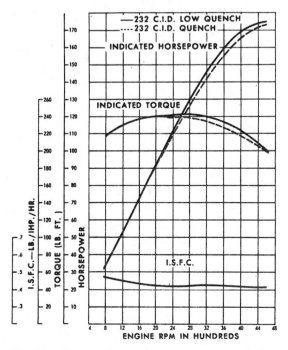

Fig. 11 - Bare engine performance characteristics

signed (dotted line in Figs. 8 and 9) as the limiting rates above which rough combustion characteristics may be expected, serious questions regarding practical application resulted. The background and contemporary knowledge concerning the relationship of chamber shape and knock control provided additional support for questioning the chamber shape considered. It was decided, however, to build prototype engines of both displacements and to establish, for the record, the exhaust emission potential of the low quench concept, to determine the magnitude of the performance problems to be encountered, and to determine whether practical solutions to these problems could be found.

ENGINE DYNAMOMETER TESTING

Power Characteristics - The low quench engine produces essentially the same horsepower as the previous quench engine within the respective displacements involved. Figs. 10 and 11 demonstrate the relative "bare engine" indicated horsepower, indicated torque, and indicated specific fuel consumption.

Fuel Consumption Characteristics - The fuel consumption characteristics at part throttle and full load were determined by the conventional "fuel fish hook" method throughout the range of speed and load. The low quench engines indicated the same fuel consumption potential as the former quench versions. Figs. 12-15 demonstrate a sampling from these data plots. Some differences in fuel flow for a given brake load are evident in these plots, but differences in engine friction levels between the engines used for these data account for the fuel flow differences. The "fuel fish hook" data are particularily useful in confirming the optimum metering level of a carburetor calibration as shown with an "X" on the curves of the referenced figures.

Ignition Requirements - The ignition timing requirements for the low quench engine are predictably different from

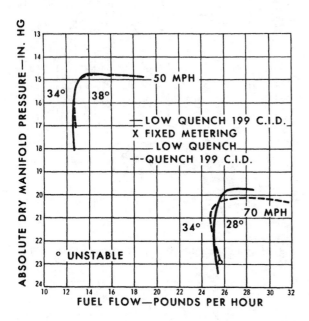

Fig. 12 - Fuel requirements at road load, 50 mph, 70 mph

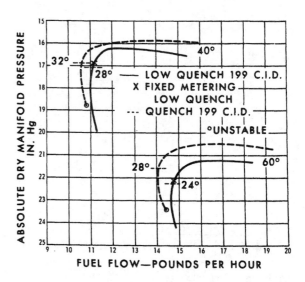

Fig. 13 - Fuel requirements at 1600 rpm, 40%, 60%

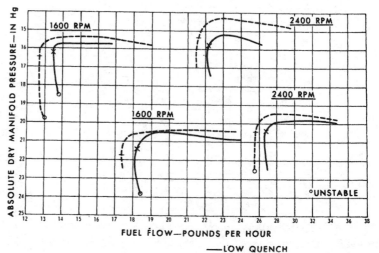

Fig. 14 - Fuel requirements - 232 CID

the quench engine. The reduction in the quenching process of the end charge in the quench engine supports the concept of faster combustion and conversely a lesser ignition advance. Fig. 16 describes the basic wide open throttle spark characteristics for both displacements of the low quench engine at a light deposit condition. For comparison, Fig. 17 describes the minimum spark advance for best torque (MBT) and "2% loss retarded" curves for a 232 engine in both combustion chamber configurations.

In-vehicle experience with the low quench engine at substantially higher deposit conditions indicated that the knock level was more sensitive than that shown in Fig. 16,

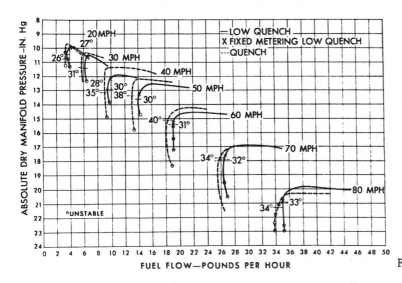

Fig. 15 - Fuel requirements at road load, 20-80 mph

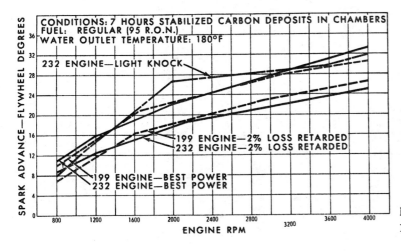

Fig. 16 - Wide open throttle spark characteristics for 199 and 232 low quench engines

Table 2 - Basic Ignition Timing

Basic Setting, deg	Displacement	Transmission	Chamber Type
10 btdc	199	Automatic	Quench
10 btdc	199	Manual	Quench
5 btdc	199	Automatic	Low quench
tdc	199	Manual	Low quench
5 btdc	232	Automatic	Quench
5 btdc	232	Manual	Quench
tdc	232	Automatic	Low quench
tdc	232	Manual	Low quench

particularily at the higher engine speed. Consequently, the centrifugal spark advance curve shown in Fig. 18 was selected. The characteristics described in this figure produce substantially the same road octane rating for their respective combustion chamber configurations. Both the quench and low quench engines use the same part throttle spark advance unit. The correction made for full load operation reflected a proper correction at part throttle, without recalibration at part throttle.

Table 2 lists the basic ignition timing settings selected for the low quench engine and the settings for the former engines for comparison. The more retarded basic settings for the emission controlled engines reflect the intention to reduce hydrocarbon level as much as reasonably possible. Also, a single distributor for both displacements of the low quench engines was desired. This was accommodated by selecting a more advanced basic setting for the 199 engine with automatic transmission. The data of Fig. 16 support a more advanced spark advance condition for the 199 engine compared to the 232 engine as shown by the "2% loss retarded" curves for both engines. The manual transmission version of the 199 low quench was not advanced to 5 deg btdc in order to meet the required exhaust emission performance.

EXHAUST EMISSION AND DRIVEABILITY PERFORMANCE

<u>Driveability Evaluation</u> - The exhaust development was a continuous process of evaluation from the date the first low quench engine was assembled. A most delicate balance between vehicle driveability quality and exhaust emission performance was the basis for carburetor development. In the early evaluation stages of driveability quality, before the final carburetor improvements were incorporated, the low quench engines were judged to be more prone to surge and appeared to require greater accelerator pump discharge volume for a driveability level equivalent to the quench engine.

In the final analysis, the low quench engines are judged to provide a driveability level comparable to the previous quench engines.

<u>Basic Exhaust Emission Performance</u> - The objectives set forth in the design section of this paper with regard to exhaust emission performance were met. The low quench engines meet the Federal requirements for emission performance. As a basis of comparison, Table 3 describes the "clean engine" emission levels for both combustion chamber configurations discussed. The data listed for the quench engines reflect nonemission controlled vehicles (nationwide vehicles prior to 1968 models) and the data for the low quench engines reflect controlled emission systems.

The "clean engine" condition is desirable for comparative purposes, primarily because it determines the potential of a given engine-vehicle combination with regard to its exhaust emission level. On the other hand, this test condition does not describe the level of emissions with normal

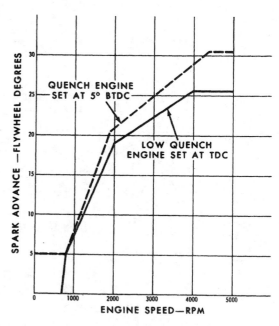

Fig. 18 - Centrifugal distributor advance characteristics

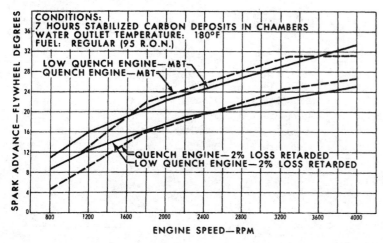

Fig. 17 - Wide open throttle spark characteristics for 232 engines - quench and low quench

deposits or the durability or deterioration characteristics of the engine-vehicle combination. Durability characteristic is defined as the reduction in all areas of engine and vehicle performance throughout the vehicle life. Deterioration refers to the increase in hydrocarbon and carbon monoxide emission throughout the life cycle of the vehicle.

Durability Testing - Durability testing of the low quench engines demonstrated that this combustion chamber provides

Table 3 - Exhaust Emission Performance - Clean Engine Condition for Vehicles Equipped with Automatic Transmissions (from Ref. 10)

Displacement	Chamber Type	HC, ppm	CO, %
199	Quench (uncontrolled)	450	2.0
199	Low quench (controlled)	190	1.0
232	Quench (uncontrolled)	350	2.3
232	Low quench (controlled)	150	1.0

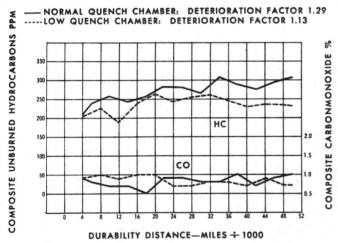

Fig. 19 - Exhaust emission deterioration with vehicle miles for the 199 and 232 CID engines

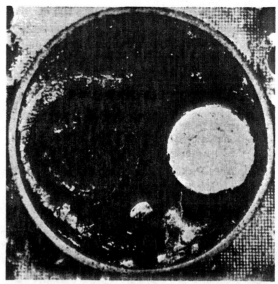

Fig. 20 - 199 CID quench chamber with 50,000 mile deposit accumulation

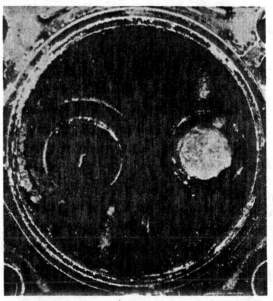

Fig. 21 - 199 CID low quench chamber with 50,000 mile deposit accumulation

Fig. 22 - 199 CID quench and low quench cylinder heads with 50,000 mile deposit accumulation

a deterioration characteristic which results in a far lower numerical slope. Fig. 19 shows the hydrocarbon deterioration factor developed for engines of both combustion chamber type as:

$$\frac{\text{emissions extrapolated to 50,000 miles}}{\text{emissions extrapolated to 4,000 miles}}$$

The quench chamber produced a deterioration factor of 1.29 and the low quench chamber produced a like factor of 1.13 (10, 14). The factors for both engines were developed using a Detroit city test route with a maximum average speed of 32 mph.

Further evidence that the low quench chamber accumulates less deposits in the moderately loaded durability schedule used is shown in Figs. 20-22. Here the cylinder head from a 199 engine with quench chamber is contrasted with the cylinder head of a 199 engine with low quench chamber. Both cylinder heads had accumulated 50,000 durability test miles. The quantity and character of the deposits shown on the quench cylinder head are several orders of magnitude greater than for the low quench cylinder head.

CONCLUSIONS

The foregoing results as described in this paper confirmed that the 199 and 232 CID engines could be modified to conform to the current Federal requirements for exhaust emission control. Although the modifications involved redesign of major components, substantial capital investments for tooling modifications and a maximum development effort for more than two years, the final results are consistent with the original objectives of the program to meet federal requirements without the aid of an external control system.

ACKNOWLEDGMENTS

The authors would like to express their appreciation to J. R. Burns for his contribution toward the carburetor development of this program and to William A. Bevill for his contribution toward the design portion of this program. The overall program was under the direction of C. E Burke, assistant chief engineer, Automotive Advanced Engineering and Research.

REFERENCES

1. "The Clean Air Act." (Public Law 88-206), December 17, 1963 and Amendment (Public Law 89-272) October 20, 1965.

2. "Standards for Ambient Air and Motor Vehicle Exhausts." The State of California, December 4, 1959.

3. D. V. Potter, G. F. Leydorf, Jr., and R. L. Lawler, "The New Rambler Six Engine - Torque Command 232." Paper 884B presented at SAE Summer Meeting, Chicago, June 1964.

4. E. W. Beckman, W S. Fagley, Jr., and J. O. Sarto, "Exhaust Emission Control by Chrysler The Cleaner Air Package." Paper 660107 presented at SAE Automotive Engineering Congress, Detroit, January 1966.

5. W. S. Fagley, Jr., Mark V. Sink, and C. M. Heinen, "Maintenance and the Automobile Exhaust - Second Report." Paper No. 486L presented at SAE National Automobile Week, Detroit, Michigan, March, 1962.

6. Earl Bartholomew, "Potentialities of Emission Reduction by Design of Induction Systems." Paper No. 660109 presented at SAE Automotive Engineering Congress, Detroit, January 1966.

7. J. H. Jones and J. C. Gagliardi, "Vehicle Exhaust Emission Experiments Using a Pre-Mixed and Pre-Heated Air Fuel Charge." Paper 670485 presented at SAE Mid-Year Meeting, Chicago, May 1967.

8. C. E. Scheffler, "Combustion Chamber Surface Area, A Key to Exhaust Hydrocarbons." Paper 660111 presented at SAE Automotive Engineering Congress, Detroit, January 1966.

9. F. W. Cook, "Antismog Carburetor Hardware and Test Equipment." Paper 660110 presented at SAE Automotive Engineering Congress, Detroit, January 1966.

10. Federal Register Volume 31, Number 61, "Control of Air Pollution From New Motor Vehicles and New Motor Vehicle Engines." Washington, D. C., March 1966.

11. H. Rabezzana, S. Kalmar, and A. Candeliste, "Combustion: An Analysis of Burning and Expansion In The Reaction Zone." Automobile Engineer, Vol. 29 (1939) pp. 347, 377.

12. D. J. Patterson and G. Van Wylen, "A Digital Computer Simulation for Spark-Ignited Engine Cycles." SAE Progress in Technology, Vol. 7, "Digital Calculations of Engine Cycles." New York: Society of Automotive Engineers, Inc., 1964.

13. A Taub, "Method and Machine for Avoiding Combustion Chamber Calculations." SAE Journal, Vol. 36, No. 4, April, 1935, p. 159.

14. J. C. Gagliardi, "The Effect of Fuel Anti-Knock Compounds and Deposits on Exhaust Emissions." Paper 670128 presented at SAE Automotive Engineering Congress, Detroit, January 1967.

15. T. A. Huls, P. S. Myers, and O. A. Uyrehara, "Spark Ignition Engine Operation and Design for Minimum Exhaust Emission". Paper 660405 presented at SAE Mid-Year Meeting, Detroit, June 1966.

APPENDIX

The determination of surface area (in.^{2}) per unit volume (in.^{3}) for a combustion chamber is, at best, a moderately difficult task to perform, especially if a fair degree of accuracy is required.

There are a number of different techniques for making this determination; however, they all require measurement of the very complicated and irregular surface which defines the combustion chamber shape. The accuracy and consistency of the determinations depend upon the areas considered to be important, the method of measurement applied to them, and the skill of the investigator.

The determinations and measurements reported in the text of this paper were all made using a technique whereby all irregular surfaces were reproduced with tape, sectioned, and planimetered. All regular surface areas were calculated from detail drawings and design layouts. All the volumes, with the exception of the gasket cutout volume which was calculated from known dimensions, were measured using the liquid displacement technique.

The areas included in the surface measurement, and those which were not, are outlined as follows.

Included -
1. Head cavity area.
2. Head flat or quench area within head gasket outline.
3. Cylinder block top surface area within head gasket outline.
4. Side area of head gasket outline.
5. Valve side areas, including cylindrical side of valve head and that part of the face projecting into the chamber.
6. Valve head surface area.
7. Piston top surface area.
8. Piston top ring land area.
9. Area of top surface of top piston ring exposed between top land diameter and cylinder bore diameter.
10. Cylinder bore surface area above top ring.
11. Spark plug cavity area.

Excluded -
1. Area behind top ring.
2. Gasket area inside first bead.
3. Chamfers less than 0.040 in.

ABSTRACT

The combination of lean mixtures to provide oxygen in the exhaust and exhaust heat conservation to enhance exhaust reactions yields significant exhaust reaction and lower hydrocarbon emissions; it requires adequately uniform mixtures, but it does not require injection of additional air into the exhaust. A vehicle with this combination and other modifications to control emissions during all types of operation exhibits low emissions and good performance and fuel economy. Full evaluation of this approach must await its application to a variety of engines, particularly smaller ones. The principle appears potentially practical and has advantages over other approaches to low emissions, but there are costs. A cost-benefit balance is required to establish its desirability for each application. The results of this experimental exploration are presented as a contribution to the multi-industry effort to provide the public with acceptable atmospheric cleanliness at minimum vehicle and fuel cost.

POTENTIALITIES OF FURTHER EMISSIONS REDUCTION BY ENGINE MODIFICATIONS

By Lamont Eltinge*, Frederick J. Marsee†, and A. Joel Warren**

INTRODUCTION

Air cleanliness is important and has taken its place with fuel economy, performance, cost, durability and reliability as a prime design criterion for engine-fuel systems. The systems that keep approximately 100 million vehicles operating in the U. S. transportation system are complex. Dr. A. M. Bueche, Vice President of Research and Development of General Electric, an observer who does not appear biased in favor of the gasoline engine, has typified the internal combustion engine as "the most highly-engineered product in human history."[1]†† Many factors must be considered for such complex and highly developed systems and advances for emission control require massive effort. That effort is being expended and progress is being made; 1968 cars have cleaner exhausts than previous models. However, if very low hydrocarbon emission levels are required, a system incorporating both low emissions from a properly modified engine and reaction of the unaugmented exhaust after it leaves the cylinder has practical interest.

Engine modifications to minimize exhaust emissions have been studied and practiced.[2-8] Bartholomew[9] reported Ethyl's earlier work in this area. Exhaust reaction has been reported and practiced,[10-15] with richer-than-stoichiometric fuel-air mixtures, air injection into the exhaust port and relatively extensive reaction in the exhaust system used to achieve low tailpipe emissions. As Ethyl continued exploration of means of obtaining and exploiting relatively uniform mixtures, we found that the *combination* of improved engines operating on leaner-than-stoichiometric mixtures, which provided oxygen in the exhaust products, and exhaust heat conservation, which increased exhaust temperatures and enhanced exhaust reactions, was a possible practical approach to very low hydrocarbon emissions. Exhaust reaction fundamentals have been studied, and a vehicle has been modified to explore the potential of this approach. The results are being reported to SAE in the hope that others may benefit from them and that, with reports of other experience, they will contribute to the joint effort of engine and fuel technologists to develop optimum solutions for the many different applications.

---
*Director, Automotive Research, Ethyl Corporation Research Laboratories, Detroit, Michigan.
†Research Engineer, Ethyl Corporation Research Laboratories, Detroit, Michigan.
**Research Associate, Ethyl Corporation Research Laboratories, Detroit, Michigan.
††Superscripts designate References at end of paper.

EXHAUST REACTION FUNDAMENTALS

The concentration of hydrocarbons in the exhaust gases entering the atmosphere is the product of:

1. The hydrocarbon concentration in the gases leaving the engine cylinders.
2. The fraction of those hydrocarbons that does not react in the exhaust system.

Normally, the unreacted fraction is high because the combination of oxygen availability, temperature and time for reaction is inadequate. Injection of air (oxygen for reaction) at the exhaust port can yield a low unreacted fraction,[11, 12, 13, 16] but the fuel-air mixture supplied to the engine has been richer than stoichiometric. The use of lean mixtures to provide the necessary oxygen has not been reported and appears to have some potential and advantages.

The effects of initial hydrocarbon concentration, temperature, pressure, reaction volume, mass flow and oxygen concentration in the exhaust are summarized in the general equation for a second-order chemical reaction of a homogeneous mixture,[10, 11] whose derivation is outlined in Appendix A.

$$C_o = C_i \, e^{-\left(\dfrac{K_r P^2 O_2 V}{K_3 \, T^2 \, W}\right)} \qquad (1)$$

where

C_o = concentration of hydrocarbons in the exhaust gases leaving the reaction volume, ppm (vol)
C_i = concentration of hydrocarbons in the exhaust gases entering the reaction volume, ppm (vol)
e = base of natural logarithms
P = exhaust pressure, psia
O_2 = oxygen concentration in the exhaust gases, % (vol)
V = volume available for exhaust gas reaction, ft^3
T = absolute temperature, °R
W = mass flow rate of air, lb. sec.$^{-1}$
K_3 = numerical constant
K_r = specific reaction rate, ft^3 lb-mol^{-1} sec^{-1}

The exponential term is the fraction unreacted; if the combination of oxygen, pressure, temperature, and time is insufficient, its argument approaches zero and its value approaches 1.0. The expression shows that the factors have interrelated effects and is a very useful aid and guide to development of systems exploiting this approach to low tailpipe emissions. Escape of hydrocarbons from a vehicle exhaust depends on 1) initial hydrocarbon concentration and oxygen availability, 2) pressure, and 3) temperature and time. Engine design and modification can affect each of them.

Initial Hydrocarbon Concentration and Oxygen Availability

This approach to cleaner exhaust requires low hydrocarbon emissions from the cylinder and 1 to 3% oxygen in the exhaust gases, as well as adequate temperature and time. The combination of low initial hydrocarbon concentration and enough oxygen in the exhaust can be attained at about 0.06 fuel-air ratio (16.7 air-fuel ratio), as illustrated by the typical results shown in Figure 1. However, good mixture preparation is clearly vital to this approach to cleaner exhaust. With poor distribution, indicated by a high maldistribution index (S_x)*,[17] some portions of the charge encounter unfavorable conditions and fail to burn. This causes surge or objectionable driveability,[18] and initial hydrocarbon concentration enough higher to offset some or all of the benefit of exhaust reactions. With uniform mixture, however, lean mixtures can burn, hydrocarbon emissions can be low, and operation can be surge-free at some conditions with fuel-air ratios leaner than 0.05.[18,19,20] The limit for a uniform mixture depends on charge dilution,[3,21] temperature,[22] combustion chamber design, ignition energy and timing, and operating conditions. Thus, relatively satisfactory combustion of adequately lean mixtures is possible. The problem is to obtain the necessary uniformity practically.

The fuel-air mixture of production engines is not always perfectly uniform. One cylinder may be 10 to 20% leaner than the average;[18,20,23,24,25] the fuel-air ratio within a cylinder may vary in one cycle or from cycle to cycle. Fuel and engine variables affect the variations. To minimize the variations, we have employed some features described by Bartholomew[9] and others developed subsequently and described in Appendix B. These include:

1. A venturi in the primary carburetor section small enough to maintain 80 to 100 fps venturi velocity at closed-throttle air flows.
2. A fuel nozzle with stepped skirt in the primary carburetor section, to enhance atomization.
3. A perforated throttle plate to minimize fuel film on the carburetor bore.
4. Idling entirely on the main fuel system of the primary carburetor, to eliminate associated stratification of the charge, "dribble" and uneven flow as fuel flow transfers from the idle system to the main fuel-metering system of the carburetor.
5. A system of primary choke-plate deflection, instead of a separate or trim idle system, to provide fuel-air ratio control at idle and deceleration conditions.
6. A ring and moat in the carburetor bore to re-entrain liquid fuel from the bore, to introduce extraneous "air" streams with minimum upset of distribution, and to promote mixing.
7. Power-jetting the transfer system of the secondary carburetor section to ensure "solid" transition at low engine speed and low manifold vacuum.
8. A system for controlled preheating of carburetor air to provide the heat required for fuel vaporization.

Even these features do not yield perfectly uniform mixture; further improvements in mixture uniformity should result from advances in the engineering understanding of fuel metering, atomization, placement, vaporization and mixing in the air stream. However, compromises of engine complexity, cost, and/or breathing may be necessary.

In addition to fuel-air ratio and mixture uniformity, initial hydrocarbon concentration reflects combustion chamber surface area,[26] ignition timing,[2,3,9,27,28] cylinder scavenging,[29] jacket temperature, mixture temperature and engine load and speed. Benefits of these factors can be combined with those from exhaust reaction; some are synergistic with it.

---

* The maldistribution index reflects nonuniformity of fuel-air ratio between cylinders, within individual cylinders, and from cycle to cycle. High values indicate reduced potential for satisfactory lean operation. Limited data suggest that when average fuel-air ratio is leaner than [(0.04 to 0.05) + (2 to 3) (S_x)], hydrocarbon emissions increase significantly; this relation seems compatible with the statistical interpretation of S_x and some lean-limit data.

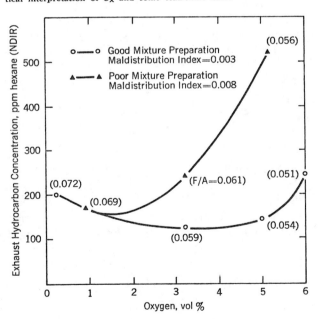

Figure 1. Effect of Mixture Preparation on Hydrocarbon Emissions at Various Oxygen Levels
389 CID, V-8, Engine R
Modif. Induction System
1200 rpm, 15 hp, 26° btc Ignition Timing

Exhaust Back Pressure

Exhaust back pressure has important effects on the scavenging and reaction processes. Raising back pressure lowers the average hydrocarbon concentration of the gases leaving the cylinder partly because it affects scavenging of the last part of the charge exhausted which has the highest hydrocarbon concentration;[29] this effect is reflected in the "minimum rate" data of Figures 2 and 3, which were taken without any exhaust insulation and, consequently, lower temperatures. Increasing back pressure in the exhaust system increases the reaction potential markedly; the effect of pressure is reflected in the P^2 term in Equation 1. When increased back pressure is accompanied by higher exhaust gas temperatures, the combined effect can reduce hydrocarbons significantly, as shown by the steeper slope of the "higher rate" curves of Figures 2 and 3.

On the negative side, raising back pressure reduces power and increases fuel consumption. Exhaust restriction can increase dilution when the throttle is closed, thereby increasing idle and deceleration emissions. Consequently, it is preferable to have a system, like that described in Appendix C, that only increases back pressure at part load. This system was designed to increase back pressure to about 3 psig during part-throttle accelerations and cruise modes. It increased fuel consumption, and implicitly exhaust mass flow, about three percent. The fuel and equipment penalties must be compared with other means of reducing hydrocarbon emissions (e.g., ignition timing) and the need for the reduction.

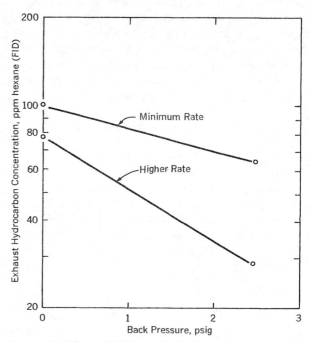

Figure 3. Effect of Exhaust Back Pressure on Emission at Different Reaction Rates
389 CID, Engine R
Simulated 40-mph Cruise
Engine Dynamometer
$\approx$ 3% O_2 in Exhaust

Temperature and Time

Reaction rate is strongly dependent on temperature, as shown in Figure 4. The increase in reaction rate reflected in the exponential term of Equation 1 far outweighs the effects of reduced densities reflected in the T^{-2} term and a 30 or 40°F increase in temperature doubles the reaction rate. Schnable et al[11] show about the same temperature dependence. The combination of volume of the exhaust system, mass flow rate and exhaust gas density establishes residence time. For typical residence times and 2 to 3% oxygen, exhaust reaction is significant if average exhaust gas temperature is about 1150°F. For minimum hydrocarbon emissions, the exhaust must be kept hot enough long enough.

Typically the temperature of exhaust gases leaving the exhaust manifold of large V-8 engines in urban driving is 600°F to 1100°F. They are this low because the heat is lost in the combustion chambers, exhaust ports, and exhaust manifolds. The temperature drops due to these heat losses depend on engine speed and load. Reduction of the heat losses leads to higher temperatures, more reaction and hence less hydrocarbon emissions.

The great temperature dependence of the reaction rate requires care in using "average" temperatures and dominates means for increasing reaction. Rates, not temperatures, must be averaged. The effective "average" temperature is complex and much closer to the highest

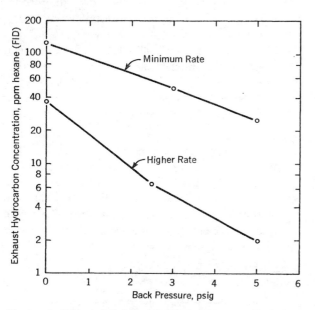

Figure 2. Effect of Exhaust Back Pressure on Emissions at Different Reaction Rates
Single-Cylinder, Engine Q
1350 rpm, 15° btc Ignition Timing
70% Full-Throttle Air Flow
$\approx$ 3% O_2 in Exhaust

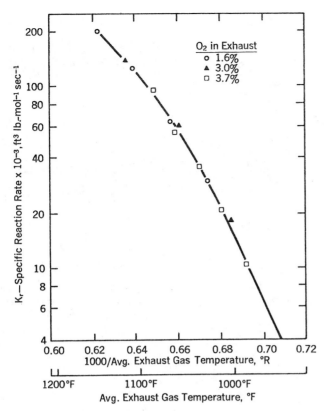

Figure 4. Effect of Temperature on Exhaust Reaction Rate
Single-Cylinder, Engine Q
1350 rpm

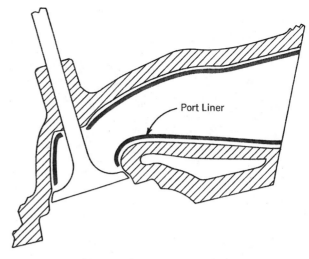

Figure 5. Exhaust Port with Liner
Liner Fabricated of 20-g. 304 s.s. or
316 and 446 s.s.

temperature than the lowest; one logical average involves a hyperbolic cosine function. In addition, it is more important to conserve heat at the highest gas temperatures.

Limited data indicate losses in the exhaust ports reduce exhaust gas temperatures by 100°F to 300°F. Stainless-steel insulating liners for exhaust ports, installed as indicated in Figure 5, increased exhaust gas temperatures about 200°F, as shown in Figure 6. Insulation of the heat cross-over passage in the cylinder heads increased temperatures 15°F to 30°F. Retarding ignition timing also increases exhaust temperatures, so with significant exhaust reaction, the usually observed benefits of spark retard [2, 9, 19] are compounded. Exhaust port insulation is most effective because exhaust gases are hottest and reaction rates are highest in the ports. Larger and shorter ports should have the same benefits.

The last part of the charge scavenged from the cylinder is not only the highest in hydrocarbons but it is also the coolest. If it had not been relatively cool all during the cycle, the reaction rates indicated by Figure 4 would have led to relatively complete reaction. This part of the charge tends to reside in the exhaust port while the exhaust valve is closed. There, mixing and thermal regeneration heat it and lead to reaction and reduction in hydrocarbon content. For this reason, too, exhaust port insulation and enlargement are particularly effective.

The combination of exhaust port insulation, insulated exhaust manifolds and enlarged and insulated exhaust pipes reduces hydrocarbon emissions. Insulating ports and enlarging and insulating pipes reduced emissions as shown in Figure 7. Doubling to tripling of exhaust manifold volume provided additional time at adequate temperature for further reaction and reduced hydrocarbon emissions another 15% in other tests. Shrouds like those shown in Figures 8 and 9 reduce heat losses from standard manifolds enough to raise exhaust gas temperatures about 40°F on the engine or chassis

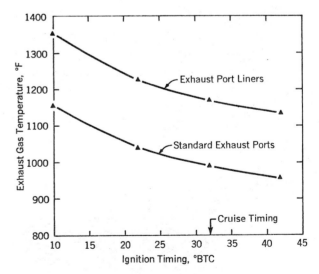

Figure 6. Effect of Port Liners on Exhaust Gas Temperature
389 CID, Engine R
Simulated 40-mph Cruise
Engine Dynamometer
≈3% O_2 in Exhaust

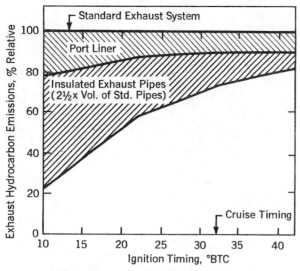

Figure 7. Effect of Exhaust Insulation and Enlargement on Hydrocarbon Emissions
389 CID, Engine R
Simulated 40-mph Cruise
Engine Dynamometer
$\approx$ 3% O_2 in Exhaust

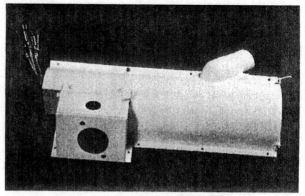

Figure 9. Complete Exhaust Shroud

dynamometer. On the road where air flow around the manifold is greater, the benefit is even greater. Insulated and enlarged exhaust pipes, too, increase reaction potential; however they are less effective because they are further downstream and cooler. Although fabricated exhaust manifolds may be desirable for thermal-stressing and quick warm-up, enlarged cast-iron manifolds, possibly with internal ceramic insulating coatings, could prove adequate.

Insulating exhaust ports reduces heat rejection to the coolant, and insulating the exhaust manifold should reduce heat rejection to the engine compartment. However, these modifications raise exhaust pipe, muffler and tailpipe temperatures at high engine speed and load. Special materials or special provision for cooling may be needed for exhaust-system durability, but we have not found them necessary in our limited experience. When high speed and load cause high manifold temperatures, blowing air through the manifold shrouds can provide desired controlled cooling. Exhaust port liners must be made of materials that will survive high temperatures, thermal shock and the corrosive effect of exhaust gases. Exhaust heat conservation increases material requirements and costs, to some extent. However, extensive heat transfer studies and design should contribute to better heat conservation and minimum material requirements and costs.

VEHICLE INCORPORATING LEAN MIXTURE AND EXHAUST REACTION

Concepts like the combination of lean mixture operation and conservation of exhaust heat to minimize hydrocarbon emissions must be incorporated in experimental vehicles to permit full evaluation. The range of conditions encountered in the official emission test and urban driving requires a complete system of controls if low total emissions are to be achieved. Emissions and performance are significantly affected by adjustments of spark timing and carburetion; comparisons must reflect equal refinement and performance. The total system must be evaluated for driveability and durability.

A standard-size passenger car with a 389 cu. in. high-compression V-8 engine having 4°BTC ported idle timing and a production dual-diaphragm distributor, 2.56 rear axle ratio and an automatic transmission was modified stepwise to include the following:

Modification Group A

1. Experimental three-barrel carburetor, with high-velocity primary mounted on an essentially production cast-iron intake manifold.
2. Choke deflection for idle mixture-ratio control and no separate idle system.
3. Ring and moat for fuel re-entrainment and mixing.
4. Modulated preheated carburetor air.

Figure 8. Exhaust Manifold and Partial Shroud

Modification Group B

1. Temperature modulation of choke qualifying position for CO minimization with good low-temperature performance.
2. Deceleration emissions control (Appendix D) and engine-braking control system, which requires uniform mixture and mixture enrichment.
3. Controlled back pressure.
4. Fuel-air mixture supplied by 3-venturi carburetor leaned to provide 2 to 3% O_2 during cruise, 1% O_2 at idle and 2 to 4% CO during deceleration.
5. Shrouds around the exhaust manifold.

Modification Group C

1. Exhaust port liners.
2. Ceramic coating inside standard cast-iron exhaust manifold, blended from two commercial products immediately prior to application.
3. Enlarged (from 1-3/4 inch I.D. to 2-7/8 inch I.D.) and externally-insulated exhaust pipe.

Successive application of these components yielded the results indicated in Table 1. They were obtained after mileage accumulation on commercial gasoline to stabilize deposits, and each value is the average of several close checks. The general comparisons of the effect of modifications are valid but the comparisons might have been a little different if all data had been obtained at the same time and at the same engine and ambient conditions. With carburetion and timing for relatively good driveability at a range of ambient temperatures (Modif. A), total cycle emissions were 211 ppm—an acceptable but not outstanding level. CO at 0.65% was somewhat better than average. Incorporation of temperature-modulated choke qualification, deceleration emissions control, back-pressure control, and exhaust manifold shrouds, and leaning carburetor settings to provide oxygen for reaction, without particular concern for excellence of driveability (Modif. B), reduced hydrocarbon emissions to 128 ppm and CO to 0.51%. Enhancing exhaust reactions by insulating exhaust ports and manifolds and insulating and enlarging the exhaust pipe (Modif. C) reduced hdyrocarbon emissions

TABLE 1
EFFECT OF MODIFICATIONS ON VEHICLE EXHAUST EMISSIONS
389 CID, V-8 Engine R, 10.5 CR,
Automatic Transmission, 2.56 Axle Ratio

	Modification Group	Cycles 1-4 HC, ppm	Cycles 1-4 CO, %	Cycles 6 and 7 HC, ppm	Cycles 6 and 7 CO, %	Total Test HC, ppm	Total Test CO, %
A	3-Venturi Carburetor (9000 deposit miles) Temp.-Modulated Choke Deceleration Control	235	1.0	199	0.35	211	0.65
B	Back-Pressure Valve 2 to 3% O_2 in Exhaust Exh. Manif. Shrouds (21,000 deposit miles) Exhaust Port Liners	157	0.79	112	0.37	128	0.51
C	Insulated Exh. Manif. Large Insulated Exh. Pipe (2500 deposit miles)	88	0.66	43	0.38	58	0.48
C'	Modif. Vacuum Advance 2.73 Axle Ratio (4000 deposit miles)	128	0.50	71	0.32	90	0.41

Individual Modes — 7th Cycle

	Weighting Factor	Corrected Hydrocarbons, ppm (NDIR) Modification Group A	B	C	C'
Idle	0.042	199	127	92	125
0-25 mph	0.244	210	129	54	72
30 mph	0.118	155	95	23	47
30-15 mph	0.062	263	99	113	145
15 mph	0.050	102	42	17	53
15-30 mph	0.455	179	125	27	50
50-20 mph	0.029	885	83	144	143

by about half to 58 ppm; the benefits, as might be expected, are greatest in the power-producing modes. Subsequently, a portion of the part-throttle ignition timing was advanced to eliminate a "flat" feeling associated with the lean mixtures, and a differential with a 2.73 ratio was installed. Ceramic insulation of the heat cross-over passage apparently scaled off at this time. (It is being replaced with a metal liner.) Emissions with those changes were about 90 ppm hydrocarbons and 0.4% CO.

Driveability is a subjective criterion. Some relatively critical engineers have driven the car, and its driveability in the first and last configurations was generally judged "commercial"; it does not exhibit holes or surge. The ability to run well on lean mixtures reflects relatively good distribution. The results indicate that adequate mixture uniformity can be achieved practically.

The experimental installation of port liners reduced port size and would be expected to reduce high-speed volumetric efficiency and therefore maximum power. In a new design this need not be true.

Applicability of this approach to a variety of cars and engines must be established by applying similar modifications to other engines; some with larger and shorter exhaust ports may be more adaptable. Those with less combustion chamber surface may have hotter exhaust and lower initial hydrocarbon concentration; vehicles and engine designs that minimize exhaust heat losses and allow more room for enlarged exhaust manifolds have an obvious potential for still lower emission levels. There is no obvious reason why the approach is not transferable, but adequate care must be taken to ensure good distribution and adequate heat conservation. Such alternate applications and the search for even lower emissions are under way, and durability is being explored.

This approach is not the only one that can be used to achieve low levels of emissions. It is encouraging that there is more than one. Comparison of approaches is complex because they must reflect equal development, equal compromises with performance, cost, driveability and economy, and equal refinement in ignition timing, carburetion and basic engine design. Several characteristics of the lean-engine and conserved-heat exhaust-reaction system may be important in such comparisons. Engines with lean mixtures should yield better fuel economy than those incorporating rich mixtures. Lean mixtures yield low CO, but the intake air and exhaust manifold must be brought near operating temperatures quickly when the engine is started to obtain maximum benefits. Since the amount of hydrocarbons and CO emitted from the cylinder is lower with lean mixtures than with rich mixtures,

1. Emission from the tailpipe is lower if there is little or no reaction.
2. There is minimal heat release in the exhaust.
3. Much less reaction and, consequently, significantly lower temperatures are required for any desired tailpipe emissions.

The lower temperature requirement should favorably affect material requirements and costs for durability; cast-iron manifolds may be practical for this approach, whereas fabricated stainless steel manifolds are used for the rich-mixture, higher-temperature approach. There is no need for or dependence on an air pump or special tubing or passages. The needed oxygen is intimately mixed with the hydrocarbons, and uneven metering of injected air is not encoutnered. There is a major requirement for good carburetion and intake manifolding. All in all, the approach seems to warrant consideration. It should be particularly interesting as new designs are undertaken.

This vehicle shows that relatively low emissions of hydrocarbons and carbon monoxide can be achieved with a modified gasoline engine; there is a potential for further improvement. However, the approach has only been tried on one engine. It is important to emphasize that before it can be fully evaluated, it must be applied to a variety of engines, particularly lower displacement engines with manual transmissions. Interpretation of the emission levels achieved must reflect the difference between engineering prototypes and similar production vehicles and effects of engine size. The modifications increase costs but appear potentially practical and attractive. The proper balance between cost and emission level must be established. Once that is done, adequate time must be allowed for evaluation, comparison with other approaches, development, application, proving and tooling for each engine and vehicle. Then the gasoline-engine system will serve the public with relatively clean atmosphere, as well as the performance, reliability and economies they have rightfully come to expect.

CONCLUSIONS

1. The combination of lean mixtures and exhaust-heat conservation can yield significant exhaust reaction and low emissions of hydrocarbons and carbon monoxide.
2. The approach requires relatively uniform fuel-air mixtures.
3. In addition, to achieve very low overall emission levels, a vehicle must incorporate other modifications designed to minimize emissions during certain types of operation.
4. An engineering prototype incorporating such an overall combination gives low emissions—about 90 ppm hydrocarbons and 0.4% CO—with good driveability and fuel economy.

5. The overall combination increases costs but appears potentially practical and attractive.

6. This approach, like others, should permit the gasoline engine to meet emission targets reflecting a reasonable balance of "costs" and benefits, while retaining essentially intact all of its other advantages to the motoring public.

REFERENCES

[1] Bueche, A. M., "Industry and the Pollution Problem." *Environmental Science and Technology*, *1*, 27 (January 1967).

[2] Jackson, M. W., Wiese, W. M. and Wentworth, J. T., "The Influence of Air-Fuel Ratio, Spark Timing and Combustion Chamber Deposits on Exhaust Hydrocarbon Emissions." *SAE Technical Progress Series*, *6*, 175 (1964).

[3] Hagen D. F. and Holiday, G. W., "The Effects of Engine Operating and Design Variables on Exhaust Emissions." *SAE Technical Progress Series*, *6*, 206 (1964).

[4] Heinen, C. M., "Using the Engine for Exhuast Control." SAE Paper 620042, presented at the SAE Meeting, November, 19, 1962, Los Angeles, California.

[5] Beckman, E. W., Fagley, W. S. and Sarto, J. O., "Exhaust Emission Control by Chrysler—The Cleaner Air Package." SAE Paper 660107, presented at SAE Automotive Engineering Congress and Exposition, January 10-14, 1966, Detroit, Michigan.

[6] Wiese, W. M., Templin, R. J. and Kline, P. C., "An Improved Device to Reduce Hydrocarbons During Deceleration." SAE Paper 620033 presented at SAE National Automobile Week, March 12-16, 1962, Detroit, Michigan.

[7] Lawrence, G., Buttivant, J. and O'Neil, C. G., "Mixture Pre-Treatment for Clean Exhaust—The Zenith 'Duplex' Carburetion System." SAE Paper 670484, presented at the SAE Mid-Year Meeting, May 15-19, 1967, Chicago, Illinois.

[8] Larborn, A. O. J., and Zackrisson, F. E. S., "Dual Manifold as Exhaust Emission Control in Volvo Cars." SAE Paper 680108, presented at SAE Automotive Engineering Congress and Exposition, January 8-12, 1968, Detroit, Michigan.

[9] Bartholomew, Earl, "Potentialities of Emission Reduction by Design of Induction Systems." SAE Paper 660109, presented at the SAE Automotive Engineering Congress and Exposition, January 11, 1966, Detroit, Michigan.

[10] Ridgway, S. L., "Homogeneous Reaction Kinetics and the Afterburner Problem." SAE Paper 590010, presented at the SAE Annual Meeting, January 12-16, 1959, Detroit, Michigan.

[11] Schnable, J. W., Yingst, J. E., Heinen, C. M. and Fagley, W. S., "Development of Flame Type Afterburner." SAE Paper 620014, presented at the SAE National Automobile Week, March 12-16, 1962, Detroit, Michigan.

[12] Cantwell, E. N. and Pahnke, A. J., "Design Factors Affecting the Performance of Exhaust Manifold Reactors." SAE Paper 650527, presented at the SAE Mid-Year Meeting, May 17-20, 1965, Chicago, Illinois.

[13] Brownson, D. A. and Stebar, R. F., "Factors Influencing the Effectiveness of Air Injection in Reducing Exhaust Emissions." SAE Paper 650526, presented at SAE Mid-Year Meeting, May 17-21, 1965, Chicago, Illinois.

[14] Steinhagen, W. K., Niepoth, G. W. and Mick, S. H., "Design and Development of the General Motors Air Injection Reactor System." SAE Paper 660106, presented at the SAE Automobile Engineering Congress and Exposition, January 10-14, 1966, Detroit, Michigan.

[15] Chandler, J. M., Struck, J. H. and Voorheis, W. J., "The Ford Approach to Exhaust Emission Control." SAE Paper 660163, presented at the SAE Automotive Engineering Congress and Exposition, January 10-14, 1966, Detroit, Michigan.

[16] Chandler, J. M., Smith, A. M. and Struck, J. H., "Development of the Concept of Non-Flame Exhaust Gas Reactors." *SAE Technical Progress Series*, *6*, 299 (1964).

[17] Eltinge, Lamont, "Determining Fuel-Air Ratio and Distribution from Exhaust Gas Composition." SAE Paper 680114, presented by title at the SAE Automotive Engineering Congress and Exposition, January 11, 1968, Detroit, Michigan.

[18] Yu, H. T. C. "Fuel Distribution Studies—A New Look at an Old Problem." *SAE Transactions*, *71*, 596 (1963).

[19] Huls, T. A., Myers, P. S. and Uyehara, O. A., "Spark Ignition Engine Operation and Design for Minimum Exhaust Emissions." SAE Paper 660405, presented at the SAE Mid-Year Meeting, June 9, 1966, Detroit, Michigan.

[20] Robison, J. A. and Brehob, W. M., "The Influence of Improved Mixture Quality on Engine Exhaust Emissions and Performance." WSCI Paper 65-17, presented at the Western States Combustion Institute Meeting, October 25-26, 1965, Santa Barbara, California.

[21] Wentworth, J. T. and Daniel, W. A., "Flame Photographs of Light-Load Combustion Point the Way to Reduction of Hydrocarbons in Exhaust Gas." *SAE Technical Progress Series*, 6, 121 (1964).

[22] Bolt, J. A. and Harrington, D. L., "The Effects of Mixture Motion Upon the Lean Limit and Combustion of Spark-Ignited Mixtures." SAE Paper 670467, presented at the SAE Mid-Year Meeting, May 15-19, 1967, Chicago, Illinois.

[23] Donahue, R. W. and Kent, R. H., Jr., "A Study of Mixture Distribution." *SAE Quarterly Transactions*, 4, 546 (1950).

[24] Cooper, D. E., Courtney, R. L. and Hall, C. A., "Radioactive Tracers Cast New Light on Fuel Distribution." *SAE Transactions*, 67, 619 (1959).

[25] Jones, J. H. and Gagliardi, J. C., "Vehicle Exhaust Emission Experiments Using a Pre-Mixed and Pre-Heated Air Fuel Charge." SAE Paper 670485, presented at SAE Mid-Year Meeting, May 15-19, 1967, Chicago, Illinois.

[26] Scheffler, C. E., "Combustion Chamber Surface Area, A Key to Exhaust Hydrocarbons." SAE Paper 660111, presented at the SAE Automotive Engineering Congress and Exposition, January 10-14, 1966, Detroit, Michigan.

[27] McReynolds, L. A., Alquist, H. E. and Wimmer, D. B., "Hydrocarbon Emissions and Reactivity as Functions of Fuel and Engine Variables." SAE Paper 650525, presented at the SAE Mid-Year Meeting, May 17-21, 1965, Chicago, Illinois.

[28] Daniel, W. A., "Engine Variable Effects on Exhaust Hydrocarbon Composition (A Single-Cylinder Engine Study with Propane as the Fuel)." SAE Paper 670124, presented at the SAE Automotive Engineering Congress, January 9-13, 1967, Detroit, Michigan.

[29] Daniel, W. A. and Wentworth, J. T., "Exhaust Gas Hydrocarbons—Genesis and Exodus." *SAE Technical Progress Series*, 6, 192 (1964).

[30] Dietrich, H. H., "Automotive Exhaust Hydrocarbon Reduction During Deceleration by Induction System Devices." *SAE Technical Progress Series*, 6, 254 (1964).

[31] Lewis, B. and von Elbe, G., *Combustion, Flames and Explosions of Gases*. Academic Press, New York and London, 1961.

[32] Perry, J. H., Chilton, C. H. and Kirkpatrick S. D., *Chemical Engineers' Handbook*, McGraw-Hill, New York, 1963.

[33] Fort, E. F., "Idle Enrichment Device," Ethyl Corporation Internal Communication, April 29, 1966.

APPENDIX A

EXHAUST REACTION THEORY

Chemical reaction of exhaust gases has significant possibilities for reducing exhaust emissions. Chemical reaction kinetics help explain the extent to which certain factors influence these reactions.

The rate of oxidation of unburned hydrocarbons in a second-order reaction is expressed by the following equation:[31]

$$\frac{-dC}{d\theta} = K_r C(O_2)^n \qquad (A\text{-}1)$$

Carbon monoxide might be involved; but with lean mixtures, which gave low CO, the instruments detected essentially no change in CO when exhaust reactions reduced unburned hydrocarbons by as much as 70%. Since the concentration of oxygen is large in relation to that of unburned hydrocarbons, its change during the reaction is negligible. Hence, integration of Equation A-1 approximates:

$$\ln \frac{C_i}{C_o} = K_r \theta\, O_2^n \qquad (A\text{-}2)$$

Replacing θ by space velocity introduces P, T, V and W, and changing O_2 from mols per cubic foot to volume percent gives:

$$\ln \frac{C_i}{C_o} = \frac{K_r V P^{n+1}(K_2 O_2)}{K_1 W T^{n+1}} \qquad (A\text{-}3)$$

The Arrhenius equation,[32]

$$\frac{d \ln K_r}{dT} = \frac{A}{RT^2} \qquad (A\text{-}4)$$

shows a relation between the specific reaction rate, K_r, and temperature. If A is assumed constant, Equation A-4 integrates to:

$$\ln \frac{K_{r_2}}{K_{r_1}} = \frac{A}{R}\left(\frac{1}{T_1} - \frac{1}{T_2}\right) \qquad (A\text{-}5)$$

Equation A-5 indicates that a plot of K_r vs $\frac{1}{T}$ on semi-log paper would give a substantially straight line of slope $-\frac{A}{R}$.

In Equation A-3, all terms except K_r and n can be measured experimentally. With n taken as unity, calculated values of K_r from Type Q single-cylinder engine data obtained at three exhaust oxygen levels fall essentially on the same curve, as shown in Figure 4. At other values of n, the data at each oxygen level fall on separate curves. Thus, the equation with n equals unity best fits the experimental data and apparently represents the reactions that occur in the exhaust system studied. The usual form of a plot such as Figure 4 is a straight line. However, this is for uniform temperature in the entire reactor. Temperature drop in the exhaust system combined with simple averaging of temperature could cause the curvature shown in Figure 4.

Substituting $n=1$ in Equation A-3 and rearranging gives an expression that shows the concentration of unburned hydrocarbons entering the atmosphere to be the product of: 1) the concentration leaving the cylinder, and 2) the fraction of those hydrocarbons that does not react in the exhaust system. It is:

$$C_o = C_i\, e^{-\left(\frac{K_r O_2 P^2 V}{K_3 T^2 W}\right)} \qquad (A\text{-}6)$$

NOMENCLATURE

C = concentration of unburned hydrocarbons, mol ft.$^{-3}$
C_o = concentration of hydrocarbons in the exhaust gases leaving the reaction volume, ppm (vol)
C_i = concentration of hydrocarbons in the exhaust gases entering the reaction volume, ppm (vol)
θ = time, sec.
O_2 = oxygen concentration in exhaust gases, mol ft.$^{-3}$ in Equations A-1 and A-2, but as volume percent in Equations A-3, A-6 and 1.
n = exponent, dimensionless
K_r = specific reaction rate, ft.^{3} lb.-mol^{-1} sec.$^{-1}$
K_1, K_2, K_3 = numerical constants
V = volume available for exhaust gas reaction, ft.^{3}
T = absolute temperature, °R
P = exhaust pressure, psia
W = mass flow rate of air, lb. sec.$^{-1}$
R = universal gas constant, ft.-lb. lb.-mol^{-1} °R^{-1}
e = base of natural logarithms, dimensionless
A = activation energy, ft.-lb. lb.-mol^{-1}

APPENDIX B

THE THREE-VENTURI CARBURETOR SYSTEM

The three-venturi carburetor shown in Figure B-1 is designed to give relatively uniform mixture and refined fuel metering and aid in the reduction of exhaust emissions. It consists of a single primary barrel for idle and light-load operation and two large secondary barrels for adequate air-capacity for good performance. The carburetor is installed on a slightly modified four-barrel manifold, Figure B-2, and fitted with a special preheat air cleaner.

The primary venturi is sized for a velocity of 80 to 100 fps at closed-throttle air flow (curb idle). Such velocity appears necessary for steady mixture discharge from the main nozzle, good atomization and accurate mixture control. A mechanism,[33] shown in Figure B-3, partially closes the primary choke plate at curb idle for controlled mixture enrichment; as the throttle valve is opened the choke plate assumes its neutral position and mixture ratio is lean. Air capacity of the primary carburetor is adequate for cruising at 40 to 50 mph.

The nozzle in the primary venturi is designed to discharge the emulsion from an annulus about its outer periphery. Fuel distribution is controlled by blocking segments of this annular channel.[9] This varies the concentration of fuel being discharged about the outer

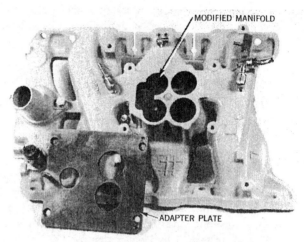

Figure B-2. Four-Barrel Manifold Modified for Three-Venturi Carburetor

edge of the nozzle. The lower section is a diverging cone with steps on its outer surface.[9] The steps create turbulence and help shear the fuel into small droplets.

In a conventional carburetor, fuel flowing from the separate curb idle discharge and/or transfer port must mix with air metered around approximately 315 degrees of the edge of the throttle plate; that arrangement does

Figure B-1. Three-Venturi Carburetor

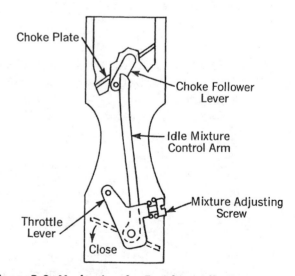

Figure B-3. Mechanism for Enriching Idle Mixture

not facilitate uniform mixtures. At curb idle the three-venturi carburetor premixes the fuel and air before they enter the manifold through a perforated throttle blade.[9] One slot in each side of the throttle, .025 inch wide and spanning 115° of a one-inch circle, provides a good discharge pattern at idle and an additional opportunity for fuel shear-atomization. The design of the throttle blade can be altered to match the closed-throttle requirements of various engines.

A by-pass duct draws *mixture* below the venturi and discharges it into a passage connected to the manifold. Engine idle speed is adjusted by a valve which increases or decreases the mixture flow in the duct. This duct empties into a moat located around the primary throttle bore. Openings between the moat and manifold are located to minimize maldistribution. This moat also provides a collection and distribution point for all unmixed air and gases, such as the crankcase ventilation and choke heat air. A lip or ring, shown in Figure B-4, protrudes into the air stream and shields the openings to form a discharge annulus below the throttle plate. It also reentrains and shear-atomizes fuel that collects on the carburetor bore.

The high velocity and good atomization in the primary venturi causes icing of the primary throttle at relatively high temperatures; ice was observed at 72°F and 28% relative humidity during a flow test. A preheat type air cleaner, however, minimizes icing and helps driveability during warmup by supplying heat needed for vaporization.

The secondary system functions at cruise speeds over 40 mph and when acceleration requires more than 25 to 30% of the total throttle travel. For a normal relationship between throttle travel and performance, the linkage opens the primary during the first 25% of travel, and the secondary during the remaining travel. Other throttle opening ratios create a feeling of over- or under-control.

The secondary metering system consists of two larger venturis with conventional main metering nozzles. Thicker throttle plates are used to minimize the air leakage at idle. The nozzles provide fuel flow and metering when air flow is high; fuel flows through "transfer slots" when air flow through the secondary venturis is too low to draw fuel through the nozzles. The "transfer slots" are located above the lower edge of the throttle plates when they are closed; fuel discharges from the slots only when the throttle plates are open.

Smooth transition between the primary and secondary, during all operating conditions, is very difficult to obtain. Lean "sag" occurs when mixtures are excessively lean at the transition between the secondary transfer slot and nozzle. The transfer slot and main discharge nozzle are calibrated to operate at lean fuel-air ratios for high-speed cruise and high-vacuum acceleration. A power enrichment system enriches the main system for heavier modes of operation. However, during low engine speed, low vacuum and low air flow conditions, the transfer slot is the only fuel supply source. Air flow is too low to induce fuel flow through the main discharge nozzles of the secondaries; so the conventional power enrichment cannot provide the additional fuel required and noticeable sag is encountered. Additional fuel must be supplied through the idle or transfer circuit. A power-enrichment device in the secondary transfer circuit, Figure B-5, functions during accelerations on the transfer port at manifold

Figure B-4. Discharge Annulus

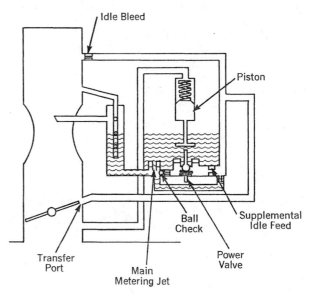

Figure B-5. Power Enrichment Device in Secondary Transfer System

vacuums below 8 inches Hg; the valve setting can vary for different engines. A fuel channel between the power jet and idle circuit intersects the idle circuit downstream from the idle feed. A restriction in this channel limits the maximum fuel-air ratio. Check valves, placed upstream of the main power jet restriction, prevent fuel feedback through the power jet during the normal idle operation.

Both primary and secondary systems of the carburetor have acceleration pumps. The primary pump is designed to deliver its shot during the last portion of the primary throttle travel. The secondary pump discharges during the initial opening of the secondary throttle. The rate and capacity of each pump are very critical for smooth transition between the two systems.

This experimental carburetor is constructed from two production carburetors, therefore each system has a float bowl. During dormant periods or operation on rough terrain, the secondary fuel level may increase until fuel drips from the nozzles. To prevent this a fuel transfer tube, similar to those used on production four-barrel carburetors, connects the primary and secondary fuel bowls.

Most conventional four-barrel carburetors lock the secondary throttles closed during the choking period. This minimizes the complexity of the mechanism and reduces backfiring during accelerations. Both the primary and secondary systems of the three-venturi carburetor require choking; the secondary cannot be locked closed, because the engine will start on the small primary system but lack drive-away performance. The secondaries are locked partially closed, but still supply adequate performance during the initial warmup.

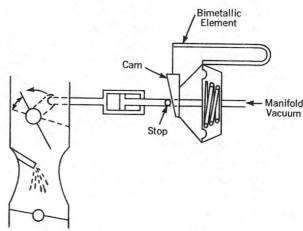

Figure B-6. Temperature-Modulated Choke

One bimetal coil supplies the temperature-sensitive motive force for both choke plates. A vacuum-actuated diaphragm opens the choke plates to the proper position when the engine begins to run. This "qualifying" position is determined by the fuel-air mixture required to sustain the engine at idle. Rich settings are required for satisfactory operation at sub-zero ambients; but they increase CO and are not needed at higher temperatures. A variable "qualifying" position stop, actuated by a separate bimetal spring that senses underhood temperature, shown in Figure B-6, varies the off-idle and idle CO from 1.5 to 5% during choking. It yields both good low-temperature operation and 0.3% lower CO emissions at normal conditions.

APPENDIX C

BACK-PRESSURE CONTROL SYSTEM

Raising back pressure lowers *average* hydrocarbon content of the gases leaving the cylinder. It also increases the residence time and temperature of the exhaust gas in the reaction zone, further reducing the hydrocarbon emissions. However, it causes partial misfire and increased hydrocarbon emission during the deceleration mode. Raising back pressure reduces volumetric efficiency and increases pumping losses. Thus, a sophisticated control of back pressure is required.

This back-pressure control system is designed to avoid the loss of maximum power while obtaining the emissions benefits; it also minimizes back pressure in the deceleration mode to avoid excess dilution and high emissions. It senses engine operating condition and adjusts back pressure appropriately. It consists of two units:

1. The back-pressure control valve, shown in Figure C-1.
2. The sensing unit, shown in Figure C-2.

The back-pressure control valve is a vacuum-operated throttle which is located 18 inches ahead of the muffler. The plate is offset and spring-loaded to an open position. A manifold-vacuum-powered diaphragm closes the valve. During low manifold-vacuum operation, the spring is strong enough to overcome the vacuum signal, thus the throttle is open and back pressure is normal. Back pressure, during part-throttle operation, is nearly constant.

The sensing unit, fabricated by the Dole Valve Company, is a combination of a pressure-responsive bellows (1) and vacuum regulator (2). Pressure from the exhaust cross-over passage acts on the bellows and exerts a force against a calibrated spring (3) which compresses the bellows (1) if pressure is too low. The vacuum-regulating valve receives a signal from a port (4) located in the carburetor. This port is located above the throttle plate and, like the port for the distributor vacuum advance, does not sense manifold vacuum when the throttle is closed. The vacuum regulator is a combination of inlet control seat (5) and bleed valve (6). If the back pressure is low, the bellows opens the regulating valve inlet and closes the bleed valve, subjecting the back-pressure control diaphragm, through regulator outlet (7), to manifold vacuum, closing the valve and thus increasing back pressure. High back pressure expands the bellows, causing the bleed control (6) to open and the inlet (5) to close, lowering the vacuum in the regulator and allowing the valve to close. Once stabilized, the system traps the proper vacuum in the unit and back-pressure regulator diaphragm until the engine operating condition changes. This control maintains back pressure at a relatively constant level.

Figure C-1. Back Pressure Control Valve

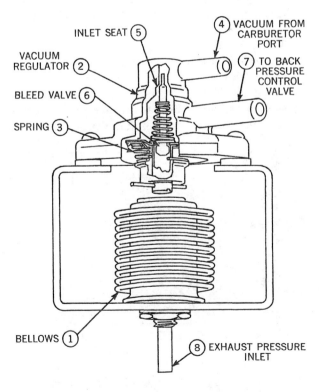

Figure C-2. Back Pressure Control Sensing Element

APPENDIX D

DECELERATION CONTROL SYSTEM

The concentration of hydrocarbons in the exhaust is highest during deceleration modes. So even though these modes are weighted much less than the acceleration modes, they contribute significantly to total vehicle emissions.

Devices designed to control deceleration emissions have been reported [6, 30]. Spark advance, manifold vacuum, fuel-air ratio, and engine operating temperature, all affect deceleration emissions.

The primary section of the three-venturi carburetor was used to conduct simulated deceleration tests with the results shown in Figure D-1. The idle was set at 500 rpm and 3.5 hp. This established the proper throttle position for the simulated test. The fuel-air ratio was set and measured. The engine was motored to higher speeds with the dynamometer. The emission levels were measured for spark advances from 6° ATC to 36° BTC. This test was repeated at different fuel-air ratios and basic ignition settings.

In contrast to the effects in power-producing modes, advancing ignition timing and increasing fuel-air ratio reduces the HC emissions at higher speeds and vacuums. Retarding ignition timing *at idle* lowers hydrocarbon emissions because it requires increased mixture flow with the throttle closed and that reduces the ratio of the dilution gas to charge. The richer mixture ratios and advanced timing reduce engine braking. Cross plots of the data, that relate the braking characteristic of the vehicle during the deceleration mode to emissions as well as ignition timing, engine speed and the amount and fuel-air ratio of the mixture, facilitate system design.

The deceleration control system, Figure D-2, increases the *mixture* flow during the high-vacuum period, retards ignition timing and increases fuel-air ratio. The mixture control valve is located in the *mixture* bypass duct of

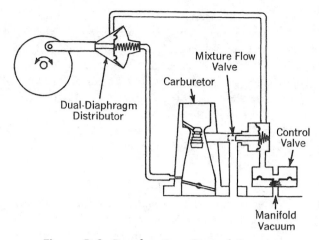

Figure D-2. Deceleration Control System

the three-venturi carburetor. The valve replaces the original idle speed adjusting screw; it is also used to adjust the curb idle speed. High manifold vacuum during closed-throttle deceleration activates a vacuum-controlled valve and opens the mixture valve to allow more mixture to enter the manifold. The choke plate of the primary carburetor is in its partially closed position, so the increased flow causes an increased pressure drop at the nozzle and mixture enrichment. The combination of increased mixture flow and richer mixture allows the ignition timing to be retarded without encountering misfire and high emissions. The vacuum-control valve admits manifold vacuum to the dual-diaphragm distributor and retards ignition timing to help restore engine braking. The system lowers deceleration mode emissions from as high as 1200 ppm to less than 300 ppm. If the retarding portion of this system is disconnected the emissions are still lower but engine braking is lost.

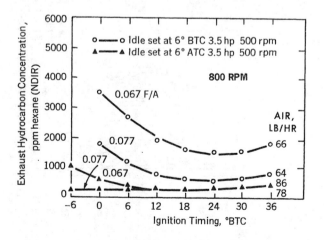

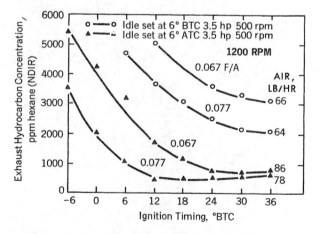

Figure D-1. Hydrocarbon Emissions During Simulated Deceleration
389 CID, V-8, Engine R
3-Venturi Carburetor

The Automobile and Air Pollution

Fred W. Bowditch
Emission Control, Engineering Staff, General Motors Corp.

AUTOMOTIVE EMISSIONS and the relationship of these emissions to ambient air pollution problems is a very complex subject. To provide the proper perspective it is necessary to have a general understanding of the different sources of emissions from the automobile, the relative levels of the emissions from these sources, the status of control systems for each of these sources, and finally, what present controls are doing for the atmospheric levels of automotive-related pollutants.

There are four sources of emissions from the automobile as shown in Fig. 1. These are the crankcase, the exhaust, the carburetor, and the fuel tank. The latter two sources are generally termed evaporative or hot soak losses, since, for the most part, these losses occur only during periods when the car is stopped. These losses amount to around 20% of all hydrocarbon losses from automobiles -- according to California Board of Health figures -- about equally divided between the carburetor and the fuel tank. Methods for controlling these losses are presently receiving considerable attention, but no systems are presently in use.

The crankcase is responsible for around another 20% of the total hydrocarbon emissions to the atmosphere. Beginning with the 1961 models in California and with 1963 models nationwide, all new cars have essentially eliminated this source of emissions by the addition of a crankcase ventilating system which returns crankcase gases to the intake system of the engine. This system is shown schematically in Fig. 2. The heart of the system is the positive crankcase valve or PCV. This valve meters the crankcase gases into the intake system as provided by carburetor calibration and therefore is carefully sized for each engine.

The exhaust accounts for the remaining approximate 60% of the automotive hydrocarbons and, incidentally, essentially all the carbon monoxide. Exhaust emission controls have been supplied in California on all new cars beginning with the 1966 models and nationwide beginning with the 1968

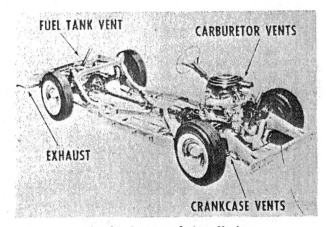

Fig. 1 - Sources of air pollution

ABSTRACT

Since the late 1940's automotive engineers and scientists have been conducting research on emission control and developing the results into practical hardware for the four sources of emission from the automobile -- the crankcase, the exhaust, the carburetor, and the fuel tank. It is estimated that 20% of all hydrocarbon losses are divided between the carburetor and the fuel tank and, at present, there is no system available for controlling these losses. The exhaust accounts for 60% of the hydrocarbons and practically all the carbon monoxide. The remaining 20% of the hydrocarbons are emissions from the crankcase. Present control systems make substantial reductions in the emissions from these two sources. Under present levels of control there is a total reduction of about 60% in both total hydrocarbon and CO emissions.

Another factor important to emission control is the need for proper maintenance. Beneficial effects of the installation of an emission control system can be negated if sufficient maintenance is not performed. Between 1968 and 1980 it is believed that research programs on gasoline engines will result in progress toward solution of the emission control problem.

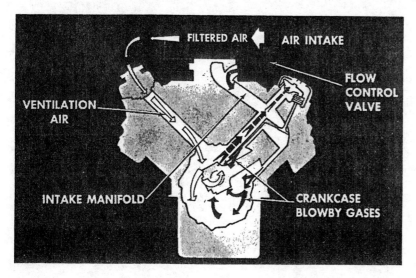

Fig. 2 - GM closed crankcase emission control system (PCV)

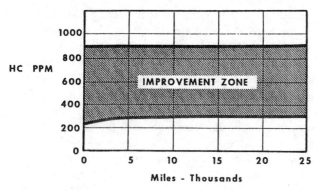

Fig. 3 - Industry improvement in emission control, hydrocarbons

models. These controls provide about 65% control of both exhaust hydrocarbons and carbon monoxide. Fig. 3 illustrates the level of exhaust hydrocarbon control. The average concentration of hydrocarbons in uncontrolled vehicles, according to a California Board of Health survey, is about 900 parts per million. Present control systems are designed to meet the federal standards of 275 ppm or a 69% reduction of exhaust levels of hydrocarbons, shown as the improvement zone in Fig. 3.

Thus, the total level of hydrocarbon control from the automobile, or the combination of crankcase and exhaust controls, provides an overall reduction of hydrocarbon emissions of about 63%.

For 1968 General Motors as well as others in the automotive industry are using two basic exhaust emission control systems, both of which have been in use in California on 1966 and 1967 cars. Fig. 4 is the air injection reactor system consisting of engine modifications and injected air as shown schematically. This system is used on Cadillacs and for all manual transmission Chevrolets.

All automatic transmission cars with the exception of Cadillac include the controlled combustion system (Fig. 5) which consists of engine modifications, carburetor and distributor changes, and an automatic temperature control of the air to maintain summer conditions at the carburetor, resulting in acceptable driveability with a leaner carburetor calibration during cold weather. To a layman, the essential difference between these systems is that A.I.R. requires an air pump while the C.C.S. has an air heater.

The exhaust control systems also serve to make substantial reductions in carbon monoxide as shown in Fig. 6. According to the same California Board of Health survey the average un-equipped car has exhaust carbon monoxide concentrations of about 3.8%. The exhaust control systems are designed to meet the federal requirements of 1.5% concentration of carbon monoxide in the exhaust, or about a 60% reduction in carbon monoxide, again shown as the improvement zone in Fig. 6.

Fig. 7 illustrates the automotive hydrocarbon control situation since the year 1940 with predictions to the year 1980. The average car with no control, up to the 1961 models, contributed about 1.25 lb per day of hydrocarbons per car. Beginning with 1961 models in California and with the 1963 models nationwide, crankcase control was added which reduced hydrocarbon emissions about 20%. Thus, each of these vehicles contributed about 1.0 lb per day to the atmosphere. Beginning with the 1966 models in California and with the 1968 models nationwide, hydrocarbon emissions were further reduced with the addition of exhaust controls so that total emissions from each of these vehicles is today about 0.46 lb per day.

Offsetting these advances in control technology is the increasing car population. Nationwide, car population has almost tripled from 1940 to the present time. We would expect this trend to continue so that by 1980 the total car population is expected to be about 110,000,000 cars or almost a four-fold increase over the 1940 vehicle population. By multiplying the factors of control levels and vehicle population, the contribution of the total vehicle population to the atmosphere can be estimated.

Such an extimate has been made for the nation and this is shown on Fig. 8. The 1940 daily total nationwide automotive hydrocarbon contribution was about 15,000 tons per

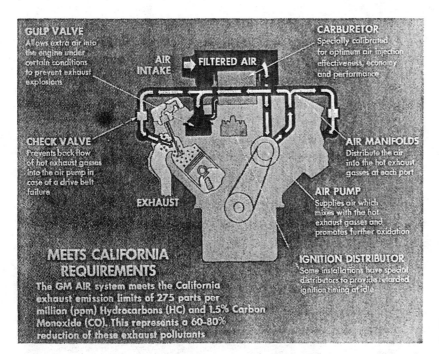

Fig. 4 - Exhaust emission control. GM air injection reactor system (A.I.R.)

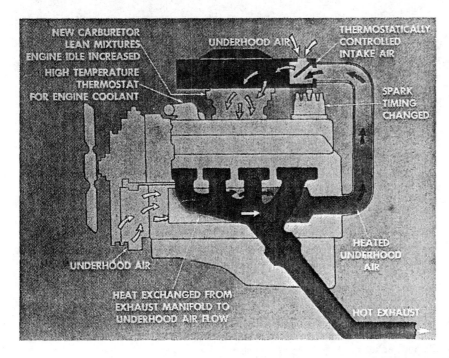

Fig. 5 - Exhaust emission control. GM controlled combustion system (C.C.S.)

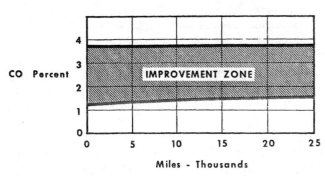

Fig. 6 - Industry improvement in emission control - carbon monoxide

day as shown on the graph. With no control measures, this level would have increased to about 110,000 tons per day by the year 1980, due primarily to the increase in car population. However, because of the hydrocarbon control systems -- both crankcase and exhaust -- total emission levels are already dropping as of today. The introduction of crankcase controls, first in California then nationwide, arrested the increasing trend in total hydrocarbon emissions and a decrease in atmospheric levels began about year-end of 1965. With no change in regulations, but because of the substitution of new equipped cars for the older unequipped cars, total emission levels by year-end 1980 will be about equivalent to the atmospheric hydrocarbon levels in 1954.

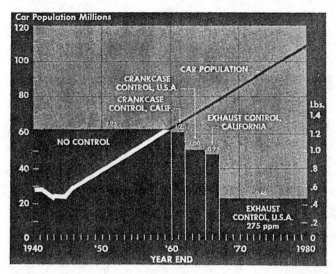

Fig. 7 - Variation of total vehicle HC emissions with model year

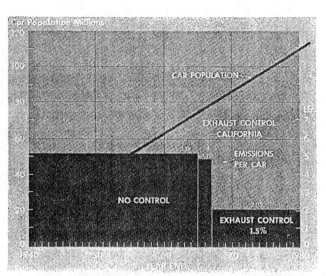

Fig. 9 - Variation of CO emissions with model year

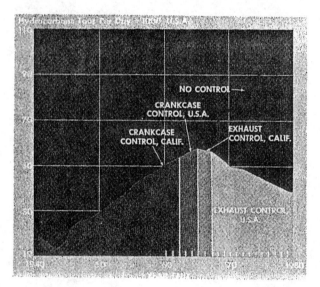

Fig. 8 - Effect of control programs on total automotive hydrocarbon contribution to the atmosphere

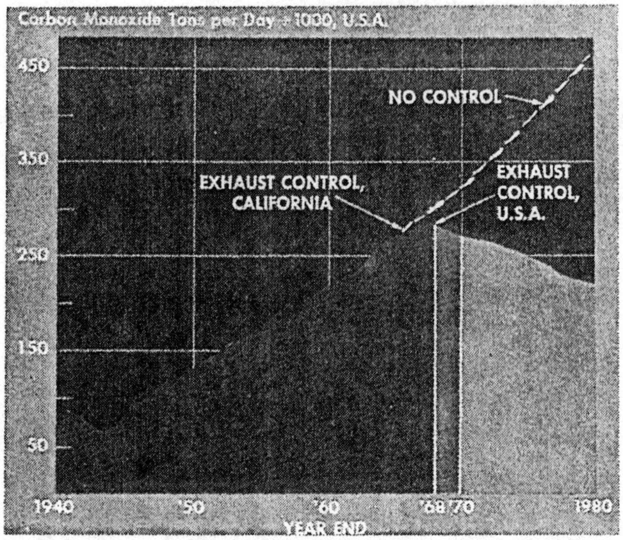

Fig. 10 - Effect of control programs on total automotive carbon monoxide contribution to the atmosphere

Similar analyses can be made of the carbon monoxide atmospheric situation as shown in Figs. 9 and 10. Thus, by 1980 carbon monoxide levels nationwide will be approximately equivalent to the average 1960 levels and reductions continue.

Much of the more recent California misunderstandings concern whether or nor control systems on the 1966 and 1967 cars are performing satisfactorily in the hands of the public. The margin of control over which the argument is centered concerns a 3% difference in passenger car control systems' effectiveness in eliminating total automotive hydrocarbon emissions; that is, a difference in exhaust concentrations of 275 ppm hydrocarbon versus 320 ppm hydrocarbon. The surveillance data accumulated by the California Automobile Club for the Motor Vehicle Control Board indicates that prior to the point of usual tune-up, 25,000 miles, the average exhaust hydrocarbon emission level is about 300 ppm. This is not quite as much as we expect to do, but we do think it is a substantial achievement. We are trying to improve as a result of what we have learned. Most of the automobile companies have set up exhaust emission laboratories both as separate laboratories as well as in-plant facilities like those shown in Fig. 11. This equipment is at least equivalent to that in the federal laboratories. These facilities provide us with information concerning how production vehicles vary from the engineering prototype vehicles. This information is termed quality audit data. It is from these data that we can say that we fully expect the 1967 California models to do better on average than the California 1966s, and the 1968 models nationwide to do better than the 1967s in California. The limited surveillance data to date supports this position at this time, and we are confident that with

Fig. 11 - In-plant facilities for exhaust emission testing

Table 1 - Exhaust Emissions with Engine Maladjustments

Typical emission control car properly adjusted

230 ppm HC   1.2% CO

Composite values after 4000 miles

Maladjustment	A.I.R. HC	A.I.R. CO	C.C.S. HC	C.C.S. CO
Intermittent miss	570	1.20	650	1.20
Low idle speed	260	1.25	280	1.20
Rich idle	230	1.80	230	1.80
Plugged PCV	280	1.55	280	1.55
Choke too rich	270	1.35	250	1.80
Advanced spark timing	280	1.20	290	1.20
A.I.R. belt off	520	3.25	---	---

sufficient data the superiority of the 1968s will be substantiated.

This quality audit data has also shown us that improvements in distributor and carburetor quality control is necessary. As a result, in the start of the 1968 model production all engines are inspected to assure proper distributor settings. Because of the particular importance of proper carburetor calibration, the automotive industry has invested millions of dollars in facilities at the carburetor plants for an audit of the flow characteristics of 100% of the carburetors to be used on 1968 models. We believe that significant additional advances in automotive emission control are being achieved through improvements in production quality control.

Still not very well understood by many people is the difficult technical job of building and maintaining cars to meet emission standards. An examination of some of the present information covering the effect of maladjustments on emission performance illustrates some of the factors affecting this problem. Table 1 itemizes this information. For this comparison a pair of cars was used which were well within the standards at 4000 miles, having emissions of 230 ppm hydrocarbons and 1.5% carbon monoxide. Both cars were equipped with a popular V8 engine and automatic transmission. One of these cars was representative of a present production air injection reactor system, and the other of a controlled combustion system. As can be seen from the tabulation, deviation from the proper setting of the various engine adjustments can create an appreciable increase in emissions. For example, with an intermittent miss, hydrocarbon levels more than doubled to 570 ppm for the A.I.R. system and 650 ppm for the C.C.S. system, but carbon monoxide was unaffected in this case. With this background on the effect of simple maladjustments, the performance of cars in the hands of the customer is of considerable interest.

Sufficient mileage has been accumulated on the 1966 models so that fairly reliable data are developing as to the major causes for the increase in emissions in the hands of the customer. Fig. 12 shows both the effect of increased mileage and of tune-up on the hydrocarbon emissions of a group of over 100 cars representative of 1966 sales in California. For the purposes of this comparison, the cars have been divided into three groups: 4000-mile group, an 8000-mile group, and an 18,000-mile group. The bar graph gives the average hydrocarbon emission data of each group of cars as received from the customer, and after restoring the car to the proper operating conditions with normal tune-up procedures. The same situation exists as far as carbon monoxide is concerned. Here again, the data in Fig. 13 show the cars as received and after the tune-up.

While the mileage at this point is low compared to the 50,000 miles durability specifications, it is significant that with proper maintenance there is practically no change in emission levels of either hydrocarbons or carbon monoxide

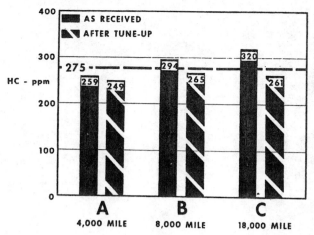

Fig. 12 - Effect of tune-up on HC exhaust emissions

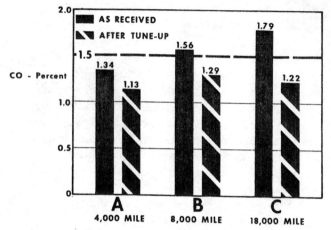

Fig. 13 - Effect of tune-up on CO exhaust emissions

with mileages from 4000 to the 18,000 mile point. The significant reduction in both hydrocarbon and carbon monoxide that occurred at the 18,000 mile level with the proper tune-up of the engine is graphic proof that a positive program of periodic maintenance is needed if the reduction of smog is going to be achieved.

These data indicate that if there is to be the control of emissions to justify the requirement for the installation of emission control systems on cars, some system of mandatory maintenance will have to be inaugurated in order to maintain acceptable emission levels in the car population.

At the present time there is no suitable test procedure and instrumentation to indicate whether the car is performing to the federal standards. This can only be determined by carefully conducted tests under controlled conditions with trained technicians and approved instrumentation. At this point the best that can be done is to follow a normal tune-up practice. The Automobile Manufacturers Association through its Vehicle Combustion Products Committee has outlined the following critical maintenance items for the two systems that are in general use at the present time; the air injection system and the engine modification system. Without going into details, the following list indicates the items to be maintained:

Air Injection Systems -
1. PCV system operational.
2. Plumbing connections tight.
3. Air pump drive belt tensioned.
4. Idle speed within limits.
5. Ignition timing within limits.
6. Idle fuel mixture within limits.

Engine Modification System -
1. PCV system operational.
2. Check for proper installation of components.
3. Vacuum distributor advance control operational.
4. Idle speed within limits.
5. Ignition timing within limits.
6. Idle fuel mixture within limits.

If this procedure is followed and normal service station diagnostic equipment used, experience has indicated that the average emission levels from the car population can be maintained within practical limits of the standards.

It has been recommended for many years that to get optimum performance from his automobile, the customer should have a tune-up at 12,000-mile intervals or annually. With the advent of exhaust emission control, the need for proper maintenance becomes a great deal more important because small factors that are not even noticeable to the customer in the majority of cases have an appreciable effect on the performance of the exhaust emission characteristics of the car. The automotive industry has stepped up to the need to modify the engines of the automobiles so that a significant reduction in air pollution would be made. The vehicle owner must be made aware of his necessary role in this program.

What of the next generation of exhaust control systems should they be required? Presently there appears to be two approaches which might be taken. Both are highly experimental at this point and neither approach at this time has the durability required regardless of expense. The first of these approaches is the use of a greatly enlarged insulated exhaust manifold in conjunction with an air pump. Research experiments with this type of system have shown exhaust control potentialities approaching 95%. However, problems of corrosion and attrition have placed severe burdens on the metallurgists for even the most exotic materials to stand up under such conditions.

The other approach is a combination of the present systems with the catalytic muffler. This approach also has a potentiality of 95% control. There are two major problems with this approach. The first is catalyst life both in terms of durability as well as attrition; neither aspect presently has an acceptable solution. The second and certainly equally important problem is that the lead additive used in present gasolines poisons most catalytic materials. The added expense of nonleaded fuels would have to be added to the hardware expense to determine total cost to the public of this control approach.

Finally, the possibility of large-scale use of alternate power sources seems quite remote at this time. While numerous power sources other than gasoline engines are being examined, we can only conclude that at the present state of the art, advanced designs of the type of gasoline-burning engine now in use offer the most practical solution for the reduction of vehicle contributions to air pollution. Reduction in vehicle emissions has become an important engine design parameter in addition to reliability, economy, and performance.

I would like to review briefly some other types of possible power sources for motor vehicles and cite some of both their strong points and weak points in relation to vehicle emission control.

Fuel injection systems, which provide measured control of fuel during engine operation, have some potential for lowering emissions but they add to the complexity of the engine, manufacturing process and maintenance procedures, as well as to greatly increased costs for the consumer; also these systems would at best provide only marginal reduction in emissions.

Stratified-charge engines, in which the fuel is injected directly into the combustion chamber, have potential for both lower hydrocarbon and carbon monoxide emissions but has not been demonstrated in actual practice. Again, this system introduces problems similar to those for fuel injection systems.

Gas turbine engines are considered to be the lowest emitter of the fuel-burning engines. Some experimental models have been developed and tested in both automobiles and heavy trucks, but some operational problems still exist. Because of the high temperatures involved, more costly materials are required than in the case of the present piston engines. This would result in higher costs for the consumer.

The low emissions of hydrocarbons and carbon monoxide from diesel engines suggests them as an appropriate automobile power source; however, problems are introduced by increased weight, noise, odor, vibration, and cost. When overloaded, improperly maintained, or improperly fueled, diesel engines also produce excess smoke and soot.

The Wankel engine, now in use in some European cars, also uses gasoline as fuel. It is a rotary rather than a reciprocating engine. Tests to date indicate that emissions would be higher than with present conventional engines.

The free piston engine has not been demonstrated as practical for broad automotive use nor is there any indication that it would produce reduced emissions.

Electric battery vehicles have the potential for eliminating the major portion of automotive emissions but present technology provides a range of only 40-80 miles before recharging is necessary. Weight and space requirements are at least double those of present powerplants. A major technical breakthrough is necessary before battery power can be considered as a practical automotive power source. It also should be pointed out that while emissions from such vehicles would be small, air pollution problems associated with the generation of electric power would undoubtedly be increased due to expanded need for power to recharge batteries.

Steam engines, which were a satisfactory source of automotive power until the early 1920s, are not now seriously considered because of weight and space requirements. They would also involve another combustion process which could introduce additional air pollution problems.

The Stirling Cycle engine would provide low emissions due to its use of a continuous combustion-type burner but in its present stage of development is not feasible for automotive use.

Fuel cells, now in use providing small amounts of auxiliary power for some space vehicles and military communication equipment, are being studied closely as a possible future source of automotive power. Major breakthroughs will be required to solve problems of weight, space, cost, and practicality before their automotive use might be demonstrated.

While the use of atomic power for automobiles has been envisioned, protection from atomic radiation would require vast amounts of lead shielding and the resulting weight problems rule out atomic automotive power at the present state of our knowledge.

The industry is continuing to examine all potential sources of power that might prove practical for motor vehicle application while curtailing or eliminating undesirable or harmful emissions. New generations of vehicle emission control equipment for present type engines are being developed at the same time that we are seeking to expand our knowledge of the basic problems.

Since the late 1940s the industry's engineers and scientists have been doing basic research on emissions and developing the results of these into practical hardware. Between now and 1980 we sincerely believe that current research and engineering development programs on our current gasoline engines will result in continued progress toward the solution of this important problem.

690503

PERFORMANCE OF A CATALYTIC CONVERTER ON NONLEADED FUEL

HAROLD W. SCHWOCHERT

Advance Product Engineering
Engineering Staff
General Motors Corporation

ABSTRACT

Catalytic converters have been combined with other emission control methods to arrive at very low emission levels. Two cars with catalytic converters successfully completed 50,000 miles on an AMA driving schedule. However, the catalyst required a nonleaded fuel for acceptable life and contained a noble metal which makes it unsuitable for large volume usage.

The search for a physically durable, chemically active, and economical catalyst, which will operate with leaded fuel, continues. In addition, durability of the catalytic converter system under the wide variety of operating conditions in customer service must be established.

INTRODUCTION

Catalytic converters have been investigated as a method of controlling automotive exhaust emissions for more than a decade. They were found to be very effective in reducing hydrocarbon and carbon monoxide emissions when new. However, the investigations also revealed a rapid decrease in catalyst activity due to poisoning by tetraethyl lead in the fuel.([1])*

Bishop and Nebel([2]) of General Motors Research Laboratories presented an SAE paper on the evaluation of a catalyst in 1959. The hydrocarbon reduction efficiency was evaluated based on a standard driving cycle which was more severe than the present exhaust emission test schedule. Average HC reduction by the catalyst had decreased from greater than 80% when new to 36% after 15,300 miles. Carbon monoxide reduction was not as severely affected by mileage accumulation.

Homfeld, Johnson and Kolbe([3]) of General Motors Engineering Staff presented an SAE paper on the development of a catalytic converter system in 1962. Again hydrocarbon control decreased rapidly with mileage accumulation. Hydrocarbon reduction as measured on the California dynamometer cycle had decreased from 75% with new catalyst to 36% with 4,000 catalyst miles. Carbon monoxide reduction was about 90% with new catalyst and decreased about one-half as much as the hydrocarbon reduction.

The effect of tetraethyl lead in commercial gasoline on catalyst activity promoted development of other techniques for emission control on production automotive vehicles. However, proposed standards for the future may exceed the ultimate capabilities of the control systems presently being used.

The catalytic converter has the potential to control automotive exhaust emissions to very low levels when combined with present emission controls. It is, therefore, being actively pursued along with other control systems such as the exhaust manifold reactor. Major incentives for catalytic converters are:

1. Design freedom in the shape and location of the converter
2. Little or no fuel economy or performance loss
3. No effect on vehicle driveability

Other incentives in comparison to exhaust manifold reactors are:

4. Lower operating temperatures
5. Minimum of expensive metals required

The major problems associated with catalytic converters have been the sensitivity of catalyst to tetraethyl lead in the fuel, use of precious metals in the catalyst, susceptibility of the catalyst to high temperature damage and the loss of catalyst due to attrition.

The objective of this evaluation was to investigate catalytic converter effectiveness and durability using nonleaded fuel and presently available catalysts and converters. Catalytic converters were combined with current production emission control methods to arrive at very low hydrocarbon and carbon monoxide levels.

The catalyst used in this evaluation contained a noble metal, a factor that makes it unsuitable for large volume usage. However, our goal was to determine the long term catalytic converter performance under favorable conditions; that is, with a nonleaded fuel and on a continuous light duty schedule on paved roadways. In addition, we hoped to stimulate interest in this approach to exhaust emission control.

Several catalyst-converter systems failed prior to arriving at the system combination that is the subject of this report. This paper gives a general description of that system, presents the test information and discusses the conditions of the evaluation.

---

*Numbers in parentheses designate references at end of paper.

DISCUSSION

DESCRIPTION OF SYSTEMS

Car A:

Car A, with a 283 in^3 V8 engine and a two speed automatic transmission, was modified by installing a carburetor tailored to 1968 specifications for vehicles equipped with controlled combustion exhaust emission control systems (CCS) and a carburetor air preheater. Carburetor calibration was not changed throughout the test interval. Typical air-fuel ratios as determined by exhaust gas analysis on the dynamometer are given in Appendix A.

A vane type air pump of 19.3 in^3 displacement was used to supply additional secondary air for oxidation in the catalytic converter. The air was introduced about two feet upstream of the converter. The converter was located in the standard muffler location (Figure 1).

Figure 1 — Catalytic converter was installed on Car A in the production muffler location.

Car B:

Car B had a production 250 in^3 six cylinder engine and a three speed manual transmission. This vehicle was equipped with an air injection reactor (AIR) system, the type produced for sale in California in 1966 and 1967, in addition to the catalytic converter. Air-fuel ratios during the durability schedule as determined by exhaust gas analysis are given in Appendix A.

The AIR system used a vane type air pump of 19.3 in^3 displacement and introduced the secondary air at the exhaust ports. A gulp valve that inducted air into the intake system upon a sudden throttle closing, such as a deceleration, was also used.

The catalytic converter was installed about three feet forward of the standard muffler location. The exhaust pipe was insulated with 3/4" of a fibrous ceramic insulation held in place by aluminum tape (Figure 2). The converter was moved forward and the pipe insulated to improve the warm-up and increase the operating temperature of the catalyst on the exhaust emission test schedule. Both of these factors improve catalytic converter effectiveness.

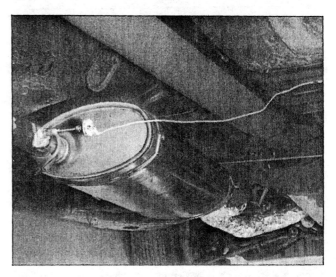

Figure 2 — The exhaust pipe of Car B was insulated and the catalytic converter was installed three feet forward of the original muffler location. Both factors improve catalytic converter effectiveness.

Converter:

The catalytic converter used in both cars was constructed of 430 stainless steel with inlet and outlet connections similar to a production muffler. The unit was of a downflow design as shown in Figure 3. The tapered inlet plenum was designed to program the flow to insure good flow distribution through the bed. The exterior of the converter was wrapped with 1/4" of a fibrous ceramic insulation as shown in the figure.

The catalyst bed volume was about 400 cubic inches and contained 4.6 pounds of catalyst. The converter had dimensions of 5-5/8" minor axis, 10" major axis by 23" long. The catalyst and container weight was 27.1 pounds.

Catalyst:

The complete chemical composition of the catalyst was not provided by the catalyst supplier. However, one of the catalytically active constituents was a precious metal. The catalyst was spherical with a nominal 1/8 inch diameter.

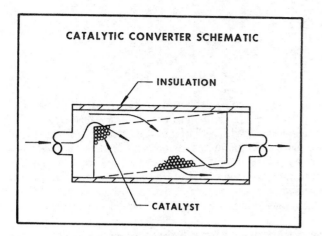

Figure 3 — The converter was of a down flow type design. A tapered inlet plenum provides good flow distribution through the bed. Insulation of the converter exterior improves the catalyst effectiveness.

CONVERTER PERFORMANCE ON AMA SCHEDULE

The driving schedule selected for mileage accumulation was the modified AMA city driving route run on the GM-Proving Ground oval (Appendix B). Conditions favorable for long catalyst life, such as light duty driving and smooth roads are inherent to this schedule.

Since this schedule is consistent with the mileage accumulation procedure specified in the Federal Register[4], the criteria for vehicle maintenance were also followed. Major tune-ups were performed prior to starting the tests and after 25,000 miles. Fuel filters and PCV valves were replaced and the PCV system was serviced at 12,000 mile intervals. Normal vehicle lubrication and other preventive maintenance services were performed at manufacturers recommended intervals.

Duplicate exhaust emission tests were run at the beginning of the schedule and after 50,000 miles, and single tests at 2500-3000 mile intervals. Tests also preceded and followed the major tune-up.

Exhaust emission tests were conducted in accordance with the Federal Register[4] except Indolene Clear test fuel was used instead of Indolene 30 (See Appendix C).

Exhaust gases were measured with nondispersive infrared analyzers. The hydrocarbon analyzers were sensitized with normal hexane. Calculations were made according to the procedure specified in the Federal Register[4].

A commercial nonleaded fuel was used for mileage accumulation (Appendix C). This fuel will be referred to as nonleaded, however, trace values of TEL of about 0.05 cc/gal were present in the fuel as received from the manufacturer.

Car A:

Commercial leaded gasoline was inadvertently added to this vehicle after about 38,500 AMA durability miles. A lead analysis of vehicle tank fuel at that time showed a TEL content of 0.5cc per gallon.

The immediate effect of this leaded fuel addition increased hydrocarbon emission 56 PPM (59%). This large increase was only temporary. However, there did appear to be a long term effect. Figure 4 shows that hydrocarbon emission was stabilized at about 97 PPM for some 30,000 miles prior to the leaded fuel error and appeared to stabilize again in the final 11,500 miles of the AMA schedule at a higher level of 107 PPM. The effect of this leaded fuel addition on hydrocarbon emission was more significant during cycles 6 and 7 of the standard exhaust emission test. The stabilized level increased from about 40 to 60 PPM HC during the hot cycles. Figure 5 shows that carbon monoxide emission was not affected.

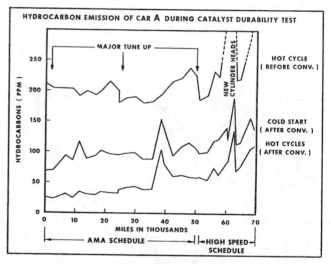

Figure 4 — Hydrocarbon exhaust emission of Car A increased steadily throughout the 69,000 test miles. The effect of an inadvertent addition of leaded fuel at 38,000 miles and of exhaust valve seat wear at 62,000 miles and at the end of the test is evident by the large increases in hydrocarbon emission.

The Federal Register[5] states that emissions at 50,000 miles will be determined by applying a best fit straight line to all applicable data points. For this report, a best fit straight line was defined as a least-squares straight line fit to all data between zero and 50,000 miles.

This technique predicts that emissions after 50,000 miles would be 117 PPM HC and 0.84% CO. Emission results for the most important test points during the AMA schedule are shown in Table 1.

One method of determining the effectiveness of a catalytic converter system is to compare the amount of pollutants oxidized by the converter to the amount entering it. This ratio expressed as a percent will hereafter be referred to as

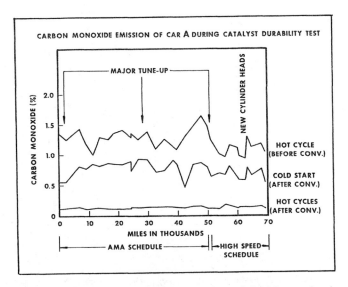

Figure 5 — Carbon monoxide emission of Car A remained relatively stable throughout the 69,000 test miles.

TABLE 1
Exhaust Emissions of Car A on AMA Schedule

	Hydrocarbons		Carbon Monoxide	
	Cold Start	Cycles 6 & 7	Cold Start	Cycles 6 & 7
Start AMA Durability*	68	26	0.56	0.12
Before inadvertent addition of Leaded Fuel (35,565 miles)	89	40	0.74	0.14
After 50,000 miles*	107	58	0.79	0.12
Projected at 50,000 miles (Best straight line fit)	117	69	0.84	0.14
*Average of two tests				

the conversion efficiency. Since only one cold start emission test was run at each mileage test point and emissions were measured after the converter, cold start conversion efficiencies are not known throughout the test. However, Table 2 shows how they decreased during the 50,000 miles of driving as determined at the start and completion of the AMA schedule.

TABLE 2
Cold Start Conversion Efficiencies for Car A on AMA Schedule

	Hydrocarbons		Carbon Monoxide	
	Cold Start	Cycles 6 & 7	Cold Start	Cycles 6 & 7
Start AMA Durability	69%	88%	68%	91%
After 50,000 miles	55%	74%	58%	92%

The hot cycle hydrocarbon conversion efficiency for the 50,000 AMA miles is shown in Figure 6. A best fit straight line to these data yields a 90% initial conversion efficiency and a decrease to 67% after 50,000 miles. An extension of this technique suggests that the hot cycle conversion efficiency would reach a level of 50% after 87,000 miles. One investigator[1] has spoken about the half life of a catalyst which in this evaluation would be predicted to be 98,000 miles.

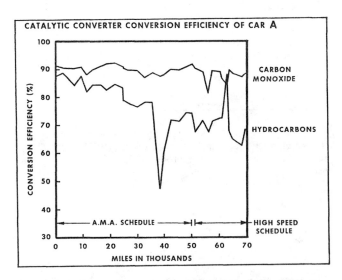

Figure 6 — The hot cycle hydrocarbon conversion efficiency of Car A decreased steadily with mileage. However, a best fit straight line to these data points predicts that a conversion efficiency greater than 50% would still be obtained after 100,000 miles. The hot cycle carbon monoxide conversion efficiency did not deteriorate.

Figure 6 also shows the carbon monoxide hot cycle conversion efficiency for the AMA mileage. Variation from one test interval to another is apparent, however, no trends are evident. A conversion efficiency of about 90% is indicated by this figure.

Figure 5 shows that the measured hot cycle CO emission for this vehicle varied between 0.1 and 0.15%. For a normal (0-10%) carbon monoxide analyzer calibration, as suggested by the Federal Register[4], the observed CO levels represent 1% of full scale. Since these levels are about the same magnitude as the accuracy of the equipment, the conversion efficiency might be represented by a band perhaps between 100 and 80%.

Figures 4 and 5 show that hot cycle emissions before the converter system were very stable throughout the 50,000 miles.

The warm-up characteristics of the catalytic converter of Car A at the start of the AMA schedule are shown in Figure 7. The importance of warm-up on the present schedule is indicated here. The change in slope of the temperature profile indicates that the catalytic reaction became exothermic between the 30 MPH modes of the second and third cycle.

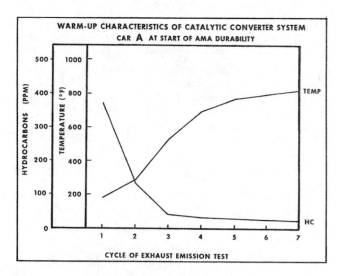

Figure 7 – Exhaust gas temperature measurements at the converter outlet of Car A during the 30 MPH mode of the exhaust emission test show that the catalytic reaction becomes exothermic between the second and third cycles. Hydrocarbon emission has nearly stabilized by the third cycle.

Car B:

The exhaust emission performance of the catalytic converter system on this vehicle is shown in Figures 8 and 9. As Figure 8 shows, the hydrocarbon emission levels into the converter system increased during the test interval. A periodic emission test at 43,457 miles revealed that the engine was missing. Hot cycle HC emission into the converter was 1047 PPM, however, even with this high input level tailpipe emission was 57 PPM. New spark plugs were installed and emission levels returned to previous values.

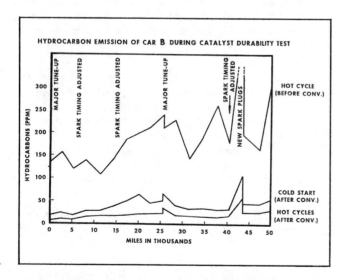

Figure 8 – Cold start hydrocarbon exhaust emission of Car B was below 50 PPM during most of the AMA schedule. Hydrocarbon emission before the converter varied considerably primarily due to poor spark (high spark plug wire resistance).

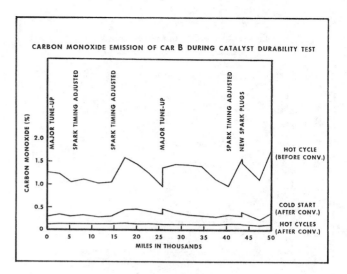

Figure 9 – Cold start carbon monoxide emission of Car B was below 0.5% throughout the 50,000 mile AMA schedule.

Near the completion of 50,000 miles, the HC emission level again was increasing and an occasional engine miss could be detected especially on a light to moderate crowd at low engine speeds. An investigation after 50,000 miles showed several spark plug lead wires to have excessive resistance. Replacement of the wires brought HC emission into the converter back to previous levels, however, tailpipe emissions were not affected as results of 55 PPM and 0.35% show.

Table 3 shows emission levels initially, after 50,000 miles and projected values at 50,000 miles using a best fit straight line to the above data points.

TABLE 3
Exhaust Emissions of Car B on AMA Schedule

	Hydrocarbons		Carbon Monoxide	
	Cold Start	Cycles 6 & 7	Cold Start	Cycles 6 & 7
Start AMA Durability*	18	6	0.30	0.12
After 50,000 miles*	55	32	0.39	0.13
Projected - 50,000 miles	59	32	0.36	0.13
*Average of two tests				

Cold start conversion efficiencies again decreased during the 50,000 miles as shown in Table 4.

Figure 10 shows the hot cycle conversion efficiencies over the 50,000 AMA miles. Little change in conversion efficiency is apparent. A discussion of half life is not in order for this reason. The high conversion efficiencies for both cold start and hot cycles after 50,000 miles suggests that

TABLE 4
Cold Start Conversion Efficiencies for Car B on AMA Schedule

	Hydrocarbons		Carbon Monoxide	
	Cold Start	Cycles 6 & 7	Cold Start	Cycles 6 & 7
Start AMA Durability	88%	96%	80%	90%
After 50,000 miles	69%	90%	70%	93%

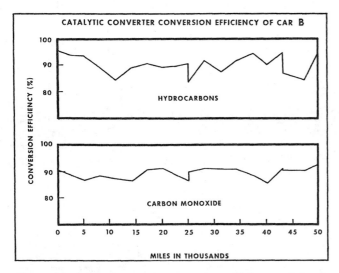

Figure 10 — The hot cycle conversion efficiency of Car B did not deteriorate. Both hydrocarbon and carbon monoxide conversion efficiencies were about 90% after 50,000 AMA miles.

low emission levels could still be expected after an extended mileage interval.

A visual check of the catalyst and container of both vehicles after 50,000 miles showed no loss of catalyst and no exterior evidence of container failure.

CONVERTER PERFORMANCE ON HIGH SPEED SCHEDULE

Upon completion of 50,000 AMA miles with two vehicles without excessive deterioration of catalytic converter performance, the effect of a higher speed schedule and rough highway conditions was questioned. Car A was chosen for mileage accumulation on a high speed schedule that included Belgian block road conditions. A tabulation summarizing this schedule is contained in Appendix D. Most accelerations in this schedule are made at wide open throttle.

As will be discussed in a later section, exhaust valve leakage due to valve seat wear became a problem. Hydrocarbon exhaust emission began to fluctuate considerably as shown in Figure 4. After 11,599 high speed durability miles, HC emission entering the converter was 1188 PPM. New cylinder heads were installed. However, after 18,265 test miles, excessive exhaust valve seat wear was again realized and the test was terminated.

Carbon monoxide emission was not affected by the exhaust valve seat wear, therefore good CO emission control was maintained throughout this high speed durability test (Figure 5). Cold start CO emission was lower during this test than during the AMA schedule, a factor which was related to carburetor choke variations. To eliminate one variable, choke adjustments were not made throughout the test period since periodic checks found them to be within manufacturers' specifications.

To determine the final effectiveness of the converter system, new cylinder heads were installed after the durability schedule was discontinued. Table 5 shows that cold start HC conversion efficiencies again decreased during this 18,265 miles of driving. Carbon monoxide emission was not affected.

TABLE 5
Cold Start Conversion Efficiencies for Car A on High Speed Schedule

	Hydrocarbons		Carbon Monoxide	
	Cold Start	Cycles 6 & 7	Cold Start	Cycles 6 & 7
Start High Speed Schedule	52%	68%	55%	90%
After 18,265 Miles of High Speed Schedule	39%	59%	57%	87%

The hot cycle conversion efficiencies for this schedule in addition to the AMA schedule are shown in Figure 6. The best fit straight line was applied to all HC test data except the high conversion efficiency points during the high speed schedule when exhaust valve leakage was apparent. Projection of this straight line indicates that more than 50% hot cycle conversion would still be obtained at 100,000 miles. A deterioration trend in the carbon monoxide conversion efficiency is not apparent.

EVALUATION OF CONTROL SYSTEM VARIABLES

Emission levels of Car B were about one-half those of Car A. Two distinct differences existed between the two systems: (1) Insulation of the exhaust pipe, and (2) Secondary air injection location. To understand the significance of these variables, additional tests were conducted on Car A. The tests were run after 69,384 catalyst miles and with new cylinder heads. Apparently due to the new heads, HC emission before the converter was lower than during the durability schedule (about 160 PPM).

Insulation of Exhaust Pipe:

The exhaust pipe of Car A was wrapped with 3/4" of fibrous ceramic insulation and aluminum foil to correspond to the system on Car B. Exhaust emissions with and without insulation are shown in Table 6.

TABLE 6
Effect of Exhaust Pipe Insulation and
Secondary Air Injection Location on Exhaust Emissions of Car A

	Hydrocarbons		Carbon Monoxide	
	Cold Start	Cycles 6 & 7	Cold Start	Cycles 6 & 7
Without Insulation	104 PPM	64 PPM	0.59%	0.14%
With Insulation	86	49	0.58%	0.14%
With Insulation and Secondary Air Injected at Exhaust Valve Ports	64	41	0.38%	0.12%

Hydrocarbon emission decreased nearly proportionately in both the cold start and hot cycles. Carbon monoxide was not affected.

Location of Secondary Air Injection:

This variable was evaluated with the exhaust pipe insulated as described. Air injection location was changed to the exhaust ports (as with production AIR systems) and backfire control was obtained with a gulp valve. Carburetion was not changed.

Exhaust emission results were improved as shown in the above tabulation. Cold start emissions, both HC and CO, were affected more than the hot cycles. This is due to oxidation of both carbon monoxide and hydrocarbons in the exhaust port location early in the exhaust emission test schedule before the catalytic converter becomes effective. The sensible heat of this reaction also improves warm-up of the catalyst.

This change in location of secondary air injection decreased hot cycle HC emission at the converter inlet by 23% and CO by 31%.

EFFECT OF CONVERTER ON PERFORMANCE AND ECONOMY

The effect of the emission control system of Car A on performance and economy was evaluated after 69,571 catalyst miles. Tests were conducted on this vehicle with the system as described and then with the catalytic converter replaced by a standard exhaust system and the secondary air pump removed.

Table 7 shows a 3% fuel economy loss in both the simulated city and highway schedules due to the catalytic converter system. The performance loss on a 0-60 mph acceleration due to this system was also 3%.

TABLE 7
Effect of Catalytic Converter System on Fuel Economy
and Performance of Car A

	With Standard Exhaust Without Secondary Air	With Catalytic Converter System
Economy, MPG		
City	14.3	13.9
Highway	16.1	15.6
30, 50, 70 mph	22.4, 19.6, 16.0	21.2, 19.0, 15.4
Performance, seconds		
0-60 mph	15.4	15.9
0-1/4 mile	20.5	20.7

The major reasons for the differences in performance and economy probably were: (1) An increase in the exhaust back pressure due to the catalytic converter and the injection of secondary air, and (2) The power requirements of the secondary air pump.

EXHAUST VALVE POUND-IN

To determine if exhaust valves and exhaust valve seats were affected by operation on nonleaded fuel, valve stem heights were monitored about every 10,000 miles.

AMA Schedule:

Exhaust valve pound-in was not a problem during this low speed driving schedule. Valve pound-in on Car A is shown in Figure 11. Pound-in on Car B was similar but slightly higher over the same mileage interval. Intake valves on both cars showed average pound-in of less than 0.003".

High Speed Schedule:

Since no exhaust valve problems were experienced in the light duty schedule, valve stem heights were measured for the first time after 11,599 miles of the high speed schedule. Figure 11 shows that five valves had pounded in more than 0.050".

New cylinder heads were installed at this time, and the test was continued. Exhaust valve seat wear was monitored at each emission test interval and gradual pound-in was observed. Again, five exhaust valves had pounded in more than 0.050" in about 5,500 miles. The test was terminated after 18,265 miles of the high speed durability schedule.

It became evident that the exhaust valves and seats were sensitive to speed and load conditions operating on nonleaded fuel. The objective of this evaluation was a catalyst

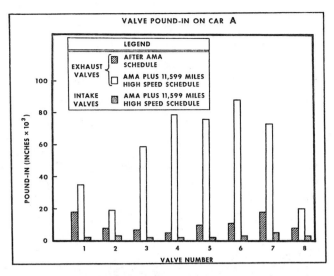

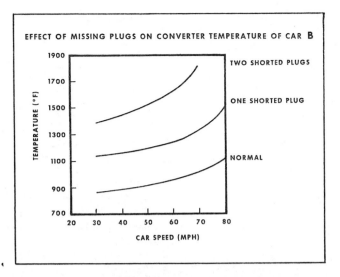

Figure 11 — Intake valves and seats were not affected by operation on nonleaded fuel. However, exhaust valve seat wear was catastrophic on a severe schedule. Exhaust valve seat wear was not a problem during light duty operation such as the AMA schedule.

Figure 12 — Maximum exhaust gas temperatures at the catalytic converter outlet during steady speed operation of Car B were: 1110°F at 80 MPH with normal ignition, 1520°F at 81 MPH (WOT) with one misfiring spark plug and 1810°F at 69 MPH (WOT) with two misfiring spark plugs.

study and not a study of exhaust valve seat wear, therefore, it was discontinued. However, it does point out that this problem exists and its significance must be recognized and evaluated.

EFFECT OF MALFUNCTIONING ENGINE ON CONVERTER TEMPERATURE

Thermal damage of the catalyst in a converter system must be of constant concern. A malfunctioning carburetor or a fouled spark plug can supply a large amount of unburned fuel that will be oxidized by the catalyst bed. Since the thermal characteristics of a catalytic converter are dependent upon the overall vehicle system, only a limited number of observations were made on Car B.

Exhaust gas temperature was measured at the converter outlet since provisions were not made for bed temperature measurements. However, temperature measurements on other systems have shown that temperature gradients in the bed of 100-150°F are possible dependent upon the vehicle operating mode. Outlet gas temperatures have also been observed to be as much as 100°F lower than bed temperatures. Therefore, catalyst bed temperatures probably were higher than those reported in the following discussion.

Figure 12 shows stabilized exhaust gas temperatures during vehicle steady speeds with normal ignition and one and two shorted spark plugs. A maximum temperature of 1520°F was observed at full throttle with one shorted plug. Two shorted plugs resulted in a maximum temperature of 1810°F, again at full throttle. The vehicle was operated at each condition just long enough (about 15 minutes) to insure a stabilized temperature had been reached.

An exhaust emission test after the temperature measurements with one shorted plug showed that the short term high temperatures did not affect the catalyst performance. Emissions of 53 PPM HC and 0.40% CO were observed. A hot cycle exhaust emission test with one shorted plug showed good emission control was obtained under an engine malfunction condition. Emissions were 60 PPM HC and 0.11% CO.

The short term effect of two shorted plugs on exhaust emissions was measurable. Cold start emission test results were 67 PPM HC and 0.55% CO after this malfunction test.

Recommended temperature limits of the catalyst by the supplier were 1700°F under continuous operation and up to about 1850°F for short periods of time on the order of 5 minutes or less.

Driving schedules that include accelerations and decelerations could cause higher temperatures than those reported for steady speed operation. However, tests on this system indicated that a large amount of malfunction was required to cause thermal damage to this catalyst.

SUMMARY

1. Two cars equipped with catalytic converters using a noble metal catalyst and operating on a nonleaded fuel successfully completed 50,000 miles on the AMA durability schedule. Cold start conversion efficiencies after 50,000 miles were: HC 55%, CO 58% for Car A; and HC 69%, CO 70% for Car B.

2. Car A operated an additional 18,265 miles on a high speed schedule with about the same low deterioration rate observed on the lower speed schedule.

3. The catalyst used in this evaluation contained a noble metal. This is not practical for large volume usage because of the high cost and limited supply of noble metals. A catalyst is needed that does not use precious metals.

4. Insulation of the exhaust pipe and location of secondary air injection both affect converter performance. Insulation of the exhaust pipe decreased cold start HC emission 17%, but did not affect CO emission of Car A. Air injection at the exhaust ports plus insulation decreased HC 38% and CO 34%.

5. The loss in simulated city and highway fuel economies due to the catalytic converter system was 3%. This system also increased the 0-60 MPH vehicle acceleration time by 3%.

6. Extreme exhaust valve seat wear was experienced on two sets of cylinder head assemblies of Car A when operated on a high speed schedule. The effect of nonleaded fuel on exhaust valve and seat life must be more definitely determined.

7. A shorted spark plug did not result in excessive exhaust gas temperatures during steady speed operation of Car B. However, more work will be required in this area once a commercially feasible catalytic system is available. Driving schedules other than steady speed operation will need evaluation.

APPENDIX A

CARBURETOR CALIBRATIONS

MODE	AIR-FUEL RATIO	
(Vehicle Speed)	Car A	Car B
Idle	13.3	11.9
15	14.4	12.3
20	14.8	12.1
25	14.6	11.9
30	15.1	12.1
40	15.0	12.8
50	14.5	13.7
0-25	14.0	13.5
15-30	14.4	14.2

APPENDIX B

MODIFIED AMA CITY DRIVING SCHEDULE RUN ON GM PROVING GROUND OVAL

This schedule consists basically of 11 laps of a 3.7 mile course. The base vehicle speed for each lap is given below:

Lap	Base Vehicle Speed, MPH
1	45
2	35
3	45
4	45
5	40
6	35
7	40
8	50
9	40
10	55
11	70

During each of the first 9 laps, there are 4 stops with 15 seconds idle. Normal accelerations and decelerations are used. In addition, there are 5 light decelerations each lap from the base speed to 20 MPH followed by light accelerations to the base speed.

Lap 10 is run at a constant vehicle speed of 55 MPH.

Lap 11 is begun with a wide open throttle acceleration from stop to 70 MPH. A normal deceleration to idle followed by a second wide open throttle acceleration to 70 MPH occurs at the mid-point of the lap.

APPENDIX C

PROPERTIES OF TEST FUELS

	INDOLENE CLEAR	COMMERCIAL NONLEADED
Pb (GM/GAL)	0.06	0.05
DISTILLATION RANGE		
IBP, °F	93	83
10% POINT, °F	140	114
50% POINT, °F	221	223
90% POINT, °F	318	312
EP, °F	410	390
SULFUR (%)	0.03	0.02
RVP	7.8	11.8
HYDROCARBON COMPOSITION		
Paraffins (%)	66	64
Olefins (%)	4	2
Aromatics (%)	30	34

APPENDIX D

HIGH SPEED DURABILITY SCHEDULE

The four types of driving conditions and the percent of distance for each condition are shown below:

HIGH SPEED

60 MPH	2%
70 MPH	10%
80 MPH	15%
90 MPH	22%

HILLS

7.2% Hill	6%
11% Hill	2%
Hill Route	14%

BELGIAN BLOCKS	5%
GENERAL (TYPICAL) DRIVING	24%

Most accelerations are wide open throttle.

REFERENCES

1. E. E. Weaver, Ford Motor Co., "Effects of Tetraethyl Lead on Catalyst Life and Efficiency in Customer Type Vehicle Operation," Paper No. 690016, presented at the SAE Annual Meeting, Detroit, Michigan, January, 1969.

2. G. J. Nebel and R. W. Bishop, General motors Coporation, "Catalytic Oxidation of Automobile Exhaust Gases -- An Evaluation of the Houdry Catalyst," Paper No. 29R presented at the SAE Annual Meeting, Detroit, Michigan, January, 1959.

3. M. F. Homfeld, R. S. Johnson and W. H. Kolbe, General Motors Corporation, "The General Motors Catalytic Converter," Paper No. 486D, presented during SAE Automobile Week, March, 1962.

4. Department of Health, Education, and Welfare, "Control of Air Pollution from New Motor Vehicles and New Motor Vehicle Engines," Federal Register, Volume 31, Number 61, Washington, D.C., March 30, 1966.

5. Department of Health, Education and Welfare, "Control of Air Pollution from New Motor Vehicles and New Motor Vehicle Engines," Federal Register, Volume 33, Number 108, Washington, D.C., June 4, 1968.

700151

The Chrysler Cleaner Air System for 1970

R. E. Goodwillie
Prod. Dev. Gen. Supvr., Engines

N. M. Jacob
Prod. Dev. Gen. Supvr., Fuel Systems

E. W. Beckman
Staff Engr., Emission Control Dev.

Engineering Office
Product Planning and Development Staff,
Chrysler Corporation

ABSTRACT

For the 1970 model passenger cars and light trucks, the Federal and California exhaust emission standards were reduced to 2.2 grams per mile hydrocarbon and 23 grams per mile carbon monoxide. This represents a reduction of approximately 33%.

This paper presents the development of the significant features of the Chrysler Cleaner Air System of exhaust emission control for 1970. Included in this development are modifications in the engine, carburetor, ignition system, and other related components. Through these modifications and by optimizing the calibration of the engine operating parameters, the new exhaust emission standards were successfully met.

BACKGROUND

The most recent statistics published by the Los Angeles County Air Pollution Control District[1]* indicate that motor vehicles contribute 68% of the hydrocarbons and oxides of nitrogen, and 98% of the carbon monoxide emitted to the Los Angeles atmosphere. These levels point out the singular importance of automotive emissions in this geographical area where the incidence of photochemical smog is many times that of other major urban areas in this country.[2]

The installation of controls on new cars and light trucks, starting on 1961 models with crankcase emission control, and 1966 models with exhaust controls, along with crankcase controls on used cars, has accounted for a downturn in total tonnage of pollutants emitted from motor vehicles into the California atmosphere.[3] This has occurred in spite of a continual increase in motor vehicle population and vehicle use during these years.

The application of the more stringent exhaust controls on 1970 models, along with the use of evaporative control systems, will accelerate this downward trend in atmospheric pollutant levels as shown in Figure 1.[4]

The automotive pollution curve for areas outside of California is similar to Figure 1, but the downward trend occurs somewhat later due to the two year timespan in the application of the emission control systems on a nationwide basis (1963 for crankcase control, and 1968 for exhaust).

The California and Federal regulations have also specified reductions in carbon monoxide levels. Although carbon monoxide does not contribute to photochemical smog, control of this material was deemed necessary. The exhaust control systems have caused a substantial drop in carbon monoxide emissions from new cars, about 70% reduction over the past decade.[5]

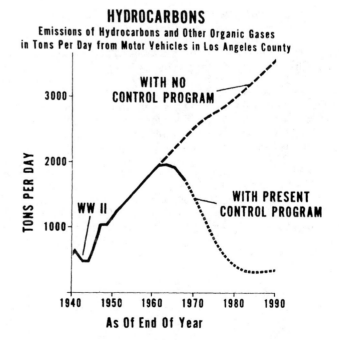

Figure 1

*The numbers refer to references listed in the Bibliography.

The vehicle emission limits that apply to 1970 model passenger cars and light trucks are shown below:[6, 7]

Source	Emission	Emission Limit
Crankcase	Blowby	None Permitted
Exhaust	HC	2.2 Grams per mile
	CO	23 Grams per mile
Evaporation (Calif. Only)	HC	6 Grams per test

The closed crankcase ventilation system, shown in Figure 2, has been used for several years and is also employed on the 1970 models. The inlet and outlet passages to the crankcase are connected to the engine induction system. Therefore, the blowby gases and ventilation air leaving the crankcase are returned to the cylinders to be burned. Thus, the closed crankcase ventilation system satisfactorily prevents emissions of engine blowby to the atmosphere.

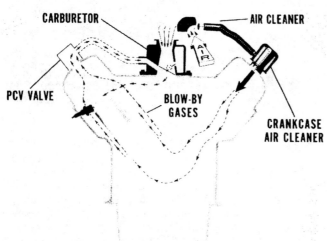

CLOSED CRANKCASE VENTILATION SYSTEM
Figure 2

Evaporative controls are being applied for the first time on 1970 model passenger cars and light trucks to be sold in California. These controls will be used nationwide on the 1971 models. The system used on Chrysler Corporation vehicles is identified as "Vapor Saver." It utilizes the engine crankcase for the temporary storage of fuel vapors emitted from the carburetor and fuel tank. These vapors are subsequently burned in the cylinders when the engine is running. The concept and development of this system is described in a companion technical paper.[8]

The above exhaust emission standards apply nationwide and are new for 1970. They represent the first reduction in emission limits since controls were first required on the 1966 models sold in California and on 1968 models nationwide. The new standards are expressed in grams per mile and are intended to limit all cars to the same weight of emissions per mile as compared to the earlier standards which were specified in concentration of pollutant (275 ppm hydrocarbons and 1.5% carbon monoxide). For the typical car produced by Chrysler Corporation, the new standards are equivalent to 180 ppm hydrocarbons and 1.00% carbon monoxide. On the heavier vehicles, however, the emission limit ranged down to 147 ppm hydrocarbons, and 0.82% carbon monoxide. These emission levels refer to a composite of emission measurements made during seven different driving modes of a test cycle that is repeated seven times.

THE CHRYSLER CLEANER AIR SYSTEM

The approach that Chrysler Corporation has taken to reduce exhaust emissions, has been to exploit all readily available engine modifications to improve combustion efficiency instead of resorting to the use of expensive and complicated add-on devices which can result in a reduction in engine reliability. A wide range of factors affecting the burning of the fuel-air mixtures has been examined to determine how improvements could be made to obtain more complete combustion and, hence, lower hydrocarbon and carbon monoxide emissions. These factors include the various components of the induction system especially the carburetor, the combustion chamber, the ignition system, and the engine factors that affect the oxidation occurring in the exhaust gas after the normal flame front has passed through.

This paper presents the development of the exhaust emission controls that are incorporated in the 1970 model passenger cars and light trucks. The major engine, carburetor, and ignition system modifications as well as other related components are described.

MODIFICATIONS FOR EXHAUST EMISSION CONTROL

New 6 Cylinder Engine

A new 198 CID engine has been developed for 1970. This engine retains the basic components of the 170 CID engine with an increase in stroke from 3.13" to 3.64" providing the increased displacement. The stroke increase resulted in a larger cylinder size, a lower combustion chamber surface area to volume ratio, and an increase in stroke bore ratio. Dynamometer emission tests had shown that each of these factors would contribute to a lower hydrocarbon emission level. The four car certification emission average for the 198 CID engine was 172 ppm hydrocarbon, confirming the dynamometer engine test results.

Extensive carburetor development to obtain good idle and part throttle fuel-air distribution was carried out on dynamometer engines during the 198 engine development program. Although all engine speeds and loads were explored, particular emphasis was placed on the driving modes that occur during typical urban operation. The requirements of low carbon monoxide emissions while maintaining good driveability have stressed the need for equal fuel-air distribution to all the cylinders of the engine. Distribution plots during the carburetion development of the 198 engine are shown in Figure 3. The fuel-air values are determined from an analysis of the exhaust gas of each cylinder.[9]

The fruits of this carburetor development work in the Laboratory in conjunction with thorough vehicle testing are reflected in an average certification CO level of .73% and fuel economy of 27 MPG in 30 MPH Proving Ground tests, without comprising our driveability standards.

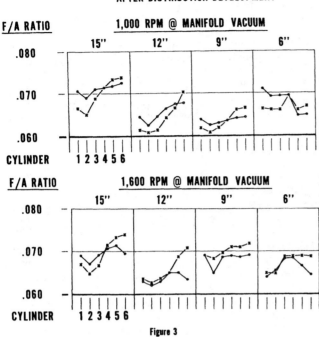

Figure 3

Combustion Chamber Design

A modification in combustion chamber design to reduce exhaust emissions was made in 1967 for Chrysler 6 cylinder engines, and in 1968 for the V-8 engines. The quench clearance in the chamber was opened up to permit more complete burning of the fuel-air mixture in the combustion chamber. The increase in quench clearance is shown in Figure 4. A substantial reduction of 25-30 PPM in HC emission level was ob-

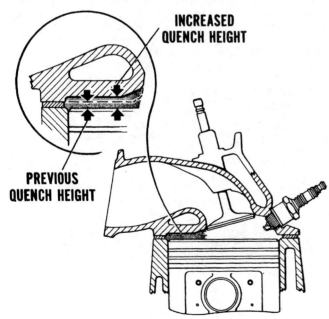

INCREASED QUENCH HEIGHT
Figure 4

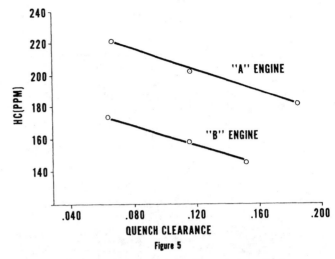

Figure 5

tained with this change. The effect of quench clearance on exhaust emission level at a constant compression ratio is shown in Figure 5.

For 1970, a reduction of .4 to .5 in compression ratio was made in the majority of the V-8 engines. This change produced a further opening up of the combustion chamber resulting in a lower surface area to volume ratio. The lower compression ratio also provides higher exhaust gas temperatures. Both of these factors contribute to a reduction in HC emission level. Dynamometer results of compression ratio vs. HC level are shown in Figure 6. Details of the engine dynamometer test schedule are presented in the Appendix.

The slight fuel economy and performance losses resulting with the reduced compression ratio are offset by fuel-air distribution improvements that have been secured and by the ability to set the spark timing closer to the maximum output spark advance of the engine at the same fuel octane requirement.

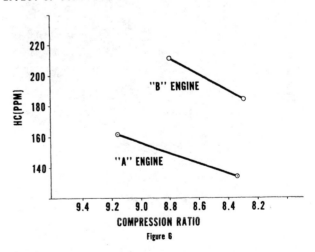

Figure 6

Engine Induction System

Heated intake air to the carburetor has been provided in almost all Chrysler Corporation 1970 vehicles. A sheet metal stove is attached to the exhaust manifold, Figures 7 and 8. Underhood air entering the stove is heated as it passes over the hot manifold. The heated air is conducted from the stove to the inlet snorkel on the air cleaner through a flexible duct. The air cleaner is designed to thermostatically control the induction air temperature at 100 ±7°F through 70 mph road load operation. At speeds above 70 mph, decreasing manifold vacuum and increasing differential pressure across the temperature control door, will cause the door to open gradually until a manifold vacuum of 5.5 in Hg. is reached. At this time, the heat control door will be in the heat off position. Under these conditions, as is true for WOT operation at all speeds, the induction air temperature will be the same as underhood air temperature.

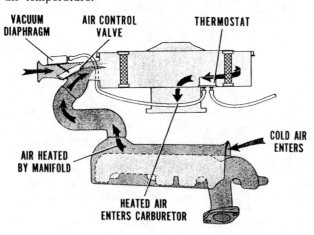

temperature of 96°F is provided on a 0° ambient day.

To provide further improvements in starting and driveability, several changes were made in the V-8 intake manifolds, Figure 10. The exhaust heat was removed from the upper portion of the manifold to provide an improvement in hot starting by decreasing the heat transfer to the carburetor during hot soak. The use of an insulating spacer between the carburetor and intake manifold is also employed to prevent an excessive temperature rise in the carburetor.

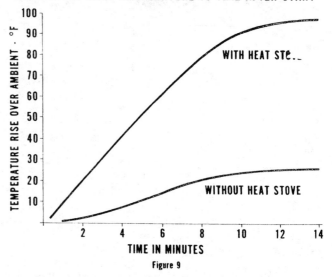

Figure 9

Temperature studies on dynamometer engines and driveability and temperature investigations in vehicles indicated the benefits of increasing the heat transfer to the upper runner of the manifold thus obtaining better side to side temperature balance. To accomplish this, pin fins were added to the floor of the upper runner in the hot spot area and the bottom floor of the exhaust heat cross-over was designed to direct the exhaust gas into the pin fin area.

Another change that has been made to insure good driveability at lean mixtures is a redesign and recalibration of the exhaust manifold heat control valve,

318 CID ENGINE
Figure 8

The use of a heated air system does not materially increase the induction air temperature during warm weather operation but does raise the intake temperature in cold weather. The decreased spread in temperature range permits the use of leaner fuel-air mixtures with satisfactory driveability. The effectiveness of the heat stove is shown in Figure 9. A carburetor inlet

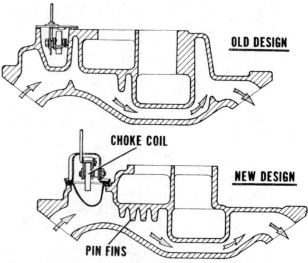

INTAKE MANIFOLD EXHAUST CROSSOVER PASSAGE
Figure 10

Figure 11. The new design provides a shield to protect the shaft bushings from direct impingement of exhaust gases. This has provided an improvement in the functional life of the valve. To increase the effectiveness of the valve, the thermostatic spring has been recalibrated to give maximum exhaust heat to the intake manifold during operation from a cold start idle to a warmed-up 30 mph road load.

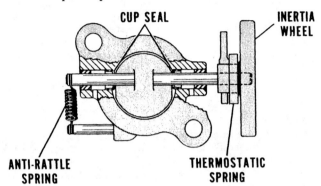

EXHAUST MANIFOLD HEAT VALVE
Figure 11

Carburetors

A number of features were added to the 1970 carburetors to accomplish the objectives of improved fuel-air distribution, decreased flow tolerances, improved carburetor manufacturing quality, and fuel evaporation loss control.

Great emphasis has been placed on attaining good geographic distribution at idle and throughout the range of part throttle operation. The improvements in distribution have resulted in lower hydrocarbon and carbon monoxide emissions and were obtained through extensive development test programs carried out on dynamometer engines. Some of the carburetor changes that were made to improve distribution are by-pass idle air systems in the throttle flanges, idle fuel feed in the by-pass air system, optimum location and direction of discharge from the crankcase ventilation system, and the use of offset main venturi.

Many carburetor models used with the 1970 Cleaner Air System have fixed by-pass idle air systems in the throttle flanges. This system is effective when the throttle valve is at or near the curb idle position. Air is taken from above the throttle valve through passages in the flange casting and discharged into the bore below the valve. The air flow quantity is fixed by the size of the passage; the direction and point of discharge is controlled by a cast slot in the lower face of the flange casting; and the velocity of discharge is controlled by the size of the cast slot. In the two and four barrel carburetors, each side of the carburetor has a separate by-pass system.

One two barrel carburetor has the additional feature of idle fuel discharged through the by-pass idle air system, Figure 12. The quantity of by-pass fuel is adjusted by conventional idle mixture needles.

In all carburetors, the location and direction of discharge from the crankcase ventilation system was selected for best distribution. In the cases of two and four barrel caburetors, a centrally located discharge is generally desired, however, the choice of front or

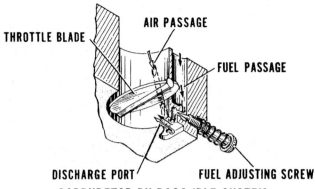

CARBURETOR BY-PASS IDLE SYSTEM
Figure 12

rear discharge in four barrel carburetors varied with the carburetor and engine. For single barrel caburetors used on six cylinder engines, considerable testing was carried out to determine the optimum location and direction.

One method of affecting distribution is to offset the main venturi with respect to the vertical centerline through the throttle bore. The amount and location of venturi offset for each side of the carburetor were determined in the dynamometer test program.

Many carburetor models have adjustable idle air bleed systems. The use of these systems result in closer tolerance flow limits in the idle and transfer ranges of the carburetor flow curve. Two methods of control are employed which differ in operating principles, Figure 13. One employs a needle which adjusts the restriction within the idle air bleed passage. The other incorporates a spring loaded ball check in an additional idle air bleed system, along with a needle adjustment. The needle adjusts the quantity of air to be bled into the idle system and the ball check, operated by vacuum within the idle system, controls the timing of the introduction of the additional air. The final adustment of these systems is made at the time of manufacture or original assembly of the carburetors and seals or cover plugs are added. It is not intended that any subsequent adjustments be made, and instructions to this effect are included in the service literature.

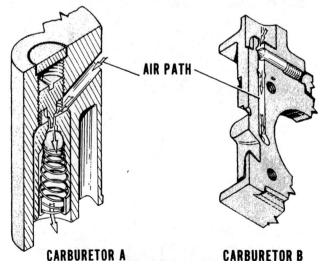

CARBURETOR A   CARBURETOR B
ADJUSTABLE IDLE AIR BLEED
Figure 13

In order to comply with the requirements for control of fuel evaporation losses, the carburetors released for use in California incorporate special fuel bowl venting. External fuel bowl vents were eliminated in the carburetors for the Vapor Saver System, therefore, it was necessary to modify all carburetors to provide vent passages and seals to prevent leakage to the atmosphere, Figure 14. The bowl vent is connected by a hose to the Vapor Saver System.

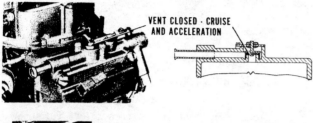
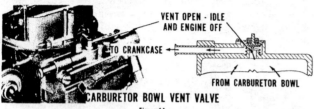

Figure 14 — CARBURETOR BOWL VENT VALVE

A new two barrel (1½ SAE) carburetor was introduced in 1970 for optional use on the 383 cu. in. engine. This carburetor has a triple venturi air section and incorporates the by-pass idle air system previously described. This carburetor has undergone thorough testing to insure satisfactory fuel-air distribution, exhaust emission control, fuel economy, performance, starting, and durability.

Chokes

A major change has been made in the mounting of the remote choke thermostatic coil units. These had previously been mounted in a cast well in a branch of the exhaust manifold (6 cylinder) or in the exhaust crossover in the intake manifold (V-8 engines). For 1970, the choke units are mounted in the same general locations, but a stamped stainless steel cup has replaced the cast well. The cup fits into a hole in the manifold and uses a gasket under the flange to seal against exhaust gas leakage, Figure 15.

The use of the stamped cup of .030" thickness in place of the cast well of .180" thickness has resulted in a faster temperature rise within the choke thermostatic coil operating space, thus making it possible for the choke to come off in a shorter distance after the cold start.

The ability to use decreased choke off distances resulted from the use of heated carburetor inlet air, improved manifold heat valve performance, and pin fins for increased intake manifold floor heat.

During the development and test program, each engine was evaluated over a wide ambient temperature range to determine the minimum choke enrichment for satisfactory cold start and warm-up performance. Peak coil temperatures during high speed operation were also monitored and component changes made as required to avoid exceeding limits imposed by available metallurgy. Each individual calibration was completed through the use of linkage geometry, thermostatic coil selection, placement depth of coil in the cup, and ventilating the upper housing. The two latter items, coil location and ventilation, were most important factors in controlling peak temperatures within the cup during severe operating conditions.

The results of these choke changes can be seen in Figures 16, 17, and 18. The choke off distance during a controlled driving sequence has been shortened by as much as 60% as shown in Figure 16. The reduction in carbon monoxide exhaust emissions is sizeable. Relating this to the first four cycles of the complete seven cycle exhaust emission test, reductions in CO of up to 35% have been attained. Figure 17 shows the average carbon monoxide reduction for the first four cycles of the emission test. Figure 18 shows the effect of the new choke system in the 225 cu. in. engine during each of the first four cycles. It should be noted that the greatest reduction in CO occurs during the second cycle. This emission benefit was made possible by the driveability gains obtained by the heated intake air and more effective intake manifold hot spot.

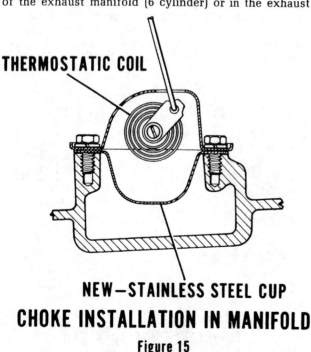

NEW—STAINLESS STEEL CUP
CHOKE INSTALLATION IN MANIFOLD
Figure 15

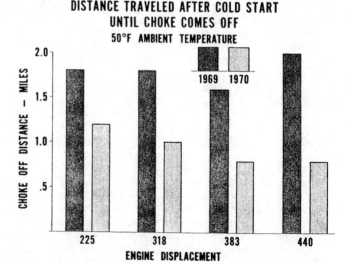

DISTANCE TRAVELED AFTER COLD START UNTIL CHOKE COMES OFF
50°F AMBIENT TEMPERATURE
Figure 16

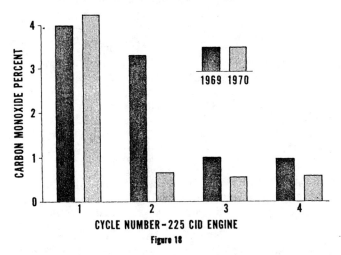

EFFECT OF EARLIER CHOKE OPENING ON CARBON MONOXIDE EMISSIONS
Figure 17

EFFECT OF EARLIER CHOKE OPENING ON CARBON MONOXIDE EMISSIONS
Figure 18

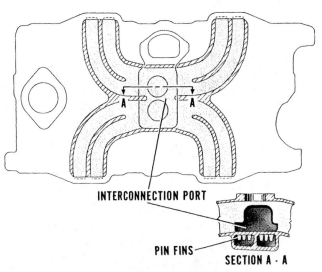

SINGLE PLANE INTAKE MANIFOLD
Figure 19

Single Plane Intake Manifold 318 CID Engine

The single plane V-8 intake manifold was first employed by Chrysler Corporation on the 1964 273 CID engine.[10] This manifold, which has all eight branches on the same level, was used to reduce engine height. A comparison of the emission results of the single plane and conventional two plane intake manifold is shown in Table 1.

TABLE I

EMISSION COMPARISON

Single Plane vs Two Plane Intake Manifold

	Vehicle Test		Engine Dynamometer Test	
	HC	CO	HC	CO
Two Plane	249	.67	108	.19
Single Plane	218	.65	96	.19

Because of the favorable emission results of these tests as well as numerous other dynamometer and vehicle tests, the single plane manifold was incorporated in the 1970 318 CID engine, Figure 19.

An explanation of the reduced HC emissions obtained with the single plane intake manifold is based on the cycle-to-cycle fuel-air delivery to the cylinders. The conventional two-level intake manifold is essentially two four-cylinder intake systems in a common casting; each fed by one-half the carburetor. As a result, the two level manifold pulsates strongly with four widely spaced impulses occurring in each level every two crank revolutions. In contrast, the single plane manifold is designed with all eight port branches feeding from a common plenum volume. This doubles the pulsation frequency but significantly reduces the magnitude. The net effect is less pulsation on the carburetor and a fuel delivery condition that results in more constant cycle-to-cycle F/A mixture to each of the engine cylinders. The single plane intake manifold also provides an improvement in engine output in the high speed range but at a sacrifice of some mid-speed torque.

Ignition System—Solenoid Control for Spark Retard at Idle

The use of retarded timing at idle has resulted in significant reductions in emissions during idle and deceleration.[11] The increased air flow provides more complete combustion, and the higher exhaust gas temperature promotes increased oxidation of the combustibles in the exhaust gas. Spark retard at idle has introduced disadvantages on some engines, however, such as rougher idle, driveability weakness off idle, and poor engine starting. To overcome some of these disadvantages, a means was sought to provide the following:
1. Normal spark advance at starting.
2. Retarded timing at idle.
3. Normal spark advance during driveaway from idle, (immediately after opening throttle).

It was desired that the retard function be directly related to a change in carburetor throttle position.

Solenoid actuation to produce the retard function was chosen because it utilized a power source unaffected by engine operating conditions. To ensure the best synchronism between throttle position and solenoid actuation, the throttle stop and idle adjustment screw were used as contacts in the throttle position switch. This switch had the advantage of having no hysteresis and perfect coincidence with any adjustment of the throttle stop.

Because of the limited available space, a solenoid built into the vacuum advance unit and working on the normal vacuum actuation arm was chosen. This is shown in Figures 20 and 21. This installation provides solenoid retard of the spark only when the carburetor throttle is closed. Normal spark timing is provided under all other conditions as shown schematically in Figure 22. For cold starting where the recommended cam start is used, the driver depresses the accelerator pedal and then releases it. This permits the idle adjusting screw to contact the fast idle cam. For hot starting, a one-third throttle opening is recommended. In either case, the throttle is partially open with the result that the spark timing is in the advanced position. During all warm engine idle operation, the idle screw is in contact with the throttle stop producing the spark retard. As soon as the throttle is opened for driving operation, the contact is broken and the spark timing is advanced, resulting in improved performance and driveability.

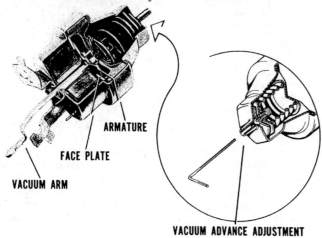

SOLENOID VACUUM UNIT
Figure 20

SOLENOID DISTRIBUTOR INSTALLED ON 1970 ENGINE
Figure 21

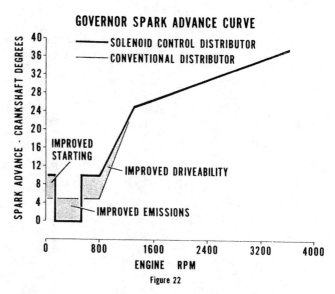

Figure 22

To overcome any possible apprehension that might be motivated by using exposed switch contacts in the vicinity of the carburetor, the current level was reduced to a very low value by means of a transistor amplifier. The two stages of this amplifier enabled sufficient gain to be obtained from inexpensive transistors and also provided the correct phasing so that the switch contacts, one of which was grounded for convenience, not only operated at low current, but also were in the low voltage part of the circuit. Thus, any minute arcing that would occur was well below the energy level required to initiate combustion.

A solenoid was needed with sufficient power to move the distributor contact plate against the forces of friction and the return spring, and as the actual movement required was relatively short (about .060), a face type armature was considered the most efficient.

Heat dissipation in the solenoid was anticipated as a problem because of the size limitations imposed. A method often used to reduce power dissipation in solenoids is to sharply reduce the current after closure of the main air gap to a level just adequate to hold the solenoid armature in the closed position. To avoid building a switch into the solenoid unit, a partial adoption of the above method was achieved by providing a feedback winding over the solenoid winding and making connections to the transistor amplifier such that a form of blocking oscillator was devised.

Frequency and wave form of the solenoid current were then adjusted so that pull-in and hold-in forces were maintained at the desired values, but power dissipation in the solenoid was reduced to acceptable levels.

The electronic unit was sealed by potting into a small plastic case which was then mounted directly to the case of the solenoid unit.

Ignition System—Distributor

Because of the importance of spark advance on driveability and exhaust emission control, a means was sought to obtain improved spark timing accuracy. Accordingly, a comprehensive design study was made of the following:

1. The concentricity and durability of the support for the distributor shaft and cam including both

the distributor and engine mounting surfaces and pilot diameters.

2. The tolerance relationship of the various distributor parts and their effect on spark timing.

3. A means to perform accurate, reproducible distributor adjustments and tests.

After extensive study, numerous drafting layouts, calculations and tests, the following design or processing changes were made in the distributor, the engine, and in the distributor test equipment used in the production plant:

Distributor bearings are now reamed in line to very close diametral and concentricity tolerances. Improved squareness and concentricity of the mounting flange and pilot diameter are obtained by machining on new equipment with greater precision. A similar program achieved improved accuracy of the mating surfaces on the engine. The weight and stop plates in the governor and the vacuum arm were retooled using special fine flow dies, thus providing greater dimensional integrity of these components.

Improved control of the breaker plate movement was obtained by two design revisions. First, a dowel was used to locate more accurately the vacuum housing to the distributor housing and lower breaker plate shown in Figure 23. Second, the hole is punched in the vacuum arm for the pin which engages the breaker plate, after assembly of the vacuum chamber.

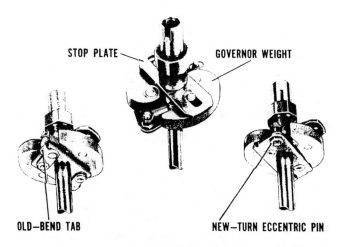

GOVERNOR SPRING ADJUSTMENT
Figure 24

The whole new concept in distributor test equipment for production was created by a combination of engineers from Chrysler's Space Division in Huntsville, Alabama, and the Power Train Group in Detroit and Indianapolis. One new machine dynamically tests and adjusts the governor mechanism when it is only a partial assembly of the cam, governor, and shaft. These components are mounted in one of the computer controlled test machines which adjusts the initial tension of the light governor spring and the loop clearance of the heavy spring while the assembly is rotated at the proper speeds for the model distributor. These springs are adjusted by rotating the eccentrically mounted spring hanger pins.

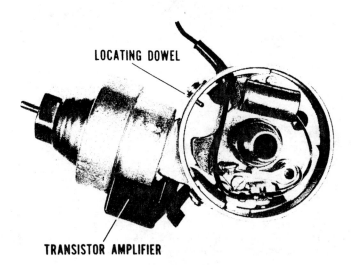

DISTRIBUTOR AND SOLENOID VACUUM UNIT
Figure 23

The governor spring adjustments are accomplished using rotatable spring loaded, eccentric spring hanger pins which can be turned from below the weight plate. This is shown in Figure 24. The vacuum chamber adjustment is accomplished by using a drawn sheet metal screw which travels through a nut thereby increasing the compression on the spring, Figure 20. The nut is held in the chamber by its shape; the screw is rotated by an Allen-type wrench inserted through the vacuum orifice in the chamber.

FINAL DISTRIBUTOR PRODUCTION TEST
Figure 25

These parts are then built into a complete distributor. Complete distributors are checked on a similar machine, shown in Figure 25, for:
  a. Governor curve conformance to specification— no adjustment can be made on this machine.

1041

b. Vacuum curve adjustment and conformance to specification.
c. Synchronism conformance to specification.
d. Contact dwell—this one adjustment is made by the operator and conformance check to engineering specifications is made by the machine.
e. Residual dwell at higher speeds conformance to specification.

All the above adjustments and inspections are performed on 100% of production. Except for dwell, all adjustments are computer controlled. The accuracy of the measurements that are made are to .1°, 1 rpm, and .1" hg. These machines can process a complete adjustment and conformance check in seconds. Because of the computer control, these machines can print out the data in many different forms for different kinds of analysis. Not only does this allow the people controlling the test equipment to more easily improve the performance of the machines, but it also quickly provides information in usable form to production people so that they may correct problems.

The result of this program is a distributor whose spark timing accuracy both initially and over long operating periods has been greatly improved for better exhaust emission control.

Coolant Temperature

Both dynamometer and vehicle emission tests indicated a reduction in HC with an increase in coolant temperature. This effect is shown in Figure 26. The increase in water temperature produces a slightly higher surface temperature in the combustion chamber. This in turn produces more complete combustion at the surface area thus reducing the HC emission level. As a result of these data, thermostats calibrated at 190° and 195°F are incorporated in our 1970 engines. These compare to 180°F thermostats of recent years. A reduction in hydrocarbon emission level of 5-10 ppm was obtained.

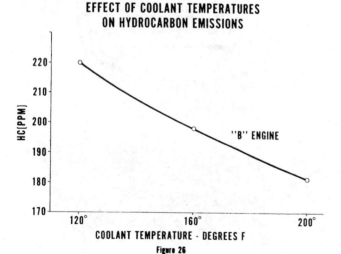

Figure 26

Engine Idle Settings

The importance of engine idle fuel-air, spark advance and rpm on exhaust emission levels has long been recognized. These idle settings determine the position of the throttle which in turn determines the air flow at idle and during closed throttle deceleration. When the intake air and fuel flow is low during deceleration, combustion is very poor with a resulting high concentration of HC. Providing an increase in idle air flow has greatly improved combustion during deceleration thus resulting in lower HC emission levels. The increase in idle air flow can be accomplished by setting a lean fuel-air mixture, a retarded spark timing or a higher idle speed. Considerable test activity has been carried out on each engine-transmission combination to determine the best idle setting compromise for exhaust emission control, engine idle quality and driveability.

To permit the use of high idle speeds on some high performance engines, a solenoid throttle stop is employed, Figure 27. The solenoid is activated when the ignition is on. The idle adjustment screw contacts the solenoid plunger, holding the throttle at a position to give the specified idle speed. When the ignition is turned off, the solenoid is deactivated, permitting the throttle to close further thus preventing engine after-run.

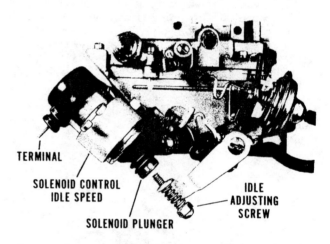

SOLENOID THROTTLE STOP
Figure 27

SERVICE REQUIREMENTS

Although numerous engine modifications have been made for emission control, including various new engine components, engine maintenance requirements have remained essentially unchanged. The positive crankcase ventilation valve (PCV) which is used to control crankcase emissions should be inspected periodically and replaced about once a year.

To assure continued control of exhaust emissions, the recommended periodic servicing of the ignition and carburetion systems should be performed. It is particularly important that the following items be properly adjusted:
1. Engine idle speed   2. Spark timing
3. Idle fuel-air ratio

CONCLUSIONS

The philosophy of promoting more complete combustion in the engine to control exhaust emissions was introduced on the 1966 model Chrysler Corporation passenger cars and light trucks sold in California. Since that time, this method of emission control has been more fully developed to provide further reduction in exhaust emission levels. Through modifications to the carburetor, engine induction system, the combustion chamber and the ignition system as well as optimizing the calibration of the engine operating parameters, the 1970 exhaust emission standards have been successfully met. The certification emission values are shown in Table 2.

While the 1970 standards represent a significant reduction in vehicle exhaust emission levels, a continuing effort is being made to effect further emission reductions in the future. Exhaust emission control is receiving top priority in advanced engine and carburetor design and development. Our goal at Chrysler is to provide vehicles at the lowest possible emission levels with the minimum compromise in performance, driveability and fuel economy, and at minimum cost to the customer.

TABLE II
1970 "F" SERIES FINAL EMISSION TEST RESULTS

Engine-Transmission	Car Model	Emission Stds.		4000 Mile Emission Results			
				Per Vehicle		Avg. Per Disp.	
		HC	CO	HC	CO	HC	CO
198-M	Dart	238	1.33	176	0.82	172	0.73
198-M	Valiant	238	1.33	184	0.75		
198-A	Valiant	214	1.19	154	0.63		
198-A	Dart	190	1.06	173	0.71		
225-M	Belvedere	212	1.18	176	1.04	151	0.83
225-A	Dart	190	1.06	126	0.77		
225-A	Satellite	190	1.06	143	0.77		
225-A	Barracuda	190	1.06	157	0.79		
318-A	Fury III	161	0.90	153	0.67	151	0.74
318-A	Fury III	161	0.90	145	0.67		
318-A	Coronet	173	0.97	170	0.87		
318-A	Barracuda	190	1.06	136	0.75		
340-M	Dart	212	1.18	206	0.88	186	0.75
340-A	Dart	190	1.06	167	0.62		
383-2A	Newport	161	0.90	158	0.83	141	0.76
383-2A	Coronet	161	0.90	137	0.82		
383-2A	Fury III	173	0.97	136	0.58		
383-4A	Road Runner	173	0.97	134	0.82		
426-M	Road Runner	193	1.07	139	0.88	147	0.77
426-A	Road Runner	173	0.97	154	0.65		
440-A	New Yorker	153	0.85	134	0.65	139	0.62
440-A	300	153	0.85	136	0.49		
440-A	Imperial	147	0.82	133	0.52		
440-A (HP)	GTX	175	0.97	153	0.81		

BIBLIOGRAPHY

1. Air Pollution Control District, Los Angeles County, "Profile of Air Pollution Control in Los Angeles County", January, 1969, p. 3.
2. Nelson, E. E., "Hydrocarbon Control For Los Angeles By Reducing Gasoline Volatility". Society of Automotive Engineers, International Automotive Engineering Congress, Detroit, Michigan, January 13-17, 1969, (#690087).
3. Air Pollution Control District, Los Angeles County, "Profile of Air Pollution Control in Los Angeles County", January, 1969, p. 35-44.
4. Ibid, p. 39.
5. Heinen, C. M., "We've Done The Job—What's Next?", Society of Automotive Engineers Metropolitan Section, Symposium on the Automobile Now and Tomorrow, New York, City, New York, April 9, 1969, (#690539).
6. Federal Register Volume 33 Number 108, Part II, "Standards for Exhaust Emissions, Fuel Evaporative Emissions, and Smoke Emissions, Applicable to 1970 and Later Vehicles and Engines", Tuesday, June 4, 1968, Washington, D. C.
7. Air Resources Board, State of California, "Exhaust Emission Standards and Test Procedures for 1970 Model Gasoline Powered Motor Vehicles Under 6001 Pounds Gross Vehicle Weight, Fuel Evaporative Emission Standard and Test Procedure for 1970 and Subsequent Model Gasoline Powered Motor Vehicles Under 6001 Pounds Gross Vehicle Weight", November 20, 1968.
8. Sarto, J. O., Fagley, W. S., and Hunter, W. A., "Chrysler Evaporation Control System—The Vapor Saver for 1970", Society of Automotive Engineers, Automotive Engineering Congress and Exposition, Detroit, Michigan, January 12-16, 1970.
9. Fagley, W. S., and Nunez, R. R., "Exhaust Gas Analysis as a Tool for Measuring Fuel-Air Ratios", Society of Automotive Engineers, Mid-Year Meeting, Chicago, Illinois, May 15-19, 1967, (#670483).
10. Weertman, W. L., and Beckman, E. W., "Chrysler Corporation's New 273 Cu. In. V-8 Engine", Society of Automotive Engineers, Automobile Week, Detroit, Michigan, March 30 - April 3, 1964, (#826A).
11. Beckman, E. W., Fagley, W. S., and Sarto, J. O., "Exhaust Emission Control by Chrysler — The Cleaner Air Package", Society of Automotive Engineers, Automotive Engineering Congress, Detroit, Michigan, January 10-14, 1966, (#660107).

APPENDIX

Engine Dynamometer Test Schedule

The engine dynamometer test procedure consisted of a series of stabilized engine operating conditions representing most of the vehicle test modes of the California cycle. Below is a comparison of the events. The operating schedule for the 318 Cu. In. engine is shown in this schedule. The speed and intake manifold vacuum values for other engines are selected to best match the conditions existing during the California cycle test modes.

Engine motoring tests are conducted separately on the engine dynamometer for evaluation of the deceleration test modes.

COMPARISON OF VEHICLE DRIVING MODE WITH ENGINE DYNAMOMETER TEST

	Weighting Factor	Engine Speed	Engine Vacuum
Idle	.042	600	—
30 MPH Cruise	.118	1200	19.5
15-30 Acceleration	.455	1200	7.0
		1600	11.0
50 MPH Cruise	.029	2000	17.0

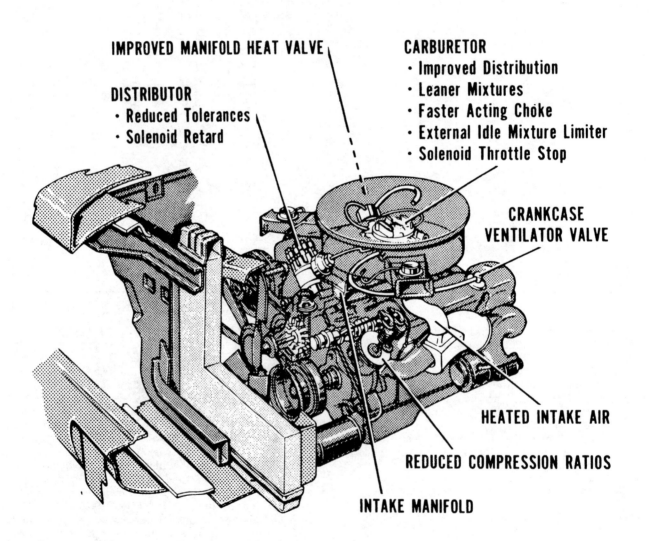

1970 CLEANER AIR SYSTEM

EFFECTIVENESS OF EXHAUST GAS RECIRCULATION WITH EXTENDED USE

ABSTRACT

Exhaust gas recirculation (EGR) was found to be an effective means of reducing automotive NO_x levels with no major unsolvable problems over 52,000 miles under city/suburban driving conditions. Compatible with air injection and engine modification systems for HC and CO control, EGR effectively reduced the NO_x levels with no decrease in reduction over the 52,000 miles. Engine wear and engine cleanliness with EGR was normal for the mileage and driving regime. However, the throttle area and exhaust valves were found to be sensitive to leaded fuels.

INTRODUCTION - Exhaust gas recirculation has been demonstrated to be an effective technique for NO_x suppression in a number of different studies. However, these studies indicated that before exhaust gas recirculation could be applied on a broad base, the following areas needed to be explored:

1. Feasibility of the recirculation technique when used in conjunction with positive crankcase ventilation and with engine modification (EM) and air injection (AIR) methods for HC and CO control.

2. Design features and parameters for a simple, dependable automatic system to provide an optimized combination of all the emission controls while maximizing economy and vehicle performance.

3. The long-term operability and effectiveness of the combination of systems and their effects on engine durability.

The recently concluded initial feasibility phase of the Esso/NAPCA research[1][2]* on the "Evaluation of Exhaust Recirculation for NO_x Control" provided some assessment of the first two areas. The feasibility program showed that exhaust gas recirculation (EGR) can be successfully combined with engine modification and air injection methods for HC and CO control in exhaust gases while retaining acceptable fuel economy and vehicle performance. Following the successful completion of the feasibility program, a durability study[3] was initiated to explore the remaining area. This program was conducted, therefore, to study the long-term operability and effectiveness of EGR as part of the complete emission control system and its effects on engine durability.

---

*Numbers in parentheses designate references at end of paper.

To aid in judging the effect of exhaust gas recycle on engine durability and control of exhaust gas emissions, four key factors were selected:

(1) Engine Wear

(2) Engine Cleanliness

(3) Vehicle Performance

(4) Effectiveness of Emission Controls with Extended Service

PROGRAM

VEHICLES - The exhaust gas recirculation system as developed in the initial feasibility program[2] was combined with the engine modification control system in three 1969 Plymouths (318 CID) and one 1969 Chevrolet (350 CID), and with the air injection control system in two 1969 Chevrolets (350 CID). Details on the vehicles are given in Table I.

MODIFICATIONS TO DISTRIBUTORS AND CARBURETORS - The EGR equipped Plymouths employed the slightly advanced Mod. 1 distributors developed in the feasibility program in place of the manufacturer's distributor. The Mod. 1 distributor has the same centrifugal advance as the manufacturer's distributor but uses a slightly altered vacuum advance response. The Mod. 1 linearly decreased the vacuum advance from 15° at 16 in. Hg. to 0° at 4 in. Hg. The manufacturer's distributor linearly decreased the vacuum advance from 15° at 16 in. Hg. to 0° at 9 in. Hg. The Mod. 1 thus had more vacuum advance at all vacuums between 16 in. Hg. and 4 in. Hg. Unmodified original equipment distributors were used for the Chevrolets with basic timing advanced five degrees (from 4° BTDC to 9° BTDC).

The three Plymouths used mean limit carburetors supplied by Chrysler Corporation. The Chevrolets used carburetors checked by Chevrolet Engineering Department to make certain that initially the carburetors were within specifications. On Chevrolets 43 and 37, however, the idle tube restrictions were enlarged to 0.046 in. and 0.042 in., respectively, to obtain better idle and off idle performance.

RECYCLE SYSTEM - The recirculation system used in this study was developed in the initial feasibility program. A detailed description of the recycle insert plate, the control valve, and the total system with pictures is given in the feasibility phase report.[2] Basically, the recycle gas is taken from the exhaust system immediately before the muffler and introduced into the carburetor just above the throttle plates. Introducing the exhaust

Table 1

TEST VEHICLES

(Manufacturers' Specifications)

	Chevrolet 37	Chevrolet 42	Chevrolet 43	Plymouth 38	Plymouth 39	Plymouth 40
Model	Impala	Impala	Impala	Fury III	Fury II	Fury II
Engine						
Bore & Stroke (in.)	4.00 x 3.48	4.00 x 3.48	4.00 x 3.48	3.91 x 3.31	3.91 x 3.31	3.91 x 3.31
Displacement (in.)	350	350	350	318	318	318
Max. Bhp at rpm	255 @ 4200	255 @ 4200	255 @ 4200	230 @ 4400	230 @ 4400	230 @ 4400
Max. Torque, lb. ft. at rpm	365 @ 1600	365 @ 1600	365 @ 1600	340 @ 2400	340 @ 2400	340 @ 2400
Compression Ratio	9.0	9.0	9.0	9.2	9.2	9.2
Emission Control	Engine Mod.	AIR	AIR	Engine Mod.	Engine Mod.	Engine Mod.
Carburetor	2 bbl.	2 bbl.	2 bbl.	2 bbl.	2 bbl.	2 bbl.
Transmission	Turbo Hydramatic	Turbo Hydramatic	Turbo Hydramatic	Torqueflite	Torqueflite	Torqueflite
Axle	2.73	2.73	2.73	2.71	2.71	2.71
Curb Weight (lb.)	3890	3890	3890	3805	3805	3805
Air Conditioning	Yes	Yes	Yes	Yes	Yes	Yes
Power Steering	Yes	Yes	Yes	Yes	Yes	Yes
Power Brakes	No	Yes	Yes	No	No	No

recycle gas above the throttle plate results in a relatively constant ratio of exhaust gas recirculated to carburetor inlet air. Recirculation is turned off below about 20 mph cruise and on deceleration. Since throttle position is used for control, exhaust gas recirculation is present under all acceleration modes. During choked operation recycle was not used because the richness of the air-fuel mixture will by itself significantly reduce NO_x.

The amount of recirculation was established by the amount needed to give an initial NO_x level below 750 ppm or by the maximum recirculation with acceptable driveability. The recirculation flow rate was adjusted to the required level by selection of the proper orifice size in the recycle line. Initial NO_x levels with EGR ranged from 610 to 720 ppm. The initial ratio of exhaust gas recirculated to carburetor inlet air is given below in Table 2.

Table 2

INITIAL RATIO OF EXHAUST GAS RECIRCULATED TO CARBURETOR INLET AIR

(Volume ratio at 75°F and 1 atm)

Vehicle	Ratio
Plymouth 38	0.15
Plymouth 39	0.12
Plymouth 40	0.13
Chevrolet 37	0.16
Chevrolet 42	0.14
Chevrolet 43	0.09

The recycle ratio was determined by comparing the recycle flow rate as measured with a one inch turbine meter in the recycle line with the carburetor air flow rate measured as a ΔP across a Meriam laminar flow meter in the air stream to the carburetor.

MILEAGE ACCUMULATION - All vehicles operated about 18 hrs./day to accumulate test mileage on the Mileage Accumulation Dynamometer (MAD)[4] in a controlled driving cycle (city-suburban) at an average speed of 32 mph.

Two different driving cycles were used. For the first 40,000 test miles, a mixed city/suburban or milk run driving schedule was used. This driving schedule is controlled by magnetic tapes recorded on the road during a 100-mile trip through a congested downtown area of a large city, suburban areas and open countryside. The tape controlled cars operate for 6 out of every 8 hours on an outdoor dynamometer which simulates level road load and cooling air conditions. The milk run schedule averages about 32 mph including extensive stop and go situations and a maximum speed of 50 mph. The driving schedule was developed for fuel and lubricant testing to replace extensive fleet testing. Ten years of experience has indicated excellent correlation with normal driving experience in connection with combustion chamber deposits, valve deposits, sludge formation, engine wear, spark plug fouling and many other items of product quality interest. At 40,000 test miles, the driving regime was changed from the milk run to a modified AMA driving schedule.[5] The AMA cycle described in the Federal Register is used for durability testing of emission control equipment for qualification purposes. The AMA schedule also averages about 32 mph. However, the AMA schedule has a period of higher (55 mph and 70 mph) speed driving along with a few wide-open-throttle accelerations to 70 mph.

Break-in mileage was accumulated following manufacturer's recommendations.

FUELS AND LUBES - Three different gasolines and two slightly different lubricating oils were used in the program. Table 3 presents the vehicle-mileage schedule for the various combinations of fuels and lubes.

Table 3

FUEL AND LUBE SCHEDULE

	Mileage Interval			
	Lube	Fuels		
Vehicles	0-52,000 miles	0-8000 miles	8000-40,000 miles	40,000-52,000 miles
Chevrolet 37	B	C	C	D
Chevrolet 42	A	C	C	D
Chevrolet 43	A	C	C	D
Plymouth 38	A	C	C	D
Plymouth 39	B	C	Unleaded*	Unleaded
Plymouth 40	A	C	C	D

Lubricating oils A and B are high quality, multi-grade premium oils that meet or exceed manufacturer's recommendations. While Oil A is a commercial oil that exceeds M.S. quality, Oil B is an experimental formulation of M.S. quality.

Fuel C is a regular grade gasoline. Fuel D is an intermediate grade gasoline. The purpose of the change from Fuel C to Fuel D was to provide a higher octane fuel to suppress part-throttle knock which developed during mileage accumulation on one vehicle. The fuel and driving schedule changes at 40,000 miles occurred simultaneously. The unleaded fuel is a commercially available unleaded gasoline. The fuel volatility was changed periodically to provide realistic weather adjusted fuels.

MAINTENANCE SCHEDULE - The manufacturer's recommended maintenance schedule was followed for all vehicles including tune-ups at 12,000 mile intervals. The oil change interval for the Plymouths was 4000 miles and 6000 miles for the Chevrolets.

---

*From 20,000 to 24,000 miles, Plymouth 39 was contaminated with Fuel C.

EMISSION MEASUREMENTS – Exhaust gas emissions were measured every 4000 miles employing modern instrumentation which meet the Federal requirements[5] for testing 1970 model vehicles. The emission testing was done on Clayton Dynamometers fitted with acceleration weights equivalent to a 4000 lb. car. Emissions were measured with recording infrared analyzers for hydrocarbons, CO, CO_2 and NO as well as recording O_2 analyzers. The data were reduced by a special type PDP-85 Digital Equipment Co. computer which is interfaced with the Beckmann analytical equipment. Beckmann programs allow the computer to accept the output from the NO, CO, CO_2, and HC analyzers and compute the corrected and weighted averages for the Federal test procedure. Indolene was used for all emissions testing.

The NO was adjusted for humidity by the equation:

$$NO\ adj. = NO\left[e^{-0.004\ (75-H)}\right]$$

where H is the humidity in grains of H_2O per lb. of air. This equation was developed in the feasibility phase for similar engines as in this program.

It will be noted that two modes of operation are indicated in the various vehicle performance tables and emissions data figures. These include "with recycle" and "without recycle." "With recycle" indicates that the vehicle has been equipped with recycle hardware and includes all modifications to accomodate recycle operation. "Without recycle" indicates that the recycle valve is closed and that no recycle gas is flowing, but all hardware and modifications are present. It is recognized that the "without recycle" is a compromise, as compared to the true no recycle case without any recycle hardware, but it is considered to be a stable reference condition that can be attained by simply opening an electrical circuit. This allowed a comparison to be obtained for hot operations over such a short period of time that external variables (humidity, etc.) would have little chance to change. The constant removal and installation of recycle hardware and engine modifications would be very time consuming, costly, and difficult to reproduce.

BASE CASE - It is recognized that since no base case - non EGR vehicles were tested at the same time, no statistical comparisons between EGR equipped and non-EGR equipped vehicles can be made. The primary purpose of this program was to obtain a first approximation of the effects of EGR. Our base case (i.e., normal wear) for comparison is then our experience with similar vehicles in taxi tests, MAD tests using the same driving schedules, etc.

TEST RESULTS

ENGINE WEAR AND CLEANLINESS - During the 52,000 test miles, the engine wear was light to normal for the driving schedules used (Table 4). For instance, the average connecting rod bearing weight loss was a low 24.1 mg. The cylinder to cylinder losses for all six vehicles ranged from 2.8 mg to 52.3 mg for the upper bearing. The average no. 1 compression ring weight loss (275 mg) was normal.

The normal oil consumption (Table 5) was another indication of normal wear. The average oil consumption for the first 40,000 test miles was 0.28 qts./1000 miles. Reflecting the higher speeds and wide-open-throttle (WOT) accelerations, the oil consumption for the AMA schedule as expected increased to 0.57 qts./1000 miles. Results of the used oil analysis were normal.[3]

Exhaust gas recirculation did not adversely affect engine cleanliness. Most varnish and sludge ratings were 9 or above (Table 6). With a perfectly clean surface defined as 10, an 8 to 9 rating is very acceptable. The rocker arm cover, push rods, valve lifter bodies, cylinder walls and the crankcase oil pan all rated varnish an average of 8.6 or higher. Sludge ratings were equally as good. The rocker arm assembly and cover, push rod chamber and cover, the valve deck, and oil screen rated 9 or above. The crankcase oil pan sludge was normal at 8.1.

The normal sludge, varnish, and wear data above indicates that a high quality oil will provide the typical high degree of protection

Table 4

SUMMARY OF ENGINE WEAR

Test Vehicle	Compression Ring Weight Loss		Connecting Rod Bearing Weight Loss		Cylinder Bore Wear Top of Ring Travel	Intake Valve Wear			Exhaust Valve Wear		
	No. 1	No. 2	Upper	Lower		Stem	Guide	Total	Stem	Guide	Total
Dimensions	mg	mg	mg	mg	10^{-4} in.	10^{-4} in.	10^{-4} in.	10^{-4} in.	10^{-4} in.	10^{-4} in.	10^{-4} in.
					Average Wear						
Plymouth 38	404.9	114.4	42.5	32.6	12.2	4.8	7.6	12.4	9.0	16.9	25.9
Plymouth 40	241.0	50.3	41.8	56.9	4.2	1.9	7.7	9.6	2.7	24.6	27.3
Chevrolet 42	146.4	57.2	13.6	6.0	9.7	7.4	8.5	15.9	17.9	29.2	47.1
Chevrolet 43	132.8	92.0	8.4	7.2	7.1	5.0	18.8	23.8	2.6	7.5	10.1
Chevrolet 37	242.2	88.2	17.4	7.0	7.1	3.2	8.5	11.7	5.0	9.5	14.5
Chevrolet 39	481.8	103.0	21.1	10.4	7.6	1.4	8.0	9.4	3.1	14.1	17.2
Average	274.9	84.2	24.1	20.0	8.0	4.0	9.9	13.8	6.7	17.0	23.7

Table 5

OIL CONSUMPTION

Car	Overall Average qts./1000 mi.	Milk Run Average qts./1000 mi.	AMA Average qts./1000 mi.
Plymouth 38	0.35	0.26	0.64
Plymouth 40	0.28	0.20	0.55
Chevrolet 42	0.30	0.23	0.51
Chevrolet 43	0.45	0.41	0.59
Chevrolet 37	0.34	0.31	0.50
Plymouth 39	0.35	0.26	0.64
Overall Average	0.35	0.28	0.57

Table 6

SUMMARY OF VARNISH, SLUDGE
AND RUST RATINGS

Varnish	Clean	\multicolumn{6}{c	}{Average for Car No.}	
		38	40	42	43	37	39	Overall*
Rocker Arm Cover	10	9.4	9.4	8.9	9.3	8.0	9.0	9.0
Push Rod Chamber Cover	10	5.0	9.3	7.8	5.0	5.0	10.0	6.4
Push Rods	10	10.0	10.0	9.0	9.5	6.5	10.0	9.0
Valve Lifter Bodies	10	9.0	9.5	9.0	10.0	9.5	9.0	9.4
Cylinder Walls (BRT)	10	8.0	9.0	9.0	9.5	7.5	9.0	8.6
Crankcase Oil Pan	10	9.0	9.0	9.5	10.0	6.5	8.0	8.8
Sludge								
Rocker Arm Assembly	10	9.7	9.6	9.7	9.4	9.5	9.8	9.6
Rocker Arm Cover	10	9.5	9.5	9.8	9.7	9.1	9.6	9.5
Push Rod Chamber Cover	10	9.6	9.7	9.7	9.5	9.6	9.9	9.6
Push Rod Chamber	10	8.8	9.0	9.7	9.2	9.1	9.8	9.2
Valve Deck	10	9.3	8.8	9.0	8.8	9.0	9.7	9.0
Crankcase Oil Pan	10	7.8	8.0	8.3	9.1	7.4	9.3	8.1
Oil Screen	10	10.00	10.00	10.00	10.00	10.00	10.00	10.00
Rust								
Valve Lifter Bodies	10	7.0	6.0	9.0	9.5	7.5	6.5	7.8
Cylinder Walls	10	9.0	8.0	9.5	9.0	9.0	8.0	8.9

*Not including Plymouth 39 using clear fuel.

against sludge and varnish formation, and engine wear with exhaust gas recirculation if the recommended oil change intervals are followed.

The first two vehicles to complete 8000 test miles, Plymouths 39 and 40, had erratic poor performance during the emission testing. Portions of the seven-mode cycle had hydrocarbon values higher than desirable. Compression pressure determinations indicated 16 psi and 70 psi on cylinders 7 and 8 for Plymouth 40 and 65 psi on cylinder 1 for Plymouth 39. Because of these low compression pressures, the head assemblies were removed. The cylinders in question had heavy deposits on the exhaust seats and valves for 8,000 miles and had indications of valve leakage. The hydrocarbon emissions of the other four test vehicles were more normal and did not indicate any problems.

New pre-measured head assemblies were installed and the durability testing continued. In order to obtain information on possible fuel additive effects, Plymouth 39 was restarted using unleaded fuel. Plymouth 40 continued using the same leaded fuel. Although compression pressures remained normal for the remainder of the test period, the vehicles occasionally indicated erratic hydrocarbon values. This short-lived erratic condition was relieved by very short periods of higher (55 mph) speed driving and was probably due to deposits from the combustion chamber causing transient valve leakage.

Because of these occasional periods of erratic hydrocarbon emissions and improvements with short periods of higher speed driving, it was decided to change the driving schedule at 40,000 miles to provide more periods of high speed operation on a regular schedule. Previous experience has shown that high speed operation and wide-open-throttle accelerations tend to reduce combustion chamber deposits (and octane requirement) due to the thermal shocking and higher gas velocities. The Federal Durability Driving Schedule[5] or AMA schedule conveniently provided the higher speed operation. This schedule averages the same as the milk run schedule, 32 mph, but provides higher speeds (70 mph vs. 50 mph) and WOT accelerations to 70 mph. Vehicle performance improved with the new schedule.

Final combustion chamber deposit ratings (Table 7) indicate that the engine deposit levels are light to normal in all areas. The combustion zone overall was normal at 1.43. The exhaust valve tops average, however, as expected was a heavy 2.75. The intake valve tops averaged a normal 0.66 demerits.

EFFECTIVENESS OF EMISSION CONTROLS - Exhaust gas recirculation remained effective for reducing NO_x levels over the 52,000 mile test period. The data is presented in Figure 1. Except for Plymouth 40, no overall trend with mileage is evident. The increase in the percent of NO_x reduction for Plymouth 40 is due to an increase in the ratio of exhaust gas recirculated to carburetor inlet air from 0.13 at 30,000 miles to 0.16 at 52,000 miles.

No overall trends with mileage are evident in the HC and CO data for the leaded fuel vehicles (Figures 2 and 3). The standard deviations for the HC data are high (10-20% vs. a normal 5-10%) probably as a result of erratic valve action (heavy ratings for exhaust valve tops) and highly transient spark plug bridging (although average spark plug cleanliness was normal). As is normal, on the average, HC emission levels remained constant after a slight increase during the first few thousand miles of operation.

Interesting comparisons can be made between Plymouth 40 on leaded fuel and Plymouth 39 on unleaded fuel. The average composite hydrocarbon emission for Plymouth 40 is 339 ppm while Plymouth 39 averages 238 ppm. The effect of lead on HC emissions was also demonstrated when the clear fuel car, Plymouth 39, was contaminated with lead from 20,000 to 24,000 miles. This was immediately evident in the hydrocarbon emissions at 24,000 miles. Additional hot cycle tests after tune-ups did not lower the HC level. After an additional 4000 miles on unleaded fuel, the hydrocarbon emissions dropped to near the pre-lead contamination level. Even after extensive mileage, the HC level remained slightly higher than levels which were realized before the lead contamination.

There was a small but statistically significant effect of recirculation on the HC and CO emissions. The HC and CO data with and

Table 7

VEHICLE DEPOSIT RATINGS

Deposit Area	Clean	Average for Car No.						Overall*
		38	40	42	43	37	39	
Piston Top	0	1.59	0.75	1.06	1.56	0.88	1.06	1.17
Cylinder Head	0	1.50	0.78	1.44	1.19	0.69	1.13	1.12
Intake Valve Top	0	0.63	0.59	0.44	1.00	0.63	0.56	0.66
Exhaust Valve Top	0	3.50	2.75	3.13	2.63	1.75	0.81	2.75
Intake Valve Undersides	0	6.56	6.13	4.38	4.34	3.29	5.09	4.94
Intake Valve Stem	0	0.34	0.50	0.25	0.44	0.31	0.69	0.37
Exhaust Valve Undersides	0	2.63	1.69	2.50	2.50	2.13	0.47	2.29
Exhaust Valve Stem	0	0.56	0.53	1.97	1.31	1.44	0.84	1.16
Intake Ports	0	1.47	1.34	1.28	2.13	1.41	1.19	1.53
Compression Ring Groove No. 1	10	6.38	7.69	8.38	6.19	7.75	7.63	7.28
Compression Ring Groove No. 2	10	8.81	9.06	8.88	8.63	8.56	8.06	8.79
Oil Groove	10	8.63	8.19	8.00	6.94	8.50	8.81	8.05
Oil Ring Slots	10	9.90	9.80	9.90	9.50	9.65	9.80	9.75
Piston Ring Zone Average	10	8.43	8.69	8.79	7.82	8.62	8.58	8.47
Ring Sticking No. 1	10	10.00	10.00	10.00	10.00	10.00	10.00	10.00
Ring Sticking No. 2	10	10.00	10.00	10.00	10.00	10.00	10.00	10.00
Piston Skirt Varnish	10	9.74	9.73	8.50	9.70	7.38	7.69	9.01
Rocker Arm Pad Wear	0	3.60	2.30	3.30	3.60	4.00	2.00	3.36
Intake Valve Tip Wear	0	2.75	2.00	4.80	4.00	4.00	2.00	3.51
Exhaust Valve Tip Wear	0	3.50	3.00	4.80	5.00	4.00	2.00	4.06

*Not including Plymouth 39 using clear fuel.

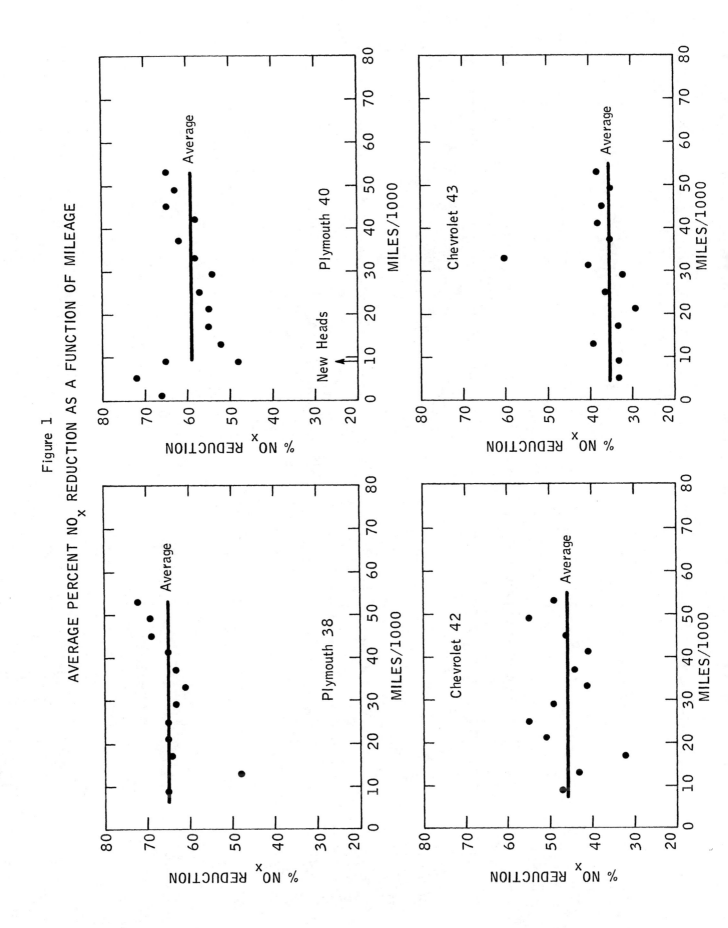

Figure 1
AVERAGE PERCENT NO$_x$ REDUCTION AS A FUNCTION OF MILEAGE

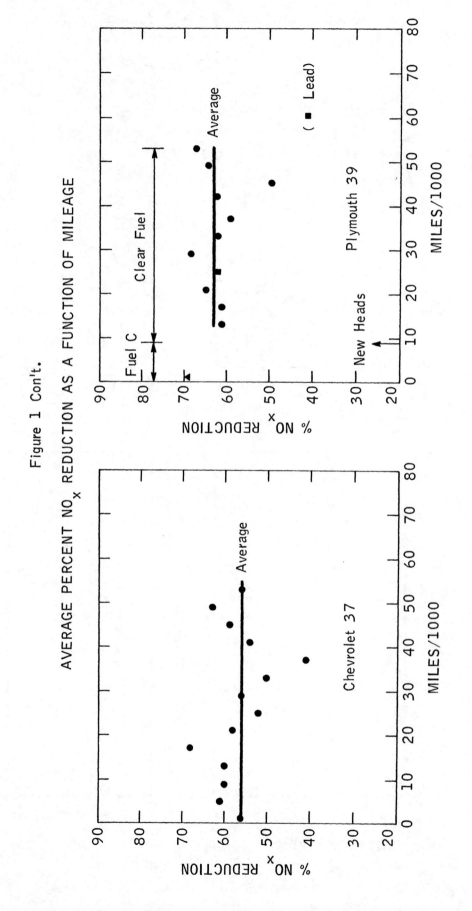

Figure 1 Con't.

AVERAGE PERCENT NO$_x$ REDUCTION AS A FUNCTION OF MILEAGE

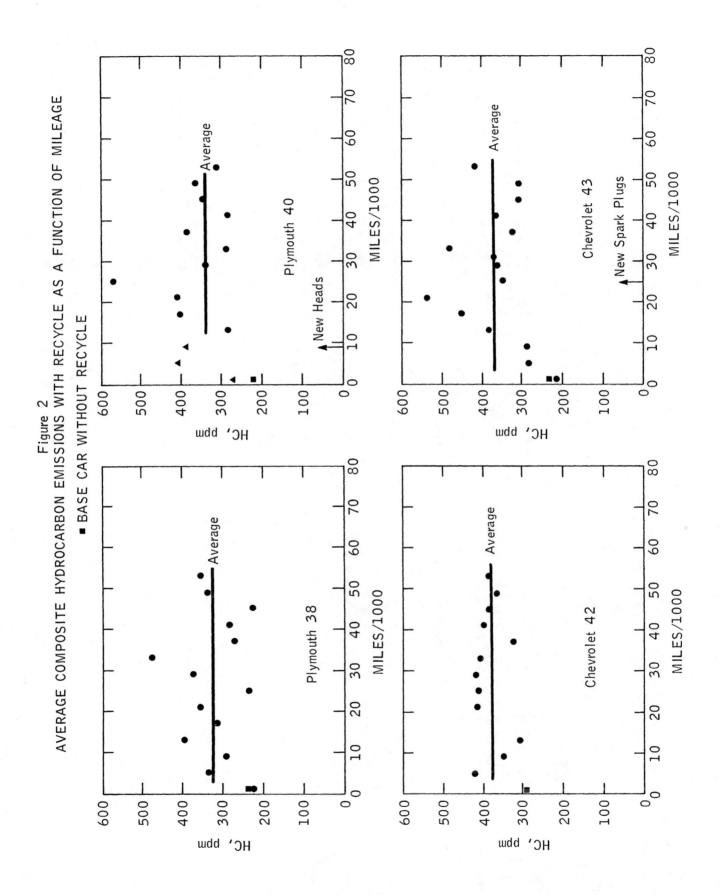

Figure 2
AVERAGE COMPOSITE HYDROCARBON EMISSIONS WITH RECYCLE AS A FUNCTION OF MILEAGE
■ BASE CAR WITHOUT RECYCLE

Figure 2 Con't.

AVERAGE COMPOSITE HYDROCARBON EMISSIONS WITH RECYCLE AS A FUNCTION OF MILEAGE

■ BASE CAR WITHOUT RECYCLE

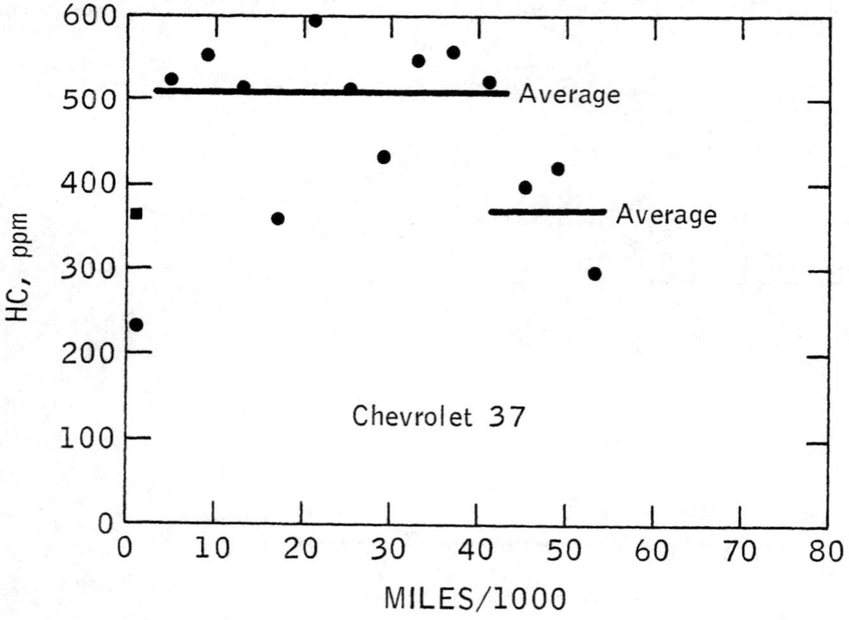

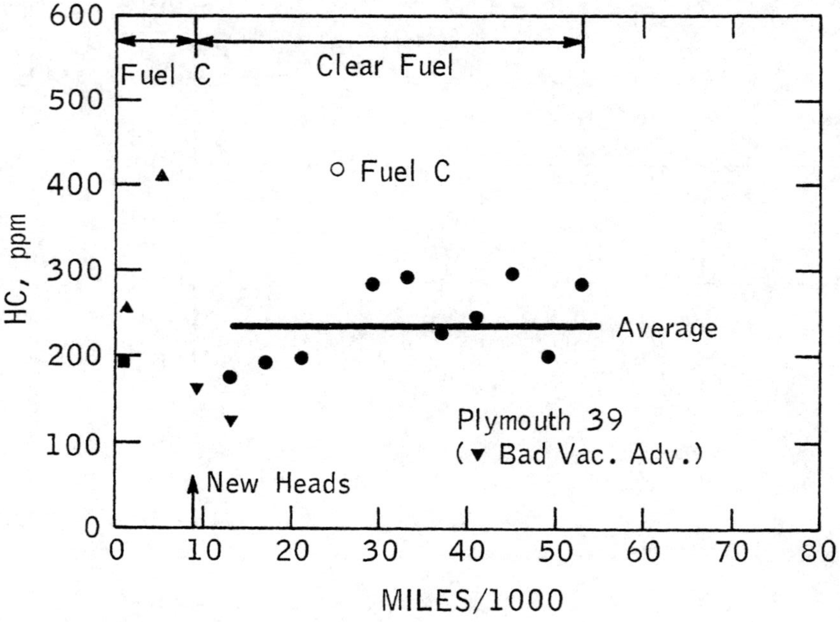

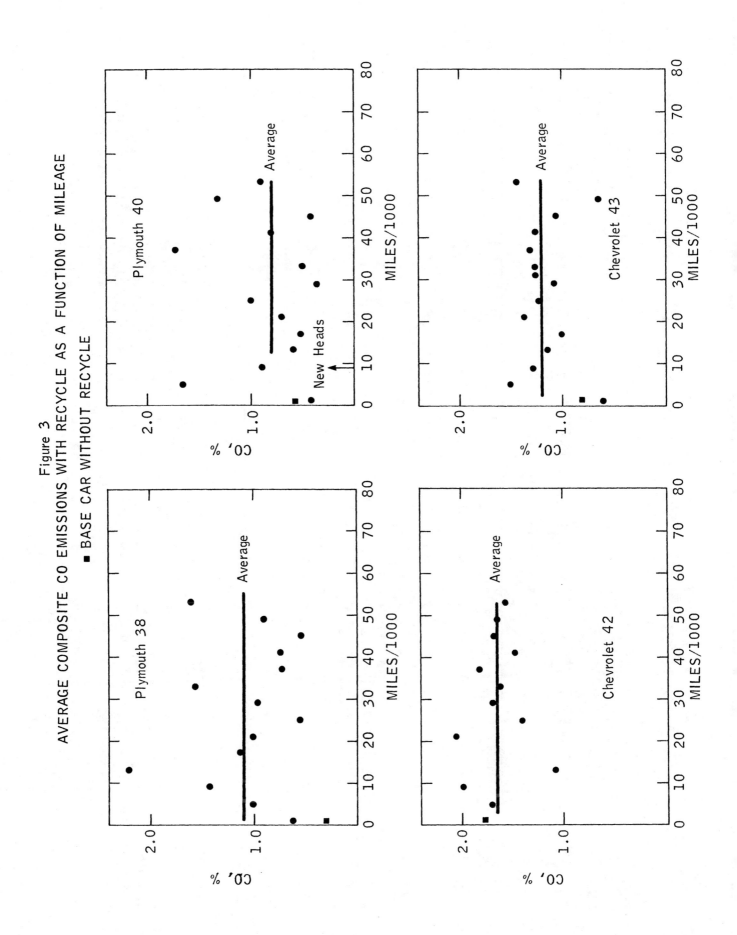

Figure 3 Con't.
AVERAGE COMPOSITE CO EMISSIONS WITH RECYCLE AS A FUNCTION OF MILEAGE

■ BASE CAR WITHOUT RECYCLE

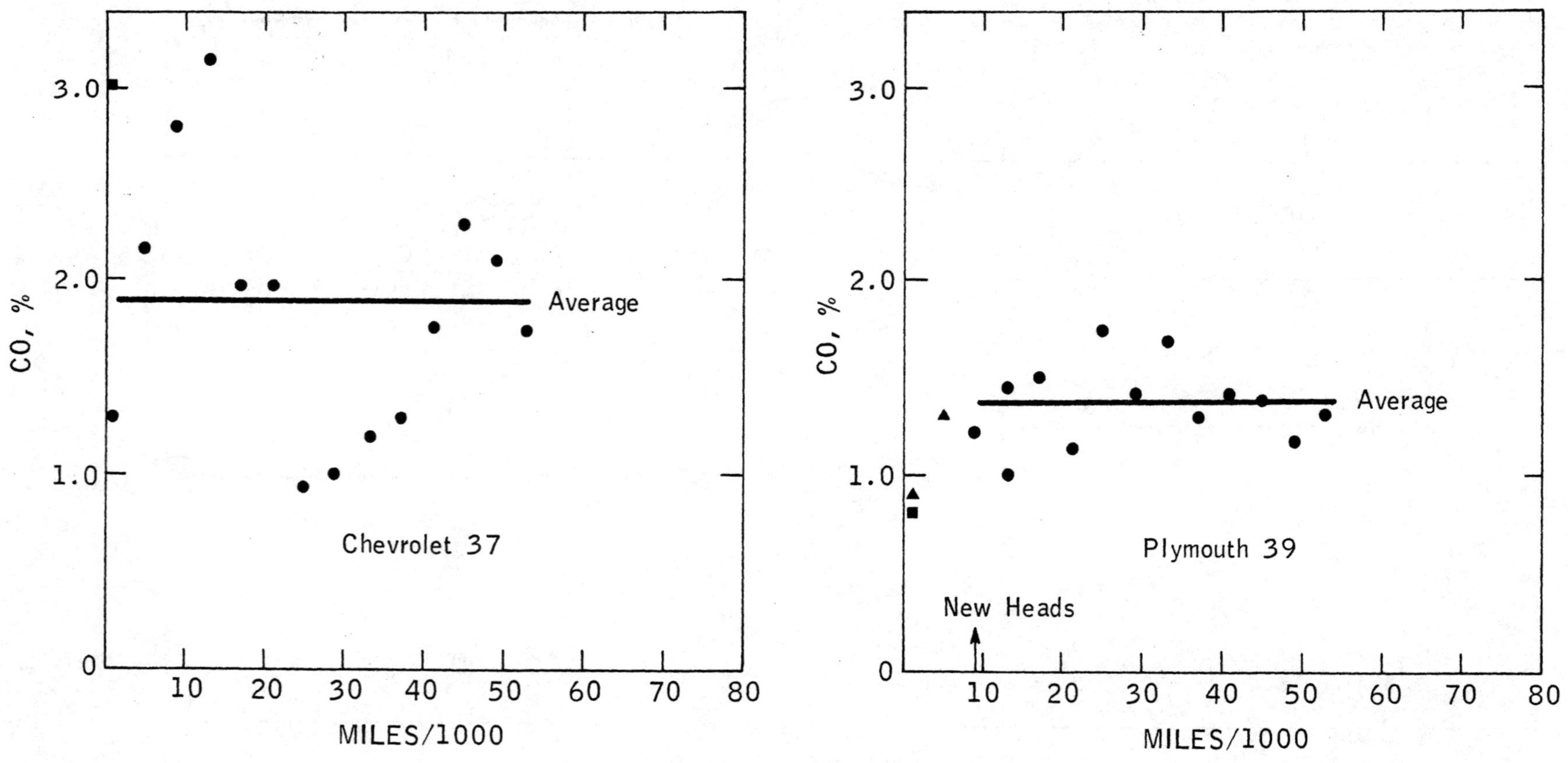

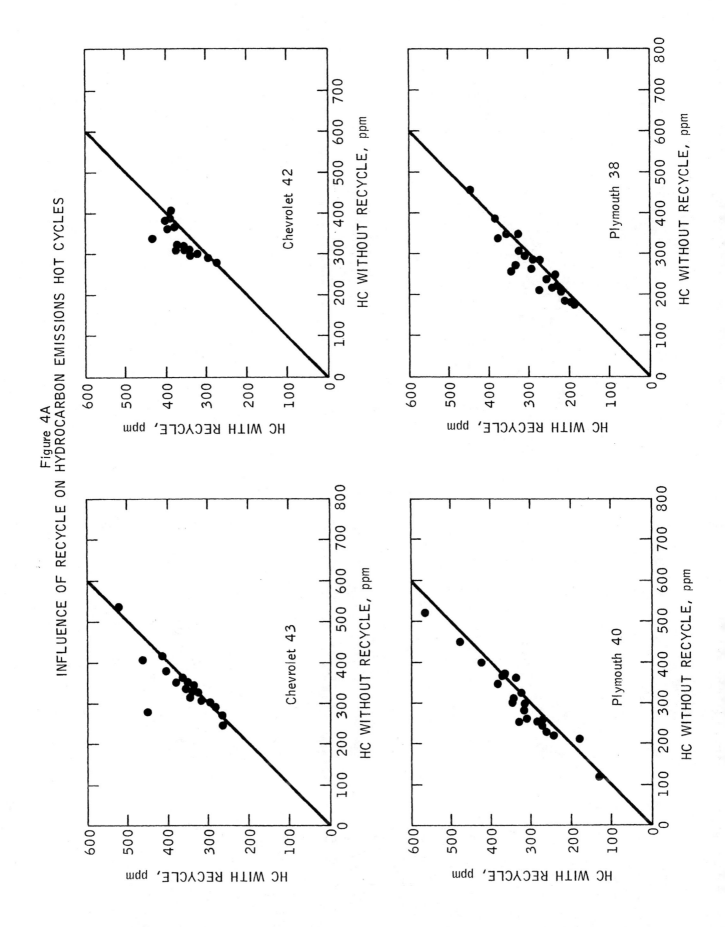

Figure 4A
INFLUENCE OF RECYCLE ON HYDROCARBON EMISSIONS HOT CYCLES

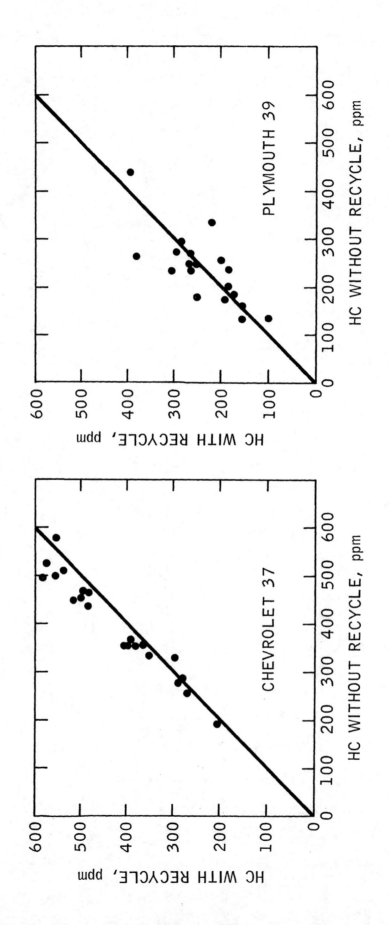

Figure 4A Con't. INFLUENCE OF RECYCLE ON HYDROCARBON EMISSIONS HOT CYCLES

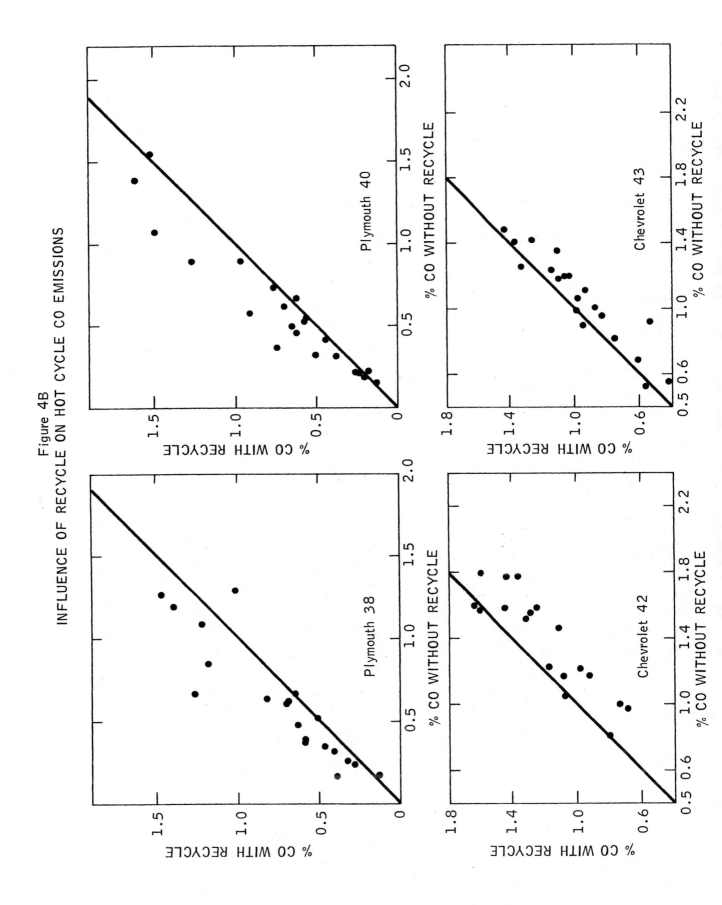

Figure 4B
INFLUENCE OF RECYCLE ON HOT CYCLE CO EMISSIONS

Figure 4B Con't.

INFLUENCE OF RECYCLE ON HOT CYCLE CO EMISSIONS

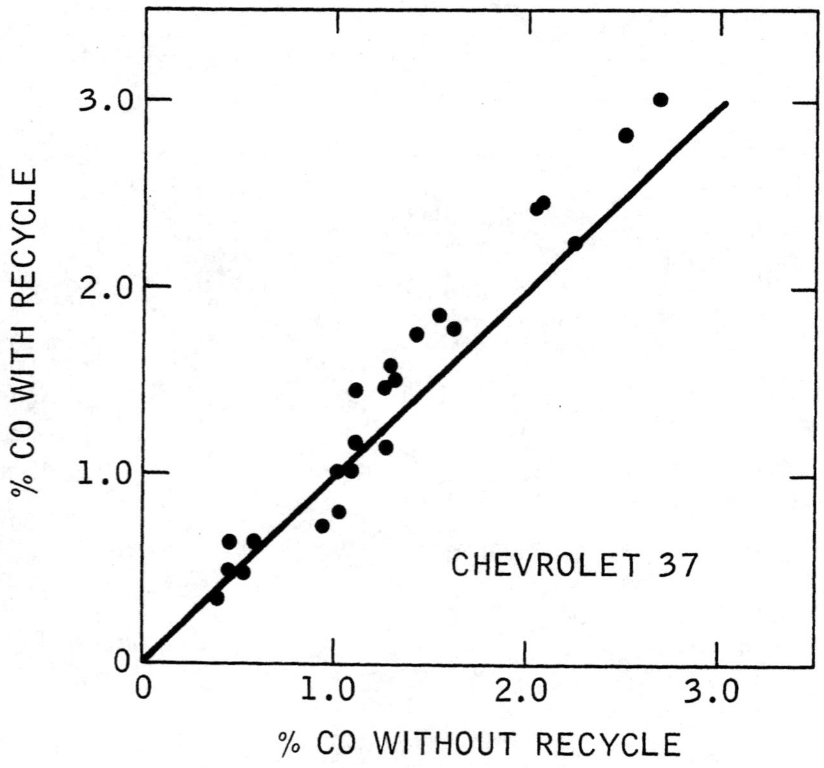

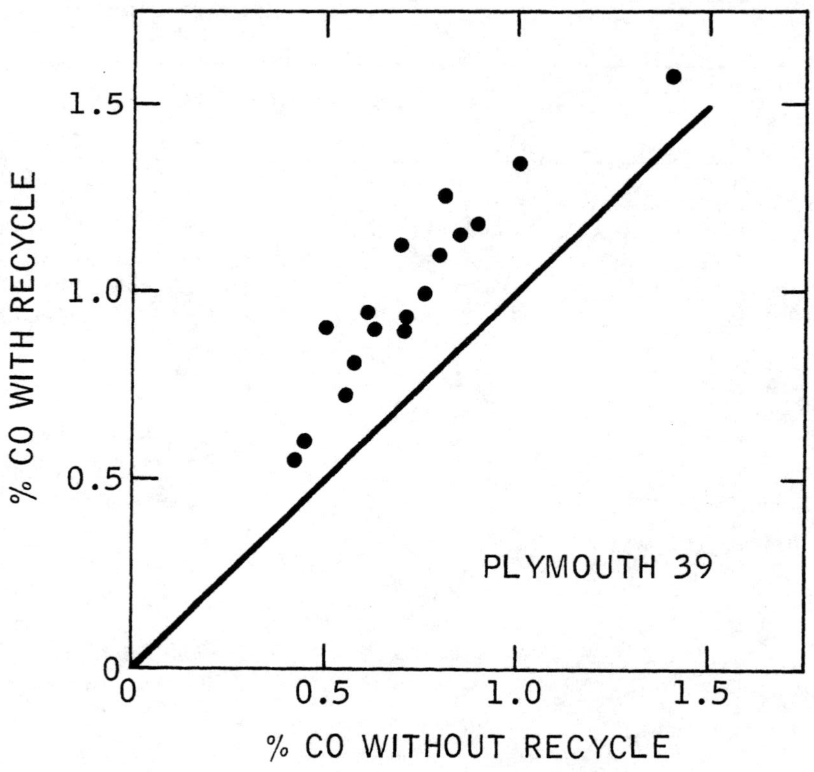

without recycle are summarized in Figures 4A and 4B. The differences are probably mainly due to spark timing as effected by the different intake manifold vacuums with and without recirculation.

The high HC emissions for the Chevrolets compared to the Plymouths are the result of a different vacuum spark advance schedule. While the Plymouths used a vacuum advance designed for EGR, the Chevrolets used a standard distributor with an additional 5° on the basic timing. This resulted in a rough idle and deceleration mode. Additional emission tests with the Chevrolets employing vacuum spark advance only when recycle was used, produced a smoother idle and deceleration modes with an accompanying reduction in HC emissions of 22% and CO emissions of 9%.

With a better spark advance, as described above, the vehicles averaged about 16% above the 1969 standards. This is about the same percentage for non-EGR vehicles after 50,000 miles.

VEHICLE PERFORMANCE - In this program, exhaust gas recirculation was present at wide-open-throttle conditions to prevent high NO_x emissions at such periods. As a result, the octane requirement of the vehicles "with recycle" was reduced from the "without recycle" case as shown below.

Figure 5

REDUCTION IN OCTANE REQUIREMENT
AS A FUNCTION OF RECYCLE

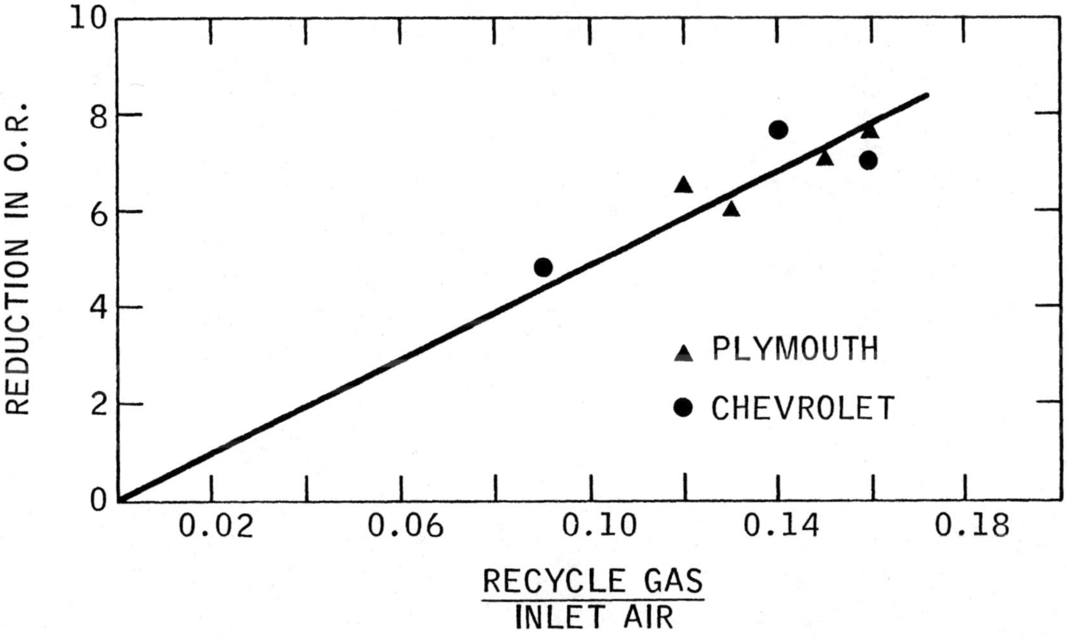

▲ PLYMOUTH
● CHEVROLET

$$\frac{\text{RECYCLE GAS}}{\text{INLET AIR}}$$

Vehicle performance and operability improved with the change in driving schedule. The WOT accelerations in the AMA schedule improved considerably the emission test results during the last 12,000 test miles. The standard deviation for the HC emissions during the AMA schedule segment was significantly lower than for tests made during the milk run segment.

The WOT acceleration times were essentially constant over the 52,000 test miles.

THROTTLE AREA DEPOSITS - At 40,000 test miles, the throttle plate deposits for the leaded fueled vehicles were as shown for Plymouth 40 in the photographs in Figure 6. In contrast, Plymouth 39 using unleaded fuel had essentially no throttle area deposits. At 52,000 test miles, the throttle plate deposits for Plymouth 40 (see photographs, Figure 7) had increased. These deposits were unexpected since vehicle operation and emission testing had improved with the new driving schedule. The clear fuel vehicle still had essentially no throttle area deposits. The amount of the deposits on the throttle plates is given in Table 8.

Table 8

CARBURETOR THROTTLE PLATE DEPOSIT WEIGHTS - mg

Vehicle	Left	Right	Total
Plymouth 38	156.1	98.0	254.1
Plymouth 40	241.5	237.2	478.7
Chevrolet 42	13.2	40.3	53.5
Chevrolet 43	24.3	28.1	52.4
Chevrolet 37	81.6	50.4	132.0
Plymouth 39	5.2	8.1	13.3

The small amount of deposit found on the throttle plates of Plymouth 39 is probably due to the leaded gasoline used from 20,000 to 24,000 test miles. The relative amount of deposits for the leaded fueled vehicles seems to be roughly related to the recycle gas - inlet air ratio. Analysis of these

Figure 6
THROTTLE PLATE AREA
AT 40,000 TEST MILES

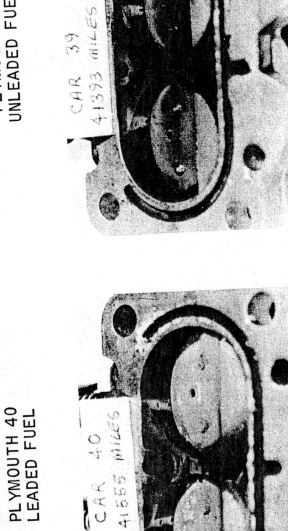

Figure 7
THROTTLE PLATE AREA
AT 52,000 TEST MILES

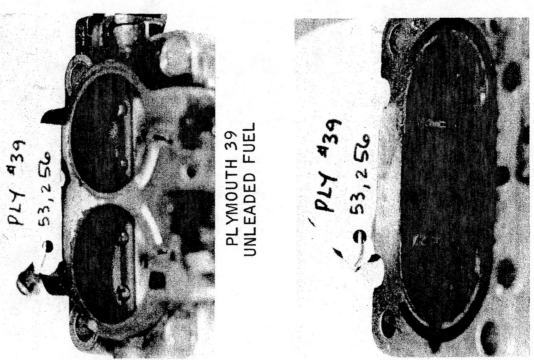
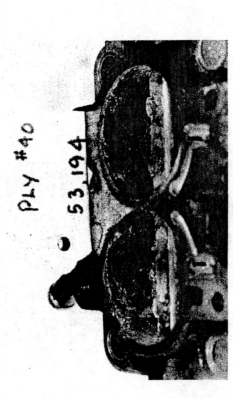

deposits and those found in the exhaust pipe indicate that the two are identical. The main component is $PbCl_2/PbBr_2$. In both the exhaust and carburetor, elemental lead accounts for the same percent by weight of the deposits. Depending upon the individual vehicle, elemental lead accounts for 25% to 45% of the deposit. Within experimental error, identical traces of lubricating oil were found in both the exhaust and carburetor.

In the area of particulate emissions, it is known that lowering the exhaust gas velocity and temperature improves the particulate trapping efficiency. Increasing the gas velocity and temperature would have the opposite effect. The switch to the AMA driving schedule increased the recycle gas velocity and temperature. The milk run driving conditions included less high speed operation. The AMA schedule has periods of not only higher cruise speeds (70 mph) but also WOT from 0 to 70 mph. Therefore, the gas velocities and temperatures in the recycle tubing increased over the last 12,000 miles of operation. The throttle plate area then was exposed to more particulates of a larger particle size. From this reasoning, the deposit analysis, and the fact that the unleaded fuel vehicle had essentially no deposits, it is our opinion that this increase in lead particulate size and amount caused the increase in throttle area deposits.

CONCLUSIONS - In summary, this study has achieved its primary purpose of obtaining a first approximation of the operability of EGR and its effects on engine durability under one set of typical driving conditions with extended mileage. This study together with previous experience indicates that for these six vehicles:

(1) Engine wear is normal with EGR over 52,000 test miles.

(2) Engine cleanliness is normal with EGR over 52,000 test miles.

(3) Vehicle performance with EGR does not decay with mileage over 52,000 test miles.

(4) The throttle plate area and exhaust valves may be sensitive to lead, particularly under certain driving conditions. The observed higher than normal standard deviation of the HC emissions were probably due to either highly transient exhaust valve leakage or highly transient spark plug bridging.

(5) EGR is compatible with air injection and engine modification systems over 52,000 test miles.

(6) EGR effectively reduces NO_x levels with no decrease in reduction over 52,000 test miles.

Therefore, our conclusion is that EGR is an effective means of reducing NO_x levels from automotive exhausts with no major unsolvable problems.

REFERENCES

1. Esso Research and Engineering Co., Final Report to NAPCA under Contract PH-86-67-25, July 1, 1969.

2. W. Glass, F. R. Russell, D. T. Wade and D. M. Hollabaugh, "Evaluation of Exhaust Recirculation for Control of Nitrogen Oxides Emissions," Paper No. 700146 presented at S.A.E. winter meeting, Detroit, Michigan, January, 1970.

3. Esso Research and Engineering Co., Final Report to NAPCA under Contract CPA-22-69-89, August 22, 1970.

4. O. G. Lewis, R. R. Risher, Jr., and J. A. Wilson, "An Eight Lane Dynamometer Highway," S.A.E. summer meeting, Atlantic City, New Jersey, June, 1959.

5. Federal Register, Vol. 33, No. 108, June 4, 1968.

ACKNOWLEDGMENTS

The work reported was partially funded by the Motor Vehicle Pollution Control Division, National Air Pollution Control Administration, Department of Health, Education and Welfare.

The work reported was also partially funded by Enjay Laboratories.

The authors also acknowledge the fine work of T. Karmilovich in obtaining all the emission tests and solving the normal day-to-day operational problems.

Methods for Fast Catalytic System Warm-Up During Vehicle Cold Starts

W. E. Bernhardt and E. Hoffmann
VOLKSWAGENWERK AG

INTRODUCTION

To achieve the emission targets prescribed by law for 1975/76 a number of emission concepts with conventional internal combustion engines and emission control systems have been examined by the automotive industry. Catalytic converters, thermal reactors and a combination of these two have been considered as emission control systems (1)*. Low emission values have been attained with these concepts when the engine is under warm working condition. However, the difficulties lie mainly in the warm-up phase during cold vehicle start-up.

To improve the over-all effectiveness of catalytic systems at vehicle start-up, extensive experimental tests were carried out during the warm-up phase on various after burning systems by the Research Department of the Volkswagenwerk AG. The intent of this paper is to illustrate the utility of improving the warm-up characteristic of catalytic emission control systems for achieving very low emission levels.

ABSTRACT

During vehicle cold start, emissions, mass flow rates, and catalytic converter space velocities vary by orders of magnitude. Therefore, catalytic exhaust control systems must be designed to operate at high efficiency almost from the moment of engine start-up. Catalysts must reach their operating temperature as quickly as possible. Therefore, the utility of different methods for improving the warm-up characteristics of catalytic systems is illustrated.

A very elegant method to speed the warm-up is the use of the engine itself as a "preheater" for the catalytic converters. High exhaust gas enthalpy to raise exhaust system mass up to its operating temperature is obtained by the use of extreme spark retard, stochiometric mixtures, and fully opened throttle. Intensive studies to investigate the effects of concurrent changes of spark timing and air/fuel mixtures on exhaust gas temperature, enthalpy, NO_x and HC emissions are discussed.

Finally, NO_x catalyst characteristics are dealt with, because the NO_x catalyst is the first in a dual-bed catalytic system. The NO_x catalyst should have high activity, low ignition temperature, and good warm-up performance. If the NO_x catalyst has a fast warm-up rate, this would result even in a significant improvement in the warm-up characteristic of the HC/CO bed.

---

*Number in ( ) indicates reference at end of paper.

WARM-UP METHODS FOR CATALYTIC SYSTEMS

Catalytic emission control systems described in this paper operate mainly with the dual-bed catalytic process. The first bed contains the reduction catalyst which reduces the oxides of nitrogen (NO_x) by carbon monoxide (CO), hydrogen (H_2), and hydrocarbons (HC) which are present in the exhaust gases. The reaction between NO_x and CO will only take place providing that the amount of oxygen (O_2) present in the exhaust gas is strictly limited to low concentrations. This oxygen limitation is met by adjusting rich fuel/air mixtures.

The second catalyst bed contains the oxidation catalyst which burns the carbon monoxide and hydrocarbons after introducing secondary air between the first and second beds. The quantity of secondary air is set high enough to ensure that there is excess oxygen for all driving conditions.

Figure 1 illustrates a dual-bed axial-flow converter. Such a concept using fresh catalysts when tested according to the CVS cold-hot test procedure gave emission values which were still twice as high as the exhaust emission standards specified for model year 1976. The results would be considerably better if the catalytic emission system could be warmed up very quickly from the moment of cold engine start-up. Methods to speed the warm-up are listed below:

- Reduce the heat capacity of the exhaust system between the engine and the dual bed converter.
- Reduce the heat capacity of the catalyst, use smaller catalyst quantities, and smaller catalyst particles.
- Mount the converter very near to the engine exhaust valves.
- Introduce secondary air in front of the first bed during the initial 120 seconds after cold engine start-up; then switch the secondary air to the connecting pipe between NO_x and HC/CO beds (staged secondary air).

Even when these features were used, the dual-bed system illustrated in Figure 1 could not reach the targets of 0.41 gm. HC/mi., 3.4 gm. CO/mi., and 0.4 gm. NO_x mi. as proposed in the Federal Register for 1976.

It is particularly difficult to fulfill the emission standard for oxides of nitrogen as the temperature in the NO_x bed increases very slowly in systems which are designed to allow an adequate residence time for the exhaust gases. In Figure 2 is illustrated the mid-bed temperature of a VW radial-flow converter with a 1.3 liter NO_x bed during the CVS cold start. It is plain to

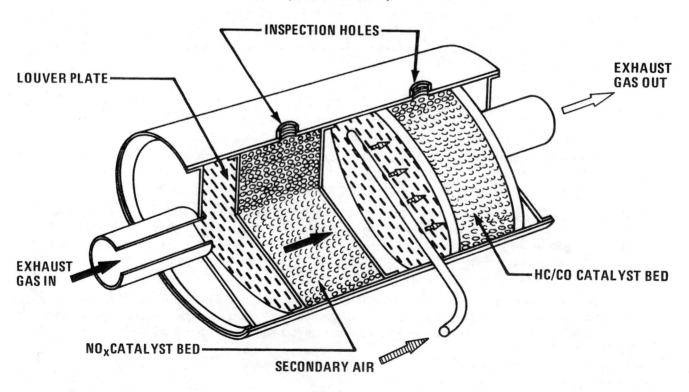

Figure 1

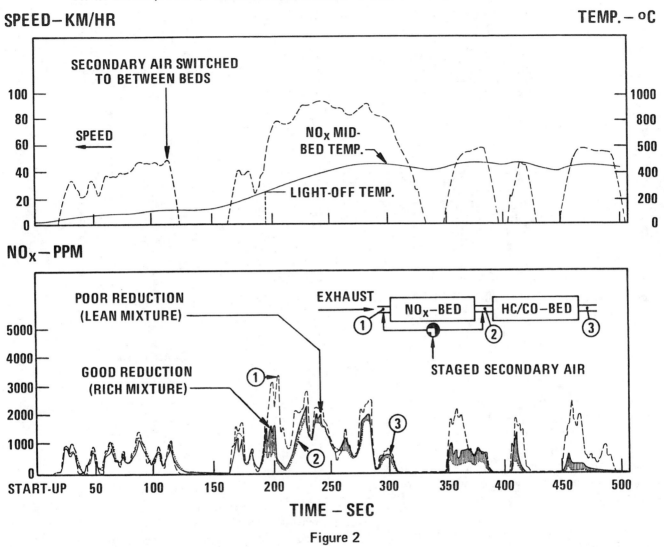

Figure 2

see that the catalyst ignition temperature of approx. 250°C was reached in the pelleted NO_x catalyst bed after 195 seconds in spite of the use of the staged secondary air feature. The ammonia problem can also be seen in Figure 2 although this is not being dealt with in this connection. For more detailed information refer to Meguerian (2).

To illustrate the problems of catalytic exhaust emission control systems Figure 3 shows the tail pipe emissions as well as the exhaust gas flow rate during the first 240 sec. of a CVS cold start test. Both the concentrations and the mass flow rate vary by orders of magnitude during the CVS cold start test procedure. Furthermore, the concentrations of CO and HC are particularly high during the first 80 sec. For this reason, catalytic emission control systems must be designed to operate with high efficiency, that means high reduction and conversion rates, as quickly as possible after the engine start-up. The catalysts should reach their operating temperatures within 20 sec, so that the emissions which are produced during the warm-up period of the engine can be controlled as quickly as possible.

To do this, further methods for improving the warm-up characteristics of dual-bed systems should be investigated.

One method which promises success is a thermal reactor acting as a "preheater" for improving catalytic converter performance. The thermal reactor is located at the cylinder heads. When starting with a rich fuel/air mixture, oxidation of carbon monoxide and hydrocarbons after adding air, ensures rapid warm-up of the

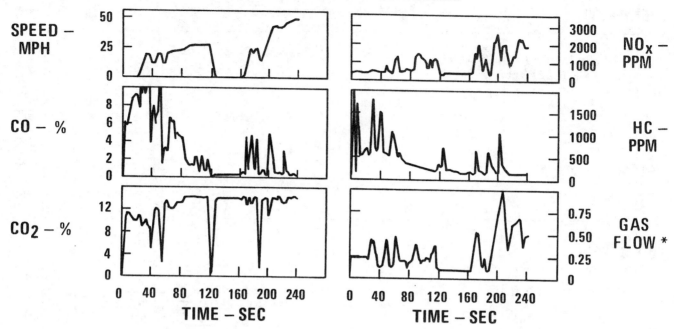

Figure 3

catalyst. In this system, secondary air is introduced in front of the thermal reactor at the cylinder head, and the reduction catalyst works as an oxidation catalyst in the starting phase. Due to the burning of high HC and CO emission levels directly after start-up a rapid warm-up of the after burning system and therefore a rapid attainment of operating temperature is ensured.

A catalytic reduction of NO_x is not necessary during the cold start phase because the engine operating temperature during this period is not high enough to produce very high NO_x emissions. After the catalysts in both beds have reached their operating temperatures (100-120 sec. after CVS cold start-up), the thermal reactor must be switched off. This is brought about by the transfer of secondary air introduction to the connecting manifold between the first and second catalyst beds.

Figure 4 illustrates a catalytic emission control system together with major hardware components which has been developed for research purposes. It consists of a monolithic dual-bed converter, series connected thermal reactor, by-pass system, exhaust gas recirculation (EGR), EGR cooler, regulating valve, and EGR filter. The efficiency of such an emission control concept can be improved by the introduction of an additional ignition system in the thermal reactor. By enrichment of the air/fuel mixture, an improvement can be obtained as can be seen in Figure 5. With 10% rich fuel/air mixtures (A/F = 13.0), the exhaust gas temperature at the thermal reactor outlet reaches 300°C five to six seconds earlier than with normal mixture strength (A/F = 16.0).

This high exhaust gas temperature increases the reactor warm-up rate, too. This means that the catalysts also reach their operating temperatures of about 300°C at least five seconds earlier.

Another method of improving the warm-up rate of the catalytic system with series connected thermal reactors is to cause an ignition failure of a single cylinder charge (which contains approx. 20,000 ppm HC) and at the same time to increase the idling speed of the engine together with wide open throttle. The technique produces an increased flow rate of exhaust gases with high unburned components during the cold start phase; the chemical energy of which can be converted into high exhaust enthalpy. This comparatively rich fuel/air mixture can be ignited in the reactor by an additional spark-plug. With an automatic control device, the ignition failure of a particular cylinder can be controlled in accordance with the firing order. After the exhaust gas has attained the operating temperature required by the catalysts the ignition system will revert to normal.

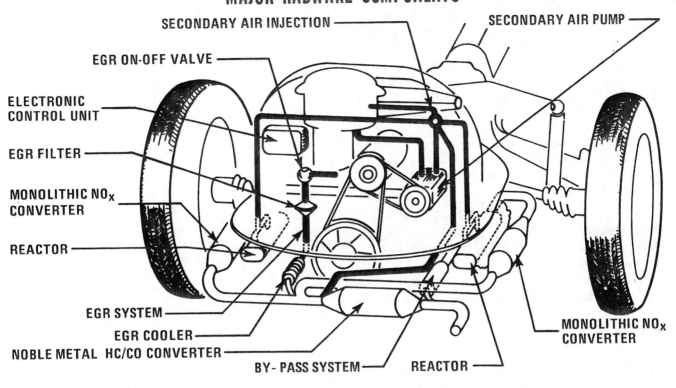

Figure 4

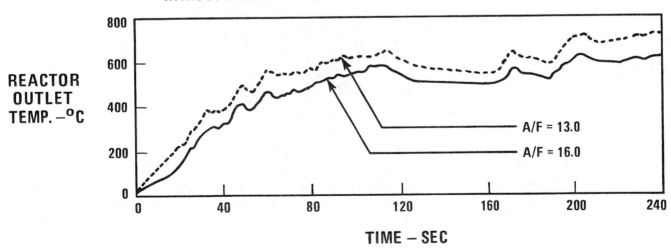

Figure 5

Figure 6 shows the results of an engine cold start test employing this warm-up system. The illustration shows the time dependence of the exhaust temperature in the reactor core and at the reactor outlet as a function of throttle opening. The test was carried out at an engine speed of 2800 rpm. The throttle valve angle was increased from 5° to 35°. A maximum speed governor controlled the ignition failure. It can be seen in Figure 6, for example, that with a throttle valve opening of 35° a reactor outlet temperature of 360°C is achieved in 10 sec. Even when considering the heat loss of the exhaust system between the thermal reactor and the catalytic system, it can be ensured that the operating temperature of the first bed can be reached in a very short interval of time.

Especially when operating under rich fuel/air conditions with thermal reactors mounted at the cylinder heads temperatures could be produced which are above the melting temperature of the monolithic materials, such as Cordierite ($Mg_2Al_4Si_5O_{18}$), Mullite ($3Al_2O_3 \cdot 2SiO_2$), and Alumina (a-Al_2O_3). In Figure 7, the monolithic catalyst reached temperatures of more than 1350°C. It can be seen by the figure that the center of the monolithic catalyst has been melted when operating together with a front mounted reactor. Tests have proved that a too high inlet concentration of HC/CO is not the cause of this high bed temperature. The cause of the thermal destruction of the material could lay in the unequal distribution of the active components on the support material. This unequal distribution of material, as for example CuO, could have led to a drastic reduction of the melting temperature from 1350 down to 975°C. Similar symptoms in the outer coating were observed during the aging process of catalysts by J. F. Roth (3).

In place of the thermal reactors, monolithic noble metal catalysts could be employed as warm-up elements because the majority of the HC and CO emissions produced by an engine are emitted in the first two minutes of the 42-min. CVS cold-hot test, while the NO_x emission in general is produced over the whole test period. For this reason, a monolithic HC/CO converter at each side of the engine which reduces the high carbon monoxide and hydrocarbon raw emissions with high efficiency could be used to improve the warm-up characteristic. Due to the good cold start performance and the low ignition temperature, the platinum monolith is particularly suitable. The ignition temperature for CO is approximately between 200 and 280°C and for HC (hexane) between 240 and 340°C.

To give a complete presentation of warm-up possibilities, other more sophisticated methods for rapid warm-up of catalytic systems should be mentioned.

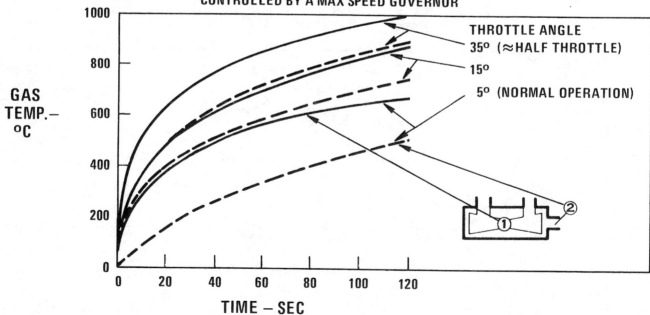

Figure 6

MONOLITHIC CATALYST FAILURE
During Rapid Warm-up Performance

Figure 7

One such approach involves the use of a gasoline heater, the other approach involves an electric heater. An electrically heated HC/CO radial-flow converter in which high temperature resistance heating rods were installed was tested within the IIEC Program (4). These heating rods were capable of heating the first layer of the catalytic bed (approx. 0.5 lb.) from the ambient temperature to 300°C within 30-40 sec. at start-up. Power requirements were available from vehicle electrical system. Similar results were obtained by the VW Research Department with pelleted catalysts in axial-flow converters which were however supplied by an external electrical system. The power available from the battery alone was not sufficient.

In most cases the gasoline heater has the disadvantage of not being able to operate against the relatively high exhaust gas back pressures. A further disadvantage is that an auxiliary heater could produce considerably high emissions. Without taking the engine emissions into consideration, the following table shows the values produced during a CVS test by a particular heating system:

HC 0.58 gm./mi.

CO 0.42 gm./mi.

NO_x 0.03 gm./mi.

It may be surprising that for HC this is more than 40% above the 1975 target, while the CO emission is approx. 12% and the NO_x emission is approx. 8% of the emission standards proposed for 1976. Fortunately, a properly designed gasoline heater operating only 100 to 120 sec. after engine start-up has considerably lower emissions.

Another method of improving the initial reaction temperature of the catalytic system is to increase the exhaust gas temperature by altering the ignition timing into the region after T.D.C. By the use of extreme retarded timing at vehicle start-up for example, the catalyst operating temperature of 250°C was reached 25 sec. earlier in the CVS test than by normal ignition timing. The influences which the ignition timing has upon the combustion process (exhaust gas temperature and exhaust gas emissions) are discussed in more detail in the next chapter. Based on extensive single cylinder measurements the utility of this warm-up technique is illustrated because of its particular importance for speeding the warm-up performance of catalytic systems.

WARM-UP TECHNIQUE BY SPECIAL ENGINE OPERATION

Only by employing after-burning systems, can the extremely low emission targets for model year 1975/76 be reached. Therefore, one of the most important tasks of the internal combustion engine is to ensure high efficiency almost immediately after engine start-up by changing the engine conditions especially for this requirement. It has been found that under appropriate operating conditions the engine itself is able to act as a preheater for the catalytic system. Warm-up spark retard and an increased idling speed of the engine with full open throttle lead to higher exhaust temperatures and thereby to a greater enthalpy of the exhaust gases, so that the after burning system could be brought rapidly up to its operating temperature.

Figure 8 shows schematically an internal combustion engine as an open thermodynamic system. If steady flow is assumed the application of the First Law of Thermodynamics gives important information about possibilities of increasing the exhaust gas enthalpy (see Figures 8 and 9).

As shown in the equations in Figure 9 the total chemical energy of the exhaust gases can be used to increase the exhaust gas enthalpy if the shaft work is zero ($W_{12} = 0$). In practice this operation condition cannot be achieved because the engine could not overcome its own mechanical friction. The maximum heat of the exhaust gases ($\dot{Q}_{12}$)$_{max}$ is therefore not at the no work condition, but at an indicated output which is appropriate to the mechanical friction of the engine. A second factor is the quantity of the exhaust gas flow rate which can be increased by opening the throttle. In an engine the condition $W_{12} = 0$ can be attained by altering the ignition timing to "retard". In this case the energy release rises very late so that the work done on the piston becomes less.

To illustrate the influence of the ignition timing on the combustion process more clearly, Figure 10 should be considered. This figure illustrates the results of a thermodynamic analysis of two combustion processes at low load. The combustion cycles differ only in the ignition timing (9° B.T.C. as opposed to normal adjustment of 27° B.T.C.) whereas other engine parameters such as air/fuel ratio and volumetric efficiency remained equal. Essential differences can be seen already in the pressure-time history (upper diagram). In the case of extreme retarded timing the maximum pressure is reduced from $16.5 \cdot 10^5$ to $7.8 \cdot 10^5$ Pa (pascal, newton/sq. meter) at 20° and 50° A.T.C., respectively. The expansion process of the working fluid can be noticed very late, runs at comparatively high pressure level and due to the opening of the exhaust valve it is cut off at a high pressure. Due to this energy loss to the ambient (that is loss of the work done on the piston) the indicated mean effective pressure was reduced from $3.90 \cdot 10^5$ to $3.05 \cdot 10^5$ Pa. This result can be taken directly from the energy release diagram (lower diagram).

APPLICATION OF FIRST LAW OF THERMODYNAMICS TO AN INTERNAL COMBUSTION ENGINE
Open Steady-Flow System

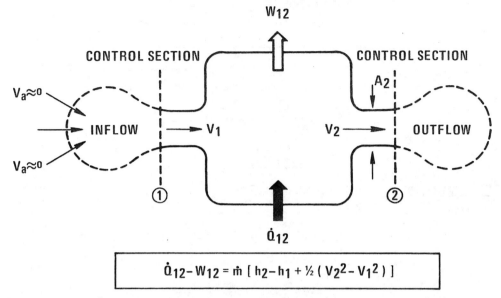

$$\dot{Q}_{12} - W_{12} = \dot{m}\,[\,h_2 - h_1 + \tfrac{1}{2}(V_2^2 - V_1^2)\,]$$

$\dot{Q}_{12}$–HEAT, W_{12}–WORK, $\dot{m}$–MASS FLOW RATE, h–ENTHALPY PER UNIT MASS, V–VELOCITY, SUBSCRIPT a–"AMBIENT".

Figure 8

APPLICATION OF FIRST LAW OF THERMODYNAMICS TO INCREASE THE SENSIBLE HEAT OF THE EXHAUST GASES

If No Work W_{12} Is Done By The Engine The Sensible Heat $\dot{Q}_{12}$ Reaches Its Maximum

$W_{12} = 0 \longrightarrow \dot{Q}_{12} = \dot{m}\,[\,h_2 - h_1 + \tfrac{1}{2}(V_2^2 - V_1^2)\,]$

$\dot{Q}_{12} = \dot{m}\,[\,h_2 + \tfrac{1}{2}V_2^2 - h_a\,]$

$h_a = h_1 + \dfrac{V_1^2}{2};\quad V_2 = \dot{m}/(A_2 \cdot \rho_2);\quad \rho$ – DENSITY

A – OUTLET AREA,

$$\dot{Q}_{12} = \dot{m}\,(h_2 - h_a) + \tfrac{1}{2}\,\dfrac{\dot{m}^3}{(A_2 \cdot \rho_2)^2}$$

$C_p \approx C_p(t);\quad$ t–TEMP; $\quad C_p$–SPECIFIC HEAT AT CONSTANT PRESSURE

$$\dot{Q}_{12} = \dot{m}\,C_p\,(t_2 - t_a) + \tfrac{1}{2}\,\dfrac{\dot{m}^3}{(A_2 \cdot \rho_2)^2}$$

(SAME SYMBOLS AS IN FIGURE 8)

Figure 9

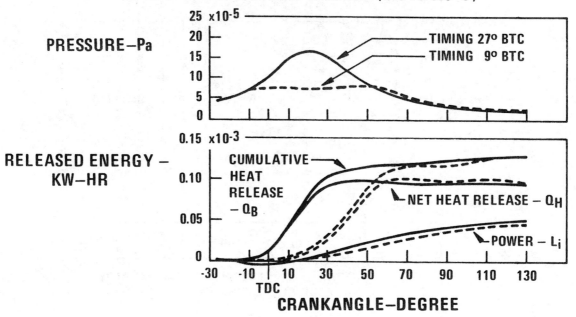

Figure 10

Due to the late energy release rise, the period in which the high temperatures are produced in the exhaust gases is considerably shortened. For this reason the heat loss to the cylinder walls during combustion is less. The same quantity of fuel is converted into energy in both cases, but in the case of extreme spark retard the piston work L_i and the heat loss to the walls $Q_w = Q_B - Q_H$ is reduced, and the internal energy of the working gases $U_i = Q_H - L_i$ is increased by the appropriate amount. The exhaust gas temperature is thereby increased by 66°C from 580 to 646°C. The exhaust gas emissions are thereby also strongly influenced; the HC emission is reduced from 132 ppm to 60 ppm by the high exhaust temperature, and the NO_x emission is suppressed from 1888 ppm to 720 ppm due to the retarded ignition timing.

For the experimental investigation a VW 1.6 liter single cylinder engine with a production type combustion chamber was used. A mechanical fuel injection system was chosen with which optimum fuel/air ratios, good mixture preparation, and an independance from the distributor setting was available.

The measurement of the exhaust gas temperature was carried out in the exhaust with an insulated thermocouple of 1.5 mm. O.D. The holding device for the thermo-element was fitted with a radiation shield.

The tests began with an increased idling speed of 2500 rpm and were later continued at 1500 rpm. The engine speeds were selected with respect to the highest possible exhaust flow rate. After adjusting the ignition timing the individual measuring points were chosen to a particular value in relation to the air flow rate, while the required air/fuel ratio was regulated by the fuel quantity. Deviations from the mean effective pressure $p_{me} = 0$ had to be balanced out by adjusting the supplied air and fuel flow rates in stages.

In addition to the measured exhaust gas temperatures Figure 11 shows the HC and NO_x emissions as well as the exhaust gas enthalpy as a function of air/fuel ratio and ignition timing. These results are in good agreement with the theory. As expected the exhaust gas temperature increased very rapidly when the ignition timing was retarded to a region after T.D.C. However, due to alteration of the ignition timing the indicated output was decreased, as already mentioned. In order to keep the engine running the cylinder charge was repeatedly increased as the timing became more retarded until finally the throttle was fully open. Therefore, the lean limit in the ignition region after T.D.C. is given as the left hand limit on the performance diagrams of Figure 11. The highest temperatures measured at the lean limit were 794°C at the engine speed of 1500 rpm and 912°C at 2500 rpm. The lean limit with spark advance is

ENGINE WARM-UP PERFORMANCE TEST

Engine Speed 1500 RPM, Steady State, No Load.

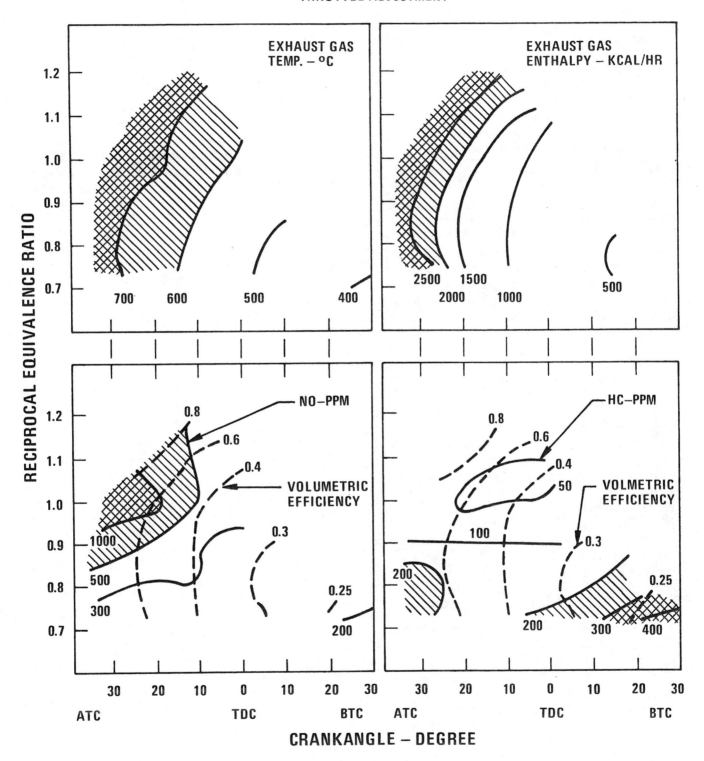

Figure 11

dependent on the air/fuel ratio in the same manner as it is with spark retard. However, the criteria for this lean limit is not the throttle opening but the breakdown of the combustion cycles which results in a rapid increase of the HC emission.

The exhaust gas temperature diagram in Figure 11 illustrates the influence of the air/fuel ratio and the ignition timing on the exhaust gas temperature. The exhaust gas temperature reaches its maximum at a reciprocal equivalence ratio of $1/\phi = 1.0$ because the combustion temperature is highest at stoichiometric mixtures.

The increase of air/fuel ratio at a constant spark advance brings the exhaust gas temperature up to a maximum at greatest possible air/fuel ratio.

This is explained by the postponement of the combustion process into the expansion stroke which is in turn caused by the lower flame speed at lean mixtures. Furthermore, the volumetric efficiency increases as the A/F ratio becomes larger. The low exhaust temperatures at rich mixtures are attributed to the cooling effect of the fuel in very rich mixtures. In combination with the volumetric efficiency, however, the largest influence on the exhaust gas temperature is exerted by the spark timing which brings about a large increase in the exhaust gas temperature.

As mentioned above, the volumetric efficiency for the desired indicated output is dependent on the spark timing and increases from about 0.25 with spark advance to a full load value of over 0.8 with spark retard, see Figure 11 (bottom). The curves of constant volumetric efficiency have the same tendency as the curves of constant exhaust gas temperature so that the desired objective of providing the hottest possible exhaust gas in largest possible quantities is achieved by the spark retard. This result is shown clearly by the enthalpy diagram. The increase in fuel flow corresponding to the increase in volumetric efficiency with spark retard raises the exhaust gas enthalpy (which is related to the enthalpy at ambient temperature) to a maximum of about 2750 kcal./hr. at 1500 rpm (Figure 11) whereas it is approximately 5600 kcal./hr. at 2500 rpm.

In the enthalpy diagram only the sensible heat portion of the exhaust energy is illustrated. The chemical energy still contained in the exhaust gas, particularly when there is a shortage of air, is not taken into account. By the use of appropriate devices (i.e. thermal reactor with secondary air injection) this energy can be used to warm-up the converters in the start-up phase.

In Figure 11 one can see that the spark timing has a very intensive influence on the composition of the exhaust gases. The HC emissions are lowered from 400 ppm to less than 50 ppm by spark retard despite the increasing amount of charge. Down to the spark retard running limit only a minimum rise can be determined. The cause of the low HC values are the high temperatures in the exhaust gas due to the delayed combustion which lead to a reaction of the still unburned hydrocarbons particularly in the presence of excess oxygen.

The NO emissions do not show the expected tendency due to the overlapping influence of the volumetric efficiency. The spark retard causes a drastic increase in NO emissions because the volumetric efficiency increases. This effect increases the temperature level in the combustion chamber due to the higher compression pressure. The NO maximum occurs in the region of $1/\phi = 1$. Furthermore the region in which the NO emissions are almost independent of the spark advance ($1/\phi \approx 0.8$) is shown in the NO emission diagram (Figure 11). Only excess oxygen in adequate quantities at lean mixtures ($1/\phi > 1$) increases the NO emissions when the spark advance is varied.

In order to ensure that hot exhaust gases in the largest possible quantities are available not only in the initial idling phase of the CVS test but also in the subsequent test procedure the possibilities of controlling the output with a wide open throttle by altering the spark timing were investigated (volumetric efficiency = constant). When doing this, the spark timing was no longer selected in conjunction with an A/F ratio and an air flow which gives the lowest fuel consumption and thus the best efficiency. Instead a timing with the largest possible exhaust gas enthalpy at least in the initial phase after start-up is used. This denotes a departure from the classical method of approach when optimizing engine characteristics.

The results of these tests are shown as a function of A/F ratio and ignition timing in Figure 12. It is also apparent here that the highest exhaust gas temperature and enthalpy values are obtained at a mean effective pressure of $p_{me} = 0$, that is at no-load condition with fully opened throttle whereby the HC and NO emissions are relatively low in this region. Therefore, it appears reasonable to run the engine in the first phase of the test procedure with extreme spark retard and permit it to develop only the output which is required for the acceleration phases of the test cycle. In this way the early desired operating temperature of the catalytic system is assured.

In Figure 12 the spark retard running limit is in good agreement with the lean limit obtained in tests with constant mean effective pressure, $p_{me} = 0$ (Figure 11). The deviation in the experimental data is within the limits of measuring accuracy. By altering the ignition timing the entire load range from no-load to full load with the throttle fully open is covered. The maximum

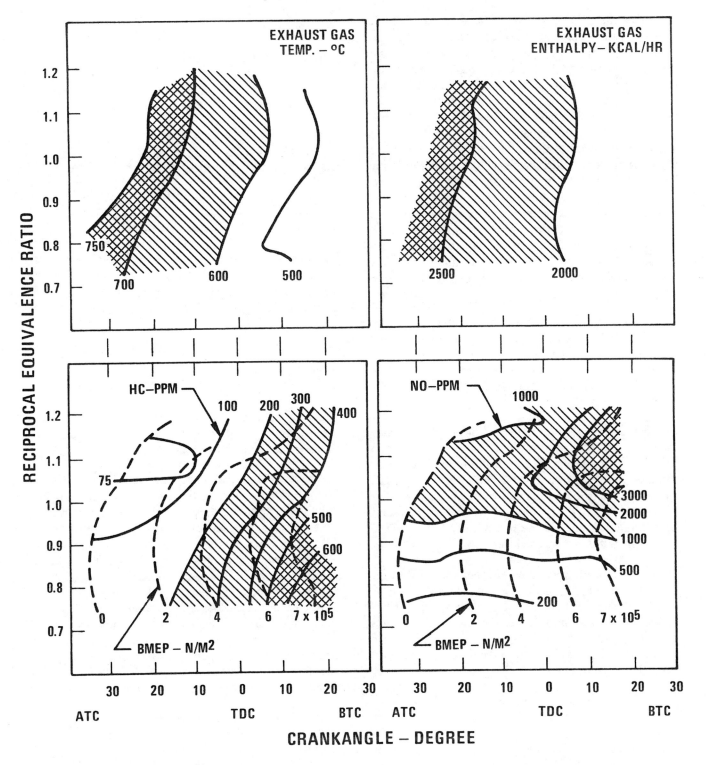

Figure 12

knock-free ignition advance point is the right hand limit of Figure 12. With constant ignition advance and by variation of the A/F ratio the maximum mean effective pressure is reached between $1/\phi = 0.85$ and $1/\phi = 0.9$. Increasing the A/F ratio causes a large drop, enrichment causes only a slight drop in the mean effective pressure.

Due to the high volumetric efficiency 70 to 90°C higher exhaust temperatures were measured over the entire investigated range as compared with the tests without load ($p_{me} = 0$) except at the spark retard running limit. Contrary to the tests at no-load conditions the exhaust temperature at rich fuel/air mixtures is dependent only on the spark advance and not on the A/F ratio. A study of the exhaust enthalpy per unit time shows the influence of the volumetric efficiency and the resultant higher exhaust gas temperatures. Between the spark retard running limit and the knock limit the exhaust enthalpy alters only by about 1000 kcal./hr., whereas with the same alteration in spark timing in Figure 11 for $p_{me} = 0$ a change of about 2000 kcal./hr. occurs. The curves of constant enthalpy show the direct dependence on spark timing which results because the fuel flow rate decreases to the same extent as the A/F ratio increases.

When compared with no load operation the available enthalpy during load variations via the spark timing is clearly larger so that this operation mode is suitable for rapid warm-up of the emission control systems even for vehicle operation during the initial phase of the CVS test.

When operating with a constant volumetric efficiency the spark advance has a similar amount of influence on exhaust emissions as during operation with constant mean effective pressure. For the curves of constant HC emissions, the tendency remains largely unchanged. For the NO emissions, the positions of the curves of constant concentration change in the region of $1/\phi > 1$ due to the constant volumetric efficiency.

The region of very low HC emission lies near $1/\phi \approx 1.1$ on the spark retard running limit (10° A.T.C. to about 25° A.T.C.). From this region the HC emissions increase particularly towards lean mixtures from 75 ppm to almost 700 ppm, that is 300 ppm more than at the comparable point in the diagram at no-load condition (Figure 11).

The NO emission diagram in Figure 12 shows the anticipated tendency with increasing emissions at lean mixtures and with an increasing output. With lean mixtures, the NO emissions increase from 1000 ppm up to 4760 ppm by spark advance. The dependence of the NO emission on the spark advance and the output exists only at A/F ratios of $1/\phi > 0.95$, that is to say when there is sufficient excess oxygen. At lean fuel/air mixtures there is no connection between NO emission and spark timing as the horizontal run of the NO concentration curves show. This demonstrates clearly the different formation mechanisms for NO. In the lean mixture region post-flame reactions have to be considered because they promote NO formation by the higher temperatures, adequate amount of excess oxygen, and sufficient time on account of the early spark timing (5). However, in the rich mixture region the NO formation must be considered as a flame reaction whereby the NO already formed is reduced again by radicals and unburned and partially burned components of the hydrocarbon oxidation reaction.

CATALYST SCREENING TESTS

Inside and outside the IIEC Program hundreds of reduction and oxidation catalysts on various supports have been developed and tested. Some of them have shown promise in the engine catalyst screening tests. In spite of properly designed catalytic systems that operate at high overall system conversion efficiencies almost from the moment of engine start-up, the tailpipe emissions were still far from IIEC engineering target levels.

At this time it is very doubtful whether a 90% emissions reduction together with the 50,000 miles requirement can be achieved because so far no catalysts are known which have good catalytic activity, chemical stability, sufficient physical durability, good long-term durability, no negative aging effects and sufficient resistance to lead, phosphorus and sulfur.

Some operating parameters are very important to the overall effectiveness of the catalytic system such as: inlet CO concentration, inlet temperature, space velocity and CO/O_2 ratio. Therefore it is necessary to make laboratory screening tests with the catalysts to find out under which conditions an optimal catalytic efficiency can be achieved. It was concluded that especially the ignition temperature and the low temperature performance of the catalyst are important items of information for the development of concept emission vehicles. Furthermore, in connection with a staged secondary air feature for fast warm-up it is important to investigate whether the catalyst is able to operate under reducing conditions as well as under oxidizing conditions.

For getting all this important information on the catalysts, the equipment shown schematically in Figure 13 was used in the laboratory. This test apparatus was similar in principle to that used within the IIEC Program (6, 7). It was designed under consideration of the following aspects:

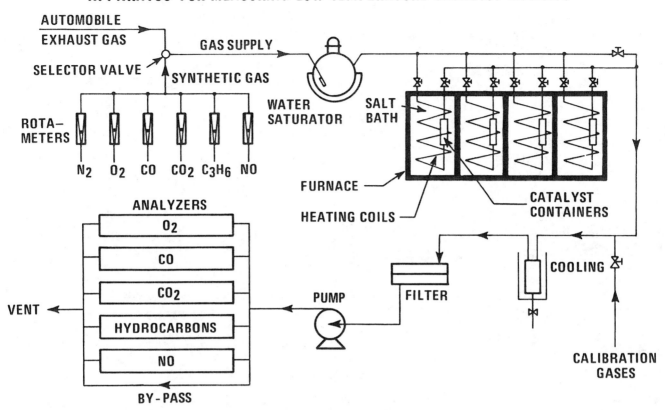

Figure 13

- The operating conditions for catalyst measurements have to represent the conditions in the catalytic system during cold engine start-up.

- The major results obtained with this test apparatus have to be comparable with those obtained in a real catalytic system. Therefore, the main requirement for the test itself is that it simulates vehicle catalytic converter space velocities.

- The operating conditions given by the following parameters: gas temperature, space velocity, mid-bed temperature, inlet concentrations, CO/O_2 ratio, secondary air flow rate, etc. have to be variable over a wide region to investigate the behavior of the catalyst very intensively under varying conditions.

The catalyst test apparatus consists essentially of four electrically heated furnaces. In this test method, a small quantity of catalyst (a few grams up to about 10 grams) is placed in a container. Vacuum pumps are used to draw a portion of synthetic gas mixture through the sample. This system enables the catalyst to be tested at different CO and O_2 levels and at different space velocities.

A detailed description of the synthetic gas compositions used is provided in Table I. For better representation of the conditions under which a catalyst must operate during cold vehicle start-up, it was decided to use engine exhaust gas. The composition of the engine exhaust gas used is summarized below:

— 500 ppm HC (sometimes propylene or hexane is added to bring the total hydrocarbon content to this level)

— 1-2 vol. % CO

— 2.5 vol. % O_2

— 10 vol. % CO_2

— 12-14 vol. % H_2O

— 1000 ppm NO_x (sometimes NO is adjusted to this level using gas cylinders).

For finding catalysts with low ignition temperatures and good low temperature performance, the pelleted type catalysts samples are supported in small 12 cc. axial-flow converters shown in Figure 14. The monolithic samples have a 1 in. O.D. and they are 3 in. long.

For improving the overall effectiveness of a catalytic system during vehicle start-up by means of using suitable catalysts it is necessary to test the aging behavior of the catalyst. Aging tests provide an estimate of catalyst deactivation in the presence of exhaust from especially unleaded fuels. It is well-known that reduction

TABLE 1

GAS COMPOSITION AND CONDITIONS FOR TESTING COLD START PERFORMANCE OF CATALYSTS (PELLETS AND MONOLITHS)

Gas Component Dry Basis	Tests For HC/CO Catalysts	Tests For NO_x Catalysts
CO – Vol. %	1.0	1.0
O_2 – Vol. %	0-1.0(a)	0-1.0(a)
HC (as C_3H_6) – ppm	350	–
CO_2 – Vol. %	12.2	12.3
H_2O – Vol. % (Wet Basis) (b)	2-3(a)	14.0
NO – ppm	500	500
Remainder	N_2	N_2

Parameter	Conditions
Volume of Catalyst – cc.	12.0 (Pellets), 38.5 (Monoliths)
Range of Space Velocity – cc./cc.-hr.(c)	8,500-90,000
Temperature Region – °C	100-750

(a) Adjustable in this range.
(b) Test gas is saturated with H_2O vapor.
(c) Gas flow rate per volume of catalyst.

CATALYST CONTAINER FOR SCREENING CATALYSTS ON HONEYCOMB SUPPORT

(Monolithic Probe: 3 in. Long, 1 in. O.D.)

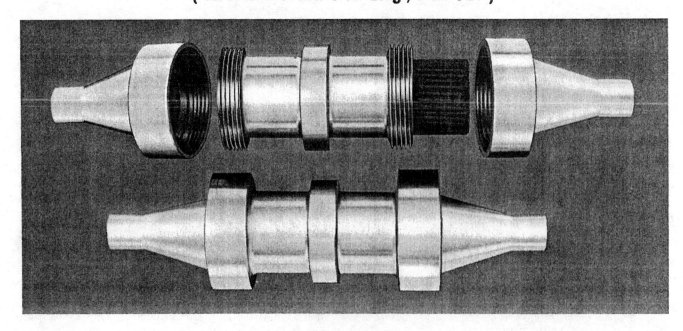

Figure 14

and oxidation catalysts lose considerable activity during aging (2, 7). To test the chemical stability and the activity as a function of aging time and of mileage, a catalyst aging test stand was developed. A schematic of the test stand is shown in Figure 15.

This test stand consists of an engine operated on an engine-dynamometer with two five-tube axial-flow converters at each side of the engine. It can be seen from Figure 15 that the NO_x catalyst containers are fitted very close to the engine, whereas the HC/CO catalyst containers are fitted in the exhaust system downstream of the NO_x converters. An engine driven secondary air pump feeds secondary air into the manifold between the NO_x and HC/CO converters in order to achieve HC/CO catalyst aging under realistic conditions. To obtain simultaneous aging of monolithic and pelleted samples one side of the exhaust system is provided for the aging of up to ten monolithic units while the other side takes up to ten pelleted catalyst samples.

The aging tests are carried out in such a manner that road load conditions are simulated. The engine is controlled by a programmed electric motor which allows the engine to operate between 1000 and 3200 rpm under engine choking conditions. The simulated average vehicle speed is about 70 kmph (44 mph).

After each 100 hours of operation (which is the equivalent of 7000 km.), samples are removed for testing in the catalyst activity test apparatus (Figure 13) in the laboratory. After activity evaluation, the samples are returned to the five-tube axial-flow converters for further aging in engine exhaust gas.

The aging tests are carried out with engine exhaust from unleaded fuels, because many of the catalysts with good low temperature performance are based on noble metals. Furthermore the sulphur content of the test fuels has to be known because catalyst activity might be affected by a small amount of sulfur in the exhaust gas. It was observed that near 750°F at a CO/O_2 ratio of 3.3 deactivation was very pronounced for sulfur contents greater than 0.01 wt. %.

Figure 16 was obtained as a typical result of measuring low temperature catalyst activity. This figure shows that the cold start performance of the NO_x catalyst investigated is promising. The ignition temperature* of this catalyst lies near 220°C. After aging only 14,000 equivalent km. in the five-tube converter, this catalyst lost much of its low temperature activity. The ignition temperature of the catalyst after the aging test is 90° higher (310°C).

---

*Ignition temperature is defined as the temperature at which 50% conversion has occurred.

CATALYST AGING TEST STAND

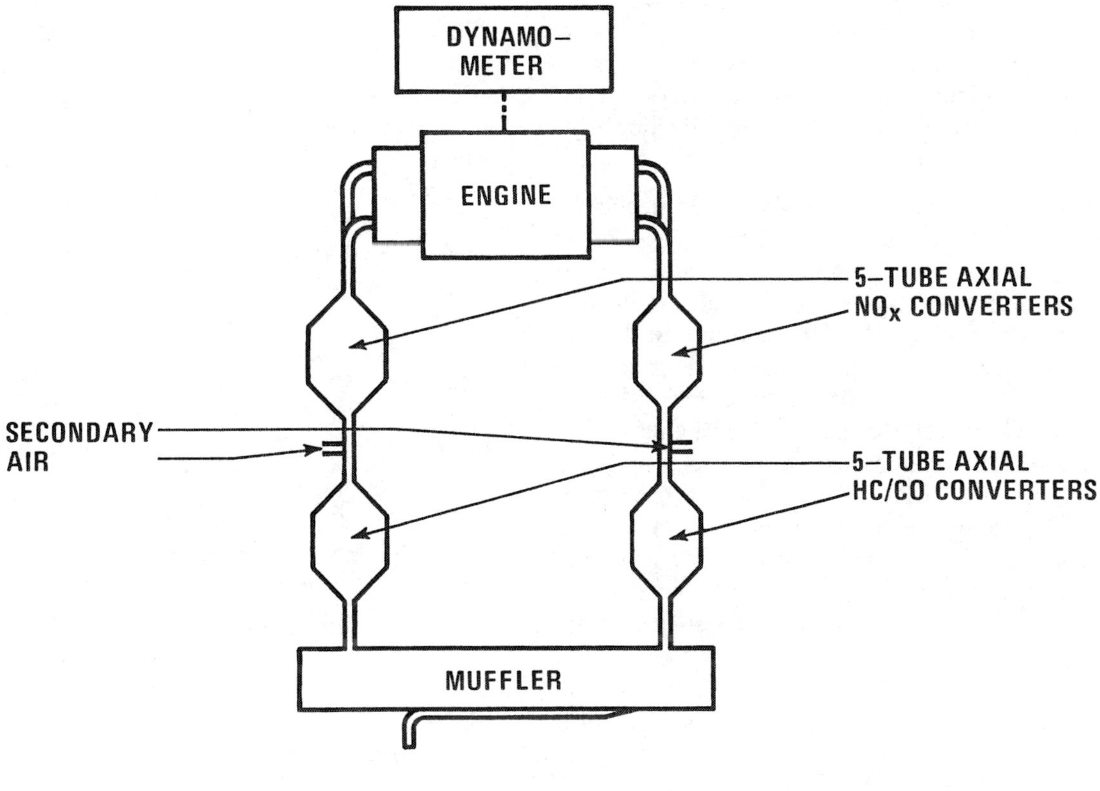

Figure 15

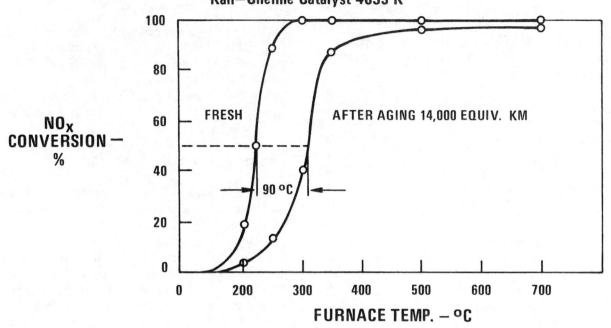

Figure 16

Figure 17 shows this relationship in more detail. Beside the distinct moving of the ignition temperature with increasing kilometers of aging it can be clearly seen that the aging influence is very stringent especially at low temperature (less than 310°C). Fortunately, at temperatures greater than 400°C the catalyst deactivation due to aging is not so severe.

The behavior of a good NO_x catalyst under reducing and oxidizing conditions is illustrated by Figure 18 for a low temperature (300°C) and by Figure 19 for a normal operating temperature (500°C). These figures show the influence of the CO/O_2 ratio due to the O_2 concentration variation on the cold start performance for NO_x and CO. For low O_2 concentrations (that is for high CO/O_2 ratios) the deactivation rate at 300°C is very high because the catalyst lost more than 50% of its effectiveness, whereas at higher temperatures (500°C) the rate of deactivation is only rather low. It is interesting to point out that in the region $O_2 < 0.5\%$ CO is consumed during NO_x conversion. For stochiometric and lean mixtures this NO_x catalyst is even in the position to achieve a complete CO conversion, in other words the investigated catalyst is able to operate under oxidizing conditions. This subject will be discussed later on.

Figure 20 shows typical results of ammonia studies conducted during the laboratory screening tests on reduction catalysts. The NH_3 concentration was measured by Draeger tubes. This figure shows the NO reduction and ammonia production over a range of temperature. It will be noted that the NH_3 production just occurs when almost 100% of the NO is reduced. At 300°C nearly all oxides of nitrogen are converted, but at about this temperature the NH_3 production reaches its peak level of about 150 ppm. With other catalysts peak levels up to 1000 ppm were measured. In this case formation of ammonia during NO_x reduction is unacceptable at 300-400°C and fortunately is very low at temperatures over 450°C. In addition, it was noticed that NH_3 production can be minimized as the space velocity* is increased (more than 50,000 hr.$^{-1}$) and that the ammonia production is a function of CO concentration and of the CO/O_2 ratio, respectively. Unfortunately, some catalysts show a considerable loss in effectiveness at high space velocities.

In connection with the staged secondary air system it is interesting that the NO_x catalyst investigated is also able to work under oxidizing conditions ($CO/O_2 < 2$). Figure 21 illustrates this behavior and shows that the monolithic NO_x catalyst is no more efficient for HC/CO

---

*Space velocity is exhaust gas flow rate at NTP per catalyst volume and has units of hr.$^{-1}$ (more exactly Standard liter per hr./liter).

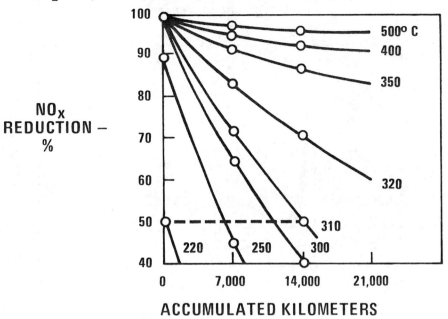

Figure 17

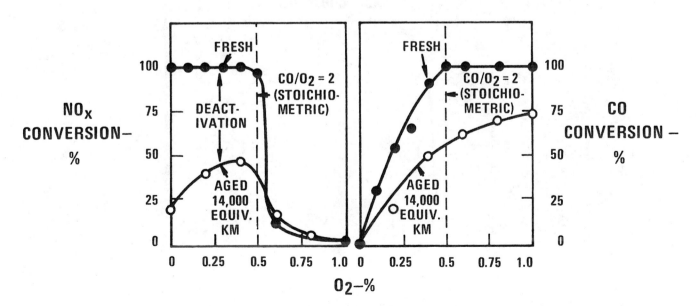

Figure 18

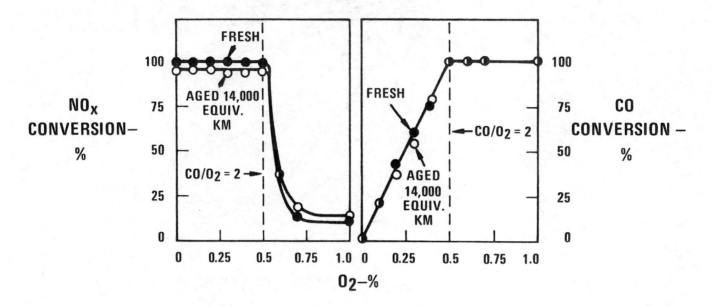

Figure 19

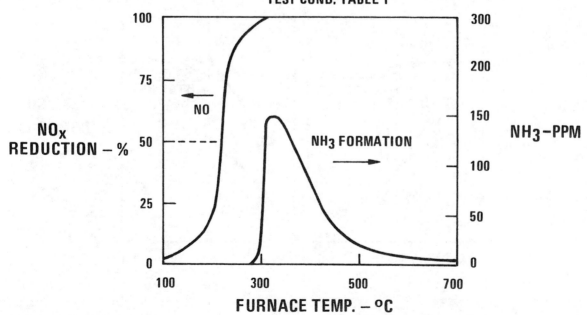

Figure 20

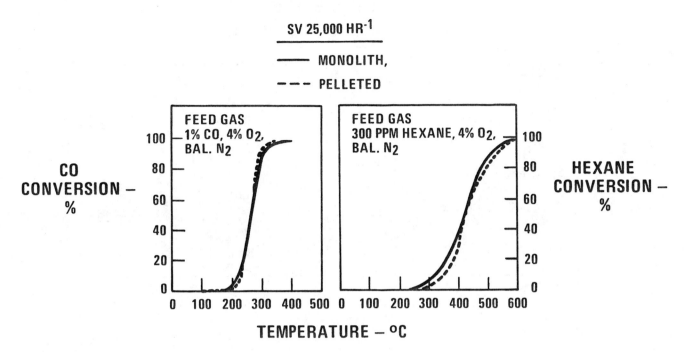

Figure 21

control than the pelleted NO_x catalyst under the chosen testing conditions in which for the HC test hexane was the only hydrocarbon constituent. In this case both catalysts have the same ignition temperatures at the conversion rate of 50%, namely 270°C for CO and 420°C for hexane conversion, because the pelleted catalyst has extremely good cold start performance.

Usually, when monolithic versus pelleted NO_x catalysts are compared with regard to conversion rate, warm-up, and back-pressure the monolithic converter shows more promise. Especially, due to the faster warm-up rate of the monolithic system (by reduced catalyst mass) the operating temperatures of the fresh NO_x and fresh HC/CO catalysts are reached 60 and 100 sec. faster, respectively, under optimal test conditions during the warm-up phase of the CVS test.

In connection with the discussions on the improvement of the warm-up characteristics of catalytic systems this paper deals mainly with the relations and problems encountered with the reduction catalysts. The reason for that is that the first catalyst in a dual-bed catalytic system is a reduction catalyst which should have a low ignition temperature and a good warm-up characteristic for reaching operating temperature as quickly as possible. Furthermore, if the NO_x catalyst mass can be reduced this would result in a significant improvement even in the warm-up rate of the HC/CO bed. But the difficulty encountered is the considerable loss in effectiveness of the NO_x catalyst as the space velocity is increased by reducing the catalyst volume. As an example, in a laboratory test with a fresh NO_x catalyst a 100% reduction of NO was measured at 24,000 hr.$^{-1}$, whereas at 90,000 hr.$^{-1}$ only a 39% reduction could be noticed ($CO/O_2 = 7.4$).

Figure 22 clearly illustrates the problems which have to be solved for getting a properly designed catalytic system with optimum operating conditions for the catalyst. It shows that the vehicle catalytic converter space velocity is continuously altering especially during the first two modes of the CVS driving cycle. It can be noted from Figure 22 that for a VW 1.7 liter engine equipped with a 67 cu. in. monolithic converter the space velocity is changing between 6000 and 60,000 hr.$^{-1}$; that is one order of magnitude. Hence, the catalyst efficiency cannot reach an optimum at any time as the space velocity, the concentrations, and the exhaust gas temperature are varying so considerably.

The catalyst and engineering technology described in the preceeding sections has been combined in research concept emission vehicles. This emission concept utilizes the same major hardware components as the concept illustrated in Figure 4 with the exception of the two

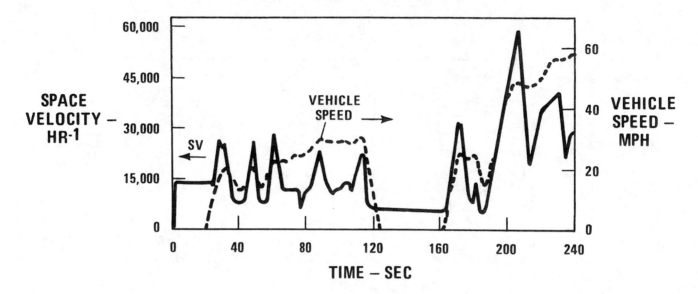

Figure 22

thermal reactors. For achieving rapid warm-up of the catalytic system the spark retard feature together with increased exhaust gas flow rate is used. A mechanical fuel injection system controls the stoichiometric fuel/air mixture between $1/\phi = 0.95$ and $1/\phi = 1.05$. The rapid warm-up resulting from the use of these components and technologies significantly reduced HC/CO mass emissions in the first 505 sec. of the CVS test. Actual cold start HC emissions measured during the "cold transient mode" were reduced by 33% from 224 to 150 ppm, whereas the CO emissions were reduced by 54% from 935 to 425 ppm with fresh monolithic catalysts. Unfortunately, the NO emissions increased during the first 505 sec. from 39 ppm up to 190 ppm by the overlapping effect of the volumetric efficiency. Furthermore, this concept has a significant fuel economy loss. When the fuel economy loss decreases, the NO_x emission level increases considerably.

While considerable success was achieved in developing research emission concepts, many major problems remain such as chemical and thermal stability of the catalyst, long catalyst life (up to 50,000 miles), sufficient physical durability of the catalyst, constant catalyst effectiveness, durability of the concept emission systems, driveability and vehicle performance, and transition from the research phase to the mass production phase.

SUMMARY

The utility of different methods for improving the warm-up characteristics of catalytic systems was illustrated and the necessity of concurrent and stringent control of NO_x and HC/CO emissions, particularly during the first two minutes after vehicle start-up was discussed.

A very elegant method to achieve a rapid warm-up is the use of extreme spark retard from the moment of engine start-up by which an increased exhaust gas flow rate with high exhaust enthalpy can be obtained. The peak level of exhaust gas enthalpy is secured when the engine operates with stochiometric mixtures ($1/\phi = 0.95 - 1.05$) and fully opened throttle. But, from the standpoint of fuel consumption it is necessary to change the warm-up spark retard as soon as possible to normal operation.

Furthermore, the possibilities of controlling the output with wide open throttle by altering the spark timing were investigated. During engine start-up the timing was no longer optimized with regard to fuel economy but to the largest possible exhaust gas enthalpy at an exactly required output during vehicle operation.

Experiments with a research concept emission vehicle which uses monolithic NO_x and HC/CO catalysts along with EGR showed considerable reduction of the emissions by using spark retard, higher exhaust gas flow rate, stoichiometric mixtures, and staged secondary air. Due to these features the warm-up rate of the catalytic system was very fast and by that a significant reduction of the emissions was achieved.

ACKNOWLEDGMENT

The authors would particularly like to acknowledge the contributions of Drs. W. Lee and A. König and Messrs. G. Ralva and G. Vogelsang for their engineering assistance; Messrs. J. Müller-Hillebrand and J. R. Hellbach who directed catalysts laboratory screening tests; Messrs R. Gospodar and D. Pickert who directed vehicle tests and engine screening programs on catalysts.

The contributions of Dr. Weidenbach of Kali-Chemie AG is particularly appreciated and hereby acknowledged.

Special thanks are extended to Prof. H. Heitland for his significant contributions to this paper.

Finally, the authors express their appreciation for the good cooperation within the IIEC Program.

BIBLIOGRAPHY

1. R. M. Campau, "Low Emission Concept Vehicles," SAE Paper No. 710294, January, 1971.

2. G. H. Meguerian and C. R. Lang, "NO_x Reduction Catalysts for Vehicle Emissions Control," SAE Paper No. 710291, January, 1971.

3. J. F. Roth, "Copper-Based Auto Exhaust Catalysts: Mechanisms of Deactivation and Physical Attrition," Paper, Monsanto Company, St. Louis, Missouri 63163, 1971.

4. R. M. Campau and R. E. Taylor, "The IIEC—A Cooperative Research Program for Automotive Emission Control," Paper No. 17-69, presented at the 34th Midyear Meeting of the API's Division of Refining, Chicago, Illinois, May 12, 1969.

5. W. E. Bernhardt, "Kinetics of Nitric Oxide Formation in Internal Combustion Engines," Paper C 149/71, presented at the Conference on Air Pollution Control in Transport Engines of the Institution of Mechanical Engineers, Solihull, England, November 9-11, 1971.

6. J. C. Kuo, H. G. Lassen and C. R. Morgan, "Mathematical Modeling of Catalytic Converter Systems," SAE Paper No. 710289, January 1971.

7. F. G. Dwyer and K. I. Jagel, "HC/CO Oxidation Catalysts for Vehicle Exhaust Emission Control," SAE Paper No. 710290, January, 1971.

Toyota Status Report on Low Emission Concept Vehicles

T. Inoue,
K. Goto,
and K. Matsumoto
TOYOTA MOTOR CO., LTD.

INTRODUCTION

This paper describes Toyota's efforts on developing low emission concept vehicles and their components.

Studies of each component were started at Toyota in 1967. From 1970, the main efforts were concentrated on finding possible combinations of the various components to meet the 1975/76 Federal Regulations.

Components studied were direct flame type afterburner, catalytic converter, lead trap for catalyst, thermal reactor, and three types of exhaust gas recirculation systems: air cleaner entry EGR, above the throttle valve EGR and manifold EGR. In parallel with component studies, some engine modifications such as choke system, camshaft overlap, combustion chamber configuration and ignition system were also investigated.

The present stage of Toyota's development of the following combination systems and their components is reported in this paper:

Package I: NO_x and HC/CO Catalytic Converter and EGR

Package II: Thermal Reactor, Catalytic Converter and EGR

ABSTRACT

The status of Toyota's development of low emission concept packages—

1. NO_x and HC/CO Catalytic Converter and EGR,

2. Thermal Reactor, NO_x and HC/CO Catalytic Converter and EGR

and their components is described.

- Variations of thermal reactor design, performance and durability characteristics are discussed.

- Above the throttle valve entry EGR has been found to have desirable flow characteristics with a simple control system.

- EGR rate over 15% brings about unacceptably poor driveability and fuel economy with smaller vehicles.

- Many types of catalytic converters for pelleted catalysts have been designed and examined for their performance and durability. A down-flow type converter has relatively good flow distribution and warm-up characteristics.

- As for HC and CO, a few prototype vehicles have met the 1975 Federal Regulation at low mileage, but the 1976 regulation for NO_x of 0.4 gm./mi. is very difficult to meet even with a significant penalty in driveability and economy for our smaller vehicles.

THERMAL REACTOR DEVELOPMENT

As a result of controlling nitrogen oxides (NO_x) by supplying a rich mixture and recycling exhaust gas, the unburned products in exhaust gas, which are carbon monoxide (CO) and hydrocarbons (HC), increase in volume and require after-burning.

Thermal reactor application was tried to improve the effectiveness of secondary air injection in the exhaust system. A DuPont-type reactor was investigated.

For the earlier thermal reactor design, shown in Figure 1, the following was considered:

1. To maintain sufficient mechanical strength under various engine operating conditions, each component was supported by the outer shell of the insulator.
2. To prevent component deformation caused by thermal shock, curved structures were employed for each part of the inner components.
3. To reduce the thermal stress, the inner parts of the insulator were isolated from the outer shell.
4. Each component was assembled with screws for experimental convenience.

The thermal reactor was insulated to minimize exhaust gas heat loss. Several insulation methods were evaluated, but the high density ceramic wool insulation was best. Figure 2 shows the test results concerning the effect of the insulation.

The insulation of the thermal reactor influences not only the reactor performance, but also the thermal condition of the surroundings, so the heat insulation should be done as perfectly as possible considering the radiation and conduction. As for the insulation, the test result of the conversion efficiency of the actual reactor is shown in Figure 3. Thus, the reaction in a thermal reactor is a function of the temperature under the condition of the fixed concentration of the unburned component and the fixed volume of the secondary air, so in case of insufficient insulation, the fuel consumption will increase to keep the temperature in the reactor sufficiently high by supplying rich mixture or retarding the ignition timing, otherwise the reaction in the reactor will not occur.

The residence time of the exhaust gas in the thermal reactor is considered to be another essential factor for its performance. The exhaust gas reaction was observed to continue from the exhaust port through the thermal reactor and into the exhaust pipe when the inner core of the thermal reactor was not large enough. Therefore, various sizes of reactors were designed and tested as shown in Figures 4 and 5 to prolong the reaction time. The smaller sized reactor requires higher reaction temperatures and

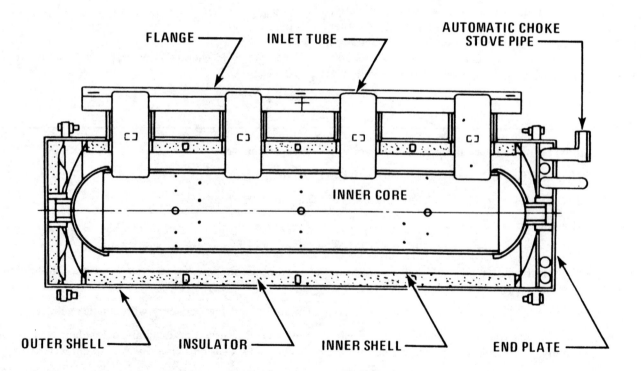

Figure 1

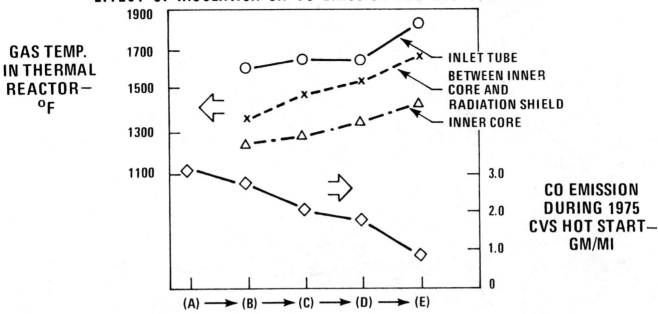

Figure 2

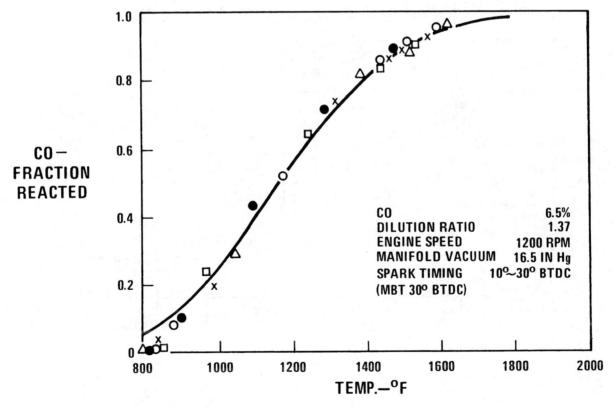

Figure 3

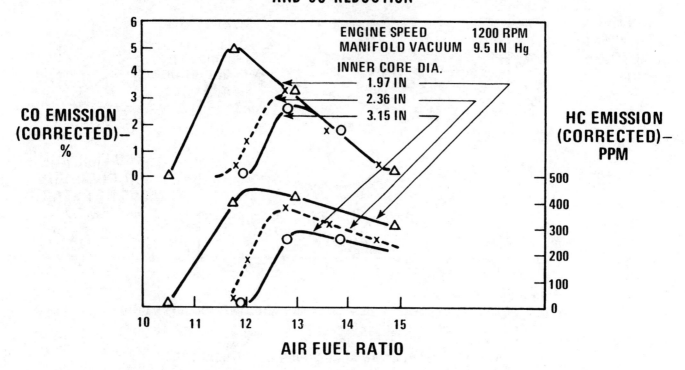

Figure 4

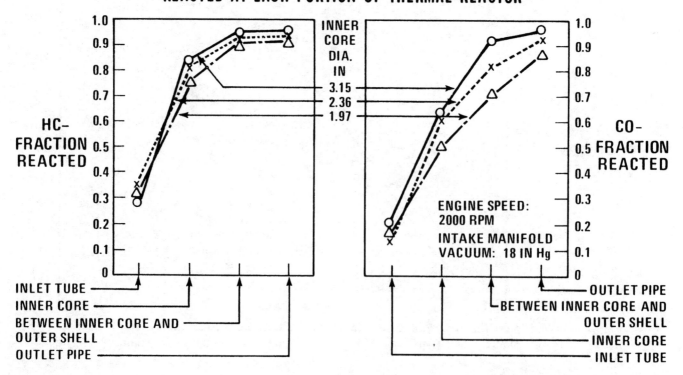

Figure 5

discharges more unburned components, so the reactor size should be as large as possible for the allowable engine space.

To promote the mixing of exhaust gas and secondary air, and to improve the reaction speed, several types of inner cores and flame holders were tested, but increases in the heat capacity made the warm-up characteristics worse.

Observing the flame propagation in the thermal reactor through a Pyrex glass window, it was found that the flame from the exhaust port was cooled out at every hole in the inner core during the cold period. In order to improve the warm-up characteristics, another type of thermal reactor as shown in Figure 6 was designed which had a large outlet port at each end of the inner core, and was intended to mix the exhaust gas and the secondary air in the free vortex and to hold the reaction in the forced vortex in the inner core. This type of reactor improved the reaction during the warm-up period.

EGR SYSTEM DEVELOPMENT

As described by many authors, lowering the maximum combustion gas temperature is an effective method to reduce nitrogen oxides. Exhaust gas recirculation (EGR), rich mixture supply, and retarding ignition timing are the most popular techniques, but deterioration of performance is a serious problem for small engines and there may be some limitations to use of these techniques in such cases.

The effects of air fuel ratio, EGR rate and ignition timing are shown in Figure 7, for seven-mode FTP hot cycle tests. Since fairly good correlation has been found between seven-mode test data and 1975 Federal test data, the same conclusions will be obtained when applying the 1975 Federal test. In the area of low nitrogen oxide emissions in Figure 7, these vehicles could not follow the test mode due to insufficient engine power. Considering the vehicle performance and the driveability, it may be concluded that 60 percent reduction of nitrogen oxides is the maximum rate obtainable for small engines by the EGR system.

Another problem is the increase of fuel consumption. The amount of nitrogen oxides in exhaust gas from the engines on which various kinds of modifications have been made was shown to be a function of the fuel consumption. Figure 8 shows the effect of EGR on fuel economy. Exhaust gas recirculation lowers NO_x levels with less fuel economy loss compared with the use of richer mixtures alone.

Several types of EGR systems were developed, but the particulate contamination in the fuel metering system

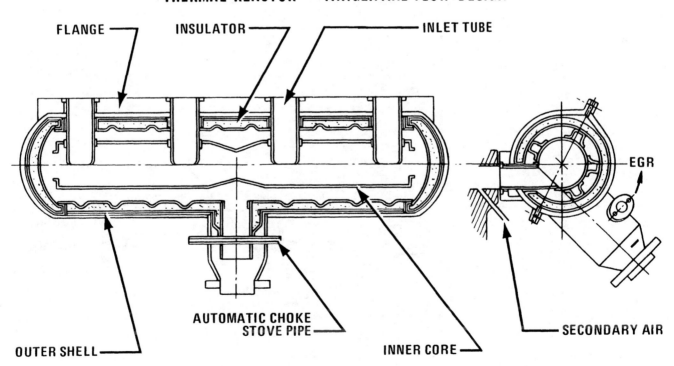

Figure 6

EFFECT OF SPARK TIMING, AIR FUEL RATIO (PRIMARY MAIN METERING JET DIA.) AND EGR RATE (EGR METERING ORIFICICE DIA.) ON NO$_x$ EMISSION
(7 MODE, MEAN OF 3 HOT CYCLES)

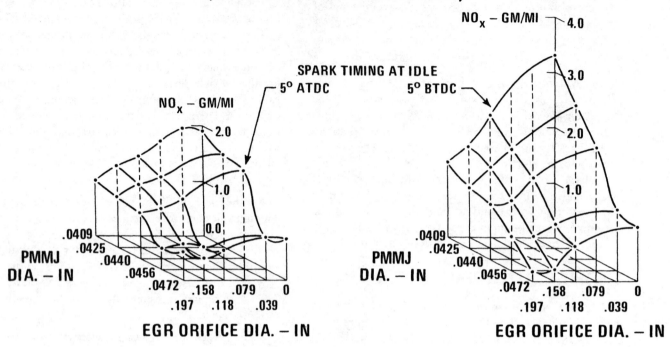

Figure 7

EFFECT OF EGR ON FUEL ECONOMY

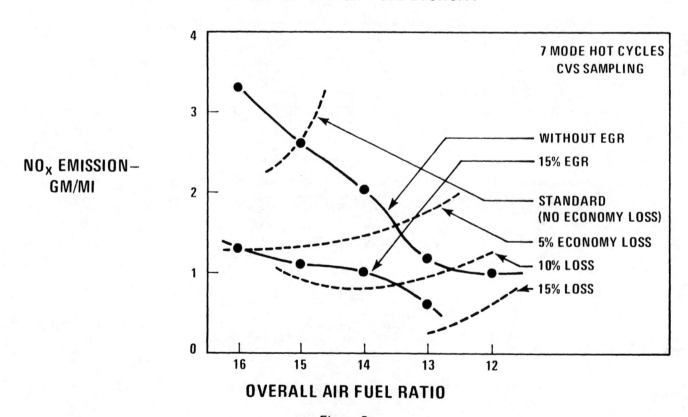

Figure 8

and the difficulty of EGR volume control were the main problems. So the above the throttle valve entry EGR system is mainly used in these tests because it is easy to control the EGR volume and the contamination of the carburetor is slight. This system is composed of an exhaust gas tapping unit, air cooling pipe, on-off control valve and a modified carburetor.

Among these components, the on-off control valve, shown in Figure 9, had many problems:

—Clogging of the metering jet with particulate matter

—Shaft sticking with particulate matter and acidic condensed water

—Burning or melting failure of the diaphragm.

Almost all of these problems were solved by optimizing the operating temperature and changing the material, but other problems like icing and corrosion of the carburetor still remain unsolved.

CATALYTIC CONVERTER DEVELOPMENT

For low emission concept packages, single and dual-bed converters were developed. Initial tests for optimizing the catalytic converter location and catalyst volume were conducted with single-bed converters.

Catalytic Converter, Location and Volume

Three locations and three single-bed converters of different volumes were selected to optimize these items. Figure 10 shows the emission test results with these converters. The vehicle tested was a 113.4 cu. in. TOYOTA Corona with manual transmission. An air injection system was installed but without an EGR system. The catalytic converters used in this test were axial-flow-type single converters with the same cross sectional area of 19.7 sq. in. Emissions were evaluated by the 1972 Federal test and the effect of location on the warm-up of the catalytic converter was determined.

Test results are as follows:

- Small volume converter must be located as close as possible to the engine, but this may cause some deterioration of catalyst due to higher temperature.
- Larger volume converter works well even at rather distant location. In this case, space and weight problems may occur.

The maximum bed temperatures were measured during the 1972 Federal test for each location and were 1035°F for 20 in. and 945°F for 45 in. from the engine. The catalyst used was a base metal extrudate and the operating temperature limit was about 1300°F. Maximum space velocities for each catalyst volume during

EGR ON-OFF VALVE

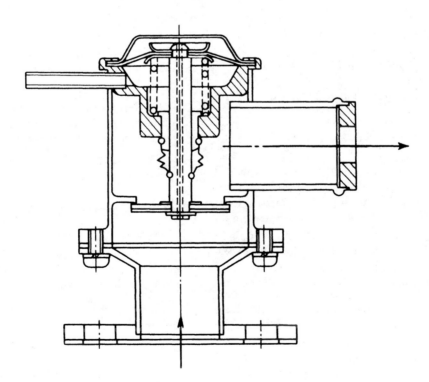

Figure 9

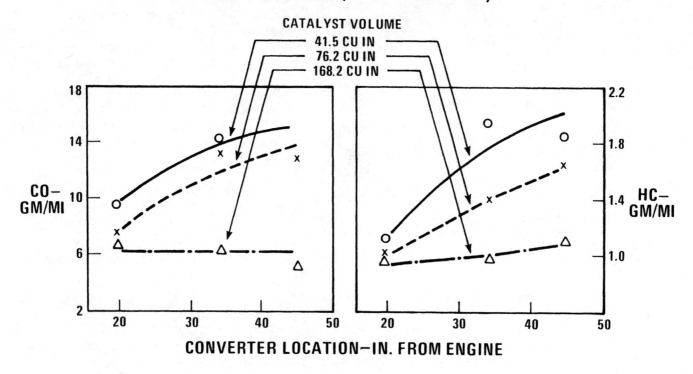

Figure 10

the test were 160,000 hr^{-1} for the 41.5 cu. in. and about 40,000 hr^{-1} for the 168.2 cu. in. converter.

Dual-Bed-Type Catalytic Converter Design

Many types of dual-bed converters were developed for the low emission concept packages I and II. Figure 11 shows the typical construction and gas flow direction of these converters.

Separated-flow type I and down-flow type I were the first stage of the development. These two converters showed fairly good conversion performance but down-flow type I was accompanied by a large pressure drop and partial deformation. Warm-up characteristics were also not so good because of the heat loss to the metallic components of the converter and to the outside. Separated-flow type I was better for pressure drop but the secondary air mixing with exhaust gas was poor and HC/CO conversion efficiency was relatively low.

Separated-flow type II was the advanced design of the type I. Heat transfer from the NO$_x$ catalyst bed to the HC/CO catalyst bed and secondary air mixing were taken into account and the warm-up performance and HC/CO conversion efficiency were improved. The durability of this converter at 1400°F. for 400 hr. was acceptable but the thickness of 4.5 in. was too large to be mounted under the front seat.

The radial-flow converter and the horizontal-flow converter were developed to improve the flow distribution. The radial-flow type showed good flow distribution but was too large and not suitable for mass production. The horizontal type also showed good flow distribution and good warm-up characteristics but back pressure was rather high.

At the present time, down-flow type II has been adopted for the packages reported in this paper. It shows good warm-up and conversion performance. The back pressure is not so high and it may not be so difficult to be mass produced.

Secondary Air Mixing and Conversion Efficiency

Because of the limited space for a dual-bed catalytic converter for pelleted catalyst, it is very important to get good mixing of the secondary air with exhaust gas within as short a distance as possible.

Figure 12 shows an example of model test results with different air mixing methods. Sample catalyst and volume were so selected that the test results clearly discriminate differences between three methods. The catalyst used was an aged base metal pelleted catalyst, and the volume was 65 cu. in. (bed thickness: 2.4 in.). Space velocity was 93,600 hr^{-1} and air flow rate was 3.5 cfm. Bed temperature was 1200°F. Air injection from the

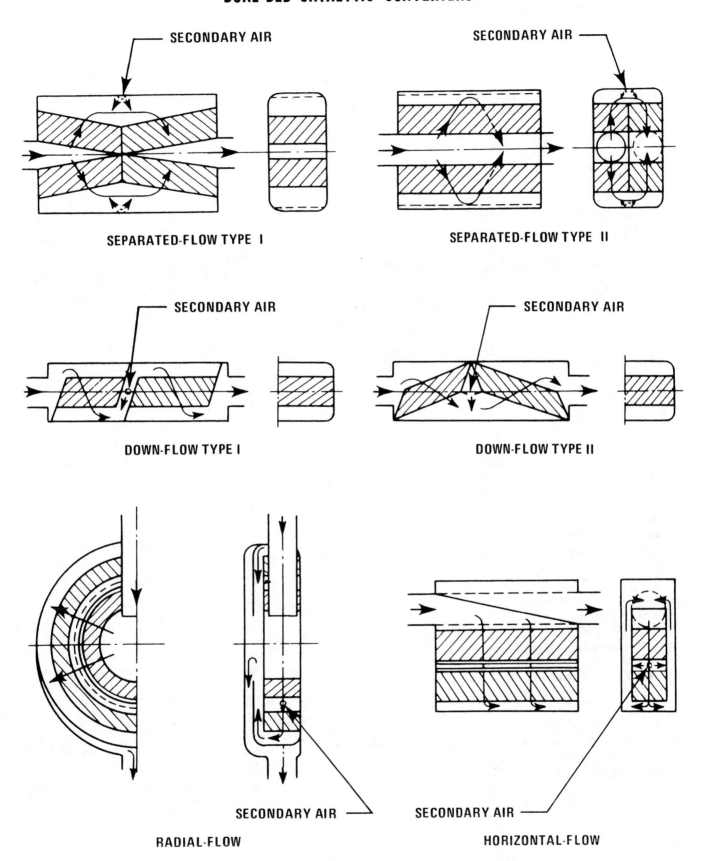

Figure 11

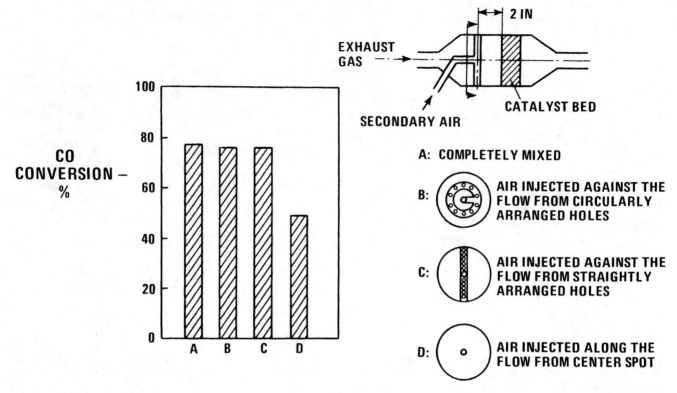

Figure 12

center spot and the relatively short distance to the catalyst bed brought about poor mixing and low conversion efficiency as shown in Figure 12, "D".

On the contrary, air injection from a circular or straight arrangement of holes improved the air mixing and the conversion efficiency was as high as that of completely mixed gas. At present, our dual-bed converter generally uses the straight pipe design "C". Gas flow and mixing with secondary air are still under study with a variety of configurations.

Over-Temperature Protection

Over-temperature of the catalytic converter system is one of the most important problems remaining unsolved with these concept packages. As mentioned by many authors, the temperature of the converter bed rises rapidly when the engine misfires. So the protection system must become effective before the bed temperature exceeds a certain temperature limit.

Over-temperature protection is even more critical on packages for smaller engines because of their higher exhaust gas temperature. As described later, a very heavy load of 2.5 in. Hg intake manifold vacuum is often used in the 1975 Federal driving cycle. Figure 13 shows the converter bed temperature at wide open throttle with and without manifold air injection. This figure also compares the converter bed temperature at two different locations: normal position A (2.5 ft. from the engine, which is about under the front seat) and rear converter position B (9 ft. from the engine, which is the main muffler location). The bed temperature difference of the two locations is about 400°F with air injection and about 300°F without air injection. After warming up converter B, bypassing converter A seems very effective for avoiding over-temperature of the converter bed under normal conditions.

But as shown in Figure 14, this is not the solution for the over-temperature problem in the case of misfire. This figure shows the temperature trace from the beginning of one cylinder misfiring under the condition of wide open throttle and without air injection. Even at 3000 rpm, the bed temperature rises to about 1730°F within 10 sec., and at 6000 rpm it becomes about 1840°F within 7 sec. The temperature limit for the HC/CO catalyst is 1560°F at the present time but it should be lowered for extended use in view of safety and durability. Therefore, some kind of over-temperature protection system must be adopted.

Figure 15 shows a schematic diagram of such a system. A thermo-sensor detects the converter bed temperature, and when it rises near the temperature limit, a controller lights the warning lamp for the driver. If it continues to

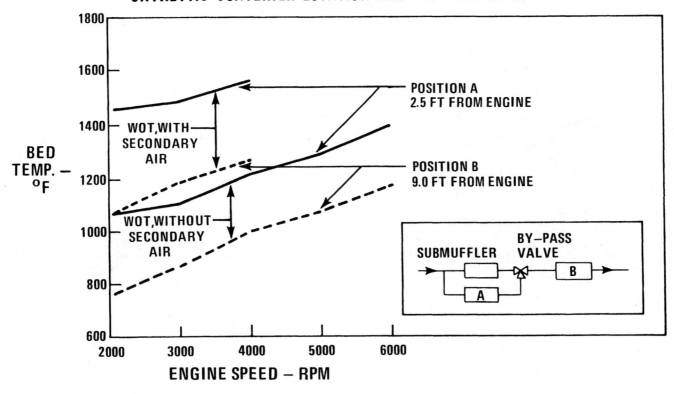

Figure 13

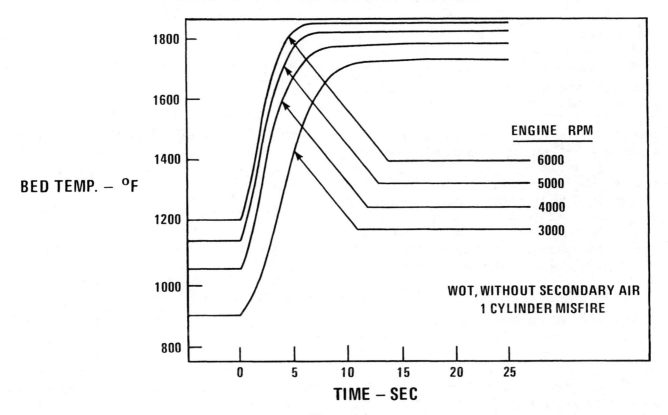

Figure 14

OVERTEMPERATURE PROTECTION SYSTEM

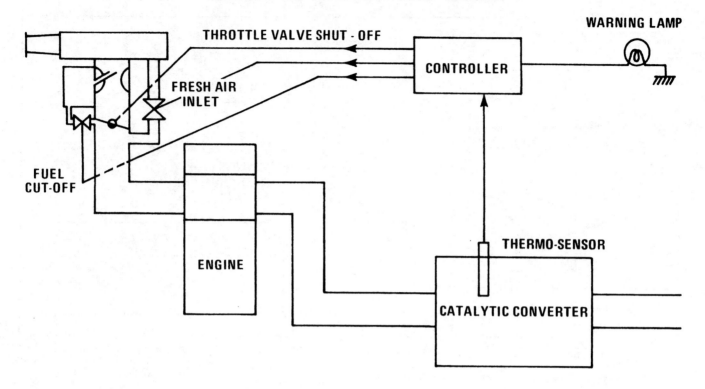

Figure 15

rise over the temperature limit, the controller works to shut the throttle valve to the idle position, the fuel of the idle system is cut-off and the fresh air inlet valve introduces air to the intake manifold and the engine blows the cooling air through the overheated catalytic converter.

Figure 16 shows the temperature trace of the actual converter bed with this system when one cylinder of the engine misfired. The converter was a 183 cu. in. separated-flow-type with a pelleted catalyst and was located at position A. The engine operating condition was wide open throttle and the engine speed of 5000 rpm was maintained constant by an engine dynamometer. After the beginning of the misfiring, the protection system begins to work when the bed temperature exceeds the temperature limit of 1560°F. The opened area of the fresh air inlet valve affects the temperature trace, and with larger opened area, the bed cools down more rapidly.

It must be noticed that after the system begins to work, the bed temperature still continues to rise for a few seconds. This temperature overshooting phenomena is caused by the residual combustible mixture in the engine and exhaust system. Because this temperature rise may partially destroy the converter bed, studies are underway to lower the temperature limit as much as possible.

Besides this protection system, it is essential to use a more reliable ignition system, such as a transistorized ignition system, improved ignition wires and spark plugs.

These ideas are still under testing, and our packages use a conventional by-pass valve at present.

LOW EMISSION CONCEPT PACKAGE I

NO_x and HC/CO Catalytic Converter and EGR

One of the low emission concept packages is the catalytic converter and EGR system. The advantage of this system is its fuel economy and cost compared to systems which additionally use a thermal reactor. However, the emission data variation from test-to-test or vehicle-to-vehicle seems to be rather high and the deterioration of the catalyst performance may affect the emission level.

Warm-up systems such as choke modification and heat riser improvement of the intake manifold and engine modifications such as combustion chamber configuration and spark advance control are being studied to decrease the raw emission level of the engine.

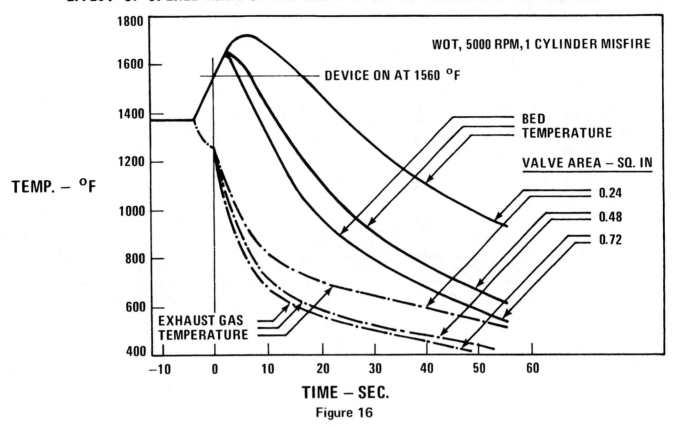

Figure 16

This package is applied to a TOYOTA Carina with 97.5 cu. in. engine and durability tests are being now conducted.

EGR Rate and Engine Performance

As shown in Figure 8, EGR greatly influences the engine performance and fuel economy. The EGR system adopted in this package is the above the throttle valve entry EGR and the average EGR rate is about 12% under the 1975 Federal driving cycle [EGR rate = gm. recirculated/(gm. recirculated + gm. air)]. Exhaust gas is extracted from the exhaust pipe just under the exhaust manifold and controlled by a vacuum actuated on-off valve depending on the engine load, vehicle speed and water temperature.

Figure 17 shows how the EGR spoils the engine performance and influences the required engine load or intake manifold vacuum to run the 1975 Federal hot cycle (0-505 sec.). The black areas in each square show the frequency of operation in each condition of engine speed and intake manifold vacuum. A full black square would mean 100 seconds driving. This figure clearly shows that with the EGR system, very heavy load ranging from 2.35 to 7.05 in. intake manifold vacuum is more frequently used compared to the system without EGR. This explains the difficulty for the smaller vehicles to meet the 1976 Federal Regulation with sufficient driveability.

System Description and Performance

Figure 18 shows the general layout of Package I. The main components for this system are:

- above the throttle valve EGR with air-cooled piping and on-off valve.
- dual-bed down-flow-type catalytic converter for NO_x and HC/CO pelleted catalysts
- by-pass valve
- secondary air pump
- air switching valve for directing air to exhaust manifold, to catalytic converter between NO_x and HC/CO catalysts, or to exhaust pipe after the converter.

The control system senses the converter bed temperatures of both the NO_x and HC/CO catalysts, engine water temperature, engine oil temperature and vehicle speed, and controls the EGR on-off valve, spark advance, air switching and by-pass valves. The EGR on-off valve is actuated by intake manifold vacuum and preloaded by a spring to shut below 2 in. Hg vacuum, thus cutting EGR at full load. Spark advance is so controlled that vacuum advance is stopped below 60 mph.

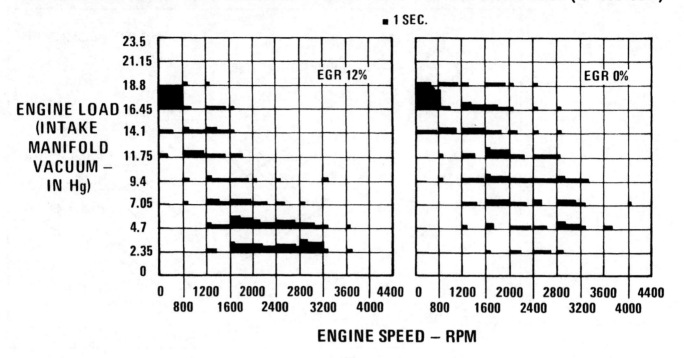

Figure 17

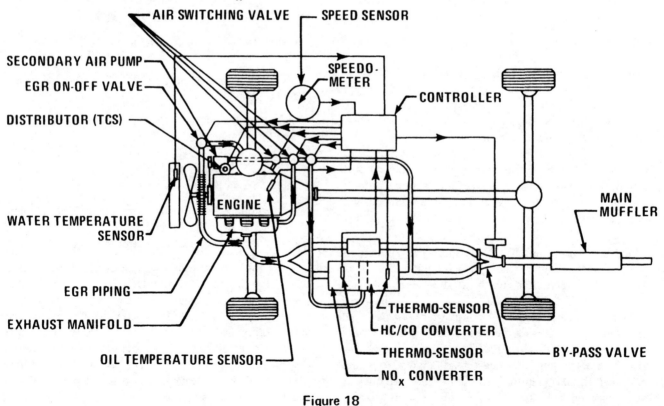

Figure 18

The by-pass valve is used to protect the catalytic converter from over-temperature at heavy load, misfire and cold starting. The controller actuates the by-pass valve when the converter bed temperature exceeds 1560°F. Secondary air is switched from exhaust manifold to converter 220 sec. after engine starting.

Table I shows the typical emission level of this concept vehicle at low mileage. Considering the interim durability test results, the emission level rapidly increases and is very sensitive to the engine tune up.

TABLE 1

LOW EMISSION CONCEPT PACKAGE I - 1975 FEDERAL TEST EMISSION DATA AT LOW MILEAGE

HC - GM/MI	0.29
CO - GM/MI	2.97
NOx - GM/MI	0.54
ECONOMY LOSS - %	12

LOW EMISSION CONCEPT PACKAGE II

This concept package consists of EGR, thermal reactor and a dual-bed catalytic converter which contains both the reduction catalyst and the oxidation catalyst. The catalytic converter itself is very helpful to control the exhaust emission, but the catalyst bed warms up very slowly due to the large heat capacity, and its deterioration is relatively high. Therefore, this package is intended to reduce the burden on the catalysts by using the thermal reactor and EGR.

While the catalyst is not reactive just after the engine starts, the secondary air is injected in each exhaust port to cause a satisfactory reaction in the thermal reactor until the temperature of the catalytic converter bed reaches sufficient level.

High temperature of the exhaust gas in the pipe may result in increasing the heat loss to the surroundings. Therefore, if preheating the catalyst is most desired, the reaction in the thermal reactor should be stopped soon after the catalytic converter is warmed up, but if the reduction of HC and CO in the thermal reactor is most desired, various control systems will be designed along driving modes.

This concept package is illustrated in Figures 19 and 20. Temperature control of the thermal reactor and the

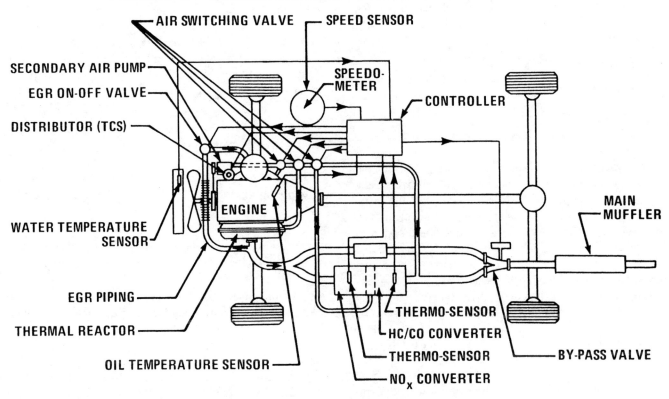

Figure 19

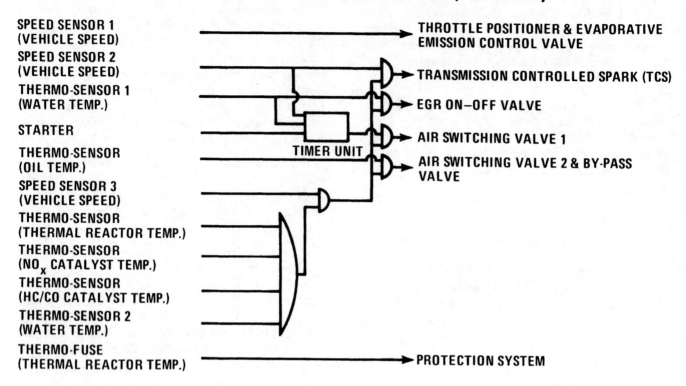

Figure 20

catalytic converter requires reliable high-temperature measurement. After studying several kinds of temperature measuring techniques, three types of thermo-sensors, as shown in Figure 21, were designed and tested, but these had the following troubles:

—Slow response

—Bad repeatability

—Corrosion of outer tube

—Snapping and abrasion of outer tube.

The thermister-type sensor was most promising because the power generated and its accuracy were relatively high, thus improving the response and the durability. This type of sensor is applied on the system. The inner space of the outer tube is evacuated and sealed to prevent corrosion of the thermister lead wire and leakage of water. The response time of the thermo-sensor may be improved by reducing the heat capacity or compensating the generated power with its differential in the electronic circuit, so both of them are being utilized. To improve the reliability of the temperature control, a thermo-fuse unit was tried but it had many troubles.

For over-temperature protection, the by-pass valve was also adopted, but its leakage reduced emission performance of the combination system, and back pressure will lower the engine power and fuel economy. (See Figure 22.) After testing several types of by-pass valves, it was found that leakage of a poppet valve was least possible, so two types of by-pass valves were designed as shown in Figures 23 and 24. The leakage of these types was lower than one percent, but to keep the sealing at closed condition, the casing should be sufficiently stiff, which causes the weight to be heavy.

Increase of the back pressure of this package is estimated from the following:

- Increase of exhaust gas volume caused by the higher temperature and the addition of the secondary air

- Increase of hydrodynamic resistance in the thermal reactor, catalytic converter, by-pass valve and additional pipe bend.

The influence of the back pressure on the engine performance and exhaust emissions was evaluated as shown in Figure 25. These data show that increase of the back pressure will decrease the volumetric efficiency, engine torque and NO_x emission. These data also show that the engine torque will be influenced by the back pressure as a result of a decrease of the volumetric efficiency and an increase in the pumping loss. Therefore, the exhaust sys-

THERMO - SENSORS FOR THERMAL REACTOR AND CATALYTIC CONVERTER

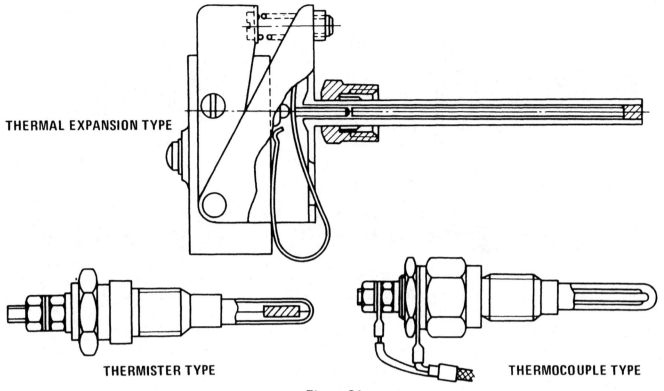

Figure 21

EXHAUST BY-PASS VALVE DESIGNS, LEAKAGE AND BACK PRESSURE

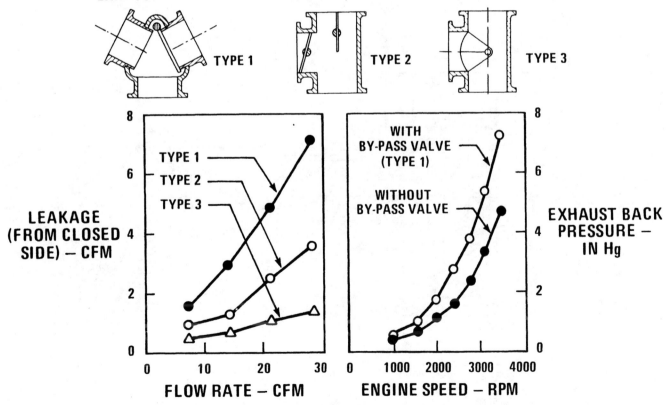

Figure 22

BY-PASS VALVE DESIGN TYPE 4

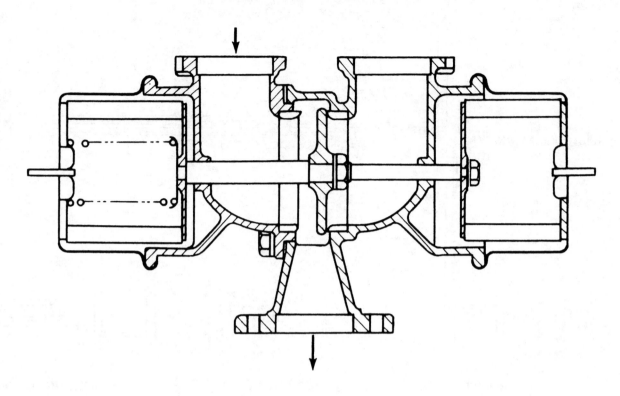

Figure 23

BY-PASS VALVE DESIGN TYPE 5

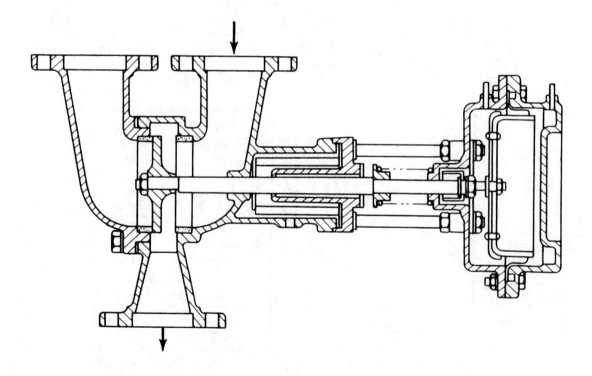

Figure 24

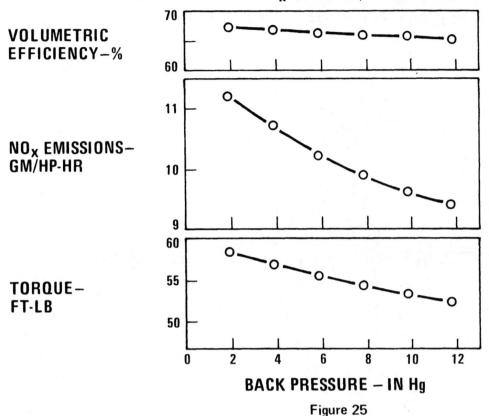

EFFECT OF BACK PRESSURE ON NO$_x$ EMISSION, TORQUE AND VOLUMETRIC EFFICIENCY

ENGINE SPEED 2400 RPM
THROTTLE ANGLE 34°
AIR FUEL RATIO 15.5

Figure 25

tem of the vehicle on which the emission control system is installed will be redesigned to match the increased exhaust volume.

Concerning the optimization of this concept package, the study of the system operation under the 1975 Federal cycle is important to get the correlation between bench data and the vehicle emission at transient mode and to improve the control system. Therefore, simultaneous measurement and real time analysis was carried out utilizing the electronic computer. The following items were measured:

—Vehicle speed

—Engine speed

—Engine intake air volume

—Engine intake manifold vacuum

—Secondary air volume to exhaust ports

—Secondary air volume to catalytic converter

—EGR volume

—CO_2, CO, HC and NO$_x$ concentration in exhaust gas.

The air volume was measured by a laminar flow element and a differential pressure sensor, and the EGR volume was measured by a calibrated metering orifice. The mixture strength of exhaust gas in each part of the exhaust system was calculated from concentrations of CO_2, CO, HC, H_2 and air volume.

These data showed good correlation with bench data and made possible the optimization of the concept package by bench tests. Figure 26 shows a part of the test chart, and Table 2 shows typical low mileage emission data on this concept package by the 1975 Federal test.

TABLE 2

LOW EMISSION CONCEPT PACKAGE II - 1975 FEDERAL TEST EMISSION DATA AT LOW MILEAGE

HC - GM/MI	0.39
CO - GM/MI	2.72
NOx - GM/MI	0.50
ECONOMY LOSS - %	14

AN EXAMPLE OF EGR RATE AND NO$_x$ EMISSION DURING 1975 FEDERAL DRIVING SCHEDULE

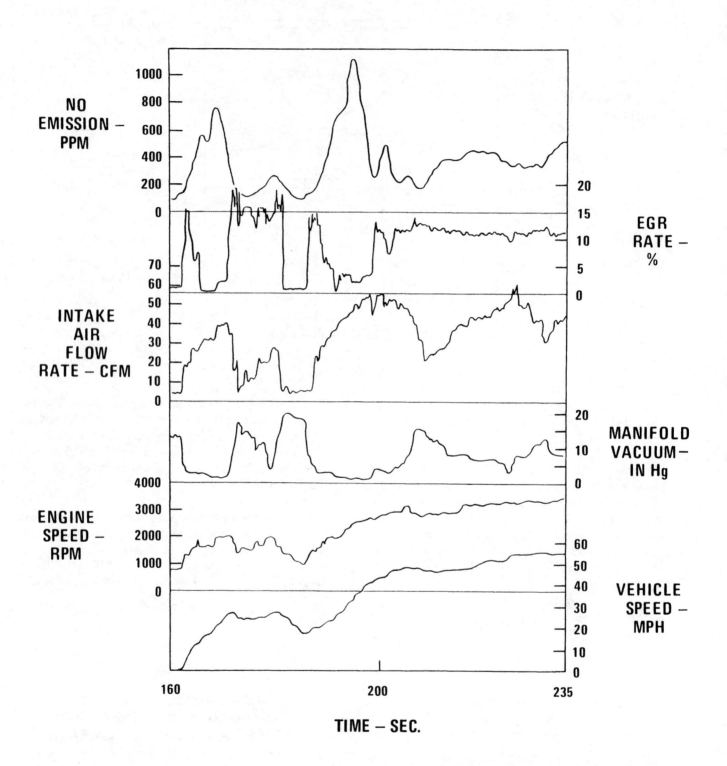

Figure 26

SUMMARY

The status of development of the low emission concept packages and their components has been described.

- The effect of heat insulation and inner core diameter for thermal reactor performance is significant. A large diameter inner core improves the conversion efficiency and extends the reactive range of air fuel ratio to the leaner side.

- A very rich mixture and large amount of EGR may attain the 1976 NO_x standard, but the penalty on the driveability and the fuel economy cannot be acceptable.

- Catalytic converter location, catalyst volume, converter bed construction and secondary air mixing method were experimentally determined for the low emission concept vehicles.

- An idea for the over-temperature protection system for the catalytic converter was examined and seemed to be effective.

- Electronic control system, especially thermo-sensor and by-pass valve designs were reported. A thermister type thermo-sensor was adopted.

- Two low emission concept packages were evaluated. At low mileage a few vehicles of these types met the 1975 HC and CO standards, but the 1976 regulation for NO_x of 0.4 gm./mi. is very difficult to meet even with significant penalty in driveability and fuel economy.

ACKNOWLEDGEMENT

The authors wish to acknowledge many co-workers in TOYOTA for their devoted contributions to develop each component and concept package, and also express their sincere appreciation to IIEC program members for their suggestions and kind discussions concerning their research work.

2/14 PREDICTING PERFORMANCE OF FUTURE CATALYTIC EXHAUST EMISSION CONTROL SYSTEMS

J. C. W. Kuo, C. D. Prater, D. P. Osterhout, P. W. Snyder and J. Wei (U.S.A.)

SYNOPSIS The general rules adopted in the development of automotive emission control catalytic converter models are to quantify all the essential physical and chemical phenomena related to the converter performance by separate laboratory experiments, and to minimize the complexity of the models by excluding from the models the less important phenomena. By adhering to these rules, predictive catalytic converter math models were constructed for both particle or monolithic support catalysts using either noble or base metal as active ingredients. The models have been used as a useful research and development tool for exploring the direction for improving total emission control systems, including catalyst development, engine calibration modification, converter location and converter design studies. The strategy in the application of the models is to maintain a continuous dialogue between test results and the math model predictions to a total vehicle-converter system that can meet stated goals. The power of the models is demonstrated by several practical examples, including identification of catalyst and engine modification required to meet U.S. 1976 emission standards, use of math models to focus on specific vehicle developments, and assist in the interpretation of emission control system test data.

INTRODUCTION

1. Mathematical models have been developed in the Inter-Industry Emission Control Program (IIEC) that predict the performance of catalytic converters used to control exhaust emission from internal combustion engines. The IIEC program is a cooperative automotive emission control research program which is described by Taylor, et al (ref. 1,2,3,4). The original emission goals of the IIEC program, set in 1967, were aimed at achieving the following emission levels as determined by the seven-cycle Federal emission test procedure (FTP, ref. 5).

Exhaust	1967 IIEC Emission Goals
Hydrocarbons	0.82 g/mile
Carbon Monoxide	7.1 g/mile
Oxides of Nitrogen	0.68 g/mile

In September 1971, the IIEC emission goals were revised to meet the requirements of the U.S. 1970 Clean Air Act Amendments for 1976 using the new non-repetitive, constant-volume-sampling hot and cold start (CVS C/H) emission test procedure (ref. 6). The new U.S. standards and the equivalent IIEC low mileage goals based on the cold start portion* of CVS C/H cycle are:

*Abbreviated CVS-III cycle (ref. 7).

	Grams/Mile	
	1976 U.S. (CVS C/H)	IIEC CVS-III Low Mileage Goals
Hydrocarbons	0.41	0.21
Carbon Monoxide	3.4	2.1
Oxides of Nitrogen	0.4	0.26

2. The catalytic converter model discussed in this paper predicts carbon monoxide and hydrocarbon emissions as well as catalyst temperature under dynamic engine operation for a point-in-the-life of pelleted or monolithic catalysts for which the catalytic activity and physical properties have been determined. This model is particularly useful for identifying what changes in engine calibration, converter design and/or catalyst properties are required for a viable emission control system. Finding the successful combination of these variables by building and testing actual converter systems on each major vehicle requires a very costly and time consuming experimental program. The predictive model can quickly screen different combinations of variables and identify the important hardware or catalyst developments required to meet desired emission levels.

3. To develop a predictive model of the kind needed, the physical and chemical phenomena that are most important in determining the performance of the converter were identified and quantified by laboratory experiments. These quantified phenomena were then used to construct a model that

represented the best information available as to "how the converter worked." Particular care was taken to make sure that the model would not become more complex than absolutely needed in order to maximize its utility. Furthermore, no arbitrary adjustment of parameters was allowed to make the model agree with actual converter performance. A search was made to identify the cause of any discrepancy. The causes of discrepancies, in descending order of occurrence, were found to be:

1. Faulty experimental data.

2. Incorrect or inprecise physical or chemical effect in the model.

The last cause, of course, required either the reevaluation of the faulty parameter or the quantification of the new effect. The constant exchange of information between experimentation and computation enables us to identify rapidly the key variables that determine the converter performance.

Description of Model

4. Catalytic converter mathematical models have been constructed for both particle and monolithic support catalysts and for using either noble or base metal catalyst. The early form of the pelleted catalyst model is described by Wei (ref. 8) and a more comprehensive version of the model is discussed by Kuo et al. (ref. 9). The pelleted catalyst model can be summarized as:

(1) For some base metal catalyst, the kinetics for both CO and hydrocarbon oxidation are first order except that they depend on oxygen concentration raised to the 0.2 power for oxygen concentration less than 2%. For noble metal catalyst, the kinetics for both CO and hydrocarbon oxidation are observed to be inhibited by the adsorption of CO, HC, and NO. The effect of O_2 concentrations is first order in the range of 0 to 10% oxygen. The total exhaust hydrocarbons are divided into two groups in these two forms of kinetics, i.e., a less easily oxidizable hydrocarbon group (primarily methane) and an easily oxidized hydrocarbon group which contains most of the remaining hydrocarbons.

(2) To account for warmup and dispersion of the highly transient inlet conditions in a packed bed, a simple cascade cell model, where each cell in a series is a completely stirred reactor, was found to adequately represent the transport processes.

(3) Gas-particle interphase heat and mass transfer is included to describe the transport resistance contributed by the stagnant gas boundary layer on the particle surface.

(4) Intraparticle heat and mass transfer is included to describe the transport resistance within the catalyst particle.

(5) Heat transfer to the catalyst support grid is included to account for the heat sink effect of the grid.

Applications

5. The strategy in applying the mathematical model for developing emission control systems is outlined in Figure 1. The fixed physical properties and oxidation kinetics are readily measured from simple reproducible laboratory tests. These measurements are obtained for catalysts fresh and aged up to the equivalent of 50,000 miles in a converter on a dynamometer stand. The transient engine exhaust flow rates, temperature and composition traces are obtained from vehicles operated under the driving cycle or condition being evaluated and are used as input to the predictive model. The model predictions indicate whether or not:

1. The proposed system meets the requirements and is ready for durability testing, or

2. An improvement is required in either the:

- catalyst properties including activity.
- converter design or location
- engine design, carburetor calibration or ignition timing.

6. If improvements are indicated, the model is used to methodically explore the benefits for modifications to the major input variables; such as, reduced catalyst density or increased activity, reduced choking time, or carburetor adjustment. When a significant improvement is identified, the hardware or catalyst modification is developed and a new set of input data obtained. After a successful combination of catalyst properties and engine performance is founded using the model, the total vehicle is tested for durability. In the event that the total vehicle does not pass the durability test, measurement of transient temperature and concentration traces for the operating conditions which caused the vehicle to fail usually combined with math model analysis usually identifies the problem area. Several examples of predictive model applications in the IIEC program that illustrate the power of this combined modeling-experimental approach follow.

Example 1 - IDENTIFICATION OF CATALYST AND ENGINE MODIFICATION REQUIRED TO MEET THE U.S. 1976 EMISSION STANDARDS

7. Vehicle emission control systems were developed in the IIEC program that would meet the original IIEC emission goals based on FTP test cycle. The CVS-III cycle, on which the new CVS C/H cycle is based, differs considerably from the old FTP cycle as shown.

Federal Test Procedures

Test	1968 FTP (1968-71 model)	CVS-III (1972-74 model)
Start of test	20 seconds after engine starts.	When starter is engaged.
Cycle	Seven repetitive.	Non-repetitive.
Test Time	979 seconds.	1372 seconds.
Measurements	Weighted average of seven discrete emission measurements per cycle.	Constant volume sampling with measurement of single diluted sample.

8. The model predicted that a vehicle equipped with a catalytic converter and tuned for optimum performance in the FTP cycle would meet the original IIEC goals with catalyst aged for the equivalent of 50,000 miles. However, as shown in Figure 2, when the same emission data are calculated by using constant-volume-sampling method, CO and HC increase by a factor of two or more and only HC meets the goal for a very short mileage. Most of the increase resulted from very high emissions produced during the first 37 seconds which are not included in the total emissions by the FTP weighting procedure while these emissions are included in the CVS measurement and contribute heavily to the total emissions. The model predictions also show that the converter inlet temperature is not high enough for the catalyst to convert the high emissions observed during the first 100 seconds, as shown in Figure 3.

9. In order to satisfy the stringent 1976 emission standards, major improvements are necessary. The model was used to explore the benefits for changing catalyst and engine characteristics. Improvement in catalyst performance was explored first. However, as shown in Figure 4, model predictions showed that even a catalyst 1000 times more active was not enough to meet the goals. Model evaluation of the benefits for reducing catalyst density showed these solutions to be unpromising.

10. Rapid preheating of the converter prior to engine ignition was also considered. The model predicted that preheating the converter to 800°F (428°C) would meet the revised goals even for relatively inactive catalyst as shown in Figure 5. However, the hardware and power required to heat converters to 428°C in a fraction of a minute were too costly to make this an attractive approach. Hence, the conclusion is reached that improvements in the engine operation during the initial minute are necessary if the more stringent CVS test requirements are to be met.

Example 2 - USE OF MATH MODELS TO FOCUS ON SPECIFIC VEHICLE DEVELOPMENTS

11. Even when both engine modification and a thermal reactor (located close to the engine) were used with a catalytic converter to reduce the emission, the revised IIEC goals were not met as the results shown below demonstrate.

Emission	IIEC Low Mileage CVS-III Goals (g/mile)	Predicted Vehicle Emissions (g/mile)
CO	2.1	7.9
HC	0.21	0.35

Model analysis of the catalytic converter performance for this vehicle using exhaust gas traces measured after the thermal reactor as inlet condition to the catalytic converter showed that a breakthrough in CO and insufficient conversion during the first minute contributed heavily to the total emissions. A systematic model evaluation of potential engine modifications are summarized in Table 1. Reduction in CO

Table 1

Predicted Benefits of Potential Engine Modifications on Outlet Emission of a Reactor-Converter System

Catalyst: Aged Mobil G (50,000 Miles on Dynamometer Stand)

Converter: Axial Flow, 60 sq. in. by 3 in.

		Cases			
		Base	1	2	3
Engine Modification					
None		X			
Modify Power Valve Acceleration (192 to 212 Seconds) Reduce CO by 50%, Increase O_2 to 1.5%			X	X	X
Lean Inlet CO During Warmup (10 to 39 Seconds) by 50%				X	X
Further Lean CO During 23 to 109 Seconds					X
All Accelerations, Max. CO = 1.5% All Other Modes, Max. CO = 0.75%					
CVS-III Emissions (G/Mile)	IIEC Goals				
Converter Inlet: CO	---	27.3	25.1	23.7	21.8
HC	---	0.93	0.93	0.93	0.93
Converter Outlet: CO	2.1	7.9	4.5	3.2	2.3
HC	0.21	0.35	0.33	0.33	0.34
Peak Catalyst Temperature (°F)	---	1690	1700	1700	1700

emission from 7.9 to 4.5 g/mile was predicted for modifying the power valve to limit CO to about 50% of the base test data. Leaning out the engine during the 10 to 39 second warmup was predicted to reduce CO from 4.5 to 3.2 g/mile. Additional leaning out during the 23-109 second period would further lower CO to 2.3 g/mile and thereby nearly meet the IIEC low mileage goals. Very high catalyst temperatures were predicted with this engine calibration and converter location as shown in Table 1. A modification of the system, such as a over-temperature protection system or relocation of the converter, to lower the peak catalyst temperature should be considered. The engine modifications were tested and produced the expected benefits; unfortunately they also caused some operability problems, such as stalling and reduced performance. Engine control developments are required to provide the lean operation needed and still maintain reasonable vehicle operability.

Example 3 - ASSIST IN THE INTERPRETATION OF EMISSION CONTROL SYSTEM TEST DATA

12. Two common problems in the performance of converter systems were evaluated with a catalytic converter. They were CO breakthrough and slow warmup. The first, CO breakthrough, was observed at the 120 second and 180 second deaccelerations of the CVS cycle and neither was predicted by the model as shown in Figure 6. A close examination of the system revealed that the air pump was bypassed when vacuum decreased on the intake manifold during deacceleration. At the same time, the oxygen analyzer was not sensitive enough to measure the brief lack of oxygen, and thus, the model used an erroneously high oxygen content and calculated complete conversion of CO and HC. The air pump bypass was eliminated and the CO breakthrough on deacceleration disappeared as the model predicted. Another breakthrough observed during the acceleration at 212 seconds (Figure 6) was not predicted by the model and was again due to insufficient air pump capacity.

13. The second problem was a slower catalyst warmup than predicted by the model as shown in Figure 7. We found that this discrepancy resulted from unknowingly drawing moist ambient air through the exhaust system as part of the safety requirements to keep exhaust fumes out of the dynamometer room. Moisture analysis revealed that the catalyst contained 8 wt % water prior to the CVS test. Desorption of this moisture during warmup required a 2000 Btu/lb of water desorbed, thus produced the predicted warmup as shown in Figure 7. Care is taken to dry converter systems prior to CVS testing.

14. The strategy of using the model to evaluate emission control systems follows the advice of the well-known scientist Sir Arthur Eddington, -- "It is not a good policy to put too much confidence in facts until they have been proven by theory."

REFERENCES

1. Taylor, R. E. and Campau, R. M., "The IIEC - A Cooperative Research Program for Automotive Emission Control," Preprint No. 17-69, Midyear Meeting, Division of Refining, American Petroleum Institute, Chicago, Illinois, May 1969.

2. Osterhout, D. P., Jagel, K. I., and Koehl, W. J., "The IIEC Program -- A Progress Report," presented at ASTM meeting, Toronto, Canada, June 1970.

3. Meisel, S. L., "Exhaust Emissions and Control," presented at symposium of Washington Academy of Sciences and American Ordnance Association, Washington, D.C., January 1971.

4. Koehl, W. J., Osterhout, D. P., and Voltz, S. E., "Accomplishments of the IIEC Program for Control of Automotive Emissions," Institute of Mechanical Engineer meeting, Solihull, England, November 1971.

5. Federal Register, Volume 33, Part II, June 4, 1968.

6. Federal Register, Volume 36, Part II, July 2, 1971.

7. Federal Register, Volume 35, Part II, November 10, 1970.

8. Wei, J., "Catalyst and Reactors," Chem. Engr. Prog. Monograph Series, Vol. 56, No. 6, 1969.

9. Kuo, J. C. W., Morgan, C. R., and Lassen, H. G., "Mathematical Modeling of Catalytic Converter System," Paper No. 710289, SAE, Detroit, Michigan, January 1971.

FIGURE 1

MATH MODEL STRATEGY
EMISSION CONTROL SYSTEM DEVELOPMENT

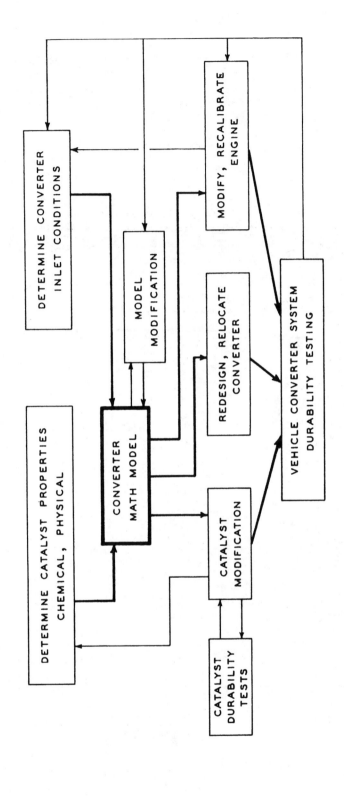

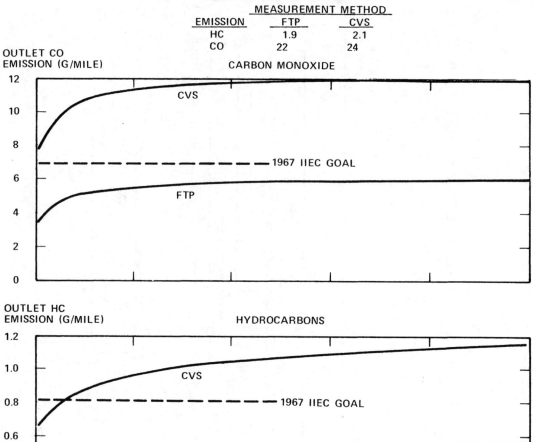

FIGURE 2

PREDICTED EFFECT OF CATALYST MILEAGE ON EMISSIONS

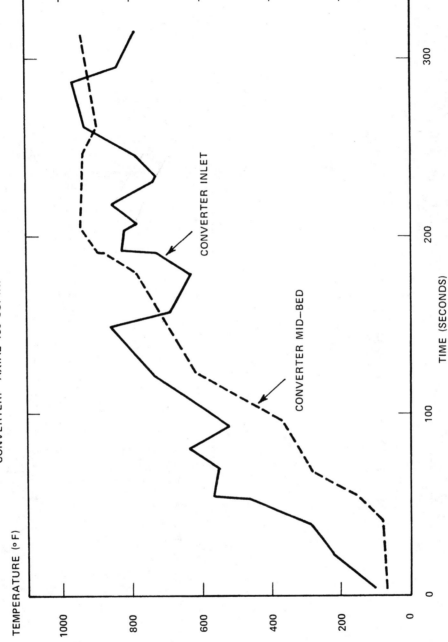

FIGURE 3
CONVERTER INLET AND MID-BED TEMPERATURE

VEHICLE: GALAXIE 289 CU. IN. ENGINE WITH THERMACTOR
CATALYST: MOBIL G (AGED 50000 MILE NON-LEADED)
CONVERTER: AXIAL 180 CU. IN.

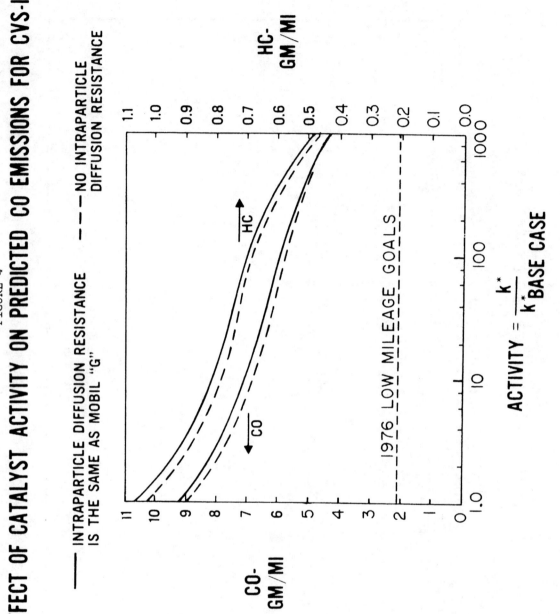

FIGURE 4
EFFECT OF CATALYST ACTIVITY ON PREDICTED CO EMISSIONS FOR CVS-III

FIGURE 5

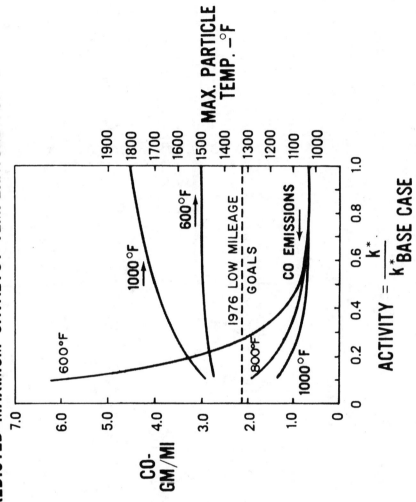

EFFECT OF INITIAL BED TEMPERATURE ON PREDICTED CO EMISSIONS AND PREDICTED MAXIMUM CATALYST TEMPERATURE FOR CVS-III

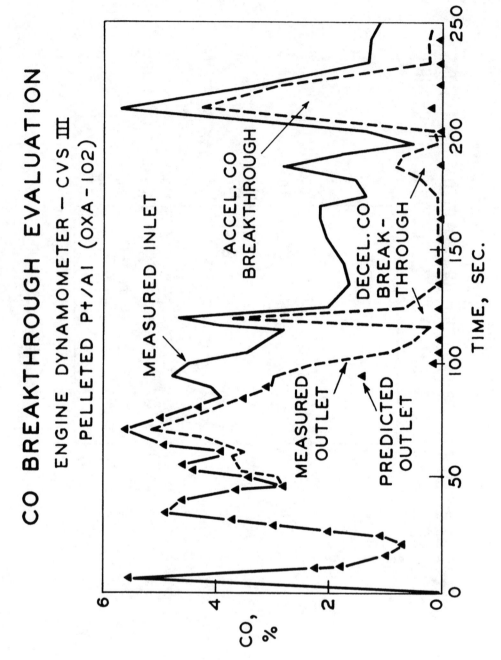

FIGURE 6

CO BREAKTHROUGH EVALUATION
ENGINE DYNAMOMETER – CVS III
PELLETED Pt/Al (OXA-102)

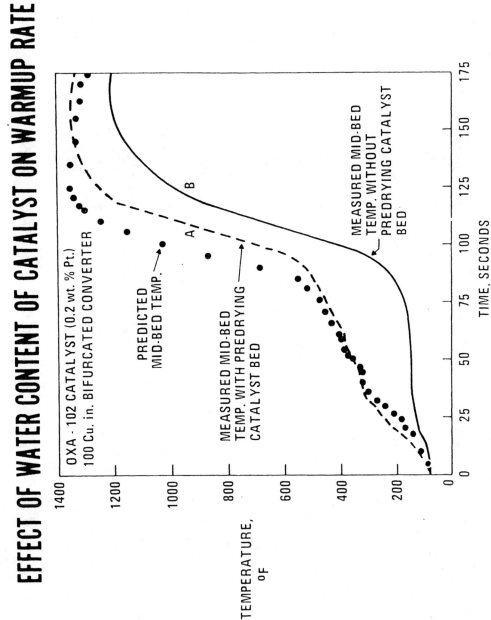

FIGURE 7

EFFECT OF WATER CONTENT OF CATALYST ON WARMUP RATE

730014

Gasoline Lead Additive and Cost Effects of Potential 1975-1976 Emission Control Systems

M. G. Hinton, Jr., T. Iura, and J. Meltzer
Aerospace Corp.

J. H. Somers
Div. of Emission Control Technology, Office of Air Programs,
Environmental Protection Agency

THE CLEAN AIR AMENDMENTS of 1970 (1)* require a marked reduction in the automotive exhaust emission species of unburned hydrocarbons (HC), carbon monoxide (CO), and oxides of nitrogen (NO and NO_2 as NO_x) for 1975-1976 model year light-duty motor vehicles. Emission standards for these exhaust pollutants are given in Table 1 (2). Proposed methods of simultaneously controlling HC, CO, and NO_x exhaust emissions from automotive spark ignition internal combustion engines to meet these stringent 1975-1976 federal standards involve not only engine system modifications, for example, improved carburetion, spark timing, compression ratio, etc., but also the combined use of additional devices, such as exhaust gas recirculation, thermal reactors, catalytic converters.

In response to a letter of inquiry of Feb. 28, 1971 from the Environmental Protection Agency (EPA) administrator concerning progress toward meeting the 1975-1976 standards, the four major United States automobile manufacturers (3-6), as well as a number of foreign manufacturers, replied that various combinations of emission control devices had been and were undergoing extensive evaluation.

The consensus was that some form of catalytic converter for

---

*Numbers in parentheses designate References at end of paper.

_____ABSTRACT_____

A study was conducted in 1971 to assess the overall effects of lead additives in gasoline on the performance, durability, and costs of emission control devices/systems which might be used to meet the 1975-1976 federal emission standards for light-duty motor vehicles. Although no system has yet demonstrated meeting the 50,000-mile emission level lifetime, all currently planned 1975-1976 emission control systems include a catalytic converter. However, lead additives are toxic to catalytic materials; they reduce catalytic activity, which results in increasing emission levels with mileage accumulation. Unleaded gasoline would be required in quantities sufficient to satisfy the demands of vehicles equipped with a catalytic converter in order to prevent catalyst activity degradation from lead additives. Implementation of such advanced emission control systems implies very high cost to the consumer, with the cost being a strong function of the required NO_x emission level. At this time, estimated overall costs to the consumer (initial, maintenance, and operating) for emission control systems being considered for the 1976 federal emission standards are $860 above average 1970 vehicle costs, over an 85,000-mile vehicle lifetime. This estimate is based on a system incorporating a dual (HC/CO, NO_x) catalytic converter, a low-grade rich thermal reactor, and exhaust gas recirculation. This system is currently considered to have the most promise to meet the 1976 NO_x standard of 0.4 g/mile.

Table 1 - Emission Standards for
1975-1976 Light-Duty
Motor Vehicles

Emission	Standards, g/mile	
	1975	1976
HC	0.41	0.41
CO	3.40	3.40
NO_x	3.10	0.4

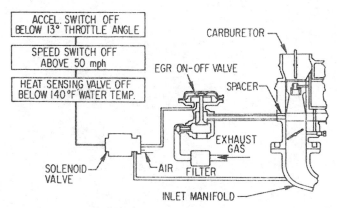

Fig. 1 - Toyo Kogyo "Mazda" EGR system, entry above throttle valve (8)

exhaust gas after-treatment was required, and that unleaded gasoline was essential to prevent excessive catalyst activity degradation over the 50,000-mile durability requirement of the federal standards (2).

The EPA administrator is empowered to regulate additives in gasoline (1). However, if such regulation is made on the basis of the effect on emission control systems, Section 211 c (2) B of the Clean Air Amendments of 1970 requires that a comparative cost-benefit analysis be made between control devices/systems needing additive regulation and those which do not.

In fulfillment of Section 211 c (2) B, an overall assessment (7) of the effects of lead additives in gasoline on the performance, durability, and costs of emission control devices/systems which may be used to meet the 1975-1976 emission standards for light-duty vehicles was made for the Office of Air Programs of the EPA. This paper (7)* is a summary discussion of the more significant results of that assessment (performed June-October 1971) in the areas of: candidate systems, specific effects of lead additives, emission levels, durability status, and potential cost impact to the consumer. Over 30 companies provided direct comments and information to supplement available published data. For direct discussions of emission control system technology, visits were made to the four major domestic automobile manufacturers, various major oil companies active in the emission control area, several catalyst manufacturers, and the two major lead additive manufacturers; in addition, three foreign automobile manufacturers were visited.

CANDIDATE SYSTEMS

An extremely broad spectrum of emission control devices/systems has been evaluated by the automotive industry and ancillary development companies. Specific emission control devices include:

1. Exhaust gas recirculation (EGR). A means of reintroducing a portion of the exhaust gas into the incoming fuel-air mixture, thereby reducing peak combustion temperature and NO formation (see Fig. 1).
2. Thermal reactor. A chamber (replacing the conventional engine exhaust manifold) into which the hot engine gases are passed. The chamber is sized and configured to increase the residence time of the gases and permit further chemical reactions, thereby reducing HC and CO concentrations (see Figs. 2 and 3).

Lean thermal reactor (LTR): air-fuel ratio (A/F) $>$ 15:1
Rich thermal reactor (RTR): A/F $<$ 15:1

3. Catalytic converter. A device containing a catalyst material which chemically decreases emissions (see Figs. 4 and 5).

HC/CO catalytic converter. A converter with a single catalyst bed for oxidizing HC and CO.

NO_x catalytic converter. A converter with a single catalyst bed for reducing NO and NO_2.

Dual catalytic converter. A converter with two beds: one for oxidizing HC and CO and one for reducing NO and NO_2.

Tricomponent catalytic converter. A converter with a single catalyst bed for simultaneously oxidizing HC and CO and reducing NO and NO_2.

Total emission control systems generally fall into the categories shown in Table 2. Figs. 6 and 7 illustrate two experimental combination systems.

EFFECT OF LEAD ADDITIVES

CATALYTIC CONVERTERS - Degradation of the performance of catalytic converters employed as pollution control devices on automobiles run on leaded and unleaded gasolines is observed to occur much more rapidly with leaded gasoline. Degradation may occur either by loss of catalytic activity, or physical attrition, or both. The lead component of gasoline thus clearly constitutes a catalyst "poison," which acts through a variety of chemical and mechanical toxicity mechanisms that are not mutually exclusive.

Even though numerous theoretical and laboratory investigations have been performed on catalyst poisoning, the complex composition of exhaust gas, the wide range and number of engine operating parameters, and the many types and configurations of catalytic materials make it very difficult to arrive at generalizations regarding the most likely mechanisms. Nevertheless, a review of these mechanisms has in-

---

*No effort has been made to update the findings or conclusions since this study was documented on Nov. 15, 1971.

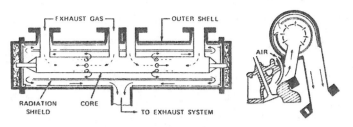

Fig. 2 - Du Pont type V thermal reactor (9)

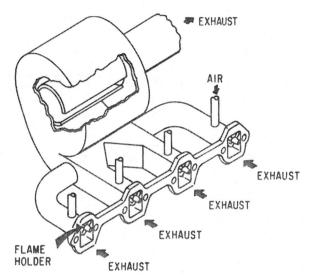

Fig. 3 - Esso rapid action manifold (RAM) reactor (10)

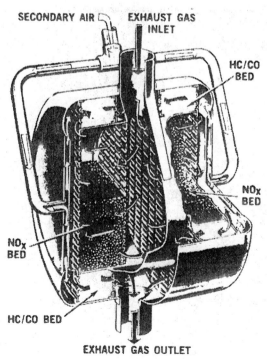

Fig. 4 - Bifurcated dual catalytic converter (11)

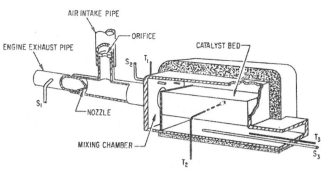

Fig. 5 - Cutaway sketch of Oxy-Catalyst converter (12)

Table 2 - Emission Control System Categories

Catalytic converter systems (no form of thermal reactor warmup device)
  HC/CO catalytic converter alone
  HC/CO catalytic converter + EGR
  Dual catalytic converter + EGR
  Tricomponent catalytic converter alone
Thermal reactor systems
  LTR + EGR
  RTR alone
  RTR + EGR
Combination systems
  LTR + HC/CO catalytic converter + EGR
  RTR + HC/CO catalytic converter + EGR
  RTR + dual catalytic converter + EGR
  RTR + NO_x catalytic converter + RTR

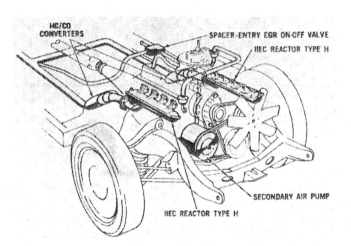

Fig. 6 - Ford combined maximum-effort/low-emission concept vehicle (13)

dicated that lead, sulfur, and phosphorus compounds would have a deleterious effect on catalysts.

Data on catalyst activity versus lifetime are available on some catalysts operating with leaded, low lead, and unleaded gasoline (lead levels of 2-3, 0.5, and 0.02-0.06 g/gal, respectively). For the lowest range, the exact amount of lead used is not clearly identified. In general, the catalyst lifetime decreases as lead content is increased. However, at very low levels, the data are not sufficient to establish a meaning-

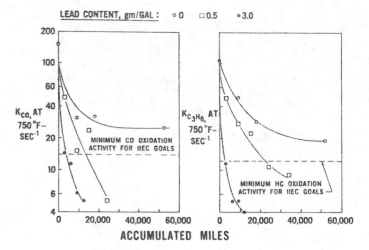

Fig. 8 - Effect of lead content in fuel on catalyst type "G" oxidation activity (14)

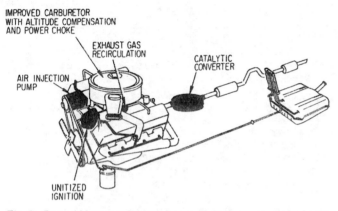

Fig. 7 - General Motors 1975 experimental emission control system (3)

ful correlation. The data do show that activity and lifetime are drastically affected with lead levels over 0.5 g/gal, and that levels of 0.02-0.06 g/gal result in significantly better performance. Figs. 8 and 9 illustrate typical mileage degradation effects.

It has been stated that a single tankful of regular leaded gasoline can destroy a catalyst. Although this cannot be substantiated, it is apparent from available data that such quantities could seriously reduce its useful lifetime. Since catalysts currently planned for automotive use are so adversely affected by lead levels in leaded gasoline, a system must be devised to prevent accidental contamination.

THERMAL REACTORS - Thermal reactor durability or effective lifetime may be significantly affected by erosive and/or corrosive deterioration caused by the presence of lead compounds in the exhaust gas. The problem has been the subject of recent intensive investigations. Symptoms of erosion are generally exhibited as a deterioration of the baffles and reactor core surface in localized areas opposite the valve ports. Du Pont (16) has analyzed the erosive behavior of a number of thermal reactor material candidates with various fuels. The results show that erosion is chemical rather than mechanical in nature and is affected by tetraethyl lead (TEL).

Du Pont also concluded that the presence of phosphorus in the fuel accelerates corrosive attack (Fig. 10). Inconel 601

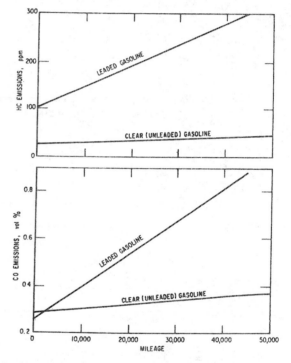

Fig. 9 - Typical vehicle emissions with a catalytic converter (15)

and Armco 18 SR were determined to be promising corrosion-resistant materials for reactor application; Uniloy 50/50 was found to be the most resistant to erosive attack.

Fuel composition effects on thermal reactor durability were investigated in an Inter-Industry Emission Control (IIEC) materials evaluation program (17). Materials tested (as core specimens in a vehicle reactor) included ferritic (nickel-free) and austenitic stainless steels, high-nickel alloys, and various coatings on low-cost materials. Tests of OR-1, a low-cost, nickel-free alloy candidate, showed that with leaded fuel, halides and phosphorus contribute heavily to metal deterioration at elevated temperatures (Fig. 11). Leaded gasoline without phosphorus showed considerably less erosion. Thus, the halides and phosphorus are major contributors to material loss. The low-cost alloy steels, however, were ultimately rejected as thermal reactor materials because of poor high-temperature strength. All of the most

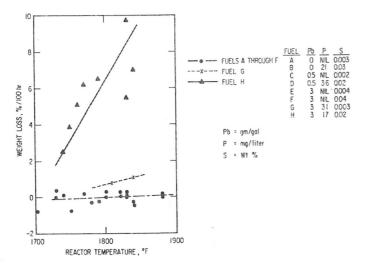

Fig. 10 - Effect of fuel additives on corrosion weight loss of Inconel 601 (16)

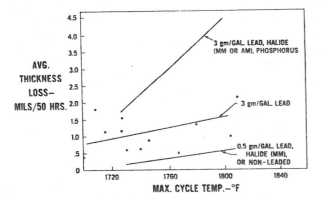

Fig. 11 - Effect of fuel variables on average thickness losses of OR-1 alloy during continuous thermal cycling (bare sandblasted specimens) (17)

promising materials were intermediate-cost nickel alloy steels.

The NASA/Lewis Research Center has been conducting a thermal reactor materials evaluation and development program for EPA (18). Coupon screening, core testing, and full-scale thermal reactor endurance testing by Teledyne-Continental* have led to the identification of two materials which appear to be suitable for rich reactor operation with leaded fuels. These are GE 1541 (15 Cr-4Al-1Y) and Inconel 601. Armco 18 SR may also be a candidate.

Endurance test results for various thermal reactor material candidates are shown in Fig. 12.

The available data on materials testing with leaded fuels indicate that corrosion effects due to lead halides and/or phosphate compounds in the exhaust are temperature related. Fig. 11, for example, suggests that at temperatures approaching 1700 F, corrosive weight loss rates are not sensitive to fuel composition. There is, therefore, a rational basis for an Ethyl Corp. claim that the lead composition of fuel has no impact on its lean thermal reactor (LTR), which

---

*Recently, the fuel used for endurance testing was switched from leaded to unleaded gasoline. No change in the weight loss of GE 1541 or Inconel 601 was observed (18).

operates at temperatures below 1700 F even under high-speed turnpike conditions (19).

Fuel lead concentrations of approximately 0.5 g/gal should have no significant detrimental effects on the better oxidation-resistant materials available. There seems to be no obvious reason (although direct data are lacking) why such materials could not function with concentrations of up to 3 g/gal of lead. However, the combined presence of lead and phosphorus additives has an accelerating influence on the corrosive deterioration of a number of different metallic alloys.

EXHAUST GAS RECIRCULATION - Discussions with the four major United States automobile manufacturers (20-23) indicated a consensus that lead could cause deposit and life problems in the EGR system. The degree of the effect would be dependent upon the EGR tap-off location and injection orifice size used; that is, the smaller the hole size the more likely clogging could or would occur. Ford (21) referred to previously published data (Fig. 13) which indicated a decrease in effectiveness to 0.15 at 6000 miles, from 0.35 initially, when a leaded fuel was used (7-10% EGR flow rate). Deposits affected flow characteristics by changing critical dimensions or by preventing control valve reseating (thereby altering the programmed rate of recycle flow and changing the NO_x reduction efficiency). All companies, however, are developing pre-1975 EGR systems for compatibility with lead additives.

The most significant reported data base for EGR system durability effects is the extended-use program conducted by Esso (25) for NAPCA (now EPA). In this test program, an EGR system previously developed was evaluated in three 1969 Plymouths and three 1969 Chevrolets over 52,000 miles under city/suburban driving conditions simulated on a tape-controlled mileage accumulation dynamometer. No major problems were reported. There was no decrease in NO_x reduction efficiency over the 52,000 miles. Engine wear and cleanliness were considered normal for the mileage and driving regime. In addition, the EGR system was found to be compatible with commonly employed air injection and engine modification systems used for HC and CO control.

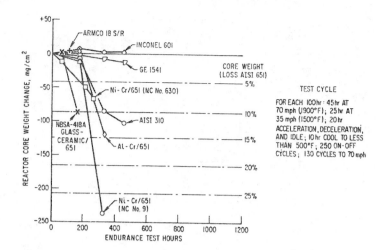

Fig. 12 - Weight change of test reactor cores in engine dynamometer endurance test (18)

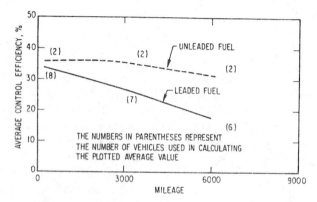

Fig. 13 - Effect of leaded fuel on the control efficiency of air cleaner EGR systems, 302 CID engine (24)

In this extended-use program, it was noted that throttle plate deposits occurred when fully leaded fuel was used. (Control tests of a vehicle with unleaded fuel indicated essentially no throttle plate deposits.) Analysis of the deposits indicated the main components were lead chlorides and lead bromides. Despite the throttle plate deposits, the NO_x control effectiveness was unaffected.

It is concluded, therefore, that lead-free or low-lead gasoline is not required for the implementation of EGR systems, per se. The presence of lead additives can result in deposits in EGR orifices, throttle plate areas, etc. The actual severity of such deposits would appear to be strongly related to the particular type of EGR system as well as to control orifice sizes used, and/or to the utilization of self-cleaning designs (plungers, specially coated surfaces, flexible snap rings, etc.) in areas susceptible to deposit buildup.

OTHER ENGINE SYSTEM PARTS - The principal deleterious effect of lead additives in gasoline on engine parts other than the emission control system, per se, is to reduce the usable lifetime of exhaust systems and spark plugs. Other reported differential effects (varnish, sludge, rust, wear, etc.), due to unleaded versus leaded gasoline, do not result in a quantifiable impact on the consumer in terms of operational considerations or cost. The use of unleaded gasoline can essentially double the exhaust system life and increase spark plug life approximately 50% in conventional (pre-1971) cars.

Ref. 26 indicates that vehicles operated on leaded gasoline have a muffler life of about 37,500 miles and a spark plug life of about 13,000 miles per set. Figs. 14 and 15 illustrate unleaded gasoline life effects based on a statistical analysis of fleet test data provided by Ethyl (27).

Similar spark plug life increases with unleaded gasoline are expected in 1975-1976 systems. If long-life exhaust systems compatible with either leaded or unleaded gasoline (for example, stainless steel) are incorporated in 1975-1976 systems, no lifetime variabilities would exist for this component.

There is considerable evidence that excessive valve seat wear can occur with the use of unleaded fuel (28-33), particularly at sustained high speed and load conditions. However, this problem can be solved at very low cost by changing to induction-hardened exhaust valve seats. One domestic manufacturer has introduced such valve seats in some 1972 models, with plans for full implementation by the end of the 1972 model year. Other manufacturers are also phasing in compatible exhaust valves and seats. All United States automobile manufacturers plan to market a system compatible with unleaded gasoline by the 1975 model year.

EMISSION LEVEL STATUS

A summary of the available emission data for the emission control system concepts identified above is presented in Table 3. Several emission control systems have met, or show promise of meeting, the federal 1975 emission standards. These systems are experimental versions, and emission data do not reflect consideration of any factor to account for variabilities in production tolerances, testing procedures, or degradation with mileage.

The following general observations are pertinent for the performance of emission control systems as shown in Table 3:

1. In general, the catalytic-converter-only systems suffer due to high emissions during the cold start portion of the constant volume sampling (CVS) test procedure, due to slow warmup. In addition, the tricomponent catalytic converter requires a precision in A/F control not adequately demonstrated to date.

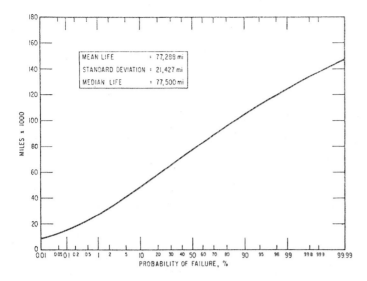

Fig. 14 - Muffler lifetime probability, operation on unleaded fuel

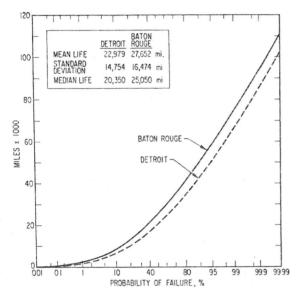

Fig. 15 - Spark plug lifetime probability, operation on unleaded fuel

2. LTR plus EGR systems have yet to demonstrate meeting 1975-1976 HC and CO standards, although improvements in thermal reactor design (flameholders, improved mixing, etc.) and fast warmup choke devices should help. NO_x levels below approximately 1.3 g/mile have not been reported for this concept.

3. The most advanced RTR plus EGR systems meet the 1975-1976 HC standard and approach the CO standard. NO_x levels below approximately 0.7 g/mile have not been demonstrated. At this NO_x level, however, the fuel economy penalty is severe (approximately 25-30 %).

4. Combination systems, that is, various combinations of thermal reactors, EGR, and catalytic converters, are judged by the automotive industry to offer the best hope for achieving minimum emission levels and are under intensive development for incorporation in 1975-1976 model cars. In these combination systems, the primary function of the thermal reactor is to warm up the catalyst bed. Therefore, it need not be a "full-size" reactor, but rather a "low-grade," less complex design.

5. Thermal reactor (LTR or RTR) plus HC/CO catalytic converter plus EGR systems are meaningful in terms of 1975 standards. However, it is generally agreed that a NO_x catalytic converter would have to be added for the lower NO_x levels required by the 1976 standards. At this time, a NO_x catalyst with the required durability (50,000 miles) has not been demonstrated. (With the NO_x catalyst, the thermal reactor is restricted to rich operation, inasmuch as all known NO_x catalysts require a reducing atmosphere.)

It is emphasized that the foregoing observations are based on experimental laboratory data only. If, as the various automobile manufacturers have suggested, levels 50% (or lower) of the 1975-1976 standards have to be achieved to account for the variation of production tolerances, test reproducibility, degradation with accumulated mileage effects, etc., then it would appear that none of the emission control systems proposed and evaluated to date will meet the 1975-1976 emission standards on a consistent basis.

DURABILITY STATUS

Because of the widespread comprehensive search for an emission control system capable of meeting 1975-1976 standards, the proliferation of concepts investigated, and the generally inadequate emission results (even at low mileage conditions), the current status of emission control system durability with respect to the 50,000-mile standard is best characterized by the following types of component/device durability data.

THERMAL REACTORS -

1. *Lean Thermal Reactor (LTR)* - One Pontiac incorporating the Ethyl LTR and EGR accumulated over 30,000 miles in various types of service, including cross-country trips. Ethyl has reported excellent durability characteristics for its LTR. Another Ethyl-equipped 1970 Pontiac is in general fleet service with the California Air Resources Board (CARB).

Table 3 - Summary of Emission Control System Emission Data

System Type — CVS Cold-Start-Emissions, g/mi Fed. Std.: 1975-1976	Single-Bag (CVS-1), Pre-July 1971			Three-Bag (CVS-3), Post-July 1971			Reference No.
	HC 0.46	CO 4.7	NO_x 3.0 / 0.4	HC 0.41	CO 3.4	NO_x 3.1 / 0.4	
Catalytic converter systems*							
HC/CO catalytic converter only, no EGR							
UOP tests							
U. S. 1971 domestic V8, normal choke	0.59-0.68	0.96-1.45	2.11-3.88	–	–	–	34
U. S. 1971 domestic V8, fast choke	0.16-0.51	1.21-2.58	4.74-5.08				–
Some foreign vehicles	0.29-1.41	0.99-3.86	1.36-2.0				
Engelhard tests							
PTX-433 catalyst, 0.2% Pt	0.70	3.80	5.0	–	–	–	35
HC/CO catalytic converter + EGR							
Ford package "B"	0.80	11.0	1.3	–	–	–	13
Chrysler	0.24	7.2	2.03	–	–	–	6
Dual catalytic converter + EGR							
Ford package "C"	0.85	10.0	0.90	–	–	–	13
Tricomponent catalytic converter							
APCO tests of UOP system, 1970 VW	1.4-2.3	10-32	0.6-1.3	–	–	–	36
Thermal reactor systems*							
LTR + EGR							
Ethyl Corp.							
Pontiac	0.64	6.4	1.52	–	–	–	37
Plymouth	0.89	8.6	1.37	0.52	6.2	1.37	37
Chrysler	0.70	7.0	1.30	–	–	–	6
RTR alone							
Modified RAM	0.07	4.2	1.89	–	–	–	10
Ford type J reactor	0.1-0.3	6-12	0.5-0.7	–	–	–	–
RTR + EGR							
RAM							
Esso tests, Chev.	0.08	3.7	0.72	–	–	–	10
EPA tests, 1971 Ford LTD							38
A	0.20	5.9	0.65	–	–	–	–
B	0.20	3.8	0.60	–	–	–	–
C	0.14	4.8	0.60	–	–	–	–
D	0.10	4.54	0.67	0.11	4.76	0.67	–
E	0.14	4.77	0.63	0.10	3.19	0.67	–
Recent Du Pont system	0.05	9.2	0.52	–	–	–	33
Ford package "A"	0.30	9.0	1.40	–	–	–	13
Chrysler	0.23	13.8	0.45	–	–	–	6
Combination systems*							
LTR + HC/CO catalytic converter + EGR							
G. M. "1975 Experimental System"**	0.54	9.2	1.0	0.40	5.5	0.95	3, 39, 40
American Motors							
Platinum-monolithic; air-injection reactor	0.40	2.35	3.0	–	–	–	–
Base metal-bead; air-injection reactor	–	–	–	0.11	3.29	3.0	–
Platinum-bead; air-injection reactor	0.45	2.96	3.0	–	–	–	–
RTR + HC/CO catalytic converter + EGR							
Ford combined concept package							
"Maximum effort" tests	0.28	3.4	0.76	–	–	–	13
"Improved fuel economy" tests	~0.3	~3.5	1.3	–	–	–	13
High-rate EGR system	–	–	–	0.25	2.95	0.55	41
RTR + dual catalytic converter + EGR							
G. M. "1975 system" + quick-heat manifold and fast choke + NO_x catalytic converter†	0.2	4.0	0.6	0.2-0.3	2.5-6.0	0.35-0.85	3, 39
Ford dual bed catalyst system	–	–	–	0.27	2.24	0.60	41, 42
AMOCO vehicle tests							
Pelletized catalysts	–	–	–	0.26	1.72	0.55	–
Monolithic catalysts	–	–	–	0.38	2.07	0.68	–
RTR + NO_x catalytic converter + RTR							
American Motors, Jeep with 360 CID; no EGR	–	–	–	0.01	2.44	0.37	40

*Laboratory, low-mileage tests.
**General Motors A.I.R. system; that is, "low-grade" LTR, lean (A/F = 15-16.5) operation.
†General Motors A.I.R. system; that is, "low-grade" RTR, rich (A/F = 14-15) operation.

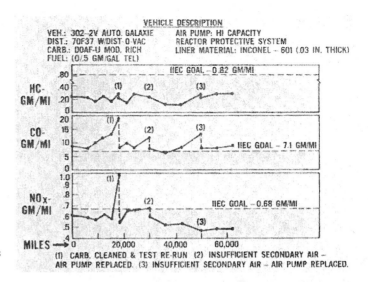

Fig. 16 - Ford type J reactor durability and cold start emissions data (38)

The total mileage of this vehicle at the last test point was 12,000 miles.

The basic nature of an LTR system would predict less difficulty in obtaining satisfactory durability than would be the case with an RTR system. This is because exhaust gas entering the reactor from the lean engine contains about 100 ppm HC, 0.1-0.4% CO, and approximately 2-4% O_2 (without an air pump). Therefore, little chemical heat is generated in the reactor, and its temperature is governed by the degree to which the sensible heat in the exhaust gas is conserved. This means that the lean reactor operates in a temperature range of 1400-1600 F, even under high-speed turnpike conditions, which is a range that good quality stainless steel should tolerate. Moreover, tests by Ethyl indicate that the LTR is not subject to destructive temperature excursions, even with a continuously misfiring spark plug. Thus, durability should not be seriously decreased by situations in which engine malfunctions might occur.

Therefore, even though a system life of 50,000 miles has not yet been demonstrated, there appears to be no fundamental reason why it could not be achieved by an LTR.

2. *Rich Thermal Reactor (RTR)* - The Ford type J reactor (no EGR) was installed in a vehicle for durability testing. As reported by Ford (38), 62,000 miles of heavy-duty operation were accumulated at the time of writing (January 1971). While failures were experienced on ancillary components as mileage was accumulated, the RTR system continued to control emissions to essentially IIEC target levels when all emission component subsystems were operational (see Fig. 16 for notation of carburetor and air-pump problems). This vehicle was equipped with a reactor overtemperature protection system, which limited peak temperatures to 1850 F.

A recent communication (43) with Ford indicated that 90,000 miles of durability operation has been completed. At test termination, a number of heat cracks in the liner were found and a hole had developed in one small area. Although a loss of performance was observed, the system was still performing relatively well.

Du Pont supplied six cars to the CARB in the fall of 1970 for evaluation in a 2-year program. The six cars, equipped with the Du Pont particulate-trapping system as well as RTR and EGR, were assigned to the state motor pool in California and were being driven in normal service by state employees. In June 1971, the average odometer reading of the six vehicles was 17,954 miles. As the vehicles had about 3000 miles prior to incorporation of the emission control system, this represents about 15,000 miles of durability testing of the emission control system.

Near the end of August 1971, a failure of a timing chain occurred in one of the six test vehicles. The failure was described as an elongation of the timing chain, which eventually rubbed a hole in the timing chain cover. The six vehicles in the control fleet were not affected.

Du Pont (44) states that similar wear was observed in three of the six CARB test cars. Symptoms of similar wear had been previously detected in three RTR-equipped vehicles tested by Du Pont. Timing chain pins, cam followers, rocker arms, and valve guides were affected. Du Pont is convinced that the wear problem is due to the lapping action of small (0.02-0.05 μ) metal oxide particles mixed in the engine oil. These small particles come from the reactor core and find their way through the EGR line to the lubrication system (presumably by entering the intake valve ports and then passing through the piston rings and/or exhaust valve guides).

Severe oxidation of the reactor core 310 stainless steel material was demonstrated in Du Pont tests of two reactors, which lost 0.5 lb of core weight (23%) after 20,000 miles of testing. Du Pont feels that the wear problem could be overcome by using a material such as Inconel 601. Since oxidation of any part of the exhaust system is a potentially similar hazard, a more complete fix would be to use an EGR gas source upstream of the reactor; the exhaust manifold crossover pipe is a possibility.

Esso tests (10) of the RAM (Rapid Action Manifold) RTR (see Fig. 3) have been demonstrative only; no durability tests have been made. If this concept were tested for durability, a better material than the 310 stainless steel used in the demonstrator model would be required.

Table 4 - EGR System Cost Data

Source	Materials Cost, $	Manufactured Cost, $	Cost to Consumer, on new car, $	Maintenance Costs, $	Description and Ref. No.
Chrysler Corp.	–	–	30.00	–	1973-1974 system (48)
Du Pont, EGR system	2.96	3.70	7.40 } 8.28	–	Cooled gas; above-throttle injection (33)
Du Pont, in-line filter	0.35	0.44	0.88		
Esso Research & Engrg.	–	~10.00	–	–	Cooled gas; above-throttle injection (49)
Universal Oil Prod.	–	–	25.00	–	(15)
Research Triangle Institute, 1970	–	–	25.00	8.00/yr	(50)
Research Triangle Inst., 1971-EGR system	–	–	20.00	–	Fixed-orifice system (51; unpublished draft)

EXHAUST GAS RECIRCULATION (EGR) - The more advanced modulating EGR system used on the Ethyl LTR car provided to the CARB was found to be free of deposits after 12,000 miles of service, and was tested successfully for 30,000 equivalent miles on the dynamometer.

Durability tests of EGR alone were conducted by Esso (45) for NAPCA (now EPA). In this test program, the EGR system was evaluated in three 1969 Plymouths and three 1969 Chevrolets over 52,000 miles under city/suburban driving conditions simulated on a tape-controlled mileage accumulation dynamometer. No major problems were reported. Engine wear and cleanliness were considered normal for the mileage and driving regime.

General Motors (46) recently reported that 300,000 miles of operation had been accumulated for the below-the-throttle type of EGR incorporated in some 1972 Buicks sold in California.

CATALYTIC CONVERTERS -

1. *HC/CO Catalytic Converter* - Engelhard (35) has reported a 50,000-mile durability test for their PTX-433 catalytic converter unit (0.2% Pt). Emission levels at the end of 50,000 miles of the AMA driving schedule are as shown in Table 3. Initial emission values were not given and, therefore, the deterioration or degradation factor cannot be stated. The test was conducted with an unleaded gasoline having a lead content of approximately 0.03 g/gal. The catalyst picked up substantial quantities of Pb, Zn, P, and Ba during the test. The Zn and Ba are contaminants which Engelhard associates with motor oil. Engelhard's present position is that the most probable reasons for PTX catalyst deterioration are metal poisons in the fuel and lubricating oils.

Other pertinent durability data, in terms of degradation with mileage, were previously shown in Figs. 8 and 9.

2. NO_x *Catalytic Converters* - Whether as a separate converter or in the form of a dual catalytic converter unit, only laboratory and low mileage data are available on NO_x catalysts. Durability tests in excess of approximately 10,000-15,000 miles have not been reported. The information is not sufficient to determine whether the durability problem is limited to activity degradation, or physical attrition, or both. Earlier Monel NO_x catalyst tests performed by Esso (47) indicated severe agglomeration problems with this bulk metal catalyst.

CONSUMER COST IMPLICATIONS

The cost implications arising from the incorporation of emission control systems in light-duty motor vehicles to meet the 1975-1976 emission standards encompass the areas of initial investment costs, maintenance-related costs, and operating costs (that is, fuel costs). In view of the fact that 1975-1976 standards have not been adequately demonstrated by automobile manufacturers and that no final selection of system components has been made, the cost estimates presented herein are limited to engineering estimates based on projections of current designs (Table 4).

The approach followed is to select a baseline vehicle having performance and fuel economy characteristics typical of cars not employing sophisticated emission control systems; that is, pre-1971 cars designed to operate within the constraints of the normal two-grade leaded gasoline supply system. Then, selected generic emission control systems (as previously defined in Table 2) are added to the baseline vehicle, and typical fuel economy penalties associated with each generic class are assessed. The fuel penalty consists of the control device effects and any compression ratio limits occasioned in the case of generic classes of emission control devices constrained to operation on unleaded gasoline. Finally, for each generic class, overall costs to the consumer (initial, maintenance, and operating costs) are summed up, based on an average vehicle operational lifetime. No costs have been included for research and development activities, compliance emission testing after car purchase, or production emission testing.

INITIAL HARDWARE COSTS - Specific control devices and their associated engine- or vehicle-related requirements were grouped in the general areas of: engine modifications, emission control system components, and exhaust system initial costs.

Engine modification and emission control system component cost penalties are based on consideration of a variety of assumptions and evaluation techniques, including cost of materials used in the device, difficulty of manufacturing, comparisons with existing automotive components, and discussions with automobile and equipment manufacturers. Every effort has been made to assure that cost levels are "comparable," both as to variation in devices within a given component class (for example, within thermal reactors) and from component class to component class (for example, from thermal reactors to catalytic converters).

In this manner, although the cost values used herein may not exactly coincide with those eventually forthcoming from the automobile manufacturers, the relative cost levels between the various generic classes of overall emission control system concepts made up of these components are meaningful on a comparative basis.

1. *Engine Modifications* -

Carburetion - All automobile manufacturers have stressed the need for improved carburetion (and/or fuel injection) for improved A/F control in engines employing complex emission control systems. Informal estimates of hardware cost increases vary widely. For purposes of the present study, a differentiation was made between "lean" and "rich" systems. A nominal increase of $13 was assessed for improved carburetion for "rich" systems. For the more difficult "lean" systems, this cost increase was selected to be $25.

Ignition/Distributor/Control Systems - All automobile manufacturers have stressed the need for some form of "unitized" ignition system for longer life and better control of ignition spark to prevent degradation of emission levels. Some have also expressed the opinion that a new engine control system may be required to provide more precise control of the variables and coordinate spark timing and carburetion. For the present study, it was assumed that all advanced emission control system concepts would incorporate features of this general type, and an estimated cost penalty of $37 was assessed all generic system classes.

Long-Life Exhaust Systems - Exhaust system components become part of or are replaced by emission control system components in nearly all advanced emission control system concepts. Also, exhaust gas temperature levels are generally increased, severely, in some cases, to provide more optimum operating conditions for certain control system components (for example, catalytic converters) or due to the operation of certain components (for example, RTR). Because of the postulated requirement for 50,000 maintenance-free miles of operation, and the interaction between control system components and normal exhaust system components, a long-life exhaust system (for example, stainless steel) was assumed to be included in every generic class of emission control system. An initial cost penalty of $60 was estimated for this system, and a cost credit of $28 allowed due to the deletion of the normal exhaust system, resulting in a net additive cost penalty of $32 for every generic class.

Exhaust Valves and Seats - As discussed previously, the use of lead-free gasoline for most cars will result in the requirement for valves and/or valve seat modifications to prevent valve recession effects. Any generic class of emission control systems operating on unleaded gasoline was assessed a $3/car cost penalty for such modifications.

2. *Emission Control System Components* -

Exhaust Gas Recirculation (EGR) Systems - A wide variety of EGR system designs exists. Similarly, there is a rather wide variation in available cost estimates of EGR systems, per se (see Table 4). For the present study, a representative value of $25 was assessed all systems incorporating EGR.

Catalytic Converters - Considering the state of development for catalytic converters, there is an understandable lack of cost data and a reluctance on the part of the catalyst and automobile manufacturers to estimate what the cost will be. However, some speculative costs have been obtained from available reports and by visits to several sources as shown in Table 5.

Since there was a great deal of disagreement on catalytic converter costs and a general lack of cost data, an independent estimate was made. This estimate considered: catalyst cost, converter material cost, manufacturing labor, overhead, and profit, installation, and sales profits. The resulting values are shown in Table 6.

Base metal and noble metal catalyst costs per vehicle are about the same, even though the cost per pound is much greater for the noble metal catalyst. The reason, of course, is that a smaller amount of noble metal catalyst is required. With the cost of the base metal and noble metal catalyst equalized, the remaining cost for metal fabrication, profit, installation, etc., also tends to equalize such that the cost to the consumer is about the same for the two types of catalytic converter.

Thermal Reactors - Du Pont (33) estimated that the cost of their RTR is $48 (two times the estimated manufacturing cost of $24). Ethyl (37) estimated the cost of their LTR to be $100 (including an upgraded exhaust pipe).

Because of these substantial variances and the general lack of cost data from the automakers, an independent cost estimate again was made. This estimate considered volumetric and materials differences between the LTR and RTR approaches and included considerations of materials cost, manufacturing labor and profit, installation costs, sales profits, and credit for standard exhaust manifolds. The resulting values are given in Table 7.

In a similar manner, the cost of a "low-grade" thermal reactor (rich or lean) was estimated to be $70/car. Such low-grade reactors would be smaller in volume, have less insulation, might not have a core liner, etc., and would approach an oversize standard exhaust manifold in configuration.

Air-Injection Pump - A nominal cost of $29 was assessed each configuration incorporating an air pump. Additional costs for plumbing, etc., were assumed to be accounted for either in the thermal reactor or catalytic converter cost estimates.

Overtemperature Protection System - Rich thermal reactor and/or catalytic converter systems may require overtempera-

Table 5 - Catalytic Converter: Consumer Cost/Car, Dollars

Source	HC/CO	NO$_x$	Dual	Tricomponent	Reference No.
UOP	150	–	–	120	15
Engelhard	50	50	–	–	52
Esso Research	–	–	84	–	53
Triangle Inst.	75	75	92	–	51
Stanford Research Inst.	130 avg.	–	–	–	54

Table 6 - Estimated Catalytic Converter Costs

Converter Type	Original Equipment Cost to Consumer, $/car
HC/CO, base or noble metal	98
Dual, base or noble metal	129
Tricomponent, noble metal	98

Table 7 - Cost Differences in Thermal Reactors

Reactor Type	Original Equipment Cost to Consumer, $/car
RTR	125
LTR	110

ture protection systems. Both the need for and the exact details of such protection systems are not established at the present time, as all automakers are searching for catalysts with higher temperature capability to avoid the system complexity introduced by the addition of protection systems.

Informal cost estimates for such systems range from $25 for the simpler approaches to approximately $100 for the more complex. For the present study, a cost penalty of $50 was assessed any generic concept incorporating either a catalytic converter or an RTR.

Summary of Initial Hardware Costs - As a primary purpose of the cost analysis effort was to provide a measure of the cost differences between the various conceptual approaches, all generic classes were assessed component or hardware cost penalties on a common basis by use of the component costs described above. Table 8 summarizes the initial hardware costs for the generic classes considered in the cost analysis. Identified are the discrete components, and their costs, of each generic class and a summation of the initial total hardware cost to the consumer, as installed in a new car.

It should be noted that systems incorporating both a thermal reactor and catalytic converter have been selected to use a low-grade thermal reactor in that its primary purpose is to warm up the catalytic converter. Therefore, it is felt that the full-size thermal reactor is unwarranted in this case.

Catalytic converter-only systems are the same as their thermal reactor plus catalytic converter counterparts, except for the deletion of the thermal reactor cost ($70). The tricomponent catalytic converter concept has not been treated, since it requires a precision in A/F ratio control not yet demonstrated. As the required A/F ratio control system has not been identified, it has not been possible to provide a reasonable cost estimate for this function.

As shown in Table 8, initial installed hardware costs range from $229 to $388. Thermal reactor systems are lowest in cost, and the dual catalytic converter-thermal reactor system is the most expensive, as would be expected. It should be noted that these values do not include the approximate $50 cost increase already incorporated in 1971 cars arising from the emission control system and engine modifications made starting in 1962-1966 model cars.

MAINTENANCE COSTS - Several maintenance cost areas affected by the emission control system concept and/or the use of leaded or unleaded gasoline were identified. These are spark plug life, maintenance of the overtemperature protection system, catalytic converter system replacement, and exhaust system replacement. An average lifetime mileage of 85,000 miles and an average automobile age of 8.4 years are used in the analysis. These values are based on the percentage of cars still registered as a function of the car age (55) and on the average miles per year as a function of the age of the car (56).

1. *Spark Plugs* - As mentioned previously, spark plugs operated with unleaded gasoline have an average longer lifetime than those operated with leaded gasoline. Although the exact lifetime levels in each case are not a priori determinable when installed in the various emission control system concepts, representative values of 13,000 miles for leaded gasoline and 20,000 miles for unleaded gasoline were selected to illustrate typical maintenance cost differences for spark plug changes in the two cases.

Each spark plug change was estimated to cost $10 (representing a combination of installations at a garage and home replacement). For the 85,000-mile baseline car life used herein, this resulted in a cost savings of $23 for cars using unleaded gasoline.

2. *Overtemperature Protection System* - As noted previously, if needed and incorporated, the overtemperature protection system is a critical part of the overall emission control system. A nominal cost value of $5/year has been assessed only for the inspection of this system; no cost for actual repair has been included. Again, based on the average 8.4 year lifetime used herein for the base line car, this results in a total inspection cost of $42 for those emission control system concepts incorporating overtemperature protection systems.

3. *Catalytic Converter* - The goal of the automobile manufacturers is to develop a catalytic converter unit capable of meeting a 50,000-mile maintenance-free and/or replacement-free requirement. Demonstration of this capability has not

Table 8 - Installed Hardware Cost Summary (Cost to Consumer in New Car)

Cost Item	Generic Concept / Thermal Reactor Systems* LTR + EGR	RTR + EGR	Combination Systems** Low-grade LTR + EGR + HC/CO C.C.	Low-grade RTR + EGR + HC/CO C.C.	Low-grade RTR + EGR + Dual C.C.
Initial costs, $					
Engine modifications					
Carburetion	25	13	25	13	13
Ignition/distributor	37	37	37	37	37
L. L. exhaust system	60	60	60	60	60
Valves, seats	–	–	3	3	3
Emission control components					
EGR	25	25	25	25	25
Thermal reactor	110	125	70	70	70
Catalytic converter, C.C.	–	–	98	98	129
Air-injection pump	–	29	29	29	29
Overtemperature protection system	–	50	50	50	50
Exhaust system credit	–28	–28	–28	–28	–28
Total installed hardware cost, $	229	311	369	357	388

*Use leaded gasoline.
**Catalytic converter + EGR are the same except for $70 decrease due to omission of low-grade thermal reactor.

been made to date. However, for the present cost assessment, it is assumed that this requirement will be met for HC/CO catalytic devices. For NO_x catalytic devices, both 25,000 and 50,000 mile replacement intervals are considered for illustrative purposes, since the demonstrated life is quite low.

Discrete costs for replacement were estimated, based on original manufactured cost, wholesaler's costs, dealer installation cost, and nominal profits. These values are given in Table 9.

It must be emphasized that these costs are for replacement of the entire unit, including both the catalyst and its container. If techniques to replace the catalyst without replacing the container become effective, these costs will be substantially reduced.

4. *Exhaust System* - As mentioned, a long-life exhaust system was a basic part of each generic class of emission control system considered. It was assumed that this system would last the life of the car. As the consumer has been penalized for this added cost, the exhaust system replacement costs normally anticipated in the base line car case must be subtracted from the overall costs to the consumer to maintain the relative cost effect of the car equipped with the new emission control system. A representative average value of $45/exhaust system replacement and an average life of approximately 37,500 miles are assumed. This results in a total cost credit of $60 for each emission control system concept for an average car lifetime of 85,000 miles.

OPERATING COSTS -

1. *Limitation and Definition* - Operating costs, as utilized

Table 9 - Catalytic Converter Replacement Costs

Converter Type	Replacement Cost, $/car
HC/CO	123
Dual	156

herein, are restricted to fuel economy cost penalties. Fuel economy cost penalty is defined as the cost of fuel for the car over its lifetime when equipped with the emission control system being evaluated, minus the car lifetime fuel cost of an average pre-1971 car. This definition gives the increase in fuel costs due to the emission control system. When added to the emission control system initial hardware and maintenance costs, the total cost of the emission control system is obtained.

The characteristics of the pre-1971 car are: Weighted average compression ratio, 9.37:1 (57, 58); miles/gal, 13.5 (58); lifetime mileage, 85,000 miles; weighted average price of fuel (leaded) for pre-1971 car, $0.3740/gal (based on regular at $0.3569/gal (58), premium at $0.3969/gal, price spread from (59), and on sales of regular at 57.4% of total gasoline sales) (58). Lifetime fuel cost for the pre-1971 car from this: $2355.

To account for the fact that some pre-1971 cars are designed to use regular gasoline and some premium, a weighted average compression ratio has been used for the baseline pre-1971 car. Similarly, since drivers of some cars designed to use regular buy premium, an average price of gasoline based on the percentage of sales of each grade has been selected as the baseline gasoline cost/gal.

2. *Fuel Economy Effects—General* - The fuel economy of a car equipped with an emission control system depends primarily on the method of reducing NO_x emissions, the NO_x emission level achieved, and whether or not the emission control system can tolerate lead in the engine exhaust (catalytic systems are lead intolerant). HC/CO emission reduction devices have only secondary effects on fuel economy, excluding unleaded gasoline effects, since engine A/F is set by the NO_x emission reduction device. These secondary effects are due to increased engine exhaust backpressure and, in some cases, to power for a secondary air pump, and are much the same for most systems. Fuel economy effects considered, therefore, consist of those due to NO_x emission reduction devices, and the use of unleaded gasoline for systems containing catalytic devices. These are considered separately and then combined to obtain an overall fuel cost penalty for several basic classes of systems which encompass the major types of systems proposed for 1976 cars.

NO_x Effects - NO_x emission reduction systems fall into two broad categories: those depending only on the use of a combination of engine A/F change and EGR, and those utilizing NO_x catalysts in conjunction with a smaller change in engine A/F and a lower EGR rate than in the first category.

Typical engine fuel economy changes with the addition of emission control systems are shown in Fig. 17. This figure is a general correlation for a number of systems examined, where the specific fuel consumption (sfc) increase (over the baseline vehicle without the specific emission control system) is shown as a function of the NO_x level achieved. This general correlation is nearly the same as the Ford/IIEC estimate (13).

Both noncatalytic converter systems and HC/CO catalytic converter systems comprise the general correlation. This is as would be expected, since the NO_x level realized is related only to conditions present in the engine cylinder (that is, A/F and percentage EGR) and not to any external device or condition.

Also shown in the figure is that relationship of sfc increase versus NO_x level estimated to occur if a NO_x catalyst at 75% efficiency were added to a system characterized by the general correlation line.

It is evident from the figure that, excluding the cost effects of their use of unleaded fuel, NO_x catalyst systems are attractive from a fuel economy standpoint. It is also evident from the figure that decreasing the required NO_x emission level for any type of system reduces fuel economy.

Unleaded Gasoline Effects - Engine efficiency improves as compression ratio increases. In general, as the compression ratio increases, however, the gasoline octane number must also increase. The addition of lead to the fuel has proved to be the least expensive way of providing high octane number fuels. Engines with emission control systems which cannot tolerate lead in the fuel (for example, catalytic converter systems) must, therefore, either operate at lower compression

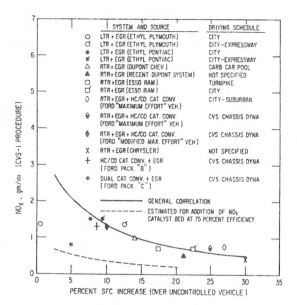

Fig. 17 - NO_x versus sfc increase

ratios with poorer efficiency than engines equipped with emission control systems which are lead tolerant, or use more expensive fuel than is required for the engines equipped with lead-tolerant emission control systems.

A 93 research octane number (RON) gasoline has been selected to be consistent with concurrent studies (EPA-funded) to determine unleaded gasoline investment and manufacturing costs (60).

The increase in average price of unleaded (clear) gasoline used versus octane number, exclusive of distribution costs, is shown in Fig. 18. For 93 RON, the cost increase is $0.20/gal. As noted on the figure, the data band corresponds to a wide range of clear-pool RONs. This band was determined from specific data points in Ref. 60 (schedules A, G, L, M, N, and O) pertinent to total added refining costs calculated for a variety of postulated lead removal schedules. However, in cases where lead was still in use in a given schedule, the cost of the lead was subtracted to reflect only the clear-pool added refining costs. It should be noted that there are three data points (60) in the region of most interest (Table 10) that have zero lead in the total pool. Also, schedule N, having very low lead content in the total pool for 1980, indicates an added refining cost of $0.10/gal at a clear-pool RON of 92.6.

To the foregoing added refining cost of unleaded gasoline (Fig. 18) was added $0.26/gal to reflect the cost of a third pump and associated storage tanks for the distribution of unleaded gasoline (60). It should be noted that this value does not incorporate any costs associated with segregated pipeline and distribution systems to ensure against contamination by leaded products; that is, only "normal" precautions were contemplated. If the trace lead-level content is required to be substantially lower than that obtainable by normal precautions, such pipeline and distribution costs would increase, perhaps substantially. The exact lead level at which this change in requirements exists is presently unknown.

There have been varying estimates from different sources as to the eventual increased cost of unleaded gasoline. Al-

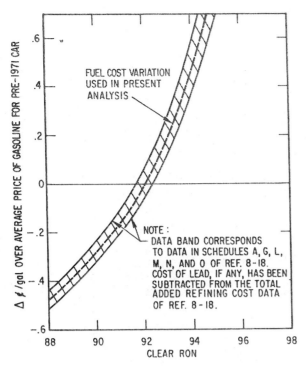

Fig. 18 - Cost of unleaded gasoline, exclusive of distribution costs

Table 10 - Clear-Pool Added Refining Costs

Schedule	Year	Clear-Pool RON	Total Added Refining Cost, $/gal
L	1980	93.5	0.34
L	1976	94.4	0.60
M	1980	94.2	0.43

though there may be differences of opinion regarding manufacturing cost increase, distribution cost effects, etc., the overall (circa 1980) cost increase of $0.46/gal used in this study for calculation purposes should be sufficiently representative to illustrate unleaded gasoline cost effects.

The 93 RON chosen is, of course, different from the 91 RON with which the United States automobile manufacturers have indicated their cars will be capable of operating, at least in the immediate future. However, several automobile manufacturers in informal discussions have indicated that they do not consider 91 RON optimum and may very well increase their RON requirement with time. Their choice of 91 RON was apparently heavily influenced by their desire to specify a fuel which could be more easily made available during the sudden transition to the use of unleaded gasoline. The long-term case is believed more appropriate for the purposes of this analysis and, hence, 93 RON has been chosen.

Both single- and three-grade (three different octane number unleaded fuels sold) unleaded gasoline cases have been analyzed, since it is anticipated that, at the initial introduction of unleaded gasoline, a single grade will be offered due to the need to retain service station pumps to sell leaded gasoline for older cars. However, as the older cars disappear from the road, the leaded fuel pumps could gradually be converted to dispense additional grades of unleaded gasoline.

It was assumed that the three-grade system would have the same clear-pool 93 RON as the single-grade system and that the overall manufacturing plus distribution cost effects ($0.46/gal over conventional leaded gasoline weighted-average price) would be the same. Although the incremental manufacturing costs for the three-grade system might be slightly different (even at the same 93 RON pool) from the $0.20/gal of the single-grade case, it is not felt that this difference would be large enough to alter significantly the results (for example, $0.01/gal gasoline cost increase is equivalent to less than $8 over 85,000 miles of operation).

The results of the analysis are shown in Table 11. As can be seen, the fuel cost penalty associated with the higher cost per gallon of unleaded fuel is not large (approximately $30). The major fuel cost penalty with the use of unleaded gasoline over leaded gasoline is due to the lowering of engine efficiency associated with reduced compression ratio required by the lower octane number of the unleaded gasoline. For the single-grade system, there is an additional $130 fuel cost penalty due to the necessity of reducing compression ratio to 8.35:1 (from 9.37:1 for the leaded fuel case; 80% knock-free technical satisfaction) with its attendant loss in fuel economy of 5.4%. In this case, then, the total fuel cost penalty attributable to the use of unleaded gasoline is $160.

The most significant effect of the three-grade 93 overall RON system is the ability to increase compression ratio (to 8.95:1 at 80% knock-free technical satisfaction). This compression ratio-octane number satisfaction relationship was determined, as was the single-grade case, in a manner similar to that developed in Ref. 61 (constant car performance, acceleration, power).

As shown in Table 11, the increase in compression ratio made possible by the three-grade system reduces the fuel economy penalty loss to $50 (for a 2% fuel economy loss compared to the leaded fuel reference case) and gives a total fuel economy penalty—compression ratio effect plus increased price/gal effect—of $80. This represents a savings of $80 over the single-grade 93 RON case shown in the table. Although not shown, a two-grade unleaded system (at the same overall 93 RON) would be expected to provide similar, but not identical, cost savings over the single-grade system.

Fig. 19 combines the NO_x emission level effects taken from Fig. 17 with the aforementioned compression ratio and higher cost of unleaded gasoline effects to obtain the total fuel cost penalty for emission control systems as a function of NO_x emission level. The following observations may be made from the figure:

1. The fuel cost penalty is sizable at the lower NO_x emission levels, regardless of the type of emission control system, due to the necessity of using low A/F and/or high EGR rates.

2. Emission control systems incorporating NO_x catalysts have more potential for lower fuel costs than other systems, particularly at low NO_x emission levels, due to their ability to

Table 11 - Cost Effects of Use of Unleaded Gasoline

Item	1-Grade	3-Grade
Calculated optimum RON	94+	94+
Pool RON used in analysis	93	93
Compression ratio, 93 RON	8.35:1	8.95:1
Percent change in fuel economy due to compression ratio change from 1970 car*	-5.4	-2
Fuel price – $Δ/gal over average fuel price for 1970 car	0.46**	0.46**
Fuel cost penalty due to price/gal of unleaded fuel, 85,000 miles;† no SFC loss, $	30	30
Unleaded gasoline fuel cost penalty over 85,000 miles;† compression ratio plus $Δ/gal effects, $	160	80

*Constant car performance (acceleration, power).
**Equivalent to $0.216/gal above leaded regular grade gasoline.
†Approximate; varies with NO_x emission level; these values assume no NO_x SFC related cost.

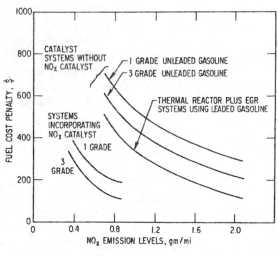

Fig. 19 - Fuel cost penalty

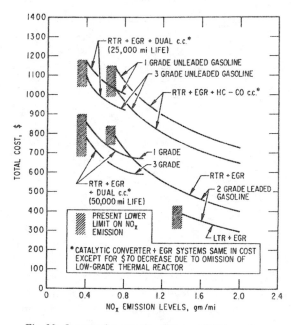

Fig. 20 - Increased consumer costs over lifetime of car

use higher A/F and less EGR. It should be noted, however, that no one has, as yet, demonstrated reasonable life with a high-performance NO_x catalyst.

3. If a durable NO_x catalyst does not become available, emission control systems which do not incorporate HC/CO catalysts are superior from a fuel cost standpoint due to their ability to use leaded gasoline and a higher compression ratio engine.

4. For emission control systems which incorporate catalytic converters and must use unleaded gasoline, multigrade gasoline systems offer significant fuel cost advantages over single-grade systems.

OVERALL CONSUMER COSTS - The overall sum of the emission control system initial hardware cost, maintenance cost, and operating cost is shown in Fig. 20 for the selected generic systems as a function of the NO_x emission level. The cost is so displayed because the various systems are not all capable of the same NO_x emission reduction, and the operating costs (fuel costs) are highly dependent on the NO_x emission reduction. Also shown on the chart are the presently demonstrated lower limits on NO_x emissions for the various systems.

A breakdown of the fuel costs from the total cost at selected NO_x emission levels is shown in Fig. 21. The emission levels selected correspond to those values which represent the presently demonstrated lower limit NO_x emission level for one or more systems.

The following cost observations and conclusions are based on the engineering cost estimates made herein:

1. All systems have high overall costs. The cost increases rapidly as NO_x emission level decreases. If the cost of an emission control system over the car lifetime were $1000, the

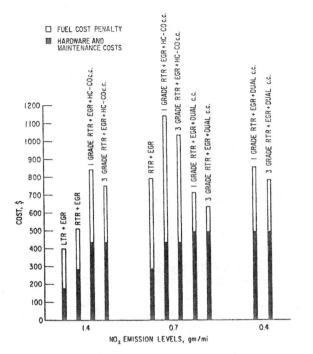

Fig. 21 - Breakdown of increased consumer costs over lifetime of car

national annual cost with all cars on the road using this system would be on the order of $10 billion.

2. Conventional spark ignition engine emission control systems incorporating a NO_x catalyst are the only systems with the potential to meet the federal standard of 0.4 g/mile by 1976. In addition, at low emission levels they are the minimum cost systems, provided that a 50,000-mile catalyst life can be achieved. The total cost of ownership in this case is approximately $860. However, severe problems in developing a durable NO_x catalyst have been encountered, and such a catalyst may not be available by 1976. If only a 25,000-mile lifetime NO_x catalyst is available, the total cost of ownership is increased by approximately $300.

3. The increased cost of unleaded gas is $0.46/gal compared to the average leaded gas cost. This amounts to about a $30 increase over the life of the car. If the lowering of the average compression ratio—and the associated fuel economy penalty—is attributed to lead removal, another $130 can be added to the cost of unleaded gasoline.

4. If the durability problems of the NO_x catalyst system cannot be resolved and its use is precluded, the RTR system, which can tolerate leaded gasoline, would be lowest in cost. Therefore, there is no cost advantage in unleaded gasoline unless a durable NO_x catalyst can be developed.

5. The lead-tolerant LTR system is attractive from a cost standpoint, but its NO_x emission levels are high.

6. In general, the lead-intolerant systems (catalyst plus thermal reactor systems) have higher initial hardware costs than those systems tolerating lead in the gasoline. This is because the catalyst plus thermal reactor systems include most of the parts of the straight thermal reactor systems, as well as the catalyst. The thermal reactor in these systems is needed to provide fast warmup of the catalyst bed. If a fast-warmup catalyst (of equivalent cost) is developed, the initial cost for the catalyst system is reduced by only $70, due to the omission of the low-grade thermal reactor.

7. Lead-intolerant systems (catalyst systems) have higher maintenance costs than those systems tolerating lead, because the cost of replacement of the catalyst bed is greater than the cost savings on spark plugs with unleaded gasoline. Muffler cost savings discussed earlier were precluded by the use of a long-life exhaust system on all 1976 systems.

8. A system incorporating a NO_x catalyst has the lowest fuel cost because it has the lowest sfc, which more than offsets the higher cost effects of unleaded gasoline.

9. A lead-intolerant HC/CO catalyst system without a NO_x catalyst has higher fuel costs than a lead-tolerant thermal reactor system. This is due primarily to the cost effects of unleaded gasoline and compression ratio, as both systems have about the same sfc for a given NO_x level.

Of course, as more definitive data become available on emission control system technology (for example, emission levels, durability, costs), the foregoing conclusions may be subject to appropriate modification.

SUMMARY AND CONCLUSIONS

Based on a review and assessment of data in the open literature and on discussions with industrial/agency sources, the more important findings with respect to the effect of lead additives in gasoline on emission control systems which may be used to meet the 1975-1976 emission standards for light-duty vehicles are as follows:

1. All emission control systems currently planned for use by major automobile manufacturers and being evaluated by them to meet the 1975-1976 federal emission standards include a catalytic converter.

2. Lead additives in gasoline are toxic to catalytic materials. Use of leaded or low-lead gasoline has demonstrated that it greatly reduces catalyst activity, thereby preventing the achievement of a 50,000-mile lifetime at low enough emission levels to meet the standards. In addition to its effect on catalyst activity, lead has been observed to degrade the structural integrity of NO_x bulk metal catalysts.

The scavengers added to gasoline to prevent the accumulation of harmful lead deposits in the engine also have detrimental effects upon catalysts. Sulfur and phosphorus have also been noted to have toxic effects.

3. Lead effects on other major emission control system components, for example, thermal reactors and EGR systems, although observed to be present to some degree, are such that materials selection and design techniques can be applied to allow lead-tolerant systems.

4. Unleaded gasoline should be made available in sufficient quantities to satisfy the demands of vehicles with catalytic converter units. The lead content should be at that level compatible with obtaining a 50,000-mile useful lifetime. However,

substantive data to establish this level precisely are not available. Most of the major automobile manufacturers and catalyst suppliers have been performing their catalytic converter development work with lead levels below 0.06 g/gal, with most of the development at levels of 0.02-0.03 g/gal. At this lead level, no automobile manufacturer has stated to date that 50,000 miles of operation at satisfactory emission levels have been achieved. It is not known whether this durability/lifetime deficiency is related to the lead level (0.02-0.03 g/gal), to other trace elements in the gasoline, or to other catalyst properties. One automobile manufacturer and one catalyst supplier have stated that a maximum lead content of 0.03 g/gal should be an adequately low level. It should be noted that this value is below the proposed ASTM specification for unleaded gasoline of 0.07 g/gal. The EPA-proposed regulations (62) specify 0.05 g/gal.

5. With regard to 1975 emission standards, both lead-tolerant systems (for example, Esso's RTR system) and lead-intolerant systems (for example, systems incorporating catalytic converters) have demonstrated approaching the standards on an experimental basis at low mileage. However, in order to meet the lower NO_x levels required in 1976, the lead-tolerant system would require the addition of a NO_x catalyst (and possibly additional components), which would render it sensitive to lead.

6. A general evaluation of emission control devices/systems envisioned by the automobile industry and ancillary development organizations has indicated that none of the systems planned for 1976 has demonstrated the capability of meeting the 1976 NO_x emission standard. However, several combination systems incorporating a NO_x catalyst have met the 1976 emission levels on an experimental basis with a new catalyst. But, at this time, a 50,000-mile lifetime has not been demonstrated. In fact, durability tests in excess of approximately 10,000-15,000 miles have not been reported.

7. At this time, estimated overall costs to the consumer (initial, maintenance, and operating) for emission control systems being considered for the 1976 federal emission standards are $860 above 1970 vehicle costs over an 85,000-mile vehicle lifetime. This estimate is based on a system incorporating a dual HC/CO, NO_x catalytic converter (with assumed replacement of the converter unit at 50,000 miles), a "low-grade" RTR, and EGR.

ACKNOWLEDGMENTS

The authors would like to acknowledge the assistance and guidance provided by G. D. Kittredge of the Environmental Protection Agency, Office of Air Programs, Division of Emission Control Technology, for whom the original study (under Contract No. F 04701-71-C-0172) was conducted.

The following technical personnel of The Aerospace Corp. participated in the study and made valuable contributions to the assessment performed under this contract: F. E. Cook, L. M. Dormant, J. A. Drake, L. Forrest, P. P. Leo, W. U. Roessler, M. J. Russi, W. M. Smalley, G. W. Stupian, K. B. Swan, and J. D. Wilson.

REFERENCES

1. Public Law 90-148, The Clean Air Act, as amended.
2. "Control of Air Pollution from New Motor Vehicles and New Motor Vehicle Engines." Federal Register, Vol. 36, No. 128, July 2, 1971.
3. Progress and Programs in Automotive Emissions Control, Progress Report to the Environmental Protection Agency, General Motors Corp., March 12, 1971.
4. American Motors Corp., letter in response to request from Administrator, Environmental Protection Agency, Subject: Regarding Requirements of the Clean Air Act, April 2, 1971.
5. H. L. Misch, Vice President - Engineering, Ford Motor Co., statement to the Environmental Protection Agency, May 6, 1971.
6. Chrysler Corp., letter in response to request from administrator, Environmental Protection Agency, Subject: Regarding Emission Control, April 1, 1971.
7. "An Assessment of the Effects of Lead Additives in Gasoline on Emission Control Systems Which Might Be Used to Meet the 1975-76 Motor Vehicle Emission Standards." Final Report. Report No. TOR-0172 (2787) - 2, The Aerospace Corp., El Segundo, Calif., Nov. 15, 1971.
8. Y. Yaneko, et al., "Small Engine-Concept Emission Vehicles." SAE Transactions, Vol. 80 (1971), paper 710296.
9. "Exhaust Manifold Thermal Reactor Development at Du Pont." Petroleum Chem. Div.: Wilmington, Del., Jan. 20, 1971.
10. R. J. Lang, "A Well-Mixed Thermal Reactor System for Automotive Emission Control." SAE Transactions, Vol. 80 (1971), paper 710608.
11. G. H. Meguerian and C. R. Lang, "NO_x Reduction Catalysts for Vehicle Emission Control." SAE Transactions, Vol. 80 (1971), paper 710291.
12. G. J. Nebel and R. W. Bishop, "Catalytic Oxidation of Automobile Exhaust Gases." Paper 29-R presented at SAE Annual Meeting, Detroit, January 1959.
13. R. M. Campau, "Low Emission Concept Vehicles." SAE Transactions, Vol. 80 (1971), paper 710294.
14. K. I. Jagel and F. G. Dwyer, "HC/CO Oxidation Catalysts for Vehicle Exhaust Emission Control." Published in SP-361, "Inter-Industry Emission Control." New York: Society of Automotive Engineers, Inc., 1971, paper 710290.
15. R. R. Allen and C. G. Gerhold, "Catalytic Converters for New and Current (Used) Vehicles." Paper presented at the Fifth Technical Meeting, West Coast Section of Air Pollution Control Assn., San Francisco, Oct. 8-9, 1970.
16. W. J. Barth and E. N. Cantwell, "Automotive Exhaust Manifold Thermal Reactors—Materials Considerations." Presented before Div. of Petroleum Chemistry, Inc., 161st Meeting of the American Chem. Soc., Los Angeles, March 28 - April 2, 1971.
17. A. Jaimee, et al., "Thermal Reactor – Design, Develop-

ment and Performance." SAE Transactions, Vol. 80 (1971), paper 710923.

18. R. E. Oldrieve and P. L. Stone, Letter Report on the Current Status of the NASA/Lewis Program for Evaluation and Development of Materials for Automobile Thermal Reactors. Materials and Structures Div., NASA/Lewis Research Center, Cleveland, April 30, 1971.

19. The Ethyl Lean Reactor System, Ethyl Corp. Research Lab., Detroit, July 1, 1971.

20. Personal discussion with representatives of General Motors Corp., June 30, 1971.

21. Personal discussion with representatives of Ford Motor Co., July 1, 1971.

22. Personal discussion with representatives of Chrysler Corp., June 29, 1971.

23. Personal discussion with representatives of American Motors Corp., Sept. 29, 1971.

24. H. L. Misch, Vice President – Engineering, Ford Motor Co., Testimony before Calif. Air Resources Board, Sacramento, March 4, 1970.

25. G. S. Musser, et al., "Effectiveness of Exhaust Gas Recirculation with Extended Uses." Paper 710013 presented at SAE Automotive Engineering Congress, Detroit, January 1971.

26. 1968/Automobile Facts/Figures Report, Automobile Mfrs. Assn., Inc., Detroit, 1968.

27. Car Maintenance Expense When Using Leaded and Nonleaded Gasoline, Ethyl Corp., Detroit, July 2, 1971.

28. W. Giles, "Valve Problems with Lead Free Gasoline." SAE Transactions, Vol. 80 (1971), paper 710368.

29. D. Godfrey and R. L. Courtney, "Investigation of the Mechanism of Exhaust Valve Seat Wear in Engines Run on Unleaded Gasoline." SAE Transactions, Vol. 80 (1971), paper 710356.

30. H. W. Schwochert, "Performance of a Catalytic Converter on Nonleaded Gasoline." SAE Transactions, Vol. 78 (1969), paper 690503.

31. Consequences of Removing Lead Antiknocks from Gasoline, A Status Report, No. AC-10, Ethyl Corp., New York, August 1970.

32. A. E. Felt and R. V. Kerley, "Engines and Effects of Lead-Free Gasoline." Paper 710367 presented at SAE Mississippi Valley Section, October 1970.

33. "Effect of Lead Antiknocks on the Performance and Costs of Advanced Emission Control Systems." Wilmington, Del.: Du Pont de Nemours & Co., July 15, 1971.

34. T. V. De Palma, "The Application of Catalytic Converters to the Problems of Automotive Exhaust Emissions." Paper presented to Interpetrol Congress, Rome, Italy, June 24, 1971.

35. Engelhard Industries, Inc., letter to Aerospace Corp., Oct. 13, 1971.

36. J. C. Thompson, "Exhaust Emissions from a Passenger Car Equipped with a Universal Oil Products Catalytic Converter." Air Pollution Control Office, Environmental Protection Agency, December 1970.

37. "The Ethyl Lean Reactor System." Ethyl Corp. Research Labs., Detroit, July 1, 1971.

38. A. Jaimee, D. E. Schneider, A. I. Rozmanith, and J. W. Sjoberg, "Thermal Reactor-Design, Development and Performance." SAE Transactions, Vol. 80 (1971), paper 710293.

39. Personal discussion with representatives of General Motors Corp., September 1971.

40. C. E. Burke, American Motors Corp., letter to Aerospace Corp., Sept. 29, 1971.

41. Conversation with B. Simpson, Ford Motor Co., October 1971.

42. J. H. Somers, Office of Air Programs, Environmental Protection Agency, Letter to Aerospace Corp., Subject: Discussion with Dr. G. Meguerian, Oct. 7, 1971.

43. B. Simpson, Ford Motor Co., letter to Aerospace Corp., Sept. 15, 1971.

44. R. C. Butler, Du Pont de Nemours & Co., letter to Office of Air Programs, Environmental Protection Agency, Aug. 30, 1971.

45. G. S. Musser, et al., "Effectiveness of Exhaust Gas Recirculation with Extended Use." Paper 710013 presented at SAE Automotive Engineering Congress, Detroit, January 1971.

46. F. W. Bowditch, Director, Automotive Emission Control, General Motors Corp., letter to Administrator, Environmental Protection Agency, Subject: Comments on Notice of Proposed Rule-Making for NO_x on 1973 New Motor Vehicles, Federal Register of 27 February 1971, Vol. 36, No. 40, Apr. 28, 1971.

47. L. S. Bernstein, et al., "Application of Catalysts to Automotive NO_x Emissions Control." SAE Transactions, Vol. 80 (1971) paper 710014.

48. Personal discussion with representatives of Chrysler Corp., June 29, 1971.

49. Personal discussion with representative of Esso Research and Engrg. Co., July 6, 1971.

50. Chapter 3: Mobile Sources, "The Economics of Clean Air." Report of Administrator of Environmental Protection Agency to Congress of United States, Senate Document No. 92-6, March 1971.

51. Personal discussion with representative of Research Triangle Inst., unpublished data, Aug. 6, 1971. The Economics of Clean Air, Annual Report of Administrator of Environmental Protection Agency to Congress of United States, February 1972.

52. Personal discussion with representatives of Engelhard Industries, Inc., July 1971.

53. L. S. Bernstein, A. K. S. Raman, and E. E. Wigg, "The Control of Automotive Emissions with Dual Bed Catalyst Systems." Presented to 1971 Central States Section of Combustion Institute, Mar. 23, 1971.

54. R. S. Yolles, H. Wise, and L. P. Berriman, "Study of Catalytic Control of Exhaust Emissions for Otto Cycle Engines." Stanford Research Inst., Final Report, SRI Project PSU-8028, April 1970.

55. Automotive News—1971 Almanac, 35th Review and Reference Ed., Detroit: Slocum Publishing Co., Apr. 26, 1971.

56. "Relationships of Passenger Car Age and Other Factors

to Miles Driven." U. S. Department of Transportation, Bureau of Public Roads, Washington, D. C.

57. "Consequences of Removing Lead Antiknocks from Gasoline. A Status Report." Detroit: Ethyl Corp., No. AC-10, August 1970.

58. "Passenger Cars: Oil, Automotive Trends." 1971 National Petroleum News Factbook Issue.

59. Personal discussion with representatives of Du Pont de Nemours & Co., regarding their winter and summer gasoline surveys, October 1971.

60. "Economic Analysis of Proposed Schedules for Removal of Lead Additives from Gasoline." Houston, Texas: Bonner & Moore Assoc., Inc., June 25, 1971.

61. E. S. Corner and A. R. Cunningham, "Value of High Octane Number Unleaded Gasolines in the United States." Presented before Division of Water, Air, and Waste Chemistry, American Chem. Soc., Los Angeles, Mar. 28-Apr. 2, 1971.

62. Regulations of Fuels and Fuel Additives, Notice of Proposed Rulemaking, Federal Register, 37, 36, 40 CFR part 80, Feb. 23, 1972.

This paper is subject to revision. Statements and opinions advanced in papers or discussion are the author's and are his responsibility, not the Society's; however, the paper has been edited by SAE for uniform styling and format. Discussion will be printed with the paper if it is published in SAE Transactions. For permission to publish this paper in full or in part, contact the SAE Publications Division and the authors.

Society of Automotive Engineers, Inc.
TWO PENNSYLVANIA PLAZA, NEW YORK, N.Y. 10001

24 page booklet.   Printed in U.S.A.

Thermal Response and Emission Breakthrough of Platinum Monolithic Catalytic Converters

Charles R. Morgan, David W. Carlson, and Sterling E. Voltz
Research Dept., Mobil Research and Development Corp.

THE WORK REPORTED in this paper was undertaken to identify the key phenomena associated with overtemperature and emission breakthrough problems in platinum monolithic converters. These studies were carried out within the Inter-Industry Emissions Control (IIEC) Program*.

During the past several years, considerable progress has been made in the development of carbon monoxide/hydrocarbon (CO/HC) oxidation catalysts for automotive emission control system (1)**. The stringent emission standards established by the 1970 amendment to the Clean Air Act and the CVS test procedure promulgated by the Environmental Protection Agency in 1970 have strongly affected the requirements for catalysts and catalytic converters. Since the CVS test procedure places considerable emphasis on cold-start performance, catalysts with lower density and higher oxidation activity at lower temperatures are required. Monolithic catalysts, which have more rapid warmup and lower pressure drop than pelleted catalysts, have been extensively evaluated in low-emission vehicles. In addition, catalytic converters are often located quite close to the engine to further increase the warm-up rate.

Catalyst durability, in this more severe environment, is one of the current critical problems. Due to the higher operating temperatures associated with locating converters close to the engines, catalyst deterioration has often been observed in 1975-76 concept vehicles (2-6). Occasionally, thermal failures (melting) have occurred during mileage accumulation runs. The specific operational modes under which melting occurred were difficult to establish on concept vehicles.

Some attempts to prevent thermal failure by the use of protection devices have been reported (3). These schemes have included bypassing the exhaust around the catalytic converter, dumping the secondary air supply, and feedback control of the air/fuel (A/F) ratio.

During certain acceleration-deceleration modes of the CVS test procedure, pulses of CO and HC are measured at the

---

*This international cooperative research effort to develop virtually emission-free automobiles was supported by American Oil Co., Atlantic Richfield Co., Fiat S.p.A., Ford Motor Co., Marathon Oil Co., Mitsubishi Motors Corp., Mobil Oil Corp., Nissan Motor Co. Ltd., Standard Oil Co. (Ohio), Sun Oil Co., Toyo Kogyo Co. Ltd., Toyota Motor Co. Ltd., and Volkswagenwerk A. G.

**Numbers in parentheses designate References at end of paper.

ABSTRACT

Stringent emission standards have strongly affected the requirement for catalysts and catalytic converters. Since considerable emphasis is placed on cold-start performance, catalysts with lower density and higher oxidation activity at lower temperatures are required. Monolithic catalysts have been extensively evaluated in low-emission vehicles. This paper identifies the key phenomena associated with overtemperature and emission breakthrough problems in platinum monolithic converters.

converter outlet, even when the catalytic converter is hot (emission breakthrough). Even though each breakthrough is relatively small compared to emissions during the cold-start portion of the CVS test procedure, their integrated contribution to the total CVS emissions can be significant.

Since both overtemperature and emission breakthrough occur when the converter is above 1000°F (and not limited by catalyst activity), these phenomena are often dependent on the rates of heat and mass transfer processes in the catalytic converters. Several theoretical studies of these processes in pelleted catalyst beds have been reported (7, 8). Kuo (9) has developed a comprehensive mathematical model for monolithic catalytic converters which includes detailed heat and mass transfer relationships. Heat and mass transfer phenomena at steady-state have been studied for several types of monolithic converter channel geometries (10-12). HC and CO emission breakthroughs associated with the secondary air supply system have also been reported. Kuo, et al. (13) reported breakthroughs resulting from insufficient oxygen during certain critical driving modes. Inoue, et al. (4) investigated the effect of secondary air distribution on catalytic converter performance.

In the present studies, an engine dynamometer facility was utilized to establish the thermal response of Engelhard PTX converters to step changes of inlet concentrations of CO, HC, and H_2. The feasibilities of sensors to indicate, and methods to prevent thermal failures of these converters were determined. Some studies of emission breakthrough were also made with this facility and were correlated with theoretical heat and mass transfer relationships. The effect of breakthrough from mass transfer as well as other causes on HC and CO emissions was investigated with a concept vehicle equipped with PTX converters.

THERMAL RESPONSE OF PLATINUM MONOLITHIC CONVERTERS ON ENGINE DYNAMOMETER FACILITY

EXPERIMENTAL APPARATUS - The experiments were performed on an engine dynamometer with two converters in series, as shown in Fig. 1. The first converter, filled with a pelleted platinum oxidation catalyst, removed most of the residual combustibles from the exhaust gas before introduction to the second converter, an Engelhard PTX-5. With this arrangement, the appropriate combustible could be more easily controlled at the desired concentration using the synthetic gas manifold system.

The feed gas for the PTX-5 converter was blended by metering the desired rates of the pure components (CO, C_3H_6, and H_2) from gas cylinders through calibrated rotameters into the engine exhaust gas. Step changes in the concentration of the combustibles were made by switching the flow of synthetic gases either to the exhaust gas or to vent with solenoid valves.

The analytical system was arranged so that a sample could be taken from either the inlet or outlet of the PTX-5 converter. Beckman NDIR analyzers were used to measure the concentrations of CO, CO_2, NO, and HCs. The oxygen was measured with a Servomex paramagnetic analyzer. The combined flow of carburetor air, PCV, and secondary air was measured with a calibrated laminar flow element.

The PTX-5 converter (laid-up type) had eleven exposed tip chromel-alumel thermocouples (0.01 in diameter), as shown in Fig. 2. This converter was modified, so that the standard inlet and outlet cones could be replaced with cones having an included angle of 30 deg, instead of 60 deg, to improve the gas flow distribution through the monolith channels. In addition, the modified inlet cone contained a flow straightener consisting of stainless steel screens installed 1 in ahead of the monolith.

THERMAL RESPONSE MEASUREMENTS - The thermal response in a PTX-5 converter to step changes of CO, C_3H_6, and H_2 was measured as a function of the concentration of the combustible species, the exhaust gas flow rate, the initial bed temperature, and the inlet and outlet cone configurations.

Some typical temperature responses to step increases in CO, C_3H_6, and H_2 are shown in Figs. 3-6. These particular temperature profiles were obtained with modified cones, a steady-state exhaust gas flow rate of 23 scfm, an inlet oxygen concentration of 7-8%, and an initial PTX-5 bed temperature of approximately 760°F. The data were obtained at this relatively low temperature so that large temperature gradients could be produced without damaging the catalyst or the thermocouples.

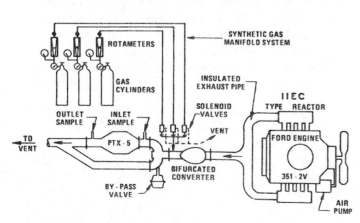

Fig. 1 - Experimental apparatus for thermal response study

A typical axial thermal response is shown in Fig. 3 for a step increase of 8.4% CO. Several points of interest are:

1. The front part of the bed, indicated by thermocouple 1A, increased 650°F in 15 s, while the rear portion of the bed, indicated by thermocouple 1C, increased 650°F in 41 s.

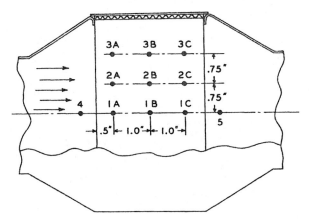

Fig. 2 - Engelhard PTX converter thermocouple locations

2. The maximum converter temperature was measured in the middle and rear portions of the bed and was 120°F higher than at the front portion of the bed, indicating that CO was still reacting after the first 1/2 in of the bed.

3. The outlet temperature increased at a much lower rate than the bed temperatures and leveled off at 1480°F, while the maximum bed temperature was 1780°F. This large drop in temperature from the catalyst to the outlet is attributed to radiation heat loss from the outlet thermocouple to the colder cone wall. Thus, a thermocouple simply placed in the outlet cone is not an adequate indicator of high-temperature excursions in a PTX-5 converter.

4. The increase in inlet temperature during the pulse of CO was attributed to radiation heat transfer from the monolith to the thermocouple.

5. When the CO feed to the exhaust gas was discontinued, the front portion of the bed cooled more rapidly than the rear portion of the bed.

The radial temperature profile during the same step increase in CO is shown in Fig. 4. A very small temperature

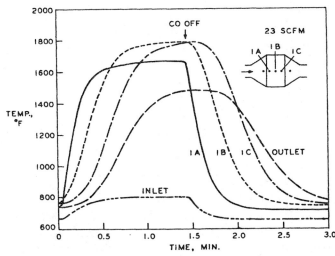

Fig. 3 - Thermal response in PTX-5 converter to a step increase of 8.4% CO

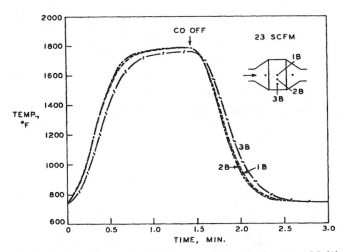

Fig. 4 - Thermal response in PTX-5 converter to a step increase of 8.4% CO

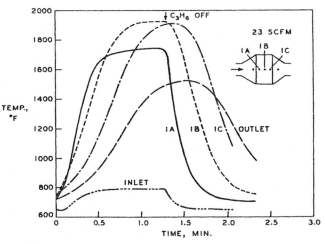

Fig. 5 - Thermal response in PTX-5 converter to a step increase of 1.4% C_3H_6

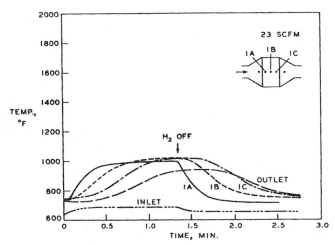

Fig. 6 - Thermal response in PTX-5 converter to a step increase of 2.1% H_2

gradient of about 30°F was observed across the bed between positions 2B and 3B. The temperatures at positions 1B and 2B were essentially the same during the entire experiment. Subsequent experiments with the standard inlet and outlet cones also showed only a small radial temperature gradient.

The axial temperature responses to step increases of C_3H_6 (Fig. 5) and H_2 (Fig. 6) were similar to that shown for CO. The maximum measured temperatures at 23 scfm were recorded at the middle and rear portions of the converter. At higher exhaust gas flow rates (40-74 scfm), the axial temperature gradient was extended through more than the first 1-1/2 in of the monolith, so that the maximum temperature was measured only at the rear portion of the bed.

The maximum measured temperature response for a wide variety of initial bed temperatures and flow rates for CO, C_3H_6, and H_2 (Figs. 7-9) are approximately equal to the calculated thermodynamic adiabatic temperature. This temperature, for each combustible species, is calculated using:

$$T_i = \frac{H_i \cdot X_i}{C_p \cdot M_g}$$

where:

T_i = adiabatic temperature increase due to oxidation of species "i," °F
H_i = reaction heat of species "i," Btu/lb-mole
M_g = gas molecular weight, lb/lb-mole
C_p = gas heat capacity, Btu/lb °F
X_i = molar concentration of species "i"

The thermodynamic adiabatic temperature does not increase exactly linearly with the concentration of the combustible species, since the heat capacity of the exhaust gas increases from 0.265 to 0.290 Btu/lb °F at 800 and 1800°F, respectively. In spite of this, the linear approximation that can be used to estimate the maximum temperature increase in a PTX-5 converter as a function of CO, C_3H_6, and H_2 concentrations is:

$$T_{max}/\% CO = 135°F$$
$$T_{max}/1000 \text{ ppm } C_3H_6 = 93°F$$
$$T_{max}/\% H_2 = 122°F$$

The maximum temperature increases for CO, C_3H_6, and H_2 are additive when the concentrations of CO, C_3H_6, and H_2 are increased in combination, as shown in Table 1, for a few experimental runs.

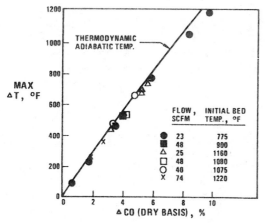

Fig. 7 - Maximum temperature response in PTX-5 to a step increase in CO concentration

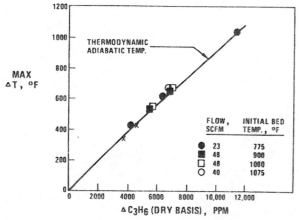

Fig. 8 - Maximum temperature response in PTX-5 to a step increase in C_3H_6 concentration

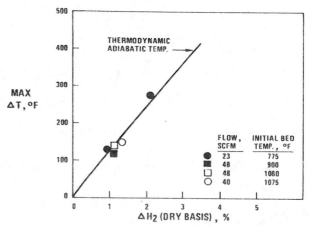

Fig. 9 - Maximum temperature response in PTX-5 to a step increase in H_2 concentration

Table 1 - Maximum Temperature Response to a Step Increase of CO, C_3H_6, and H_2

Exhaust, Flow, scfm	Δ%, CO	Δppm, C_3H_6	Δ%, H_2	ΔT_{max}, °F	
				Measured	Calculated
48	3.40	2090	1.07	775	784
40	4.06	2499	1.28	950	936
74	2.27	243	0	515	549

The rate of thermal response in a PTX-5 converter provides an estimate of the maximum available response time for the operation of a catalyst overtemperature protection device. Previously, the temperature in the first 1/2 in of the monolith was shown to increase at a faster rate than in the middle or rear portion (Fig. 3). Also, the maximum rate of temperature increase in this portion of the bed occurred at the beginning of the pulse, and was approximately linear with time. Since the thermal response to CO was at least as fast as those for the other combustibles (C_3H_6 or H_2), the initial rate of temperature increase with a step increase of CO has been evaluated as a function of initial bed temperature, flow rate, and concentration of CO.

The initial rate of temperature increase was directly proportional to the CO concentration, for a wide range of gas flow rates (Fig. 10). The effect of the exhaust gas flow rate on thermal response was evaluated by plotting the slope of the lines in Fig. 10 with the flow rate as shown in Fig. 11. This initial thermal response increases approximately with the square root of the flow rate. The thermal response at 1200°F initial bed temperature and 70 scfm flow rate was approximately 16°F/(s) (%CO). At these flow conditions, a sudden pulse of 8% CO would cause the bed temperature to increase no more than 512°F in 4 s. Thus, an overtemperature protection system with a response time of up to 4 s would be adequate to prevent the catalyst temperature from exceeding 1700°F in this particular case.

THERMAL FAILURE OF PTX-5 CONVERTER

A PTX-5 converter was tested on the engine dynamometer facility at conditions that normally result in high catalyst temperatures. It was located in place of the bifurcated converter shown in Fig. 1. The converter was equipped with 2 1/16 in diameter chromel-alumel thermocouples located in the center of the entrance and exit cones of the PTX unit, and a 0.02 in chromel-alumel thermocouple located inside a monolith channel about 1/2 in from the entrance. A fuse to indicate thermal failure (that is, temperature near the melting point of the monolith) during operation was constructed by inserting a 24 gage nichrome wire (2550°F melting point) through one of the center monolith channels. Both ends of the wire were extended through insulated openings in the inlet and outlet cones of the PTX-5 converter. The external leads of the wire were connected to a circuit which indicated when the fuse wire had melted.

The engine was then operated at various cruise conditions, while different spark plug wires were shorted out to provide the temperatures shown in Table 2. As expected, the outlet temperature increased in all cases, as the number of nonfiring spark plugs were increased. The catalyst temperature is not reported because the small thermocouple became inoperable. Based on the axial thermal response measurements (Fig. 5), the interior monolith temperature may have been much higher than the outlet temperature; however, no melting occurred. At idle operation, the inlet temperature increased as the number of nonfiring plugs were increased, which suggest that some of the HCs from the nonfiring cylinders were apparently preburned homogeneously in the exhaust system. The higher exhaust flow rates at 5 and 30 mph apparently reduced this preburning of combustibles, since the inlet temperature remained relatively constant as the number of nonfiring plugs was increased.

The effect of ignition failure was also investigated by turning off the ignition at 30 mph cruise with the throttle open. Within 8 s, the converter inlet temperature exceeded

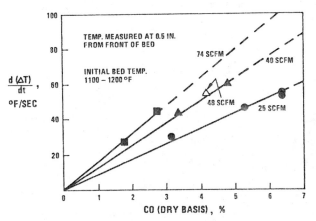

Fig. 10 - Initial rate of temperature increase in PTX-5 with a step increase of CO

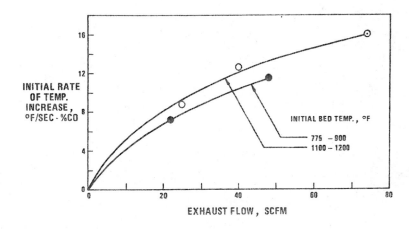

Fig. 11 - Initial rate of temperature increase in PTX-5 with a step increase of CO

2000°F and the nichrome fuse had melted. Subsequent inspection of the interior of the PTX-5 converter revealed that the monolith had melted about 1/2 in from the entrance, as shown in Fig. 12. It should be noted that a measurement of the converter outlet temperature is not adequate to indicate failure, since this temperature only reached 1775°F during this operation. This same mode of operation also caused thermal failure of a bifurcated converter containing a particle platinum catalyst.

THERMAL PROTECTION SCHEMES

The control of catalyst overtemperature excursions requires a sensor for either the combustible concentrations in the exhaust gas and/or the catalyst temperature. When the signal from a sensor exceeds a predetermined value, an electronic command can be given to operate a device to limit the catalyst temperature. For example, some converter protection systems that have been reported are:

1. Bypass valve that diverts exhaust around the catalyst at high loads and speeds, or at high catalyst temperatures detected by thermocouples installed in the catalyst bed (1).
2. Feedback control system which provides a desired A/F (2).

TOTAL IGNITION FAILURE - Since total ignition failure at moderate-to-high cruise conditions caused actual melting of the catalyst, some methods to prevent thermal failure during this mode of operation were investigated. In order to establish the effect of decreasing the amount of unburned fuel pumped to the engine after ignition failure, the transmission on the engine dynamometer was simultaneously shifted to neutral. This significantly reduced the number of engine revolutions after ignition loss (Fig. 13). With the automatic transmission in drive, the engine takes about 210 revolutions and 16 s to stop from an initial cruise speed of 30 mph. With a simultaneous shift to neutral and ignition loss at 30 mph, the number of engine revolutions is decreased 210-15, and the time for complete stop is 2 s. As previously described, a total ignition loss at 30 mph resulted in temperature excursions in the PTX converter of at least 1400°F increase and melting of the catalyst. With a simultaneous shift to neutral, this temperature increase was reduced to only 120°F at 30 mph and 350°F at 50 mph (Table 3), thereby preventing thermal damage to the catalyst on the engine dynamometer.

Other methods to decrease the amount of unburned fuel pumped by the engine to the catalyst after total loss of ignition were then evaluated. The following methods were

Table 2 - PTX-5 Temperature During Failure of Spark Plugs to Fire

Speed, mph	Exhaust Flow, scfm	No. of Nonfiring Spark Plugs	Temperature, °F	
			Inlet	Outlet
Idle	26	0	1100	1150
↓	↓	1	1225	1225
		2	1375	1375
		3	1550	1525
5	38	0	980	1030
↓	↓	1	1040	1385
		2	1000	1600
		3	900	1825
30	48	0	970	1000
↓	↓	1	1080	1350
		2	1000	1575
		3	900	1775

Fig. 12 - Thermally damaged PTX-5 converter

activated manually on the engine dynamometer at the moment of total ignition failure:

Bypass - A bypass protection system was operated in response to a loss of ignition at 30 and 50 mph, and prevented thermal damage by limiting the catalyst temperature increase to 248 and 470°F, respectively (Table 3). The time required for the bypass valve to move from the fully closed to fully opened position was 2 s.

Vacuum-Throttle Off - Instead of bypassing the gas, the amount of unburned fuel pumped to the catalyst was decreased by simultaneously closing the throttle and opening the intake manifold to the atmosphere upon ignition loss. The intake manifold was opened to the atmosphere by a valve installed at an opening in the spacer plate between the carburetor and engine. When this valve was opened at the time of ignition loss, the increase in catalyst temperature was limited to 550°F (Table 3).

Throttle Off - The closure of the throttle without opening the intake manifold to atmosphere at ignition loss was not sufficient to protect the catalyst from overtemperature. The bypass had to be opened as the catalyst temperature increased 2200-2300°F. A tighter seal or a separate valve to isolate the carburetor from the intake manifold would improve the efficiency of this method.

Vacuum Off - The opening of the intake manifold to the atmosphere, without closure of the throttle at ignition loss, did not protect the catalyst from overtemperature. The bypass again had to be opened at a catalyst temperature of 2200-2300°F. However, the effectiveness of this method might be improved, if the restrictions in the opening to the intake manifold were decreased.

Based on these preliminary results, the automatic vacuum-throttle shut-off appears to be the most feasible method of preventing thermal failure, among those evaluated in this study.

EMISSION BREAKTHROUGH

MEASUREMENTS ON DYNAMOMETER FACILITY - The emission breakthrough in a fresh PTX-5 converter with excess oxygen was evaluated on the engine dynamometer facility shown in Fig. 1. The bifurcated converter was replaced with a section of exhaust pipe, so that the entire concentration of combustibles in the exhaust gas did not have to be supplied from the synthetic gas manifold system.

During CVS testing, emission breakthrough often occurs during sharp acceleration modes when there is a sharp peak in the flow rate of the exhaust gas. This mode of operation was simulated on the engine dynamometer system by acceleration of 20-40 mph. The exhaust gas flow rate increased 42-112 scfm, and then declined to 64 scfm, as shown in the top portion of Fig. 14. The unconverted CO (bottom of Fig. 14) increased from zero to a maximum of 7.4% at the peak flow rate, and then returned to zero. In order to determine

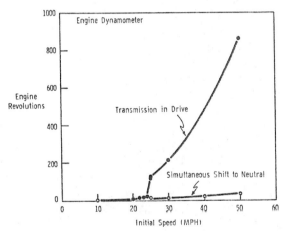

Fig. 13 - Engine revolutions after ignition loss with automatic transmission

Table 3 - Maximum Increase in PTX-5 Converter Temperatures After Total Loss of Ignition

Operation	Temperature Increase After Ignition Failure	
	30 mph	50 mph
No protection	>1400°F	>1400°F
Shift-to-neutral	120	350
Bypass	248	470
Vac-throttle off	547	510
Throttle off	1165*	—
Vacuum off	1150*	—

*Bypass opened at 2200-2300°F to protect catalyst.

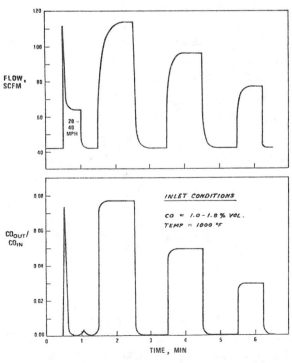

Fig. 14 - CO breakthrough in a PTX-5 converter (modified entrance and exit cones)

whether this amount of breakthrough could be sustained at this peak flow rate under steady-state operation, an approximate step increase in the exhaust flow rate was accomplished by rapidly increasing the load on the engine. When the flow rate sharply increased to approximately the peak flow rate (114 scfm), the unconverted CO increased to 7.7%, and was sustained at this steady-state flow rate. The sharp decrease in the flow rate produced a corresponding decrease in the unconverted CO. Step changes in flow rate to 96 and 76 scfm, respectively, resulted in less severe breakthroughs, as shown in Fig. 14.

These breakthrough experiments were conducted at the following conditions:

Initial converter inlet temperature = 1000°F

Inlet CO = 1.0-1.8%

Inlet HC (NDIR) = 70 ppm as C_6

Inlet O_2 = 4-6%

Inlet NO = 1500-1900 ppm

The midbed temperatures in the PTX-5 converter at the maximum flow rates were 1260-1520°F.

The effects of catalyst temperature, exhaust gas flow rate, and inlet HC concentration on the CO breakthrough are shown in Fig. 15. At 64 scfm, the unconverted CO is about 0.6% at an inlet HC concentration of about 10 ppm (NDIR-C_6), and increases to about 2.2% at 480 ppm HCs.

When the exhaust gas flow rate is increased to 110 scfm, unconverted CO increases to 7.7% at 1460°F, and decreases significantly as the catalyst temperature increases. Part of the conversion at the high temperatures is probably due to the homogeneous gas phase oxidation of CO. An increase in the inlet HCs slightly increases the unconverted CO.

The HC breakthrough with propylene (C_3H_6) was higher than that measured for CO at 1200-1600°F. As with CO, the fraction of unconverted C_3H_6 also increased with flow rate and decreased with temperature (Fig. 16).

The conditions for these CO and C_3H_6 breakthrough experiments (Figs. 15 and 16) are summarized:

Converter inlet temperature = 1150°F

Inlet CO = 0.9-3.2%

Inlet HC (NDIR) = 5-500 ppm as C_6

Inlet O_2 = 8.5%

The degree of CO breakthrough in the fresh PTX-5 converter was not affected by use of the standard or modified inlet flow straightener cones (Fig. 17). However, with aged PTX-5 converters where intrinsic catalytic activity is rate-limiting, breakthrough may possibly be reduced with modified inlet cones that provide better flow distribution. At exhaust gas flow rates below 40 scfm, no CO breakthrough was detected; at flow rates above 40 scfm, the fraction of unconverted CO increased with increasing flow rates, as shown in Fig. 17.

The breakthrough with methane is compared to that of propylene in Fig. 18. At lower temperatures, the fraction of

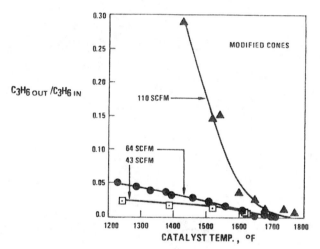

Fig. 16 - C_3H_6 breakthrough in a PTX-5 converter

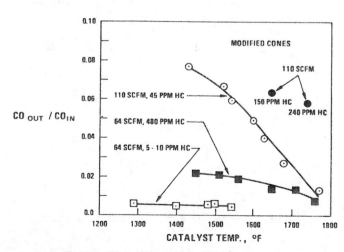

Fig. 15 - CO breakthrough in a PTX-5 converter

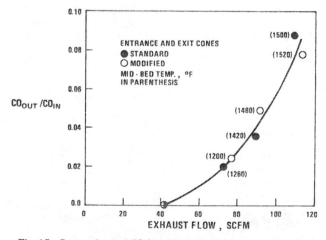

Fig. 17 - Comparison of CO breakthrough between standard and modified cones

unconverted methane was much larger than that of propylene. Essentially, no methane was converted below 1150°F. As the catalyst temperature was increased above 1150°F, the fraction of unconverted methane decreased significantly.

In summary, these data have indicated the maximum CO, C_3H_6, and CH_4 conversions obtained in a fresh PTX-5 converter at high flow rates and with greater than 1% excess oxygen. If the exhaust gas flow rate is 64 scfm/converter, the maximum conversion would be about 97% for CO and 94% for C_3H_6. In terms of mass emissions, a vehicle equipped with dual fresh PTX-5 converters, and operated at 128 scfm and 5% inlet CO, would produce a maximum of 6.82 g CO/min. For a 5000 lb inertia weight vehicle with 351 in^3 displacement engine, the gas flow rate reaches or exceeds 128 scfm for about 20 s during the CVS test cycle. Based on this analysis, the CO emissions breakthrough would contribute about 0.3 g CO/mile to the total test emissions. These breakthrough emissions would probably increase in an aged PTX-5 converter where the chemical reaction rate is limiting. Experimental evaluation of breakthrough on aged converters was not included in this work.

EFFECT OF MASS TRANSFER LIMITATIONS ON BREAKTHROUGH - The emission breakthrough observed in the dynamometer facility (Figs. 14-18) was generally attributed to mass transfer limitations in the monolith. At typical converter operating temperatures (1000-1600°F) after warmup, the chemical reaction rate at the channel wall is very fast, unless the catalyst has been severely deactivated. For relatively fresh catalysts, the slowest step in the overall reaction rate, therefore, becomes the rate of diffusion between the bulk gas stream and the channel wall. This rate is determined by the geometry of the structure and the hydrodynamics of the flow.

When mass transfer is the rate controlling step, the CO conversion in the monolith at 1000°F may be expressed by

$$\left(\frac{x_{out}}{x_{in}}\right)_{CO} = \exp -\left[(4.01 \times 10^{-6})\left(\frac{S^2}{\epsilon}\right)\left(\frac{A_c L_c}{G}\right)(N_{Sh}) + (0.128)(N_{Sh})\right] \quad (1)$$

where:

- S = superficial monolith surface area, ft^2/ft^3
- ϵ = void fraction of the monolith
- A_c = cross-sectional area, in^2
- L_c = length, in
- G = gas flow rate, scfm
- N_{Sh} = Sherwood number, dimensionless, proportional to mass transfer rate

For several monolithic structures with the same exterior dimensions, the most significant variables from Eq. 1 are S and N_{Sh}. While the superficial area is primarily affected by the number of channels/square inch, the Sherwood number is affected by the geometry of these channels. The key properties reported for several available monolithic supports are listed in Table 4.

To test Eq. 1, the theoretical fraction of CO unconverted was plotted against $1/G$, using the data for American Lava 8 corrugations/inch support in Table 3, with $N_{Sh} = 2.0$ (10). Another line for $N_{Sh} = 2.4$ (11) was also plotted against $1/G$. The experimental data obtained with PTX-5 on the

Table 4 - Some Properties of Monolithic Supports

Monolith	Shape	N_{Sh}	S, ft^2/ft^3	ϵ
American lava				
8 corrugations/in		2.0*	636	0.64
		2.4**		
12 corrugations/in			816	0.64
Corning W-1		3.0†	720	0.70
DuPont Torvex				
1/8 in		3.66**	384	0.60
1/16 in			460	0.50

*Ref. 10.
**Ref. 11.
†Ref. 12.

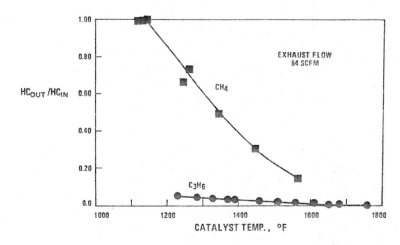

Fig. 18 - Comparison of HC breakthrough between CH_4 and C_3H_6

dynamometer facility was consistent with these plots (Fig. 19). Experimentally, therefore, an average $N_{Sh} = 2.2$ appeared to best fit the PTX-5 converter data.

Using Eq. 1 and the properties in Table 4, the expected fraction of CO unconverted was plotted against flow rate for several monoliths with 5 in diameter and 3 in length (Fig. 20). The advantages of increasing the number of cells/in, and of using square channels instead of the sinusoidal channel geometry, are significant. A slight advantage was predicted for segmenting a 3 in length piece into three spaced 1 in sections. This is a result of the entrance effect, where heat and mass transfer rates are higher until laminar flow is fully developed.

BREAKTHROUGH MEASUREMENTS WITH CONVERTERS ON A CONCEPT VEHICLE - CO and HC emission breakthrough problems have been reported previously in the development of catalytic converter systems on concept vehicles (4, 6).

A 1971 Ford Galaxie with a 351 CID engine and dual exhaust equipped with the following emission control hardware was used for emission breakthrough studies:

1. Low thermal inertia reactors (without secondary air).
2. PTX-5 converters, located at a comparable distance from the engine in each exhaust line.
3. Secondary air supply to the catalytic converters from a 19 in^3 air pump, 1.8:1 drive ratio.
4. Programmed EGR for NO_x control, with manual on-off override.
5. Engine was calibrated for 2% CO in the exhaust gas at idle with the EGR on.

Thermal response and emission measurements were usually obtained over the cold-start CVS cycle. Occasionally, a modified cycle was used to evaluate breakthrough on acceleration/deceleration modes.

In the current work, breakthrough frequently appeared during hard acceleration modes and sharp deceleration modes of the CVS cycle after warmup (Fig. 21). The high breakthrough peaks in this run were much larger than the emissions predicted by mass transfer limitations. Apparently, they were caused by momentary periods of insufficient secondary air supply during the sharply fluctuating driving pattern. The breakthrough peaks were substantially reduced by the constant addition of 8 scfm of compressed air together with 4-15 scfm air normally supplied to each converter from the secondary air pump (Fig. 21). Further studies were made with acceleration/deceleration driving cycles to simulate problem areas in the CVS cycle. These studies showed unusual breakthroughs occurring after sharp speed correc-

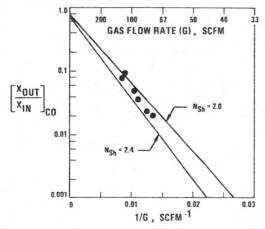

Fig. 19 - CO breakthrough in PTX-5 with high intrinsic catalytic activity

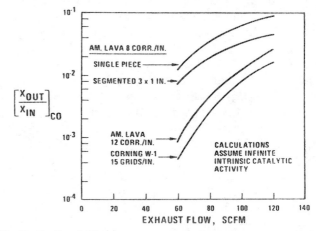

Fig. 20 - Predicted CO conversion for monoliths of 5 in diameter × 3 in length at high flow rates

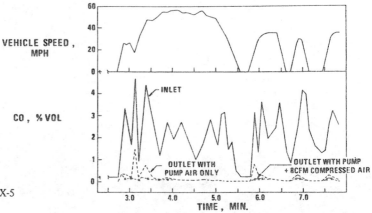

Fig. 21 - Breakthrough during CVS cold start after warmup (PTX-5 units with standard inlet cone)

tions (closing or near-closing of throttle) following acceleration modes. Again, a boost in the secondary air supply alleviated the problem (Fig. 22).

Inoue, et al. (4) related breakthrough to insufficient radial mixing of secondary air with exhaust gas. However, their work was applied strictly to the problem of injecting and distributing air immediately before the catalyst entrance face. In the present study, the secondary air was injected through specially designed distributors a full 28 in upstream of the catalyst to provide good radial mixing. The unusual breakthroughs were, therefore, probably related to axial mixing.

To improve backmixing, a perforated plate baffle was installed approximately 4 in upstream of each converter. The baffles substantially reduced the amount of breakthrough observed on deceleration in the special acceleration/deceleration driving cycle (Fig. 23). These results suggest that a temporary oxygen-deficient "slug" of exhaust gas was created under the conditions of a sudden drop in air pump speed under high load. The baffles provided sufficient backmixing to overcome this condition at low exhaust flow rates.

During CVS tests, the most significant breakthroughs occurred with accelerations. The baffles had much less effect on acceleration breakthroughs, probably because the residence time may be too short to provide adequate mixing at the higher gas flow rates. More baffle plates may have helped, but they would also be expected to slow down the converter warmup rate, adding to cold-start emissions.

SUMMARY AND CONCLUSIONS

THERMAL RESPONSE

1. Thermal response behavior of an Engelhard PTX-5 converter from pulses of high concentrations of combustibles was established for a variety of flow conditions.
2. The maximum temperature increase in the converter bed was approximately equal to the thermodynamic adiabatic temperature increase.
3. The maximum rate of temperature increase occurred at the front portion of the converter. At steady-state, the peak temperatures were in the middle to rear portion of the bed.
4. Generally, the rate of converter temperature increase was sufficiently slow to allow time for activating an overtemperature protection device.
5. Converter thermal failure (melting) was experienced upon total loss of ignition with the throttle open. Melting due to ignition failure can be prevented by various methods that decrease the unburned fuel pumped to the converter.

EMISSIONS BREAKTHROUGH - The engine dynamometer studies with the PTX-5 converter have established that emissions breakthrough occurs at greater than 40 scfm/converter and increases with flow rate; decreases with increasing catalyst temperature; increases for CO with increasing HCs; and is the total for CH_4 below 1150°F at 64 scfm, and generally is much greater for CH_4 than C_3H_6.

Vehicle studies with similar monolithic converters have confirmed these results, and show that more severe breakthrough problems may be associated with the secondary air supply system. These conditions may be controlled by increased secondary air, from larger, more powerful air pumps; and perforated plate baffles installed ahead of the converter, behind the point of secondary air injection.

From the standpoint of mass transfer limitations, breakthrough with monolithic converters can generally be decreased by larger size converters, either greater length or cross-sectional area; and the use of monolithic forms with high external surface area and improved channel geometry more conducive to rapid gas-solid heat and mass transfer.

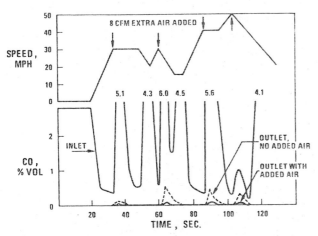

Fig. 22 - CO breakthrough on acceleration/deceleration cycles (Ford concept vehicle with PTX-5 converters)

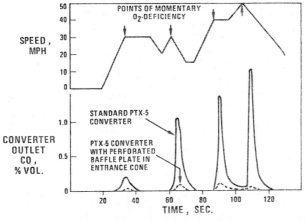

Fig. 23 - CO breakthrough on acceleration/deceleration cycles (Ford concept vehicle with PTX-5 converters)

ACKNOWLEDGMENTS

The authors are grateful to their many co-workers in the IIEC program who contributed to this effort. Special thanks are due H. G. Lassen and D. A. Watt, Ford Motor Co., who provided the Engelhard PTX converter equipped with thermocouples, and assisted in the engine calibration of the Ford concept vehicle. Appreciation is also extended to Dr. J. C. W. Kuo and R. C. Murphey, Mobil Research and Development Corp., for their assistance in this effort.

REFERENCES

1. P. W. Snyder, W. A. Stover, and H. G. Lassen, "Status Report on HC/CO Oxidation Catalysts for Exhaust Emission Control." Paper 720479 presented at SAE National Automobile Engineering Meeting, Detroit, May 1972.

2. W. E. Bernhardt and E. Hoffman, "Methods for Fast Catalytic System Warm-Up During Vehicle Cold Starts." Paper 720481 presented at SAE National Automobile Engineering Meeting, Detroit, May 1972.

3. K. Tanaka, M. Akutagawa, K. Ito, Y. Higashi, and K. Kobayashi, "Toyo Kogyo Status Report on Low Emission Concept Vehicles," Paper 720486 presented at SAE National Automobile Engineering Meeting, Detroit, May 1972.

4. T. Inoue, K. Goto, and K. Matsumoto, "Toyota Status Report on Low Emission Concept Vehicles." Paper 720487 presented at SAE National Automobile Engineering Meeting, Detroit, May 1972.

5. R. M. Campau, A. Stefan, and E. E. Hancock, "Ford Durability Experience on Low Emission Concept Vehicles." Paper 720488 presented at SAE National Automobile Engineering Meeting, Detroit, May 1972.

6. Chrysler Corp.: "Application to EPA for Suspension of 1975 Motor Vehicle Emission Standards." March 1972.

7. J. C. W. Kuo, C. R. Morgan, and H. G. Lassen, "Mathematical Modeling of CO and HC Catalytic Converter Systems." SAE Transactions, Vol. 80 (1971), paper 710289.

8. J. L. Harned, "Analytical Evaluation of a Catalytic Converter System." Paper 720520 presented at SAE National Automobile Engineering Meeting, Detroit, May 1972.

9. J. C. W. Kuo, unpublished work.

10. D. F. Sherony and C. W. Solbrig, "Analytical Investigation of Heat and Mass Transfer and Friction Factors in a Corrugated Duct Heat or Mass Exchanger." Int. Jrl. Heat Mass Transfer, Vol. 13 (1970), pp. 145-159.

11. R. D. Hawthorne, "Afterburner Catalysts—Effects of Heat and Mass Transfer Between Gas and Catalyst Surface." Paper 29c presented at AIChE 71st National Meeting, February 1972.

12. W. Kays and A. L. London, "Compact Heat Exchangers." New York: McGraw-Hill Inc., 1964.

13. J. C. W. Kuo, D. P. Osterhout, C. D. Prater, P. W. Snyder, and J. Wei, "Predicting Performance of Future Catalytic Exhaust Emission Control Systems." Paper 2-14 presented at XIV International Automobile Technical Congress of FISITA, London, England, June 1972.

Society of Automotive Engineers, Inc.
TWO PENNSYLVANIA PLAZA, NEW YORK, N.Y. 10001

This paper is subject to revision. Statements and opinions advanced in papers or discussion are the author's and are his responsibility, not the Society's; however, the paper has been edited by SAE for uniform styling and format. Discussion will be printed with the paper if it is published in SAE Transactions. For permission to publish this paper in full or in part, contact the SAE Publications Division and the authors.

16 page booklet.   Printed in U.S.A.

LIST OF SYMBOLS

A/F = Air/Fuel ratio of the conventional engine or overall ratio of the pre-chamber engine

α_t = Overall air/fuel ratio of the pre-chamber engine

α_m = Main chamber air/fuel ratio of the pre-chamber engine

α_a = Pre-chamber air/fuel ratio

λ = Pre-chamber to main chamber mass air flow ratio

β = Pre-chamber to total fuel flow ratio

α_{ai} = Pre-chamber theoretical air/fuel ratio at the time of spark ignition

α_{mi} = Main chamber theoretical air/fuel ratio at the time of spark ignition

EGR = Exhaust gas recirculation

MBT = Minimum spark advance for best torque

Copyright ©Society of Automotive Engineers, Inc. 1974
All rights reserved.

Pre-Chamber Stratified Charge Engine Combustion Studies

By
Egils A. Purins
FORD MOTOR COMPANY

INTRODUCTION

Many claims have been made recently about the stratified charge engine, including the pre-chamber engine, as a solution to the automobile emissions problem. In the past the pre-chamber engine was investigated for good fuel economy. The advent of emission control has revised interest in the pre-chamber engine utilizing a lean mixture as an alternative to the conventional engine.

This investigation was initiated to determine if the pre-chamber engine concept has inherent emission advantage over the conventional, to identify its disadvantages and to gain understanding of the pre-chamber engine combustion process.

Before looking at the test data a review of the pre-chamber engine operation is in order. Figure 1 illustrates that engine combustion at lean air-fuel ratios (A/F) forms a minimum of carbon monoxide (CO) and nitrogen oxide (NO) and discharges low amounts of unburned hydrocarbon (HC). Therefore, from a steady-state emission standpoint lean combustion is very desirable. But the problem is that the lean mixture is not readily ignitable by conventional means and, if ignited in a conventional combustion chamber, the combustion process procedes at such a slow rate that the efficiency suffers drastically. Providing a richer mixture at the spark plug than in the rest of the combustion chamber eliminates the ignitability problem. In addition, some form of induced combustion turbulence would improve the combustion rate. The pre-chamber concept does all of this. A rich mixture is introduced into a separate enclosure (pre-chamber) basically for ignition. In this manner the rich mixture is mechanically prevented from mixing completely with the leaner portion. This not only maintains a richer mixture in the pre-chamber for a positive ignition, but also results in some degree of stratified combustion which is beneficial for low NO emissions (1).* An orifice between the rich pre-chamber and the lean main combustion chambers is used to induce the required amount of combustion turbulence to obtain efficient combustion of the lean mixture.

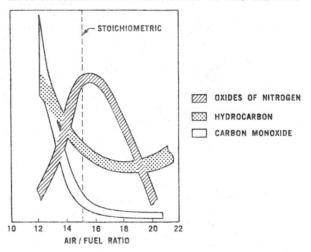

EMISSION CONCENTRATION RELATIONSHIP TO AIR/FUEL RATIO

Figure 1

* Indicates reference source.

ABSTRACT

Single-cylinder experiments were conducted with a 3-valve, carbureted pre-chamber stratified charge engine in comparison with a conventional engine. The pre-chamber engine operation is governed by many design and operating variables. This investigation was limited to determining the effect of overall air/fuel ratio, ignition timing and EGR on emissions and fuel economy at a single road load test condition.

It was found that, as for the conventional engine, these operating variables are also significant for the pre-chamber engine and that a compromise must be made between good fuel economy and low emissions. The main virtue of the pre-chamber engine was found to be the ability to operate at leaner overall air-fuel ratio. This resulted in lower nitrogen oxide (NO) emissions than the conventional engine without EGR. The unburned hydrocarbons (HC) were found to be higher for the pre-chamber engine up to the conventional engine lean misfire A/F ratio.

Exhaust gas introduced into the pre-chamber was found to reduce NO emissions significantly without a large corresponding increase in HC emissions as observed with the conventional engine.

Only at very low NO emissions with severely retarded spark timing and/or high EGR rate did the pre-chamber engine show a fuel economy advantage over the conventional engine.

As the test program was limited to one load and speed, the results should not be construed to be typical of all modes of operation.

PRE-CHAMBER STRATIFIED CHARGE ENGINE

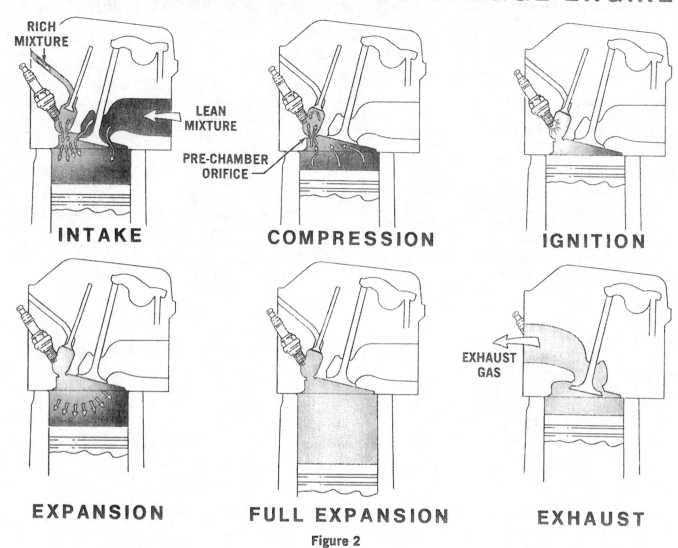

Figure 2

Figure 2 describes the pre-chamber engine process. Note that the rich and lean mixtures do not stay in their respective chambers. On the intake stroke, if more than one pre-chamber volume of the rich mixture (α a) is inducted, then the excess will flow through the orifice and either partially or fully mix with the leaner main chamber mixture (α m). On the compression stroke the reverse takes place. The leaner main chamber mixture and some of the rich mixture is pushed into the pre-chamber and dilutes the original rich charge (α a). The mixture strength in the pre-chamber at the time of ignition (α ai) depends on the ignition timing, the degree of mixing taking place in the main chamber on the intake stroke and the degree of pre-chamber overfill. Figure 3 provides the approximate α ai for a chosen ignition timing at the air flow (λ) and air-fuel ratio (α a, α m) conditions used for these tests. It is possible to obtain the same α ai by different combinations of α a, α m, λ and ignition timing the effect of which was not thoroughly investigated. It was the intent of this study to learn only about the operating parameters of α t, ignition timing and EGR in comparison with a conventional engine.

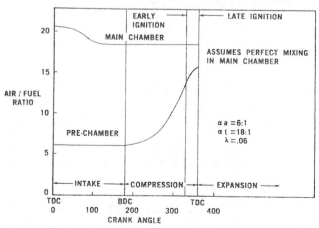

Figure 3

TEST EQUIPMENT AND INSTRUMENTATION

TEST ENGINE — Experiments were conducted with a 50 cu. in. single cylinder engine built on a 400 CID V-8 engine block with pistons only in the front two cylinders. Only the #1 cylinder was used for combustion while the #5 piston in the opposite bank served as counterbalance. The stock crankshaft was modified for best dynamic balance for this two cylinder operation.

A specially designed pre-chamber single cylinder head was used on the firing cylinder with the remaining cylinders blocked to complete the cooling system. A 1973 production 400 CID engine cylinder head was also tested to obtain comparative baseline data.

The head designs are illustrated in Figures 4 and 5 with specifications covered in Appendix A. Figure 4 shows the designed locational relationships between the pre-chamber, the orifice and the main chamber, the spark plug location relative to the orifice and the pre-chamber intake valve, and the proximity of the main chamber intake valve to the pre-chamber orifice. The spark plug was so located as to receive minimum of fuel impingement from the pre-chamber intake valve or the pre-chamber orifice. The pre-chamber cup was installed into a machined recess in the cylinder head with clearance calculated to give contact between cup and head for heat transfer when expanded at the higher operating temperatures. This design was to minimize spark plug fouling and to provide a heat sink for the vaporization of fuel particles in the fuel-rich charge entering the pre-chamber. The main chamber and its intake port were designed for minimum mixture turbulence prior to combustion.

Figure 5 shows the open wedge design (squishless) conventional combustion chamber which was run for comparison purposes.

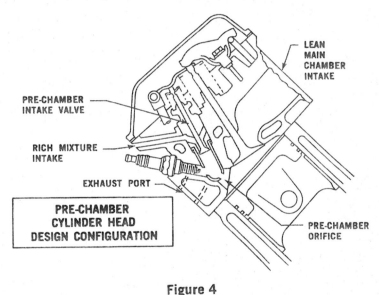

Figure 4

CONVENTIONAL COMBUSTION CHAMBER CONFIGURATION

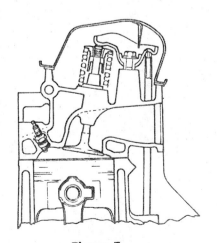

Figure 5

The exhaust system design permitted manual control of exhaust gas heat to the intake systems by changing the exhaust gas flow path as diagramed in Figure 6. Figures 6 and 7 illustrate the engine set-up and further describe the intake and exhaust system. The exhaust system was insulated with approximately .25 in. thick asbestos wrap to conserve heat energy for the intake charge pre-heating. The insulated exhaust system acted as a HC/CO reactor with the main exhaust pipe volume equal to one cylinder displacement.

A .94 inch diameter venturi Bendix 1V carburetor was used to meter fuel to the main combustion chamber and a specially constructed .25 inch diameter venturi carburetor was used for the pre-chamber with independent and manually controlled throttles and needle valve fuel adjustments for both carburetors.

Because of an unsolvable engine vibration problem which caused fuel handling difficulties the pre-chamber carburetor was isolated from the engine through a length of wire-reinforced silicone rubber hose, suspended over the engine and isolated from the engine mounting bed plate.

Indolene Clear fuel was used throughout the testing to minimize the combustion deposit formation.

The ignition system used was a conventional breaker-type automotive 12V system with a .035 inch gapped resistor spark plug.

INSTRUMENTATION — The engine was coupled to an electric dynamometer equipped with a strain gage load read-out.

The gas analysis equipment used included Beckman and Intertech analyzer for CO, CO_2, O_2 and HC analysis and a chemiluminescent NO analyzer. The HC readings were taken as N-hexane. The exhaust was sampled downstream from a mixing chamber to minimize measurement errors resulting from the single cylinder exhaust gas inhomogeneity.

The carburetor air flow measurements were made with Meriam laminar element flow meters of 20 and 4 cfm flow capacity.

Fuel flow rates were calculated from burette timed samples. The burette fuel measuring system was designed to maintain a constant pressure fuel supply during measurements.

Combustion pressure measurements were made using a tranducer installed in the main combustion chamber and displayed and analyzed with a Digital Processing Oscilloscope.

A more detailed description of the instrumentation used is included in Appendix D.

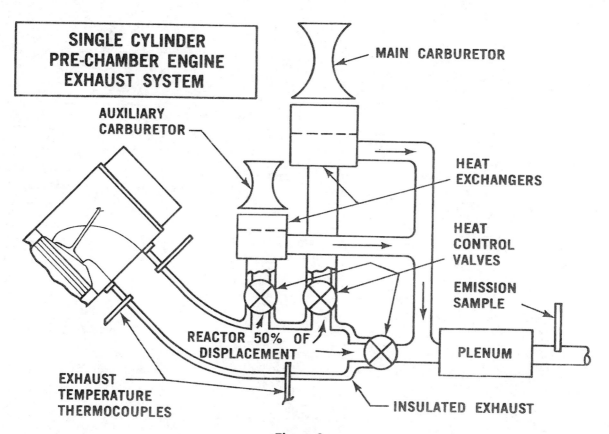

Figure 6

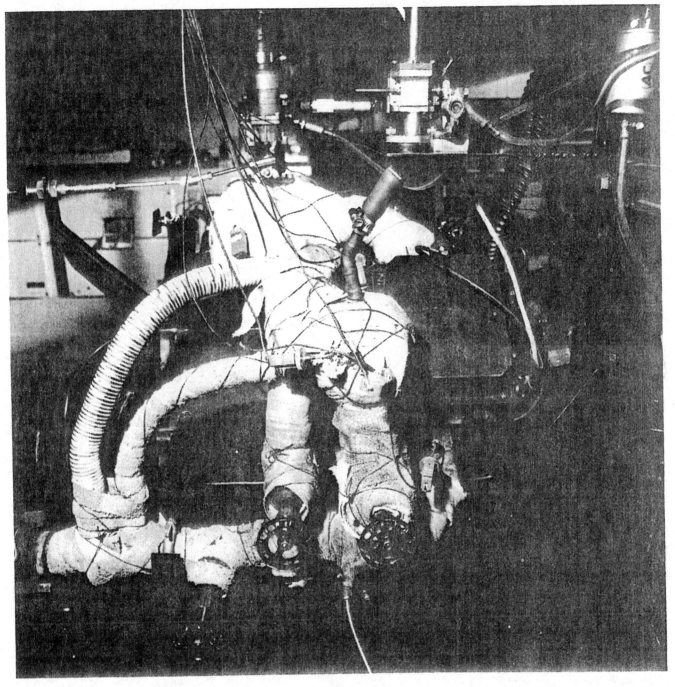

Figure 7

TEST PROCEDURE

Except as indicated, the investigation was conducted at steady state 1500 rpm and 60 psi IMEP. This test condition can be related to a light 40 mph acceleration of a standard size vehicle, a condition which frequently occurs on a CVS vehicle emission test. Both the pre-chamber head and the conventional head were run on the same basic engine block at comparable conditions.

Engine operating conditions were manually adjusted and only one condition varied at a time. The operating conditions investigated were as follows:

1. The overall air/fuel ratio (αt)

2. Spark advance relative to minimum advance required for best torque (MBT)

3. Exhaust temperature effect on HC/CO emissions

4. Exhaust Gas Recirculation (EGR)

A pre-chamber volume with 8.9% of the total clearance volume and a 12mm diameter circular orifice was used for all pre-chamber engine tests.

To describe the true effects at a given constant operating condition the emissions are reported on a specific mass basis as described in Appendix B.

The overall air/fuel ratio was obtained by exhaust analysis using the Spindt Method (3) and by fuel and air flow measurements. The two methods were required to agree within .5 A/F. The overall air/fuel ratio data reported is based on exhaust analysis. The reported pre-chamber and main chamber α_a and α_m are, of necessity, calculated from fuel and air flow measurements.

Intake charge mixture temperatures were maintained in the range of 170-190°F for the main intake system (and the conventional engine) and 370-400°F for the pre-chamber intake charge measured near the intake valves.

The exhaust backpressure was maintained at a normal test cell level minimum and did not exceed 1.0 inch Hg measured at the exhaust port outlet.

The exhaust temperatures were measured at the exhaust port outlet and/or in the reactor as shown in Figure 6.

Determination of burn time was accomplished using the method of Rassweiler and Withrow (8). Pressure rise due to combustion is analogous to the charge burning. In brief, the difference between observed pressure change and calculated pressure change due to piston motion (ΔPc) is calculated over a small time interval and corrected to a constant volume. The fraction burned at any point during combustion is the sum of the ΔPc from ignition to that point divided by the total sum of ΔPc from ignition to the point where combustion is complete. Complete combustion has occurred when the observed pressure change equals the change due to piston motion.

The exact points where significant burning begins and burning ends are difficult to establish reliably. Therefore, the time to burn 10 to 90% of the charge was taken to compare the effect of operating parameters on ignition lag and burn time. Each burn and lag time calculation was made using the average of 512 consecutive p-t traces.

AIR-FUEL RATIO REQUIREMENTS

To obtain the effects of varied overall air-fuel ratio (αt) for the pre-chamber engine, as compared to a conventional combustion chamber, tests were conducted with αt leaned from 13:1 to lean limit at MBT spark advance where a substantial HC increase was observed. Other test conditions were maintained constant as shown in Table 1. The pre-chamber air-fuel ratio (αa) was maintained at a constant 6:1 and the mass air flow ratio of pre-chamber-to-main chamber (λ) was held constant at .06 which results in approximately 5 pre-chamber volumes of rich charge overflowing into the main chamber on the intake stroke. The overall air-fuel ratio was adjusted by varying the main intake A/F.

Table 1
AIR / FUEL RATIO REQUIREMENT TEST CONDITION FOR FIGURES 8 TO 14

Pre-Chamber Head (Ref. Test #22-07)	Conventional Head (Ref. Test #12-06)
1500 RPM	1500 RPM
60 IMEP	60 IMEP
6:1 αa	8.6 CR
.06 λ	MBT spark
MBT spark	No EGR
12 mm dia. orifice	
8.9% pre-chamber volume	
8.5 CR	
No EGR	

The most noticeable difference in the results is the lower NO emissions at A/F richer than 20:1 for the pre-chamber as shown in Figure 8. The lower NO is thought to be due to stratified combustion. With the combustion initiated in the rich mixture zone of the pre-chamber, the rich condition results in low peak temperatures in the pre-chamber and a lack of excess oxygen serving to minimize NO formation. The bulk of the main combustion chamber mixture is lean, likewise producing minimum of NO with only a fraction of the total mixture consumed at high NO forming air/fuel ratios.

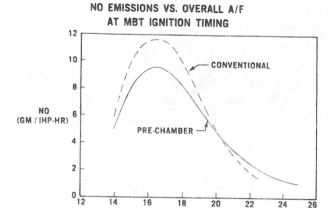

Figure 8

Shown in Figure 9, pre-chamber operation is accompanied by approximately a 3% fuel consumption penalty over the conventional chamber in the operating range of 16:1 to 19:1 αt. After 20:1 A/F combustion in the conventional chamber deteriorates rapidly due to misfires.

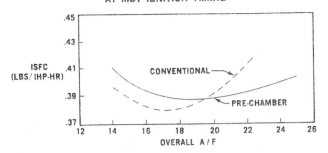

Figure 9

The difference in MBT ignition timing requirement for the conventional and pre-chamber is shown in Figure 10. The difference is attributed mainly to increased ignition lag time for the conventional chamber. Calculations made from combustion chamber pressure measurements show in Figure 11 that the ignition lag increases rapidly with lean A/F ratio at ignition although the burn time may be less. The higher spark advance is required for the conventional chamber because of low mixture turbulence in the vicinity of the spark plug and a relatively leaner A/F ratio which results in high ignition lag time. The burn time for the pre-chamber condition was longer by approximately 10% throughout the A/F ratio range but may decrease with the use of smaller orifice diameter or a larger pre-chamber volume or both.

The difference in HC emissions shown in Figure 12 at A/F ratio below 19:1 is mainly due to the difference in exhaust temperature indicated in Figure 13 and a 20% higher combustion chamber surface-to-volume ratio for the pre-chamber engine. At A/F over 19:1, the conventional chamber HC emissions increase rapidly, indicating misfires.

The lack of significant differences in CO emissions in Figure 14 is mainly due to a lack of significant CO conversion in the exhaust at the exhaust temperatures observed. The increase in CO above 19:1 A/F ratio is a result of increased mass flow due to the leaner operation.

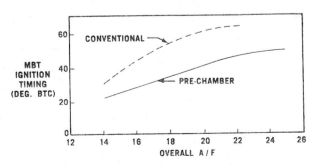

Figure 10

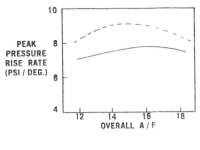

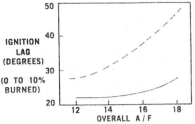

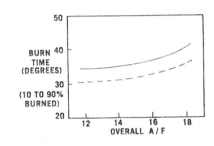

Figure 11

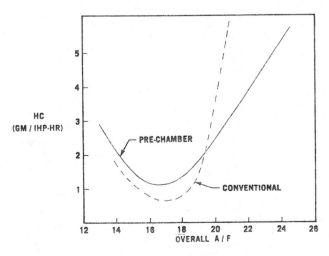

Figure 12

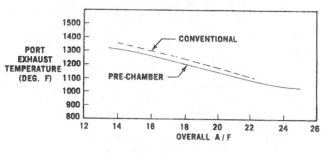

Figure 13

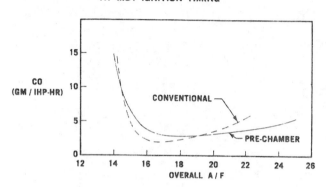

Figure 14

IGNITION TIMING

For a conventional engine retarded ignition timing has proved to be the one variable which significantly reduces HC, CO and NO emissions. This investigation was conducted to determine its significance for the pre-chamber engine. It should be noted that any change in ignition timing for the pre-chamber engine also is accompanied by a change in the pre-chamber air-fuel ratio at the time of ignition (α ai). Theoretically, at constant αa and αm, the αai will become leaner with retarded ignition timing. Referring again to Figure 3, a decrease in spark advance by 20 degrees from MBT (from 25° to 5° BTC) will only increase α ai by about one air-fuel ratio. (assuming perfect mixing). The significance of the change in α ai alone was not investigated.

To determine the total effect, the ignition timing was incrementally decreased from a maximum advance of MBT + 5° for both the pre-chamber and the conventional engine. All other engine conditions were maintained constant as shown in Table 2. Overall air-fuel ratios of 18:1 for the pre-chamber and 16:1 for the conventional chamber were chosen as typical for this road load condition.

The results shown in Figures 15 and 16 point out that ignition timing is critical on NO emissions for both the pre-chamber and conventional combustion chamber; but after retarding a few degrees from MBT, a rather large fuel consumption penalty is received. Because of the initial NO emission advantage of the pre-chamber engine, the conventional engine must be run more retarded from MBT to obtain the same NO emission level as the pre-chamber.

Table 2
IGNITION TIMING REQUIREMENT TEST CONDITIONS FOR FIGURES 15 TO 22

Pre-Chamber Head (Ref. Test #22-07)	Conventional Head (Ref. Test #12-06)
1500 RPM	1500 RPM
60 IMEP	60 IMEP
6:1 αa	8.6 CR
18:1 αt	16:1 A/F
20.5:1 αm	No EGR
.06 λ	
8.5 CR	
12mm dia. orifice	
8.9% pre-chamber volume	
No EGR	

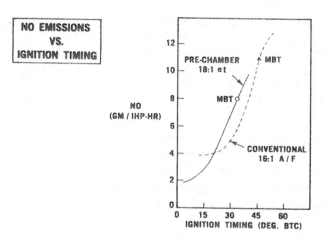

Figure 15

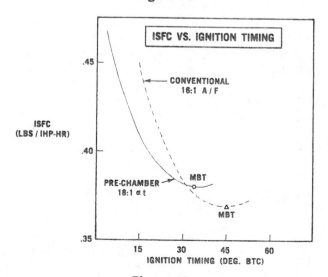

Figure 16

The HC emissions in Figure 17 show a large decrease with spark retarded for both engines. This is due to increase in exhaust temperature shown in Figure 19 which results in better post reaction. The pre-chamber has 30% higher HC emissions at MBT. At about 30° retarded, both engines have about equal HC because the exhaust temperatures approach about 1400°F with good post reaction.

The CO emissions increase for both engines to about 10 degrees retarded ignition and decrease when retarded further. This phenomenon, shown in Figure 18, is thought to be due to HC to CO conversion in the reactor with the temperature insufficiently high to complete the conversion to CO_2. The CO peaks for both are at about the same 1220°F reactor temperature, shown in Figure 19. The CO increase was studied separately and will be discussed later in the text.

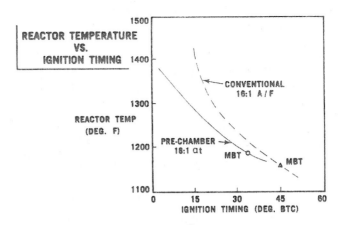

Figure 19

Figures 20 and 21 illustrate the NO/ISFC and HC/ISFC trade-offs with spark advance. It shows that in order to get a low NO emissions with retarded spark, a large sacrifice in fuel economy must be made. At the higher NO level the conventional chamber has a fuel economy advantage. This advantage is lost, however, where spark is retarded to obtain NO levels of 4 gm/IHP-HR or less. At the 4 gm NO level the conventional engine specific fuel consumption begins to deteriorate very rapidly. By retarding spark alone, the conventional engine cannot reach the low NO level obtainable with the pre-chamber. There is a similar trade-off in HC emission control except that the pre-chamber engine at the same NO level condition shows higher HC emissions than the conventional.

The degree that combustion characteristics are affected by ignition timing is shown in Figure 22. The peak pressures are almost identical for the pre-chamber and conventional engine and decrease in the same proportion with retarded spark.

The peak pressure-rise-rate which is an indicator of the relative rate of burn is slightly lower (11%) for the pre-chamber engine measured in the main combustion chamber. The 10% to 90% burn time is calculated to be 17% longer for the pre-chamber engine at MBT (41 degrees vs. 34 degrees) and increases at a slower rate with retarded spark than the conventional.

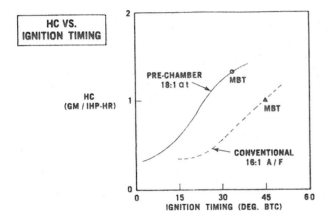

Figure 17

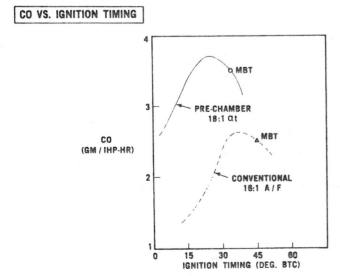

Figure 18

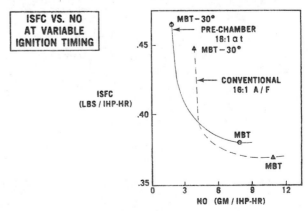

Figure 20

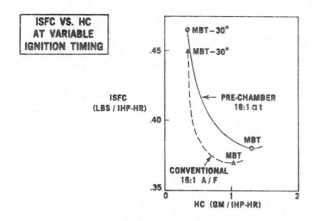

Figure 21

The ignition lag time at MBT is 43% lower for the pre-chamber than the conventional (28 vs. 48 degrees). The reason for this difference is due to the difference in A/F ratio and mixture turbulence near the spark gap at the time of ignition as explained under A/F effects. At equal spark retard the difference in lags is maintained. Because of this ignition lag difference MBT ignition timing cannot be used as an indicator of relative burn rates between stratified and conventional combustion chambers. Instead, the pressure-rise-rate at MBT spark is a better indicator of the relative burn rates between different engines.

EXHAUST SYSTEM HC/CO CONVERSION

Because the tailpipe HC/CO emissions are greatly affected by the degree of post reaction in the exhaust system, it was of interest to measure the HC/CO emission concentrations leaving the exhaust port. To prevent the exhaust manifold reaction, the exhaust gas was cooled to 800°F with N_2 gas injected at the port. This procedure was used because an abnormally high HC concentration is measured at the port as reported by other investigators (4, 2). Therefore, a simple procedure of sampling and comparing HC and CO concentration at port and after the exhaust manifold could not be used.

It was assumed that no change in mass NO emissions would result from the N_2 injection and the NO concentration was used as a tracer gas to determine the degree the exhaust sample was diluted for the correction of measured HC/CO concentration.

The engine was run at three constant exhaust mass flow rates using different speeds and ignition timing to obtain an exhaust temperature range of 1000° to 1700°F before cooling with N_2.

To determine the amount of HC/CO conversion taking place at different reactor temperatures, the mass emissions at 800°F reactor temperature were compared to emissions without cooling. Figure 23 illustrates the changes taking place in the exhaust system at different exhaust temperatures. Only small amounts of HC are converted below 1000°F with conversion reaching 50% at 1300°F. Surprisingly, the CO is shown to increase in the exhaust manifold with increasing exhaust temperature between 800°F and 1400°F with maximum increase at 1200°F. The data suggest that HC is converted to CO in the exhaust system faster than CO is converted to CO_2 at temperatures below 1200°F. Thus, until a sufficiently high exhaust temperature is reached for adequate conversion of CO, an increase in tailpipe CO emissions will take place. This means that when working with lean burn engine emissions, one must be careful not to create a CO problem when post oxidizing HC in the exhaust system. CO increases to approximately 1200°F exhaust temperature and then decreases with a break-even point at approximately 1400°F.

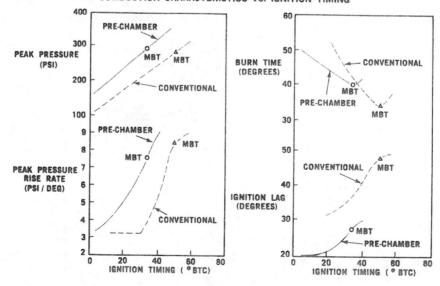

Figure 22

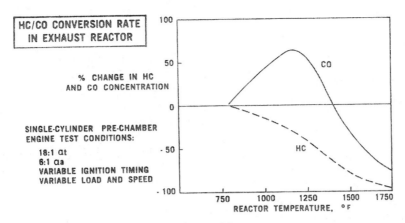

Figure 23

A rough accounting of HC converted and the increase in CO showed good correlation up to 1300°F after which CO conversion becomes significant.

Table 3 makes comparison of port emission for the conventional and the pre-chamber engine at the same overall A/F. It is evident that the pre-chamber engine emits exhaust with 30 to 40% higher HC emission concentration from the combustion chamber at the same test conditions with nearly the same exhaust temperature. The 22% larger combustion chamber surface-to-volume ratio (S/V) is thought to be the main contributor to the higher HC emissions.

EXHAUST GAS RECIRCULATION

The pre-chamber combustion process initiates in a fuel rich mixture in the pre-chamber and then passes through the orifice into the main combustion chamber which may or may not have a uniform air/fuel mixture. If some stratification does exist in the main chamber, it is likely that a richer mixture will be found at the pre-chamber orifice. This will be consumed next after the pre-chamber with ignition finally passing into the lean region. At the air/fuel ratios used for this investigation ($\alpha a = 6:1$ and $\alpha m = 20.5:1$), MBT ignition occurred theoretically at about $\alpha ai = 14:1$ assuming perfect mixing in the main chamber.

It is apparent that the early combustion taking place at rich A/F contributes heavily to the overall NO emissions. It seemed, therefore, that EGR would be effective in reducing NO without a significant increase in HC emissions (7) if introduced into this rich initial burning charge.

This investigation was conducted at constant 1500 rpm and 60 IMEP load with $18:1\ \alpha t$ for the pre-chamber engine. For comparison, the conventional head was tested richer, at 16:1 A/F, because this has been shown to be typical when EGR is used.

The EGR and ignition timing were both variable. EGR was incrementally added to the intake charge up to the maximum rate supporting good combustion at MBT. At each EGR setting, ignition timing was varied from MBT +5 to MBT -30 degrees.

For the pre-chamber head EGR was first introduced only in the main combustion chamber intake system. In a second test, the EGR was introduced

Table 3
COMPARISON OF PORT EMISSIONS FROM PRE-CHAMBER HEAD AND CONVENTIONAL HEAD
HC/CO MEASURED AFTER PLENUM WITH N_2 COOLING

Conditions: $\alpha t = 18:1$, $\lambda = .06$, $\alpha a = 6:1$, MBT Ignition Timing

	PRE-CHAMBER				CONVENTIONAL			
RPM	IMEP psi	Exh. Temp. °F	HC ppm	CO Mole %	IMEP psi	Exh. Temp. °F	HC ppm	CO Mole %
1000	55	1110	352	.11	55	1094	199	.08
1500	67	1263	262	.11	66	1272	154	.11
2000	79	1402	193	.13	80	1363	131	.08

only into the pre-chamber intake manifold. The effect of EGR concurrently in both manifolds was not investigated. EGR for the conventional head was also introduced into the intake manifold.

The EGR rate was calculated from CO_2 measurements as described in Appendix C and is defined as follows: % EGR = (CO_2 intake - CO_2 background) / (CO_2 exhaust - CO_2 intake).

For the test conditions in Table 4, the effects of EGR at MBT spark are shown in Figures 24-33. As indicated, the maximum EGR rate for the pre-chamber was approximately 7%, for the main combustion chamber 20% and 17% for the conventional. The latter two were maximum values because of erratic combustion and a large increase in HC emissions shown in Figure 27. At 7% pre-chamber EGR, an HC increase was not observed, but the engine load surged sufficiently to be declared a maximum rate. This surging was probably due to the extensive dilution of the pre-chamber mixture with exhaust gas causing large variations in the ignition lag time. The 7% maximum overall rate shown is a 123% EGR rate for the pre-chamber mixture. This means that the mixture inducted in and through the pre-chamber consisted of one part of 6:1 air/fuel mixture and 1.23 parts of exhaust gas. The actual dilution of pre-chamber charge at the time of ignition was not measured in this study but from other reported data (7) it is estimated that gas/fuel ratios in the range of 17-19:1 existed in the pre-chamber at ignition.

Table 4
EGR EVALUATION (REF. TEST #22-07)
TEST CONDITIONS FOR FIGURES 24 TO 37

Pre-Chamber Head	Conventional Head
1500 RPM	1500 RPM
60 IMEP	60 IMEP
6:1 α a	8.6 CR
20.5:1 α m	16:1 A/F
18:1 α t	
.06 λ	
8.5 CR	
12mm dia. orifice	
8.9% pre-chamber volume	

Figure 24 illustrates the degree that ignition timing must be advanced for MBT condition with increasing EGR at a constant air/fuel ratio. Like trends are shown for all three tests — increasing MBT spark requirement with EGR due to increased burn time and increased ignition lag. The pre-chamber EGR condition shows an exceptionally rapid increase in MBT requirement. Calculations from pressure-time measurements taken for the combustion cycle show that the major cause for the increase in the pre-chamber MBT spark requirement is ignition lag with burn duration remaining relatively constant. Figure 25 shows the calculated ignition lag and the burn duration for 10% to 90% of the combustion chamber mass. The high ignition lag measured indicates a large amount of EGR present at ignition resulting in a lean gas/fuel ratio.

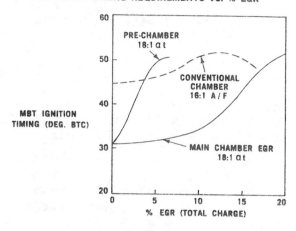

Figure 24

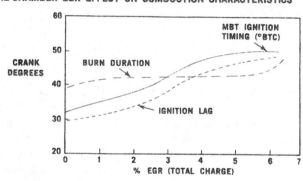

Figure 25

Comparison of reactor temperatures, presented in Figure 26, show basically that at this test condition the reactor for the conventional engine was doing a better job, increasing in temperature as higher HC concentrations are received at sufficiently high temperature for HC to CO conversion. HC and CO data in Figures 27 and 28 also show this and agree with the earlier discussion of data in Figure 23.

The HC emissions increase rapidly for main chamber EGR and the conventional chamber with EGR due to incomplete combustion of the lean end gas diluted by the EGR.

Figure 29, NO emission results, shows the extreme effectiveness of pre-chamber EGR in reducing NO emissions as postulated at the beginning of the text. In comparison, large amounts of EGR are required for the main chamber and the conventional chamber resulting in high HC emissions at a comparable low NO level.

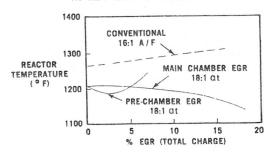

Figure 26

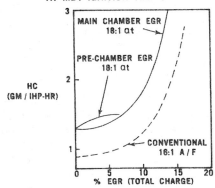

Figure 27

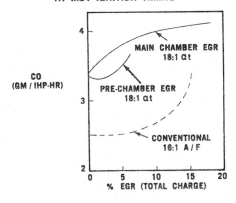

Figure 28

Figure 30 shows that specific fuel consumption is oppositely affected. The pre-chamber EGR with the largest NO decrease shows the fastest increase in ISFC with increasing EGR. A cross plot in Figure 31 shows that a definite trade-off exists for decrease in NO emissions to an increase in fuel consumption with increasing EGR at MBT spark. It is interesting to note that the same relationship exists both for pre-chamber and main chamber EGR, both tested at 18:1 αt and shown as one line. The conventional run at 16:1 A/F shows the same trend at low EGR rates and the curves displaced only by the initially higher NO level due to the difference in A/F and the stratification effect discussed earlier.

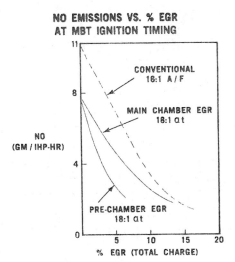

Figure 29

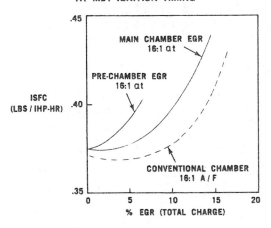

Figure 30

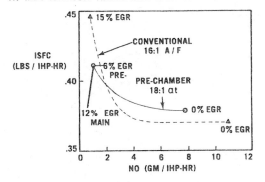

Figure 31

Figure 32 expresses the NO and HC trade-off with EGR on a percent basis. At MBT spark and a more than 50% reduction in NO with EGR, the HC increase is more rapid for the conventional than the pre-chamber.

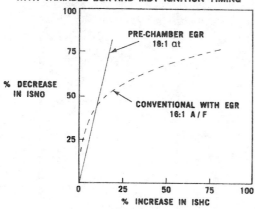

Figure 32

In NO vs. ISFC trade-off with EGR at MBT spark, shown in Figure 33, the pre-chamber has a lower percentage loss in ISFC at over 85% decreased NO level. At a smaller NO decrease, the conventional has a smaller increase in ISFC. Figures 34 through 37 are crossplots of data at variable EGR rates and retarded spark. To this point only their individual effects have been demonstrated. The following analysis will extract the best compromise combination.

At a constant 2 gm/IHP-HR NO level, in Figure 34, HC emissions and ISFC increase rapidly as more EGR is introduced in conjunction with less ignition retard. At this test condition it is to HC advantage to operate with less EGR and more spark retard. At the higher NO levels, the pre-chamber engine is at a disadvantage because it cannot obtain the low ISFC that is possible with the conventional. The pre-chamber engine is 3% higher in minimum ISFC which is at about the 1.3 gm HC level.

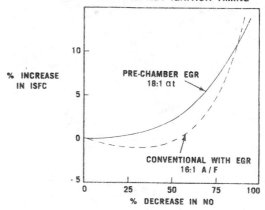

Figure 33

At 1 gm/IHP-HR NO emission level, shown in Figure 35, the positions are reversed and the pre-chamber engine has a 6% advantage over the conventional in minimum ISFC, again occurring at about the 1.3 gm HC level.

Figure 36 shows that the crossing over point where both engines are equal in ISFC occurs at about the 1.5 NO level. At lower than 1.5 NO the pre-chamber engine has lower ISFC but at a higher level the conventional is lower in ISFC.

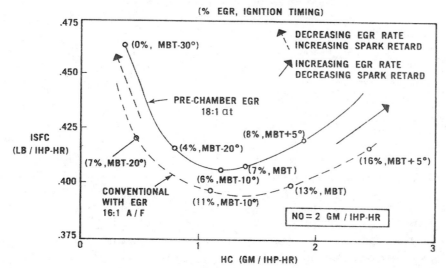

Figure 34

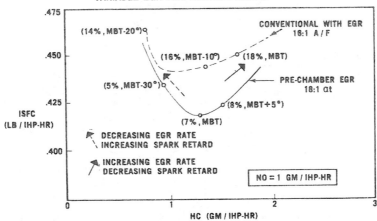

Figure 35

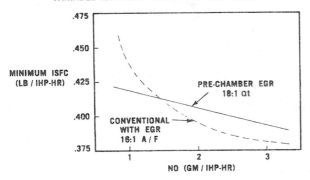

Figure 36

Figure 37 shows how HC emissions compare when both engines are operated at the lowest ISFC possible for a given NO level with varied spark and EGR. The difference in ISFC is not great in the low 1 to 2 ISNO range.

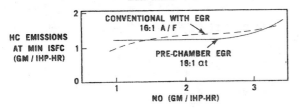

Figure 37

In summary, EGR introduced into the pre-chamber is more effective than EGR introduced in the main chamber because much less EGR is required to reach a given NO level and results in a minimum increase in HC. The pre-chamber engine, in combination with EGR introduced in the pre-chamber and retarded spark, has lower ISFC, at less than 1.5 ISNO emission level and the same HC emissions, than a conventional engine.

SUMMARY AND CONCLUSIONS

A pre-chamber engine and a conventional chamber engine were compared from the emission and economy standpoint on a single cylinder test at one "road load" operating condition. It was found that both engines respond similarly to the operating variables of A/F ratio, spark advance and exhaust gas recirculation. The main difference found was that the pre-chamber engine can be operated at a leaner overall A/F ratio than the conventional engine which results in lower NO emissions without exhaust gas recirculation. An added bonus of the leaner combustion is the presence of an adequate amount of O_2 in the exhaust gas for post oxidation of HC and CO in the exhaust system without secondary air injection.

From the limited single-cylinder tests conducted on this particular design of pre-chamber engine, in comparison to a conventional production engine, the following conclusions are reached:

1. The pre-chamber engine was found to have the ability to operate at 3 to 4 overall air/fuel ratios leaner than the conventional engine. The conventional engine would misfire at ratios leaner than 19:1, whereas the pre-chamber engine operated satisfactorily at 23:1.

2. At MBT spark timing in the air/fuel range of 14 to 19, the pre-chamber engine shows lower NO emissions than the conventional engine but higher levels of HC emissions and poorer fuel economy.

3. With both the conventional and pre-chamber engines, only residual CO is present in the exhaust gas at leaner than 16:1 A/F ratio. At operating conditions where exhaust gas temperature is at about 1250°F CO is formed from post oxidation of HC in the exhaust system and, because of the low initial concentration level, a large percentage increase in tailpipe CO is observed.

4. EGR can be effectively utilized in reducing NO emissions from the pre-chamber engine. When EGR is introduced into the pre-chamber, a substantial drop in NO emission results without any significant increase in HC emissions.

5. At the low NO emission levels of <1.5 GM/IHP-HR the pre-chamber engine with EGR has lower fuel consumption than the conventional engine also with EGR. At the higher NO emission levels, the conventional engine has lower fuel consumption.

ACKNOWLEDGMENT

The author thanks associates William E. Redinger, James H. Rush and John S. Siddall for obtaining the data and to Ethyl Corporation Research and Development Department, especially Randall H. Field, for their efforts in carrying out this test program.

REFERENCES

1. A. Simko, M. A. Choma and L. L. Repko, "Exhaust Emission Control by the Ford Programmed Combustion Process — PROCO." SAE 720052

2. T. A. Huls and H. A. Nickol, "Influence of Engine Variables on Exhaust Oxides of Nitrogen Concentrations from a Multicylinder Engine." SAE 670482

3. R. S. Spindt, "Air-Fuel Ratio from Exhaust Gas Analysis." SAE 650507

4. W. A. Daniel and J. T. Wentworth, "Exhaust Gas Hydrocarbons — Genesis and Exodus." SAE 486B

5. B. A. D'Alleva, "Procedure and Charts for Estimating Exhaust Gas Quantities and Compositions." GMP-372 Research Laboratories Publication, General Motors.

6. Ather A. Quader, "Effects of Spark Location and Combustion Duration on Nitric Oxide and Hydrocarbon Emissions." SAE 730153

7. James J. Gumbelton, Robert A. Bolton and H. Walter Lang, "Optimizing Engine Parameters with Exhaust Gas Recirculation." SAE 740104

8. G. M. Rassweler and L. Withrow, "Motion Pictures of Engine Flames Correlated with Pressure Cards." SAE Trans., Vol. 42, No. 5, 1936 Pg. 185.

APPENDIX A

ENGINE SPECIFICATIONS

CONVENTIONAL ENGINE

Bore	4.00 in.
Stroke	4.00 in.
Compression Ratio	8.6:1
Displacement	50.26 cu. in.
Combustion Chamber Surface-to-volume ratio at TDC	5.58 in.$^{-1}$
Intake Valve Dia.	2.05 in.
Exhaust Valve Dia.	1.66 in.
Valve Lift	.43 in.
Camshaft Timing	256° Duration
	17° BTC I.O.
	21° ATC E.C.
	38° Overlap

PRE-CHAMBER ENGINE

Same as conventional except:

Total Combustion Chamber Surface-to-volume ratio	6.73 in.$^{-1}$
Pre-Chamber Cup Volume	.598 cu. in. (8.9%)
Pre-Chamber Orifice Dia.	.472 in. (12mm)
Pre-Chamber Cup Wall Thickness	.08 in.
Pre-Chamber Cup I.D.	.70 in.
Exhaust Valve Dia.	1.63 in.
Main Intake Valve Dia.	1.93 in.
Pre-Chamber Intake Valve Dia.	.57 in.

APPENDIX B

CALCULATION OF SPECIFIC MASS EXHAUST EMISSIONS

Indicated data is used to allow comparison of engine combustion chamber performance between engines with different mechanical efficiencies. Since the single cylinder pre-chamber engine has higher friction than ⅛ of that of an 8-cylinder engine, the performance only on indicated basis is meaningful.

Expressing emissions on a specific basis allows for comparison of emission levels among different engines.

The specific emissions are calculated using exhaust concentration (HC, CO or NO) times its molecular weight times the exhaust flow rate and divided by the horsepower.

$$\text{HC, GM/IHP-Hr} = \frac{(1.30343 \times 10^{-3}) \text{ (HC, as a hexane, PPM) (Air Flow \#/Hr) (Moles Exh./Mole Air)}^*}{\text{IHP}}$$

$$\text{CO, GM/IHP-Hr} = \frac{(4.3860) \text{ (CO\%) (Air Flow \#/Hr) (Moles Exh./Mole Air)}^*}{\text{IHP}}$$

$$\text{NO, GM/IHP-Hr} = \frac{(7.2036 \times 10^{-4}) \text{ (NO, PPM) (Air Flow \#/Hr) (Moles Exh./Mole Air)}^*}{\text{IHP}}$$

* From Ref. (5)

APPENDIX C

EGR RATE CALCULATIONS

The formula used for calculation of "Percent EGR" is:

$$\% \text{ EGR} = \frac{CO_2 \text{ intake} - CO_2 \text{ background}}{CO_2 \text{ exhaust} - CO_2 \text{ intake}} \tag{1}$$

Where:

CO_2 intake = Mole percent of CO_2 in the intake manifold with recirculated exhaust.

CO_2 background = Mole percent of CO_2 in the intake manifold **without** recirculated exhaust. This includes the effects of ambient CO_2 and CO_2 from valve overlap effects.

CO_2 exhaust = Mole percent of CO_2 in the exhaust gas to be recirculated.

For applications where secondary air is introduced before the exhaust to be recirculated is picked up, the "Percent EGR" should be corrected by the formula.

$$\% \text{ EGR} = \frac{R(1 - O_2/21)}{1 + R(O_2/21)}$$

Where:

R = Percent EGR calculated by formula (1)

O_2 = Mole percent of oxygen in the recirculated exhaust gas

This definition gives the relationship between EGR flow and fresh fuel/air mixture. It is important to note that EGR rates higher than 100% are possible with this definition. This condition exists whenever the EGR mass flow rate exceed the mass flow of fresh fuel and air.

APPENDIX D

DESCRIPTION OF TEST INSTRUMENTATION

COMBUSTION MEASUREMENTS

The Tektronix Digital Processing Oscilloscope System used for compiling the combustion data consisted of the following components shown.

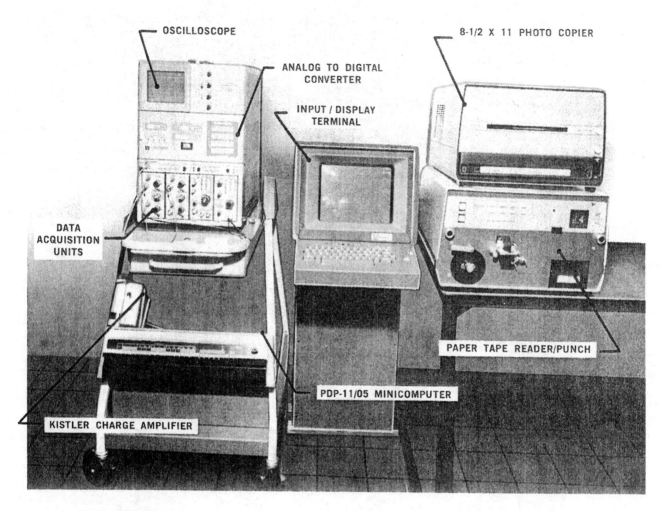

770YA Oscilloscope
P7001 Processor
119-0465-00 Data Acquisition Processor
4010-10 Terminal
4610 Hard Copier

4911 Tape Perforator and Reader
021-0117-00 Interface
7818 Amplifier
7B53A Time Base

The D.P.O. System has the capability of inputting, digitizing and sorting in the computer four waveforms simultaneously. The A-D converter has a digitizing rate of ~ 6 s/data point and can store consecutive cycles from an engine below 2000 rpm.

The computer can be user programmed in a variety of ways. The language is DEK BASIC modified to include commands for data input from the oscilloscope and also special routines for waveform analysis.

The turn around time for data analysis is about 10 minutes for the combustion calculations. This allows for on-the-spot interpretation and validation of data while engine is running.

The cylinder pressure measurement was made with a Kistler Model 618-A5 pressure transducer installed in the main combustion chamber.

SAE 750414

THE EFFECTS OF EXHAUST GAS RECIRCULATION AND RESIDUAL GAS ON ENGINE EMISSIONS AND FUEL ECONOMY

by

Yasuo Kaneko, Hiroyuki Kobayashi and Reijiro Komagome
Mitsubishi Motors Corporation

ABSTRACT

Three exhaust emissions, hydrocarbons, carbon monoxide and oxides of nitrogen, from the automotive spark-ignition engines are presently subject to regulatory control. Of these harmful pollutants, NOx emissions are the hardest to control under current status of emission control technology. Accordingly, exhaust gas recirculation (EGR) has been receiving continued efforts as one of promising NOx control.

This paper reports the effects of EGR on the mechanism of NOx reduction and engine fuel economy, on the basis of research made in the following areas:

(1). NOx formation in a combustion vessel.

(2). Studies on EGR effects in a single-cylinder engine.

(3). Effects of EGR on NOx and HC emissions and fuel economy.

COPYRIGHT 1975
SOCIETY OF AUTOMOTIVE
ENGINEERS, INC.

INTRODUCTION

The major task of automotive emission control is basically to seek the best compromise between exhaust emission control and the two principal requirements of the automotive engine; higher thermal efficiency and power output per CID.

Generally, NOx reduction for the automotive engine is contradictory with fuel economy improvement, but EGR is an effective means of controlling NOx generated in the combustion process without significant adverse effect on fuel economy.

Although the NOx reduction attainable by EGR alone may not satisfy the regulatory requirement for automotive exhaust NOx emissions, EGR can still be useful as an auxiliary NOx control measure.

In this paper, NOx formation and its reduction potential by EGR are reported from the following three preliminary studies:

(1). NOx formation in a combustion vessel. Formation of NOx in the combustion process, at varied ratios of oxygen and nitrogen, was observed using a closed-type combustion vessel. From this laboratory study, major factors affecting NOx formation and those of EGR efficiency have been clarified.

(2). EGR efficiency in a single-cylinder engine. The NOx reduction efficiency of an EGR system, in which a portion of the exhaust gas is recycled into the intake manifold, was studied using a single-cylinder engine. This experiment has revealed that the NOx reduction efficiency of EGR in the induction system is substantially the same as that of residual gas in the cylinder or that of exhaust gas blown back into the intake manifold during the intake- and exhaust-valve overlaps.

(3). Effects of EGR on HC and NOx emissions and fuel economy. In this experimental study, the interrelationships of EGR flow rate, ignition timing and air fuel ratio with consequent effects on HC, CO and NOx emissions and fuel economy are clarified.

EXPERIMENTAL APPARATUS AND METHODS

Experiments to determine NOx concentration in the combustion were conducted using a closed-type combustion vessel and a single-cylinder engine.

EXPERIMENT BY A COMBUSTION VESSEL - A spherical-combustion vessel of 524 cc displacement (100 mm dia.) was used in this experiment, and the mixture in the vessel was ignited at the spherical center.

The sectional view of the combustion vessel is shown in Figure 1. In such a vessel, the combustion takes place by flame-propagation in a non-swirling mixture, and it is different from the swirl combustion that generally takes place in the combustion chamber of internal combustion engines.

However, the experiment of NOx formation in the combustion vessel is desirable when analyzing factors affecting NOx formation such as mixture strength and combustion gas temperature.

Two ion-probes were provided in the vessel as shown in Figure 1 to confirm the spherical flame-propagation in progress. In those cases where flame-propagation time differed between the two probes, the applicable test data were discarded.

Methane or propane was used as a test fuel, and a fuel mixture, blended at different O_2/N_2 ratio, was burned in the combustion vessel to determine NOx formation. The experimental apparatus, shown in Figure 2, consists of the combustion vessel, associated sub-systems for mixture preparation, blended gas charging, and dilution gas preparation for NOx analysis.

The blended gas composition was determined from the respective partial pressures of the ingredient gases and the blended gas charged into the combustion vessel at the atmospheric pressure and temperature. The gas was ignited under stabilized condition.

In order to attain higher measurement accuracy of NOx concentration in the exhaust gas, it was necessary to solve the problem of NOx absorption into condensed water in the sampling apparatus.

Accordingly, the exhaust gas was introduced into a dilution-gas tank under vacuum conditions where the condensed water was to be vaporized. The exhaust gas was diluted with algon gas to bring it to atmospheric pressure. Then the diluted gas was collected into a sampling bag and analyzed for NOx concentration by a Chemi-luminescence analyzer.

The NOx concentration generated in the combustion vessel was computed from the product of the NOx concentration of diluted gas and the dilution factor determined from the partial pressure reading of the manometer C.

APPARATUS AND EXPERIMENTAL METHODS USED IN THE SINGLE-CYLINDER ENGINE TEST - Major specification of the single-cylinder engine used for this test and the operating conditions are shown in Table 1. The operating conditions were selected to be representative of typical driving conditions, where NOx emissions would be expected.

Table I

Engine Specifications and Operating Conditions

Type of Engine	Water-cooled, verical, 4-cycle gasoline engine
Valve mechanism	OHC
Type of combustion chamber	Semi-spherical
Bore x Stroke	74 x 75 mm
Displacement	322.6 cc
Compression ratio	8.1
Operating condition	2000 rpm, 60% in charging efficiency

Figure 3 shows the single-cylinder engine and related test apparatus.

The air/fuel ratio of the mixture was determined from the fuel supply, adjusted by using a variable jet in the carburetor [2], and intake air flow measured with the laminar-flow meter [4].

To determine the contained inert gas, the mixture under compression stroke was sampled through the electro-magnetic sampling valve [11]. The CO_2 concentration in the mixture [hereinafter referred to $(CO_2)e$] was analyzed with an improved NDIR [13], which is capable of CO_2 analysis with a small amount of sample gas circulated repeatedly.

The recirculated gas from the exhaust pipe was dehydrated by the heat exhanger [7], and then introduced into the intake manifold via a flow-control valve [9]. The EGR rate was measured with the circular nozzle.

RESULTS OF EXPERIMENTS AND DISCUSSION

NOx FORMATION IN THE COMBUSTION VESSEL - The results of NOx formation over a wide range of O_2/N_2 ratios has clarified major factors leading to NOx formation and also suggested promising areas in NOx control.

Figures 4 and 5 show NOx concentration curves obtained at varied rates of fuel, oxygen, and nitrogen when using propane and methane as fuel. Excess oxygen coefficient λ is determined from the formula below.

$$\lambda = \frac{\text{oxygen introduced in the vessel}}{\text{oxygen required to completely burn the fuel introduced}}$$

For example, the point A in Figure 4 shows the NOx concentration, which will be formulated when the ratio of propane, oxygen and nitrogen is 4:40:60. In the case of 4% propane, "$\lambda = 1$" corresponds to 20% oxygen. Therefore, it is assumed that approximately 20% out of 40% oxygen is used for the propane fuel combustion and the remaining 20% oxygen in the post flame is considered to be the effective O_2 concentration for NOx formation.

As can be seen in this Figure, NOx concentration increases as the oxygen ratio from point A becomes higher and it reaches the peak concentration at the ratio of 60% oxygen and 40% nitrogen. But NOx concentration decreases with further increases in oxygen ratio.

It is also mentioned that the peak NOx concentration, when keeping the ratio of propane or methane constant, takes place where the product of effective oxygen concentration in the post flame and nitrogen concentration becomes the maximum value.

The line B-B drawn in Figure 4 connects the point of 40% O_2 and 60% N_2 and other points where the product of effective O_2 and N_2 are maximum.

Figure 5 indicates that a similar trend in NOx concentration, as in the case of propane fuel, can also be observed with methane fuel.

NOx formation area, as shown in Figures 4 and 5, are provided under the following conditions:

(1). In an exceedingly over-rich mixture, the flame cannot be propagated, and accordingly no NOx is formed.

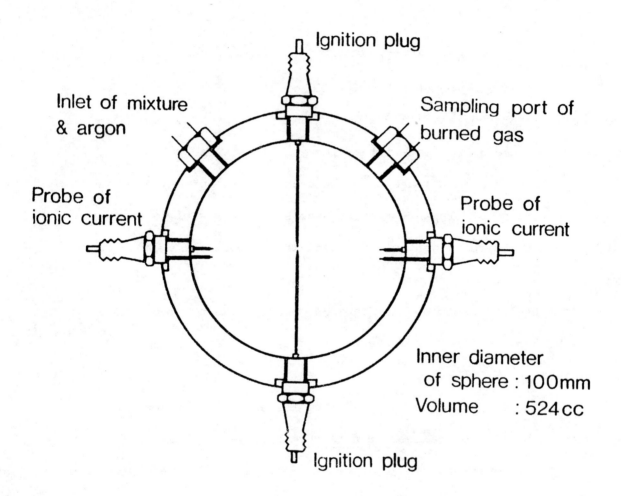

FIG. 1

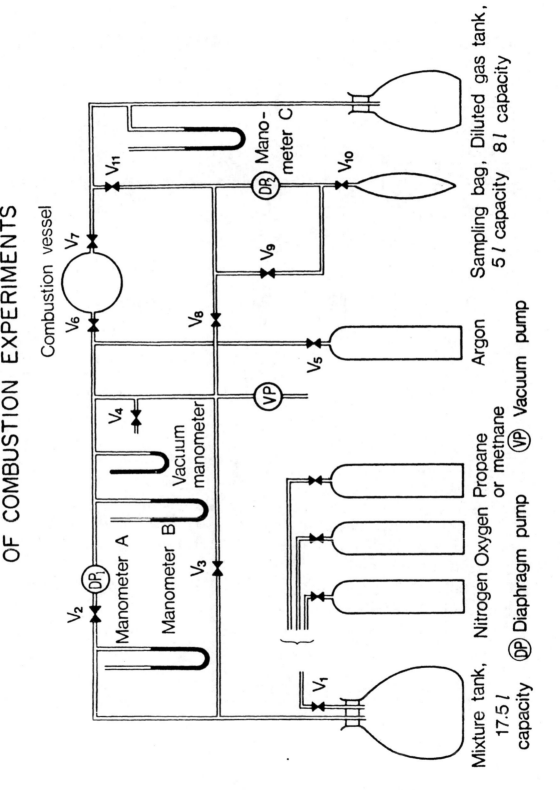

FIG. 2

SCHEMATIC DIAGRAM OF EXPERIMENTAL APPARATUS

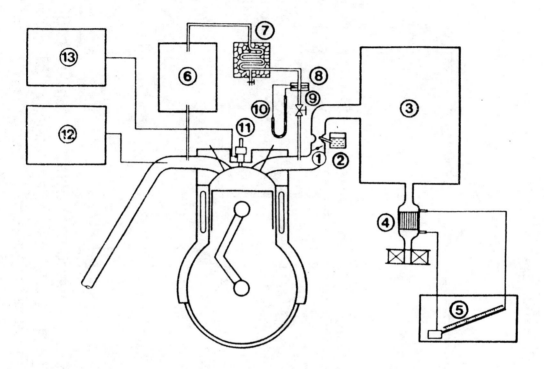

① Throttle valve
② Carbureter
③ Surge tank
④ Laminar flow meter
⑤ Inclined-tube manometer
⑥ Surge tank
⑦ Heat exchanger
⑧ Circular orifice
⑨ Flow control valve
⑩ U-tube
⑪ Gas-sampling valve
⑫ Non-dispersive type infrared gas analyzer (NDIR)
⑬ Improved NDIR

FIG. 3

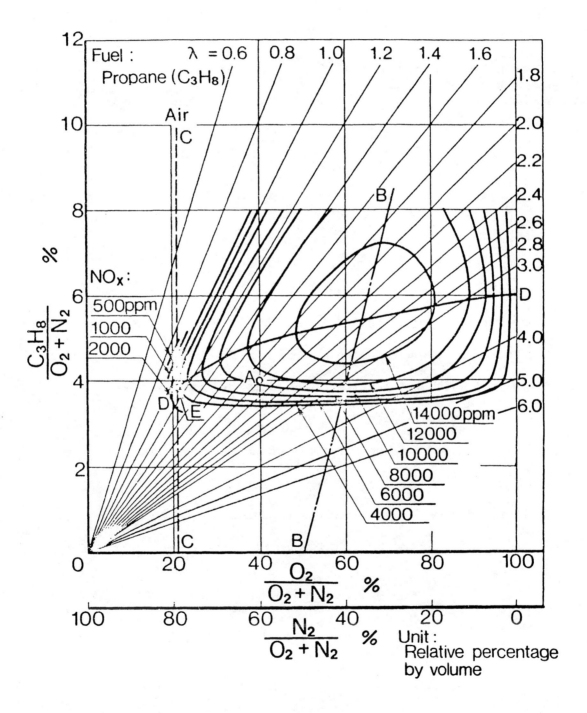

FIG. 4

The value λ of an over-rich mixture at the combustible limit differs by fuels and, in the case of propane fuel burned in a spherical combustion vessel, it was found to be approximately 0.7.

(2). Similarly, in an exceedingly over-lean mixture the flame cannot be propagated either.

The combustible limit of over-lean mixture depends on a certain amount of fuel concentration and is not determined by a specific λ value. Even within the combustible limit, therefore, the value λ becomes larger as O_2 concentration increases and, at nearly 100% O_2 concentration, λ equals 5 or 6 respectively with methane or propane.

(3). When N_2 concentration is zero, obviously no NOx is generated.

As indicated by both figures, in the case where oxygen concentration on the abscissa is constant, NOx concentration increases as the fuel becomes richer and reaches the peak concentration at a certain value of λ. After going through the peak NOx concentration decreases with the richer fuel, although the maximum combustion gas temperature is further raised at such a mixture strength range.

The D-D curve in both figures connects those peak points plotted for each oxygen concentration. Therefore, the cross-point of D-D curve and B-B line indicates the maximum NOx concentration. At such point, O_2 concentration and λ become 65% and 2.3 respectively, regardless of any fuel used. The C-C line shows the case of 21% O_2 and 79% N_2, which is close to air quality.

Figure 6 indicates the relation between O_2 concentration and λ over D-D curves in Figure 4 and Figure 5. There is little difference by methane or propane fuel. This suggests that the λ, where NOx concentration becomes maximum, can be constant for each O_2 concentration irrespective of the fuel used. The λ at E point on C-C line is approximately 1.1 and it nearly corresponds to the air/fuel ratio 16:1 for gasoline fuel. It is interesting to note that this λ value tends to become smaller when decreasing O_2 concentration from E along D-D line, namely keeping NOx concentrations maximum.

Figure 7 may serve to clarify this tendency and there are drawn NOx concentration curves and D-D curve by choosing $(O_2)3$ concentrations on the abscissa and the mixture strength on the ordinate. Let us consider the case where NOx concentration is reduced from 3,500 ppm at point E to 1,700 ppm at point G on D-D line, by decreasing both $(O_2)e$ and fuel concentration. The NOx reduction effects can be considered from the following two aspects: One is the NOx reduction between points E and H, where $(O_2)c$ is constant and the fuel concentration is slightly lower at point H than point E, which in turn results in reduced NOx formation due to lower combustion gas temperature. Another is the NOx reduction between points H and G, where fuel concentration is the constant and $(O_2)e$ is reduced from H to G. The combustion gas temperature remains nearly constant and NOx is reduced by lower $(O_2)e$ concentration.

In general, NO formation in the spark-ignition engine can be reduced when the mixture combustion is progressed in the presence of inert gas introduced into the combustion chamber. The NOx reduction may be affected by both of the two factors; reduced combustion gas temperature and lower $(O_2)e$ concentration. In the case where NOx concentration shifts from point E to point I keeping λ constant, this NOx reduction is mainly due to lower combustion gas temperature and is little affected by reduced $(O_2)e$ concentration.

In Figure 8, the relationship between λ and NOx concentration is indicated using some CO_2 gas concentrations added to the mixture, where CO_2 gas is representative of EGR effect. The relation of λ and NOx concentration, at 0% CO_2 in the figure, corresponds to that along the C-C line in Figure 4. NOx concentrations are lower with increasing CO_2 concentration and the λ, for maximum NOx concentration under constant CO_2%, becomes proportionately smaller. This tendency is similar to the trend of the ratio λ decreasing with smaller $(O_2)e$ concentration along D-D as shown in Figure 7.

Such a tendency is also apparent in the spark-ignition engine operated by gasoline with EGR. In that case, the NOx concentration is the highest at a air/fuel ratio of approximately 16:1 (or $\lambda = 1.09$). As EGR rate increases, the air/fuel ratio at which NOx becomes the peak concentration tends to shift to richer side, as in Figure 11.

In either case, NOx formation is reduced by EGR and this NOx control is achieved primarily by lower combustion gas temperature and leaner $(O_2)e$ concentration, as a result of EGR.

Therefore, NOx concentration can be controlled by applying either of the two effects aforementioned, predominantly depending on the mixture composition.

STUDIES FOR EGR EFFECTS USING A SINGLE-CYLINDER ENGINE - EGR for NOx control can be further classified into two different types according to the mechanism.

External EGR . . . . . Effect by recirculation of a portion of the exhaust gas.

Internal EGR . . . . . Effect by residual gas in the combustion chamber and by exhaust gas flown back into the combustion chamber during the valve overlap interval.

In order to analyze NOx reduction results by the respective EGR effects, the total EGR (E_T) and external EGR (E_o) are defined as follows:

$$E_o = \frac{\text{External EGR weight}}{\text{Dry intake air weight}} \qquad (1)$$

$$E_T = \frac{\text{Gas weight by external and internal EGR}}{\text{Dry intake air weight}} \qquad (2)$$

Where the recycle gas by external EGR in this experiment can be regarded as dry combustion gas, because it was dehydrated as mentioned before, while the internal EGR deals with wet combustion gas having the same composition as the exhaust gas. The symbols are defined as follows:

C_d - CO_2 concentration in dry combustion gas

C_w - CO_2 concentration in wet combustion gas

C_m - H_2O concentration in wet combustion gas

m_f - Molecules in fuel

m_r - Molecules in internal EGR gas

m_e - Molecules in external EGR gas

ψ - $\dfrac{m_e}{m_r}$

M_d - Average molecules in dry combustion gas

M_w - Average molecules in wet combustion gas

ϕ - Air/fuel ratio

$(CO_2)e$, defined as CO_2 concentration under compression stroke, can be expressed as;

$$(CO_2)e = \frac{m_r(C_w + C_d\psi)}{(1 + m_f) + m_f[(1 - C_m) + \psi]} \quad (3)$$

from (1),

$$E_o = \frac{m_e M_d}{28.83} \quad (4)$$

from (3) and (4), we have;

$$\psi = \frac{\dfrac{28.83}{M_d} E_o \left[\dfrac{C_w}{(CO_2)e} - (1 - C_m) \right]}{\left(1 + \dfrac{0.307}{\phi}\right) + \dfrac{28.83}{M_d} E_o \left[1 - \dfrac{C_d}{(CO_2)e}\right]} \quad (5)$$

can be expressed as function of ψ as follows;

$$E_T = \frac{\psi + \dfrac{M_w}{M_d}}{\psi} E_o \quad (6)$$

$$E_T = \frac{M_w\left(1 + \frac{0.307}{\phi}\right)}{28.85\left[\dfrac{C_w}{(CO_2)e} - (1 - C_m)\right]} \qquad (7)$$

C_d, C_m, C_w, M_d and M_w can be obtained by analyzing the exhaust gas composition or can be calculated, assuming possible thermodynamical equilibriums between combustion gas composition. If $(CO_2)e$ and E_o are attained, E_T can be calculated through (5), (6) and (7).

Figure 9 shows the relation between $(CO_2)e$ and E_T, for the case of $\phi = 14$, using E_o as a parameter. The curves shown in the figure are drawn from calculations which assumed thermodynamical equilibriums between combustion gas compositions. When keeping $(CO_2)e$ constant, E_T varys a little depending on different values of E_o. Therefore, when considering NO reduction, $(CO_2)e$ cannot be related directly with the reduction while E_T may be immediately connected. Namely, E_T should be derived from ϕ, $(CO_2)e$ and E_o first and then the NO reduction can be examined corresponding to E_T obtained.

Figure 10 shows the relation between air/fuel ratio ϕ and $(CO_2)e$ using different percentages of external EGR and a specific valve timing as a parameter. Furthermore, the value for three other valve timings, without external EGR, are added to show the effect of internal EGR. The maximum $(CO_2)e$ point for each operating condition is always situated at the approximate stoichiometric air/fuel ratio.

Figure 11 shows the relation between air/fuel ratio and specific NO emission and the data represented here correspond to those in Figure 10. It is revealed from the figure that the maximum point of specific NO emission shifts to richer air/fuel ratios with increased EGR rate. This phenomenon is considered to be similar to that mentioned concerning NO formation in the combustion vessel.

Figure 12 shows the relation between E_T and specific NO emission where E_T is calculated from $(CO_2)e$, in Figure 10, and air/fuel ratio. Different air/fuel ratios are adopted as parameters and the respective data for each air/fuel ratio makes a smooth curve. In other words, it can be said that in both the case of E_O introduction by small valve timing overlap and the case where E_T is increased up to 10 to 12% by varying valve overlap or valve timing completely identical effects are experienced.

RELATIONSHIPS AMONG NOx, HC, FUEL ECONOMY, AND EGR

Figures 13, 14 and 15 show how fuel economy and NOx, HC and CO emissions vary and their dependence on air/fuel ratios, retarded spark timings, EGR, or the absence of EGR. As far as Figure 13 is concerned, the following comments can be made:

(1). The use of EGR is effective in control of NOx emission with less penalty on fuel economy.

(2). The EGR used in combination with the leanest possible mixture and the ignition timing set at MBT brings satisfactory NOx control.

The following can be also said from Figure 16:

(1). The external EGR increases HC emission, which can be reduced by leaner mixture and delayed ignition timing.

(2). CO emission is scarcely affected by EGR or spark retard and depends mostly on air/fuel ratio. Accordingly, if CO emissions at a given air/fuel ratio are assumed to be constant, increased EGR rate will bring better fuel economy.

The foregoing findings suggest a preferable direction in applying engine modifications for emission control. In this paper, the discussion is particularly concentrated on NOx reduction which is difficult to achieve by after-treatment devices in the exhaust system. As an example, the determination of the optimum EGR rate, under the following conditions, is demonstrated below.

(1). NOx reduction target . . . . . . 50 % of current engine

(2). HC increase . . . . 20% at maximum

(3). CO level . . . . current level

(4). Fuel economy penalty . . . . . . 10% or less

In Figure 16, which is cross plotted from Figures 13 through 15, those ranges meeting the above requirements are illustrated by hatched area, which may be summarized as follows:

An optimum EGR rate of about 10%, the leanest possible air/fuel ratio, and an ignition timing of 10° delayed from MBT are desirable to satisfy the above requirements. However, such conditions are subject to those transient characteristics which govern vehicle driveability.

CONCLUSIONS

The experiments and studies conducted at our laboratory using combustion vessel and single-cylinder engine have revealed the following facts:

(1). The use of EGR is an effective means of controlling NOx with less penalty on fuel economy.

(2). Although NOx reduction by EGR alone may not satisfy regulatory requirements, EGR used in combination with leanest possible mixture and MBT ignition timing may provide satisfactory NOx control.

(3). External EGR increases HC emissions which can be reduced with lean mixtures and delayed ignition timing.

(4). EGR can be used to provide better fuel economy at any given A/F ratio without influencing CO emissions.

(5). In the case of constant-volume combustion, where the rates of fuel O_2 and N_2 concentrations are varied in a wide range, O_2 and N_2 concentrations in the combustion gas along with mixture strength are the major factors which determine NOx formation.

(6). NO reduction effect by EGR is caused partly by combustion gas temperature drop and partly by O_2 concentration decrease in the post flame.

(7). When reducing O_2 concentration and proportionately increasing N_2 concentration, the excess oxygen coefficient, at which the maximum NOx concentration is attained, becomes smaller.

(8). The total EGR flow, inclusive of both external and internal EGR, can be determined by calculating from $(CO_2)e$; namely, CO_2 concentration under engine compression stroke, air/fuel ratio, and external EGR ratio.

(9). In the case of a constant total EGR flow, NOx reduction is identical irrespective of the ratio of internal and external EGR involved in the total EGR.

ACKNOWLEDGEMENT

The authors wish to acknowledge the helpful advice provided by Professor Kumagai, Mechanical Engineering Department, Tokyo University; Assistant Professor Sakai of the same department and also by IIEC members.

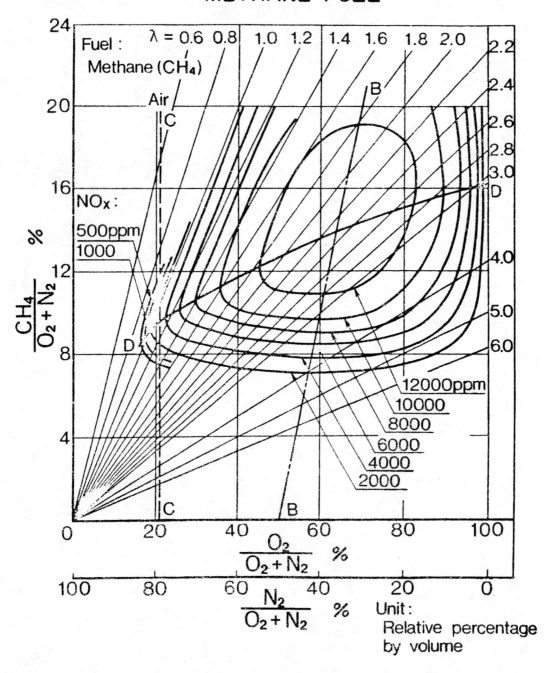

FIG. 5

RELATIONSHIP BETWEEN OXYGEN CONCENTRATION AND EXCESS OXYGEN COEFFICIENT ON MAXIMUM NO_x CONCENTRATION

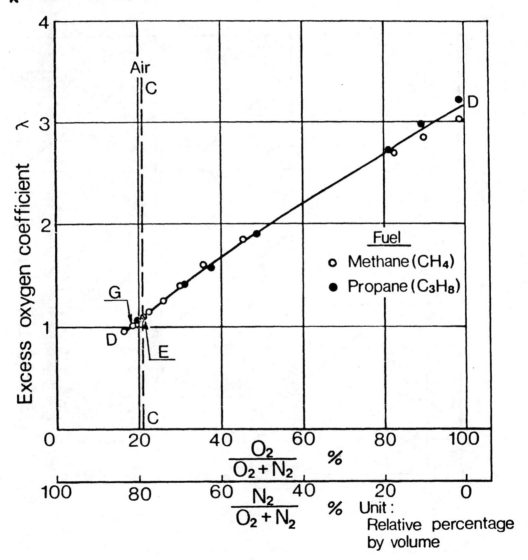

FIG. 6

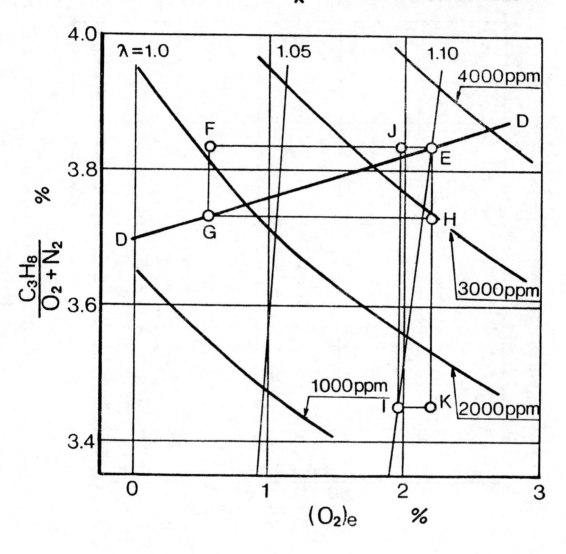

FIG. 7

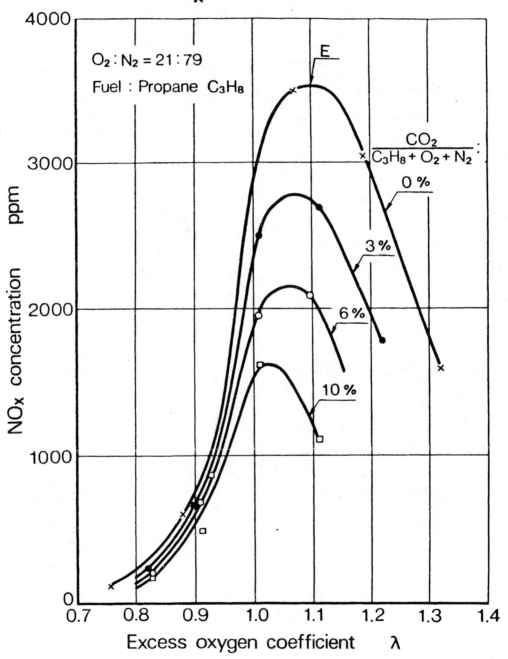

FIG. 8

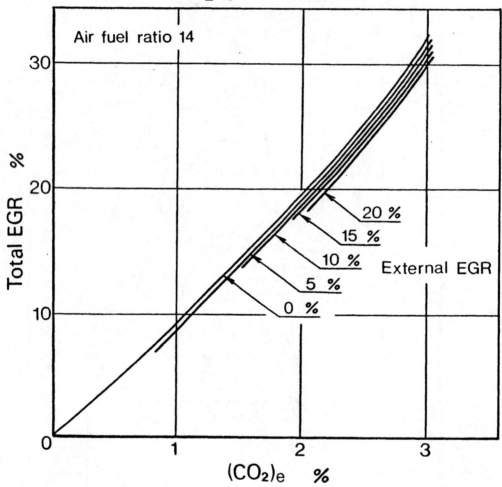

FIG. 9

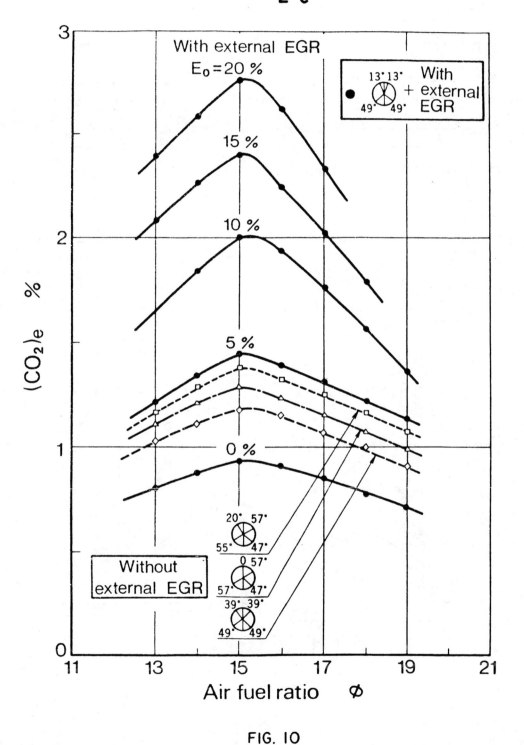

FIG. 10

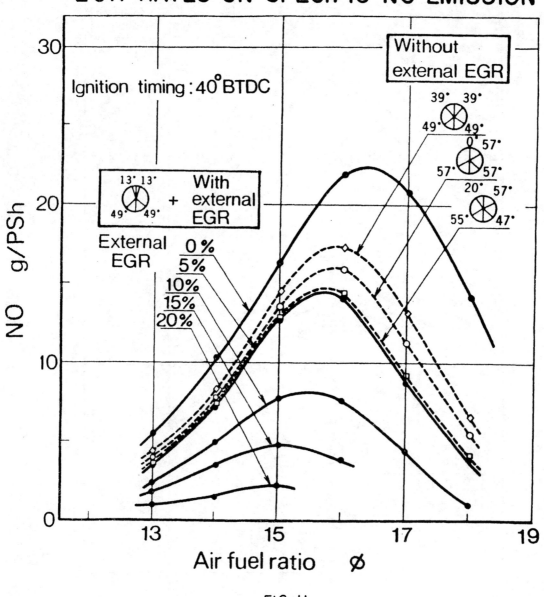

FIG. II

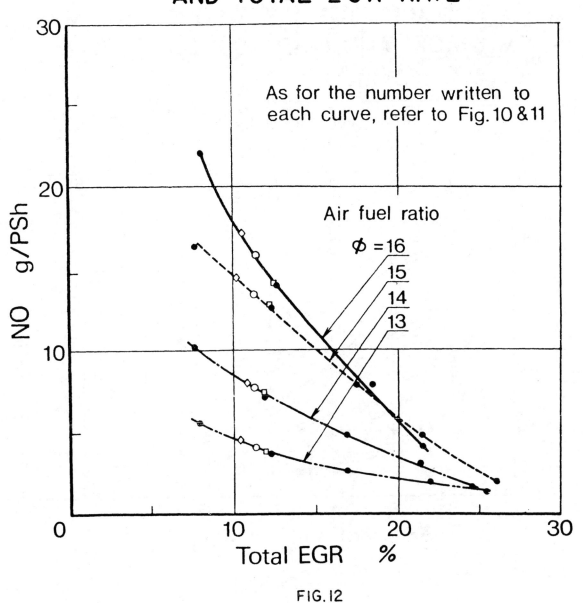

FIG. 12

EFFECTS OF EXTERNAL EGR RATE, AIR-FUEL RATIO, AND IGNITION RETARD ON NO MASS EMISSION AND FUEL CONSUMPTION

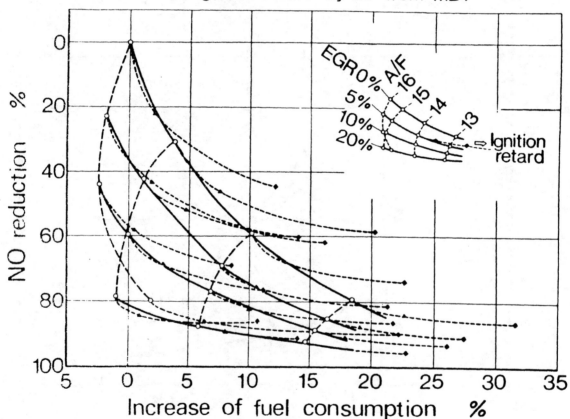

FIG.13

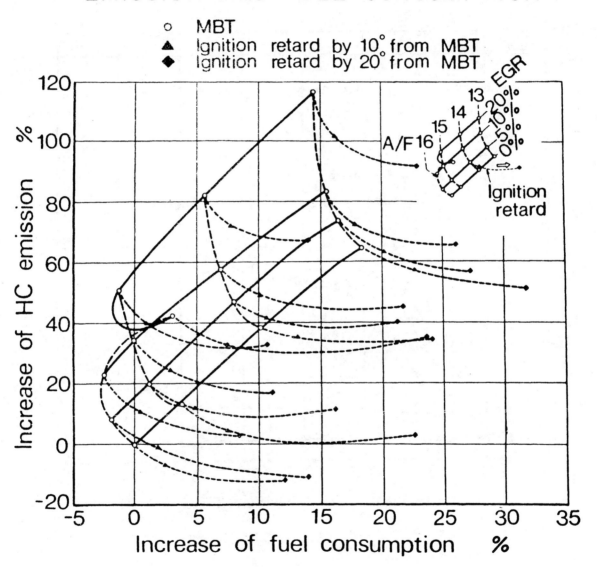

FIG. 14

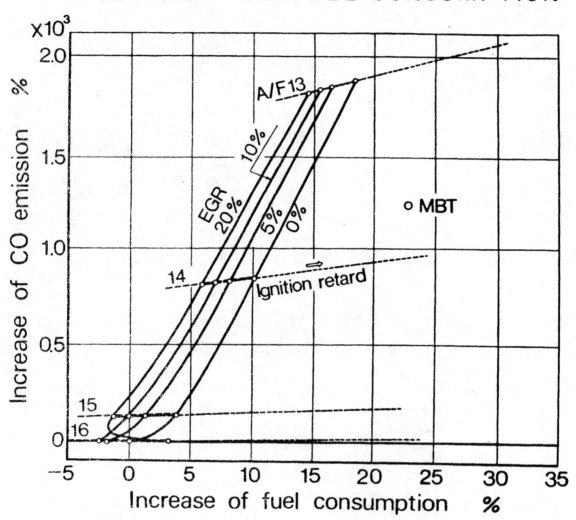

FIG. 15

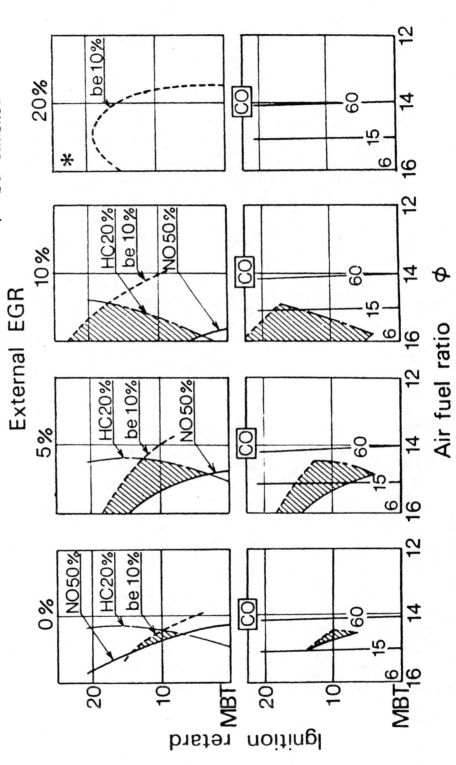

FIG. 16

ERRATA

SAE PAPER 750414

"THE EFFECTS OF EXHAUST GAS RECIRCULATION
AND RESIDUAL GAS ON ENGINE EMISSIONS AND
FUEL ECONOMY"

by

Y. Kaneko, H. Kobayashi, and R. Komagome
Mitsubishi Motors Corporation

Page 11:

The statement, "Then, $\psi = 0$ in case external EGR = 0, and E_T becomes as;" should be included following formula (6) and prior to formula (7) on the next page.

750178

Reliability Analysis of Catalytic Converter as an Automotive Emission Control System

K. Matsumoto, T. Matsumoto, and Y. Goto
Toyota Motor Co., Ltd.

THE CATALYTIC CONVERTER SYSTEM was introduced into the market in an effort to meet the stringent 1975 automotive emission standards. Never before has Toyota tried to introduce, on such a large scale, such a completely new concept as the catalytic converter without sufficient field experience. For this reason, it is very important to anticipate possible problems and failures which may occur in the field before its introduction, so that we can prepare countermeasures to minimize converter troubles.

The purpose of this study is to pick up any failure mode which may possibly affect the life of the catalytic converter and to determine its criticality in order to minimize converter troubles in the field.

Converter troubles are caused mostly by over-temperature in the catalyst bed. Factors which cause the converter over-temperature are the concentration of unburned gas, the flow rate of gas, and the gas temperature at the converter inlet. These factors, as parameters of operations of the engine and exhaust emissions control system, determine the converter temperature. Therefore, failure modes which may cause the converter over-temperature are chosen from those which affect these three factors.

In this paper, a fault analysis is performed on the catalytic converter which uses a precious metal pelletted catalyst.

Determination of the failure modes and their criticality which could cause melting and/or other thermal damage to a catalytic converter are discussed in this paper. Our approach consists of the following two steps:

1. An application of the failure mode effect analysis techniques to predict all possible factors which could contribute to melting and other thermal damage of a catalytic converter.
2. Confirmatory vehicle tests to determine the actual effects of component failure modes which were judged to have greater influence on converter problems.

The component failure modes and their effects are discussed further on.

OUTLINE OF THE CATALYTIC CONVERTER

The major specifications of the catalytic converter, which are discussed below, are shown in Table 1. A cross section of the converter is also shown in Fig. 1. The converter is mounted under the vehicle floor beneath the driver's seat. A schematic layout of the emission control system is shown in Fig. 2. A secondary air injection system which injects air into the engine exhaust ports has an air switching mechanism to prevent the converter from over-temperature. A warning circuit is incorporated which detects exhaust gas temperature in the converter bed. A warning light comes on when the bed temperature exceeds the predetermined 920°C. The canister

---

ABSTRACT

A reliability analysis was conducted by means of the failure mode effect analysis technique on the catalytic converter which has been adopted in the 1975 Toyota models. A series of vehicle tests was conducted to determine the failure modes criticality. The study was performed on a precious metal pelleted catalyst. Critical durability failure modes to the durability of the catalytic converter and their criticality were classified and substantiated through this study. The results are presented and discussed.

The results obtained by this analysis are useful for improving the reliability of the catalytic converter.

Copyright © Society of Automotive Engineers, Inc. 1975
All rights reserved.

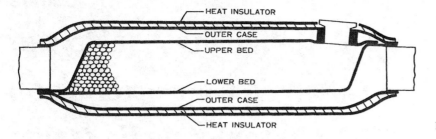

Fig. 1 - Cross section of catalytic converter

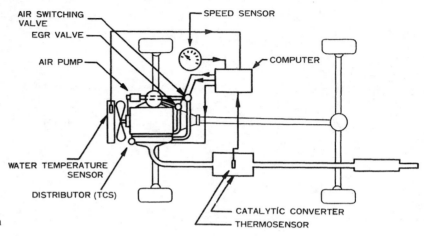

Fig. 2 - Schematic layout of emission control system

Table 1 - Major Specifications of the Converter

Item	Specification
Catalyst	Pt and Pd
Substrate	Alumina pellet
Converter	Down-flow
Volume	2 litre
Location	Under-floor

Table 2 - Major Specifications of the Vehicle

Vehicle: Toyota Celica Hard Top 1975 California model
Curb weight: 2482 lb
4-speed manual transmission
Engine: Type: 4-stroke gasoline engine
Cylinder: 4-cylinder in-line
Bore stroke: 88.5 × 89 mm
Displacement: 2.2 litre

is made of stainless steel so as to endure high temperature. The vehicle on which we have conducted the study was a prototype Toyota Celica for the 1975 California market. The major specifications of the car are shown in Table 2.

CATALYTIC CONVERTER DURABILITY - Durability of the catalytic converter depends on the temperature history (see Fig. 3). When all systems operate normally, temperature in a catalytic converter bed remains below 850°C. With this normal operation, it has been confirmed that our catalytic converter lasts over 50,000 miles. If some problems occur in the engine or the emissions control system, the temperature in a catalytic converter may rise above 850°C.

Durability problems under this over-temperature condition have been recognized as follows. Stainless steel, of which the converter consists, will melt at or over 1300°C. If the bed temperature exceeds 1000°C for an extended duration, the converter will be damaged due to the material strength deterioration which is caused by high temperature corrosion and fatigue, and due to thermal stress. Concerning the catalyst itself, if they are exposed to a temperature higher than 1100°C for an extended duration, the converter efficiency is decreased,

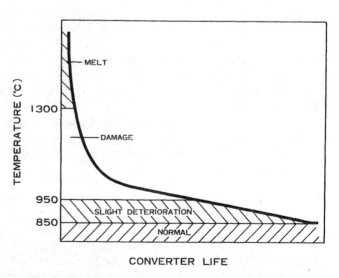

Fig. 3 - Durability of catalytic converter versus temperature

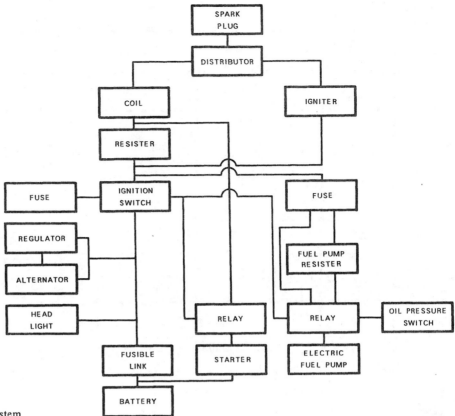

Fig. 4 - Block diagram of engine electrical system

catalysis tend to have a higher attrition rate due to decreased crushing strength of the substrates, and the substrates shrink.

Here, melting is defined as the condition where the bed plate and/or catalysts are melted away into the exhaust pipe and lose the function of catalytic action in a relatively short time. There are two conditions in converter melting. One is the condition where the upper bed plate of the converter melts away first due to the explosive combustion of unburned exhaust gases at the converter inlet when the converter is supplied with a high volume of unburned gas. In this case, hot catalysts work as an igniter. If this abnormal condition continues, the lower bed plate may also melt away. The other condition is when the lower bed plate melts away when an excessive oxidation reaction occurs in the bed as the exhaust gas passes through the converter. In the latter case, the volume of unburned gas is less than the former one. In both cases, the converter outer case with the heat insulator prevents the catalysts or broken bed debris from scattering over the ground.

The canister damage is defined as cracking and/or distortion of the bed plate. If the damage increases, the catalyst pellets move out of the bed and then gradually lose their function. Minor damages may not have an apparent adverse effect on conversion efficiency but may cause noise problems from the converter.

FAULT ANALYSIS REGARDING CONVERTER MELTING AND DAMAGE

To analyze faults which may cause converter problems of melting and damage, the failure mode effect analysis (FMEA) techniques (1-4)* were applied. Major engine troubles which require engine overhaul for repair were not taken into consideration in this study because such troubles are far more crucial than converter failure.

SYSTEM RELIABILITY BLOCK DIAGRAM - Logic block diagrams which show the functional interrelationship among subsystems and parts were prepared. Effects of component failure were traced and estimated for converter failure with this diagram. A diagram example is shown in Fig. 4 with regard to the engine electrical system. In addition to the example, similar diagrams were also prepared for other subsystems such as the fuel system, secondary air injection system, EGR system, exhaust system, and regulating system.

LIST OF FAILURE MODE - For each component, any possible failure mode and its effect on the converter system was listed for analysis. Since the purpose of this analysis involves melting and resultant damage to a catalytic converter, listing of the failure effects were limited to over-temperature of the catalytic converter. Failure modes of the components and their effects were listed by completing the FMEA format, an example of which is shown in Fig. 5. The listed failure modes exceeded 1000 items.

DETERMINATION OF CRITICAL FAILURE MODES - The criticality of each failure mode was classified by the following categories. Failure modes which were classified in category 1 were defined as critical failure modes.

1. Modes which cause excessive temperature in the catalytic

---

*Numbers in parentheses designate References at end of paper.

SUBSYSTEM	IGNITION					
DEVICE	DISTRIBUTOR		FAILURE MODE EFFECT ANALYSIS		DATE	OCT. '73
					BY	M. ANDO

ITEM IDENTIFICATION		FUNCTION	FAILURE MODE	FAILURE EFFECT ON		CATEGORY OF OVER-TEMPERATURE
NAME	I.D. NUMBER			SUBSYSTEM	ENGINE	
CONTACT BREAKER	M-4-1	SWITCHING THE SIGNAL CURRENT	FAIL TO CLOSE	ABILITY TO SPARK IS LOST	WHOLE CYLINDER MISFIRING	1
			FAIL TO OPEN	↑	↑	1
			INTERMITTENT OPERATION	ABILITY TO PROPER SPARK IS LOST	INTERMITTENT WHOLE CYLINDER MISFIRING	1
TERMINAL	M-4-2	TRANSMITTING THE SIGNAL CURRENT	DROP OFF	ABILITY TO SPARK IS LOST	WHOLE CYLINDER MISFIRING	1
			LOOSE	ABILITY TO PROPER SPARK IS LOST	INTERMITTENT WHOLE CYLINDER MISFIRING	1

Fig. 5 - Format for failure mode effect analysis; engine electrical failures

converter or which relate to the malfunction of secondary air switching.

2. Modes which need the air switching operation to maintain temperature below 850°C. (In normal conditions, the bed temperature is controlled by switching the secondary air supply.)

3. Modes which are considered to have no effect or an insignificant effect on the bed temperature.

There were approximately 300 items defined as critical failure modes. Fig. 6 shows the top level fault tree on critical failure modes at this stage of the analysis. This figure also represents the parts which have critical failure modes.

CRITICALITY ANALYSIS - Criticality is determined by the following formula for each failure mode which has been classified in category 1.

$$Cr = T \cdot D \quad (1)$$

where:

Cr = criticality for a catalytic converter
T = damage intensity
D = correction factor for prompt repair expectation

Cr shows the probability of converter trouble due to component failure. Generally speaking, it is important that the above formula includes the probability parameter of failure modes in order to determine the priority of corrective actions. However, it is more important to clarify the design inadequacies in order to complete corrective engineering action within the limited time frame. Thus, a probability estimation was excluded from this study.

Damage Intensity - In the case of melting, the damage intensity is defined simply with a specific temperature level. However in the case of converter damage, the damage intensity must be determined with both parameters of temperature level and thermal stress history. Therefore, the damage intensity was classified as shown in Table 3 considering temperature levels and stress duration. This classification was based upon experiences with our catalytic converters while they were tested for thermal cycle durability on the engine dynamometer and by actual vehicle durability test runs. Damage intensity cannot be determined on paper. Vehicle tests are necessary to determine the damage intensity of each failure mode. Even under the same kind of failure mode, damage intensity shown in Table 3 is variable depending on vehicle operating conditions. To correct this problem, the damage intensity of a specific failure mode was determined under the most severe thermal conditions.

Correction Factor for Repair Expectation - Even in the case of a failure mode with a high degree of damage intensity, early corrective action to the engine or emission control system may save the converter from melting and damage. Poor drivability and the operation of the warning lamp, due to system failure, can alert the driver to take necessary corrective action. Prompt and proper corrective action is expected and therefore the effect of thermal stress will be minimized if a component failure is accompanied by poor vehicle drivability which alerts the vehicle driver of a malfunction. If deteriorated drivability does not accompany a failure, it may take a long time until the failure is recognized and suitable repair is performed, and it will subject the converter to thermal stress for an extended duration. Some failure modes may affect the converter too quickly and too heavily to take any corrective action based on poor driveability or activation of the warning light. Taking these cases into consideration, the correction factor for repair expectation was classified as shown in Table 4. A correction factor was determined for each failure mode based on vehicle tests by which actual temperature characteristics and drivability were evaluated.

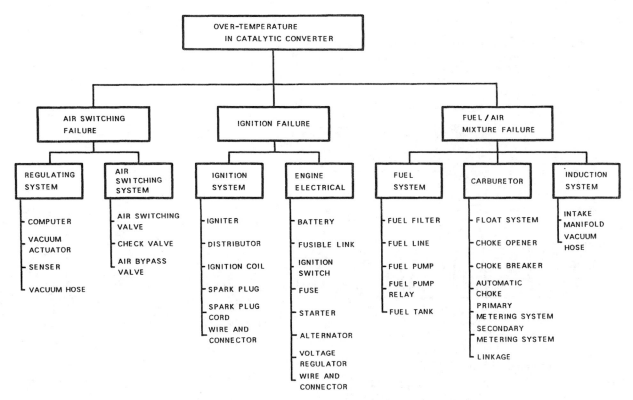

Fig. 6 - Simplified fault-tree for over-temperature of catalytic converter

Table 3 - Damage Intensity

	Bed Temperature (t) Classification	
Damage Intensity (T)	Extended Exposure, °C	Peak Allowable Temperature, °C
10: rapid melting	t ≥ 1400	
8: melting	1400°C > t ≥ 1300	
6: damage in a short time	1300°C > t ≥ 1100	t < 1350
4: damage	1100°C > t ≥ 900	t < 1300
1: slight functional deterioration	950°C > t ≥ 850	t < 1100
0: same as normal operation	t < 850	

Table 4 - Correction Factor for Repair Expectation

Correction factor (D)	Remarks	Explanation
10	No corrective action in time or no corrective action expected	Any corrective action is too late to prevent converter from destruction, or component failure does not accompany significant drivability deterioration
8	Detectable but usable	Drivability deterioration detectable but can be operated without repair
3	Detectable and prompt corrective action expected	Normal operation difficult
1	Vehicle does not run without repair	Impossible to start or to continue running

VEHICLE TESTS FOR CRITICALITY AND CORRECTION FACTOR DETERMINATION

The purpose of vehicle tests are:
1. To determine the damage intensity of the classified critical failure modes.
2. To determine the correction factor for failure modes through driveability evaluation.

TEST ITEMS - There are two test items to consider, the converter bed temperature characteristics and vehicle drivability.

Bed temperature characteristics for typical failure modes were determined. The temperature was recorded at four positions in the bed in order to determine the highest bed temperature (see A to D points in Fig. 7), since it is known that the temperature distribution in the bed varies widely with the concentration of unburned gases and the gas flow rate. In addition to these points, exhaust gas temperature at the inlet of the converter was checked (see E and F points in Fig. 7), since the temperature of the upper bed plate depends on the inlet gas temperature except when gas flow is extremely low.

Vehicle drivability for failure modes was evaluated to determine the correction factor.

VEHICLE DRIVING PATTERN - A typical vehicle driving pattern was established to represent various vehicle operations in the field. This pattern included high-speed driving, heavy acceleration and deceleration to simulate various operating modes in a short testing time (see Fig. 8). Constant speed

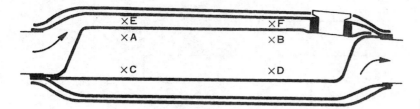

Fig. 7 - Temperature measuring points

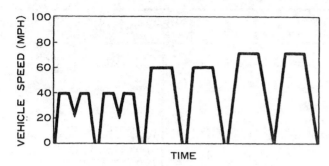

Fig. 8 - Driving pattern

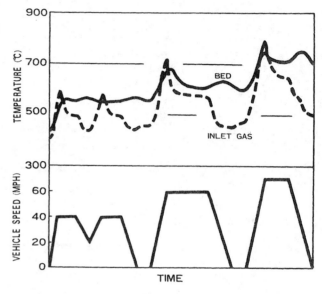

Fig. 9 - Converter temperature under normal operation

modes and long idle modes were added to accommodate some failure modes. Most of the tests were conducted on a chassis dynamometer to get good repeatability. It was confirmed that there was no significant difference in the catalyst bed temperature between vehicle operation on a test track and on the chassis dynamometer. In addition, a series of confirmation tests was conducted on city/suburban roads, highways, and in mountain areas.

SIMULATION PROCEDURES REPRESENTING COMPONENT FAILURE - Critical component failures were classified into several groups, each of which was judged to have the same effect on the subsystem. A series of tests was conducted on each group. For instance, although there are many component failure modes on the engine electrical system which relates to one cylinder misfire, all of these modes were classified as one cylinder misfire and the test was represented only by the disconnection of the spark plug cord of the number 1 cylinder.

Tests were conducted on several degrees of subsystem failure, when converter temperatures varied with the degree of the failure. In addition, the effect of secondary air switching on the converter temperature was also checked during these tests in combination with subsystem failures.

TEST RESULTS

Temperature characteristics in the converter and vehicle drivability during testing are discussed in this section.

NORMAL CONDITIONS - For reference, Fig. 9 shows the bed temperature characteristics when all system operations are normal. Bed temperature was in the range between 500-750°C.

ONE CYLINDER MISFIRING - One cylinder misfiring was represented by one spark plug cord disconnection. The bed temperature reached 900-1100°C during the pattern driving. The temperature exceeded 1000°C, even at lower vehicle speed—about 40 mph in this case—and increased by about 100

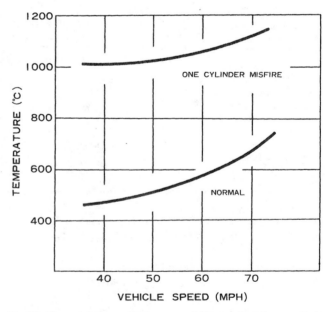

Fig. 10 - Converter temperature versus vehicle speed with one cylinder misfiring

deg at higher vehicle speeds (see Fig. 10). Therefore, the converter will be damaged with one cylinder out of four misfiring if the vehicle is used without any corrective action. The bed temperature did not change during steady speed operation, but dropped by some 100 deg during acceleration by switching the secondary air from the exhaust ports to the air cleaner.

The top speed that the car could reach with one cylinder

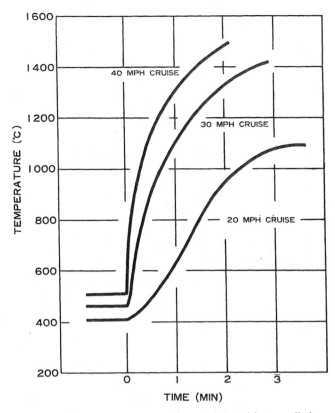

Fig. 11 - Converter temperature characteristics with two cylinder misfiring

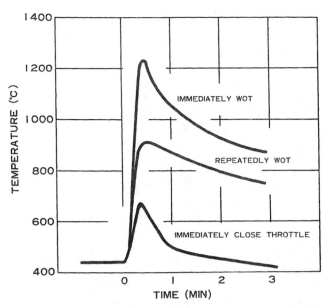

Fig. 12 - Converter temperature characteristics at 40 mph cruise with whole cylinder misfiring

misfiring was approximately 90 mph. But tests were conducted only up to 70 mph. Vehicle drivability was not severely deteriorated, even though slight surge and hesitation was observed at lower speeds. It was judged that many drivers would continue to drive the car without any corrective action, which would result in converter damage. The warning system is expected, in this case, to work effectively as a subsystem malfunction indicator before the catalytic converter has suffered serious damage.

TWO CYLINDER MISFIRING - The bed temperature exceeded the critical 1300°C at any speed during constant speed operation except below 30 mph as shown in Fig. 11. The temperature became even higher during pattern driving which included acceleration and deceleration. The lower bed plate melted in this case, but the gas temperature at the converter inlet decreased with two cylinders misfiring.

Vehicle drivability was extremely poor so that any driver could recognize the trouble. Engine idle was very rough and the engine was likely to stall. However, any corrective action would be too late because the temperature rises so sharply and extremely even at low vehicle speed unless the trouble is found and corrected during idle before vehicle operation. The warning system also works too late. Air switching does not improve the situation. This is one of the most fatal modes for the converter.

WHOLE CYLINDER MISFIRING - The failure mode effect is divided into two types in this case. One is the condition that whole cylinders misfire continuously so that the engine stalls and continued driving is impossible. The other is the condition that misfire occurs intermittently so that continued driving is possible in spite of whole cylinder misfiring. In the former case, the bed temperature level is closely related to the driver's behavior with the accelerating pedal when the engine misfires and also to the vehicle speed when misfire occurs.

Fig. 12 shows the bed temperature characteristics when whole cylinder misfire happens at 40 mph cruising. When a driver released the accelerating pedal to close the throttle valve immediately after misfire, the temperature increased sharply by as much as 200 deg and dropped sharply too as the engine stalled and the vehicle lost speed. No damage would occur in this case. When a driver pushed the accelerating pedal to the floor immediately after misfire and kept the throttle valve wide open or pushed the pedal to the floor repeatedly, the temperature reached 900-1200°C quickly but also dropped quickly. Therefore, no significant damage would happen unless this type of trouble happens repeatedly. Fig. 13 shows the bed temperature characteristics at 60 mph. In this case, the bed temperature reached just above 900°C if the driver released the pedal immediately after the misfire. The temperature exceeded 1400°C if the driver pushed the pedal to the floor repeatedly, which resulted in the converter melting. It should be noted that the gas temperature at the inlet of the converter exceeded 1300°C caused by explosive combustion in the exhaust pipe so that the upper bed of the converter was melted by this high temperature gas. Since the normal driver's response when whole cylinders misfire is believed to be that of pressing the accelerating pedal to the floor in order to recover decreased vehicle speed, there will be converter melting when whole cylinder misfire occurs at vehicle speeds higher than 40 mph. Drivability is the worst when whole cylinders misfire since the engine stalls and cannot be restarted. Therefore, any driver would recognize the trouble and call the repair shop. However, corrective action will be too late in most cases. This failure mode is judged as very critical. In the case

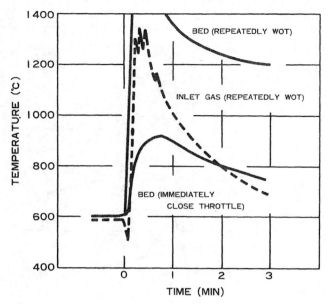

Fig. 13 - Converter temperature characteristics at 60 mph cruise with whole cylinder misfiring

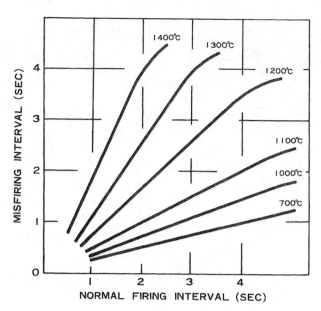

Fig. 14 - Converter temperature versus whole cylinder misfiring rate at 50 mph

of intermittent whole cylinder misfire under which the vehicle would be able to continue to run, the bed temperature increased roughly in proportion to the misfire rate and to the vehicle speed (see Figs. 14 and 15). The converter receives fatal damage in many cases with the higher misfire rate and higher vehicle speed since excessive temperature is maintained because the vehicle continues to run. The vehicle drivability is very poor during misfire, but it easily returns to normal when misfiring ceases. Therefore, there is a high probability of fatal converter damage when the vehicle is operated at a higher speed since the driver may continue to drive the car even though there is intermittent poor drivability.

CHOKE VALVE STICKING, EXCESSIVELY RICH MIXTURE - The bed temperature increased as the choke valve was stuck at a lower opening angle as shown in Fig. 16. If the choke valve was stuck at 20 deg or less opening angle, the bed temperature easily exceeded 1100°C. The temperature became even higher if the choke sticking was accompanied by an air switching malfunction which continued to inject air into the exhaust ports. Concerning vehicle drivability, the vehicle was not operative with the choke valve at a fully closed position since the engine always stalled just after starting. At the opening angle of 10 deg, the vehicle was difficult to operate even with a high degree of driving technique, and the engine stalled at idle and low speed operation due to spark plug fouling. At the opening angle of 20 deg, the vehicle had very poor drivability at lower speeds with hesitation and surging. However, the vehicle showed rather smooth operation at higher speeds. There was no deterioration in vehicle drivability with choke sticking at or over 30 deg of opening angle. Reviewing these test results, it can be expected that the vehicle will receive service to correct the trouble before the converter is seriously damaged. However, the conversion performance of the catalyst may be deteriorated at 1100°C or higher temperature. The warning system will help to protect the cata-

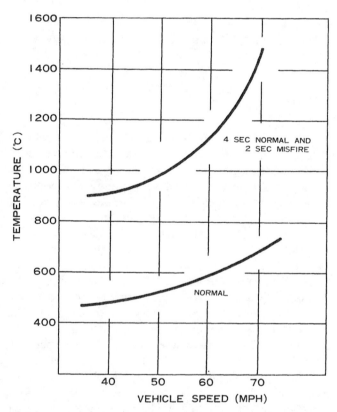

Fig. 15 - Converter temperature versus vehicle speed with whole cylinder misfiring at rate of one third

lytic converter from over-temperature stress generated by the choke valve sticking.

CARBURETOR FLOODING, EXCESSIVELY RICH MIXTURE - Tests were conducted to determine the effects of the following conditions as typical examples of excessively rich mixture:

1. Flooding carburetor.

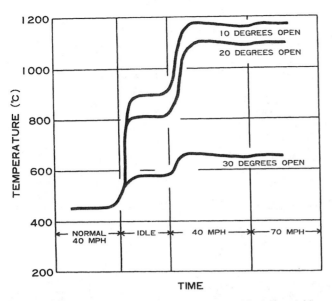

Fig. 16 - Converter temperature characteristics with choke sticking

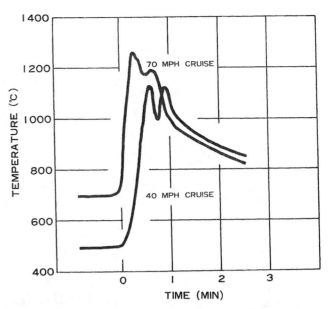

Fig. 17 - Converter temperature characteristics with carburetor flooding

2. Primary main jet falls out.
3. Secondary main jet falls out.
4. Plugged air bleeder.

Among these conditions, plugged air bleeder and secondary main jet fall-out showed a relatively mild bed temperature increase which only reached 920°C. The other two items showed higher but similar bed temperature characteristics. Therefore, the temperature characteristics with carburetor flooding is discussed here as the typical example (Fig. 17). When the carburetor is flooded during engine stop, the engine cannot be started and thus there is no effect on the converter. Tests were conducted on carburetor flooding during vehicle operation. The bed temperature increased sharply immediately after flooding and the peak temperature was higher at higher vehicle speeds as shown in Fig. 17. The peak temperature exceeded 1200°C but did not reach the melting temperature even at 70 mph. The vehicle speed also began to drop immediately after the carburetor flooding even though the driver pushed the accelerating pedal to the floor. The bed temperature decreased as the vehicle speed decreased. In most cases, the engine stalled soon after the carburetor flooded. Therefore this failure mode is not considered to be fatal to the converter although the catalyst efficiency will be deteriorated when this failure occurs at a higher speed vehicle operation.

INSUFFICIENT FUEL SUPPLY, EXCESSIVELY LEAN MIXTURE - Insufficient fuel supply was represented by inserting various sizes of orifice in the fuel line which has 5 mm inside diameter (ID). With an orifice of 0.8 mm ID or more, both the converter temperature and the drivability were normal. In the case of inserting an orifice of 0.6 mm ID, the converter temperature and the drivability were also normal at vehicle speeds below 60 mph. But the temperature increased to 1100°C and the vehicle had heavy surge, as shown in Fig. 18, when the vehicle was operated over 70 mph. With a 0.4 mm orifice, even engine idle was impossible. If there is insufficient fuel supply for an extended duration, equivalent to the effect of a 0.6 mm orifice which is the worst case, it will cause

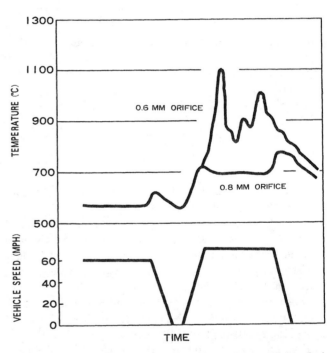

Fig. 18 - Converter temperature characteristics with insufficient fuel supply

converter damage. However, the vehicle will not be operated for long periods under this condition since the bed temperature beyond 1000°C accompanies heavy vehicle surge. Therefore, it is expected that the failure will be corrected before the converter is damaged and that this failure mode has only minor effects on the converter reliability. There is no significant effect from air switching.

FUEL RUN-OUT - Preliminary tests indicated that the converter was exposed to the most stringent circumstances when the vehicle was operated at a higher speed and when the accelerating pedal was pushed to the floor immediately after the

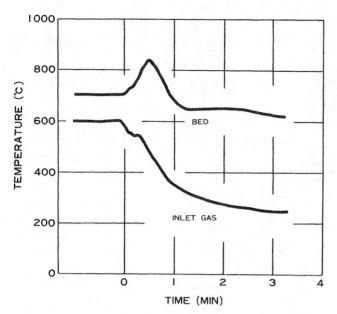

Fig. 19 - Converter temperature characteristics with fuel run-out

Table 5 - Evaluation of Criticality

Subsystem Failure	Damage Intensity	Correction Factor	Criticality
One cylinder misfiring	4	10	40
Two cylinder misfiring	10	10	100
Continuously whole cylinder misfiring	10	10	100
Intermittent whole cylinder misfiring	10	10	100
Choke valve sticking	6	3	18
Carburetor flooding	6	1	6
Primary main jet fall-out	4	1	4
Secondary main jet fall-out	1	8	8
Plugged air bleeder	1	8	8
Insufficient fuel supply	4	3	12
Fuel run-out	0	1	0

fuel run-out. Then tests were conducted at higher speed and also by fully opening the throttle valve immediately after the fuel run-out. The bed temperature decreased after a slight temperature peak which was 900°C, as shown in Fig. 19. Therefore, this type of failure had no adverse effects on the converter.

SUMMARY AND CONCLUSIONS

All possible component failure modes which would adversely affect the catalytic converter reliability were listed according to the failure mode effect analysis technique, and critical failure modes were investigated. A series of vehicle tests was conducted to determine the criticality of these critical failure modes. In order to conduct vehicle tests effectively, these critical component failure modes were classified into groups which were represented as the same subsystem failure, such as one cylinder misfiring, whole cylinder misfiring, choke sticking, and excessive fuel supply. Criticality was determined on these subsystem failures by means of vehicle tests and are shown in Table 5. There were approximately 130 items which were finally judged to be critical component failure modes. The ignition system was proven to have the most critical effects on the converter. Following are conclusions of this study:

1. Catalytic converter melting is caused by ignition system component failures which cause two cylinder or more frequent misfire.
2. Failure modes which have high potential to cause converter damage such as bed cracking or deformation are:
   a. Ignition system component failures which cause one cylinder or more frequent misfire.
   b. Choke component failures which cause slightly open choke valve sticking.
3. Failure modes which may not cause serious damage to the converter but have the potential to deteriorate the conversion efficiency of the catalyst are: carburetor component failures which supply the engine with excessively rich mixture; fuel system and carburetor component failures which cause insufficient fuel supply.
4. The warning system is ineffective to prevent the converter from melting since any corrective action is too late. However, the warning system often works effectively to protect the converter from thermal damage if prompt and suitable action is taken based upon the activation of the warning system.
5. The secondary air switching system has the potential to decrease the thermal influence of component failures on converter damage and on deterioration of the catalyst performance.

The results of this study are considered to be comprehensive as far as the general effects on converter reliability is concerned although this study was conducted on only one small car. However it is advised that the following items be taken into consideration when the results of this study are applied to other vehicles concerning the criticality to the converter:

1. Type of catalytic converter and catalyst volume.
2. Engine displacement.
3. Characteristics of the engine and emission control system.
4. Power train.

ACKNOWLEDGMENTS

This study involved the contributions of many people of Toyota Higashi-Fuji Technical Center. The authors would particularly like to acknowledge the contributions of K. Oishi and H. Miyagi who conducted the basic study on catalytic converter temperature characteristics.

The authors also thank M. Ando and H. Hasegawa for their close cooperation in conducting the analysis and vehicle tests.

REFERENCES

1. J. Mateyka, R. Danzeisen and D. Weiss, "Fault-Tree Applications to the Automobile Industry." Paper 730587 pre-

sented at SAE National Automobile Engineering Meeting, Detroit, May 1973.

2. SAE Aerospace Recommended Practice, Design Analysis Procedure for Failure Mode, Effects and Criticality Analysis—ARP 926.

3. E. Karmiol and S. Greenberg, "Reliability Factors in the Design Process." Annals of Assurance Sciences, Proceedings of the 1970 Annual Symposium on Reliability, February 1970.

4. W. Jordan, "Failure Modes, Effects and Criticality Analyses." Annals of Assurance Sciences, Proceedings of 1972 Annual Reliability and Maintainability Symposium, January 1972.

760199

NO_x Catalytic Converter Development

Y. Kaneko, T. Ohinouye, H. Kobayashi, and S. Abe
Mitsubishi Motors Corp. (Japan)

THE DUAL-BED CATALYTIC CONVERTER SYSTEM is now one of the most promising systems to cope with future stringent exhaust emission regulations and, at the same time, to minimize fuel economy loss accompanied with usual emission control systems.

In this system, the engine is operated at air-fuel ratios slightly on the fuel-rich side of stoichiometry in order to keep the first bed in reducing atmosphere throughout actual vehicle operations. After the exhaust passes through the first bed where reduction of NO_x takes place, secondary air is injected into the exhaust between the two catalyst beds and the exhaust containing excessive air is passed through the second bed where HC and CO are oxidized.

One of the main problems in this system is the considerable amount of ammonia produced in the first, reduction catalyst bed, which is converted back to NO in the second, oxidation catalyst bed, thus reducing the net NO_x reduction efficiency of the system.

Three approaches to the solution of this problem have been considered: find a catalyst that selectively reduces NO_x to N_2 with minimal ammonia formation; convert ammonia produced to N_2 by adding an ammonia catalyst bed between the two beds; and adjust the exhaust gas conditions to minimize ammonia on a given catalyst.

Since the presentation of our initial dual-bed system (1)* at the 1972 SAE meeting, we have continued to study ways to optimize our dual-bed system, using various promising NO_x catalysts.

This paper discusses ammonia forming characteristics of particular NO_x catalysts and the effects of engine exhaust gas parameters and clarifies the effect of bleed air introduced ahead of NO_x catalyst to adjust the CO/O_2 ratio of the exhaust.

EXPERIMENTAL

SYSTEM DESCRIPTION - A four-cylinder production engine with air control devices was used. The specifications of the engine and related emission control devices are shown in Table 1.

The layout of the dual-bed catalytic converter system is illustrated in Figure 1. The figure shows: the air switching valve for cold start; the air control valve to prevent overheating at full load conditions; and the air bleed orifice which adjusts the CO/O_2 ratio of the exhaust to minimize ammonia formation in the NO_x converter. A diameter of air bleed orifice corresponding to the main passage area was selected so that a maximum NO_x reduction efficiency can be attained during CVS-III (hot) mode operation as shown in Figure 2.

TEST APPARATUS AND NO_x CATALYSTS - Figure 3 shows the arrangement of the test apparatus. A heat exchanger is incorporated in the upstream of the NO_x catalyst to control the exhaust temperature and an extra oxidizing converter is used in series to assure complete oxidation of ammonia.

The intake air, measured by means of a laminar flow meter, passes through a carbure-

*Numbers in parentheses designate References at end of paper.

———ABSTRACT

The ammonia forming characteristics of two NO_x catalysts have been thoroughly studied under steady-state conditions after defining the ranges of such parameters as intake air flow, inlet gas temperature, inlet exhaust emissions and inlet CO/O_2 under the CVS-III (hot) mode.

A four-cylinder 1.6 liter engine equipped with a dual-bed catalytic converter system was used. Two catalysts having the same catalytic materials and loadings were prepared on pelleted and monolithic supports. The test results show how and why the bleed air, added ahead of the NO_x catalyst, is effective in reducing ammonia formation and improving the net NO_x reduction efficiency of the system.

Table 1 - Engine Specifications

Model	Mitsubishi 4G32
Type	water-cooled, vertical, 4-cycle gasoline engine
No. of Cylinders	4, in-line
Bore x Stroke	76.9 x 86 mm
Displacement	1597 cc
Compression ratio	8.5
Carburetor	2-barrel, down-draft, Stromberg type, specially metered for dual catalytic converter system
EGR device	Manifold entry type with vacuum-controlled valve
Secondary air devices	* Saginaw type air pump (220 cc/rev.) * Air switching valve for cold phase operation * Split air bleed orifice to NO_x reducing catalytic converter

Table 2 - NO_x Catalytic Converter Details

CATALYST IDENTIFI.	MDY-922	AM-1404S
CONVERTER CONSTRUCTION x ---TEMPERATURE MEASURING POINT		
CATALYST TYPE	PELLET	MONOLITH
CATALYST VOLUME	0.85 LITER	1.03 LITER
CATALYST BULK DENSITY	0.7	0.61
CATALYST MATERIAL	* MIXED (NICKEL–RHODIUM) * SAME METAL LOADING FOR BOTH TYPES	

tor with variable jets where the air-fuel ratio is controlled to adjust the CO level in the exhaust. The exhaust NO_x level is regulated through the variable EGR valve or by changing the spark timing.

The secondary air to the oxidizing converter and the bleed air to the reducing converter, shown in the figure, are manually controlled.

The concentrations of exhaust emissions are corrected for dilution by the use of the following factor, k:

$$k = \frac{14.7}{CO_2 + 0.5\ CO + HC.\ \times 10^{-4}} \quad (1)$$
$$(\%) \quad\ (\%) \quad (ppmc)$$

The exhaust gas is sampled at points ①~④ as shown in the figure and NO_x reduction efficiencies and ammonia formation are calculated as follows:

$$\text{Gross } NO_x \text{ red. eff.} = \frac{NO_x① - NO_x②}{NO_x①} \times 100\% \quad (2)$$

$$\text{Net } NO_x \text{ red. eff.} = \frac{NO_x① - NO_x③(=NO_x④)}{NO_x①} \times 100\% \quad (3)$$

$$NH_3 \text{ formation} = \frac{NO_x③ - NO_x②}{NO_x①} \times 100\% \quad (4)$$

where $NO_x①$~④ are corrected NO_x concentrations measured at ①~④ and $NO_x④$ is used to monitor the deterioration of the regular oxidizing catalyst.

Two kinds of NO_x catalysts, a pelleted (MDY-922) and a monolithic (AM-1404S), as shown in Table 2, were tested. Though cata-

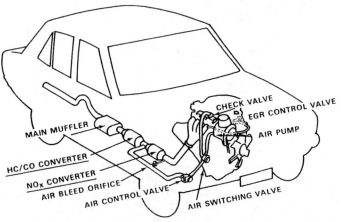

Fig. 1 - Dual catalytic converter system layout

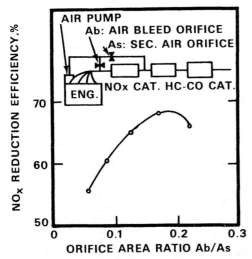

Fig. 2 - Effect of orifice area ratio on NO_x reduction efficiency during CVS-III (hot) mode

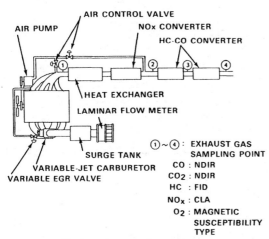

Fig. 3 - Test apparatus arrangement

lyst substrates were different in shape, the same catalyst materials were used at equivalent loadings for comparison purposes.

EVALUATION OF PARAMETERS

For the evaluation of a candidate NO_x catalyst under steady-state conditions, it is important to properly select values or ranges of such parameters as intake air flow, inlet gas temperature, inlet exhaust emissions and inlet CO/O_2 ratio.

Figure 4 shows the inlet NO_x mass emission distribution based on NO_x measurements made every minute during CVS-III (hot) cycle. It can be observed that the low NO_x mass emission portions indicate rather enhanced rates because of their high frequencies. Average intake air flow, NO_x concentration and inlet gas temperature for each specified range are tabulated above the figure.

These average values are plotted in Figure 5 and all possible combinations of these parameters that occur during CVS-III (hot) cycle are shown by the shapes of the domains.

In principle, the standard activity evaluation point, marked by the solid circle, was selected to fall near the center of the domains. However, the temperature and the input NO_x concentration selected were slightly lower than those indicated by the contours of the domains to better distinguish factors affecting ammonia formation.

Figure 6 shows NO_x reduction efficiencies plotted versus inlet CO concentrations for three different inlet NO_x concentrations. The lower NO_x concentration clearly causes a wider difference between gross and net efficiencies, namely, causes more ammonia to form. It is preferable, therefore, to evaluate NO_x catalysts under low NO_x concentrations to clearly distinguish their ammonia formation characteristics.

RESULTS AND DISCUSSION

The effect of intake air flow on the relation between net NO_x reduction efficiency and inlet CO/O_2 ratio are shown in Figure 7 for AM-1404S and MDY-922 NO_x catalysts at the temperature and the NO_x concentration corre-

AVERAGE FOR EACH RANGE

	WHOLE RANGE	RANGE ①	RANGE ②	AREA Yht	AREA Yst
INTAKE AIR FLOW (m^3/Hr)	30	83	91	38	26
NOx CONC. (PPM)	369	1607	2119	588	243
INLET GAS TEMP. (°C)	473	575	597	407	464

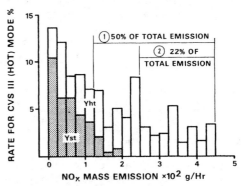

Fig. 4 - NO_x emission distribution during CVS-III (hot) cycle

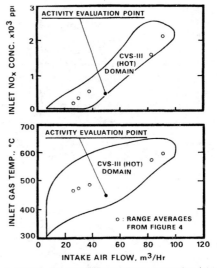

Fig. 5 - Relationship between inlet NO_x concentration, inlet gas temperature and intake air flow during CVS-III (hot) cycle

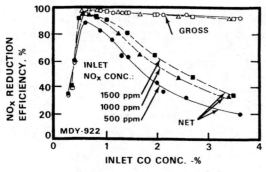

Fig. 6 - Effect of inlet NO_x concentration on NO_x reduction efficiency

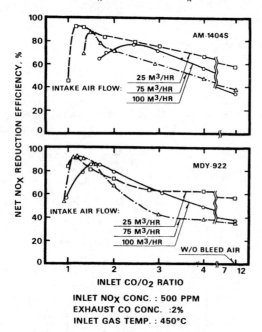

Fig. 7 - Effect of intake air flow on the relation between net NO_x reduction efficiency and inlet CO/O_2 ratio

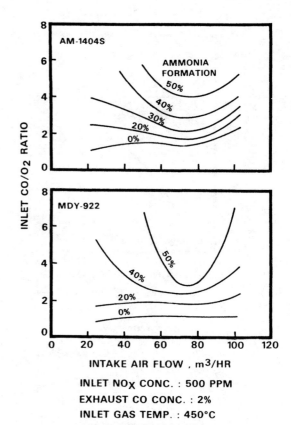

Fig. 8 - Effect of intake air flow and inlet CO/O_2 ratio on ammonia formation

sponding to the standard evaluation point. For AM-1404S, the peak NO_x efficiency drops as the intake air flow is increased and the peak moves to the richer side of CO/O_2 ratio. This tendency is less pronounced with MDY-922 as shown in the lower figure. For both catalysts, the net NO_x efficiency passes through a minimum when the air flow is varied at constant CO/O_2 ratio; this occurs, however, only on the rich side of the peak.

The effects of the inlet CO/O_2 ratio and intake air flow on ammonia formation are shown in Figure 8 for both catalysts. Ammonia formation appears to become more sensitive to inlet CO/O_2 ratios when the intake air flow is increased; also in the case of MDY-922, the closeness of contour lines at about 75 m³/hr indicates that ammonia production is most sensitive at this air flow rate.

Figure 9 shows ammonia formation contours for AM-1404S defined by inlet gas temperature and intake air flow at three different CO/O_2 ratios; the corresponding CVS-III (hot) domain is also shown by dashed curve. There is a

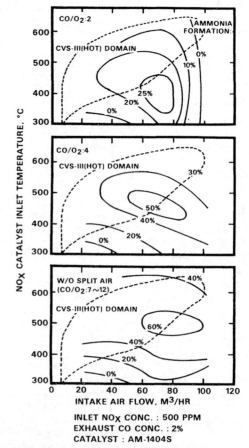

Fig. 9 - Effect of intake air flow, CO/O_2 ratio and inlet gas temperature on ammonia formation

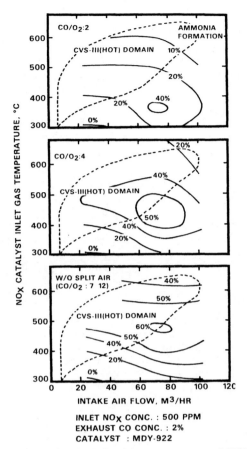

Fig. 10 - Effect of intake air flow, inlet gas temperature and CO/O_2 ratio on ammonia formation

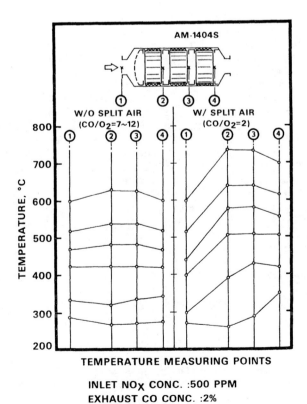

Fig. 11 - Temperatures at measuring points in AM-1404S catalys

peak island at intake air flow around 70 to 80 m³/hr in each of the three plots and it is interesting that, as the CO/O_2 ratio increases, the ammonia formation islands move toward areas of higher temperatures with higher peak values.

The correlations between the CVS-III (hot) domain and the peak islands show that ammonia formation distributed over the domain can be effectively reduced by decreasing CO/O_2 ratio through addition of bleed air. In other words, the bleed air system is quite effective in reducing ammonia formation during the CVS-III (hot) mode and consequently increasing net NO_x reduction efficiency of the system.

Similar diagrams for MDY-922 are shown in Figure 10 with almost similar trends as in Figure 9, except that a relatively higher peak appears in the upper figure for MDY-922 as compared with that for AM-1404S.

When the CO/O_2 ratio is adjusted to 2.0 by adding bleed air to the exhaust, the inlet CO concentration is in fact lowered to about 1% from the pre-adjusted exhaust CO concentration of 2%; and this appears to cause reductions in the peak value of ammonia formation.

Furthermore, in order to explain why the peak island moves along the temperature axis in each of these figures, the temperatures in the catalytic converters were measured and are shown in Figure 11. It is remarkable that the temperature rises in the converter with use of bleed air, while it remains almost constant without the use of bleed air.

If the inlet gas temperatures in Figure 9 were replaced with the corresponding converter temperatures, which more accurately represent temperatures of actual chemical reactions, all peak islands would remain around 450 to 500°C, where ammonia is preferentially produced.

Though the data are omitted here, the same argument applies to MDY-922 in Figure 10. In this case the converter temperature was measured at the point shown in Table 2.

In order to clarify the ammonia formation trend within the catalyst, four different volumes of AM-1404S catalysts were compared at three intake air flows. The test results are shown in Figure 12. Ammonia formation passes through a maximum with increasing volume of catalyst at the low intake air flow, while it continuously increases at the higher intake air flow. These results indicate that ammonia formed in the front part of the catalyst can react within the catalyst when the intake air flow is low enough to assure space velocities below some small value. This may be taken as further proof for ammonia intermediacy in NO_x reduction (2).

Although CO/O_2 ratio adjustment by adding bleed air through an orifice ahead of the NO_x catalyst is quite effective in improving NO_x reduction efficiency of the system, further improvements may be still possible.

Figure 13 shows CO/O_2 distributions over the CVS-III (hot) domain for three different orifice ratios; the data are plotted on planes

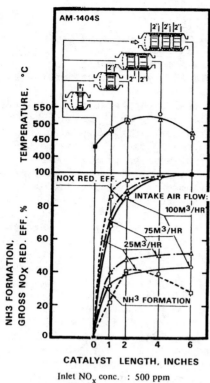

Inlet NO_x conc. : 500 ppm
Exhaust CO conc.: 2%
Inlet CO/O_2 : 4

Fig. 12 - Effect of catalyst length and intake air flow on ammonia formation and gross NO_x reduction efficiency

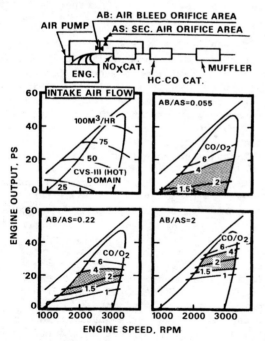

Fig. 13 - Effect of orifice area ratio on CO/O_2 ratio during CVS-III (hot) cycle

defined by engine output versus engine speed. The figure shows that, with a fixed orifice ratio, CO/O_2 ratio can be adjusted to the preferred shaded range of 1.5 to 4, only over a part of the domain. Therefore, the orifice ratio should be made so that it can vary continuously or stepwise to allow the effective adjustment of CO/O_2 ratios over the entire domain. For instance, since an air flow near 75 m^3/hr has been found to be most sensitive for ammonia formation, it may be more effective to put emphasis on this specific point. Such optimization practice appears very promising but needs further detailed study.

SUMMARY AND CONCLUSIONS

A dynamometer mounted four-cylinder engine equipped with a dual-bed catalytic converter system was used to study ammonia formation at a suitable combination of parameters selected to represent the CVS-III (hot) mode operation. A pelleted and a monolithic NO_x catalyst having the same active ingredients and approximately the same loadings were used.

1. There was no substantial difference in the behavior of the two catalysts except for a small difference in peak ammonia formation at low CO/O_2 ratios.

2. A peak area of ammonia formation exists in the plane defined by the inlet gas temperature and intake air flow; this peak island stays on a specific intake flow rate of about 70 to 80 m^3/hr irrespective of the CO/O_2 ratios. Though the peak island seems to move toward higher temperature zones as CO/O_2 ratio increases actually they remain within a specific catalyst temperature zone between about 450 and 500°C.

3. Maximum ammonia formation at the peak island drops as the CO/O_2 ratio is decreased by addition of bleed air because the inlet CO concentration is reduced considerably.

4. Thus, it can be concluded that the bleed air addition is very effective in reducing ammonia formation on NO_x catalysts and in improving net NO_x reduction efficiency of the system under CVS-III (hot) mode because it pushes the area of maximum ammonia formation out of the CVS-III (hot) domain and lowers the inlet CO concentration.

5. Test results with catalysts having different volumes and lengths show that ammonia formation goes through a maximum with increased length of catalyst at low intake air flows.

6. The results of this study suggest that the air bleed system can be further improved by effectively controlling the orifice area ratio.

REFERENCES

1. Y. Kaneko and Y. Kiyota, "Mitsubishi Status Report on Low Emission Concept Vehicles", SAE 720483 (1972).

2. R. L. Klimisch and K. C. Taylor, "Ammonia Intermediacy as a Basis for Catalyst Selection for Nitric Oxide Reduction" Environmental Science and Technology, vol. 7 (1973).

780006

The Fast Burn with Heavy EGR, New Approach for Low NO$_x$ and Improved Fuel Economy

H. Kuroda, Y. Nakajima, K. Sugihara,
Y. Takagi, and S. Muranaka
Nissan Motor Co., Ltd.
(Yokosuka/Japan)

IMPROVEMENT IN EXHAUST EMISSIONS and fuel economy is strongly required for automotive engines hereafter. Especially, substantial reduction of NO$_x$ emissions from spark ignition engines is one of the most difficult tasks. In order to reduce NO$_x$ emissions substantially, there will be two approaches, i.e the reduction in cylinder and also the decomposition by catalyst. The first approach could be realized by EGR, retarded spark timing, rich or lean air-fuel ratio, etc. Among them, EGR is the most effective measure considering fuel economy and driveability, as revealed in past studies of the authors and others (1–3).*

With conventional engines, however, a large amount of EGR impairs engine operating stability so that marked reduction of NO$_x$ could not be achieved.

In the present work, to seek a marked reduction of NO$_x$ just by EGR, analytical study deals with combustion around the practical engine operating stability limit of four cylinder conventional S.I engine.

---

*Numbers in parentheses designate References at end of paper.

ABSTRACT

In the way to seek a marked reduction of NO$_x$ just by EGR which was proved to be the best answer from the past studies, this analytical study deals with combustion around the practical engine stability limit.

The results show that practical engine operating stability limit with the four cylinder engine is determined by the occurrence of several percent of slow burn.

The combustion patterns giving the practical engine stability limit were investigated by the simultaneous analysis of cylinder pressure traces and multi-ionization signals.

As the result, four kinds of combustion patterns with the use of EGR have been classified. It was concluded that the practical engine stability limit was determined by the occurrence of several percent of slow burn, and short combustion duration would be an answer to improve the engine stability.

The concept derived from the above mentioned study pointed the way to achieve both combustion stability with heavy EGR and improvement of fuel economy together with low NO$_x$ emission level.

EXPERIMENTAL APPARATUS AND PROCEDURE

A schematic diagram of the experimental apparatus for the combustion analysis is shown in Figure 1. A Datsun in-line 4 cylinder 1.8 litre engine was used for these experiments. Its detailed specifications are given in Table 1. The ignition system is current production type. Test fuel is our standard unleaded gasoline silver-N of which specifications are given in Appendix A.

Experiments applying short combustion duration (fast burn) showed that the fast burn in combination with heavy EGR improves engine stability and has a potential to achieve substantial reduction in NO$_x$ emission along with some improvement in fuel economy. This potential was also directed by mathematical model study.

This fast burn concept was embodied in the newly developed Nissan Z Fast Burn Engine.

0148-7191/78/0227-0006$02.50
Copyright © 1978 Society of Automotive Engineers, Inc.

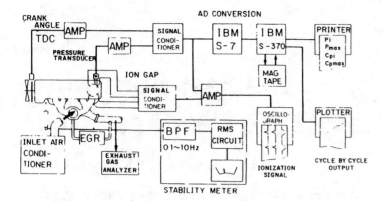

Fig. 1 - Schematic diagram of test apparatus

Table 1 — Specifications of test engine

Model	Datsun L18
Number of cylinders	4
Bore	85 mm
Stroke	78 mm
Displacement	1770 cc
Compression ratio	8.7 to 1
Combustion chamber shape	Wedge
Carbureter	Modified Stromberg type
Spark energy	50 mJ
Spark plug gap	1.1 mm

A piezoelectric pressure transducer was installed in the No. 4 cylinder for the cylinder pressure measurements. The on-line data processing of the output signal was done by an IBM 370 computer. Both cycle-by-cycle and statistical calculation were carried out. The indicated mean effective pressure (P_i) and heat release rate in the cylinder for each cycle, the standard deviation and the fluctuation rate of P_i for 400 consecutive cycles were calculated by the computer. The mass fraction of the charge burned was also computed from the average pressure trace.

A special objective of this experiment is to understand the correlation of P_i levels and flame propagation. Four ion gaps were installed in the same No. 4 cylinder to detect the flame front (Figure 2). In order to achieve the objective, ionization signals and cylinder pressure traces recorded on the oscillograph were compared with the corresponding P_i values calculated by the computer.

Engine stability has been judged by human sensing so far. For more precise measurement with higher reproducibility, an engine stability meter was newly developed and used for the experiments. This stability meter consists of a sensor detecting the engine's transverse displacement, a low pass filter and a root mean square circuit. The output of this meter and the stability scale by human sensing were almost proportional as shown in Figure 3.

By operating an engine on the dynamometer at the same conditions under which the engine in a vehicle shows an acceptable surge limit, it was experimentally confirmed that the practical stability limit of the engine corresponds to the human-sensing stability scale of 3 on the dynamometer.

Most test data were acquired for two speed-torque sets, 1400 rpm-3kgm and 1600 rpm-5kgm which are representative engine operating conditions under Japanese 10 mode test cycle for cars of 1250 kg I.W. class. In each test condition, air-fuel ratio, EGR rate and spark timing were varied.

To minimize the effect of air temperature and humidity changes on the emission data, especially for NO_x, an inlet air conditioner was used. Temperature and relative humidity were controlled to 25±1°C and 55±2% respectively.

Exhaust gas was sampled at each exhaust port and tail pipe to measure CO, CO_2, HC, NO and NO_x concentration. CO_2 concentration in all four intake manifold branches was measured to calculate average and cylinder-by-cylinder EGR rates by the following equation.

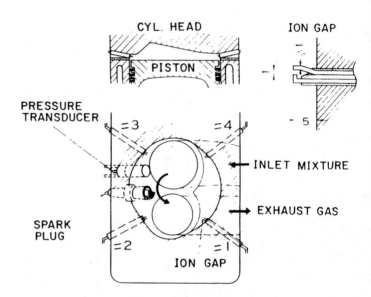

Fig. 2 - Instrument for the measurement of ionization signals

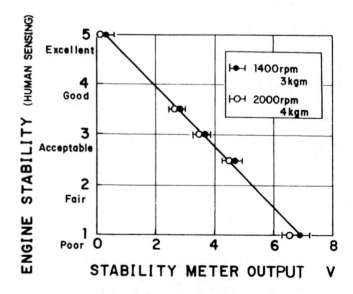

Fig. 3 - Correlation between engine stability by human sensing and the output of stability meter

$$\text{EGR rate (\%)} = \frac{\text{EGR Gas Volume}}{\text{Intake Air Volume}} \times 100 \, (\%)$$

$$= \frac{[CO_2]_{IN} - [CO_2]_a}{[CO_2]_{EX} - [CO_2]_{IN}} \times 100 \, (\%)$$

[CO_2] IN : CO_2 concentration in the intake manifold
[CO_2] a : CO_2 concentration in the air
[CO_2] EX : CO_2 concentration in the exhaust manifold

Since cylinder-to-cylinder distribution range of air-fuel ratio was ±0.1 to the set A/F and that of EGR rate was ±1% throughout the tests, data obtained from No. 4 cylinder was considered to have generality.

RESULTS

ANALYSIS OF COMBUSTION PHENOMENA UNDER HEAVY EGR

Judgement of Engine Stability by the Fluctuation Rate of P_i — When combustion fluctuation is discussed, fluctuation rates of P_{max} or $(dP/d\theta)_{max}$ have been used as indices so far because they were easy to obtain. But they are not appropirate indices since they vary with spark timing and do not always correspond to car surge and deterioration of emissions which are the final result of combustion fluctuation. Among many indices obtained from cylinder pressure traces, P_i seems to be the most appropriate index as cycle-by-cycle engine torque is directly represented by P_i (4) (5). Consequently, in this study, quantitative analysis of engine stability was tried using P_i fluctuation rate as an index.

Under the steady state operation on the engine dynamometer, heavier EGR rate increases P_i fluctuation and deteriorates engine stability as shown in Figure 4. Without EGR, P_i distribution falls in very narrow range and engine stability is excellent. At 20% EGR, engine stability reaches acceptable limit and P_i spread becomes a wide band. At 28% EGR, engine stability is poor and cycles of zero P_i appear. This is further exemplified by the photographs of pressure traces. Some cycles show the same traces as that of motoring cycle.

The relationships among EGR rate, engine stability and P_i fluctuation rate (C_{pi}) at three different A/F levels are shown in Figure 5.

C_{pi} is defined as follows:

$$C_{pi} = \frac{\sigma p_i}{\overline{P_i}} \times 100\%,$$

where P_i : indicated mean effective pressure
C_{pi} : fluctuation rate of P_i
σp_i : standard deviation of P_i
$\overline{P_i}$: mean P_i

As seen in this figure, richer A/F gives better engine stability. EGR rates for the acceptable stability limit are 15, 20 and 25% with A/F of 16, 14.5 and 12.5 to 1 respectively. From the standpoint of fuel economy, however, richer A/F is not advisable. At each A/F, C_{pi} level of about 10% was found to correspond to the acceptable

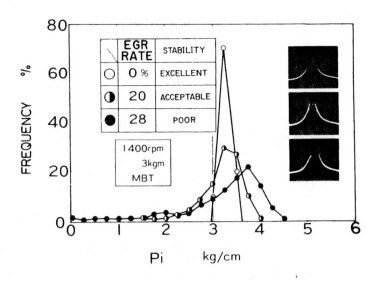

Fig. 4 - Effect of EGR rate on the cylinder pressure traces and the distribution of indicated mean effective pressure

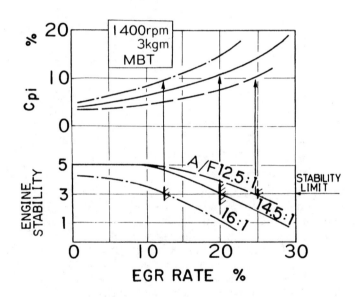

Fig. 5 - Effect of EGR rate on engine stability and P_i fluctuation rate

stability limit. Experiments under another speed-torque condition (1600 rpm-5 kgm) confirmed this value. From this, 10% C_{pi} was considered to give the stability limit in this study.

The above discussions are based on the test data at MBT spark timing. Relationships among EGR rate, P_i fluctuation and engine stability with different spark timings will be discussed in Appendix B.

Correspondence of Ionization Signals and P_i Level — As shown before, P_i fluctuation increases with heavy EGR. In order to examine the difference in the combustion process corresponding to the various levels of P_i and to

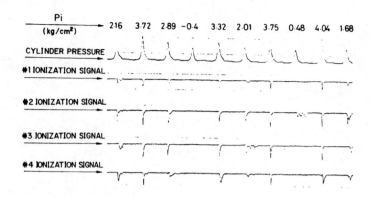

Fig. 6 - Sample data from simultaneous measurement of cylinder pressures and ionization signals

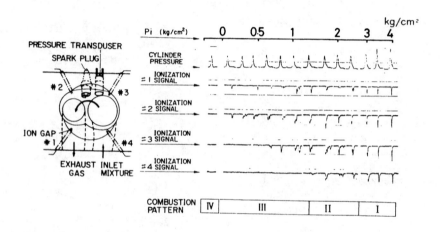

Fig. 7 - Relationship between P_i and ionization signals

classify the combustion, thermodynamic calculations of combustion and the analysis of ionization signals were carried out. Figure 6 shows P_i, cylinder pressure traces and ionization signals under the test conditions of 1400 rpm-3 kgm, 14.5 to 1 A/F and 28% EGR. P_i values were calculated by the on-line computer. The cylinder pressure traces and the ionization signals at four locations were recorded on multi-channel oscillograph. The cycle-by-cycle correspondence between P_i and the outputs on the oscillograph was made possible by shortcircuitting the output of pressure transducer for several cycles. Levels of the ionization signals from four ion gaps had been adjusted to the same value at the engine operating condition where combustion is very stable and flame propagation speed is maximum. This adjustment is required to compensate for the difference in the ion gap and signal line impedance.

From this figure, a relationship between the level of P_i and the occurrence and level of ionization signals was studied. Some cycles show high P_i and sharp ionization signals for all four locations while others show very low or negative P_i values and ionization signals of which are weak and irregular or totally lacking. It is also observed that a cycle of relatively high P_i follows a cycle of very low or negative P_i. This fact can be explained by the unburned charge of the prior cycle being carried over to the following cycle as residual gas. Ionization signals of irregular shape are supposed to correspond to very flexuous flame front and the size of flame front irregularity is comparable with the gap size (1 mm) (6).

To clarify the relationship between P_i levels and ionization signals, each combustion cycle was arranged in the order of P_i level. As seen in Figure 7, the number and the level of ionization signals increased with higher P_i levels. Considering ionization signals (at four locations) and P_i levels, four types of combustion pattern were identified:

I. Ionization signals occur at all four locations. At the same time the signal level is high and its shape is sharp.
2.8 kg/cm² < P_i < 4 kg/cm²

II. Ionization signals occur at all four locations, but the signal level is comparatively low and its shape is not neat.
1.5 kg/cm² < P_i < 2.8 kg/cm²

III. Ionization signals at one to three locations are lacking.
0 < P_i < 1.5 kg/cm²

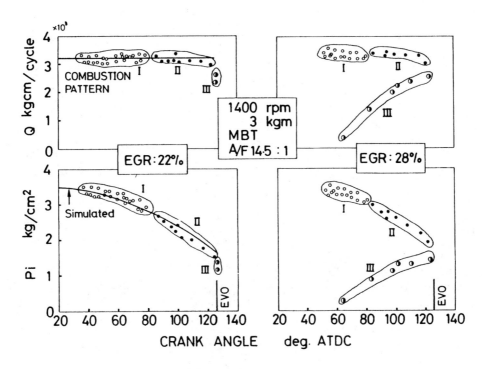

Fig. 8 - Relationships among combustion period, P_i level and heat release

IV. Ionization signals do not appear at any location.
$P_i < 0$

Another interesting fact found in this figure is the occurence of ionization signals at a specific location with different P_i levels. Corresponding to the increase of P_i value, ionization signals occur in the order of #2, #1, #3 and #4, while the distance between the spark plug and each ion gap is in the order of #2, #3, #1 and #4 from short to long. This difference can be explained by the counterclockwise mixture motion caused by combustion chamber configuration.

The Relationships among P_i level, Combusion Period and Heat Release — From the cylinder pressure traces for each combustion pattern, the combustion period (crank angle at combustion termination) and the heat release were calculated and compared with the P_i levels. The results for 22 and 28 percent EGR are given in Figure 8. The combusion period is defined as the crank angle where the calculated heat release stops rising. From Figure 8, it is observed that the combustion terminates early for the Type I cycles, while for the cycles of Type II, combustion is slow and in some cases lasts until the exhaust valve opens. P_i level of Type I is higher than that of Type II but the calculated heat release is almost the same for both patterns. From this fact and the ionization signals shown in Figure 7, the flame is assumed to have completely propagated in the cylinder in the case of Type II combustion as well. This assumption is confirmed by a mathematical model study. In this study, triangular shape heat release pattern was used, where the total heat release and the combustion initiating timing was kept constant while the combustion duration was varied.

Solid line in Figure 8 shows the calculated result with the models. It is evident that the experimental data with Type I and II cycles correspond well with the calculation.

Since cycles of Type III show low P_i level, low heat release, large variation in combustion duration and lack of ionization signals, it is believed that the combustion of these cycles has not been completed before the exhaust valve opens or the flame has been extinguished during the expansion stroke, leaving some unburned portion.

In addition to this, because the friction mean effective pressure (P_f) is slightly higher than 1 kg/cm² at this operating condition, the brake mean effective pressure (P_e), which has more direct relation to engine stability than P_i, is more sensitive to the change of combustion duration.

Classification of Combustion Pattern — Combining these observations, combustion patterns under 0 to 30% EGR are conveniently classified into four categories as shown in Table 2. They are Normal Burn, Slow Burn, Partial Burn and Misfire. Their boundaries, however, are not quite clear.

Under a hypothetical condition where cycle-by-cycle fluctuation does not exist, the combustion pattern would shift from Type I through Type IV continuously with the increase of the EGR rate.

Effect of EGR on Combustion Pattern Distirbution — In order to examine how the frequency of occurrence of the four kinds of combustion patterns varies with the EGR rate, classification of cycles at various EGR levels was carried out, applying the standards of Table 2. Figure 9 shows the results. From this figure it can be said that;

(1) With the increase of EGR, slow burn appears first. Then the engine stability deteriorates with the in-

Table 2 – Classification of combustion pattern

1400 rpm, $\overline{P_i}$ = 3.3 kg/cm²

COMBUSTION PATTERN		REMARKS	P_i kg/cm²	CRANK ANGLE AT COMBUSTION TERMINATION	IONIZATION SIGNAL	
					NUMBER	SHAPE
I	NORMAL BURN	Short Combustion Duration Complete Flame Propagation	$4 > P_i > 2.8$	30 ~ 80 deg. ATDC	All 4	Sharp & High
II	SLOW BURN	Long Combustion Duration Complete Flame Propagation	$2.8 > P_i > 1.5$	80 ~ EVO *)	All 4	Flexuous & Low
III	PARTIAL BURN	Incomplete Flame Propagation	$1.5 > P_i > 0$	Large Variation	3 ~ 1	Flexuous & Low
IV	MISFIRE	———	$0 > P_i$	———	0	———

* Exhaust valve opening

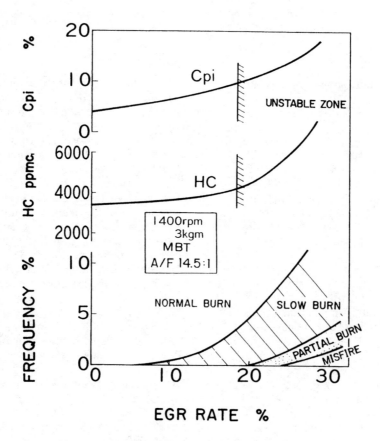

Fig. 9 – Effect of EGR on the distribution of combustion patterns, Cpi and HC emissions

crease of slow burn and at 5% slow burn the engine reaches the acceptable stability limit. In other words, combustion at the stability limit contains 5% of slow burn and there is neither misfire nor partial burn at this condition.

(2) With further increase of EGR rate beyond the engine stability limit, partial burn begins to appear. Once partial burn appears, unburned charge is carried over to the next cycle as residual gas. This further increases P_i fluctuation and the engine stability deteriorates rapidly. At the same time, because of partial burn, unburned HC in the exhaust and also fuel consumption sky-rocketed.

(3) Misfire cycles appear at a much higher EGR rate, where partial burn appears at the frequency of several percent. At this condition, because of the higher frequency of slow burn, deterioration in engine stability, fuel economy and HC emission is very great.

AN APPROACH TO IMPROVED COMBUSTION WITH HEAVY EGR

It was revealed that the factor limiting the practical engine stability limit is neither misfire nor partial burn but the appearance of slow burn at a frequency of about 5%. With increased EGR, the reduced flame temperature decreases burning velocity. Longer combustion duration tends to be more susceptible to the factors causing combustion fluctuation. Thus the slow burn appears and increases.

In order to avoid engine instability caused by slow burn, short combustion duration (fast burn) was conceived to be a good answer.

Although past literatures predicted higher NO_x emission with the fast burn (7) (8), the potential of the combination of the fast burn and heavy EGR was apparently overlooked. In order to confirm whether the fast burn increases or decreases NO_x emission under heavy EGR, a demonstration of fast burn concept in the engine was carried out as the next step. The relationship between NO_x formation level and combustion duration with heavy EGR was also studied by utilizing a mathematical model. The results of this study are presented in Appendix C.

An Approach to Achieve Fast Burn in the Engine — There are two approaches to burn the mixture rapidly in the cylinder:

(1) Utilization of swirl and squish which intensify mixture turbulence.

(2) Decrease of flame propagation distance by such means as the optimization of combustion chamber shape and the location of spark plug and/or the adoption of multi-point ignition.

A dual spark plug system as shown in Figure 10 was adopted because the second approach enables faster combustion in any engine under any operating condition and has the potential to further shorten the combustion duration resulting from the first approach which increases the combustion speed by swirl and squish. Combustion chamber shape and spark plug location were optimized experimentally to equalize the flame propagation from two spark plugs. The displacement and the compression ratio of this dual plug engine are the same as those in Table 1.

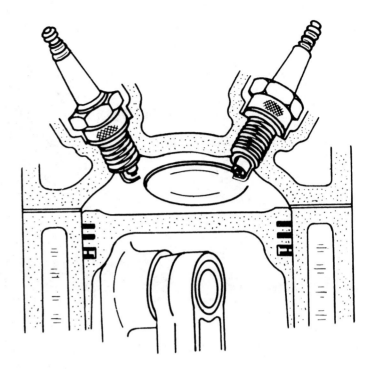

Fig. 10 – Configuration of dual spark plugs

The Effect of Dual Spark Plug Fast Burn On Combustion with Heavy EGR — Combustion duration and P_i fluctuation of the conventional and the dual spark plug fast burn engine were compared. Figure 11 shows the combustion duration of the two processes with zero and 20% EGR. The mass fraction of cylinder charge burned for each process was calculated by the average pressure trace of 400 consecutive cycles. With fast burn, combustion duration showed a marked decrease. With 20% EGR, combustion duration of fast burn was almost the same as that of conventional engine with no EGR.

P_i frequency distribution was compared for both combustion processes at 14.5 to 1 air-fuel ratio and 20% EGR. As shown in Figure 12, the fluctuation rate of P_i (C_{pi}) reaches 11% for conventional engine, which is over the limit for engine stability, while, for fast burn, C_{pi} is as small as 4%. This C_{pi} level is the same as that of conventional engine with no EGR.

It is also clear from the cylinder pressure traces shown in Figure 12 that combustion with heavy EGR was improved by the fast burn, and the cycle-by-cycle fluctuation of cylinder pressure decreased greatly.

A high speed movie technique was used to observe the combustion in the cylinder. Details of the equipment used was previously presented in another paper by Nissan (9). Several photographs of the fast burn at 800 rpm are shown in Figure 13.

The flame propagation from the two spark plugs is seen to be almost identical. Similar results were also obtained at 1400 rpm.

Engine Performance Improvement by Fast Burn — Engine performance for various EGR rates was compared between the fast burn and the conventional engine (Figure 14). As stated above, since combustion with heavy EGR

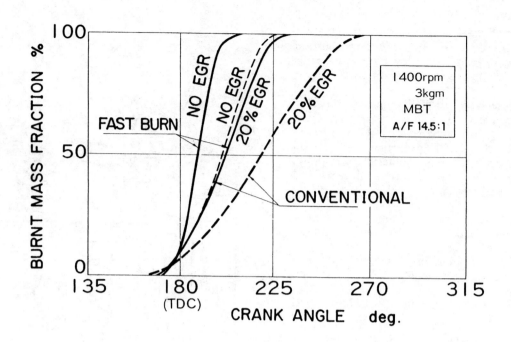

Fig. 11 - Effect of fast burn on combustion duration

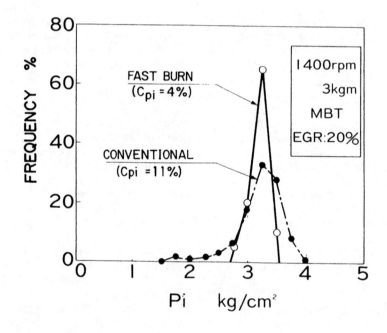

Fig. 12 - Improvement in P_i distribution by fast burn

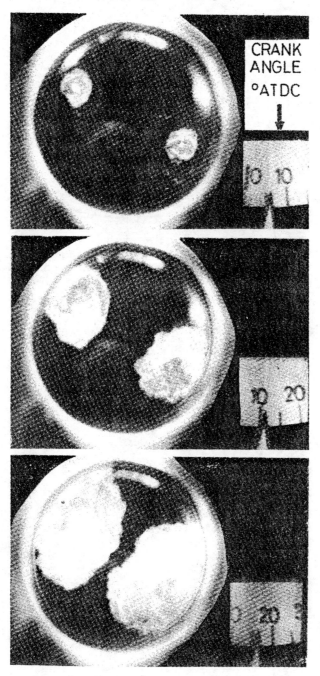

Fig. 13 - Highspeed photographs of flame propagation in a dual spark plug engine

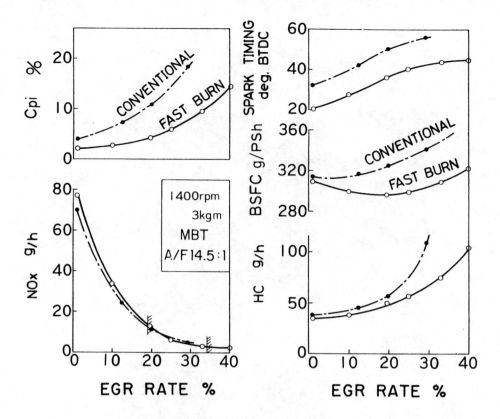

Fig. 14 - Improvement in emissions and fuel economy achieved by fast burn engine

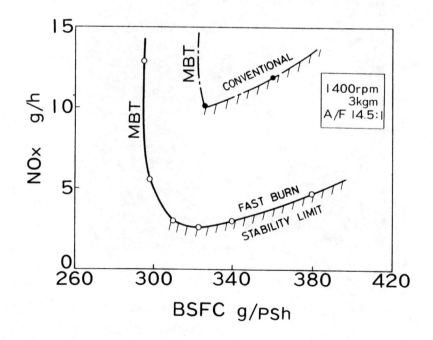

Fig. 15 - Improved fuel economy combined with low NO_x emission achieved by fast burn engine

Table 3 — Japanese 10 mode test data

Emission Control System;
    EMS (Z Engine) + Heavy EGR + Secondary Air
    + Oxidation Catalyst
Engine   ; Nissan Z Fast Burn Engine
    4 Cylinder in line, 1.8 liter
Transmission ; Four Speed, Manual
Inertia Weight ; 1,250 kg

	HC g/km	CO g/km	NO_x g/km	Roll Economy km/l
Test Data at Japanese MOT	0.12 ~ 0.20	0.03 ~ 0.04	0.13 ~ 0.16	13.0 ~ 13.3
Japanese 1978 Standards	0.25	2.1	0.25	—

Table 4 — U.S. LA-4 test data (CVS-CH)

Vehicle   ; Datsun 810
Engine   ; Nissan Z Fast Burn Engine,
    4 cylinder, 2.0 liter
Transmission ; Four Speed, Manual
Inertia Weight ; 2.750 lbs

	HC g/mile	CO g/mile	NO_x g/mile	Roll Economy MPG
In house Test Data	0.15 ~ 0.19	1.2 ~ 2.5	1.2 ~ 1.4	26.3 ~ 26.8
U.S. Federal St'ds for 1980	0.41	7	2.0	—

is greatly improved by the fast burn, the EGR rate at the engine stability limit with the fast burn is 33%, which is 15% higher than the value for the conventional engine. At EGR rates lower than the limit for the conventional engine, NO_x from the fast burn is higher than that of conventional engine. At higher EGR rates, the fast burn can largely reduce NO_x emissions. At the same time, stabilized combustion decreases HC emissions compared with the conventional engine. Also, MBT spark timing is delayed over 10 degrees because of shortened combustion duration.

Fuel economy was markedly improved by the fast burn, the difference becoming larger with higher EGR rate. Improved combustion by the fast burn with heavy EGR plus the reduction of pumping losses by EGR improve fuel economy. Appendix D comments further on the factors affecting fuel economy improvement by EGR. Thus, reduced NO_x and improved fuel economy can be compatible.

Figure 15 shows the relationship between NO_x emission and brake specific fuel consumption for the conventional and the fast burn engine. Vertical parts of the two L-shaped curves were drawn from the test results with MBT spark timing at various EGR rates, while the hatched parts correspond to the data with various spark timings retarded from MBT and EGR rates at the engine stability limit. Engines can be operated in the zone above the L-shaped line. From this figure, it can be seen that the fast burn can achieve a marked reduction in NO_x emission while maintaining good fuel economy.

Vehicle Emission Data

In order to meet the 1978 Japanese Emission Standards, the development of an emission control system was carried out. Specifications of the system are listed in Table 3. NO_x target of 0.25 g/km was attained just by EGR. Spark timing was set close to MBT to improve fuel economy. Emission levels of a Datsun 810 (inertia weight 1250 kg) in the certification test at Japanese MOT are shown in Table 3. This fuel economy level is so far the highest among the models of 1250 kg inertia weight class for 1978 standards. Also, 10 to 15% improvement in fuel economy is achieved for the suburban and the highway driving conditions compared to 1976 model.

The development of emission control system for the U.S. standards has been initiated with the same concept. As seen from recent test results of this system given in Table 4, this system has demonstrated superior roll economy compared to the present model.

CONCLUSION

1. A new concept simultaneously achieving low NO_x and improved fuel economy has been demonstrated. The new concept involves fast burn and heavy EGR, the fast burn being achieved by dual spark plug ignition in a sophisticated combustion chamber configuration.

2. Based on experimental work presented herein, combustion pattern under heavy EGR has been classified into four types. They are normal burn, slow burn, partial burn and misfire. With the increase of EGR rate, slow burn appears first and increases its percentage in the total cycles. Partial burn and misfire appear at higher EGR rate. The engine reaches its practical stability limit at the EGR rate where several percent of slow burn exists and neither partial burn nor misfire is yet present.

3. Fast burn has been shown to overcome the slow burn limitation of conventional engines and to greatly extend the stable combustion range under heavy EGR conditions, with resulting marked reduction in NO_x emission and improved fuel economy. Mathematical model study has also indicated the advantage of the fast burn with heavy EGR.

4. The Nissan Z Fast Burn Engine which embodies the fast burn concept was installed in its domestic model for 1978 standards. This production model is further proof that gains in fuel economy and driveability can be achieved with a marked reduction in NO_x emissions.

ACKNOWLEDGEMENT

The authors wish to thank Messrs. Leonard Raymond and T. Saito for their instructive advice in preparing this paper. We also thank the personnel of Central Engineering Laboratories of Nissan Motor Company for their help in conducting the experiments and processing the data.

REFERENCES

1. S. Ohigashi, H. Kuroda, Y. Nakajima, Y. Hayashi and K. Sugihara, "Heat Capacity Changes Predict Nitrogen Oxides Reduction by Exhaust Gas Recirculation." SAE Paper 710010
2. H. Kuroda, Y. Nakajima, Y. Hayashi and K. Sugihara, "An Approach to the Low Emission Engine with the Consideration of Fuel Economy." SAE Paper 750416
3. Y. Kaneko, "The Effects of Exhaust Gas Recirculation and of Residual Gas on Engine Emission and Fuel Economy." SAE Paper 750414
4. T. Mizunuma, T. Sato and I. Matsuno, "An Approach to Combustion Control for Low Pollution Engines by two Staged Combustion Process." 16th International Congress of FISITA, May 1976
5. S. Sanda, T. Toda, H. Nohira and T. Konomi, "Statistical Analysis of Pressure Indicator Data of an Internal Combustion Engine." SAE Paper 770882
6. S. Ohigashi, Y. Hamamoto and A. Kizima, "Effects of Turbulence on Flame Propagation in Closed Vessel." Bulletin of the Japan Society of Mechanical Engineers Vol. 14 No. 74 1971
7. Athor A. Quader, "Effects of Spark Location and Combustion Duration on Nitric Oxide and Hydrocarbon Emissions." SAE Paper 730153
8. Y. Hosho, Y. Oyama, T. Yamauchi and T. Nishimiya, "Exhaust Emission Reduction with Lean Mixture (2)" NAINENKIKAN Vo. 13 No. 152 1974
9. Y. Sakai, K. Kunii, M. Sasaki, N. Kakuta and H. Aihara, "Combustion Characteristics of a Torch Ignited Engine — Analytical Measurements of Gas Temperature and Mixture Formation" Conference on Stratified Charge Engines. London Nov. 1976
10. A. A. Quader, "What Limits Lean Operation in Spark Ignition Engines — Flame Initiation or Propagation?" SAE Paper 760760
11. S. Matsuoka, T. Yamaguchi and Y. Umemura, "Factors Influencing the Cyclic Variation of Combustion of S. I. Engine." SAE Paper 710586
12. Y. Kim, "Mechanism of Cyclic Combustion Variation." Preprint of the Japan Society of Mechanical Engineers, No.740-5, April 1974

APPENDIX A

Table A-1 — Fuel Specifications

Designation	:	Silver N Gasoline
Specific gravity	:	0.753
RON	:	90.6
Distillation °C	:	IBP 33.0
		10% 55.5
		50% 103.0
		90% 155.5
		EP 184.5
RVP kg/cm² (at 37.8°C)	:	0.620
Lead ml/l	:	0.001 (−)
Sulfur wt %	:	0.001 (−)
Aromatic vol %	:	40.5
Olefin vol %	:	0.5

APPENDIX B

Relationship between acceptable engine operating stability limit and P_i fluctuation rate (C_{pi}) at spark timing different from MBT.

As seen in Figure B-1, the dashed line representing the human sensing stability scale of 3 almost falls in with the line for C_{pi} of 10% at spark timing retarded from MBT. In this region, P_i distributions for different spark timings at the same stability level also coinside well with each other as shown in Figure B-2. Accordingly, it can be said that the engine stability limit is determined by P_i distribution or C_{pi} at the spark timing of MBT or retarded from MBT. At the spark timing advanced from MBT, however, equi-C_{pi} lines do not coincide with equi-stability lines as seen in Figure B-1. This seems to indicate that the factors dominating the engine stability limit at the advanced side are different from those at the retarded side. At the advanced side, very few misfire or close-to-misfire cycles will hurt the engine stability very much, while the combustion in most of the other cycles is fairly good (Figure B-3). This result corresponds well to the result of A. A. Quader using a mono cylinder engine (10).

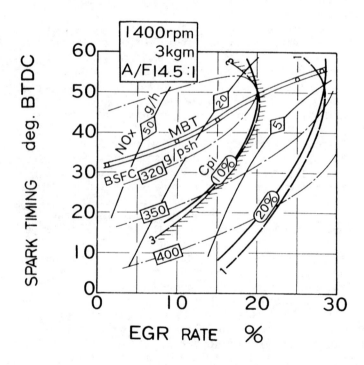

Fig. B-1 - Relationships among EGR rate, spark timing, and engine stability limit

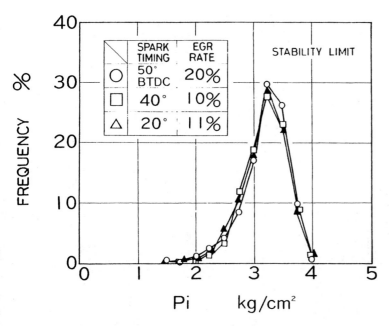

Fig. B-2 - P_i distribution at acceptable engine stability limit

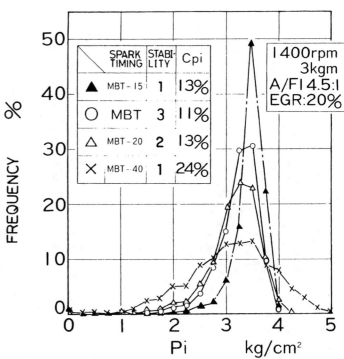

Fig. B-3 - Effects of spark timing on P_i distribution and engine stability

APPENDIX C

From the studies on combustion fluctuation in the past, when the problem of exhaust emissions was not a concern for worry as it is today, it was suggested that the reduced combustion duration was effective for decreasing the combustion fluctuation (11) (12). The effect of reduced combustion duration on the emissions, however, has not been examined yet. In addition, present experiments have proven that the combination of the fast burn and heavy EGR would be effective in a marked reduction of NO_X and fuel economy. From the reasons mentioned above, a mathematical model study was carried out to examine the relationship between NO_X emission and fuel economy with the fast burn and heavy EGR. The relationship between ISFC and formed NO in the cylinder was calculated according to the block diagram as shown in Figure C-1. In this calculation the pattern of heat release rate was given as a triangular shape, and the P-V diagram was calculated from this heat release rate.

Equilibrium concentration of combustion products such as CO_2, H_2O, CO, H_2, N_2, H_2, OH and O are calculated by using the following equations.

$$CnHm + \frac{1}{\phi}(n + \frac{m}{4})(O_2 + 3.76N_2) + \psi(aCO_2 + bH_2O + cCO + dH_2 + eO_2 + fN_2 + gOH + hO) = aCO_2 + bH_2O + cCO + dH_2 + eO_2 + fN_2 + gOH + hO$$

$$2CO + O_2 \rightleftarrows 2CO_2$$
$$2H_2 + O_2 \rightleftarrows 2H_2O$$
$$4OH \rightleftarrows 2H_2O + O_2$$
$$O_2 \rightleftarrows 2O$$

where ϕ is the equivalence ratio and ψ is the factor derived from the sum of the EGR rate and residual gas. Equilibrium constants were taken from the JANAF Gas Table.

In this procedure the effects of EGR were considered for the reduction of combustion gas temperature due to the increase of heat capacity and gas concentration in the cylinder.

In order to calculate NO concentration, the kinetic equation derived from the following extended Zel'dovich reaction formulae were used.

$$N_2 + O_2 \underset{K_2}{\overset{K_1}{\rightleftarrows}} NO + N$$

$$O_2 + N \underset{K_4}{\overset{K_3}{\rightleftarrows}} NO + O$$

$$N + OH \overset{K_5}{\rightarrow} NO + H$$

And the reaction rate constant as shown in table C-1 was utilized for this calculation.

The results are given in Figure C-2. Short combustion duration (fast burn) yields higher NO_X emission around MBT spark timing, while at slightly retarded timing combustion duration does not effect on the relation between NO_X and ISFC.

Combining the above mentioned past studies and this result, at the same NO_X and ISFC level, faster combustion would give better stability with an actual engine. In other words, Figure C-2 could be understood to show that combustion stabilization with heavy EGR realized by the fast burn made a marked reduction in NO_X emission possible.

Table C-1 — Reaction rate constants for the calculation of NO emissions

Reaction rate constant	A	b	E
K_1	7.0×10^{13}	0	75500
K_2	1.55×10^{13}	0	0
K_3	13.3×10^9	1	7080
K_4	3.2×10^9	1	39100
K_5	4.0×10^{13}	0	0

$$K = AT^b \exp(-E/RT) \quad cm^3/mol \cdot S$$

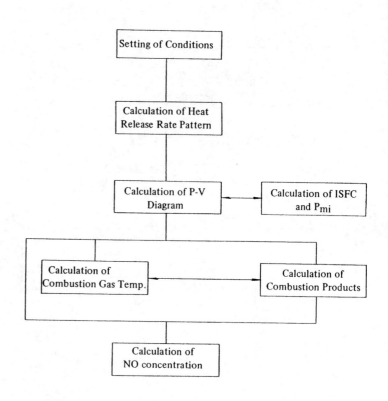

Fig. C-1 - Block diagram for the calculation of NO concentration

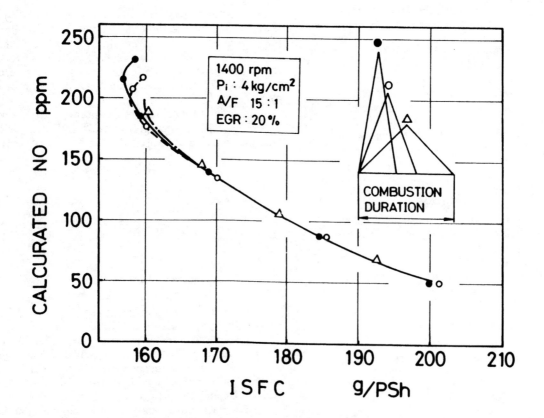

Fig. C-2 - Relationship between calculated ISFC and NO emission for various combustion duration with heavy EGR

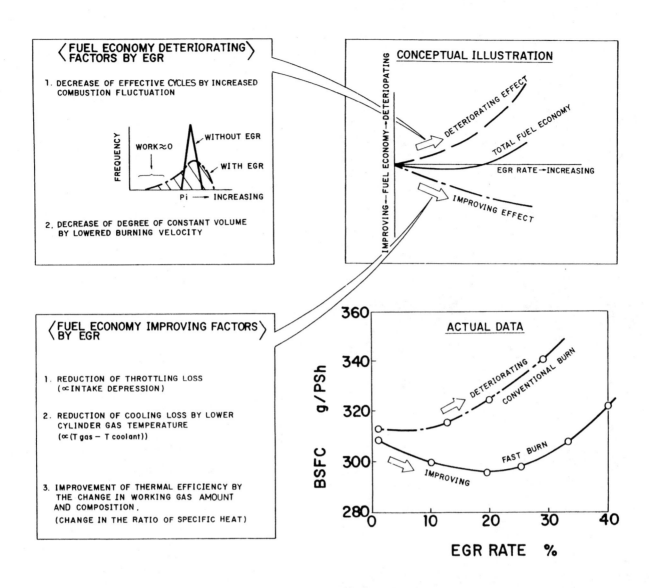

Fig. D-1 - How EGR effects fuel economy

APPENDIX D

FACTORS EFFECTING FUEL ECONOMY IMPROVEMENT BY EGR.

As shown in Figure D-1, EGR can either increase or decrease fuel economy. Actual gain or loss is determined by the relative strength of respective factors which is the function of combustion condition. Since the increase of combustion fluctuation and the decrease in thermal efficiency appear at relatively low EGR rates in the conventional engine and the reduction of pumping loss is small in this range, deterioration in fuel economy begins at low EGR rate.

With fast burn, however, the factors improving fuel economy can be fully utilized because the combustion is stable even at high EGR rates.

780203

Ford Three-Way Catalyst and Feedback Fuel Control System

Robert E. Seiter and Robert J. Clark
Ford Motor Co.

THE PURPOSE OF THIS PAPER is to describe the Ford three-way catalyst and feedback fuel control system (TWC/FB), its function and performance, and discuss the developmental rationale behind it. The TWC is so named because it promotes reactions of all three regulated pollutants; HC, CO and NOx. At the present time, Ford also uses a conventional oxidation catalyst (COC) in its TWC/FB system to further reduce HC and CO emissions. As a 1978 pilot program, the system will be installed on approximately 30,000 Pinto and Bobcat vehicles for sale in California. This pilot program was undertaken to gain experience with TWC/FB systems in anticipation of expanded usage to meet future emission standards. To enable direct comparison with conventional California emission control systems, the emission targets selected for the program were the 1978 California standards of 0.41 gm./mi. HC, 9.0 gm./mi. CO and 1.5 gm./mi. NOx rather than proposed future NOx levels of 1.0 or 0.4 gm./mi. Secondary program objectives were to support initial 1978 production, improve fuel economy over 1977 California levels, and incur no depreciation in other vehicle functional areas such as driveability and performance.

This paper is organized into the following divisions:
- Background of Early Ford TWC Research
- Description of the Production System
- Performance of the Production System
- Discussion of Major Developmental Issues
- Summary
- Prognosis for Future Applications

BACKGROUND

Early TWC testing at Ford[1,2,3]* began in 1968 with evaluations of the NOx reducing capability of conventional oxidizing catalyst (COC) formulations when operated near stoichiometry.

---
* Numbers in parentheses designate references at end of paper.

ABSTRACT

The objective of this paper is to describe the Ford Motor Company (Ford) approach of meeting exhaust emission regulations with a three-way catalyst and feedback control system. A pilot program was initiated to gain production experience with three-way catalyst systems in anticipation of expanded usage to meet future emission standards.

The Ford system consists of a three-way catalyst with feedback control monitoring the exhaust oxygen concentration and controlling the fuel flow to produce a stoichiometric exhaust mixture. Mixture control is critical since catalyst NOx conversion efficiency is diminished when the exhaust mixture deviates from stoichiometry. Briefly, the control loop consists of zirconium dioxide exhaust sensor to indicate oxygen concentration, an electronic control unit, a vacuum regulator to proportion a vacuum signal to the carburetor, and a feedback controlled carburetor with vacuum modulated main fuel system.

Highlights of the development program and system details are described.

0148-7191/78/0227-0203$02.50
Copyright © 1978 Society of Automotive Engineers, Inc.

Steady state dynamometer testing indicated that, although peak COC efficiency for NOx reduction was relatively high at stoichiometry, small changes in air/fuel (A/F) ratio either rich or lean from stoichiometry caused a dramatic decrease in efficiency. Even with early TWC formulations, it appeared that the A/F ratio would have to be controlled within ± 0.1 A/F of stoichiometry to achieve good simultaneous conversion of HC, CO and NOx.

It was subsequently discovered[4] that cycling the feedgas A/F during dynamometer testing to simulate the control of a carbureted system significantly widened the apparent window of A/F in which acceptable simultaneous conversion occurs. Cycling the feedgas had the effect of reducing the peak efficiency at a stoichiometric mean A/F while increasing the efficiency at other mean A/F ratios close to stoichiometry (Figure 1). Dynamometer data indicated that acceptable NOx conversion could be obtained over an A/F range great enough to make carburetor control feasible.

Favorable TWC durability results along with the availability of a feasible carburetor and control system design prompted Ford to embark on a production pilot program in early 1976. The rationale behind the pilot TWC/FB program was to gather valuable certification, production, and field experience for future designs.

DESCRIPTION OF THE PRODUCTION SYSTEM

The production system can be described in three sub-systems: catalyst; closed loop fuel control; Thermactor secondary air controls. After describing each sub-system, the composite operation of the TWC/FB system can be more readily understood.

CATALYTIC CONVERTER - The catalytic converter assembly consists of two separate and unique catalytic substrates (Figure 2). The front substrate is the TWC which treats HC, CO and NOx emissions. The TWC contains platinum and rhodium in the ratio of 11:1 or 9:1 depending on EPA-emission family, with a loading rate of 50 gm./ft.^{3}. The catalyst formulation was developed by Engelhard Industries. In addition to oxidizing HC and CO into CO_2 and H_2O, when operated lean of stoichiometry or even when operated rich of stoichiometry for short periods of time, the TWC reduces NO into N_2 and CO_2 when operated slightly rich of stoichiometry. Some of the NO also combines with H_2 to produce ammonia (NH_3).

The rear substrate is the COC which treats HC and CO emissions when operated in an atmosphere having an excess of oxygen. The substrate coating contains platinum and paladium in the ratio of 3:2 or 2:1 depending on EPA-emission family, with a loading rate of 25 gm./ft.^{3}. To assure sufficient oxygen under all warmed-up operating conditions, secondary air is injected into the catalyst assembly just ahead of the COC substrate via a venturi ring which assures even distribution of air across the catalyst face. The COC oxidizes HC and CO into

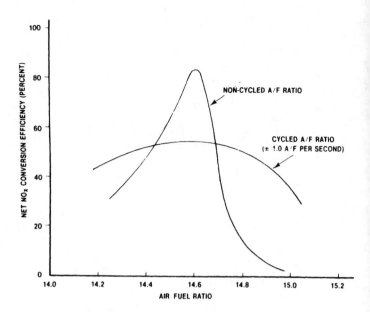

Fig. 1 - TWC/COC net NO_x conversion efficiency versus A/F ratio with and without A/F cycling

CO_2 and H_2O, but also causes some of the NH_3 produced by the TWC to be reoxidized into NO and H_2O.

The NO_x reformation due to the addition of a clean-up COC catalyst has a considerable negative effect on overall NO_x efficiency (Figure 3). The dual substrate catalyst was an essential compromise for obtaining simultaneous conversion of HC, CO and NOx over a wide A/F ratio range.

CLOSED LOOP FUEL CONTROL - To maintain the A/F within the limits for efficient catalysis, a feedback controlled fuel metering system is used. A standard carburetor offers the range of A/F control shown in the large band shown in Figure 4, which is unsatisfactory for TWC applications. A feedback controllable carburetor was developed to maintain the A/F within the narrower band shown in Figure 4. The use of a feedback controlled carburetor is one factor which makes the Ford system different from earlier industry TWC systems which used fuel injection [5,6,7].

The feedback control loop senses the composition of exhaust gas, engine speed and coolant temperature, and responds by continuously modulating the carburetor fuel flow towards a stoichiometric A/F mixture. Feedback control inherently compensates for other factors that affect A/F such as temperatures, altitude, and fuel composition. The components comprising the control system include an exhaust gas oxygen (EGO) sensor, electronic control unit (ECU) and vacuum regulator-solenoid (VRS). Prior to further description a definition of the two system operating modes, closed loop and open loop, will be made.

Closed Loop: In the closed loop mode the feedback system is operational and continuously corrects the A/F toward a stoichiometric mixture. Figure 5 shows that a continuous chain of sensing and actuating components form a loop. Each component is

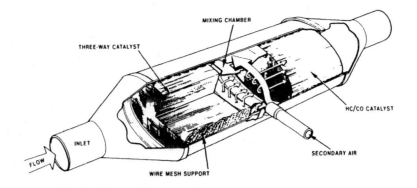

Fig. 2 - TWC/COC dual catalytic converter assembly

sensitive to the performance of the upstream component and responds by communicating to the downstream component.

Open Loop: Carburetor fuel flow is not corrected by the control system in the open loop mode. The open loop mode is activated whenever a stoichiometric A/F is not desired. Thus, open loop operation is activated during engine warm-up when the EGO sensor is below operating temperature or where the feedback correction would interfere with effective choke operation. It is also activated at idle and closed throttle when the idle fuel system is the primary fuel metering circuit.

Feedback Carburetor - As mentioned earlier, a feedback carburetor is used to maintain the air-fuel ratio within the limits for efficient catalysis in the TWC. Fuel flow is varied in proportion to commands from the control loop to a feedback fuel metering valve within the carburetor. The commands are received by the carburetor in the form of a vacuum signal ranging between 0-5 inches Hg.

Figure 6 illustrates the feedback fuel metering valve and its operation. The feedback fuel metering valve consists of a tapered metering rod in an orifice. The relative position of the rod in the orifice limits fuel in the feedback circuit. The position of the rod is varied through a piston and diaphragm assembly and in proportion to a vacuum signal from the vacuum regulator-solenoid (hence ECU command). Fuel from the feedback fuel circuit supplements fuel entering through the main metering system. The feedback fuel metering valve enables control of fuel flow in the feedback fuel circuit, and thereby trims the A/F mixture reaching the engine. The carburetor has the capability to make mixture adjustments of approximately $\pm$ 2 A/F ratios. Figure 7 correlates the signal given to the carburetor with the fuel metering responses that occur.

In all aspects, except the feedback fuel circuit, the FB carburetor is similar to the non-feedback carburetor used on other 2.3L engine applications. In the feedback version, the power valve circuit has been eliminated and the space used to accommodate the feedback fuel circuit. Sufficient power enrichment is provided within the feedback control system for satisfactory driveability. During the full throttle accelerations the control system directs

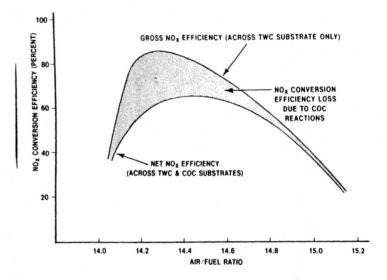

Fig. 3 - Gross versus net NO_x conversion efficiency

the carburetor to provide richer mixtures, rather than correcting toward stoichiometric mixtures.

A conventional idle circuit is used in the feedback carburetor. The idle fuel mixture is not feedback controlled and the control system is in the open loop mode at idle.

Exhaust Gas Oxygen (EGO) Sensor - The EGO sensor, which was designed and developed by Robert Bosch GmbH [8] is mounted in the exhaust manifold and provides an electrical signal to the ECU as a function of the oxygen level in the exhaust gas. The EGO Sensor has the unique property of producing an abrupt change in voltage output at a stoichiometric A/F which is precisely the operating condition required for a TWC (Figure 8). An output signal of 0.6-1.0 volt is generated when the exhaust gas mixture is rich while the output signal generated in the presence of a lean exhaust gas mixture is 0.2 volt or less.

The EGO Sensor is basically a galvanic device with a zirconium dioxide (ZrO_2) solid electrolyte and porous platinum electrodes. The sensor is constructed such that the outer electrode is exposed

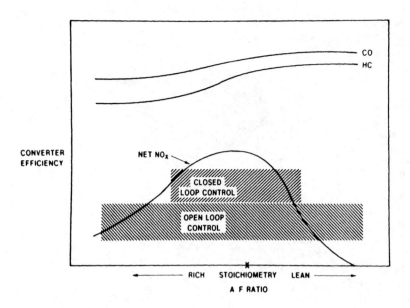

Fig. 4 - Typical TWC/COC conversion characteristics

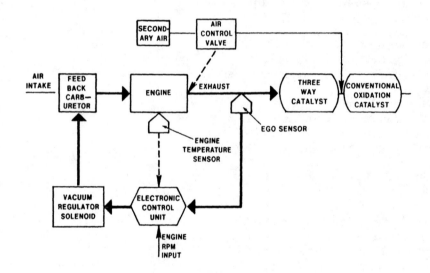

Fig. 5 - Three-way catalyst/feedback control system block diagram

to the exhaust gas while the inner electrode is exposed to atmosphere (Figure 9). When hot (300°C), the electrolyte allows the passage of oxygen ions from one electrode to the other while blocking the passage of others. The resulting oxygen pressure differential produces an electrical potential. The abrupt change in the partial pressure of oxygen which occurs at stoichiometry produces the switch-like change in electrical output which is so critical to the design of any closed loop TWC control system.

As used in the 1978 Ford TWC/FB system, maintenance or replacement of the EGO sensor will normally not be required over 50,000 miles. This is a significant difference from other closed loop TWC applications which require EGO sensor replacement at 15,000 mile intervals.

Electronic Control Unit (ECU) - The ECU, which resembles the Ford Dura Spark electronic ignition module in outward appearance, is the information processing component of the feedback control system. It receives inputs from the EGO sensor and other engine operating condition signaling devices to be discussed later in the paper. The ECU interprets the various inputs and then transmits the proper output signal to the vacuum regulator-solenoid.

During closed loop operation, the ECU output consists of a varying average electrical output which will constantly adjust the vacuum regulator-solenoid duty cycle (percent on time) to maintain stoichiometric operation.

Open loop ECU output consists of a constant average electrical output which will produce a fixed level of the vacuum regulator-solenoid duty cycle. Thus, the position of the carburetor feedback diaphragm is fixed when closed loop operation is not desired.

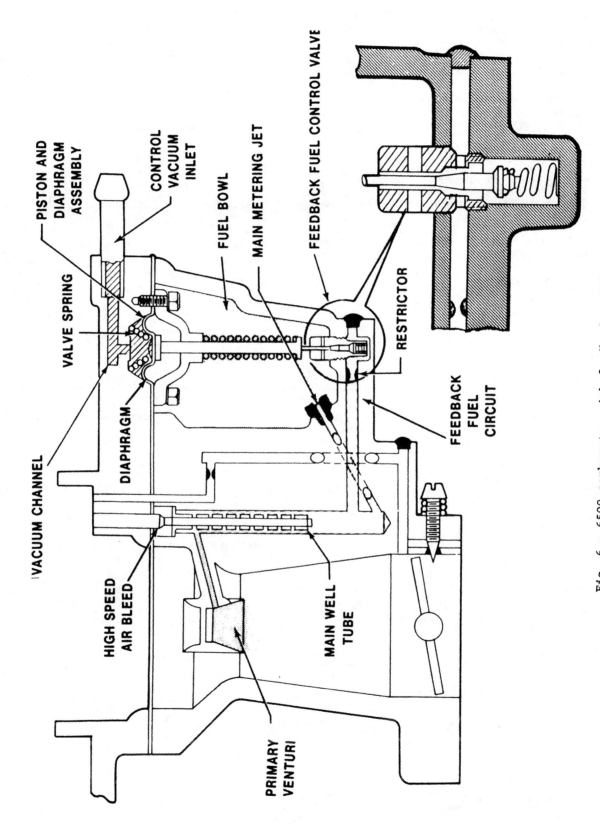

Fig. 6 - 6500 carburetor with feedback modulated fuel metering system

Calibration flexibility items provided for in the ECU are as follows including present specifications:

- **Base Ramp Rate:**
  The speed at which the A/F correction signal is changed. It is expressed in the amount of time required to change from a full lean to a full rich correction signal and is rated at 1000 rpm. Presently, two different ramp rate functions are available; a slow ramp rate with a 15-30 second calibration range and a fast ramp rate having a 7.5-15 second range. Only the slow ramp rate is presently used and is set at 20 seconds.

- **Engine Speed Dependence:**
  The ECU can be programmed to provide ramp rate revisions automatically as an inverse of engine rpm or to make no adjustment as the calibration dictates. The present system is engine speed dependent.

- **Open Loop Duty Cycle:**
  Two different open loop outputs or duty cycles may be selected between 0-100 percent vacuum regulator-solenoid duty cycle. Only one of these is used for 1978 and it is set at 50 percent.

- **Bias:**
  By selecting different speeds for rich to lean and lean to rich correction, the A/F ratio to which the system controls can be adjusted to be slightly rich or lean of stoichiometry. This feature was not used for 1978.

Vacuum Regulator-Solenoid - The combination vacuum regulator and solenoid (VRS) unit both pulsates and modulates the vacuum signal supplied to the feedback carburetor as a function of the electrical output from the ECU. Manifold vacuum is applied to the lower portion of the unit (Figure 10), containing the vacuum regulator which reduces manifold vacuum to a constant 5 inches of Hg. The armature of the electric solenoid which is located in the upper portion of the unit is fitted with a conical tip at each end. In the current off condition (as illustrated) the output vacuum port is exposed to atmosphere through the vent while the 5 inch Hg vacuum signal from the vacuum regulator is blocked producing a 0 inch Hg carburetor vacuum signal. In the current on condition, the armature rises to

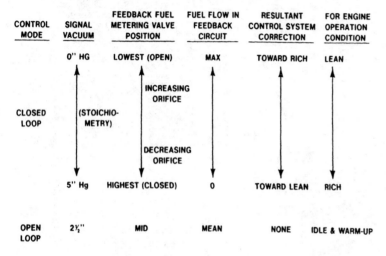

Fig. 7 - Feedback carburetor signal vacuum versus fuel metering response

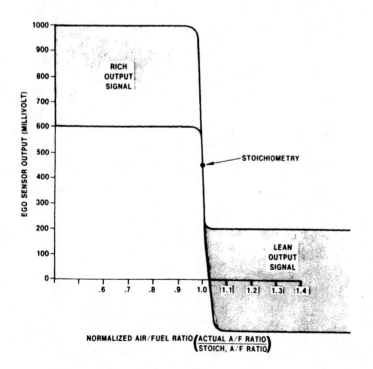

Fig. 8 - EGO sensor voltage output range versus normalized air/fuel ratio

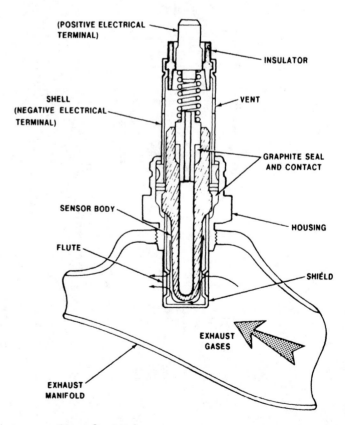

Fig. 9 - Exhaust gas oxygen sensor

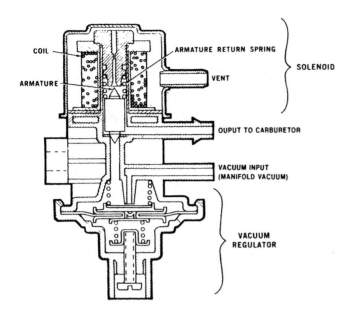

Fig. 10 - Vacuum regulator - solenoid

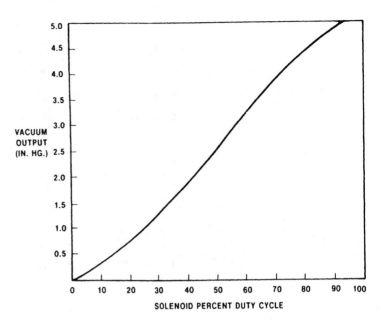

Fig. 11 - Vacuum regulator solenoid percent duty cycle versus vacuum output

block off the vent and allow the full 5 inch Hg vacuum signal to be applied to the carburetor.

To achieve vacuum signals between 0 and 5 inches of Hg the current is supplied and removed by the ECU at a constant rate of 10 cycles per second while the length of time the armature rests in the on (full 5 inch Hg vacuum) or off (0 inch Hg vacuum) position determines the average vacuum supplied to the car= buretor. Vacuum output level as a function of duty cycle appears in Figure 11.

Other Control Loop Inputs - The ECU receives two other inputs from vacuum electric switches which indicate when open loop operation is required. A normally closed "idle vacuum" switch, with vacuum signal input from the carburetor spark port, is used to indicate idle and closed throttle conditions. Spark port vacuum is zero at closed throttle, the electrical switch is closed, a completed circuit to the ECU is formed, and the ECU is structured to provide open loop operation at that point. As the throttle is opened spark port vacuum rapidly rises above the switch point resulting in an open circuit and subsequent closed loop operation.

A similar scheme is used to determine if engine coolant is above predetermined temperature. A combination ported vacuum signal (PVS) valve and a normally open "cold start" switch are used for this function. The PVS valve ports manifold vacuum to the vacuum electric switch only below 125°F coolant temperature. This furnishes an "on-off" signal as a function of temperature. Therefore, the normally open vacuum switch provides a closed circuit (vacuum applied) during operation below 125°F. The ECU is thereby signaled to provide open loop operation until the EGO sensor has reached its operating temperature and the choke cycle has been completed.

THERMACTOR SECONDARY AIR CONTROL - The requirement for stoichiometric exhaust mixture at the TWC has necessitated management of the secondary air (Thermactor) system. Secondary air is injected at the exhaust ports during warm-up for control of feedgas HC and CO plus promotion of rapid catalyst light-off (Figure 12). During this mode the TWC functions as an oxidation catalyst. The air injection points at the exhaust ports are identical to those used for other 2.3L engines equipped with Thermactor. After EGO sensor and catalyst operating temperatures are reached, the secondary air flow is diverted from the exhaust ports to enter the exhaust stream after the TWC and just in front of the COC (Figure 13). The Thermactor components, such as the air pump and bypass valve, are the same as those used on the non-TWC applications. An air control valve and a new air supply tube have been added to the system to accomplish the necessary air switching function.

COMPOSITE SYSTEM OPERATION - The concurrent operation of the three sub-systems described above produces a controlled A/F ratio and simultaneous catalysis of all three regulated pollutants. Figure 14 is an overall system schematic diagram.

During engine warm-up, the system is operated in the open loop mode without feedback control. A conventional choke system provides cranking and warm-up enrichment. Secondary air is injected at the exhaust ports.

When the engine coolant temperature reaches 125°F, the catalyst and EGO sensor have reached their operating temperatures, the choke system has cycled and the engine begins closed loop (feedback controlled) operation. The secondary air supply is routed downstream of the TWC. Undiluted exhaust gas flows past the EGO sensor which responds in proportion to the exhaust gas oxygen level. The feedback loop and carburetor correspondingly alter the fuel flow toward a stoichiometric A/F ratio. Thus, the TWC is presented with an exhaust mixture suitable for

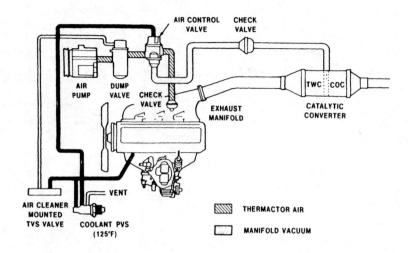

Fig. 12 - Thermactor air control system - Engine coolant **temperature** lower than 125°F

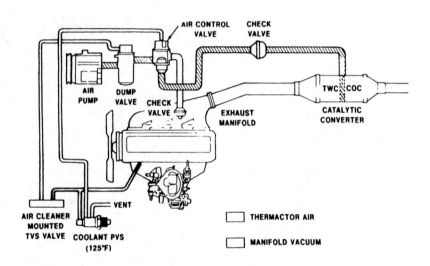

Fig. 13 - Thermactor air control system - Engine **coolant** temperature higher than 125°F

efficient operation. Figure 15 gives two examples of the chain of action that occurs to continuously drive the A/F ratio toward a stoichiometric mixture.

The resultant effect of closed loop fuel control is illustrated in Figure 16. During hot CVS emission testing, continuous Ricardo instrument A/F measurements were made with closed loop fuel control and with a conventional carburetor. Representative portions of those A/F traces are shown in Figure 16.

All other components installed with this system are conventional and have been used in other 2.3L engine calibrations. Those components include ported exhaust gas recirculation (EGR). Catalytic control of NOx did provide the opportunity for reduction of EGR with inherent improvement in driveability and fuel economy.

SYSTEM PERFORMANCE

EMISSION - Two vehicles successfully completed the Federal 50,000 mile emission durability test sequence. The significant difference between the cars was catalyst composition, that is platinum to rhodium (Pt/Rh) ratio and platinum to pladium (Pt/Pd) ratio. Emission levels and deterioration factors over 50,000 miles are summarized in Figure 17.

One 50,000 mile vehicle meets the necessary criteria to qualify this TWC/FB system against the 1981 emission standards of 0.41 gm./mi. HC, 3.4 gm./mi. CO and 1.0 gm/mi. NOx. Requirements in 1981 for SHED, end-of-line surveillance test, etc., are not addressed here but will have significant impact on engine calibrations for that model year.

Certification of 4,000 mile vehicles was completed in time to support initial 1978 production. Three different calibrations were certified for the following Pinto and Bobcat applications:

- Manual Transmission   2.73 Axle Ratio   2750 Inertia Weight   Sedans
- Manual Transmission   3.18 Axle Ratio   3000 Inertia Weight   Station Wagon (Optional Axle for Sedan)
- Auto. Transmission   3.18 Axle Ratio   2750 & 3000 Inertia Weight   Sedan and Station Wagon

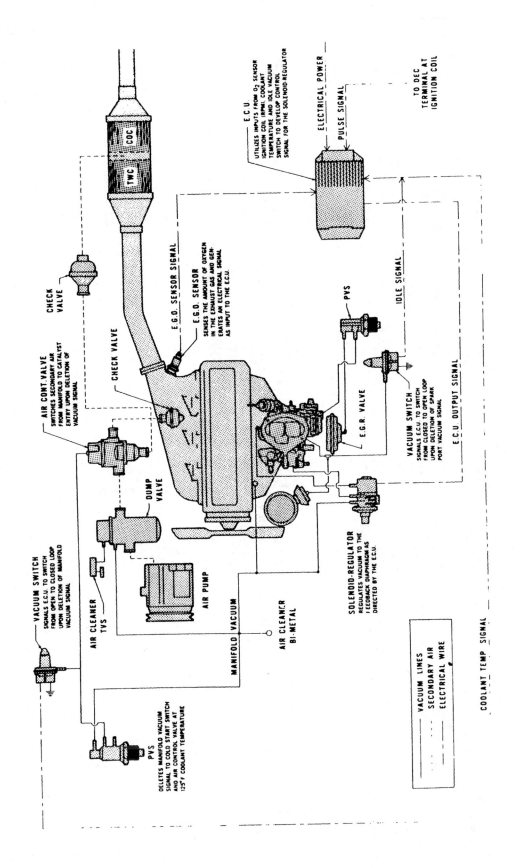

Fig. 14 - "TWC+COC" system (with feedback carburetor)

ENGINE OPERATING CONDITION	EGO SENSOR OUTPUT	ECU	VACUUM SOLENOID REGULATOR	VACUUM SIGNAL TO CARBURETOR F.B. CIRCUIT	POSITION OF METERING ROD	FUEL FLOW	RESULTANT CORRECTIONS
RICH OF STOICHIOMETRY	HIGH OUTPUT VOLTAGE (<.35 VOLT)	DIRECTS VAC. SOL.-REG. TO GREATER "ON TIME" (50→100%)	INCREASED DUTY CYCLE RESULTS IN HIGHER OUTPUT VACUUM (>2.5 IN. Hg.)	>2.5 IN. Hg SIGNAL PULLS METERING ROD UP (SMALLER ORIFICE)	HIGHER POSITION FOR DECREASE IN F.E. FUEL FLOW	DECREASED	TOWARD LEAN
LEAN OF STOICHIOMETRY	LOW OUTPUT VOLTAGE (>1 VOLT)	DIRECTS VAC. SOL.-REG. TO LESS "ON TIME" (50%→0)	DECREASED DUTY CYCLE RESULTS IN LOWER OUTPUT VACUUM (<2.5 IN. Hg)	<2.5 IN. Hg SIGNAL MOVES METERING ROD DOWN (LARGER ORIFICE)	LOWER POSITION FOR INCREASED F.B. FUEL FLOW	INCREASED	TOWARD RICH

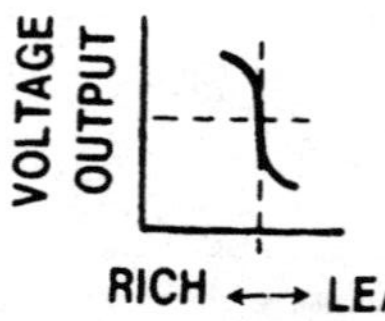

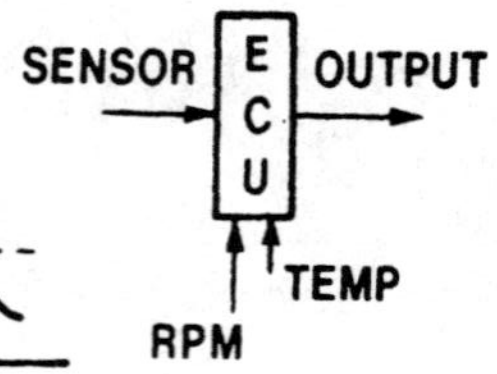

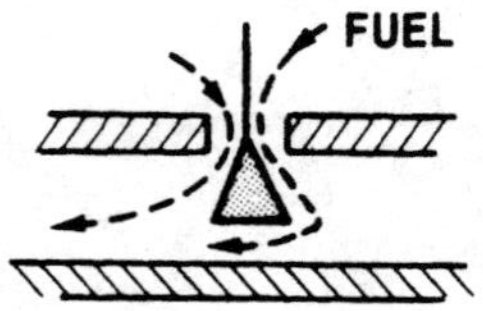

Fig. 15 - Control loop operating sequence

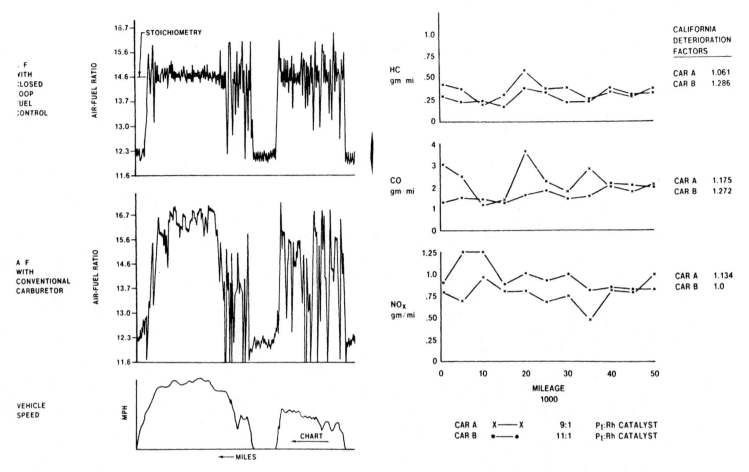

Fig. 16 - Comparison of closed loop and conventional fuel control

Fig. 17 - 50 000 mile TWC/FB emission durability

Table 1 - Summary of emission data from 4,000 mile TWC/FB vehicles

CALIBRATION	APPLICATION	DATA SOURCE	FEEDGAS			TAILPIPE			CATALYST EFFICIENCY			VEHICLES/TESTS
			HC	CO	NOx	HC	CO	NOx	HC	CO	NOx	
8-2T-R0	2750 INERTIA WT. MANUAL TRANSMISSION	EPA FORD	— 2.94	— 31.23	— 1.67	.40 .30	1.60 1.95	.66 1.04	— 90%	— 94%	— 38%	1/1 3/14
8-2P-R0	3000 INERTIA WT. MANUAL TRANSMISSION	EPA FORD	— 3.05	— 28.47	— 1.82	.37 .26	2.20 1.47	.66 1.04	— 91%	— 95%	— 43%	1/1 2/8
8-1P-R1	2750 INERTIA WT. AUTOMATIC TRANSMISSION	EPA FORD*	— 2.01	— 22.25	— 1.90	.24 .15	3.00 1.51	.68 .91	— 93%	— 93%	— 52%	1/1 1/4
8-1P-R1	3000 INERTIA WT. AUTOMATIC TRANSMISSION	EPA FORD*	— 1.93	— 21.37	— 1.71	.24 .18	5.00 2.23	.83 .91	— 91%	— 90%	— 47%	1/1 1/3

Table 1 summarizes the emission data from 4,000 mile certification vehicles as well as from some representative Ford development vehicles. Feedgas emission data (before catalyst) are indicated for the development vehicles. Typically, between 38 and 52% NOx conversion efficiency was found with the TWC/FB system. Simultaneously, 90 to 93% HC and 90 to 95% CO conversion efficiencies were maintained.

FUEL ECONOMY The largest group of data available to compare the TWC/COC feedback system to a COC non-feedback system is the developmental data for each of the systems. These data are judgmentally comparable even though vehicle differences existed. A comparison of these data (Table 2) is useful to describe the gross effects as follows:

. The TWC/COC feedback system has an impressive 7 to 14% fuel economy advantage over a COC non-feedback system when calibrated to the same 0.41 gm./mi. HC, 9.0 gm./mi. CO, 1.5 gm./mi. NOx 1978 California emission standards.

Table 2 - Fuel economy comparison

	TWC/COC FUEL ECONOMY ADVANTAGE EPA CYCLE M-H AVERAGE	
TWC/COC FEEDBACK CALIFORNIA VS COC NON-FEEDBACK CALIFORNIA	MILES/GALLON	Δ%
MANUAL TRANSMISSION CALIBRATION	+1.5	+7
AUTOMATIC TRANSMISSION CALIBRATION	+3.0	+14
TWC/COC FEEDBACK CALIFORNIA VS COC NON-FEEDBACK FEDERAL		
MANUAL TRANSMISSION CALIBRATION	-1.0	-4
AUTOMATIC TRANSMISSION CALIBRATION	0	0

Fuel economy of vehicles developed against Federal Standards of 1.5 gm./mi. HC, 15 gm./mi. CO, 2.0 gm./mi. NOx was better than fuel economy attained with TWC systems (against California standards). Even though the TWC/FB system provided the capability to obtain higher fuel economy than conventional systems, the more stringent California emission standards still accounted for a loss of 1.0 mpg fuel economy from Federal levels for manual transmission vehicles.

The fuel economy advantage of the automatic transmission (A/T) TWC/COC feedback system is twice that of the manual transmission (M/T) when compared to the corresponding COC non-feedback California system. This difference is explained largely by the fact that the EGR rate of the COC non-feedback A/T calibration was significantly higher than that of the M/T.

VEHICLE DRIVEABILITY - The drive quality of vehicles equipped with TWC/COC feedback system is improved over that of vehicles with the conventional California emission systems at similar emission levels principally because the EGR rates are significantly reduced. As earlier illustrations indicated, there are many part throttle operating modes in which the A/F is actually leaner than a non-feedback system. The driveability is, however, generally improved since the EGR rate reductions more than offset the A/F change. The exception to this is that part throttle response, although still acceptable, was compromised due to the deletion of the power valve.

Wide open throttle (WOT) performance is equal to that of conventional systems. Since the feedback vacuum source is manifold vacuum and there are no vacuum reservoirs in the system, there is no vacuum available to effect a lean correction during a WOT acceleration. This full rich condition is calibrated to satisfy WOT fuel flow requirements of being between LBT (least fuel for best torque) and LBT + 6% as are the conventional 2.3L calibrations.

Idle quality is not depreciated over conventional systems since there is no feedback control on the idle system. Early development with a feedback controlled stoichiometric idle A/F yielded unacceptable idle quality particularly on manual transmission vehicles. Subsequent testing indicated that feedback idle was not required since NOx formation at idle was considered insignificant. The feedback idle feature was deleted in favor of idle quality.

The feedback metering rod is positioned in a fixed neutral (50% duty cycle) open loop position during cold start until engine coolant temperature reaches $125^{\circ}F$. This cold start open loop strategy results in choke calibrations that are similar to those of conventional carburetor systems.

MAJOR DEVELOPMENTAL ISSUES

IDLE SYSTEM CONTROL-Early system development included a closed loop idle system since it was estimated that the idle circuit contribution to the total A/F would have to be controlled to support proper catalyst function. The idle system was similar to the main feedback system in that a diaphragm and rod assembly was employed and it received the same signal as the main system. The idle feedback mechanism, however, controlled an air bleed rather than fuel flow. The first problem to surface with this system was that the idle quality at stoichiometry was marginal on manual transmission vehicles. The second problem was that during a deceleration the feedback system would make a lean correction that sometimes produced stalls on idles immediately following a rapid deceleration.

Open loop idle operation combined with an idle A/F richer than stoichiometry solved the idle quality and stalls after decel problems. Subsequently, however, careful monitoring of test vehicles during the course of development showed that the idle A/F ratio drifted rich with time. Carburetor idle diaphragm distortion was the cause of the problem. Since initial emission tests indicated that an idle feedback system was not required, a conventional idle system was adopted, thus deleting the feedback idle diaphragm.

With a conventional non-feedback idle system, stoichiometric idle quality and decel stalls were no longer an issue, but open loop control was still required. Since all closed throttle operating modes were significantly richer than stoichiometry, the main system feedback diaphragm assembly would be driven to a full lean correction position during these modes. This produced severe hesitation or stumbles on accelerations from an idle or deceleration mode.

The problem was that accelerations from a closed throttle condition began with the carburetor in a full lean correction condition. Open loop closed throttle operation at a 50% duty cycle was, therefore, retained so that the transition to open throttle would begin from a neutral feedback condition.

As a further refinement to the calibration, the richer idle open loop operation was ultimately extended to conditions slightly off idle, to correct a M/T driveability problem under light load stoichiometric operating conditions. This was achieved by raising the vacuum solenoid switch point from 1 inch Hg to 3 inch Hg of spark port vacuum. The idle system was then calibrated so that under low speed cruise operating conditions near the switch point, the transition between open and closed loop control due to slight throttle position changes would not be noticeable.

FEEDBACK CARBURETOR DIAPHRAGM DEVELOPMENT—One of the most significant problems encountered during the development of the feedback carburetor was that of obtaining a control diaphragm having a combination of low hysteresis, stability over time, and good durability. A long series of problems began when it was found that the first diaphragms did not retain flexibility over time with exposure to fuel and heat. This caused the carburetor A/F to drift rich. A development program ensued that would evaluate more than twenty designs involving various elastomer materials and suppliers, manufacturing techniques and dimensional changes.

Originally the rich shift appeared to take place and rapidly stabilize, but additional evaluations showed that the drift continued with time. It was eventually concluded that as a result of the diaphragm being formed from a semi-cured flat sheet, the material tended to return to its original flat shape. This caused diaphragm distortion and disruption of the balance of forces required for proper operation. Molding rather than forming the diaphragm was adopted to solve this problem.

Unfortunately, the drift rich problem was encountered again. Additional testing meanwhile showed that improved materials could be developed. Batch control was improved and design revisions were also incorporated such as; removal of mating surface sealing beads, deletion of the diaphragm gasket, shallower diaphragm depth, larger radii that would conform to the carburetor casting, and the reduction of fastener torque. The revisions were so successful that the durability was improved to a level that was three times the objective. The 50,000 mile durability vehicles not only completed durability successfully, but obtained outstanding deterioration factors.

EXHAUST GAS OXYGEN (EGO) SENSOR Use of the Bosch exhaust gas oxygen sensor has been previously reported in trade publications (7). Other researchers (5) have stated that the sensor mounting location was a critical parameter to both sensor durability and performance. Criteria for sensor location were short warm-up time, minimal thermal shock effects, short response time for detection of A/F ratio changes and using a location where multi-cylinder exhaust flow is well mixed.

Sensor location did not appear to pose a serious problem in the Ford 2.3L engine. The sensor was positioned in the exhaust manifold near the outlet, where exhaust flow is well mixed, and response time was found to be satisfactory. Control system logic is structured to insure the sensor is at operating temperature before closed loop operation begins.

As mentioned earlier, the 1978 Ford TWC/FB system will not require maintenance or replacement of the EGO sensor over 50,000 miles, while other systems currently specify sensor replacement at 15,000 mile intervals. Two factors encouraged the use of EGO sensors for 50,000 miles: (1) Sensor aging characteristics were known; (2) The specific sensor requirements of the Ford control system over 50,000 miles were also known. The Ford TWC/FB system includes EGR, a Thermactor secondary air system, and a dual substrate (TWC + COC) catalyst. These features allowed more latitude in A/F control while maintaining tailpipe emissions below objective. Other systems, on the other hand, have employed only a TWC catalyst which required greater accuracy of A/F control. Effectively, the additional components in the Ford system reduce the overall sensitivity of the system to A/F ratio control. Although the sensor aging characteristics known to occur during 50,000 miles were acceptable for the 1978 Ford TWC/FB

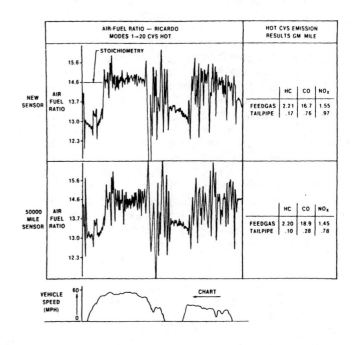

Fig. 18 - Comparison of new and 50 000 mile EGO sensors

system, future emission control systems may require sensor replacement. The sensor maintenance decision will vary as a function of the application and emission standards.

Figure 18 illustrates that aged EGO sensor operating characteristics did not affect A/F control to the point where emissions were significantly altered. Consecutive hot CVS emission tests were conducted with a new and a 50,000 mile aged sensor. Continuous A/F measurements were made with Ricardo A/F measuring equipment. There was a slight shift in mean A/F ratio and loss in accuracy of control when the 50,000 mile sensor was used. Sensor voltage output was diminished with the aged sensor, but it still provided an A/F switching function about the 350 mv. stoichiometric condition which is critical to the control system. The salient point from Figure 18 is that the emission levels (feedgas and tailpipe) are not markedly different when the new or 50,000 mile sensor are used.

RAMP RATE Again, base ramp rate is defined as the time required for the system to move from one correction extreme to the opposite extreme (rich to lean, or lean to rich). Base ramp rate is engine speed dependent in the Ford system. Selection of a suitable ramp rate was accomplished by fabricating electronic control units with capacity for varying the base ramp rate. Although faster ramp rates were desired for more precise A/F control, the limiting factor for ramp rate was driveability. Subjective drive evaluations were conducted at various ramp rates. A base rate of 20 seconds was found to be satisfactory for emission control as well as driveability.

HC CONTROL—Since the stoichiometric A/F required for TWC is controlled at the carburetor, secondary air could no longer be introduced into the exhaust ports. Reductions in HC emissions that normally occurred in the exhaust manifold were lost. The desired stoichiometric A/F also results in feedgas HC levels greater than those typically found at leaner A/F mixtures. The TWC strategy facilitated NOx control but aggravated HC control, which was particularly difficult on M/T applications.

Improvement in catalyst light-off time (time to reach 50% conversion efficiency or 500°F substrate temperature) tended to reduce HC emissions. Light-off time has an important effect since much of the total HC comes from pre-light off modes. An increase in choke cycle high cam engine speed from 2300 rpm (normal for a COC system) to 2700 rpm provided a relatively consistent light-off.

Due to the inherent HC sensitivity of M/T applications, additional development of these calibrations was pursued as a contingency action. Deletion of vacuum spark advance during warm-up modes produced faster and more consistent light-off with slightly reduced HC. The fuel economy effect was insignificant since vacuum advance was deleted for only a short time during cold start warm-up. The calibration strategy that produced the most significant HC reduction over the base calibration was to provide secondary air at the exhaust ports during idles and decelerations, when NOx formulation is low.

This feature gave excellent HC results and posed no control loop difficulty since an open loop closed throttle strategy was already employed.

Another HC control strategy that is still being developed is modulation of both carburetor fuel flow and secondary air to control the A/F ratio presented to the TWC. The strategy has been called "Managed Secondary Air." Early attempts at utilizing secondary air to adjust the A/F of the exhaust mixture were abandoned due to poor fuel economy. To assure stoichiometry at the TWC, the engine had to be operated richer than stoichiometry since there was no capability to correct lean operation. The strategy developed to address this problem was to utilize carburetor modulation for rich corrections and secondary air modulation for lean corrections. Such a system would allow conventional power enrichment beyond stoichiometry for engine response while maintaining stoichiometry at the TWC.

COMPARISON OF ANALOGOUS SYSTEMS—Fig. 19 highlights the composition of the Ford 1978 California TWC/FB system and contrasts it with the 1977 California Volvo (5) system. This comparison illustrates that these systems are similar only in the utilization of TWC technology. Significant differences exist in application of that technology into a total system strategy.

SUMMARY

As measured against each of the established objectives, the TWC/COC catalyst system with closed loop fuel control represents a significant achievement.

FORD AND VOLVO — THREE-WAY CATALYST SYSTEM COMPARISON

	1978 FORD TWC/FB SYSTEM	1977 VOLVO SYSTEM
ENGINE, APPLICATION	2.3L 4 CYLINDER	2.1L 4 CYLINDER
FUEL SYSTEM	FEEDBACK CARBURETOR HOLLEY-WEBER 6500	MECHANICAL FUEL INJECTION BOSCH K-JETRONIC
A/F CONTROL RANGE	± 2 A/F	± .25 A/F (ESTIMATED)
CATALYST SIZE	80 IN^3 TWC + 80 IN^3 COC	102 IN^3 TWC
• TWC Pt:Rh	9:1	5:1
• Pt Rh LOADING RATE	50 GM./FT^3	50 GM./FT^3
• FORMULATION SUPPLIER	ENGELHARD	ENGELHARD
• TYPE	MONOLITH	MONOLITH
• LOCATION FROM EXHAUST FLANGE	48 IN.	38 IN.
• SCHEDULED REPLACEMENT INTERVAL	NONE	NONE
SECONDARY AIR SUPPLY	BETWEEN TWC & COC	NONE
EGR	3—8 CFM	NONE
EGO SENSOR	BOSCH	BOSCH
• SCHEDULED REPLACEMENT INTERVAL	NONE	15,000 MILES
NO$_x$ EFFICIENCY	40-50%	80% (ESTIMATED)

Fig. 19 - Ford and Volvo - three-way catalyst system comparison

EMISSION CERTIFICATION—Not only did the system pass the 50K emission and durability testing, but also had excellent deterioration factors. The 1981 Federal emission standards of 0.41 gm./mi. HC, 3.4 gm./mi. CO and 1.0 gm./mi. NOx were met with one vehicle, but production feasibility for other applications and other conditions has not been established. Many requirements such as evapotive loss control and end-of-line surveillance must be considered before any engine system is viable for use in 1981. Four vehicles were required for 4K calibration certification and all passed the 1978 California emission standards with a comfortable margin. Certification was received in time to support initial production.

- FUEL ECONOMY—The fuel economy as compared to other conventional California emission systems for 1978 was improved (14% Metro-Highway advantage for automatic transmission applications and 7% advantage for manual transmission applications) as measured on development vehicles. Fuel economy of the California TWC/FB system is still reduced (by as much as 4%) from economy of conventional systems calibrated for the less severe Federal standards.

- DRIVEABILITY—Part throttle drive quality of the automatic transmission vehicles improved slightly and the manual transmission drive quality is overall the best of any 2.3L system offered in California since this engine's introduction in 1974. Maximum performance was maintained since full throttle calibrations satisfied dynamometer fuel and spark requirements.

- ENGINEERING & MANUFACTURING EXPERIENCE—The pilot program is highly effective in raising the level of TWC/FB system expertise throughout associated engineering and manufacturing activities. The result was to provide a stronger corporate base for future three-way catalyst development.

PROGNOSIS FOR FUTURE APPLICATIONS

Three-way catalysts and closed loop control systems will of necessity play an important role in future emission systems. The competitive nature of the marketplace will lead to the most cost effective systems for each application. Feedback controlled carburetors and high Pt:Rh ratios are likely to be prevalent on applications having a lesser NOx task while feedback controlled fuel injection or electronic fuel metering systems and low Pt:Rh ratios will be utilized on the more difficult packages. Due to the mine ratio for Pt:Rh of 19:1 being significantly higher than anticipated usage ratios, considerable effort will no doubt be expended to develop an alternate material for rhodium.

TWC/FB systems represent a major advance in our ability to meet future emissions standards, but they are not the whole answer. The 0.41 gm./mi. HC standard which is to be introduced nationwide in 1980 will present an extremely difficult emission task. Additionally, the task of certifying and maintaining acceptable end-of-line and Selective Enforcement Audit HC levels will be greatly aggravated. Since most of the CH-CVS HC and CO emissions are now produced in the first minutes of operation prior to catalyst light-off, the most significant task ahead will be to develop techniques and associated hardware required to reduce catalyst light-off time.

In summary, the refinement of TWC formulations and the development of the necessary closed loop control systems has resulted in a production feasible means of applying the after treatment philosophy to the control of NOx as well as HC and CO emissions. This development is a significant step forward in the evolution of emission control systems for the future.

REFERENCES

1. M. E. Hyde, R. G. DeLosh, "Three-way Catalyst Design Parameter Seminar." Ford Motor Company Internal Publication.

2. H. G. Lassen, E. E. Weaver, "Engine Dynamometer Screening Tests of Catalysts Capable of Controlling Nitric Oxide." Ford Motor Company Internal Publication, November 1968.

3. J. H. Jones, J. T. Kummer, K. Otto, M. Shelef and E. E. Weaver, "Selective Catalytic Reaction of Hydrogen with Nitric Oxide in the Presence of Oxygen." Environmental Science and Technology, 5 (790-798), 1971.

4. M. K. Adawi, et al, "Method of Improving the Operational Capacity of Three-Way Catalysts." U. S. Patent No. 4,024,706. May 24, 1977.

5. G. T. Engh, S. Wallman, "Development of the Volvo Lambda Sond System." SAE Paper 770295, presented at International Automotive Engineering Congress, Detroit, Michigan, 1977.

6. J. G. Rivard, "Closed Loop Electronic Fuel Injection Control of the Internal Combustion Engine." SAE Paper 730005, presented at International Automotive Engineering Congress, Detroit, Michigan, 1973.

7. "Lambda-Sond: Complete Emission Control?" Automotive Engineering, 85 (45-51), 1977.

8. E. Hamann, H. Manger, L. Steinke, "Lambda-Sensor with Y2O3 Stabilized ZrO2 Ceramic for Application in Automotive Emission Control System." SAE Paper 770401, presented at International Automotive Engineering Congress, Detroit, Michigan, 1977.

Effect of Catalytic Emission Control on Exhaust Hydrocarbon Composition and Reactivity

Marvin W. Jackson

Environmental Activities Staff
General Motors Corporation

INTRODUCTION

The composition of the hydrocarbon mixture in automobile exhaust gases can be profoundly influenced by the type of emission control. A previous paper (1)* reported compositional differences between the General Motors experimental and production Air Injection Reactor (AIR) control systems and experimental Controlled Combustion Systems (CCS). Some authors (2, 3 and 4) reported results for 1970-3 production AIR-type and CCS-type controls and experimental oxidation and dual bed catalytic converter systems. Others (5 and 6) reported results for experimental and production oxidation catalytic converters, an experimental dual bed converter, a production stratified charge engine and experimental and production lean burn engines. Some of the earlier papers did not report results using the 1975-8 Federal Test Procedure (7), since it had not been finalized when the tests were conducted.

This paper describes the efforts of the General Motors Research Laboratories in determining the effects of various emission controls on exhaust hydrocarbon composition and calculated photochemical reactivity. Exhaust gases from fourteen 1970-4 model and twenty 1975-7 model General Motors cars and six experimental model cars were collected during tests conducted according to the 1975-8 Federal Test Procedure and analyzed by gas chromatography. Numerous individual hydrocarbons were determined, but for brevity data for only seventeen hydrocarbons will be presented. Calculated photochemical

*Numbers in parentheses designate references at end of paper.

ABSTRACT

Exhaust gases from fourteen 1970-4 model and twenty 1975-7 model General Motors cars were collected during 1975-8 Federal Test Procedure tests and analyzed by gas chromatography. Hydrocarbon reactivity was calculated from the chromatographic analyses, using several reactivity scales.

The use of oxidation catalytic converters on the 1975-7 model cars greatly changed the exhaust hydrocarbon composition in comparison to 1970-4 model cars. In general, such use caused individual paraffins to increase in carbon percent and individual olefins and acetylene to decrease. For example, the methane carbon percent was 5.0 for 1970-4 model cars (nonconverter cars) and 14.5 for converter cars; ethylene percent was 12.5 for nonconverter cars and 7.4 for converter cars; propylene was 6.5 for nonconverter cars and 2.9 for converter cars; and acetylene was 7.9 for nonconverter cars and 2.2 for converter cars. Because of these large changes in hydrocarbon composition, each of the reactivity scales evaluated indicated that converter cars produced exhaust hydrocarbon mixtures that were less reactive than those of nonconverter cars. The reductions in reactivity per gram ranged from about 10 to 35 percent.

reactivities for the hydrocarbon emissions will be presented using six different reactivity scales.

EXPERIMENTAL

TEST CARS - The forty test cars and their test fuels are described in Table I. A variety of recent production model cars, one 1970 model, one 1972 model, three 1973 models, nine 1974 models, fourteen 1975 models, two 1976 models, four 1977 models and six experimental models, was selected for testing in the "as received" condition. Cars K and S were borrowed from Research employees; the remaining twenty-eight 1970-6 model cars were selected from the Research Garage fleet. All of the 1970-6 model cars were driven in customer-type service. The four 1977 model cars were either back-up certification data or fuel economy cars or development cars which were started on test prior to the start of production for the 1977 model year. These 1977 model cars were provided by three of the General Motors passenger car divisions and were driven on the Milford Proving Ground. The six experimental model cars were provided by Engineering Staff.

Eleven of the cars were California-production models, the remaining twenty-three 1970-7 model cars were Federal-production models. Fifteen of the cars were equipped with AIR, twenty-one of the cars were equipped with CCS, twenty of the cars were equipped with oxidation catalytic converters, two were equipped with three-way catalytic converters, two were equipped with dual bed catalytic converters (three-way and oxidation converters) and one was equipped with a lean reactor. Except for cars A, B and MM, all of the cars were equipped with Exhaust Gas Recirculation (EGR). Car B was equipped with a high valve overlap camshaft for exhaust nitrogen oxide (NOx) control. Car MM was equipped with an experimental stratified charge engine and car NN employed lean combustion.

One car was fueled with a commercial leaded gasoline, twelve were fueled with unleaded

Table I   Car Descriptions and Test Fuels.

Year	Car	Make	Fed. or Calif.	Eng. CID	Carb. bbl	Emission Control[a]				Highest Test Mileage	Unleaded Fuel	No. of Tests
1970	A	A	Fed.	455	2	CCS				3600	Sun. 230[b]	2
1972	B	B	Calif.	350	2	CCS	HVO			42105	Indolene	1
1973	C	B	Fed.	350	2	AIR	EGR			21833	Chevron	1
1973	D	C	Fed.	455	4	CCS	EGR			20576	Indolene	1
1973	E	C	Fed.	455	4	CCS	EGR			16998	Chevron	1
1974	F	D	Fed.	455	4	CCS	EGR			13442	Amoco 91	2
1974	G	E	Fed.	350	2	CCS	EGR			14138	Amoco 91	1
1974	H	B	Calif.	350	4	AIR	EGR			2538	Indolene	2
1974	I	B	Fed.	350	4	AIR	EGR			2955	Amoco 91	1
1974	J	B	Fed.	400	2	AIR	EGR			5132	Amoco 91	1
1974	K	F	Fed.	350	4	CCS	EGR			9296	Amoco 91	2
1974	L	G	Fed.	350	2	AIR	EGR			12634	Amoco 91	1
1974	M	C	Fed.	455	4	CCS	EGR			8146	Amoco 91	1
1974	N	C	Fed.	400	2	CCS	EGR			14173	Amoco 91	1
1975	O	D	Fed.	455	4	CCS	EGR	OC		12317	Amoco 91	3
1975	P	H	Fed.	500	4	CCS	EGR	OC		8741	Amoco 91	2
1975	Q	E	Fed.	231	2	CCS	EGR	OC		20180	Amoco 91	4
1975	R	B	Fed.	350	2	CCS	EGR	OC		17705	Amoco 91	3
1975	S	F	Fed.	350	4	CCS	EGR	OC		2942	Total	2
1975	T	I	Fed.	262	2	AIR	EGR	OC		18140	Shell	4
1975	U	C	Fed.	400	2	CCS	EGR	OC		22317	Shell	3
1975	V	J	Fed.	250	1	CCS	EGR	OC		20907	Marathon	4
1975	W	K	Fed.	140	2	CCS	EGR	OC		2200	Amoco 91	3
1975	X	D	Calif.	455	4	CCS	EGR	OC		20142	Amoco 91	9
1975	Y	B	Calif.	350	4	AIR	EGR	OC		20146	Amoco 91	9
1975	Z	F	Calif.	350	4	CCS	EGR	OC		19995	Amoco 91	8
1975	AA	I	Calif.	250	1	AIR	EGR	OC		19333	Amoco 91	8
1975	BB	K	Calif.	140	1	AIR	EGR	OC		19982	Amoco 91	8
1976	CC	B	Fed.	350	4	CCS	EGR	OC		2510	Amoco 91	1
1976	DD	I	Fed.	305	2	CCS	EGR	OC		3925	Amoco 91	1
1977[c]	EE	D	Calif.	231	2	AIR	EGR	OC		9808	Indolene	3
1977[c]	FF	F	Calif.	403	4	AIR	EGR	OC		9901	Indolene	3
1977[c]	GG	I	Calif.	250	1	AIR	EGR	MOC	OC	9903	Indolene	4
1977[c]	HH	I	Calif.	305	2	AIR	EGR	OC		9824	Indolene	5
Exp.	II	K	--	140	EFI	TWC	EGR			1446	Indolene	1
Exp.	JJ	K	--	140	EFI	TWC	EGR			50000	Indolene	1
Exp.	KK	B	--	350	4	AIR	EGR	TWC	OC	547	Indolene	1
Exp.	LL	B	--	350	4	AIR	EGR	TWC	OC	10000	Indolene	1
Exp.	MM	B	--	350	1	SCE				12448	Cetron	2
Exp.	NN	G	--	350	2	LC	EGR	LR		1990	Indolene	1

a CCS = Controlled Combustion System, HVO = High Valve Overlap, AIR = Air Injection Reactor, EGR = Exhaust Gas Recirculation, OC = Oxidation Converter, MOC = Manifold Oxidation Converter, TWC = 3-Way Converter, SCE = Stratified Charge Engine, LC = Lean Combustion, LR = Lean Reactor.
b Leaded.
c Back-up data or fuel economy car or development car.

Indolene (an Amoco Oil Company physical-property-controlled-gasoline), and the remaining cars were fueled with one of six commercial unleaded gasolines.

EMISSION TEST PROCEDURE - Excluding the 1977 model cars, exhaust emissions were measured according to the 1975-8 FTP, but without the vehicle conditioning and evaporative emission measurement portions of the official procedure(7). Instead, the test car was driven about seven miles on either the Technical Center roads or a chassis dynamometer, the day before the test. (Occasionally, a 1975-8 FTP emission test was used to condition the car, the day before the test.) Also, the canister of the evaporative emission control system was disconnected from the carburetor and a purged canister was put in its place during the test. These steps replaced the regimented steps of the official procedure and were taken to control the condition of the test car prior to the emission test. The 1977 model cars were tested using the full 1975-8 FTP including the vehicle conditioning, but not evaporative emission measurement.

Besides measuring individual hydrocarbons during each 1975-8 FTP, as will be discussed later, the total hydrocarbon emissions were measured in the usual manner.

The tests on the 1977 model cars were conducted in the Milford Proving Ground Vehicle Emissions Laboratory and the remaining tests were conducted in the Research Vehicle Emissions Laboratory.

SAMPLE COLLECTION PROCEDURE FOR INDIVIDUAL HYDROCARBONS - Three exhaust sample collection methods were used. For the 1972-6 and the experimental model cars, a constant-volume sampler (CVS) bag-sampling method was used. With this method, all of the exhaust gas was diluted with air in the usual manner by the CVS, and a small portion of the diluted exhaust gas was pumped into a small Tedlar bag. A bag of dilution air was simultaneously collected. Both the diluted exhaust gas and dilution air bags were filled at the following flow rates:
a. for cycles 1-5:   2.0 liters per minute
                      (4.3 cubic feet per hour)
b. for cycles 6-18: 4.7 liters per minute
                      (10.0 cubic feet per hour)
c. for cycles 19-23: 2.7 liters per minute
                      (5.7 cubic feet per hour).

Thus, both bags were filled in proportion to the cold-hot weighting factors (0.43, 1.00 and 0.57) specified for the three portions of the 1975-8 FTP. This variable fill-rate collection scheme eliminated the need to collect three diluted exhaust gas samples and three dilution air samples, as specified in the 1975-8 FTP, and greatly simplified the experimental work. The diluted exhaust gas bag was analyzed by gas chromatography immediately after completion of the emissions test. The dilution air bag was analyzed after the exhaust gas sample.

For the 1977 model cars, a second type of CVS bag-sampling method was used. With this method, the six CVS bags (three diluted exhaust gas and three dilution air) were collected in the usual manner. Then, a combined diluted exhaust gas bag was prepared by sampling from the three diluted exhaust gas bags collected during cycles 1-5, 6-18 and 19-23. A similar combined dilution air bag was also prepared. The second sampling step was conducted in a fixed-flow-rate varying-sampling-time manner so that the cold-hot weighting factors and the varying durations of the three portions of the 1975-8 FTP were reflected in the filling of the combined bags. The combined diluted exhaust gas bag was again analyzed first by gas chromatography. As with the first CVS bag-sampling method, Tedlar bags were used.

The total bag-sampling method had been used for the 1970 model car. With this method, all of the exhaust gas was cooled to about 17°C in a large water-cooled heat exchanger in order to remove water from the exhaust, and was collected in a large 8500 liter (300 cubic foot) polyethylene bag. Because of the limited size of the bag, only a portion of the exhaust gas generated during the 1975-8 FTP could be collected each day. On the first day, the bag was filled with exhaust gas from cycle 1; on the next day, the exhaust from cycles 2-5; on another day, cycles 6-11 and cycles 12-18 (in separate bags); and, finally, on another day, cycles 19-23 of the 1975-8 FTP. Each bag sample was analyzed by gas chromatography immediately after it was filled. This method was very time consuming and was abandoned in favor of the variable fill-rate CVS bag-sampling method described above.

GAS CHROMATOGRAPHIC ANALYSIS - Since the experiments reported have been conducted over a period of years and in two different laboratories, the same gas chromatographic separation method was not used for all tests. In fact, three chromatographic methods were used. It is believed that the three methods would give equivalent results. In the first method, the contents of each bag were analyzed for individual hydrocarbons with a Perkin-Elmer Model 800 gas chromatograph, using the separation method described by McEwen(8). This method employed a sodium hydroxide deactivated alumina (Alcoa F-10) packed column in a series with a silicone fluid (Dow Corning-200) coated capillary column (Perkin-Elmer). A single gas sample, alumina column flow direction reversing and temperature programming of both columns were used.

The second separation method employed a Perkin-Elmer Model 900 gas chromatograph with an oxypropionitrile bonded on porous silica (Waters Durapak OPN/Porafil C) packed column and a Versilube F-50 fluid coated capillary column (Perkin-Elmer). Two gas samples and temperature programming of both columns were used.

With the first two separation methods, a subtractor column (mercuric perchlorate on Coast Engineering GC-22 packing) could be used to remove olefins, acetylene and aromatics from any sample in order to obtain a paraffin only chromatogram. This provided a means to determine the amount of olefin (by subtraction) in a paraffin-olefin peak on a chromatogram.

Aromatics could also be determined in paraffin-aromatic peaks. In almost every case, a possible paraffin-olefin peak was found to be completely paraffinic for converter equipped cars.

The third method was similar to the second method and employed a Perkin-Elmer Model 900 with a silica gel (Davison-58) packed column and a Versilube F-50 fluid coated capillary column (Perkin-Elmer). Two samples and temperature programming of the capillary column were used.

A Research Laboratories-written computer program was used to calculate the "ppm C" hydrocarbon concentrations, "ppm" concentrations, various measures of hydrocarbon reactivity in units of reactivity per gram, average carbon number and average molecular weight from the areas on the chromatograms.

For the CVS bag-sampling methods diluted exhaust gas concentrations were corrected for the concentrations determined for the dilution air, in order to calculate the net concentration for each individual hydrocarbon. The dilution air correction was proportioned, for the amount of dilution air passing through the CVS, as specified in the 1975-8 FTP.

For the total bag-sampling method which had been used for the 1970 model car, the concentrations determined for each bag were weighted by the measured volumes of each bag and also by the cold-hot weighting factors (0.43, 1.00, and 0.57) of the 1975-8 FTP, in order to calculate the average concentration for each individual hydrocarbon.

With each gas chromatographic analysis, the carbon-percent-of-total-hydrocarbon was calculated for each individual hydrocarbon by dividing its net or average (depending upon the sampling method being used) concentration in "ppm C" by the net or average total hydrocarbon concentration in "ppm C" and multiplying by 100. These values of carbon-percent-of-total-hydrocarbon will usually be called "carbon percents."

RESULTS

The results of the chromatographic analysis are summarized in Tables II through V. The Appendix lists the results for each car. For the 1975 and 1977 model cars, the results were linearly regressed with mileage and the resulting equation was evaluated at one-half of the final test mileage. Only these mid-point values have been tabulated.

Table II lists the results averaged for each model year. Table III lists some statistical information about the averages for all fourteen nonconverter cars and Table IV lists similar information about the averages for all twenty converter cars. Table V lists the results for the experimental cars compared to the nonconverter and converter car results.

INDIVIDUAL HYDROCARBON PERCENTS - Near the top of Tables II through V are listed the individual hydrocarbon results, expressed as carbon-percent-of-total-hydrocarbon, for the five "nonreactive" hydrocarbons that EPA requested information about in 1974 (9). The classification of "nonreactive" does not mean that the five hydrocarbons listed are completely nonreactive or that they do not react to form photochemical smog or oxidant. All five hydrocarbons react to some degree. However, methane is by far the least reactive hydrocarbon and, perhaps, it is the only hydrocarbon that should be called "nonreactive." The other four hydrocarbons are the next least reactive hydrocarbons.

In each of these Tables, the nonreactive and reactive hydrocarbons are listed in order of decreasing carbon percent, based on the averages for all fourteen nonconverter cars (see Table III). Combining the nonreactive and reactive hydrocarbon results for nonconverter cars, the individual exhaust hydrocarbon percents in the order of decreasing concentration are as follows

Table II Summary of Results by Model Year.

	1970	1972	1973	1974	1975	1976	1977
No. of Calif. Cars	-	1	-	1	5	-	4
No. of Fed. Cars	1	-	3	8	9	2	-
Mileage	3600	42105	19802	9162	16075	3218	9859

CARBON PERCENT OF TOTAL HYDROCARBON

Nonreactive Hydrocarbons	1970	1972	1973	1974	1975	1976	1977
Acetylene	7.4	10.7	7.8	7.7	2.5	3.6	0.3
Methane	6.2	6.3	4.9	4.7	11.3	9.8	28.0
Benzene	6.4	3.8	3.4	4.9	3.7	4.8	1.6
Ethane	1.2	0.9	1.0	1.0	2.6	3.0	4.5
Propane	0.1	0.0	0.1	0.1	0.3	0.3	0.3
Nonreactive Total	21.4	21.7	17.1	18.4	20.4	21.6	34.7
Reactive Hydrocarbons							
Ethylene	10.1	15.5	12.8	12.4	7.4	7.2	7.8
Toluene	11.7	11.4	9.1	7.5	6.5	7.2	7.7
Xylenes	9.7	2.3	4.1	7.7	5.9	7.3	2.0
Propylene	8.0	9.6	6.3	6.1	3.1	3.9	1.5
Trimethlpentanes	2.1	3.0	5.5	4.0	6.4	5.0	15.6
n-Butane	1.6	7.0	2.0	4.7	5.3	3.9	7.2
i-Pentane	4.4	5.2	4.5	3.6	5.4	4.5	5.9
Butenes	6.0	2.3	2.9	4.0	2.5	4.2	0.4
Methylpentanes	2.0	1.2	2.9	2.0	3.0	3.2	1.5
n-Pentane	0.6	2.1	1.4	1.8	2.1	2.1	0.8
Ethylbenzene	1.7	0.6	1.1	1.7	1.4	1.9	0.6
i-Butane	0.8	2.3	0.8	0.6	1.0	0.8	0.3
Other Hydrocarbons	19.8	15.6	29.7	25.6	29.6	27.2	14.0
Reactive Total	78.6	78.3	82.9	81.6	79.6	78.4	65.3
Hydrocarbon Classes							
Total Parafins	23.9	34.5	36.8	33.2	52.9	46.4	74.3
Total Olefins	32.0	31.1	25.7	25.8	16.4	17.0	9.9
Acetylene	7.4	10.7	7.8	7.7	2.5	3.6	0.3
Total Aromatics	36.7	23.7	29.7	33.3	28.2	33.0	15.5

HYDROCARBON REACTIVITY PER GRAM

Molar-Based Reactivity Scales	1970	1972	1973	1974	1975	1976	1977
NO_2 Formation Rate	0.0542	0.0511	0.0467	0.0485	0.0366	0.0380	0.0222
Altshuller	0.0571	0.0556	0.0499	0.0518	0.0362	0.0392	0.0234
Methane Exclusion	0.0167	0.0190	0.0166	0.0165	0.0132	0.0139	0.0109
Dimitriades	0.1612	0.1773	0.1548	0.1560	0.1210	0.1232	0.1078
Carbon-Based Reactivity Scales							
Methane Exclusion	0.9380	0.9370	0.9510	0.9530	0.8870	0.9020	0.7200
Cal. Air Res. Bd.	0.4808	0.4469	0.4160	0.4215	0.3300	0.3443	0.2456

RELATIVE REACTIVITY PER GRAM (NONCONVERTER CAR AVERAGE = 100.0)

Molar-Based Reactivity Scales	1970	1972	1973	1974	1975	1976	1977
NO_2 Formation Rate	111.3	104.9	95.9	99.6	75.2	78.0	45.6
Altshuller	109.8	106.9	96.0	99.6	69.6	75.4	45.0
Methane Exclusion	100.0	113.8	99.4	98.8	79.0	83.2	65.3
Dimitriades	101.3	112.5	98.2	99.0	76.8	78.2	68.4
Carbon-Based Reactivity Scales							
Methane Exclusion	98.7	98.6	100.1	100.3	93.4	94.9	75.8
Cal. Air Res. Bd.	112.8	104.8	97.6	98.9	77.4	80.7	57.6

AVERAGE HYDROCARBON PROPERTIES

	1970	1972	1973	1974	1975	1976	1977
Molecular Weight	47.27	42.47	49.95	50.06	47.99	47.91	34.33
Carbon Number	3.42	3.04	3.59	3.61	3.39	3.40	2.35
H/C Ratio	1.80	1.94	1.89	1.85	2.12	2.04	2.60
Nonmethane HC MW	55.68	48.76	57.02	56.60	66.27	63.91	63.56
Nonmethane HC C No.	4.07	3.53	4.13	4.11	4.76	4.62	4.51
Nonmethane HC H/C	1.66	1.79	1.77	1.74	1.89	1.82	2.07

EXHAUST EMISSIONS - GRAM PER MILE

	1970	1972	1973	1974	1975	1976	1977
Hydrocarbon	3.43	1.36	3.95	1.96	0.71	0.95	0.27
Methane	0.21	0.06	0.20	0.09	0.08	0.09	0.08
Nonmethane HC	3.22	1.30	3.75	1.87	0.63	0.86	0.19

Table III Summary of Average Statistics for Nonconverter Cars (1970-4 Models).

	95% Lower Limit on Average	Average	95% Upper Limit on Average
Mileage		13546	
CARBON PERCENT OF TOTAL HYDROCARBON			
Nonreactive Hydrocarbons			
Acetylene	6.5	7.9	9.4
Methane	4.1	5.0	5.8
Benzene	4.0	4.6	5.2
Ethane	0.9	1.0	1.1
Propane	0.0	0.1	0.1
Nonreactive Total	16.3	18.6	20.8
Reactive Hydrocarbons			
Ethylene	11.6	12.5	13.5
Toluene	7.2	8.4	9.7
Xylenes	5.4	6.7	7.9
Propylene	5.6	6.5	7.4
Trimethylpentanes	2.8	4.1	5.4
n-Butane	3.0	4.1	5.1
i-Pentane	3.3	4.0	4.6
Butenes	3.1	3.8	4.4
Methylpentanes	1.8	2.2	2.6
n-Pentane	1.3	1.6	2.0
Ethylbenzene	1.3	1.5	1.8
i-Butane	0.5	0.8	1.1
Other Hydrocarbons	22.2	25.3	28.5
Reactive Total	79.1	81.4	83.7
Hydrocarbon Classes			
Total Paraffins	31.1	33.4	35.8
Total Olefins	25.2	26.6	28.1
Acetylene	6.5	7.9	9.4
Total Aromatics	29.5	32.1	34.7
HYDROCARBON REACTIVITY PER GRAM			
Molar-Based Reactivity Scales			
NO$_2$ Formation Rate	0.0473	0.0487	0.0501
Altshuller	0.0502	0.0520	0.0538
Methane Exclusion	0.0161	0.0167	0.0172
Dimitriades	0.1527	0.1576	0.1626
Carbon-Based Reactivity Scales			
Methane Exclusion	0.9420	0.9500	0.9590
Cal. Air Res. Bd.	0.4137	0.4264	0.4340
RELATIVE REACTIVITY PER GRAM (NONCONVERTER CAR AVERAGE = 100.0)			
Molar-Based Reactivity Scales			
NO$_2$ Formation Rate	97.1	100.0	102.9
Altshuller	96.5	100.0	103.5
Methane Exclusion	96.4	100.0	103.0
Dimitriades	96.9	100.0	103.2
Carbon-Based Reactivity Scales			
Methane Exclusion	99.2	100.0	100.9
Cal. Air Res. Bd.	97.0	100.0	103.0
AVERAGE HYDROCARBON PROPERTIES			
Molecular Weight	46.48	49.30	52.11
Carbon Number	3.34	3.55	3.76
H/C Ratio	1.83	1.86	1.89
Nonmethane HC MW	53.71	56.06	58.42
Nonmethane HC C No.	3.90	4.07	4.24
Nonmethane HC H/C	1.71	1.74	1.77
EXHAUST EMISSIONS - GRAM PER MILE			
Hydrocarbon	1.63	2.45	3.27
Methane	0.07	0.12	0.17
Nonmethane HC	1.55	2.33	3.11

Table IV Summary of Average Statistics for Converter Cars (1975-7 Models).

	95% Lower Limit on Average	Average	95% Upper Limit on Average
Mileage		13548	
CARBON PERCENT OF TOTAL HYDROCARBON			
Nonreactive Hydrocarbons			
Acetylene	1.5	2.2	2.8
Methane	10.7	14.5	18.3
Benzene	3.0	3.4	4.1
Ethane	2.5	3.0	3.4
Propane	0.2	0.3	0.4
Nonreactive Total	19.9	23.4	26.8
Reactive Hydrocarbons			
Ethylene	6.7	7.4	8.2
Toluene	6.2	6.8	7.4
Xylenes	4.3	5.3	6.2
Propylene	2.4	2.9	3.3
Trimethylpentanes	6.1	8.1	10.0
n-Butane	4.8	5.5	6.3
i-Pentane	4.7	5.4	6.2
Butenes	1.7	2.2	2.8
Methylpentanes	2.3	2.7	3.1
n-Pentane	1.5	1.8	2.2
Ethylbenzene	1.1	1.3	1.5
i-Butane	0.6	0.8	1.0
Other Hydrocarbons	23.2	26.4	29.6
Reactive Total	73.2	76.6	80.1
Hydrocarbon Classes			
Total Paraffins	51.8	56.5	61.3
Total Olefins	13.4	15.1	16.9
Acetylene	1.5	2.2	2.8
Total Aromatics	23.0	26.2	29.4
HYDROCARBON REACTIVITY PER GRAM			
Molar-Based Reactivity Scales			
NO$_2$ Formation Rate	0.0308	0.0339	0.0370
Altshuller	0.0311	0.0340	0.0368
Methane Exclusion	0.0123	0.0128	0.0134
Dimitriades	0.1140	0.1186	0.1231
Carbon-Based Reactivity Scales			
Methane Exclusion	0.8170	0.8550	0.8930
Cal. Air Res. Bd.	0.2954	0.3146	0.3338
RELATIVE REACTIVITY PER GRAM (NONCONVERTER CAR AVERAGE = 100.0)			
Molar-Based Reactivity Scales			
NO$_2$ Formation Rate	63.2	69.6	76.0
Altshuller	59.8	65.4	70.8
Methane Exclusion	73.7	76.6	80.2
Dimitriades	72.3	75.3	78.1
Carbon-Based Reactivity Scales			
Methane Exclusion	86.0	90.0	94.0
Cal. Air Res. Bd.	69.3	73.8	78.3
AVERAGE HYDROCARBON PROPERTIES			
Molecular Weight	42.07	45.25	48.43
Carbon Number	2.95	3.19	3.43
H/C Ratio	2.10	2.21	2.31
Nonmethane HC MW	63.98	65.49	67.01
Nonmethane HC C No.	4.58	4.70	4.81
Nonmethane HC H/C	1.87	1.92	1.96
EXHAUST EMISSIONS - GRAM PER MILE			
Hydrocarbon	0.54	0.65	0.75
Methane	0.07	0.08	0.09
Nonmethane HC	0.46	0.57	0.67

(the nonreactive hydrocarbons are underlined): ethylene, toluene, acetylene, xylenes, propylene, methane, benzene, trimethylpentanes, n-butane, i-pentane, butenes, methylpentanes, n-pentane, ethylbenzene, ethane, i-butane, and propane. Propane is not the 17th most concentrated hydrocarbon in nonconverter car exhaust. Many other reactive hydrocarbons have carbon percents greater than 0.1, but they have been excluded from this discussion for brevity.

Based on all fourteen nonconverter cars, the average methane carbon percent was 5.0 (see Table III). At the 95 percent confidence level, the lower and upper limits of the average methane percent were 4.1 and 5.8 (see Table III). These are the limits that the average for another group of similar 1970-4 model (nonconverter) cars would lie between with 95 percent probability. The average methane percent is an important value in setting a nonmethane hydrocarbon emission standard. For example, if EPA chose to convert the 0.41 gram-per-mile total-hydrocarbon standard to a nonmethane-hydrocarbon standard, the new value would be 0.39 (0.41 times 0.95) gram per mile.

HYDROCARBON CLASS PERCENTS - Near the middle of Tables II through V, carbon percents for the paraffins, olefins, acetylene, and aromatics are listed. As indicated in Table IV, the paraffin percent was 56.5 (averaged for all twenty converter cars), the aromatic percent was 26.2, the olefin percent was 15.1 and the acetylene percent was 2.2.

HYDROCARBON REACTIVITIES - Hydrocarbon reactivity-per-gram values are shown just below the middle of Tables II through V. The Tables

Table V Comparison of Emission Control Systems.

	Nonconverter	Oxidation Converter	3-Way Converter	Dual Bed Converter	Stratified Charge Engine	Lean Comb. With Reactor
Mileage	13546	13548	25723	5274	12448	1990
No. of Cars	14	20	2	2	2	1

CARBON PERCENT OF TOTAL HYDROCARBON

	Nonconverter	Oxidation Converter	3-Way Converter	Dual Bed Converter	Stratified Charge Engine	Lean Comb. With Reactor
Nonreactive Hydrocarbons						
Acetylene	7.9	2.2	3.4	3.2	7.1	8.2
Methane	5.0	14.5	7.8	33.7	5.9	5.3
Benzene	4.6	3.4	2.9	2.4	2.8	5.9
Ethane	1.0	3.0	2.4	4.4	1.1	1.9
Propane	0.1	0.3	0.2	0.2	0.1	0.1
Nonreactive Total	18.6	23.4	16.4	43.9	17.0	21.4
Reactive Hydrocarbons						
Ethylene	12.5	7.4	5.2	8.0	20.5	19.4
Toluene	8.4	6.8	10.0	6.4	6.7	13.3
Xylenes	6.7	5.3	4.0	2.3	5.3	4.8
Propylene	6.5	2.9	4.6	3.0	10.7	9.1
Trimethylpentanes	4.1	8.1	8.9	6.9	5.2	4.4
n-Butane	4.1	5.5	1.6	1.7	1.2	2.2
i-Pentane	4.0	5.4	2.4	1.8	2.6	1.2
Butenes	3.8	2.2	4.2	3.2	4.2	4.5
Methylpentanes	2.2	2.7	1.6	0.6	2.4	1.2
n-Pentane	1.6	1.8	0.6	0.5	0.6	0.3
Ethylbenzene	1.5	1.3	1.4	0.7	0.7	1.0
i-Butane	0.8	0.8	0.2	0.4	1.2	0.1
Other Hydrocarbons	25.3	26.4	39.1	20.6	21.8	17.0
Reactive Total	81.4	76.6	83.6	56.1	83.0	78.6
Hydrocarbon Classes						
Total Paraffins	33.4	56.5	43.6	56.2	26.6	22.8
Total Olefins	26.6	15.1	17.6	14.5	39.5	36.1
Acetylene	7.9	2.2	3.4	3.2	7.1	8.2
Total Aromatics	32.1	26.2	35.4	26.1	26.8	32.8

HYDROCARBON REACTIVITY PER GRAM

	Nonconverter	Oxidation Converter	3-Way Converter	Dual Bed Converter	Stratified Charge Engine	Lean Comb. With Reactor
Molar-Based Reactivity Scales						
NO_2 Formation Rate	0.0487	0.0339	0.0416	0.0262	0.0595	0.0562
Altshuller	0.0520	0.0340	0.0398	0.0304	0.0682	0.0640
Methane Exclusion	0.0167	0.0128	0.0127	0.0107	0.0188	0.0192
Dimitriades	0.1576	0.1186	0.1164	0.1144	0.2028	0.1940
Carbon-Based Reactivity Scales						
Methane Exclusion	0.9500	0.8550	0.9220	0.6630	0.9410	0.9470
Cal. Air Res. Bd.	0.4264	0.3146	0.3603	0.3116	0.5308	0.5080

RELATIVE REACTIVITY PER GRAM (NONCONVERTER CAR AVERAGE = 100.0)

	Nonconverter	Oxidation Converter	3-Way Converter	Dual Bed Converter	Stratified Charge Engine	Lean Comb. With Reactor
Molar-Based Reactivity Scales						
NO_2 Formation Rate	100.0	69.6	85.4	53.8	122.2	115.4
Altshuller	100.0	65.4	76.5	58.5	131.2	123.1
Methane Exclusion	100.0	76.6	76.0	64.1	112.6	115.0
Dimitriades	100.0	75.3	73.9	72.6	128.7	123.1
Carbon-Based Reactivity Scales						
Methane Exclusion	100.0	90.0	97.1	69.8	99.1	99.7
Cal. Air Res. Bd.	100.0	73.8	84.5	73.1	124.5	119.1

AVERAGE HYDROCARBON PROPERTIES

	Nonconverter	Oxidation Converter	3-Way Converter	Dual Bed Converter	Stratified Charge Engine	Lean Comb. With Reactor
Molecular Weight	49.30	45.25	54.94	29.96	43.78	43.45
Carbon Number	3.55	3.19	3.93	2.04	3.14	3.14
H/C Ratio	1.86	2.21	1.84	1.95	1.92	1.81
Nonmethane HC MW	56.06	65.49	72.41	59.21	49.51	48.91
Nonmethane HC C No.	4.07	4.70	5.24	4.28	3.59	3.57
Nonmethane HC H/C	1.74	1.92	1.79	1.81	1.77	1.68

EXHAUST EMISSIONS - GRAM PER MILE

	Nonconverter	Oxidation Converter	3-Way Converter	Dual Bed Converter	Stratified Charge Engine	Lean Comb. With Reactor
Hydrocarbon	2.45	0.65	0.48	0.62	1.06	1.04
Methane	0.12	0.08	0.04	0.21	0.06	0.06
Nonmethane HC	2.33	0.57	0.44	0.41	1.00	0.98

also list relative reactivity-per-gram values based on setting the averages for all fourteen nonconverter cars equal to 100.0. Six reactivity values were calculated for each car, each value based on the gas chromatographic analysis for that car and one of six reactivity scales. These scales are simply measures or estimates of how various individual hydrocarbons react in the photochemical smog reaction. Six scales were used because, at this time, there is no consensus of opinion as to which is the best. There may never be a consensus of opinion.

The molar-based reactivity scales were:
a. the NO_2-formation-rate scale (1)
b. the Altshuller scale (10)
c. the methane-exclusion scale
d. the Dimitriades scale (11)

The carbon-based reactivity scales were:
e. the methane-exclusion scale
f. the California Air Resources Board (CARB) scale (12)

A molar-based reactivity scale is one that uses multiplying values based on molar concentrations (ppm). A carbon-based scale uses multiplying values based on carbon concentrations (ppm C). For a molar-based scale, the hydrocarbon concentration in ppm is multiplied by the reactivity value and for a carbon-based scale, the concentration in ppm C is multiplied by the reactivity value. The detailed reactivity calculation method has been previously described (1 and 13).

The Altshuller scale was modified by assigning reactivity-per-mole values of 1 to the C_4 and C_5 paraffins, instead of zero. This modification was made because the C_4 and C_5 paraffins react in the photochemical smog reaction. The CARB scale was modified by including acetylene with the paraffins instead of the olefins. This modification was made because acetylene reacts more like paraffins than olefins in the smog reaction. (The CARB method of analysis includes acetylene with the olefins.) The NO_2-formation-rate and Dimitriades scales were not modified. The methane-exclusion scales assign a reactivity-per-mole (ppm) or reactivity-per-ppm C of zero to methane and 1 to all other hydrocarbons.

The reporting of reactivity per gram was selected because a factor to convert from the emissions in gram per mile to reactivity per mile was desired. Other reactivity units are not as convenient to use.

As indicated in Tables II through V, there is considerable variation in the magnitude of the reactivity-per-gram value calculated for each reactivity scale. Based on all fourteen nonconverter cars (see Table III), the reactivity-per-gram averages varied from 0.0167 to 0.9500. The magnitudes of these values are not significant by themselves. They become important only when they are compared to similar values from other types of cars or when used in setting an exhaust hydrocarbon reactivity standard. For example, if EPA chose to convert the 0.41 gram-per-mile total hydrocarbon standard to a hydrocarbon reactivity standard using the molar-based methane exclusion scale, the new value for such a standard would be 0.0068 (0.41 times 0.0167) reactivity per mile.

There was relatively little car-to-car variation in the reactivity values within the nonconverter or the converter car classes. For example, the 95 percent confidence limits for the six reactivity averages (see Table III) were plus or minus less than 4 percent for nonconverter cars. The 95 percent confidence limits were plus or minus less than 9 percent for converter cars.

AVERAGE HYDROCARBON PROPERTIES - Near the bottom of Tables II through V, values for the average molecular weight, average carbon number, and average hydrogen-carbon ratio of the exhaust hydrocarbons are listed. Also, similar values for nonmethane hydrocarbons are listed. These values are reported because they are useful in interpreting reactivity data and in calculating the hydrocarbon density that should be used to calculate "true" hydrocarbon mass emissions in a

CVS-type test.

For example, the reactivity-per-gram values can be converted to reactivity-per-mole values by multiplying by the average molecular weight. With the NO_2-formation-rate scale, the corresponding reactivity-per-mole value is 2.40 (reactivity per gram, 0.0487, times molecular weight, 49.30). A value of 2.40 indicates that the average hydrocarbon mixture from the nonconverter cars is about 85 percent as reactive as ethylene(1).

The average hydrogen-carbon ratio of 1.86 (see Table III) is very close to the EPA estimated value of 1.85 used in the 1975-8 FTP for hydrocarbon mass emission and fuel economy calculations. The hydrocarbon density in gram per cubic foot is based on the average hydrogen-carbon ratio. The density equation is 1.177 [12.011 + (H/C) (1.008)].

EXHAUST EMISSIONS - At the bottom of Tables II through V, values for hydrocarbon, methane and nonmethane hydrocarbon emissions in gram per mile are listed. As shown in Table II, the methane emissions for nonconverter cars varied by a factor of 3.5 times (0.21 divided by 0.06) over the various model years, whereas the methane fraction varied by a factor of only 1.3 times (6.3 divided by 4.7).

The methane, nonmethane hydrocarbon, and total hydrocarbon emissions were calculated using the hydrocarbon density specified in the 1975-8 FTP (16.33 gram per cubic foot based on a hydrogen-carbon ratio of 1.85) so that the sum of the methane and nonmethane hydrocarbon emissions would equal the total hydrocarbon emissions. If a density of 18.88 gram per cubic foot (corresponding to a hydrogen-carbon ratio of 4.00) had been used for methane, then the sum of the methane and nonmethane hydrocarbon emissions could only be made to equal the total hydrocarbon emissions by using the correct density values for both the nonmethane and total hydrocarbon emissions. For this paper, the correct density values could have been used. However, the correct values are usually not known since these values can only be determined from a complete gas chromatographic analysis.

DISCUSSION

SELECTION OF REACTIVITY SCALES - A wide selection of reactivity scales has been provided. Such a selection is provided so that the wide range of opinion regarding which reactivity scales is the most appropriate might be satisfied. Some reactivity scales are molar-based and some are carbon-based. The molar-based scales assume that photochemical smog manifestations are proportional to the hydrocarbon concentrations in ppm. The carbon-based scales assume that manifestations are proportional to the concentrations in ppm C.

There are three scales which assign high measures of photochemical reactivity to the olefins on a molar basis--the NO_2 formation rate, the Altshuller, and the Dimitriades scales. Another scale assigns high measures of reactivity to the olefins on a carbon basis--the California Air Resources Board scale. The other two scales assign zero reactivity to methane and treat all other hydrocarbons equally on either a molar or carbon basis--the two methane exclusion scales.

Scales which assign high measures of the reactivity to the olefins, which react the most rapidly in the photochemical smog reaction, probably best approximate smog formation near the sources of hydrocarbon pollution, that is, in urban areas. The scales which exclude methane probably best approximate the end result for all hydrocarbon emissions, that is, all hydrocarbons, except methane, react if given enough time.

Readers may choose the reactivity scale that best approximates their opinion as to the fate of hydrocarbon emissions in the photochemical smog reaction.

COMPARISON BY MODEL YEAR - There are some significant differences between the 1970 through 1974 model years--the nonconverter car model years. However, most of these differences can be considered minor in comparison to the dramatic differences between the 1974 and 1975 model year cars (see Table II). Acetylene was much lower in the exhaust from the 1975 model cars in comparison to the 1974 model cars; methane and ethane were much higher; and ethylene, propylene and the butenes were much lower.

The percent total paraffins was higher for the 1975 model cars and the percent total olefins and percent total aromatics were lower in comparison to the 1974 model cars. Also, all measures of photochemical reactivity showed that the 1975 models produced hydrocarbon mixtures with lower reactivity per gram than the 1974 models.

These dramatic differences also occurred between the 1973 and 1975 models, the 1974 and 1976 models and the 1974 and 1977 models.

Further discussion of these differences can be meaningless when considering the number of cars tested for each model year. Perhaps, the best approach is to average and compare the 1970-4 model cars, the nonconverter cars, to the 1975-7 model cars, the converter cars.

COMPARISON OF NONCONVERTER AND CONVERTER CARS - Tables III and IV summarize comparisons of the data for nonconverter cars and that for converter cars. Figures 1 through 4 show the more important differences between these two types of cars. As shown in Figure 1, the 95% confidence limits on the average acetylene percent for the nonconverter cars did not overlap similar limits for the converter cars. Therefore, acetylene was significantly lower in the exhaust from converter cars in comparison to nonconverter cars, at the 95% confidence level. Methane and ethane percents were significantly higher for the converter cars.

The ethylene, propylene and butenes percents were all significantly lower (see Figure 2) in converter car exhaust than in nonconverter car exhaust at the 95% confidence level. Because of these three differences, total olefins were significantly lower (see Figure 3) in converter car exhaust. Figure 3 also shows that total paraffins were significantly higher in converter

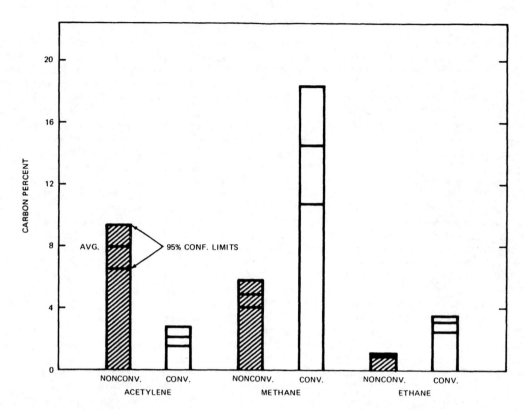

Figure 1   Comparison of Average Acetylene, Methane and Ethane Carbon Percents for Nonconverter Cars (1970-4 Models) and Those for Converter Cars (1975-7 Models).

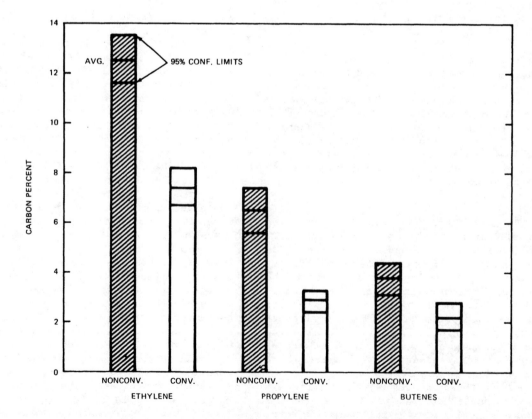

Figure 2   Comparison of Average Ethylene, Propylene and Butenes Carbon Percents for Nonconverter Cars (1970-4 Models) and Those for Converter Cars (1975-7 Models).

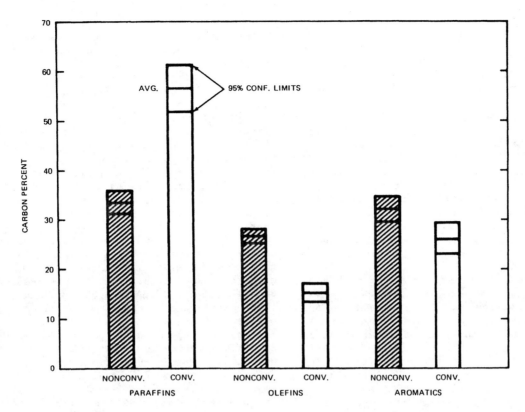

Figure 3 Comparison of Average Paraffin, Olefin and Aromatic Carbon Percents for Nonconverter Cars (1970-4 Models) and Those for Converter Cars (1975-7 Models).

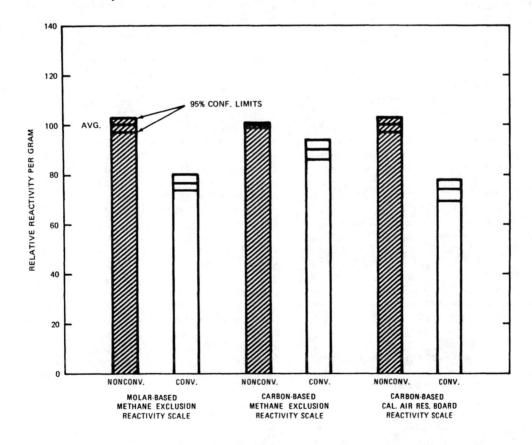

Figure 4 Comparison on Average Relative Reactivity-Per-Gram Values for Nonconverter Cars (1970-4 Models) and Those for Converter Cars (1975-7 Models).

car exhaust and total aromatics were significantly lower.

Because of the differences shown in Figures 1 through 3, it is not suprising that all six measures of photochemical reactivity showed that the hydrocarbon mixture in the exhaust from converter cars was significantly lower in reactivity per gram than that from nonconverter cars. The results for three of the six reactivity scales are shown in Figure 4. Both methane exclusion scales and the California Air Resources Board scale showed lower reactivity per gram for the converter cars. Thus, converter-equipped cars achieved a greater control of total photochemical reactivity (emissions in gram per mile times reactivity per gram) than that indicated by the reduction in hydrocarbon mass emissions alone.

Much of the difference using the molar-based methane exclusion scale was caused by a difference in average molecular weight. For example, considering two cars with equal hydrocarbon emissions in gram per mile, the car that produced the exhaust hydrocarbon mixture with the highest average molecular weight or highest average carbon number would most likely produce the least reactive exhaust. Such differences in average molecular weight and average carbon number are shown in Tables III and IV. The nonmethane hydrocarbon average molecular weight and average carbon number were significantly higher for the converter cars in comparison to nonconverter cars.

Tables III and IV also show average hydrogen-carbon ratio results. This ratio for the nonmethane hydrocarbon emissions was significantly higher (1.92 versus 1.74) for converter cars in comparison to nonconverter cars. Based on these data, if EPA converted to a nonmethane hydrocarbon standard, a density of 16.41 gram per cubic foot should be used to calculate nonmethane hydrocarbon emissions in FTP-type tests for similar converter equipped cars. The present FTP procedure uses 16.33 gram per cubic foot.

EFFECT OF MILEAGE ACCUMULATION - The five 1975 California production model cars were systematically tested as a function of mileage. Each of the five cars was tested new and at 2000, 4000, 8000, 12000, 16000 and 20000 miles. Only average results for all five cars at each mileage are shown in Figures 5, 6 and 7. A less systematic study was conducted with the four 1977 California production model cars. Each of the four cars was tested between 4000 and 5000 miles and at 6000 and 10000 miles. Two cars were also tested at 8000 miles. Because of the variety of test mileages, results for each car are shown in Figures 8, 9 and 10.

For the 1975 California production model cars (see Figures 5 and 6), the methane and ethane percents decreased with increasing mileage; acetylene percent increased slightly and ethylene, propylene and the butenes percents increased. For the 1977 California production model cars (see Figures 8 and 9), the methane and ethane percents decreased with increasing mileage, acetylene (not shown) was very low and decreased slightly, ethylene increased, propylene increased slightly and the butenes (not shown) were very low and also increased slightly.

Since for both the 1975 and 1977 California production model cars the ethylene, propylene and the butenes percents increased with increasing mileage, the reactivity-per-gram values would be expected to also increase with increasing mileage. This is generally the case as shown in Figures 7 and 10. The only exception was for the 1977 model cars using the California Air Resources Board scale. With this scale, the relative reactivity per gram decreased with increasing mileage.

All of the trends discussed in this section can be explained by the following: (a) methane emissions in gram per mile increased slightly with increasing mileage (from a regression value of 0.076 at 4000 miles to 0.081 at 20000 miles for 1975 California production model cars), (b) ethane emissions increased slightly (from a regression value of 0.017 at 4000 miles to 0.020 at 20000 miles for the same cars) and (c) nonmethane hydrocarbon (reactive hydrocarbon) emissions increased considerably (from a regression value of 0.517 at 4000 miles to 0.801 at 20000 miles for the same cars). These facts explain the observations that: (a) the methane percent decreased with increasing mileage, (b) the ethane percent decreased with increasing mileage, (c) the ethlyene, propylene and butenes percents increased with mileage and (d) reactivity-per-gram values also increased.

COMPARISON OF EMISSION CONTROLS - Table V compares the results with experimental emission controls to nonconverter car (1970-4 models) and oxidation converter car (1975-7 models) results. Figures 11 through 14 show some of the highlights of such a comparison. The lean combustion engine with lean reactor and the stratified charge engine produced exhausts with methane percents about equal to that of nonconverter cars (see Figure 11) and ethylene and propylene percents higher than those of nonconverter cars (see Figures 12 and 13). These ethlyene and propylene percents were much higher than those of converter cars.

In comparison to the converter cars, the three-way converter cars were lower in methane and ethylene and somewhat higher in propylene. Dual bed converter cars, in comparison to converter cars, were much higher in methane and about equal in ethlyene and propylene.

Based on the above compositional differences, it would be expected that the exhausts from the lean combustion engine with lean reactor and stratified engine would tend to be high in reactivity. Three-way and dual converter cars would be expected to be low in reactivity. Except for the carbon-based methane exclusion scale, these expectations proved to be correct (see Table V). Using the carbon-based methane exclusion scale, the relative-reactivity-per-gram values for the three-way converter, the lean combustion engine with lean reactor and the stratified charge

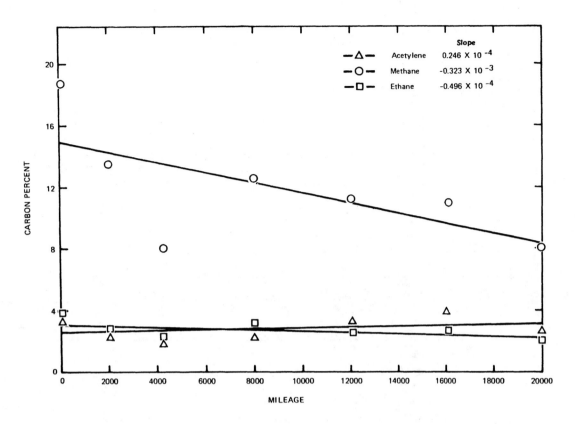

Figure 5 Effect of Mileage Accumulation on Average Acetylene, Methane and Ethane Carbon Percents for Five 1975 California Production Model Cars.

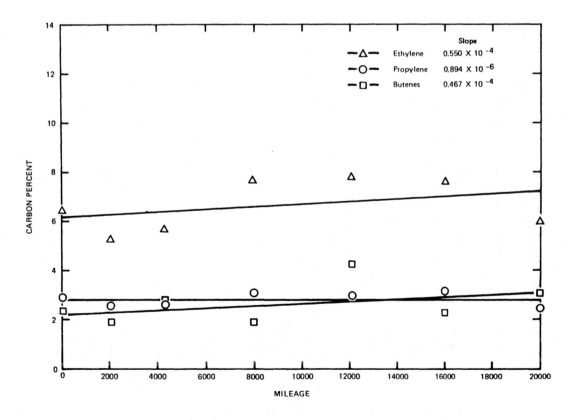

Figure 6 Effect of Mileage Accumulation on Average Ethylene, Propylene and Butenes Carbon Percents for Five 1975 California Production Model Cars.

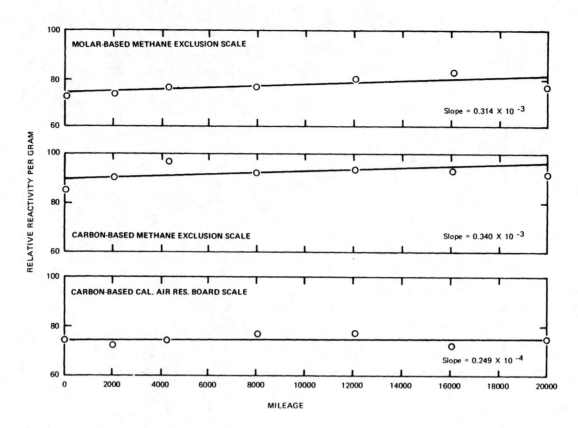

Figure 7 Effect of Mileage Accumulation on Average Relative Reactivity-Per-Gram Values for Five 1975 California Production Model Cars.

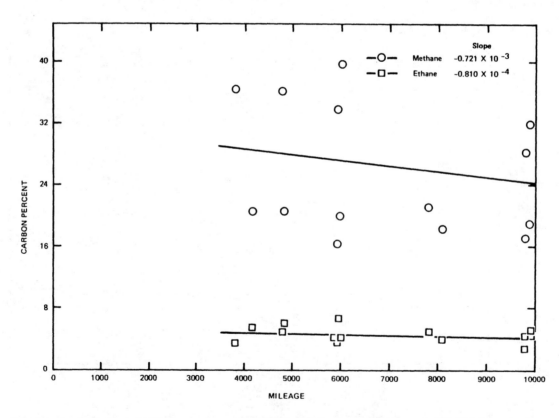

Figure 8 Effect of Mileage Accumulation on Average Methane and Ethane Carbon Percents for Four 1977 California Production Model Cars.

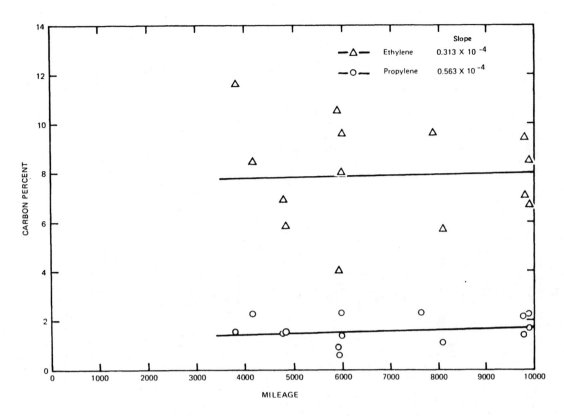

Figure 9 Effect of Mileage Accumulation on Average Ethylene and Propylene Carbon Percent for Four 1977 California Production Model Cars.

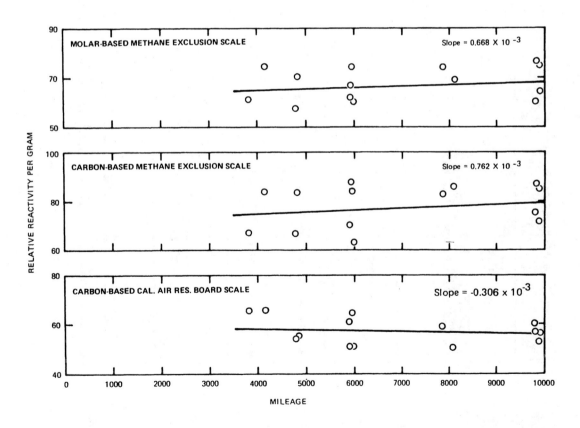

Figure 10 Effect of Mileage Accumulation on Average Relative Reactivity-Per-Gram Values for Four 1977 California Production Model Cars.

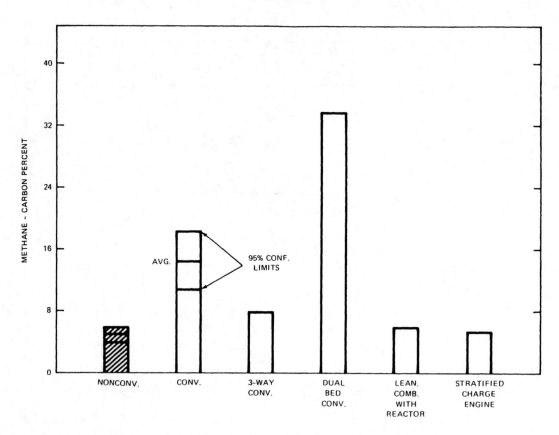

Figure 11 Comparison of Average Methane Percents for Various Types of Emission Control.

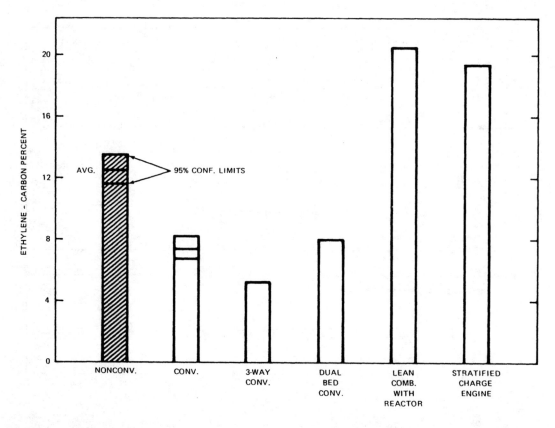

Figure 12 Comparison of Average Ethylene Percents for Various Types of Emission Control.

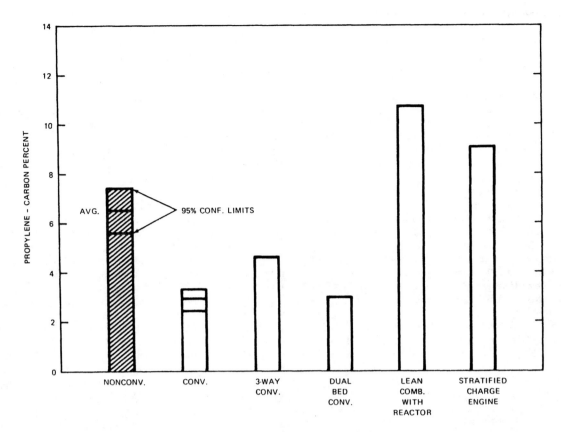

Figure 13 Comparison of Average Propylene Percents for Various Types of Emission Control.

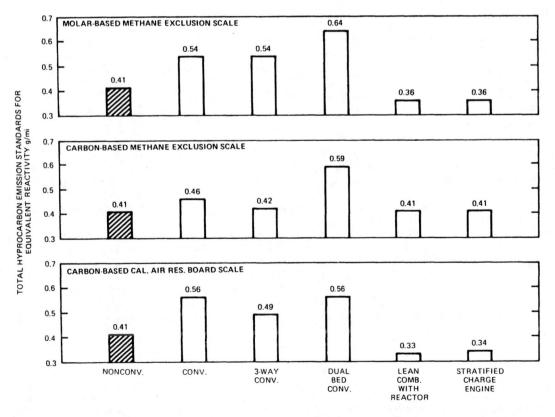

Figure 14 Comparison of Average Total Hydrocarbon Emission Standards for Equivalent Reactivity for Various Types of Emission Control.

engine cars were about equal in reactivity to that of the nonconverter cars. Using the same scale, the dual converter cars were lower in reactivity.

One way to compare these control systems on a reactivity basis is to compare total hydrocarbon mass emission standards for equal reactivity. Such a comparison is shown in Figure 14. This figure shows results for the molar-based and carbon-based methane exclusion scales and the California Air Resources Board scale. As shown in the figure, the highest equal reactivity mass emission standards, using all three reactivity scales, were for the dual bed converter cars. The lowest equal reactivity mass emission standards were for the lean combustion engine with lean reactor and stratified charge engine cars. The equal reactivity standards for the oxidation converter and three-way converter cars were higher than 0.41 gram per mile.

The results discussed in this section can be explained by the following observations: (a) all automotive-type catalysts are ineffective in oxidizing methane, (b) the same catalysts are effective in oxidizing nonmethane hydrocarbons, in particular ethylene, acetylene, propylene and the butenes and (c) lean engine operation cracks some of the fuel into large percents of low-boiling-nonfuel-type-hydrocarbons (except methane)(1). In this respect, the stratified charge engine is a type of lean combustion engine.

<u>PREDICTION OF REACTIVITY PER GRAM FROM LIMITED GAS CHROMATOGRAPHIC ANALYSES</u> - The gas chromatographic analyses and reactivity calculations described in this paper are tedious and time-consuming to perform. It would be convenient if reactivity-per-gram values could be accurately predicted empirically from a simplified gas chromatographic analysis limited to, say, the C_1 to C_3 hydrocarbons. (The analysis time for this range of hydrocarbons is believed to be about the maximum practical limit that might be considered for a hydrocarbon reactivity standard.) In order to determine if accurate predictions of reactivity are possible, linear multiple-regression analyses of all the data were performed using LIMREG, a stepwise regression package on the Dartmouth Time-Sharing System. The results are summarized in Tables VI and VII.

In these calculations, the reactivity-per-gram values for five of the six reactivity scales were regressed against the methane, ethane, ethylene, propylene, and acetylene carbon percents. Prediction of reactivity using the carbon-based methane exclusion scale is unnecessary since this reactivity value can be calculated by simply measuring methane and total hydrocarbon concentrations in ppm C. Of course, total hydrocarbon concentrations would be measured with a flame-ionization analyzer using the method described in the 1975-8 FTP (7). The regressions were tried in all combinations of up to the five hydrocarbons.

A complete gas chromatographic analysis for all of the individual hydrocarbons is <u>not</u> needed to determine the carbon percents used in these regressions. These percents can be determined from the C_1 to C_3 hydrocarbon concentrations in ppm C and the total hydrocarbon concentration in ppm C. Carbon percents were selected instead of mole or weight percents because such percents can only be determined from a complete gas chromatographic analysis.

As indicated in Table VI, the indexes of determination (R^2) for methane alone are poor and vary from 0.12 to 0.43. However, for all five hydrocarbons, the indexes are quite good and vary from 0.83 to 0.96. The highest indexes and the lowest standard errors (best correlations) occurred for the molar-based methane exclusion and Dimitriades scales.

One way to determine the accuracy of our regression equations is to use our equations and predict reactivity using EPA chromatographic data (5 and 6) for only the five hydrocarbons. The predicted reactivity would then be compared to the value calculated using EPA data for all of the hydrocarbons. The first step of such a study was to convert the individual hydrocarbon chromatographic data reported by EPA in gram per mile to values proportional to the area on EPA's original chromatograms. This conversion made use of EPA's reported average molecular weight for each individual hydrocarbon peak. The proportional-to-area values could then be used in our computer program and the carbon percents and the various reactivities could be calculated. By definition, area percent is equal to carbon percent.

Figures 15 and 16 compare calculated and predicted reactivity values using our equations and EPA data. As shown in the figures, most values fall close to the equivalency lines. Generally, EPA's reactivity-per-gram values can be predicted well. The two points farthest away from the equivalency line on Figure 15, are for a dual converter car and car with a converter on only one bank of a V-8 engine.

No physical significance should be attached to the sign or magnitude of the coefficients and constants listed in Table VII.

COMPARISON WITH LITERATURE

There are at least five papers in the literature (2 through 6) that deal with detailed individual hydrocarbon composition and reactivity of exhaust gases from cars similar to those reported in this paper. The oldest of these papers (2 and 3) reported results using the 1972 FTP. Their hydrocarbon composition data was given in mole percents and it was impossible to convert their results to carbon percents. Also, their reactivity data was reported in a fashion so that reactivity per gram could not be calculated. No further discussion of these two papers will be made.

The Bureau of Mines (presently called the Energy Research and Development Administration) study (4) reported results for ten 1970-3 model cars, three cars equipped with experimental oxidation converters, one experimental dual bed converter car and one experimental lean reactor car. All of these cars were fueled and tested

Table VI   Regressions Used to Predict Reactivity Per Gram.

	Methane	Ethane	Methane Plus Ethane	Ethylene	Propylene	Acetylene	Index of Determination	Standard Error As Pct. of Average
			MOLAR-BASED REACTIVITY SCALES					
	NO_2 Formation Rate							
a.	X						0.43	18.9
b.			X	X	X	X	0.84	10.1
c.	X	X		X	X	X	0.84	10.2
	Altshuller							
a.	X						0.35	21.7
b.			X	X	X	X	0.91	8.1
c.	X	X		X	X	X	0.91	8.2
	Methane Exclusion							
a.	X						0.28	12.4
b.			X	X	X	X	0.88	5.1
c.	X	X		X	X	X	0.95	3.4
	Dimitriades							
a.	X						0.12	16.9
b.			X	X	X	X	0.96	3.5
c.	X	X		X	X	X	0.96	3.5
			CARBON-BASED REACTIVITY SCALES					
	Cal. Air Res. Board							
a.	X						0.28	16.6
b.			X	X	X	X	0.82	8.3
c.	X	X		X	X	X	0.83	8.3

Table VII   Regression Equation and Terms Used to Predict Reactivity Per Gram.

Reactivity = A(Methane) + B(Ethane) + C(Ethylene) + D(Propylene) + E(Acetylene) + Constant

	Equation Coefficients $\times 10^4$					Equation Constant
	Methane	Ethane	Ethylene	Propylene	Acetylene	
	MOLAR-BASED REACTIVITY SCALES					
	NO_2 Formation Rate					
c.	-4.02	-4.28	4.08	24.85	-1.71	0.0316
	Altshuller					
c.	-3.60	-4.95	8.36	24.73	-0.33	0.0279
	Methane Exclusion					
c.	-1.30	-3.03	2.72	1.60	2.19	0.0108
	Dimitriades					
c.	-3.49	-2.99	38.62	48.19	2.11	0.0847
	CARBON-BASED REACTIVITY SCALES					
	Cal. Air Res. Board					
c.	-17.08	-40.65	50.44	156.04	12.01	0.2720

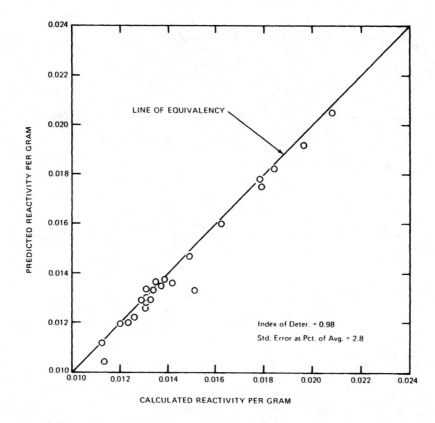

Figure 15 Correlation Between Calculated and Predicted Reactivity-Per-Gram Values (Molar-Based Methane Exclusion Scale) Using GM Empirical Equations and EPA Chromatographic Data.

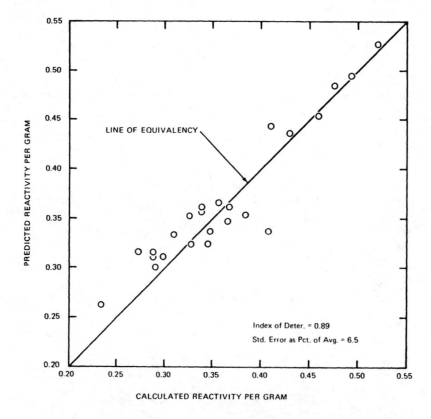

Figure 16 Correlation Between Calculated and Predicted Reactivity-Per-Gram Values (Carbon-Based California Air Resources Board Scale) Using GM Empirical Equations and EPA Chromatographic Data.

with both Indolene and one of two special gasolines containing either 35 or 40 percent aromatics. For simplicity, their gas chromatographic composition data was mainly given in groups so that data for only methane, ethlyene, acetylene and benzene could be extracted. Their composition data was given in weight percents and it could not be converted to carbon percents. However, they did report reactivity-per-gram values using the NO_2 formation rate (1) and Altshuller (10) scales.

For the nonconverter cars reported by the Bureau of Mines, the average methane value was 7.0 weight percent in comparison to our value of 5.8 weight percent (same as 5.0 carbon percent, see Table III). Their ethylene value was 10.9 weight percent versus our value of 12.6 weight percent. Their reactivity-per-gram values were 0.0488 versus our value of 0.0487 for the NO_2 formation rate scale and 0.0500 versus our value of 0.0520 for the Altshuller scale. For brevity, no additional comparisons using the other Bureau of Mines data will be made. However, in general the Bureau of Mines data is similar to our data.

Our results are compared to the two EPA papers (5 and 6) in Tables VIII and IX. These two EPA studies reported results for one 1972 model car and five 1975, two 1976 and eight 1977 model cars equipped with production oxidation converters. Results were also reported for one three-way converter car, one dual bed converter car, one stratified charge engine car and four lean burn engine cars. For each of these cars, they listed complete individual hydrocarbon composition values in gram per mile. As previously mentioned, their composition data was converted to values proportional to area and then, the carbon percents for all of the hydrocarbons and reactivity-per-gram values were calculated using our computer program.

As shown in Tables VIII and IX, there are many similarities between the two EPA studies and our paper, but there are also many differences. The following discussion will be general in nature, since the two tables provide the detailed comparisons. Both EPA and our paper rank (a) the dual bed converter cars the highest in methane; (b) the nonconverter cars, stratified charge engine cars and lean burn or lean combustion engine cars the lowest in methane and (c) the oxidation converter cars intermediate in methane.

Both studies rank the stratified charge engine cars and the lean burn or lean combustion engine cars (a) the highest in ethylene, (b) the highest in total olefin and (c) the highest in reactivity per gram (except the carbon-based methane exclusion scale). Both rank the dual bed converter cars the lowest in average reactivity.

There are differences in the fuel-type hydrocarbons--probably caused by fuel composition differences. Also, it should be pointed out that neither of these two EPA references mention correcting their results for the hydrocarbons in the dilution air that passes through the CVS. These "dirty air" corrections are significant for at least methane. It is believed that these corrections were made in both studies, but the authors simply failed to mention it.

When the complexity of a gas chromatographic analysis and a reactivity calculation are considered, our data, in general, agrees quite well with that reported by the Bureau of Mines and EPA.

SUMMARY

Exhaust gases from fourteen 1970-4 model and twenty 1975-7 model General Motors cars and six experimental model cars were collected during 1975-8 Federal Test Procedure tests and analyzed by gas chromatography. Numerous individual hydrocarbons were determined, but data for only the sixteen most concentrated hydrocarbons will be discussed below. The average composition and reactivity results for the forty cars are summarized below.

1. For the fourteen nonconverter (1970-4) model cars, individual exhaust hydrocarbons in the order of decreasing carbon percent of total hydrocarbon were: ethylene, toluene, acetylene, xylenes, propylene, methane, benzene, trimethylpentanes, n-butane, i-pentane, butenes, methylpentanes, n-pentane, ethylbenzene, ethane and i-butane.
2. The use of oxidation catalytic converters greatly changed the exhaust hydrocarbon composition. In general, such use caused individual paraffins to increase in carbon percent and individual olefins and acetylene to decrease.
3. For the twenty oxidation converter (1975-7) model cars, individual hydrocarbons in the order of decreasing carbon percent were: methane, trimethylpentanes, ethylene, toluene, n-butane, i-pentane, xylenes, benzene, ethane, propylene, methylpentanes, acetylene, butenes, n-pentane, ethylbenzene and i-butane.
4. The methane percent was 5.0 for nonconverter cars and 14.5 for converter cars, ethylene percent was 12.5 for nonconverter cars and 7.4 for converter cars and propylene percent was 6.5 for nonconverter cars and 2.9 for converter cars.
5. For nonconverter cars, the paraffin carbon percent was 33.4, the aromatic percent was 32.1, the olefin percent was 26.6 and the acetylene percent was 7.9. For converter cars, the paraffin percent was 56.5, the aromatic percent was 26.2, the olefin percent was 15.5 and the acetylene percent was 2.2.
6. Because of these large changes in hydrocarbon composition, all of the reactivity scales evaluated indicated that converter cars produced exhaust hydrocarbon mixtures that were less reactive than those of nonconverter cars. The reductions in reactivity per gram ranged from about 10 to 35 percent. Thus, overall reduction in total photochemical reactivity (emissions in gram per mile times reactivity per gram) of the exhaust hydrocarbon was greater for converter cars than indicated by the percent

Table VIII. Comparison with Literature.

	Nonconverter		Oxidation Converter	
	EPA	This Paper	EPA	This Paper
Mileage	---	13546	---	13548
No. of Cars	1	14	15	20

CARBON PERCENT OF TOTAL HYDROCARBONS

	EPA	This Paper	EPA	This Paper
Nonreactive Hydrocarbons				
Acetylene	12.3	7.9	3.4	2.2
Methane	5.7	5.0	15.1	14.5
Benzene	2.7	4.6	2.4	3.4
Ethane	1.1	1.0	2.8	3.0
Propane	0.0	0.1	0.0	0.3
Nonreactive Total	21.8	18.6	23.7	23.4
Reactive Hydrocarbons				
Ethylene	13.4	12.5	7.3	7.4
Toluene	14.2	8.4	12.8	6.8
Xylenes	4.3	6.7	4.1	5.3
Propylene	7.0	6.5	3.8	2.9
Trimethylpentanes	4.6	4.1	8.0	8.1
n-Butane	4.6	4.1	4.6	5.5
i-Pentane	2.7	4.0	5.2	5.4
Butenes	3.6	3.8	3.1	2.2
Methylpentanes	0.5	2.2	0.8	2.7
n-Pentane	2.0	1.6	4.0	1.8
Ethylbenzene	0.9	1.5	0.9	1.3
i-Butane	0.8	0.8	0.3	0.8
Other Hydrocarbons	19.6	25.3	21.4	26.4
Reactive Total	78.2	81.4	76.3	76.6
Hydrocarbon Classes				
Total Paraffins	34.7	33.4	54.4	56.5
Total Olefins	25.4	26.6	16.4	15.1
Acetylene	12.3	7.9	3.4	2.2
Total Aromatics	27.6	32.1	25.8	26.2

HYDROCARBON REACTIVITY PER GRAM

Molar-Based Reactivity Scales	EPA	This Paper	EPA	This Paper
NO$_2$ Formation Rate	0.0431	0.0487	0.0360	0.0339
Altshuller	0.0493	0.0520	0.0355	0.0340
Methane Exclusion	0.0179	0.0167	0.0132	0.0128
Dimitriades	0.1586	0.1576	0.1234	0.1186
Carbon-Based Reactivity Scales				
Methane Exclusion	0.9430	0.9500	0.8490	0.8550
Cal. Air Res. Bd.	0.4095	0.4264	0.3249	0.3146

RELATIVE REACTIVITY PER GRAM (GM NONCONVERTER CAR AVERAGE = 100.0)

Molar-Based Reactivity Scales	EPA	This Paper	EPA	This Paper
NO$_2$ Formation Rate	88.5	100.0	73.9	69.6
Altshuller	94.8	100.0	68.3	65.4
Methane Exclusion	107.2	100.0	79.0	76.6
Dimitriades	100.6	100.0	78.3	75.3
Carbon-Based Reactivity Scales				
Methane Exclusion	99.3	100.0	89.4	90.0
Cal. Air Res. Bd.	96.0	100.0	76.2	73.8

AVERAGE HYDROCARBON PROPERTIES

	EPA	This Paper	EPA	This Paper
Molecular Weight	41.80	49.30	43.03	45.25
Carbon Number	3.00	3.55	3.03	3.19
H/C Ratio	1.91	1.86	2.17	2.21
Nonmethane HC MW	50.52	56.06	63.10	65.49
Nonmethane HC C No.	3.68	4.07	4.54	4.70
Nonmethane HC H/C	1.70	1.74	1.87	1.92

EXHAUST EMISSIONS - GRAM PER MILE

	EPA	This Paper	EPA	This Paper
Hydrocarbon	1.15	2.45	0.45	0.65
Methane	0.07	0.12	0.07	0.08
Nonmethane HC	1.08	2.33	0.38	0.57

Table IX. Comparison with Literature.

	3-Way Converter		Dual Bed Converter		Stratified Charge Engine		Lean Burn		Lean Comb. with Reactor
	EPA	This Paper	EPA	This Paper	EPA	This Paper	EPA	This Paper	This Paper
Mileage	---	25723	---	5274	---	12448	---		1990
No. of Cars	1	2	1	2	1	2	4		1

CARBON PERCENT OF TOTAL HYDROCARBON

	EPA	This Paper	EPA	This Paper	EPA	This Paper	EPA	This Paper
Nonreactive Hydrocarbons								
Acetylene	5.4	3.4	1.5	3.2	8.3	7.1	9.0	8.2
Methane	18.6	7.8	25.5	33.7	4.3	5.9	3.3	5.3
Benzene	3.2	2.9	1.4	2.4	3.9	2.8	4.0	5.9
Ethane	1.0	2.4	3.9	4.4	1.0	1.1	0.8	1.9
Propane	0.0	0.2	0.0	0.2	0.0	0.1	0.0	0.1
Nonreactive Total	28.2	16.4	32.3	43.9	17.5	17.0	17.1	21.4
Reactive Hydrocarbons								
Ethylene	5.5	5.2	4.2	8.0	16.1	20.5	17.0	19.4
Toluene	13.7	10.0	6.9	6.4	14.9	6.7	13.4	13.3
Xylenes	5.3	4.0	1.4	2.3	5.3	7.4	5.0	4.8
Propylene	5.3	4.6	1.8	3.0	6.7	10.7	8.1	9.1
Trimethylpentanes	7.0	8.9	7.2	1.6	8.8	5.2	4.3	4.4
n-Butane	3.5	1.6	8.8	1.7	2.9	2.6	3.1	2.2
i-Pentane	3.0	2.4	10.3	1.8	1.4	4.2	2.3	1.2
Butenes	2.4	4.2	1.8	3.2	4.0	2.4	4.8	4.5
Methylpentanes	0.6	1.6	1.8	0.6	0.0	0.6	1.0	1.2
n-Pentane	2.6	0.6	7.1	0.5	1.4	0.7	2.1	0.3
Ethylbenzene	1.3	1.4	0.4	0.7	1.1	1.2	0.7	1.0
i-Butane	0.6	0.2	0.6	0.4	0.9	0.7	0.7	0.3
Other Hydrocarbons	21.0	39.1	21.0	20.6	19.0	21.8	19.6	17.0
Reactive Total	71.8	83.6	67.7	56.1	82.5	83.0	82.9	78.6
Hydrocarbon Classes								
Total Paraffins	48.6	43.6	77.6	56.2	28.9	26.6	27.9	22.8
Total Olefins	15.0	17.6	9.3	14.5	29.7	39.5	33.0	36.1
Acetylene	5.4	3.4	1.5	3.2	8.3	7.1	9.0	8.2
Total Aromatics	31.0	35.4	11.6	26.1	33.1	26.8	30.1	32.8

HYDROCARBON REACTIVITY PER GRAM

Molar-Based Reactivity Scales	EPA	This Paper	EPA	This Paper	EPA	This Paper	EPA	This Paper
NO$_2$ Formation Rate	0.0351	0.0416	0.0261	0.0262	0.0507	0.0595	0.0566	0.0562
Altshuller	0.0351	0.0398	0.0218	0.0304	0.0561	0.0682	0.0604	0.0640
Methane Exclusion	0.0123	0.0127	0.0113	0.0107	0.0178	0.0188	0.0188	0.0192
Dimitriades	0.1167	0.1164	0.1011	0.1144	0.1733	0.2028	0.1834	0.1940
Carbon-Based Reactivity Scales								
Methane Exclusion	0.8140	0.9220	0.7450	0.6630	0.9570	0.9410	0.9670	0.9470
Cal. Air Res. Bd.	0.3258	0.3603	0.2319	0.3116	0.4579	0.5308	0.4791	0.5080

RELATIVE REACTIVITY PER GRAM (GM NONCONVERTER CAR AVERAGE = 100.0)

Molar-Based Reactivity Scales	EPA	This Paper	EPA	This Paper	EPA	This Paper	EPA	This Paper
NO$_2$ Formation Rate	72.1	85.4	53.8	53.4	104.1	122.2	116.2	115.4
Altshuller	67.5	76.5	41.9	58.5	107.9	131.2	116.2	123.1
Methane Exclusion	73.7	76.0	67.7	64.1	106.6	112.6	112.6	115.0
Dimitriades	74.0	73.9	72.6	64.1	110.0	128.7	116.4	123.1
Carbon-Based Reactivity Scales								
Methane Exclusion	85.7	97.1	78.4	69.8	100.7	99.1	101.8	99.7
Cal. Air Res. Bd.	76.4	84.5	54.4	73.1	107.4	124.5	112.4	119.1

AVERAGE HYDROCARBON PROPERTIES

	EPA	This Paper	EPA	This Paper	EPA	This Paper	EPA	This Paper
Molecular Weight	39.34	54.94	34.83	29.96	47.77	43.78	47.76	43.45
Carbon Number	2.77	3.93	2.38	2.04	3.45	3.14	3.45	3.14
H/C Ratio	2.17	1.95	2.60	2.52	1.82	1.92	1.82	1.81
Nonmethane HC MW	64.21	72.41	63.85	59.21	53.32	49.21	51.76	48.91
Nonmethane HC C No.	4.66	4.28	4.51	4.28	3.88	3.59	3.76	3.57
Nonmethane HC H/C	1.75	1.79	2.13	1.81	1.72	1.77	1.74	1.68

EXHAUST EMISSIONS - GRAM PER MILE

	EPA	This Paper	EPA	This Paper	EPA	This Paper	EPA	This Paper
Hydrocarbon	0.16	0.48	0.31	0.62	0.28	1.06	0.74	1.04
Methane	0.03	0.04	0.08	0.21	0.01	0.06	0.02	0.06
Nonmethane HC	0.13	0.44	0.23	0.41	0.27	1.00	0.72	0.98

1294

reduction in hydrocarbon mass emissions alone.

7. Analyses performed on experimental three-way converter cars and dual bed converter cars indicated that exhausts from such cars were higher in paraffins, lower in olefins and lower in reactivity than similar values for nonconverter cars. The reductions in reactivity per gram, using five of the six scales evaluated, ranged from about 15 to 46 percent.

8. Analyses performed on a stratified charge engine car and a lean combustion engine car indicated that exhausts from such cars were lower in paraffins, higher in olefins and higher in reactivity than similar values for nonconverter cars. The increases in reactivity per gram, using five of the six scales evaluated, ranged from about 13 to 31 percent.

ACKNOWLEDGEMENT

The author wishes to thank the following Research Laboratories employees: G. J. Morris and J. R. Collins for conducting the emission tests and P. A. Mulawa for conducting the last few chromatographic analyses on the 1975 California production model cars. Acknowledgement is given to A. S. Prostak and R. L. Williams, of the Milford Proving Ground, for obtaining the chromatograms for the 1977 model cars.

REFERENCES

1. M. W. Jackson, "Effects of Some Engine Variables and Control System on Composition and Reactivity of Exhaust Hydrocarbons," SAE Vehicle Emissions, Part II, PT-12, p. 241, 1968.
2. E. E. Wigg, R. J. Campion, and W. L. Petersen, "The Effect of Fuel Hydrocarbon Composition on Exhaust Hydrocarbon and Oxygenate Emissions," presented at SAE meeting, Deroit, MI Jan. 1972. (SAE Paper No. 720251).
3a. Wigg, E. E., "Reactive Exhaust Emissions from Current and Future Emission Control Systems," presented at SAE meeting, Detroit, MI, Jan. 1973. (SAE Paper No. 730196).
3b. E. E. Wigg, "Fuel-Exhaust Compositional Emission Control Systems," Proc. Amer. Petrol. Inst., Div. of Ref., Vol. 52, p. 964, 1972.
4. "Aldehyde and Reactive Organic Emissions from Motor Vehicles," Parts I and II, report prepared for EPA by Bureau of Mines, Bartlesville, OK, March 1973. (EPA-IAG-0188 and MSPCP-IAG-001).
5. F. M. Black and R. L. Bradow, "Patterns of Hydrocarbon Emissions from 1975 Production Cars," presented at SAE meeting, Houston, TX, June 1975. (SAE Paper No. 750681).
6. F. Black and L. High, "Automotive Hydrocarbon Emission Patterns in the Measurements of Nonmethane Hydrocarbon Emission Rates," presented at SAE meeting, Detroit, MI, Feb.-March 1977. (SAE Paper No. 770144).
7a. Federal Register, Vol. 35, No. 219, p. 17288, Nov. 10, 1970.
7b. Federal Register, Vol. 17, No. 221, P. 24250, Nov. 15, 1972.
7c. Federal Register, VOL. 42, No. 124, p. 32906, June 28, 1977.
8. D. J. McEwen, "Automobile Exhaust Hydrocarbon Analysis by Gas Chromatography," Anal. Chem., Vol. 38, No. 8, p. 1047, June 1966.
9. Federal Register, Vol. 39, No. 92, p. 16905, May 10, 1974.
10. A. P. Altshuller, "An Evaluation of Techniques for the Determination of the Photochemical Reactivity of Organic Emissions," J. APCA, Vol. 16, No. 8, p. 257, May 1966.
11. B. Dimitriades, "The Concept of Reactivity and Its Possible Application in Control," Proceedings of the Solvent Reactivity Conference, Research Triangle Park, NC, Nov. 1974. (EPA-650/3-74-010).
12. Status Report on Catalytic-Device-equipped Test Fleets, presented at California Air Resources Board Meeting, San Francisco, CA, Jan. 9, 1974.
13. M. W. Jackson and R. L. Everett, "Effect of Fuel Composition on Amount and Reactivity of Evaporative Emissions," SAE Vehicle Emissions, Part III, PT-14, p. 802, 1971.

Appendix Table A-I Summary of Individual Nonconverter Car Results.

	1970 Car A Make A	1972 Car B Make B	1973 Car C Make B	1973 Car D Make C	1973 Car E Make C	1974 Car F Make D	1974 Car G Make E
Engine	455-2	350-2	350-2	455-4	455-4	455-4	350-2
Calif. or Fed.	Fed.	Calif.	Fed.	Fed.	Fed.	Fed.	Fed.
Mileage	3600	42105	21833	20576	16998	13442	14138

CARBON PERCENT OF TOTAL HYDROCARBON

Nonreactive Hydrocarbons							
Acetylene	7.4	0.7	7.8	9.6	6.0	11.1	7.3
Methane	6.2	6.3	5.0	5.6	4.0	6.4	4.3
Benzene	6.4	3.8	4.2	2.1	3.8	4.5	5.3
Ethane	1.2	0.9	0.9	1.1	0.9	1.1	0.9
Propane	0.1	0.0	0.1	0.0	0.3	0.0	0.2
Nonreactive Total	21.4	21.7	18.0	18.4	15.0	23.1	17.9
Reactive Hydrocarbons							
Ethylene	10.1	15.5	14.2	13.1	11.0	14.3	13.2
Toluene	11.7	11.4	8.0	10.4	8.9	5.3	6.4
Xylenes	9.7	2.3	4.8	3.1	4.4	7.2	7.8
Propylene	8.0	9.6	5.7	7.7	5.4	7.2	7.1
Trimethlpentanes	2.1	3.0	2.3	11.0	3.1	4.8	3.4
n-Butane	1.6	7.0	2.0	2.7	1.3	5.2	3.8
i-Pentane	4.4	5.2	3.8	2.1	7.0	2.9	3.5
Butenes	6.0	2.3	3.1	2.1	3.5	4.9	3.2
Methylpentanes	2.0	1.2	3.2	1.8	3.8	1.8	1.9
n-Pentane	0.6	2.1	1.0	1.4	1.7	1.5	2.1
Ethylbenzene	1.7	0.6	1.1	1.2	1.1	1.6	2.1
i-Butane	0.8	2.3	0.9	0.1	1.3	0.5	0.9
Other Hydrocarbons	19.8	15.6	31.9	24.4	32.7	19.7	26.7
Reactive Total	78.6	78.3	82.0	81.6	85.0	76.9	82.1
Hydrocarbon Classes							
Total Parafins	23.9	34.5	32.9	39.7	37.8	32.9	33.7
Total Olefins	32.0	31.1	25.6	27.3	24.3	26.8	27.6
Acetylene	7.4	10.7	7.8	9.5	6.0	11.1	7.3
Total Aromatics	36.7	23.7	33.7	23.5	31.9	29.2	31.4

HYDROCARBON REACTIVITY PER GRAM

Molar-Based Reactivity Scales							
NO$_2$ Formation Rate	0.0542	0.0511	0.0461	0.0475	0.0466	0.0490	0.0496
Altshuller	0.0577	0.0556	0.0516	0.0498	0.0483	0.0542	0.0532
Methane Exclusion	0.0167	0.0190	0.0168	0.0171	0.0158	0.0156	0.0170
Dimitriades	0.1612	0.1773	0.1588	0.1577	0.1480	0.1666	0.1604
Carbon-Based Reactivity Scales							
Methane Exclusion	0.9380	0.9370	0.9500	0.9440	0.9600	0.9360	0.9570
Cal. Air Res. Bd.	0.4808	0.4469	0.4228	0.4174	0.4079	0.4156	0.4318

AVERAGE HYDROCARBON PROPERTIES

Molecular Weight	47.27	42.47	49.05	47.37	53.44	43.72	49.82
Carbon Number	3.42	3.04	3.54	3.39	3.84	3.14	3.59
H/C Ratio	1.80	1.94	1.83	1.95	1.89	1.90	1.85
Nonmethane HC MW	55.68	48.76	56.08	54.74	60.24	50.68	55.97
Nonmethane HC C No.	4.07	3.53	4.08	3.96	4.36	3.68	4.06
Nonmethane HC H/C	1.66	1.79	1.72	1.80	1.79	1.75	1.76

EXHAUST EMISSIONS - GRAM PER MILE

Hydrocarbon	3.43	1.36	1.90	6.72	3.22	2.42	1.70
Methane	0.21	0.06	0.10	0.38	0.13	0.16	0.07
Nonmethane HC	3.22	1.30	1.80	6.34	3.09	2.26	1.63

Appendix Table A-II Summary of Individual Nonconverter Car Results.

	1974 Car H Make B	1974 Car I Make B	1974 Car J Make B	1974 Car K Make F	1974 Car L Make G	1974 Car M Make C	1974 Car N Make C
Engine	350-4	350-4	400-2	350-4	350-2	455-4	400-2
Calif. or Fed.	Calif.	Fed.	Fed.	Fed.	Fed.	Fed.	Fed.
Mileage	2538	2955	5132	9296	12634	8146	14173

CARBON PERCENT OF TOTAL HYDROCARBON

Nonreactive Hydrocarbons							
Acetylene	8.2	11.0	5.3	5.4	11.2	3.5	6.4
Methane	5.5	6.2	5.1	2.7	6.0	1.6	4.7
Benzene	5.7	4.6	5.4	3.8	4.6	4.9	5.3
Ethane	1.1	1.4	0.8	0.8	1.3	0.6	1.0
Propane	0.0	0.0	0.0	0.0	0.0	0.0	0.2
Nonreactive Total	20.6	23.2	16.5	12.7	23.1	10.8	17.7
Reactive Hydrocarbons							
Ethylene	12.8	11.7	13.4	13.2	11.4	10.1	11.2
Toluene	10.2	6.9	5.8	6.8	6.4	9.0	10.5
Xylenes	6.9	8.2	6.5	7.7	8.1	8.3	8.3
Propylene	6.2	7.3	4.3	7.0	7.1	4.0	4.4
Trimethlpentanes	5.4	4.1	3.4	2.7	3.8	5.3	3.4
n-Butane	4.1	3.5	4.8	6.1	3.8	3.7	4.6
i-Pentane	3.5	4.0	7.3	2.6	4.1	4.2	3.7
Butenes	3.5	4.4	4.3	4.1	4.1	4.0	4.7
Methylpentanes	1.8	2.0	3.1	2.0	1.9	2.8	1.8
n-Pentane	1.3	1.5	2.4	1.3	1.5	2.9	2.2
Ethylbenzene	1.5	0.8	2.0	1.6	1.8	2.2	1.3
i-Butane	0.4	0.4	1.4	0.6	0.4	0.4	0.7
Other Hydrocarbons	21.9	22.1	28.0	31.7	22.5	32.4	25.5
Reactive Total	79.4	76.8	83.5	87.3	76.9	89.2	82.3
Hydrocarbon Classes							
Total Parafins	32.2	32.2	37.7	27.0	32.2	37.3	33.3
Total Olefins	26.4	25.4	25.1	28.1	25.9	22.7	24.6
Acetylene	8.2	11.0	5.3	5.4	11.2	3.5	6.4
Total Aromatics	33.2	31.4	31.9	39.5	30.7	36.5	35.7

HYDROCARBON REACTIVITY PER GRAM

Molar-Based Reactivity Scales							
NO$_2$ Formation Rate	0.0484	0.0481	0.0473	0.0529	0.0478	0.0469	0.0467
Altshuller	0.0512	0.0518	0.0497	0.0578	0.0509	0.0480	0.0494
Methane Exclusion	0.0170	0.0172	0.0162	0.0166	0.0175	0.0151	0.0160
Dimitriades	0.1574	0.1574	0.1533	0.1648	0.1522	0.1433	0.1490
Carbon-Based Reactivity Scales							
Methane Exclusion	0.9450	0.9380	0.9490	0.9730	0.9400	0.9840	0.9530
Cal. Air Res. Bd.	0.4248	0.4142	0.4109	0.4601	0.4169	0.4026	0.4162

AVERAGE HYDROCARBON PROPERTIES

Molecular Weight	47.70	45.97	50.34	54.11	45.81	61.64	51.42
Carbon Number	3.44	3.31	3.61	3.92	3.30	4.45	3.71
H/C Ratio	1.84	1.86	1.92	1.78	1.86	1.83	1.83
Nonmethane HC MW	55.08	53.78	58.05	58.58	53.22	65.11	58.93
Nonmethane HC C No.	4.01	3.91	4.20	4.26	3.87	4.72	4.31
Nonmethane HC H/C	1.71	1.73	1.80	1.73	1.73	1.77	1.65

EXHAUST EMISSIONS - GRAM PER MILE

Hydrocarbon	1.30	1.70	1.08	2.43	1.82	2.96	2.27
Methane	0.07	0.10	0.06	0.07	0.11	0.05	0.11
Nonmethane HC	1.23	1.60	1.02	2.36	1.71	2.91	2.16

Appendix Table A-III Summary of Individual Converter Car Results.

	1975 Car O Make D	1975 Car P Make H	1975 Car Q Make E	1975 Car R Make B	1975 Car S Make F	1975 Car T Make I	1975 Car U Make C
Engine	455-4	500-4	231-2	350-2	350-4	262-2	400-2
Calif. or Fed.	Fed.	Fed.	Fed.	Fed.	Fed.	Fed.	Fed.
Mileage	12317	8741	20180	17705	2942	18140	22317
CARBON PERCENT OF TOTAL HYDROCARBON							
Nonreactive Hydrocarbons							
Acetylene	2.8	2.3	2.6	1.5	1.6	2.6	3.5
Methane	12.1	14.5	9.9	11.7	12.3	7.8	9.2
Benzene	2.2	3.6	5.4	7.4	5.4	2.5	3.8
Ethane	2.7	2.3	2.7	2.1	3.0	2.3	2.6
Propane	0.4	0.3	0.4	0.2	0.0	0.2	0.1
Nonreactive Total	20.2	23.0	19.6	22.9	22.4	15.4	19.2
Reactive Hydrocarbons							
Ethylene	4.8	8.5	6.68	8.6	9.5	6.9	7.9
Toluene	6.0	7.4	5.8	8.4	5.6	7.6	8.0
Xylenes	7.1	6.3	5.4	6.9	5.4	4.6	4.9
Propylene	3.0	3.2	3.0	3.5	3.9	2.5	3.4
Trimethylpentanes	5.8	5.6	5.2	4.8	7.4	7.6	8.1
n-Butane	7.2	3.5	7.3	4.4	6.6	5.8	6.3
i-Pentane	4.4	3.5	6.5	4.4	3.5	10.4	6.7
Butenes	2.6	1.8	2.6	2.9	2.4	2.4	2.3
Methylpentanes	2.8	2.6	3.3	2.1	2.0	3.7	2.7
n-Pentane	2.3	2.0	2.9	2.2	1.2	2.1	2.7
Ethylbenzene	1.7	1.6	1.3	1.3	1.3	1.3	1.2
i-Butane	1.1	1.2	1.6	0.9	0.8	0.8	1.4
Other Hydrocarbons	31.0	29.6	28.7	26.6	27.6	29.0	27.4
Reactive Total	79.8	77.0	80.4	77.1	77.6	84.6	80.8
Hydrocarbon Classes							
Total Parafins	52.9	50.7	55.8	46.3	50.5	59.5	53.2
Total Olefins	17.5	17.8	15.6	18.3	18.3	13.9	16.3
Acetylene	2.8	2.3	2.6	1.5	1.6	2.6	3.5
Total Aromatics	26.8	29.2	26.0	33.9	29.6	24.0	27.0
HYDROCARBON REACTIVITY PER GRAM							
Molar-Based Reactivity Scales							
NO$_2$ Formation Rate	0.0378	0.0384	0.0364	0.0385	0.0389	0.0334	0.0364
Altshuller	0.0354	0.0375	0.0349	0.0393	0.0393	0.0330	0.0364
Methane Exclusion	0.0128	0.0128	0.0137	0.0132	0.0136	0.0136	0.0140
Dimitriades	0.1151	0.1243	0.1193	0.1277	0.1302	0.1174	0.1233
Carbon-Based Reactivity Scales							
Methane Exclusion	0.8790	0.8550	0.9010	0.8830	0.8770	0.9220	0.9080
Cal. Air Res. Bd.	0.3393	0.3448	0.3168	0.3516	0.3456	0.2993	0.3253
AVERAGE HYDROCARBON PROPERTIES							
Molecular Weight	48.71	43.52	48.67	47.34	45.01	52.45	49.38
Carbon Number	3.44	3.07	3.44	3.34	3.18	3.70	3.50
H/C Ratio	2.13	2.15	2.12	2.06	2.13	2.15	2.08
Nonmethane HC MW	67.55	65.46	65.41	65.77	63.48	67.16	64.16
Nonmethane HC C No.	4.85	4.72	4.68	4.77	4.57	4.80	4.60
Nonmethane HC H/C	1.90	1.84	1.95	1.76	1.86	1.96	1.92
EXHAUST EMISSIONS - GRAM PER MILE							
Hydrocarbon	0.63	0.69	0.65	0.80	0.83	0.85	0.76
Methane	0.08	0.10	0.06	0.09	0.10	0.06	0.07
Nonmethane HC	0.55	0.59	0.59	0.71	0.73	0.79	0.69

Appendix Table A-IV Summary of Individual Converter Car Results.

	1975 Car V Make J	1975 Car W Make K	1975 Car X Make D	1975 Car Y Make B	1975 Car Z Make F	1975 Car AA Make I	1975 Car BB Make K
Engine	250-1	140-2	455-4	350-4	350-4	250-1	140-1
Calif. or Fed.	Fed.	Fed.	Calif.	Calif.	Calif.	Calif.	Calif.
Mileage	20907	2200	20142	20146	19995	19333	19982
CARBON PERCENT OF TOTAL HYDROCARBON							
Nonreactive Hydrocarbons							
Acetylene	2.8	1.2	2.5	3.7	1.7	1.6	4.4
Methane	10.8	12.2	16.9	10.8	8.7	9.9	11.7
Benzene	2.4	5.8	4.2	2.3	2.9	2.3	2.5
Ethane	2.6	2.3	3.4	2.6	2.8	2.9	1.5
Propane	0.2	0.2	0.4	0.3	0.3	0.5	0.6
Nonreactive Total	18.8	21.7	27.4	19.7	16.4	17.2	20.7
Reactive Hydrocarbons							
Ethylene	9.0	7.5	8.8	6.3	8.0	4.9	5.4
Toluene	5.4	7.3	5.7	6.2	5.4	5.5	6.3
Xylenes	6.3	8.4	4.5	5.4	5.6	5.1	7.1
Propylene	4.0	3.3	4.2	2.5	3.2	2.2	2.0
Trimethylpentanes	5.5	5.2	5.8	8.0	6.9	7.8	5.4
n-Butane	5.0	4.4	4.4	4.5	5.0	6.5	3.3
i-Pentane	4.9	3.4	5.4	5.6	6.2	6.9	4.5
Butenes	2.6	1.9	3.9	2.4	2.6	2.4	1.7
Methylpentanes	3.6	2.4	2.7	3.2	3.6	3.8	3.1
n-Pentane	1.2	1.9	2.1	2.0	2.7	3.2	1.9
Ethylbenzene	1.3	1.8	1.2	1.5	1.6	1.3	1.5
i-Butane	0.8	0.7	0.9	1.1	0.8	1.4	0.7
Other Hydrocarbons	31.6	30.1	23.0	31.6	32.0	31.8	36.4
Reactive Total	81.2	78.3	72.6	80.3	83.6	82.8	79.3
Hydrocarbon Classes							
Total Parafins	50.6	46.6	54.5	55.0	54.9	60.7	49.3
Total Olefins	18.9	17.4	20.0	14.3	17.6	11.9	11.2
Acetylene	2.8	1.2	2.5	3.7	1.7	1.6	4.4
Total Aromatics	27.7	34.8	22.5	27.0	25.8	25.8	35.1
HYDROCARBON REACTIVITY PER GRAM							
Molar-Based Reactivity Scales							
NO$_2$ Formation Rate	0.0405	0.0374	0.0381	0.0336	0.0393	0.0327	0.0377
Altshuller	0.0405	0.0379	0.0384	0.0336	0.0378	0.0309	0.0324
Methane Exclusion	0.0138	0.0126	0.0134	0.0131	0.0136	0.0126	0.0123
Dimitriades	0.1308	0.1225	0.1304	0.1140	0.1252	0.1076	0.1057
Carbon-Based Reactivity Scales							
Methane Exclusion	0.8920	0.8780	0.8310	0.8920	0.9130	0.9010	0.8830
Cal. Air Res. Bd.	0.3537	0.3497	0.3477	0.3149	0.3378	0.2886	0.3056
AVERAGE HYDROCARBON PROPERTIES							
Molecular Weight	46.91	47.41	40.01	49.20	51.04	52.97	49.54
Carbon Number	3.32	3.36	2.80	3.48	3.61	3.74	3.52
H/C Ratio	2.10	2.08	2.26	2.11	2.11	2.14	2.05
Nonmethane HC MW	64.04	68.53	60.25	67.38	60.61	70.41	71.56
Nonmethane HC C No.	4.60	4.95	4.33	4.84	4.77	5.03	5.18
Nonmethane HC H/C	1.90	1.82	1.89	1.90	1.94	1.97	1.79
EXHAUST EMISSIONS - GRAM PER MILE							
Hydrocarbon	0.70	0.59	0.82	0.72	0.74	0.71	0.49
Methane	0.08	0.07	0.14	0.08	0.06	0.07	0.06
Nonmethane HC	0.62	0.52	0.68	0.64	0.68	0.64	0.43

Appendix Table A-V Summary of Individual Converter Car Results.

	1976 Car CC Make B	1976 Car DD Make I	1977 Car EE Make D	1977 Car FF Make F	1977 Car GG Make I	1977 Car HH Make I
Engine	350-4	305-2	231-2	403-4	250-1	305-2
Calif. or Fed.	Calif.	Fed.	Calif.	Calif.	Calif.	Calif.
Mileage	2510	3925	4808	9901	9903	9824

CARBON PERCENT OF TOTAL HYDROCARBON

Nonreactive Hydrocarbons						
Acetylene	2.8	4.5	0.0	0.0	0.4	0.7
Methane	9.3	10.3	34.7	38.2	18.7	20.2
Benzene	5.7	3.9	1.7	1.4	1.5	1.9
Ethane	3.1	2.9	3.4	5.5	5.1	4.1
Propane	0.3	0.4	0.0	0.0	0.9	0.2
Nonreactive Total	21.2	21.9	39.8	45.1	26.6	27.1
Reactive Hydrocarbons						
Ethylene	7.4	7.0	11.0	7.3	4.9	8.2
Toluene	7.3	7.0	8.5	5.2	9.3	7.7
Xylenes	8.3	6.4	1.8	1.6	2.9	1.5
Propylene	3.4	4.4	1.1	1.4	1.1	2.5
Trimethylpentanes	5.4	4.6	12.8	13.5	18.3	17.8
n-Butane	3.1	4.7	5.4	6.9	8.0	8.3
i-Pentane	4.7	4.4	4.2	4.9	7.1	7.3
Butenes	3.6	4.7	0.0	0.0	0.4	1.1
Methylpentanes	3.5	2.9	1.5	0.9	1.8	1.7
n-Pentane	2.0	2.2	0.6	0.6	0.9	0.9
Ethylbenzene	2.2	1.6	0.5	0.8	0.9	0.5
i-Butane	0.5	1.0	0.4	0.2	0.3	0.2
Other Hydrocarbons	27.4	27.0	12.4	11.9	17.6	14.9
Reactive Total	78.8	78.1	60.2	54.9	73.4	72.9
Hydrocarbon Classes						
Total Paraffins	45.3	74.4	71.5	79.8	72.9	73.1
Total Olefins	14.9	19.0	12.2	8.9	6.3	12.3
Acetylene	2.8	4.5	0.0	0.0	0.4	0.1
Total Aromatics	37.0	29.1	16.3	11.3	20.4	13.9

HYDROCARBON REACTIVITY PER GRAM

Molar-Based Reactivity Scales						
NO$_2$ Formation Rate	0.0358	0.0400	0.0222	0.0190	0.0210	0.0265
Altshuller	0.0361	0.0403	0.0255	0.0198	0.0213	0.0268
Methane Exclusion	0.0136	0.0142	0.0103	0.0097	0.0114	0.0122
Dimitriades	0.1201	0.1264	0.1186	0.1022	0.0954	0.1150
Carbon-Based Reactivity Scales						
Methane Exclusion	0.9070	0.8970	0.6530	0.6180	0.8130	0.7980
Cal. Air Res. Bd.	0.3338	0.3548	0.2680	0.2243	0.2276	0.2623

AVERAGE HYDROCARBON PROPERTIES

Molecular Weight	49.38	46.44	29.57	28.34	41.32	38.10
Carbon Number	3.52	3.29	2.01	1.90	2.86	2.64
H/C Ratio	2.00	2.09	2.68	2.88	2.42	2.40
Nonmethane HC MW	65.68	62.14	60.21	60.74	69.58	63.73
Nonmethane HC C No.	4.75	4.48	4.28	4.30	4.95	4.51
Nonmethane HC H/C	1.80	1.84	2.04	2.10	2.03	2.10

EXHAUST EMISSIONS - GRAM PER MILE

Hydrocarbon	0.90	1.00	0.30	0.30	0.15	0.33
Methane	0.08	0.10	0.10	0.11	0.03	0.07
Nonmethane HC	0.82	0.90	0.20	0.19	0.12	0.26

Appendix Table A-VI Summary of Individual Experimental Car Results.

	Car II Make K	Car JJ Make K	Car KK Make B	Car LL Make B	Car MM Make B	Car NN Make G
Engine	140-EFI	140-EFI	350-4	350-4	350-1	350-2
Mileage	1446	50000	547	10000	12448	1990

CARBON PERCENT OF TOTAL HYDROCARBON

Nonreactive Hydrocarbons						
Acetylene	2.8	3.9	3.2	3.1	7.1	8.2
Methane	8.7	6.8	32.0	35.4	5.9	5.3
Benzene	2.6	3.2	1.7	3.1	2.8	5.9
Ethane	2.3	2.4	4.4	4.4	1.1	1.9
Propane	0.2	0.1	0.2	0.2	0.1	0.1
Nonreactive Total	16.5	16.3	41.4	46.4	17.0	21.4
Reactive Hydrocarbons						
Ethylene	3.6	6.7	4.7	11.3	20.5	19.4
Toluene	9.4	10.7	6.1	6.8	6.7	13.3
Xylenes	4.3	3.6	3.2	1.4	5.3	4.8
Propylene	3.1	6.0	2.7	3.4	10.7	9.1
Trimethylpentanes	11.0	6.8	7.6	6.2	5.2	4.4
n-Butane	1.6	1.5	1.9	1.5	1.2	2.2
i-Pentane	2.3	2.4	1.9	1.7	2.6	1.2
Butenes	1.5	7.0	3.2	3.2	4.2	4.5
Methylpentanes	1.8	1.5	2.5	0.7	2.4	1.2
n-Pentane	0.7	0.5	0.4	0.6	0.7	0.3
Ethylbenzene	1.2	1.6	0.6	0.8	0.7	1.0
i-Butane	0.2	0.1	0.4	0.3	1.2	6.1
Other Hydrocarbons	42.9	35.3	25.4	15.8	21.8	17.1
Reactive Total	83.5	83.7	58.6	53.6	83.0	78.6
Hydrocarbon Classes						
Total Paraffins	49.2	38.1	56.3	56.1	26.6	22.8
Total Olefins	12.1	23.1	10.8	18.3	39.5	36.1
Acetylene	2.8	3.9	3.2	3.1	7.1	8.2
Total Aromatics	35.9	34.9	29.7	22.5	26.8	32.8

HYDROCARBON REACTIVITY PER GRAM

Molar-Based Reactivity Scales						
NO$_2$ Formation Rate	0.0360	0.0473	0.0242	0.0283	0.0595	0.0562
Altshuller	0.0327	0.0469	0.0247	0.3333	0.0682	0.0640
Methane Exclusion	0.0115	0.0139	0.0099	0.0115	0.0188	0.0192
Dimitriades	0.0996	0.1333	0.0998	0.1291	0.2028	0.1940
Carbon-Based Reactivity Scales						
Methane Exclusion	0.9130	0.9320	0.6800	0.6460	0.9410	0.9470
Cal. Air Res. Bd.	0.3141	0.4065	0.2895	0.3336	0.5308	0.5080

AVERAGE HYDROCARBON PROPERTIES

Molecular Weight	56.64	53.25	31.25	27.45	43.78	43.45
Carbon Number	4.04	3.82	2.16	1.91	3.14	3.14
H/C Ratio	1.99	1.91	2.44	2.61	1.92	1.81
Nonmethane HC MW	78.60	66.22	65.44	52.98	49.51	48.91
Nonmethane HC C No.	5.68	4.80	4.73	3.83	3.59	3.57
Nonmethane HC H/C	1.81	1.77	1.81	1.81	1.77	1.68

EXHAUST EMISSIONS - GRAM PER MILE

Hydrocarbon	0.32	0.63	0.64	0.59	1.06	1.04
Methane	0.03	0.04	0.20	0.21	0.06	0.06
Nonmethane HC	0.29	0.59	0.44	0.38	1.00	0.98

780672

Performance and Emission Predictions for a Multi-Cylinder Spark Ignition Engine with Catalytic Converter

P. C. Baruah, R. S. Benson, and H. N. Gupta
Univ. of Manchester Institute of Science and Technology

NOMENCLATURE

a	Speed of sound
a_A	Speed of sound after isentropic change of state to reference pressure
a_{ref}	Reference speed of sound
A	Non-dimensional form of a, $\left(\dfrac{a}{a_{ref}}\right)$
A_a	Non-dimensional form of a_A, $\left(\dfrac{a_A}{a_{ref}}\right)$
A_{chem}	Pre-exponential factor depending upon catalyst type and species being oxidized
C	Surface roughness constant
C_p	Specific heat of gas at constant pressure
D_A	Diffusivity of gas, $\left(\dfrac{m^2}{s}\right)$
D_e	Equivalent diameter
E_a	Activation energy

ABSTRACT

A mathematical model is developed to represent an oxidizing catalytic converter in the exhaust system of a spark ignition engine in which the flow is non steady. By using the basic mass transfer, heat transfer and chemical reaction rate equations on the path lines the heat generated at the catalyst surface and the friction factor are allowed for in the generalized non steady flow relations using the method of characteristics. The model is included in a multi-cylinder engine simulation program. Secondary air injection into the exhaust system is represented by a simple mixing process without chemical reaction. A series of tests were carried out on a four cylinder two litre engine with a carbon monoxide and hydrocarbon oxidizing converter and secondary air injection. Comparison of results between experiments and computer calculations shows excellent agreement when the converter is new, but that if the catalyst surface is poisoned or aged the hydrocarbon prediction deteriorates. The carbon monoxide predictions, however, remain fairly good. The tests showed that the CO levels entering the exhaust pipe are dependent on the engine condition. Prediction of the overall engine performance and emission levels is very good. The comprehensive simulation program offers an excellent tool for examining practical locations of catalytic converters.

0148-7191/78/0605-0672$02.50
Copyright © 1978 Society of Automotive Engineers, Inc.

f	Friction factor $\dfrac{\tau_w}{\frac{1}{2}\rho u^2}$
f_{CO}	Scale factor for CO formation
f_g	Pressure drop coefficient of carburettor $\left(\dfrac{\Delta p}{\frac{1}{2}\rho u^2}\right)$
FF	Darcy friction factor (4f)
G	Mass velocity of gas, $\left(\dfrac{kg}{m^2 s}\right)$
h_q	Heat transfer coefficient, $\left(\dfrac{kJ}{m^2 sK}\right)$
H_L	Heat loss factor
κ	Specific heat ratio
k	Boltzmann's constant
k_{chem}	Chemical rate constant (m/s)
k_{mt}	Mass Transfer coefficient (m/s)
k_q	Thermal conductivity $\left(\dfrac{kJ}{s\,mK}\right)$
$\bar{k}_s$	Overall rate constant, (m/s)
ℓ	Distance m
ℓ_{ref}	Reference length m
L	Non-dimensional length $\left(\dfrac{\ell}{\ell_{ref}}\right)$
L_{cv}	Catalytic converter length (active part) m
M	Mols
M_w	Molecular weight
N_{Nu}	Nusselt number, $\left(\dfrac{h_q D_e}{k_q}\right)$
N_{Pr}	Prandtl number, $\left(\dfrac{\mu C_p}{k_q}\right)$
N_{Re}	Reynolds number, $\left(\dfrac{D_e G}{\mu}\right)$
N_{Sc}	Schmidt number, $\left(\dfrac{\mu}{\rho D_A}\right)$
N_{Sh}	Sherwood number, $\left(\dfrac{k_{mt} D_e}{D_A}\right)$
p	Pressure
q_{HR}	Rate of heat transfer per unit mass per unit time
Q_{HR}	Heat of reaction
R	Gas constant
S	Total surface area of the catalyst
$(Sh)_{lim}$	Limiting Sherwood number
t	Time
T	Gas temperature
T_{ig}	Initial gas temperature
T_{fg}	Final gas temperature
T_g	Gas temperature
T_s	Surface temperature of the catalyst
u	Particle velocity
U	Non-dimensional form of u, $\left(\dfrac{u}{a_{ref}}\right)$
v	Total volume of the catalytic reactor block
V_g	Velocity of gas
x_i	Mol fraction of ith specie
X_i	Concentration of ith specie $\left(\dfrac{kg\,mol}{m^3}\right)$
y	Fractional concentration
Z	Non-dimensional time $\left(\dfrac{a_{ref} \cdot t}{\ell_{ref}}\right)$
λ	Riemann variable $\left(A + \dfrac{k-1}{2} U\right)$
β	Riemann variable $\left(A - \dfrac{k-1}{2} U\right)$
ϵ	Bed void fraction (void volume/total volume)
ϵ_o	Minimum energy of attraction
μ	Viscosity of gas $\left(\dfrac{kg}{ms}\right)$
ρ	Gas density
σ	Intermolecular distance when $\phi(r) = 0$ (Lennard-Jones force constant). (Angstrom)
Ω_D	Collision integral
τ_w	Wall shear stress
ΔL_{cv}	Converter step length m

THE INITIAL WORK ON MODELLING SPARK IGNITION Engines was confined to the power cycle and to some steady flow models of catalytic converters. More recently (1)(2)(3)* models have been presented for completely integrated systems which include the cylinder gas

---

* Numbers in parentheses designate References at end of paper.

exchange process and the intake and exhaust systems. These papers described basic engines with carburettors but no emission control devices. In a companion paper (4) the application of the gas dynamic model for an EGR valve (5) to a multi-cylinder engine is described. As part of the continuing research programme at UMIST on modelling of spark ignition engine systems this paper describes some further work on oxidizing catalytic converters. A mathematical model is presented to allow for the influence of the catalytic converter on the non-steady flow systems in the engine as well as predicting the performance of the converter. The basic theory of non-steady flow applied to chemically reacting systems described in reference 1 is modified to include the oxidation process in the converter. To limit the length of the paper only the theory relevant to non-steady flow in the catalytic converter is presented. For the full development of the mathematical expressions for non-steady flow with chemically reacting systems the reader is referred to reference 1 and for the numerical techniques in their solution, to reference 6.

A number of mathematical models have been presented for oxidation catalytic converters notably by Kuo et al (7), Harned (8), Bauerle and Nobe (9). None of these include wave action. There is an extensive literature of engine tests with these and other types of converters as well as methods for analysing their performance, in this paper we presented a mathematical model applied to the non-steady flow situation based on the general consensus of the mode of operation of these converters presented in earlier papers (7) (8) (9).

Since for successful operation of oxidation converters secondary air injection in the exhaust system is required for low air-fuel ratios, this is included in the modelling.

The work is wholly confined to monolithic converters for oxidation of hydrocarbon and carbon monoxide.

THEORETICAL CONSIDERATIONS

The catalytic converter may be represented as a duct of constant cross-sectional area. Within this duct the flow is non-steady and there are chemical reactions. The flow can be described by three Riemann variables (1), namely λ, β and A_a. The first two (λ, β) represent the wave lines and the third (A_a) the path line. The identity of a fluid element is preserved along a path line. The path line chemical composition is defined by the molar concentrations. In the present case we use some thirteen species (1) H_2O, (2) H_2, (3) OH, (4) H, (5) N_2, (6) NO, (7) N, (8) CO_2, (9) CO, (10) O_2, (11) O, (12) Ar, (13) C_3H_8. The latter representing hydrocarbons. In the analysis we consider a time (Z) - distance (L) plane. At a given time, Z_1, (figure 1), a path line (A_a) will be located at a point L_1 and after a time step ΔZ the path line will be located at L_2. For the catalytic converter we consider the chemical reactions on each path line during the time step. The scheme used in this work is basically the same as that reported by Rolke et al (10). The reaction rate will be controlled by the catalyst type, the gas chemical composition, pressure and temperature at time Z_1 and these may be different for each path line. Furthermore, the distance travelled by the fluid element ΔL_{cv} will also be different as well as the temperature of the catalyst surface at the point L_i. All these will be taken into account in the model.

Owing to the chemical reactions there will be heat generated in the converter and this will affect the magnitude of the Riemann variables. Along the wave lines the changes in λ, β may be represented by

$$d\lambda = (d\lambda)_F + (d\lambda)_{A_a} + (d\lambda)_q + (d\lambda)_f \quad (1)$$

$$d\beta = (d\beta)_F + (d\beta)_{A_a} + (d\beta)_q + (d\beta)_f \quad (2)$$

and along the path line

$$dA_a = (dA_a)_q + (dA_a)_f \quad (3)$$

The suffix F refers to area change, A_a to entropy change, q to heat generation and transfer, f to friction. Except for numerical changes due to the nature of the flow in the converter the only terms which have a significant algebraic difference from the

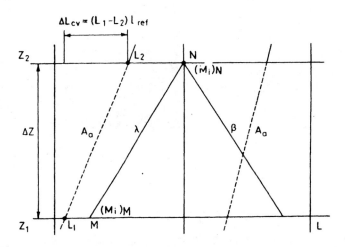

Fig. 1 - Characteristic mesh diagram for catalytic converter

basic expressions for flow with chemical reaction described in reference 1 are the heat transfer terms which we can write as

$(d\lambda)_q = (d\lambda)_{\text{heat transfer}} + (d\lambda)_{\text{heat generated by reaction}}$

$(d\beta)_q = (d\beta)_{\text{heat transfer}} + (d\beta)_{\text{heat generated by reaction}}$

$(dA_a)_q = (dA_a)_{\text{heat transfer}} + (dA_a)_{\text{heat generated by reaction}}$

for simplicity we will use subscripts HR for the second terms.

The first term in each case is related to convective heat transfer and the second term to the heat generated in the chemical reaction. In normal calculations (without heat generation) we consider the flow to be fully developed turbulent flow and Reynolds analogy is used for heat transfer. In the present case we will replace this term by a heat loss fraction, H_L, related to the heat generated in the reaction so that

$$(d\lambda)_q = (1-H_L)(d\lambda)_{HR} \quad (4)$$

$$(d\beta)_q = (1-H_L)(d\beta)_{HR} \quad (5)$$

$$(dA_a)_q = (1-H_L)(dA_a)_{HR} \quad (6)$$

where H_L lies between 0 and 1.

The change in the Riemann variables due to the heat released in the catalytic converter will then be

$$(d\lambda)_q = (d\beta)_q = \frac{(\kappa-1)^2}{2}(1-H_L)\frac{q_{HR}\ell_{ref}}{a_{ref}^3}$$

$$\frac{2}{(\lambda+\beta)}dZ \quad (7)$$

$$(dA_a)_q = \frac{2(\kappa-1)A_a}{(\lambda+\beta)^2}(1-H_L)\frac{q_{HR}\ell_{ref}}{a_{ref}^3}dZ \quad (8)$$

ℓ_{ref} and a_{ref} are reference length and speed of sound, respectively, q_{HR} is the heat released by the catalytic reaction per unit mass per unit time (J/Kgs).

The flow in the converter channels is normally laminar. In the present theory we represent this by a friction factor f, defined in the normal manner

$$f = \frac{\tau_w}{\frac{1}{2}\rho u^2} \quad (9)$$

The channel calculations, discussed later will consider developing laminar flow and use a friction factor FF which is 4f.

The cross-sectional area of the exhaust pipe entering the converter and the corresponding area of the tail pipe are less than the overall cross-section of the converter (figure 2). The effect of the flow expansion and contraction is to produce an increase in total pressure drop across the converter above the loss in the channels. To represent this loss in the calculation an additional boundary is added at the exit from

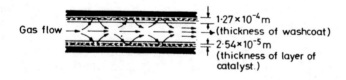

a) CATALYTIC REACTOR CHANNEL GEOMETRY

b) MOUNTING OF CHANNELS (3.1×10^5 CELLS/m²)

■ Non-porous ceramic material (Silica alumina called Cordierite)
▨ A thin layer of porous γ-alumina (wash coat)
··· Dispersed catalyst within porous γ-alumina as microcrystallites

Fig. 2 - Essential features of catalyst element

the converter. This is a gauze boundary (11), represented by

$$f_g = \frac{\Delta p}{\frac{1}{2}\rho_1 u_1^2} \qquad (10)$$

where Δp is the pressure drop and ρ_1, u_1 the upstream conditions to the "gauze" namely, the exit from the channels. The coefficient f_g may be a constant or a function of the Reynolds number. In practice the addition of the gauze boundary has no significance on the predicted results for the converter conversion efficiency, but does significantly effect the pressure level if an EGR system is included.

In order to carry out a non-steady flow calculation for the converter we therefore only require to determine the heat released in the converter, q_{HR}, and the corresponding friction coefficient. These will depend on the processes occurring within the converter and we will now turn to these. The calculation procedure is a step-by-step method so it is assumed that at any time Z_1 all the information is known. The problem is to find the state at the next time Z_2. In the same manner as we have considered chemical reaction along path lines to be "quasi-steady" over the time step ΔZ so we will consider "quasi-steady" processes taking place along a path length ΔL_{cv}. We will consider in this analysis a fresh catalyst whose reactivity has not been reduced by poisoning or surface deposits. Its extension to calculations for aged, poisoned or suppressed catalyst would be straightforward, but would require significantly more information than is currently available concerning both the nature and the quantitative extent of catalyst fouling.

In a catalytic converter, reactants flow through the bed or matrix and reaction products are carried away by the flow of a waste stream. The reactions which convert reactants to product molecules, however, take place at discrete catalytic sites distributed throughout a thin, porous substrate layer coated on the surface of the matrix. The overall process of converting reactants to products in the flowing stream involves a number of sequential steps including transport processes between gas stream and catalytic sites as well as the actual steps in the reaction at the catalyst sites. These steps are:

(1) Mass transfer of reactants to the external surface of the catalyst (gas-particle interphase diffusion).

(2) Diffusion of reactants into the pores of the substrate layer (intraparticle diffusion).

(3) Adsorption of reactants on the active catalytic sites.

(4) Reaction at the active sites.

(5) Desorption of combustion product molecules from the catalyst sites.

(6) Diffusion of combustion products through the porous substrate to the external surface (intraparticle diffusion).

(7) Mass transfer of combustion products from the catalyst surface into the flowing waste stream (interphase diffusion).

The reactions taking place in step (4) are highly exothermic, liberating considerable heat at the active sites. This in turn must be transferred to the waste stream by thermal conduction through the porous layer and by convective or conductive transport from the catalyst surface to the flowing gas stream.

The qualitative effects of these sequential processes are that, concentrations of combustibles are lower within the catalyst, at the active sites, than in the gas stream flowing over its surface, and that the combustion product concentrations and the temperature are higher within the catalyst. The concentration and temperature differences which develop, provide the driving forces for the various diffusion and heat and mass transfer steps in the overall sequence.

A complete description of the rate of the overall process would require a knowledge and description of each of the seven steps in the sequence. The present state of knowledge and available data on catalytic converters does not permit such a detailed breakdown. However, sufficient data are available to allow independent description of the rates of convective mass transfer processes (Steps 1 and 7) taking place external to the catalyst, and of the neat process taking place within the porous catalyst structure (Steps 2 through 6).

Rolke et al (10) reported that under normal design rates for oxidation catalysts, the effective reaction rate is impeded by slow diffusion of reactants through the porous substrate to such an extent that only that fraction of the catalyst near the surface of the thin porous substrate layer (0.005 inch in thickness) is being used effectively to catalyse the oxidation reactions. It is assumed in the present work that the active catalytic material is concentrated on the outer layer of the particle. Thus a pragmatic description of the chemical rate is possible without the need for separate calculations of effectiveness factors to account for the intraparticle diffusion, adsorption and desorption.

Heat transfer between catalyst and gas stream is important because it determines the internal temperature of the catalyst, at which chemical reactions take place at the active sites.

It is observed that the rate of oxidation of a given species over the catalyst is proportional to its concentration in the gas stream and independent of the concentration of oxygen in the gas stream provided there is sufficient oxygen available than required for complete oxidation of gaseous combustibles. For oxidation of waste streams containing several oxidizable species it is udually assumed that each species is oxidized independently with its own rate.

The rates of oxidation for similar catalysts prepared on support of different geometric surface areas are proportional to the geometric surface area of the matrix rather than the volume of the catalyst.

The above observations are consistent with the following expression for the local rate of reaction at a point within a catalytic converter bed.

$$\text{Rate}\left(\frac{\text{mols}}{\text{sec m}^3}\right) = -\frac{V_g}{\epsilon}\frac{dX_i}{dL_{cv}} = \bar{k}_{s_i}\frac{S}{\epsilon v}X_i \quad (11)$$

In this expression V_g is the gas velocity m/s, ϵ the bed void fraction (void volume/total volume), X_i the mol concentration (kg mol/m^3) for reacting species i, L_{cv} the active part of the catalytic converter length, m, v the total volume of the catalytic converter, m^3, S the total surface area of the catalyst m^2 and $\bar{k}_{s_i}$ is the overall rate constant m/s.

If we consider that the effects of temperature and the flow conditions through the catalyst and the catalyst channel may be represented by two rate limiting processes <u>in series</u>, namely, mass transfer of reactants from the waste stream to the catalyst surface and an effective chemical reaction within the catalyst localized near its external surface, then the overall rate constant $\bar{k}_{s_i}$ can be represented in terms of the rates of the two processes, thus

$$\frac{1}{\bar{k}_{s_i}} = \frac{1}{k_{mt_i}} + \frac{1}{k_{chem_i}} \quad (12)$$

where k_{mt_i} is the mass transfer coefficient and k_{chem_i} is the chemical reaction rate constant for species i.

The rate equation (11) can be expressed in terms of the flow rate $G(kg/m^2s)$, the gas stream density ρ and the mol fraction x_i of species i in the form

$$\frac{dx_i}{dL_{cv}} = -\frac{\bar{k}_{s_i}}{G}\frac{S}{v}x_i\rho \quad (13)$$

If we fix the variables G and ρ at time Z_1 and let $y_i = x_i/x_i^o$ is the mol fraction at Z_1 then integration of (13) in the time step ΔZ will yield.

$$y_{2_i} = y_{1_i}\exp\left[-\bar{k}_{s_i}\frac{S}{v}\frac{\rho}{G}\Delta L_{cv}\right] \quad (14)$$

where subscripts 1, 2 refer to the relative concentrations of species i at Z_1 and Z_2 and ΔL_{cv} is the distance traversed by the path line in time ΔZ as shown in figure 1.

The ratio S/v is a function of the converter geometry, the parameter ρ, G and ΔL_{cv} can be determined from the state (λ, β, A_a) at time Z_1. To obtain the new concentration of the species at time Z_2 a knowledge of $\bar{k}_{s_i}$ is required. We will examine the two components of $\bar{k}_{s_i}$ in expression (12), namely, k_{chem_i} and k_{mt_i}.

The chemical rate constant, k_{chem}, which represents the rate of the net processes taking place within the porous substrate layer, is determined by the chemical composition and microstructure of the active metal crystallites distributed in that layer and by the microstructure of the porous substrate. These factors are in turn determined by the method used in preparing the catalyst. The chemical rate constant depends also upon the species being oxidized. It has a relatively strong dependence upon temperature, which can be represented over a working range in respect of I.C. engine exhaust gas temperature by Arrhenius equation.

$$k_{chem} = A_{chem}\exp\left(-\frac{E_a}{RT_s}\right) \quad (15)$$

where E_a is the activation energy, T_s the catalyst surface temperature and A_{chem} is a constant depending on the catalyst type and the species being oxidized. At low temperatures the behaviour of the overall rate constant, $\bar{k}_s$, is limited by k_{chem} and its value approaches that of k_{chem}.

The rate of mass transfer of reactants from the waste stream to the catalyst surface, k_{mt}, is determined by the geometry of the catalyst bed and by the hydrodynamics of flow of the waste stream through the catalyst bed. It has a relatively slight dependence upon temperature and upon the nature (molecular size and shape) of the reactants. The overall rate constant, $\bar{k}_s$, approaches the value of k_{mt} as a limit at high temperature when the value of k_{chem} has become large relative to k_{mt}.

k_{mt} is calculated by using the analogy between heat and mass transfer. The results of heat transfer studies on a particular matrix geometry can usually be expressed in a correlation of the form

$$N_{Nu} = fn(N_{Re}, N_{Pr}) \quad (16)$$

where N_{Nu} is the Nusselt number, N_{Re} is the Reynolds number and N_{Pr} the Prandtl number as defined in the nomenclature.

The analogy between heat and mass transfer, as used here, utilizes the above heat transfer expression to develop the mass transfer correlation, thus

$$N_{Sh} = fn\ (N_{Re}, N_{Sc}) \quad (17)$$

where N_{Sh} is the Sherwood number and N_{Sc} is the Schmidt number as defined in the nomenclature and the functional relationships are identical. From the Sherwood number the mass transfer coefficient K_{mt} can be determined.

The calculation procedure is that generally used for catalytic reaction with matrix structures (10). In the catalytic converter discussed in this paper a honeycomb type matrix was used for catalyst support (figure 2). As the gases flow through these tubes, laminar flow develops and in the limiting case a fully developed laminar flow is achieved. Analytical and numerical solutions for the Sherwood number has been developed by Kays and London (12) and Hawthorne (13).

Tabulated values for the limiting Sherwood number for fully developed laminar flow for a number of channel sections are given in Table 1. For developing flow Hawthorne (13) gives the expression

$$N_{Sh} = (Sh)_{lim} \left(1 + CD_e N_{Re} \frac{N_{Sc}}{\Delta L_{cv}}\right)^{0.45} \quad (18)$$

where the limiting Sherwood number $(Sh)_{lim}$ is given in Table 1 for a given section and C is a roughness constant (0.078 for smooth surfaces, 0.095 for catalytic monoliths (10)) and D_e is the equivalent diameter of the channel. Using the heat-mass transfer analogy (10), the Nusselt number is

$$N_{Nu} = (Sh)_{lim} \left(1 + CD_e N_{Re} \frac{N_{Pr}}{\Delta L_{cv}}\right)^{0.45} \quad (19)$$

For the monolithic catalytic converter used in this investigation the mass transfer coefficient for the species i is then given by Sherwood Number relation,

Table 1

Limiting Sherwood Numbers and Friction Factors for Fully Developed Laminar Flows in Conduits of Various Geometries

Channel Geometries	$(Sh)_{lim}$	$(FF \cdot N_{Re})_{lim}$
1. Oblong (For Johnson & Matthey Catalytic Converter)	3.790	62.000
2. Hexagonal	3.621	60.216
3. Circular	3.980	64.000
4. Square	3.236	56.908
5. Triangular with equal sides	2.686	53.332

$$k_{mt_i} = \frac{D_{A_i}}{D_e} (Sh)_{lim} \left(1 + 0.095\ D_e N_{Re} \frac{N_{Sc}}{\Delta L_{cv}}\right)^{0.45} \quad (20)$$

D_{A_i} is the diffusivity (m^2/s) or diffusion coefficient of the species i.

The diffusion coefficient D_A for the species i must be calculated from data provided in Tables of Data of Physical Properties. These data are based on the methods of statistical thermodynamics. We will only give the appropriate expressions.

For one component diffusing in a mixture of n components Wilke (14) gives the diffusion coefficient D_{1m} of the component 1 in a multi-component gas as

$$D_{1m} = \frac{1 - x_1}{\sum_{j=2}^{j=n} \left(\frac{x_j}{D_{1j}}\right)} \quad (21)$$

where x_1 is the mol fraction of gas 1, x_j the mol fraction of gas j and D_{1j} the binary gas-diffusion coefficient for gases 1 and j. The binary gas-diffusion coefficient, D_{12}, is given by Reid and Sherwood (15) for gas 1 diffusing through a mixture of gases 1 and 2 at temperature T, pressure p as

$$D_{12} = \frac{(0.00217 - 0.0005B)BT^{3/2}}{Cp} \quad (22)$$

The parameter B is related to the molecular weights M_w of the two species by

$$B = \left(\frac{M_{w1} + M_{w2}}{M_{w1} M_{w2}}\right)^{\frac{1}{2}} \quad (23)$$

The parameter C is related to the driving force of the diffusion process by

$$C = \left(\frac{\sigma_1 + \sigma_2}{2}\right)^2 \Omega_D \quad (24)$$

σ_1, σ_2 are the Lennard-Jones potentials for gases 1 and 2, typical values of which are given in Table 2, (16) Ω_D is the collision integral for the two gases and is related to the parameter kT/ϵ_0, where k is the Boltzmann constant and ϵ_0 is the minimum energy of attraction. For a gas pair 1,2

$$\left(\frac{\epsilon_0}{k}\right)_{12} = \left(\left(\frac{\epsilon_0}{k}\right)_1 \left(\frac{\epsilon_0}{k}\right)_2\right)^{\frac{1}{2}} \quad (25)$$

Hirschfelder et al (17) give tables relating kT/ϵ_0 and Ω_D a short abstract of which is given in Table 3.

Table 2 - Lennard-Jones Potentials as Determined from Viscosity Data*

Molecule	Compound	$\sigma(\text{Å})$	ϵ_o/k (K)
Ar	Argon	3.542	93.3
CO	Carbon monoxide	3.690	91.7
CO_2	Carbon dioxide	3.941	195.2
C_3H_8	Propane	5.118	237.1
H_2	Hydrogen	2.827	59.7
H_2O	Water	2.641	809.1
NO	Nitric oxide	3.492	116.7
N_2	Nitrogen	3.798	71.4
O_2	Oxygen	3.467	106.7

* Reference 16

Table 3 - Typical Values of the Collision Integral Ω_D Based on the Lennard-Jones Potential*

kT/ϵ_o	Ω_D
0.30	2.6620
0.50	2.0660
0.75	1.6670
1.00	1.4390
1.25	1.2960
1.50	1.1980
2.00	1.0750
3.00	0.9490
4.00	0.8836
5.00	0.8422
10.00	0.7424
50.00	0.5756
100.00	0.5130
400.00	0.4170

* Reference 17.

The diffusion coefficient D_{A_i} in expression (20) corresponds to D_{1m} in expression (21), where gas 1 corresponds to the species i. In the present analysis we consider only CO and C_3H_8 as the reacting gases. In this case we consider for CO the binary mixtures, $CO-CO_2$, $CO-H_2$, $CO-H_2O$, $CO-N_2$, $CO-O_2$, and for C_3H_8 the binary mixtures, $C_3H_8-CO_2$, $C_3H_8-H_2$, $C_3H_8-H_2O$, $C_3H_8-N_2$, $C_3H_8-O_2$. It will be seen that the parameters B and $(\epsilon_o/k)_{12}$, in expressions (23) and (25), can be calculated as input data. The pressure p and temperature T and the mol fraction x_i at the beginning of the time step, Z_1 at location L_1 (Figure 1), enable the diffusion coefficient to be evaluated and hence the mass transfer coefficient k_{mt_i} (i = CO, C_3H_8) from (20). The chemical reactions taking place on the catalyst surface are considered to be

$$CO + \tfrac{1}{2}O_2 = CO_2 \qquad (26)$$

$$C_3H_8 + 5O_2 = 3CO_2 + 4H_2O \qquad (27)$$

The second reaction represents the hydrocarbon reaction.

The constants E_a, A_{chem} for these reactions are given in Table 5, and the catalyst surface temperature T_s at time Z_1 and location L_1 is known. Using these data the chemical rate constant k_{chem} at location (Z_1, L_1) can be obtained for each reaction using the Arrehnius equation (15). Hence the overall coefficient $\bar{k}_{s_i}$ can be calculated from (12) for each reaction and the relative concentration y of CO and C_3H_8 at the time Z_2 evaluated from (14).

Thus for CO

$$(x_2)_{CO} = (y_2)_{CO} (x_1)_{CO} \qquad (28)$$

and for C_3H_8

$$(x_2)_{C_3H_8} = (y_2)_{C_3H_8} (x_1)_{C_3H_8} \qquad (29)$$

If there is inadequate oxygen present then the reaction will not go to completion according to (14).

For the purpose of the calculation the available oxygen $(x_1)_{O_2}$ in the catalyst at Z_1, L_1 is subdivided into two parts, one part for each reaction (26), (27).

For reaction (26) the available oxygen is

$$\left[(x_1)_{O_2}\right]_{CO} = \left(\frac{(x_1)_{CO}}{(x_1)_{C_3H_8} + (x_1)_{CO}}\right)(x_1)_{O_2} \qquad (30)$$

and for reaction (27) the available oxygen is

$$\left[(x_1)_{O_2}\right]_{C_3H_8} = \left(\frac{(x_1)_{C_3H_8}}{(x_1)_{C_3H_8} + (x_1)_{CO}}\right)(x_1)_{O_2} \qquad (31)$$

For the reactions to proceed to (28) and (29) the molar concentrations of CO and C_3H_8 should satisfy the following relations:

$$(x_1)_{CO} \lesssim 2.0 \left[(x_1)_{O_2}\right]_{CO} \qquad (32)$$

$$(x_1)_{C_3H_8} \lesssim 0.2 \left[(x_1)_{O_2}\right]_{C_3H_8} \qquad (33)$$

If either $(x_1)_{CO}$ or $(x_1)_{C_3H_8}$ exceeds the above limits the reaction will not proceed to completion, since there will be inadequate oxygen present, and expressions (28) and (29) are then replaced by

$$(x_2)_{CO} = (y_2)_{CO} \times 2.0 \left[(x_1)_{O_2}\right]_{CO}$$

$$+ \left[(x_1)_{CO} - 2.0 \left[(x_1)_{O_2} \right]_{CO} \right] \quad (34)$$

$$(x_2)_{C_3H_8} = (y_2)_{C_3H_8} \times 0.2 \left[(x_1)_{O_2} \right]_{C_3H_8}$$
$$+ \left[(x_1)_{C_3H_8} - 0.2 \left[(x_1)_{O_2} \right]_{C_3H_8} \right] \quad (35)$$

Having obtained the final mol fractions of CO and C_3H_8 at the time Z_2 and location L_2 we can calculate the heat generated q_{HR} at the catalyst bed during the time step. This is obtained directly from the heat of reaction at constant pressure, Q_{p_i}, for the two reactions (26 and (27) at the gas temperature T_1 at time Z_1. The normal procedure (18) is used for the calculations of $(Q_p)_{C_3H_8}$, $(Q_p)_{CO}$; then

$$Q_{HR} = \frac{(x_2-x_1)_{C_3H_8}(Q_p)_{C_3H_8} + (x_2-x_1)_{CO}(Q_p)_{CO}}{M_w} \text{kJ/kg} \quad (36)$$

where M_w is the mean molecular weight of the gas mixture.

The heat generated per unit mass per unit time q_{HR} along the path line is then

$$q_{HR} = \frac{Q_{HR} G}{\rho \Delta L_{cv}} \text{ kJ/kgs} \quad (37)$$

This value is used to calculate the state change in (8).

The friction factor for the channel, FF, is calculated by methods developed by Kays and London and Hawthorne (12) (13). For developing flow

$$FF = \frac{(FF\, N_{Re})_{lim}}{N_{Re}} \left[1 + 0.0445\, N_{Re} \frac{D_e}{\Delta L_{cv}} \right]^{0.5} \quad (38)$$

The parameter $(FF\, N_{Re})_{lim}$ is the product of the limiting friction factor and Reynolds number for fully developed laminar flow. Typical values for different channel geometries based on (12) are given in Table 1.

The friction factor f used in the wave equation is taken as FF/4.

The catalyst surface temperature, T_s, is calculated from the mean gas temperature during reaction at the surface of the catlysts. From a simple energy balance we can define:

$$T_{fg} = T_{ig} + \frac{Q_{HR}}{C_p} \quad K \quad (39)$$

and

$$T_g = 0.5(T_{ig} + T_{fg}) \quad K \quad (40)$$

The temperature T_{ig} corresponds to the temperature in the gas stream at time Z_1 at location L_1 and is calculated from the Riemann variables

$$T_{ig} = T_1 = \left(\frac{\lambda_1 + \beta_1}{2}\right)^2 \frac{a_{ref}^2}{(\kappa R)_1} \quad K \quad (41)$$

where κ, R are the ratio of the specific heats and gas constant (kJ/kg K) at Z_1, L_1.

The surface temperature T_s, is then calculated from the expression

$$q = -h_q S \frac{\Delta L_{cv}}{L_{cv}} (T_g - T_s) \quad kJ/s \quad (42)$$

where q is the heat transfer associated with the heat released Q_{HR}

$$q = \frac{Q_{HR} Gv}{L_{cv}} \quad kJ/s \quad (43)$$

Now

$$k_q = \frac{\mu C_p}{N_{Pr}} \quad (44)$$

and h_q is obtained from the Nusselt relationship through the analogy of heat and mass transfer.

Equation (19) with $C=0.095$ gives the Nusselt relationship

$$h_q = \frac{k_q}{D_e} (Sh)_{lim} \left[1 + 0.095\, D_e\, N_{Re} \frac{N_{Pr}}{\Delta L_{cv}} \right]^{0.45} \quad (45)$$

and then we have

$$T_s = T_g + \frac{G\, Q_{HR}}{h_q\, \Delta L_{cv} (S/v)} \quad K \quad (46)$$

For the "converter pipe" the rate equations for the NO and other reactions are not included. Therefore, the calculation proceeds in a straightforward manner. Since the active length ΔL_{cv} is required before the state along the path line the uncorrected values of λ_1, β_1 at time Z_1 are used to determine the value, thus

$$\Delta L_{cv} = \left(\frac{\lambda_1 - \beta_1}{\kappa - 1}\right) \ell_{ref} \Delta Z \quad m \quad (47)$$

For the λ, β characteristics a slightly different procedure is used to calculate q_{HR} for expression (7). Initially the molar concentrations of all the species are calculated at the mesh points (figure 1) by linear interpolation from the path lines. At time Z_1 the molar concentrations are $(M_i)_1$ and at time $Z_2, (M_i)_2$. By linear interpolation between the two mesh points at Z_1 the value of $(M_i)_M$ at M can be obtained. Hence the change in the number of mols of species i along the characteristic MN is then $(M_i)_N - (M_i)_M$. The heat released for the C_3H_8 and CO reactions is then

$$Q_{HR} = \left[\left[M_{C_3H_8}\right]_N - \left[M_{C_3H_8}\right]_M\right]\left(Q_p\right)_{C_3H_8} + \left[\left[M_{CO}\right]_N\right.$$
$$\left. - \left[M_{CO}\right]_M\right]\left(Q_p\right)_{CO} \qquad \text{kJ} \qquad (48)$$

If M_w is the mean molecular weight of the mixture at mesh point N and the total number of mols at M is M_T then

$$q_{HR} = \frac{Q_{HR}}{M_T M_w} \cdot \frac{1}{\Delta t} \qquad \text{kJ/kgs} \qquad (49)$$

The time Δt is related to ΔZ by

$$\Delta t = \frac{\ell_{ref} \Delta Z}{a_{ref}} \qquad (50)$$

Hence

$$q_{HR} = \frac{Q_{HR} \, a_{ref}}{M_T M_w \ell_{ref} \Delta Z} \qquad (51)$$

For the secondary air injection a simple junction model is used with mixing <u>without</u> chemical reaction.

EXPERIMENTAL APPARATUS AND TEST PROGRAMME

The experimental investigation was carried out on a four cylinder Vauxhall Victor 2000 cc engine. Full details of the engine are given in Table 4. The general arrangement of the engine and test bed is shown in figure 3. A full description of the data processing equipment and the dynamic pressure measuring techniques have been given in previous papers (3) (4). Transient pressures were measured in the cylinder (in the head and in the liner at midstroke), in the exhaust pipe ahead and downstream of the catalytic converter and in the intake system downstream of the carburettor. The pressure records were recorded on a Bell and Howell VR-3700 magnetic tape recorder in analog form. These were then processed through an ADC on a DEC PDP 15 computer. The mean of 32 consecutive records was used for analysis. The exhaust gas sampling instrumentation, on-line through stainless steel tubes, comprised an Analytical Development Co NDIR gas analyser for NO, CO and CO_2, a Taylor Servomax Oxygen Analyser type OA72, for O_2 an Analysis Automation model 523 Flame Ionisation Detector for hydrocarbon analysis and a British Oxygen Chemiluminescence Luminox 201 NO/NO_x analyser for NO/NO_2.

The secondary air supply was obtained from the laboratory low pressure air supply via a regulator valve and a rotameter for flow measurement, to an injection tube mounted just downstream of the pipe junction. The injection tube was inserted normally into the pipe and bent in the flow direction.

The catalytic converter was supplied by Johnson Matthey Chemicals Ltd. A drawing of the essential features of the element is given in figure 2 and the properties and typical design parameters in Table 5. The catalytic element comprised a cylindrical block 15 cm long, 10 cm diameter with a large number of channels forming a honeycomb structure. The geometric shape of the honeycomb was formed from a dense, non-porous ceramic material - silica alumina - called Cordierite. On the surface of the geometric matrix a thin wash coat of porous γ-alumina was applied. The thickness of the wash coat was approximately 0.0125 cm, one gram covering about 1000 m^2 area. The active noble metals (Platinum and Palladium) were dispersed within the thin layer of porous alumina as micro crystallites. The active block of the converter was canned in a stainless steel pipe.

In the first set of experiments (series 1 to 3) the engine operated on conventional five star petrol containing lead additive. This caused poisoning of the catalyst, a second element was then used and the engine run with a special minimum lead fuel. Because extensive running with this special fuel would damage the exhaust valve these runs were of limited duration. In between the two sets of runs there was an engine failure which

Table 4 - Dimensions of the Test Engine

Name of the engine	Vauxhall Victor 2000 cc
Type of Carburettor	Constant Vacuum Type Zenith Stromberg 175 CD-2S
Cycle	Four Stroke
Number of Cylinders	Four cylinders in line inclined at 45 degrees
Cylinder bore	95.25 mm
Stroke	69.24 mm
Connecting rod length	136.50 mm
Cylinder volume	1975 cc
Compression ratio	8.5
Angle of Ignition	9 degrees BTDC at idling
Idling engine speed	750 r.p.m.
Spark plug position	33.5 mm from the nearest cylinder edge
Mean inlet valve diameter	42.0 mm
Mean exhaust valve diameter	38.0 mm
Valve timing, EVO	114.57° ATDC
EVC	393.43° ATDC
IVO	326.57° ATDC
IVC	605.43° ATDC
Maximum Power (nominal)	90 HP at 5500 r.p.m.
Firing order	1-3-4-2
Fuel	Conventional Five Star Petrol

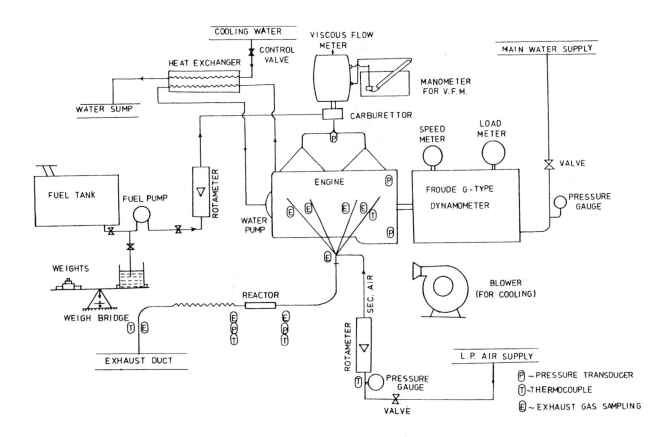

Fig. 3 - General arrangement of experimental rig

Table 5 - Properties and Design Parameters for Catalytic Converter

Length of the active part of the converter	= 0.15 m
Diameter of the converter	= 0.10 m
Surface to volume ratio	= 933.0
Void fraction	= 0.695
Hydraulic radius	= 0.0745 cm
Effective diameter	= 0.00298 m
Number of cells per square inch	= 200
Equivalent open diameter	= 0.0834 m
Thickness of a layer of catalyst	= .00254 cm
Thickness of wash coat	= .0127 cm
Wash coat used	= γ-alumina
Catalyst used	= Platinum-Palladium
Pre-exponential factor of chemical rate equation for CO	= $1.7600 \times 10^8 (sec^{-1})$*
Pre-exponential factor of chemical rate equation for HC	= $0.3500 \times 10^8 (sec^{-1})$*
Ratio of activation energy to gas constant for CO and HC	= 8944 (K)*

*Supplied by Johnson Matthey

required replacement of the pistons and liners. This had a significant effect on the prediction of the CO emissions as reported later.

The remaining part of the test bed was standard. Fuel measurement was by weight balance rather than flow meter.

All the experiments were carried out at 3000 r.p.m. at full throttle over a range of air fuel ratios from about 12:1 to about 17:1. The following groups of test were carried out. The secondary air injection flow is expressed as a percentage of the air inlet mass flow rate.

SERIES 1 - Bare exhaust system with no air injection or catalytic converter.

SERIES 2 - Bare exhaust system with secondary air injection and dummy pipe replacing catalytic converter.

SERIES 3 - Complete exhaust system with secondary air injection and catalytic converter.

SERIES 4 - Series 3 repeated with a new catalytic converter element with trace-lead 97 octane fuel with 13 mg lead per litre.

In series 2 and 3 the secondary air injection rate was varied from zero to a maximum quantity to ensure adequate oxygen for oxidation of CO before entry to the converter (in the case of Series 2 the dummy converter) plus some oxygen for oxidation of any free hydrocarbon present. At air-fuel ratios of 15.08 and above there was adequate oxygen for oxidation of CO and hydrocarbons.

In Series 4, two ignition timings were tested and repeat tests carried out to assess catalyst ageing effects.

COMPUTER CALCULATIONS

As far as possible these covered the same range of air-fuel ratios and secondary air rates as the experiments. The system is shown in figure 4 and the pipe dimensions are given in Table 6. The catalytic converter was represented by pipes VIII, IX, X. The pipe bore was the same as the connecting pipe and the pipe length selected so that the volume of the pipes corresponded to the converter total flow volume between the flanges. The active part of the converter, pipe IX, had the same length as the catalyst element. Pipe XIII was a flexible pipe. The carburettor was represented by the same model as described by Benson et al (19).

The main objective in the calculations was to assess the modelling of the secondary air injection and the catalytic converter. Thus the emission levels at exit from the cylinder and entry to the exhaust system are the base points to commence the calculation.

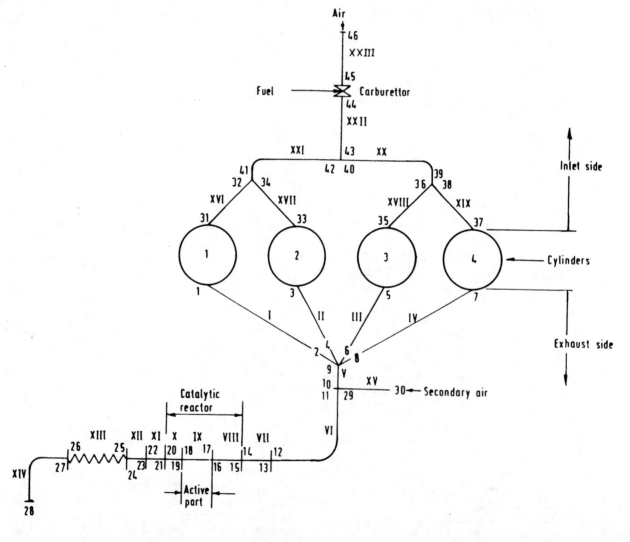

Fig. 4 - Diagram showing pipe and pipe end No's for computational analysis with secondary air and catalytic converter

Table 6 - Pipe Arrangements with Secondary Air and Catalytic Converter*

Pipe No.	Length (m)	No. of Meshes	Diameter (m)	Friction Factor	Wall Temperature (K)
Exhaust Pipes					
I	0.3610	3	0.0350	0.0050	770.0
II	0.3250	3	0.0350	0.0050	770.0
III	0.3250	3	0.0350	0.0050	770.0
IV	0.3610	3	0.0350	0.0050	770.0
V	0.1000	1	0.0510	0.0050	770.0
VI	0.4660	5	0.0510	0.0050	770.0
VII	0.1000	1	0.0510	0.0050	770.0
VIII	0.1250	1	0.0640	0.0500	770.0
IX	0.1500	2	0.0834	0.0050	770.0
X	0.0940	1	0.0686	0.0500	770.0
XI	0.1000	1	0.0510	0.0050	770.0
XII	0.0910	1	0.0510	0.0050	770.0
XIII	0.3800	4	0.0760	0.0050	770.0
XIV	0.3800	4	0.0760	0.0050	770.0
Air Pipes					
XV	0.2450	3	0.0063	0.0075	320.0
XVI	0.2220	2	0.0340	0.0050	320.0
XVII	0.2220	2	0.0340	0.0050	320.0
XVIII	0.2220	2	0.0340	0.0050	320.0
XIX	0.2220	2	0.0340	0.0050	320.0
XX	0.1700	2	0.0400	0.0050	310.0
XXI	0.1700	2	0.0400	0.0050	310.0
XXII	0.0500	1	0.0460	0.0050	310.0
XXIII	0.0500	1	0.0400	0.0050	310.0

* Correspond to figure 4.

It was shown in an earlier paper (1) that the NO predictions are strongly related to the flame speed in the engine cylinder. In the present calculations the flame speed is calculated by multiplying the laminar flame speed by a flame factor to allow for turbulence. For the Vauxhall engine it was shown (3) that a flame factor of 6.5 gave reasonable results when comparing the predicted with the measured maximum cylinder pressure. This value was used in the present tests and slight adjustment to 6.0 when required. The mean NO levels at the pipe inlet agreed favourably with measurements.

It was shown that for this engine (3) the mean CO levels lay between the quilibrium value and some value based on the freezing of CO. Attempts to use rate controlled expressions to predict CO were not successful and this was considered to be due to non-equilibrium concentrations of OH and H being present. A simple model was therefore suggested based on the equilibrium value of CO, CO_{eq}, and the peak value, CO_{peak}. The instantaneous CO was then taken as

$$x_{CO} = x_{CO_{eq}} + f_{CO} \left(x_{CO_{peak}} - x_{CO_{eq}} \right) \quad (52)$$

with f_{CO} lying between 0 and 1. For the previous tests on this engine a value of f_{CO} equal to 0.5 had been shown to hold over the whole air fuel ratio (3) (4). When the preliminary results from the present experiments were analysed it was observed that the factor f_{CO} was no longer constant. At rich mixtures it approached zero, implying equilibrium conditions. There was, however, no fixed trend as shown in Table 7. These values were therefore used in series 1, 2 and 3. However, when the engine was rebuilt after the engine failure with new pistons and liners the value of f_{CO} returned to 0.5 and was constant over the whole air fuel ratio range for all the tests. No explanation can be advanced at this stage, except to point out that series 1, 2 and 3 were carried out after extensive testing of the engine at full throttle in previous research projects.

The calculation of the CO levels in the exhaust pipe assumes that the CO was frozen

Table 7 - Values of f_{CO} at Different Air/Fuel Ratio

Air/Fuel Ratio	f_{CO}
12.29	0.0
12.44	0.0
12.80	0.0
13.46	0.23
13.81	0.36
14.42	0.20
15.08	0.133
15.72	0.10
16.56	0.20

at the concentration at the pipe inlet, except in the presence of the catalyst. For the hydrocarbon emissions no model was available for the prediction of hydrocarbon levels at inlet to the exhaust pipe. The experimental <u>average</u> value for each cylinder was therefore considered on the boundary condition. The concentration was assumed to be frozen except in the presence of the catalyst. The nitric oxide emissions in the pipe were rate controlled.

It was noted in earlier paper (3) that this engine suffered considerable maldistribution of fuel in the intake system. For the present calculation the air fuel ratio at each cylinder was set at a value corresponding to the computed air fuel ratio based on the CO, CO_2 and HC readings at the cylinder exhaust using an Eltinge chart (20).

To compare the emission calculations with the measurements at the sample points the average mass fraction of species i, $(x_i)_{av}$, was calculated from the instantaneous mass fraction, x_i, and the instantaneous flow rate $\dot{m}$ by the expression.

$$(x_i)_{av} = \frac{\Sigma \dot{m} \, x_i \cdot dt}{\Sigma \dot{m} \, dt}$$

The calculations were carried out for some 10 cycle (20 engine revolutions) until the average concentration $(x_i)_{av}$ was steady. The average computing time was 715 seconds on a CDC 7600.

A full set of results is given by Gupta (21). In figures 5 to 12 a representative selection is presented.

DISCUSSION OF RESULTS

SERIES 2 (SECONDARY AIR INJECTION INTO EXHAUST MANIFOLD ONLY) - In figure 5 representative results are given for a rich mixture (5(a), 5(b), 5(c)) and a lean mixture (5(d), 5(e), 5(f). In all cases there is reasonable agreement between experiment and calculation. The mixing model assumes no chemical reaction for CO and CO_2 so that the mass flow g/s for these gases should be independent of the secondary air flow rate. Allowing for small variations from test to test the experimental results confirm this assumption. Thus the reduction in the percentage by volume of these gases with increase in secondary air flow is due to dilution of the exhaust gases. For nitric oxide concentration the experimental results as well as the calculations indicated that the NO concentration was effectively frozen. The decrease on a volume basis with increase in secondary air flow rate is due to dilution. The results at the pipe junction ahead of the injection point, show, as would be expected, little change with secondary air flow rate. The computer calculation had no implied flow direction to allow for the injection pipe which was directed downstream. Since, however, the main flow velocities were in this direction the program automatically allowed for the pipe direction.

From these and other results the simple mixing model at the point of secondary air injection and the assumption of frozen flow for CO, CO_2 (and hence HC) are shown to be reasonable.

SERIES 3 AND 4 (SECONDARY AIR INJECTION AND CATALYTIC CONVERTER)- The results of both these groups of tests will be discussed together although we are effectively dealing with two separate engine systems due to the change in pistons and liners as well as the catalytic converter.

In figure 6 the measured and computed CO_2, CO, HC along the exhaust system are shown with air injection and catalytic converter for series 3. The calculations give the same trends but the predictions for the CO changes are rather better than the hydrocarbons. The wide variation in the emission levels at the cylinders at the low air fuel ratios (figure 6(a)) should be noted. The calculations show that the simple mixing model at the pipe junction gives good results for the prediction of the emission in the main exhaust pipe.

In figure 7 the measured and computed concentrations of CO, CO_2 and HC upstream and

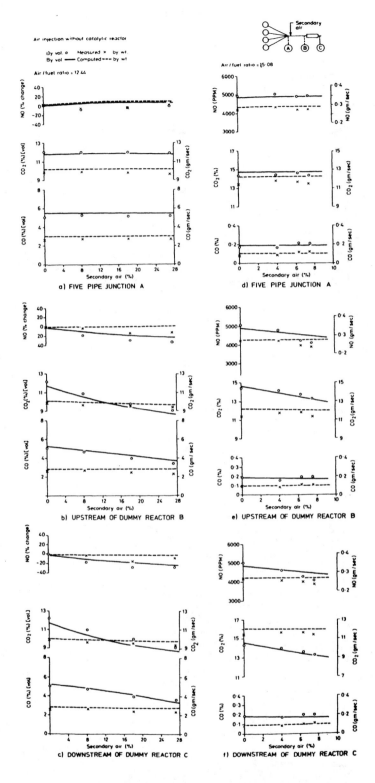

Fig. 5 - Comparison of measured and computed emissions with secondary air injection in exhaust pipe and dummy catalytic converter

downstream of the catalytic converter with different secondary air flow rates are shown for series 3 and 4. The results for CO and CO_2 for the original converter (series 3) are very good, however, for HC the predicted reductions in HC were much less than the measured reductions. This was due to the poisoning of the catalyst for the HC reactions. With a new catalytic converter and a trace leaded fuel the run times were limited; tests were therefore only carried out at zero and maximum secondary air injection (series 4). Here we now see good agreement between the predicted and measured results for the HC emissions.

In figure 8 the results are summarised for series 4 over the whole air fuel ratio range. For rich mixtures air injection was necessary for the converter to operate successfully; for weak mixtures there was adequate air available for oxidation of CO and HC. The predicted reductions for CO and HC are acceptable for fresh catalytic converters (figures 8(a) and 8(b)). After 10 hours operation (figure 8(c)) the measured HC emissions at exit from the converter are clearly higher than the predicted values. Thus to allow for catalysis ageing some modification to the rate constants must be included in the model. The prediction of the overall performance of the converter with air fuel ratio and secondary air obtained by integrating the catalytic converters in the comprehensive simulation program is very satisfactory considering the complexity of the overall system. We can summarise these results in the form of the conversion efficiencies (figure 9). In this figure the results are presented for both series (3), with converter poisoning, and series (4), fresh converter and trace leaded fuel. Of interest here is that for no air injection and consequently inadequate levels of oxygen at low air fuel ratios the predictions for CO were better for the poisoned catalyst than the fresh catalyst. However, these results are of academic interest, in practice secondary air will be used; the predictions are not unreasonable over the whole air fuel ratio range (one must be very careful to note that one is dealing with the differences of small numbers when assessing efficiency). For the hydrocarbons, with inadequate oxygen, the predictions are very poor. Once again, however, with a fresh converter and adequate quantity of air in the exhaust (figure 9(b)) the predictions are very good. Also shown in figure 9(b) are the results with an aged converter. It will be seen that except for one air fuel ratio the CO conversion efficiency is not greatly affected by ageing up to 8 hours, however, the hydrocarbon efficiency is.

Space precludes the presentation of detailed cyclic results for the various tests; it is hoped that these will be given in another paper. In figure 10 the calculated transient gas temperatures and CO concentrations are shown into and out of the catalytic converter. The four peaks in temperature correspond to the gas flows from each

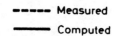

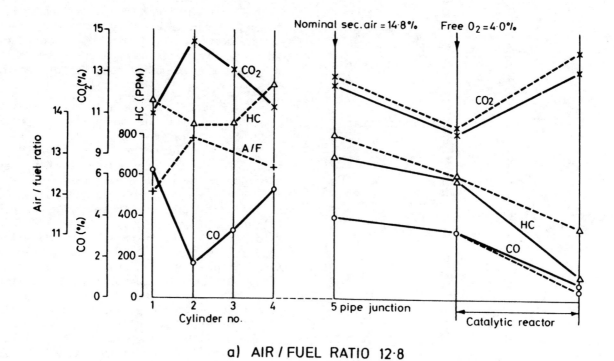

a) AIR/FUEL RATIO 12.8

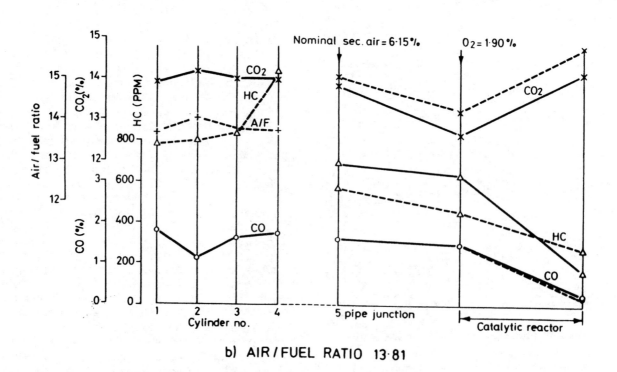

b) AIR/FUEL RATIO 13.81

Fig. 6 - Comparison of measured and computed emission concentrations along the exhaust system with air injection and catalytic converter - series 3

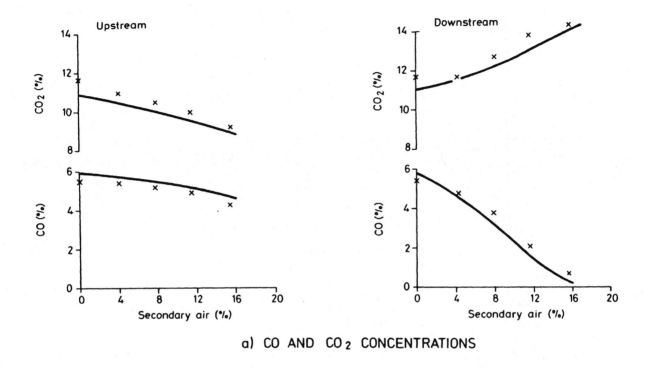

a) CO AND CO₂ CONCENTRATIONS

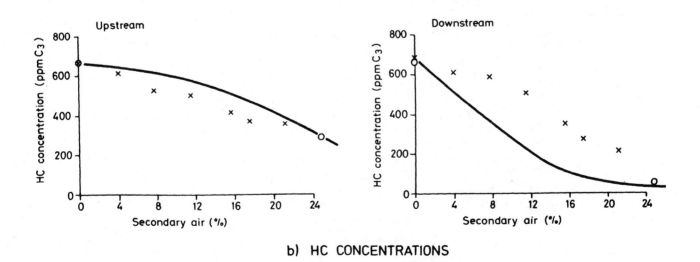

b) HC CONCENTRATIONS

Fig. 7 - Comparison of measured and computed concentrations of CO, CO_2 and HC upstream and downstream of the catalytic converter with different secondary air flow rates

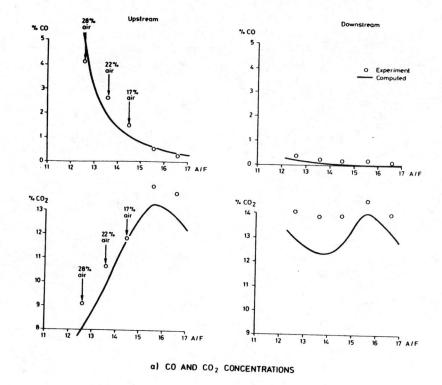

a) CO AND CO$_2$ CONCENTRATIONS

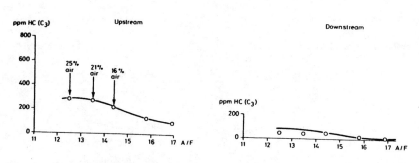

b) HC CONCENTRATIONS (REACTOR AGE 0 HOURS)

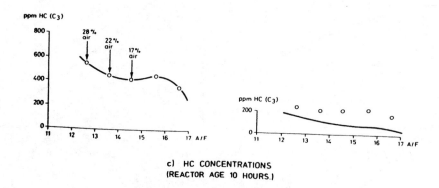

c) HC CONCENTRATIONS (REACTOR AGE 10 HOURS)

Fig. 8 - Comparison of computed and measured CO, CO$_2$ and HC concentrations upstream and downstream of catalytic converter with secondary air injection and air-fuel ratio (series 4)

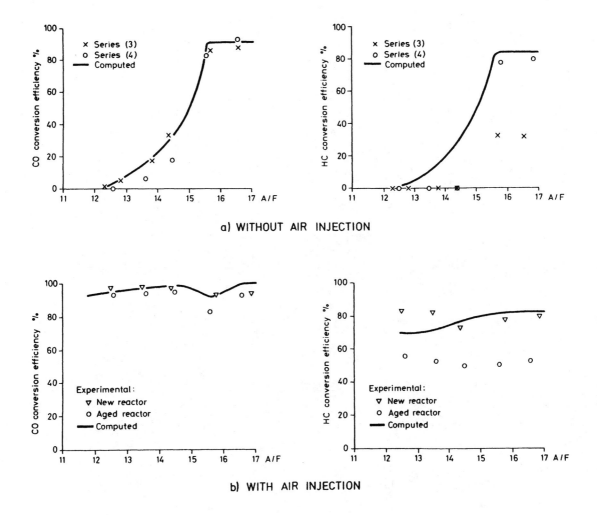

Fig. 9 - Comparison of computed and measured catalytic converter conversion efficiencies with and without air injection

cylinder. There is little difference in the peak and minimum temperature into and out of the converter, but there is a temperature rise of the order of 75K, for the bulk of the flow for the test conditions indicated. This rise is due to the catalytic reaction. The CO concentrations fluctuate marginally with crank angle. This is not unexpected as during the exhaust stroke the cylinder temperature is nearly constant and hence the CO concentration is almost constant with crank angle for a particular cylinder. There is some mixing at the pipe junction and this causes the fluctuations shown. Typical transient pressure diagrams are shown in figure 11. Here there is a good correlation between prediction and experiment.

The final group of results, shown in figure 12, give the overall engine performance with and without catalytic converter. The major effect of the catalytic converter will be to increase the back pressure in the exhaust system which will, in turn, effect the conditions at the commencement of the compression stroke and pumping work. The former effects are seen in the maximum cylinder pressure which is higher with a catalytic converter than without a converter. The net result is an increase in the computed indicated mean effective pressure. The difference is small and it is doubtful whether one should read too much into the small differences in the results shown in figure 12. From the point of the modelling of the complete engine system the trends calculated are in broad agreement with the experiments.

CONCLUSIONS

The simulation of an oxidizing catalytic converter by an equivalent pipe and using the quasi-steady mass transfer, heat transfer and chemical rate data on the path lines gives excellent results for the prediction of emissions in an exhaust system in a general computer simulation program of a spark ignition engine. The non-steady flow

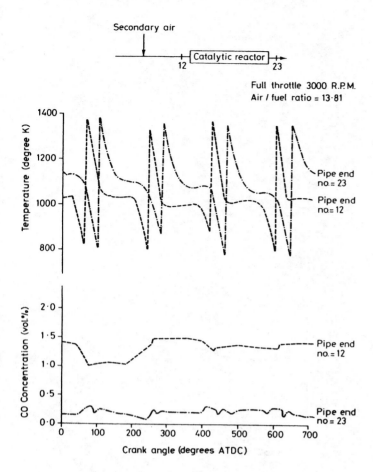

Fig. 10 - Variation of gas temperature and CO concentration with crank angle at inlet to and exit from catalytic converter at 3000 rpm, A/F = 13.81

equations include the heat generated at the catalyst surface expressed in quasi-steady relations and the friction factor under developing laminar flow conditions.

The accuracy of the prediction of the conversion efficiency of the converter depends on the state of the catalyst. For a fresh catalyst the efficiency can be predicted within 5 per cent for CO and within 10 per cent for HC. For the converter used in these experiments the ageing of the HC catalyst reaction was fairly fast but the CO reaction was stable over the tests with only a slight deterioration, it should be noted that even the special fuel had a trace of lead.

The experiments indicated that further work is required to calculate the CO emissions from the cylinder in a more consistent manner. The evidence from the tests reported here indicates that the value of the concentration lies between the equilibrium and the cycle peak value of CO but the constant f_{CO} included in the simple expression (equation (52)) given in an earlier paper (3), does not hold as the engine cylinder conditions deteriorate. However, when the liners and pistons

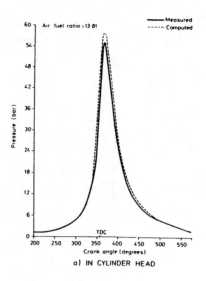

a) IN CYLINDER HEAD

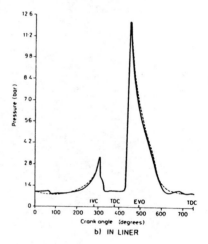

b) IN LINER

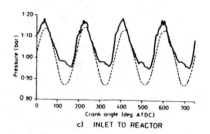

c) INLET TO REACTOR

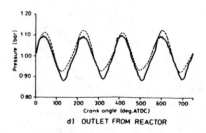

d) OUTLET FROM REACTOR

Fig. 11 - Comparison of measured and computed pressure diagrams with catalytic converter

were replaced the constant f_{CO} reverted to its original value.

The analysis and experiments showed that for this engine the CO concentration in the exhaust pipe on a mass basis is effectively constant even when secondary air is injected into the exhaust but that the mass concentration drops in the presence of a suitable catalyst. Thus using the methods described in this paper the most effective practical location can be determined. The same remarks apply in the hydrocarbon emissions except that at the present time the prediction of the hydrocarbon levels at entry to the exhaust pipe is not included in the simulation. This is currently under way using methods which have been fully developed elsewhere.

The predictions of the pressure diagrams in the exhaust system and cylinder give good agreement with experiments. For this engine the catalytic converter influences the pressure levels in the exhaust system which in turn control the trapped conditions in the cylinder and hence the maximum cylinder pressure and the indicated mean effective pressure.

Further work to allow for ageing of the catalyst and a better representation of the CO reaction in the cylinder are required. The methods should also be extended to include three-way converters.

ACKNOWLEDGEMENTS

The authors wish to acknowledge with thanks the funds received from the Science Research Council (Grant No. B/RG/3876) for the investigation. They also wish to thank Johnson Matthey for the provision of catalytic converters and the technical information concerning the catalysts. Dr. H.N. Gupta held a British Commonwealth Scholarship during the course of the investigation.

REFERENCES

1. R.S. Benson, W.J.D. Annand and P.C. Baruah, "A Simulation Model Including Intake and Exhaust Systems for a Single Cylinder FourStroke Cycle Spark Ignition Engine." Int.J.Mech.Sci. Vol.17, No.2. pp.97-124, 1975.

2. R.S. Benson, P.C. Baruah and B. Whelan, "A Simulation Model for a Crankcase Compression Two-Stroke Spark Ignition Engine Including Intake and Exhaust Systems". Proc.I.Mech.E., Vol.189, 7/75, 1975.

3. R.S. Benson and P.C. Baruah, "Performance and Emission Predictions for a Multi-Cylinder Spark Ignition Engine." Proc.I.Mech.E. Vol.191, 32/77, 1977.

4. P.C. Baruah, R.S. Benson and S.K. Balouch, "Performance and Emission Predictions of a Multi-Cylinder Spark Ignition Engine with

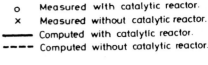
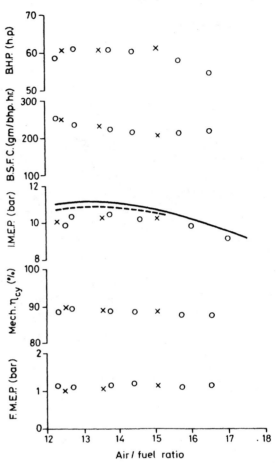
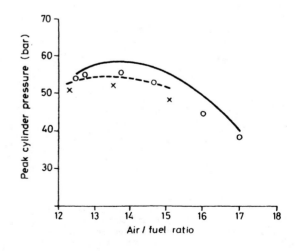

Fig. 12 - Engine performance with and without catalytic converter

Exhaust Gas Recirculation." To be presented to SAE Passenger Car Meeting, June 1978.

5. R.S. Benson, P.C. Baruah and P.G. Evans, "Steady and Non-Steady Flow Through an Exhaust Gas Recirculating Valve." Thermofluids Conference 1974, Institution of Engineers, Australia, National Conference Publication No.74/7, pp.81-85.

6. R.S. Benson, R.D. Garg and D. Woollatt, "A Numerical Solution of Unsteady Flow Problems." Int.J.Mech.Sci. Vol.6, No.1, pp.117-144, 1964.

7. J.C.W. Kuo, C.R. Morgan and H.G. Lassen, "Mathematical Modelling of CO and HC Catalytic Converter Systems." SAE Paper No.710289, 1971.

8. J.L. Harned, "Analytical Evaluation of a Catalytic Converter System." SAE Paper No. 720520, 1972.

9. G.L. Bauerle and K. Nobe, "Two-Stage Catalytic Converter. Transient Operation." Ind. Eng. Chem. Process Des. Develop. Vol.12, No.4, 1973.

10. R.W. Rolke, R.D. Hawthorne, C.R. Garbett, E.R. Slater, T.T. Phillips and G.D. Towell, "Afterburner System Study", Shell Development Co., Emeryville, California, 1972.

11. R.S. Benson and P.C. Baruah, "Non-Steady Flow Through a Gauze in a Duct."J.Mech.Eng.Sci. Vol.7, No.4, pp.449-459, 1965.

12. W.M. Kays and A.L.London, "Compact Heat Exchangers", 2nd Ed., McGraw-Hill Book Co., New York, 1964.

13. R.D. Hawthorne, "Afterburner Catalysts - Effect of Heat and Mass Transfer Between Gas and Catalyst Surface." Presented at 71st AIChE National Meeting, Dallas, Feb.1973.

14. C.R. Wilke, "Diffusional Properties of Multicomponent Gases". Chemical Engineering Progress, Vol.46, No.2, 1950.

15. S.C. Reid and T.K. Sherwood, "The Properties of Gases and Liquids - Their Estimation and Correlation." McGraw-Hill Book Co., 1966.

16. R.A. Svehla, NASA Tech. Report R-132, Lewis Research Center, Cleveland, Ohio.

17. J.O. Hirschfelder, C.F. Curtiss and R.B. Bird, "Molecular Theory of Gases and Liquids." John Wiley and Sons Inc., New York, 1954.

18. R.S. Benson, "Advanced Engineering Thermodynamics." Pergamon Press, 1967.

19. R.S. Benson, P.C. Baruah and R. Sierens, "Steady and Non-Steady Flow in a Simple Carburettor." Proc.I.Mech.E., Vol.188, 53/74, 1974.

20. L. Eltinge, "Fuel-Air Ratio and Distribution from Exhaust Gas Composition." SAE Paper No.680114, 1968.

21. H.N. Gupta, "Theoretical and Experimental Investigation of Multi-Cylinder Spark Ignition Engine with Air Injection and Catlytic Reactor." Ph.D. Thesis, Faculty of Technology, University of Manchester, 1977.

780843

Emission Control at GM

R. J. Schultz
Emission Control Systems Project Center

General Motors has developed a computer controlled catalytic converter system. The C-4 system is the primary element in the GM effort to meet the twin goals of higher vehicle fuel economy and lower exhaust emissions. Major system components include a catalytic converter, electro-mechanical carburetor, exhaust oxygen sensor, coolant temperature sensor and electronic control module. The primary function of these components is to provide a precisely controlled air-fuel ratio to the catalytic converter for efficient reduction of regulated exhaust constituents.

In August, 1977, General Motors formed an Emission Control System Project Center to analyze and resolve the complex problems of meeting legislated exhaust emission and fuel economy standards.

In the development of a highly sophisticated emission control system, each component such as the electronic control module, carburetor and converter is interdependent with every other component in the system. This necessitates a high degree of coordination between car divisions, component divisions and outside suppliers.

The Project Center has the overall responsibility for the coordination of the design and development of control systems to meet the new emission and fuel economy standards. Of equal importance, the systems must also provide levels of vehicle performance, driveability and reliability that are satisfactory to our customers.

The federally legislated emission levels that we must meet have become increasingly stringent since the initial standards in 1968. The 1981 standards represent a 97% reduction in exhaust hydrocarbons, a 96% reduction in carbon monoxide and 76% reduction in oxides of nitrogen when compared to uncontrolled 1960 models. The impact of these reductions is compounded by the more recent fuel economy standards.

The 20 miles per gallon legislated fleet average of 1980 represents a 67% increase from the actual 1974 GM fleet average of 12 miles per gallon. Further increases must occur in subsequent years, as shown in Figure 1.

C-4 SYSTEM

The engine control system being developed has been designated the C-4 system. C-4 stands for "Computer Controlled Catalytic Converter". This system is the primary element in the GM effort to meet the twin goals of higher fuel economy and lower emissions. A phased implementation program has been initiated for the new system.

In model year 1978, in California, a limited number of 2.5 litre L4 and 3.8 litre V6 engines with closed loop carburetor systems were merchandised.

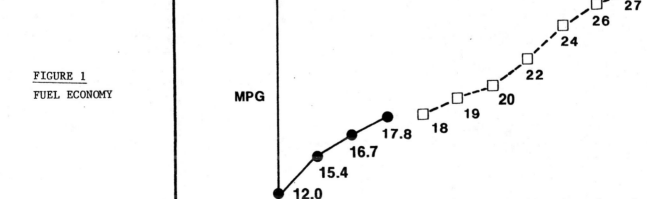

FIGURE 1
FUEL ECONOMY

In model year 1979, in California, C-4 systems will be installed on the 2.5 litre L4 and 2.8 and 3.8 litre V6 engines offered in most of GM's small and mid-size cars.

In 1980, closed loop system usage is planned to include all cars sold in California. This experience will provide the background for the installation of C-4 systems nationwide in 1981.

Obviously, any system that will meet the specific emission and fuel economy requirements of all GM's 4, 6 and 8 cylinder engines -- engines with displacements from 1.6 to 6 litres -- must incorporate considerable flexibility. The system that evolves must be economically adaptable to allow the development engineers from each car division to tailor the system to their specific engine/vehicle combinations while still achieving the desired levels of emission control, fuel economy and driveability.

The system, like the 1978 Pontiac and Buick systems, is based on the knowledge that if the air-fuel mixture is held very close to a stoichiometric mixture (14.6:1), then the oxygen in the exhaust, including that produced by reducing NOx will be just the right amount to oxidize HC and CO.

Figure 2 shows the location of major components of the 1979 closed loop system:

- The coolant temperature sensor in the engine block
- The electro-mechanical carburetor that houses the air-fuel solenoid
- The electronic control module in the passenger compartment
- The diagnostic light in the cluster
- The catalytic converter under the floor
- The oxygen sensor in the exhaust manifold

The interaction of principal elements of a basic closed loop system are shown in Figure 3. In operation, the control loop begins with the exhaust oxygen sensor, as it checks for the presence or absence of oxygen in the exhaust stream. This information is analyzed and combined with other signals by the electronic control module which commands the carburetor to enrich or lean out the fuel mixture. In this way, the catalytic converter can then act as a more efficient after-treatment device.

That is a very general description of the C-4 system. Let me be a little more specific.

ELECTRONIC CONTROL MODULE

The schematic diagram shown in Figure 4 in the dotted block is the electronic control module. The signal, received from the exhaust oxygen sensor, is evaluated by a comparator circuit to determine if the exhaust gas is too rich or too lean. Because the signal from the sensor lags the air-fuel mixture at the carburetor by as much as 1.2 seconds at idle, the electronic control module must compensate for this delay. It does this in two

FIGURE 2   COMPUTER CONTROLLED CATALYTIC CONTROL SYSTEM

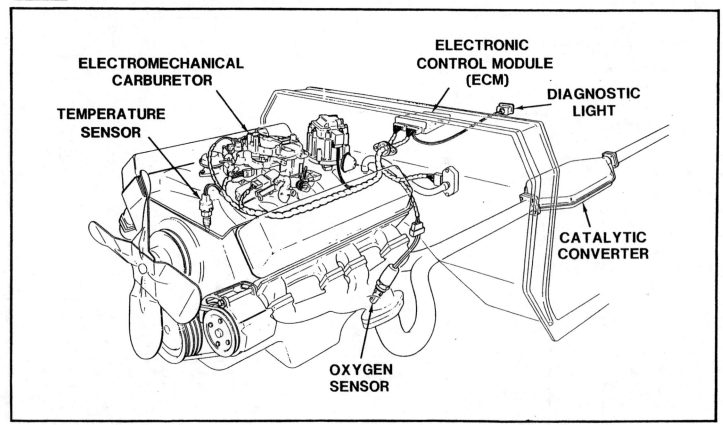

ways: one is a quick correction, or proportional control; and the other is a correction over a time period, or integral control.

Figure 5 illustrates this concept. The oxygen sensor output voltage is as indicated. As long as the sensor voltage indicates the exhaust is lean, the integral correction drives the carburetor increasingly rich. At the instant the oxygen sensor voltage crosses the reference voltage, which is the horizontal dotted line, indicating the exhaust is now rich, the electronic control module makes an instant proportional correction to bring the carburetor mixture back to stoichiometry. The mixture then continues to change as an integral correction until the next time the output crosses the reference voltage.

The left hand side of Figure 4 shows the open loop control or cold start logic. This is necessary because for a period of time during engine start up the sensor is too cold to provide air-fuel ratio information and the catalytic converter too cold to provide effective after-treatment of the exhaust gas. The rest of the time, the system operates "closed loop". The open loop carburetor operating mode can be programmed with a pre-selected air-fuel ratio, or can be varied as a function of manifold vacuum, coolant temperature, rpm, etc. or a combination of signals.

The electronic control module chooses between open loop or closed loop control based on programmed decision criteria such as coolant temperature, time, sensor impedance, voltage output, or a combination of signals.

EXHAUST OXYGEN SENSOR

The performance of the closed loop system is dependent on the accuracy and response of the exhaust oxygen sensor. It consists of a cone-shaped zirconia dioxide ceramic body coated on the inside and outside with platinum. The outer surface of platinum is coated with an inert protective coating and the entire element is placed inside a louvered metal shield to protect the cell from the highly errosive exhaust gases, while still allowing gas circulation around the element. The zirconia and platinum form an electro-chemical cell when heated to operating temperature by the exhaust gas stream. The voltage produced in the cell is dependent on the ratio of the partial pressures of atmospheric oxygen inside the cell and exhaust gas oxygen on the outside. Sensor voltage is also affected by temperature.

Figure 6 shows the response curve of the sensor. It is very sharp at the stoichiometric air-fuel ratio, causing the sensor to be seen by the electronic control module as a switch, with high voltage output when the air-fuel ratio is rich, and low output when the air-fuel ratio is lean. Voltage output and time response are affected by exhaust temperature and aging.

ELECTRO-MECHANICAL CARBURETOR

The mechanism in the carburetor which controls the fuel flow is show, in section, in Figure 7. The solenoid plunger in the carburetor drives the metering rods, and adjusts the air-fuel ratio by cycling between the rich and lean stops. The percentage of each cycle spent at the lean stop is referred to as the duty cycle. The air-fuel ratio is controlled by varying the duty cycle from a minimum of 10% to a mximum of 90%.

FIGURE 3   C-4 SYSTEM OPERATION

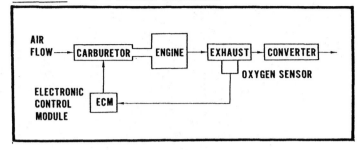

FIGURE 4   ELECTRONIC CONTROL MODULE SCHEMATIC

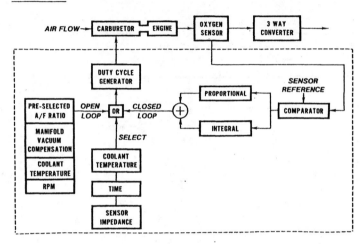

FIGURE 5   AIR/FUEL RATIO CORRECTION

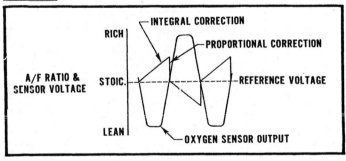

FIGURE 6 EXHAUST OXYGEN SENSOR CHARACTERISTIC CURVE

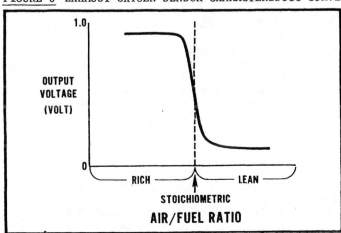

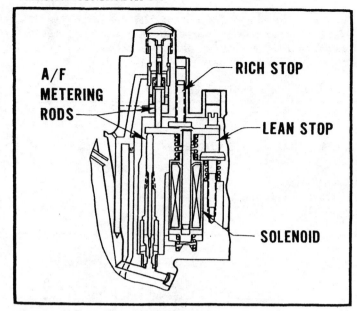

FIGURE 7 CARBURETOR SOLENOID

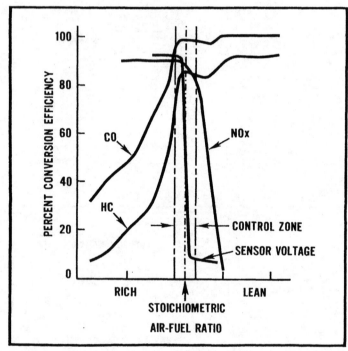

FIGURE 8 EFFICIENCY OF THREE-WAY CATALYST AND AIR-FUEL RATIO SENSOR OUTPUT

CONVERTER

The three-way converter plays a very important role in the C-4 system. The new three-way converter differs from today's oxidizing converter in that it contains rhodium. The rhodium is required to reduce oxides of nitrogen.

Figure 8 is an actual strip chart recording showing converter efficiency on the vertical axis versus air-fuel ratio on the horizontal axis. HC conversion efficiency is shown in red, CO in green, and NOx in blue. Note that each horizontal division represents one-tenth of an air-fuel ratio. Also, note the rapid decrease in efficiencies as the air-fuel ratio is operated away from the stoichiometry.

The converter can operate at maximum efficiency only while the air-fuel mixture is maintained at a stoichiometric level; and, therefore requires the precise control made possible with integrated electronics.

SELF DIAGNOSTICS

The C-4 system also includes a limited self-diagnostic capability. The diagnositc light is located in the instrument panel and will be standard on closed loop cars with digital electronic control modules in the 1979 model year.

In the event of certain system malfunctions, the "check engine" light will glow and the driver will be alerted to the need for service. Taking the vehicle to a service center, the mechanic need only ground the test point and the instrument panel light will flash a coded number to indicate the specific circuit malfunction. This code represents a specific malfunction. By referring to the service manual, the mechanic is directed to a diagnostic service procedure. Following the steps in the procedure will lead to the proper corrective action. When a driver complaint is not accompanied by a "check engine" light, the mechanic utilizes both the diagnostic procedure plus conventional diagnostic tools to correct the complaint.

Except for the scheduled replacement of the exhaust oxygen sensor, the emission system requires no periodic maintenance or adjustment for the first 50,000 miles. However, tune-ups, oil changes, etc. are still recommended at prescribed intervals.

FUTURE CONSIDERATIONS

The entire C-4 system is being designed in such a way as to make it expandable in future design generations to include control of other functions where needed.

Some of the functions we are considering are programmed spark timing and idle speed control. Which of these are actually put into production will be determined after sufficient development work has been done to determine the value of each function, as well as the specific needs of each engine.

ELECTRONIC SPARK TIMING

The electronic spark timing is similar to the Misar system used on some 1977 and 1978 Oldsmobiles. All of the spark advance engine calibration data is stored in memory. By sensing the engine speed and manifold pressure, the system selects the proper spark advance for each engine operating condition. This results in more flexible and accurate control of spark timing than is possible with the conventional centrifugal and vacuum advance mechanisms.

The items highlighted in Figure 9 are those required to implement electronic spark timing. These are: a manifold pressure sensor, additional signal lines to the distributor, the distributor, and the electronic control module.

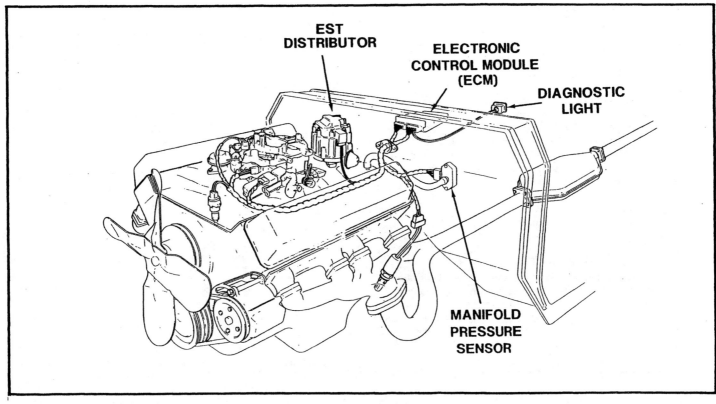

FIGURE 9   ELECTRONIC SPARK TIMING

IDLE SPEED CONTROL

The purpose of idle speed control is to automatically adjust the throttle to maintain a constant idle speed, independent of loads on the engine.

The items highlighted in Figure 10 are required to implement idle speed control. These are the electric motor actuator which positions the throttle, and the electronic control module. This system would replace the presently used idle stop solenoid.

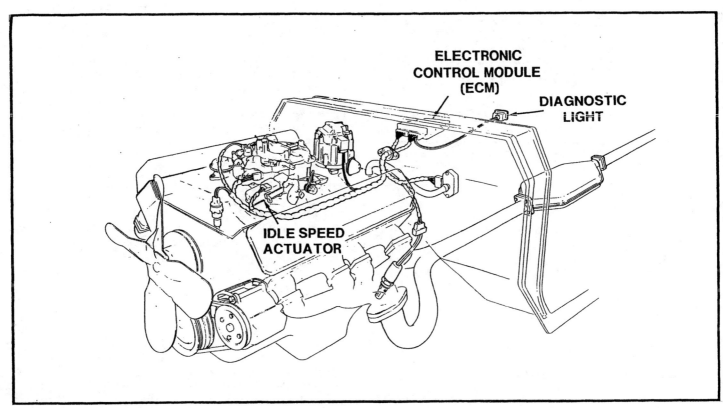

FIGURE 10   IDLE SPEED CONTROL

EMISSION CONTROL AT GM

THROTTLE BODY INJECTION

The electronics required for the closed loop carburetor system have permitted the development of a new fuel injection concept referred to as throttle body injection. This concept replaces the carburetor with an assembly which contains injectors similar to those required for Seville's electronic fuel injection.

Many of the components used in the throttle body injection system are common with the closed loop carburetor system. New components required are the air-fuel meter assembly, the electronic control module, the fuel pump, and the fuel filter.

The air-fuel meter assembly shown in Figure 11, contains one or two injectors, depending on the fuel and air quantities required by the engine. Injectors feed fuel directly into the throttle body. The amount of injected fuel is electronically controlled just as in port fuel injected systems. In addition to TBI, other future developments in engine and emission control being investigated are: EGR control, choke control, air control, evaporative emission purge systems, knock limiter, and transmission controls.

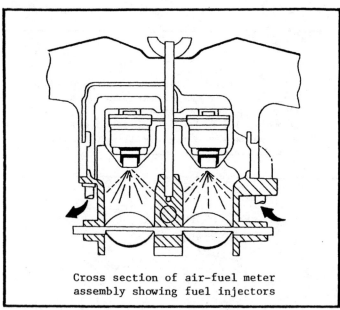

Cross section of air-fuel meter assembly showing fuel injectors

FIGURE 11 THROTTLE BODY INJECTION

EGR CONTROL

As we begin to better understand the closed loop engine EGR requirements, programmed EGR may prove beneficial. By employing engine speed, engine temperature, manifold pressure, and atmospheric pressure, the electronic control module can provide the logic signals to program the desired exhaust gas recirculation for each engine operating condition.

CHOKE CONTROL

Cold start-up accounts for nearly 70 percent of total HC and 60 percent of CO emissions during the federal test procedure. During this phase, the engine is operating in a controlled open loop mode. The use of a programmed choke system with computer control could provide added flexibility in attempting to control these start-up emissions. Using this approach, engine speed, manifold pressure, barometric pressure, and coolant temperature provide input data to the electronic control module. Based on this data, the desired start-up logic can be generated to position the choke.

AIR CONTROL

Air control can be achieved with programmed air switching. The objective is to inject air into the exhaust only when required for emission reduction and to divert the air to the air cleaner the remainder of the time. While in the bypass mode, both the pump and engine backpressure are reduced.

EVAPORATIVE EMISSION PURGE SYSTEM

During the evaporative emission test, fuel vapors are stored in a charcoal canister. Uncontrolled purging of these vapors can adversely affect the amount of HC measured in the subsequent exhaust emission test. By programming the electronic control module to permit canister purge only when engine and converter conditions will permit efficient burning and after-treatment of the fuel vapors, most of the adverse effect on the exhaust emissions can be eliminated.

KNOCK LIMITER

A knock limiter device, introduced by General Motors on the Buick turbocharged V6 engine in 1978, is one alternative means of optimizing spark control.

This device detects a sudden increase in noise level above the background noise indicating engine knock. The electronic control module sends a command signal to the distributor to retard the spark. The amount of retard depends on several factors including the severity of the knock. The spark is then restored to normal timing at a predetermined rate until the next occurrence of knock.

TRANSMISSION CONTROLS

Lock up clutch controls on automatic transmissions appear desirable in future model years. The availability of the on-board microcomputer can provide the control system required for up shift, lock up and down shift points.

SUMMARY

While the systems I've described are still in the development stages, we are already looking ahead to the future to assess the impact of the on-board electronic controller on our future products. This will be an ongoing study. However, it is already evident that electronics are a powerful and useful tool in making our products cleaner and more efficient in the years to come.

The Effects of Varying Combustion Rate in Spark Ignited Engines

R. H. Thring
Ricardo Consulting Engineers Ltd.

IT HAS BEEN SHOWN by calculation (1,2)* that, for given engine operating conditions, there is an optimum rate of combustion for minimum NOx emissions from spark ignited engines. In essence, this is because if the combustion rate is too slow, there is plenty of time for NOx formation, while if it is too fast, charge temperature becomes very high and more NOx is formed. It was considered that rates of combustion higher than those attained in current production spark ignition engines might be desirable, firstly because current rates of combustion were believed to be slower than the optimum for minimum NOx emissions and secondly because previous research had indicated that fast burn improved the lean mixture and EGR tolerances (3). Also, there were indications that fast burn improved economy.

Mayo (3) reported an experimental programme aimed at examining the effects of squish, spark gap position, swirl and charge velocity on combustion rate, emissions and fuel economy, using two V8 engines. A number of limitations were encountered with the work, and this programme was initiated as a continuation project. A single cylinder engine was used, and the effects of swirl and number of ignition points on combustion rate, emissions and fuel economy were examined. An improved data acquisition system was used enabling 300 pressure diagrams to be collected over a period of about five minutes.

OBJECTIVES

To study the effects of varying combustion rate by swirl and number of ignition points over the largest possible range of engine speed, load, equivalence ratio, EGR, octane number, compression ratio and spark timing, with a view to the reduction of engine baseline emissions, especially NOx, and fuel consumption.

DETAILS OF EQUIPMENT AND TECHNIQUES

THE ENGINE - The project was a fundamental one, and one which would require sophisticated instrumentation. Also it was going to study changes that might be small, and therefore would require a high degree of reproduci-

---

*Numbers in parentheses designate References at end of paper.

ABSTRACT

It has been shown by calculation that, for given engine operating conditions, there should be an optimum rate of combustion for minimum NOx emissions from spark ignited engines. This paper gives experimental results from a single cylinder engine which confirm the theory, and show that, for a particular engine, the normal combustion rate needed reducing at zero EGR and increasing at high EGR rates, in opposition to its natural tendency to decrease. The effect on economy was a small loss at zero EGR, but an appreciable improvement at high EGR.

Cyclic variation and octane requirement studies are also included.

bility. It was therefore decided that a single cylinder research engine should be used. Compression ratio was an important variable, and it was required to use multiple spark plugs in the combustion chamber, and also a shrouded intake valve. For these reasons, a Ricardo E6 single cylinder variable compression ratio research engine was chosen. It was found to be possible by means of using a brass plate sandwiched between the cylinder head and cylinder block to arrange for four 10 mm access holes to the combustion chamber in addition to the two 14 mm access holes already existing. Two alternative combustion chambers were used; a bathtub type of chamber designed to have no squish, and an open disc chamber (Figs. 1 and 2). The four 10 mm access holes in each chamber were used for the installation of four 10 mm spark plugs, each of which could be switched in or out independently, to vary the rate of burning of the charge. There was no phase shift facility between the different spark plugs; those that were switched on fired simultaneously. The engine was already designed to incorporate the option of a shrouded intake valve. The engine specification is given in Table 1.

The two access holes in the cylinder head were used for a Ricardo balanced disc indicator and a Kistler piezo pressure transducer. The engine was equipped with electronic inlet port fuel injection, and an inlet mixing and damping chamber which had a volume equal to eight times the swept volume of the engine. An exhaust thermal reactor was used, with secondary air injection close to the exhaust valve.

A slotted wheel was mounted on the crankshaft and a chopper disc on the camshaft. The signals from the pick-ups were used to trigger the data acquisition system at the crank angles required for the recording of the cylinder pressures.

Data Acquisition and Processing - The instrumentation block diagram is given in Fig. 3. The hub of the system was an Intertechnique Histomat S. The analogue to digital converter, core store and display of the machine were used. Since the core store of the machine was too small to store more than six pressure diagrams, it was programmed to keep a running total of pressure diagrams until 300 were acquired. Variance was also calculated. The total diagram was punched on

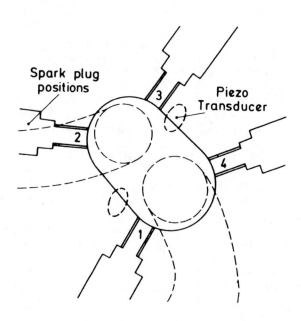

Fig. 1 - No-squish bathtub type combustion chamber

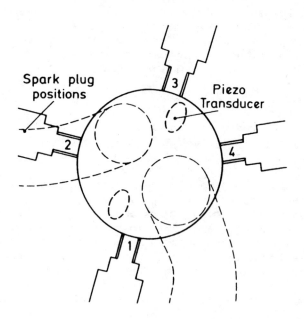

Fig. 2 - Open disc type combustion chamber

Table 1 - Ricardo E6 Brief Engine Specification

Bore	3 in. (76 mm)
Stroke	4⅜ in. (111 mm)
Swept Volume	31 in³ (.506 litre)
Ratio con rod/crank radius	4.343
Inlet Valve Timing	9/38 crank degrees
Exhaust Valve Timing	13/38 crank degrees
Compression Ratio	up to 11 to 1, in this form

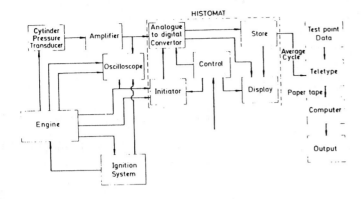

Fig. 3 - Instrumentation block diagram

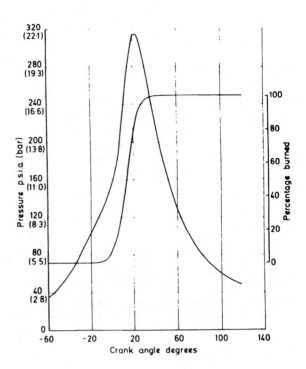

Fig. 4 - Typical computer output

to paper tape, and then fed into a Data General "Eclipse" computer. The reading from the balanced disc indicator was used to fix the absolute level of the pressure diagram, not obtainable from the piezo signal because of the finite time constant of the amplifier/transducer. The computer programme was used to calculate such results as maximum pressure rise rate, polytropic indices, etc., and also to calculate curves of per cent of charge burned versus crank angle; Fig. 4 is a typical example. The method used was that of Rassweiler and Withrow (4). The benefit of this method is that it uses no empirical correlation (such as the Eichelberg heat transfer correlation which is often used), the information required being entirely derived from the pressure data. Basically, the method is as follows: over one crank angle increment the pressure change measured is compared to the pressure change expected if there were no combustion. These differences are normalised to constant volume, and summed over the whole curve. The fraction burned at any point is the sum of the differences up to that point divided by the total. To calculate the expected pressure rise over an increment in the absence of combustion, a value is needed for the polytropic indices of the charge during combustion. Polytropic indices during compression and expansion were obtained by linear regression on the straight parts of the log P log V plot, immediately on each side of the burn region. The polytropic indices during the combustion period are different from these, so experiments were carried out to evaluate the effect of using the compression index, expansion index and a linear interpolation between the two, as values for the polytropic index during combustion. The resulting difference in burn time* was negligible so it was concluded that the lack of a genuine polytropic index during combustion was not a serious difficulty.

In order to check the absolute levels of the mean pressure diagrams, two balanced disc indicator readings were taken, one before and one after the pressure peak. The absolute levels of the mean diagrams were set up using the first reading, and the second reading compared with the value so obtained from the piezo transducer trace. It was found that the pressure value obtained from the balanced disc indicator was usually 1 to $3\frac{1}{2}$ lb/in^2 (.07-.24 bar) greater than the value obtained from the piezo signal, and this notwithstanding the fact that the balanced disc indicator signal was corrected for disc and gas column acceleration. Therefore some data were processed with an arbitrary 5 lb/in^2 (.35 bar) pressure error, and although the effect on the calculated values of the polytropic indices was large, the effect on the burn time was small.

RESULTS - ANALYSIS AND DISCUSSION

THE EFFECT OF NUMBER OF SPARK PLUGS - Table 2 shows some results obtained with five spark plugs in the engine, using the no-squish bathtub combustion chamber. The spark plug

---

*The expression "burn time" is used throughout this paper to mean the number of degrees crank angle required for combustion to progress from 10% to 90% of the mass of the charge burnt. This parameter was used as recommended by Mayo (3) because of ripples in the burn curve at the start and finish of the burn.

numbers correspond to those given in Fig. 1. The per cent change shown is based on plug number 5, which was in the standard position in the cylinder head. It can be seen that the spark plug positions near the exhaust valve were bad positions under these conditions, presumably because they were not well scoured by fresh charge during the intake stroke. This combustion chamber was found to have too low an HUCR (highest useful compression ratio) to carry out the required testing, and the remainder of the work was carried out with the open disc chamber (Fig. 2).

SPAN OF VARIABLES - In order to gain an appreciation of the effects of adjusting the input variables on the values of the output variables some experiments were carried out with the two extreme values of the input variables speed, load and number of ignition points (Table 3). The burn time was nearly halved by using four spark plugs while the BSNOx was increased by 13-113% depending on speed and load. Fuel consumption was reduced 7-28%.

Table 4 shows the results of varying equivalence ratio, manifold pressure, spark timing, speed and EGR rate over their extremes of range, using one spark plug and operating at spark timings fixed in terms of crank degrees rather than a fixed per cent load loss from best torque.

It can be seen that changing equivalence ratio had most effect on burn time in the range .9 to 1.0, while cyclic irregularity had a minimum at 1.1. Increasing load had very little effect on burn time or cyclic irregularity, and retarded ignition gave increased burn time and cyclic irregularity. The burn time nearly doubled and the cyclic irregularity nearly trebled over the speed range, but EGR had little effect.

EFFECT OF AIR/FUEL RATIO - Figure 5 shows some of the results of a survey over a range of air/fuel ratios. These results were all obtained at a constant BSNOx level of 12.8 g/bhp.h (17.2 g/kW.h), EGR being applied as required. The purpose of these experiments was to see if there was still an improvement in economy with fast burn even when the NOx emissions were brought back to their original levels by EGR. It can be seen that there was very little difference between the resulting fuel economies for the slow and fast burn conditions. There appeared to be a significant reduction in CO emissions at air/fuel ratios leaner than about 16/1, but HC emissions were worse except at air/fuel ratios leaner than about 17½/1.

The secondary purpose of these experiments was to see if there was an optimum exhaust oxygen concentration for minimum HC emissions across the air/fuel ratio range. The results indicated that the optimum oxygen level was not critical to within ± ½%, but it appeared that 4% oxygen was a reasonable overall level to use for minimum hydrocarbon emission.

STUDY OF COMBUSTION RATES -

Preliminary - Figures 6-9 show the results of a study of the effects of varying the rate of combustion in stages from fastest to slowest, at two different manifold pressures, 18 in and 26 in Hg (457 and 660 mm Hg). The rate of combustion was varied by the use of different combinations of the four spark plugs, the use of ordinary or extended electrode spark plugs, and the use of a shrouded intake valve. These tests were carried out at 1500 rev/min (25 rev/s), 7:1 compression ratio, zero EGR

Table 2 - Minimum Fuel Consumptions at 2000 rev/min (33.3 rev/s) 25 lb/in² (1.7 bar) BMEP

Spark Plug Number	BSFC lb/bhp.h	(g/kW.h)	% Change in BSFC
1	.882	(537)	+ 2
2	.970	(590)	+ 12
3	.975	(593)	+ 12
4	.923	(561)	+ 6
5	.869	(529)	0
all five	.820	(499)	- 6

Table 3 - Span of Burn Times with Plain Inlet Valve

C.R. 7:1
Fuel 96 RON unleaded indolene
Zero EGR
Spark Timing retarded for 2% torque loss
MAP Manifold Air Pressure, in.Hg (mm Hg)

10-90% BURN TIME, CRANK DEGREES

		Four Plugs		One Plug	
Speed rev/min	(rev/s)	MAP 14 in. (356 mm)	MAP 26 in. (660 mm)	MAP 14 in. (356 mm)	MAP 26 in. (660 mm)
3000	(50)	18.6	18.3	34.1	34.1
1000	(16.7)	15.8	14.1	30.9	25.6

BSNOx g/bhp.h (g/kW.h)

		Four Plugs		One Plug	
Speed rev/min	(rev/s)	MAP 14 in. (356 mm)	MAP 26 in. (660 mm)	MAP 14 in. (356 mm)	MAP 26 in. (660 mm)
3000	(50)	56.2 (75.4)	31.7 (42.5)	26.4 (35.4)	21.3 (28.6)
1000	(16.7)	20.2 (27.1)	27.4 (36.7)	17.9 (24.0)	21.7 (29.1)

BSFC lb/bhp.h (g/kW.h)

		Four Plugs		One Plug	
Speed rev/min	(rev/s)	MAP 14 in. (356 mm)	MAP 26 in. (660 mm)	MAP 14 in. (356 mm)	MAP 26 in. (660 mm)
3000	(50)	1.658 (1009)	.619 (377)	2.282 (1388)	.680 (414)
1000	(16.7)	.749 (456)	.496 (302)	1.035 (630)	.534 (325)

BMEP lb/in² (bar)

		Four Plugs		One Plug	
Speed rev/min	(rev/s)	MAP 14 in. (356 mm)	MAP 26 in. (660 mm)	MAP 14 in. (356 mm)	MAP 26 in. (660 mm)
3000	(50)	12.8 (0.883)	68.3 (4.71)	9.14 (0.630)	61.0 (4.21)
1000	(16.7)	29.6 (2.04)	89.6 (6.18)	21.6 (1.49)	85.2 (5.87)

Table 4 - Span of Variables

Compression Ratio 8:1
Plug Number   3
Fuel   100 RON
Standard Extended Electrode Spark Plug

Equiv Ratio	MAP psi (bar)	Spark Timing deg.C.A.	Speed rev/min (rev/s)	EGR %	ISFC lb/ihp.h (g/kW.h)	ISNO g/ihp.h (g/kW.h)	σ Pmax %	10-90% Burn Time Crank Degrees
0.8(lean)	8 (.55)	-35	1500 (25)	0	.351(214)	12.17(16.3)	10.96	26.6
0.9	" "	-30	" "	"	.364(221)	16.05(21.5)	10.33	25.1
1.0	" "	-30	" "	"	.362(220)	13.17(17.7)	7.05	20.9
1.1	" "	-30	" "	"	.408(248)	7.51(10.1)	6.85	19.7
1.2(rich)	" "	-30	" "	"	.448(273)	3.25(4.4)	6.99	19.5
1.0	6 (.41)	-30	1500 (25)	0	.370(225)	9.58(12.9)	8.70	25.3
"	11 (.76)	-20	" "	"	.387(235)	14.29(19.2)	9.82	22.9
"	14 (.97)	-15	" "	"	.412(251)	17.64(23.7)	10.09	27.0
1.0	8 (.55)	-45	1500 (25)	10	.388(236)	9.57(12.8)	9.26	22.4
"	" "	-15	" "	0	.403(245)	9.38(12.6)	10.48	28.3
"	" "	0	" "	0	.524(319)	4.30(5.3)	10.32)	-
1.0	8 (.55)	-30	750 (12.5)	0	.437(266)	12.18(16.3)	3.33	17.1
"	" "	-30	2250 (37.5)	"	.372(226)	11.77(15.8)	9.12	26.6
"	" "	-30	3000 (50)	"	.364(221)	10.61(14.2)	9.88	28.6
1.0	8 (.55)	-30	1500 (25)	10	.380(231)	6.63(6.2)	11.56	30.5
"	" "	-30	" "	20	.440(268)	0.69(0.9)	12.39	28.0
"	" "	-30	" "	25	.439(267)	0.51(0.7)	10.48	27.6

and a nominal air/fuel ratio of 16/1. The ignition timing was retarded from MBT to give 1½% IMEP loss.

ISNOx Results - It can be seen in Fig. 6 that reducing the burn time gave increased NOx emissions for all loads, fuels and combustion systems used in this part of the study. The reproducibility errors were at worst ± 5%. Theoretical predictions have shown that there should be a minimum in the curve of NOx versus combustion rate, so a comparison was made between theory and practice, and the results are shown in Fig. 7. The theoretical curve was taken from Reference 2. The experimental results shown in Fig. 7 are those given in Fig. 6 replotted to make them compatible. The Figures are not strictly comparable because some of the engine conditions were different, but it is clear that the reason no minimum was apparent in the experimental results was that the range of burn times explored was too narrow, and that the burn rates used were too rapid for these particular engine conditions.

It appeared from the results that the ISNOx emissions were not a unique function of burn rate, even at a given engine condition and correcting for any errors in setting the ignition timing. For example, it appeared that for a given burn time, spark plug number 2 gave significantly more NOx than the others. This was consistent with the expected result in that theoretical models show that a large proportion of the total NOx emitted comes from the first part of the charge to be burned, due to its continued heating by the combustion of the rest of the charge. Spark plug number 2 was the nearest one to the hot exhaust valve (Fig. 2).

Also, the ISNOx emissions were reduced at the light load condition by the addition of the shrouded intake valve. The reason for this was probably that the increased turbulence within the cylinder increased the gas to wall heat transfer coefficients enough to significantly reduce the sum of the time-temperature histories of many of the portions of the charge, where the charge density was relatively low, but that the reduction at the higher charge density was not significant. The Appendix gives a discussion of the results of static blowing tests on the cylinder head with both plain and shrouded intake valves.

The addition of the extended electrode spark plugs appeared in some cases to increase the NOx produced. The reason for this could be that bringing the points of ignition out into the chamber moved the position of the first part of the charge to be burnt (which is where a large proportion of the total NOx is produced) further away from the relatively cold walls of the combustion chamber, hence increasing the time-temperature history of this portion of the charge. It was partly because of these discoveries that it was decided to carry out a comparison between the use of multiple spark plugs and the use of a shrouded intake valve to attain a given burn rate over a range of engine speeds, loads and EGR rates.

ISFC Results - The ISFC results are given in Fig. 6. The reproducibility errors were at worst at about ± 5%. The ISFC at 26 in Hg (660 mm Hg) was 4 to 5% higher than at 18 in Hg (457 mm Hg) MAP possibly due to reduced combustion efficiency which was reflected in the CO and hydrocarbon emissions. An alternative explanation could lie in the friction values

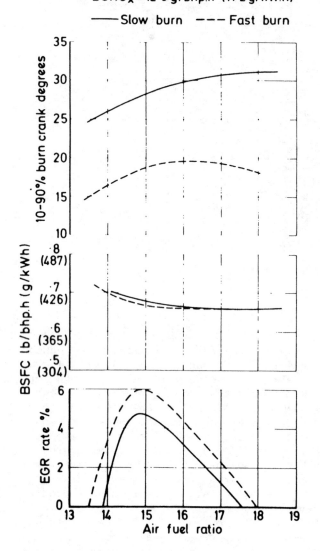

Fig. 5 - Burn time and BSFC versus air/fuel ratio

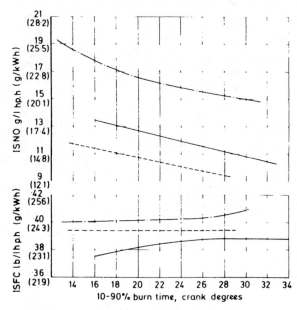

Fig. 6 - ISNOx and ISFC versus burn time

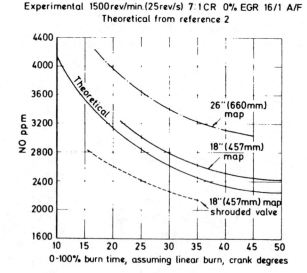

Fig. 7 - Experimental and theoretical NO

used, which were 8% different at the two manifold pressures. The discrepancy between dynamometer and directly measured IMEP's is discussed in Reference 6. The ISFC with the shrouded valve was increased by 2 to 3% at the light load and unchanged at the high load. The explanation could be that the heat transfer to the combustion chamber walls per unit mass of charge was significantly increased at the lower charge density while the change at the higher charge density was not enough to show up in the ISFC results. This would be because the heat transfer per unit mass of charge decreased with increasing charge density.

The ISFC figures showed very little change across the range of burn rates examined, although there appeared to be a slight improvement as the burn rate was increased. The change was small because the range of burn rates studied was limited, for example, the calculations in Reference 2 show an indicated thermal efficiency falling by only 5% across a range of 0-100% burn time of 60 degrees crank angle. The experimental results shown in Fig. 6 are in good agreement with this.

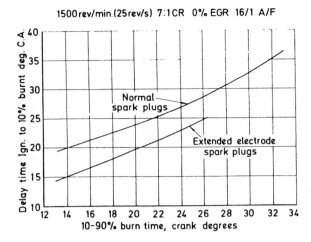

Fig. 8 - Delay time versus burn angle

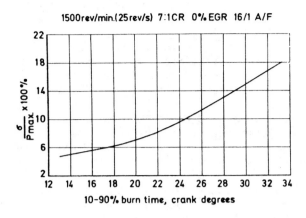

Fig. 9 - Normalised standard deviation at the peak pressure point

Delay Time and Cyclic Irregularity - Figures 8 and 9 show the results of delay time and cyclic irregularity. There was no distinct difference in delay time between any of the engine conditions tested during the study of combustion rates, except for the extended electrode spark plug results, where the delay time was reduced by 25-30% (Fig. 8). This reduction was reflected in the ignition timing. A possible reason for this reduction could be that the electrodes of the extended electrode spark plugs were running at a considerably higher temperature at the moment when the arc occurred, and therefore that the gas temperatures in the gap were higher during the combustion initiation period, hence increasing reaction rates and reducing delay time.

The cyclic irregularity results are shown in Fig. 9, which shows the standard deviation of the 300 cylinder pressures recorded at the crank angle where the pressure peak occurred on the mean diagram, normalised by dividing by the mean peak pressure and expressed as a percentage. The maximum standard deviation of cylinder pressure was found to occur at a crank angle slightly in advance of the peak pressure, the difference being virtually constant at about 6 degrees crank angle. Subject to scatter, the normalised standard deviations, both the maximum value and the value at the peak pressure point, were unique functions of burn time.

The reduction in cyclic irregularity with burn time could be because reduced burn time required later ignition timing, which meant that the combustion initiation period, which is the source of cyclic irregularity (7,8,9,10) occurred at a time when the charge pressure and temperature were higher than with a longer burn time. This theory is supported by the fact that a comparison between the standard deviation results and the ignition timing results revealed that over-advanced ignition timings yielded high values of standard deviation.

COMPARISON BETWEEN THE USE OF MULTIPLE SPARK PLUGS AND THE USE OF A SHROUDED INTAKE VALVE TO ATTAIN A GIVEN BURN TIME -

Object - The object of these experiments was to find out whether the use of two different methods of reducing burn time gave similar results. The implication was that if the results were similar, then burn time would be a good parameter to assist in the judgement of combustion chambers, while if the results were dissimilar, then combustion rate could not be considered without attention to the means of achieving it.

Method - Tests were carried out at a burn time in the centre of the range available, namely 23 degrees crank angle for 10-90% burn. Three combustion chamber configurations were chosen that gave burn times close to this figure:
(1) masked inlet valve and extended electrode spark plug number 1
(2) masked inlet valve and flush spark plug number 3
(3) plain valve and flush spark plug numbers 1 and 3

The tests were all carried out at 7:1 compression ratio and 16/1 air/fuel ratio. The ignition timing was retarded from MBT to give 1½% IMEP loss.

They were carried out in two parts:
Part 1 - The load was varied from 14 in to 25 in Hg (356 mm to 635 mm Hg) manifold pressure at 1500 rev/min (25 rev/s). The speed was then varied from 1000 rev/min to 3000 rev/min (16.7 rev/s to 50 rev/s) at a constant 18 in Hg (457 mm Hg) manifold pressure. This part was all carried out with zero EGR.

Part 2 - The results of Part 1 were extended by the addition of EGR in 5% stages up to a maximum level of the misfire tolerance minus 2%. This work was carried out at the same speed and load combinations as used in Part 1.

ISNOx Results - The differences in ISNOx between the three engine configurations tested were not consistent over the range of load, speed and EGR used. At high speeds there was little difference in ISNOx between the configurations, while at low speeds the plain valve configuration gave more NOx than the others. The explanation for this could be that at high speeds the turbulence levels were high enough both with and without the shrouded valve for the extra effect of the shrouded valve to make little difference, while at low speeds the shrouded valve made a considerable difference. It would follow from this that the NOx levels would also diverge as the load was reduced, and this was found to be the case.

The curves of ISNOx versus EGR rate were almost unchanged over the speed range and, although the absolute levels were lower, the curve shapes were very similar over the load range (e.g. 60% reduction in NOx for 10% EGR).

ISFC Results - There was no difference apparent in ISFC between the three configururations tested across the range of speed, load and EGR rate studied.

EXAMINATION OF THE EFFECT OF VARYING THE COMBUSTION RATE AT DIFFERENT EGR RATES -

Object - Theoretical work had shown that although ISNOx levels rose as the combustion rate was increased at zero EGR, the slope of the curve decreased as EGR was applied, and at an EGR rate of 25% a cross-over occurred with increased combustion rates causing a reduction in ISNOx levels at EGR rates greater than 25%. It was therefore considered desirable to carry out an experimental investigation to find out whether the predictions were borne out in practice.

Method - The results from this investigation can be divided into four parts:

Part 1 - This work was carried out at 1500 rev/min (25 rev/s), 71 lb/in^2 (4.9 bar) IMEP, 7:1 compression ratio, 16/1 air/fuel ratio, using a plain inlet valve and 100 RON unleaded gasoline. The ignition timing was retarded to give 1½% IMEP loss. The combustion rate was varied by changing the number of spark plugs in operation, the combinations used being chosen to give a reasonably uniform spacing between the fastest and slowest rates. The results of this work are shown in Fig. 10.

Part 2 - This work was carried out at 1500 rev/min (25 rev/s), 100 lb/in^2 (6.9 bar) IMEP, 7:1 compression ratio, 16/1 air/fuel ratio, using a plain inlet valve and 100 RON unleaded gasoline. The combustion rate was varied using the same spark plug combinations as were used in Part 1, and the EGR rates used were 20 and 25%. These values were chosen because the results from Part 1 had indicated that at high EGR rates it did not matter what the value of the combustion rate was, and therefore investigations were needed to find out whether there was an advantage to fast burn under different conditions. Some of the results of this work are given in Table 5.

Part 3 - This work was carried out at 10:1 compression ratio, the other conditions being the same as Part 1. Some of the results of this work are given in Table 6.

Part 4 - This work was carried out at stoichiometric air/fuel ratio and 18/1 air/fuel ratio, the other conditions being the same as Part 1. Some of the results of this work are given in Table 7.

Results - Figure 10 shows that NOx increased with increasing combustion rate over the range examined here at low EGR rates, while at high EGR rates the NOx decreased with increasing combustion rate. The cross-over point depended on load, and was about 16% EGR at 71 lb/in^2 (4.9 bar) IMEP and about 22% at 100 lb/in^2 (6.9 bar) IMEP at the engine conditions given in Fig. 10. The explanation for this is as follows: NOx formation depends on the time-temperature history of each element of the gas in the charge. At very short burn times the charge temperatures are high, giving high NOx levels. At very long burn times there is enough time for high NOx levels to form. The curve of NOx versus burn time therefore has a minimum value at some intermediate burn

Table 5 - ISNOx as a function of burn time at 100 lb/in^2 (6.9 bar) IMEP
1500 rev/min (25 rev/s)
16/1 A/F Plain Inlet Valve

EGR %	Burn Time	ISNOx g/ihp.h (g/kW.h)
20	23.0	4.0 (5.4)
20	26.8	4.3 (5.8)
20	32.5	3.5 (4.7)
20	34.0	4.0 (5.4)
25	25.4	2.2 (3.0)
25	27.6	2.6 (3.5)
25	29.0	2.3 (3.1)
25	31.0	3.6 (4.8)

Table 6 - Effect of Compression Ratio on Burn Time
1500 rev/min (25 rev/s)
71 lb/in^2 (4.9 bar) IMEP
16/1 Air/Fuel Ratio

C.R.	Zero EGR		20% EGR	
	4 Ignition Points	1 Ignition Points	4 Ignition Points	1 Ignition Points
7:1	15.3	28.3	22.0	35.0
10:1	14.4	28.5	19.5	29.3
% reduction	6%	-1%	11%	16%

Table 7 - ISNOx as a Function of Burn Time at Different Air/Fuel Ratios
1500 rev/min (25 rev/s) 7:1 C.R. 70 lb/in² (4.8 bar) IMEP

Air/Fuel Ratio	EGR Rate %	Burn Time Crank Degrees	ISNOx g/ihp.h (g/kW.h)
Stoichiometric	0	13.5	13.0 (17.4)
"	"	16.4	11.8 (15.8)
"	"	24.1	10.5 (14.1)
"	"	25.5	10.8 (14.5)
"	20	20.1	3.0 (4.0)
"	"	24.5	2.5 (3.4)
"	"	32.0	2.0 (2.7)
"	"	35.3	2.0 (2.7)
18:1	0	17.9	8.1 (10.9)
"	"	21.1	8.8 (11.8)
"	"	29.0	7.4 (9.9)
"	"	30.8	6.6 (8.9)
"	20	27.0	1.0 (1.3)
"	"	29.5	0.7 (.94)
"	"	32.3	1.0 (1.3)
"	"	33.3	1.1 (1.5)

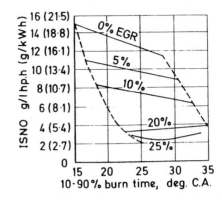

Fig. 10 - ISNOx versus burn time

time. The shape of the curve, its absolute levels, and the position of its minimum point depend upon engine conditions, including the EGR rate. As EGR is increased, the position of the minimum on the curves moves towards shorter burn times. This is because the EGR reduces charge temperatures within the cylinder, and since the net rate of formation of NOx is exponentially dependent upon temperature, shorter burn times are required before the temperature is raised high enough for the increase in temperature with burn time to compensate for the reduction in reaction time available. Following from this argument, it would be expected that an increase in load would shift the minimum points for given EGR rates towards the longer burn times. This was confirmed by experiment; for example, at 20% EGR rate the minimum NOx point occurred at a 10-90% burn time of less than 23 degrees crank angle at 71 lb/in² (4.9 bar) IMEP, while at 100 lb/in² (6.9 bar) IMEP it occurred at a burn time of greater than 34 degrees crank angle. It was not possible to determine exactly where the minimum points did lie because the range of burn time available was not great enough.

When the compression ratio was increased to 10:1, the NOx values at zero EGR were very similar to those obtained at 7:1, but as the EGR rate was increased the NOx levels became progressively higher than the corresponding levels at 7:1 compression ratio at the same burn rate. The expected increase in NOx due to the higher compression ratio was evidently masked by the increased surface to volume ratio. This effect would be greater at zero EGR when charge temperatures were higher.

It was also observed that increased compression ratio gave decreased burn time, especially at high EGR rate (Table 6). This would be due to the associated increase in charge temperature and pressure, which would have more effect at high EGR where the baseline temperature would be lower.

The results at different air/fuel ratios (Table 7) indicate that the cross-over EGR value above which shorter burns reduce NOx was over 20% at 18/1 air/fuel ratio. This would be because of the higher charge temperatures at the richer air/fuel ratio giving the same effect as was observed when the load was increased.

OCTANE REQUIREMENT STUDY -

Object - It was considered that the rate of burning of the charge in an engine would have an effect upon the octane requirement of the engine. In view of the apparent incompatibility between the present trend towards low octane gasolines and the requirement for improvements in fuel economy, it was considered desirable to carry out experimental work to quantify this effect.

Method - The experiments carried out can be divided into two parts:

Part 1 - Experiments were carried out at 1500 rev/min (25 rev/s) and 16/1 air/fuel ratio using a plain inlet valve. Three fuels were used and two EGR rates.

Part 2 - Part 1 showed that the dependence of HUCR* on combustion rate was very much less than expected, particularly with 80 RON gasoline. Further experiments were therefore undertaken at the same speed, 1500 rev/min (25 rev/s), and air/fuel ratio 16/1, also with a plain inlet valve. Two fuels were used, and the engine was run at 26 in Hg (660 mm Hg) manifold pressure and zero EGR. The number of spark plug combinations was increased from the four used in Part 1 to every possible combination of the four spark plugs.

---

*HUCR (highest useful compression ratio) was determined by gradually raising the compression ratio of the engine while it was running at the desired test condition until knock was just audible.

Results - Figure 11 shows some of the results of HUCR versus burn time, using three different fuels. It can be seen that HUCR increased as the burn time was reduced, due to the reduced time available for pre-knock reactions to occur in the end-gas. However, this effect decreased as the RON of the fuel was decreased, and at 80 RON the effect had almost disappeared. The most likely explanation is as follows:

At high compression ratios, with the four peripheral spark plugs in a disc chamber the ratio of flame surface to cold wall surface area in the end-gas region was lower than it was at low compression ratios. Therefore the tendency to knock with four spark plugs operating was greater at the lower compression ratios associated with the lower octane fuels than it was at the higher compression ratios associated with the higher octane fuels.

The results at 26 in Hg (660 mm Hg) manifold pressure showed similar trends.

SUMMARY AND CONCLUSIONS

GENERAL - A Ricardo E6 single cylinder research spark ignition engine was converted to allow the use of four spark plugs in the combustion chamber. Cylinder pressure transducers were also fitted, and a data acquisition and processing system devised to enable such parameters as 10-90% burn time to be calculated. Experimental work was carried out to examine the effect of changing the combustion rate both by altering the number of spark plugs and varying the swirl. Parameters varied were speed, load, air/fuel ratio, EGR, octane number, compression ratio and spark timing.

It was found that at low EGR rates there was no benefit in fast burn. However at high EGR rates some reduction in NOx was gained with fast burn, especially at light load, associated with slight reductions in fuel consumption and CO.

COMBUSTION RATE - Multiple spark plugs increased the rate of combustion of the charge. Different spark plug positions gave different combustion rates. The spark plug position giving the slowest burn was not necessarily the one that was worst scoured during the intake stroke. The use of a shrouded intake valve increased the combustion rate. The use of extended electrode spark plugs sometimes increased the combustion rate, depending upon the spark plug position. Increased compression ratio decreased burn time, more significantly at high EGR rate conditions.

NOx EMISSIONS - ISNOx increased with combustion rate at zero EGR rate, and decreased with combustion rate at high EGR rates, over the range of combustion rates attainable with the combustion system used. This is in agreement with theoretical predictions. The changes found were in the region of 25% at zero EGR

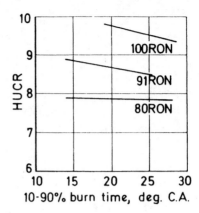

1500 rev/min.(25rev/s) 16/1 A/F, plain inlet valve
Number of ignition points varied from 1 to 4
Zero EGR, 18" (457mm.) Hg map

Fig. 11 - Highest useful compression ratio versus burn time

rate and 20% at 20% EGR rate. To obtain minimum NOx emissions at increasing EGR rates, it was desirable to increase the combustion rate in opposition to its natural tendency to decrease.

ISNOx emissions were not a unique function of combustion rate, some results being as much as 15% different from the mean value for a given combustion rate. In particular, combustion initiation near the exhaust valve gave high NOx values at high load, and the use of a shrouded inlet valve gave low values of NOx at light load.

The curves obtained of ISNOx versus EGR rate were almost unchanged over the speed range and although the absolute levels were different the curve shapes were very similar over the load range (e.g. 60% reduction in NOx for 10% EGR).

Ignition retard had a much greater effect on NOx at 18/1 air/fuel ratio than at stoichiometric.

ISNOx results as functions of burn time were found to follow similar trends at 10:1 compression ratio as at 7:1, but the absolute values were greater at high EGR rates.

FUEL CONSUMPTION - ISFC changed by 3-5% across the range of combustion rates studied, which is in good agreement with theory.

When operating at constant load, ISFC was found to decrease with increasing EGR rate, to give a minimum at about 20% EGR of up to 5% below the value at zero EGR.

HYDROCARBON EMISSIONS - ISHC was unchanged by increased combustion rate at low load, but increased by 20-25% at the high load, due to reduced exhaust temperature.

ISHC increased with increasing EGR rate when operating at constant load, probably due to the fall in exhaust temperature.

CO EMISSIONS - ISCO decreased with increasing EGR rate when operating at constant load, probably due to the longer burn time and the fact that the recirculated CO had a second chance to be oxidised.

IGNITION DELAY TIME RESULTS - Ignition delay time was found to decrease with decreasing burn time, there being no distinct differences between any of the engine conditions tested, except for the results obtained with extended electrode spark plugs, where the delay time was reduced by 15-30%.

Delay time was found to be insensitive to ignition timing. It would appear that the speed of the initiation reactions depended more upon the temperature of the surrounding surfaces (i.e. the spark plug electrodes) than upon the bulk temperature and pressure of the charge.

CYCLIC IRREGULARITY RESULTS - Cyclic irregularity, expressed as a normalised standard deviation of cylinder pressure at the peak pressure point, was a unique function of combustion rate. It decreased with increasing combustion rate. The maximum standard deviation occurred 4-8 degrees before the peak pressure.

There was a threshold level of standard deviation above which the engine became very unstable. At 1500 rev/min (25 rev/s), 70 lb/in^2 (4.8 bar) IMEP, the value of this threshold level was about 15%.

HIGHEST USEFUL COMPRESSION RATIO RESULTS - Increased combustion rate gave a higher HUCR although the effect was diminished with low ON fuels. This was thought to be the result of the lower surface to volume ratio of the combustion chamber at the lower compression ratios.

ACKNOWLEDGEMENTS

The author would like to thank the Directors of Ricardo Consulting Engineers for permission to publish this work, the Ford Motor Company for their sponsorship of the project, and also Mr. M.T. Overington and Mr. R.A. Haslett of Ricardo Consulting Engineers for many fruitful discussions throughout the duration of the project.

REFERENCES

1. P. Blumberg and J.T. Kummer, "Prediction of NOx Formation in Spark Ignited Engines - An Analysis of Methods of Control." Combustion Science and Technology, Vol. 4, 1971, p. 73.

2. P. Eyzat and J.C. Guibet, "A New Look at Nitrogen Oxides Formation in Internal Combustion Engines." Paper 680124 presented at SAE Automotive Engineering Congress, Detroit, January 1968.

3. J. Mayo, "The Effect of Engine Design Parameters on Combustion Rate in Spark Ignited Engines." Paper 750355 presented at Automotive Engineering Congress and Exposition, Detroit, February 1975.

4. G.M. Rassweiler and L. Withrow, "Motion Pictures of Engine Flames Correlated with Pressure Cards." SAE Transactions, Vol. 42, No. 5, 1936, p. 185.

5. R.A. Haslett and R.H. Thring, "Polynomial Expressions for the Ratio of Specific Heats of Burnt and Unburnt Mixtures of Air and Hydrocarbon Fuels." Ricardo DP 19417.

6. R.H. Thring, "The Discrepancy between Dynamometer and Directly Measured IMEP's." Ricardo DP 76/616.

7. S. Curry, "A Three-Dimensional Study of Flame Propagation in a Spark Ignition Engine." SAE Transactions, Vol. 71, 1963.

8. Sir Harry R. Ricardo, "The High Speed Internal Combustion Engine." Blackie, 1964.

9. E.S. Starkman, F.M. Strange and T.J. Dahm, "Flame Speeds and Pressure Rise Rates in Spark Ignition Engines." Pre-print for SAE Meeting, August 10-13, 1959.

10. R.E. Winsor and D.J. Patterson, "Mixture Turbulence, A Key to Cyclic Combustion Variation." SAE 730086.

APPENDIX

DISCUSSION OF STEADY STATE FLOW RIG TEST RESULTS

Air was blown through a clean and dirty inlet port, with and without the masked valve, and measurements of flow rate and swirl made for various valve lifts.

Ricardo swirl ratio is the predicted ratio of swirl to engine crankshaft speed. In D.I. diesel engines this value is typically 2.0, so in the E6 the masked valve gave a relatively high swirl (3.5) and the plain valve a relatively low swirl (0.65).

The values of the gulp factor obtained indicated that the restrictiveness of the cleaned port both with and without the mask was unlikely to reduce volumetric efficiency significantly at maximum load and speed. The carbon deposits in the port before cleaning affected port properties significantly.

GM Micro-Computer Engine Control System

R. A. Grimm,
R. J. Bremer,
and S. P. Stonestreet
GM Emission Control Project Center
Flint, MI

GENERAL MOTORS INTRODUCED a microprocessor based emission control system on a limited number of 1979 California and all 1980 California passenger vehicles with spark ignition engines. The system was introduced as the Computer Controlled Catalytic Converter (C-4) system to emphasize the primary function of the system: air-fuel control which allows efficient utilization of three-way catalytic converters (single or dual bed). An expanded version of the system will be introduced to allow GM engine families to meet the stringent 1981 emission standards while attaining effective utilization of the engine's fuel economy potential consistent with good performance and driveability.

The purpose of this paper is to review the components and implementation of the carbureted C-4 system. The 1979 and 1980 systems will first be reviewed followed by an in-depth look at the system capabilities.

SYSTEM EVOLUTION

The C-4 System has followed an evolution which has allowed a programmed introduction of the system into the total automotive environment. The total automotive environment ranges from such areas as engine family certification and vehicle assembly to dealer servicing.

The basic function included in the system is a closed loop carburetor control (CLCC) system to maintain the air-fuel ratio near stoichiometry so that the three-way converter can oxidize or reduce the exhaust gas constituents. The system has been implemented with both single-bed and dual-bed catalytic converters.

Other major functions that are included in the first systems are secondary air management control (SAMC). Expanded system capabilities include electronic spark timing (EST), idle speed control (ISC), controlled canister purge (CCP), torque converter clutch control (TCC), exhaust

ABSTRACT

The General Motors microprocessor based Computer Controlled Catalytic Converter (C-4) system used on 1980 California vehicles and nationwide in 1981 is presented. The basic system function is to maintain engine exhaust air-fuel ratio control so that oxidation and reduction characteristics of three-way catalytic converters can be effectively utilized.

Control strategies and components used in the Closed Loop Carburetor Control (CLCC) function are reviewed. Other system components and functions including secondary air management control, electronic spark timing, idle speed control, torque converter clutch control, controlled canister purge, exhaust gas recirculation control, early fuel evaporative control and system self diagnostics are presented.

0148-7191/80/0225-0053$02.50
Copyright © 1980 Society of Automotive Engineers, Inc.

gas recirculation control-on/off (EGR), early fuel evaporation control - on/off (EFE) and system self-diagnostics (SSD).

An open loop throttle-body fuel injector system (OLTBI) has been introduced in 1980 by Cadillac. Future systems will include closed loop throttle body fuel injection (CLTBI).

The function evolution is depicted in Fig. 1.

BASIC SYSTEM OPERATION

A number of closed loop A/F control systems used with three way converters have been described. (1-2)* The three-way converter plays a very important role in the C-4 system. To control all three emissions in a single catalyst is not simple. Catalytic control of HC and CO emissions requires an oxidizing atmosphere in the exhaust whereas catalytic control of NOx emissions requires a reducing atmosphere in the exhaust. Fortunately, there are available a variety of catalysts that can control all three emissions simultaneously--but, in order to do so, the constituents of the exhaust must be maintained at a very precise value close to the chemically correct air-fuel ratio mixture or stoichiometric mixture. Fig. 2 shows a typical single-bed three-way catalyst performance characteristic.

The system has been implemented with dual-bed three-way converters which use the first converter bed for exhaust gas reduction (NOx conversion) and a second bed with injected air to provide oxidation capabilities. The dual-bed converter provides improved oxidation capabilities over the single-bed converter while trading off reducing capabilities. The dual-bed converter utilizes a more extensive air management control system with air being switched between the converter and engine exhaust manifold.

The components associated with the basic C-4 control system include the carburetor, exhaust oxygen sensor and electronic control module (ECM), Fig. 3. The exhaust oxygen sensor indicates to the ECM whether the exhaust mixture is rich or lean of stoichiometry, Fig. 4. In closed loop operation the ECM generates the appropriate control correction signal to the carburetor to maintain the exhaust mixture near stoichiometry. As long as a rich or lean mixture is indicated, the

---
*Numbers in parentheses designates Reference at end of paper.

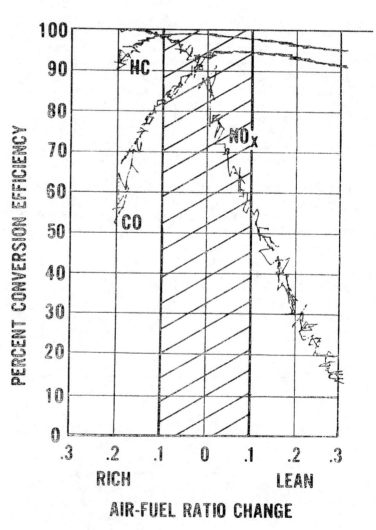

Fig. 2 - Three-way catalyst performance characteristics

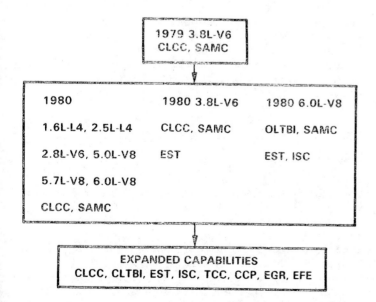

Fig. 1 - C-4 system evolution

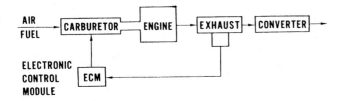

Fig. 3 - Computer controlled catalytic converter (C-4) system

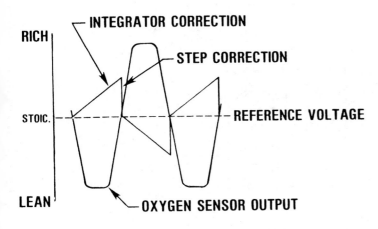

Fig. 5 - Integrator and step correction - carburetor control signal

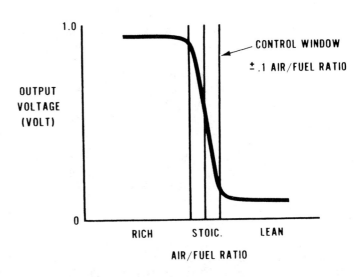

Fig. 4 - Exhaust oxygen sensor characteristic

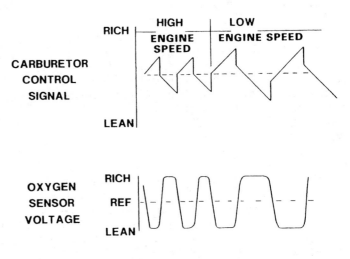

Fig. 6 - Gain scheduling feature

carburetor control signal integrates in the corrective direction at a fixed rate. When the exhaust oxygen sensor indicates a mixture change, the control signal steps in the opposite direction of the previous signal and begins to integrate in the corrective direction as illustrated in Fig. 5.

A single set of integrator and step calibrations is generally insufficient to satisfy all of the system needs. The inclusion of several inputs to the system allows increased flexibility in the selection of the integrator rate and step values with engine operating conditions. Two signals included in the system to expand the capability are engine speed and engine load. The engine speed allows idle calibrations to differ from part throttle calibrations. Manifold pressure or throttle position is used to distinguish between various engine load levels allowing variations in calibrations. The C-4 system includes the capability for two levels of gain scheduling (integrator and step) with engine speed (Fig. 6) with expansion up to 16 levels as a function of engine speed and load inputs.

A further enhancement of the closed loop operation is the addition of an adaptive control feature. The adaptive control feature is closely related to the gain scheduling feature in that the same engine speed and load divisions are used to define engine operating areas. The adaptive control algorithm stores a time weighted function of the carburetor control signal in ECM memory, while the engine is operating in a defined area. When the engine operating area changes via an engine speed or load change, the adaptive memory assigned to the new operating point is used to initialize the integrator and thereby the carburetor control signal. This concept is illustrated in Fig. 7.

In summary, the basic features associated with closed loop operation are:
 o gain scheduling (integrator and step)
 o adaptive control

This form of closed loop control produces a carburetor control signal oscillation or limit cycle about stoichiometry. The amplitude and frequency of the oscillation is dependent on engine transport delay time and system calibration. Engine transport delay is dependant on manifold and cylinder residence times which vary with speed and load. The summarized closed loop control features allow system gain variation at different engine speed and load conditions so that the limit cycle amplitude can be minimized.

The system must operate in several other modes to satisfy normal engine requirements. Basic added features are open loop, enrichment and inhibit. The inhibit feature is used during engine crank conditions to turn off all ECM control outputs. This allows the carburetor to remain at a rich mixture during engine crank. The enrichment feature generates an enriched mixture at high engine loads.

The open loop feature provides a transition between crank - start conditions (inhibit feature) and closed loop operation. The feature provides carburetor control signal scheduling as a function of coolant temperature, engine load and speed. The coolant temperature along with exhaust oxygen sensor voltage and time since engine start are used to disable the open loop feature and allow closed loop operation. The mode selection decision flow chart is shown in Fig. 8.

An additional ECM capability provided is continuously powered memory (CPM). This allows carburetor control signals calculated by the adaptive control feature to be retained after the ignition is turned off because a limited portion of ECM

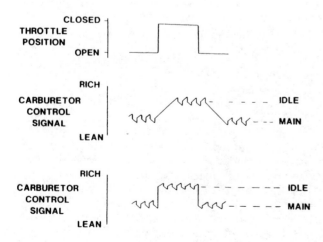

Fig. 7 - Adaptive control feature

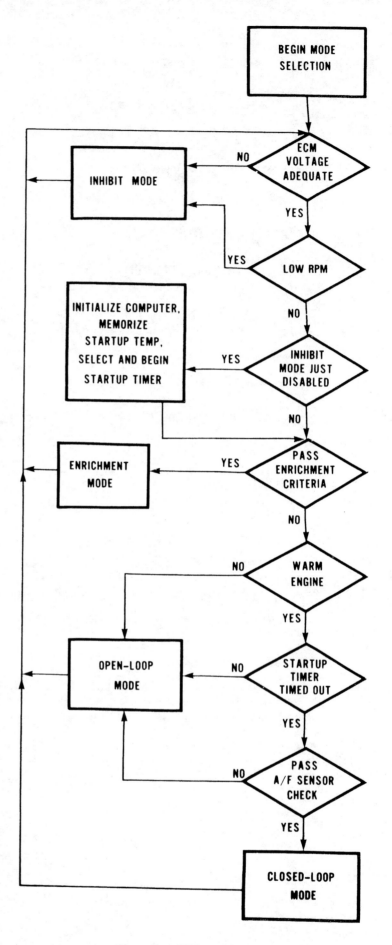

Fig. 8 - Mode selection

memory is powered directly from the vehicle battery. The values stored in the CPM are then used to compute the open loop carburetor control signal and initialize the adaptive memories. This feature allows open loop mode compensation for carburetor/engine variations which has been identified by the adaptive feature.

At each engine start the CPM data is checked to insure the stored values are valid. If for some reason the battery power has been removed (example-battery disconnected), nominal carburetor calibration values are substituted for the CPM values. New adaptive values will be learned as the system operates.

BASIC SYSTEM COMPONENTS

With the fundamental system operation defined, the components can be more fully described. The basic system components are as follows:
- o Electronic Control Module (ECM)
- o Carburetor
- o Exhaust Oxygen Sensor
- o Engine Coolant Temperature Sensor
- o Engine Speed Sensor
- o Engine Load Sensor (Options)
    - manifold pressure switch(es)
    - manifold pressure sensor
    - throttle position switch(es)
    - throttle position sensor

The C-4 systems have included the ECM capability to use the above input sensors and switches. The 1979 and some 1980 systems use manifold vacuum and throttle switches as ECM inputs with other 1980 systems using manifold pressure and throttle position sensors.

ELECTRONIC CONTROL MODULE (ECM)

The C-4 ECM is a microprocessor based control module (Fig. 9) which can be divided into three major operating areas:
Input
Processing
Output

Operation of these areas have been presented (3) and will only be reviewed in this paper. The input area of the ECM is associated with the transformation of the sensor signals into a format understood by the processing section of the ECM. The input section must interface discrete (on/off) and analog signals with the processor area. Discrete signals require level shifting obtained with discrete components and integrated circuits, while the analog signals need buffering along with an analog to digital converter (ADC).

The processor area consists primarily of the following:
Central Processor Unit (CPU)
Program Memory - Read Only (ROM)
Calibration Memory - Programmable
  Read Only (PROM)
Operational Memory - Random Access
  Memory (RAM)

The central processor operates on instructions stored in the Program Memory. Input data and intermediate results are stored in the Random Access Memory, as directed by the CPU. User specified calibration constants are stored in the Calibration Memory for use by the CPU. In the C-4 system the program memory or software is coordinated among the users to provide one basic ECM with only user calibration memory variations.

The 1980 ECM's were introduced with a Motorola 6800 series chip set and limited use of the General Motors Custom Microcomputer (GMCM) chip set. Available bytes of memory in the two 1980 systems are as follows:

	Without EST	With EST
Program Memory	4096	8192
Operation Memory	128	256
Calibration Memory	256	1024

The output section of the ECM primarily uses discrete outputs (on/off) with the exception of the pulse width modulated carburetor control signal and the electronic spark timing signal in the expanded systems. All outputs use saturated driver transistors with a mix of Darlington and single stage devices to satisfy load current requirements.

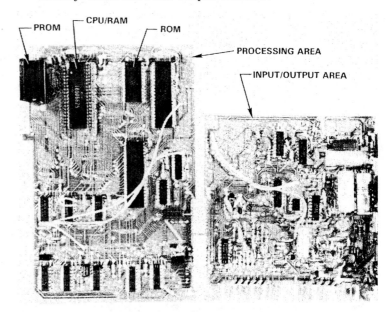

Fig. 9 - Electronic control module

CARBURETORS

The C-4 system carburetor is a modification of previously available open loop carburetors. The addition to the carburetor includes a pulse width modulated solenoid which controls the carburetor fuel metering system.

As the pulse-width modulated control signal drives the solenoid from stop to stop an average A/F ratio is generated based on the ratio of the solenoid on time to the total pulse width period. A typical carburetor performance characteristic is shown in Fig. 10.

Incorporated with the carburetor is an integral throttle position sensor. The sensor is linkage driven and is mounted in various locations on the carburetor. An illustration of the sensor is shown in Fig. 11.

EXHAUST OXYGEN SENSOR

The detailed description of the exhaust oxygen sensor operation has been previously discussed (4-5). The exhaust oxygen sensor generates a voltage in a reduced oxygen environment (rich) which is higher than produced when the sensor is placed in an excess oxygen environment (lean). The sensor behaves very much like a switch. (Fig. 12).

The sensor output voltage and internal impedance are temperature sensitive. These characteristics must be considered when locating the sensor.

ENGINE COOLANT TEMPERATURE SENSOR

The engine coolant temperature sensor is a thermistor which produces a nonlinear resistance variation with temperature. The ECM interface and software is used to linearize the coolant temperature measurement for use in various control decisions and algorithms.

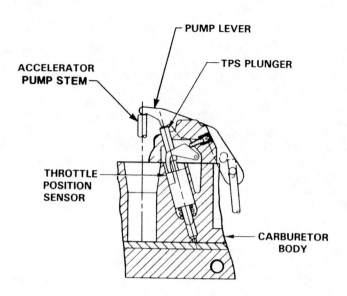

Fig. 11 - Throttle position sensor

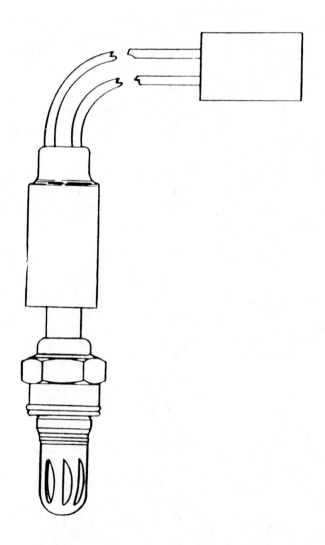

Fig. 12 - Exhaust oxygen sensor

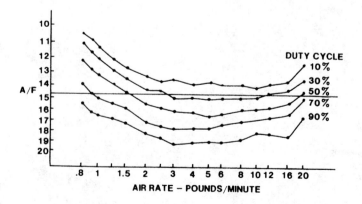

Fig. 10 - Typical carburetor performance characteristics

ENGINE SPEED SENSOR

Engine speed is calculated in the ECM by measuring time between ignition coil pulses in the basic system configuration. In systems with EST an engine position pulse is generated within the distributor and used to compute engine speed. This information is used for the gain scheduling and adaptive features.

MANIFOLD PRESSURE AND THROTTLE POSITION SWITCHES

Manifold pressure and throttle position switches are used in some systems to provide information for the adaptive and enrichment features. Switch calibrations are established per the users requirements.

MANIFOLD PRESSURE SENSORS

Two basic types of manifold pressure transducers can be utilized by the system - differential (manifold vacuum) and absolute, Fig. 13. Both sensors utilize integrated electronics for output signal amplification and buffering. The absolute sensor is implemented as a strain gage type transducer and a capacitive type transducer with the sensor output characteristics meeting the same specifications.

The differential sensor is constructed only as a strain gage type transducer.

ADDITIONAL BASIC FUNCTIONS

The basic system provides control of secondary air sources used in the converter system. The basic system allows secondary air to be directed to the engine exhaust manifold during open loop operation via an ECM output to a control valve. During closed loop operation the air is directed to the dual-bed converter (when used) or into the air cleaner for silencing.

In addition to providing basic engine/emission controls the system includes a self-diagnostic capability. This feature when used in conjunction with the system service procedure allows a mechanic to isolate the malfunction and repair the system. The feature incorporates an instrument panel lamp, labeled "CHECK ENGINE", and a diagnostic lead from the ECM. When the ECM detects a system fault the "CHECK ENGINE" lamp is turned on.

With the diagnostic lead grounded the "CHECK ENGINE" lamp flashes a code to the mechanic. The code then indicates the general area of the system malfunction, i.e. coolant measurement, carburetor control signal, etc. The service procedure is then used to isolate the fault with available test tools, including dwell meter and test light. An example of the diagnostic procedure is included in Fig. 14.

EXPANDED SYSTEM CAPABILITY

Expanded systems include Electronic Spark Timing (EST) control. The major component added to support this function is the EST distributor. Fig. 15. The distributor contains an engine crankshaft position sensor which is conditioned by an integrated electronic module to provide a reference position signal to the ECM. The integrated electronic module also provides

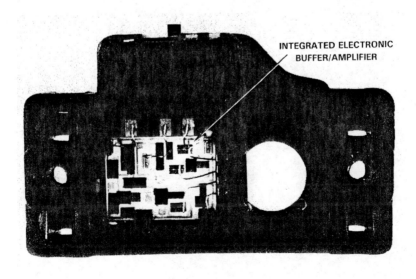

Fig. 13 - Pressure sensor

a limited spark timing control capability for use during engine crank and as back-up if an ECM malfunction occurs.

During normal operation the reference signal is fed to the ECM where engine speed and spark timing is computed. At the computed time a signal is issued to the distributor which triggers the ignition event.

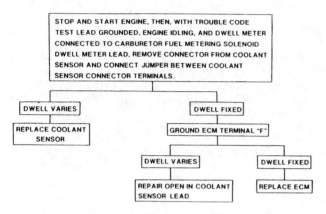

Fig. 14 - Typical service procedure

A barometric pressure sensor can be included in the system. This sensor provides barometric compensation for EST and other functions. The sensor is constructed using the same techniques as the manifold absolute pressure sensor.

Expanded engine control systems are expected to include the capability to perform an idle speed control (ISC) function. The primary addition to the system is a DC motor throttle actuator (Fig. 16) which operates the throttle lever. The ISC function provides engine load compensation which is typically required at low idle speeds. Lowering the idle speed provides improved fuel economy.

Additional functions being considered to replace existing mechanical hardware include torque converter clutch (TCC) control, exhaust gas recirculation (EGR) control, charcoal canister purge (CCP) control and early fuel evaporation (EFE) control.

A vehicle speed sensor is included in expand systems for use in various functions.

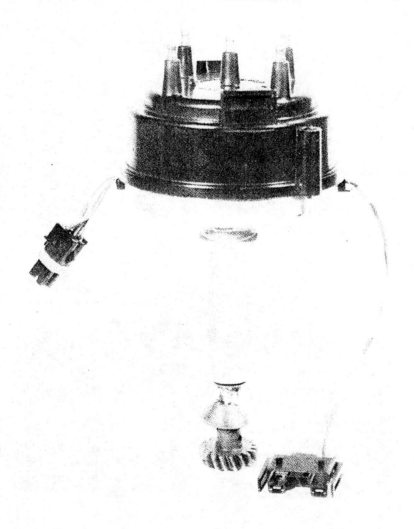

Fig. 15 - Electronic spark timing distributor

Fig. 16 - Idle speed throttle actuator

SYSTEM ALGORITHMS

A review of the system algorithms will further identify system operation and capabilities. The algorithms are incorporated in the ECM program memory (ROM). The calibrations used in the algorithms are stored in the calibration memory (PROM).

CLOSED LOOP CARBURETOR CONTROL

The discussion of closed loop carburetor control covers five areas. Table Initialization, Mode Selection, Start-Up Enrichment Mode, Open Loop Mode and Closed Loop Mode.

However, before we discuss these five areas that describe the operation, the concept of a Universal Load Axis needs to be presented, Fig. 17. There are a number of look-up tables used in the control of fuel, spark, etc. Generally one axis of these tables represents engine load in the form of throttle position or manifold pressure. For each algorithm table there is a calibration option to assign one of five inputs to be the "load" variable. The five options are:

Vacuum (differential pressure)
MAP (manifold absolute pressure)
2 ATM MAP (for turbo charged
 applications)
Throttle position
Throttle position corrected by
 barometer

The load variables are scaled so that the normal range of the load variable represents 0 to 255 counts of an 8 bit computer word. Increasing load values

- UNIVERSAL LOAD IS CONSIDERED TO BE A DIMENSIONLESS NUMBER WITH A RANGE 0 TO 255.

- FOR EACH ALGORITHM TABLE THERE IS A CALIBRATION OPTION TO ASSIGN 1 OF 5 INPUTS TO BE THE LOAD VARIABLE.

- THE FIVE LOAD OPTIONS ARE INDICATED BELOW:

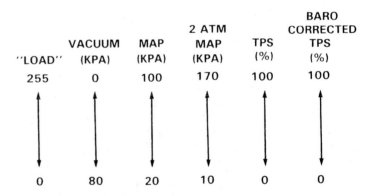

- INCREASING "LOAD" VALUES CORRESPOND TO INCREASING LOAD.

Fig. 17 - Universal load axis definition

(computer counts) always correspond to increasing load.

TABLE INITIALIZATION

Fig. 18 represents the flow chart used to initialize the computer when the ignition switch is turned on. The computer has the capability to memorize carburetor commands required to hold the A/F ratio near stoichiometry. These are

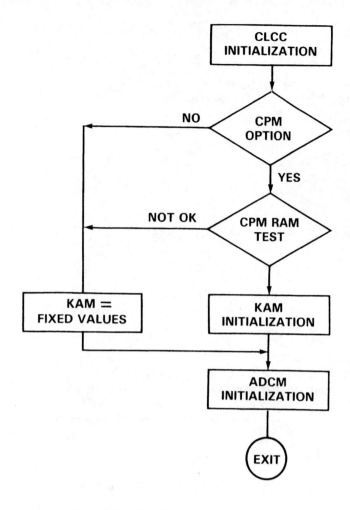

Fig. 18 - Table initialization

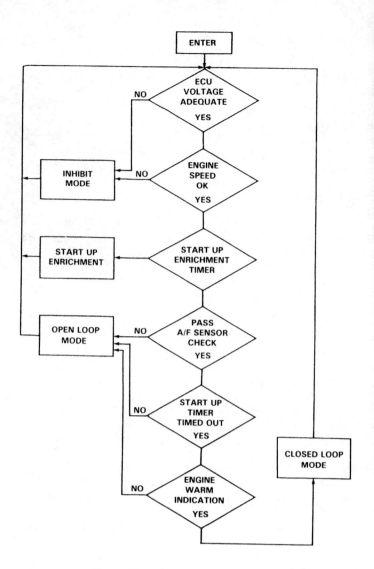

Fig. 19 - Mode selection

memorized as a function of engine operating conditions, LOAD and RPM, and can be memorized when the engine is operating "closed loop" for use when the engine is operating "open loop".

CPM, continuously powered memory as previously discussed, is memory used to store information when the ignition switch is turned off. The use of CPM is an option. KAM represents "Keep Alive Memory" used to accumulate the information when the engine is operating. KAM is stored in CPM when the ignition is turned off.

ADCM is Adaptive Duty Cycle Memory. This is the working memory used for control when the engine is operating. Carburetor commands (duty cycles) are memorized as a function of LOAD and RPM in a 4 x 4 memory matrix and continually used for carburetor control.

As the flow chart indicates, when the ignition switch is turned on, if CPM is selected and if a test indicates the memory is "good", the CPM memory is transferred to KAM memory. If CPM is not selected or if the CPM is "not good", fixed calibration values are transferred to the KAM memory.

The next step is to transfer KAM memory to the ADCM table. The ADCM is a sixteen value table with LOAD and RPM axis.

The next decision made is to establish the correct mode of operation, Fig. 19, inhibit, start-up enrichment, open loop or closed loop. This selection flow chart indicates the requirements for each mode of operation. The inhibit mode is selected if the ECM voltage is too low or the engine speed is too low. In the inhibit mode there is no current supplied to any output device except the check engine lamp.

After the inhibit mode, the start-up enrichment mode is enabled for a calibrated time. The duty cycle is a function of coolant temperature.

After the start up enrichment has "timed out" the system can go to the open loop or closed loop modes. To get to the closed loop mode the exhaust oxygen sensor must pass a "ready" test, a start up timer must be "timed out" and the engine coolant temperature must be above a specific level. If these conditions are not satisfied, the system goes into the open loop mode.

The open loop flow chart, Fig. 20, indicates the sequence of events that determines the carburetor control duty cycle. The first check is to determine if the engine requires enrichment. This is determined by testing for a minimum LOAD at the operating RPM. If this minimum load is exceeded an enrichment duty cycle is output to the carburetor. This enrichment duty cycle is determined by using the ADCM duty cycle value and subtracting a fixed duty cycle. If the minimum load is not exceeded the computer makes an "open loop" duty cycle calculation. This duty cycle is determined by taking the ADCM value and adding another duty cycle that is taken from a table set up to be a function of LOAD and COOLANT TEMPERATURE. This open loop duty cycle is further modified if the cold start program modifier (CSPM), a long term timer, has "timed-out". If this occurs a duty cycle bias will be added to the previously determined open loop duty cycle. Before any calculated duty cycle is output to the carburetor, it is checked to assure it does not exceed maximum and minimum limits.

CLOSED LOOP MODE

When the three closed loop conditions are met the system is allowed to go into the closed loop mode.

The closed loop flow chart, Fig. 21, indicates the sequence of events that determines the carburetor control duty cycle. The first check is to determine if engine requires enrichment. The test for enrichment is identical to that used in the "open loop mode". When enrichment is called for, the amount is determined in one of two ways. First, is to determine a duty cycle based on barometric pressure. A second option is to pick up the ADCM value when the load level equals that required to enable enrichment. As the load further increases, the duty cycle is richened from the ADCM value, at a linear rate.

If the conditions for enrichment are not met, the computer determines in which of the 16 ADCM cells or regions the engine is operating. If a new cell has just been entered, the output to the carburetor will be the cell duty cycle value. The duty cycle thereafter will be modified based on

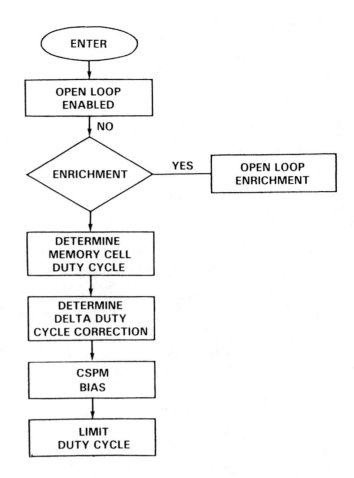

Fig. 20 - Open loop mode

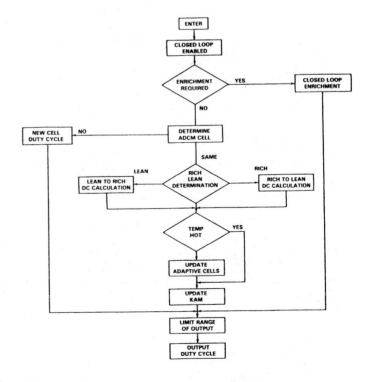

Fig. 21 - Closed loop mode

the reading from the air-fuel sensor. Step/integral control is used to move the duty cycle back to the level representing the stoichiometric mixture. There is a single step table and two integral tables. They all have 16 values based on LOAD and RPM and they have the same boundaries as the ADCM table. Thus, when there is a rich air-fuel indication and the engine has been operating in a cell (under a particular LOAD-RPM condition) a lean step command will be issued (based on the LOAD-RPM conditions) and a rich-to-lean integral rate will be commanded (based on LOAD-RPM conditions). The same control applies when a lean air-fuel ratio is indicated.

As the flow chart indicates, there are temperature limits above and below which the ADCM memory is not allowed to "learn". Finally, the output duty cycle is used to "learn" the KAM values. The output duty cycle range is limited as in the open loop mode.

This represents the basic control philosophy for the closed loop carbureted control system.

ELECTRONIC SPARK TIMING

Fig. 22 represents the flow chart for spark control. There are two modes of operation. First, is "by-pass mode." This mode of operation is for cranking the engine and for backup should a computer malfunction occur. The by-pass mode test is used to assure the conditions are satisfied to allow computer control of spark advance.

When the conditions are satisfied, the computer will calculate the proper time in the engine cycle for ignition to occur. Spark advance is the algebraic sum of advances from five tables. These are:

Main Advance - A table of spark advance based on LOAD and RPM.
Start-Up-Advance - A table of spark advance based on a delta coolant temperature. Delta temperature is the current coolant temperature minus the start-up coolant temperature. This factor is deleted when a calibrated coolant temperature is exceeded.
Supplemental Spark Advance - A table of spark advance based on LOAD. To enable this advance mode, a set of steady state conditions (RPM, LOAD, and temperature) must be satisfied for a period of time. If any of the conditions are not satisfied, the mode is disabled without delay.
Coolant Advance - A table of spark advance based on LOAD and COOLANT TEMPERATURE.
Altitude Advance - A table of spark advance based on LOAD and BAROMETRIC PRESSURE.

EXHAUST GAS RECIRCULATION

Fig. 23 represents the flow chart for control of EGR. EGR is enabled based on three inputs; engine running, coolant temperature and load. If the engine is not running, or the computer fails to operate, the EGR is enabled. After the engine starts, EGR will be enabled if the coolant temperature is greater than a specific level and if the load is less than a specific level. If the coolant temperature is too cold or the load, too great, EGR can be disabled.

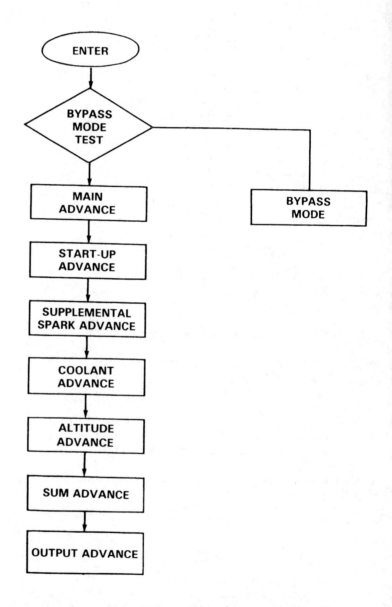

Fig. 22 - Electronic spark timing

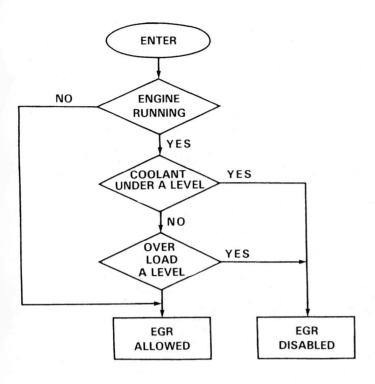

Fig. 23 - EGR control

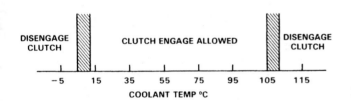

Fig. 24 - TCC engagement as a function of coolant temperature

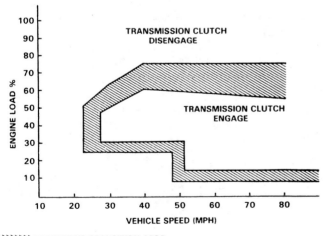

Fig. 25 - TCC engagement as a function of vehicle speed and load

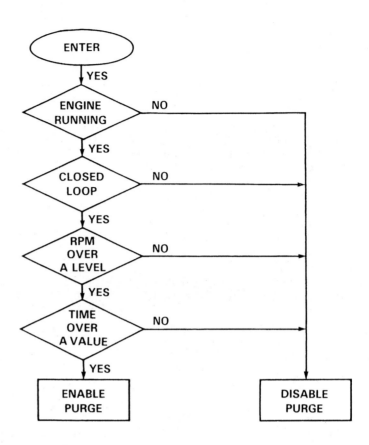

Fig. 26 - Controlled canister purge

TRANSMISSION CLUTCH CONTROL

Figs. 24 and 25 represent the control regions for the torque converter clutch. First, coolant temperature must be within a high and low limit before clutch engagement will be allowed. However, there are other conditions that also must be satisified to engage the transmission clutch. These conditions are represented by a table based on LOAD and VEHICLE SPEED. The table indicates conditions required to allow engagement of the transmission clutch. There are two such tables, one for third gear and one for second gear operation if required.

CONTROLLED CANISTER PURGE

Fig. 26 is a flow chart indicating the conditions required to enable purge. To enable purge requires that the closed loop fuel control system be operating in the closed loop mode, the RPM be greater than a specific level and the time from engine start has exceeded a specific value. If any of these three conditions are not met, the purge is disabled.

AIR CONTROL

Fig. 27 is a flow chart indicating the condition required for port air, converter air, or diverted air. If the engine is not running, the air will be diverted. There are five other sets of conditions that will result in diverted air. They are:
- o The computer has been commanding enrichment for a period of time.
- o The exhaust oxygen sensor has been indicating rich for a period of time.
- o There is a high load condition determined from LOAD and RPM.
- o There is a steady state condition detemined from LOAD and RPM.
- o There has been an ECM malfunction.

If none of these conditions are satisfied, the air will be directed to either the engine exhaust ports or a dual-bed converter, based on coolant temperature and first open loop condition. If the coolant temperature is below a particular value and the closed loop fuel system is operating in the open loop mode for the first time, the air will be directed to the ports. If either of these two conditions is not met, the air will be directed to the converter.

IDLE SPEED CONTROL

Fig. 28 is a flow chart representing idle speed control. As indicated, the desired idle speed is a function of coolant temperature with modification for the park/neutral (P/N) condition and for the low battery voltage condition.

The commands, to close the loop, are pulse widths issued to the idle speed motor. These pulse widths are determined based on how far above or below the desired RPM the actual RPM is. There are five discrete RPM error correction values (pulse widths) as a function of RPM error with the capability of including a "dead zone" around the desired RPM. All

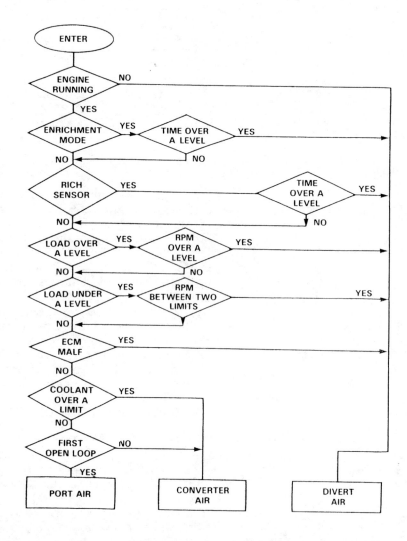

Fig. 27 - Air management

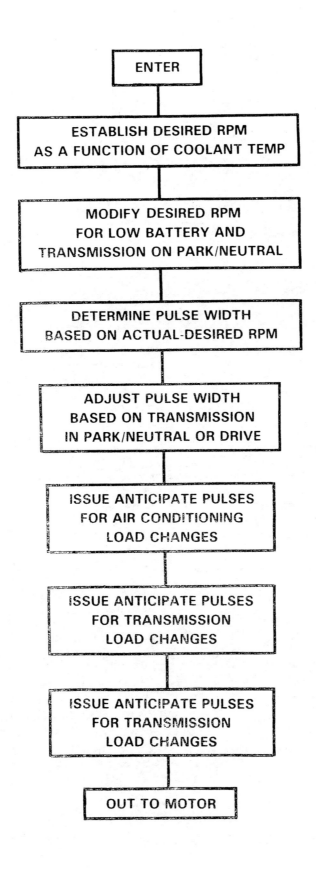

Fig. 28 - Idle speed control

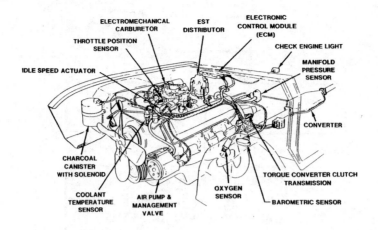

Fig. 29 - Fully expanded C-4 system

correction values are reduced when the transmission is in the P/N state.

In addition to the closed loop control of RPM, there are open loop corrections for specific operating conditions. One condition is decel throttle control. The throttle is controlled during vehicle deceleration utilizing available system information. Another condition is anticipation of the air conditioning and park/neutral to drive load changes. When there is an electrial indication that these loads are being applied to the engine, an anticipate pulse is issued to compensate for the effect of the change in load on RPM.

SUMMARY

The General Motors microcomputer based emission control system, as applied to carbureted engines, has been presented. The primary function of this system is to provide effective utilization of the three-way catalytic converter's oxidation and reducing capabilities by maintaining engine exhaust air-fuel control.

The control strategy used in the closed loop carburetor control system was reviewed. Additionally, expanded system functions (Fig. 29), including electronic spark timing, idle speed control, and torque converter clutch control were presented.

REFERENCES

1. W. D. Creps, R. A. Spilski, "Closed Loop Carburetor Emission Control System." SAE Paper No. 750368, February, 1975.

2. S. Akaeda, et al, "Development of the Nissan Electronically Controlled Carburetor System." SAE Paper No. 780204, February, 1978.

3. G. H. Miller, "Interfacing to the Microprocessor." SAE Paper No. 790235, February, 1979.

4. D. S. Eddy, et al, "Sensor for On-Vehicle Detection of Engine Exhaust Gas Composition." SAE Paper No. 730575, February, 1973.

5. E. Hamann, et al, "Lambda-Sensor with Y_2O_3 Stabilized ZrO_2 - Ceramic for Application in Automotive Emission Control Systems." SAE Paper No. 770401, February, 1977.

865033

ECONOMY INCREASE FOR OTTO ENGINES BY INTRODUCING EXHAUSTS INTO CARBURETTOR

S. Veinović, Faculty of Mechanical Engineering Kragujevac
M. Kalajžić, IPM Belgrade
K. Golec, Politechnika, Krakow
E. Hnatko, VZŠ Zagreb

ABSTRACT Preparation of fuel by introducing exhausts into carburettor emision pipe creates conditions for total evaporation of petrol and more equal distribution of micture among cylinders. This, associated with some other measures, generates conditions necessary for engine optimization in terms of economy and toxicity.

Applying this procedure to a number of locally built engines we have reached 10% lower fuel consumption. Leaner and better prepared mixture results in better economy, depending on the initial design of the original engine.

1. INTRODUCTION

Finding unique solution for antagonistic demands: lower fuel and energy consumption, on one hand, and environmental protection on the other hand, requires a complex analysis of relations between vehicle-engine-fuel-traffic-environment.

Most of the existing low class passenger cars are equipped with carburettors featuring many disadvantages, such as nonuniform cycles and supply to multi-cylinder engines, and low portion of vapour components in A/F mixture.

European trend towards cheaper cars with up to 1.4l engines is a chance for sophisticated systems burning lean mixtures.

This paper presents a possibility for improving the quality and quantity of mixture in single-throat carburettors by introducing exhausts into emulsion tube. This will be called GEYSER carburettor hereafter.

2. OPERATION OF "GEYSER" CARBURETTORS

Depression produced by cylinders attracts fuel from carburettors. Portion of easy-evaporating gasoline components evaporates along the way and mixture made of air, vapours, drops and fuel film enters cylinder.

Fuel evaporation may be intensified in many ways, but emulsification is a common process for all kinds of carburettors.

There are several advantages of using exhausts:

- they have three-fold influence on fuel: thermal, kinetic and catalytic: thermal, and.
- they are warm immediately after a successful start-up.

Figure 1 is a presentation of the GEYSER operation principle. Conventional elements are: 1-throat, 2-fuel inlet, 3-emulsion tube, 4-emulsification air jet, 5-main fuel jet, 6-throttle.

Our expirience with a number of different engines indicates the necessity of having one or more regulators: 7-emulsified mixture flow control, and 8-exhaust control.

With conventional carburettors, fuel vapor bubbles are intiated by a very small equantity of air. With the GEYSER type, warm exhausts are rapidly foroced into the carburettor, atomizing liquid gasoline and causing intensive evaporation.

This mixture is of better quality, which is important for more uniform distributi-

S. Veinović, Mašinski fakultet, YU - 34000 Kragujevac, Yugoslavia

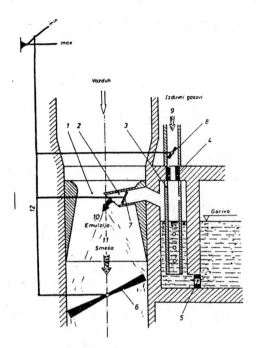

Fig. 1 Operation principle: exhausts inlet into carburettor emulsion pipe

on among cylinders and leaner A/F mixture. For this reason, it is resonable to expect improvements in economy and toxicity without risking driving performance.

3. TEST RESULTS

Most of the experiments were conducted with single-throat carburettors, only due to practical reasons. Multi-throat carburettors impose major design changes, not to mention more complicated way of exhaust regulation and flow.

With 128A.064 engine, produced localy for "Z-101" car, 1116cc, the following results were obtained:

CARBURETTOR	Fuel consumption l/100 km		
	ECE R 15	90 km/h	120 km/h
Serial	12.4	7.9	11.0
Geyser	10.6	7.0	10.0

It should be mentioned that 128A.064 series engines can be further optimized, and leaner mixture and better distribution among cylinder are direct outcomes of the GEYSER application.

"126p" car engine features a 28 IMB 10 carburettor of simplified design. To apply GEYSER carburettor on this engine, many changes had to be made.

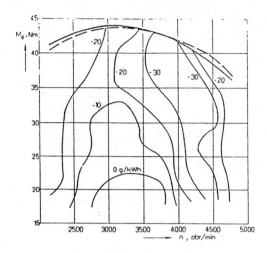

Fig. 2 Specific fuel consumption with GEYSER and conventional carburettors of "126p" car

Figure 2 shows relative change of specific fuel consumption in the entire engine operation range.

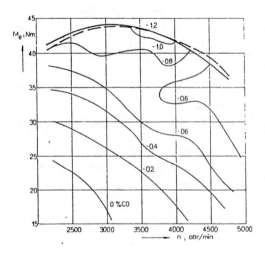

Fig. 3 CO emission: GEYSER and conventional carburettors of "126p" car

With low loads, the GEYSER efects is lost reaching 10% in the rest of the load range.

To retain good performance at high revs and loads, carburettor is adjusted for rich mixture.

Figure 3 presents change of CO emission indicating that with GEYSER carburettors maximum power is not significantly reduced by making the mixture leaner.

4. CONCLUSIONS

(1) Any system producing a higher quality A/F mixture by increasing vapour components portion creates conditions for reducting fuel consumption and emission in S.I. engines.

(2) Main features of the GEYSER carburettors are leaner mixture and more unform distribution between cylinders. Our measurements revealed economy improvement of up to 10% with lower CO emission.

REFERENCES

(1) VEINOVIĆ, S., HNATKO, E., KALAJŽIĆ, M., Karburator za intenzivnu pripremu goriva, Patentna prijava 18797/80.

(2) GOLEC, K., VEINOVIĆ, S., Gaznik glokovego silnika spalinowego z zaplonem iskrowym, Swiadectwo Autorskie o dokonami wynalazku Nr. 200 100 Warszawa, 1984.

(3) VEINOVIĆ, S., KALAJŽIĆ, M., HNATKO, E., GOLEC, K. Intensive fuel evaporation in carburettor with the aim to increase the engine economy JUMV-Science and motor vehicles'83, Opatija, april 1983., pp.96-98.

(4) GRUDEN, D., RICHTER, H., KORTE, V., Möglichkeiten zur Verbesserung der wirtschaftlichkeit von Ottomotoren Automobil-Industrie, 2/84.

(5) VEINOVIĆ, S., RADONJIĆ, D., ZAVADA, J. Karburatori automobilskih motora, "Naučna knjiga", Beograd, 1985., pp.237-246 and 370-372.

C354/87

Implications of the future European emissions scene for the motor industry—effect on overall industry costs and complexity

P E JEE, BSc
Ford Motor Company Limited, Basildon, Essex

SYNOPSIS

This paper, which is based upon the experience of Ford companies in Europe, sets out to review the business implications of the European emissions situation as it has affected the indigenous motor industry as a whole over the past 4 years, and how it will continue to be affected.

BACKGROUND

Prior to 1985, two basic emission standards (ECE 15 and A10) covered the whole of Western Europe. Countries with no standard in effect also received vehicles complying with ECE 15.03 emissions regulations.

Since then, however, mainly in response to environmental damage which is still largely unexplained, several European countries have developed stricter emission standards either individually, or within the framework of the European Community. This led to no fewer than 6 unique emission standards, namely ECE 15.03, ECE 15.04, A10, 1977/79 US, EEC Draft 5th Amendment and 1983 US. Of these, all but the 1977/79 US standard, which operated in Switzerland between October 1, 1986 and October 1, 1987, are still in force. Longer term, the number of unique standards will reduce to 3, when ECE 15.03 is superseded by ECE 15.04 and when A10 is withdrawn in Sweden around December 1988. 15.04 application could be very limited. It deserves mention that in many instances, the initial implementation of these standards has been encouraged by tax incentives.

Under this scenario, and assuming widespread enforcement of the EEC draft 5th Amendment levels, our gasoline engined car volumes will be split between EEC draft 5th Amendment (80-85%) and 83 US standards (15-20%).

Running in parallel with the overall emissions debate, and affected by it, has been the move, initially to reduce the lead content of gasoline, but now focussed on deleting lead from petrol.

RATIONALISING THE EMISSION STANDARDS

To try to contain the burgeoning engineering workload resulting from the diversity of emission regulations, it has been necessary to rationalise the range of potential emissions derivatives, and ensure that wherever possible, closely equivalent vehicle specifications cover more than one emissions role.

For example, the EEC 5th Amendment level for cars over 2.0 liter is effectively as stringent as 83 US requirements. Ford therefore uses a single derivative of the 2.9 V6 engine equipped with a three way catalytic converter to comply with both standards. Similarly, in the fuel injected role for engines between 1.4 liters and 2.0 liters, a single derivative with three way catalyst has been developed to cover 83 US and EEC Draft 5th Amendment markets. In this case however, the emission standards are more divergent than for the larger engines (EEC Draft 5th Amendment being less severe), the similar hardware level resulting from the use of fuel injection for power/image reasons, means that the commonisation cost penalty in this case is much reduced.

Because of the differences in severity between the two standards for engines below 2.0 liters, we also market carburetted engines with unregulated catalyst systems where compliance with the EEC 5th Amendment levels only is required. The purpose being to increase the product offering at lower cost where the standards will permit. This does have the disadvantage of increasing complexity.

ENGINEERING TECHNIQUES FOR REDUCING EMISSIONS

A review of the engineering techniques which are employed to reduce exhaust emissions, is necessary to understand the impact of emission regulations on the motor industry.

In essence, there are two different basic approaches. One is to clean-up the exhaust after it leaves the combustion chamber by means of add-on devices, the other is to modify the actual combustion process to ensure that it is as efficient and clean as possible.

Various devices are available to reduce the level of pollutants in the exhaust gas after combustion, and these are sometimes used either in isolation or in combination.

C354/87 © IMechE 1987

1. Catalytic conversion, which at the moment is most prominent in the public's mind. There are two types:

    - The oxidising catalyst which reduces the CO and HC content in the exhaust gas. This operates in conjunction with an open loop fuel metering system and is equally applicable to carburetted and fuel injected engines. This hardware can meet EEC 5th Amendment standards below 2.0 liters, but not 83 US.

    - Three way catalyst which reduces the CO, HC and NOx content in the exhaust simultaneously. For maximum conversion efficiency it is necessary that the engine operates at stoichiometric combustion conditions, which requires closed loop fuel metering systems featuring fuel injection, exhaust gas oxygen sensor and electronic engine controls. Such an arrangement is necessary to comply with 83 US standards.

2. Another approach is pulse air. This system uses exhaust gas pulsations to operate valves to admit air (excess oxygen) into the exhaust stream to assist in the oxidation of hydrocarbons (HC) and carbon monoxide (CO). Pulse air is at its most effective when admitted into the exhaust port where the combustion gases are hottest. Alternatively, the air can be introduced into the exhaust down pipe, or into the catalyst to assist "light off" and oxidation.

3. Exhaust gas recirculation (EGR) is often employed to reduce the production of oxides of nitrogen (NOx). It achieves this by introducing a proportion of inert exhaust gas into the intake charge to act as a heat sink thereby reducing peak combustion and slowing down the rate of change of temperature. EGR is usually controlled by an electro-vacuum system or by the electronic engine control module.

The other main approach to emission control, that of improving the combustion process itself, is in certain ways more difficult and in terms of investment more expensive, as it generally involves redesign and retooling of the cylinder head. With present levels of technological understanding, this approach is also not capable of achieving the most severe emission reduction tasks. However, it does avoid major price increases to the customer, is robust to poor servicing/misfuelling, and also offers the prospect of significant fuel economy gains due to the lean air/fuel mixture used.

Achievement of the necessary combustion process control involves careful development of engine combustion chamber and intake port shape, piston crown design and intake manifold arrangement. Particular attention must also be paid to the two fundamental calibration parameters -- ignition timing and fuel metering.

The next generation of lean burn engines will benefit from advanced 3D electronic ignition systems which will control the spark flexibly as a function of engine speed and load, free from the constraints of a mechanical/vacuum system. This is necessary for the very wide ranging spark demands of a lean burn engine. Fuelling can be provided by modern generation carburettors with flexible calibration capability, able to handle the transition from cold to normal operating temperatures, from lean fuelling at part load cruising, to richer fuelling for transients (driveability) and full load for power and durability.

Alternatively, fuel injection can be utilised for high power variants, or used as part of a total emission control system which is explained in the following section of this paper.

HARDWARE APPLICATION

Our current thinking on the application of this hardware to the various emission standards is as follows:

- 15.04 engines basically require no significant add-on componentry. Careful calibration of the fuel and ignition systems is generally sufficient.

- The EEC 5th Amendment requires different hardware levels for the different displacements.

    - Below 1.4 liter (Stage I for Homologations from October 1990) we foresee lean burn technology as the prime route, assisted by additional carburettor features and possibly pulse air, particularly in heavier vehicle classes.

        Stage 2 (October 1992 Homologation) is more problematic since we do not know what the levels will be. We are currently engineering to 15% less than Stage 1 for the tax incentives available in Germany and the Netherlands. At this level, even with additional ignition and carburettor features, and pulse air, the trade-off in fuel economy may not be attractive. An alternative approach, with benefits from a reduced engineering workload and manufacturing complexity view point, is simply to insert an oxidation catalyst in the exhaust system. We are currently evaluating both routes on our small engine ranges. If Stage 2 is tougher than a 15% reduction, EGR will assist with NOx control if necessary, but it is conceivable that the best solution would be a richer calibration with an unregulated 3 way catalyst. Depending on severity, the indicative manufactured cost increase over a 15.04 calibration would be between 55 and 135 European Currency Units. The price to the customer would, of course, be significantly greater to cover total costs including engineering, supply, transportation and inventory.

- For engines in the 1.4-2.0 category (October 1991 Homologation) a non catalyst approach is in the general case unlikely in the medium term, and lean burn plus oxidation catalyst is the prime route. This will be supported by pulse air and EGR if necessary. Indicative manufactured cost effect is 130-230 ECU's and fuel consumption could suffer by around 5% by comparison with that achieved from a comparable 15.04 calibration.

- For engines over 2.0 liters (October 1988 Homologation), the standards are similar in severity to 83 US and the same technology will be used -- full electronic engine management system controlling spark and fuel (multi-point fuel injection) with a closed loop 3 way catalyst and pulse air and EGR as appropriate. Compared with a 15.04 calibration carburetted model, the additional indicative manufactured cost of the 1983 US system is 500-600 ECU's and the fuel consumption penalty is around 8-10%.

• 1983 US markets require the same hardware level as the over 2.0 liter EEC 5th Amendment variants, irrespective of displacement. However, in some cases, where high performance is not a premium requirement, a more cost-effective single-point central fuel injection system will be used, rather than a multi-point fuel injection.

DEVELOPMENT PROCESS

The Development Process of any low emission engine is a complex task involving iterative testing and analysis to balance and attempt to optimise many conflicting parameters. For example:

• Spark retard reduces emissions but harms fuel economy, drive quality and performance.

• EGR reduces NOx emissions but can adversely affect fuel economy, and drive quality.

• Compression ratio reductions reduce HC and NOx emissions but harm fuel economy and performance.

• Lean fuelling reduces CO and NOx emissions but may worsen HC emissions and driveability if carried to extremes.

It can often be relatively simple to demonstrate acceptable emission compliance on one specially prepared engine and vehicle. The task is, however, to demonstrate that compliance can be achieved by all engines under production tolerance and assembly line conditions consistent with maintaining the required high standards of performance, engine starting, vehicle driveability and fuel economy and an overall excellent standard of total vehicle quality as perceived by the customer. These are the factors that the customer sees, not the emission results.

To accomplish this task with its conflicting objectives takes time and resources. A typical development plan takes 2½ years from Approval to Production. This is a deliberately compressed plan designed to achieve 83 US calibrations in the shortest possible time.

WORKLOAD IMPACT

The increase from 2 standards to 5 stricter standards has had a substantial impact on the European Automotive Industry. Obviously, I can only talk about our own experience on emissions, but press reports indicate that our major European competitors have gone through a similar process.

Our emissions program is costing around 200 million US dollars and has occupied up to 400 people. Hundreds of man years effort has been expended. To put this into context, we are looking at a degree of engineering effort and level of expenditure roughly equivalent to an all new engine program.

Executing these programs has been complicated by the uncertainty that has surrounded development of the emission legislation. Indeed, even at the time this paper is written, the EEC agreement is still only a draft, with acceptance blocked by Denmark. It should also be remembered that Austria gave virtually no warning of its decision to mandate 83 US emission standards and that Switzerland has just introduced emission standards for which fundamental technical details were not published until as little as one year ago. There is also the uncertainty, as to which EEC countries will enforce the 5th Amendment, and when.

Calibration for 83 US standards or for 5th Amendment engines over 1.4 liter requires a much greater degree of engineering and investment than 15.04 engines. Given the relatively smaller proportion of manufactured vehicles that 83 US markets represent for us, this obviously leads to an imbalance between effort and returns. In a world of strictly finite engineering resource (calibration engineers, dynamometers, emission rolls, prototype engine build capacity) and funding, it inevitably dilutes effort on the key task of remaining competitive.

Against the background of ever-increasing Japanese product competitiveness and technical innovation, the impact on the European industry of major programs delayed or cancelled as a result of the diverted emissions program, has been serious.

Further, in the most severe emissions markets, the substantial customer cost increases of the added emissions hardware could result in medium term reductions in market size.

This could exacerbate the manufacturing over-capacity problems already existing in Europe.

MANUFACTURING/SERVICE RELATED PRESSURES

Over and above the direct workload impact of the current emissions situation, there are other, less obvious, but nevertheless important effects:

- In-plant complexity. By 1988, Ford will be producing roughly twice as many vehicle/engine/emission combinations as in 1985. Each combination requires unique parts and assembly operations, causing inefficiency in-plant (thus higher costs). Our research has also shown that increasing complexity has a direct, adverse impact on product quality.

- Aftermarket Parts complexity. Every new part in every new model generates a requirement to stock replacement parts for in-service repair/replacement. As a rough guideline, we retain these parts in our parts depots for up to 10 years after the model has ceased production. Our dealers are also affected by this as they have to maintain stocks of commonly used parts. In markets where a large number of different emission derivatives are sold due to the existence of both voluntary and mandatory emission levels, this obviously adds to dealer stock cost and control problems.

- Servicing. Every new model requires full Service support involving in-depth technical training of Service technicians in all the dealers of the country where the model is sold. Again, this has a major impact on dealer costs and efficiency.

DISTORTION

However, possibly the most serious impact of the way the various emission regulations have been introduced from an overall industry viewpoint has been the distortions that have been created. The European Community is supposed to be about harmonisation, the recent history of exhaust emission legislation has been the exact opposite with the following results:

- Beween competitors. Due to the different engineering and market supply philosophies between manufacturers, the demands placed on their resources and therefore their capabilities to respond quickly to a rapidly changing emissions position in Europe varied significantly.

- Within markets. In some markets 2 or more emission standards are in effect, with the more severe standards encouraged by financial incentives. As an example, Germany is officially a 15.04 market. However, some areas of the market demand EEC Draft 5th Amendment calibrations to obtain incentives. Government fleets by contrast, insist on cars meeting 83 US standards. The introduction of smog alerts that differentiate between 5th Amendment and 83 US cars has also led to market distortion.

- Versus the Japanese. Japanese manufacturers already have to meet severe emission standards in their home market and are also major exporters to the US. Given their significant cost advantage, they were ideally placed to capitalise on the European emission situation without having to embark upon lengthy design and development programs.

SUMMARY

This paper has attempted to cover the impact that the emissions programs have had on the European car industry. It has not been the intention to denigrate the need for stricter emission standards and the paper should not be interpreted this way. As a company, we wholeheartedly support the introduction of rational, stricter emission standards. What has hurt the European industry, and continues to cause problems, is the proliferation of different standards, the uncertainty surrounding development of these standards and the extremely short time between their announcement and their de facto introduction in major markets.

© IMechE 1987 C354/87

About the Editor

Daniel J. Holt holds a Masters of Science degree in Mechanical Engineering and a Masters of Science degree in Aerospace Engineering. He is currently the Editor-at-Large for SAE's Automotive Engineering International magazine. For 18 years Mr. Holt was the Editor-in-Chief of the SAE Magazines Division where he was responsible for the editorial content of Automotive Engineering International, Aerospace Engineering, Off-Highway Engineering, and other SAE magazines.

He has written numerous articles in the area of safety, crash testing, and new vehicle technology.

Prior to joining SAE Mr. Holt was a biomedical engineer working with the Orthopedic Surgery Group at West Virginia University. He was responsible for developing devices to aid orthopedic surgeons and presented a number of papers on crash testing and fracture healing.

Mr. Holt is a member of Sigma Gamma Tau and a charter member of West Virginia University's Academy of Distinguished Alumni in Aerospace Engineering. He is also a member of SAE.

ERAU-PRESCOTT LIBRARY